The elements shown in color
are those that commonly
undergo bonding with carbon.

Noble Gases 0

						4.0026
						$1s^2$
						He 2
						Helium

III A	IV A	V A	VI A	VII A	
10.81	12.011	14.0067	15.9994	18.9984	20.179
2.0	2.5	3.1	3.5	4.1	[He]$2s^2 2p^6$
[He]$2s^2 2p$	[He]$2s^2 2p^2$	[He]$2s^2 2p^3$	[He]$2s^2 2p^4$	[He]$2s^2 2p^5$	
B 5	**C** 6	**N** 7	**O** 8	**F** 9	**Ne** 10
Boron	Carbon	Nitrogen	Oxygen	Fluoride	Neon
26.9815	28.0855	30.97376	32.06	35.453	39.948
1.5	1.7	2.1	2.4	2.8	[Ne]$3s^2 3p^6$
[Ne]$3s^2 3p$	[Ne]$3s^2 3p^2$	[Ne]$3s^2 3p^3$	[Ne]$3s^2 3p^4$	[Ne]$3s^2 3p^5$	
Al 13	**Si** 14	**P** 15	**S** 16	**Cl** 17	**Ar** 18
Aluminum	Silicon	Phosphorus	Sulfur	Chloride	Argon

	I B	II B	III A	IV A	V A	VI A	VII A	
58.70	63.546	65.38	69.72	72.59	74.9216	78.96	79.904	83.80
1.8	1.8	1.7	1.8	2.0	2.2	2.5	2.7	
[Ar]$3d^8 4s^2$	[Ar]$3d^{10}4s$	[Ar]$3d^{10}4s^2$	[Ar]$3d^{10}4s^2 4p$	[Ar]$3d^{10}4s^2 4p^2$	[Ar]$3d^{10}4s^2 4p^3$	[Ar]$3d^{10}4s^2 4p^4$	[Ar]$3d^{10}4s^2 4p^5$	[Ar]$3d^{10}4s^2 4p^6$
Ni 28	**Cu** 29	**Zn** 30	**Ga** 31	**Ge** 32	**As** 33	**Se** 34	**Br** 35	**Kr** 36
Nickel	Copper	Zinc	Gallium	Germanium	Arsenic	Selenium	Bromine	Krypton
106.4	107.868	112.41	114.82	118.69	121.75	127.60	126.9045	131.30
1.4	1.4	1.5	1.5	1.7	1.8	2.0	2.2	
[Kr]$4d^{10}$	[Kr]$4d^{10}5s$	[Kr]$4d^{10}5s^2$	[Kr]$4d^{10}5s^2 5p$	[Kr]$4d^{10}5s^2 5p^2$	[Kr]$4d^{10}5s^2 5p^3$	[Kr]$4d^{10}5s^2 5p^4$	[Kr]$4d^{10}5s^2 5p^5$	[Kr]$4d^{10}5s^2 5p^6$
Pd 46	**Ag** 47	**Cd** 48	**In** 49	**Sn** 50	**Sb** 51	**Te** 52	**I** 53	**Xe** 54
Palladium	Silver	Cadmium	Indium	Tin	Antimony	Tellurium	Iodine	Xenon
195.09	196.9665	200.59	204.37	207.2	208.9804	(209)	(210)	(222)
1.4	1.4	1.5	1.4	1.6	1.7	1.8	2.0	
[Xe]$4f^{14}5d^9 6s$	[Xe]$4f^{14}5d^{10}6s$	[Xe]$4f^{14}5d^{10}6s^2$	[Xe]$4f^{14}5d^{10}6s^2 6p$	[Xe]$4f^{14}5d^{10}6s^2 6p^2$	[Xe]$4f^{14}5d^{10}6s^2 6p^3$	[Xe]$4f^{14}5d^{10}6s^2 6p^4$	[Xe]$4f^{14}5d^{10}6s^2 6p^5$	[Xe]$4f^{14}5d^{10}6s^2 6p^6$
Pt 78	**Au** 79	**Hg** 80	**Tl** 81	**Pb** 82	**Bi** 83	**Po** 84	**At** 85	**Rn** 86
Platinum	Gold	Mercury	Thallium	Lead	Bismuth	Polonium	Astatine	Radon

158.9254	162.50	164.9304	167.26	168.9342	173.04	174.967
1.1	1.1	1.1	1.1	1.1	1.1	1.1
[Xe]$4f^9 6s^2$	[Xe]$4f^{10}6s^2$	[Xe]$4f^{11}6s^2$	[Xe]$4f^{12}6s^2$	[Xe]$4f^{13}6s^2$	[Xe]$4f^{14}6s^2$	[Xe]$4f^{14}5d 6s^2$
Tb 65	**Dy** 66	**Ho** 67	**Er** 68	**Tm** 69	**Yb** 70	**Lu** 71
Terbium	Dysprosium	Holmium	Erbium	Thulium	Ytterbium	Lutetium
(247)	(251)	(254)	(257)	(258)	259	260
≈1.2	≈1.2	≈1.2	≈1.2	≈1.2		
[Rn]$5f^9 7s^2$	[Rn]$5f^{10}7s^2$	[Rn]$5f^{11}7s^2$	[Rn]$5f^{12}7s^2$	[Rn]$5f^{13}7s^2$	[Rn]$5f^{14}7s^2$	[Rn]$5f^{14}6d 7s^2$
Bk 97	**Cf** 98	**Es** 99	**Fm** 100	**Md** 101	**No** 102	**Lr** 103
Berkelium	Californium	Einsteinium	Fermium	Mendelevium	Nobelium	Lawrencium

ORGANIC CHEMISTRY

Reproduction of a woodblock print, in the authors' collection, from a book published in 1497 in Basel, Switzerland, depicting an alchemist (standing at the right) and his two assistants, one working at the "fume hood" and the other taking a sample from the cask. The alchemist is holding a retort, an all-in-one distillation apparatus in which the long snout serves as the condenser. (Another retort is in use in the fume hood, and a third one is on the floor.)

ORGANIC CHEMISTRY

Marye Anne Fox

James K. Whitesell

The University of Texas
Austin, Texas

▲

Jones and Bartlett Publishers

Boston   *London*

Editorial, Sales, and Customer Service Offices
Jones and Bartlett Publishers
One Exeter Plaza
Boston, MA 02116
617-859-3900
800-832-0034

Jones and Bartlett Publishers International
7 Melrose Terrace
London W6 7RL
England

Library of Congress Cataloging-in-Publication Data

Fox, Marye Anne, 1947-
    Organic chemistry / Marye Anne Fox, James K. Whitesell.
      p.   cm.
    Includes bibliographical references and index.
    ISBN 0-86720-207-6
    1. Chemistry, Organic.   2. Biochemistry.   I. Whitesell, James K.
II. Title.
QD251.2.F69   1994
547—dc20   93-43800
                                                  CIP

Acquisitions: Arthur C. Bartlett, David E. Phanco
Developmental Editor: Patricia Zimmerman
Production Editor: Judy Songdahl
Manufacturing Buyer: Dana L. Cerrito
Design: Nancy Blodget
Illustrations: Sarah Mittelstadt Bean
Typesetting: The Clarinda Company
Cover Design: Marshall Henrichs
Printing and Binding: Rand McNally
Cover Printing: John P. Pow Company

Cover: Opiate drugs such as morphine are effective in relieving pain because they
bind to the same site in the central nervous system as do the enkephalins. The image
on the cover is of crystals of an enkephalin, viewed between crossed polarizers so
as to bring out the vivid rainbow display of colors. (Photograph © Dr. Dennis
Kunkel/Phototake NYC)

Printed in the United States of America
98 97 96 95   10 9 8 7 6 5 4 3

Brief Contents

Chapter 1   **Structure and Bonding in Alkanes**   *1*

Chapter 2   **Alkenes, Arenes, and Alkynes**   *27*

Chapter 3   **Functional Groups Containing Heteroatoms**   *59*

Chapter 4   **Chromatography and Spectroscopy**   *109*

Chapter 5   **Stereochemistry**   *155*

Chapter 6   **Understanding Organic Reactions**   *193*

Chapter 7   **Mechanisms of Organic Reactions**   *227*

Chapter 8   **Nucleophilic Substitutions at sp^3-Hybridized Carbon**   *267*

Chapter 9   **Elimination Reactions**   *303*

Chapter 10   **Electrophilic Addition to Carbon–Carbon Multiple Bonds**   *331*

Chapter 11   **Electrophilic Substitution of Aromatic Molecules**   *369*

Chapter 12   **Addition and Substitution by Heteroatomic Nucleophiles at sp^2-Hybridized Carbon**   *401*

Chapter 13   **Addition and Substitution by Carbon Nucleophiles at sp^2-Hybridized Carbon**   *451*

Chapter 14   **Skeletal-Rearrangement Reactions**   *477*

Chapter 15   **Multistep Syntheses**   *505*

Chapter 16   **Polymeric Materials**   *541*

Chapter 17   **Structures and Reactions of Naturally Occurring Compounds Containing Oxygen Functional Groups**   *581*

Chapter 18   **Structures and Reactions of Naturally Occurring Compounds Containing Nitrogen Functional Groups**   *613*

Chapter 19   **Noncovalent Interactions and Molecular Recognition**   *651*

Chapter 20   **Catalyzed Reactions**   *681*

Chapter 21   **Cofactors for Biological Redox Reactions**   *713*

Chapter 22   **Energy Storage in Organic Molecules**   *745*

Chapter 23   **Molecular Basis for Drug Action**   *789*

Contents

Preface   *xxi*

Acknowledgments   *xxv*

Chapter 1
Structure and Bonding in Alkanes   *1*

1-1 **Atomic Structure**   *6*

1-2 **sp^3 Hybridization at Carbon**   *8*

1-3 **Covalent Bonding**   *9*

1-4 **Ionic Bonding**   *10*

1-5 **Alkanes**   *12*

    Sigma Bonding   *13*

    Drawing Three-Dimensional Structures   *14*

    Structural Isomers   *15*

1-6 **Cycloalkanes**   *16*

    Structure and Formula   *16*

    Ring Strain   *17*

1-7 **Nomenclature**   *18*

1-8 **Alkane Stability**   *22*

    Combustion   *22*

    Heats of Formation   *23*

1-9 **Physical Properties of Hydrocarbons**   *24*

    Conclusions   *25*

    Review Problems   *26*

Chapter 2
Alkenes, Arenes, and Alkynes   27

2-1 Alkenes   27
sp^2 Hybridization   27
Pi Bonding   28
Structures of Alkenes   29
Molecular Orbitals   30
Higher Alkenes   34
Isomerism in Alkenes   34
Nomenclature for Alkenes   34
Alkene Stability   37

2-2 Dienes   42

2-3 Aromatic Hydrocarbons   43
Resonance Structures   44
Stability   45
Aromaticity and Hückel's Rule   46
Arenes   48

2-4 Alkynes   51
sp Hybridization   51
Higher Alkynes   52

2-5 Physical Properties of Hydrocarbons   54
Conclusions   55
Summary of New Reactions   56
Review Problems   56

Chapter 3
Functional Groups Containing Heteroatoms   *59*

3-1 Compounds Containing sp^3-Hybridized Nitrogen   59
Ammonia   59
Amines   60

3-2 Polar Covalent Bonding in Amines   62
Dipole Moments   63
Hydrogen Bonding   64
Solvation   65
Nucleophiles and Electrophiles   66
Bond Cleavages   67

3-3 Arrow Pushing to Indicate Electronic Motion   69

3-4 Acidity and Basicity of Amines   69

3-5 Compounds Containing sp^2-Hybridized Nitrogen   71
Double Bonding at Nitrogen   71
Oxidation Levels   74

3-6 Compounds Containing sp-Hybridized Nitrogen: Triple Bonding at Nitrogen   77

3-7 Compounds Containing sp^3-Hybridized Oxygen in an O–H Bond   77
Water   77

Alcohols: R—OH   *78*

Homolytic Cleavages: Bond Energies and Radical Structure   *79*

Heterolytic Cleavage of C–OH Bonds: Formation
of Carbocations   *84*

Ordering Alcohol Reactivity by Class   *85*

Conjugation in Radicals and Cations   *86*

3-8   **Ethers: R—O—R**   *88*

3-9   **Carbonyl Compounds: R₂C=O**   *89*

3-10   **Carboxylic Acids and Derivatives**   *90*

3-11   **Sulfur-containing Compounds**   *94*

3-12   **Aromatic Compounds Containing Heteroatoms**   *95*

Heteroaromatic Molecules   *95*

Heteroatom-substituted Arenes   *98*

3-13   **Alkyl Halides**   *99*

3-14   **Nomenclature**   *101*

Conclusions   *102*

Summary of New Reactions   *105*

Review Problems   *105*

Chapter 4

Chromatography and Spectroscopy   *109*

4-1   **The Use of Physical Properties to Establish Structure**   *109*

4-2   **Chromatography**   *111*

Liquid Chromatography on Stationary Columns   *112*

Paper and Thin-Layer Chromatography   *115*

Gel Electrophoresis   *116*

Gas Chromatography   *117*

4-3   **Spectroscopy**   *118*

Nuclear Magnetic Resonance Spectroscopy   *119*

Infrared Spectroscopy   *135*

Ultraviolet and Visible Spectroscopy   *140*

Mass Spectroscopy   *147*

Conclusions   *151*

Review Problems   *152*

Chapter 5

Stereochemistry   *155*

5-1   **Geometric Isomerization: Rotation about Pi Bonds**   *155*

5-2   **Conformational Analysis: Rotation about Sigma
Bonds**   *159*

Ethane   *159*

Butane   *161*

Gauche and *Anti* Conformers   *164*

5-3   **Cycloalkanes**   *165*

5-4   **Cyclohexanes**   *167*

Cyclohexane   *167*

Monosubstituted Cyclohexanes   *169*

Multiply Substituted Cyclohexanes   171

Fused Six-membered Rings   172

5-5   **Chirality**   173

5-6   **Absolute Configuration**   177

5-7   **Polarimetry**   179

5-8   **Designating Configuration**   181

A Single Center of Chirality   181

Multiple Centers of Chirality   182

Meso Compounds   184

Fischer Projections   185

5-9   **Optical Activity in Allenes**   186

5-10   **Stereoisomerism at Heteroatom Centers**   187

Conclusions   188

Summary of New Reactions   190

Review Problems   190

Chapter 6
Understanding Organic Reactions   193

6-1   **Reaction Profiles (Energy Diagrams)**   193

6-2   **Characterizing Transition States: The Hammond Postulate**   196

6-3   **Relative Rates from Reaction Profiles**   198

6-4   **Reactive Intermediates**   198

Carbanions   199

Carbenes   200

Radical Ions   201

6-5   **Thermodynamic Factors**   203

Enthalpy Effects   203

Entropy Effects: The Diels-Alder Reaction   205

6-6   **Chemical Equilibria**   206

Relating Free Energy to an Equilibrium Constant   206

Acid-Base Equilibria   207

6-7   **Acidity: Quantitative Measure of Thermodynamic Equilibria**   208

Electronegativity   209

Bond Energies   209

Inductive and Steric Effects   209

Hybridization Effects   212

Resonance Effects   212

Enolate Anion Stability   214

Aromaticity   215

6-8   **Kinetic and Thermodynamic Control**   216

6-9   **Reaction Rates: Understanding Kinetics**   219

Unimolecular Reactions   220

Bimolecular Reactions   222

Conclusions   222

Summary of New Reactions   223

Review Problems   224

Chapter 7
Mechanisms of Organic Reactions   *227*

7-1 **Classification of Reactions**   *227*

    Addition Reactions   *228*

    Elimination Reactions   *228*

    Substitution Reactions   *229*

    Condensation Reactions   *229*

    Rearrangement Reactions   *230*

    Isomerization Reactions   *231*

    Oxidation and Reduction Reactions   *231*

7-2 **Bond Making and Bond Breaking: Thermodynamic Feasibility**   *232*

    Energy Changes in Homolytic Reactions   *233*

    Energy Changes in Heterolytic Reactions   *234*

7-3 **How to Study a New Organic Reaction**   *236*

7-4 **Mechanism of a Concerted Reaction: Concerted Nucleophilic Substitution (S_N2)**   *239*

7-5 **Mechanism of Two Multistep Heterolytic Reactions: Electrophilic Addition and Nucleophilic Substitution (S_N1)**   *244*

    Electrophilic Addition of HCl to an Alkene   *245*

    Multistep Nucleophilic Substitution (S_N1): Hydrolysis of Alkyl Bromides   *248*

7-6 **Mechanism of a Multistep Homolytic Cleavage: Free Radical Halogenation of Alkanes**   *251*

    Energetics of Homolytic Substitution in the Cholorination of Ethane   *252*

    Steps in a Radical Chain Reaction   *253*

    Relative Reactivity of Halogens   *255*

    Regiocontrol in Homolytic Substitution   *257*

7-7 **Synthetic Applications**   *260*

    Conclusions   *260*

    Summary of New Reactions   *261*

    Review Problems   *263*

Chapter 8
Nucleophilic Substitutions at *sp^3*-Hybridized Carbon   *267*

8-1 **Review of Mechanisms of Nucleophilic Substitution**   *267*

8-2 **Functional-Group Transformations through S_N2 Reactions**   *270*

    Williamson Ether Synthesis   *271*

    Reaction of Alkyl Halides with Nitrogen Nucleophiles   *273*

    Phosphines as Nucleophiles   *276*

8-3 **Preparation of Carbon Nucleophiles**   *277*

    sp-Hybridized Carbon Nucleophiles: Cyanide and Acetylide Anions   *277 sp²*- and *sp³*-Hybridized Carbon Nucleophiles: Organometallic Reagents   *278*

Protonation as a Limitation   *283*

8-4   Reactivity of Carbon Nucleophiles in S$_N$2 Reactions   284

S$_N$2 Reactions by *sp*-Hybridized Carbon Nucleophiles   *284*

Alkylation of Other Organometallics   *285*

S$_N$2 Opening of Ethylene Oxide by Grignard Reagents   *285*

Alpha-Halogenation of Enolate Anions   *286*

Enolate Anion and Enamine Alkylation   *288*

Ester Enolate Anion Alkylations   *289*

Alkylation of Beta-Dicarbonyl Compounds   *290*

8-5   Synthetic Methods: Functional-Group Conversion   294

Conclusions   296

Summary of New Reactions   297

Review Problems   300

Chapter 9
Elimination Reactions   *303*

9-1   Mechanistic Options for Eliminations   *304*

9-2   E1 versus E2 Elimination Reactions: Dehydrohalogenation of Alkyl Halides   *306*

Regiochemistry   *306*

Leaving Groups   *312*

Nucleophiles   *313*

Stereochemistry   *314*

9-3   Elimination of HX from Vinyl Halides   *316*

9-4   Elimination of HX from Aryl Halides   *319*

9-5   Dehydration of Alcohols   *320*

9-6   Elimination of X$_2$ *321*

9-7   Oxidations of Alcohols   *322*

9-8   Oxidation of Hydrocarbons   *325*

9-9   Synthetic Methods   *326*

Conclusions   326

Summary of New Reactions   327

Review Problems   328

Chapter 10
Electrophilic Addition to Carbon–Carbon Multiple Bonds   *331*

10-1   Electrophilic Addition to Alkenes   *331*

10-2   Addition of HX   *333*

Gas-Phase Reactivity   *333*

Solution-Phase Reactivity: The Effect of Water on HX Addition   *334*

Hydration   *336*
Regiochemistry   *337*
Addition to Dienes   *338*
Stereochemistry   *339*
Rearrangements   *340*
HX Addition to Alkynes   *341*

10-3   Reversing Markovnikov Regiochemistry   *344*
Radical Addition of HBr   *344*
Hydroboration-Oxidation   *347*

10-4   Other Electrophiles   *349*
Oxymercuration-Demercuration   *349*
Addition of X_2 *351*
Carbenes   *355*
Epoxidation   *356*
Ozonolysis   *358*
Carbocations as Electrophiles   *360*

10-5   Radical Additions   *361*

10-6   Synthetic Methods   *362*
Conclusions   *362*
Summary of New Reactions   *363*
Review Problems   *365*

Chapter 11
Electrophilic Substitution of Aromatic Molecules   *369*

11-1   Mechanism of Electrophilic Aromatic Substitution   *369*

11-2   The Introduction of Groups by Electrophilic Aromatic Substitution: Activated Electrophiles   *371*
Halogenation   *372*
Nitration   *373*
Sulfonation   *374*
Friedel-Crafts Alkylation   *374*
Friedel-Crafts Acylation   *376*

11-3   Reactions of Substituents and Side Chains   *378*

11-4   Substituent Effects   *380*
Electron Donors and Acceptors   *382*
An Exception: Electrophilic Substitution of Halogen-substituted Aromatics   *386*
Multiple Substituents   *389*
Using Substituent Effects in Synthesis   *390*

11-5   Electrophilic Attack on Polycyclic Aromatic Compounds   *393*

11-6   Synthetic Applications   *395*
Conclusions   *396*
Summary of New Reactions   *396*
Review Problems   *398*

Chapter 12
Addition and Substitution by Heteroatomic Nucleophiles at sp^2-Hybridized Carbon   *401*

12-1	**Nucleophilic Addition to Carbonyl Groups**	*402*
12-2	**Complex Metal Hydride Reductions**	*404*
	Aldehydes and Ketones *404*	
	Reduction of Derivatives of Carboxylic Acids *406*	
	Relative Reactivity of Carbonyl Compounds toward Hydride Reducing Agents *408*	
12-3	**Biological Hydride Reductions** *410*	
12-4	**Nonhydride Chemical Reductions** *410*	
	Catalytic Hydrogenation *410*	
	Dissolving-Metal Reductions *413*	
12-5	**Anions as Nucleophiles** *415*	
12-6	**Addition of Oxygen Nucleophiles** *417*	
	Addition of Water: Hydrate Formation *417*	
	Hydroxide Ion as a Nucleophile: The Cannizzaro Reaction *419*	
	Addition of Alcohols *420*	
12-7	**Addition of Nitrogen Nucleophiles** *423*	
	Amines *423*	
	Other Nitrogen Nucleophiles *426*	
12-8	**Nucleophilic Acyl Substitution of Carboxylic Acids and Derivatives** *427*	
	Hydrolysis of Carboxylic Acid Derivatives *429*	
	Interconversion of Carboxylic Acids and Esters *431*	
	Transesterification *433*	
	Amide Hydrolysis *433*	
	Substitution Reactions of Acid Chlorides *434*	
	Reactions of Acid Anhydrides *436*	
	Formation of Carboxylic Acids from Nitriles *437*	
12-9	**Sulfonic Acid Derivatives** *438*	
12-10	**Phosphoric Acid Derivatives** *440*	
12-11	**Synthetic Applications** *441*	
	Conclusions *441*	
	Summary of New Reactions *444*	
	Review Problems *446*	

Chapter 13
Addition and Substitution by Carbon Nucleophiles at sp^2-Hybridized Carbon   *451*

13-1	**Reaction of Carbon Nucleophiles with Carbonyl Groups** *452*	
	Cyanide *452*	
	Grignard Reagents *453*	
	Organolithium Reagents *455*	

Conjugate Addition   *456*
The Wittig Reaction   *459*

13-2   **Enolates and Enols as Nucleophiles: The Aldol Condensation**   *460*

Base-catalyzed Condensation of Aldehydes   *461*
Acid-catalyzed Condensation of Aldehydes   *463*
Aldol Condensations of Ketones   *464*
Crossed Aldol Condensations   *465*
Intramolecular Aldol Condensation   *467*

13-3   **The Claisen Condensation**   *468*

Base-induced Claisen Condensation   *468*
Crossed Claisen Condensations   *469*
Reformatsky Reaction   *470*
Dieckmann Condensation   *470*

13-4   **Synthetic Applications**   *472*

Conclusions   *472*
Summary of New Reactions   *473*
Review Problems   *475*

Chapter 14
Skeletal-Rearrangement Reactions   477

14-1   **Carbon–Carbon Rearrangements**   *477*

Cation Rearrangements   *477*
An Anionic Rearrangement   *481*
A Pericyclic Rearrangement: The Cope Rearrangement   *482*

14-2   **Carbon–Nitrogen Rearrangements**   *487*

The Beckmann Rearrangement   *487*
The Hofmann Rearrangement   *489*

14-3   **Carbon–Oxygen Rearrangements**   *491*

The Baeyer-Villiger Oxidation   *492*
A Pericyclic Rearrangement: The Claisen Rearrangement   *493*

14-4   **Synthetic Applications**   *494*

Conclusions   *500*
Summary of New Reactions   *501*
Review Problems   *502*

Chapter 15
Multistep Syntheses   *505*

15-1   **Grouping Chemical Reactions**   *506*
15-2   **Retrosynthetic Analysis: Working Backward**   *509*
15-3   **Complications: Reactions Requiring both Functional-Group Transformation and Skeletal Construction**   *512*
15-4   **A Multistep Example**   *514*
15-5   **Selecting the Best Synthetic Route**   *517*
15-6   **Criteria for Evaluating Synthetic Efficiency**   *519*

15-7   **"Real World" Examples: Functional-Group Compatibility**   *522*

15-8   **Protecting Groups**   *527*

Protecting Groups for Aldehydes and Ketones   *527*
Protecting Groups for Alcohols   *529*
Protecting Groups for Carboxylates   *529*
Protecting Groups for Amines   *530*

15-9   **Practical Examples**   *532*

Ibuprofen and Ketoprofen   *532*
Valium   *535*
Conclusions   *537*
Summary of New Reactions   *538*
Review Problems   *539*

Chapter 16
Polymeric Materials   *541*

16-1   **Linear and Branched Polymers**   *543*
16-2   **Types of Polymerization**   *545*
16-3   **Addition Polymerization**   *545*

Radical Polymerization   *546*
Ionic Polymerization   *547*
Cross-Linking   *551*
Heteroatom-containing Addition Polymers   *553*

16-4   **Condensation Polymers**   *556*

Polyesters   *556*
Polysaccharides   *557*
Polyamides   *560*
Polypeptides   *561*

16-5   **Extensively Cross Linked Polymers**   *564*
16-6   **Three-Dimensional Structure**   *567*

Polypropylene   *567*
Polypeptides   *568*
Cellulose and Starch   *575*
Conclusions   *576*
Summary of New Reactions   *578*
Review Problems   *578*

Chapter 17
Structures and Reactions of Naturally Occurring Compounds Containing Oxygen Functional Groups   *581*

17-1   **Lipids**   *581*

Fats and Waxes   *581*
Saponification   *584*
Micelles   *585*

Bilayer Membranes *587*

Terpenes *590*

Steroids *594*

17-2 **Terpene Biosynthesis** *595*

17-3 **Carbohydrates** *600*

17-4 **Classification of Sugars** *601*

Trioses *601*

Aldotetroses *602*

Aldopentoses *602*

Aldohexoses *604*

Ketoses *606*

17-5 **Oligomeric Carbohydrates** *607*

Conclusions *609*

Summary of New Reactions *610*

Review Problems *610*

Chapter 18
Structures and Reactions of Naturally Occurring Compounds Containing Nitrogen Functional Groups *613*

18-1 **Methods for Forming Carbon–Nitrogen Bonds: A Review** *613*

Amines *614*

Amides *614*

Imines *615*

Nucleophilic Addition to Imines *616*

Mannich Condensation *616*

Nitrile Reduction *617*

Beckmann Rearrangement of Oximes *618*

18-2 **Amino Acids** *619*

Hydrophobic and Hydrophilic Properties *621*

Acidic and Basic Properties *621*

Zwitterionic Character of Amino Acids *622*

18-3 **Peptides** *623*

18-4 **Peptide Synthesis** *624*

Merrifield Solid-Phase Peptide Synthesis *626*

18-5 **Alkaloids** *631*

18-6 **Nucleic Acids** *634*

18-7 **Aminocarbohydrates** *638*

18-8 **Abiotic Synthesis** *639*

Adenine *640*

Carbohydrates (Ribose) *644*

18-9 **Synthetic Methods for Preparing Nitrogen-containing Compounds** *645*

Conclusions *646*

Summary of New Reactions *647*

Review Problems *648*

Chapter 19
Noncovalent Interactions and Molecular
Recognition   *651*

19-1   **Nonpolar (Hydrophobic) Interactions**   *651*
19-2   **Polar Interactions: Dipole–Dipole Interactions**   *653*
19-3   **Polar Interactions: Hydrogen Bonds**   *655*
19-4   **Polar Interactions: Metal–Heteroatom Bonds**   *659*
19-5   **Multiple Hydrogen Bonds in Two Dimensions**   *664*
19-6   **Genetic Coding, Reading, and Misreading**   *665*
19-7   **Molecular Recognition of Chiral Molecules**   *671*
        Three-Point Contacts Are Necessary for Chiral Recognition   *671*
        Resolution   *674*
        Biological Significance of Chirality   *676*
        Conclusions   *677*
        Review Problems   *678*

Chapter 20
Catalyzed Reactions   *681*

20-1   **General Concepts of Catalysis**   *682*
        Transition-State Stabilization   *683*
        Effect of Solvation on S_N2 Reactions   *684*
20-2   **Avoiding Charge Separation in Multistep Reactions**   *687*
20-3   **Distinction between Catalysis and Induction**   *692*
20-4   **Base Catalysis**   *694*
20-5   **Intermolecular and Intramolecular Reactions
        Compared**   *695*
20-6   **Transition-Metal Catalysis**   *696*
20-7   **Catalysis by Enzymes**   *700*
        Enzyme-Substrate Binding   *700*
        Catalysis by the Enzyme Chymotrypsin   *701*
20-8   **Enzymes and Chiral Recognition**   *704*
20-9   **Artificial Enzymes: Catalytic Antibodies**   *706*
        Conclusions   *709*
        Review Problems   *710*

Chapter 21
Cofactors for Biological Redox Reactions   *713*

21-1   **Molecular Recognition**   *714*
21-2   **Recycling of Biological Reagents**   *715*
21-3   **Cofactors: Chemical Reagents for Biological
        Transformations**   *716*
21-4   **Pyridoxamine Phosphate: Reductive Amination of Alpha-
        Ketoacids as a Route to Alpha-Amino acids**   *718*

21-5   NADPH: Hydride Reduction of Beta-Ketoacids   *721*

21-6   FADH$_2$: Electron-Transfer Reduction of an Alpha,Beta-
Unsaturated Thiol Ester   *723*

21-7   Acetyl CoA: Activation of Carboxylic Acids (as Thiol
Esters) toward Nucleophilic Attack   *724*

21-8   Thiamine Pyrophosphate and Lipoic Acid: Decarboxylation
of Alpha-Ketoacids   *726*

21-9   Mimicking Biological Activation with Reverse Polarity
Reagents   *730*

21-10   Tetrahydrofolic Acid: A One-Carbon Transfer Cofactor
for the Methylation of Nucleic Acids   *736*

Conclusions   *741*
Summary of New Reactions   *741*
Review Problems   *743*

Chapter 22
Energy Storage in Organic Molecules   745

22-1   Reaction Energetics   *745*

Catalysis   *748*
Multistep Transformations   *749*

22-2   Complex Reaction Cycles   *749*

22-3   Energy Storage in Anhydrides   *750*

22-4   Energy Storage in Redox Reactions   *752*

22-5   Energy Storage in Fatty Acid Biosynthesis   *754*

Carbon–Carbon Bond Formation   *754*
Reduction   *757*
Synthesis of Longer Chains   *758*

22-6   Energy Release in Fatty Acid Degradation   *759*

22-7   The Krebs Cycle   *760*

22-8   Controlling Heat Release   *767*

22-9   Energy Release from Carbohydrates through
Glycolysis   *769*

Isomerization of Glucose to Fructose   *770*
Cleavage of Fructose into Three-Carbon Fragments   *772*
*Conversion of the Three-Carbon Fragments into Acetic Acid
Derivatives*   *776*

22-10   Biological Reactions in Energy Storage and
Utilization   *780*

Conclusions   *784*
Review Problems   *785*

Chapter 23
Molecular Basis for Drug Action   789

23-1   Chemical Basis of Disease States   *790*

23-2   Intact Biological Systems as Chemical Factories   *793*

23–3   **Beta-Blockers: Modern Antacids**   *794*

23–4   **Beta-Phenethylamines: Peptide Mimics**   *795*

23–5   **Blocking Tetrahydrofolic Acid Synthesis**   *797*

23–6   **Antibiotics Affecting Membrane Structure and Ion Balance across Membranes**   *801*

23–7   **Disruption of Bacterial Cell Walls**   *803*

23–8   **Drugs Affecting Nucleic Acids Synthesis**   *809*

   Conclusions   *810*

   Review Problems   *812*

Appendix   *813*

Glossary   *819*

Index   *851*

Preface

Each year, most of the thousands of students who finish a first course in organic chemistry clearly express their dissatisfaction with what they have learned. They convey their displeasure both vocally and, even more persuasively, by "voting with their feet"—that is, by not enrolling in other advanced science courses. Ask a typical group of such students what was wrong with their course and you will hear the same answer that this query draws from deans of medical schools, from educational psychologists who specialize in the instruction of mathematics and science, from university administrators, and even from many instructors of the courses: all say that a typical organic chemistry text contains too much information, much of which is excruciatingly detailed, disconnected from "real life," irrelevant to other parts of a technical or liberal education, and just plain boring. Now, even among the strongest students, many emerge from a year of organic chemistry without a good picture of what a practicing organic chemist does.

Adopting a "less is more" philosophy for an introductory undergraduate course, we have tried in this text to address each of these common criticisms in an intellectually demanding year-long course.

- ▶ First, the content of this rigorous course is presented in less than a thousand pages.

- ▶ Second, the course is developed as a "story," with each chapter containing only those topics and reactions that are needed to understand the intellectual roots of organic chemistry as it is currently practiced.

- ▶ Third, specific examples are included at each stage to illustrate familiar, concrete uses of the chemistry under discussion.

- ▶ And, fourth, the story that we tell is intended to enhance the student's appreciation of the significance of chemistry in other science and pre-professional courses, in undergraduate research in a modern organic chemistry laboratory, and in industrial and biomedical research.

In attempting to accomplish these objectives, we have had to take a substantially different approach from that in virtually all other currently available organic texts. Like most synthetic chemists, we began by "working backward." We first asked ourselves what topics a well-informed organic student should understand after a one-year course in organic chemistry. We consulted extensively with health-profession faculty and with chemists of every stripe, both in the United States and abroad (industrial and academic, synthetic and mechanistic, material and biological). These conversations confirmed our own initial supposition that polymer chemistry, naturally occurring compounds, energy conversion and storage within organic molecules, molecular recognition and information transfer, modes of action of natural and artificial catalysts, and design criteria for new materials and biologically active molecules are of key importance in describing the contributions of organic chemistry to civilization. Most currently available texts, if they treat these topics at all, do so only as brief subsidiary applications rather than as intrinsic intellectual goals of the course.

Greater coverage of these topics, however, meant that something else would have to go, if we were to adhere to our first objective of concise presentation.

▶ First, we have tried to remove redundancy, believing that it is unnecessary, for example, to treat the complex metal hydride reductions of aldehydes, ketones, esters, and amides as four separate, seemingly unrelated reactions. This approach has required that we move away from the functional-group organization that has been widely used since the early sixties as a means of tabulating reactions, reasoning that this organization has become unwieldy, owing to the ongoing development of large numbers of new reagents.

▶ Second, we have tried to exercise restraint in choosing which chemical topics and reactions to include. Only those reactions that recur in the book's unfolding chemical story are retained, along with closely related ones that illustrate basic chemical principles and mechanisms for these essential reactions. We reasoned that good pedagogy should inhibit us from feeling obliged to include every chemical topic and detail known by either author. Rather, we sought to identify those topics absolutely required to reach our objective of giving the student sufficient information to understand the principles and practice of modern organic chemistry.

These goals led to an organizational structure that begins with seven chapters that deal primarily with the three-dimensional structures of various organic functional groups (Chapters 1 through 5) and the relation between structure and reactivity, both from a thermodynamic point of view and from a kinetic one (Chapters 6 and 7). As soon as the student has been exposed to the range of organic functional groups, spectroscopy is introduced (Chapter 4) to facilitate work in the laboratory. The next seven chapters (Chapters 8 through 14) deal with specific reaction types, each organized by common mechanism rather than by functional group, and are followed by an integrative chapter (Chapter 15) that incorporates these reactions into strategies for planning the synthesis of new compounds. Finally, Chapters 16 through 23 illustrate how the structural features considered in the first part of the text, together with the specific reactions covered in the second part, can be sources of insight into the chemical structure and function of important naturally occurring and manufactured materials: polymers, proteins, and natural and artificial enzymes. How these materials accomplish specific chemical conversions in biological systems by molecular recognition, catalysis, and energetic coupling with cofactor conversions is shown by example, ultimately describing the function of pharmaceutical agents in the last chapter.

This textbook presupposes only the knowledge of chemistry typically attained in a high school course or in the first semester of standard college chemistry. For curricula that so require, the self-contained course presented in this book can be offered in the freshman year, without the prerequisite quantitative development of a one-year college general chemistry course. The topics covered here afford a solid basis for a description of common natural organic phenomena, which might effectively instill in students a greater enthusiasm for the more-abstract topics of introductory physical chemistry.

Apart from organizing the text itself in what we think to be a better way, we have included a number of learning aids and motivational stimulants.

▸ First, each chapter contains exercises for testing immediate mastery of the concepts in a section, as well as end-of-chapter problems that help to integrate the concepts in the chapter as a whole. Both the exercises and the problems cover a range of difficulty, progressing from those that provide basic reinforcement of a concept to those that require the student to apply the concept to a different situation. We have written detailed answers for all the exercises and problems, preparing the *Study Guide and Solutions Manual* ourselves to ensure correspondence of the explanations given in the manual with the presentation of concepts in the text.

▸ Second, each chapter contains boxed material—examples of the practical utility of the reactions and materials being considered.

▸ Third, each chapter includes a narrative summary (Conclusions) of the principal ideas of importance in the chapter. These summaries, together with a list of Important Topics in the *Study Guide and Solutions Manual* are intended to help the student recognize, and learn, the main concepts presented therein. Most chapters contain a list of reactions that are new to the chapter, and Chapters 7 through 18 also include tables that regroup the reactions considered according to what they accomplish as synthetic transformations.

▸ Fourth, the book's index is comprehensive, listing pages on which chemicals and topics are covered for easy access to a given topic, if needed for reinforcement when it is discussed again in a new context in a later chapter.

▸ Fifth, a complete alphabetized glossary of key definitions is included in the text, together with citations to the chapter and section in which the term is introduced and developed. A chapter-by-chapter glossary is given in the *Study Guide and Solutions Manual* to assist the student in preparing for examinations; the definitions constitute an additional means of reviewing the concepts developed.

▸ Sixth, the publisher has made it possible to supply a "Value Package" that includes a plastic model-making kit for the construction of three-dimensional molecules. This special package includes the textbook, the *Study Guide and Solutions Manual*, and the model kit included at no extra cost when all three are ordered as a package (ISBN: 0-86720-882-1).

▸ Finally, a set of fifty full-color transparencies is available to qualified adopters.

We hope that students will enjoy and benefit from the experience of learning modern organic chemistry as it is presented in this book. We will be grateful indeed to our readers for their evaluation of our work.

Marye Anne Fox
James K. Whitesell

Acknowledgments

Preparing an organic chemistry text that pedagogically departs so markedly from the traditional approach of the past three decades has been a fascinating experience that has been significantly aided by the very useful and detailed criticisms of a number of reviewers, whose names are given below. We are indeed grateful to each of them. Their comments were universally helpful; any errors or deviations from their advice are our own responsibility. We are also deeply grateful for the highly professional editing of Patricia Zimmerman, for the financial and moral support of Art Bartlett and Dave Phanco, and for the technical assistance of Susie Pruett, Michael Fox, Matthew Fox, and Charlotte Hicks.

Steven W. Baldwin
Professor of Chemistry, Duke University

Eric Block
Professor of Chemistry, State University of New York, Albany

John I. Brauman
Professor of Chemistry, Stanford University

William D. Closson
Professor of Chemistry, State University of New York, Albany

Dennis P. Curran
Professor of Chemistry, University of Pittsburgh

William P. Dailey
Professor of Chemistry, University of Pennsylvania

William P. Jencks
Professor of Biochemistry, Brandeis University

George L. Kenyon
Professor of Chemistry and Pharmaceutical Chemistry,
University of California, San Francisco

John L. Kice
Professor of Chemistry and Dean of the Faculty of Science,
Mathematics, and Engineering, University of Denver

Doris Kimbrough
Professor of Chemistry, University of Colorado, Denver

David M. Lemal
Professor of Chemistry, Dartmouth College

Maher Mualla
Assistant Professor of Chemistry, Adrian College

C. Dale Poulter
Professor of Chemistry, University of Utah

David A. Shultz
Assistant Professor of Chemistry, North Carolina State University

Christopher T. Walsh
Professor of Biochemistry, Harvard Medical School

Chapter 1

Structure and Bonding in Alkanes

Organic chemistry—what is it and why do we have a full year course devoted to the subject? Chemistry itself is the study of the properties and transformations of matter, and organic chemistry is a subset of the discipline, concentrating on compounds that contain the element carbon. As we will learn during this course, the chemistry of carbon is far richer than that of any of the other elements. This richness results in part from the number of different types of strong bonds that carbon readily forms and in part from the ease with which many carbon atoms join together to form long chains. This diversity is apparent even in the forms of carbon itself, such as diamond and graphite. Diamond is hard and colorless, but graphite is soft and black. However, above all else is the fascination that goes with the study of organic chemistry, which is, indeed, the chemistry of life. The life forms on this planet—from algae to fish and ultimately to mammals, including human beings—are components of a complex web that is amazingly diverse in form and structure, but all of it is based on organic chemistry. Indeed, though distinct in detail, the fundamental chemistry that imparts life to algae is the same as that operating within human cells.

Today, we have a good understanding of the complex chemistry of living systems, as a result of an explosive growth in knowledge in the twentieth century. But how did we get to this point in organic chemistry? We can trace the origins of organic chemistry to the period before Christ, with the production of soap (from animal fats and plant oils) and of wood tar, a resin prepared from charcoal and used as an important article of trade. Sucrose, a crystalline sugar obtained from sugar cane, and plant extracts to be used as flavorings and perfumes also figured prominently as valued items in ancient civilization. Among pure compounds of interest to elite members of the Egyptian and Roman empires are two other classes: dyes with which to color their clothes and poisons with which to kill their enemies. Organic compounds were obtained from natural sources to address both needs: purple dyes from plants and a red dye from an insect; and extracts from the highly toxic foxglove and nightshade plants to be used in the complicated political intrigues that arose among the nonworking upper classes.

Morphine

The struggles of the alchemists (like the fellows in the frontispiece of this book) to make gold from less-valuable metals diverted attention from compounds of carbon until the sixteenth century when scientists began to turn their attention to practical (and less greedy) endeavors. In particular, Philippus Paracelsus, a German-Swiss holding a chair in medicine at the University of Basel, became convinced that drugs could be found that would relieve the suffering of the masses. Indeed, he was the first to recognize that opium, an extract of the poppy plant, could be used as a pain reliever. Yet even though it was recognized that naturally occurring compounds could be useful, organic chemistry did not flourish. The structures of organic compounds were not understandable until John Dalton advanced his atomic theory in 1803. At that time, many chemists began to focus their attention on organic compounds obtained from nature, and morphine, the active constituent of opium, was isolated in 1804 by a French chemist. Nonetheless, it was not until 1847 that the empirical formula of morphine was determined and another three-quarters of a century lapsed before the correct structure (shown in the margin) was proposed in 1925.

Indeed, it was not until Friedrich Wöhler's synthesis of urea in 1828 that it was realized that the molecules found in nature are composed of atoms and can be described and handled in the same way as minerals and metals. What an astounding generalization: that the hand holding this book, a green leaf, and the shell of a turtle, for example, have common chemical characteristics that in some ways resemble those found in rocks and ores. It was from this discovery that organic chemistry was born.

The lack of detailed structures for organic compounds did not prevent the chemists of the eighteenth and nineteenth centuries from applying the knowledge that they had to the preparation of new and much less costly dyes and to the isolation of compounds from plants for medical purposes. For example, the extracts of the toxic plant foxglove previously used as poisons were turned to beneficial use as a heart stimulant, and the important structural features of naturally occurring large molecules such as those in cotton, silk, and wool were recognized. As the number of useful compounds from nature increased, so did interest in organic chemistry.

Thanks to the curiosity and tireless drive of chemists in the twentieth century, we now have a detailed understanding of the inner workings of some cells and a fairly complete chemical picture of the organic chemistry behind the complex operations essential for multicell animals. With this knowledge, chemists have synthesized sophisticated compounds with properties that enhance the quality of life. The variety of uses for organic compounds is truly amazing, ranging from the natural and synthetic polymers that are the basis of many of the materials for clothing, housing, equipping, and transporting our populace to the modern wonder drugs, such as the penicillins, for the treatment of many human diseases.

The objective of this course, then, is to develop sufficient knowledge of organic chemistry that the structures and reactions of seemingly complicated systems such as organic polymers and penicillin antibiotics become understandable. The very complexity imparted to organic chemistry by the unique bonding states available to carbon demands that we begin slowly, developing a firm understanding of the nature of covalent bonds (and especially the multiple bonds) that are readily formed to carbon.

Like the early chemists, we must proceed step by step, learning the structures of various kinds of organic molecules and how they are determined before studying a variety of typical reactions that take place with these classes of compounds. When we have a good grasp of organic structure and reactivity, we can then see how modern chemistry is practiced: we will learn how syntheses of new compounds and materials are planned, about the properties of synthetic and naturally occurring polymers, and about the structure and function of natural products containing oxygen and nitrogen. We will then be ready to understand how organic chemistry works in nature through the use of cofactors as biological reagents, various substituent groups to help two reactants to recognize each other, and enzymes to control reaction rates. When these concepts have been mastered, we will be able to understand how organisms use the organic reactions presented in the earlier part of the book to store and release energy in fats and sugars, to transfer information in replication and reproduction, and to avoid disease states. Although the journey that you are about to undertake may seem long, and sometimes tedious, it leads to a fascinating goal and will provide rich intellectual satisfaction for the traveller who has worked hard along the way.

Organic chemistry, then, is the chemistry of carbon compounds. All organic compounds (including the compounds present in living organisms) are composed of atoms, as are the metals and salts used by the alchemists. But, unlike minerals, most of which contain metals, organic compounds are especially rich in carbon and hydrogen, and many also contain small amounts of oxygen, nitrogen, sulfur, phosphorus, and the halogens. The central role of carbon in organic chemistry follows directly from its position in the **periodic table** (which is inside the front cover of this book). The numerous types of molecules found in living matter are possible only because many different compounds can be formulated from a small number of elements by the joining of the component atoms into chains and rings to produce many different structures. This is most easily achieved with elements near the center of a row of the periodic table because such atoms require the addition or removal of several electrons to attain a stable, filled valence shell, and this is accomplished through the formation of chemical bonds to neighboring atoms.

For biological efficiency, naturally occurring compounds must also be as light as possible and must derive from common, easily accessible atoms. The lightest atom near the center of the periodic table is carbon, and the earth's atmosphere is a rich source of this element (now, from carbon dioxide and, in prebiotic times, from methane). Even silicon, from the same column of the periodic table as carbon, is much less versatile and has a much less rich chemistry than carbon. Thus, the observation that organic compounds are largely composed of carbon—bound to hydrogen, other carbon atoms, or other light elements such as nitrogen and oxygen—makes chemical sense.

If we are to understand the chemistry of the molecules of nature, we must first understand the chemistry of simple carbon compounds. In studying organic chemistry, we will explore the various modes of bonding of carbon in organic compounds. We will begin with the simplest compounds: those that contain only carbon and hydrogen and are thus called **hydrocarbons.** In this chapter, we consider the chemistry of a family of hydrocarbons, the alkanes, that contain only single bonds between carbon atoms.

Hydrocarbons are familiar to us in our everyday life: the natural gas that we burn to cook our food, the liquid gasoline used to power our vehicles, and the bottles for soft drinks, cooking oils, and shampoo—all are hydrocarbons. Clearly, different hydrocarbons exist at room temperature as gases, liquids, or solids. Their structures at the molecular level, about which we will learn in this chapter, are similarly diverse. We are quite aware of the sharply contrasting properties of two forms of carbon itself: diamond is a brilliant, extremely hard crystal, whereas graphite (like that in pencils) is extremely soft, leaving a trail of carbon particles as it is drawn across the surface of a sheet of paper. These physical differences are due to the contrasting three-dimensional cross-linked net of chemical bonds in diamond and the less highly interconnected array of planes of atoms in graphite. Another form of carbon has recently been discovered: isolated from soot, this form comprises a family of compounds called fullerenes because of their structural resemblance to the geodesic domes first designed by Buckminster Fuller. This is a family of nearly spherical clusters, one of which is shown with the structures of carbon in diamond and graphite in Figure 1-1.

In this chapter and the next one, we will learn how changes in bonding can dramatically alter the shapes of hydrocarbons. We will learn about the hybridization of carbon and its consequences for the structure of organic molecules. We will learn how atomic and molecular orbitals combine to form different kinds of chemical bonds: covalent and ionic, sigma (σ) and pi (π) bonds. We will learn how to draw the structures of molecules in three dimensions and to calculate sites of formal charge. We will learn how to recognize functional groups and structural isomerism in hydrocarbons and to correlate physical properties with the functionality. We will also learn how to rank various hydrocarbons according to relative stability, to recognize particularly stable molecules from their structures and chemical formulas, and to unambiguously name specific molecules.

Before we consider bonding in hydrocarbons in detail, it is essential that we understand two very important principles. The first is that a molecule exists because of chemical bonds; that is, because the favorable attraction of its negatively charged electrons for its positively charged nuclei exceeds the electrostatic repulsions arising from the interactions of nucleus

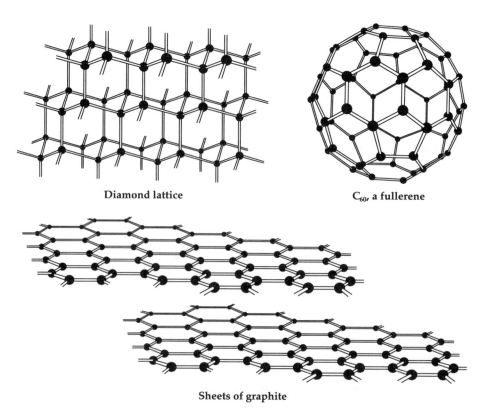

Diamond lattice C_{60}, **a fullerene**

Sheets of graphite

▲ FIGURE 1-1

Three-dimensional representations of subunits of three forms of carbon.

with nucleus and electron with electron. The balance of these forces produces a chemical bond and determines the distance between adjacent bonded nuclei in a molecule. In a chemical bond, a pair of electrons mutually attracted to both nuclei is found near the "line" connecting the atomic nuclei.

The second principle concerns the repulsive forces (electrons versus electrons) that dominate the interactions between nonbonded atoms in a molecule. This interaction is between the electron cloud of one atom and that of another. The nucleus-electron $(+/-)$ attractive interactions are weaker than the electron-electron $(-/-)$ repulsive interactions when atoms are separated by distances longer than those of typical chemical bonds. These repulsions fix the relative positions of nuclei that are not connected by a chemical bond. To minimize repulsion, the bonds emanating from a single atom are directed as far as possible from each other.

These principles lead to specific molecular shapes. For example, when a carbon atom is bound to four other atoms (as in **methane,** CH_4, a hydrocarbon composed of carbon surrounded by four hydrogens), the molecule adopts an arrangement in which the four hydrogen nuclei (each located at a fixed distance from the carbon atom) are as far from each other as possible. This produces a tetrahedron-shaped molecule (Figure 1-2) with an HCH angle of about 109.5°. When carbon has only three neighboring atoms, minimal electrostatic repulsion again places the three neighbors as far from each other as possible. This placement is achieved when all three bonded atoms lie in the same plane as the carbon atom to which they are bound (a trigonal planar arrangement like that shown in Figure 1-2 for formaldehyde), with an HCH angle of about 120°. When carbon has but two closest neighbors placed as far as possible from each other, a linear molecule with a bond angle of 180° results, as is found, for example, in carbon dioxide.

These simple principles have become progressively more quantitative through the years. Theoretical chemists have given not only numerical justification for their use, but also a description that permits a more-detailed understanding of the location of electrons in molecules. Although we do not deal with their quantitative aspects in this book, these simple principles of molecular attraction and repulsion are very helpful in describing chemical bonds in organic compounds. Because the bonds in molecules are between atoms, we must first understand the atomic orbitals from which the molecular orbitals used in bonding are constructed.

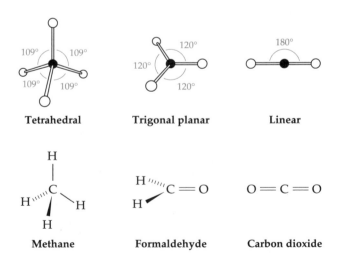

▲ FIGURE 1-2
Simple representative tetrahedral, planar, and linear carbon compounds.

1-1 ▸ Atomic Structure

Atomic orbitals describe probability surfaces within which an electron is likely to be found. Precise calculations have been made to describe the shapes of the atomic orbitals of hydrogen because it is the simplest element: only a single proton need be accommodated in that atom's nucleus and a single electron located in its possible atomic orbitals. These shapes, calculated for hydrogen, are assumed to apply equally in describing the atomic orbitals of heavier elements. These calculations produce the different shapes for the hydrogen atomic orbitals shown in Figure 1-3: we will be most concerned with the spherical *s* orbital and the three propeller-shaped *p* orbitals directed along the *x*, *y*, and *z* axes of a molecule. The electrons in elements in the first row of the periodic table (hydrogen and helium) can be accommodated by *s* orbitals (1*s*), but those in the second row also require *p* orbitals (1*s*, 2*s*, and 2*p*). Five spatially more complex *d* orbitals must be considered for elements appearing in the third or subsequent rows of the periodic table but need not be considered in describing first- or second-row atoms.

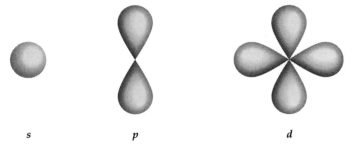

s p d

▲ FIGURE 1-3
Shapes of hydrogenic atomic orbitals.

Complete occupancy of any set of these orbitals (for example, the 1*s* orbital, the 2*s* and 2*p* orbitals, the 3*s*, 3*p*, and 3*d* orbitals) leads to a spherical distribution about the central atom. This is easy to grasp in considering the *s* orbitals, but it also holds for the three equivalent *p* orbitals with their propeller axes disposed along three orthogonal directions (Figure 1-4) and for the *d* orbitals. Thus, the geometric sum of completely filled p_x, p_y, and p_z orbitals is roughly a sphere. At the center of this sphere (at the nucleus), the probability of encountering an electron is negligible. The nucleus of the atom is therefore said to be at a **node** of each *p* or *d* suborbital, a position at which electron density is zero. Although the atomic orbitals of elements in each row of the periodic table have approximately these same shapes, the average radius of the *s* orbital and the length of the lobes of the *p* or *d* orbitals become larger as the number of electron shells increases in the progression down the table.

According to the **Pauli Exclusion Principle,** each electron must have a distinct set of principal, secondary, azimuthal, and spin **quantum numbers:** that is, each electron must be unique. The first three quantum numbers define the orbital; for example, $2p_x$. The last defines the relative spin of the electron in the orbital: sometimes this is indicated by an arrow or by a plus or a minus sign, but, because the absolute spin is arbitrary, these labels are often omitted. Because there are only two possible spin quantum numbers for an electron, an orbital (or suborbital) is completely filled by two electrons of opposite spin. Thus, an *s* orbital can accommodate exactly two electrons, and each of the three *p* suborbitals can accommodate two electrons

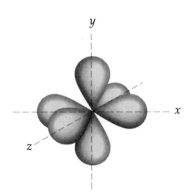

▲ FIGURE 1-4
A three-dimensional representation of mutually orthogonal *p* orbitals. The p_x orbital is directed along the *x* axis; the p_y along the *y* axis; and the p_z along the *z* axis.

for a total of six. The periodic table describes, in each row, the number of electrons needed to completely fill each of the orbital types encountered in that row: two electrons fill the valence shell of a first-row element; eight more are needed for a second-row element; eighteen more are required for a third-row element; and so forth. When each orbital type is completely filled, an atom is said to have a complete, or filled, **valence shell.**

As mentioned earlier, first-row elements accommodate electrons within a $1s$ orbital: here, the number 1 refers to the principal quantum number for these electrons, which is identical with the number of the row in which the atom is found, and the letter refers to the shape of the atomic orbital, s for spherical atomic orbitals. Hydrogen has one electron in this $1s$ orbital, and so we describe hydrogen's atomic **electron configuration** as $1s^1$, in which the superscript specifies the number of electrons in the $1s$ orbital. Similarly, helium's electron configuration is $1s^2$. Because a first-row element has only one orbital ($1s$), the two electrons of helium completely fill its $1s$ valence shell.

Second-row elements have electrons in $1s$, $2s$, and $2p$ orbitals. As in the first-row elements, each s orbital can hold two electrons. Similarly, each of the p suborbitals (directed respectively along the x, y, and z axes) can hold two electrons. A filled second-row valence shell is therefore attained when an atom has ten electrons, two in the $1s$ orbital of the first shell and eight in the second shell (two in the $2s$ and six in the three $2p$ orbitals).

Because carbon is so important to organic chemistry, our primary focus is the **atomic structure** of carbon. Its atomic number (6) tells us that a neutral carbon atom has six electrons. Using hydrogenlike atomic orbitals in the conventional way for describing electronic orbital occupancy, we place these six electrons in the energetically lowest lying orbitals. These are the spherical $1s$ and $2s$ orbitals and the propeller-shaped $2p$ orbitals.

To clearly see how these electrons are accommodated in carbon, let us compare carbon's atomic structure with those of other elements in the first and second rows of the periodic table. **Valence electrons** are those present in the last, incomplete valence shell. Hydrogen and helium contain only s electrons. Thus, the hydrogen atom can be described as having a single valence electron in a $1s$ orbital. The helium atom has the same orbital doubly occupied, which means that it has a filled electronic shell and has no need for access to additional electrons for a completed valence configuration. For helium to take on additional electrons would require the use of orbitals of much higher energy. Helium, therefore, does not enter into chemical bonding with other atoms and is described as an **inert gas.** Any molecule or ion with a completely filled valence shell—that is, with the electronic configuration of an inert gas—is particularly stable.

The next heavier elements of the periodic table must accommodate electrons in the $2s$ and $2p$ orbitals. For example, lithium has the electronic configuration given in Figure 1-5. Two electrons are accommodated in its $1s$ orbital and a third electron is in its $2s$ orbital. A $2s$ orbital has spherical symmetry, but it has a larger radius than the $1s$ orbital. The electron in the $2s$ orbital of lithium is a valence electron because it participates in bonding. Lithium metal is neutral because the number of electrons (3) is exactly equal to the number of protons in its nucleus (3). An uncharged atom is sometimes called atomic, metallic, or elemental. If the $2s$ electron of lithium is removed, a positively charged ion (Li^+) results. Because Li^+ has a completely filled valence configuration ($1s^2$), it is relatively stable in this ionic state.

The next element (beryllium, atomic number 4) can accommodate a second electron in the $2s$ orbital, but boron, atomic number 5, must place its fifth electron in a $2p$ orbital. In the progression from boron to neon (at the

H: $1s^1$

He: $1s^2$

Li: $1s^2 2s^1$

Be: $1s^2 2s^2$

B: $1s^2 2s^2 2p^1$

C: $1s^2 2s^2 2p^2$

N: $1s^2 2s^2 2p^3$

O: $1s^2 2s^2 2p^4$

F: $1s^2 2s^2 2p^5$

Ne: $1s^2 2s^2 2p^6$

▲ **FIGURE 1-5**
Electronic configurations of first- and second-row elements. The number preceding each letter is the principal quantum number that defines the valence shell, the letter designates the orbital shape, and the superscript specifies the number of electrons in the orbital or suborbital.

right-hand side of the second row of the periodic table), six electrons are sequentially added to these orthogonal $2p$ orbitals. Carbon's atomic structure can thus be written: $1s^2, 2s^2, 2p^2$.

EXERCISE 1-A

Specify the atomic orbitals (using $1s$, $2s$, $2p$, $3s$, $3p$, $3d$, etc., orbitals) and their occupancy to define the electronic configuration of each of the following atoms or ions:

(a) atomic boron

(d) elemental phosphorus

(b) metallic magnesium

(e) S^{2-}

(c) Mg^{2+}

1-2 ▸ sp^3 Hybridization at Carbon

We are now ready to specifically consider the electronic configuration of carbon. As mentioned in Section 1-1, carbon has six electrons: two in the filled $1s$ shell and four in the only partly filled $2s/2p$ shell. As we know, the outer shell, which can be partly or completely filled, is the valence shell; accordingly, the electrons in that shell are valence electrons. You probably learned several theories in your first chemistry course that suggest how valence electrons are distributed among the s and p orbitals. For example, we can formulate carbon as in Figure 1-5: that is, with two core electrons in its filled $1s$ orbital, with two spin-paired electrons (as required by the Pauli Principle) in its filled $2s$ orbital, and with the remaining two valence electrons singly distributed in two of the three $2p$ orbitals in accordance with **Hund's Rule.** This rule states that, when possible, electrons tend to singly occupy orbitals of identical energy. Alternatively, the four valence electrons of carbon can be distributed singly among the four possible suborbitals available to second-row elements ($2s$, $2p_x$, $2p_y$ and $2p_z$), as in Figure 1-6. In either arrangement, carbon's electronic configuration is far from a filled-shell, inert-gas configuration. Four electrons would have to be either lost or gained for carbon to have a completely filled valence shell.

A hydrogen atom has one electron (in a $1s$ orbital). By the paired association of the four electrons of four hydrogen atoms with one carbon, the electronic configuration requirements of each atom can be met. In CH_4, carbon is associated not only with its own four electrons, but also with each of the single valence electrons contributed by each of the four hydrogens, giving carbon access to a total of eight electrons and filling carbon's valence shell. In like fashion, each hydrogen atom has access to one of carbon's electrons, providing the two electrons needed to fill its valence requirement. The driving force for this favorable association of hydrogen atoms with carbon is the electrostatic attraction of the electrons of each atom for the nucleus of its partner in the chemical bond. Carbon is almost always limited to four bonding partners because the addition of a fifth partner would require an electron at the much higher energy level of a $3s$ orbital.

Each C–H bond in methane is equivalent to the others. This can be understood with a model based on the electronic configuration formulated as at the top of Figure 1-6. In this model, there are no positions at which four hydrogen atom $1s$ orbitals can overlap effectively with the carbon orbitals. In particular, it would be harder for a hydrogen atom to approach the

C: $1s^2$ $\underbrace{2s^1 2p_x^{\,1} p_y^{\,1} p_z^{\,1}}$

Four sp^3-hybrid orbitals

▲ **FIGURE 1-6**
Electronic configuration of carbon. Four atomic orbitals are mixed to form four sp^3-hybrid orbitals.

smaller-radius 2*s* orbital than to approach the more-elongated 2*p* orbitals. Furthermore, the two lobes of a 2*p* orbital have equal electron density on each side of the nucleus; thus, at best, only half of this electron density could be used in bonding if the 2*s* or 2*p* orbital character does not change as the bond is formed.

This problem can be solved, however, if the carbon orbitals are mixed to form **hybrid orbitals.** For example, the 2*s* and the three 2*p* orbitals can be mixed to form a new type of orbital referred to as an sp^3-**hybrid.** The mixing of four atomic orbitals (one *s* and three *p* orbitals) produces four sp^3-hybrid orbitals: $1s^2(2sp^3)^4$. The hybrid orbitals must maintain the overall symmetry of the atomic orbitals from which they were composed. Because these hybrid orbitals must occupy separate regions in space, they are directed as far as possible from each other. Simple geometry tells us that this is best accomplished if the hybrid orbitals point toward the corners of a pyramid, creating the tetrahedral geometry illustrated in the margin. The directionality problem is then solved, because each hybrid orbital points in a defined direction to a region where good overlap with a hydrogen 1*s* orbital is possible. With the hybrid orbitals maximally separated, a 109.5° HCH angle is formed between each of the fully equivalent C–H bonds.

Because these hybrid orbitals are composed substantially (three parts out of four) of *p* orbitals, they are elongated, but the fractional *s* contribution (1/4) fattens them. The *s* character of the hybrid orbital gives it finite electron density at the nucleus. The larger the fraction of *s* character, the more electronegative is the hybrid orbital. Thus, the shape of these sp^3-hybrid orbitals is as shown in the margin.

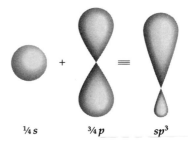

¼ *s*   ¾ *p* sp^3

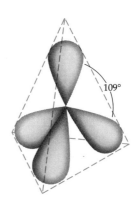

109°

1-3 ▸ Covalent Bonding

We now allow an electron in each hybrid carbon orbital to interact with a hydrogen 1*s* electron to form a **chemical bond.** Because bonding consists of the sharing of electrons through **orbital overlap,** a spherical hydrogen atom binds at the terminus of each of the four directional sp^3-hybrid orbitals of carbon (Figure 1-7). The four electrons available from the four hydrogen atoms satisfy the electronic requirement of carbon and give the resulting molecule (CH_4) an electronic stability comparable to that of an inert gas.

The attainment of a filled-valence-shell configuration is both a goal and a limitation. For more electrons to be associated with carbon, they would have to be placed in third-level (3*s*, 3*p*, or 3*d*) orbitals. Although it is energetically favorable for electrons to be associated with the positively charged nucleus, accommodating them in an orbital beyond the valence shell would place them farther from the nucleus, which is thermodynamically costly.

In an alternative depiction of the structure of CH_4, two dots placed between two atoms represent the shared electrons in the overlapping orbitals. The resulting picture, called a **Lewis dot structure,** is shown in Figure 1-8 (on page 10). In the Lewis dot structure of methane, each pair of dots represents one of the valence electrons of carbon and the valence electron of hydrogen. Because these electrons are shared by the two atoms (carbon and hydrogen), they form what is called a **covalent bond.**

If either atom in a covalent bond has a greater tendency to attract the shared electrons than the other, the electrons in the covalent bond shift (are polarized) toward the more-electronegative atom. **Electronegativity** measures the tendency of a particular atom to attract electrons. The most-electronegative atoms are at the top and at the right of the periodic table. In the progression from left to right, electronegativity increases. The order of

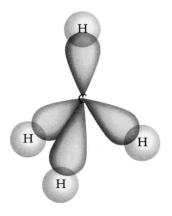

▲ FIGURE 1-7
A three-dimensional (tetrahedral) representation of methane. Four equivalent C–H bonds are formed, each by the overlap of a carbon sp^3-hybrid orbital with a hydrogen 1*s* orbital. The bonds are directed as far from each other as possible to minimize electron repulsion.

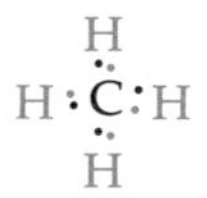

$$H \cdot \overset{\displaystyle H}{\underset{\displaystyle H}{\overset{\cdot\cdot}{C}}} \cdot H$$

▲ FIGURE 1-8
Lewis dot structure of methane. Four of the electrons are shown in color to emphasize the concept that the covalent bonds of methane are formed by the sharing of four electrons of carbon with the single valence electron of four hydrogen atoms. Note that no geometry is implied by a Lewis dot structure.

electronegativity of second-row elements is carbon < nitrogen < oxygen < fluorine. Electronegativity also increases in the progression from the bottom to the top of a column. For example, among the halogens, fluorine is most electronegative; that is, in order of electronegativity, iodine < bromine < chlorine < fluorine. These trends result from the greater nuclear (positive) charge as the atomic weight increases across a given row of the periodic table and from the greater distance of the valence electrons from the nucleus as atomic weight increases down a column.

Carbon and hydrogen have very similar electronegativities, and so the sharing of the electrons in a C–H covalent bond (Figure 1-8) is nearly equal. Thus, C–H covalent bonds in hydrocarbons are **nonpolar;** there is very little charge polarization associated with these bonds. When a highly electronegative atom such as fluorine is attached to carbon, however, the electrons in the C–F covalent bond are not shared equally. Instead, a partial shift of electrons occurs, placing a partial negative charge on fluorine and leaving carbon partially positively charged. The periodic table can be used to predict trends in electronegativity and hence when **polar covalent bonding**—that is, unequal sharing of the electrons in a covalent bond connecting two different atoms—is likely. The chemical and physical consequences of bond polarization will be considered in more detail in Chapter 3.

EXERCISE 1-B

Based on the relative electronegativities of the relevant atoms, choose the more-polar bond in each of the following pairs of compounds:

(a) $HO-H$ or H_2N-H

(b) CH_3-H or CH_3-F

(c) H_3C-OH or H_3C-NH_2

(d) CH_3-OH or CH_3-SH

(e) H_3C-OH or H_3C-Br

1-4 ▸ Ionic Bonding

Because they have nearly equal electronegativity, hydrogen and carbon equally share the electrons in a covalent bond connecting them. The alternative formulation shown in reaction 1, where electrons are not equally shared, must therefore be incorrect.

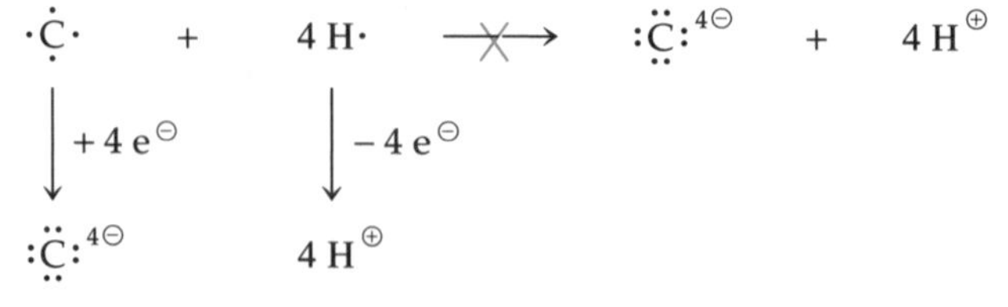

$$\cdot \dot{\underset{\cdot}{C}} \cdot \quad + \quad 4\,H\cdot \quad \xrightarrow{\;\;\times\;\;} \quad :\overset{\cdot\cdot}{\underset{\cdot\cdot}{C}}:^{4\ominus} \quad + \quad 4\,H^{\oplus} \tag{1}$$

$$\Bigg\downarrow +4\,e^{\ominus} \qquad\qquad \Bigg\downarrow -4\,e^{\ominus}$$

$$:\overset{\cdot\cdot}{\underset{\cdot\cdot}{C}}:^{4\ominus} \qquad\qquad 4\,H^{\oplus}$$

The ions shown at the right-hand side of the reaction are those that would be formed if the valence requirement of carbon were satisfied by the addition of four electrons taken from four hydrogen atoms, thus forming four protons. This carbon would bear four negative charges, having acquired four extra electrons. Its formal charge is obtained by comparing the number of valence electrons in the neutral atom with the number of electrons borne by the ion.

The **formal charge** (FC) of any atom is calculated by comparing the number of valence electrons in the neutral atom with the sum of the number of unshared electrons plus half the number of shared electrons. This is because the number of valence electrons is equal to the number of positive charges in the nucleus not offset by electrons in the filled levels of lower quantum number.

Formal charge = number of valence electrons of atom
                     − number of unshared electrons
                     − 1/2 number of shared electrons

Thus, the formal charge of carbon in $::C::$ and in CH_4 can be calculated:

$::C::$ FC = 4 valence electrons − 8 unshared electrons
                − 0/2 shared electrons = −4

     H
H:C̈:H   FC = 4 valence electrons − 0 unshared electrons
     Ḧ   − 8/2 shared electrons = 0

EXERCISE 1-C

Calculate the formal charge of each second-row atom in the following Lewis dot structures.

      H
(a) H:C̈:C:::N:
      H

(c) Ö::Ö:Ö:

      H H
(e) H:N̈:C̈:H
      H H

    :O:
(b) :Ö:C:Ö:

       H :O: H
(d) H:C̈:C:C̈:H
       H   H

The formation of ions, such as that proposed in reaction 1, results from **electron transfer** from the less- to the more-electronegative atom. This transfer fails to occur in C–H bonds because the two participating atoms, carbon and hydrogen, have similar electronegativities. On the other hand, electron transfer does occur in bonds between atoms that differ more significantly in electronegativity. For example, as shown in reaction 2, chlorine (a halogen atom) combines with sodium (a first-column element); by electron transfer, two ions are formed. Chlorine, lacking only one electron to achieve an inert-gas configuration, has a high electronegativity and actively seeks to acquire this last electron, becoming the negatively charged chloride ion. Similarly, atomic sodium has one extra electron beyond an inert-gas configuration: the loss of this electron to form a sodium cation is therefore easy. (Note that the electronic configuration of Cl^- in reaction 2 is analogous to C^{4-} in reaction 1 and that the sodium ion configuration in reaction 2 is parallel to that of a proton in reaction 1.)

$$:\ddot{C}l\cdot \quad + \quad Na\cdot \quad \longrightarrow \quad :\ddot{C}l:^{\ominus} \quad + \quad Na^{\oplus} \qquad (2)$$

$$\downarrow +e^{\ominus} \qquad\qquad \downarrow -e^{\ominus}$$

$$:\ddot{C}l:^{\ominus} \qquad\qquad Na^{\oplus}$$

The significant difference between these two reactions, which causes reaction 1 to fail and reaction 2 to proceed, is that the difference in electronegativity (see the periodic table) is much greater between sodium and chlorine than between carbon and hydrogen.

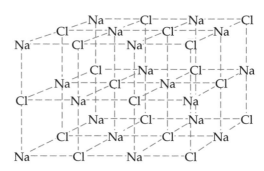

▲ FIGURE 1-9
Crystal structure of sodium chloride.

Because the interaction of sodium and chlorine atoms produces ions of opposite charge, there is a strong **electrostatic attraction** between them. In a solid composed of these ions, Na^+ and Cl^- are present in a 1:1 ratio. The electrostatic forces holding an ionic crystal together do not define a particular bond, however, and the electrons are not shared between two atoms (as they are in the bonding of methane). Instead, each positively charged sodium ion is surrounded by six negatively charged chloride ions (and each chloride is surrounded by six sodium ions), resulting in strong electrostatic attraction. The electrostatic attraction between two oppositely charged ions in an ionic crystal is called an **ionic bond** (Figure 1-9).

In contrast, nonionic organic molecules are held together in molecular crystals by much weaker **van der Waals attractions,** which result from the attraction of the bonded electrons of one molecule to the nuclei of another. The amount of energy required to disrupt the strong electrostatic forces in an ionic crystal is much greater than that needed to interfere with van der Waals forces in an organic molecular solid. The melting points of (ionic) salts are thus often very high (NaCl: mp 801 °C), whereas many organic compounds have sufficiently low melting points that they exist as liquids or even as gases at room temperature.

Because of the position of carbon near the center of the periodic table and its relatively low electronegativity, ionic bonds to carbon are only infrequently encountered in organic chemistry. Covalent bonding—that is, the nearly equal sharing of electrons between two atoms—is the norm. Polar covalent bonds are encountered when shifts in electron density occur; that is, when hydrogen is replaced by a more- (or sometimes less-) electronegative atom in the more-complex structures that we will consider later in this book.

1-5 ▸ Alkanes

Organic chemistry would be simple indeed, and in fact quite boring, if methane were the only hydrocarbon that could be constructed from carbon and hydrogen. Luckily, carbon bonds readily to itself, forming structures that can contain tens or hundreds or even millions of carbon atoms.

Suppose that two sp^3-hybridized carbon atoms approach each other for bonding. Because hybrid orbitals are directional, only one orbital can point directly toward the other atom. As long as the two atoms are sp^3-hybridized, a single covalent bond is formed as the result of overlap of the hybrid orbitals. When two carbon atoms form a covalent bond, three other, vacant bonding sites are available at each carbon atom (Figure 1-10). When hydrogen atoms are covalently bound to carbon at these sites, the stable

▲ FIGURE 1-10

A covalent bond formed by overlap of sp3-hybridized carbon orbitals.

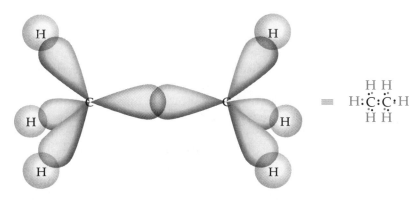

▲ FIGURE 1-11

Covalent bonding in ethane, shown by overlapping orbitals at the left and by the Lewis dot structure at the right.

molecule shown in Figure 1-11, H_3CCH_3 **(ethane),** is formed. This structure satisfies the valence-shell requirements of each carbon and hydrogen atom, as the Lewis dot structure shows.

Hydrocarbon structures can become increasingly more complex, forming a family of related compounds called **alkanes,** which contain only C–C and C–H single bonds. With the formation of each new carbon–carbon covalent bond, two hydrogen atoms are needed per carbon, plus an additional hydrogen atom at each end of the chain. This family therefore has an overall molecular formula of C_nH_{2n+2}.

Sigma Bonding

Each C–C and C–H bond consists of a nearly equal sharing of electrons between two atoms of similar or identical electronegativity. The bond is directional, with the region of orbital overlap between the two atomic orbitals located along the line (axis) that connects the nuclei of the two atoms. This bond is referred to as a **σ** (sigma) **bond;** that is, one in which electron density is arranged symmetrically along the axis connecting the two bonding atoms. Because electron density lies along this axis, the extent of orbital overlap is not significantly affected if the atoms rotate about the internuclear axis. Thus, **free rotation** can occur about a σ bond without appreciably affecting its bond strength because of the very low barrier to this rotational motion. For example, holding one of the carbons of ethane fixed while rotating the other need not break the C–C or any C–H bond. Because all bonds in ethane are σ bonds, the participating atoms can rotate freely about each of them so that an infinite number of three-dimensional structures differing by this rotation are available.

The σ bond connecting two carbon atoms is formed by the overlap of two sp^3 orbitals, one from each carbon. In ethane, the three σ bonds that connect each carbon atom with three hydrogen atoms are formed by sp^3–1s

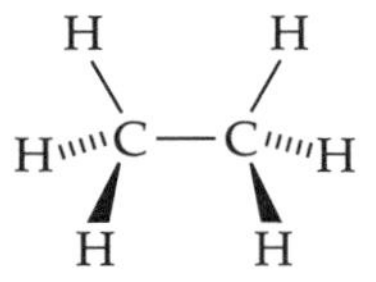

▲ **FIGURE 1-12**
A three-dimensional sawhorse
representation of ethane.

overlap; these bonds are shorter (1.10 Å) than the sp^3–sp^3 σ bond between carbons (1.54 Å).

Different molecular arrangements constituted from the same atoms are called **isomers.** In particular, **rotational isomers** are one class of **stereoisomers,** which differ only in the way that atoms are arranged in space. We will present stereoisomerism and related topics in Chapter 5.

Drawing Three-Dimensional Structures

We can depict the three-dimensional character of the sp^3-hybridized atoms in alkanes by using single lines to represent σ bonds lying in the plane of the page, solid wedges to indicate those coming toward the observer, and hatched lines to indicate those going away from the observer. One of many possible arrangements of the hydrogen atoms in ethane is shown in Figure 1-12. The arrangement of wedges and hatched lines resembles that of the legs of a sawhorse in this three-dimensional structure. This method for depicting structure is therefore called a **sawhorse representation.** It is useful at this stage to construct a three-dimensional model of the structure of ethane to assure yourself that a tetrahedral geometry can be maintained at each carbon atom even though there is free rotation about the C–C bond. You will often find that making such three-dimensional molecular models helps you to visualize the structure of molecules much more clearly than can be done only by reading the text. You are encouraged to do so whenever a new type of molecule is described.

EXERCISE 1-D

With the three-dimensional structure of ethane in mind, replace one of the hydrogen atoms with a chlorine atom. Draw at least six three-dimensional representations that place the chlorine atom in at least six different regions in space. With a space-filling model in hand, orient the molecule to correspond to the sawhorse structures that you have drawn. Notice that all six representations are pictures of the same molecule.

Hydrocarbon skeletons can also be represented by a line notation in which each line segment represents a carbon–carbon bond, as shown at the right in Figure 1-13. No C–H bonds are shown; their presence is inferred as needed to meet the valence requirement of carbon. When a compound contains other atoms besides carbon and hydrogen, all such atoms are drawn in specifically. This convention does not show three-dimensional structure: it is merely a useful way for depicting structural isomers, which are the subject of the next section. Although this is a convenient shorthand, it often conveys less graphic information than the space-filling representation of Figure 1-12 but does clearly indicate the attachment of one atom to another, or **connectivity,** in a molecule.

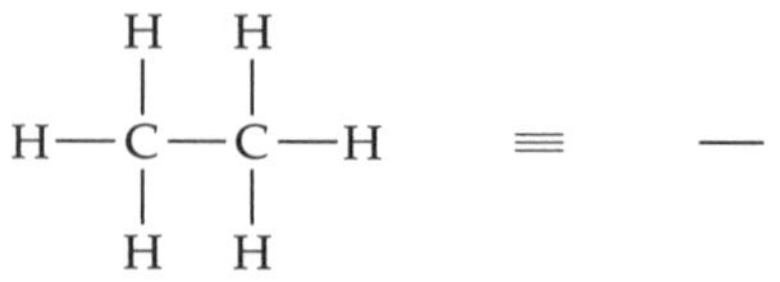

▲ **FIGURE 1-13**
The line notation at the right is a shorthand method for indicating bonding in ethane.

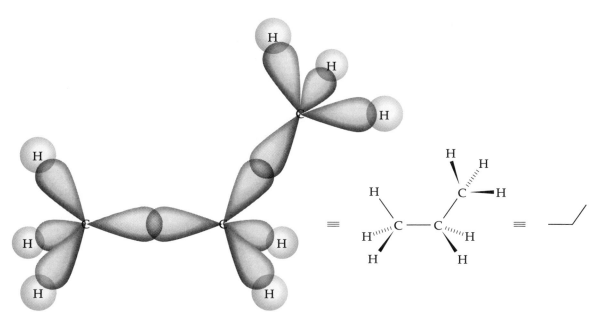

▲ FIGURE 1-14

Three structural representations of propane.

Structural Isomers

It is a relatively simple matter to conceive of higher analogs of the hydrocarbons discussed so far. For example, if we were to add a carbon atom and two hydrogens to ethane, we would obtain the structure shown in Figure 1-14, which represents only one of the many possible stereoisomers of **propane,** C_3H_8.

By adding a fourth carbon atom (and two more hydrogens), we obtain C_4H_{10}, **butane.** There are two ways in which four carbon atoms can be bound to each other:

$$C_4H_{10} \qquad C_4H_{10}$$

The C_4H_{10} isomer at the left (in which all the carbons are arranged in a row) is derived from propane by the addition of a carbon atom to one end of the skeleton, whereas the isomeric structure shown at the right, also C_4H_{10}, is obtained by the addition of a carbon atom to the central carbon of propane.

These two molecules do not have the same sequence of chemical bonds no matter how they are oriented in space. This is easily seen if we consider that one of the carbon atoms of the structure at the right is attached to three other carbon atoms, whereas each carbon atom in the structure at the left is attached to, at most, two other carbon atoms. Although the normal tetrahedral bond angles and lengths are maintained, and although the overall formulas are identical, these two compounds are different molecules: they are **structural isomers** in which the carbon backbones differ.

The number of possible structural isomers increases as the number of carbon atoms increases. It is left as an exercise for you to determine the number of such structural isomers that can be obtained with five, six, or seven carbons.

EXERCISE 1-E

Draw all possible isomeric carbon skeletons with the following overall formulas:

(a) C_5H_{12} (b) C_6H_{14} (c) C_7H_{16}

(Hint: It will help greatly if you draw the skeletal structures first. Then add the number of hydrogens necessary to complete the valence of each carbon.)

1-6 ▸ Cycloalkanes

Structure and Formula

An alternative mode of bonding with three or more carbon atoms is also possible. For example, the structure of ethane can be modified by the binding of an additional carbon atom not to just one carbon (as in Figure 1-14) but to two. This bonding produces the structure shown in Figure 1-15, a cyclic, three-carbon compound called **cyclopropane.** Close inspection reveals a difficulty with this structure. Because three points determine a plane, the three carbon nuclei in cyclopropane must be coplanar. Simple geometry requires the sum of the CCC angles within this cyclic structure to be 180°, and so each CCC angle must be 60°. However, all three carbon atoms are formally sp^3-hybridized and would prefer the normal tetrahedral bonding angle of about 109°. This deviation of 49° from the normal bonding angle at an sp^3-hybridized atom confers appreciable **ring strain** on this molecule and destabilizes it. This strain causes cyclopropane to have a higher potential energy content than it would otherwise have: the "extra" energy released when cyclopropane is burned, called its strain energy, is about 28 kcal/mole.

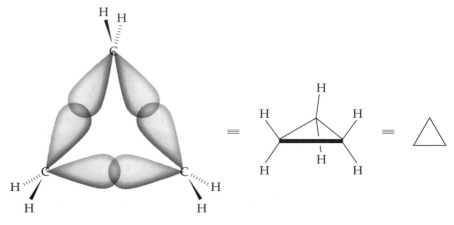

▲ FIGURE 1-15
Three structural representations of cyclopropane.

A CYCLOPROPANE–CONTAINING INSECTICIDE

Few naturally occurring compounds contain cyclopropane rings. An exception is the pyrethrins, illustrated by pyrethrin II. These compounds are found in marigolds, a class of flowers also known as pyrethrums (from which the name is derived), and are potent insecticides. Although these compounds are isolated from natural sources, they cause severe allergic dermatitis and systemic allergic reactions in some people.

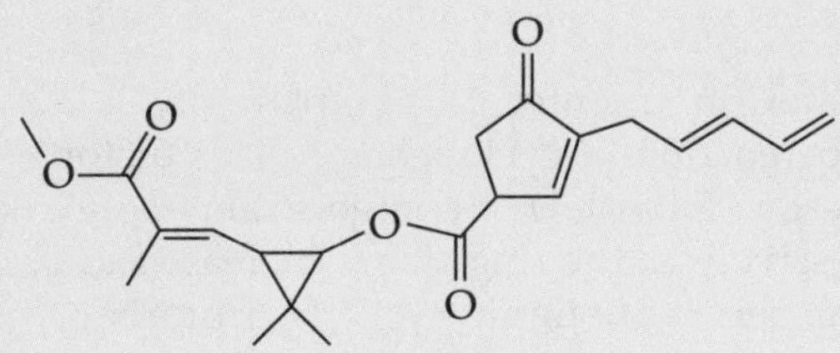

Pyrethrin II

Notice that constraining three carbons to a ring affects the atomic formula of cyclopropane. It is now C_3H_6 instead of C_3H_8 as in the straight-chain hydrocarbon propane. Each time a ring is formed, two fewer hydrogens are needed to satisfy carbon's valence requirement. As mentioned in Section 1-5, the overall formula for isomeric hydrocarbons composed only of sp^3-hybridized atoms is C_nH_{2n+2}. A cycloalkane with one ring has the formula C_nH_{2n}. Any deviation from the overall formula for molecules containing only σ bonds must be due to the introduction of rings: each time that a ring is formed, two fewer hydrogens are needed. Thus, we can immediately recognize from the formula of an alkane whether it includes one or more rings.

EXERCISE 1-F

Assuming that each of the following formulas represents a hydrocarbon containing only σ bonds, predict whether the compound has zero, one, or two rings and draw at least three possible carbon skeletons that correspond to your prediction.

(a) C_5H_{10} (b) C_5H_8 (c) C_6H_{12} (d) C_6H_{10} (e) C_7H_{14}

(Hint. It will be easiest to start with the largest ring possible and then to successively make the ring smaller and smaller.)

Ring Strain

Let us consider ring strain in cycloalkanes with four carbons. Two saturated cyclic C_4 isomers, **methylcyclopropane** and **cyclobutane,** along with the five-carbon **cyclopentane,** are shown in Figure 1-16. Methylcyclopropane has the same ring structure as that of cyclopropane itself, and the same geometric considerations apply. With cyclobutane, geometric arguments predict that the bond angles will be 90° if all four carbon atoms are in a plane and all C–C bond lengths are equal. In fact, however, because three (not four) points determine a plane, the carbon atoms in cyclobutane can deviate from planarity to prevent unfavorable spatial interactions (which we will

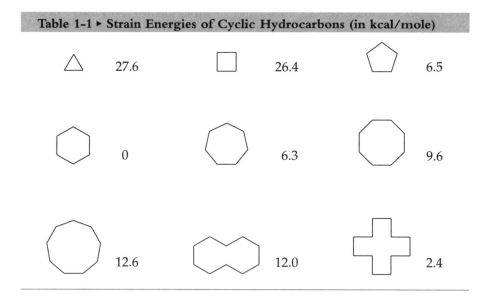

▲ FIGURE 1-16

Ring conformations in cycloalkanes: planar methylcyclopropane, folded cyclobutane, and folded cyclopentane.

consider again later, in Chapter 5). Nonplanarity is also possible in the larger, five-membered ring of cyclopentane. This distortion reduces the resulting strain energy (Table 1-1): the large strain energy of cyclopropane is decreased progressively as the ring size is increased to cyclobutane and cyclopentane. In the series of simple cyclic hydrocarbons, as the size of the ring increases beyond four atoms, distortion from planarity becomes very easy. Cyclohexane, for example, is quite nonplanar and can easily attain the desired tetrahedral angle of 109° at each atom. The three-dimensional structures of cyclic hydrocarbons are discussed in more detail in Chapter 5.

Table 1-1 ▸ Strain Energies of Cyclic Hydrocarbons (in kcal/mole)		
△ 27.6	☐ 26.4	⬠ 6.5
⬡ 0	⬣ 6.3	⯃ 9.6
⬟ 12.6	⬡⬡ 12.0	✚ 2.4

1-7 ▸ Nomenclature

Because of the great number of compounds containing carbon and because of the range of complexity of their skeletons, it is often necessary, and practical, to refer to a specific compound by a unique name. The International Union of Pure and Applied Chemistry (IUPAC) has provided a set of rules for naming organic compounds in an exact way. In accord with the **IUPAC rules,** a compound is specifically identified by a root, which establishes the number of carbon atoms in the longest continuous chain, and a suffix, which describes the kind of bonds present in the molecule. Prefixes are used to indicate where additional carbons or other substituents are attached. In Chapter 2, we will learn about hydrocarbons in which carbon atoms are

Table 1-2 ▸ IUPAC Nomenclature for Simple Hycrocarbons		

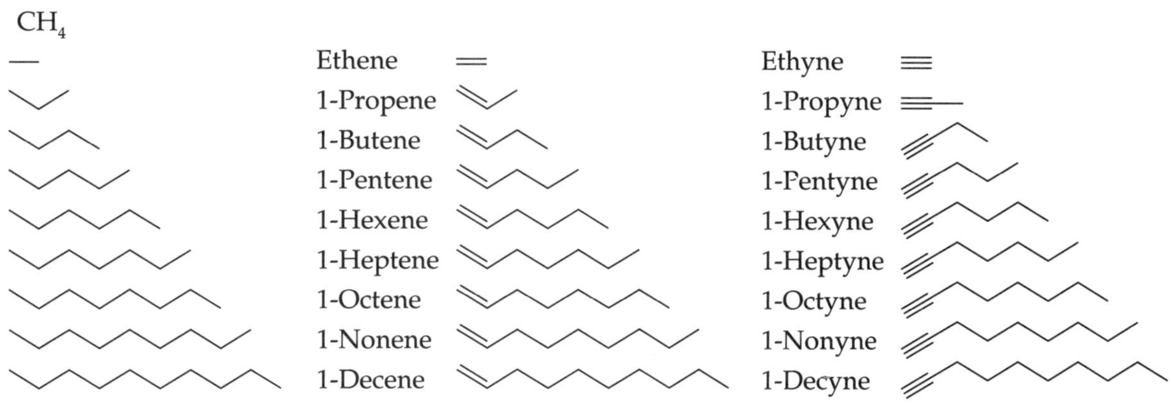

Methane	CH_4				
Ethane	—	Ethene	=	Ethyne	≡
Propane	⋀	1-Propene	⋀	1-Propyne	≡—
Butane	⋀⋀	1-Butene	⋀⋀	1-Butyne	⫽⋀
Pentane	⋀⋀⋀	1-Pentene	⋀⋀⋀	1-Pentyne	⫽⋀⋀
Hexane	⋀⋀⋀⋀	1-Hexene	⋀⋀⋀⋀	1-Hexyne	⫽⋀⋀⋀
Heptane	⋀⋀⋀⋀⋀	1-Heptene	⋀⋀⋀⋀⋀	1-Heptyne	⫽⋀⋀⋀⋀
Octane	⋀⋀⋀⋀⋀⋀	1-Octene	⋀⋀⋀⋀⋀⋀	1-Octyne	⫽⋀⋀⋀⋀⋀
Nonane	⋀⋀⋀⋀⋀⋀⋀	1-Nonene	⋀⋀⋀⋀⋀⋀⋀	1-Nonyne	⫽⋀⋀⋀⋀⋀⋀
Decane	⋀⋀⋀⋀⋀⋀⋀⋀	1-Decene	⋀⋀⋀⋀⋀⋀⋀⋀	1-Decyne	⫽⋀⋀⋀⋀⋀⋀⋀

connected not only by σ bonds, but also by additional chemical bonds. The same rules are used to name these multiply bound compounds.

A molecule that contains only sp^3-hybridized carbon atoms (an alkane) is designated by the suffix **-ane,** which indicates the absence of any groups other than C–C and C–H σ bonds. If a double bond is introduced into a molecule, the ending is changed to **-ene,** and such compounds are known as **alkenes.** If a triple bond is introduced into a molecule, the compound is called an **alkyne,** the **-yne** ending indicating the presence of the triple bond. The roots are derived from Greek and Latin; except for those of the first four members of the series, the root designates the number of carbon atoms in the molecule. Table 1-2 gives the skeletons and names of hydrocarbons containing from one to ten carbon atoms; comparable nomenclature rules apply to larger systems. When a cyclic structure is present, the prefix **cyclo-** is inserted before the root that designates the number of carbons. For example, the cycloalkanes shown in Table 1-1 are named cyclopropane, cyclobutane, cyclopentane, and so forth.

For branched hydrocarbons, IUPAC nomenclature rules dictate that the longest continuous carbon chain be identified as the root, with branching groups named as **alkyl substituents.** An alkyl group can be considered to be an alkane from which one hydrogen has been removed. The name is derived by replacing the -ane suffix with **-yl** as a suffix. Thus, CH_3 is a methyl group, C_2H_5 an ethyl group, and so forth. The position in the main chain to which an alkyl group is attached is designated by number: the numbering of carbon atoms in the chain starts at the end closest to the attachment of the substituent, so that the lower of two possible numbers can be assigned to that carbon, as shown in Figure 1-17 for several C_6 hydrocarbons.

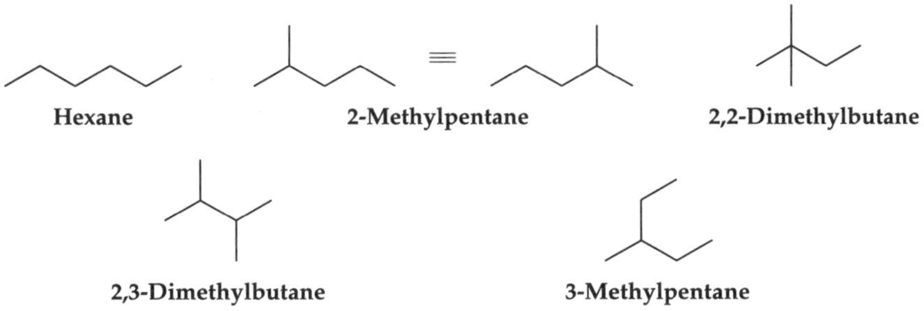

Hexane   **2-Methylpentane**   **2,2-Dimethylbutane**

2,3-Dimethylbutane   **3-Methylpentane**

▲ FIGURE 1-17
Names of the isomeric hexanes (C_6H_{14}).

The presence of more than one alkyl group along the hydrocarbon chain is designated by a Greek prefix (di-, tri- tetra-, penta-, etc.). In this case, *each* alkyl group must be assigned a number (as low as possible) to indicate its position. Note that the same compound can be drawn in several ways, as shown for 2-methylpentane. But, whether we number from the right or the left so as to assign the methyl group the lowest possible position along the carbon chain, we obtain the same unique name for either representation. The two structures shown are really the same compound. (Use your molecular models to convince yourself that they are indeed identical.) Figure 1-17 also illustrates two isomeric dimethylbutanes. Note that the isomer shown at the lower right is properly named 3-methylpentane rather than 2-ethylbutane. (Recall that the IUPAC rules stipulate that the longest continuous carbon chain is the root in the name of a compound.)

Alkyl groups with more than two carbons are often designated by common names, and it is also necessary to designate the point on the alkyl group at which it is attached to the main chain. Table 1-3 lists the structures of some common alkyl groups. The prefix ***n-*** (normal) refers to a straight-chain alkyl group, the point of attachment being at a **primary carbon**—that is, one bonded to only one other carbon. The prefix **iso-** describes an alkyl group in which the point of attachment is at the end of a carbon chain that bears a methyl group at the second carbon from the opposite end. (This is easier to picture than to describe: see the series isopropyl, isobutyl, isopentyl in Table 1-3.) The term ***s-*butyl** designates an alkyl group whose point of attachment is at carbon-2 of a C_4 straight chain. Here, ***s-*** is read as secondary, indicating attachment at a **secondary carbon**—that is, one attached to two carbons. The term ***t-*butyl** indicates attachment at the group's central carbon: $—C(CH_3)_3$. Here, ***t-*** is read as tertiary to indicate attachment

Table 1-3 ▸ Some Common Alkyl Group Substitutents and Their Common Names

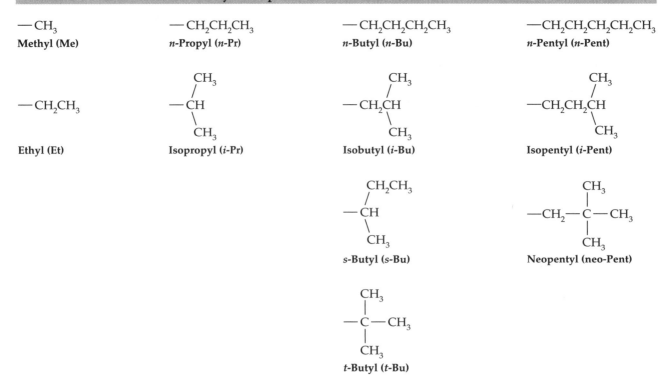

at a **tertiary carbon**—that is, one bound to three other carbons. The prefixes *s*- and *t*- are used for butyl chains because longer alkyl groups often contain more than one type of secondary carbon, and so the *s*- and *t*- designations would not be unique. The prefixes *n*- and *iso*- are used for longer chains. Unlike *s*- and *t*-, which refer to points of attachment, the prefix **neo-** is used almost exclusively in neopentyl, —$CH_2C(CH_3)_3$. The alkyl groups shown in Table 1-3 are encountered frequently and should be studied until they are easily recognized.

EXERCISE 1-G

Write acceptable names for each of the following hydrocarbons:

(a) 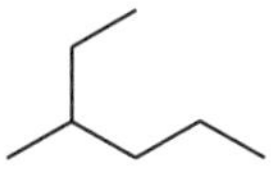   (c) 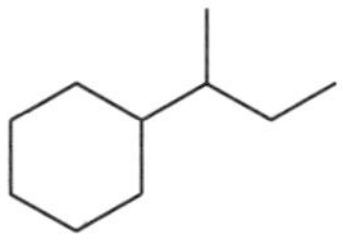   (e)

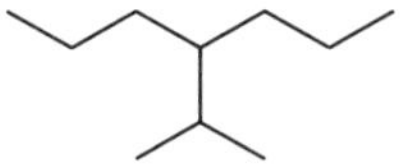

(b)   (d)

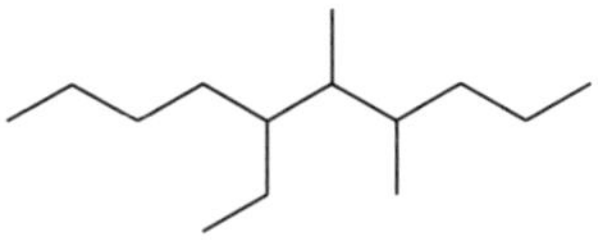

Because the carbon atoms in a cycloalkane are sp^3-hybridized, alkyl substituents bound to the ring carbons in a cycloalkane are directed either above or below the atoms of the ring. If only one group is present, the designations "above" and "below" are arbitrary because the ring can be easily viewed from its opposite side. If two or more groups are present, however, the relative positions of the groups are fixed: if two groups are on the same side of the ring, they are said to be in a *cis* arrangement; if they are on opposite sides of the ring, they are said to be in a *trans* arrangement. Because *cis* and *trans* isomers cannot be interconverted by free rotation about a sigma bond, these compounds are isomeric. In naming compounds in which this kind of geometric isomerism is possible, it is therefore necessary to designate the relation of multiple substituents on a cycloalkane as being either *cis* or *trans*. For example, the upper structure shown in the margin is correctly named 2,6-dimethyl-4-*t*-butyldecane and the lower one is *trans*-1-*s*-butyl-3-isopropylcyclobutane.

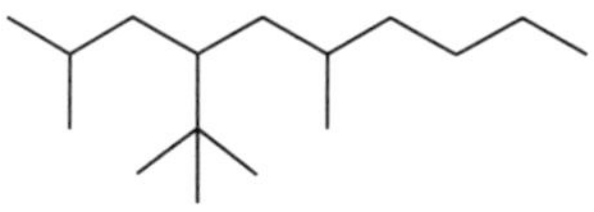

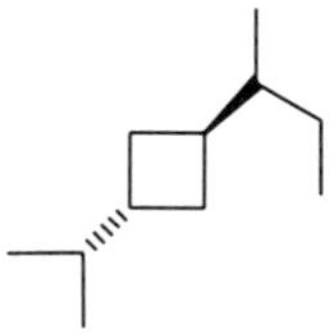

To develop your skills in naming alkanes, assign names to the isomers that you drew for Exercises 1-E and 1-F.

EXERCISE 1-H

Draw acceptable structures to correspond to each of the following IUPAC names:

(a) 3,3,4-trimethyloctane

(b) *n*-propylcyclopentane

(c) 3-ethyl-2-methylhexane

(d) *cis*-1,2-dimethylcyclopentane

(e) *trans*-1,4 dimethylcyclohexane

1-8 ▶ Alkane Stability

Isomeric alkanes usually have slightly different energies. The order of stability of isomers can be determined by measuring the heat of combustion (ΔH°_c), the amount of heat released upon conversion of the isomers into a common product.

Combustion

Alkanes burn in air, producing water and carbon dioxide. The reactions are:

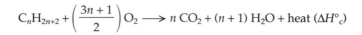

$$C_nH_{2n+2} + \left(\frac{3n+1}{2}\right)O_2 \longrightarrow n\,CO_2 + (n+1)\,H_2O + \text{heat}\,(\Delta H^\circ_c)$$

When a given alkane is completely burned to carbon dioxide and water, a characteristic amount of heat is given off. The greater the amount of heat given off (per mole of CO_2 released), the higher was the alkane's energy content and the less stable was the hydrocarbon. In a calorimeter, which measures the amount of heat released, a closed small vessel containing a measured quantity of the alkane to be burned is immersed in a liquid. The heat released in the chemical reaction warms the liquid and the resulting change in temperature is noted. From the known heat capacity of the liquid, the heat released in the reaction is calculated. This heat is then converted into a molar basis to obtain the **heat of combustion.**

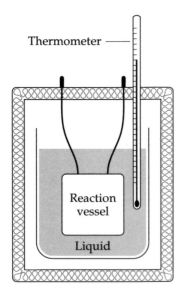

Thermometer

Reaction vessel

Liquid

EXERCISE 1-I

Calculate the heats of combustion for each of the following compounds, given the following raw data. (The heat capacity of water is 1.0 cal/g °C; assume that no heat is lost during the measurement.)

(a) 1.0 g of C_3H_6 produces enough heat to warm 1000 g of water by 12 °C.

(b) 1.0 g of C_6H_{12} warms 250 g of water by 45 °C.

HYDROCARBON BRANCHING AFFECTS GASOLINE QUALITY

The degree of branching of hydrocarbons also affects how rapidly they undergo combustion. Too-rapid burning of gasoline in an internal combustion engine leads to "preignition," or "knocking" or "pinging," because the explosion occurs before the piston has reached the top of the cylinder and is ready to start back down. Isooctane (a "trivial" or non-IUPAC name for 2,2,4-trimethylpentane) is used as a standard against which other hydrocarbons and mixtures of hydrocarbons are rated: it is arbitrarily assigned the value of 100 on the "octane scale," with *n*-heptane, originally used as a fuel, representing zero octane. Measurements of the rate of combustion are done both in the laboratory and in actual engines, and the average of these two methods is used as the octane rating for gasoline.

Isooctane

Table 1-4 ▸ Heats of Combustion of Several Alkanes

Alkane	$\Delta H°*$ Combustion (kcal/mole)	Alkane	$\Delta H°$ Combustion (kcal/mole)
CH_4	212.9	△	499.8 (166.6/C)[†]
CH_3CH_3	373.0		
$CH_3CH_2CH_3$	530.4	□	656.3 (164.1/C)
$CH_3CH_2CH_2CH_3$	687.8		
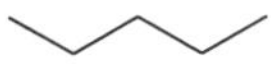 ∨∨∨	845.0	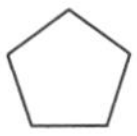 ⬠	793.6 (158.7/C)
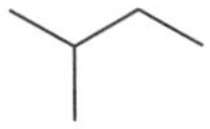 (branched)	843.4	 ⬡	944.7 (157.4/C)
(branched)	840.0	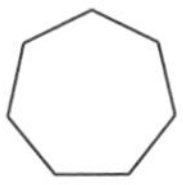 ⬡ (heptagon)	1108.3 (158.3/C)

*$\Delta H°$ (combustion) is the heat released when one mole of compound is completely oxidized to CO_2 and water under standard conditions (1 atm O_2 at 0 °C).
[†]Values in parentheses are obtained by dividing $\Delta H°$ (combustion) by the number of carbon atoms.

The values shown in Table 1-4 represent heats released (on a molar basis) for various alkanes. Several trends are clear. The greater the number of carbon atoms in an alkane, the greater is its molar heat of combustion because more molecules of CO_2 and H_2O are produced, releasing energy as they are formed. In a series of isomeric hydrocarbons, linear alkanes (for example, *n*-pentane) have a higher heat of combustion than do more highly branched isomeric alkanes (for example, neopentane, 2,2-dimethylpropane). Cyclopropane and cyclobutane have higher heats of combustion, per carbon, than larger ones because of ring strain.

Heats of Formation

A second measure of the relative stabilities of isomeric alkanes is found in their heats of formation. The **heat of formation** ($\Delta H°_f$) is a theoretical number that describes the energy that would be released if a molecule were formed from its component elemental atoms in their standard states.

$$n\,C + (n + 1)\,H_2 \longrightarrow C_nH_{2n+2} + \text{heat } (\Delta H°_f)$$
(graphite)   (gas)

Thus, the heat of formation of an alkane represents a measure of the amount of heat that would be released if the alkane were formed from elemental carbon and hydrogen. These heats of formation are very often calculated from the heat of combustion by the sequence shown at the top of the next page. (The heats of formation of carbon dioxide and water are constant irrespective of the material from which they are formed.) The observed heats of formation provide the same information concerning relative stability that is obtained from heats of combustion.

$$C_nH_{2n+2} + \left(\frac{3n+1}{2}\right)O_2 \quad \xrightarrow{\Delta H^\circ_c \text{ (alkane)}}$$

$$\uparrow \Delta H^\circ_f \text{ (alkane)} \qquad\qquad\qquad\qquad\qquad\qquad n\,CO_2 + (n+1)\,H_2O$$

$$n\,C + (n+1)\,H_2 + \left(\frac{n+1}{2}\right)O_2 \quad \Delta H^\circ_f \text{ (}CO_2 + H_2O\text{)}$$

1-9 ▸ Physical Properties of Hydrocarbons

Because hydrocarbons contain only atoms of similar electronegativity, they are quite nonpolar. Because there is minimal charge polarization in the bonds of a hydrocarbon, polar intermolecular interaction between these molecules is weak. If there were no attractive interactions between molecules at all, hydrocarbons would exist as gases at all temperatures. Although several of the low-molecular-weight members of these series are indeed gases at room temperature, all of them form liquids or solids at lower temperatures.

The major intermolecular interaction in hydrocarbons is van der Waals attraction, in which the electrons of one molecule are attracted to the nuclei of another. These van der Waals attractions are relatively weak and are easily disrupted. Clearly, the more atoms in a given molecule, the greater is the sum of the van der Waals attractions for another molecule of its kind. Thus, as molecular weight increases, van der Waals attractions generally increase, producing stronger intermolecular interactions and higher melting points and boiling points. These trends are apparent in Table 1-5.

Because these intermolecular attractions are weak, they have a less-significant effect on these physical properties than do the stronger electrostatic intermolecular forces in more-polar molecules. Thus, polar liquids, which are strongly attracted to each other by electrostatic forces, interact more strongly with themselves than with a nonpolar hydrocarbon. As a result, hydrocarbons are not as soluble in polar liquids (such as water) as they are in other nonpolar liquids, which interact only through weak intermolecular attractive forces. This is an example of "like dissolves like," an old adage of organic chemistry.

Table 1-5 ▸ Physical Properties of Alkanes

Name	Formula	Boiling Point (°C)	Melting Point (°C)
Methane	CH_4	−164	−182
Ethane	C_2H_6	−89	−183
Propane	C_3H_8	−42	−190
Butane	C_4H_{10}	−0.5	−138
2-Methylpropane (Isobutane)	C_4H_{10}	−12	−159
Hexane	C_6H_{14}	69	−95
Cyclohexane	C_6H_{12}	81	6
Octane	C_8H_{18}	126	−57
2,2,4-Trimethylpentane (Isooctane)	C_8H_{18}	99	−107

EXERCISE 1-J

Of the following pairs of compounds, which member of each pair is more likely to fit the given description?

(a) higher boiling point: pentane or octane

(b) higher melting point: octane or decane

(c) higher solubility in hexane: water or heptane

Conclusions

Hydrocarbon structure is based on a tetravalent carbon atom—that is, carbon participates in covalent bonding in which its valence requirement as a group IV element in the periodic table is satisfied by forming four bonds with other atoms. That the valence electronic configuration of carbon is met can be checked by drawing a Lewis dot structure, which specifies the position of shared electrons between atoms and accounts for nonbonded electrons.

Three types of hybridized atomic orbitals can be formed at carbon. Four equivalent sp^3-hybridized orbitals are directed toward the corners of a tetrahedron in order to minimize electron repulsion. Hydrocarbons containing only sp^3-hybridized atoms are called alkanes and are considered to be saturated. The bond angles at an sp^3-hybridized carbon are approximately 109° and the bond lengths are about 1.54 Å for a carbon–carbon bond and 1.10 Å for a carbon–hydrogen bond. The carbon skeleton is held together by sigma (σ) bonds.

Alkanes that lack rings (acyclic) have an overall formula of C_nH_{2n+2}. For each ring present, a cycloalkane requires two fewer hydrogen atoms than does the open-chain alkane with the same number of carbons. Ring formation of small cycloalkanes introduces strain.

Alkanes are named in accord with the IUPAC rules. Each compound is named by combining a root, which specifies the number of carbons in the longest continuous chain of carbon atoms, with a suffix (-ane) which specifies the chemical family of the compound as an alkane. Branches are named as alkyl groups, and their positions are indicated by a prefix that specifies the position along the longest chain to which the group is bound. Alkanes containing rings are named by inserting the prefix cyclo- before the root descriptor.

Relative stabilities of alkanes are determined by measuring the amount of heat either that is released when the alkane is completely burned to water and CO_2 (its heat of combustion) or that would be released if the alkane were formed from elemental carbon and hydrogen. Linear alkanes have higher heats of combustion (that is, are less stable) than their more highly branched isomers.

Because hydrocarbons are nonpolar, they associate only through weak intermolecular actions (dominated by van der Waals attractions). As a group, these compounds have relatively low melting points and boiling points. They are most soluble in other nonpolar liquids.

Review Problems

1-1 Calculate the formal charges on each atom in each of the following compounds and ions.

(a) tetrafluoroborate, BF_4^-

(b) ammonium, NH_4^+

(c) methane, CH_4

(d) hydronium, H_3O^+

(e) molecular hydrogen, H_2

1-2 Draw a Lewis dot structure for each of the following molecules:

(a) CO

(b) CO_2

(c) acetylene, $HC \equiv CH$

(d) carbonate, CO_3^{2-}

(e) formaldehyde, $H_2C = O$

1-3 Draw structures that correspond with each of the following names:

(a) 5-*s*-butylnonane

(b) *trans*-1,3-diethylcycloheptane

(c) 4-isopropyl octane

(d) *cis*-1-*t*-butyl-3-methylcyclopentane

1-4 Provide the IUPAC name for each of the following structures:

(a)

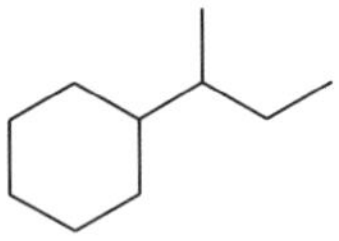

(c)

(b)

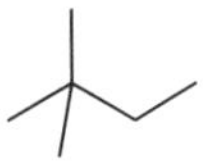

(d)

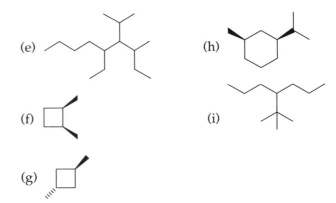

(e)

(f)

(g)

(h)

(i)

1-5 As we will see in Chapter 3, it is possible to replace individual hydrogen atoms in an alkane by halogen atoms, and isomers are formed if nonequivalent hydrogens are thus substituted. We can therefore look for equivalence or nonequivalence of various hydrogens in an alkane if we consider whether halogenated derivatives of the alkane exist as isomers. Which isomeric pentane skeleton has exactly:

(a) one monofluoro derivative?

(b) three different monofluoro derivatives?

(c) four different monofluoro derivatives?

1-6 Which isomer of the following pairs has the higher heat of combustion?

(a) 2-methylhexane or heptane

(b) 2,2-dimethylpropane or 2-methylbutane

(c) octane or *cis*-1,2-dimethylcyclohexane

Chapter 2

Alkenes, Arenes, and Alkynes

The alkanes, discussed in Chapter 1, are hydrocarbons that contain only sp^3-hybridized carbon atoms. Because the atoms in an alkane are connected only by C–C and C–H sigma (single) bonds, the alkanes constitute the simplest class of organic compounds. However, if we are to understand the chemical and physical properties of other classes of organic compounds, we must study the effects caused by the introduction of carbon atoms that are not sp^3-hybridized and of elements other than carbon and hydrogen along the carbon backbone of an alkane.

In this chapter, we will study the structures of hydrocarbons that have double bonds **(alkenes)** and triple bonds **(alkynes)** between carbon atoms, as well as those that have several such multiple bonds. We will learn that these multiple bonds give rise to a characteristic geometry at the atoms participating in these bonds and that the multiple bond constitutes an area of special reactivity in the molecule. We will also see that interactions between groups of multiple bonds cause some of these compounds (aromatic compounds) to have unusual stability. In the next chapter, we will learn about the structures and properties of different classes of organic compounds that contain second- and third-row elements other than carbon.

2-1 ▸ Alkenes

When a carbon atom forms σ bonds to three (rather than four) atoms, its valence requirement is satisfied by the formation of a double bond. In this section, we will learn about the geometry and chemical character of such double bonds (composed of one σ and one π bond) and how carbon must rehybridize (from the sp^3 hybridization found in alkanes) in order to participate in π bonding.

sp^2 Hybridization

So far, we have considered hybridization in which the 2s orbital is mixed with all three 2p orbitals of carbon to form four sp^3-hybrid orbitals. If, in-

$$\text{C: } 1s^2 \quad \underbrace{2s^1 2p_x{}^1 p_y{}^1}_{\substack{\text{Three} \\ sp^2\text{-hybrid} \\ \text{orbitals}}} \quad \underbrace{2p_z{}^1}_{\substack{\text{One } p \\ \text{orbital}}}$$

▲ FIGURE 2-1
Orbital mixing in the formation of sp^2-hybridized carbon.

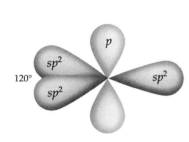

stead, the *s* orbital is mixed with only two *p* orbitals (Figure 2-1), leaving one *p* orbital unhybridized, three equivalent **sp^2-hybrid orbitals** result. As with the sp^3-hybridized orbitals, these sp^2-hybridized orbitals are directed as far as possible both from each other and from the remaining *p* orbital not taking part in the hybridization. This arrangement is best achieved when the three sp^2-hybridized orbitals are in a plane and are directed toward the vertices of a triangle (with 120° angles) and the remaining *p* orbital is perpendicular to this plane, as shown in the margin. A carbon–carbon σ bond can now be formed by the overlap of two sp^2-hybridized orbitals, one from each of two adjacent carbon atoms, pointed toward each other as shown in Figure 2-2. (As we will see in the next subsection, this σ bond also permits *p* orbitals on the two bound carbon atoms to interact.) Although an sp^2-hybridized orbital has less *p* character than does the sp^3-hybrid used for σ bonding in the alkanes, the sp^2–sp^2 σ bond present in alkenes exhibits many characteristics similar to those of sp^3–sp^3 σ bonds.

There is one significant difference between sp^2-hybridized carbons and sp^3-hybridized carbons. Because the sp^2 orbital has a larger fraction of *s* character, it has greater electron density near the nucleus. As a consequence, the electrons in an sp^2-hybridized orbital are held more tightly by the nucleus than are electrons near an sp^3-hybridized orbital. When forming a bond, an sp^2-hybridized atom behaves as if it were more electronegative than if it were sp^3-hybridized. We will see in later chapters how this difference influences the relative stability of anions formed at carbons of different hybridization.

Pi Bonding

In the three-dimensional arrangement shown in Figure 2-2, the *p* orbitals on adjacent sp^2-hybridized carbon atoms are coplanar. They are then in correct geometric positions to interact above and below the molecular plane (as shown by the color) so that a second bond can be formed. In this type of bonding, referred to as π (pi) **bonding,** electron density is not along the axis connecting the two bonded atoms (as in σ bonds); rather, it is above and below the axis.

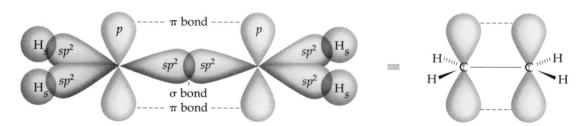

▲ FIGURE 2-2
Sigma and pi overlap between sp^2-hybridized atoms. For clarity, the *p* orbitals participating in π bonding are in color.

A hydrocarbon that contains a π bond has two fewer hydrogens than does a molecule in which the atoms are sp^3-hybridized. Lacking the full number of hydrogens that would be present in an alkane with no rings but with the same number of carbon atoms, an alkene is said to be **unsaturated.** An alkane, in contrast, with no π bonds, is referred to as **saturated.**

Pi-bond formation is always accompanied by σ-bond formation, and so the bonding connecting the two carbon atoms is referred to as a **double bond.** As mentioned earlier, hydrocarbons containing double bonds are referred to as alkenes to differentiate them from the alkanes, which lack this multiple bonding. Alkenes have the overall formula C_nH_{2n}; that is, an alkene requires two fewer hydrogens (because of the double bond) than the corresponding alkane.

The introduction of a double bond has the same effect on a hydrocarbon's molecular formula as does constraining an alkyl chain into a ring. The difference between the number of hydrogens, m, present in an alkene from that expected for a straight-chain alkane $(2n + 2)$ with the same number of carbons (n) therefore defines the number of multiple bonds or rings present—that is, $[(2n + 2) - m]/2 =$ the number of rings plus the number of double bonds. This value is called the **index of hydrogen deficiency,** or the **degree of unsaturation.**

EXERCISE 2-A

For the following molecules, predict whether zero, one, or two double bonds or ring structures are present. Draw three carbon skeletons that correspond with your prediction.

(a) C_5H_{10}

(b) C_5H_8

(c) C_6H_{12}

Structures of Alkenes

Each sp^2-hybridized carbon participating in π bonding requires two additional σ bonds (beyond the C–C σ and C–C π bonds) to satisfy its valence electron requirement. Two carbon–hydrogen bonds can be formed by overlap of a $1s$ hydrogen orbital with an sp^2-hybridized carbon orbital, as shown in Figure 2-2. The molecule constructed in this way, **ethene** (also called **ethylene**), can be represented by the Lewis dot structure and line model shown here.

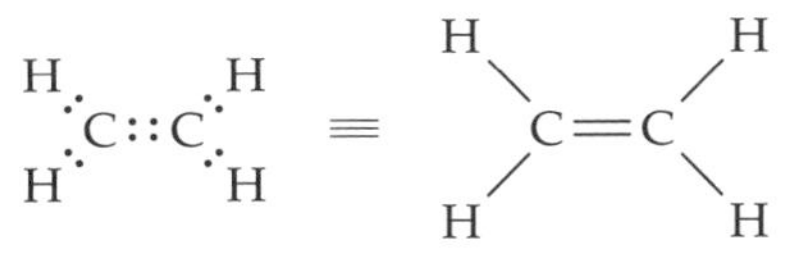

Because each of the carbons in ethene has access to a full complement of valence electrons, its electronic configurational demand is satisfied and this molecule (C_2H_4) is a stable species.

A subtle but significant consequence of the presence of two bonds (rather than one) between atoms is that the atoms are held more closely to each other. The distance between two carbons connected by a single bond is typically 1.54 Å, whereas that between carbons joined by a double bond is 1.34 Å.

ETHENE: A NATURAL RIPENING AGENT

Ethene, commonly called ethylene, occurs naturally in many plants. Indeed, this gas has been shown to be a natural ripening agent for fruits and vegetables. This is why the ripening of fruits is hastened by putting them in a closed bag. The ripening can be even further accelerated by including a lemon or lime, fruits that produce comparatively large amounts of ethylene. Fruits are ripened commercially, for fresh delivery to grocery stores, by placing them in an ethylene atmosphere.

The existence of a π bond between carbon atoms in ethene requires the overlapping p orbitals to be coplanar. This makes rotation about the C–C σ bond impossible without disruption of the π bond. For example, if the aligned geometry shown in Figure 2-2 were altered by a 90° rotation about the carbon–carbon bond to one in which the p orbitals were perpendicular, as in Figure 2-3, overlap between the p orbitals would be reduced to zero. Because overlap is necessary for π bonding, this rotation effectively breaks the π bond and makes it impossible for the two π electrons to be shared between the two atoms. Although the structure at the left in Figure 2-3 can be represented by a Lewis dot structure in which four electrons are shared between the carbon atoms, that at the right has only two electrons shared in the σ bond between carbons, with a single electron being localized on each carbon.

A structure bearing a single unpaired electron at a carbon atom is called a **radical.** Because the structure at the right in Figure 2-3 has two non-interacting radical centers, it is called a **biradical.** The valence requirement of neither carbon atom is satisfied in the biradical, and the orthogonal (perpendicular) geometry is expected to be unstable compared with its isomeric structure at the left. Breaking the π bond upon rotation costs energy, and so free rotation (like that in molecules containing only σ bonds) is not possible about a C=C π bond.

Molecular Orbitals

Bonding between atoms occurs by the overlap of atomic orbitals. These bonds can be separated into two classes: σ bonds (resulting from direct

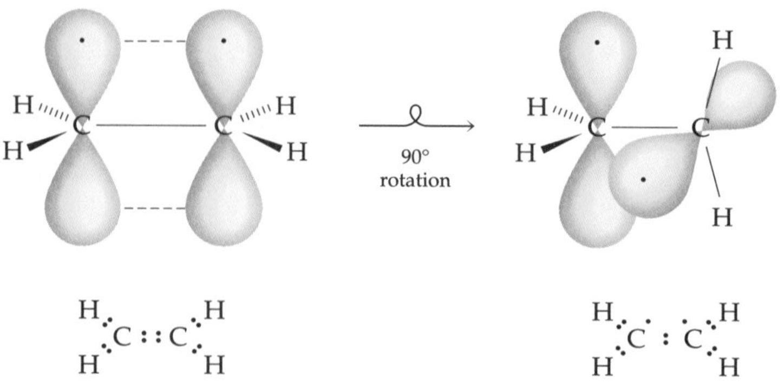

▲ FIGURE 2-3
Twisting about a carbon–carbon π bond. When the p orbitals are perpendicular, as at the right, only one pair of electrons is shared between the atoms in a covalent bond. The electronic requirement of neither carbon atom is then met.

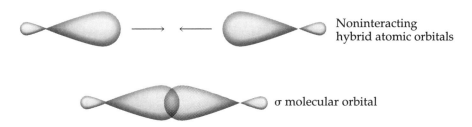

▲ FIGURE 2-4
Overlap of two σ atomic orbitals to form a σ molecular orbital.

overlap of hybrid orbitals having some *s* character) and π bonds (resulting from the overlap of *p* orbitals). It is useful to view these bonds as orbitals themselves, referred to as **molecular orbitals** to differentiate them from simple and hybrid atomic orbitals.

Let us consider the formation of a σ bond between sp^3-hybridized orbitals. First, we start with completely noninteractive carbon atoms (Figure 2-4). As these carbons and their orbitals move closer together, the orbitals increasingly overlap, and bonding ensues. Because bonding is energetically favorable, the molecule is more stable than the separated atoms from which it derived, and the total energy content of the molecule is lower than that of the isolated atoms.

We can visualize the energy of this interaction (Figure 2-5) if we define the noninteractive hybrid orbitals as lying at an arbitrary zero point of energy. Overlap of these hybrids results in a more-stable, **bonding molecular orbital.** The formation of a σ bond by overlap of these hybrid orbitals is shown at the bottom of Figure 2-5, in which the bonding character of the molecular orbital is indicated by showing the two overlapping lobes in the same color. (These molecular orbitals, like the atomic orbitals from which they are constructed, are mathematical surfaces that describe the likely positions of electron density.) Because we started at zero energy and have formed a combination that is energetically more favorable, there must exist, in principle, a way of combining the σ orbitals that is equivalently unfavorable. This **antibonding molecular orbital** is shown at the top of Figure 2-5, in which the unfavorable interaction is indicated by shading the orbitals in different colors. That two molecular orbitals should result from the overlap of the two atomic hybrid orbitals should not surprise us because we have already seen that hybridization produces exactly the same number of hybrid orbitals as atomic orbitals from which they were formed.

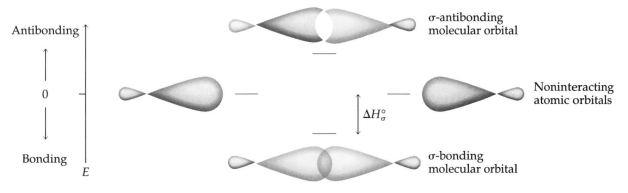

▲ FIGURE 2-5
Bonding and antibonding σ molecular orbitals.

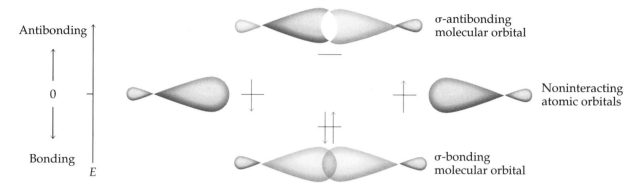

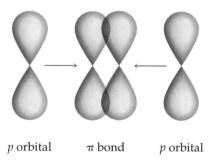

▲ FIGURE 2-6
Electron filling of the σ-bonding molecular orbital. The arrows in color represent electrons of opposite spin. The single electrons from the two noninteracting atomic orbitals fill the σ-bonding molecular orbital, leaving the σ-antibonding molecular orbital vacant.

Figure 2-5 was drawn without consideration of the number of electrons occupying the interacting orbitals. Generally, one electron is available from each of the hybrid atomic orbitals, and these two available electrons are inserted into the energetically more favorable, σ-bonding orbital (Figure 2-6). That this bonding molecular orbital is filled is shown by the two arrows in color (representing electrons with opposite spins). This leaves the σ-antibonding orbital unpopulated and explains why the sharing of valence electrons produces favorable chemical bonding. Only the bonding orbital need be considered because it is the only one that contains electrons.

An analogous pair of π molecular orbitals can be constructed from p atomic orbitals (Figure 2-7). The interaction of the p orbitals (of sp^2-hybridized atoms) also forms a bonding and an antibonding combination (Figure 2-8). Because each carbon contributes one electron, these two electrons occupy the π-bonding orbital (Figure 2-9). The degree of overlap of hybrid orbitals is greater in a σ geometry than in a π geometry, and the energy stabilization gained in σ bonding (Figure 2-5) is usually greater than that in π bonding (Figure 2-8).

Both σ and π bonding and antibonding orbitals maintain energetic symmetry about an arbitrary zero of energy, at which the atomic orbitals that combine to produce the molecular orbital do not interact at all. In both σ and π molecular orbitals, bonding results from the overlap of atomic orbitals with the same phase (represented by color), with interacting hybrid atomic orbitals producing σ molecular orbitals and interacting p orbitals producing π molecular orbitals. Conversely, antibonding molecular orbitals result when the interacting atomic orbitals have opposite phasing. Al-

p orbital   π bond   *p* orbital

▲ FIGURE 2-7
Interaction of p orbitals to form a π bond.

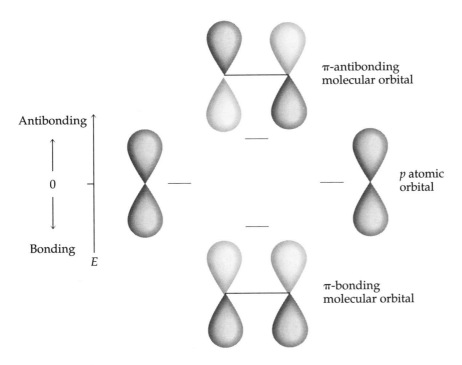

▲ FIGURE 2-8
Bonding and antibonding π molecular orbitals.

though this energetic symmetry of molecular orbitals is also maintained in most molecules with more-extensive π-bonding systems, the analysis becomes somewhat more complicated as the number of atomic orbitals increases.

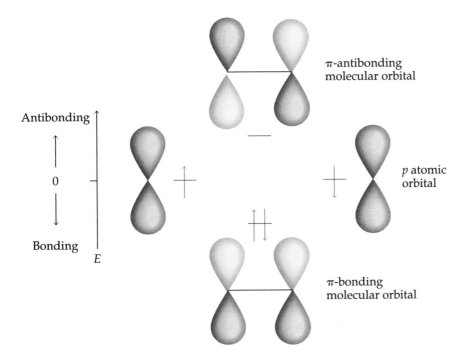

▲ FIGURE 2-9
Electron filling of the π-bonding molecular orbital. The single electrons from the two noninteracting *p* atomic orbitals fill the π-bonding molecular orbital, leaving the π-antibonding molecular orbital vacant.

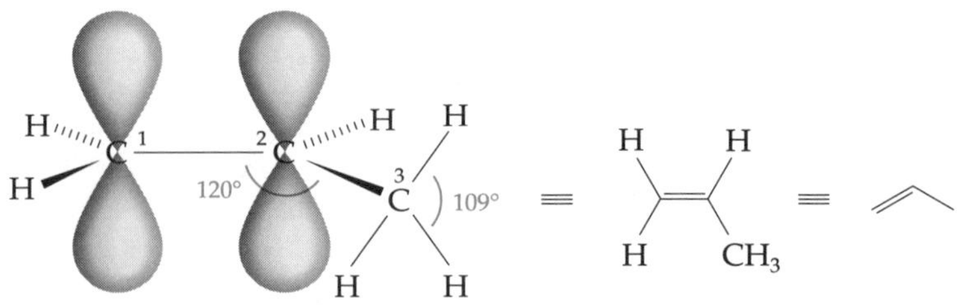

▲ **FIGURE 2-10**
Structural representations of propene.

Higher Alkenes

Pi bonds are also encountered in hydrocarbons containing more than two carbons. For example, by replacing a hydrogen of ethene with a methyl group to form propene, we obtain the structure shown in Figure 2-10. The double bond can be represented by stick notation or by the hatch/wedge (sawhorse) notation. The angle defined by the two sigma bonds connecting the sp^2-hybridized central carbon atom to the other sp^2-hybridized atom and to the sp^3-hybridized atom (C-1—C-2—C-3) is 120°, as is consistent with the hybridization of the central atom. The angle at C-3 (C-2—C-3—H-3) is 109°, owing to the sp^3-hybridization of C-3. The σ-bond length between C-2 and C-3 (between sp^2- and sp^3-hybridized carbons) is only slightly shorter (1.50 Å) than that of a typical sp^3–sp^3 carbon–carbon σ bond (1.54 Å) but is longer than the sp^2–sp^2 carbon–carbon double bond (1.34 Å).

Isomerism in Alkenes

1-Butene

trans-2-Butene

cis-2-Butene

In a C_4 straight-chain hydrocarbon containing a multiple bond, there are two options for isomerism: positional and geometric. In a **positional isomer,** the sequence in which atoms are connected differs because the position of the functional group has been shifted. In a **geometric isomer,** the atomic connectivity along a chain is unchanged, but the relative disposition of one or more groups about a bond with restricted rotation differs.

In butene, the position of the double bond can be between the first and second atoms of the chain (1-butene) or between the second and third (2-butene). Because there is restricted rotation about the carbon–carbon double bond, there are two possible geometries for 2-butene: one in which the two hydrogens are on the same side of the double bond and the other in which they are on opposite sides. If the same groups are on the same side, the molecule is referred to as a *cis* isomer; if they are on opposite sides, it is called a *trans* isomer. The barrier for interconversion of *cis*- and *trans*-2-butene, which requires a rotation that breaks the π bond, can be used as a measure of the energy of the π bond. An average value for the energy of a carbon–carbon π bond is about 63 kcal/mole, a value well above that available to molecules at room temperature.

Nomenclature for Alkenes

Many specific chemical reactions can take place at the site of a double bond in a hydrocarbon. Hydrocarbons that contain such multiple bonds constitute families (alkenes) that differ from the alkanes from the point of view of chemical reactivity. They contain a molecular feature (a **functional group**— in this case, the multiple bond) that undergoes characteristic and selective

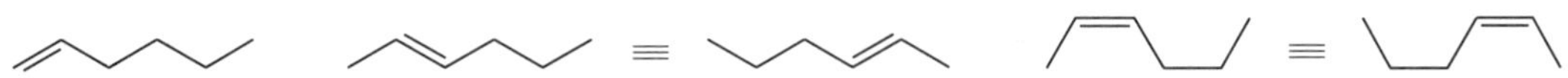

1-Hexene   *trans*-2-Hexene   *cis*-2-Hexene

trans-3-Hexene   *cis*-3-Hexene   4-Methyl-1-pentene

▲ FIGURE 2-11

Names of some C_6H_{12} hydrocarbons.

chemical reactions. In the naming of an alkene, the position of the functional group is indicated by a number immediately before the designated ending. Numbering of the longest carbon chain containing the double bond starts at the end closest to the functional group, which means that this group (the multiple bond) is assigned the lowest possible number in the chain. Figure 2-11 shows how some isomeric alkenes are named. Notice that, in the naming of 4-methyl-1-pentene, the lower of the two possible numbers is assigned to the functional group (the alkene), not to the substituent methyl group. The numbering sequence along the longest chain starts at the end nearest the functional group; that is, the position of the functional group takes precedence over that of the alkyl group.

In summary, use the following steps to apply the IUPAC rules for naming hydrocarbons:

1. Determine the longest continuous carbon chain that contains the functional group (if there is one) and name it with the appropriate root and suffix.

2. Assign the functional group the lowest possible number.

3. Name substituent branches as alkyl groups with their positions indicated by numbers.

4. Indicate multiple substituents with Greek prefixes.

EXERCISE 2-B

Write acceptable names for each of the hydrocarbons shown below:

(a) 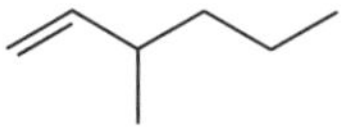   (c)

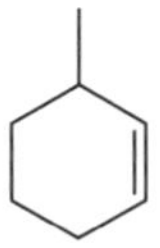

(b) 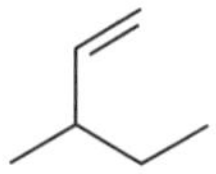   (d)

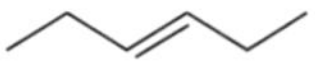

EXERCISE 2-C

Draw acceptable structures to correspond to each of the following IUPAC names:

(a) 2-cyclopropyl-1-hexene

(b) 3-ethyl-1-octene

(c) 2-isobutyl-1-heptene

The *cis* and *trans* designations (for specifying the isomeric structures of alkenes) are clear in simple cases such as 2-butene in which there are two identical substituents at each end of the double bond. This designation becomes ambiguous, however, if the substituents are different. For example, consider whether 3-methyl-2-pentene is a *cis* or *trans* isomer.

(E)-3-Methyl-2 pentene   **(Z)-3-Methyl-2 pentene**

To assist in resolving such ambiguity, IUPAC has adopted a method for specifying such isomers uniquely. This method consists of establishing group priorities at each end of the double bond and then specifying whether the groups of highest priority at each end are on the same or opposite sides of the double bond.

At each of the carbons participating in double bonding, we assign priorities to the two attached groups according to the atomic number of the attached atom. For example, in 3-methyl-2-pentene, a hydrogen atom and a carbon atom are attached to C-2. Carbon, being of higher atomic number, has priority over hydrogen, and the methyl group therefore has priority at C-2. At C-3, a methyl and an ethyl group are attached: there is no atomic number difference at the first atom attached. We then move out along each chain until we find an atomic number difference. In both the methyl and the ethyl groups, a CH_2 (methylene) group is attached to C-3; however, the methyl then has a hydrogen, whereas ethyl has a carbon substituent. Because carbon has a higher atomic number than hydrogen, ethyl takes priority over methyl at C-3.

Having established priorities, we then designate the spatial relation between the groups taking priority by using the German designations **E** (*entgegen*, opposite) or **Z** (*zusammen*, together). In *(E)*-3-methyl-2-pentene, the ethyl group at C-3 is on the opposite side of the double bond from the methyl group at C-2. Thus, this structure is the *E* isomer. In the *Z* isomer, the ethyl and methyl groups are on the same side.

Some alkenes need no stereochemical designator: 2-methyl-2-butene has no *cis* or *trans* isomers because it has two identical groups on one of the double-bond carbons.

2-Methyl-2 butene

Recall that *cis* and *trans* isomers are also encountered in multiply substituted cycloalkanes. When these substituents are on the same face of a ring, the isomer is *cis;* when on opposite sides, it is *trans.* Two examples are shown.

cis-**1,2-Dimethyl-**
cyclopropane

trans-**1,4-Dimethyl-**
cyclohexane

For more practice, you may want to return to Exercise 2-A and name the isomeric hydrocarbons that you drew.

EXERCISE 2-D

Write an acceptable names for each of the following hydrocarbons:

(a)

(c)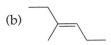

(b)

(d)

EXERCISE 2-E

Draw a structure to correspond with each of the following IUPAC names:

(a) *(E)*-2-octene   (c) *cis*-2-octene

(b) *(Z)*-3-octene   (d) *trans*-3-octene

Alkene Stability

Like alkanes, isomeric alkenes usually have slightly different energies. As for the alkanes, the order of stability of the isomers of alkenes can be determined by measuring how much heat is released upon conversion of the isomers into a common product.

Combustion. Both alkanes and alkenes burn in air. Only the reaction stoichiometry differs.

$$C_nH_{2n+2} + \left(\frac{3n+1}{2}\right) O_2 \longrightarrow n\, CO_2 + (n+1)\, H_2O + \text{heat } (\Delta H^\circ_c)$$

$$C_nH_{2n} + \left(\frac{3n}{2}\right) O_2 \longrightarrow n\, CO_2 + n\, H_2O + \text{heat } (\Delta H^\circ_c)$$

One can measure the heat of combustion of various alkenes (Table 2-1, on page 38). As in alkanes, the greater the number of carbon atoms in an alkene, the greater its molar heat of combustion. An alkane, containing more hydrogen atoms than its corresponding alkene, has a higher heat of combustion than the sum of those of the corresponding alkene and hydrogen. Within a series of alkenes, the *cis* isomer, which is destabilized by the interaction of two large groups on the same side of the double bond, has a higher heat of combustion than does the *trans* isomer. In the butenes, for example, *cis*-2-butene has a higher heat of combustion than does *trans*-2-butene and that of 1-butene is higher still. Thus, the order of stability of isomeric butenes is *trans*-2-butene > *cis*-2-butene > 1-butene.

EXERCISE 2-F

From the generalizations outlined in this section, predict which member of the following isomeric pairs will have the higher heat of combustion:

(a) 1-hexene or *(E)*-2-hexene   (c) octane or 2,5-dimethylhexane

(b) *(E)*-2-hexene or 2-methyl-2-pentene   (d) *(Z)*-2-pentene or *(E)*-2-pentene

Table 2-1 ▸ Heats of Hydrogenation and Combustion of Several Alkenes

Hydrocarbon	$\Delta H°^a$ Hydrogenation (kcal/mole)	$\Delta H°^b$ Combustion (kcal/mole)	Hydrocarbon	$\Delta H°$ Hydrogenation (kcal/mole)	$\Delta H°$ Combustion (kcal/mole)
$H_2C{=}CH_2$	−32.8		(cyclohexene)	−28.6	
(propene)	−30.1				
(1-butene)	−30.3	−649.5	(cross/methylenecyclobutane)	−20.7	
(3-methyl-1-butene)	−30.3	−805.2			
(3,3-dimethyl-1-butene)	−30.3		(cross/methylenecyclopentane)	−24.0	
(cis-2-butene)	−28.6	−647.8		−57.1	−607.4
(trans-2-butene)	−27.6	−646.8		−54.1	−761.7
(2-methylpropene)	−28.4	−645.4		−60.8	−768.8
(2-methyl-2-butene)	−28.5	−803.4		−60.5	
(2-methyl-2-pentene)	−26.9	−801.8			
(2,3-dimethyl-2-butene)	−26.6				

$^a\Delta H°$ (hydrogenation) is the heat released when one mole of the alkene is completely hydrogenated under standard conditions (1 atm H_2 at 0 °C).
$^b\Delta H°$ (combustion) is the heat released when one mole of compound is completely oxidized to CO_2 and water under standard conditions (1 atm O_2 at 0 °C).

Hydrogenation. Isomeric alkenes can also be ranked in order of stability by measuring the heat released upon the addition of hydrogen to generate a common alkane. For example, as shown in Figure 2-12, 1-butene and *cis*- and *trans*-2-butene all produce butane when hydrogen is added across the double bond. This reaction requires the presence of a **catalyst,** a compound that is not directly involved in the stoichiometry of the reaction but is necessary in order for the reaction to proceed at a reasonable rate. Platinum is often used as this catalyst, and the general reaction shown in Figure 2-12 is thus referred to as a **catalytic hydrogenation.** This reaction is also called an **addition reaction** because two simple molecules combine to form a product of higher molecular weight. Organic chemists commonly use a short-cut notation in writing such equations: they write the reagents necessary for the chemical reaction on either side of an arrow connecting the reactant and product without indicating the stoichiometry of the reaction.

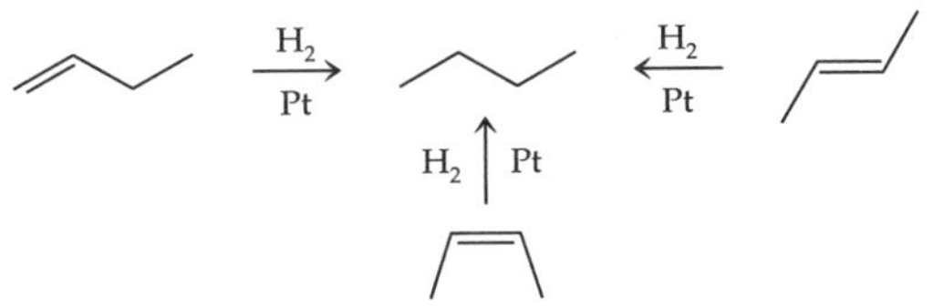

$$C_4H_8 \ + \ H_2 \ \longrightarrow \ C_4H_{10} \ + \ \text{heat}$$

▲ FIGURE 2-12

Catalytic hydrogenation of isomeric butenes.

This reaction specifically involves the double bond. The heat of hydrogenation therefore describes not the overall stability of the molecule, but rather the relative stabilities of the reactive part of the molecule, the carbon–carbon double bond. (Recall that the π bond of an alkene constitutes a functional group, a subset of atoms in a molecule at which a particular reactivity is centered or a specific reaction takes place.)

Heats of hydrogenation for several alkenes are given in Table 2-1. Many of the same trends that emerged from the combustion analysis are also apparent in calorimetric measurements of hydrogenation: 1-butene is less stable than *cis*-2-butene, which is less stable than *trans*-2-butene. With regard to the relevant stability of alkenes, we can classify heats of hydrogenation in sets according to the number of alkyl substituents on the double bond. In the isomeric hexenes, for example (Figure 2-13), 2,3-dimethyl-2-butene has four alkyl groups attached to the double-bond carbons, 3-methyl-2-pentene has three alkyl groups on the double bond, 2-hexene has two alkyl groups on the double bond, and 1-hexene has only one alkyl group attached to one of the double-bond carbons. The more-stable alkene is the one in which the double bond is more highly substituted. This is revealed in the lower heat of hydrogenation of the more-branched (more highly substituted) alkene.

As mentioned earlier, an sp^2-hybridized atom is more electronegative than an sp^3-hybridized atom. Because alkyl groups are much more polarizable than is a hydrogen atom, they more readily satisfy the electron demand of the sp^2-hybridized carbons of a π bond. Therefore, the replacement of hydrogens on a double bond by alkyl groups stabilizes the alkene and accounts for the observed order of stability.

EXERCISE 2-G

Rank the following groups of compounds in order of decreasing heats of hydrogenation (that is, by increasing stability):

(a) 1-heptene, 3-heptene, 2-methyl-2-hexene

(b) 1-methylcyclooctene, 3-methylcyclooctene, 1,2-dimethylcyclooctene

(c) 3-ethyl-1-octene, 2-ethyl-1-octene, 3-ethyl-2-octene

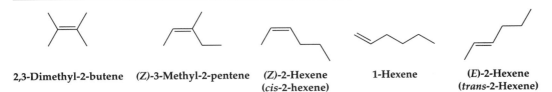

2,3-Dimethyl-2-butene   **(Z)-3-Methyl-2-pentene**   **(Z)-2-Hexene**   **1-Hexene**   **(E)-2-Hexene**
                                                          (*cis*-2-hexene)   (*trans*-2-Hexene)

▲ FIGURE 2-13

Structures of isomeric hexenes (C_6H_{12}).

ALKENES AS HIGH-QUALITY GASOLINE COMPONENTS

Alkenes have a higher octane rating than do the corresponding alkanes. It is common practice in petroleum refineries to treat petroleum
mixtures with catalysts at high temperatures, at which the equilibrium
between alkane and alkene plus H_2 favors the latter. The hydrogen
produced is a valuable by-product that can be used in other processes.

Hydrogenation also allows one to distinguish the contributions of double bonds and rings to the index of hydrogen deficiency. By exhaustive hydrogenation, two hydrogens are added to each double bond but not to saturated rings (except highly strained cyclopropanes). Any remaining
hydrogen deficiency after catalytic hydrogenation must be caused by the
presence of rings rather than multiple bonds.

Hyperconjugation. An orbital description of the stabilizing effect of
alkyl groups on adjacent π bonds derives from hyperconjugation. Illustrated in Figure 2-14 for propene, **hyperconjugation** results from the donation of electron density from an adjacent σ bond to a π-antibonding molecular orbital. Because there is free rotation about the carbon–carbon σ bond
connecting the C-2 carbon to C-3 in propene, one possible three-dimensional representation places one of the C–H bonds in the methyl group in
good alignment with the p orbital of the double bond, as shown at the left
in Figure 2-14. The alignment of the electrons in this σ bond with the empty
antibonding (π^*) orbital weakly stabilizes the molecule. This spatial
arrangement permits a favorable electronic interaction that does not occur if
the σ bond is positioned in the plane of the double bond, as shown in Figure 2-15.

Each time an alkyl group replaces a hydrogen on a carbon–carbon double bond, the number of such possible hyperconjugative interactions increases. For example, in Figure 2-16, 2-butene has two possible interactions

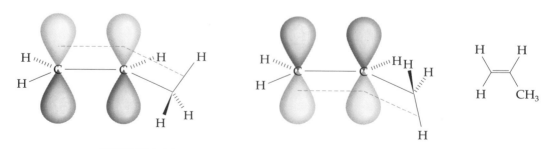

▲ FIGURE 2-14
Required geometries for hyperconjugation in propene.

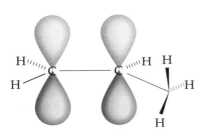

No hyperconjugation is possible from the carbon–hydrogen bond of the methyl group in this three-dimensional arrangement.

(in which any of the three hydrogens on the alkyl group can so overlap), whereas there is only one hyperconjugative group in 1-butene. The more alkyl groups on the double bond, the greater is the number of atoms that can hyperconjugate and the more stable is the double bond. This stability reveals itself in a lower heat of hydrogenation for the more highly substituted double bond (Table 2-1).

EXERCISE 2-H

Draw three-dimensional representations that illustrate hyperconjugative stabilization of the double bond of each of the following compounds:

(a) 1-butene   (b) *trans*-2-butene   (c) 2,3-dimethyl-2-butene

---

Heats of Formation. The relative stabilities of isomeric alkenes can also be ranked according to their heats of formation. As defined in Chapter 1, the heat of formation is a theoretical number that describes the energy that would be released if a molecule were formed from its component elemental atoms in their standard states.

$$n\,\text{C} \quad + \quad n\,\text{H}_2 \longrightarrow \text{C}_n\text{H}_{2n} \quad + \quad \text{heat } (\Delta H^\circ_f)$$
$$\text{(graphite)} \qquad \text{(gas)}$$

As a means of ordering alkene stability, heats of formation provide the same information as heats of combustion. In any case, we are most interested in the *differences* between the heats of combustion (or hydrogenation or formation) of isomers and not in the absolute values.

EXERCISE 2-I

Describe how a measured heat of hydrogenation of an alkene, together with the heat of combustion of H_2 and the heat of combustion of its hydrogenation product, can give the heat of combustion of the alkene.

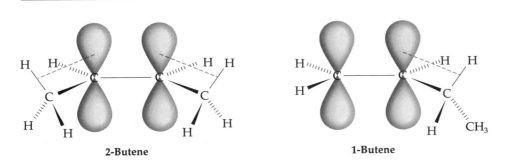

2-Butene   1-Butene

▲ FIGURE 2-16
Hyperconjugative interactions in 1- and 2-butene.

2-2 ▸ Dienes

As we know, a double bond in a molecule is a site of special reactivity, a functional group. More than one such group can exist in a molecule, and compounds containing two double bonds are called **dienes.**

Dienes can have several different positional relations, as shown for the two double bonds in the C_5 skeleton in Figure 2-17. In 1,3-pentadiene, there are p orbitals on four adjacent atoms, whereas the array of p orbitals in 1,4-pentadiene is interrupted by an sp^3-hybridized carbon atom at C-3. The interaction of the p orbitals in these two systems is therefore different. As in simple alkenes, each of the positions of the double bonds is indicated by a number. In 1,3-pentadiene, the double bonds interact directly and are referred to as **conjugated.** In 1,4-pentadiene, the two double bonds do not interact directly and are therefore referred to as **isolated.** The introduction of a second double bond into a molecule to form a diene (with the removal of two hydrogens from adjacent carbons) changes the overall formula from C_nH_{2n} (for the alkenes) to C_nH_{2n-2}. In 1,2-pentadiene, the two double bonds are abutting and are therefore referred to as **cumulated.**

Pi-orbital overlap between aligned double bonds stabilizes the molecule. For example, 1,3-pentadiene is more stable than 1,4-pentadiene and is most stable in the geometry shown in Figure 2-17 in which all the atoms lie in one plane (and, hence, in which the p orbitals are perfectly aligned). The molecule prefers this geometry to one in which rotation about the C-2–C-3 σ bond puts the π bonds in perpendicular planes.

EXERCISE 2-J

Write acceptable names for each of the following hydrocarbons:

(a)   (d)

(b)

(c)   (e)

EXERCISE 2-K

Draw acceptable structures to correspond with each of the following IUPAC names:

(a) 2-(1-cyclobutenyl)-1-hexene   (c) 1,2-pentadiene
(b) *trans*-1,4-heptadiene   (d) *(E)*-3-methyl-1,3-pentadiene

More-extended conjugated systems also can be constructed. β-Carotene (Figure 2-18), the compound responsible for the yellow-orange color of carrots, and vitamin A, a compound needed for light sensitivity in human vision, contain long conjugated systems that make them sensitive to light in the visible region. These compounds are called **polyenes** because of the presence of several (*poly* is Greek for many) double bonds. The dependence of light absorption on extended conjugation is discussed further in Chapter 4.

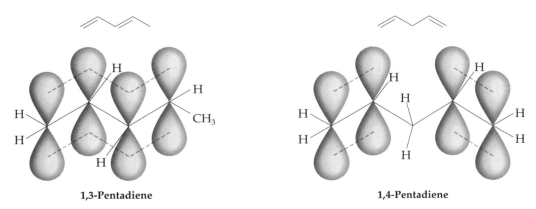

▲ FIGURE 2-17
Conjugated and isolated dienes.

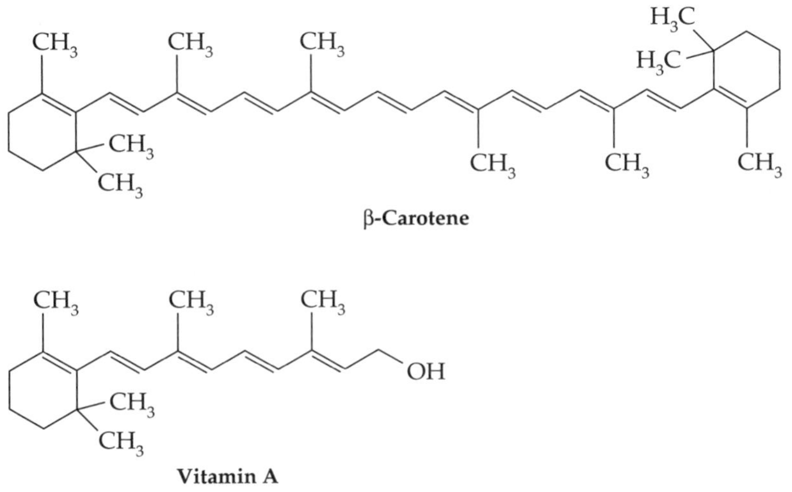

β-Carotene

Vitamin A

▲ FIGURE 2-18
Two naturally occurring polyenes.

2-3 ▸ Aromatic Hydrocarbons

Planar, conjugated, cyclic, unsaturated molecules constitute another important class of sp^2-hybridized hydrocarbons. Some of these compounds have unusual stability and are referred to as **aromatic hydrocarbons,** originally because of their characteristic odor. The parent compound of this family is benzene (Figure 2-19). Benzene, C_6H_6, consists of an array of six sp^2-hybridized carbons, each attached by a σ bond to a hydrogen atom. The

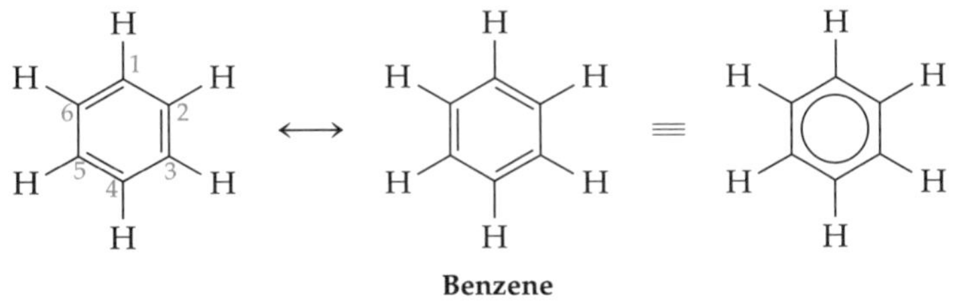

Benzene

▲ FIGURE 2-19
Kekulé structures (resonance contributors) of benzene.

Benzene

Cyclohexatriene

formulation of its structure as a planar cyclic arrangement of CH units is one of the classic tales of organic chemistry, having been imagined in a dream by the German chemist August Kekulé as a snake biting its tail.

Benzene is known to have equivalent carbon–carbon bond lengths at each position around the ring, and so a description of benzene as cyclohexatriene (with alternating single and double bonds) must be wrong. If benzene existed as the cyclohexatriene shown in the margin, the double bonds would be shorter than the single bonds, with an **alternation** in bond length from one atom to the next around the ring.

Resonance Structures

We have identified a fundamental problem with the way in which we draw chemical structures: we cannot represent the structure of benzene properly with a single structure using only single and double bonds. The bond alternation implied by this structure can be avoided, however, if we recognize that there are two possible arrangements for the π bonds in benzene. This is indicated in the two representations in Figure 2-19 in which a double bond can be placed either between C-1 and C-2 or between C-1 and C-6. There is no reason for an energetic preference for one or the other of these two representations, which are called **Kekulé structures.** Benzene is better represented as a combination of each of these structures.

Although these Kekulé structures, which depict benzene as having localized double bonds, do not exist, they serve as a convenient shorthand for allowing us to count multiple bonds and electrons. By convention, we use a double-headed arrow to indicate that these structures are **resonance contributors,** or **resonance structures,** differing only with respect to the formal localization of electrons and *not* with respect to positions of atoms. Often, another notation, in which a hexagon (representing the ring carbons of benzene) encloses a circle (representing the conjugated triene units), is used to indicate the equal contributions of these resonance structures. This representation is frequently used because of our inability to draw descriptive structures with conventional bond representations.

EXERCISE 2-L

Draw an alternative resonance structure to each of those shown here:

(a) 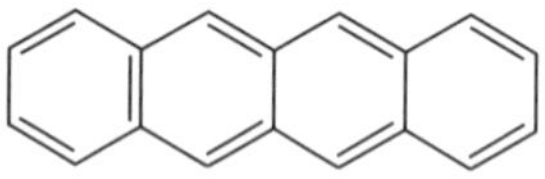 **Tetracene**

(b) 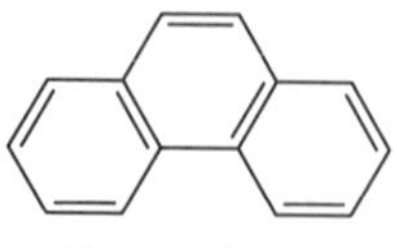 **Phenanthrene**

(c) 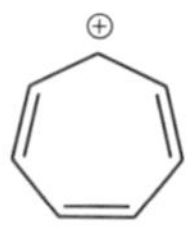 **Cycloheptatrienyl (tropylium) cation**

(d) 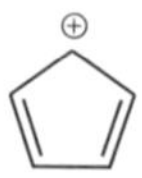 **Cyclopentadienyl cation**

Many other possible resonance contributors can be drawn to depict possible electron distributions in benzene. Resonance structures are drawn by keeping the position of each atom fixed while shifting electrons within the π system. In these drawings, two electrons from a π bond are localized on a single carbon, making that atom negatively charged and some other atom of the conjugated system positively charged.

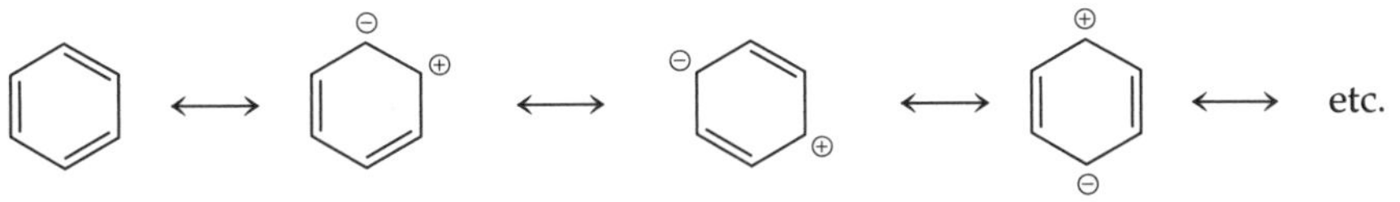

These structures, which are neutral overall but contain equal numbers of locally charged (plus and minus) centers, are called **zwitterions.**

These zwitterionic resonance contributors are of higher energy than the uncharged Kekulé forms shown in Figure 2-19. Not only does it cost energy to create charged centers from a neutral species, but these structures also have one fewer covalent π bond than do the Kekulé contributors. Thus, although we can write such structures, their importance in describing the electron distribution in benzene is minor, and we can usually neglect such structures in our thinking. Much the same can be said about biradical contributors, in which single electrons are localized on two of the atoms of the conjugated systems.

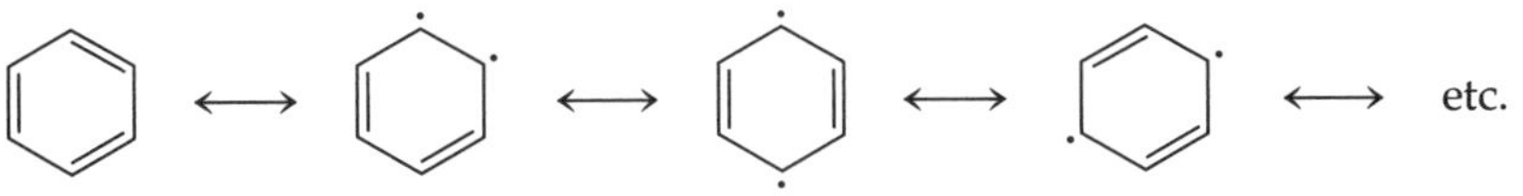

(Recall the twisted structure of ethylene discussed earlier.) Unlike the zwitterionic contributors, these biradical structures do not have charge-separated states; however, a biradical contributor lacks one of the π bonds that is intact in a Kekulé contributor.

Stability

The unusual stability of benzene (and related structures) is seen both in its heat of hydrogenation and in its chemical reactivity, which differ appreciably from those usually observed in conjugated alkenes, dienes, or trienes. The differences in reactivity will be discussed further in Chapter 11, but we consider the hydrogenation data here. More-severe conditions of temperature and pressure must be employed to add hydrogen catalytically to benzene to generate cyclohexane than to analogous alkenes.

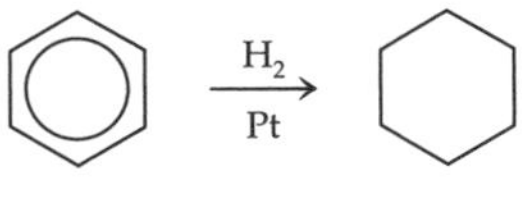

$$\Delta H° = -49.3 \text{ kcal/mole}$$

The stoichiometry of the reaction tells us that three moles of hydrogen are taken up per mole of benzene. If the double bonds were noninteractive (as in a hypothetical cyclohexatriene), the heat of hydrogenation of benzene would be approximately three times that of cyclohexene. The difference between the heat of hydrogenation of benzene (49.3 kcal/mole) and three times that of cyclohexene (3 × 28.4 kcal/mole) is approximately 36 kcal/mole.

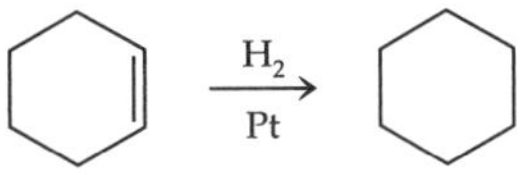

$$\Delta H° = -28.4 \text{ kcal/mole}$$

This difference cannot be due to a simple conjugation effect because the difference between the heat of hydrogenation of 1,3-cyclohexadiene (54.9 kcal/mole) and twice the heat of hydrogenation of cyclohexene (2 × 28.4 kcal/mole) is small (approximately 2 kcal/mole).

The larger difference seen with benzene must therefore have a different origin, and the special stability afforded by a planar cyclic array of *p* orbitals containing six electrons is known as **aromaticity.**

Other evidence that aromaticity is not a simple conjugation effect can be seen in the differing stabilities of benzene and cyclic hydrocarbons containing four and eight CH units, respectively. Despite the fact that cyclobutadiene can be written with two resonance contributors analogous to the Kekulé structures of benzene, as shown in the margin, it is an exceedingly unstable molecule that can be prepared and studied only at very low temperature under special conditions. Its preparation, in fact, was not achieved until the late 1960s, whereas benzene was isolated early in the nineteenth century and can be easily stored at room temperature. When benzene does react, much more rigorous conditions are required than with simple unsaturated hydrocarbons; that is, reactions of aromatic compounds are induced only with some difficulty.

Cyclooctatetraene has been shown to exist not as a planar hydrocarbon but rather as the tub-shaped unsaturated molecule shown in the margin. In this geometry, the *p* orbitals are not well aligned for interaction.

Aromaticity and Hückel's Rule

Both cyclobutadiene and cyclooctatetraene contain multiples of four electrons; that is, four and eight electrons, respectively. Benzene, on the other hand, contains two electrons more than a multiple of four (that is, $6 = 4n + 2$, in which *n* is an integer—in this case, $n=1$). This distinction was recognized in 1938 by Erich Hückel who generalized this observation into what has come to be known as **Hückel's Rule:** any planar, cyclic, conjugated system containing $4n + 2$ π electrons (in which *n* is an integer) experiences unusual aromatic stabilization, whereas those containing $4n$ π electrons do not.

For aromatic molecules to be stabilized by orbital interaction as predicted by Hückel's Rule requires that the *p* orbitals (Figure 2-20) be aligned

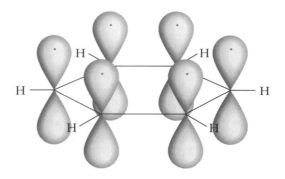

▲ **FIGURE 2-20**

Benzene as a Hückel six-electron system, with the contribution of one electron from each carbon being represented by a dot in each *p* orbital.

CYCLOOCTATETRAENE: A NONAROMATIC CONJUGATED CYCLIC HYDROCARBON

Cyclooctatetraene was produced in large quantity in World War II by German chemists seeking a way to convert acetylene (ethyne) into benzene. Acetylene is a low-boiling hydrocarbon that can spontaneously explode when stored under pressure. It can be readily prepared from calcium carbide, which, in turn, is prepared by heating $CaCO_3$ and coal. Trimerization of acetylene yields benzene, whereas the joining of four molecules of acetylene leads to cyclooctatetraene.

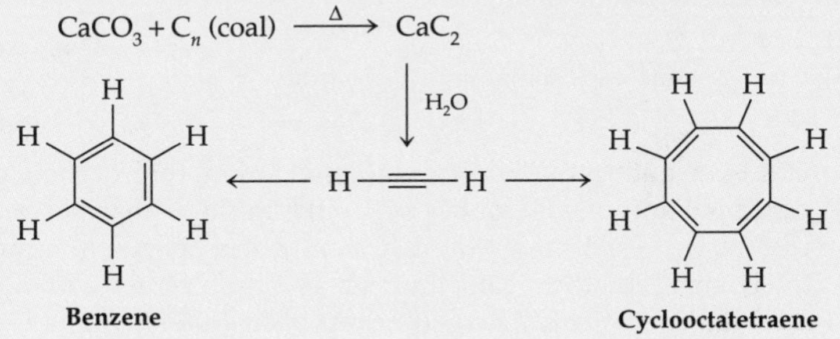

$$CaCO_3 + C_n \text{ (coal)} \xrightarrow{\Delta} CaC_2$$

Benzene   **Cyclooctatetraene**

in a planar geometry. Larger or smaller cyclic structures that maintain this alignment also are subject to Hückel's Rule. For example, a cyclic, planar conjugated array containing $4n + 2$ electrons in a smaller ring is found in the cyclopropenyl cation and in the cyclopentadienyl anion (Figure 2-21). Notice that, in the cyclopropenyl cation, one orbital is vacant and formally bears a (+1) charge. We can write resonance structures for this cation by shifting electrons in the π bond from the position between C-1 and C-2 to the position between C-2 and C-3. All of the atoms remain in the same position in each resonance structure. Only the position of the electrons is changed.

In the cyclopentadienyl anion, one p orbital is doubly occupied and bears a formal (−1) charge. By shifting electrons in the π system as we did in the cyclopropenyl cation, we can write five equivalent resonance contributors, of which only two are shown in Figure 2-21. We can count the number of electrons in such systems by recognizing that each p orbital in a formal π bond is populated by a single electron. In the cyclopropenyl cation, two of the p orbitals (those participating in the π bond) contain one electron. The

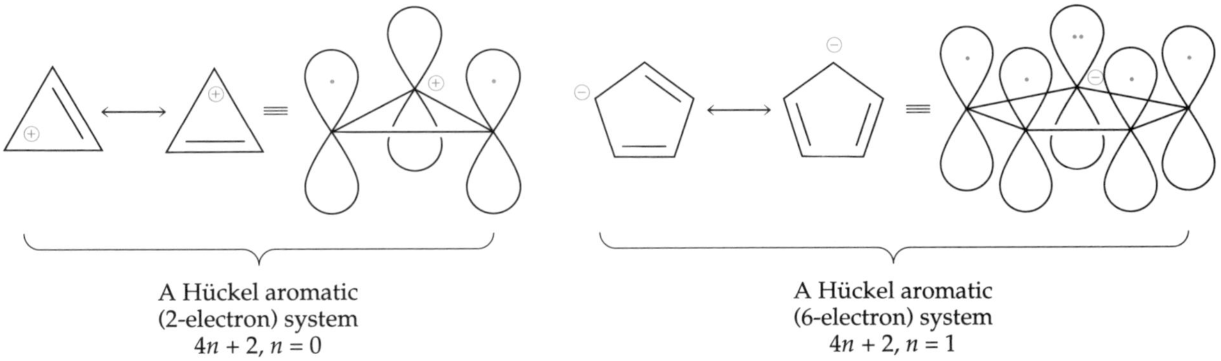

A Hückel aromatic
(2-electron) system
$4n + 2, n = 0$

A Hückel aromatic
(6-electron) system
$4n + 2, n = 1$

▲ FIGURE 2-21
Hückel aromatic rings containing $4n + 2$ electrons.

▲ **FIGURE 2-22**

A Hückel antiaromatic
(4-electron) system
$4n$, $n = 1$

A Hückel antiaromatic
(4-electron) system
$4n$, $n = 1$

Two anti-Hückel systems, each containing $4n$ electrons.

third p orbital is vacant. In the cyclopentadienyl anion, four of the p orbitals are singly occupied and one is doubly occupied. In these analyses, a double bond contributes two electrons to the π system, a center with a vacant p orbital (positively charged atom) contributes zero, and a center with a doubly occupied p orbital (negatively charged atom) contributes two, as is consistent with our earlier discussion of formal charge calculation. Thus, cyclopropenyl cation is a two-electron Hückel system $[(4 \times 0) + 2]$ and cyclopentadienyl anion is a six-electron Hückel molecule $[(4 \times 1) + 2]$.

Both of the oppositely charged ions shown in Figure 2-22—that is, the cyclopropenyl anion and the cyclopentadienyl cation—contain four electrons ($4n$, in which $n = 1$) and therefore lack the aromatic stabilization characteristic of a Hückel system. These structures are known to be so unstable that they have in fact been called **antiaromatic.**

EXERCISE 2-M

Using Hückel's Rule, predict which of the following hydrocarbons, one of whose resonance contributors is shown below, will exhibit aromatic stabilization.

(a)   (b)   (c)   (d)   (e)

Arenes

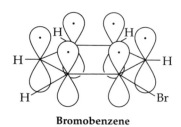

Bromobenzene

Derivatives of benzene obtained by the replacement of hydrogen by other groups or by the fusion of additional rings are called **arenes**. The delocalized π orbitals of benzene shown in Figure 2-20 are perpendicular to any substituents that may be attached to the carbon atoms. For example, a three-dimensional representation of bromobenzene shows that the carbon–bromine bond is in the atomic plane and is thus completely orthogonal to the aromatic π array. Substituted benzenes maintain a Hückel number of electrons in the π system and, to a first approximation, exhibit the same aromaticity as the parent benzene.

Because substituted benzenes are quite common, a special terminology is used to refer to the relative positioning of the two substituents in disub-

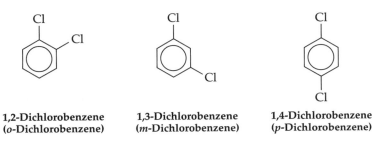

1,2-Dichlorobenzene
(*o*-Dichlorobenzene)   **1,3-Dichlorobenzene**
(*m*-Dichlorobenzene)   **1,4-Dichlorobenzene**
(*p*-Dichlorobenzene)

▲ FIGURE 2-23
Disubstituted chlorobenzenes.

Toluene   *o*-**Xylene**   *m*-**Xylene**   *p*-**Xylene**

▲ FIGURE 2-24
Common names of mono- and dimethylbenzenes.

stituted structures. For example, the 1,2 isomer of dichlorobenzene (Figure 2-23) is alternatively referred to as *ortho* substituted (*o*-), the 1,3 isomer is called *meta* substituted (*m*-) and the 1,4 isomer is called *para* substituted (*p*-). (Note that a "1,5" isomer is identical with a 1,3 and that a "1,6" isomer is identical with a 1,2 isomer.) Because methyl substituted benzenes (Figure 2-24) also are common reagents, they have common names. Methylbenzene is commonly referred to as toluene and the dimethylbenzenes are referred to as xylenes.

EXERCISE 2-N

Write acceptable names for each of the following hydrocarbons:

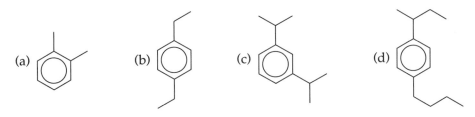

(a)   (b)   (c)   (d)

EXERCISE 2-O

Draw structures to correspond with each of the following names:

(a) *ortho*-dichlorobenzene

(b) *meta*-dibromobenzene

(c) *para*-diiodobenzene

The particular stability of a benzene ring can also be extended to fused cyclic aromatic hydrocarbons. These **polycyclic aromatic hydrocarbons** exhibit chemical stability similar to that of benzene. A representative sample,

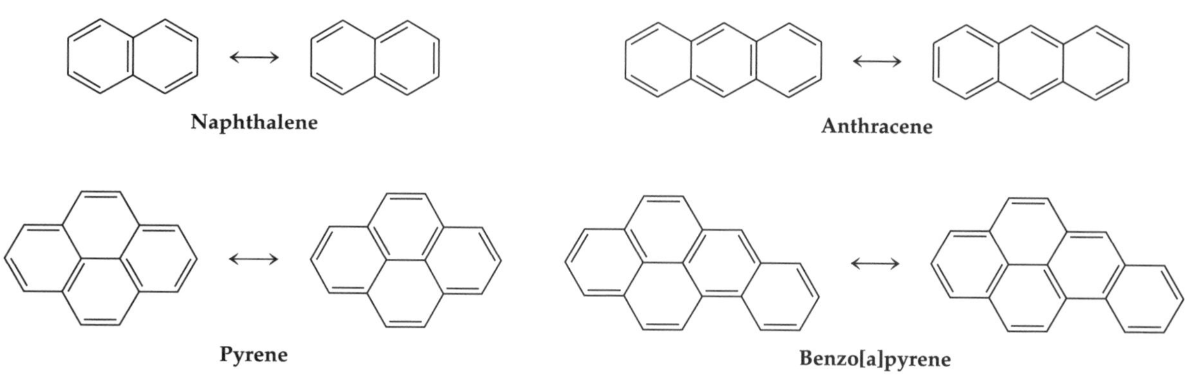

Naphthalene

Anthracene

Pyrene

Benzo[a]pyrene

▲ FIGURE 2-25

Some fused-ring (polycyclic) aromatic hydrocarbons.

together with their common names, is shown in Figure 2-25. Like monocyclic aromatic compounds, these polycyclic aromatic compounds have several important resonance contributors. Bear in mind that each compound shown can have more than one important contributing resonance structure. Benzo[a]pyrene was one of the first clearly identified **carcinogens,** or cancer-inducing agents. This compound, found in soot formed from the partial combustion of wood, was shown in the nineteenth century to be responsible for chemically induced scrotal cancer in chimney sweeps in London.

If a benzene ring is a substituent on a carbon chain, the C_6H_5 fragment is called **phenyl** (Figure 2-26); and, if a (generic) arene is present, it is called an **aryl group.** An alkene substituent is called a **vinyl group** when the attachment is to one of the carbons in the double bond. When a three-carbon alkenyl chain is attached at the atom adjacent to the double bond, the substituent is called an **allyl group.**

EXERCISE 2-P

Write acceptable names for each of the following hydrocarbons:

(a) 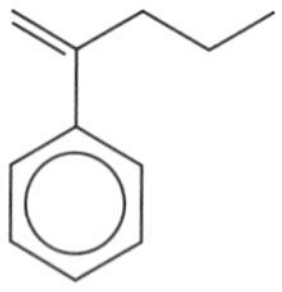   (b) 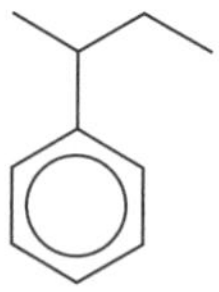   (c) 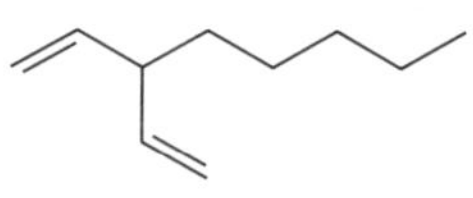   (d)

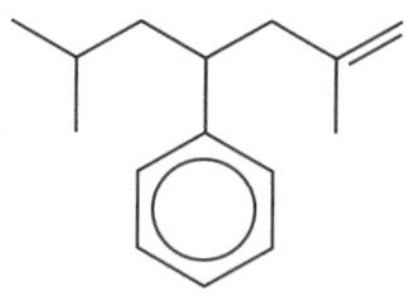

ALLYL: A GROUP FOUND IN GARLIC

Many common names of organic compounds and groups derive from the botanical names of the plants from which they were first isolated. The allyl group is present in a key amino acid isolated from garlic, and the term "allyl" derives from the botanical name for garlic (*Allium sativum*), which comes from the Latin word *allium*, from a Celtic word meaning pungent.

Garlic supposedly protects against stroke, coronary thrombosis, and hardening of the arteries. Its extracts also have antibacterial and antifungal activity—and allegedly repel vampires.

CARCINOGENICITY OF BENZO[A]PYRENE

A derivative of benzo[a]pyrene, rather than the hydrocarbon itself, is the real culprit in inducing cancer. Because benzo[a]pyrene is a large hydrocarbon with a very low solubility in water, it collects in the liver, which is composed in part of fats—hydrocarbon-rich molecules that will be discussed in Chapter 17. There are many enzymes in the liver that carry out oxidation reactions on waste products of metabolism, as well as on unneeded materials consumed in the diet. These oxygenated materials have a higher solubility in water and can thus be excreted. Unfortunately, the oxidation product of benzo[a]pyrene, shown below, interacts with DNA and results in abnormal cell growth (cancer).

▲ **FIGURE 2-26**

Some unsaturated substituent groups. In the aryl group, X is an undefined substituent for which the point of attachment can be at the *ortho, meta,* or *para* position. It is common for the aromatic ring to have multiple substituents.

Phenyl	Vinyl	Aryl	Allyl
$-C_6H_5$	$-C_2H_3$	$-C_6H_{(5-n)}X_n$	$-C_3H_5$

2-4 ▸ Alkynes

When a carbon atom σ-bonds to two (rather than three or four) atoms, its valence requirement is satisfied by the formation of two π bonds. When both π bonds are directed to the same atom, a triple bond (composed of one σ and two π bonds) is formed. In this section, we will learn about the geometry and chemical character of triple bonds and how *sp* hybridization (rather than the *sp*³ hybridization in alkanes or the *sp*² hybridization in alkenes) is required in order to form triple bonds.

sp Hybridization

In the third fundamental type of carbon hybridization, an *s* orbital mixes with one of the *p* orbitals to produce two hybrid orbitals, as shown in Figure 2-27. As before, these two hybrid orbitals are directed as far from each other as possible—in this case, producing a 180° bond angle at the *sp*-**hybridized** atom. The remaining 2*p* orbitals (which do not participate in hybridization) are orthogonal to each other and to the hybrid orbitals. The geometry of orbitals at an *sp*-hybridized carbon atom is therefore that shown in the margin.

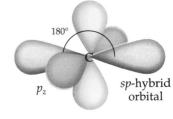

▶ FIGURE 2-27

Orbital mixing in *sp*-hybridized carbon.

$$C: 1s^2 \; \underbrace{2s^1 2p_x^1}_{\substack{\text{Two} \\ sp\text{-hybrid} \\ \text{orbitals}}} \; \underbrace{2p_y^1 p_z^1}_{\substack{\text{Two } p \\ \text{orbitals}}}$$

When two such *sp*-hybridized orbitals overlap to form a carbon–carbon σ bond (Figure 2-28), the additional valence electronic requirement of carbon is met by the formation of an additional σ bond (by overlap of the remaining *sp*-hybrid orbital of each carbon) to another substituent in a geometry colinear with the C–C sigma bond. When this substituent is hydrogen, the resulting structure is **ethyne,** also called **acetylene.** This geometry also allows for optimal overlap between the aligned p_y and p_z orbitals on adjacent carbons above and below, and in front of and behind, the σ bond. (The directional subscripts, p_y and p_z, are arbitrary.) This overlap thus forms one σ and two π bonds between the carbons. The σ bond connecting triply bound carbons is completely surrounded by a cloud of π-electron density; its bond length (1.20 Å) is less than that of a single or a double bond. Rewriting the Lewis dot structure with line notation indicates a triple bond between the carbon atoms.

$$\text{H:C:::C:H} \quad \equiv \quad \text{H}\!-\!\text{C}\!\equiv\!\text{C}\!-\!\text{H}$$
$$\text{C}_2\text{H}_2$$

Two of the electrons shared between the two carbons are accommodated in a σ bond and four are accommodated in two π orbitals.

Because the carbon–carbon σ bond at an *sp*-hybridized atom has less *p* character than those at sp^2- or sp^3-hybridized atoms, the *sp*-hybrid orbitals are less elongated and the σ bond formed from them is shorter, whether the *sp*-hybridized carbon is bound to an *sp*-, sp^2-, or sp^3-hybridized atom. A summary of the dependence of bond lengths and bond angles on hybridization is given in Table 2-2 for several hydrocarbons.

Higher Alkynes

As mentioned earlier, a hydrocarbon containing a triple bond is called an alkyne. Having formally lost two hydrogens in forming each π bond, an alkyne has an overall formula of C_nH_{2n-2}; that is, four fewer hydrogens than the corresponding alkane. Thus, the presence of one triple bond is equivalent to two units of unsaturation. As with the alkanes and the alkenes, alkynes can be ranked in order of stability by determining their heats of combustion or heats of hydrogenation. As we observed for substituted alkenes in Table 2-1, alkyl substitution stabilizes the multiple bond, making, for example, 2-butyne more stable than 1-butyne. Because of the linearity of an alkyne, no possibility for *E-Z* isomerism exists in this family.

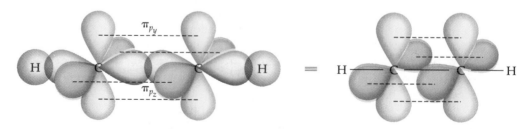

▲ FIGURE 2-28

A three-dimensional view of C–C σ and π bonding in ethyne.

Table 2-2 ▸ Bond Lengths and Angles in Representative Hydrocarbons

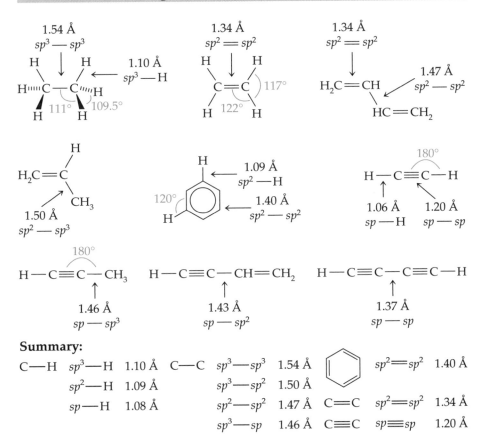

Summary:

C—H	sp^3—H	1.10 Å	C—C	sp^3—sp^3	1.54 Å		sp^2=sp^2	1.40 Å
	sp^2—H	1.09 Å		sp^3—sp^2	1.50 Å			
	sp—H	1.08 Å		sp^2—sp^2	1.47 Å	C=C	sp^2=sp^2	1.34 Å
				sp^3—sp	1.46 Å	C≡C	sp≡sp	1.20 Å
				sp^2—sp	1.43 Å			
				sp—sp	1.37 Å			

Alkynes are named according to the same IUPAC rules as those discussed in Section 2-1 for alkenes, with a **-yne** suffix indicating the presence of a triple bond. As before, we must (1) find the longest chain that contains a triple bond; (2) begin numbering the chain so as to assign the functional group the lowest possible number; (3) name branches as alkyl groups; and (4) use Greek prefixes to indicate multiple substituents.

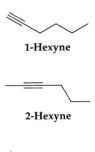

1-Hexyne

2-Hexyne

3-Hexyne

EXERCISE 2-Q

Write acceptable names for each of the following hydrocarbons:

(a)   (b)   (c)   (d)

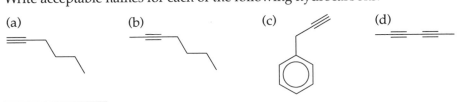

EXERCISE 2-R

Draw acceptable structures to correspond with each of the following IUPAC names:

(a) 2-heptyne   (c) 2-methyl-3-hexyne

(b) 3-hexyne   (d) 1,5-octadiyne

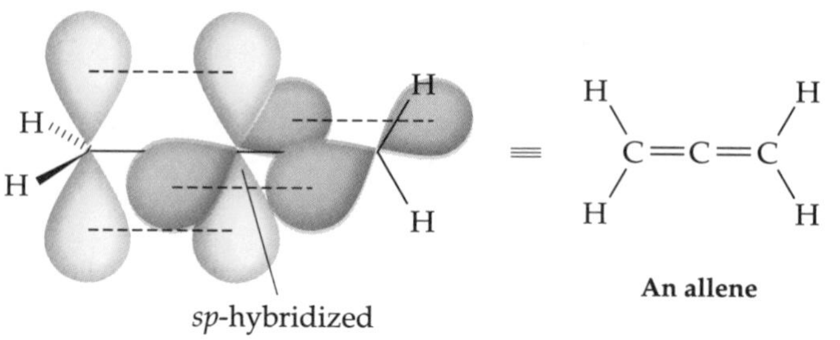

▲ FIGURE 2-29

A three-dimensional view of π bonding in an allene (1,2-propadiene).

Although the most-common functional group that incorporates *sp*-hybridized carbon atoms is the alkyne—that is, a compound with a carbon–carbon triple bond—it is also possible to have *sp*-hybridized atoms in molecules that have two double bonds emanating in opposite directions from a common carbon atom. These compounds are referred to as **allenes,** one of which is shown in Figure 2-29.

The central *sp*-hybridized carbon of an allene has a σ bond and a π bond to each of the two neighboring carbons. However, unlike the alkyne in Figure 2-28, the two π bonds of 1,2-propadiene are directed toward different carbon atoms. The orthogonality of the *p* orbitals on the *sp*-hybridized atom dictates the geometry for the rest of the molecule. Thus, the hydrogens and carbon at one end of an allene are in a plane that is perpendicular to the plane containing the two hydrogens and carbon at the other end of the molecule.

An allene is also referred to as a cumulated diene in that the double bonds share a common carbon atom. Because the two π bonds are orthogonal, they cannot interact as they would in a conjugated diene. Therefore, a cumulated diene is less stable than a conjugated diene.

EXERCISE 2-S

Using sawhorse designations, draw the two possible three-dimensional structures of 2,3-pentadiene. Translate these two drawn structures into molecular models, and convince yourself that the two isomers are not identical.

Like the carbon–carbon double bond, the carbon–carbon triple bond is a functional group. In parallel to the structures presented in Section 2-2 for dienes, it is also possible to have conjugated diynes, conjugated enynes, isolated diynes, or isolated enynes.

1,3-Pentadiyne	Penta-3-ene-1-yne	1,4-Pentadiyne	Penta-4-ene-1-yne
(a conjugated diyne)	(a conjugated enyne)	(an isolated diyne)	(an isolated enyne)

2-5 ▸ Physical Properties of Hydrocarbons

As is true of alkanes, unsaturated hydrocarbons (alkenes, dienes, aromatic hydrocarbons, and alkynes) are relatively nonpolar and interact intermolecularly primarily through van der Waals attraction. Accordingly, these families have relatively low melting points and boiling points (Table 2-3) and are soluble in nonpolar solvents.

PHYSIOLOGICALLY ACTIVE ALKYNES

Several naturally occurring compounds containing alkynes have been isolated from microbes. Among them are the dynemicins, illustrated by the structure of dynemicin A, which contains two alkynes and an alkene (as well as other functional groups). These compounds have potent antibacterial and anticancer activity but, unfortunately, are probably too toxic to mammals to be used as pharmaceutical agents.

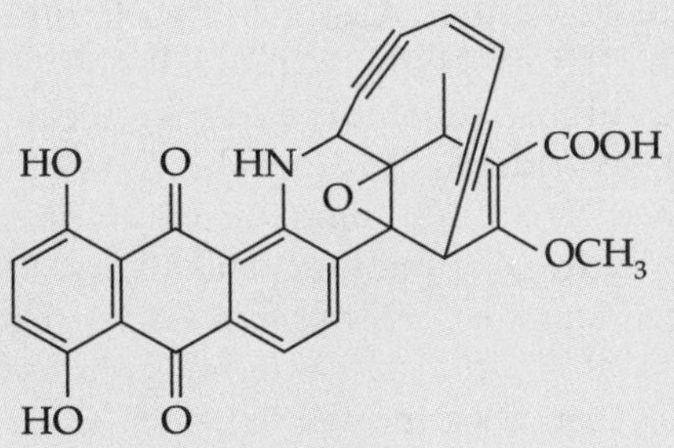

Dynemicin A

Table 2-3 ▸ Physical Properties of Some Unsaturated Hydrocarbons

Name	Formula	Boiling Point (°C)	Melting Point (°C)
Ethene	C_2H_4	−104	−169
Ethyne	C_2H_2	−84	−81
Propene	C_3H_6	−48	−185
Propyne	C_3H_4	−23	−102
1-Butene	C_4H_8	−6	−185
cis-2-Butene	C_4H_8	+4	−139
trans-2-Butene	C_4H_8	+1	−105
1-Butyne	C_4H_6	8	−126
2-Butyne	C_4H_6	27	−32
Benzene	C_6H_6	80	5.5

Conclusions

In hydrocarbons bearing sp^2-hybridized carbons, the alkenes, maximum stability is achieved when the bond angles are about 120° and the p orbitals are aligned, resulting in a planar structure. The carbon skeleton is held together by sigma (σ) and pi (π) bonds. The length of a carbon–carbon double bond is shorter (1.34 Å) than that of a carbon–carbon single bond. If a molecule contains more than one double bond, the groups can be conjugated, isolated, or cumulated. In conjugated systems, delocalization stabilizes the molecule.

Molecules composed of sp^2-hybridized atoms in planar, cyclic conjugated arrays display aromatic properties if they contain $4n + 2$ electrons. The unusual stability of aromatic molecules can be predicted empirically by Hückel's rule. Aromatic systems usually have two or more dominant resonance contributors, which together describe the delocalized electron density of the molecule.

In hydrocarbons bearing adjacent *sp*-hybridized orbitals (the alkynes), the hybrid orbitals are directed at 180° to minimize electron repulsion, with the atoms bound to the *sp*-hybridized carbon atoms being colinear. The triple bond, composed of one σ bond and two π bonds orthogonal to the σ system, is approximately 1.20 Å in length, shorter than a double or single bond.

The bond energy of a σ bond is greater than that of a π bond. Pi bonds are thus more reactive and form the basis for localized chemical reactivity. A multiple bond can then be considered a functional group. Relative stabilities of isomers can be ranked on the basis of heats of combustion, heats of hydrogenation, or heats of formation. A more highly substituted alkene is more stable than a less highly substituted isomer because of the higher electronegativity of an sp^2-hybridized atom and because of hyperconjugation.

The introduction of a double bond into a hydrocarbon backbone reduces the number of hydrogens in an acyclic alkane (C_nH_{2n+2}) by two, and so an acyclic alkene has an overall formula of C_nH_{2n}. The introduction of additional double bonds also reduces the number of hydrogens by two per multiple bond. The presence of the triple bond in an alkyne reduces the number of hydrogens from the alkane formula by four. Thus, an acyclic alkyne has the overall formula C_nH_{2n-2}. All hydrocarbons can be named according to IUPAC rules.

Summary of New Reactions

Catalytic Hydrogenation

$$C_nH_m \xrightarrow[\text{Pt}]{H_2} C_nH_{m+2}$$

Hydrocarbon Combustion

$$C_nH_m \xrightarrow[\Delta]{O_2} n\,CO_2 + m/2\,H_2O$$

Review Problems

2-1 Draw structures that correspond with each of the following names:

(a) *cis*-1,3-pentadiene

(b) 4-methylcyclopentene

(c) 3-*t*-butyl-1-hexene

(d) *m*-bromotoluene

(e) octa-1-ene-4-yne

(f) 1,4-dihydroxynaphthalene

2-2 For each of the following pairs, determine whether catalytic hydrogenation (where the molar equivalent of hydrogen uptake is followed) can be used to distinguish between the compounds.

(a) cyclohexane and 1-hexene

(b) 1-hexene and (Z)-2-hexene

(c) cyclohexane and methylcyclopentane

(d) cyclohexane and cyclohexene

(e) 1-butene and 1-butyne

(f) 1-butene and 1-pentene

2-3 Where possible, assign *E* or *Z* configuration to the following alkenes:

(a) 2-pentene

(c) 3-methyl-2-pentene

(b) 2-methyl-2-butene

(d) 4-methyl-2-pentene

(e) 2-methyl-1-butene   (f) 1,4-hexadiene

2-4 Calculate the index of hydrogen deficiency in each of the following naturally occurring hydrocarbons. From this value, calculate the number of double bonds, given the indicated number of rings for each compound.

(a) limonene (responsible for "citrus" odor), $C_{10}H_{16}$, one ring

(b) acenaphthene (in coal tar and sauna mud), $C_{12}H_{10}$, three rings

(c) benzo[a]pyrene (a carcinogen in soot), $C_{20}H_{12}$, five rings

(d) β-pinene (in pine needles and bark), $C_{10}H_{16}$, two rings

(e) caryophyllene (oil of cloves), $C_{15}H_{24}$, two rings

(f) β-cadinene (produces odor of cedar), $C_{15}H_{24}$, two rings

2-5 Determine whether *cis-trans* isomerism is possible in each of the following compounds, drawing the structure of the geometric isomers where possible.

(a) 1-hexene

(b) 2-pentene

(c) 2-methyl-1-butene

(d) 2-methyl-2-butene

(e) 2-fluoro-2-butene

(f) 1,2-dichlorocyclohexane

(g) 1,2-dimethylcyclobutene

(h) 3,4-dimethylcyclobutene

2-6

(a) Arrange the following alkenes in order of their relative stabilities: *trans*-3-heptene; 1-heptene; 2-methyl-2-hexene; *cis*-2-heptene; 2,3-dimethyl-2-pentene

(b) For which pairs of compounds in part *a* can relative stabilities be determined by comparing heats of hydrogenation?

2-7 Which compound would you expect to have the larger heat of hydrogenation: (a) *cis*-cyclooctene or *trans*-cyclooctene; (b) *cis*-2-hexene or *trans*-2-hexene? Explain.

2-8 When 1,2-deuterocyclopentene reacts with hydrogen in the presence of finely divided platinum, a single 1,2-dideuterocyclopentane is formed. Which geometric isomer is it? [Assume that catalytic hydrogenation delivers both hydrogens to the same face of the double bond, and recall that deuterium (atomic weight = 2) is an isotope of hydrogen (atomic weight = 1).] What predominant isomer would be expected from the reaction of cyclohexene and deuterium (D_2) in the presence of finely divided platinum?

2-9 1-Methylcyclohexene and methylidenecyclohexane exist in equilibrium when dissolved in strong aqueous acid. Assuming that the stability of the alkene controls the equilibrium, which alkene is present at the higher concentration?

2-10 α-Phellandrene, $C_{10}H_{16}$, a naturally occurring product found in wormwood, produces the odor of bitter fennel. Upon treatment of α-phellandrene with an excess of hydrogen in the presence of a platinum catalyst, a compound with the formula $C_{10}H_{20}$ is produced.

(a) What is its index of hydrogen deficiency?

(b) How many rings does α-phellandrene have?

2-11 Assign the hybridization of each carbon in the following compounds:

(a) 1,2,6-heptatriene

(b) 3-phenyl-1-propyne

(c) vinylcyclopropane

(d) *m*-xylene

2-12 Which of the following pairs represent resonance contributors?

(a)

(b)

(c)

(d)

(e)

2-13 Draw a significant resonance contributor that describes an electron distribution different from that shown in *a*, *b*, and *c* on the next page.

(a) allyl radical

(b) a cyclic pentadienyl cation

H H

⊕

(c) pentadienyl anion

⊖

2-14 Bonding molecular orbitals are formed by the overlap of atomic orbitals that have the same phasing, whereas antibonding molecular orbitals result from the overlap of atomic orbitals with opposite phasing. (Indeed, filled orbitals that are out of phase contribute as much to the destabilization of a collection of atoms as filled orbitals that are in phase do to stabilize it). Explain from the point of view of orbital phasing why no net contribution to the stability of the molecule results from *p* orbitals rotated at right angles.

2-15 The primary reaction that takes place in human and mammalian vision is initiated when geometric isomerization about one of the double bonds in rhodopsin produces the *trans*-isomer after the absorption of light. This change in geometry triggers a chemical reaction that results ultimately in a signal, which the brain interprets as a visual event.

Consider, as a model for vision, a simple light-induced *cis*-to-*trans* isomerization; for example, *cis*-2-butene → *trans*-2-butene. Light interacts with the π bond by promoting an electron from the bonding π molecular orbital to the antibonding molecular orbital, forming what is known as an excited state.

(a) From what you know about molecular orbitals, compare and contrast the relative magnitude of the barriers encountered upon rotation about the C-2—C-3 bond of *cis*-2-butene in the ground state and in the excited state.

(b) The geometric isomerization from a *cis* to a *trans* isomer requires that the *p* orbitals change from an aligned to an orthogonal to an aligned geometry. On the basis of your answer to part

a, explain why geometric isomerization results upon interaction of an alkene with light.

(c) Predict whether photoexcitation should allow for the conversion only of *cis* into *trans* isomers, only *trans* into *cis* isomers, or of both *cis* into *trans* and *trans* into *cis* isomers.

2-16 Predict whether pentalene (C_8H_6) should be stable as a planar hydrocarbon. Explain your reasoning.

2-17 Unlike most hydrocarbons, azulene ($C_{10}H_8$) is highly colored (deep blue). Although its isomer naphthalene does not have significant zwitterionic character, azulene does. For example, azulene dissolves in aqueous acid, but naphthalene does not. (Hint: Consider resonance contributors in which the two electrons of a π bond are moved to a single *p* orbital, producing a formal negative charge in that orbital and a formal positive charge in another vacant *p* orbital.)

(a) Draw a resonance structure of azulene in which the five-membered ring is anionic and the seven-membered ring cationic.

(b) Can azulene be considered to be aromatic?

(c) Azulene has an appreciable dipole moment. What does this observation imply about the relative importance of the resonance structure drawn in part *a*?

(d) Can a similar charge-separation argument explain the properties of pentalene (see Problem 2-16)? Why or why not?

2-18 Although cyclopentene is a stable compound that has chemical reactivity like that of a typical alkene, cyclopentyne is much less stable than a typical acyclic alkyne and cannot be stored at room temperature. Explain this difference in stability on the basis of what you know about hybridization and the preferred geometries for alkenes and alkynes.

Chapter 3

Functional Groups Containing Heteroatoms

Most of the concepts of bonding developed for hydrocarbons in Chapters 1 and 2 also apply to molecules that contain other second- or third-row atoms. To differentiate them from carbon, which is a constituent atom of all organic compounds, all other atoms (except hydrogen) are designated as **heteroatoms.** The chemical reactivity of an organic compound containing one or more heteroatoms usually differs significantly from that of an analogous compound that lacks these atoms. To emphasize this difference, that part of the molecule containing the heteroatom is referred to as a *functional group.*

In this chapter, we will consider how structure is altered by the replacement of a carbon atom by another atom in the second row of the periodic table. Specifically, we will compare the structure of CH_4 (methane) with those of NH_3 (ammonia), H_2O (water), and HF (hydrofluoric acid) and consider how the presence of the heteroatom affects the structures and reactivities of the organic derivatives of ammonia, water, and hydrofluoric acid.

3-1 ▸ Compounds Containing sp^3-Hybridized Nitrogen

When the carbon atom of methane is replaced by nitrogen, the result is **ammonia** (NH_3), the simplest member of another family of compounds. Derivatives of ammonia in which one or more hydrogen atoms are replaced by alkyl or aryl groups are called **amines.**

Ammonia

Nitrogen has five electrons in its valence shell. Mixing the 2s and 2p suborbitals, as we did with carbon in Chapter 1, results in four sp^3-hybridized orbitals (Figure 3-1). In a neutral nitrogen atom, however, five electrons must be accommodated in these hybrid orbitals, and so one hybrid orbital is doubly occupied. The two electrons accommodated in this filled orbital must

N: $1s^2\ \underbrace{2s^22p_x{}^1p_y{}^1p_z{}^1}$

Four
sp^3-hybrid
orbitals

▲ **FIGURE 3-1**
Mixing of atomic orbitals in nitrogen to form sp^3-hybridized orbitals.

have opposite spins: these electrons are referred to as a **lone pair** because they are associated only with one atom and do not take part in a covalent bond. The single electron in each of the three remaining sp^3-hybridized orbitals participates in covalent bonding with another atom. In this way, an octet electronic configuration is achieved, satisfying nitrogen's valence requirement. As in an sp^3-hybridized carbon, each of these hybrid orbitals is directed toward the corner of a tetrahedron so as to minimize electron-pair repulsion, and bond angles of approximately 109° are observed. Thus, in ammonia (NH_3), the tetrahedral geometry of methane is roughly preserved.

When each of the partly filled hybrid orbitals of nitrogen overlaps with the 1 s orbital of a hydrogen atom, the structure for ammonia, shown in the margin, is obtained. Because the electron density of the nonbonded electron pair is closer to nitrogen's nucleus than is that in an N–H σ bond, the lone pair exerts a somewhat greater repulsive force toward the electrons in the σ bonds than do the σ-bonding electrons for each other. As a result, the HNH angle is slightly smaller than the expected tetrahedral angle because of the slightly larger angle formed by H, N, and the lone pair. Ammonia has three-fold symmetry, with an HNH bond angle of 107°. (The angle formed by an H–N bond and a vector directed toward the lone pair is slightly larger than 109°.) Because there are only three groups attached to nitrogen in ammonia and similar compounds, this spatial arrangement is referred to as **pyramidal** (rather than tetrahedral) because the four atoms (nitrogen and three hydrogens) are located at the corners of a pyramid.

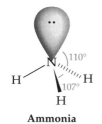

Ammonia

Amines

Methylamine, CH_3NH_2, has a methyl group (which also is sp^3-hybridized, see Chapter 1) bound to nitrogen by a σ bond; this bond replaces one of the $N(sp^3)$–$H(s)$ σ bonds of ammonia. Approximately the same geometry is found at nitrogen as that in ammonia and at carbon as that in an alkane. The nitrogen atom, specifically the lone pair of electrons, is a site of high chemical reactivity and is therefore the functional group in methylamine.

Alkyl substituents other than the methyl group of methylamine also can be attached to nitrogen. The resulting compounds belong to the family referred to as **amines.** An amine can be named either as an alkyl derivative of ammonia (as in methylamine) or in accord with the IUPAC system as a nitrogen derivative of an alkane; that is, with the NH_2 **(-amino)** group as a substituent of an alkane (as in aminomethane). The amino group can be placed at various positions along a hydrocarbon chain. Figure 3-2 shows two amines, one in which the amino group is placed at the second carbon of a six-atom fragment (at the top) and the other in which the amino group is

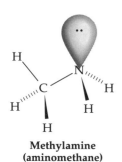

**Methylamine
(aminomethane)
CH_3NH_2**

at the first carbon. These compounds are referred to as 2-aminohexane and 1-aminohexane, respectively. Methylamine and both 1- and 2-aminohexane are referred to as **primary amines** because nitrogen is connected to only one carbon substituent. A common convention (to emphasize the functional group, rather than the alkyl chain) is to designate a primary amine as RNH_2 in which R represents any alkyl group.

Replacement of two of the hydrogens of ammonia with alkyl groups results in a **secondary amine.** With three alkyl groups and no hydrogens, we have a **tertiary amine.** Specific examples of these groups are dimethylamine and trimethylamine. Attachment of a fourth alkyl group to nitrogen requires that both of the electrons of the lone pair be used to form a covalent bond. As a result, the nitrogen is positively charged and such cations are referred to as **quaternary ammonium ions,** as in the tetramethylammonium cation.

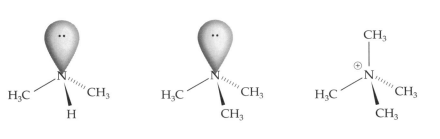

NH₂

2-Aminohexane
(a primary amine)

1-Aminohexane
(a primary amine)

▲ FIGURE 3-2
Two primary amines.

| Dimethylamine (a secondary amine) | Trimethylamine (a tertiary amine) | Tetramethylammonium cation (a quaternary ammonium ion) |

Notice that the designation primary, secondary, or tertiary for an amine characterizes the degree of substitution at nitrogen, not at carbon. Thus, both 1-aminohexane and 2-aminohexane are primary amines, despite the

---

SOME SIMPLE, NATURALLY OCCURRING AMINES

Compounds that contain nitrogen are pervasive in nature. Many of them have well-defined biological functions that are important, and even essential, to life. As living materials decompose, many of these complex structures decompose to simple amines. For example, the odor of "old" fish is due to a mixture of amines. Putrescine, a diamine, is produced by bacteria during the decomposition of animal tissue. Its name provides an excellent description of its odor.

$$H_2N - CH_2 - CH_2 - CH_2 - CH_2 - NH_2$$
Putrescine

$$H_2N - (CH_2)_3 - NH - CH_2 - CH_2 - CH_2 - CH_2 - NH - (CH_2)_3 - NH_2$$
Spermine

Very commonly, people associate bad odors with toxicity: things that smell bad are bad for you. In some cases, such as rotting flesh, this assessment is correct. However, putrescine occurs naturally in all cells, and compounds such as spermine are believed to be essential to cell division. Spermine, a tetramine with its own unique aroma, was first isolated as its acid-base salt with phosphoric acid from semen in 1678 by Anton von Leeuwenhoek, a Dutch chemist.

fact that the amino group is attached to a primary carbon in the first and to a secondary carbon in the second compound (Figure 3-2).

EXERCISE 3-A

Classify each of the following structures as a primary, secondary, or tertiary amine or as a quaternary ammonium salt.

(a)

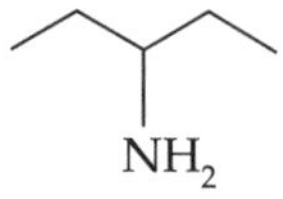

(d)

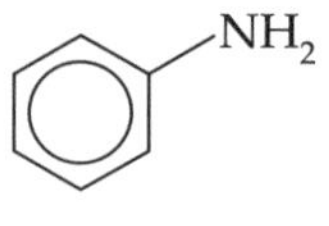

(g)

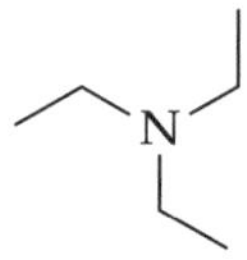

(b)

(e) Et_4NCl

(c)

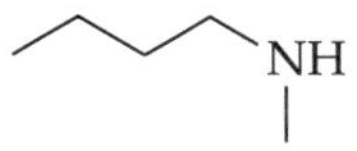

(f)

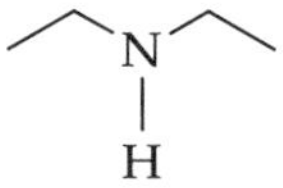

(h)

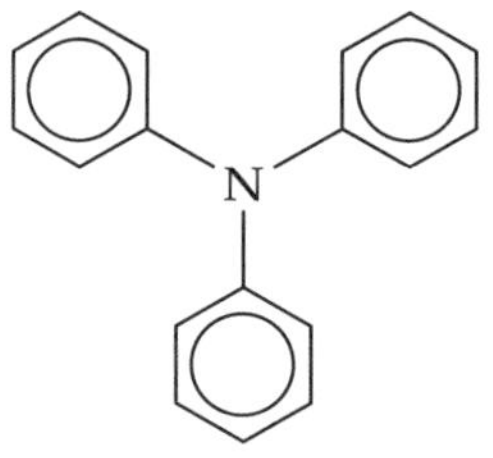

Trimethylamine and the tetramethylammonium ion can be represented by the Lewis dot structures shown in Figure 3-3. As we did for carbon in Chapter 1, we calculate the formal charge on nitrogen in trimethylamine by comparing the number of valence electrons (5) with the sum of half the number of shared electrons ($6/2 = 3$) and the number of unshared electrons (2): FC(N) = 5 – (3 + 2) = 0. Thus, the nitrogen atom in trimethylamine bears a formal charge of zero. Tetramethylammonium ion can be represented by the Lewis dot structure shown at the right in Figure 3-3 and the formal charge on nitrogen and carbon can be similarly calculated: FC(N) = 5 – (8/2 + 0) = +1. Thus, nitrogen bears a formal charge of +1 in the quaternary ammonium ion.

► FIGURE 3-3
Lewis dot structures and formal-charge calculation for trimethylamine and tetramethylammonium.

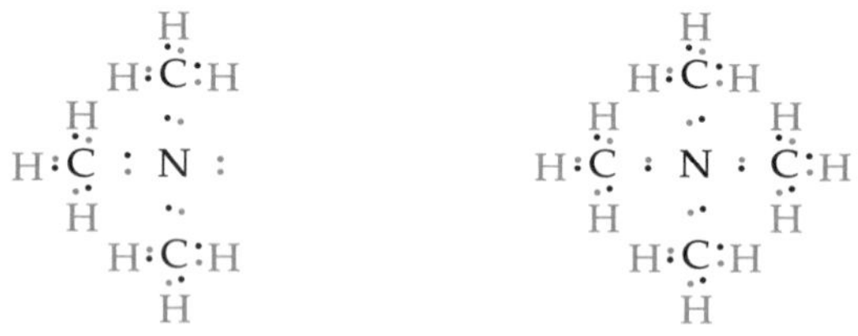

$$\text{Formal charge: at N} = 5 - \left(\frac{6}{2} + 2\right) = 0 \qquad \text{at N} = 5 - \left(\frac{8}{2} + 0\right) = +1$$

$$\text{at C} = 4 - \left(\frac{8}{2} + 0\right) = 0 \qquad \text{at C} = 4 - \left(\frac{8}{2} + 0\right) = 0$$

3-2 ► Polar Covalent Bonding in Amines

The replacement of a carbon atom of an alkane by a nitrogen atom makes the connecting σ bonds much more polar. The uneven charge distribution between atoms that are connected by a polar covalent bond causes amines to have different physical properties from those of the alkanes having similar structures and molecular weights.

Dipole Moments

Let us now consider the chemical changes caused by the introduction of an amino group into a hydrocarbon. Hydrocarbons have only nonpolar covalent bonds, whereas the covalent bonding in amines results in regions of partial positive or negative charge within the molecule. This partial charge separation is usually expressed as a molecular **dipole moment,** which is the resultant of the individual bond polarities projected into three dimensions.

What structural features in amines produce a dipole moment? First, amines have a lone pair that does not participate in bonding, which by itself induces a dipole moment. Furthermore, the electronegativities of carbon and nitrogen are unequal. Nitrogen is more electronegative than carbon, as indicated by its position in the periodic table. (Electronegativity increases in the progression from the left to the right along a row of the periodic table.) The greater electronegativity of nitrogen means that it more strongly attracts electrons toward itself and, hence, away from carbon. Therefore, the carbon–nitrogen bond in methylamine is polarized so that electron density in the bond is shifted toward nitrogen, further enhancing the molecular dipole induced by the lone pair.

This shift can be indicated in several ways, as shown in Figure 3-4. In one method, the direction of the shift of electron density in the σ bond is indicated by an arrow pointed toward the center of partial negative charge. Alternatively, lowercase delta (δ) is used to indicate the development of partial positive and negative charges on the atoms constituting the polar covalent bond.

The presence of a dipole moment has important consequences for both a molecule's physical properties and its chemical reactivity. For example, because of the significant electron density on nitrogen, the negative end of a carbon–nitrogen dipole is attracted to the positive end of a dipole in another molecule. Intermolecular forces are greater, therefore, in methylamine than in a hydrocarbon of similar molecular weight. As a result, the boiling points of amines are higher than those of the hydrocarbons of the same molecular weight (Table 3-1, on page 64). Alcohols and ethers, which are discussed in later sections, also have higher boiling points and melting points than do structurally similar alkanes of the same molecular weight.

EXERCISE 3-B

For each of the following pairs, choose the molecule that is likely to have the larger dipole moment. Explain your reasoning.

(a) NH_3 or NF_3

(b) trimethylamine or 2-methylpropane

(c) triphenylamine or triphenylmethane

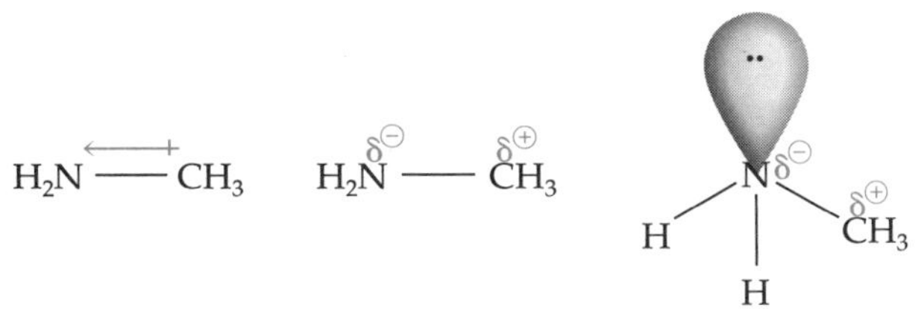

▲ FIGURE 3-4
Methods for indicating partial charge separation in a polar covalent bond.

Table 3-1 ▸ Boiling Points (in °C) of Selected Hydrocarbons and Heteroatom-containing Compounds

Hydrocarbon	bp	Amine	bp	Alcohol or Ether	bp
CH_4	−164	NH_3	−33	H_2O	100
CH_3CH_3	−89	CH_3NH_2	−6	CH_3OH	65
$CH_3CH_2CH_3$	−42	CH_3NHCH_3	7	$CH_3OCH_2CH_3$	11
		$CH_3CH_2NH_2$	16	CH_3CH_2OH	78
$CH_3CH_2CH_2CH_3$	−0.5	$CH_3CH_2CH_2NH_2$	48	$CH_3CH_2CH_2OH$	97
		$CH_3NHCH_2CH_3$	37	$CH_3CH(OH)CH_3$	82
		$CH_3CH(NH_2)CH_3$	33	$(CH_3)_3COH$	82
		$(CH_3)_3N$	3	$CH_3(CH_2)_3OH$	117
$CH_3(CH_2)_3CH_3$	36	$CH_3(CH_2)_3NH_2$	78	$CH_3CH_2OCH_2CH_3$	35
				$CH_3(CH_2)_4OH$	138

$$\overset{\delta\ominus}{X} - \overset{\delta\oplus}{H} \cdots\cdots \overset{\delta\ominus}{X} - \overset{\delta\oplus}{H}$$

A hydrogen bond

Hydrogen Bonding

Like the carbon–nitrogen bond, the nitrogen–hydrogen bond is polar be-cause hydrogen, like carbon, is less electronegative than nitrogen. Polar co-valent bonds are formed between hydrogen and highly electronegative atoms (those in the fifth, sixth, and seventh columns of the periodic table). The hydrogen of a polar H–X σ bond also often participates in further asso-ciation by hydrogen bonding. Hydrogen bonding causes a lengthening of the polar heteroatom–hydrogen σ bond. The partially positively charged hydrogen atom associates with a partially negatively charged center in an-other molecule. The weak attraction of a hydrogen atom bonded to an elec-tronegative atom X for a lone pair of electrons on another electronegative atom Y (X—H···Y) is a **hydrogen bond.**

Let us consider the interaction between two ammonia molecules. The partial positive charge on hydrogen induced in the polar N–H bond is at-tracted to the high electron density (partial negative charge) of the nitrogen lone pair of another molecule (Figure 3-5). Thus, this hydrogen is linked both with the nitrogen to which it is formally bound by a covalent bond and, more weakly, with the lone pair on another ammonia molecule through a hydrogen bond. A network is set up throughout the entire vol-ume of a sample of liquid ammonia in which many such hydrogen bonds

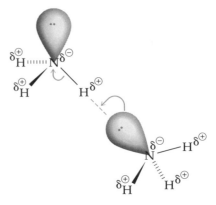

▲ FIGURE 3-5
Intermolecular hydrogen bonding in ammonia.

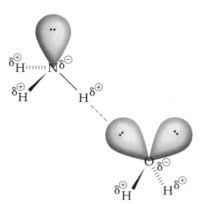

▲ FIGURE 3-6
Intermolecular hydrogen bonding between ammonia and water.

link many individual ammonia molecules together. When hydrogen bonds connect separate molecules, they are referred to as **intermolecular hydrogen bonds;** when they connect groups within the same molecule, they are called **intramolecular hydrogen bonds.**

Intermolecular hydrogen bonding is important wherever hydrogen is covalently attached to such highly electronegative atoms as nitrogen, oxygen, sulfur, or a halogen, with bonds to nitrogen, oxygen, or fluorine being strongest. Because the hydrogen bond is not needed to fulfill the valence requirement of either the heteroatom or the hydrogen atom, a hydrogen bond is much weaker than a normal covalent σ bond. Typically, hydrogen-bond strengths vary from about 1 to 5 kcal/mole.

Solvation

Hydrogen bonding is not restricted to interactions between identical molecules; it also occurs in mixtures of heteroatom-containing molecules. For example, consider the interaction of ammonia with water (H_2O), whose structure is considered in more detail in later sections of this chapter. If a water molecule is oriented as in Figure 3-6, an N—H···O hydrogen bond can be formed. Both the nitrogen and oxygen atoms in this hydrogen-bonded pair bear other hydrogens and lone pairs not associated with the hydrogen bond of interest. They are therefore free to form other hydrogen bonds, resulting in a network of N—H···O and O—H···N bonds throughout the liquid. This enhanced hydrogen bonding is the reason for the high solubility of ammonia (a polar molecule) in water (also a polar molecule).

EXERCISE 3-C

For each of the following pairs, choose the molecule that is likely to be more soluble in water. Explain your reasoning.

(a) ammonia or triethylamine

(b) methylamine or *n*-octylamine

(c) trimethylamine or *n*-propylamine

In a general description of the **solvation** of methylamine, polar solvent molecules orient themselves around the bond dipoles present at various atoms in methylamine. By representing a polar solvent molecule by an

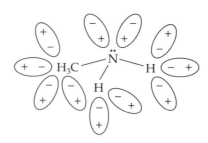

▲ FIGURE 3-7
Orientation of polar solvent molecules around methylamine.

ellipsoid bearing plus and minus charges, we obtain the picture shown in Figure 3-7 in which different regions of methylamine associate differently with polar solvent molecules. Solvation near the nitrogen–hydrogen σ bond entails N–H hydrogen bond donation (N—H···X), whereas that near the carbon–nitrogen bond is mainly through electrostatic interaction if the solvent itself lacks a polar X–H bond (in an **aprotic solvent**) or through hydrogen-bond acceptance (N···H—X) if the solvent has a polar X–H bond (in a **protic solvent**).

Nucleophiles and Electrophiles

The partial charge separation in methylamine (Figure 3-4) allows us to make predictions about its chemical reactivity. Because the carbon end of the dipole is partially positively charged, electron-rich reagents react at that site. Electron-rich reagents are referred to as **nucleophiles** because they seek centers of positive charge: the word nucleophile derives from the Greek *nucleo,* nucleus, and *philos,* loving. Similarly, the partially negatively charged end of the dipole (at nitrogen) tends to interact with reagents that are electron deficient. Electron-deficient reagents are referred to as **electrophiles:** electrophile is derived from the Greek words *electros,* electron, and *philos,* loving.

The concept of **electrostatic interaction** (nucleophiles interacting with positive centers and electrophiles interacting with centers of electron density) is one of the fundamental principles of organic chemistry. You should begin to recognize which reagents can effectively act as nucleophiles and which ones as electrophiles. Table 3-2 lists some species that are commonly used. Nucleophiles either are negatively charged or at least bear an unshared pair of electrons. Electrophiles either are positively charged or are at least electron deficient by virtue of attachment to a more-electronegative group. Some reagents appear in both lists: for example, the oxygen atom of water can act as a nucleophile, whereas the protons of water are electrophilic.

Nucleophilicity can be defined as the tendency of an atom, an ion, or a group of atoms to donate electrons toward an atom (usually carbon), and **electrophilicity** as the tendency of an atom, an ion, or a group of atoms to accept electron density from some atom. A negatively charged ion (an **anion**) is thus more nucleophilic than the corresponding neutral compound, and a positively charged ion (a **cation**) is a better electrophile than the corresponding neutral reagent. We will make extensive use of the concept of electrophiles and nucleophiles when we consider reactions in later chapters.

Nucleophiles can also act as **Lewis bases** (electron-pair donors), whereas electrophiles can act as **Lewis acids** (electron-pair acceptors).

Table 3-2 ▸ Some Common Nucleophiles and Electrophiles

Nucleophiles:

NH_3 $^{\ominus}NH_2$ RNH_2 $RNH^{\ominus}$ R_2NH $R_2N^{\ominus}$ R_3N $^{\ominus}N_3$

H_2O $^{\ominus}OH$ ROH $RO^{\ominus}$

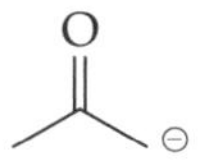

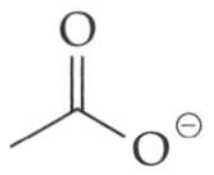

$^{\ominus}F$ $^{\ominus}Cl$ $^{\ominus}Br$ $^{\ominus}I$

HSH $^{\ominus}SH$ $S^{2\ominus}$ RSH $RS^{\ominus}$ R_3P $^{\ominus}C \equiv N$

Electrophiles:

$H_3O^{\oplus}$ $H^{\oplus}$ H_2O ROH BF_3 $^{\oplus}NO_2$ Cl_2 Br_2 I_2 $^{\oplus}R$

Bond Cleavages

The polarization of the carbon–nitrogen bond also has an effect on its **bond-dissociation energy.** The bond-dissociation energy of the C–N bond in methylamine defines the quantity of heat consumed when the carbon–nitrogen bond is cleaved to generate a methyl radical and an NH_2 radical.

Homolytic Cleavage

$$H_3C \overset{\frown}{-} NH_2 \longrightarrow H_3C\cdot + \cdot NH_2 \qquad \Delta H° = 85 \text{ kcal/mole}$$

This reaction is a **homolytic cleavage,** in which the two electrons shared in a σ bond become unpaired as the bond is broken, so that one bonding electron is distributed to each atom, resulting in two free radicals. (Recall from Chapter 2 that a radical is a species containing an atom bearing an unpaired electron and hence one too few electrons for a full valence shell.) We indicate the motion of *one* electron toward an originally bonded atom in this homolytic cleavage with a half-headed curved arrow.

Because the carbon–nitrogen bond is polar, electron density must be shifted against the electronegativity gradient in order to distribute electrons equally in this cleavage. Because the bond-dissociation energy is a measure of the endothermicity of a homolytic cleavage, polar covalent bonds often have higher bond-dissociation energies than do nonpolar ones, if other factors are comparable. This increase is due to the fact that the separation of charge requires energy. For example, an O–H bond is stronger than a similarly positioned C–H bond. The stronger bond is usually shorter because of stronger overlap in the σ bond, and its bond-dissociation energy is higher. This can be seen in the representative examples of bonds whose typical energies and lengths are listed in Table 3-3 (on page 68). The bond lengths and bond-dissociation energies for most of the common bonds of organic chemistry are given in Tables 3-4 and 3-6 (pages 72 and 80), which are also reproduced inside the back cover of this book.

An alternative to homolytic cleavage is **heterolytic cleavage,** in which the two electrons initially shared in a covalent bond are distributed unequally; that is, both electrons of the σ bond go to one of the atoms. This process is easier in a σ bond between atoms of unlike electronegativity than in a nonpolar bond. However, spontaneous heterolytic cleavage is difficult because it requires not only the energy to break the σ bond, but also the energy necessary to completely separate the positive and negative charges produced. In a polar covalent bond, part of this additional energy cost has

Bond	Bond Length (in Å)	Bond Energy (in kcal/mole)
Table 3-3 ▸ Typical Bond Lengths and Bond Energies		
$-\overset{\displaystyle \vert}{\underset{\displaystyle \vert}{C}}-H$	1.10	93–105
$-\overset{\displaystyle \vert}{\underset{\displaystyle \vert}{C}}-\overset{\displaystyle \vert}{\underset{\displaystyle \vert}{C}}-$	1.54	84–90
$-\overset{\displaystyle \vert}{N}-H$	1.01	91–103
$-\overset{\displaystyle \vert}{\underset{\displaystyle \vert}{C}}-\overset{\displaystyle \vert}{N}-$	1.47	82–85
$-O-H$	0.97	102–109
$-\overset{\displaystyle \vert}{\underset{\displaystyle \vert}{C}}-O-$	1.43	80–94
$F-H$	0.92	136
$-\overset{\displaystyle \vert}{\underset{\displaystyle \vert}{C}}-F$	1.40	107–108

already been paid in polarization of the σ bond; but, in the absence of other factors, heterolytic cleavage of a particular bond still requires more energy than does homolytic cleavage. We will learn later how this order can be reversed in the presence of polar solvents that help to stabilize the ions produced.

There are two modes by which heterolytic cleavage can occur in polar molecules. In reaction 1, the two electrons in the C–N bond in methylamine are shifted toward nitrogen, producing $^+CH_3$ and $^-NH_2$.

Possible Heterolytic Cleavages

$$H_3C \overset{\frown}{-} NH_2 \longrightarrow H_3C^{\oplus} + {}^{\ominus}NH_2 \tag{1}$$

$$H_3C \overset{\frown}{-} NH_2 \longrightarrow H_3C^{\ominus} + {}^{\oplus}NH_2 \tag{2}$$

We indicate that *two* electrons are shifted by using a curved full-headed arrow. The two electrons of the σ bond are no longer in a bonding orbital; rather, they are localized on nitrogen.

In the alternative mode of heterolytic cleavage shown in reaction 2, the two electrons in the C–N σ bond are shifted to carbon to produce $^-CH_3$ and $^+NH_2$. Again, the full-headed curved-arrow notation shows the flow of two electrons from the σ bond to carbon. The heterolytic cleavage in reaction 2 is more difficult than that in reaction 1 because it opposes the inherent electronegativity tendency in the polar bond; that is, in reaction 1, heterolytic cleavage simply completes the shifting of electrons from carbon toward the more-electronegative nitrogen atom, whereas reaction 2 reverses this flow of electrons and thus requires more energy.

EXERCISE 3-D

Using your knowledge of electronegativity, show the preferred direction of electron flow with a curved full-headed arrow when each of the following bonds is cleaved heterolytically:

(a) $CH_3–SCH_3$ (b) $CH_3S–H$ (c) $CH_3–OH$ (d) $CH_3O–CH_3$

3-3 ▶ Arrow Pushing to Indicate Electronic Motion

The use of half- and full-headed arrows to indicate the motion of electrons is a recurrent motif in the remainder of this book as we develop an understanding of how reactions of organic molecules occur. Whenever we use a full-headed arrow, we are indicating that two electrons are being moved from the position at the tail of the arrow to a position near the head. In contrast, a half-headed arrow shows the movement of only one electron. We always use half-headed arrow in pairs when bonds are broken homolytically. Heterolytic cleavages of neutral molecules give rise to ionic products (one cation and one anion), whereas homolytic cleavages produce two neutral species (radicals), each of which bears a single unpaired electron.

3-4 ▶ Acidity and Basicity of Amines

A heteroatom bearing both a bond to hydrogen and a lone pair can act either as an acid or as a base. The more commonly used definition of acids and bases is that originally suggested by Johannes Brønsted: an acid acts by transferring a proton to an acceptor. Therefore, a **Brønsted acid** is defined as a proton donor. For example, when HCl reacts with water to form H_3O^+ and Cl^-, a proton (H^+) is donated from HCl to water. Similarly, a **Brønsted base** is a proton acceptor. In the ionization of water ($2\ H_2O \rightleftharpoons H_3O^+ + {}^-OH$), one water molecule acts as a Brønsted acid, donating a proton, whereas the other acts as a Brønsted base, accepting a proton.

These same reactions can be thought of in the Lewis sense in that Cl^- accepts the electrons in the H–Cl bond freed by the interaction of H with the base (water). In the Lewis-base sense, water acts as an electron donor as one of the lone pairs changes from being a nonbonding pair in water to participating in an O–H σ bond in the hydronium ion. Thus, Lewis basicity parallels nucleophilicity and Lewis acidity parallels electrophilicity.

The concepts of Brønsted and Lewis acidity are equally useful for mineral acids (such as HCl or H_2SO_4); but, in organic molecules, acids often include bonds other than X–H. Here, the Lewis-acid concept can be used, whereas Brønsted acidity is not relevant. For example, when the lone pair of electrons on nitrogen of trimethylamine interacts with an electron acceptor such as boron trifluoride,

$$NMe_3 + BF_3 \rightleftharpoons Me_3N^+\!\!-\!\!BF_3^-$$

the amine acts as an electron donor (a Lewis base) and BF_3 acts as an electron acceptor (a Lewis acid). Thus, a Lewis acid-base reaction takes place in this example even in the absence of an X–H bond in the reactants or the product.

When the lone pair on nitrogen in an amine interacts to form a covalent bond by donation of its electrons to an electrophile, E^+, nitrogen is a Lewis basic site.

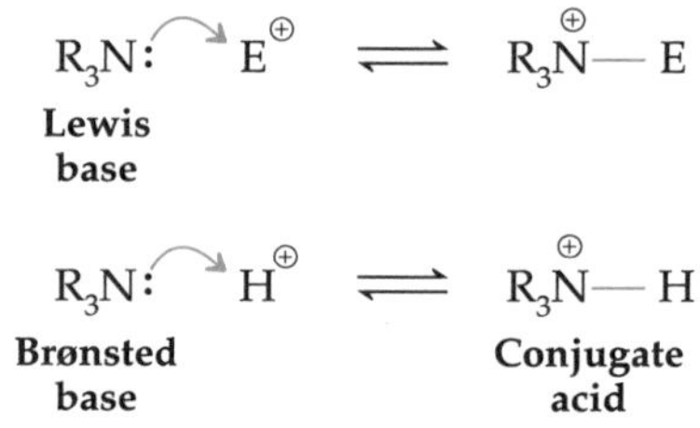

$$R_3N: \quad E^\oplus \quad \rightleftharpoons \quad R_3\overset{\oplus}{N}\!-\!E$$

**Lewis
base**

$$R_3N: \quad H^\oplus \quad \rightleftharpoons \quad R_3\overset{\oplus}{N}\!-\!H$$

**Brønsted   Conjugate
base   acid**

When the electrophile is a proton (H^+), nitrogen acts as a Brønsted base. Brønsted acid-base reactions are usually reversible. When the electrophile is a carbon atom, a new carbon–nitrogen bond is formed, often irreversibly. The presence of a lone pair of electrons (or a π bond) is both necessary and sufficient for a compound to act as a Brønsted or Lewis base.

EXERCISE 3-E

From your knowledge of the periodic table, predict whether each of the following compounds can act as Lewis base. Why or why not?

(a) Mg^{2+}

(b) aluminum trichloride

(c) boron trifluoride

(d) triethylamine

(e) tin tetrachloride

EXERCISE 3-F

For each of the following compounds, determine whether a Lewis basic site exists. If so, write the product that would be formed by interaction with a protic acid such as HCl.

(a) butane

(b) 1-butanol ($CH_3CH_2CH_2OH$)

(c) methyl ether (CH_3OCH_3)

(d) tertiary butanol [$(CH_3)_3COH$]

(e) methyl sulfide (CH_3SCH_3)

Because a heteroatom–hydrogen bond is highly polar, organic compounds bearing X–H bonds are also acids. In a primary amine, for example, the N–H bond is polarized as shown in Figure 3-8, and a nitrogen lone pair from a second amine acts as a base to induce the transfer of the proton from the positive end of the N–H dipole: $RNH_2 + RNH_2 \rightleftharpoons RNH^- + RNH_3^+$. In this way, the amine shown at the left acts as an acid. This reaction is reversible. In the reaction in Figure 3-8, the equilibrium lies far to the left, toward the two neutral amines. However, stronger bases, which will be dis-

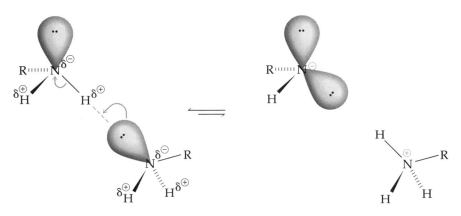

▲ FIGURE 3-8

An acid-base reaction ($2\,RNH_2 \rightleftharpoons RNH^- + RNH_3^+$) of a primary amine.

cussed in Chapter 5, can remove a proton from an amine so that the equilibrium lies far to the right: $RNH_2 + B \rightleftharpoons RNH^- + BH^+$.

The tendency of an —XH group to act either as an acid or as a base is referred to as **ambiphilicity.** The requirements for ambiphilicity are a polarized X–H bond and a nonbonded lone pair on X. This dual acid-base reactivity is a property not only of amines, but also of alcohols (ROH), thiols (RSH), and phosphines (RPH_2), which are considered in later sections. (Notice, for example, that water and alcohols appear as both electrophiles and nucleophiles in Table 3-2.)

3-5 ▸ Compounds Containing sp^2-Hybridized Nitrogen

Like carbon, nitrogen can participate in double and triple bonds. When a nitrogen atom is sp^2-hybridized, a double bond (one σ bond and one π bond) is formed by overlap with another sp^2-hybridized atom.

Double Bonding at Nitrogen

The same mixing of atomic orbitals to form hybrid orbitals described in Chapter 2 for carbon compounds can be applied to nitrogen to construct three sp^2-hybrid orbitals, leaving one p orbital unhybridized (Figure 3-9). This sp^2-hybridized nitrogen can bond with an sp^2-hybridized carbon, forming a planar structure in which overlap of the p orbitals orthogonal to the molecular plane forms a π bond. The carbon–nitrogen and nitrogen–hydrogen σ bonds and the nitrogen lone-pair hybrid orbital are coplanar, producing a structure much like that of the alkenes.

N: $1s^2$ $\underbrace{2s^2 2p_x^1 p_y^1}$ $2p_z^1$

Three   One p
sp^2-hybrid   orbital
orbitals

▲ FIGURE 3-9

Mixing of atomic orbitals of nitrogen to form sp^2-hybridized orbitals.

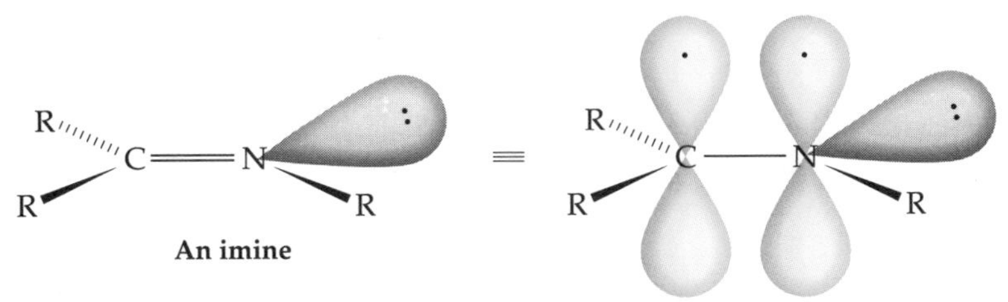

An imine

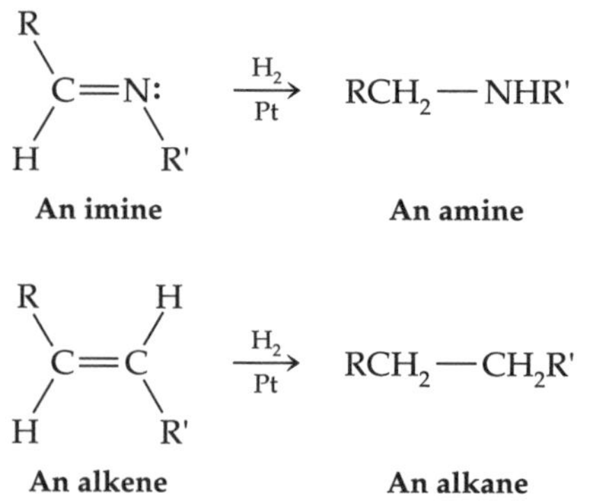

$$\underset{\substack{H}}{\overset{\substack{R}}{\diagdown}}C=N:\diagup R' \quad \xrightarrow[\text{Pt}]{H_2} \quad RCH_2-NHR'$$

An imine   An amine

$$\underset{\substack{H}}{\overset{\substack{R}}{\diagdown}}C=C\underset{\substack{R'}}{\overset{\substack{H}}{\diagup}} \quad \xrightarrow[\text{Pt}]{H_2} \quad RCH_2-CH_2R'$$

An alkene   An alkane

▲ FIGURE 3-10

The catalytic hydrogenation of an imine to an amine is analogous to that of an alkene to an alkane.

As a result, the geometries of these C=N double-bond–containing compounds, known as **imines,** bear close analogy to those of alkenes. For example, an imine contains two fewer hydrogen atoms than the corresponding amine, just as an alkene contains two fewer hydrogens than the structurally similar alkane. There is restricted rotation about a C=N bond in an imine, just as there is about a C=C bond in an alkene. Like carbon–carbon π bonds, the π bond of an imine undergoes catalytic hydrogenation, producing an amine (Figure 3-10).

It is useful to compare the strengths of single and multiple bonds between various atoms. Table 3-4 is a summary of average bond energies for various types of bonds obtained from heats of formation. Thus, for example, the entry listed for C–H is obtained as the heat required to convert methane into carbon and hydrogen—that is, of $CH_4 \rightarrow C + 4\,H$, $\Delta H°/4 = 99$ kcal/mole. The entry for C–C is then obtained by measuring the heat of formation of ethane and subtracting the bond energies of the six C–H bonds. The other bond energies listed are obtained by similar estimations. The

Table 3-4 ▶ Average Bond Energies (kcal/mole)

Example: $CH_4 \rightarrow C + 4\,H$, $\Delta H°/4 = 99$ kcal/mole

C—H	C—C	C=C	C≡C	
99	83	146	200	
N—H	C—N	C=N	C≡N	
93	73	147	213	
O—H	C—O	C=O	C≡O	O=C=O
111	86	179	257	225 (each)
H—H	N—N	N=N	N≡N	
104	39	100	226	
H—F	O—O	^{3}(O=O)		
136	35	119		
H—Cl	C—Cl	Cl—Cl		
103	81	58		
H—Br	C—Br	Br—Br		
87	68	46		
H—I	C—I	I—I		
71	51	36		

QUININE: AN ALKALOID

Naturally occurring compounds with basic nitrogen atoms are referred to as alkaloids. Quinine is an alkaloid isolated from the bark of the cinchona tree, which grows only in tropical regions such as the South Pacific.

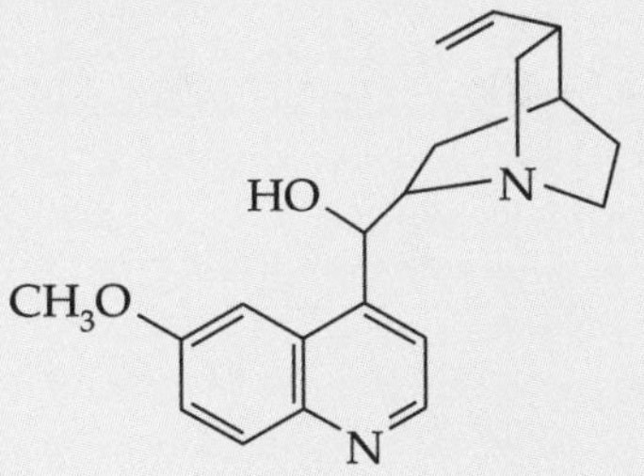

Quinine

Quinine was the first compound found to be effective for the treatment of malaria, a complicated disease state caused by a parasite. The control of many of the Pacific islands by the Japanese during World War II raised concerns about supplies of quinine to treat American troops operating in the Pacific. Two American chemists, Robert B. Woodward (Nobel Prize in Chemistry, 1965) and William von Eggers Doering, were the first to prepare quinine in the laboratory. This synthesis was the first preparation of such a complicated molecule and set the stage for a major revolution in how chemists viewed their ability to prepare molecules found in nature.

process of determining average bond energies reduces all atoms to their atomic states. Thus, these averaged values do not change from one molecule to another.

As we can see from Table 3-4, the energy required to break a double bond between two given atoms is greater than that needed to break a single bond and is greater still if the atoms are linked by a triple bond. We can also see that an H–X bond becomes weaker in the progression down a column of the periodic table: H–F > H–Cl > H–Br > H–I. This trend is also observed when these halogens are bound to carbon or to themselves.

Let us use these energies to analyze the feasibility of catalytic hydrogenation. With these bond energies in hand, we can predict whether a given transformation is energetically feasible. For example, in the catalytic hydrogenation of an alkene, the $C=C$ bond is converted into a $C-C$ bond in the alkane product: an $H-H$ bond in hydrogen gas is broken as two $C-H$ bonds are formed.

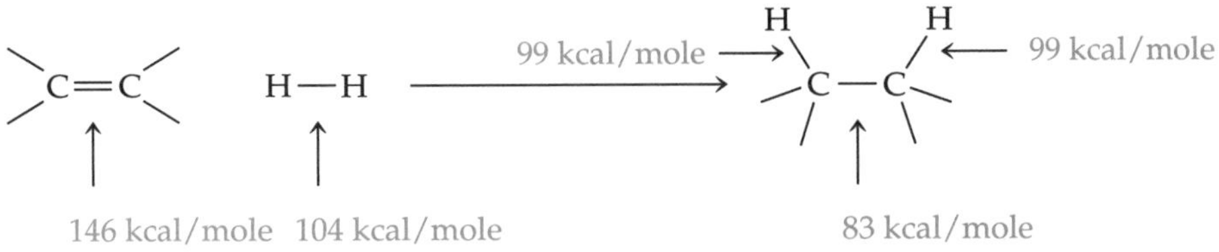

By assigning positive values to the energies that must be supplied to break bonds and negative values to the energies released when bonds are made, we can calculate $\Delta H°$ for this reaction as:

$$\Delta H° = +146 + 104 - 83 - (2 \times 99) = -31 \text{ kcal/mole}$$

The negative value of this sum tells us that energy is released; that is, this is a favorable reaction.

In the same way, we can also calculate $\Delta H°$ for the catalytic hydrogenation of an imine as:

$$\Delta H° = +147 + 104 - 99 - 93 - 73 = -14 \text{ kcal mole}$$

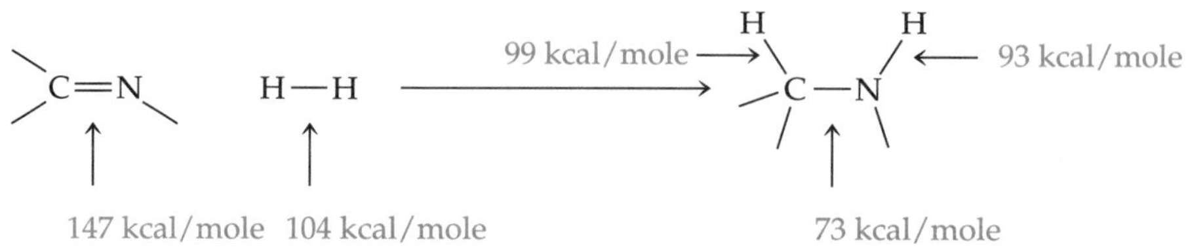

Again, this reaction is predicted to be energetically favorable, as is consistent with our observation that amines are produced by catalytic hydrogenation. Table 3-4 is thus extremely valuable as a rough predictive tool for determining the feasibility of proposed reactions.

EXERCISE 3-G

Using the bond energies in Table 3-4, predict whether or not the following proposed conversions are thermodynamically feasible.

(a) 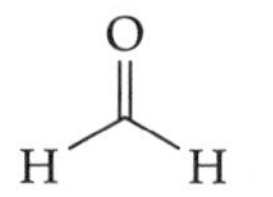 $\xrightarrow[\text{Catalyst}]{\text{H}_2}$ CH_3OH

(b) N_2 $\xrightarrow[\text{Catalyst}]{\text{H}_2}$ NH_3

(c) $H_3C-C\equiv N$ $\xrightarrow[\text{Catalyst}]{\text{H}_2}$ $CH_3CH_2NH_2$

Oxidation Levels

It is useful to have a method of "electron bookkeeping" in describing conversions in which the number of multiple bonds or the number of bonds to heteroatoms is changed. In applying this method, we compare the oxidation levels of the atoms participating in such reactions. This, in turn, enables us to know what type of reagent to use (and how much) to accomplish a particular chemical transformation. Reagents that can induce an oxidation or a reduction are called **redox reagents. Reduction** entails the addition of electrons and is always accompanied by the oxidation of a reaction partner. **Oxidation** is the loss of electrons from a compound.

Consider, for example, the catalytic hydrogenation of ethene. We can assign a formal oxidation level to the carbons in this neutral molecule by assuming that the usual oxidation level of hydrogen is +1. Thus, the two carbons in C_2H_4 must together bear a formal -4 oxidation level in order for ethene to exist as a neutral molecule. Because ethene is symmetrical, each carbon atom must have the same formal oxidation level, which we calculate as -2.

If we conduct the same analysis for ethane (a product of catalytic hydrogenation—see Figure 3-10), we find that each of the carbon atoms must bear a formal −3 oxidation level. Thus, the formal oxidation level of carbon has been reduced (made more negative) by the addition of hydrogen to the molecule. In our example, H_2 is oxidized (losing an electron) from H(0) to a formal H^+ state; that is, the hydrocarbon is reduced as molecular hydrogen is oxidized.

If we repeat the same analysis with ethyne, we find that the carbon atoms in ethyne (C_2H_2) are at a higher oxidation level than in ethene (C_2H_4). We conclude, as shown in Figure 3-11, that the higher the number of multiple bonds at a given carbon atom, the higher is its oxidation level. To introduce more multiple bonds, we must employ an oxidizing reagent; to have fewer multiple bonds, we must use a reducing reagent.

EXERCISE 3-H

Calculate the formal oxidation level of the carbons in ethyne.

The same considerations also apply to heteroatom-containing compounds. The catalytic hydrogenation of an imine to the corresponding amine is similar to the reduction of an alkene to an alkane. As with an alkene, an imine carbon is at a higher oxidation level than in the amine product formed by catalytic hydrogenation.

We can recognize a more highly oxidized functional group as being one with more multiple bonds or with more bonds to heteroatoms. For example, a $C\equiv N$ triple bond (whose structure is developed in more detail in the next section) is at an even higher oxidation level than a $C=N$ double bond, which is more highly oxidized than a $C-N$ single bond. Thus, the compounds listed in Figure 3-11 go from a more highly oxidized state to a more highly reduced one in the progression from the left to the right side of the figure.

The same procedure used for calculating oxidation level in hydrocarbons can be applied to heteroatom-containing organic compounds. We begin by assuming a usual oxidation state of −1 for halides, −2 for oxides and sulfides, −3 for nitrogen, and +1 for hydrogen. We can then calculate the formal charge on carbon required to produce molecular neutrality. Let us apply the procedure to ethylamine, $C_2H_5NH_2$. The net formal oxidation level for the two carbons must be −4 = −[7 hydrogens (+1) + 1 nitrogen (−3)], the same level that we calculated for ethene. Similarly, for ethanol, C_2H_5OH, in which an OH group replaces the NH_2 group of ethylamine, or in ethyl chloride, C_2H_5Cl, in which Cl replaces NH_2, the same formal level

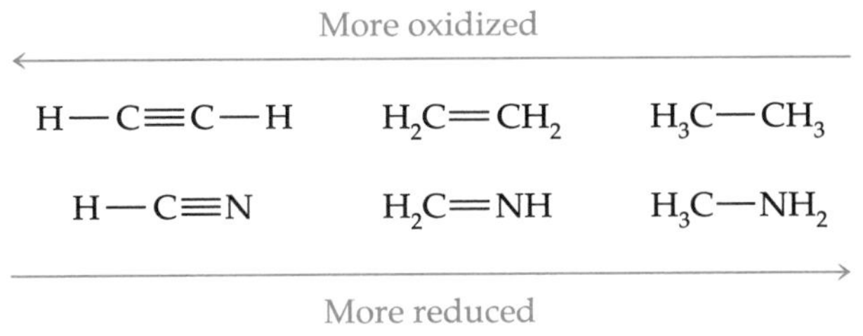

More oxidized

$$H-C\equiv C-H \qquad H_2C=CH_2 \qquad H_3C-CH_3$$

$$H-C\equiv N \qquad H_2C=NH \qquad H_3C-NH_2$$

More reduced

▲ FIGURE 3-11
Oxidation levels of some hydrocarbons and nitrogen-containing compounds. The greater is the degree of multiple bonding, the higher is the formal oxidation level of a compound.

Table 3-5 ▸ Some Common Oxidizing and Reducing Agents

Reducing Agent	Oxidizing Agent
H$_2$/M (catalytic hydrogenation)	Cr(VI) (chromate), often as CrO_4^{2-} or $Cr_2O_7^{2-}$
NaBH$_4$ (sodium borohydride)	Mn(VII) (permanganate), often as MnO_4^-
LiAlH$_4$ (lithium aluminum hydride)	Cu(II) (cupric)
NaB(CN)H$_3$ (sodium cyanoborohydride)	Os(VIII) (osmate)
	Fe(III) (ferric)

is obtained: (C_2H_5OH: $-4 = -$ [6 hydrogens (+1) + 1 oxygen (−2)]; C_2H_5Cl: -4 = −[5 hydrogens (+1) + 1 chlorine (−1)]). Thus, a bond to a heteroatom has the same effect on the oxidation level of carbon as does a π bond.

Once we recognize whether a given transformation of one compound to another is an oxidation or a reduction, we can choose among possible reagents to effect the desired reaction (Table 3-5). In many common reagents used for reduction, hydrogen is present at an oxidation level lower than its usual +1 state; that is, as hydrogen gas, H$_2$ (0), or formally as hydride, H$^-$, in complex metal hydride reducing agents such as NaBH$_4$ and LiAlH$_4$ that we will learn about in Chapter 12.

EXERCISE 3-I

Determine for each of the following pairs of compounds whether any of the carbon atoms are at different oxidation levels. If so, determine which structure—that at the left or that at the right—is the more-oxidized species.

(a) 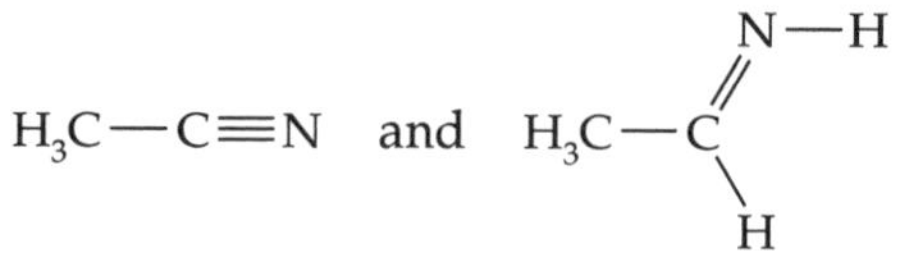 H$_3$C—C≡N   and   H$_3$C—C(=N—H)(H)

(b) 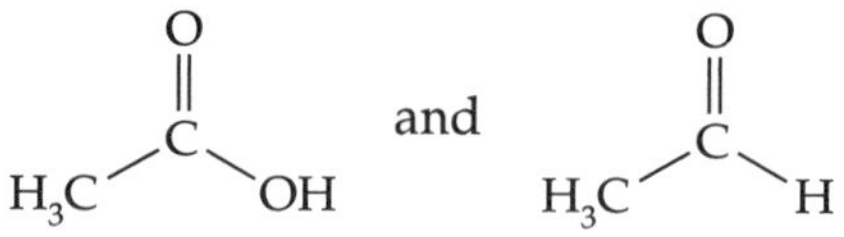 H$_3$C—C(=O)—OH   and   H$_3$C—C(=O)—H

(c) CH$_3$CH$_2$NH$_2$ and   H$_3$C—C≡N

(d) 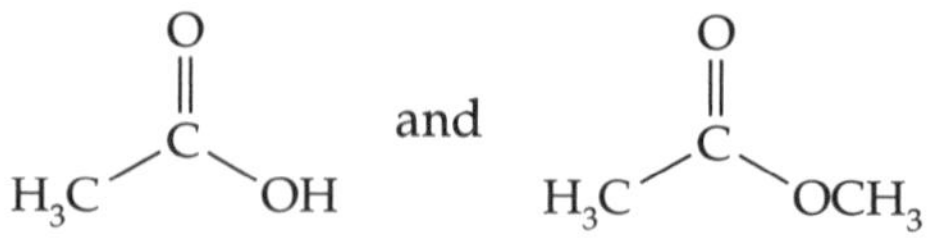 H$_3$C—C(=O)—OH   and   H$_3$C—C(=O)—OCH$_3$

(e) 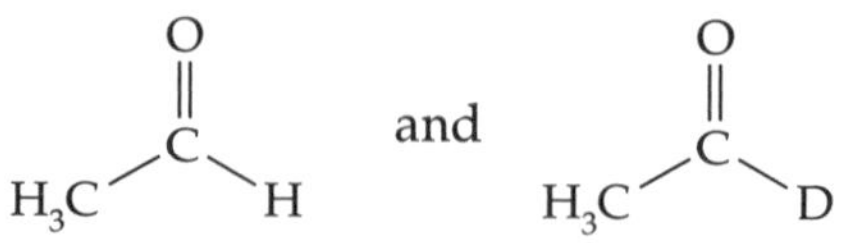 H$_3$C—C(=O)—H   and   H$_3$C—C(=O)—D

Oxidation reactions are accomplished with reagents that can change their formal oxidation levels by taking on additional electrons, thus undergoing reduction. Many oxidation reagents therefore contain a transition metal having two or more relatively stable oxidation states. For example, chromate, Cr(VI), permanganate, Mn(VII), cupric, Cu(II), osmate, Os(VIII), and ferric, Fe(III), ions are very often used. Several of these reagents are particularly convenient because their colors change when they are reduced, thereby allowing the reaction to be followed readily. For example, chromate reagents are red orange in the +6 oxidation state but become green chromium (+3) salts when they are reduced upon inducing the oxidation of an organic substrate. Manganese is purple as the permanganate ion [Mn(VII)] in $KMnO_4$ and red brown when reduced to Mn(IV) as MnO_2.

3-6 ▸ Compounds Containing *sp*–Hybridized Nitrogen: Triple Bonding at Nitrogen

The final mode for hybridization of nitrogen, shown in Figure 3-12, is the mixing of a 2*s* and a 2*p* atomic orbital to form two *sp*-hybridized orbitals, leaving two 2*p* orbitals unmixed. These remaining orthogonal *p* orbitals can participate in the same type of bonding as in alkynes. An *sp*-hybridized nitrogen atom, when linked to carbon by a triple bond, constitutes a **nitrile** functional group. The C≡N group is also called a **cyano** group or **cyanide ion** when it exists by itself as a negatively charged ion (⁻CN) or when found in inorganic reagents. As in the alkynes, the *sp*-hybrid orbitals participating in σ bonding are directed at 180° relative to each other. The *p* orbitals are orthogonal to the σ bond, overlapping above and below, and in front of and behind, the atomic plane.

N: $1s^2$ $\underbrace{2s^2 2p_x^1}$ $\underbrace{2p_y^1 2p_z^1}$
      Two   Two *p*
  sp-hybrid orbitals
   orbitals

▲ FIGURE 3-12
Mixing of atomic orbitals of nitrogen to form *sp*-hybridized orbitals.

    Cyanide ion   **Nitrile**

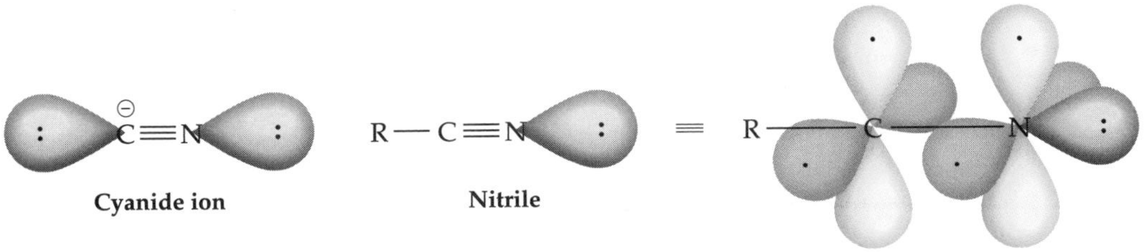

3-7 ▸ Compounds Containing *sp³*-Hybridized Oxygen in an O–H Bond

When the carbon atom of methane is replaced by oxygen, **water** (H_2O) is obtained. Derivatives of water in which one of the hydrogen atoms is replaced by an alkyl or an aryl group are called **alcohols.** Compounds in which both of the hydrogen atoms of water are replaced are called **ethers.**

Water

The same structural features that characterize bonding in organic compounds containing nitrogen atoms are also encountered in oxygen-containing compounds. Let us first consider the structure of water, H_2O. Figure 3-13 shows the electronic configuration of oxygen used in forming *sp³*-hybrid orbitals. The 2*s* and 2*p* orbitals can be combined to form four *sp³*-hybrid orbitals. Because these orbitals must accommodate six electrons, oxygen can form only two covalent bonds before it reaches a filled-valence-shell electronic configuration. When these bonds are formed with hydrogen atoms, we obtain water, whose Lewis dot structure and geometry are shown in Figure 3-14. Here, the six valence electrons of oxygen are shown as black dots and the two electrons contributed by two hydrogen atoms are shown

O: $1s^2$ $\underbrace{2s^2 2p_x^2 p_y^1 p_z^1}$
    Four
  sp³-hybrid
   orbitals

▲ FIGURE 3-13
Mixing of atomic orbitals of oxygen to form *sp³*-hybridized orbitals.

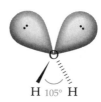

▲ FIGURE 3-14
Lewis dot structure and a three-dimensional representation of water.

as orange dots. The valence requirement of each atom is satisfied, and the sp^3-hybridized orbitals are roughly directed from each other at approximately a tetrahedral angle. The lone-pair–lone-pair repulsion in water is slightly larger than that between a lone pair and a σ bond (as in ammonia), which, in turn, is larger than the bond–bond repulsion in methane. As a result, the angle between the lone pairs is expanded slightly and the HOH angle is somewhat compressed (to about 105°) from the ideal tetrahedral angle.

Alcohols: R—OH

Replacement of one of the O–H bonds in water with an O–C bond to an alkyl group results in an alcohol. The simplest alcohol, in which a methyl group has replaced a hydrogen, is methanol, shown in the margin. Notice the close similarity between the structure of methanol and that of methylamine (Section 3-1). Compounds bearing the O–H functional group are known as *alcohols*.

As in the amines shown in Figure 3-2, the O–H group can be at various positions in an alcohol. Examples are the three isomeric six-carbon alcohols shown here.

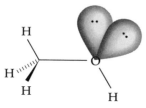

**Methanol
(hydroxymethane)**

CH_3OH

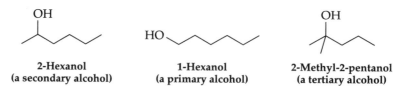

| **2-Hexanol** | **1-Hexanol** | **2-Methyl-2-pentanol** |
| (a secondary alcohol) | (a primary alcohol) | (a tertiary alcohol) |

Nomenclature for these compounds is in accord with the IUPAC rules for hydrocarbons (see Chapter 2), except that the ending **-anol** is used to indicate the presence of an O–H group. Thus, the number preceding the name of the compound indicates the position along the chain at which the O–H group is attached.

Alcohols are subclassified as primary, secondary, and tertiary according to the nature of the carbon to which the O–H group is attached. Unlike amines, in which one, two, or three carbons can be bound to nitrogen, an alcohol can have only one carbon atom attached to oxygen and retain its functional-group identity. Therefore, the designation of the subclass of an alcohol derives from the type of carbon to which the O–H group is attached. Thus, 2-hexanol is a **secondary alcohol** because the O–H group is attached to a secondary carbon; that is, the carbon to which the O–H is attached is itself bound to two other carbon atoms. 1-Hexanol is a **primary alcohol** because the carbon to which the O–H group is attached is primary—that is, attached to one other carbon atom—and 2-methyl-2-pentanol is a **tertiary alcohol** in that the carbon bearing the O–H group is attached to three other carbons.

For each of the following compounds, classify the alcohol as primary, secondary, or tertiary.

(a) [structure with OH]

(b) [structure with OH]

(c) [structure with OH]

(d) [structure with OH, phenol]

(e) [structure with OH]

(f) [structure with OH]

Like nitrogen, oxygen is more electronegative than either carbon or hydrogen: this leads to partial charge separation in the covalent bonds between carbon and oxygen and between oxygen and hydrogen.

$$HO\overset{\longleftarrow}{}CH_3 \qquad \overset{\delta^\ominus}{H}O\overset{\delta^\oplus}{-}CH_3 \qquad \overset{\delta^\oplus}{H}\overset{\delta^\ominus}{-}OCH_3$$

Methanol also hydrogen bonds with itself (Figure 3-15) and with other polar molecules such as water (Figure 3-16). Accordingly, the boiling point of an alcohol is significantly higher than that of a hydrocarbon of a similar molecular weight (see Table 3-1), and low-molecular-weight alcohols are significantly soluble in water and other polar solvents. As the molecular weight of the alcohol increases, the polar O–H group becomes a smaller fraction of the total molecular volume and the solubility of the alcohol in polar solvents decreases, whereas solubility in nonpolar solvents increases.

Homolytic Cleavages: Bond Energies and Radical Structure

As with the amines, the cleavage of the carbon–oxygen or oxygen–hydrogen bond of an alcohol can be, in principle, either homolytic or heterolytic. Let us consider the structural factors that might determine the preferred mode of cleavage.

The principal measure of the ease of homolytic cleavage is the bond-dissociation energy. Table 3-6 lists specific bond-dissociation energies for

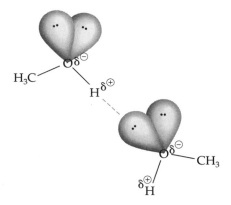

▲ FIGURE 3-15
Hydrogen bonding of methanol to methanol.

▶ FIGURE 3-16
Hydrogen bonding of
methanol to water.

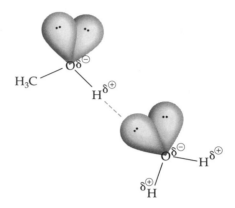

some of the common bonds encountered in organic chemistry. These values are not the same as those in Table 3-4 (average bond energies) because the values in the two tables are arrived at in different ways. The bond-dissociation energies in Table 3-6 describe the energy required to break a specific bond in a particular molecule, whereas the average bond energies in Table 3-4 are typical for various single or multiple bonds and are derived from measurements of heats of formation of molecules containing such bonds. We will find uses for both values: bond-dissociation energies provide an accurate assessment of the energy required to break a particular bond in a homolytic fashion; average bond energies can be used to estimate roughly the overall change in energy from one stable species to another.

A given bond type (for example, a carbon–hydrogen bond) has a somewhat different bond-dissociation energy in different molecules. For example, the entry in Table 3-6 for a C–H bond in methane (105 kcal/mole) differs from the average bond energy for a typical C–H bond (99 kcal/mole) in Table 3-4. Clearly, the cleavage of CH_3—H → CH_3 + H is more energetically costly than that of an average C–H bond. Bond-dissociation energies also vary with the degree of substitution of the atoms taking part in the bond being broken. Thus, the energy required to break a primary carbon–hydrogen bond (100 kcal/mole) is greater than that of a secondary one (96 kcal/mole), which, in turn, is greater than that of a tertiary one (93 kcal/mole), as shown in Figure 3-17. Because these homolytic cleavages produce one fragment (the hydrogen radical) in common, the decreasing difference in bond-dissociation energy in this series may be a result of the increasing stability of the carbon radical formed or the ground-state weakening of the C–H bond or both.

Table 3-6 ▸ Bond Dissociation Energies (kcal/mole)

Bond	X =	H	F	Cl	Br	I	OH	NH₂	CH₃
Ph—X		111	126	96	81	65	111	102	101
CH₃—X		105	108	85	70	57	92	85	90
CH₃CH₂—X		100	108	80	68	53	94	84	88
(CH₃)₂CH—X		96	107	81	68	54	94	84	86
(CH₃)₃C—X		93		82	68	51	93	82	84
PhCH₂—X		88		72	58	48	81		75
H₂C=CH—CH₂—X		86		68	54	41	78		74
H—X		104	136	103	87	71	119	107	105
X—X		104	38	59	46	36	51	66	90

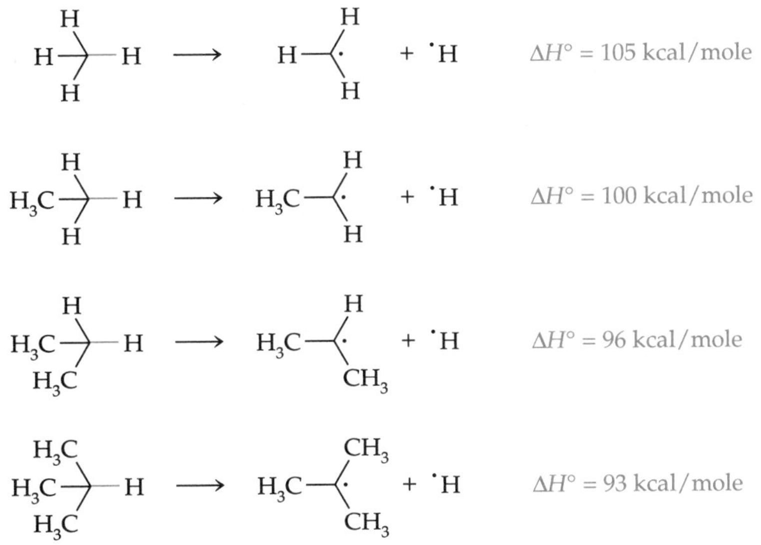

▲ FIGURE 3-17

Bond-dissociation energies for several simple hydrocarbons.

In a relatively nonpolar bond (such as that between carbon and hydrogen or between carbon and carbon), the radical stabilization energy is important in determining the order of bond-dissociation energies for different kinds of C–H or C–C bonds. We conclude that **radical stability** follows the order tertiary > secondary > primary and influences the ease of homolytic cleavage. (As in the subclassification of alcohols, the subclassification of radicals is based on the number of carbon atoms attached to the carbon bearing the unpaired electron.)

In contrast, bond-dissociation energies for bonds between carbon and more-electronegative atoms such as oxygen and the halogens do not differ as greatly with the degree of substitution (Figure 3-18). The same 1°, 2°, and 3° carbon radicals are formed upon cleavage of a carbon–oxygen bond in this series as are produced in the breaking of a carbon–hydrogen bond in the series shown in Figure 3-17. However, the bond strength between carbon and oxygen increases as the degree of substitution increases,

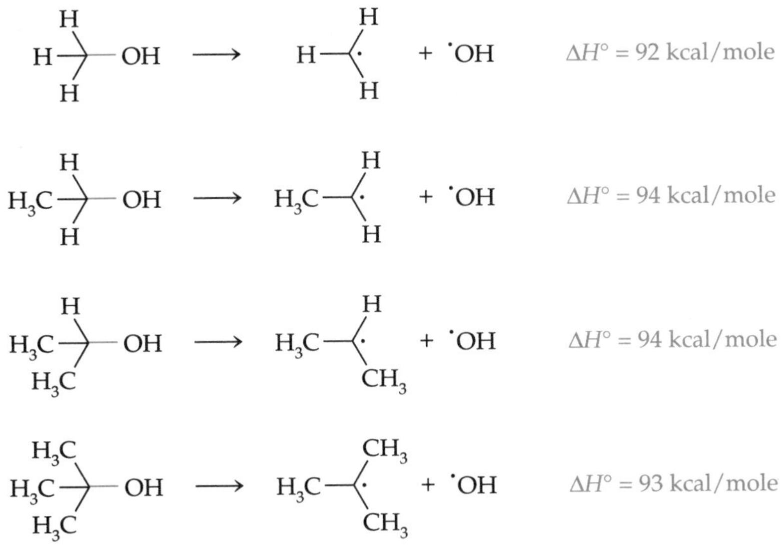

▲ FIGURE 3-18

Bond-dissociation energies of C–O bonds in several simple alcohols.

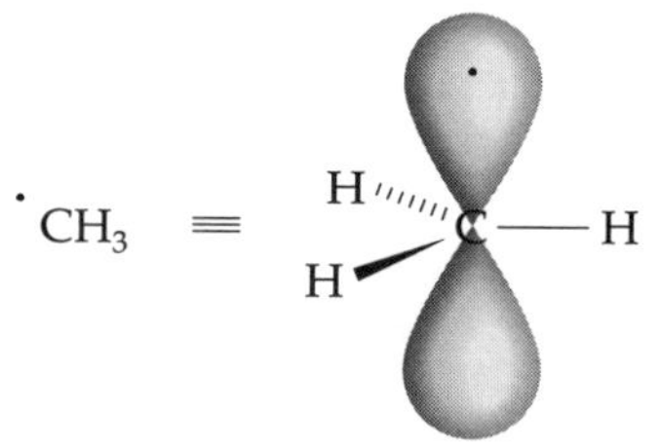

▲ FIGURE 3-19
A three-dimensional representation of methyl radical.

counteracting the effect of radical stability. The bond-dissociation energies of C–X bonds therefore do not easily conform to a degree-of-substitution trend, and individual entries in Table 3-6 must be used for this family.

Let us consider the structure of a carbon radical in order to understand the order of radical stability. The structure of a methyl radical is shown in Figure 3-19. In the methyl radical ($CH_3 \cdot$), there are three equivalent C–H bonds and one nonequivalent p orbital bearing a single electron: in this configuration, the carbon atom has only seven valence electrons. Lacking one electron from a filled valence shell, radicals are electron deficient and highly reactive.

A radical has one unpaired electron. Because an s orbital is closer to the nucleus than is a p orbital, the lowest energy arrangement of the methyl radical has as many electrons as possible in hybrid orbitals (with s character). This is achieved when the single electron is held within a p orbital and the three hybrid orbitals are doubly occupied. The methyl radical is therefore sp^2-hybridized, bearing three σ C–H bonds and a singly occupied p orbital. In this hybridization, the HCH angle is 120° and the three C–H bonds are coplanar. Thus, when a methyl radical is formed in the homolytic cleavage of a CH_3–X bond, the carbon undergoes a geometric change from tetrahedral to planar and a rehybridization from sp^3 to sp^2.

By replacing one of the hydrogens of the methyl radical by a methyl group, we obtain the ethyl radical shown in Figure 3-20. An alkyl group is more polarizable than a hydrogen atom and can better satisfy the high electron demand of the electron-deficient sp^2-hybridized radical center. Furthermore, one of the carbon–hydrogen bonds of the CH_3 group can be coplanar with the singly occupied p orbital so that **hyperconjugative stabilization** parallel to that seen in alkenes in Chapter 2 further stabilizes the ethyl radi-

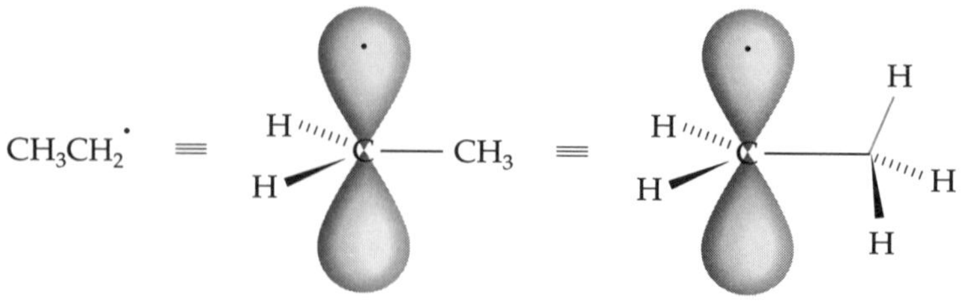

**Ethyl radical
(a primary radical)**

▲ FIGURE 3-20
A three-dimensional representation of an ethyl radical. Note that it is possible to align one of the C–H bonds on the adjacent carbon with the singly occupied p orbital, thus permitting hyperconjugative stabilization of the radical.

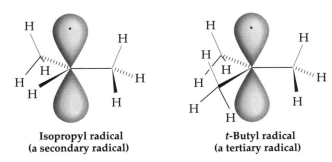

Isopropyl radical
(a secondary radical)

t-Butyl radical
(a tertiary radical)

▲ FIGURE 3-21
Hyperconjugation in the isopropyl and t-butyl radicals.

cal. Any primary radical (like the ethyl radical in Figure 3-20) has this hyperconjugative interaction, which is not possible in the simple methyl radical shown in Figure 3-19. Thus, a primary radical (like the ethyl radical) is more stable than the methyl radical.

As we replace additional hydrogen atoms in the methyl radical by alkyl groups, giving, for example, the isopropyl radical and the tertiary butyl radical shown in Figure 3-21, there are additional possibilities for polarization and hyperconjugative stabilization in the interaction with the electron-deficient radical center. Thus, a tertiary radical is more stable than a secondary radical, which, in turn, is more stable than a primary radical. This order of radical stability (3° > 2° > 1°) is consistent with that observed from bond dissociation energies for carbon–hydrogen bonds at various types of carbon.

EXERCISE 3-K

For each of the following pairs of compounds, choose the one that requires less energy for heterolytic cleavage of the C–X bond:

(a) ![structure] or ![structure with I]

(b) ![phenethyl alcohol] or ![1-phenylethanol]

(c) ![structure with I] or ![structure with I]

(d) Br~~~~ or Br~~~~

(e) ![cyclohexane with Br] or ![cyclohexane with Br Br]

Heterolytic Cleavage of C–OH Bonds: Formation of Carbocations

As in amines, heterolytic cleavage of a carbon–oxygen bond can occur, in principle, in two ways that differ in the direction of the flow of electrons.

$$H_3C\overset{\frown}{-}OH \longrightarrow {}^{\oplus}CH_3 \quad {}^{\ominus}OH$$

$$H_3C\overset{\frown}{-}OH \longrightarrow {}^{\ominus}CH_3 \quad {}^{\oplus}OH$$

Of these two ways, the top reaction is more favorable because the atom with the higher electronegativity (oxygen) takes on negative charge.

Spontaneous heterolytic cleavage of a carbon–oxygen bond is very difficult by either route because of the high energy cost of cleaving bonds and forming ions, but the reaction can be assisted significantly by the addition of a proton (from an acid) to one of the lone pairs of oxygen. (For now, we will indicate this addition simply as a proton and will reconsider this reaction in more detail in later chapters.) For example, in the protonation of methanol, one of the lone pairs on oxygen must be shared between oxygen and the incoming proton. This is indicated by the curved-arrow notation in Figure 3-22, in which we obtain a protonated alcohol having a formal positive charge on oxygen. In a **protonated alcohol,** oxygen bears three σ bonds; this intermediate is called an **oxonium ion** and is formally charged +1. The ending *-onium* indicates positive charge and the prefix *ox-* indicates that the charge resides substantially on oxygen. This positive charge induces even further electronic polarization of the carbon–oxygen and hydrogen–oxygen bonds toward oxygen. Heterolytic cleavage of the carbon–oxygen bond in the second step of Figure 3-22 thus produces a methyl cation and a neutral water molecule. Because the oxonium ion is already positively charged, this cleavage does not require further charge separation and is therefore much more easily accomplished than is heterolytic cleavage without acid.

In general, the ease with which water is lost **(dehydration)** from an oxonium ion depends on the character of the alcohol. In analogous reactions of different alcohols, the loss of water becomes progressively easier in the order from methanol to a primary alcohol to a secondary alcohol to a tertiary alcohol (Figure 3-23). Because water is the common product, this order is largely governed by the stability of the resulting alkyl cation, called a **carbocation** or **carbonium ion.** This order of stability of carbocations is the same as that of radicals. Because a carbocation has only six valence electrons, it is even more electron deficient than a radical.

The structural factors that control radical stability and geometry also apply to carbocations. Thus, as shown in the margin, rehybridization from the sp^3-hybrid in the alcohol to the sp^2-hybridized carbocation produces a planar cation with a vacant *p* orbital orthogonal to the atomic plane. Because a carbon with a vacant *p* orbital is even more electron deficient than a

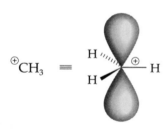

$${}^{\oplus}CH_3 \quad \equiv$$

$$H_3C-\overset{..}{\underset{|}{O}}\overset{H}{\overset{\oplus}{\frown}} \longrightarrow H_3C\overset{\frown}{-}\overset{\overset{H}{|}}{\underset{|}{\overset{\oplus}{O}}}\overset{H}{\underset{H}{}} \longrightarrow {}^{\oplus}CH_3 \; + \; :\overset{\overset{H}{|}}{\underset{|}{O}}\overset{}{\underset{H}{}}$$

**Protonated alcohol
(oxonium ion)**

▲ FIGURE 3-22

Protonation and dehydration of methanol.

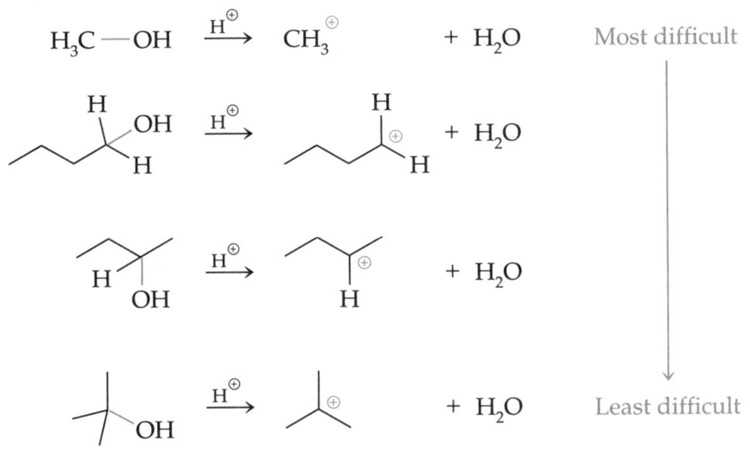

$H_3C-OH \xrightarrow{H^{\oplus}} CH_3^{\oplus} + H_2O$ Most difficult

(structure) $\xrightarrow{H^{\oplus}}$ (structure) $+ H_2O$

(structure) $\xrightarrow{H^{\oplus}}$ (structure) $+ H_2O$

(structure) $\xrightarrow{H^{\oplus}}$ (structure) $+ H_2O$ Least difficult

▲ FIGURE 3-23
Dehydration of methanol and three isomeric butanols.

radical (which bears a single electron in the *p* orbital), the stability afforded by hyperconjugation (Figure 3-24) is even greater. Because hyperconjugation is stabilizing, the *t*-butyl cation is more stable than the isopropyl cation, which is more stable than the ethyl cation, which is more stable than the methyl cation. The order of stability of carbocations, like that of radicals, is therefore $3° > 2° > 1° > CH_3^+$.

Ordering Alcohol Reactivity by Class

The greater stability of tertiary than secondary than primary cations can be affirmed experimentally because the type of cation correlates with the type of alcohol from which it is derived; that is, a tertiary cation is formed from a tertiary alcohol, and so forth. The order of stability for cations discussed in the preceding section therefore predicts a corresponding order in the ease of their formation from tertiary, secondary, and primary alcohols. An example of this reactivity sequence can be found when alcohols are treated with the **Lucas reagent,** a mixture of Brønsted and Lewis acids, in a reaction in

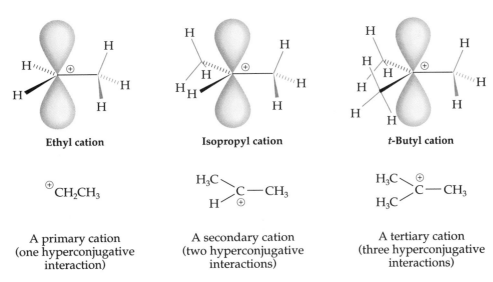

Ethyl cation

$^{\oplus}CH_2CH_3$

A primary cation
(one hyperconjugative
interaction)

Isopropyl cation

$H_3C \diagdown C-CH_3$ with H and $\oplus$

A secondary cation
(two hyperconjugative
interactions)

t-**Butyl cation**

$H_3C \diagdown \overset{\oplus}{C}-CH_3$ with H_3C

A tertiary cation
(three hyperconjugative
interactions)

▲ FIGURE 3-24
Hyperconjugation in a primary, secondary, and tertiary cation.

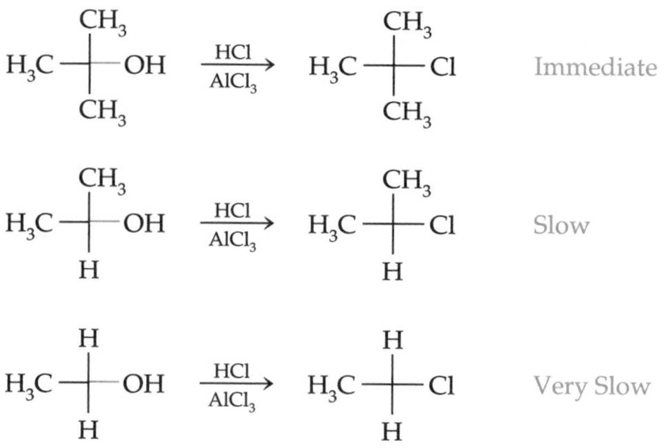

▲ FIGURE 3-25
Rates for appearance of an insoluble layer in a Lucas test.

which the alcohol undergoes heterolytic cleavage to ultimately form the corresponding alkyl chloride, as shown in Figure 3-25. This conversion proceeds through the intermediacy of a carbocation.

In the **Lucas test,** a layer appears because the product alkyl halide does not have a polar O—H bond that can participate in hydrogen bonding. The product is therefore much less soluble in water than is the starting alcohol. How fast this layer appears describes how fast the alcohol reacts. As shown in Figure 3-25, this layer appears immediately with tertiary alcohols, slowly with secondary alcohols, and virtually not at all (no reaction after as long as 5 or 10 minutes) with primary alcohols. Notice that in this reaction the —OH group is replaced by —Cl. It is therefore called a **substitution reaction,** a class of reactions that we will consider in much more detail in Chapter 7.

EXERCISE 3-L

For each of the following alcohols, predict whether the reaction with the Lucas reagent (HCl/AlCl$_3$) would be immediate or slow or would not take place at all.

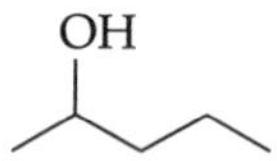

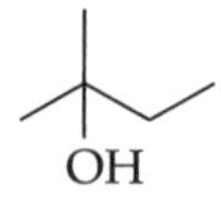

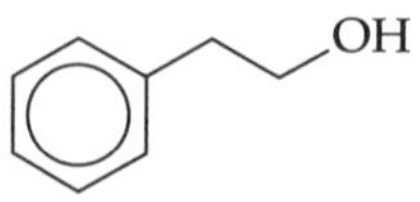

Conjugation in Radicals and Cations

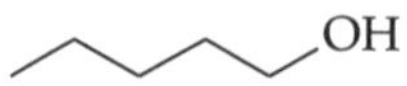

Let us now consider the interaction of the vacant p orbital of a carbocation with an adjacent double bond. Overlap of the carbocationic center with the π bond results in the conjugated orbital array shown in Figure 3-26. The presence of easily polarized π electrons adjacent to a vacant p orbital allows for resonance interaction. This disperses positive charge to the atoms at opposite ends of this conjugated, three-carbon-atom system. This shift of electron density is represented by the resonance structures at the right and left in Figure 3-26, with the curved-arrow notation in the center showing the flow of electrons. (Recall from Chapter 2 that a resonance structure is a rep-

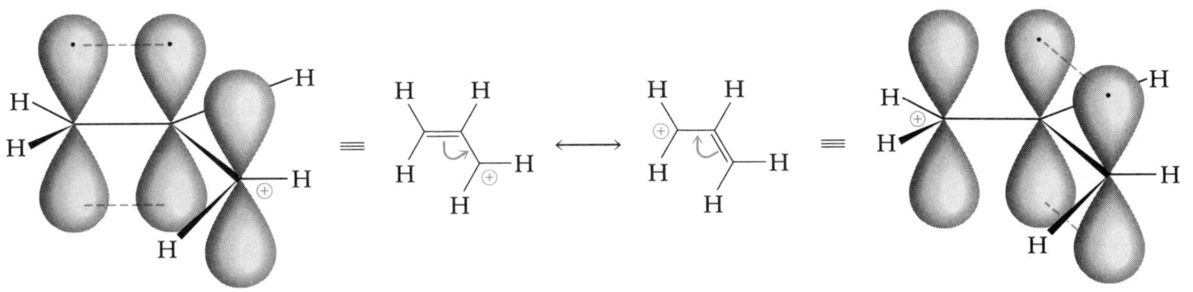

▲ FIGURE 3-26

Three-dimensional representations of significant resonance contributors in an allyl cation.

resentation of the electron distribution in a molecule in which atomic positions are fixed.) Whenever we can write two or more reasonable resonance structures, the molecule or intermediate being represented is unusually stable because of delocalization. In this case, partial positive charge is distributed over two carbons rather than being localized on one. This primary **allyl cation,** containing two electrons in three adjacent *p* orbitals, is almost as stable as a secondary alkyl cation.

A similar interaction takes place between a carbocationic center and a phenyl substituent. The resulting **benzyl cation,** which has a vacant *p* orbital adjacent to an aryl ring (Figure 3-27), also has significant resonance stabilization and is about as stable as a tertiary cation. Thus, the order of cation stability presented earlier for carbocations should be expanded: benzylic ~ 3° > allylic ~ 2° > 1° > CH_3.

The same factors that are important in stabilizing the allyl and benzyl cations apply to the allyl radical and the benzyl radical shown in Figure 3-28 (on page 88); that is, because we can describe the electronic arrangement with several important resonance structures, these radicals are more stable than simple primary radicals. A more-complete order of radical stability, inferred from the C–H bond dissociation energies in Table 3-6, is: allylic > benzylic > 3° > 2° > 1° > CH_3.

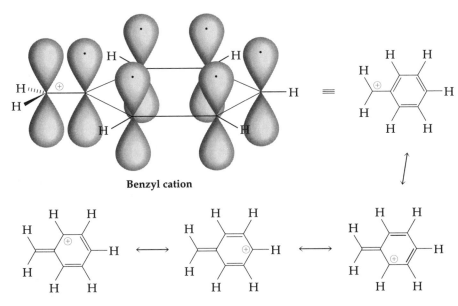

Benzyl cation

▲ FIGURE 3-27

Resonance contributors for a benzyl cation.

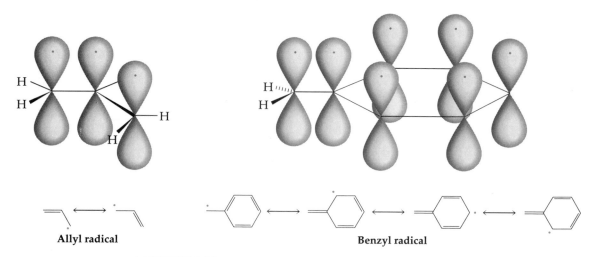

▲ FIGURE 3-28

Resonance contributors for the allyl and benzyl radicals.

EXERCISE 3-M

Rank the following isomers in order of facility for acid-catalyzed dehydration: fastest first, slowest last. (Hint: Acid-catalyzed dehydration proceeds through a carbocation.)

(a)

(b)

(c)

3-8 ▸ Ethers: R—O—R

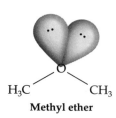

Methyl ether

Functional groups other than alcohols contain oxygen. Unlike alcohols, **ethers** have two carbon atoms bound to an oxygen atom, as in methyl ether. A near tetrahedral geometry is maintained at the sp^3-hybridized oxygen atom, and the carbon–oxygen bonds are polarized so that the electron density on oxygen is higher than that on carbon. Because of this molecular dipole, ethers are more polar than hydrocarbons and are more soluble with polar solutes and other polar solvents.

Because ethers do not have an OH group, they cannot donate a hydrogen bond to themselves or to other molecules. The polarized carbon–oxygen bond enhances van der Waals attractions between ether molecules (compared with hydrocarbons); but, without hydrogen bonding, ethers have weaker intermolecular interactions than do alcohols. Therefore, the boiling points of ethers are comparable to those of hydrocarbons of similar molecular weights but lower than those of the analogous alcohols.

The bond-dissociation energies of the carbon–oxygen bonds in ethers are similar to those in alcohols, but heterolytic cleavage of the C–O bond in an ether generally requires a very strong acid and rigorous heating. In general, ethers are very unreactive and, as a result, find utility in organic chemistry as solvents. An ideal **solvent** is one that interacts with solutes well enough to dissolve them but remains chemically inert. Ethers are good solvents because the polar carbon–oxygen bonds participate in dipole–dipole interactions with other polar molecules. In addition, the lone pair of electrons on oxygen in an ether participates in hydrogen bonding with compounds bearing polar X-H groups. However, ethers are resistant to heterolytic cleavage reactions because they do not have acidic X-H groups themselves and thus do not serve as proton donors, a reaction characteristic of alcohols. Although ethers are polar, they lack an acidic proton on a heteroatom. They are therefore *aprotic* solvents.

Ethers usually derive their names from the alkyl groups attached to oxygen, as in methylethyl ether, $CH_3OCH_2CH_3$ and ethyl ether, $CH_3CH_2OCH_2CH_3$. Notice that we do not use a prefix (such as di) to indicate two identical groups in ethyl ether: when only one group is specified, the compound is assumed to be a symmetrical ether.

3-9 ▸ Carbonyl Compounds: $R_2C{=}O$

Oxygen atoms also participate in multiple bonding. The removal of two hydrogen atoms (one bonded to a carbon atom and the other to an adjacent oxygen atom) from an alcohol results in a $C{=}O$ double bond (Figure 3-29). These units are referred to as **carbonyl groups.** A carbonyl carbon that bears a hydrogen and an alkyl group is called an **aldehyde,** whereas one that bears two alkyl groups is called a **ketone.** An aldehyde is produced by the oxidation of a primary alcohol, whereas the oxidation of a secondary alcohol gives a ketone. We can recognize the conversion of an alcohol into an aldehyde or a ketone as an oxidation by counting the number of bonds to heteroatoms. In an alcohol, there is one bond to oxygen; after oxidation,

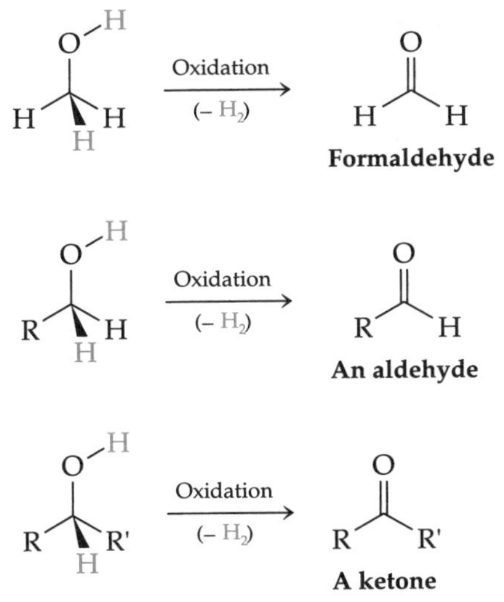

▲ FIGURE 3-29

Carbonyl compounds produced by alcohol oxidation.

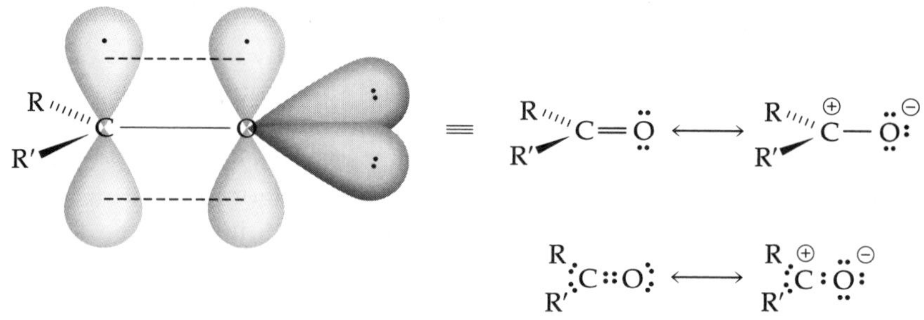

▲ FIGURE 3-30
Several representations of bonding in a ketone.

there are two. An aldehyde is named by adding an -**anal** suffix; a ketone is named by adding an -**anone** suffix.

The carbon atom in a carbonyl group is bonded to only three atoms and is thus trigonal and sp^2-hybridized. The geometries of carbonyl compounds are therefore similar to those of imines, which contain C=N double bonds. Thus, as shown in Figure 3-30, a carbonyl compound is planar, with a CCO bond angle of about 120° and with a π orbital formed by overlap of *p* orbitals above and below the carbon–oxygen atomic plane. Like imines, carbonyl compounds can be catalytically hydrogenated to alcohols. The higher stability of the C=O π bond makes the reaction more difficult, however, and much more rigorous experimental conditions are required. Alkyl groups stabilize C=O double bonds even more than C=C double bonds.

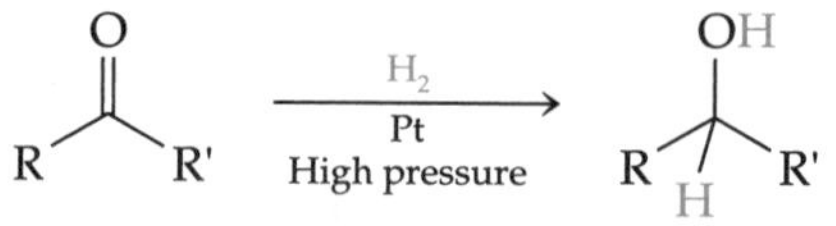

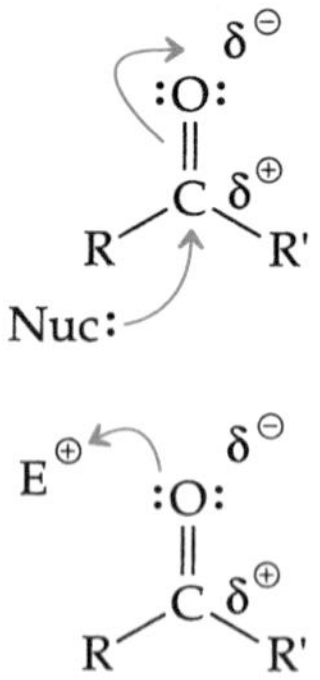

▲ FIGURE 3-31
Nucleophilic attack at a carbonyl carbon, and electrophilic attack at a carbonyl oxygen.

Because oxygen is more electronegative than carbon, it attracts the electrons in the double bond more strongly. The electrons in a π orbital are held less tightly than those in a σ bond, and thus the degree of polarizability is greater for a π bond than for a σ bond. We can write a resonance structure in which the electrons initially shared between carbon and oxygen in the π bond are shifted completely to oxygen. The right-hand resonance contributor in Figure 3-30, when regarded as a Lewis dot structure, meets the valence requirement of oxygen, but not of carbon. Furthermore, there is one covalent bond fewer and a formal charge separation in this structure, with carbon bearing a positive charge and oxygen a negative charge. The structure at the right therefore contributes less to the real structure of the carbonyl group than does that at the left. However, because it does contribute to some degree, the carbonyl group is polarized, and so the carbon end of the carbon–oxygen double bond is electron deficient and the oxygen end is electron rich. Therefore, the carbon atom of a carbonyl group is easily attacked by nucleophiles, as is the oxygen atom by electrophiles (Figure 3-31). These reactions will be covered in more detail in Chapters 12 and 13.

3-10 ▸ Carboxylic Acids and Derivatives

Because an oxygen atom needs only two covalent bonds to fulfill its valence requirement, it does not ordinarily participate in triple bonding. A carbon atom, however, can bond to two different oxygen atoms. Replacement of

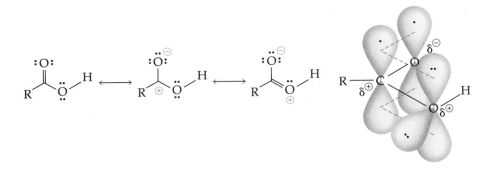

▲ FIGURE 3-32
Resonance contributors of a carboxylic acid.

the hydrogen of an aldehyde with an OH group produces a class of compounds known as **carboxylic acids.** A carboxylic acid (RCO_2H) is easily named by the addition of an **-anoic acid** suffix to the root designating the length of the appropriate hydrocarbon. The carbon atom of a carboxylic acid is sp^2-hybridized and therefore trigonal, with a formal π bond to oxygen forming a carbonyl group, as shown at the left in Figure 3-32. There are two additional resonance structures shown for the carboxylic acid, both with negative charge on the carbonyl oxygen. One has positive charge on carbon, whereas the other has this positive charge delocalized to the adjacent oxygen by the formation of a π bond between carbon and that oxygen. The electron density of the oxygen–hydrogen bond is thus shifted even further toward oxygen than it is in an alcohol, the partial positive charge thus facilitating the loss of a proton. Deprotonation of carboxylic acids is easier than that of alcohols: that is, carboxylic acids are more acidic than alcohols. We will see in Chapter 6 how resonance stabilization of the anion resulting from the loss of a proton from a carboxylic acid is also important in enhancing the acidity of the O–H group.

The ability to donate a proton (that is, to act as an acid) characterizes much of the chemistry of carboxylic acids. The carbon atom in a carboxylic acid is at a higher oxidation level (three bonds to oxygen) than that in an aldehyde. Indeed, this carbon has an oxidation level of +3, the same as that in a nitrile, in which carbon is triply bound to nitrogen.

There are other functional groups like carboxylic acids in which the carbon atom of the carbonyl group bears three bonds to heteroatoms. All of them have the same +3 oxidation level for the carbonyl carbon and are considered derivatives of carboxylic acids. Several of these functional groups are shown in Figure 3-33. You should become familiar with the name of

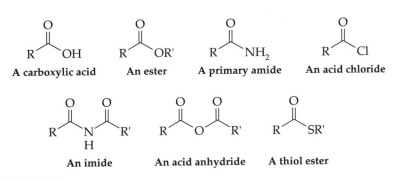

▲ FIGURE 3-33
Derivatives of carboxylic acids.

each functional group. **Esters** are named as alkyl derivatives with the suffix **-anoate,** amides as **-anoamides, anhydrides** as **-anoic anhydrides,** and **acid chlorides** as **-anoyl chlorides.** We will consider their chemistry in detail in Chapters 12 and 13, but for now it is sufficient to notice that the carbonyl carbons in these functional groups are at the same oxidation level and that each bears three bonds to heteroatoms. Resonance structures similar to those of the parent carboxylic acid in Figure 3-32 can be written for each of these derivatives.

Contributions from resonance structures analogous to the middle and right-hand structures in Figure 3-32 significantly influence the physical properties of carboxylate derivatives. For example, rotation about the carbon–nitrogen bond in an amide is much more difficult than rotation about the C–N bond in an amine. This **restricted rotation** is caused by the partial double-bond character of the carbon–nitrogen bond in the amide; that is, it is due to a contribution from the zwitterionic resonance contributor at the right.

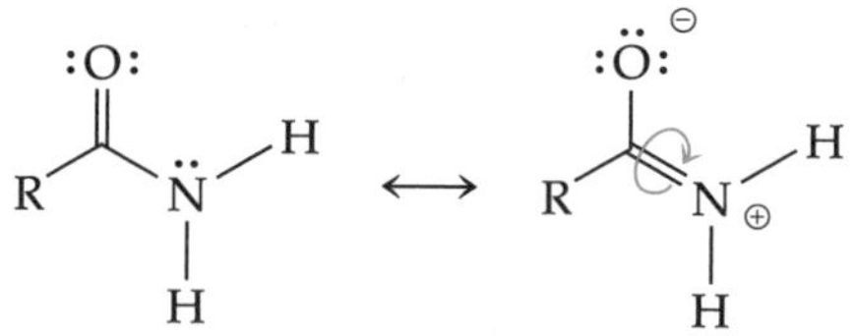

The overlap between nitrogen and the carbonyl carbon in an amide is even greater than that with oxygen in a carboxylic acid: in amides, the barrier to rotation is approximately 18 kcal/mole. Furthermore, to the extent that this resonance structure is important, less positive charge is localized on carbon; therefore, attack on carbon by a nucleophile, like that shown in Figure 3-31, is more difficult. We will see in Chapter 16 that this partial double-bond character has important consequences for the structures of peptides and proteins, which contain many such amide groups.

EXERCISE 3-N

Classify the functional group of each of the following compounds:

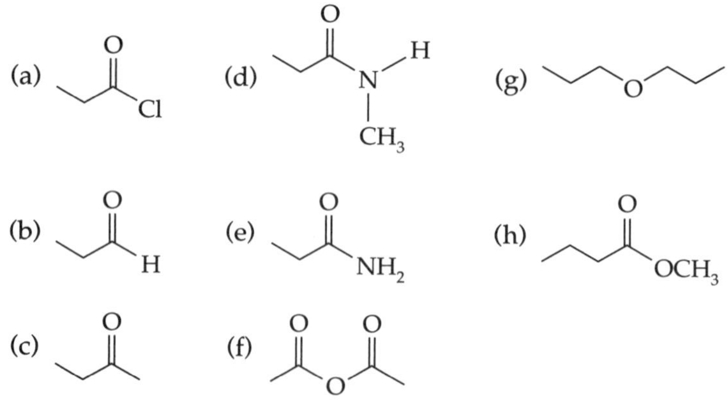

The ease with which a nucleophile attacks a carbonyl carbon is related to the amount of partial positive charge at that site. The contribution of resonance structures like those shown below causes relatively less positive charge to be localized on carbon.

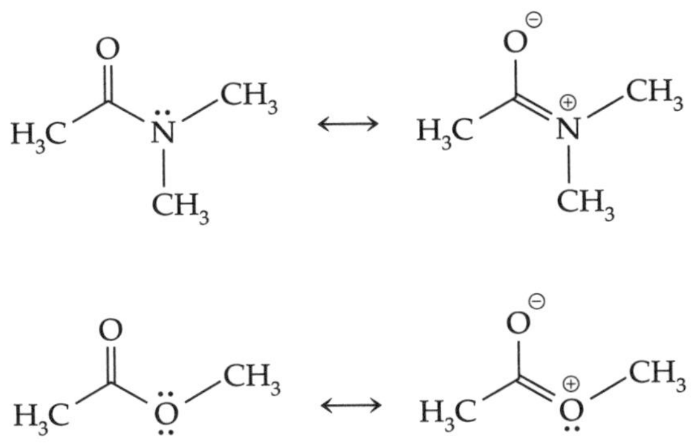

Because a carbonyl carbon in an acid derivative is less electron deficient than it is in a simple aldehyde or ketone, amides and esters are attacked by nucleophiles less readily. Because nitrogen is less electronegative than oxygen, it can more readily release charge to the carbonyl oxygen; so amides are less reactive toward nucleophiles than are esters.

EXERCISE 3-O

Choose the compound in each of the following pairs that would be more easily attacked by a nucleophile at the carbon end of the C=O double bond. Explain your reasoning.

(a) CH_3CHO or CH_3COCH_3

(b) CH_3CHO or $CH_3CO_2CH_3$

(c) CH_3COCH_3 or $CH_3CON(CH_3)_2$

(d) CH_3CONH_2 or $CH_3CO_2CH_3$

We can summarize the oxidation levels of compounds bearing oxygen as shown in Figure 3-34. Consider the oxidation levels of the two different carbons in ethanal (CH_3CHO) and the two identical carbons in ethyne (H—C≡C—H). In ethanal, the methyl group is at an oxidation level of −3 and the carbonyl carbon at +1. In ethyne, both carbons are at a −1 oxidation level. If we add the values for ethanal and ethyne individually, we find the same total, −2. This correspondence is of chemical consequence because it means that the interconversion of these two compounds involves neither oxidation nor reduction overall. Indeed, we will see in detail in Chapter 10

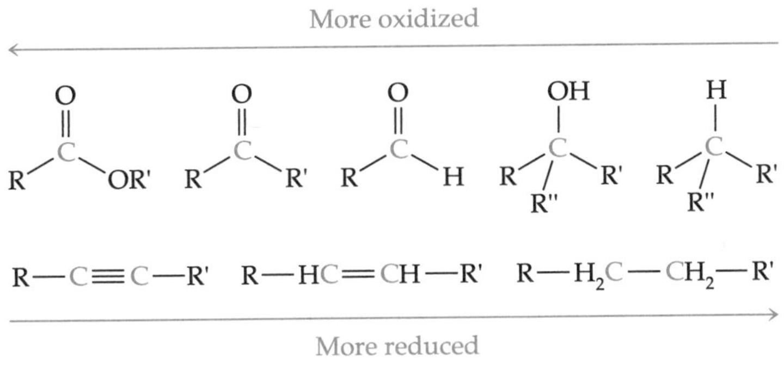

▲ FIGURE 3-34
A comparison of the oxidation levels (for the carbons in color) of oxygen-containing compounds with those of hydrocarbons.

that this reaction takes place by hydrolysis, the addition of water, not by treatment with a redox reagent.

The addition of water across a carbon–carbon double bond results in an alcohol. This addition reaction does not shift the oxidation levels of hydrogen or oxygen in water at all. However, a hydrogen atom is added to one carbon of the alkene (as a consequence, that carbon is reduced), whereas addition of the OH group to the other carbon is an oxidation. These two processes in the same molecule, reduction and oxidation, exactly balance one another, and no overall change in oxidation level of the molecule occurs when water is added to an alkene.

$$\text{H}-\text{OH}$$

$$\bigg\rangle\!=\!\bigg\langle \quad \xrightarrow{\text{Acid}} \quad \overset{\text{H}}{\bigg\rangle}\!-\!\overset{\text{OH}}{\bigg\langle}$$

These two examples show, as we have seen before, that the presence of a multiple bond between carbon atoms means that the carbons are at a higher oxidation level than in an alkane and that a π bond changes the oxidation level of a molecule to the same extent as does the introduction of a single bond to a heteroatom.

3-11 ▸ Sulfur-containing Compounds

A thiol

$$R^{\diagdown S \diagup} R'$$

A thioether

$$R \overset{\overset{\displaystyle O}{\|}}{\diagup\diagdown} SR'$$

A thiol ester

Oxygen and sulfur are in the same column of the periodic table and therefore have similar valence electronic requirements. Like oxygen, sulfur forms hybrid orbitals that participate in covalent bonding. For example, when sulfur is bound to one carbon, in analogy with the oxygen in an alcohol, a **thiol** is formed. When sulfur is bonded to two alkyl or aryl carbons (in analogy to bonding to oxygen in an ether), the functional group is called a **thioether** or alkyl sulfide. A **thiol ester** is a compound in which an SR group replaces the OR group in an ester.

DIMETHYL SULFOXIDE: A VERSATILE SOLVENT

Dimethyl sulfoxide (DMSO) is an odorless dipolar aprotic organic solvent with unusual properties. By virtue of its highly polarized sulfur–oxygen bond, DMSO is miscible with water but also quite soluble in other, less-polar organic solvents. Furthermore, it passes readily through the skin and will even carry with it other organic molecules. It was considered a possible way to deliver drugs to the blood stream for compounds that are destroyed in the digestive system. This application of DMSO has not been commercialized, in part because of concern about possible toxic side effects of the solvent itself. Another complication is that DMSO is reduced in the body to methyl sulfide, a compound with a highly disagreeable odor.

$$\overset{\overset{\displaystyle O}{\|}}{H_3C \diagup S \diagdown CH_3} \quad \xrightarrow{\text{Reduction}} \quad H_3C \diagup S \diagdown CH_3$$

**Dimethyl sulfoxide
(DMSO)**   **Methyl sulfide**

Many of the chemical properties of thiols and alcohols and of thioethers and ethers are similar. Most of the significant differences between these functional groups result because sulfur's valence shell is the third level. Thus, $3s$ and $3p$ atomic orbitals are used to form sulfur's hybrid atomic orbitals. These third-level orbitals are significantly larger than those of the second level, and a size mismatch in the overlap between carbon and sulfur results in a weaker carbon–sulfur covalent bond than that between carbon and oxygen. In addition, sulfur's electronegativity is significantly lower than that of oxygen (2.5 versus 3.5), and it is more polarizable. Finally, because sulfur, in the third row of the periodic table, has access to $3d$ orbitals, its valence shell can be expanded beyond eight electrons: sulfur often participates in bonding with more than four atoms. As a result, the chemistry of sulfur compounds is somewhat more complex than that of oxygen. For example, **sulfonic acids** and their derivatives have no analogy in oxygen chemistry. This expanded valence capability also makes possible amidelike derivatives of sulfonic acids. Some **sulfonamides** are potent antibacterial substances known as the sulfa drugs, which are frequently used in medicine.

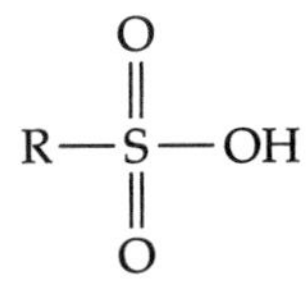

$$
\begin{array}{c}
O \\
\parallel \\
R-S-OH \\
\parallel \\
O
\end{array}
$$

A sulfonic acid

$$
\begin{array}{c}
O \\
\parallel \\
R-S-NH_2 \\
\parallel \\
O
\end{array}
$$

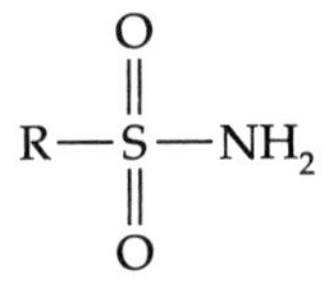

A sulfonamide

EXERCISE 3-P

A thiol ester is analogous to an ester except that the singly bonded oxygen is replaced by sulfur. Would you expect a thiol ester to be more or less active than a simple ester toward nucleophilic attack at the carbon end of the $C{=}O$ double bond? Explain your reasoning.

3-12 ▸ Aromatic Compounds Containing Heteroatoms

We know from Chapter 2 that planar, cyclic, conjugated molecules containing $4n + 2$ electrons (n = an integer) are aromatic compounds that have unusual stability. Aromatic molecules in which one or more carbon atoms are replaced by heteroatoms are **heteroaromatic compounds.** These compounds have a stability similar to that of their all-carbon analogs.

Heteroaromatic Molecules

Many aromatic molecules contain nitrogen, oxygen, sulfur, or other heteroatoms in the ring. As a family, they are called **heterocyclic aromatics,** or heteroaromatics, because the heteroatom is one of the component atoms of the cyclic array. These compounds have common, rather than systematic, names. Three such compounds that contain five ring atoms are shown in Figure 3-35 (on page 96): **furan, pyrrole,** and **thiophene.** Each of these heterocycles can be represented by the cyclic array shown at the left in Figure 3-35, in which one lone pair of electrons on the heteroatom is held in a p orbital perpendicular to the molecular plane and, hence, aligned for conjugative interaction with the p orbitals of the carbon–carbon double bonds. These structures are analogous to the cyclopentadienyl anion (Figure 2-20) because each contains six electrons in a planar cyclic delocalized π system, making them Hückel aromatics. In this geometry, the nitrogen–hydrogen bond of pyrrole must project in the atomic plane and be orthogonal to the π system. This position is occupied by a lone pair in both furan and thiophene, as shown in the left-hand structure in Figure 3-35.

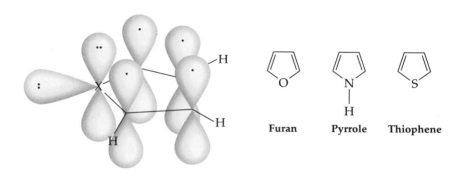

▲ FIGURE 3-35
Representative heteroaromatic molecules.

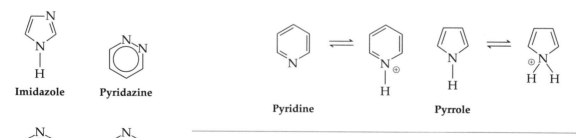

▲ FIGURE 3-36
Several representations of pyridine.

As shown in Figure 3-36, there are also six-membered heteroatomic aromatics. For example, we can write Kekulé-like contributors for the structure of **pyridine.** In these structures, we find a cyclic, delocalized, six-electron system just like that in benzene. In contrast with pyrrole, in which the lone pair is part of the π system, the lone pair on nitrogen in pyridine is contained in an sp^2 orbital in the plane of the six ring atoms and is orthogonal to the p orbitals.

EXERCISE 3-Q

Both pyridine and pyrrole have lone pairs on nitrogen that can be protonated in an acid-base reaction. In view of Hückel's rule, which protonation will be easier? That is, will pyrrole or pyridine be the stronger base? Explain your reasoning.

A heterocyclic aromatic can contain more than one heteroatom, and each structure shown in the margin represents a five- or a six-membered ring containing two nitrogen atoms. Three biologically important bases,

CAFFEINE: A HETEROAROMATIC STIMULANT

Caffeine, a cyclic compound containing nitrogen, is present in both tea and coffee. It has a dramatic stimulating effect on people, and both tea and coffee have been consumed for centuries for this effect. In this century, caffeine has been marketed by itself and in combination with other ingredients for use as a stimulant by those who do not like coffee or when drinking a beverage is not convenient. Until recently, the caffeine sold in this way was prepared by adding a methyl group to theobromine, a related compound obtained from cocoa fruits. (Theobromine is also present in tea.)

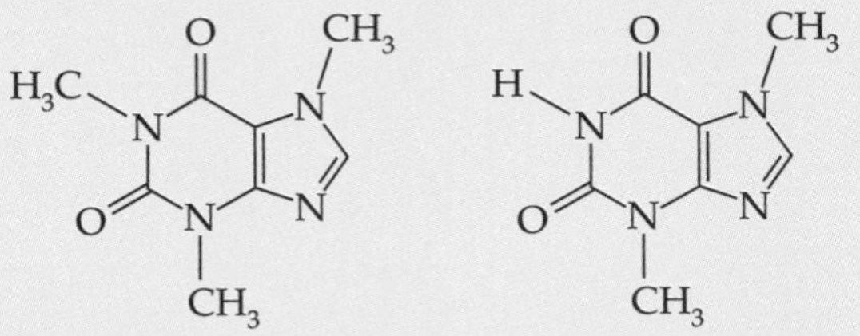

Caffeine   Theobromine

However, the relatively large demand for decaffeinated coffee has resulted in large quantities of caffeine being available by extraction from coffee beans. At first, halogenated organic solvents were used to remove the caffeine from the bean, but concern about possible adverse effects of these solvents on health has stimulated the development of an alternate process that uses steam or supercritical carbon dioxide.

uracil, thymine, and **cytosine,** are hydroxy or amino derivatives of **pyrimidine** (Figure 3-37).

Heteroatoms are also found in fused-ring molecules (structurally similar to the polycyclic aromatic hydrocarbons). Figure 3-38 shows three common fused structures—**quinoline, pteridine,** and **purine**—as well as two

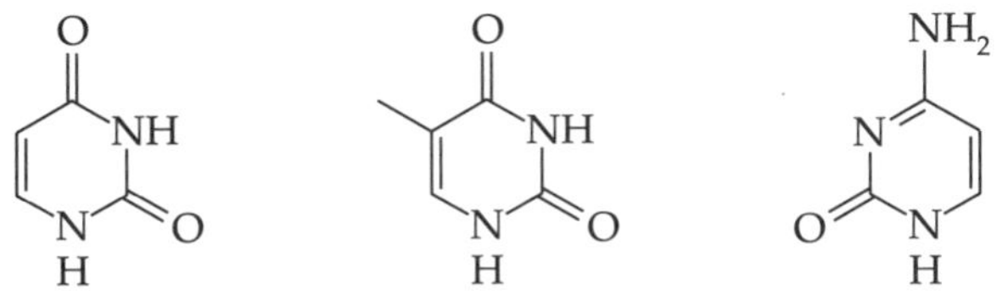

Uracil   Thymine   Cytosine

▲ FIGURE 3-37
Biologically important derivatives of pyrimidine.

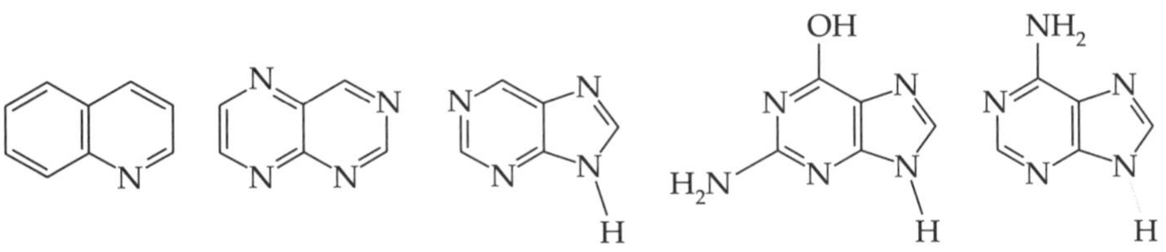

Quinoline   Pteridine   Purine   Guanine   Adenine

▲ FIGURE 3-38
Representative fused-ring heteroaromatics.

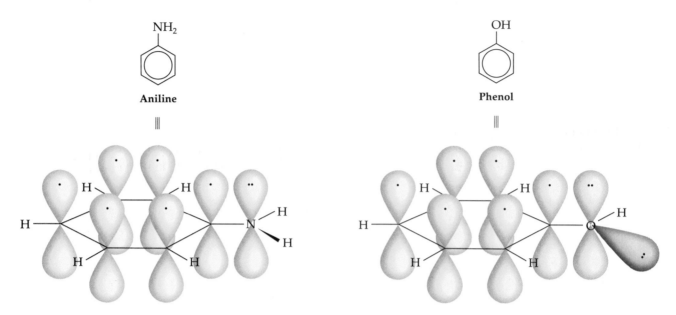

▲ FIGURE 3-39

Three-dimensional representations of orbital interaction in aniline and phenol with sp^2-hybridized heteroatoms.

purine derivatives, guanine and adenine. Purines are subunits of biologically important systems. **Guanine** and **adenine,** together with uracil, thymine, and cytosine, are aromatic bases and are components of nucleotides, which constitute the chemical basis for genetic coding—a subject discussed in Chapter 22. Many different derivatives of pteridine have been isolated from insects and are responsible for the bright and varied colors in butterfly wings.

Heteroatom-substituted Arenes

In addition to the heteroaromatics, which have a heteroatom present in the ring, a number of important compounds have a heteroatom attached to an all-carbon aromatic ring. Because the ring contains only carbon atoms, such compounds are not called heteroaromatics. For example, as shown in Figure 3-39, sp^2 hybridization of nitrogen and oxygen in **aniline** and **phenol** would produce the optimal geometry for the overlap of a lone pair in a heteroatomic p orbital with the aromatic array of p orbitals on carbon in the ring, although overlap of electrons from the heteroatom with the ring π system can also take place even without rehybridization. We shall see the effect of this extended conjugation on the chemical reactivity of aromatic compounds in Chapter 10.

From Chapter 2, we know that aromatic rings can have alkyl substituents, as in toluene. It is also possible for these alkyl substituents to bear heteroatoms. Several of these compounds are encountered frequently and are usually referred to by common names: for example, benzoic acid, benzaldehyde, acetophenone, and anisole.

Benzoic acid   **Benzaldehyde**   **Acetophenone**   **Anisole**

Will contribution by the zwitterionic resonance structures for aniline shown below make aniline a stronger or a weaker base than it would be if its structure could be described simply by the resonance contributor at the left? Explain your reasoning.

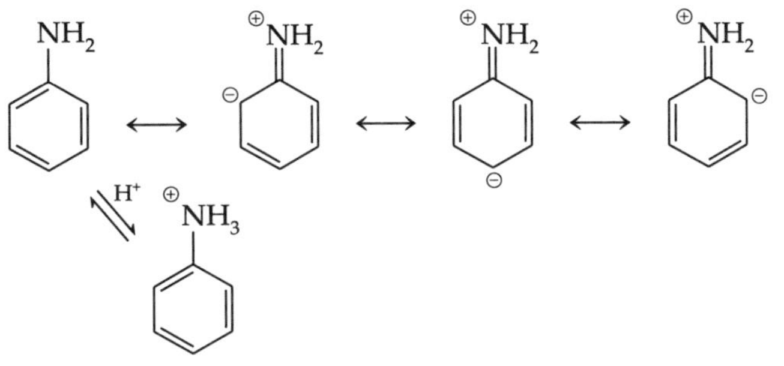

3-13 ▶ Alkyl Halides

The next-to-last column on the right-hand side of the periodic table (before the inert gases) contains the halogens. The atomic structure of fluorine is shown in Figure 3-40. Fluorine requires only one σ bond to satisfy its valence requirement. In the structure of hydrofluoric acid, seven valence electrons are contributed by fluorine and a single electron is contributed by hydrogen to satisfy the valence requirements of both atoms, making H—F a stable molecule.

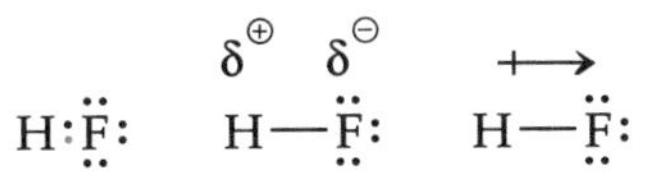

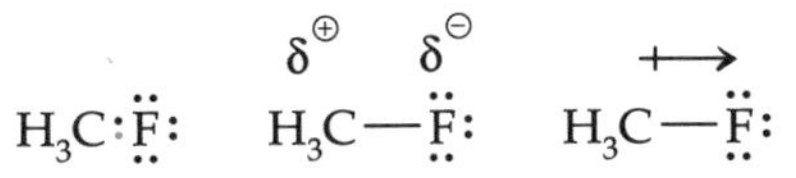

$$\text{F: } 1s^2\ \underbrace{2s^2 2p_x^2 p_y^2 p_z^1}_{\substack{\text{Four}\\ sp^3\text{-hybrid}\\ \text{orbitals}}}$$

▲ FIGURE 3-40
Mixing of atomic orbitals in fluorine to form sp^3-hybridized orbitals.

The σ bond is nonetheless highly polarized because there is a substantial difference in electronegativity between hydrogen and fluorine.

Carbon–fluorine bonds are otherwise similar to carbon–nitrogen or carbon–oxygen bonds, as in methyl fluoride.

CHEMICALLY INERT CARBON–FLUORINE BONDS

An important distinction between the bond between fluorine and carbon and other bonds to carbon is its much greater strength. For example, a C–F bond is 25% stronger than a C–H bond. As a result, fluorocarbons are unusually stable, and polymers such as Teflon, in which there are only C–F and C–C bonds, are almost completely inert to chemical reaction, except with strong reducing agents. They can therefore be used for applications where other organic materials are degraded, such as in coatings for heating utensils or as seals for containers of corrosive liquids.

▶ FIGURE 3-41
Isomeric butyl fluorides.

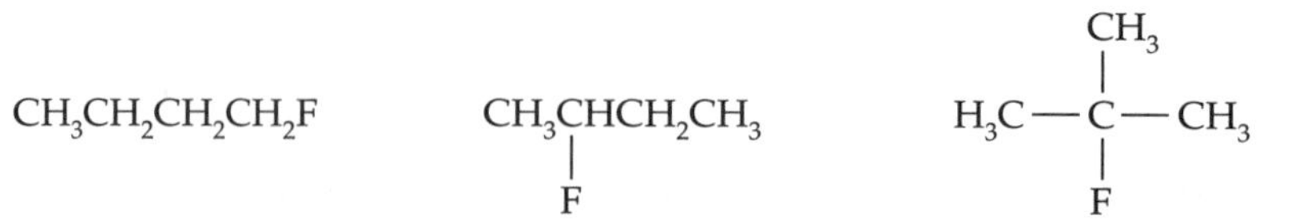

$$CH_3CH_2CH_2CH_2F \qquad CH_3CHCH_2CH_3 \qquad \underset{F}{\overset{CH_3}{H_3C-C-CH_3}}$$
$$\qquad\qquad\qquad\qquad \overset{|}{F}$$

1-Fluorobutane   **2-Fluorobutane**   **2-Fluoro-2-methylpropane**
(a primary alkyl halide) **(a secondary alkyl halide)** **(a tertiary alkyl halide)**

Like alcohols, alkyl fluorides can be primary, secondary, or tertiary (Figure 3-41).

Other **alkyl halides** have carbon bonded to one of the other halogen atoms (chlorine, bromine, or iodine) and have structures similar to those of alkyl fluorides. They can be named either as halogenated alkanes (for example, bromoethane) or as alkyl halides (for example, ethyl bromide). From the bond-dissociation energies of alkyl fluorides, chlorides, bromides, and iodides (while keeping the alkyl group constant), we conclude that the σ bond becomes weaker as the difference in size between carbon and the halogen increases (Figure 3-42). In the progression from the top to the bottom of the periodic table, the electronegativity of the halogen decreases, whereas the size and hence its ability to respond to charge demand (polarizability) increases. The stability of the anion (with the negative charge on I^- spread over a much larger area than on F^-) also is important. Therefore, heterolytic cleavage within a series of alkyl halides also becomes easier in the progression from alkyl fluorides to alkyl iodides, and the rates of C–X heterolytic cleavage increase as the R–X bond becomes weaker (Figure 3-43), in parallel to the acidities of HX.

EXERCISE 3-S

Remembering that the dipole moment of a molecule is the vector sum of its bond dipoles, predict whether a dipole moment will exist in any of the following multihalogen-substituted compounds and draw (in three dimensions) the direction of the dipole:

(a) CCl_4 (b) $CHCl_3$ (c) CH_2Cl_2 (d) CH_3Cl (e) CBr_4

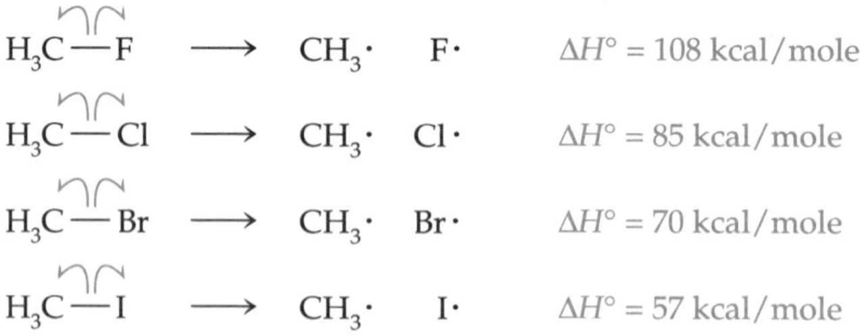

$$H_3C-F \longrightarrow CH_3\cdot \quad F\cdot \qquad \Delta H° = 108 \text{ kcal/mole}$$

$$H_3C-Cl \longrightarrow CH_3\cdot \quad Cl\cdot \qquad \Delta H° = 85 \text{ kcal/mole}$$

$$H_3C-Br \longrightarrow CH_3\cdot \quad Br\cdot \qquad \Delta H° = 70 \text{ kcal/mole}$$

$$H_3C-I \longrightarrow CH_3\cdot \quad I\cdot \qquad \Delta H° = 57 \text{ kcal/mole}$$

▲ FIGURE 3-42
Bond-dissociation energies in methyl halides.

▶ FIGURE 3-43
Relative rates of heterolytic
cleavage in alkyl halides.

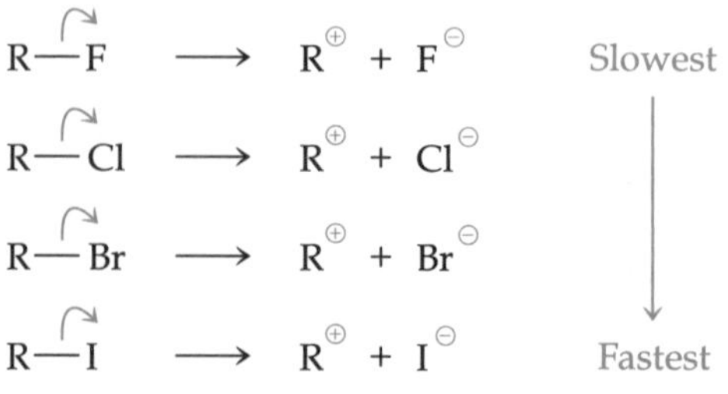

$$R-F \longrightarrow R^{\oplus} + F^{\ominus} \qquad \text{Slowest}$$

$$R-Cl \longrightarrow R^{\oplus} + Cl^{\ominus}$$

$$R-Br \longrightarrow R^{\oplus} + Br^{\ominus}$$

$$R-I \longrightarrow R^{\oplus} + I^{\ominus} \qquad \text{Fastest}$$

3-14 ▸ Nomenclature

Each heteroatom-containing functional group considered in this chapter is named in accord with the IUPAC rules presented in Chapter 1, except that the suffix is changed to identify the functional group. Table 3-7 (pages 102 and 103) is a summary of the suffixes used to name these families. This table also includes the minimal representation needed to characterize each of the functional groups considered. Like hydrocarbons, these compounds are named by locating the longest continuous carbon chain that contains the functional group. The root designates the number of carbons and the suffix designates the functional group. Substituents are assigned numbers to indicate their positions along the carbon skeleton.

The frequent occurrence of low-molecular-weight members of some functional groups has made it convenient to use common names, rather than the IUPAC nomenclature, to name them. For example, formaldehyde (HCHO), acetaldehyde (CH_3CHO), acetyl for CH_3CO—, acetic acid (CH_3COOH), and acetone (CH_3COCH_3) are used almost to the exclusion of conventional IUPAC names.

EXERCISE 3-T

Write acceptable names for each of the following compounds:

(a)

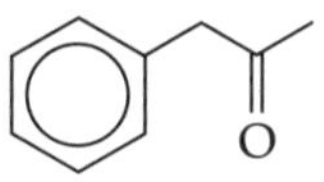

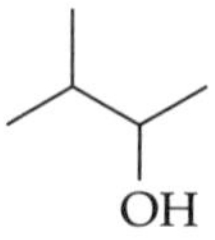

(b)

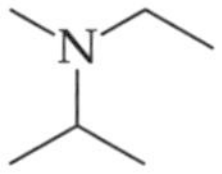

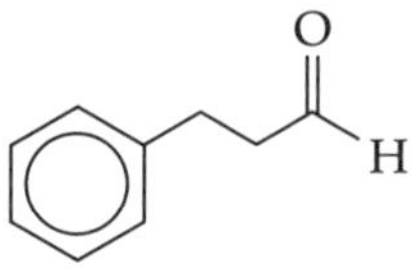

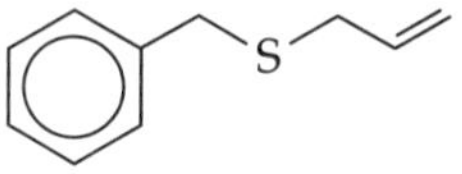

(c)

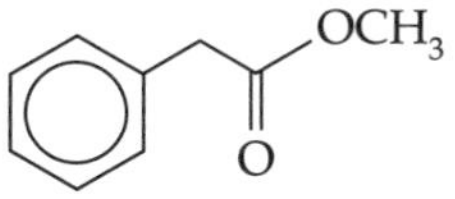

(d)

(e)

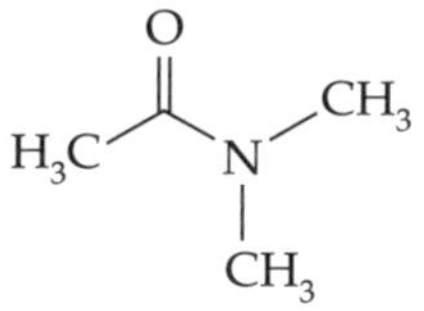

(f)

(g)

(h)

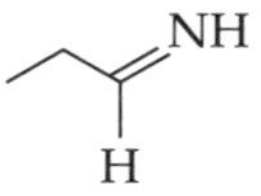

(i)

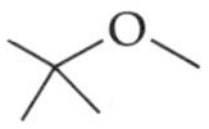

EXERCISE 3-U

Draw acceptable structures corresponding to each of the following IUPAC names:

(a) butanone

(b) 2-hexanone

(c) 3-pentanone

(d) 4-methylpentanal

(e) 2-chloropropanoic acid

(f) methyl propanoate

(g) dimethylamine

(h) propanoamide

(i) butanoyl chloride

(j) ethyl 2-bromopropanoate

Table 3-7 ▶ Nomenclature of Various Functional Groups

Functional Group	Composition	IUPAC Suffix	Example
Alcohol	R—OH	-anol	CH_3CH_2OH **Ethanol**
Ether	R—O—R'	alkyl ether	CH_3—O—CH_3 **Methyl ether**
Aldehyde	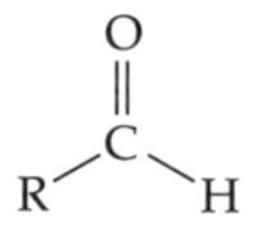 (structure) R—C(=O)—H	-anal	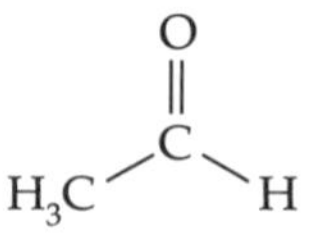 (structure) H_3C—C(=O)—H **Ethanal**
Ketone	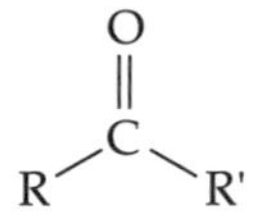 (structure) R—C(=O)—R'	-anone	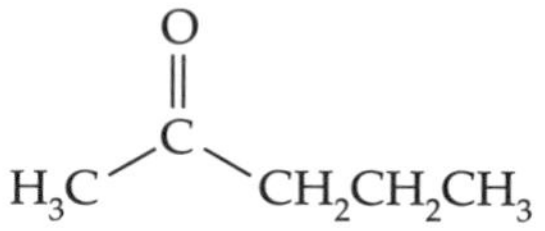 (structure) H_3C—C(=O)—$CH_2CH_2CH_3$ **2-Pentanone**
Carboxylic acid	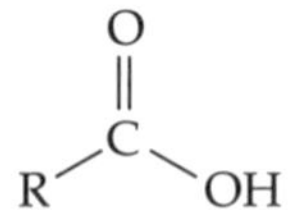 (structure) R—C(=O)—OH	-anoic acid	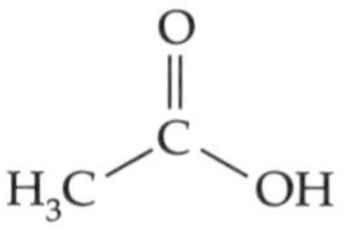 (structure) H_3C—C(=O)—OH **Ethanoic acid**
Acid chloride	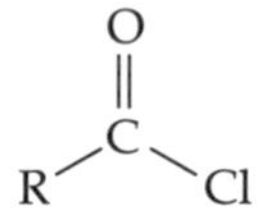 (structure) R—C(=O)—Cl	-anoyl chloride	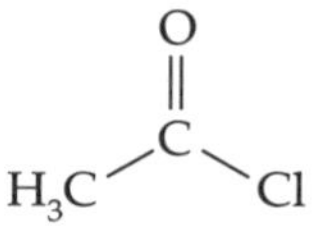 (structure) H_3C—C(=O)—Cl **Ethanoyl chloride**
Amide	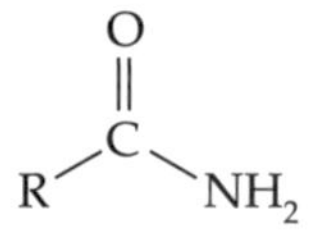 (structure) R—C(=O)—NH_2	-anoamide	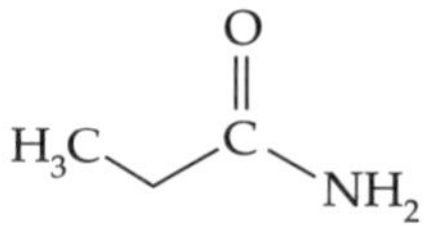 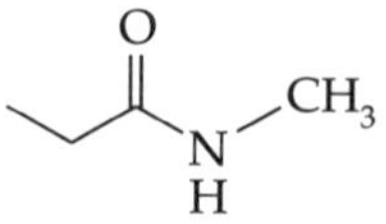 (structure) H_3C—CH_2—C(=O)—NH_2 **Propanoamide** (structure) —C(=O)—N(H)—CH_3 ***N*-Methyl propanoamide**

Conclusions

By considering the electronic structures of nitrogen, oxygen, and fluorine (and atoms in the same columns of the periodic table), we can make important predictions about their derivatives with respect to bond strength, molecular geometry, and reactivity with nucleophiles and electrophiles. This analysis helps us to understand the reactivity of these functional groups and to recognize oxidation and reduction levels among organic compounds containing heteroatoms.

The classification of heteroatom-containing molecules into subgroups is based upon the degree of substitution of the heteroatom-bearing carbon, except in amines for which the level of substitution on nitrogen is used instead. Thus, the terms primary, secondary, and tertiary applied to alcohols,

Table 3-7 ▸ Nomenclature of Various Functional Groups *(continued)*

Functional Group	Composition	IUPAC Suffix	Example
Anhydride	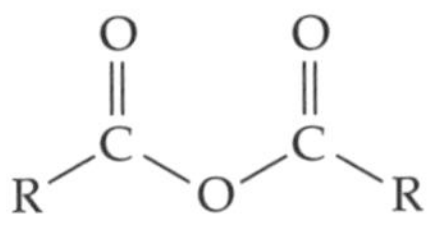	-anoic anhydride	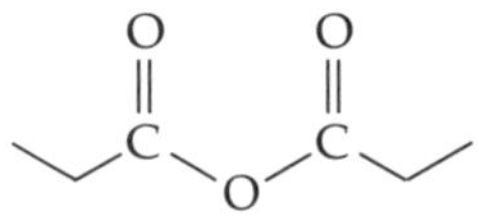 **Propanoic anhydride**
Ester	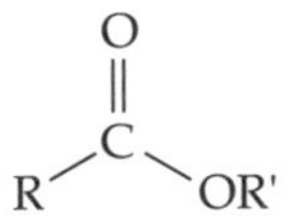	alkyl -anoate	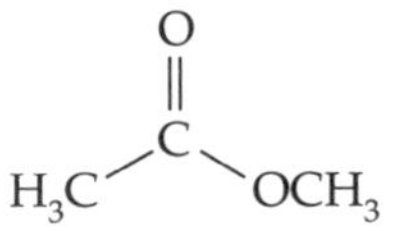 **Methyl ethanoate**
Amine	RNH_2, R_2NH, R_3N	alkylamine	$H_3C-NH-CH_2CH_3$ **Methyl ethylamine**
Nitrile	$R-CN$	-anonitrile	CH_3CH_2CN **Propanonitrile**
Imine	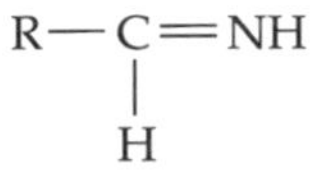 $R-\overset{\underset{\mid}{H}}{C}=NH$	-anal imine	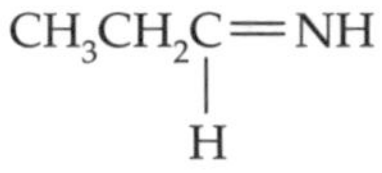 $CH_3CH_2\overset{\underset{\mid}{H}}{C}=NH$ **Propanal imine**
Thioether	$R-S-R$	alkyl thioether	$CH_3SCH_2CH_3$ **Methyl ethyl thioether** **(methyl ethyl sulfide)**
Alkyl halide	$R-X$	haloalkane or alkyl halide	CH_3CH_2F **Fluoroethane** **(ethyl fluoride)**

ethers, and alkyl halides refer to the number of carbon substituents on the carbon bearing the oxygen or halogen substituent. When applied to amines, these same designations refer to the number of alkyl groups attached to nitrogen.

Heteroatoms alter the structure of carbon compounds because the presence of one or more lone pairs of electrons on an atom of higher electronegativity induces significant partial charge separation within the molecule. The heteratom usually functions as a locus for chemical activity; that is, as the functional group in a molecule. Reactions of heteroatom-containing compounds usually take place at bonds to or near the heteroatom.

The difference in electronegativity between carbon and a heteroatom (X) to which it is bound results in polarization of the C–X σ bond. In many cases, the vectorial sum of such polar covalent bonds causes a net molecular dipole moment, which has consequences for the molecule's physical properties (greater reactivity toward charged reagents, higher melting points and boiling points, higher solubility in polar solvents, and so forth). The presence of a dipole moment within a molecule makes that molecule subject to attack by nucleophiles and electrophiles. Nucleophiles attack molecules at centers of partial positive charge, whereas electrophiles attack at centers of partial negative charge.

A heteroatom bound to both carbon and a hydrogen can participate in hydrogen bonding. This interaction derives from polarization of the X–H bond so that hydrogen is attracted to a lone pair of a heteroatom in another molecule or at another site within the same molecule. Hydrogen bonding has an important effect on the three-dimensional structure of a molecule (if intramolecular) and on solvation and intermolecular association (if intermolecular).

Bond cleavages in organic molecules can be accomplished in homolytic or heterolytic pathways. In a bond homolysis, the two electrons initially shared between the two atoms in a covalent bond are partitioned equally to the two radical fragments. In the alternative pathway, heterolytic cleavage, both electrons of the covalent bond are transferred to one of the participating atoms, leaving the other atom with none of the electrons of the bond. Radicals result from homolytic cleavage, whereas ions result from heterolytic bond cleavage.

Multiple bonding between carbon and nitrogen is possible, as in imines (double bond) and nitriles (triple bond). Double bonding between carbon and oxygen is found in aldehydes, ketones, carboxylic acids, esters, amides, or other derivatives of carboxylic acids. Because oxygen's valence shell is filled with two bonds (and two lone pairs of electrons), triple bonds to oxygen are not found in stable molecules, except in carbon monoxide.

Alcohols are functional groups bearing oxygen σ bound both to carbon and to hydrogen. A characteristic reaction of alcohols is the acid-catalyzed loss of water, the first step of which generates an oxonium ion from which water is lost to form a carbocation. The reactivity of an alcohol is dependent on the character of the carbon atom to which the OH group is attached. Carbocations (formed by heterolytic cleavage) and radicals (formed by homolytic cleavage) follow the same relative order of stability: benzyl ~ tertiary > secondary ~ allyl > primary > methyl. This order of stability arises from the greater polarization of a highly substituted carbon and from resonance stabilization and greater hyperconjugation by alkyl groups. The Lucas test can be used to chemically distinguish primary, secondary, and tertiary alcohols by the rates at which they undergo conversion into alkyl halides.

Ethers lack the OH group of alcohols and are therefore much less reactive than alcohols. Their primary use in organic chemistry is as polar aprotic inert solvents. Carbonyl groups are highly polarized so that carbon bears appreciable partial positive charge, making it a potential site for nucleophilic attack.

Sulfur-containing compounds are similar to those containing oxygen. The chemistry of thiols is similar to that of alcohols, and thioethers are similar to ethers. Thiol esters have some of the features of carboxylic acid esters, but because of mismatch in orbital size between the sulfur and the adjacent carbon, the carbon–sulfur π bond in the zwitterionic resonance contributor is considerably weaker than the carbon–oxygen π bond. Enhanced reactivity toward nucleophilic attack at carbon results from this weaker interaction.

Aromaticity is maintained in the presence of heteroatoms, and the same considerations regarding the number of electrons delocalized in a stabilized aromatic ring apply. Several heteroaromatic compounds containing more than one heteroatom are important in nucleic acid chemistry.

Alkyl halides contain highly polar carbon–X bonds. Often such compounds have high dipole moments and are readily attacked by nucleophilic reagents at the partially positively charged carbon.

Nomenclature of heteroatom-containing compounds follows the IUPAC rules discussed in Chapter 2: roots designate the number of carbon atoms in the longest chain containing the functional group; suffixes designate the identity of the functional group; and numbers and positions of substituents are represented by prefixes and Arabic numerals.

Summary of New Reactions

Protonation of Amines

$$R_3N + HA \rightleftharpoons R_3\overset{\oplus}{N}{-}H + A^-$$

Alcohol Oxidation

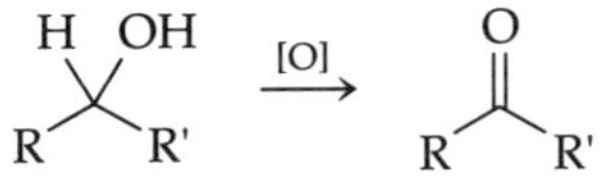

Alcohol Substitution: Lucas Test

$$R{-}OH \xrightarrow[ZnCl_2]{HCl} R{-}Cl \qquad 3° > 2° > 1°$$

Alcohol Dehydration

$$R{-}OH \xrightarrow[Heat]{H^\oplus} \overset{\oplus}{R} + H_2O \qquad 3° > 2° > 1°$$

alkene

Catalytic Hydrogenation of Aldehydes, Ketones, and Imines

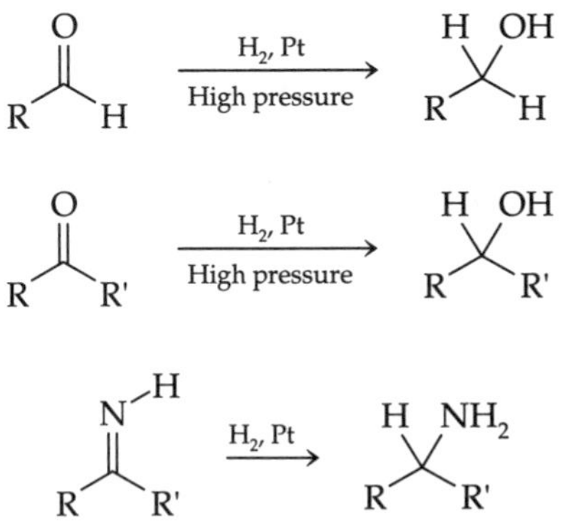

Review Problems

3-1 Like alkenes, imines can exist as geometric isomers. Draw the *cis* and *trans* isomers of ethanal imine. Would you expect interconversion of these isomers to be easier or harder than the *cis–trans* isomerization of 2-butene?

3-2 Classify the following alcohols and amines as primary, secondary, or tertiary. Name each compound according to the IUPAC rules.

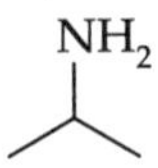

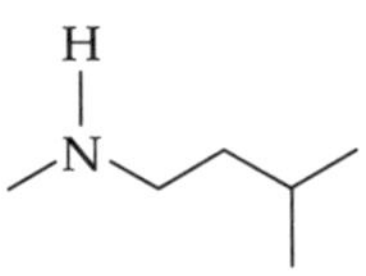

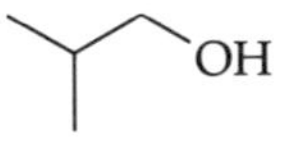

(e) 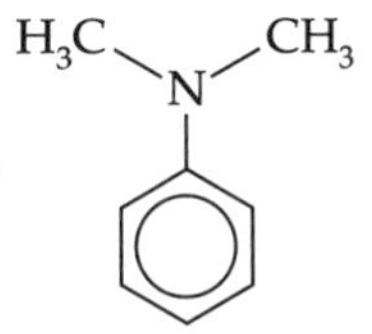   (f)

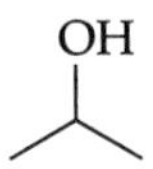

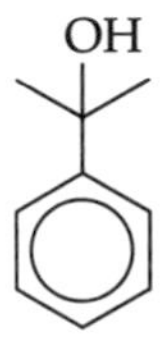

3-3 Ethers, esters, aldehydes, and thioethers dissolve in concentrated sulfuric acid. Why?

3-4 Environmentalists are greatly concerned about an atmospheric ozone hole centered on Antarctica and thought to be caused in part by the presence of fluorochlorocarbons in the atmosphere. Ozone, O_3, absorbs high-energy ultraviolet light (which is dangerous to plant and animal life), and its absence imperils these species.

(a) Draw a Lewis dot structure of ozone, O_3, being sure to indicate formal charge on each atom.

(b) By drawing a resonance structure, explain how the two O–O bonds in ozone are of equivalent length.

(c) From the hybridization of the oxygen atoms in the structure that you have drawn, predict whether ozone is linear or bent.

3-5 The acidity of a sulfonic acid is due to the high stability of the conjugate base derived by deprotonation of the acid. Write significant resonance structures for the monoanion of benzene sulfonic acid ($C_6H_5SO_3H$), and use them to explain why the acidity of sulfonic acids is higher than that of carboxylic acids.

3-6 Explain why iodomethane has a smaller dipole moment ($\mu = 1.62$ D) than fluoromethane ($\mu = 1.85$ D).

3-7 Use resonance structures to explain why formaldehyde has a larger dipole moment than methanol.

3-8 Calculate the formal oxidation level of carbon in

(a) ethyne

(b) acetonitrile (CH_3CN)

(c) ethyl amine

Explain why the catalytic hydrogenation of ethyne to ethane is called a reduction.

3-9 Although ethyl ether has a substantially higher molecular weight than ethanol, ethanol has the higher boiling point. Explain.

3-10 Derivatives of butane can be obtained by the replacement of C–H bonds with C–Cl bonds when butane is exposed to chlorine gas in the presence of ultraviolet light. Determine (a) how many different monochlorobutanes are possible? (b) how many dichlorobutanes? (c) how many trichlorobutanes?

3-11 Draw structures of all geometric isomers of

(a) 1,1,2-trichlorocyclopentane

(b) 1,2,3-trichlorocyclopentane

(c) 1,2,4-trichlorocyclopentane

3-12 From what you know about intermolecular interactions, decide which compound in each of the following pairs has the higher boiling point.

(a) pentane, C_5H_{12}, or octane, C_8H_{18}

(b) ethyl alcohol, CH_3CH_2OH, or methyl ether, CH_3OCH_3

(c) ethylene glycol, CH_2OHCH_2OH, or ethyl alcohol, CH_3CH_2OH

3-13 Write structural formulas that correspond to the following descriptions:

(a) four esters with the formula $C_4H_8O_2$

(b) two aldehydes with the formula C_4H_8O

(c) a secondary alcohol with the formula C_3H_8O

(d) three ketones with the formula $C_5H_{10}O$

(e) a tertiary amine with the formula $C_4H_{11}N$

(f) a tertiary alkyl bromide with the formula C_4H_9Br

3-14 Dimethyl sulfoxide (H_3CSOCH_3, often called DMSO), methylene chloride (CH_2Cl_2), dimethylformamide [$HCON(CH_3)_2$, called DMF], methanol, ethyl ether ($CH_3CH_2OCH_2CH_3$, often simply called ether), and tetrahydrofuran [$—(CH_2)_4O—$, called THF] are common organic solvents. Classify each of these solvents as dipolar aprotic, polar protic, or nonpolar. Identify the structural feature in each molecule from which its solvent classification derives.

3-15 Identify each of the following reagents as a nucleophile or an electrophile:

(a) triethylamine

(b) hydroxide ion

(c) Fe^{3+}

(d) methanethiol, CH_3SH

3-16 Identify the functional group in each of the following compounds. Does the molecule act as a Lewis acid or base?

(a)

(b) $H_3C—\overset{\overset{O}{\|}}{\underset{\underset{O}{\|}}{S}}—OH$

(c)

(d)

3-17 What is the relation between the members of the following pairs of structures? Are they identical? positional (or structural) isomers? geometric isomers? or resonance contributors?

(a)

(b)

(c)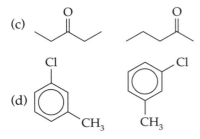

(d)

3-18 Secondary structure in deoxyribonucleic acid (DNA) has as its basis specific hydrogen bonding between guanine and cytosine and between adenine and thymine, all of which are bound through a nitrogen atom to a sugar of the backbone. Draw structures to show how guanine can form three hydrogen bonds to cytosine and how adenine can form two hydrogen bonds to thymine.

3-19

(a) The aromatic ring in phenol (C_6H_5OH) behaves as if it is particularly electron rich. Draw resonance structures in which the electrons in one of the lone pairs on oxygen is shifted to another position to explain this observation.

(b) Assuming that parallel resonance contributors to those drawn in answer to part *a* control the electron density of other aromatic rings, rank the following compounds in order of decreasing ring electron density (most electron rich first).

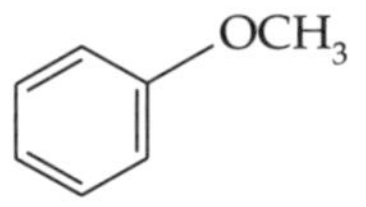

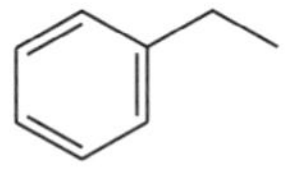

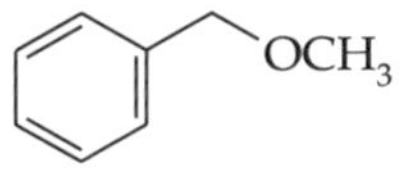

3-20 Provide IUPAC names for each of the following structures:

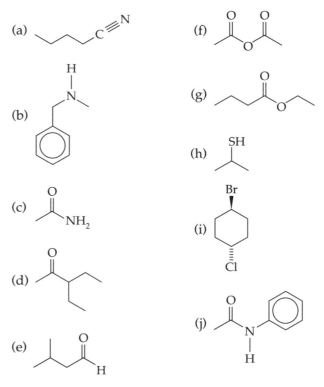

3-21 Draw structures that correspond to each of the following IUPAC names:

(a) 2-aminobutane
(b) methyl ethyl ether
(c) *t*-butyl allyl thioether
(d) methyl pentanoate
(e) benzoyl chloride
(f) *N*-phenyl ethanoamide
(g) *p*-bromoiodobenzene
(h) 2-methylthiophene
(i) *N*-methylaniline
(j) propanoic acid

Chromatography
and
Spectroscopy

In Chapters 2 and 3, we learned how to identify the functional groups of organic chemistry and how they are assembled from carbon atoms and heteroatoms. We also saw how specific functionality can influence typical physical properties. Although what we learned about the relation between structure and functional group is logical, we were asked to accept much of the structural information without independent proof. In this chapter, we will learn how physical techniques can be used to gather evidence about structure. We will also learn how to identify products correctly and to follow reactions with confidence as we examine the properties and reactions of these functional groups in the remainder of the text.

Among the interesting questions to be addressed in this chapter are: How do we know whether a given sample is a pure compound or a mixture? How do we assign structure to a compound? What characteristics of the individual functional groups assure us that such assignments are likely to be correct?

4-1 ▶ The Use of Physical Properties to Establish Structure

When a student learns a new chemical reaction, the identities of the reactants, reagents, and products are given. The practicing chemist working in a laboratory, however, usually knows only the structure of the reactant and the reagents with which it is treated and must demonstrate the structure of the product, even if it follows logically. In this chapter, we will describe how a practicing chemist assigns structure by establishing the identity of products, and thus determines whether a desired reaction has been successful.

Physical properties such as melting points or boiling points can be used to help assign structure to a compound, provided that the compound has previously been prepared and these physical properties have been measured and recorded. The greater the number of physical properties that correspond to those of the known compound, the greater is the chemist's confidence that an assignment of structure is correct. For example, if a compound thought to be 2-octanol exhibits a boiling point corresponding to that listed in reference books for this compound, it would be reasonable to hypothesize that this assignment is possible. However, it is difficult to establish boiling points to closer than within 2 or 3 degrees, and a perusal of the compounds listed in even a simple reference such as the *Handbook of Chemistry and Physics* will quickly convince you that many compounds have the same boiling-point range within 2 or 3 degrees. If a second physical property (such as the melting-point range) also corresponds to that listed for 2-octanol, the structural assignment can be made with greater confidence because far fewer candidates will match both properties.

You might be even more assured if your compound's chemical reactivity also corresponds to that of 2-octanol. For example, suppose that you treat the compound with a strong oxidizing agent (say, chromic acid) and that you obtain a product mixture that can be distilled to give a clear liquid that boils at 173 °C. Because this result corresponds to that which you would get from the expected oxidation product, 2-octanone, evidence is accumulating that the original assignment was correct.

Chemists working in the nineteenth century and first half of the twentieth century spent a great deal of their time not only conducting the reaction of interest, but also transforming the product into other known compounds simply to show in a more-convincing way that the suggested structural assignment for the product was correct.

A principal reason that research productivity has dramatically increased in the past 40 years is the availability of new instrumentation and techniques for the isolation of single compounds from mixtures and for the structural identification of organic compounds. Many of these methods provide direct evidence for the presence of functional groups and their spatial arrangements, thereby greatly facilitating the assignment of structure. The most common of these techniques are subjects of this chapter. They are grouped into two general classes: chromatography and spectroscopy.

Chromatography is the principal means by which components of mixtures like those formed in chemical reactions are separated. It is used both to obtain pure individual components of a mixture and to determine the ratio of components present. In chromatography, molecules are **partitioned** between two different phases, and separation is directly related to the difference in solubility that different molecules show in each of these phases.

In addition to being a powerful separation tool, chromatography is a method by which to assess the magnitude of intermolecular, noncovalent interactions. Furthermore, because compounds differ in mobility, chromatography can be used to demonstrate a correspondence between a compound and a reference sample of known structure.

Spectroscopy constitutes a set of techniques that measure the response of a molecule to the input of energy. The resulting spectrum is a series of bands that show the magnitude of the interaction of a compound as a function of the incident energy. The energy source can be optical photons—as in ultraviolet spectroscopy, visible spectroscopy, and infrared spectroscopy—or radio-frequency energy—as in nuclear magnetic resonance spectroscopy.

A somewhat different technique, known as mass spectroscopy, determines the mass of ions formed when molecules are bombarded with high-energy electrons.

In this chapter, we consider how these spectroscopic techniques are used to assist in the identification of organic products, although not at the level of detail necessary for general application in the laboratory. (Such details are absorbed naturally as you use these techniques in a more-specialized course or as a practicing chemist.) Instead, we deal with the physical basis of each of these methods and how characteristic spectra are interpreted to obtain structural information.

4-2 ▸ Chromatography

At some time in our lives, most of us have had an undesirable encounter with chromatography: for example, the ink on a neatly written homework paper becomes rain soaked, causing the ink to "run"; that is, to disperse into the component colors as they dissolve and flow at different rates across the paper surface. In fact, the word chromatography was first suggested by the Russian chemist Mikhail Tswett at about the turn of the century to describe the separation of pigments as "colored writing." As in this simple (perhaps undesired) experiment, chromatographic separations are usually accomplished by introducing organic compounds onto a **stationary phase** (the paper) and then allowing a **mobile phase** (the water) to flow past the mixture. Each component interacts with (**adsorbs on**) the stationary phase and dissolves in the mobile phase to a different extent. Those that are bound less tightly to the stationary phase and are more soluble in the mobile phase travel farther than other components of the mixture. The various methods of chromatography differ with respect to the mobile phase (a liquid or a gas), the stationary phase (paper, gels, or solid packings), and the driving force for the motion of the mobile phase (pressure, gravity, or an electrical field).

Modern chromatographic techniques make use of the difference in solubility of different molecules in a moving phase relative to a stationary phase. In **gas chromatography,** the mobile phase is a **carrier gas** and the stationary phase is either a solid or a solid coated with a nonvolatile liquid. In **liquid chromatography,** the mobile phase is a liquid and the stationary phase is a solid composed of small particles around which the liquid phase can flow. All of these techniques depend on differences in the strength of interaction of the various components of a mixture with the solid phase. Thus, it is important that the surface area of this phase be as large as possible. Very fine particles of stationary phase are often used for the most-demanding separations because the ratio of surface area to volume increases as the size of the particles decreases. The differences in interaction with the stationary phase (**adsorption**) and with the mobile phase result in different **mobilities** as a mixture of compounds passes over a solid support.

Chemists routinely use a simple example of selective partitioning when they extract organic molecules into an organic phase (such as ether) from water containing inorganic salts. In **extraction,** the separation of organic and inorganic compounds takes place because organic compounds are generally more soluble in ether than in water, whereas the inorganic materials are more soluble in water. These differences in solubility are usually very large, and it is often necessary to extract an aqueous layer only once to obtain most, if not all, of an organic material.

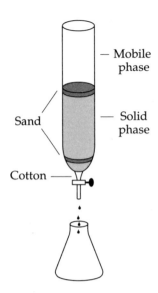

▲ FIGURE 4-1
Column chromatography. A mixture of compounds placed at the top of a solid support slowly moves down the column as a liquid mobile phase flows over the solid support, partitioning the components of the mixture differently. The components therefore elute separately as the eluent flows under the influence of gravity.

Liquid Chromatography on Stationary Columns

A simple liquid chromatography column is constructed by packing a solid stationary phase (typically, alumina or silica gel) as a slurry into a burette. Both alumina and silica gel are **polymers**—arrays of large molecules composed of simple, repeating subunits (in this case, Al_2O_3 and SiO_2, respectively). As shown in Figure 4-1, a plug of cotton is usually inserted into a burette (or other glass column with a restriction at one end) so that the stopcock does not become clogged with small particles of the solid phase. A layer of sand is added to form a more even surface, and then the solid phase is added. A second layer of sand is added at the very top of the column so that the solid phase is not disrupted as the mobile phase is added. A concentrated solution of the mixture of compounds to be separated is applied at the top of the solid phase; then the mobile phase, or **eluent,** is added and allowed to flow toward the bottom under the influence of gravity.

Those components of a mixture that adhere more tightly to the solid phase move more slowly down the column, requiring a greater volume of eluent before they reach the bottom. Thus, as the chromatography proceeds, a series of bands develops and the mixture of compounds is separated into its components. This motion of solute and solvent through the solid phase is called **elution.** The difference in the volume of solvent required for two different compounds to pass through a column is a measure of the degree to which each interacts with the solid phase. The ratio of solvent volumes required for elution represents the degree of separation of two components and is referred to as **alpha** (α). The time that a compound takes to pass through the column is its **elution time.** As long as the rate of flow of the solvent remains constant, the separation factor α can also be determined from the ratio of elution times.

Both alumina and silica gel are polar materials that have surfaces covered with hydroxyl groups. As a result, molecules with higher polarity adsorb more strongly to these highly polar, metal oxide supports (recall "like prefers like"). Correspondingly, for a mixture of compounds differing in polarity, the less-polar components elute faster.

EXERCISE 4-A

Which compound in each of the following pairs would be more likely to flow first from an alumina column eluted with ethyl acetate?

(a) acetone or 2-propanol

(b) benzene or cyclohexane

(c) acetic acid or methyl acetate

(d) cyclohexyl chloride or cyclohexylamine

Liquid chromatography conducted in an open chromatographic column such as that shown in Figure 4-2 is often referred to as **column chromatography.** The degree of separation indicates the relative ease of elution of two components. The ease of separation, or **resolution,** in chromatography depends not only on the degree of separation but also on how much each component has spread while passing through the column. Band spreading decreases as the average size of the particles constituting the stationary phase is made smaller; but, with very small particle sizes, the flow of solvent induced by gravity all but stops. This problem is overcome in **high performance liquid chromatography,** or **high pressure liquid chro-**

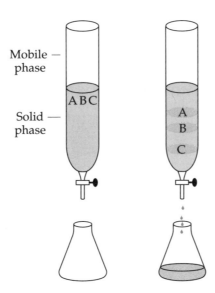

▲ FIGURE 4-2
Separation of compounds A, B, and C by elution from a chromatographic
column.

matography (HPLC), in which the mobile phase is driven through a sealed
column by a mechanical pump. The same principles apply in HPLC as in
simple column chromatography but, because smaller particles can be used,
the degree of separation of compounds is better in HPLC.

Let us consider how chromatography can be used for the separation of
a mixture of compounds A, B, and C, resolved into bands as in Figure 4-2
and flowing from the column as in Figure 4-3. The least-polar compound
(C) elutes first. As additional solvent flows through the column, compounds A and B continue to move and are ultimately eluted, in turn, from
the column. Each of these components can be collected in a separate flask.
Thus, **chromatographic separation** is achieved. Because each component is

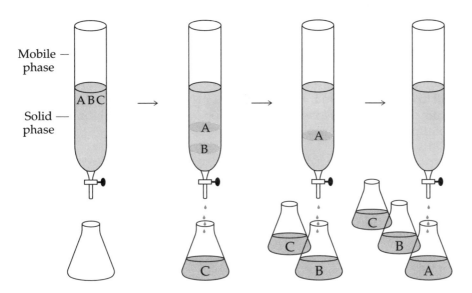

▲ FIGURE 4-3
Elution of compounds A, B, and C as separate samples after chromatographic
resolution.

obtained as a single compound, chromatography is also a method for purification. When solvent is removed from the eluent in each of the flasks in Figure 4-3, the individual components of the original mixture, A, B, and C, are obtained in relatively pure form.

To accomplish this separation, however, it is necessary for us to know when a component is eluting from the column. For this reason, liquid chromatographic columns are often coupled with a **detector** that responds to a change in some physical property when an additional compound is present in the eluting solvent. Thus, the simple arrangement shown in the column of Figure 4-1 is modified so that the effluent flows first through a detector before being collected in a flask. One very useful chromatographic detector is the **refractive index detector,** a so-called universal detector because it responds to virtually all compounds. The basis of its operation is that a path of light bends as it passes from one medium to another and the degree of bending is related to the difference in the refractive indices of the two media. As light passes from one compartment containing the solvent and sample to another containing only the solvent, it is bent by an amount that depends on the difference in refractive index between the two liquids (Figure 4-4). The change in the path of the light is detected by comparing the intensity of the light received by two photocells positioned so that they "see" equal light when there is no difference in refractive index and thus no bending of the light. Indeed, the **refractive index** is defined as the ratio of the degree of bending of a light beam when it passes through a material to that when it passes through air. The refractive index of a solution differs from that of a pure solvent, and thus the eluent (containing a dissolved component) emerging from a chromatographic column has a different refractive index from that of the pure solvent itself. When the refractive index of the liquid phase is continually monitored as it flows through such a detector, a change in refractive index indicates the presence of an additional component in the liquid. A plot of the detector response as a function either of the volume of effluent flowing through the column or of the elution time, called a **chromatogram,** will show peaks for each component (A, B, and C in our example) as each separately flows from the column and through the detector. An added advantage to the use of a detector to monitor the elution of samples is that the ratio of the areas under each peak (its **integration**) is proportional to the ratio of concentrations of the individual molecules composing the original mixture. In the chromatogram shown in Figure 4-5, the ratio of the area under peak A to that under peak B to that under peak C is approximately 4:1:3. Assuming that the detector's response to each component is the same, this ratio is proportional to the relative amounts of A, B, and C present. Most commercial HPLC instruments are equipped with detectors for observing (as a chromatogram) the elution of a purified compound.

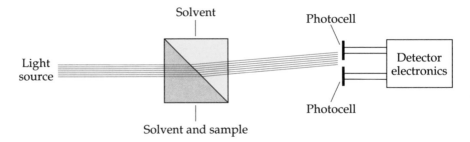

▲ FIGURE 4-4

A refractive index detector. The magnitude of bending of the path of incident light indicates the presence of a compound having a refractive index that differs from that of the solvent.

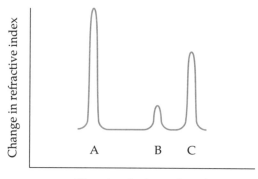

▲ FIGURE 4-5

A chromatogram showing the elution of compounds A, B, and C from a chromatographic column. Because the refractive index returns to the level of pure solvent, this chromatogram provides evidence that A has been completely separated from B, which is completely separated from C. The absence of overlap of peaks indicative of the elution of components of a mixture is called a base-line separation.

Other changes in physical properties can also be used to detect the presence of a sample in the eluting solvent. Other commercially available chromatographic detectors employ ultraviolet absorption spectroscopy, fluorescence spectroscopy, or electrochemical conduction as a means of registering the presence of a compound as it passes from the column and through the detector. Irrespective of the physical characteristic that is measured, however, such detectors are designed to indicate when the composition of the eluent has changed.

Paper and Thin–Layer Chromatography

Another variant of liquid chromatography uses sheets of support (Figure 4-6) rather than the cylindrical columns employed in both column chromatography and HPLC. In **paper chromatography,** the mixture of compounds to be separated is applied as small drops of a solution near one edge of a sheet of chromatographic paper. This edge is immersed in a solvent that acts as the mobile phase, or eluent, which is pulled up the paper by capillary action. Alternatively, in **thin-layer chromatography,** a flat solid support such as a sheet of glass, plastic, or aluminum foil is coated with a

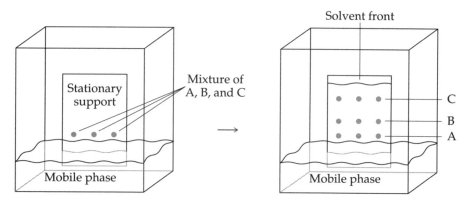

▲ FIGURE 4-6

Chromatographic separation by thin-layer chromatography. The eluent moves up over a dry stationary support by capillary action.

thin layer of silica gel or alumina. As in paper chromatography, the solvent moves upward through the solid support by capillary action. This movement achieves the same separation of a mixture as that accomplished by gravity in column chromatography.

The *R*_f **value** is the ratio of the distance migrated by a substance compared with the farthest point reached by the solvent (the **solvent front**). The *R*_f value usually parallels the ratio of elution times observed in liquid chromatography. It is common practice to employ thin-layer chromatography first in order to find a solvent (or mixture of solvents) that will separate a mixture before proceeding to HPLC.

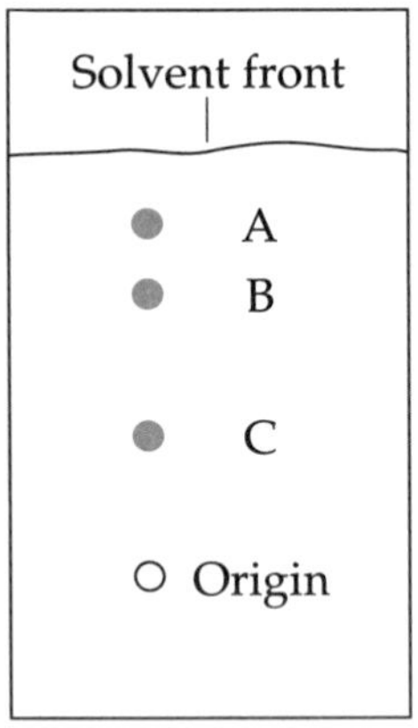

EXERCISE 4-B

Calculate R_f values for the three compounds separated in the thin-layer chromatogram shown at the left.

Large biological molecules often have many polar functional groups and bind too tightly to silica gel and alumina for effective chromatographic resolution. For these types of compounds, a modified silica gel is used as a stationary phase in which a nonpolar organic molecule has been chemically bonded to the surface of the particles. This stationary phase, which is now nonpolar, binds the less-polar compounds most tightly. The more-polar compounds are carried more rapidly through the column by the solvent, which is often a mixture of a hydrocarbon (hexane) and a small amount of an alcohol such as 2-propanol. Because the normal order of elution is reversed (with the more-polar compounds eluting first), this technique is referred to as **reverse phase chromatography.** In contrast, the use of unmodified silica gel or alumina is sometimes called **normal phase chromatography.**

Gel Electrophoresis

We will see in later chapters that biological molecules often have many charged centers and are therefore sometimes called **polyelectrolytes.** Such a molecule can bear both positive and negative charges at various sites, giving either a net positive or a negative charge or, if the charges are exactly balanced, no overall charge. (The zwitterions presented in Chapters 2 and 3 exemplify one class of compounds that are neutral overall but have sites of both positive and negative charge.) Even in a neutral molecule, these ionic centers interact strongly with a stationary support. To separate molecules of this kind, a polar organic polymer like polyacrylamide (properties of polymers will be discussed in more detail in Chapter 16) is used as the support. The polymer is saturated with water, causing it to swell, and ionic compounds move through this stationary phase under the influence of an electric field. Thus, negatively charged ions migrate toward the positive pole when an electric field is applied, whereas cations migrate in the opposite direction. Molecules with a higher charge-to-mass ratio migrate faster, thus effecting separation of a mixture as shown in Figure 4-7.

The migration of an ion (or multi-ion) under the influence of an electric field is known as **electrophoresis.** The use of an electric field to induce the movement of polyelectrolytes through a gel as a separation technique is referred to as **gel electrophoresis.** This technique is used extensively for the separation and purification of biological macromolecules where the relative ease of migration depends on both the charge and the size: smaller molecules and those with higher charge move proportionally faster. The relative

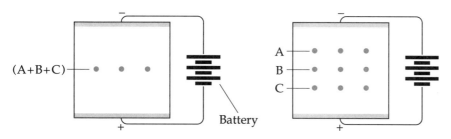

▲ FIGURE 4-7

In gel electrophoresis, charged polyelectrolytes migrate across a gel under the influence of an electric field. Negatively charged ions migrate toward the positive electrode; positively charged ions migrate toward the negative electrode; neutral (including zwitterionic) molecules do not migrate. Shown at the left are three identical mixtures of A, B, and C that separate into components under the influence of an electric field. The separation shown at the right would be attained if A were positively charged, B were neutral, and C were negatively charged.

ease of migration on these gels is often used to compare an unknown with reference samples of known composition.

Gas Chromatography

In gas chromatography, a carrier gas (nitrogen or helium) sweeps a sample from a heated injector block onto and through a long chromatographic column heated within an oven. The gaseous effluent flows over a detector that registers the passage of each compound. Two common types of detectors are used for detecting the presence of compounds in the gas steam. A **thermal conductivity detector** measures the difference in thermal conductivity between the carrier gas alone and the gaseous sample coming from the column. A **flame ionization detector** senses the presence of ions that are generated as the effluent from the column is burned in a hydrogen flame. A gas chromatograph is shown schematically in Figure 4-8. Gas chromatography is used both in research laboratories and in routine analyses. It is the method of choice for analysis of trace amounts of compounds such as pesticides or illicit drugs present in body fluids.

As in the chromatograms obtained by liquid chromatography, gas chromatographic analysis gives rise to a series of compounds, displayed as peaks in a gas chromatogram, eluting at various times after injection onto the column. The stationary phase in gas chromatography can be the walls of

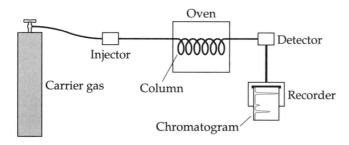

▲ FIGURE 4-8

A schematic description of a gas chromatograph. An inert carrier gas moves under pressure over a column lined or filled with solid adsorbent. The mixture injected at the head of the column is fractionated and detected as the effluent gas passes through a detector.

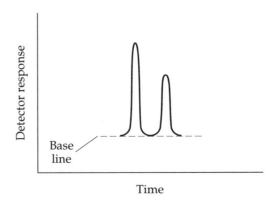

▲ FIGURE 4-9
A typical gas chromatogram illustrating base-line separation of two
injected components.

an empty column or solid packing within a column or a polymeric liquid
that coats either the wall or the porous solid packing. Typically, much
longer columns (10–100 meters) are used in gas chromatography than in liquid column chromatography.

Retention times in gas chromatography are influenced by the strength
of the covalent interactions of the compounds being separated with the stationary phase, just as in liquid chromatography. Roughly, these interactions
can be considered to be governed by the effects of polarity and van der
Waals interactions, which also influence boiling points. It is common in the
use of gas chromatography to find an order of elution that approximately
follows the boiling points of the compounds separated. Because very long
columns are used in gas chromatography, **base-line separation** is often
achieved for chemically similar compounds (Figure 4-9).

─────────

EXERCISE 4-C

Use your knowledge of how molecular structure influences physical properties to predict which of the compounds in each of the following pairs will
emerge first from a gas chromatography column.

(a) ⌃⌄ or ⌃⌄⌃⌄⌃   (c) [structure] or [structure with OH]

(b) [structure with O, N] or H₂N [structure with OH]   (d) CH_4 or CCl_4

─────────

4-3 ▸ Spectroscopy

After chromatography has been used for separation and purification, spectroscopic techniques are employed to identify the components of a mixture.
For the remainder of the chapter, we will consider different types of energy
probes that can be used to establish the chemical structure of pure samples.
Simple examples will illustrate how these techniques can be used to identify the products of chemical reactions. We will consider several common
spectral techniques in the order of increasing energy of the spectroscopic
probe.

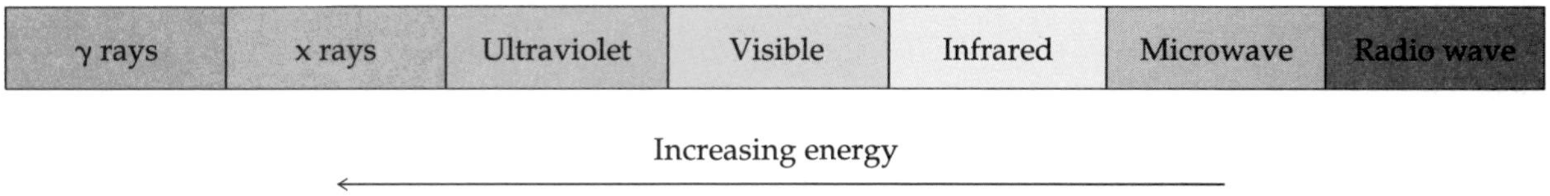

γ rays	x rays	Ultraviolet	Visible	Infrared	Microwave	Radio wave

Increasing energy

←

▲ FIGURE 4-10

Energetic order in the electromagnetic spectrum.

Many spectroscopic techniques rely on the interaction of a compound with **electromagnetic radiation,** which can be considered either a particle (called a photon) or a wave traveling at the speed of light. When regarded as a wave, light can be described by its wavelength (λ) or frequency (ν). Wavelength is the distance of one complete wave cycle, and frequency is the number of wave cycles that pass a fixed point in a defined time. (One **hertz,** Hz, equals one cycle per second.)

For some spectroscopic techniques, we will use wavelength to define energy content. For others, we will use frequency. These quantities are directly related because the product of wavelength and frequency equals the speed of light, c (3×10^{10} cm/sec). Thus:

$$\lambda = \frac{c}{\nu} \qquad \text{and} \qquad \nu = \frac{c}{\lambda}$$

and the energy of a photon can be easily calculated:

$$\varepsilon = h\nu = \frac{hc}{\lambda}$$

in which h = Planck's constant (6.6×10^{-34} joules/sec). (One joule = 4.186 calories. One calorie is the heat required to raise the temperature of 1 gram of water by one degree Celsius.) The energy of the photon increases with its frequency and is inversely proportional to its wavelength. Thus, high frequency means high energy, as does short wavelength. The various regions of the electromagnetic spectrum are shown in Figure 4-10.

In subsequent sections, we will consider the interaction of organic molecules with electromagnetic waves of increasing energy: radio frequencies in nuclear magnetic resonance spectroscopy; infrared photons in infrared spectroscopy; visible photons in visible absorption spectroscopy; and ultraviolet (UV) photons in UV absorption spectroscopy. Finally, we will consider the interaction of high-energy electrons with organic compounds in mass spectroscopy.

Nuclear Magnetic Resonance Spectroscopy

Atomic nuclei with odd mass numbers have angular momentum and behave as if they were spinning about an axis. Because the nucleus is positively charged, this spinning motion causes the nucleus to behave as if it were a tiny magnet that can therefore be aligned by interaction with an applied directional magnetic field. The magnetic fields created by the nuclear spins of these atoms will align either with or against this applied field, denoted by H_0 in Figure 4-11 (on page 120). Because it is slightly more favorable for these nuclear spins to align with the magnetic field than against it, the number of molecules in the parallel alignment will be slightly in excess of those in the antiparallel alignment. In the absence of a magnetic field, these spins are oriented completely randomly.

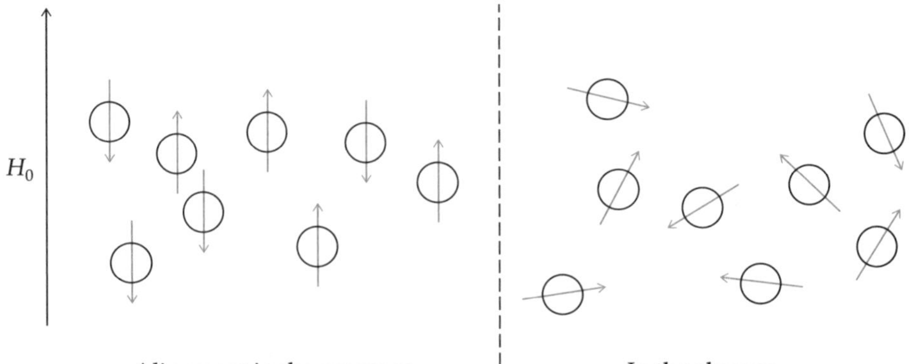

▶ FIGURE 4-11
Parallel and antiparallel alignment of nuclear spins under an applied magnetic field, H_0.

| Alignment in the presence of a strong magnetic field | In the absence of a magnetic field |

Electromagnetic energy that matches the energy difference between the parallel and antiparallel spin states of the nucleus can be absorbed by the molecule being studied, causing its spin to "flip" from the lower-energy, parallel state to the higher-energy, antiparallel state. When this occurs, the spin is said to be in **resonance** with the applied electromagnetic radiation, giving rise to a signal on a detector. A plot of this signal intensity against the energy of the absorbed radiation describes the amount of energy needed to have a spinning nucleus (most commonly ^1H or ^{13}C in organic molecules) come into resonance with the applied radiation field. Because this spin is a property of the nucleus of the atom, this technique is called **nuclear magnetic resonance (NMR).** (Note that in this context the term *nuclear* has nothing to do with radioactivity.)

The frequency of the energy required to induce spin-state flipping of nuclei varies directly with the magnitude of the applied magnetic field. The larger the field, the larger the difference between parallel and antiparallel spin states and the higher the energy of the signal required to induce the change. Commercial spectrometers have very large magnets that employ superconducting wires to produce the magnetic field. With these field strengths, electromagnetic energy in the radio-frequency range is required. Spectrometers are classified by the frequency used to change the spin state of magnetically active nuclei. The highest-field machines currently available from commercial instrument manufacturers operate at 750 MHz (1 megahertz, MHz, equals one million cycles per second), although instruments using from 100 to 300 MHz signals are much more common.

These frequencies actually correspond to very little energy: 100 MHz corresponds to only about 1×10^{-5} kcal/mole. This radio-frequency energy can be taken up only by magnetic nuclei and, though ^1H is magnetically active, other nuclei of interest to organic chemists, such as ^{12}C and ^{16}O, are not. Carbon-13 is magnetically active but is present only to the extent of approximately 1% in normal samples, and very sensitive instruments must be used to observe the spin-state changes of this nucleus. All nuclei with an odd number of protons or an odd mass have magnetic properties, making (in addition to ^1H and ^{13}C) ^{19}F, ^{31}P, ^{15}N, ^{17}O, ^2H, and ^{14}N amenable to NMR analysis.

In NMR spectroscopy, the **effective field** (H_{eff}) felt by the nucleus differs from the **applied field** (H_{app}) by the tiny **local magnetic field** (H_{loc}) set up by the circulating electron cloud surrounding the nucleus.

$$H_{eff} = H_{app} - H_{loc}$$

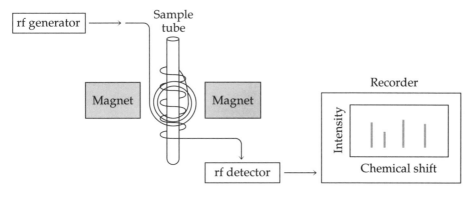

▲ FIGURE 4-12

Schematic representation of the key features of a nuclear magnetic resonance spectrometer (rf = radio frequency).

The electron density of each specific atom in a molecule varies with the nature of the surrounding atoms and is slightly different for each nonequivalent atom in a molecule. Each unique set of hydrogens and each unique set of carbons in a molecule resonate at a different frequency and, hence, give rise to a unique NMR signal. The shift of this signal from that expected from the applied field is described as a **shielding** of the nucleus of interest from the applied field. The NMR spectrum is a plot of signal intensity against the amount of energy required to bring a given proton (or ^{13}C nucleus) into resonance. This value is directly related to the degree of shielding of the atom.

As shown schematically in Figure 4-12, an NMR spectrometer is operated by placing a sample in a thin tube held within the poles of a magnet. The sample is then held precisely within a coil and the sample is irradiated with radio-frequency (rf) energy. The amount of energy required to bring a given nucleus into resonance depends both on the strength of the magnetic field and on the identity of the nucleus being observed. A detector coil is wrapped around the sample. In early instruments, the absorption of energy by the sample was detected by a decrease in the intensity of the energy received by the detector coil. Modern instruments operate differently by detecting the signal emitted by nuclei as they give off energy in going from the higher to the lower spin state. In either case, the difference in energy is related directly to the frequency of this energy (either absorbed or emitted).

A plot of signal intensity versus frequency is called a **spectrum,** in which the unique signal produced by each nucleus is displayed as a peak at a specific resonance frequency. A relative frequency scale, called the **chemical shift,** is used to describe the magnitude of the change of the observed resonance energy for a given nucleus from that observed for a standard. Tetramethylsilane [$(CH_3)_4Si$, often called TMS] is commonly used as a standard for both proton (^1H) and carbon (^{13}C) spectroscopy, and most protons and carbons resonate at lower frequencies than those of this standard. The high field position of TMS results from the attachment of the four methyl groups to silicon, a less-electronegative element. Active nuclei that resonate at frequencies only slightly below TMS are said to appear in the **upfield** region; those shifted to much lower frequencies are said to be **downfield.**

What can we learn about molecular structure from an NMR spectrum? From the number of signals, we establish how many unique types of magnetically active nuclei are present. From the chemical shifts, we learn what kind of chemical environment surrounds each of the nuclei.

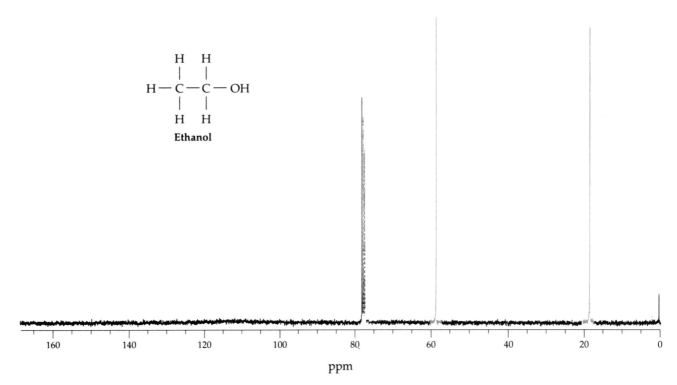

▲ FIGURE 4-13

A ^{13}C NMR spectrum of ethanol. The signal at 0 ppm is that for tetramethylsilane. That at 17.9 ppm is from the CH_3 carbon and that at 57.3 ppm is from the CH_2 carbon that is attached to oxygen. The three peaks centered at 77 ppm are from solvent ($CDCl_3$).

From the splitting of the observed signal, we can also learn how many magnetically active nearest neighbors each proton or carbon has.

For example, the ^{13}C and ^1H NMR spectra of ethanol are shown in Figures 4-13 and 4-14. Three groups of signals in the ^1H spectrum represent the three kinds of hydrogens in ethanol, and two sharp peaks in the ^{13}C spectrum indicate the presence of two different carbons. Note that the signals in the carbon spectrum are recorded as a series of single sharp lines, whereas those in the proton spectrum are split into symmetrical patterns. As we will see in the following sections, this splitting describes the number of active nuclei on an adjacent atom.

The delta (δ) scale in which

$$\delta = \frac{(\omega_{\text{standard}} - \omega_{\text{signal}})}{\omega_{\text{standard}}} \times 10^6$$

is used to express the frequency (ω) of a signal. Thus, signals at lower frequency (and lower field) than that of the standard have positive δ values. These δ values are also often referred to as parts per million (ppm) downfield from the standard. The majority of proton signals range between 0 and 12 ppm, whereas the range for carbon is larger, from 0 to 250 ppm. These values, as either δ or ppm, represent the chemical shift of the signal; that is, the difference from the standard. Because the range in ^1H spectroscopy is smaller than that for ^{13}C, the accidental overlap of two nonequivalent signals is more likely to be found in a proton spectrum than in a carbon spectrum.

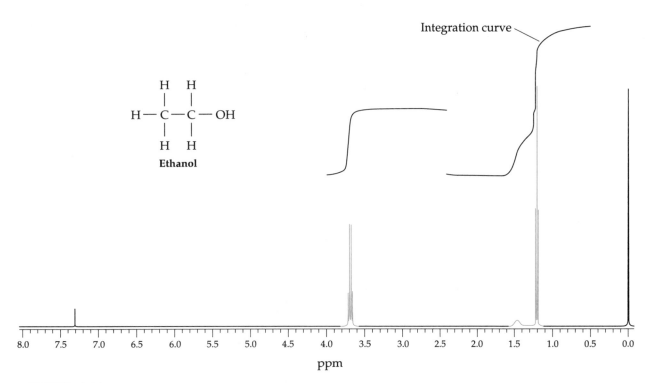

▲ FIGURE 4-14

A 360 MHz ^{1}H NMR spectrum of ethanol. The signal at 0 ppm is that for tetramethylsilane. The broad singlet at 1.5 ppm represents the OH proton. The signal at about 1.2 ppm is split into a three-peak pattern (called a triplet) and is from the CH_3 group. The area under this peak is three times the area under the peak at 1.5 ppm. The signal at about 3.7 ppm is split into a four-peak pattern (called a quartet) and is from the CH_2 protons. The area under this peak is twice that under the peak at 1.5 ppm. Splitting results directly from protons on adjacent atoms. (The proton of the OH group is moving from the oxygen of one molecule to that of another; as a result, it is not usually split by other protons and does not itself contribute to splitting.) The integration curve is explained on page 130. The small peak at about 7.3 ppm is from $CHCl_3$, present as an impurity in the $CDCl_3$ solvent.

^{13}*C NMR.* Because ^{13}C spectra are often simpler than ^{1}H spectra, we will consider them first. Under the usual instrumental conditions, each non-equivalent carbon-13 nucleus is recorded as a distinct, sharp signal.

The sharpness of the signal in ^{13}C NMR spectroscopy is very important for two reasons. First, a sharp signal indicates that the absorption is concentrated in a narrow frequency range, and the peak is therefore higher than those of random noise signals produced by the electronic circuitry. Second, the narrower the signal, the closer in frequency two absorptions can be and still be distinguishable. As mentioned earlier, the natural abundance of ^{13}C is only about 1%. This heavier isotope is randomly distributed among the "normal" (^{12}C) carbons and, thus, it is very improbable that both of two adjacent carbons will be ^{13}C.

Notice that the two distinct carbons in ethanol are recorded as signals with different chemical shifts in Figure 4-13. In fact, as shown in Table 4-1, (on page 124), the chemical shifts of most distinct carbons are different. Many ^{13}C signals are characteristic of the type and number of attached atoms other than hydrogen. In this table, an α substituent is an atom directly attached to the carbon being observed, a β substituent is an atom one carbon removed down the chain, and a γ substituent is an atom attached to

Table 4-1 ► General Effects of Substituents on the Carbon-13 NMR Shifts

For any sp^3 carbon,

$$C—\alpha—\beta—\gamma$$

add to 0.0 for each:

α or β substituent	8.0
presence in a five- or six-membered ring	−5.0

In addition, add for each:

α oxygen substituent	38.0
α nitrogen substituent	22.5
α *trans* C=C	2.5
α *cis* C=C	−2.5
α ester or acid	−2.5
α ketone	7.5
α aldehyde	15.0
γ carbon	−2.0
γ oxygen	−5.0

For 3° carbons, add for each:

β substituent	−1.5

For 4° carbons, add for each:

β substituent	−3.5

For an sp^2 carbon of a C=C bond,

$$\begin{array}{cc} \beta—\alpha & \alpha'—\beta' \\ \diagdown & \diagup \\ C=C \\ \diagup & \diagdown \\ \beta—\alpha & \alpha'—\beta' \end{array}$$

add to 121.0 for each:

α or β carbon substituent	8.0
α' carbon substituent	−8.0

a β substituent. For example, the methyl carbon of ethanol has one α substituent (the carbon of the methylene group) and one β substituent (the oxygen). Notice that we consider *only* nonhydrogen groups in this analysis.

The carbon that bears the OH group in ethanol is shifted significantly downfield from the CH_3 group because of the presence of the electronegative oxygen. The carbon-13 spectrum allows us not only to count the number of different carbons in a molecule of unknown structure, but also to have some idea of the immediate environment of each of the carbon atoms present. For example, using the values in Table 4-1, we can predict that the methyl group (CH_3—) of ethanol would resonate at about 16 δ because it has one α and one β substituent (8 + 8), whereas the methylene (—CH_2—) carbon should appear near 54 δ because it has two α substituents, one of which has an added effect because it is oxygen (8 + 8 + 38). Both of these predictions are close to the observed values for the carbons in ethanol: δ 17.9 and δ 57.3, respectively. Although the effect of α substituents varies with the chemical nature of the attached atom, the β effect does not.

Although it is possible to count the number of different *types* of carbons present in a molecule from its ^{13}C spectrum, the intensity of the signal is only roughly related to *how many* carbon atoms are present. The difference in peak size is caused by differences in the rate at which irradiated carbons relax to the equilibrium distribution of their two energy states in the presence of a magnetic field, which in turn is influenced by the proximity of a given nucleus to other spin centers in the molecule. For our purposes, it is not necessary to define these factors in greater detail here, but you should realize that ^{13}C NMR signal intensity does not correlate accurately with the number of carbon atoms responsible for a given resonance.

EXERCISE 4-D

For each of the following compounds, use symmetry properties to predict how many distinct carbon signals are observed in its ^{13}C NMR spectrum.

(a)   (c)   CO$_2$H   (e)   OH

(b)   O   (d)   Br / Br   (f)   Br   (g)   Br

EXERCISE 4-E

Using the values in Table 4-1, predict chemical shifts expected for each carbon in the following molecules. Then associate each of your predictions with one of the observed signals so as to arrive at the smallest average error between prediction and experiment. What is the average error for the four predictions that you have made for part *a* and the five predictions for part *b*?

(a) δ 36.6, 29.5, 18.9, 11.6

(b)   OH δ 67.0, 41.6, 23.3, 19.1, 14.0

---

^{1}H NMR. Like those from carbons in ^{13}C NMR, the signals from protons in ^{1}H NMR spectra are recorded as separate absorption peaks for non-equivalent nuclei. A proton NMR spectrum provides four important pieces of information: the number of unique signals, the chemical shift, the splitting pattern, and the integrated signal intensity.

To interpret a ^{1}H NMR spectrum, we must first recognize the number of signals with distinct chemical shifts. For example, in Figure 4-14, there are three distinct signals corresponding to the three kinds of protons present in ethanol. The broad single peak that appears at 1.5 ppm represents the proton on oxygen. The signal for the CH$_2$ group at 3.7 ppm is split into four lines and that for the CH$_3$ at highest field (1.2 ppm) is split into three lines.

The possibility of accidental overlap in ^{1}H spectroscopy is reduced by using a high-field-strength instrument (for example, 300–500 MHz) rather than a lower-field instrument (60–100 MHz) because different NMR absorptions are more widely separated at the higher field strength. For example, two ^{1}H spectra of linalool are shown in Figures 4-15 and 4-16 (on pages 126 and 127), at 90 MHz and 360 MHz, respectively. The signal at about 5 ppm is quite broad and very hard to interpret in the spectrum at 90 MHz. In contrast, that taken at 360 MHz exhibits sharp peaks and, although the patterns are complex, they can be readily interpreted.

The center of each of ethanol's three signals (at 1.2, 1.5, and 3.7 ppm) defines the chemical shift for each type of hydrogen. The chemical shift

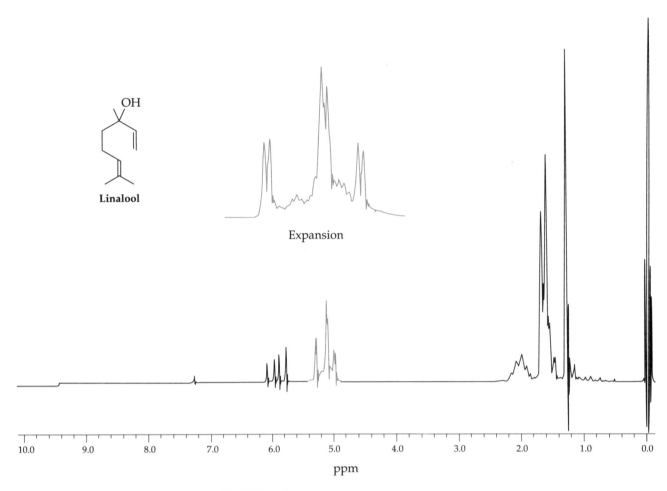

Expansion

▲ **FIGURE 4-15**
A 90 MHz ^{1}H NMR spectrum of linalool. The signal in the range from 4.9 to
5.3 ppm is expanded to show the significant peak overlap.

describes the environment of the magnetically active nucleus. For example,
the spectrum of ethanol in Figure 4-14 shows three different kinds of protons; but, in contrast with the sharp lines observed in the carbon spectrum
(Figure 4-13), the signal for each type of proton has a more-complex pattern.
It is possible to correlate the chemical shift of each proton signal with its
molecular environment (Table 4-2, on page 128) in somewhat the same way
that we did earlier for ^{13}C NMR spectroscopy. It is also possible to correlate
proton chemical shifts with functional-group type, although not in as quantitative a way. For example, the protons closer to the more-electronegative
oxygen atom in ethanol are shifted farther downfield, as is the case for the
carbon bearing oxygen in the ^{13}C NMR spectrum of ethanol.

Next, we note that each signal is split into a complex pattern. This signal splitting can provide valuable information about the structure of the
molecule. The splitting of the signal for a proton into multiple lines, forming a **multiplet,** is the direct result of interaction with neighboring protons.
The number of lines in a multiplet can be interpreted to reveal the number
of hydrogens on adjacent carbons and even the dihedral angle relations.
The splitting patterns observed in these signals are caused by the interaction of the magnetic spin of the nucleus with neighboring nuclei. This interaction is referred to as **coupling.** Compare the spectrum of ethanol in Figure

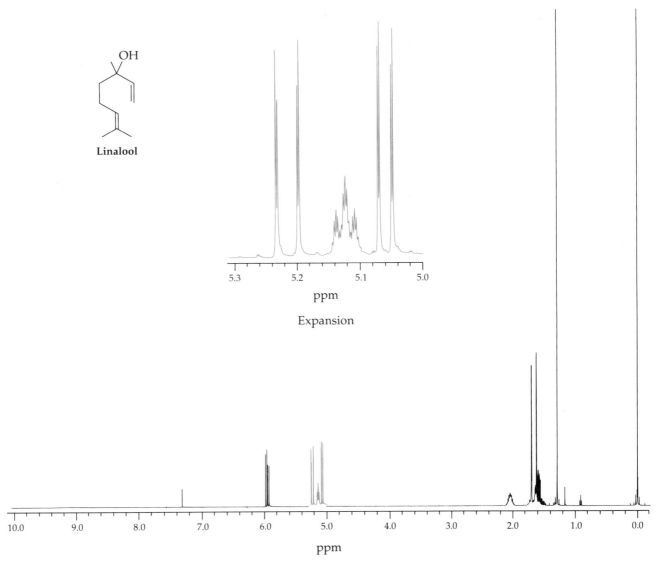

▲ FIGURE 4-16

A 360 MHz ^{1}H NMR spectrum of linalool. The expansion of the signal
between 5.0 and 5.3 ppm shows that this region can be resolved into two pairs
of doublets and a triplet.

4-14 with that of ethyl bromide in Figure 4-17 (page 129). Both compounds
have ethyl groups and exhibit the same general pattern, although the chemical shifts observed are different. In both cases, the upfield CH$_3$ signal is
split into a three-line multiplet (a **triplet**), whereas the methylene (CH$_2$)
group is split into a four-line multiplet (a **quartet**).

Let us first examine the CH$_3$ groups in both compounds. The methyl
hydrogens interact with the two hydrogens of the adjacent methylene unit.
As illustrated at the right in Figure 4-18 (page 129), when the interaction
with the methylene hydrogens on the adjacent carbon atom is blocked, a
sharp singlet is seen. To understand the observed methyl splitting pattern,
we must consider the number of ways in which the interacting spins on adjacent hydrogens can align with or against the applied magnetic field. (The
hydrogens being observed are shown in color in Figure 4-18; the hydrogens
splitting the observed signals are in boxes). The two methylene hydrogens

Table 4-2 ▸ Representative ^{1}H Chemical Shifts

Type of Proton	Chemical Shift (δ)	Type of Proton	Chemical Shift (δ)
—CH$_3$	0.7–1.3	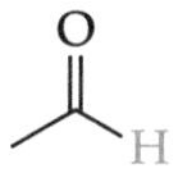	
—CH$_2$—	1.2–1.4	(aldehyde H, C=O with H)	9.5–10.0
—C—H	1.4–1.7	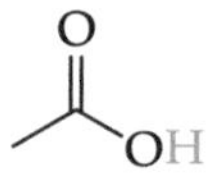 (carboxylic acid, C=O with OH)	10.0–12.0
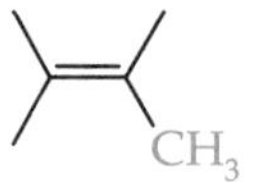 (alkene with CH$_3$)	1.5–2.5	—C—OH	1.0–6.0 (changes with solvent)
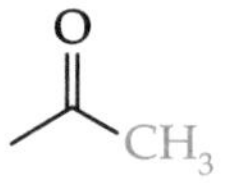 (ketone with CH$_3$, C=O)	2.1–2.6	O—C—H	3.3–4.0
Ar—CH$_3$	2.2–2.7	Cl—C—H	3.0–4.0
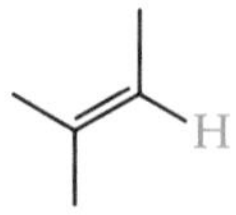 (alkene with H)	4.5–6.5	Br—C—H	2.5–4.0
Ar—H	6.0–9.0	I—C—H	2.0–4.0
C≡C—H	2.5–3.1		

can be oriented in three possible ways: both aligned with, both against, and one with and one against the applied field. There are two possible arrangements of identical energy for the last combination, and the arrows below the decoupled signal represent the alignments possible. The relative abundance of each type of alignment gives the observed pattern: a quartet from three interacting neighboring hydrogens, a triplet from two interacting neighboring hydrogens.

The field experienced by the hydrogens of the methyl group is the sum of the applied field from the magnet and the small magnetic field of these neighboring hydrogens. When the alignment of each of the neighboring nuclei is in the same direction as the applied field, the effective field is larger. When their alignment is against the field, the effective field is smaller. If one nucleus is aligned with and the other against the applied field, the effects cancel. Thus, the hydrogens on the methyl group are in fact exposed to three different fields: one larger, one equal, and one smaller than the applied field. Each of these slightly different environments exhibits its own unique resonance frequency for the methyl hydrogens. Thus, instead of seeing a sharp singlet for the methyl group, we see a multiplet resulting from the contribution of the spins of the methylene group. One-fourth of the signal is upfield and another one-fourth is downfield from the expected position; one-half of the signal, twice as intense, is at the center. This 1:2:1 ratio produces the three-line pattern constituting a triplet.

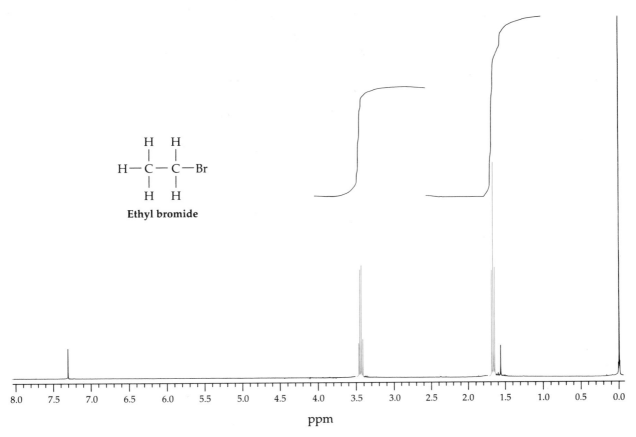

▲ FIGURE 4-17

A 360 MHz ^{1}H NMR spectrum of ethyl bromide. The sharp singlet at 0 ppm is a TMS standard. The triplet from the CH$_3$ group appears at 1.7 ppm and the CH$_2$ group quartet is at 3.4 ppm. (The small peak at 1.55 ppm is due to contamination of the sample by H$_2$O and that at 7.3 ppm is due to HCCl$_3$.)

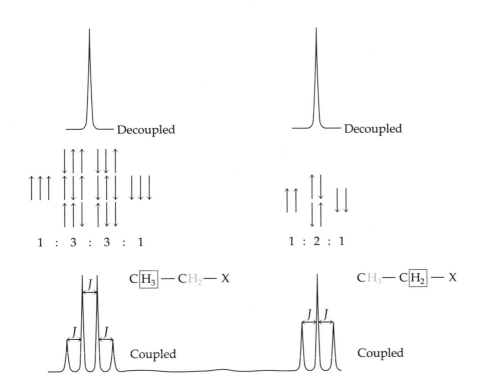

◄ FIGURE 4-18

Coupling with protons on adjacent carbons is responsible for the observed splitting pattern (J = coupling constant). The protons under observation are in color; those splitting the observed signals are in boxes. The sets of arrows describe the possible alignments of adjacent nuclear spins with or against the applied magnetic field. The statistical abundance of each type determines the ratio of peak heights in the coupled spectrum. (J = coupling constant.)

Let us compare this triplet splitting pattern of the methyl group with that of the methylene ($-CH_2-$) hydrogens at the left in Figure 4-18. The methylene hydrogens are influenced by the three neighboring hydrogen nuclei on the methyl group. These three neighbors can have eight different orientations in relation to the applied field, distributed in a $1:3:3:1$ ratio. In turn, this distribution of spins results in four different effective field strengths experienced by the methylene group hydrogens, which give rise to a four-line pattern, or quartet.

The splitting by neighboring hydrogen nuclei can be selectively removed by a process called **spin-spin decoupling.** Indeed, this is normally done in obtaining ^{13}C spectra to remove the effect of hydrogens attached to the carbons. (This spectral simplification, by a process called **proton decoupling,** is the reason that single-line ^{13}C NMR spectra can be obtained. In this routine technique, the entire proton region is irradiated while the ^{13}C NMR spectrum is recorded.) By applying a large radio-frequency signal that equals the resonance frequency of the methylene group, these protons are induced to flip rapidly from one spin state to the other. The protons on the methyl group then experience an average of all three of the possible orientations of the methylene protons and resonate at a single frequency. Thus, the $1:2:1$ triplet in the coupled spectrum at the right in Figure 4-18 is changed to a singlet in the decoupled spectrum.

The degree of separation between the lines in a multiplet varies directly with the effect of the neighboring hydrogens on the magnetic field. In general, the closer in space the hydrogens are, the larger the splitting. The magnitude of this splitting is referred to as the **coupling constant,** which is often designated simply as J. The magnitude of the interaction of two types of nuclei on each other will be equal. As a result, the coupling constants will be identical in the related multiplets: that is, the peak separations in the triplets in Figures 4-14 and 4-17 are exactly equal to the peak separations in the quartets. The J values for the methyl and methylene hydrogens of both ethanol and ethyl bromide are 7 Hz.

The **multiplicity** of the signal (that is, the number of peaks into which a signal is split) observed in a coupled ^1H NMR spectrum provides valuable information about the number of hydrogens on adjacent positions. Figure 4-19 shows the shapes of simple multiplets that are derived by equal interaction with different numbers of nearest-neighbor nuclei. A doublet results when one magnetically active neighboring nucleus is present, a triplet when there are two, a quartet with three, and a pentet with four.

Unlike that of ^{13}C NMR, the relative intensity of a signal in ^1H NMR spectroscopy *is* proportional to the number of protons contributing to the signal. The smooth curve superimposed on the signals in the spectrum in Figure 4-14 is an **integration curve;** it is a measure of the area under each split signal of the curve. The vertical rise in the "stair step" of an integration curve can be used to calculate the ratio of the number of hydrogens responsible for each signal. Thus, in Figure 4-14, the $1:2:3$ ratio of the three peaks is proportional to the number of hydrogens responsible for each of the three different chemical shifts in ethanol. In Figure 4-17, the ratio of the integrals of the two peaks is $2:3$ (or $1:1.5$), corresponding to the ratio of two methylene to three methyl hydrogen atoms. Because the absolute area under an integration curve depends on instrument sensitivity, not the number of hydrogens, the integral gives a ratio of numbers of hydrogens, not the absolute number of each type.

Finally, from the number of unique resonance signals, both ^1H and ^{13}C NMR spectroscopy provide easy methods for establishing symmetry in a molecule. For example, consider the isomeric alcohols shown in Figure 4-20.

Pattern	Multiplicity	Number of nearest neighbors	Examples
1 : 1	Doublet	1	CH_3 — $CHBr_2$
1 : 2 : 1	Triplet	2	CH_3 — CH_2Br
1 : 3 : 3 : 1	Quartet	3	CH_3 — CH_2Br
1 : 4 : 6 : 4 : 1	Quintet	4	Br — CH_2 — $CHBr$ — CH_2 — Br

◄ FIGURE 4-19
The splitting patterns in ^1H NMR spectra of several compounds. The nuclei for which the patterns are illustrated are highlighted in color.

Only three signals are observed in the ^{13}C NMR spectrum for the five carbons of 3-pentanol, whereas each of the five carbons in 2-pentanol gives rise to its own unique signal. Notice that there is symmetry in 3-pentanol that makes the two methyl groups, at C-1 and C-5, equivalent, as well as the two methylene groups, at C-2 and C-4. As a result of this symmetry, both the

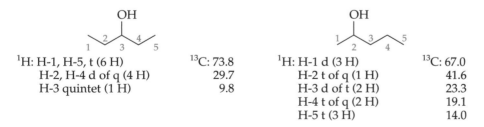

^1H: H-1, H-5, t (6 H) ^{13}C: 73.8
    H-2, H-4 d of q (4 H)   29.7
    H-3 quintet (1 H)   9.8

^1H: H-1 d (3 H) ^{13}C: 67.0
    H-2 t of q (1 H)   41.6
    H-3 d of t (2 H)   23.3
    H-4 t of q (2 H)   19.1
    H-5 t (3 H)   14.0

▲ FIGURE 4-20
Observed splitting patterns and integrals in ^1H NMR spectra and chemical shifts in ^{13}C NMR spectra. The signal for H-1 (and H-5) of 3-pentanol is a triplet, split by two adjacent hydrogens, and that for H-2 (and H-4) is a doublet of quartets, split by the single hydrogen on C-3 and by three hydrogens on C-1 (and C-5).

carbons and the hydrogens at C-1 and C-5 in 3-pentanol are identical and give rise to one signal in the ^{1}H and ^{13}C NMR spectra. The same can be said for the hydrogens and carbons at C-2 and C-4. Unlike the methylene groups of ethanol and ethyl chloride, however, the C-2 (and C-4) hydrogens are split not only by the hydrogens on C-1 (and C-5), but also by the single hydrogen at C-3. The coupling constants of the C-2 (and C-4) hydrogens with these two different sets of hydrogens are not the same, resulting in a pattern that is a doublet (coupling with the C-3 hydrogen) of quartets [coupling with the three C-1 (and C-5) hydrogens]. The signal for the hydrogen on C-3 appears as a quintet, being coupled to the four equivalent hydrogens on C-2 and C-4, and is downfield from the signals for the other hydrogens because of the oxygen on C-3.

See if you can explain the ^{13}C and ^{1}H signals and splitting pattern in 2-pentanol shown at the right of Figure 4-20. Clearly, the chemical shifts of hydrogens on carbons 1, 3, 4, and 5, though not identical, are not dramatically different. Instead of the simple pattern of 3-pentanol, a broad signal, a multiplet, is observed. Even though the proton spectrum of 2-pentanol is hard to interpret, its very complexity makes it distinguishable from its symmetrical isomer 3-pentanol. This assignment of isomers is also completely unambiguous from the number of signals in the ^{13}C NMR spectra.

EXERCISE 4-F

Each of the ^{1}H spectra shown here and on pages 133 and 134 corresponds to one of the isomers in parts *a* through *d*. Choose between the alternative compounds and give reasons for your assignment.

(a)

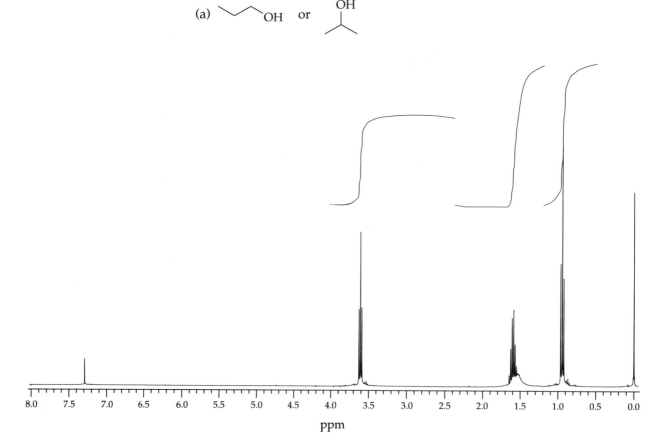

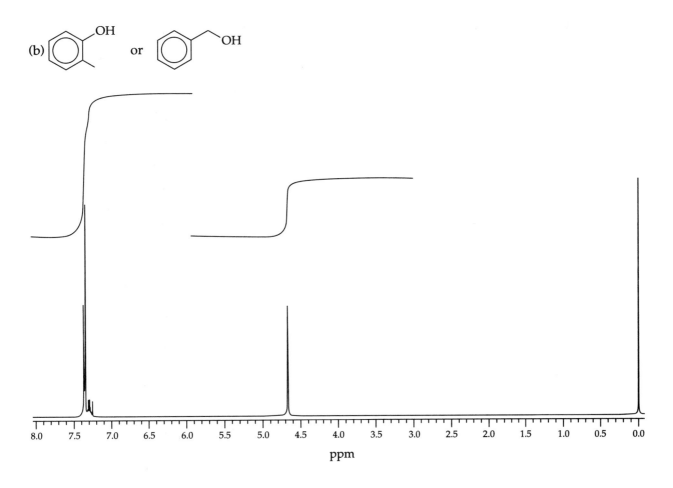

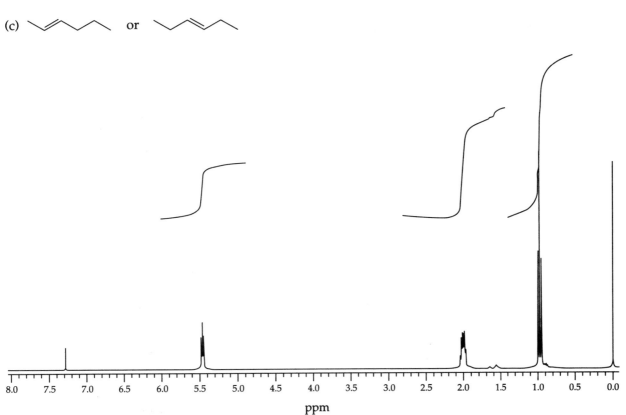

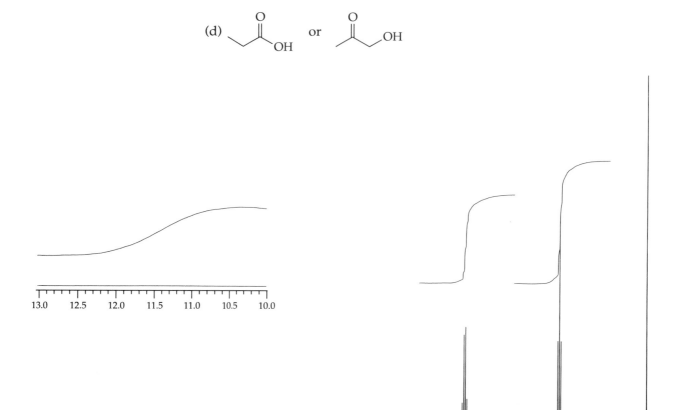

ppm

EXERCISE 4-G

Predict the approximate ^{1}H NMR spectrum (number of signals, approximate chemical shift, multiplicity, and integration) for each of the following compounds.

(a) 1-butanol   (c) 2-butanol

(b) 1-butanal   (d) 2-butanone

In conclusion, ^{1}H NMR spectra can be interpreted on the basis of the number of signals, the chemical shifts, the splitting patterns, and integration. These spectral features provide valuable structural information about the nature of attached atoms (from chemical-shift positions) and the number of neighboring hydrogens (from splitting and integration). ^{1}H NMR spectroscopy is thus a very useful complement to ^{13}C NMR spectroscopy for assigning structure.

Medical Applications of NMR Spectroscopy. NMR spectroscopy is used not only for identifying pure compounds, but also for detecting differences in relative abundances of magnetically active nuclei in solid samples and water-filled tissues. Very large NMR spectrometers are now available for medical and biological research applications, and plant matter, animals, or even a whole human body can be inserted in the magnet and thus analyzed. Such a spectrometer is shown in Figure 4-21.

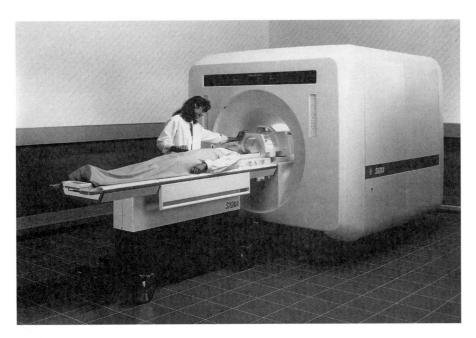

▲ FIGURE 4-21
NMR spectrometer used for three-dimensional imaging of human bodies.
(Courtesy of GE Medical Systems.)

This spectrometer produces a proton spectrum from which a three-dimensional map of water concentration is made of an organ or other object of interest. Deviations from normal water concentrations and distributions are indicative of medical anomalies such as a tumor. Representative three-dimensional images of human brains obtained with such a spectrometer are shown in Figure 4-22 (page 136). Differences from "normal" patterns observed for healthy people can indicate disease. The same instrument can be used to search for weak points or irregularities in man-made objects. Figure 4-23 (page 137) includes a picture of polystyrene tubing taken with an optical camera and a three-dimensional NMR image of a cross section of the tubing. This method is called noninvasive imaging by medical personnel and nondestructive testing by material scientists.

The technique is considerably more sensitive (and at present more expensive) than x-ray imaging and does not damage tissue. Physicians usually use the expression **magnetic resonance imaging** (MRI) to differentiate this technique from x-ray imaging. (The word nuclear is omitted in order to avoid alarming that segment of the public who might otherwise connect the term—incorrectly—with nuclear fusion and fission.)

Infrared Spectroscopy

The infrared (IR) region of the electromagnetic spectrum, as the name implies, is just beyond that section of the electromagnetic spectrum that the human eye perceives as red light. Conventional infrared spectrometers record spectra using light of wavelengths ranging from 2,000 to 15,000 nm, though it is more common to use a scale based on wave number, or cm^{-1} (700–5,000 cm^{-1}). The cm^{-1} scale is a designation of frequency; that is, of how many waves of a given wavelength are completed within one centimeter.

The key principle of **infrared spectroscopy** is that the absorption of infrared radiation occurs when there is a match between the radiant energy and the frequency of a specific molecular motion, usually bond bending or

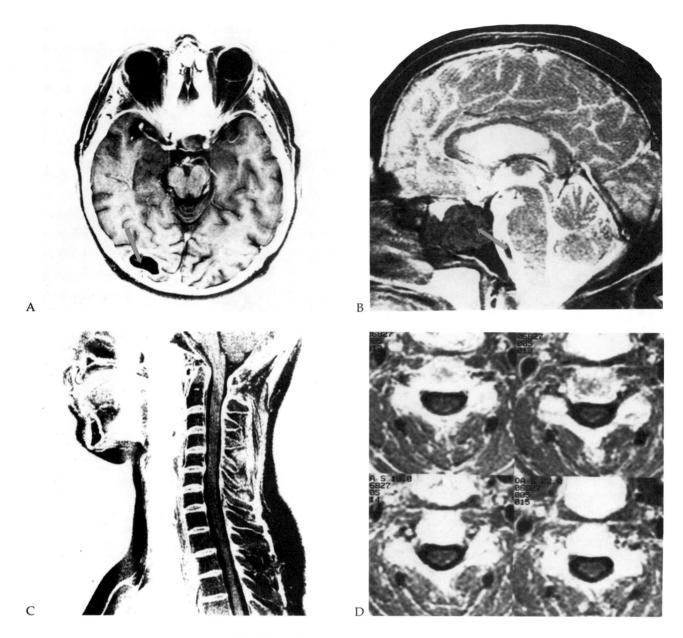

▲ FIGURE 4-22

Three-dimensional NMR imaging (MRI) representations of various parts of
the human body: (A) a horizontal layer of the brain, clearly showing such
features as the eyes, as well as a blood clot (arrow); (B) a vertical layer from
the head of a different patient with an enlarged pituitary gland (arrow); (C) a
vertical layer of the spine; and (D) four horizontal slices. The images shown in
parts B and C are similar to what would be shown in x-ray of the same areas
of the body. On the other hand, the cross-sectional slices (A and D) are unique
and allow a physician to examine internal body structures that cannot be
visualized by x-ray techniques. (Courtesy of Matthew C. Ohlson, Central
Texas Imaging Center, Austin.)

stretching. An average C–H bond length, for example, describes the average
position about which atoms vibrate back and forth at a defined frequency,
stretching and compressing as if they were held together by a spring. Absorption of infrared energy results in the stretching, as well as the bending,
of bonds when these motions result in changes in the dipole moment of a
molecule; thus, signals in the infrared spectrum are indicative of the pres-

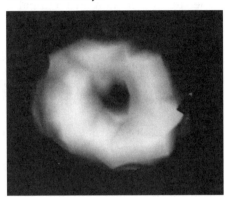

◄ FIGURE 4-23
Optical photograph (left) and three-dimensional NMR cross-sectional image (below) of a piece of polystyrene tubing. (Reprinted with permission from S. L. Dieckman et al. 1992. *J. Am. Chem. Soc.* 114:2717. Copyright 1992 American Chemical Society.)

ence of unsymmetrical or polar bonds. (Rotations about bonds require even less energy; such absorptions are in the microwave region of the electromagnetic spectrum.)

The great utility of infrared spectroscopy in assigning structures to organic molecules comes from the observation that specific functional groups exhibit characteristic infrared absorptions. In essence, the "springs" that hold the atoms together have characteristic strengths that are predominantly influenced only by the immediate neighboring atoms. Table 4-3 lists

Table 4-3 ► Typical IR Bands for Specific Functional Groups		
Functional Group	**Band (cm^{-1})**	**Intensity**
C—H	2850–2960	Medium
=C—H	3020–3100	Medium
C=C	1620–1680	Medium
≡C—H	3300–3350	Strong
R—C≡C—R′	2100–2260	Medium (R ≠ R′)
Ar—H	3000–3030	Medium
 ⬡	1600, 1500	Strong
RO—H	3400–3650	Strong, broad
—C—O—	1050–1150	Strong
C=O	1640–1780	Strong
R$_2$N—H	3300–3500	Medium, broad
—C—N—	1030, 1230	Medium
—C≡N	2210–2260	Medium
RNO$_2$	1540	Strong

characteristic infrared absorptions for frequently encountered organic functional groups.

Although the frequencies of IR bands are reasonably characteristic, the specific intensity of the absorption, as well as the width of the band, varies from one compound to another. Most bands characteristic of a functional group appear at frequencies higher than 1200 cm^{-1}. In the region from about 700 to 1200 cm^{-1}, a series of complex bands characteristic of a specific molecule (rather than a functional-group type) is usually observed in what is called the **fingerprint region.** A comparison of the infrared spectrum of an unknown compound with a library of infrared spectra of known compounds can often enable the unambiguous identification of both the functional group(s) present and the specific structure. In essence, there are so many absorption bands in the fingerprint region that correspondence of the spectra of two different compounds in all details (including intensity) is improbable.

You should learn to recognize the characteristic infrared absorptions for a small number of functional groups. For example, a $C=O$ absorption of a carbonyl compound almost always appears as a strong band between 1640 and 1780 cm^{-1}, whereas $O-H$ and $N-H$ stretches are recorded as broad bands in the range from 3200 to 3600 cm^{-1}. By using the information in Table 4-3, you can identify the presence of many of the common functional-group types and, conversely, can eliminate from consideration those that are not present.

Consider how you can use infrared spectroscopy to distinguish 2-octanone from 2-octanol and thus to determine whether the oxidation reaction discussed in Section 4-1 proceeded as expected. The starting material, 2-octanol, shows an absorption characteristic of the OH functional group at about 3600 cm^{-1} (and a weaker bond at about 3400 cm^{-1} assigned to a hydrogen-bonded OH stretch), whereas this absorption is absent in the product, 2-octanone, having been replaced by the strong stretching absorption of the carbonyl group ($\sim 1700 \text{ cm}^{-1}$), as shown in Figure 4-24. These spectral characteristics distinguish these compounds from other conceivable products. For example, the infrared spectrum of octanoic acid, which has both a $C=O$ and an $-OH$ group, shown in Figure 4-25 (page 140), differs from both the alcohol and the ketone.

CHARACTERIZING MOLECULES IN INTERSTELLAR SPACE
BY INFRARED SPECTROSCOPY

The utility of infrared spectroscopy is not limited to the chemistry laboratory: it has also been used in heat sensors and for remote sensing in which the observer is at a physically distant location from the sample being analyzed. For example, infrared spectrometers aboard both *Voyager I* and *Voyager II* (unmanned spacecrafts that have been exploring the solar systems for more than ten years) detected six simple hydrocarbons (ethyne, ethene, ethane, propane, propyne, and butadiyne) and three carbon-containing nitriles (HCN, $N\equiv C-C\equiv N$, and $HC\equiv C-C\equiv N$) in the atmosphere of Titan, a large moon of Saturn. Infrared spectrometry was sufficiently sensitive that C_2N_2 was detected at the parts-per-billion level. It is interesting to speculate why these compounds are uniquely produced in an atmosphere whose major constituents are N_2 and CH_4.

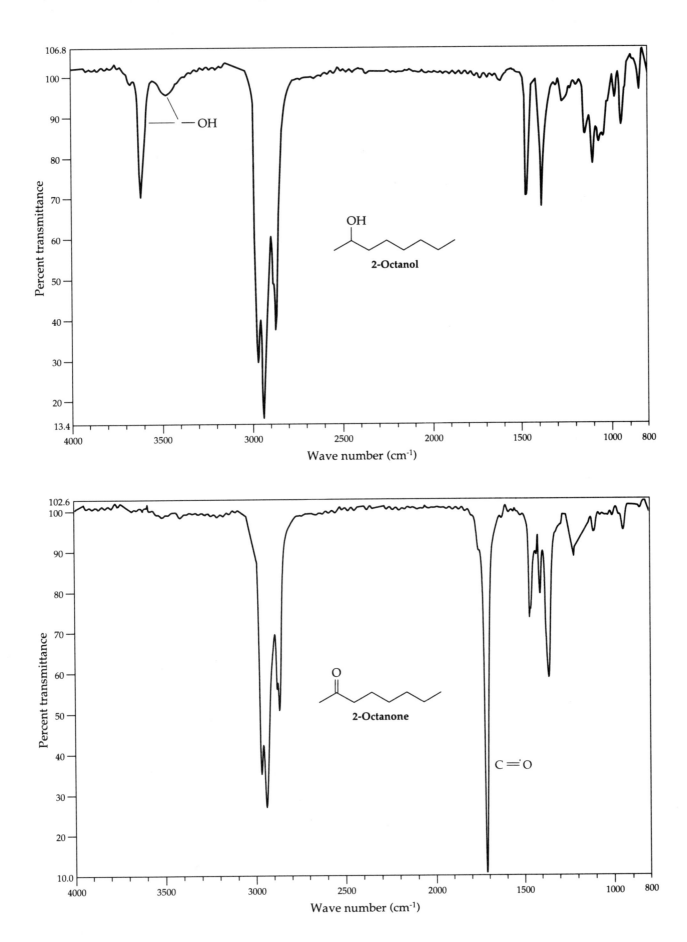

▲ FIGURE 4-24
Infrared spectra of 2-octanol (upper) and 2-octanone (lower).

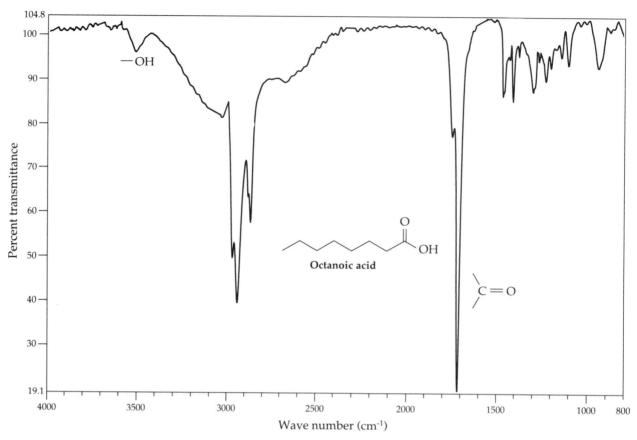

▲ FIGURE 4–25
Infrared spectrum of octanoic acid.

EXERCISE 4-H

Which region of the infrared spectrum might be used to distinguish between the pairs of isomers shown?

(a) [structure] and [structure OCH₃]   (c) $CH_3CH_2NH_2$ and $CH_3C{\equiv}N$

(b) [structure OH] and [structure O, H]   (d) [benzene ring] and [cyclohexane ring]

Ultraviolet and Visible Spectroscopy

As we proceed from the infrared to the visible and ultraviolet regions of the spectrum, the amount of energy of the photon increases. The energy in the ultraviolet region is large enough to perturb the electronic structures of many organic molecules. Even in the lower-energy visible region, some organic molecules can be excited.

In Chapters 1 and 2, we learned that the electronic structure of molecules can be represented by electrons located in molecular orbitals. The bonding (π) and antibonding (π^*) orbitals formed, for example, by the interaction of two p atomic orbitals are equally split about the zero point of energy. The antibonding character of the π^*orbital can be recognized by the

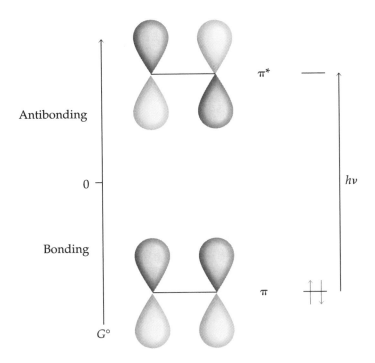

Antibonding

0

Bonding

$G°$

π^*

π

$h\nu$

▲ FIGURE 4-26
Photoexcitation in the visible or ultraviolet regions consists of the absorption of a photon, which promotes an electron from a filled molecular orbital to a vacant one.

node between the two carbons; this node results from the inversion of sign in the orbital phasing of the contributing atomic *p* orbitals. Two electrons located in the π orbital confer net bonding on the molecule. When a photon (hv) of sufficient energy interacts with a molecule, it is absorbed, promoting one of the electrons from a bonding to an antibonding orbital (Figure 4-26). After this process, called **photoexcitation,** the π and π^* orbitals are each singly occupied (Figure 4-27). Because an electron has moved from a π to a π^* molecular orbital, this change is called a π,π^* (or $\pi{\rightarrow}\pi^*$) **transition** (read as pi-to-pi-star). The combined effect of one electron in a bonding molecular orbital and one electron in an antibonding molecular orbital is that there is no net bonding interaction between the atomic *p* orbitals. We will see in Chapter 5 how this change in bonding forms the chemical basis for vision.

This electronic transition can occur only when the energy of the absorbed photon equals that required to raise a bonding electron to an antibonding orbital. A visible or ultraviolet spectrum is simply a plot of the intensity of light absorption as a function of wavelength (corresponding to

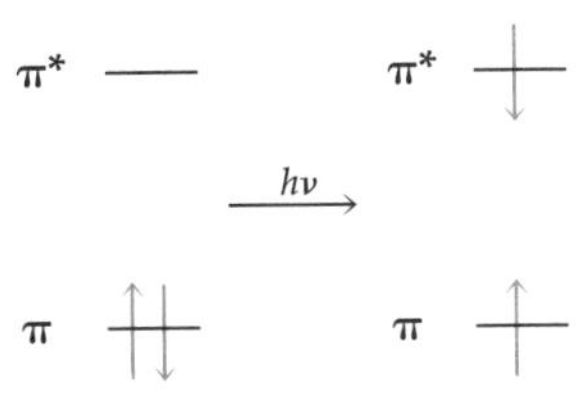

π^* —— π^*

$\xrightarrow{h\nu}$

π π

▲ FIGURE 4-27
After photoexcitation, one electron is left in the bonding (π) orbital and the second electron is located in the antibonding (π^*) orbital. This state (shown at the right) is then called a π,π^* excited state.

MOLECULAR OXYGEN: A STABLE TRIPLET MOLECULE

The lowest-energy arrangement for molecular oxygen (O_2) is a triplet electronic state in which there is only one bond between the atoms and there are two unpaired electrons, one on each atom.

$$:\!\overset{\displaystyle .}{\underset{\displaystyle .}{O}}\!-\!\overset{\displaystyle .}{\underset{\displaystyle .}{O}}\!: \quad\xrightarrow{\text{Sensitizer}}\quad :\!\overset{\displaystyle ..}{O}\!=\!\overset{\displaystyle ..}{O}\!:$$

$^3O_2$ **Triplet oxygen** $^1O_2$ **Singlet oxygen**

By interacting with other excited molecules (sensitizers), triplet oxygen can be converted into the singlet state, a very unstable compound that reacts rapidly with organic compounds. Virtually no organic molecules are stable for long in the presence of singlet oxygen. In fact, these oxidation reactions have been implicated in aging and in other modes of cell damage. However, long-chain polyenes having ten or more conjugated double bonds (many of which, like vitamin E, are found in nature) interact with singlet oxygen in a special way that results in the conversion of both species into the triplet state.

$$\left(\!\!\bigwedge\!\!\right)_n + {}^1O_2 \longrightarrow \left[\left(\!\!\bigwedge\!\!\right)_n\right]^3 + {}^3O_2$$

The triplet state of the polyene reverts rapidly to the stable singlet state. Thus, polyenes serve as deactivators of $^1O_2$ and, by their presence, protect other molecules from oxidation.

excitation energy). Thus, in absorption spectroscopy, transitions between filled and vacant orbitals are measured as a function of wavelength. For ethylene, this transition requires high energy and the absorption appears at about 171 nm (at the high energy end of the ultraviolet spectrum). We can calculate the energy (in kcal/mole) of light of this wavelength using the formula:

$$E = \frac{2.86 \times 10^4 \text{ kcal} \cdot \text{nm} \cdot \text{mole}^{-1}}{\text{wavelength (nm)}}$$

Thus, 200 nm corresponds to 143 kcal/mole, 300 nm to 95 kcal/mole, and 400 nm, the break between the ultraviolet and visible regions, to 72 kcal/mole.

Let us now consider the π-orbital picture for butadiene. Because there are four molecular orbitals, we must differentiate the π bonding and antibonding orbitals further. We use the designations ψ_1, ψ_2, ψ_3, and ψ_4 to indicate orbitals of decreasing bonding (or increasing antibonding) character. Interaction of the four atomic p orbitals results in four molecular orbitals, with two bonding orbitals (ψ_1 and ψ_2) located below and two antibonding orbitals (ψ_3 and ψ_4) above zero energy in Figure 4-28. The arrangement of lowest energy is that in which bonding is continuous along a chain of carbon atoms (ψ_1). The next level of increasing energy, bonding orbital ψ_2, has a node at the center of the chain, between C-2 and C-3. [Recall that, at a

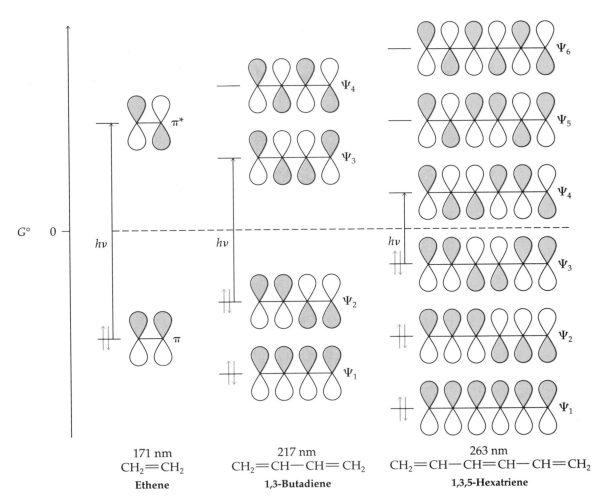

171 nm
$CH_2\!=\!CH_2$
Ethene

217 nm
$CH_2\!=\!CH\!-\!CH\!=\!CH_2$
1,3-Butadiene

263 nm
$CH_2\!=\!CH\!-\!CH\!=\!CH\!-\!CH\!=\!CH_2$
1,3,5-Hexatriene

▲ FIGURE 4-28
Molecular orbitals of ethylene, butadiene, and hexatriene.

node, orbital phase inverts; thus, orbital phasing (indicated by color) is reversed from one side of the node to the other.] In the two antibonding orbitals, ψ_3 and ψ_4, bonding in ψ_3 is maintained only between C-2 and C-3, and there is no bonding between adjacent carbons in ψ_4.

In the ground state (the lowest-energy arrangement) of butadiene, the four π electrons populate ψ_1 and ψ_2. Because ψ_2 of butadiene lies at a higher level of energy than the π orbital of ethene and ψ_3 of butadiene lies at a lower level than π* of ethene, the energy difference between the highest occupied and lowest unoccupied molecular orbitals of butadiene is smaller than that for ethene. Thus, 1,3-butadiene absorbs light of longer wavelength and lower energy (217 nm) than does ethene (171 nm).

In general, the greater the extent of conjugation, the smaller is the energy difference and the farther the absorption is shifted to longer wavelengths. Figure 4-28 also illustrates the π-molecular orbitals for 1,3,5-hexatriene. As for butadiene, the six molecular orbitals are arranged energetically with increasing numbers of nodes. We can see that the energy required for electronic excitation of hexatriene is even lower than that for 1,3-butadiene, and the absorption for this triene is at 263 nm.

Conjugative effects are also important in aromatic rings. For example, benzene has an absorption maximum at 256 nm (not far from that for 1,3,5-hexatriene), whereas naphthalene absorbs at 286 nm. Some characteristic absorption maxima of other conjugated molecules are given in Table 4-4.

Table 4-4 ▸ Representative Absorption Maxima for Typical UV-absorbing Compounds

Compound	λ_{max} (nm)	Compound	λ_{max} (nm)
$CH_2{=}CH_2$	171	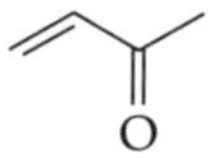	(π, π^*) 210
	182		(n, π^*) 315
	217		256
	226		
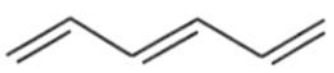	263	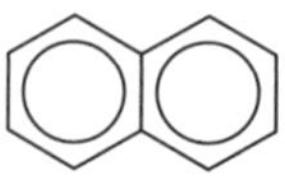	286
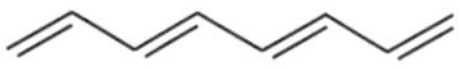	290		
	(π, π^*) 188 (n, π^*) 279	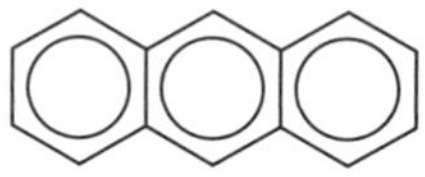	375
		$-N{=}N-$	~350

Extended conjugation increases the observed absorption maximum to longer and longer wavelengths. Ultimately, the absorption maximum shifts from the ultraviolet into the visible spectrum. A compound that removes some wavelengths of light from white light by absorption is perceived by the human eye as having color. For example, as shown in Figure 4-29, β-carotene has a long conjugated hydrocarbon skeleton with many conjugated π bonds. The absorption maximum is between 450 and 500 nm, which means that the energy difference between its highest filled and lowest unfilled molecular orbitals is small. This absorption maximum corresponds to blue light and, as a result of this absorption, β-carotene is perceived as having a bright yellow orange color. This is the same pigment that will appear on your hands if you peel fresh carrots. Similar compounds, such as lycopene, are responsible for the color of other vegetables, such as tomatoes.

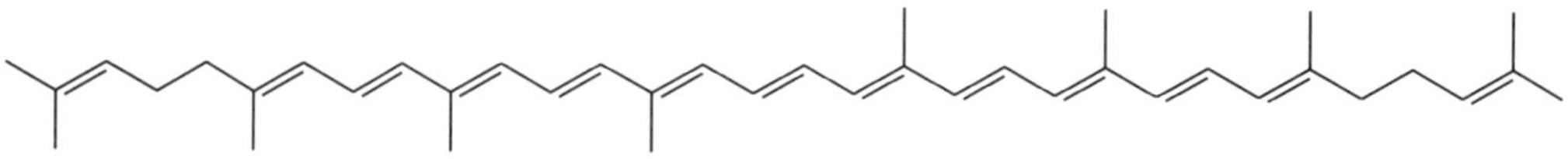

Lycopene (red)
20 mg/kg of tomatoes

EXERCISE 4-1

Which compound in each of the following pairs will exhibit an electronic transition at the longer wavelength?

(a) 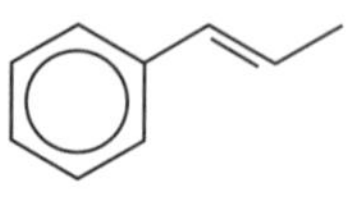   or

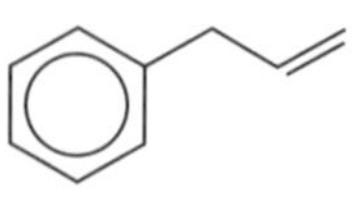

(b) 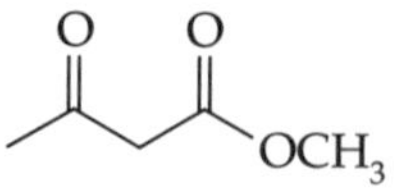   or

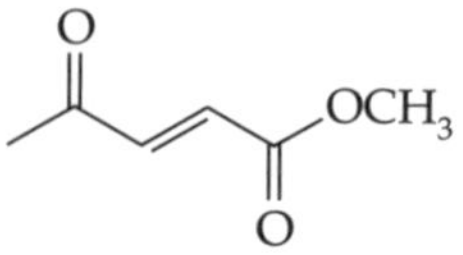

(c) 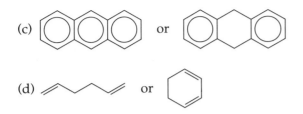 or

(d) 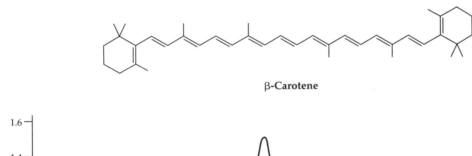 or

In addition to unsaturated hydrocarbons, other functional groups absorb ultraviolet light. For example, ketones and conjugated enones show weak absorption spectra in which the absorption maxima are shifted to longer wavelengths than would be expected for C=C bonds. In addition to the promotion of an electron from the π to the π^* orbital in such compounds (as we have seen for hydrocarbons), their carbonyl groups also show absorption resulting from the promotion of an electron from one of the nonbonded, lone pairs of electrons on oxygen to the π^* orbital. In acetone, this so-called **n,π^* (or n $\rightarrow$ π^*) transition** (read as n-to-pi-star) results in a weak absorption band at about 279 nm.

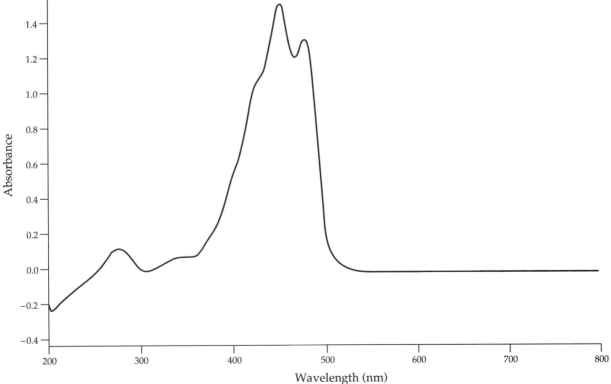

β-Carotene

▲ FIGURE 4-29

An absorption spectrum of β-carotene in the ultraviolet and visible regions.

CONJUGATED OXYGEN–CONTAINING FLOWER PIGMENTS

Anthocyanins are oxygen heterocycles that absorb strongly in the visible region of the spectrum. They occur naturally in plants and are responsible for purple, mauve, and blue colors. Although highly colored, these compounds have not been used commercially to any extent as dyes because they degrade (fade) upon prolonged exposure to ultraviolet light.

Cyanidine
(an anthocyanin)

The position of the n,π* transition also has a molecular-orbital basis. The energy levels of the π and π* orbitals of the carbonyl group are equidistant from zero energy (Figure 4-30). In contrast, the nonbonding electron pair on oxygen is located in an orbital that is only slightly below the zero of energy. The energy of the n,π* transition is therefore lower than that of the π,π* transition; the absorption bands of n,π* transitions are usually at longer wavelengths (lower energies) than those of π,π* transitions and are usually less intense. As we have seen earlier in relation to hydrocarbons, further conjugation of the carbonyl group also results in a shift in the absorption band to longer wavelengths. For 3-butene-2-one, the n,π* absorption band

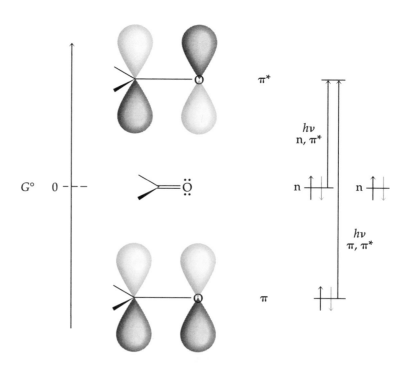

▲ FIGURE 4-30
Pi and nonbonding molecular orbitals in formaldehyde.

CHLOROPHYLL: A CONJUGATED NITROGEN–CONTAINING PIGMENT

Chlorophylls are a group of intensely green pigments found in plants. These complex molecules, represented by chlorophyll *a*, are central to the conversion of light energy into chemical energy in the process known as photosynthesis. Upon absorption of a photon of light, an electron in the extended π system of chlorophyll is promoted to an antibonding orbital, producing an excited state. In photosynthesis, an electron is transferred along a chain of molecules and ultimately effects the reduction of carbon dioxide. This process of reduction of carbon dioxide is called "carbon fixation" and results in incorporation of the carbon into carbohydrates.

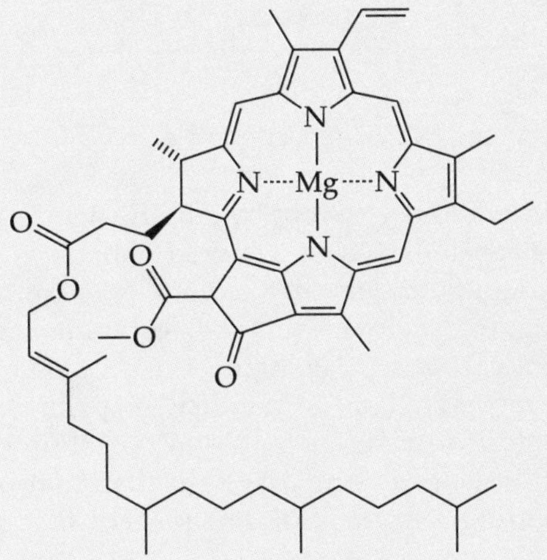

Chlorophyll *a*

occurs at 315 nm and is accompanied by a stronger π,π* band at 210 nm; that is, at a wavelength very similar to that seen for butadiene.

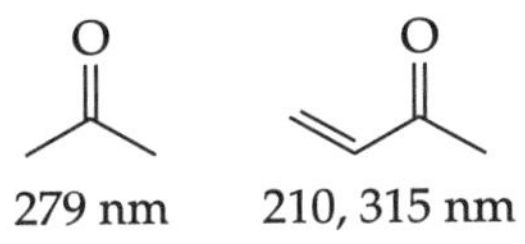

279 nm   210, 315 nm

Mass Spectroscopy

Energies much higher than those required for electronic transitions can cause the expulsion of an electron from a molecule. In a technique called **mass spectroscopy** (MS), molecules are bombarded with high-energy electrons, resulting in the ejection of electrons. In the most-common type of mass spectrometer (shown schematically in Figure 4-31, page 148), an ionizing electron beam passes through a gaseous sample. The high-energy electron strikes a molecule, dislodging an electron from a bonding orbital and producing a cation radical. The cations formed are therefore highly energetic and often break into smaller cationic fragments.

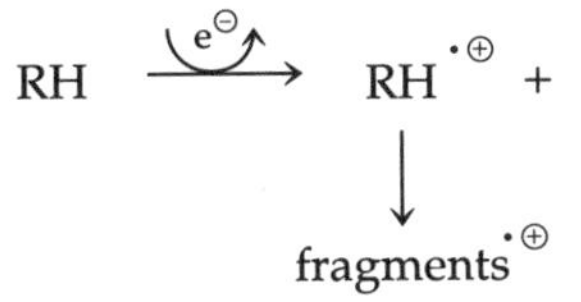

$$RH \xrightarrow{e^{\ominus}} RH^{\cdot\oplus} + e^{\ominus}$$

$$\downarrow$$

$$fragments^{\cdot\oplus}$$

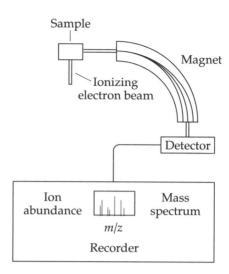

▲ **FIGURE 4–31**
A schematic representation of a mass spectrometer.

The resulting mixture of ions is swept by a high vacuum into a strong magnetic field. The ions are deflected in a curved path according to their mass-to-charge ratio (m/z) and are then directed to a detector that determines the m/z ratio. The magnetic field is scanned, bringing ions of successively higher m/z ratio to the detector. The highest-molecular-weight peak, that of the **molecular ion,** represents an unfragmented (parent) ion with the same mass as that of the starting material. A plot of ion abundance, as a function of m/z, of ions with lower molecular weight than that of the parent ion is thus measured, yielding a **fragmentation pattern** that is characteristic of that specific molecule.

The accuracy with which the m/z ratio can be measured is determined by the quality of the magnet (especially its uniformity) and the sophistication of the detector. In a low-resolution mass spectrum, m/z ratios are determined to about plus or minus 0.2 mass units. For some purposes, this accuracy is sufficient for assigning molecular weights to the ions (assuming, as is generally the case, that the charge is one).

High-resolution mass spectrometers are highly sophisticated instruments that have an accuracy of a small fraction of a mass unit. With this level of accuracy, it is possible to distinguish between combinations of atoms that differ only slightly. For example, carbon dioxide and propane have, to the nearest integer, the same molecular weight (44). However, this weight is based upon naturally occurring distributions of ^{12}C and ^{13}C, as well as ^{16}O, ^{17}O, and ^{18}O isotopes. In a mass spectrum, the molecular ion peak is that associated with those isotopes of highest abundance (in this case, ^{12}C and ^{16}O). With these isotopes, CO_2 is calculated to have a mass of 43.9898 and propane (C_3H_8) to have a mass of 44.0626. These two masses can be readily distinguished by high-resolution mass spectrometry, which generally has an accuracy of plus or minus 0.0001 mass units. There is usually only a single (reasonable) combination of atoms that corresponds to any given mass. It is possible therefore to calculate the formula for the ion from the exact mass of a parent ion found in a high-resolution mass spectrum.

Mass spectroscopy examines individual ions and determines their mass and does not measure the average properties of a bulk sample. Thus, the atomic masses that we must use in calculating an exact mass value are those of the individual isotopes (for example, 12.0000 for ^{12}C instead of

12.011), not those that we typically use—that is, atomic masses that represent the effect of all isotopes weighted for their relative abundances. Because we are observing individual ions in a mass spectrometer, it is also very easy to recognize the presence of elements such as chlorine, which exists as two abundant isotopes (^{35}Cl and ^{37}Cl). Thus, each fragment will be represented by two ions that are two mass units apart.

An example of a low-resolution mass spectrum of hexane is shown in Figure 4-32. The highest-molecular-weight peak appears at $m/z = 86$. This peak represents the parent ion, obtained by the simple loss of an electron without fragmentation. The much smaller peak that appears at $m/z = 87$ represents that fraction of molecules containing a higher-weight isotope (either ^{13}C or ^{1}H): the ratio of intensities of the parent and (parent + 1) ions is defined by the natural abundance of ^{13}C, ^{2}H, and so forth, and hence by the probability of incorporating one of the higher-mass isotopes into the ionized molecule. Lower-molecular-weight peaks also appear in this spectrum at $m/z = 71$, 57, 43, and 29. These peaks represent the sequential loss of a

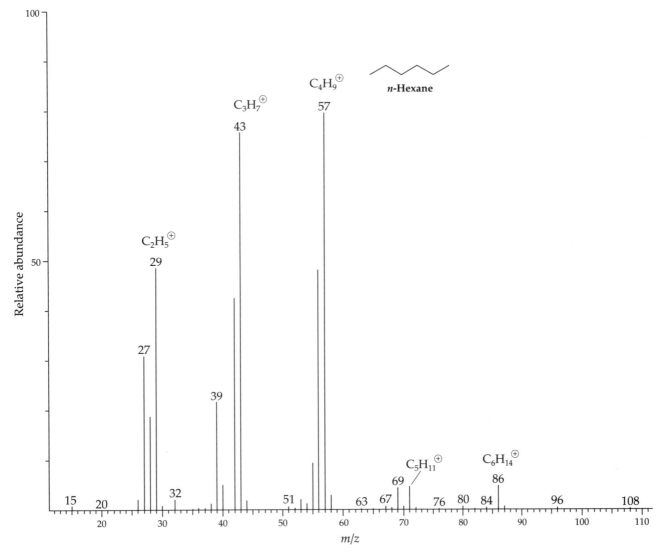

▲ FIGURE 4-32
A low-resolution mass spectrum of *n*-hexane.

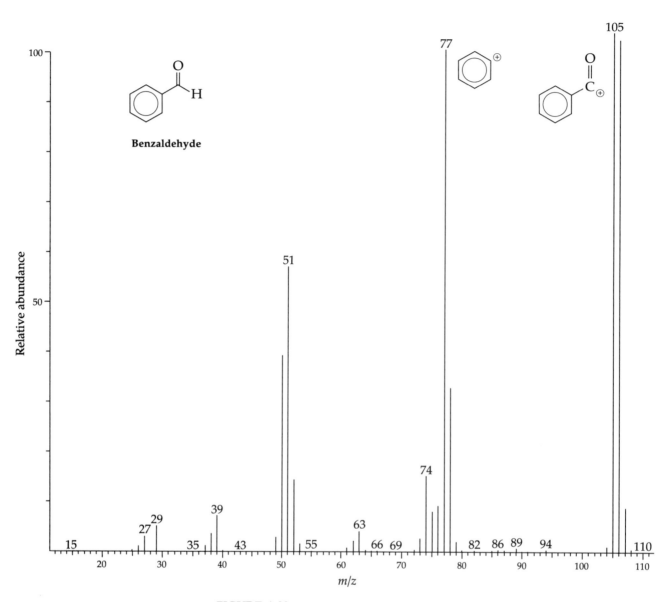

▲ FIGURE 4-33
A low-resolution mass spectrum of benzaldehyde.

methyl group and then methylene groups along the straight chain of the parent ion. The most-intense peak, at $m/z = 57$, is referred to as the **base peak.** Often, mass spectral data are reported as a series of peaks whose intensities are given as a fraction of this most-intense base peak. Notice that, even in this simple compound, some bonds are cleaved more readily than others. Thus, the base peak at $m/z = 57$ results from cleavage of the bond between carbons 2 and 3, not from cleavage of the more-abundant C–H bonds or one of the carbon–carbon bonds to the two methyl groups.

A knowledge of which bonds are most readily cleaved can assist in the interpretation of the fragmentation patterns observed in mass spectroscopy. For example, as shown in the mass spectrum of benzaldehyde in Figure 4-33 (106), the peak with the highest m/z value, corresponding to the molecular weight, is the parent peak. Other strong peaks are for ions that result from cleavage on either side of the carbonyl group, with loss of H or C_6H_5. These fragmentations are represented by more-significant peaks than are those resulting from cleavage of C–H or C–C bonds in the aryl ring.

Mass spectroscopy is thus very helpful in assigning structure. Detection of the parent molecular ion provides a determination of molecular weight and, at high resolution, can provide a unique formula for each ion produced. The identity of lower-molecular-weight ions in the spectrum produced by fragmentation of the molecular ion also can be used to help establish structure.

EXERCISE 4-J

For each of the following compounds, calculate the parent peak and predict the base peak (major fragmentation) in the mass spectrum.

(a) $CH_3CH_2CH_2CH_2OH$

(b) $CH_3COCH_2CH_3$

(c) CH_3CH_2CHO

(d) $CH_3CHClCH_3$

Conclusions

Structures of organic compounds are established by a comparison of their physical properties with those of model compounds. In addition to simple physical properties such as melting points, boiling points, and chemical reactivity, specific structural features in the molecule can be definitively characterized from evidence obtained through the use of instrumentation.

Spectroscopic techniques are most effective when used to analyze pure compounds. Chromatography is used to separate mixtures of compounds into individual components. Chromatographic separations take place by the differential adsorption of a mixture of compounds on a solid or liquid stationary support while a fluid mobile phase flows over this fixed support, eluting the organic compounds. Column chromatography employs a solid support (usually alumina or silica gel) through which a liquid solvent flows under the influence of gravity. In paper and thin-layer chromatography, sheets of adsorbent (paper or a thin layer of silica gel spread on a glass support) through which solvent moves by capillary action are used. In gel electrophoresis, a polar polyacrylamide gel is swollen with solvent and subjected to an electrical field. Polyelectrolyte molecules migrate under the influence of this external electrical field, producing separation. In gas chromatography, the mixture to be analyzed passes through a very long column either packed with a solid support or lined with a liquid adsorbent. An inert gas acts as a carrier to move the organic molecules through the column.

The time required for the elution of a desired compound is directly related to the degree of adsorption of that compound on the stationary support. Strong polar interactions and van der Waals attractions assist in this adsorption, causing longer elution times and higher adsorptivity. The retention time is therefore a rough indicator of the polarity and size of a given molecule. Chromatography can thus be used not only as a purification technique, but also as an analytical method for the characterization of mixtures because retention times represent yet another characteristic physical property.

Spectroscopic techniques entail the interaction of an energy source with molecules, producing a spectrum that can be interpreted to reveal the presence of characteristic groups. Nuclear magnetic resonance spectroscopy detects the nuclear spin flipping, induced by radio-frequency energy, of a molecule placed in a high magnetic field. Because ^{13}C and ^{1}H are magnetically active nuclei, NMR spectroscopy can provide important structural information about organic molecules. In ^{13}C NMR spectroscopy, the chemical shifts for each unique carbon can be predicted for a proposed structure. In ^{1}H NMR spectroscopy, chemical shifts, as well as the multiplicity and integration of signals, provide information on the type of protons, their number of nearest neighbors, and the number of protons responsible for an observed signal.

Infrared spectroscopy is a method for observing characteristic stretching and bending of bonds. The infrared spectra of many common organic functional groups have characteristic absorptions. In addition, the fingerprint region of the infrared spectrum often provides a unique pattern for each molecule. Structural assignments can often be made by a comparison of the spectrum of an unknown compound with the infrared spectra of known structures.

Ultraviolet, as well as visible, absorption spectroscopy probes electronic transitions from filled to vacant molecular orbitals. The absorption maxima in these bands provide information regarding the degree of conjugation and the types of electronic transitions possible within a compound.

In mass spectroscopy, high-energy electrons collide with a given molecule in the gas phase. The high-energy electrons effect ionization, producing a parent ion (a radical cation) and facilitating the determination of an accurate molecular weight. From isotopic abundances, high-resolution mass spectra also provide a molecular formula. Fragmentation patterns observed in the mass spectrum provide valuable information concerning the structure of a molecule.

Review Problems

4-1 Which compound in each of the following pairs would have the longer chromatographic retention time?

(a) 2-butanol and butanal

(b) octadecane and octanoic acid

(c) cyclohexanol and benzene

(d) pyridine or guanine

4-2 For each of the pairs of compounds in Problem 4-1, determine whether they could be separated best by gas chromatography, high-pressure liquid chromatography, or gel electrophoresis. Explain.

4-3 For each of the following compounds, a ^{1}H NMR spectrum and a listing of characteristic infrared bands are given, together with an atomic formula. Propose one or more structures consistent with each set of data. If the data listed are permissive of isomers, suggest another spectroscopic method that might be used to distinguish them.

(a) $C_2H_3Cl_3$: δ 3.95 (d, 2 H), 5.77 (t, 1 H); 2950 and several below 850 cm^{-1}

(b) C_2H_4O: δ 2.20 (d, 3 H), 9.80 (m, 1 H); 1730 cm^{-1}

(c) $C_2H_4O_2$: δ 2.10 (s, 3 H), 11.37 (s, 1 H); broad band at 3200 and strong band at 1710 cm^{-1}

(d) $C_2H_4O_2$: δ 3.77 (br s, 3 H), 8.08 (br s, 1 H); 1745 and 1250 cm^{-1}

(e) C_2H_6O: δ 1.22 (t, 3 H), 2.58 (br s, 1 H), 3.70 (q, 2 H); broad band at 3600 cm^{-1}

(f) $C_3H_5ClO_2$: δ 1.73 (d, 3 H), 4.47 (q, 1 H), 11.22 (s, 1 H); broad band at 3200, strong band at 1710, and several below 850 cm^{-1}

(g) C_3H_5NO: δ 3.47 (s, 3 H), 4.20 (s, 2 H); 2250 and 1100 cm^{-1}

(h) C_3H_6O: δ 2.72 (quintet, 2 H), 4.73 (t, 4 H); 1120 cm^{-1}

(i) C_3H_6O: δ 3.58 (s, 1 H), 4.13 (m, 2 H), 5.13 (m, 1 H), 5.25 (m, 1 H), ca. 6.0 (m, 1 H); broad band at 3600, 3050, 2980, 1420 cm^{-1}

(j) C_6H_6ClN: δ 3.60 (s, 2 H), 6.57 (d, 2 H), 7.05 (d, 2 H); broad bands at 3520 and 3400, 3050, 1490, 1590, 910 cm^{-1}

4-4 Give a structure for a compound with a formula $C_{10}H_{12}O_2$ that is consistent with the following spectra. Would the assignment of structure be unambiguous from the carbon spectrum alone?

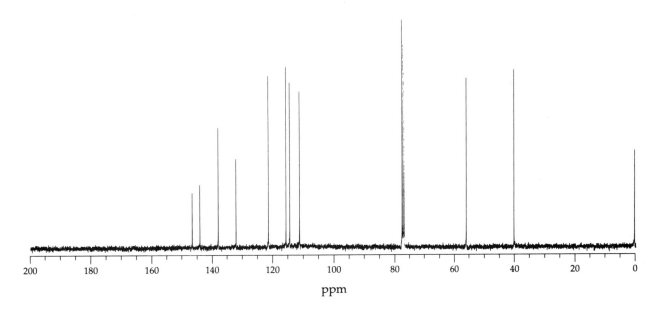

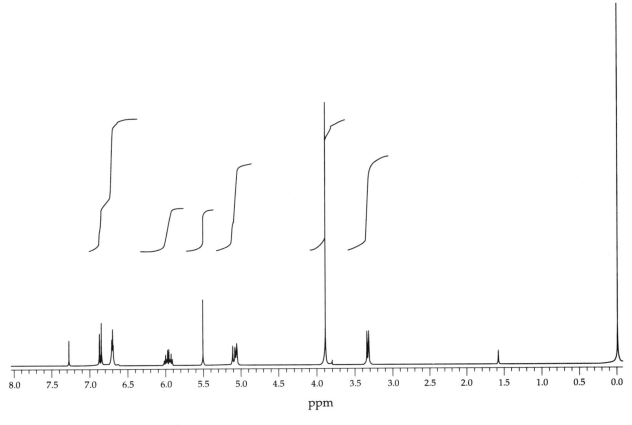

4-5 For each of the following compounds, how many peaks you would expect to find in the ¹H NMR spectrum? What would you expect their splitting and integration to be? How many peaks would be expected in the ¹³C NMR spectrum? What characteristic peaks would you expect to find in its infrared spectrum?

(a) ethyl bromide

(b) 2-propyne

(c) 2-propyne-1-ol

(d) allyl bromide

(e) 2-nitropropane

(f) *N,N*-dimethylformamide

(g) methyl ethyl sulfide

(h) 2-propanol

(i) vinyl acetate

(j) 2-bromobutanoic acid

(k) 2-butanone

(Continued on next page.)

(l)   butanal

(m) 3-methoxybutanol

(n)   toluene

4-6 Suggest a spectroscopic method that can easily distinguish between members of the following pairs of compounds. Describe exactly what you will see for each compound in your chosen method.

(a)   *n*-butylamine and *t*-butylamine

(b)   *s*-butylamine and *t*-butylamine

(c)   methylene cyclopentane and 1-methylcyclohexene

(d)   *m*-methylphenol (*m*-cresol) and benzyl alcohol

(e)   *p*-methylphenol (*p*-cresol) and anisole ($C_6H_5OCH_3$)

(f)   styrene oxide ($C_6H_5CHCH_2O$) and acetophenone

(g)   acetophenone and *p*-methoxybenzaldehyde (*p*-anisaldehyde)

(h)   *m*-xylene and *p*-xylene

4-7 Propose a structure for hydrocarbon X, whose mass spectrum shows a parent peak at $m/z = 86$ with a major peak at P–15 (parent – 15 mass units) and whose ^{13}C NMR spectrum shows two peaks: δ 19.5 and 34.3, respectively. Explain how other isomeric structures can be definitely eliminated by this data.

4-8 Isomers A and B have a molecular formula of C_6H_{12}. Isomer A has a ^{13}C NMR spectrum with peaks at 13.7, 17.8, 23.1, 35.0, 124.9, and 131.6 δ; isomer B has peaks at 12.7, 13.7, 23.1, 29.2, 123.9, and 130.7. How many double bonds are present in these compounds? Propose possible structures for isomers A and B, excluding structures that are specifically eliminated by the ^{13}C data. What additional information, if any, would be of use in making an unambiguous spectral assignment?

4-9 From the ultraviolet absorption data in Table 4-4, roughly predict the wavelength region in which each of the following common solvents would absorb. Would these be acceptable solvents for measuring the absorption spectrum of naphthalene?

(a)   cyclohexane

(b)   tetrahydrofuran

(c)   toluene

(d)   acetone

4-10 Predict the parent ion mass and the major mass spectral fragments to be expected for each of the following compounds:

(a)   pentanal

(b)   acetophenone

(c)   ethanol

(d)   ethylether

Chapter 5

Stereochemistry

Now that we understand the composition of the principal functional groups of organic chemistry, we are ready to think about three-dimensional structure in more detail. Some of the examples of compounds presented in Chapters 1 through 3 are referred to as **constitutional isomers,** in which the atoms in two molecules having the same molecular formula are attached in different sequences; for example, ethanol and methyl ether. In this chapter, we will learn about two classes of **stereoisomers** (those differing not in atomic connectivity but only in how the atoms are held in space). These stereoisomers are: **conformational isomers** (those that can be interconverted by rotation about a σ bond) and **configurational isomers** (those that can be interconverted only by the breaking and reforming of bonds). Two important subclasses of configurational isomers are **geometric isomers** (those in which restricted rotation in a ring or at a multiple bond determines the relative spatial arrangement of atoms) and **optical stereoisomers** (those that differ spatially only at one or more atoms lacking a plane of symmetry). We will also learn about energy differences between different isomers within these classes and how chemical and physical properties are affected by the spatial arrangement of atoms.

5–1 ▸ Geometric Isomerization: Rotation about Pi Bonds

As we learned in Chapter 2, π bonding requires the overlap of two (or more) *p* orbitals on adjacent atoms. Because of the barrier to rotation about a double bond, geometric isomers such as *cis-* and *trans-*2-butene, for example, differ in how groups on adjacent carbons of the alkene are arranged relative to each other. In both isomers, the geometry of the π bond forces the two sp^2-hybridized carbons of the double bond to be coplanar with the four attached substituent atoms. Carbons 1, 2, and 3 of 2-butene define a plane, and the dihedral angle made between this plane and that containing C-4 (that is, the C-2–C-3–C-4 plane) defines the degree of twisting about the π

cis-2-Butene

trans-2-Butene

bond. In *cis*-butene, this dihedral angle is 0°, and in *trans*-butene it is 180°. In both of these arrangements the overlap of the *p* orbitals of the π bond is at a maximum. In contrast, when this angle is 90°, there is no π-bonding character.

In a rotation of this dihedral angle from 0° to 180°, the π bond is broken (which requires energy) and is then reformed. When we plot this energy change as a function of the dihedral angle (Figure 5-1), we find that an infinite number of structures are possible, the least stable of which is the one at 90°. At this point, the *p* orbitals on C-2 and C-3 are orthogonal; there is no π bonding and the difference in energy between the geometries at 0° and 180° and that at 90° is a rough measure of the π-bond strength.

Plots such as that in Figure 5-1 also describe net reaction energetics. The energy difference between a reactant and a product is defined as the **heat of reaction,** $\Delta H°$. Here the product, *trans*-2-butene, lies at a lower energy than the reactant, *cis*-2-butene, for the reasons discussed in Chapter 2. Therefore, it is energetically unfavorable for the reaction to proceed from the *trans* to the *cis* isomer. The energy difference for reactions ($\Delta G°$) has contributions from both **enthalpy** ($\Delta H°$, heat content) and **entropy** ($\Delta S°$, disorder).

$$\Delta G° = \Delta H° - T\Delta S°$$

For isomerization reactions, it is usual to consider only the change in enthalpy.

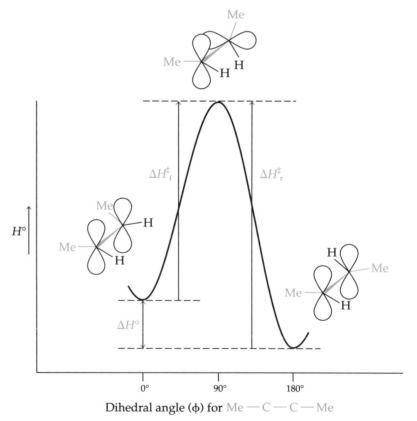

▲ FIGURE 5-1

Energy changes induced by rotation about the C=C bond of 2-butene (Me = methyl group). Activation energies for the forward (*cis* → *trans*: $\Delta H^{\ddagger}_f$) and back reactions (*trans* → *cis*, $\Delta H^{\ddagger}_b$) are indicated by the vertical arrows.

The sign of $\Delta H°$ tells us if an isomerization is endothermic (+) or exothermic (−): in Figure 5-1, $\Delta H°$ is negative and thus the reaction is exothermic. Even so, $\Delta H°$ provides no direct information on the **energy barrier**, $\Delta H^{\ddagger}$; that is, how much energy is required to reach the most unfavorable point along the path followed in the conversion of the reactant into product. To describe the route by which a reaction takes place, we must know not just the energy difference between the reactants and products, but also the energy of the least-favorable arrangement through which the atoms must pass as the reaction proceeds. This point of highest energy is called the **transition state.**

In the geometrical isomerization in Figure 5-1, the transition state is that arrangement in which there is no stabilization from π bonding. This transition state (with p orbitals at a dihedral angle of 90°) can collapse to either the *cis* or the *trans* isomer without any additional energy barrier. Because the transition state is the least-stable species in a reaction pathway, no additional energy is required when it is converted into a more-stable species.

The energy difference between the reactant and this high-energy transition state is referred to as the **energy of activation,** or **activation energy** ($\Delta H^{\ddagger}$ or E_{act}). In this specific transformation, the energy of activation for the forward reaction (*cis* → *trans*) is slightly less than that of the reverse reaction (*trans* → *cis*). Activation energies for the forward and reverse reactions always differ by precisely the value of $\Delta H°$ of the reaction.

EXERCISE 5-A

For each of the following isomerizations, draw the transition state required for the conversion of the structure at the left into the structure at the right. Can you estimate from what you know about bond energies from Chapters 2 and 3 whether the activation barrier in each of these reactions is larger or smaller than that for the isomerization of *cis*-2-butene to *trans*-2-butene?

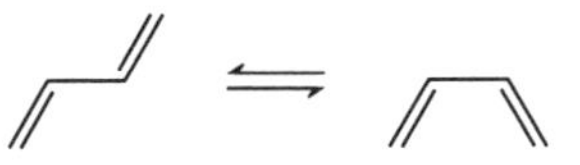

The activation energy required for the interconversion of *cis* and *trans* isomers is larger than can be easily provided at typical reaction temperatures (<300 °C). However, we learned in Chapter 4 that absorption of a photon of ultraviolet light by an alkene, diene, or triene is accompanied by the promotion of a bonding electron to an antibonding orbital. In the simple case of an alkene with only one double bond, the resulting arrangement with one electron in both the bonding and the antibonding π-molecular orbitals leads to essentially no net bonding between the atomic p orbitals. In this excited state, rotation about the carbon–carbon bond occurs readily, even at very low temperatures. Ultimately, the electron in the energetically higher, antibonding orbital returns to its original position in the bonding orbital, releasing most of the absorbed energy as heat or light.

The relatively free rotation about the carbon–carbon bond that occurs upon absorption of a photon has significant biological consequences because *cis–trans* isomerization forms the chemical basis for mammalian

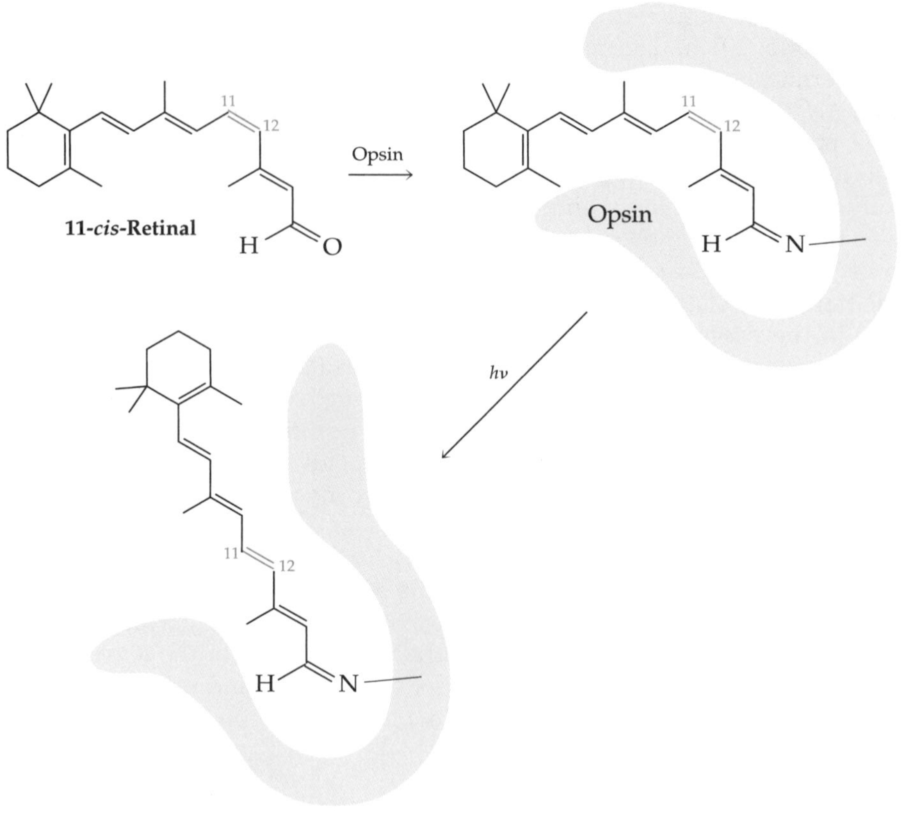

▲ FIGURE 5-2
Geometric isomerism in rhodopsin induced by the absorption of light.

vision. Light-sensitive receptor cells in the eye contain **11-*cis*-retinal** chemically bound to the protein **opsin** (through an imine functional group), forming **rhodopsin.** Absorption of visible light (notice the extended conjugation) by the 11-*cis* isomer results in an excited state that readily undergoes isomerization to the 11-*trans* isomer (Figure 5-2).

The *trans* isomer is a more-extended molecule than the *cis*, and this isomerization requires a change in the shape of the opsin protein. This change, in turn, initiates the release of calcium ions. This increase in ion concentration triggers a nerve impulse that is interpreted by the brain as vision. Thus, this electronic transition, whose energy is defined by the energy difference between the filled and vacant orbitals, provides a way for inducing new reactivity in the excited state.

The isomerization of *cis* and *trans* alkenes can also be accomplished in the laboratory. The irradiation of an alkene with ultraviolet light of sufficient energy to promote a π electron effectively breaks the π bond, resulting in free rotation (Figure 5-3). At wavelengths of light at which the *cis* isomer

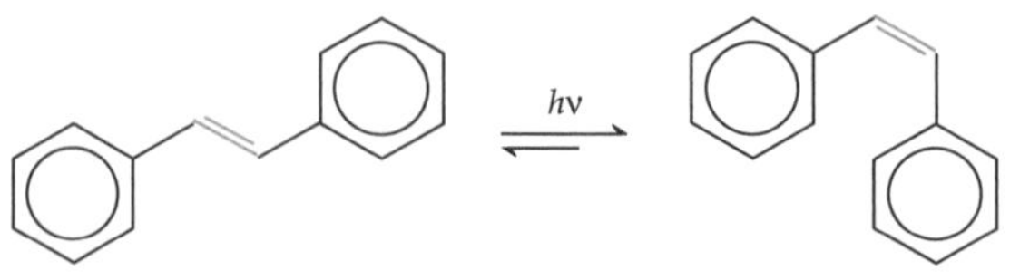

trans-Stilbene   *cis*-Stilbene

▲ FIGURE 5-3
Geometric isomerization in an alkene induced by the absorption of light.

absorbs light to a lesser extent than does the *trans* isomer, the rate of conversion of the *trans* isomer into the *cis* is greater than that of *cis* into *trans*, and the *cis* isomer dominates the equilibrium. This reaction can be used, therefore, to produce *cis* alkenes from the more thermodynamically stable *trans* isomers.

5-2 ▸ Conformational Analysis: Rotation about Sigma Bonds

Isomers that have the same skeletons (that is, with component atoms attached in the same sequence) but differ from each other with respect to the relative positions of some atoms in three-dimensional space by virtue of rotation about σ bonds are called **conformational isomers,** or **conformers.** A quantitative description that relates relative atomic positions to the changes in potential energy during rotation about a σ bond describes the energetics of conformational interconversion, a process known as **conformational analysis.**

We know from Chapters 1 and 2 that rotation about a single bond does not require bond cleavage and is therefore faster than rotation about a multiple bond, which requires the addition of significant energy to break the π bond. Thus, changes in energy in the interconversion of conformational isomers are relatively small, and it is often said that there is "free" rotation about a σ bond.

Ethane

Let us consider rotation about the carbon–carbon bond of ethane. Although there are an infinite number of conformations, there are only two extremes (Figure 5-4). (We recommend that you build a model of this simple molecule so that you can better understand the analysis that follows.) In the left-hand structure, called an **eclipsed conformation,** each of the carbon–hydrogen bonds at C-1 is aligned with those at C-2 (dihedral angle = 0°). In the right-hand structure, called a **staggered conformation,** each C–H bond at C-1 is fixed at a 60° dihedral angle so that it is exactly between two C–H bonds at C-2. In both structures, tetrahedral geometry at carbon is maintained. The left-hand structure can be converted into the right-hand one by simple rotation about the C–C σ bond by 60°.

If, instead of looking at the C–C bond as it is depicted in Figure 5-4, we were to view these same conformational isomers of ethane from the end, with the C–C bond directed away from us as shown in Figure 5-5, C-2 and the carbon–carbon bond would be hidden behind C-1. We can regard the circle in these representations as the electron density of the σ bond, with the front carbon implied at the junction of the three bonds and with the back carbon hidden. These representations, called **Newman projections,** show the orientation of the hydrogens on the front carbon relative to those on the back carbon and are quite useful for conformational analysis. (This terminology is in recognition of the chemist Melvin Newman, of the Ohio State

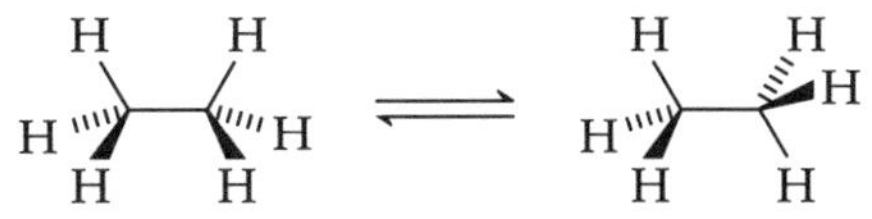

▲ FIGURE 5-4
Rotation about the C–C bond of ethane.

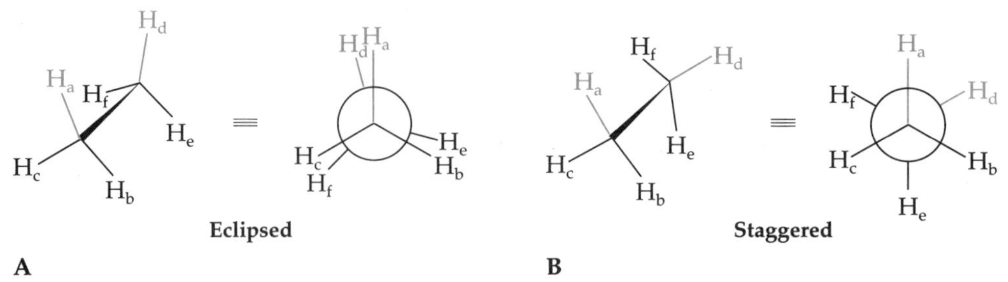

▲ FIGURE 5-5

Two sawhorse representations of ethane and their corresponding Newman projections. (The eclipsed substituents are drawn slightly displaced from planarity for easier visualization.)

University, who first showed the utility of these representations in conformational analysis. As we progress in organic chemistry, we will encounter many scientists whose names are associated with theories and the discovery of reactions.)

Keep in mind that the carbons are not flat and that, indeed, the molecules represented in the Newman projections are the same as those shown in the sawhorse notations. In the Newman projection in Figure 5-5A, all the hydrogen atoms of ethane are aligned with each other in the eclipsed conformation. If we identify each of the hydrogen atoms with a subscript, we see that the $C–H_a$ and the $C–H_d$ bonds are coplanar and on the same side of the carbon–carbon internuclear axis. The σ bonds to H_b and H_e and to H_c and H_f also are coplanar. Rotation about the carbon–carbon σ bond of the eclipsed conformation in Figure 5-5A by 60° (moving the substituents on the back carbon) changes the relative positions of the hydrogens in these pairs. In the resulting conformation, shown in Figure 5-5B, the bonds to the front carbon are exactly between those of the back carbon in a staggered conformation.

The change in the relative positions of the bonding electrons between the eclipsed and staggered conformations results in an energy difference as well. The electrons of the carbon–hydrogen bonds to the front and back carbons are closer to each other in the eclipsed conformation, resulting in greater electron–electron repulsion. Thus, the eclipsed conformation is energetically less favorable than the staggered conformation. The increase in electron–electron repulsion upon rotation from a staggered to an eclipsed conformation is referred to as **torsional strain.**

EXERCISE 5-B

Draw Newman projections to illustrate each of the following descriptions:

(a) an eclipsed conformation of 2,2,3,3-tetramethylbutane, viewed down the C-2–C-3 bond

(b) the staggered conformation of 2,2,3,3-tetramethylbutane, viewed down the C-2–C-3 bond

(c) the staggered conformation viewed down the C-1–C-2 bond of propane

(d) the eclipsed conformation viewed down the C-1–C-2 bond of propane

We can draw a profile for rotation about the σ bond of ethane that relates the relative potential energy (degree of torsional strain) to the dihedral angle between a pair of hydrogens, one on the front and one on the back carbon. In the Newman projection shown in Figure 5-5A, H_a, C-1, and C-2

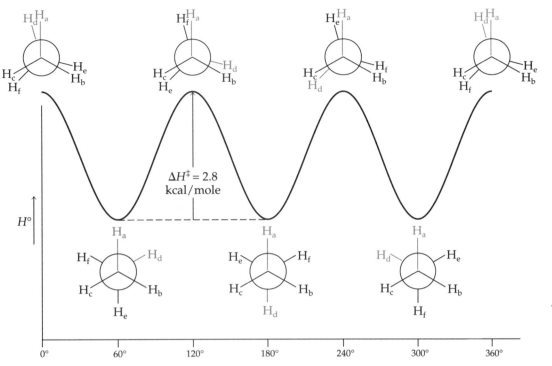

Dihedral angle (Φ) between the H_a, C-1, C-2 plane and the C-1, C-2, H_d plane

▲ FIGURE 5-6

Energy changes induced by rotation about the C–C bond of ethane.

define a plane and the dihedral angle Φ is defined as the angle made between that plane and one defined by C-1, C-2, and H_d. We thus begin with a high-energy conformation with the molecule in an eclipsed arrangement, $\Phi = 0$. As we keep the front carbon stationary and rotate the back carbon clockwise by 60°, we obtain a staggered conformation, which is of lower energy because torsional strain is relieved (Figure 5-6). As we continue the rotation through another 60°, we come to another eclipsed conformation in which H_d now aligns with H_b. However, because all of the hydrogens are identical, we have arrived at an eclipsed conformation that is identical in energy with the first. Thus, the relative energy increases during rotation until it reaches the original value. A series of sequential 60° rotations provides a smooth energy profile for the interconversion of the eclipsed and staggered conformations until H_a has returned to its original position.

The energy difference between the eclipsed and the staggered conformations is about 2.8 kcal/mole (about 0.9 kcal/mole for each H–H torsional interaction). The staggered conformation is at the bottom of an energy well, whereas the eclipsed conformation is at the top. Indeed, no additional energy is required for rotation of the eclipsed conformation. A structure that exists at the bottom of a **potential energy well** (as do these staggered conformations) has a real-time existence and is either a stable compound or, if this energy well is at a high energy relative to a reactant or product, a **reactive intermediate.** Species at the top of a potential energy curve are **transition states** and have only a fleeting existence.

Butane

Newman projections are also useful for the analysis of more-complex hydrocarbons. For example, rotation about the central carbon–carbon bond of butane results in several different eclipsed and staggered arrangements that

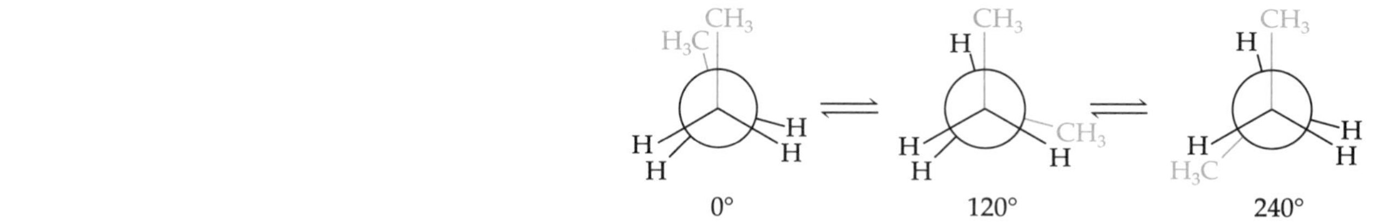

▲ FIGURE 5-7
Newman projections of eclipsed conformers of *n*-butane.

also differ in energy. In Figure 5-7, the left-hand conformation (in which two methyl groups are eclipsed) is energetically unfavorable both because of torsional interactions between the bonds and because of the repulsive interaction resulting from two large methyl groups being in the same region of space. Destabilization resulting from groups that are close to each other is referred to as a **steric effect,** or **steric strain,** and is caused by van der Waals repulsion. Thus, there is both torsional and steric strain in the left-hand conformation, known as the *syn* **eclipsed conformer.**

In the other two eclipsed conformations shown in Figure 5-7, the methyl groups are eclipsed with hydrogens. Because a hydrogen atom is smaller than a methyl group, there is less steric strain and the total destabilization resulting from both torsional and steric strain is only slightly larger than that for ethane. These two conformations have identical energies, and both are more stable than the *syn* eclipsed structure.

The three staggered isomers of butane are represented by Newman projections in Figure 5-8. Torsional strain is at a minimum in these staggered conformations. In the middle structure, the dihedral angle between the methyl groups is 180°; whereas, in the left- and right-hand structures, this angle is 60°. There is some steric interaction between the methyl groups in the left- and right-hand structures. As a result, they are higher in energy than is the middle one. Isomers bearing substituents near each other in a staggered conformation (that is, separated by a 60° dihedral angle) are referred to as *gauche* **conformers.** Those in which substituents are separated by a 180° dihedral angle are referred to as *anti* **conformers.** The two *gauche* isomers are mirror images of each other and cannot be superimposed without rotation about the C-2–C-3 bond. (We will see other compounds that have this relation later in the chapter.)

We can now construct an energy profile for rotation about the central carbon–carbon bond of butane based on energetic considerations, as shown in Figure 5-9. The energy difference between *gauche-* and *anti*-butane is 0.9 kcal/mole, and the energy barrier for the conversion of the *gauche* into the *anti* conformer (by way of the eclipsed conformer that represents the transi-

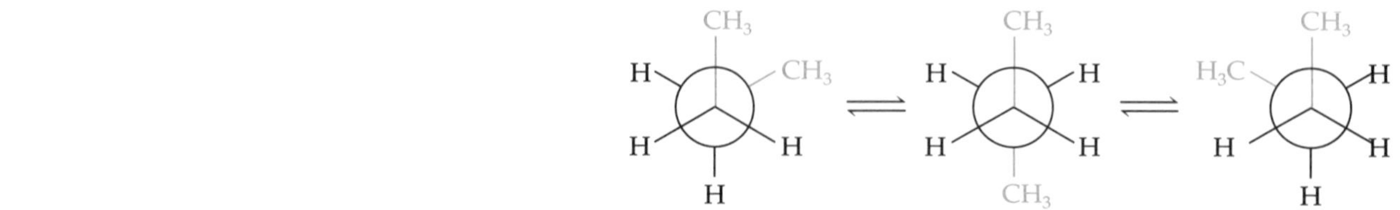

▲ FIGURE 5-8
Newman projections of staggered conformations of *n*-butane.

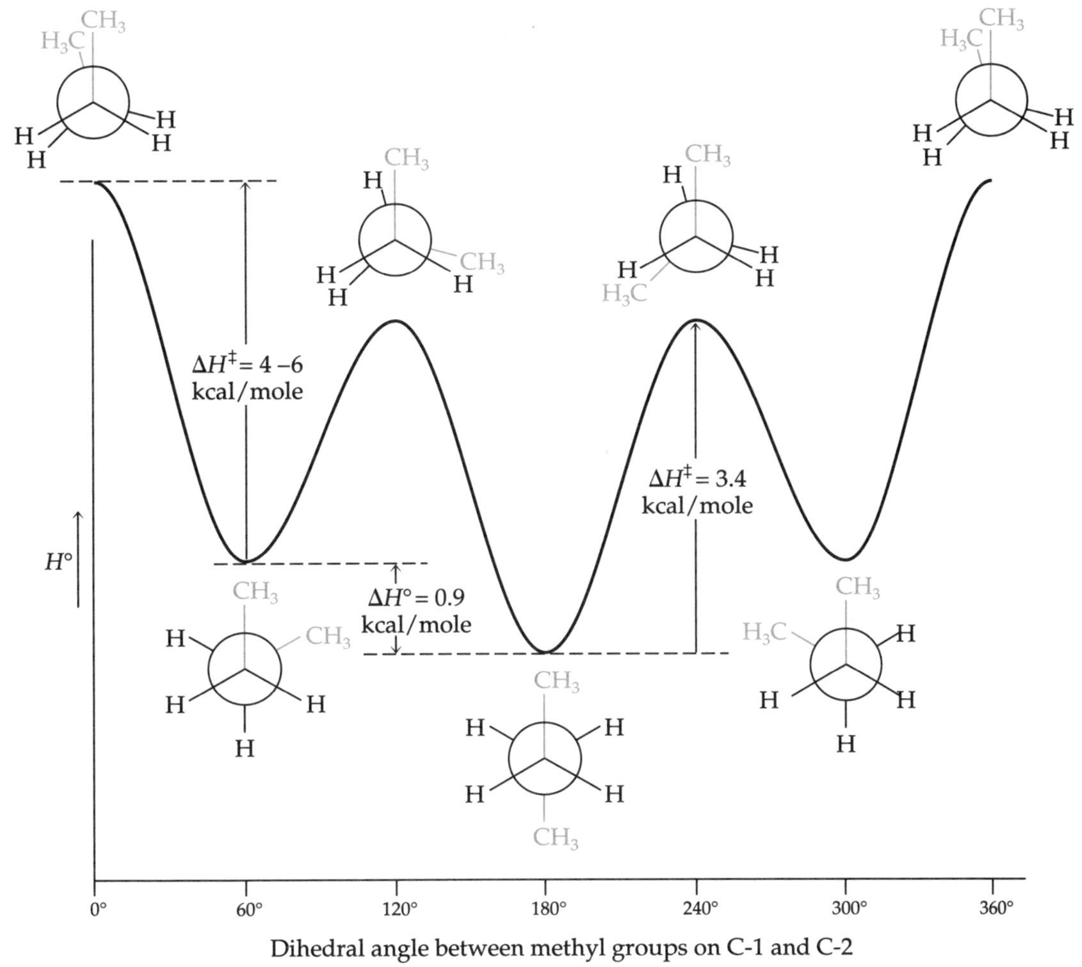

▲ FIGURE 5-9
Energy changes induced by rotation about the C-2–C-3 bond of *n*-butane.

tion state) is about 3.4 kcal/mole. The energy of the *syn*-eclipsed conformer, in which the methyl groups are aligned, is difficult to measure accurately but has been estimated to be between about 5 and 7 kcal/mole higher than that of the *anti* isomer.

The magnitude of the energy cost for each of the conformational interactions in ethane and butane is summarized in Table 5-1. These values can also be assumed to be reasonable approximations for the indicated interactions in other molecules. The value of the CH_3-CH_3 *gauche* steric strain is

Table 5-1 ▸ Approximate Energy Costs for Steric Interactions

Type of Interaction	Energy (kcal/mole) ($\Delta H°$)
H–H steric strain (*gauche*)	0
CH_3–CH_3 steric strain (*gauche*)	0.9
H–H torsional strain (eclipsed)	0.9
CH_3–H steric and torsional strain (eclipsed)	1.25
CH_3–CH_3 steric and torsional strain (eclipsed)	~ 3–5

obtained as the energy difference between the *anti* and the *gauche* isomers of butane. That for the CH_3-H eclipsed steric and torsional strain is derived from $\Delta H°$ between *anti* butane and the eclipsed conformer at a 120° (or 240°) dihedral angle, corrected for H-H eclipsing and partitioned equally to the two *gauche* interactions. The estimate for the CH_3-CH_3 interaction energy is derived from $\Delta H°$ for the conformers at 0° and 180°, corrected for the two eclipsing H-H interactions

EXERCISE 5-C

Suppose that, instead of viewing butane down its C-2–C-3 bond, you viewed hexane down its C-3–C-4 bond. Draw the energy profile that you would obtain and qualitatively compare the magnitude of the hills and valleys in your profile with that of Figure 5-9.

Gauche and *Anti* Conformers

$$A \underset{k_b}{\overset{k_f}{\rightleftharpoons}} B$$

The free energy difference between two isomers determines their relative abundance at equilibrium, and an equilibrium constant can be calculated from the difference in free energy of the reactants and the products. In the equilibration of a reactant A and product B, the relative amounts of A and B present at equilibrium depends directly on the size of the energy difference, $\Delta G°$, between them. The equilibrium constant (K) can also be expressed as a ratio of the rate constants for the forward (k_f) and reverse (k_b) reactions:

$$K = \frac{k_f}{k_b}$$

These rates can be calculated through the use of the **Arrhenius equation:**

$$k_f = A e^{-\Delta H^{\ddagger}_f/RT}$$

and

$$k_b = A e^{-\Delta H_b/RT}$$

in which T is temperature in kelvins (K) and R is the ideal gas constant (1.987 cal/K). This equation tells us that the rate is greater when the activation energy barrier is smaller. We can find the equilibrium constant, K, by calculating the ratio of these rates if we assume that the preexponential factors, A, for the forward and reverse reactions are similar (because the difference in entropy is small) and thus cancel.

$$K = \frac{k_f}{k_b} = \frac{A_f e^{-\Delta H^{\ddagger}_f/RT}}{A_b e^{-\Delta H_b/RT}} \cong \frac{e^{-\Delta H^{\ddagger}_f/RT}}{e^{-\Delta H_b/RT}} = e^{-(\Delta H^{\ddagger}_f - \Delta H^{\ddagger}_b)/RT} \cong e^{-(\Delta G°)/RT}$$

Here, $\Delta G°$, the change in free energy, is approximated by the difference in activation enthalpies if it can be assumed that entropy differences are small. Notice that the dependence of the equilibrium constant on $\Delta G°$ (the difference in free energy between starting material and product) is exponential. Thus, small differences in $\Delta G°$ result in much larger changes in K. We can also solve for $\Delta G°$, and find a logarithmic dependence:

$$\ln K = - \frac{\Delta G°}{RT}$$

in which $\Delta G° = \Delta H° - T\Delta S°$. The enthalpy difference determines relative abundance at equilibrium assuming that contributions from entropy differ-

Table 5-2 ▸ Conformational Equilibrium Ratios as a Function of Energy Differences

$\Delta H°$ (kcal/mole)	Percentage of More-Stable Isomer[a] (25 °C)
0.0	50.0
0.65	75.0
1.3	90.0
1.7	95.0
2.7	99.0
4.1	99.9

[a]Calculated from $K = e^{-\Delta H°/RT}$.

ences are negligible. Thus:

$$\ln K \cong - \frac{\Delta H°}{RT}$$

We can use this equation to calculate the ratio of the more-stable to the less-stable conformer as a function of the energy difference between them, leading to the values shown in Table 5-2. Either by interpolating from the values in this table or by solving the Arrhenius equation, we find that the 0.9 kcal/mole energy difference between the *gauche* and *anti* forms of butane corresponds to an equilibrium in which the *anti* conformation is favored by a factor of about 4 over each of the *gauche* isomers. Because there are two *gauche* conformations and only one *anti* form, the *anti:gauche* ratio is approximately 2:1; that is, the equilibrium mixture is composed of about 66% of the *anti* conformer and 34% of the two *gauche* conformers. The high energy *syn* eclipsed conformation ($\Delta H° \simeq 5$–7 kcal/mole) does not significantly contribute to the conformational equilibrium: the ratio of the staggered (*anti*) to the eclipsed conformer is greater than 1000/1.

EXERCISE 5-D

Using the Arrhenius equation or the data in Table 5-2, estimate the energy difference required to give the following equilibrium distributions of a more-stable conformer A with a higher-energy conformer B at 25 °C:

(a) a 55:45 mixture of isomers A and B

(b) a 70:30 mixture of A and B

(c) a 99.99:0.01 mixture of A and B

5-3 ▸ Cycloalkanes

The factors that control the conformational equilibrium of butane (namely, torsional and steric strain) can also be applied to cycloalkanes. The factors that favor staggered over eclipsed conformers (eclipsing interactions that induce torsional and steric strain) also play a role in how small- and large-ring saturated hydrocarbons exist in three dimensions.

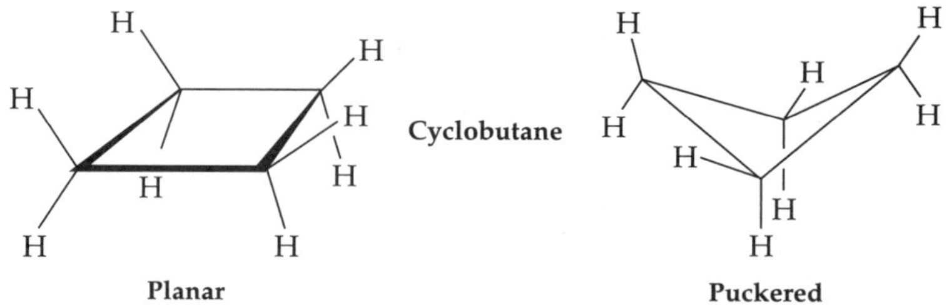

▲ FIGURE 5-10

Three-dimensional views of planar and puckered cyclobutane. Because of
minimization of torsional, steric, and angle strain, the puckered form
dominates the conformational equilibrium of cyclobutane.

Three factors control conformational preference in small rings: torsional and steric strain caused by the eclipsing of C–H (or other) bonds;
and angle strain caused by distortion from the angle dictated by the hybridization of the ring atoms. Because three points determine a plane, the
carbon atoms of cyclopropane are coplanar; and, in fact, its C–H bonds are
eclipsed. The ring strain in cyclopropane is therefore caused both by angle
strain (as described in Chapter 1) and by torsional strain. Similarly, planar
cyclobutane would be destabilized by angle strain caused by bond angles
(90°) that are smaller than the ideal tetrahedral angle. Furthermore, additional destabilization results from the eclipsing of substituents on all four
bonds at each carbon in the planar form (Figure 5-10). This strain is reduced
in the **puckered form** in which one atom is moved out of the plane of the
other three. Because the puckered form has less complete eclipsing and thus
lower torsional strain, cyclobutane exists in a conformation that is substantially distorted from planarity. Likewise, a completely planar conformation
of cyclopentane would have all ten carbon–hydrogen bonds eclipsed (Figure 5-11), whereas a puckered form in which one of the carbons lies out of
the plane defined by the remaining ring carbons has less of this torsional
strain. (Of nearly equal energy to the puckered form, another conformation
of cyclopentane has two atoms out of the plane of the other three, one above
and one below. See if you can find it by using your molecular models.)

For cyclic systems, the conformations in which all carbon atoms are in
one plane have C–C–C bond angles dictated by the geometry of regular
polyhedra, and any deviation from planarity always reduces some or all of
these angles. For a six-membered planar ring, this angle is 120°, clearly
larger than the tetrahedral angle of 109° of sp^3-hybridized carbons. The nonplanar form of cyclohexane, however, has angles close to 109°, an angle
much smaller than that of the planar form. Thus, a change from the planar
form is accompanied by relief of *both* torsional *and* angle strain. Similar nonplanar, flexible conformations also exist for even larger cycloalkane rings.

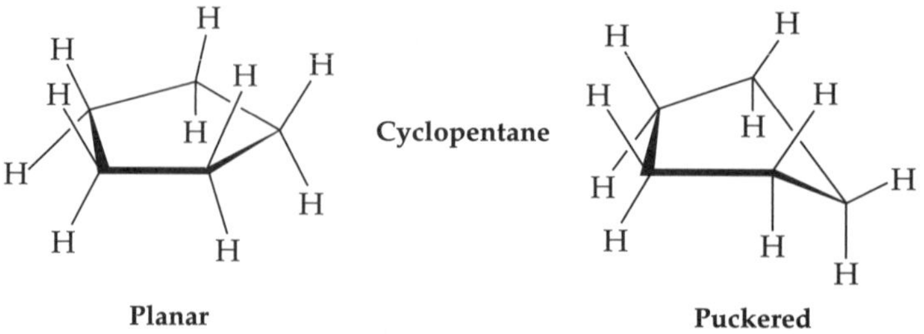

▲ FIGURE 5-11

As for cyclobutane, the puckered form of cyclopentane is the preferred
conformation because strain is minimized.

For five- and four-membered rings, the planar conformers have angles (108° and 90°) that are *smaller* than tetrahedral. Deviation from the plane thus decreases angle strain for large rings but increases it for small rings. For six-membered and larger rings, the deviation from planarity results in a conformation that is lower in energy because of a reduction both in angle strain and in torsional strain. For five- and four-membered rings, deviation from planarity results in a conformation that balances a decrease in torsional strain with an increase in angle strain.

As a result of this balance between increasing angle strain and decreasing torsional strain, five- and four-membered rings must adopt conformations that are not ideal. Furthermore, the differences in energy between these low-energy conformations and the planar form are much smaller than that for cyclohexane, and interconversions between the various conformations of cyclopentane and cyclobutane are much more rapid than are those of cyclohexane.

We can compare the effects of bond-angle strain and eclipsing interactions for different ring systems by examining the heats of combustion per CH_2 group listed in Table 5-3. This value decreases with increasing ring size and reaches a minimum for cyclohexane (in which torsional and angle strain are relieved in a nonplanar conformation). The difference between the heats of combustion per CH_2 group for cyclobutane and cyclohexane is 7 kcal/mole. Because there are four CH_2 groups in cyclobutane, there is an expenditure of 28 kcal/mole (4×7 kcal/mole) of ring strain in a four-membered ring, the sum of the energetic costs of bond-angle distortions from the ideal 109.5° and torsional strain caused by partial eclipsing. The same calculation for cyclopropane yields a value of 30 kcal/mole of ring strain.

Table 5-3 ▸ Heat of Combustion per CH_2 in Cycloalkanes	
Compound	**Heat Released per CH_2 (kcal/mole)**
Cyclopropane	167
Cyclobutane	164
Cyclopentane	159
Cyclohexane	157
Cycloheptane	158
Cyclooctane	159

5-4 ▸ Cyclohexanes

In this section, we use the foregoing principles of conformational analysis to estimate the relative stabilities of several possible three-dimensional arrangements in cyclic alkanes. Again, you may find it very useful to prepare molecular models to help you visualize the various conformations in three dimensions.

Cyclohexane

We can draw two double Newman projections of cyclohexane by viewing simultaneously down the C-2–C-3 and C-6–C-5 bonds. When the bonds to C-1 and C-4 are staggered, we obtain the Newman projections shown in

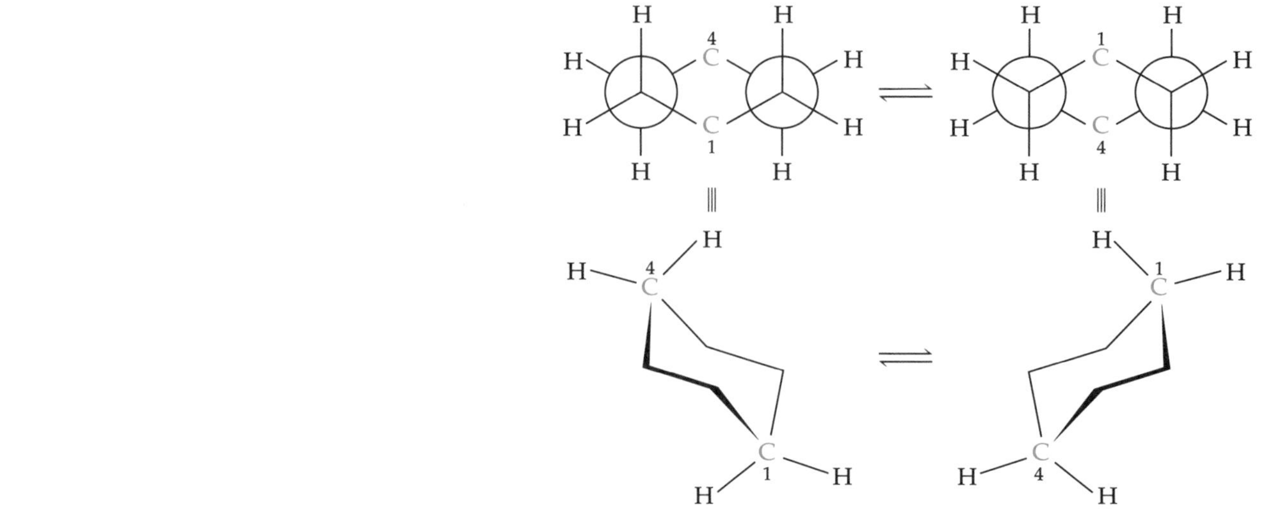

▲ FIGURE 5-12

Cyclohexane in a chair conformation visualized in three dimensions as a Newman projection and a line drawing. The equivalent structures at the left are converted into the equivalent structures at the right by ring flipping.

Figure 5-12. By viewing cyclohexane from the side and omitting some of the hydrogen atoms, we obtain the carbon skeletons shown in Figure 5-12. Because these drawings look roughly like the back, seat, and footrest of a chair, the staggered conformations of cyclohexane are called **chairs.** Alternatively, an eclipsing interaction between C-1 and C-4, as shown in Figure 5-13, results in a much less stable isomer whose carbon skeleton roughly resembles a **boat;** therefore, this conformation is referred to as boat cyclohexane.

With your molecular models, try to flip from one chair conformation to another and from one boat form to another. As you work with your model, it will soon become clear that it is possible to convert one chair into another by proceeding through an intermediate boat conformation. (Convince yourself, using your model, that an *anti* relation between C-1 and C-4 is not possible.) You will also soon find another, less energetically demanding, twisted conformation of the boat, which is difficult to draw (because the two C–C bonds are no longer parallel) but easy to see when manipulating a model. Chair conformations interconvert very readily, proceeding through this **twist-boat conformation.**

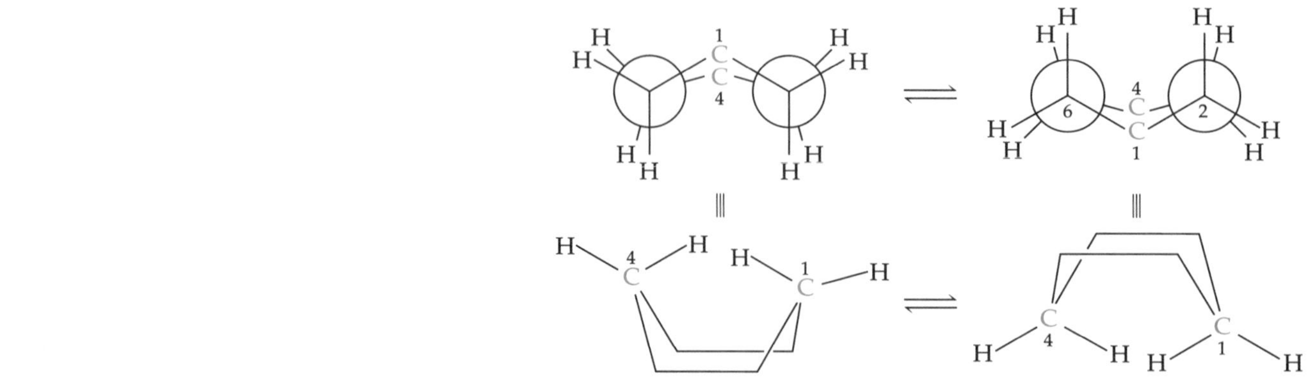

▲ FIGURE 5-13

Newman projections of two eclipsed conformations of cyclohexane visualized down C-2–C-3 and C-6–C-5, with the corresponding boat conformations.

The energies of the boat and the twist-boat conformations are significantly lower than that of a **half-chair conformation,** which represents the transition state that you obtain as you convert a chair into a boat. The energy difference between the chair and the boat conformations of cyclohexane is about 7 kcal/mole; this value represents the energy difference between the staggered and eclipsed conformations at each of the six bonds around the ring. The barrier to interconversion of these conformations is approximately 11 kcal/mole, the difference in energy between the chair and half-chair conformations. This amount of thermal energy is available at room temperature, and chair-to-chair ring flipping readily takes place.

Monosubstituted Cyclohexanes

Let us now consider the various conformations of a cyclohexane ring with a methyl group attached to C-1. There are two unique chair conformations of methylcyclohexane, and they do not have the same energy. In the conformation at the upper left in Figure 5-14, the methyl group points away from the "seat" of the chair; whereas, in the upper right-hand conformation, the methyl group is roughly coplanar with the seat. These two positions are **axial** and **equatorial,** respectively, and the two chair conformations can be interconverted simply by flipping the atoms of the ring back and forth. (Use a model to convince yourself that a ring flip converts each axial substituent into an equatorial position and changes each equatorial substituent into an axial one.)

By looking down the C-1–C-6 bond of methylcyclohexane in the Newman projection shown at the lower left in Figure 5-14, we see that, when the methyl group is in the axial position, there is a *gauche* butanelike interaction with a hydrogen on C-5, as well as with one on C-3 (not shown but can be envisioned in a Newman projection visualized along the C-1–C-2 bond) of

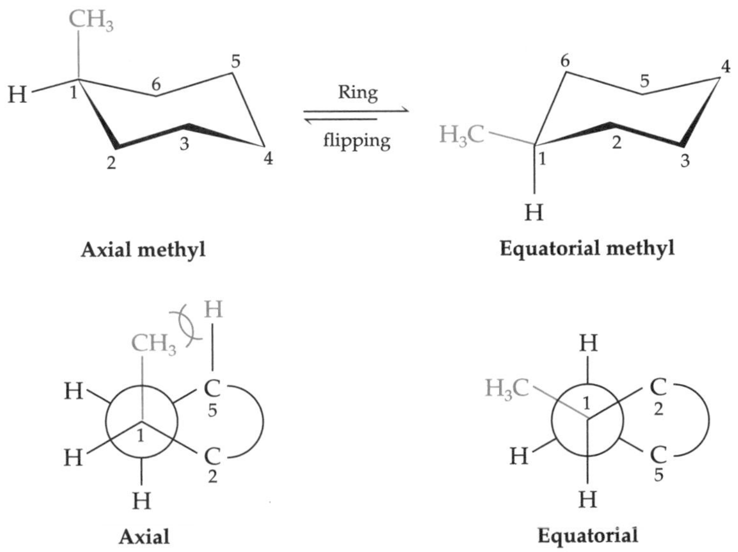

▲ FIGURE 5-14

Three-dimensional representations of the ring-flipped chairs of methylcyclohexane. The Newman projections in the lower half of the figure are obtained by visualizing down the C-1–C-6 bond, with C-1 shown in the front and C-6 hidden from view by C-1. The overlapping arcs between the methyl group and the hydrogen at C-5 represent repulsive interaction between these groups.

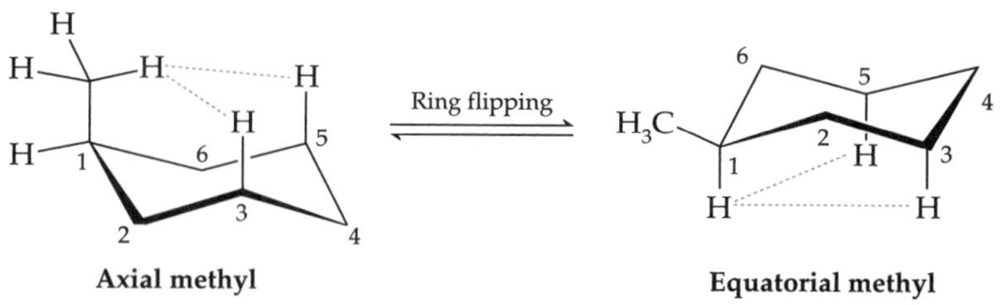

▲ FIGURE 5-15

Diaxial interactions in chair methylcyclohexane.

the ring. Conversely, when the methyl group occupies the equatorial position, its relative orientation to C-5 (and C-3) is *anti*. The axial isomer is therefore destabilized by about twice the energy of a *gauche* butane interaction because of the two steric interactions, which are referred to as **1,3-diaxial interactions** (Figure 5-15). Indeed, the equatorial isomer is more stable than the axial isomer by about 1.8 kcal/mole, an experimental value that is very close to that which we would predict for two *gauche* butanelike interactions (2×0.9 kcal/mole). (Again, the use of molecular models will be helpful here. Be sure to notice that a 1,3-diaxial interaction is the same interaction that destabilizes *gauche* butane.)

Neither chair conformation of methylcyclohexane has the strong destabilizing eclipsing interactions of the boat and twist-boat conformations. In the boat isomer shown at the left in Figure 5-16, there is costly steric interaction between the C-1 methyl group and one of the hydrogens on C-4. Even in the alternative boat conformation, there is still steric interaction between C-1 and C-4.

Chair cyclohexanes in which substituents occupy equatorial positions are therefore more stable than those with axial substituents. In general, either of these chair conformations is more stable than the other conformational possibilities; that is, the boat and twist-boat conformations.

The energy difference between a conformation having a substituent in an axial position and one having the substituent in an equatorial position depends on the steric requirement, or size, of the substituent. The larger the substituent, the greater is the steric strain resulting from 1,3-diaxial interactions. Table 5-4 lists values that represent the contribution of each 1,3-diaxial interaction to the relative destabilization of the axial conformation. (Recall from Table 5-2 how these energy differences affect the conformational equilibrium.) Some substituents are so large that they effectively act as **conformational anchors,** or locks. For example, the energy difference between an axial and equatorial conformation for a *t*-butyl group is very large. Although ring flipping between the chair conformations is still rapid, the equilibrium is so strongly dominated by the conformer bearing the large group in the equatorial position that we can consider the conformation "locked," keeping all other substituents in axial or equatorial positions as appropriate.

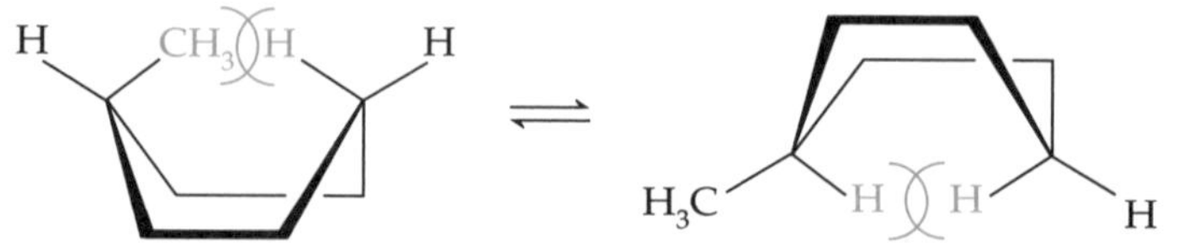

▲ FIGURE 5-16

Flagpole interactions in the boat conformations of methylcyclohexane.

Table 5-4 ▸ Energy Cost of 1,3-Diaxial Interactions

1,3-Diaxial Interaction	Energy (kcal/mole)
H and CH_3	0.9
H and CH_2CH_3	1.0
H and isopropyl	1.1
H and phenyl	1.5
H and *t*-butyl	2.7

Multiply Substituted Cyclohexanes

We can apply the ideas developed for the conformations of monosubstituted cyclohexanes to multiply substituted cyclohexanes. In *cis*-1,4-dimethylcyclohexane, one of the methyl groups is in an equatorial position and the other is axial. Chair–chair ring flipping changes the position of each of these substituents, converting each axial position into an equatorial one and moving each equatorial substituent to an axial position. This ring flipping results in a conformation that is identical in all respects with the first (Figure 5-17A, page 172).

The situation is quite different for *trans*- 1,4-dimethylcyclohexane (Figure 5-17B) because both of the methyl groups must be either equatorial or axial. Thus, we would expect the energy difference between these two chair conformations to be large. The conformation with two equatorial methyl groups is clearly the more stable. In *cis*- and *trans*-1,3-dimethylcyclohexanes, the *trans* isomer has one axial and one equatorial substituent in both ring-flipped chair conformations (Figure 5-18A). Conversely, the *cis* isomer has a conformation in which both methyl groups are equatorial. As a result the diequatorial conformer is more stable than the alternative, in which both methyl groups are axial (Figure 5-18B). The pattern of axial and equatorial substituents is completely inverted upon a ring flip, as illustrated in this multisubstituted cyclohexane:

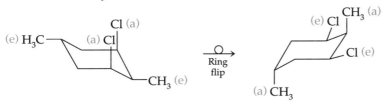

For each of the following isomeric pairs, indicate which isomer is the conformationally more stable form and draw its preferred conformation as a three-dimensional representation:

(a) *cis*-1-*t*-butyl-2-methylcyclohexane or *trans*-1-*t*-butyl-2-methylcyclohexane

(b) *cis*-1,4-diisopropylcyclohexane or *trans*-1,4-diisopropylcyclohexane

(c) *cis*-1,3-dibromocyclohexane or *trans*-1,3-dibromocyclohexane

(d) *cis*-1-*t*-butyl-3-ethylcyclohexane or *trans*-1-*t*-butyl-3-ethylcyclohexane

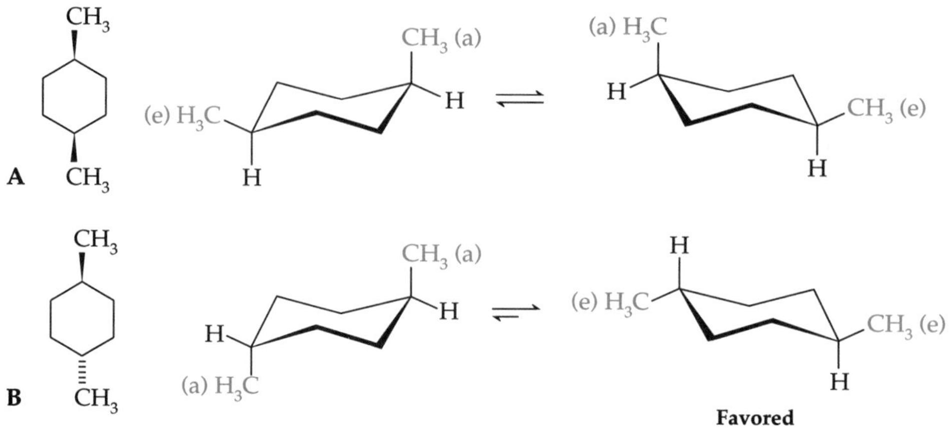

▲ FIGURE 5-17

Chair conformations of (A) *cis*- and (B) *trans*-1,4-dimethylcyclohexane.

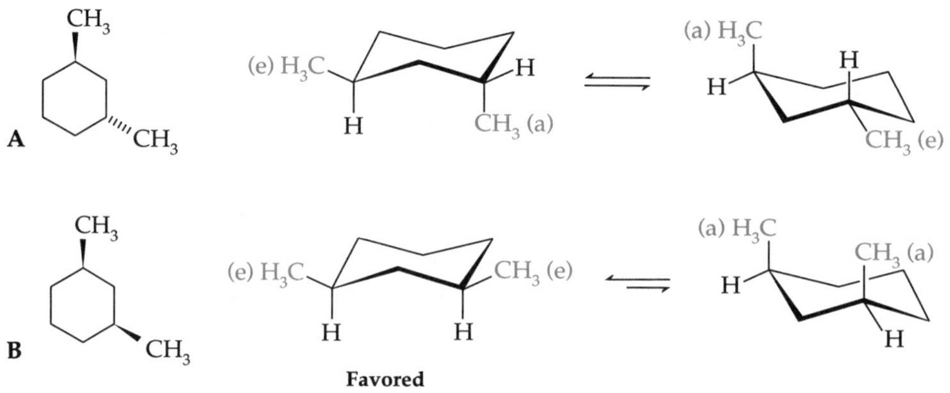

▲ FIGURE 5-18

Chair conformations of (A) *trans*- and (B) *cis*-1,3-dimethylcyclohexane.

Fused Six-membered Rings

The conformations of *cis*- and *trans*-1,2-dimethylcyclohexanes are useful in the conformational analysis of fused, saturated rings. Thus, we can imagine *trans*-decalin, a hydrocarbon in which two cyclohexane rings have two carbon atoms in common, as deriving from the most-stable conformation of *trans*-1,2-dimethylcyclohexane by mentally extending the two methyl groups, with two additional carbons, into another ring.

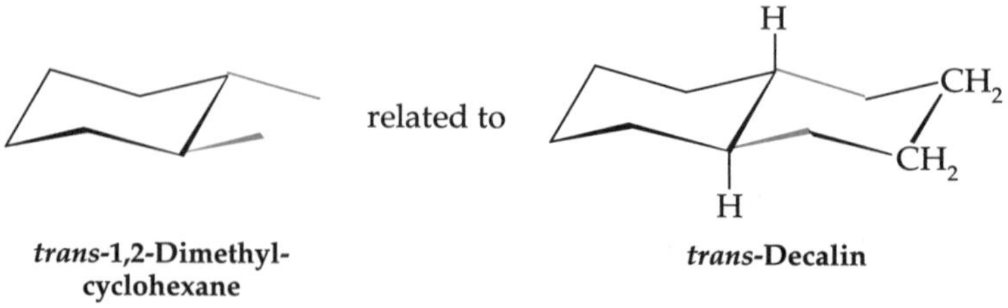

trans-1,2-Dimethyl-
cyclohexane

related to

trans-Decalin

The *trans* ring fusion is clearly indicated by the relative positions of the hydrogens at the **bridgehead positions;** that is, at the carbon atoms that are common to both rings. Unlike dimethylcyclohexane, however, *trans*-decalin cannot flip to another, stable chair-chair form. Although the two carbons added to form the second ring of *trans*-decalin can reach the two equatori-

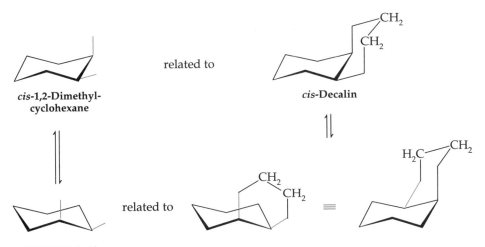

▲ FIGURE 5-19

A comparison of the conformation of *cis*-decalin with the chair form of
cis-1,2-dimethylcyclohexane.

ally oriented carbons without strain, they cannot form sufficiently long
links with two adjacent, axial methyl groups. (Use models to convince
yourself that this is the case.) As a result, ring flipping is blocked in *trans*-
decalin. The additional two carbons needed to close the ring, however, provide a chain long enough to reach between the axial and equatorial methyl
groups of *cis*-1,2-dimethylcyclohexane, forming *cis*-decalin (Figure 5-19). In
this case, one of the methyl groups is axial and the other is equatorial; and
chair–chair ring flipping can, and indeed does, take place quite readily. This
ring flip can be visualized as shown at the right-hand side of Figure 5-19.
(Again, use models to convince yourself that the lower right-hand structure
is another representation of the upper right-hand, ring-flipped structure.)

EXERCISE 5-F

Draw in three dimensions *cis*-decalin and its isomers bearing methyl groups
at C-2 and C-3 (using the skeletal numbering scheme shown in the margin).
For each of the isomers that you have drawn, let both rings undergo a ring
flip and draw the ring-flipped isomer. In each case, decide which of the
ring-flipped structures is the conformationally more stable alternative.

cis-Decalin

5-5 ▸ Chirality

A final type of stereoisomerism is that found in otherwise identical molecules that differ only in the way in which atoms are arranged in three dimensions but that cannot be interconverted by bond rotation. This type of
stereoisomerism deals with **chirality,** or molecular handedness.

Chiral molecules are those that are not superimposable on their mirror
images and are very often encountered in nature. One way of recognizing
chirality in objects is to look for handedness. Your left and right hands are
clearly different even though each of the component parts (for example, the
thumbs) appear to be the same. A hand is chiral because one of its ends (the
fingers) is different from the other end (the wrist); its thumb is different
from its little finger; and its back is different from its palm. Thus, although
your left hand is very similar in some ways to your right hand, they can not

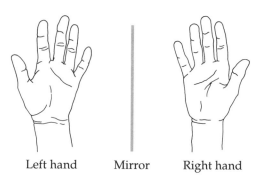

▶ FIGURE 5-20
Left and right hands: representative chiral objects that are
mirror images of each other.

Left hand   Mirror   Right hand

be superimposed (Figure 5-20). Furthermore, right- and left-handed gloves
also are chiral, and thus different: a right-handed glove fits your right hand,
not your left. The difference between your hands is maintained no matter
how they are oriented—for example, with the thumbs up or down. Although your two hands are not the same, they are related: they are **mirror
images** of each other. Only if a glove is completely flat (one that has a mirror plane), like those used by children, can it be used on either hand: even
then, the face of the glove that is the palm when worn on one hand is on the
back of the hand when worn on the other.

The presence of a mirror plane through an object assures that it will be
superimposable on its mirror image. A **mirror plane** is a plane through
which each part of an object on one side of the plane is mirrored by an identical one on the opposite side. If a human hand were completely flat and the
back identical with the palm, the hand would have a mirror plane through
the palm and fingers. With hands like this, the same glove would fit both
the right and left hands identically. By analogy, then, if there is no mirror
plane through any conformation of a molecule, the molecule can be considered to be "handed"—that is, **chiral.**

Chirality is a property of molecules (or, more generally, of objects).
Molecules (objects) are either chiral or not. There is no in between. A molecule that has conformational freedom is considered to be **achiral** (not chiral)
as long as at least one of its conformations has a mirror plane of symmetry,
even if the others lack such a plane.

If an atom has substituents oriented in three-dimensions such that
there is no mirror plane through this atom, then (except in very special
cases) it will not be possible to find a mirror plane for the molecule as a
whole. Because of its **center of chirality,** then, the molecule is chiral. Such
atoms are sometimes referred to as **chiral atoms** or **chiral centers,** but chirality is a property only of a complete object, not of its parts. Thus, we use
the expression *center of chirality* to emphasize that an atom contributes to the
overall handedness of the molecule.

EXERCISE 5-G

Locate one or more mirror planes in the following molecules:

(a) ethylene   (b) benzene   (c) *s-cis*-butadiene   (d) propyne

Any atom that is *sp*- or *sp²*-hybridized has a mirror plane (that containing the bound atoms). Therefore, the carbons of an alkene or alkyne are not
centers of chirality, irrespective of substituents. On the other hand, an *sp³*-
hybridized carbon does have the three-dimensionality necessary to impart
chirality to a molecule.

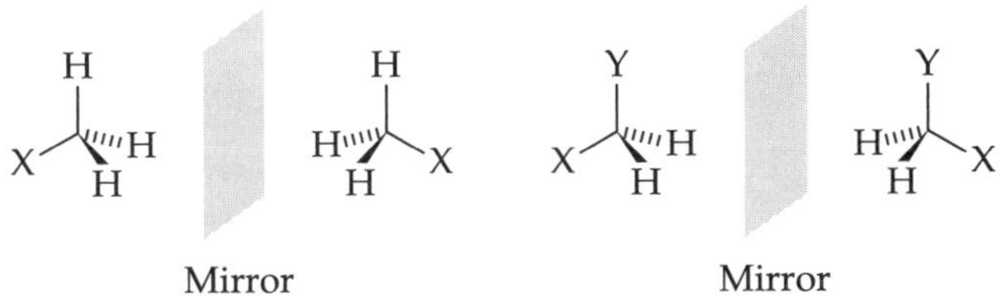

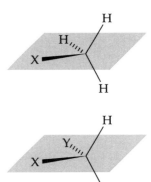

▲ FIGURE 5-21
Mirror images of mono- and disubstituted methanes.

Let us consider mono- and disubstituted carbon atoms: examples of such compounds are shown in three dimensions in Figure 5-21 as pairs of mirror images. In both cases, the molecule at the right of the mirror can be superimposed on that of the left by rotation such that we lay the X groups on top of each other and superimpose the carbons as well. We say that two molecules are superimposable if a conformation exists so that each of four substituents can be placed upon each other and are thus oriented in exactly the same direction in space. (Use molecular models to confirm that this is so.) It is because of the presence of a mirror plane (defined by the H–C–X plane in the monosubstituted carbon and by X–C–Y in the disubstituted one) that this superimposability is possible. With respect to these planes, the hydrogen coming toward the observer mirrors the hydrogen going back.

Any sp^3-hybridized carbon atom that bears two identical substituents has a mirror plane through which one of these substituents is a reflection of the other. That plane contains the tetrahedral carbon and the two other substituents, X and Y. Thus, for an sp^3-hybridized carbon atom to be a center of chirality, it is necessary and sufficient that four *different* groups be bound to carbon. A generalized example is shown in Figure 5-22 with four substituents, A, B, C, and D, arranged at a tetrahedral center of chirality. The two representations show two unique ways that these four groups can be oriented in three dimensions. They are mirror images of each other and cannot be superimposed. For example, if we place the two A's and B's on top of each other, the C and D groups are at the wrong positions. Conversely, if we orient the molecules so as to superimpose the C and D groups, then the substituents A and B are not at the same places. These two molecules are therefore stereoisomers, differing only in the way in which these four substituent groups are oriented in space. Stereoisomers that are related to each other as nonsuperimposable mirror images are referred to as **enantiomers.**

Enantiomers can be interconverted only by switching the positions of two substituents, a process that requires the breaking and reforming of σ bonds at the center of chirality. Specifically, enantiomers are *not* interconverted by rotations about σ bonds and therefore differ significantly from conformational isomers. Thus, in analyzing a molecule for the presence of a center of chirality, we are free to use any conformational representation to compare structures (for example, eclipsed or staggered) without worrying about configurational interconversion at the center of chirality.

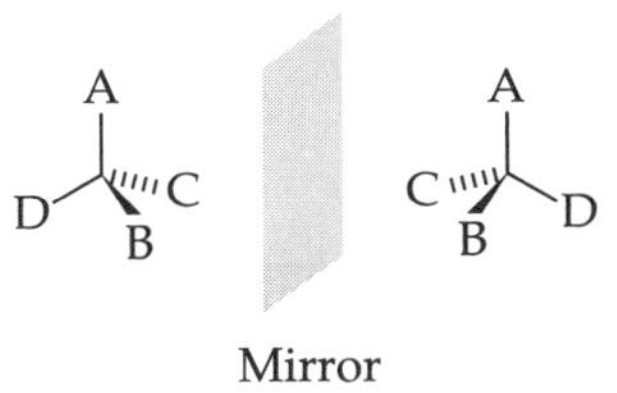

▲ FIGURE 5-22
Mirror images of tetrahedral carbon bearing four different substituents.

STEREOCHEMISTRY IN ODOR RECOGNITION

Human olfactory receptor sites are very sensitive to the shape of gaseous molecules. For example, a floral odor is caused by molecules that are spherical at one end and elongated at the other, somewhat like a miniature guitar, whereas a peppermintlike odor is produced by molecules that are ellipsoidal in shape. For many components of perfume, shape seems to be more important than chemical composition: (+)-camphor, hexachloroethane, and cyclooctane have nearly the same odor, even though they have quite different molecular and structural formulas. However, all of these compounds are roughly bowl-shaped, which allows a reasonable fit with the receptor site for what perfumers call "camphoraceous" molecules.

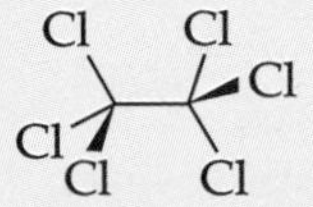

Hexachloroethane 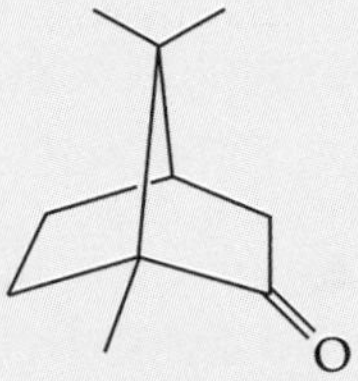   Camphor 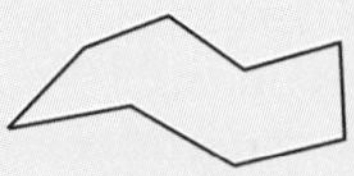   Cyclooctane

The ability to detect the presence of small quantities of molecules at various receptor sites varies with each person: those highly skilled for such a task are well respected (and highly paid) by the perfumers in the south of France and by wineries throughout the world.

Because stereochemistry significantly affects the shape of a molecule, a molecule's absolute configuration also strongly affects its odor. Enantiomers, for example, can elicit quite different responses: the characteristic aromas of oil of spearmint and oil of caraway are due to the separate enantiomers of carvone.

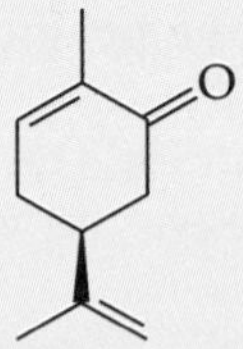

(+)-Carvone   (−)-Carvone
(from oil of caraway)   (from oil of spearmint)

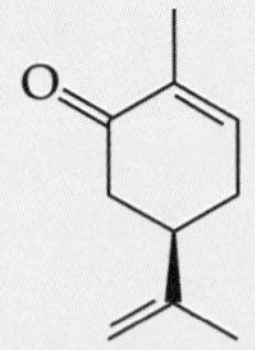

EXERCISE 5-H

In each of the following molecules, indicate the presence of a center of chirality with an asterisk.

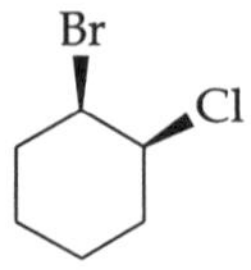

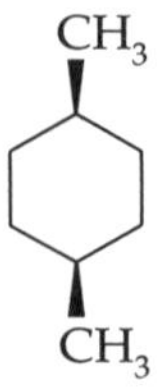

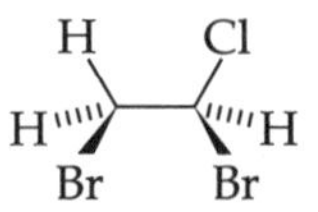

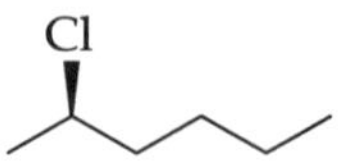

5-6 ▸ Absolute Configuration

Because there are two different stereoisomers for a molecule that has a center of chirality, we need to be able to refer uniquely to one or the other of the enantiomeric pair in exactly the same way that we need to specify, for example, a left or right shoe. An unambiguous method for specifying **absolute stereochemistry** was developed by three chemists and is known as the Cahn-Ingold-Prelog rules. (Vladimir Prelog was awarded the Nobel Prize in 1975 for his contributions to our understanding of organic stereochemistry.) In contrast, **relative stereochemistry** refers only to the relation between two molecules; for example, it may be said of two enantiomers that they are mirror images of each other, but which one is which is not specified.

The specification of absolute stereochemistry makes use of the same priority rules employed to describe *E* and *Z* isomerism in Chapter 2. In accord with these rules, priority is assigned to each of the four groups attached to the center of chirality, which are then uniquely defined as 1, 2, 3, and 4. Let us assume that the substituents A, B, C, and D have priorities that decrease in that order. We then view the molecule by looking down the bond between carbon and D, putting this lowest-priority substituent as far as possible from the observer. For the left-hand isomer in Figure 5-22, this means that we must be positioned in the plane of the page and to the upper right of the molecule, visualizing (as indicated by the arrow in Figure 5-23) along the bond from the center of chirality to substituent D. In viewing the molecule in this way, we see a picture similar to a Newman projection in which the D substituent is hidden, as at the right in Figure 5-23. (Make a model to convince yourself that this is so.) The assignment of priorities from highest to lowest to the remaining substituents then proceeds in either a clockwise or a counterclockwise direction. Because A has the highest and B the next-highest priority here, the direction is counterclockwise, designated **S** (from the Latin *sinister*, left), and the structure shown in Figure 5-23 is therefore referred to as an *S* isomer and as having the *S* configuration at its center of chirality.

To assign configuration to the right-hand structure in Figure 5-22, we must again position ourselves so as to visualize down the bond from the center of chirality to substituent D. This would produce the projection shown in Figure 5-24. Here, the direction from A to B to C is clockwise, and the isomer is assigned the stereochemical designation *R* (from the Latin *rectus*, right).

Let us consider a specific example and assign the absolute configuration to the isomer of 2-bromobutane shown in the margin. Our first task is

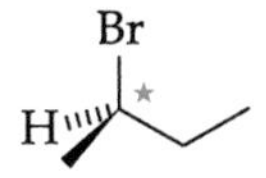

(R)-2-Bromobutane

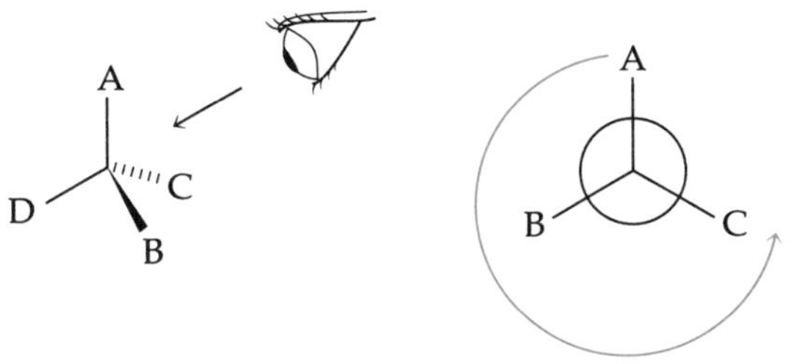

▲ FIGURE 5-23

Orienting a center of chirality to assign absolute configuration (*S* if the priorities are A > B > C > D).

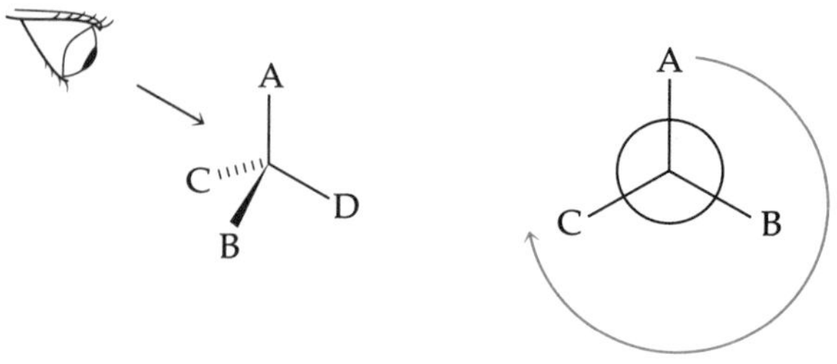

▲ FIGURE 5-24
Orienting a center of chirality to assign absolute configuration (*R* if the priorities are A > B > C > D).

to identify the center of chirality in the molecule. Carbon-1 bears three hydrogens and cannot be a center of chirality. Carbon-2, which bears four different substituents, is a center of chirality, as indicated by the asterisk. Carbon-3 bears two hydrogens and is therefore not a center of chirality; nor is carbon-4, which bears three hydrogens. Using the atomic-number rule to assign priority, we find that the substituents at carbon-2 have the priority $Br > CH_3CH_2 > CH_3 > H$ (recall the rules discussed in Chapter 2). Because hydrogen is of lowest priority, we must visualize the molecule by looking down the C–H bond; that is, from C-2 toward hydrogen. From this orientation, the direction from bromine to ethyl to methyl is clockwise; so the isomer shown is the *R* enantiomer.

Now that we have a three-dimensional representation of (*R*)-2-bromobutane, it is easy to draw the *S* isomer. We simply exchange any two substituents of the *R* configuration, as shown in Figure 5-25 in which there are six views of (*S*)-2-bromobutane. Because the interconversion of enantiomers requires the breaking of σ bonds, it would be quite difficult in most cases to accomplish this switching in practice.

It is important for you to be able to assign absolute configuration to a given center of asymmetry, as this will aid in your understanding of the relation between structures drawn in various ways.

EXERCISE 5-I

Assign absolute stereochemistry to each center of chirality in the following molecules by applying the Cahn-Ingold-Prelog rules.

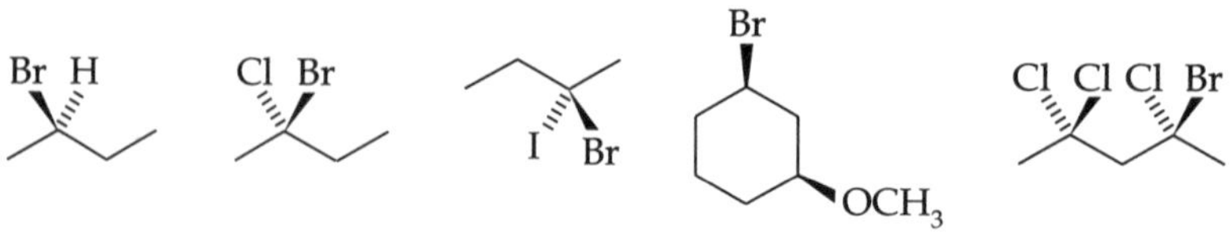

(a)   (b)   (c)   (d)   (e)

EXERCISE 5-J

Draw three-dimensional representations of each of the following stereoisomers.

(a) (*R*)-2-bromopentane

(b) (*S*)-3-bromo-3-chlorohexane

(c) (*R*)-2-fluoro-2-chlorobutane

(d) (*R*)-1-bromo-(*S*)-2-fluorocyclohexane

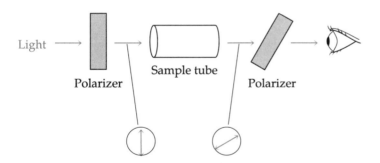

(S)-2-Bromobutane

◄ FIGURE 5-25
Six three-dimensional
representations of the center
of chirality in 2-bromobutane.

5-7 ▸ Polarimetry

Because all the chemical bonds in a chiral molecule are also present in its enantiomer, one might expect two enantiomers to have identical physical properties. This is generally true except when chiral molecules interact with other chiral objects. The circularly polarized components of plane-polarized light are chiral, and the plane of polarization is rotated to the right by one enantiomer and to the left by the other. Figure 5-26 schematically outlines the operation of a **polarimeter** capable of measuring this optical effect.

Ordinary light is an electromagnetic wave that oscillates in all directions perpendicular to the path of propagation. However, as a light beam passes through a set of polarizers, the photons that are not directionally aligned with the polarizer are absorbed. The light beam that emerges from the polarizer is composed of photons that have all electric and magnetic oscillations in the same plane. As this plane-polarized light passes through a chiral medium, the asymmetric nature of the chiral molecule causes the plane to rotate from its original position. A polarizer placed behind the

Light →

Polarizer

Sample tube

Polarizer

▲ FIGURE 5-26
A schematic representation of a polarimeter. Light is composed of rays that have electric fields oscillating from a point source in all directions. However, after passing through a polarizer, it is constrained to a single plane. The direction of alignment of the polarizer inserted between the light source and the sample tube (indicated by a double-headed arrow in a circle) defines the incident plane of light. This plane is rotated during electronic interaction with chiral molecules present in the sample, and the plane of polarized light emerging from the sample tube has rotated from its original plane of polarization. A second polarizer (again indicated by a double-headed arrow in a circle) placed between the sample-tube exit and the observer must be rotated by some angle from alignment with the first polarizer in order to compensate for the rotation induced by the optically active sample. The amount of rotation of this plane is determined by the specific rotation of the chiral sample.

sample must be rotated by some angle to compensate for the rotation of the plane.

Only objects that are chiral can rotate a plane of polarized light, and thus chiral molecules are often referred to as being **optically active.** The extent of rotation observed in the experiment depends on the magnitude and asymmetry of the sample's electric field (which is a sensitive function of the particular molecule being measured), the wavelength of the light, and the number of optically active molecules in the sample. The extent to which a given molecule rotates a plane of polarized light (on a molar basis) is referred to as its **specific rotation,** $[\alpha]$. Often the yellow light emitted by a sodium lamp (the D-line) is used as the polarized light. The specific rotation induced by this light is called $[\alpha]_D$. The observed rotation is the product of the specific rotation and those factors relating to the number of molecules present—that is, to the concentration in the sample compartment and the path length of the sample cell.

$$\text{Observed rotation} = \alpha = [\alpha] \times c \times l$$

$[\alpha]$ = specific rotation (at the wavelength of the polarized light)

c = concentration (in g/ml)

l = path length (in dm)

Enantiomers differ only with respect to the *sign* of the specific rotation; that is, the direction of rotation of a plane of polarized light. A pure sample of one enantiomer of a pair rotates a plane of polarized light to a degree exactly equal to that of the other member but with the opposite sense. A sample that is a 50:50 mixture of enantiomers thus does *not* show optical activity because the two compounds present have effects of equal magnitude but of opposite direction.

A sample that contains only a single enantiomer is optically pure. **Optical purity** defines the excess of one enantiomer over the other: a 9:1 mixture of a compound and its enantiomer is considered to be $(0.9 - 0.1) \times 100\%$, or 80% optically pure. Thus, it is possible to gauge optical purity by

---

SEPARATION OF ENANTIOMERS IN THE LABORATORY

The first resolution (separation of enantiomers constituting a racemic mixture) to be conducted in a laboratory was the work of Louis Pasteur, whose contributions to microbiology (fermentation, pasteurization of milk, sterilization of surgical instruments, development of a rabies vaccine, and many others) are as well known as those to chemistry. Pasteur decided to investigate the crystals that form in wine barrels during fermentation and aging (and on the corks of wine bottles). These crystals are called racemic acid from the Latin word for grapes, *racemus*. All the crystals were found to have the same chemical composition, as a mixed sodium ammonium salt of tartaric acid ($HO_2CCH(OH)CH(OH)CO_2H$); but, when viewed carefully under a microscope, they appeared to constitute two different sets, differing in the three-dimensional shape of the crystals. Indeed, these shapes were mirror images of each other. With tweezers, Pasteur painstakingly separated the two different types and showed that they had exactly the same physical and chemical properties, except that their solutions rotated a plane of polarized light in opposite directions.

comparing the observed degree of rotation of a sample with the rotation obtained for a sample containing only one enantiomer. For an exactly equimolar mixture of two enantiomers, the rotations induced by each enantiomer exactly cancel and the resulting mixture is **optically inactive** (0.5 − 0.5 × 100% = 0% optical purity). Such a 50:50 mixture of enantiomers is referred to as a **racemic mixture,** as a **racemic modification,** or simply as a **racemate.** By definition, a racemic mixture (a 1:1 enantiomeric mixture) is optically inactive and its rotation is always 0°.

EXERCISE 5-K

Calculate the optical purity of an enantiomeric mixture

(a) in which specific rotation of one enantiomer is +100° and the observed rotation of the mixture is +10°.

(b) in which the specific rotation is +200° and the observed rotation is +50°.

5-8 ▸ Designating Configuration

Enantiomers are a stereoisomeric pair of molecules related as nonsuperimposable mirror images. The members of this pair are therefore separate compounds whose disposition in three-dimensional space must be individually defined. In this section, we will learn how to specify the absolute configuration of centers of chirality in molecules containing one or more such centers.

A Single Center of Chirality

As we have seen, the absolute configuration of a chiral molecule can be specified by applying the IUPAC rules to designate the configuration as *R* or *S*. This method specifies absolutely the direction of groups in space without relation to physical properties. An alternate way of referring uniquely to one member of the enantiomeric pair is to use the sign of its specific rotation. Thus, the enantiomer that rotates a plane of polarized light in a clockwise direction (when the observer looks at the light) is called the **(+) isomer,** whereas its mirror image, which rotates the plane of polarized light in

---

DETERMINATION OF ABSOLUTE CONFIGURATION IN CHIRAL NATURAL PRODUCTS

Only in the 1930s did it become possible to determine the actual arrangement of atoms in three-dimensional space about a center of chirality by x-ray crystal analysis. Long before, however, chemists were drawing three-dimensional representations of molecules based upon whether they were related to D- or L-glyceraldehyde. A molecule was said to belong to a series if it could be converted into another compound already in that series by reactions that were not expected to change stereochemistry. The original assignment of the arrangement for the D and L series was made arbitrarily but ultimately was shown to be correct (a 50:50 chance).

a counterclockwise direction, is referred to as the **(−) isomer.** This (+)-(−) designation does not specify how the groups are arranged spatially but simply relates one structure to the sign of its specific rotation. It is important to note that no simple relation exists between the sign of the optical rotation (±) and the absolute configuration (*R,S*) of an enantiomer.

An equivalent representation is to use lowercase *d* (dextrorotatory, from the Greek for "right rotating") to indicate the (+) enantiomer and lowercase *l* (levorotatory, from the Greek for "left rotating") to indicate the (−) enantiomer. A third representation for relative configuration at centers of chirality (D and L, based on correspondence with naturally occurring glyceraldehyde, which is dextrorotatory) will be introduced in Chapter 17, where carbohydrates and other bioorganic molecules are described.

Multiple Centers of Chirality

A molecule can have more than one center of chirality. Let us consider the structure of 2-bromo-3-chlorobutane (Figure 5-27). Note that this molecule contains two centers of chirality: one at C-2 and one at C-3. We can draw four different three-dimensional representations of this molecule in which the carbon–halogen bonds are held in the plane of the paper. In accord with the IUPAC rules of nomenclature, these drawings represent the 2*R*,3*S*, 2*R*,3*R*, 2*S*,3*S*, and 2*S*,3*R* stereoisomers. With every additional asymmetric center in a molecule, the number of possible stereoisomers doubles. Thus, the maximum possible number of stereoisomers for *n* centers of chirality is equal to 2 raised to the *n*th power, or 2^n.

Note that, in Figure 5-27, the 2*R*,3*R* and 2*S*,3*S* isomers are an enantiomeric pair and that the 2*R*,3*S* and 2*S*,3*R* isomers are another enantiomeric pair, but the 2*R*,3*R* is not an enantiomer of the 2*R*,3*S* isomer, nor are the 2*S*,3*S* and the 2*S*,3*R* isomers enantiomers of each other. Stereoisomers that are not mirror images are referred to as **diastereomers.** Thus, in Figure 5-27, the relation between the 2*R*,3*R* and the 2*R*,3*S* isomers is diastereomeric, as is the relation between the 2*S*,3*S* and 2*S*,3*R* isomers. Although enantiomers have identical physical properties (melting points, boiling points, and so forth) except for the sign of their specific rotation, diastereomers have different physical properties.

It is tedious and very difficult to separate enantiomers that have identical melting and boiling points and solubilities by the use of recrystallization or chromatographic columns that are not themselves chiral. However, because diastereomers do not have identical physical properties, they can be separated by various physical methods, including chromatography and recrystallization. We can make use of the physical differences between diastereomers to effect the separation of enantiomers by first converting the mixture into diastereomers by reaction with a single, optically active enantiomer as a reagent. For example, the acid-base reaction of a racemic mixture of 2-chloropropanoic acid with one enantiomer of α-phenethylamine produces a mixture of diastereomeric salts, as shown in Figure 5-28. In this acid-base reaction, the configuration does not change at the center of chiral-

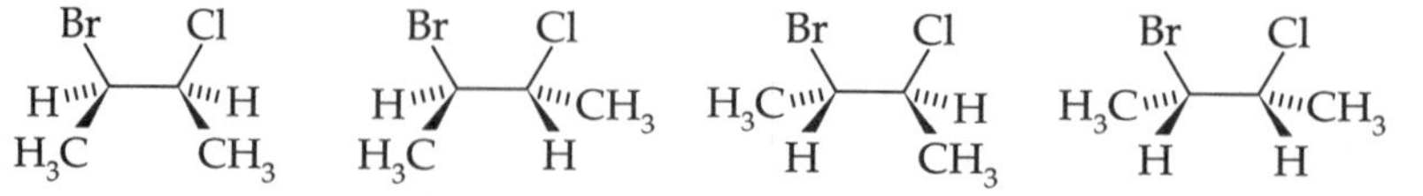

▲ **FIGURE 5-27**
Possible configurations for 2-bromo-3-chlorobutane.

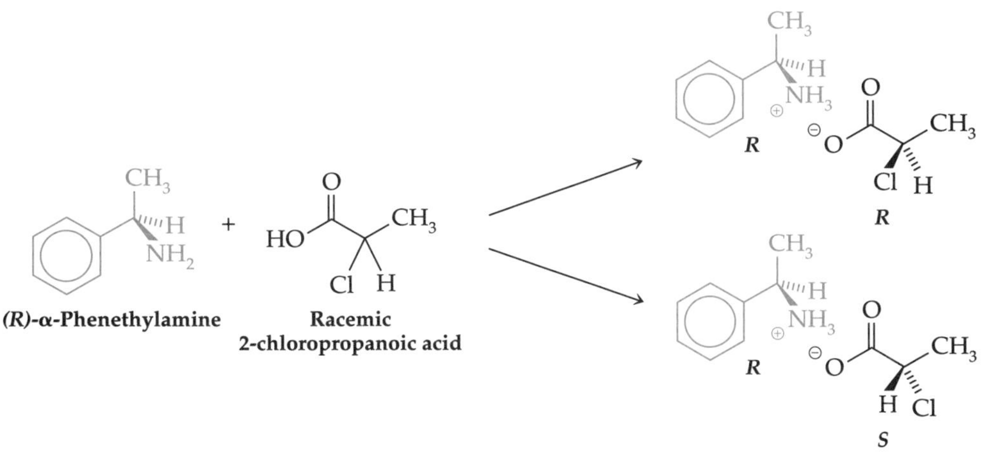

▲ FIGURE 5-28
Formation of a diasteromeric salt by interaction of a chiral base with a racemic acid.

ity in either the acid or the amine because bonds are neither broken nor made at these carbon atoms. Thus, the salts formed from the racemic starting material retain the original configuration at each center of chirality.

The two salts (*R,R* and *R,S*) are not mirror images and are therefore diastereomers. At this stage, they can be separated because diastereomers have different physical properties. Here, repeated recrystallization of the salts can yield a single diastereomer as a pure, crystalline solid. The separated diastereomers can then be reconverted into their components, the carboxylic acid and the amine. (How could you accomplish this process in the laboratory?)

This method of **resolution** (namely, any technique for separating enantiomers) by forming and then separating diastereomers, followed by regeneration of the original reactant, makes use of a fundamental difference between stereoisomeric pairs that are diastereomers and those that are

THALIDOMIDE: DISASTROUS BIOLOGICAL ACTIVITY
OF THE "WRONG" ENANTIOMER

A dramatic and unfortunate consequence of absolute stereochemistry is found in the story of thalidomide, a drug produced as an antidepressant. Because of the keen insight of Frances Kelsey, a researcher at the Federal Drug Administration, thalidomide was never approved for use in the United States. However, this prescription drug was already in use in the 1950s in Canada and in European countries and, despite strong warnings against prescribing thalidomide for pregnant women or even women likely to become pregnant, it was being used to treat "morning sickness." Unfortunately, thalidomide was marketed as a racemate. As the story unfolded, it became clear that the antidepressant activity was due to one enantiomer; the other was both a mutagen and an antiabortive. The net result of its use was the birth of many very seriously deformed children, often having vestigial arms and legs. Curiously, the observation that Kelsey had used to hold back approval of thalidomide was that it caused abortions at high doses in rats. Clearly, human beings differ from rats in more ways than just size.

enantiomers. Diastereomers can be separated because they have different physical properties; enantiomeric pairs do not (except for the sign of optical rotation).

In an alternate method for resolving a racemic mixture into individual enantiomers, diastereomers are formed by adsorption on a chiral chromatography column. One of the enantiomers interacts more strongly with the chiral support than the other; the less strongly adsorbed enantiomer elutes from the column first.

Meso Compounds

The rule that there are 2^n stereoisomers for a compound with n centers of chirality does not hold when two (or more) identically constituted centers are present. For example, let us consider 2,3-dibromobutane (Figure 5-29). We can draw three-dimensional representations of 2,3-dibromobutane that are analogous to those shown in Figure 5-27 for 2-bromo-3-chlorobutane. The 2*R*,3*R* and the 2*S*,3*S* stereoisomers are nonsuperimposable mirror images; that is, they are enantiomers and, individually, they are optically active. They are therefore referred to as a *d,l* pair and together constitute a racemate. However, the two representations shown at the bottom of Figure 5-29 [analogous to (2*R*,3*S*)- and (2*S*,3*R*)-2-bromo-3-chlorobutane] are not different isomers because they are superimposable and thus represent only a single stereoisomer. (Make a model to convince yourself that this is so.) This stereoisomer of 2,3-dibromobutane has a mirror plane of symmetry in the center of the molecule. Through this mirror plane (perpendicular to the C-2–C-3 bond), each of the centers of chirality is reflected, *R* to *S* and *S* to *R*. As a consequence of this symmetry, there is no distinction between the designations 2*R*,3*S* and 2*S*,3*R* because they describe the same molecule, differing only in the end of the carbon chain from which numbering begins.

Because the molecule contains this plane of symmetry, it is optically inactive, despite the presence of centers of chirality. The term **meso compound** is used to designate such a stereoisomer. One can recognize a meso compound by the presence of a mirror plane or center of symmetry interrelating centers of chirality in the molecule. A meso compound is, by definition, optically inactive and is a diastereomer of the enantiomers of the *d,l* pair, from which it can be separated without resolution. Members of the *d,l* pair, however, are enantiomers and can be separated only by resolution. For compounds with meso stereoisomers, the number of possible stereoisomers is reduced from the maximum of 2^n.

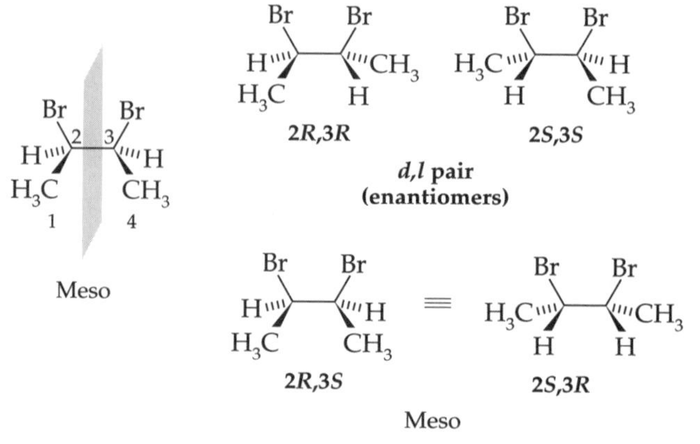

▲ FIGURE 5-29
Possible configuration of 2,3-dibromobutane.

EXERCISE 5-L

For each of the molecules in Exercise 5-H, calculate the number of possible stereoisomers.

EXERCISE 5-M

Of the following compounds, identify those that are optically active and those that are meso compounds. Methyl and ethyl groups are abbreviated Me and Et, respectively.

(a)   (d) 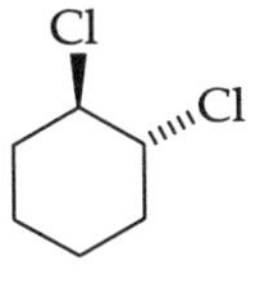   (g)

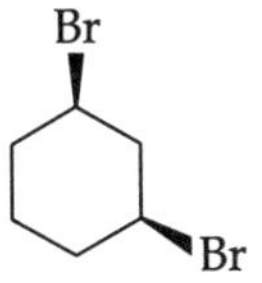

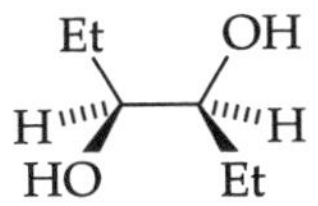

(b) 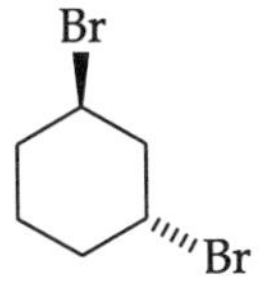   (e)

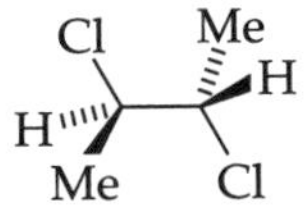

(c) 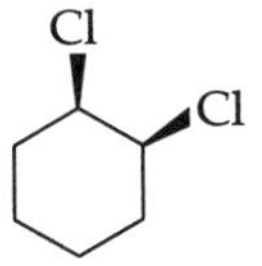   (f)

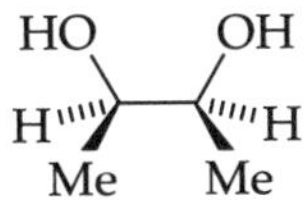

EXERCISE 5-N

Draw Newman projections of each stereoisomer in parts *e*, *f*, and *g* of Exercise 5-M in which the hydrogens at the centers of chirality are eclipsed.

Fischer Projections

Stereoisomers that contain more than one center of chirality can often be more easily recognized and compared through the use of a stick notation, called a **Fischer projection,** to indicate absolute configuration. In a Fischer projection, the intersection of two lines indicates the position of a chiral carbon. By convention, the horizontal lines indicate substituents that are directed toward the observer, and the vertical lines indicate substituents directed away from the observer. A prototype center of chirality bearing substituents A, B, C, and D is shown as a Fischer projection in the left-hand diagram of Figure 5-30. The Fischer projection is equivalent to the hatch/wedge representation at the right.

The carbon skeleton in a Fischer projection is usually arranged vertically and the substituents are arranged horizontally with C-1 at the top. The three stereoisomers of 2,3-dibromobutane depicted in Figure 5-29 can be represented by the Fischer projections shown in Figure 5-31. With the Fischer notation, it is easy to see the plane of symmetry between carbons 2 and 3 in the meso isomer.

Notice that in a Fischer projection all of the bonds in sequential centers of chirality are eclipsed. The Fischer projection thus represents an unstable conformation, which is nonetheless useful for recognizing configurational isomers.

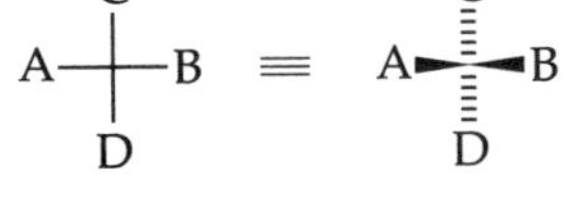

▲ FIGURE 5-30
The Fischer projection of the enantiomer at the left is equivalent to that shown in the hatch/wedge representation at the right.

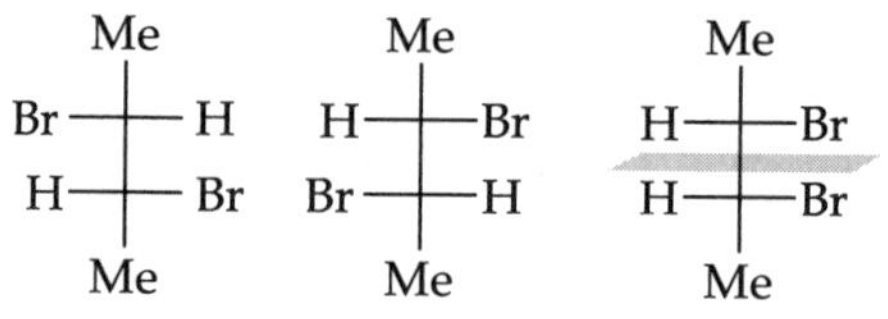

2*R*,3*R* **2*S*,3*S*** **Meso**

▲ FIGURE 5-31
Three configurational isomers of 2,3-dibromobutane. The two structures at the left and in the middle are enantiomers constituting a *d,l* pair. Each of these stereoisomers is chiral and is diastereomeric to the achiral meso compound at the right. The plane of symmetry in the meso compound is shown.

EXERCISE 5-O

Determine whether the members of each of the following pairs of compounds are enantiomers, diastereomers, constitutional isomers, or identical.

(a) 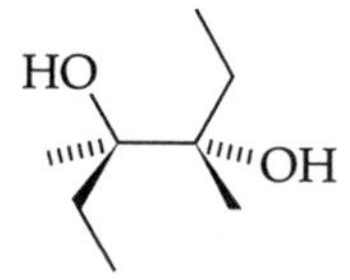 and 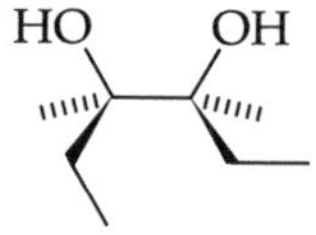

(b) 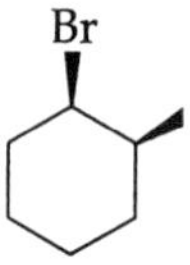 and 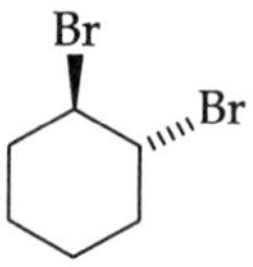

(c) 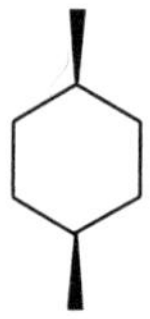 and 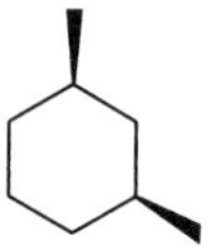

(d) 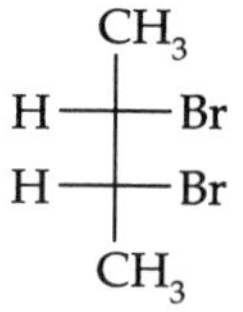 and 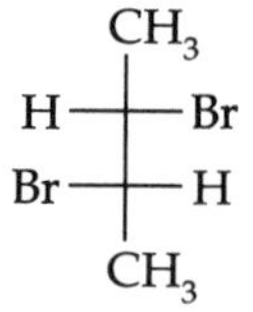

(e) 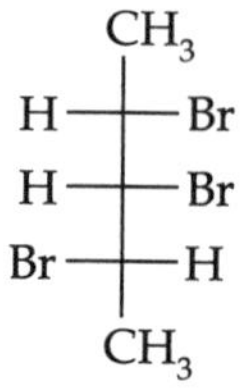 and 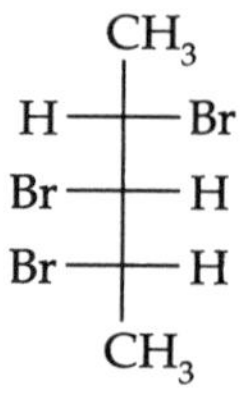

EXERCISE 5-P

Draw Fischer projections that represent the compounds shown in parts *e, f,* and *g* of Exercise 5-M.

5-9 ▸ Optical Activity in Allenes

In meso compounds, centers of chirality are present in a molecule that is itself achiral and optically inactive. Conversely, it is possible, although unusual, for molecules that lack chiral tetrahedral carbon atoms to be chiral and optically active. For example, two isomers of 2,3-pentadiene (Figure 5-32) are not superimposable and, hence, are enantiomers and are optically active. A necessary condition for chirality, and hence for optical activity, is the absence of a molecular mirror plane of symmetry, a feature that this molecule clearly lacks. In principle, therefore, one should look first at the symmetry properties of a molecule to determine whether it is optically active, recognizing that potential centers of chirality are most frequently encountered at tetrahedral carbon atoms.

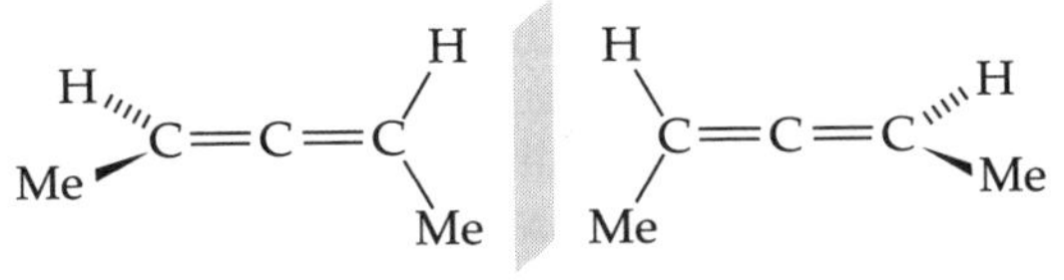

▲ FIGURE 5-32
Enantiomers of 2,3-pentadiene.

5-10 ▸ Stereoisomerism at Heteroatom Centers

We should not conclude that centers of chirality exist solely at carbon. Consider the structure of methylethylphenylamine.

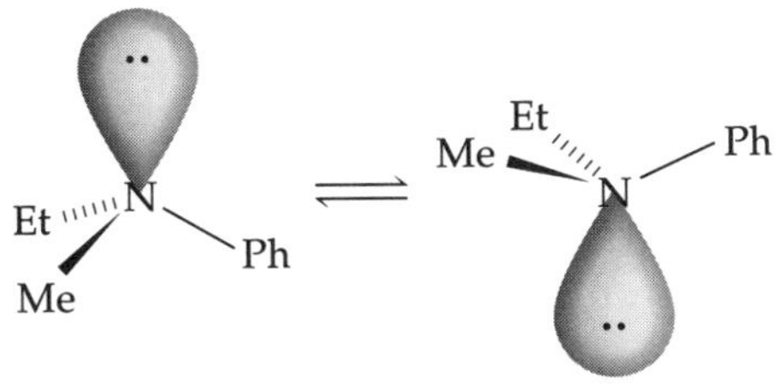

We know from Chapter 3 that a lone pair of electrons occupies a bonding site—in this case, making nitrogen tetrahedral. In this arrangement, nitrogen is indeed a center of chirality because there is no mirror plane of symmetry through this atom. If this arrangement were rigid, we would be able

---

WHY YOUR MOTHER TELLS YOU TO EAT YOUR BROCCOLI

The body is a marvelous chemical "factory." A vast array of chemical transformations required for life are constantly being performed. In addition, the body must deal effectively with unwanted and unneeded chemicals that are consumed, a task taken on in major part by the liver, where a complex series of oxidations and hydrolyses convert relatively nonpolar, lipophilic molecules into much more water soluble products that can be easily excreted in the urine. Many different enzymes catalyze these reactions, but some of them are specifically responsible for degrading carcinogenic compounds. Recently, a compound known as sulforaphane was isolated from broccoli and has been shown to induce increased activity by these detoxification enzymes. Notice that sulforaphane has a sulfoxide group and that there are four different groups arranged about the sulfur (the oxygen, a four-carbon chain terminated by isothiocyanate, a methyl group, and the lone pair). Thus, the sulfur in this compound is chiral and only one enantiomer (*R*) is found in the plant.

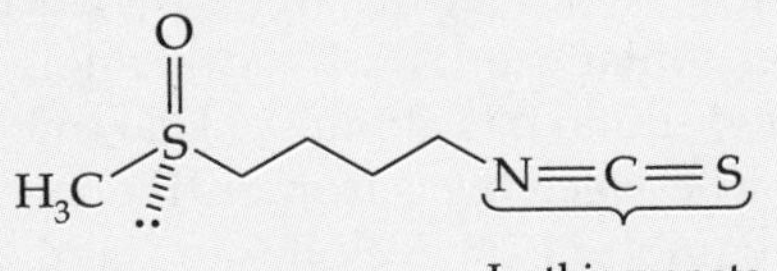

Isothiocyanate

Sulforaphane

to resolve such amines into two nonsuperimposable mirror images and each alone would be optically active. However, the unshared electron pair does not maintain its position on one side of the molecule but moves rapidly from one side to the other even at low temperature. This redisposition of the electron pair, known as **nitrogen inversion,** has the effect of converting one enantiomer into the other. The fast rate of nitrogen inversion makes it impossible to obtain neutral amines in optically active form except in very unusual cases.

However, inversion at nitrogen can be stopped by a simple chemical reaction. When, for example, the lone pair of electrons participates in the formation of a hydrogen–nitrogen bond, an ammonium salt is obtained. Because this salt does not have a nonbonded lone pair, it cannot undergo inversion.

Enantiomeric pair

Thus, the enantiomers of the ammonium salt cannot interconvert, and the nitrogen atom constitutes a center of chirality. Inversion is slower for third-row elements, and chiral phosphines ($R_1R_2R_3P$) and sulfoxides (R_1R_2SO) are known.

Conclusions

Stereoisomers are isomers that differ not in atom connectivity but rather in the three-dimensional disposition of atoms. The stereoisomers described in this chapter are: geometric isomers, conformational isomers, and configurational isomers.

The energy difference between geometric isomers is estimated as an enthalpy difference, $\Delta H°$. The energy required to accomplish an interconversion of geometric isomers is that required to overcome a barrier to reaching the highest energy point, the transition state, along the reaction coordinate: the energy of activation ($\Delta H^{\ddagger}$) defines the energy difference between the starting state and the transition state.

Conformational isomers are interconverted by rotations about σ bonds. Eclipsed and staggered conformations differ with respect to torsional strain caused by electron repulsion between aligned bonds. The staggered isomer of ethane is more stable than the eclipsed isomer by about 2.8 kcal/mole. The most-stable staggered conformer of butane is favored by about 5-to-7 kcal/mole over the least-stable, eclipsed conformation. The energies of various staggered conformers may differ because of different steric interactions resulting from van der Waals repulsion. The actual energy of a conformation depends on the dihedral angle separating bulky substituents. A *gauche* isomer is one containing a 60° dihedral angle between carbon substituents, whereas an *anti* isomer contains a 180° dihedral angle. The energy difference between *gauche*- and *anti*-butane is about 0.9 kcal/mole, and the energy barrier for interconversion between these isomers is about 3.4 kcal/mole.

Conformational equilibria in cyclohexane and other cyclic saturated compounds also are governed by torsional and steric interactions. For cyclohexane, torsional strain is minimized in the chair conformation. The alternative boat and twist-boat have fully or partially eclipsed bonds and thus include appreciable torsional strain. Ring flipping from one chair conformation of cyclohexane to another has the effect of converting each axial substituent into an equatorial one, and vice versa. The axial substituents undergo 1,3-diaxial interactions, whereas equatorial groups do not. Therefore, those conformers whose large substituents are in the equatorial position are most stable.

Chirality is a characteristic of molecules that lack a mirror plane of symmetry and usually have a center of chirality. An sp^3-hybridized atom bearing four different substituents constitutes a center of chirality, although a small number of compounds such as allenes are chiral and lack mirror symmetry even in the absence of an sp^3-hybridized center of chirality.

Stereoisomers that are mirror images of each other are called enantiomers, and stereoisomers that are not mirror images are referred to as diastereomers. Enantiomers differ with respect to the direction of rotation of a plane of polarized light and how they interact with other chiral molecules, but otherwise they have identical physical and chemical properties. Diastereomers have different chemical and physical properties.

Whereas diastereomers can be separated by physical and chromatographic methods, enantiomers are usually resolved by a process that converts the enantiomers into diastereomers, which are then separated. Reversal of the reaction used to form the diastereomers then regenerates the starting materials in optically pure form.

A chiral molecule is optically active; that is, it rotates a plane of polarized light by a value characteristic of that particular compound. This specific rotation can then be used to gauge optical purity in a sample by comparison with that calculated from the observed rotation.

There are several methods for uniquely identifying an enantiomer. The absolute configuration of a specific enantiomer is specified by the use of IUPAC rules. The designations (+) and (−), as well as *d* and *l*, can be used to indicate the direction of rotation of the plane of polarized light for a given compound. However, there is no direct connection between either the (+), (−) designations or the *d,l* designations and the absolute arrangement specified by the *R,S* designations.

A stereoisomer bearing centers of chirality symmetrically disposed within the molecule is called a meso compound and is optically inactive by virtue of the existence of a mirror plane or center of symmetry interrelating the centers of chirality. For a molecule with *n* centers of chirality, there are in principle 2^n stereoisomers minus the number of meso compounds. (Some theoretically possible stereoisomers are geometrically impossible.)

The two methods for the three-dimensional representation of molecules introduced in this chapter are Newman projections and Fischer projections. Newman projections are used to show conformational relations (those resulting from rotation about σ bonds) and can effectively illustrate the dihedral angles between substituent groups about σ bonds. The use of Fischer projections enables us to easily recognize the existence of mirror planes within molecules; they are a convenient means of representing the stereochemical relations of molecules containing multiple centers of chirality.

Although optical activity is encountered at atoms other than sp^3-hybridized carbon (for example, in allenes or in quaternary ammonium salts), its occurrence is relatively rare.

Summary of New Reactions

Photochemical *trans–cis* Isomerization

$$\text{（structure）} \underset{\text{hv}}{\overset{}{\rightleftharpoons}} \text{（structure）}$$

Acid–Base Reaction

$$RCO_2H + R'NH_2 \longrightarrow RCO_2^- + R'NH_3^+$$

Review Problems

5-1 Crotonic acid, $CH_3CH=CHCO_2H$, a compound found in Texas clay and formed by the dry distillation of wood, exists as the *E* isomer. Draw this molecule as the correct geometric isomer.

5-2 Draw Newman projections, visualizing down the C-2–C-3 bond, of the most-stable and least-stable conformations of 2,3-dimethylbutane. Using the values in Table 5-1, calculate the energy difference between these conformers and estimate the ratio of the most stable to the least stable at equilibrium.

5-3 Draw an approximate potential-energy diagram to describe a 360° rotation about the C-2–C-3 bond of 2,2,3,3-tetramethylbutane.

5-4 The preference of a methyl group for an equatorial rather than an axial position on cyclohexane relates to the number of *gauche* interactions in these two isomers. Draw, in three dimensions, 1-methylcyclohexane in its preferred conformation and in the conformation attained by flipping the ring. Then draw a Newman projection for each of these conformers, visualizing down the C-1–C-2 bond. From the structures, count the number of *gauche* and *anti* interactions that are like those considered in the chapter for butane.

5-5 Explain why the indicated isomer is the more stable of the following pairs of compounds:

(a) *trans*-1,2-dimethylcyclohexane is more stable than the *cis* isomer

(b) *cis*-1,3-dimethylcyclohexane is more stable than the *trans* isomer

(c) *trans*-1,4-dimethylcyclohexane is more stable than the *cis* isomer

5-6 For the following substituted cyclohexanes, label each substituent as axial or equatorial. If a ring flip will produce a more-stable conformation,

draw the expected more-stable conformation in three dimensions.

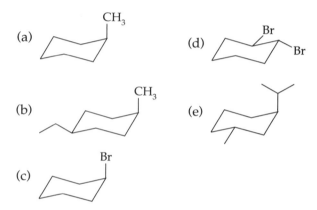

5-7 Despite the usual preference for substituents on chair cyclohexanes to occupy an equatorial position, 2-hydroxypyran exists predominantly in a chair conformation with the hydroxyl group axial. Draw this conformation, indicating the directionality of the lone pairs on oxygen.

2-Hydroxypyran

5-8 For each center of chirality in the following molecules, assign an *R* or *S* configuration according to the IUPAC rules.

(a)

**Ascorbic acid
(vitamin C)**

(b)

α-D-Glucose

(c) 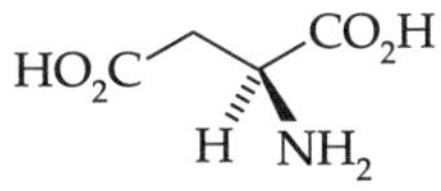 HO₂C⎓⎓⎓CO₂H
H NH₂

Aspartic acid

(d)

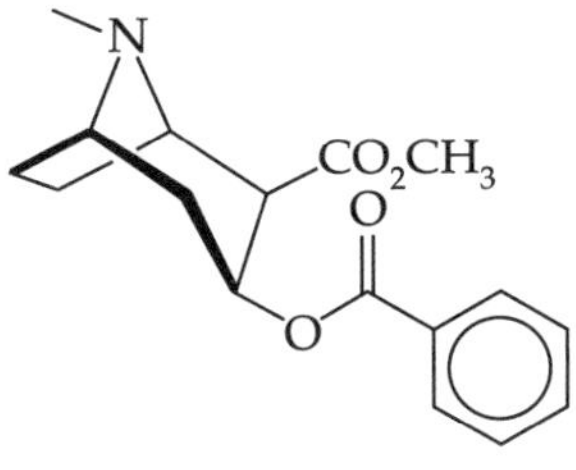

Cocaine

(e)

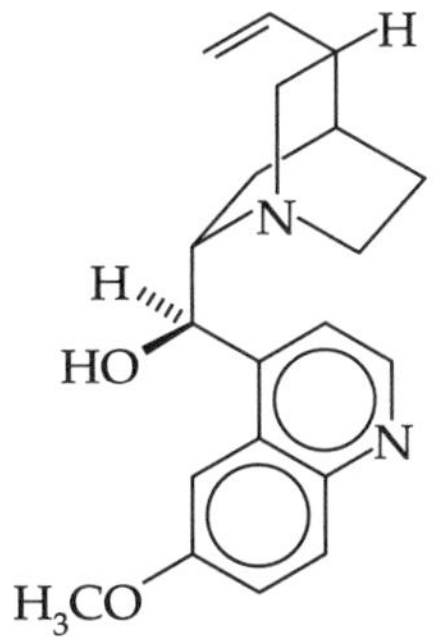

H,,,
HO

H₃CO

Quinine

(f)

CHO
H⎓⎓OH
HO⎓⎓H
H⎓⎓OH
CH₂OH

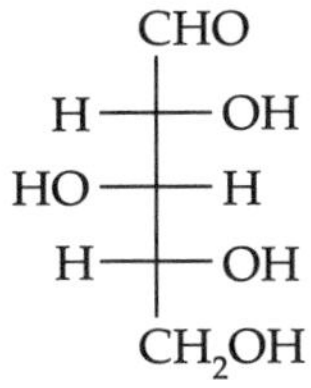

Xylose

5-9 Draw Fischer projections that represent the following:

(a) (*S*)-2-pentanol

(b) (*R*)-serine, HOCH₂CH(NH₂)COOH

(c) (*S*)-glyceraldehyde (2-hydroxypropanal)

(d) (*R*)-3-methylheptane

(e) (2*R*,3*R*)-dihydroxybutane

(f) meso-2,3-dihydroxybutane

5-10 For each of the following pairs of structures, identify the relation between them. Are they enantiomers, diastereomers, structural isomers, or two molecules of the same compound?

(a)

(b)

(c)

(d) 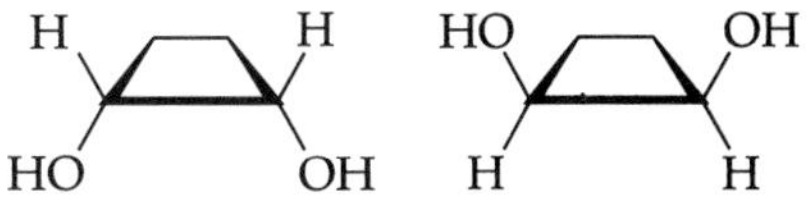 H⎓⎓H   HO⎓⎓OH
HO   OH   H   H

(e) 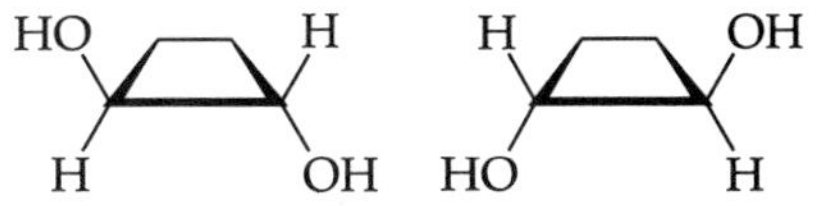 HO⎓⎓H   H⎓⎓OH
H   OH   HO   H

(f)

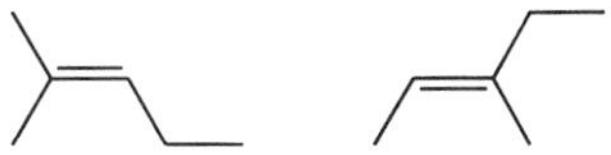

(g)

(h)

(i) 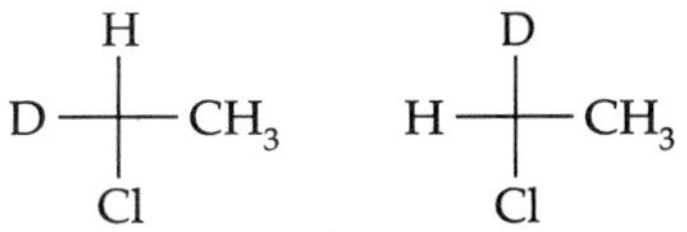 D⎓⎓CH₃   H⎓⎓CH₃
    H   D
    Cl   Cl

(j) 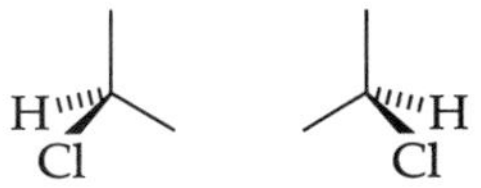 H,,,   ,,,H
   Cl   Cl

(k) 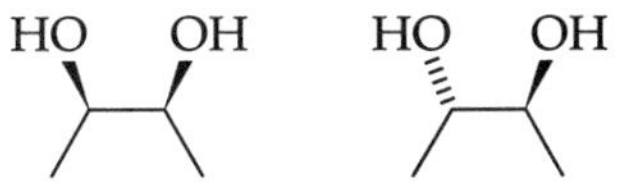 HO OH   HO OH

(l) 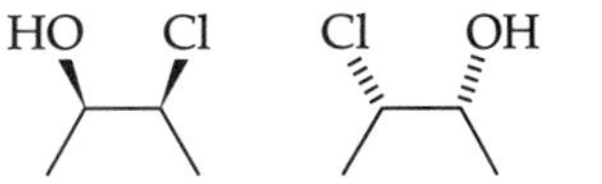 HO Cl   Cl OH

(m)

(n) 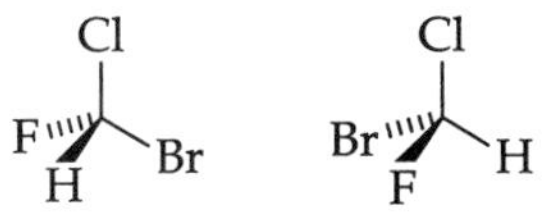 Cl⎓⎓⎓⎓⎓⎓Cl
F,,, Br Br,,, H
 H   F

(o) 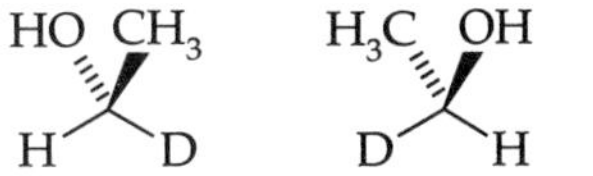 HO CH₃   H₃C OH
,,,   ,,,
H   D   D   H

5-11 The structure of cholesterol, the principal sterol found in all mammals, is shown here without stereochemistry. Identify any centers of chirality, and calculate the number of possible stereoisomers of cholesterol.

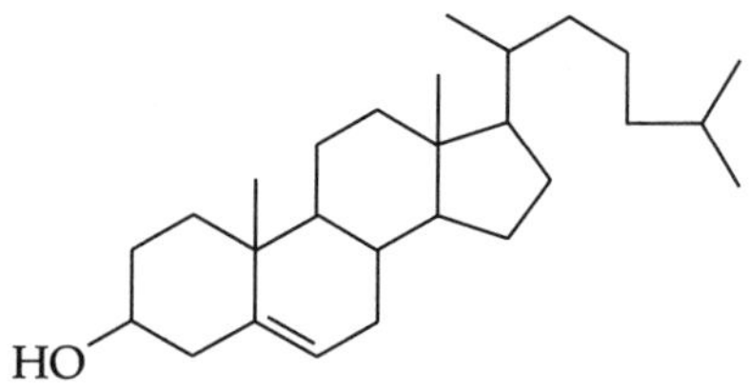

Chapter 6

Understanding
Organic
Reactions

With our knowledge of the structure of organic molecules, we are ready to begin a consideration of their reactions—why they proceed, what controls their rates, and how the structure of a reactant affects its reactivity. To do so, we will learn how to recognize when a reaction proceeds smoothly from reactant to product and when it takes place by several steps through a reactive intermediate. We must also learn how energy content changes as a reaction takes place and how energy changes and chemical structure affect equilibria (thermodynamics) and reaction rates (kinetics).

The energy content of a molecule is usually expressed as its free energy, $\Delta G°$. **Free energy** is a measure of the potential energy of a molecule or group of molecules, and it is partitioned between *enthalpy* ($\Delta H°$) and *entropy* ($\Delta S°$), equation 1:

$$\Delta G° = \Delta H° - T\Delta S° \qquad (1)$$

Enthalpy relates to bonding, whereas entropy relates to the degree to which a molecular system is ordered. The entropy contribution to free energy depends on temperature because $\Delta S°$ is multiplied by the temperature T (in kelvins). As the temperature increases, the $T\Delta S°$ term becomes larger and can sometimes dominate over the $\Delta H°$ term.

6-1 ▸ Reaction Profiles (Energy Diagrams)

The progression from reactant to product can be followed on a **potential energy surface** by plotting the change in potential energy as the reaction proceeds, just as we did in the conformational interconversions in Chapter 5. Although free energy ($\Delta G°$) has both **enthalpic** ($\Delta H°$, bond energies) and **entropic** ($\Delta S°$, disorder) components, most organic reactions proceed with only very small entropy changes when the number of moles of products equals the number of moles of reactants. Therefore, potential energy changes can often be approximated by changes in enthalpy. A **reaction profile** (also called an **energy diagram**) describes the changes in enthalpy in the course of a reaction. The measure of how far a reaction has proceeded is

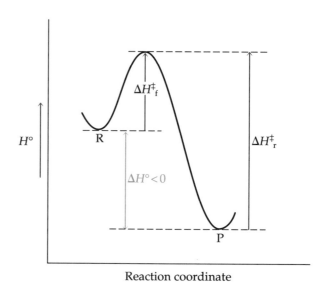

▲ FIGURE 6-1
Potential energy diagram for an exothermic reaction.

called the **reaction coordinate.** For example, the reaction coordinate can follow how far a critical bond that is breaking in the reaction has stretched or
how much a bond angle has expanded as an atom rehybridizes from sp^3 to
sp^2 as the reaction occurs.

When free energy is released in a reaction, the conversion is said to be
exergonic (Figure 6-1). If entropy changes are negligible (as is common in
many organic reactions), we can use the term **exothermic** to describe a reaction in which energy is released (as one in which the total bond energy content of the products is lower than that of the reactants). When free energy
must be supplied to drive a reaction, the conversion is said to be **endergonic** (Figure 6-2). Similarly, the term **endothermic** can be used to describe
one in which the bond energy content of the products is higher than that
of the reactants. A **thermoneutral reaction** is one in which the reactants and

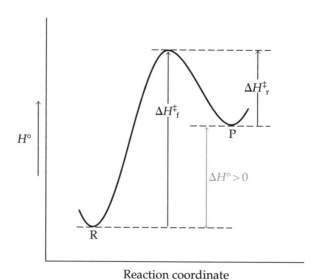

▲ FIGURE 6-2
Potential energy diagram for an endothermic reaction.

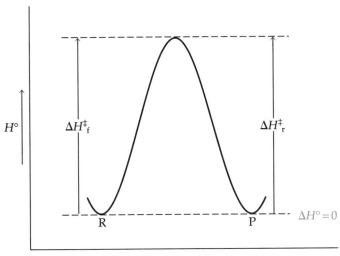

▲ FIGURE 6-3
Potential energy diagram for a thermoneutral reaction.

the products have the same energy content (Figure 6-3). If a reaction is **reversible,** it can proceed rapidly back and forth with similar ease many times: that is, from reactant to product and from product back to reactant. In principle, all chemical reactions are reversible; but, in practice, some reactions are referred to as **irreversible** because the conversion of product back into reactant is extremely slow under certain reaction conditions.

A reaction can proceed directly from reactant to product (a **concerted reaction**) or it can occur through a sequence of steps that includes the formation of one or more intermediates. An intermediate appears in a reaction profile as a minimum, or *energy well,* in the potential energy curve. Because the reactant is converted directly into product without the formation of any intermediates in a concerted reaction, no energy minima except the wells describing the reactant and product are observed. Thus, the energy diagrams in Figures 6-1 through 6-3 describe concerted reactions.

As stated in Chapter 5, the top of the smooth curve that connects the reactants and products represents the transition state of the reaction. The transition state is the structure of highest energy along the reaction pathway. Because the transition state is at an energy maximum (rather than at a minimum), it is very unstable and has only a transient existence.

The energy required to reach the transition state from the reactant energy minimum is defined as the **activation energy** (sometimes written as E_{act}), which can often be approximated by the enthalpy of activation ($\Delta H^{\ddagger}$) encountered in reaching the transition state in reactions in which entropy changes are small. In the conversion of a reactant into a product, we must supply enough energy to reach the transition state and hence to overcome the barrier separating these species. The higher the activation energy of a reaction, the more difficult it is to reach the transition state and the slower is the reaction. The barrier is thus often called the **activation energy barrier,** an expression used interchangeably with activation energy.

In a thermoneutral reaction, the activation energies for the forward and reverse processes are identical, as is apparent in Figure 6-3. In an exothermic reaction (Figure 6-1), the activation energy barrier for the forward reaction, $\Delta H^{\ddagger}_{f}$, is lower than that for the reverse reaction, $\Delta H^{\ddagger}_{r}$ whereas, in an endothermic reaction (Figure 6-2), the reverse is true.

A step-by-step (nonconcerted) reaction, showing transition states 1 and 2.

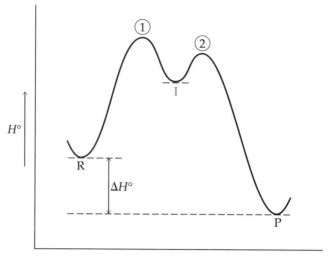

Reaction coordinate

As mentioned earlier, a reaction that takes place not in a single step, but rather in several stages, includes the formation of one or more **reactive intermediates**—species that have a real-time existence. Unlike a transition state, an intermediate must overcome some energy barrier, however small, before a reaction can proceed either forward or backward. It is this barrier that gives the intermediate a measurable lifetime. For this reason, an intermediate appears in a reaction profile as an energy minimum.

Consider the profile shown in Figure 6-4, in which a reactant, R, is converted into a product, P, through an intermediate, I, in a transformation that is exothermic overall. The enthalpy change of the reaction, $\Delta H°$ (roughly described by the changes in bond energies in the reactant and product in a reaction with a negligible entropy change), can be read from this profile in the same way as was done in Figure 6-1. However, an intermediate, I, of higher energy than either the reactant or the product is formed en route. Because the entire reaction (R → P) takes place in two steps (first R → I and then I → P), we can dissect this reaction (Figure 6-4) into two simpler reactions: an endothermic step for the R → I conversion and an exothermic reaction for the I → P transformation. Note that each of these stages of the complete reaction individually represents a concerted transformation.

EXERCISE 6-A

Draw a reaction profile to represent each of the following situations:

(a) an exothermic reaction with a small activation barrier

(b) an exothermic reaction with a large activation barrier

(c) an endothermic reaction with a small activation barrier

(d) a thermoneutral reaction with a large activation barrier

6-2 ▶ Characterizing Transition States: The Hammond Postulate

By definition, a transition state has only a transitory existence and can be characterized only by inference; that is, by relating the transition state's structure to a stable (or metastable) species that it closely resembles both

CALCULATING STRUCTURES OF UNOBSERVABLE MOLECULES

To fully understand a reaction, we must completely understand the structure and energy of the transition state that leads to product. However, because transition states have no real-time existence, they cannot be studied directly in the laboratory, and any experimentally derived inferences about transition states are indirect; for example, by the use of the Hammond Postulate, as discussed in this chapter. Theory, however, is not so limited, because it is possible to use quantum mechanics to calculate the energies even of nonexistent molecules and transition states. Most calculations are based on the theory developed by Erwin Schrödinger in 1925 to describe the hydrogen atom; but, unfortunately, it is not yet possible to rigorously solve the same equations for larger molecules. Because there are too many charged particles interacting in large molecules, a simplification of the calculation that ignores some of the interactions is required so that approximate structures can be calculated with currently available computers. These calculational methods have been used to predict the existence of new molecules before they were prepared, to describe "renegade molecules" (such as compounds containing square planar carbon or π bonds with perpendicular p orbitals) whose structures and reactivities do not correspond to normal patterns, and to quantify activation energies and the amount of bond bending or stretching required to form a transition state.

geometrically and energetically. According to the **Hammond Postulate,** *a transition state most closely resembles the stable species that lies closest to it in energy.* Thus, the Hammond postulate asserts that, in an endothermic reaction, the transition state is more similar to the product than to the reactant; whereas, in an exothermic reaction, the transition state more closely resembles the reactant. Respectively, these transition states are referred to as **late** (productlike) and **early** (reactantlike) to indicate how far along the reaction coordinate the highest point lies.

The rationale for the Hammond Postulate is based on ideas of how bonding changes in approaching the transition state. Because a transition state must be of higher energy than either the reactant or the product, partial bond breaking must take place as the transition state is approached. The greater the extent of bond breaking, the more energetically costly it is to reach that geometry. In an endothermic reaction like that represented in Figure 6-2, the progress from reactant, R, to product, P, requires energy. The activation energy of an endothermic reaction must be at least as large as the $\Delta H°$ of the reaction (by definition), and the degree to which bonds are broken and reformed in the transition state is similar to that in the product.

EXERCISE 6-B

Consider the reaction of an alcohol with a hydrogen halide, HX: ROH + HX $\rightarrow$ RX + H_2O. From the bond energies given inside the back cover of this book, calculate the energy change encountered for the reaction when the alcohol is ethanol and X = Cl, Br, I. For each, state whether the reaction is endothermic, exothermic, or thermoneutral and whether the transition state is more reactantlike (early) or productlike (late), assuming that the reaction is concerted.

Let us consider how the geometries of the two transition states in a two-step reaction like that depicted in Figure 6-4 might resemble either the reactant or the product. We start by dividing the whole reaction into two halves: the R → I and the I → P reactions. Because the R → I conversion is endothermic, the transition state between R and I most closely resembles the "product"; that is, intermediate I. Because the second step is highly exothermic, the Hammond Postulate tells us that its transition state more closely resembles the "reactant" of the second step—that is, the intermediate, I—than the final product, P. Thus, if we wish to understand the step-by-step conversion of R → P, we must characterize I, because both transition states in this two-step reaction more closely resemble the intermediate, I, than either the reactant, R, or the product, P.

6-3 ▸ Relative Rates from Reaction Profiles

In a reaction that takes place in more than one step, the step in which the transition state is of highest energy determines the overall rate of the conversion. This slowest step is called the **rate-determining,** or rate-limiting, **step** and is the "bottleneck" in a sequence. The success of a multistep transformation requires the transition state of this rate-determining step to be reached as the reaction proceeds.

In the profile shown in Figure 6-4, the formation of transition state 1 is the rate-determining step. According to the Hammond Postulate, that transition state can be approximated most closely by intermediate I, because, as we have seen, the transition state lies close in energy to this intermediate. If a similar reaction proceeded through the same general pathway, but through a more-stable intermediate, the transition state would be lower in energy. Thus, the corresponding activation barrier would also be lower and the reaction would proceed more rapidly.

EXERCISE 6-C

Identify the rate-determining step in each of the following energy diagrams:

(a)   (b)

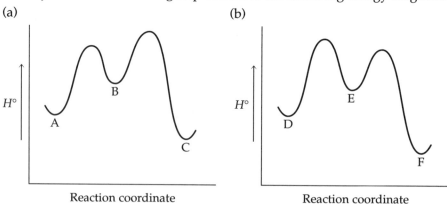

6-4 ▸ Reactive Intermediates

Because intermediates are so important in understanding multistep reactions, let us consider the structural characteristics of the most-common organic intermediates. Knowledge of the molecular and electronic structures of a number of reactive intermediates helps us to understand how reactions occur.

In some multistep reactions, the intermediate is sufficiently stable to be isolated. More frequently, however, the intermediate is highly energetic and reactive. If it cannot be isolated, the intermediate must be characterized indirectly. Common intermediates encountered in organic chemistry include carbocations, radicals, carbanions, carbenes, and radical ions.

Recall from Chapter 3 that both a carbocation and a radical contain a carbon atom bearing three substituents in a trigonal arrangement. The sp^2-hybridized atom in a carbocation or radical is electron deficient (compared with carbon's valence-shell requirement), and the stability of these intermediates is increased by substitution with alkyl groups: for both carbocations and radicals, the observed order of stability is: $3° > 2° > 1° > CH_3$.

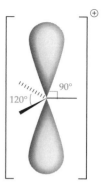

Carbocation
(sp^2-hybridized)

Carbanions

Like carbocations and radicals, a **carbanion** often has three σ bonds, but it also bears an unshared electron pair (Figure 6-5) and is *not* electron deficient because its valence shell is filled. [One method for preparing carbanions is to remove a proton from a C–H bond (**deprotonation**) by treating the compound with a strong base. The electrons originally in the C–H bond then become a nonbonded lone pair on carbon.] Although both the electron-deficient carbocation and the radical are sp^2-hybridized, simple carbanions are pyramidal. With three σ bonds and a lone pair, the carbanion is electronically similar to an amine: the carbanion and amine are therefore said to be **isoelectronic.** Like an amine, a simple alkyl carbanion is sp^3-hybridized with a doubly occupied sp^3-hybrid nonbonding orbital directed at approximately a tetrahedral angle away from the bonding orbitals. Because the trigonal carbon of a carbanion bears a formal charge of −1, it is often strongly associated (**ion paired**) with a positively charged metal counterion. When the carbanionic carbon is adjacent to a π system, it can rehybridize to sp^2, with negative charge being distributed by resonance throughout the π orbital array. Resonance structures are then used to describe the electron distribution in the resulting conjugated anion.

Radical
(sp^2-hybridized)

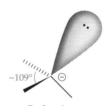

Carbanion
(sp^3-hybridized)

▲ FIGURE 6-5
Comparison of the three-dimensional structures of a carbocation, a radical, and a carbanion.

EXERCISE 6-D

Draw the structures of all significant resonance contributors for the anions formed by deprotonation of each of the following compounds. Show the location of formal charge in each ion.

(a) ![structure a]   (b) ![structure b]   (c) ![structure c]   (d) ![structure d]

EXERCISE 6-E

Draw three-dimensional structures for the following reactive intermediates:

(a) allyl cation

(b) cyclopropyl anion

(c) cyclopropyl cation

(d) allyl anion

Because simple carbanions are sp^3-hybridized, the carbanionic carbon can be a center of chirality, whereas the carbon center of a carbocation or a radical, having a symmetry plane through the atom, cannot be a center of chirality. In most cases, however, inversion of configuration of carbanions is rapid and results in a racemic mixture.

Carbenes

As already noted, carbocations and radicals are trigonal, having three σ bonds at carbon, and carbanions are pyramidal, with three σ bonds and a fourth hybrid orbital occupied by a lone pair of electrons. **Carbenes** are another class of neutral reactive intermediates that bear only two σ bonds. A carbon atom with two σ bonds is called digonal; with three, trigonal; and with four, tetrahedral. For a digonal carbon to be neutral, two nonbonded electrons must be present in addition to those participating in the two covalent bonds.

EXERCISE 6-F

Calculate the formal charge (see Chapter 1) of the carbenic carbon in:

(a) $:CH_2$

(b) $:CCl_2$

A neutral digonal intermediate, a carbene, is quite electron deficient (despite its neutrality) because it lacks two electrons from the octet needed for an inert-gas configuration. Like cations and radicals, a singlet carbene (in which the nonbonded electrons are spin-paired) is sp^2-hybridized. In this hybridization, a p orbital is oriented perpendicular to the plane containing the two σ bonds and an additional sp^2-hybridized orbital (Figure 6-6). The electron pair is generally in this hybrid orbital (rather than in the p orbital). The greater s character (33%) in an sp^2 hybrid places the electrons closer to the positively charged nucleus, making that arrangement more stable.

In a **singlet** molecule, all electrons in the molecule are paired, and generally two electrons of opposite spin are paired in each molecular orbital. A **singlet carbene** has its electrons paired (Figure 6-6), producing both an electrophilic center (the vacant p orbital) and a nucleophilic center (the doubly occupied sp^2-hybridized orbital). Carbenes are therefore reactive in an **ambiphilic** sense; that is, toward both electron-rich and electron-poor reagents.

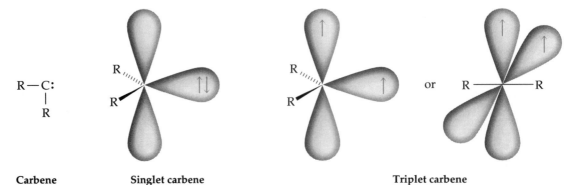

| Carbene | Singlet carbene | Triplet carbene |

▲ FIGURE 6-6
Orbital occupancy in a singlet and a triplet carbene.

Although carbenes are highly reactive intermediates, they are sufficiently stable for spectroscopic study when isolated from each other and from other molecules with which they might react. For example, they can be characterized in the laboratory by absorption or infrared spectroscopy when trapped in a frozen matrix of solid argon or other inert species. The parent carbene (methylene, $:CH_2$) has been found in outer space where very low temperatures and the low density of molecules inhibit its intermolecular trapping.

When the electrons are not spin-paired, the carbene is in a **triplet state.** In a triplet, the two electrons of the same spin must be accommodated in two different orbitals. If the carbenic carbon remains sp^2-hybridized, a **triplet carbene** has one electron in p and sp^2-hybrid orbitals; if it rehybridizes to a linear sp hybrid, the two electrons are held in orthogonal p orbitals. Because of this structure, a triplet carbene has radicallike reactivity.

Radical Ions

Reactive intermediates can also be formed by the addition or removal of an electron. For example, if an electron is removed from the π bond of an alkene, the resulting π orbital contains only one electron. Because the resulting structure has one electron fewer than needed for neutrality, it is positively charged; and, because it contains a single electron, it is also a radical. We can write resonance contributors for this **radical cation** that localize the odd electron on either of the two atoms of the π system, with positive charge being borne formally at the other atom (Figure 6-7). The π system thus has unpaired-electron character (and therefore behaves as a radical) and is electron deficient (like a carbocation). It can therefore also act as an electrophile.

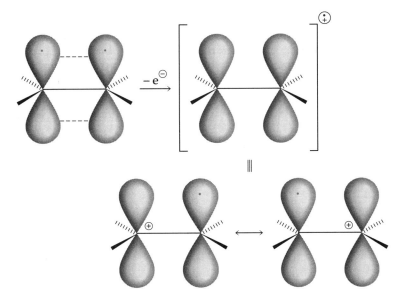

▲ FIGURE 6-7
Electron distribution in an alkene radical cation.

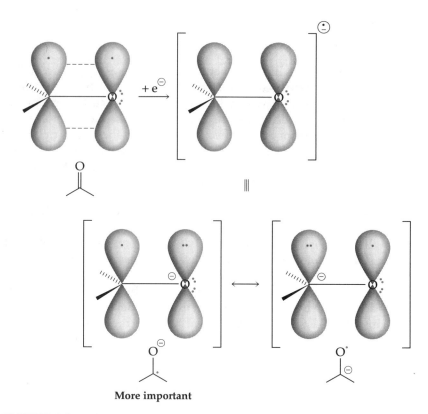

▲ **FIGURE 6-8**
Electron distribution in a ketyl radical anion.

Similarly, the addition of an electron to a π system generates a **radical anion,** a structure in which three electrons are accommodated in the π bonding and antibonding orbitals. Because the bonding orbital is of lower energy than the antibonding orbital, it contains two electrons, whereas the antibonding orbital accommodates only one. Therefore, there still is net bonding in the radical anion's π system.

When the π system to which the electron is added is a carbonyl group, the resulting radical anion is known as a **ketyl.** Resonance structures can localize two electrons in one of the *p* orbitals (for example, on oxygen) and one electron on carbon, or vice versa (Figure 6-8). Because oxygen is the more-electronegative atom, these two contributors are not of equal energy, and so the electron density in this ketyl is strongly polarized toward oxygen. As a result of this unequal contribution of resonance structures, the oxygen atom has anionic character and the carbon atom has radical character in this radical anion. As we shall see in many of the later chapters of the book, a clear understanding of the structures and relative reactivity of these various intermediates is critical in understanding how reactions proceed.

EXERCISE 6-G

Draw all significant resonance contributors for:

(a) benzene cation radical

(b) naphthalene anion radical

(c) acetone anion radical

6-5 ▸ Thermodynamic Factors

How can we predict when a given reaction will occur? Two factors, kinetics and thermodynamics, are important. When we speak about **kinetics,** we are concerned with the rate at which reactions proceed, whereas **thermodynamics** describes the relative energies of the reactants and products. In this chapter, we consider thermodynamic equilibria first and reaction rates later. For a reaction to be practical, thermodynamics must favor the desired product and its rate must be fast enough for the reaction both to be completed within a reasonable time and to dominate over other competing reactions.

Regardless of the reaction under consideration, the principle of **microscopic reversibility** tells us that the transition state in the forward direction is the same as that in the reverse direction, both in concerted and in multistep reactions. Furthermore, the difference between the activation energies for the forward and reverse reactions is always given by $\Delta H°$ (if the contribution of entropy can be ignored). In this way, thermodynamics is often intimately related to the observed kinetics (rates of reactions).

We know that free energy has both enthalpic (bond energies) and entropic (disorder) components. Enthalpy changes are almost always important in chemical reactions, but entropy changes are usually significant in organic reactions only when the number of product molecules differs from the number of reactant molecules. Let us first consider examples that typify each of these components.

Enthalpy Effects

We can often gauge the thermodynamic feasibility of a simple reaction by contrasting the strengths of the bonds that must be broken with those that are formed as the reactant is converted into product. If the relevant bond energies are known, the calculation of the heat of reaction is straightforward. If the new bonds formed in the product(s) are stronger than those that must be broken in the reactant(s), the reaction is exothermic and, consequently, thermodynamically favorable. The same calculation can be used for multistep reactions because only the relative energies of the starting material(s) and final product(s) determine the overall thermodynamics of the reaction. We cannot learn from these calculations anything about reaction rates, which are governed by transition-state energies rather than by the energies of reactants or products.

Let us consider a specific example. Upon treatment with base, a proton can be removed from the α position of a carbonyl compound (Figure 6-9). In the **enolate anion** formed, two chemically different sites bear partial negative charge, as indicated by the resonance contributors at the right in Figure 6-9. It is therefore possible to reprotonate at either of these sites (that is, at

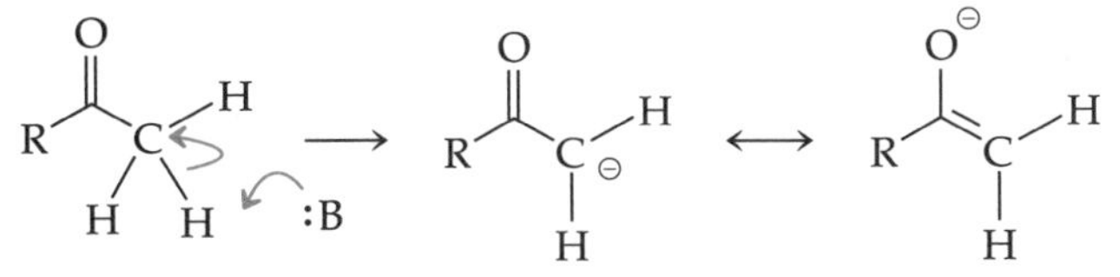

▲ FIGURE 6-9

Formation of an enolate anion by deprotonation of a carbonyl compound that has an α-hydrogen. An alkoxide is often used as the base. The enolate anion has two important resonance contributors that differ in the site at which negative charge is localized: on carbon or on oxygen.

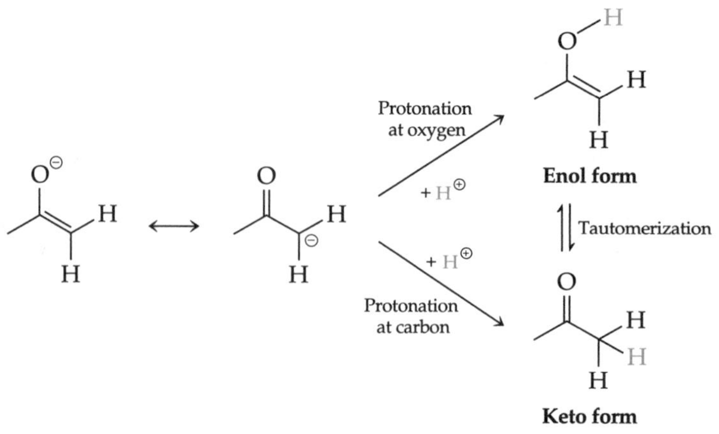

▶ FIGURE 6-10

Protonation of an enolate anion
forms either an enol (by
delivering the proton to
oxygen) or a carbonyl
compound (by delivering the
proton to carbon). The
interconversion of a ketone and
its enol form is called
tautomerization.

carbon or at oxygen) to form two distinct products (Figure 6-10). When reprotonation is at carbon, the original carbonyl compound is formed again.
When it is at oxygen, an **enol** is generated, so named because of the presence of a hydroxyl group attached to an alkenyl group. An isomerization
proceeding by deprotonation at the C–H bond adjacent to a carbonyl group
and protonation at the carbonyl oxygen is called a **keto-enol tautomerization.** The keto and enol forms shown in Figure 6-10 are **tautomers** because
they differ only with respect to the position of an acidic hydrogen. The
process by which they are interconverted is called **tautomerization.** A tautomerization can be induced by treatment with acid or base. With base, the
α carbon is deprotonated in the first step, producing an enolate anion that is
reprotonated on oxygen in a second step. With acid, the carbonyl oxygen is
protonated in the first step, with removal of the α-hydrogen occurring in
the second step.

Let us use what we know about bond strengths to determine whether
the keto or enol form dominates this tautomeric equilibrium. The ketone
has a carbon–oxygen double bond, whereas the enol has a carbon–oxygen
single bond (179 versus 86 kcal/mole, Table 3-4); the ketone has a
carbon–carbon single bond, whereas the enol has a carbon–carbon double
bond (83 versus 146 kcal/mole); and the ketone has a carbon–hydrogen
bond, whereas the enol has an oxygen–hydrogen bond (99 versus 111
kcal/mole). In total, the bonds of the ketone are stronger than those of the
enol by 18 kcal/mole ([179 + 83 + 99] − [86 + 146 + 111] = +18 kcal/mole).
Thus, the ketone is more stable and is the dominant species at equilibrium.

EXERCISE 6-H

Write the structure of a tautomeric isomer for each of the following compounds:

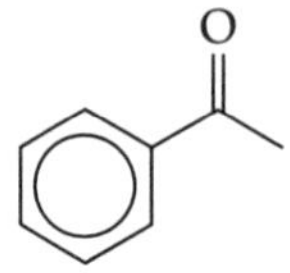

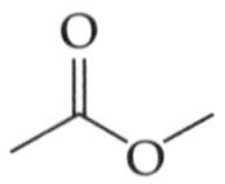

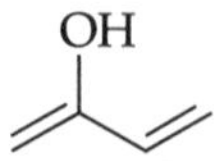

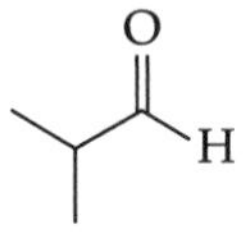

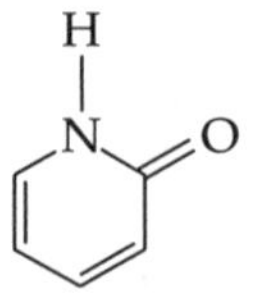

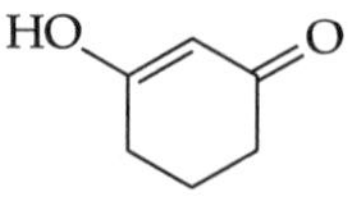

Refer to the bond strengths in Table 3-6 or at the back of the book, and predict whether the following reactions are endothermic or exothermic:

(a)  [structure with Br] + HCl ⟶ [structure with Cl] + HBr

(b) 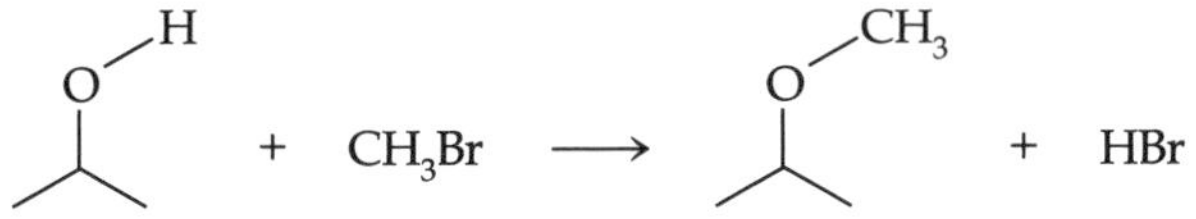 [structure with O–H] + CH₃Br ⟶ [structure with O–CH₃] + HBr

(c) 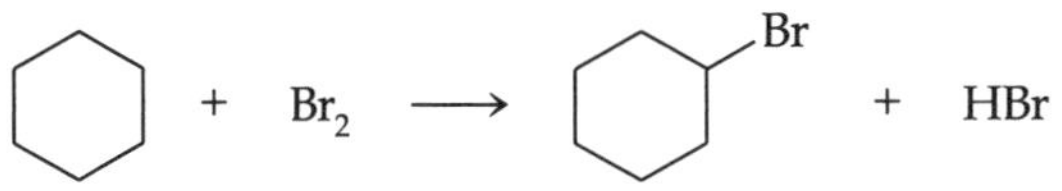 [cyclohexane] + Br₂ ⟶ [cyclohexyl bromide with Br] + HBr

Entropy Effects: The Diels–Alder Reaction

Most organic reactions have small entropy changes, but sometimes a single product is formed from several simpler reactants. For example, in the **Diels-Alder reaction,** two hydrocarbon reactants combine to form a cyclic product without proceeding through any intermediate that can be isolated. In its simplest form, the Diels-Alder reaction combines a diene with an alkene (Figure 6-11) to form a cyclohexene. We can follow how this reaction proceeds by drawing *curved, full-headed arrows* to indicate the motion of each electron pair as the π electrons of each of these reactants are delocalized in a cyclic, conjugated transition state. The six electrons from the two π systems of the reactants are delocalized in the transition state. The rules about aromaticity explained in Chapter 2 also apply to transition states, and so the Diels-Alder reaction has an unusually low activation energy. Because $4n + 2$ (6) electrons participate in the cyclic transition state, the reaction proceeds rapidly.

The reactants have a total of four σ bonds (three in butadiene and one in ethylene) and three π bonds (two in butadiene and one in ethylene). The product has six σ bonds in the carbon skeleton and one π bond. Therefore, two π bonds have been converted into two σ bonds. Because the average bond strength of a carbon–carbon π bond is lower than that of a σ bond by 20 kcal/mole (63 versus 83 kcal/mole), enthalpy—that is, bond energies—dictates that this reaction proceed from the left to the right and that $\Delta H° = -40$ kcal/mole. This is indeed observed with most Diels-Alder reactions. If this difference in bond energy ($\Delta H°$) between reactants and products is negative, the reaction is exothermic; if positive, endothermic.

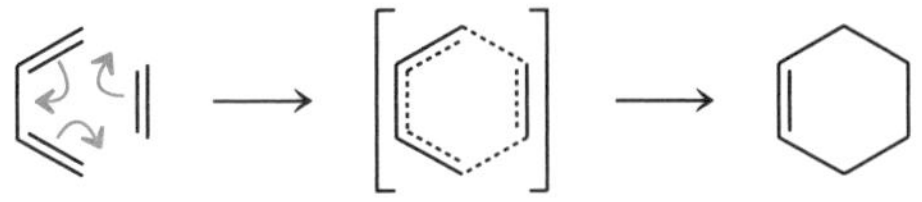

▲ FIGURE 6-11

The Diels-Alder reaction. Partial bonds being formed and broken in the transition state are indicated by dashed lines. Because the Diels-Alder reaction is concerted, it does not include the formation of an intermediate. The structure in brackets, between the reactants and product, is a transition state that has no real time existence.

Notice, however, that there are two reactants and only a single product in the Diels-Alder reaction. A single product is more highly ordered than the two reactant molecules because a single molecule has fewer translational and rotational degrees of freedom than do two. Thus, unlike most organic reactions in which the number of reactant and product molecules is the same, product formation in the Diels-Alder reaction should be disfavored by entropy—that is, a measure of disorder.

Because entropy favors the reactants in the Diels-Alder reaction, conditions in which entropy dominates cause the product to revert to starting material. The **retro-Diels-Alder reaction,** the reverse of the reaction considered in Figure 6-11 (which therefore has the same transition state), almost always occurs only at high temperatures, at which the effects of entropy begin to become important. Whether the equilibrium lies with the products or reactants in Diels-Alder reactions therefore depends on temperature.

Near room temperature, changes in the entropy contribution ($T \Delta S°$) to the reaction free energy are generally smaller than those in enthalpy ($\Delta H°$) in most organic reactions (especially in those having the same number of reactant as product molecules), and they can often be ignored. In many reaction profiles, therefore, the y axis (potential energy) is enthalpy, or bond energies, and thus does not include entropy.

EXERCISE 6-J

Calculate the entropy change (in calories per K) that would be necessary to reverse a reaction $A + B \rightarrow C$ at room temperature in which $\Delta H°$ is:

(a) –1 kcal/mole   (b) –10 kcal/mole   (c) –20 kcal/mole

6-6 ▸ Chemical Equilibria

Relating Free Energy to an Equilibrium Constant

Changes in free energy control chemical reactions: that is, the more exergonic a reaction is (the more negative $\Delta G°$ is), the larger its equilibrium constant, K, is (shifted farther toward product).

$$\Delta G° = -RT \ln K = \Delta H° - T\Delta S° \tag{2}$$

in which R is the ideal gas constant and T is the temperature in kelvins. Because the contributions of enthalpy are often more important than those of entropy to free energy changes in organic reactions, we can often relate changes in bond energies to chemical equilibrium constants. Specifically, $\Delta G°$ or $\Delta H°$ can also be related to the equilibrium constant K by equation 2.

For a reversible reaction between A and B to produce C and D (equation 3) an equilibrium constant, K, can be written either in terms of concentrations of reagents at equilibrium (equation 4) or as a ratio of the forward (k_1) and reverse (k_{-1}) reaction rate constants (equation 5).

$$A + B \underset{k_{-1}}{\overset{k_1}{\rightleftharpoons}} C + D \tag{3}$$

$$K = \frac{[C][D]}{[A][B]} \tag{4}$$

$$K = \frac{k_1}{k_{-1}} \tag{5}$$

EXERCISE 6-K

For a general reaction $A + B \rightleftharpoons C + D$ taking place at room temperature, calculate K for each of the following conditions:

(a) $k_1 = 10^{10}(\text{mole/l})^{-1} \cdot \text{sec}^{-1}$; $k_2 = 10^8(\text{mole/l})^{-1} \cdot \text{sec}^{-1}$

(b) initial concentration of A = initial concentration of B and the final concentration of C = 0.5 initial concentration of A

(c) $\Delta G° = -1\,\text{kcal/mole}$

(d) $\Delta G° = -10\,\text{kcal/mole}$

(e) $\Delta G° = -30\,\text{kcal/mole}$

Acid–Base Equilibria

One of the principal types of chemical equilibria is related to **acid–base** chemistry, in which a proton is transferred from an acid to a base. Thus, cleavage of an H–X bond in an organic acid (to generate an anionic intermediate bearing a lone pair of electrons on X^-) forms the **conjugate base** (the deprotonated form X^-) of the acid. Such anionic intermediates are much more reactive than their protonated precursors toward other functional groups, and the initiation of reaction sequences by X–H deprotonation is a critical step in many organic transformations.

Bond cleavage of a general acid, HA, to generate an anion requires a **base,** a species known to be active either as a proton acceptor or as an electron-pair donor. The flow of electrons in the reaction of a base with a hydrocarbon acid is shown in equation 6.

$$B:\!\curvearrowright\!H\!-\!A \;\;\rightleftharpoons\;\; \overset{\oplus}{B}\!-\!H \;+\; :A^{\ominus} \tag{6}$$

To be more specific, one can write an equilibrium for a general acid, HA, with water acting as a base, as in equation 7.

$$H_2O:\!\curvearrowright\!H\!-\!A \;\;\rightleftharpoons\;\; H_3O^{\oplus} \;+\; :A^{\ominus} \tag{7}$$

The equilibrium constant, K, is defined in the usual way (equation 8; compare with equation 4); and, because the concentration of water is constant in aqueous solution, one can define another equilibrium constant, K_a, as K times the water concentration, in equation 9. For example, applying this equation to acetic acid, CH_3CO_2H (equation 10) gives $K_a = 10^{-5}$.

$$K = \frac{[A^-][H_3O^+]}{[HA][H_2O]} \tag{8}$$

$$K_a = K[H_2O] = \frac{[A^-][H_3O^+]}{[HA]} \tag{9}$$

$$K_a(CH_3CO_2H) = \frac{[CH_3CO_2^-][H_3O^+]}{[CH_3CO_2H]} = 10^{-5} \tag{10}$$

Because of the ability to cover order of magnitude changes in a simple way, a convention has been adopted to use a negative logarithm scale, in which **pK_a** is defined as the negative logarithm of K_a (equation 11), to describe acidity.

$$pK_a = -\log K_a;\; pK_a(CH_3CO_2H) = 5 \tag{11}$$

Instead of saying that the K_a of acetic acid is 10^{-5}, we say that the pK_a of acetic acid is 5. Thus, a small K_a corresponds to a large pK_a. The acid dissociation constant, K_a, of acetic acid is quite small, but the K_a values for most organic acids are even smaller. The smaller the K_a of an acid is, the larger the positive value of its pK_a and the less acidic the corresponding compound.

EXERCISE 6-L

Calculate the pK_a values of acids having the following acid dissociation equilibria:

(a) $K = 4 \times 10^{-6}$ (d) $K = 1$

(b) $K = 3 \times 10^{-40}$ (e) $K = 5$

(c) $K = 1250$

6-7 ▸ Acidity: Quantitative Measure of Thermodynamic Equilibria

A convenient way to describe the relative acidity of two organic compounds is to order them according to pK_a. A ranking of pK_a values to describe the deprotonation of a number of functional groups is informative because this value defines how easily heterolytic cleavage of an X–H bond can occur. On this scale, the less positive (or more negative) the pK_a, the stronger the acid and the easier it is to cleave the H–X bond. The more stable the anion X$^-$, the more acidic the X–H bond. Thus, because acidity is directly related to the stability of the anion generated, this scale can be used to interrelate the relative reactivity of various anions as bases.

Mineral acids typically have negative pK_a values, suggesting that their acid dissociation equilibria lie far to the right in equation 7 and that they are, therefore, very strong acids. Different functional groups have different characteristic pK_as: carboxylic acids typically have pK_as of about 5, phenols (aromatic alcohols) have pK_as of about 10, and aliphatic alcohols have pK_as of about 16. More values are listed in Tables 6-1 and 6-2 (pages 210–211).

In the following sections, we consider electronic and structural features that influence the pK_a values of various molecules. The strength of an acid (its pK_a value) depends directly on the extent to which a proton is transferred from the acid to a base, which in turn is determined by the stability of the resulting anion. First, we consider how acidity is influenced by changing the atom to which the acidic proton is attached. Then we examine more-subtle effects, by keeping the atom to which the acidic proton is attached constant. By doing so, we can minimize the effects of electronegativity and see how several other factors significantly influence acidity: bond energies, inductive and steric effects, hybridization, resonance stabilization, and aromaticity.

EXERCISE 6-M

Which member of the following pairs of compounds has the lower pK_a (use Table 6-1)?

(a)

(b)

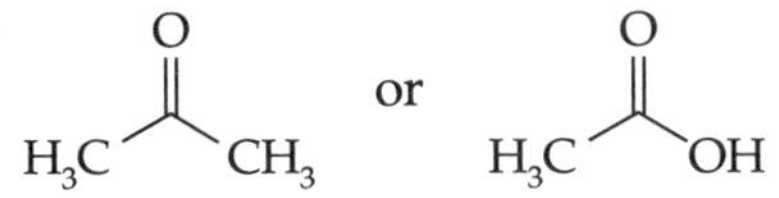

(c)

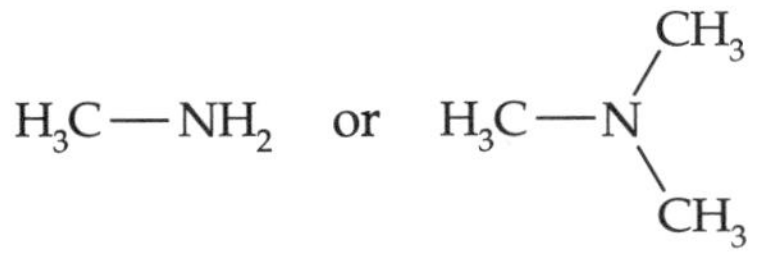

H_3C-NH_2 or $H_3C-N\overset{CH_3}{\underset{CH_3}{}}$

(d)

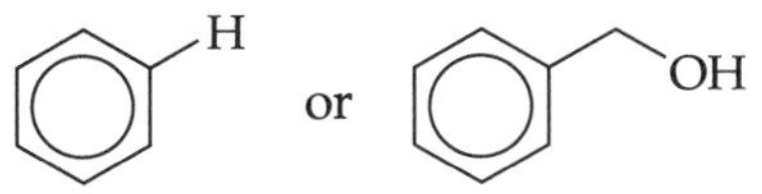

Electronegativity

The chemical bond connecting a proton to an electronegative atom is highly polar, usually making the molecule a strong acid. For example, the acidities of compounds containing second-row elements, CH_4 (pK_a about 48), NH_3 (38), H_2O (16), and HF (3), increase steadily as the atom to which the proton is attached becomes more electronegative. (Remember that pK_as are logarithms, so that HF is about 10^{45} times as acidic as methane.)

Bond Energies

Within a single column of the periodic table, the trend in acidity is opposite that expected solely from electronegativity. As the atomic weight increases down a column, the ability of the atom to bear negative charge increases because of its larger size, even though its electronegativity decreases. The H–X (or C–X) bond-dissociation energy also decreases in the progression down the periodic table from HF to HI because of the increasingly mismatched orbital sizes. Thus, the acidity of HI ($pK_a \approx -10$) is greater than HBr ($pK_a \approx -9$), which is greater than HCl ($pK_a \approx -7$), which, in turn, is greater than HF ($pK_a \approx +3$).

EXERCISE 6-N

Predict which member of each of the following pairs of compounds is more acidic and give reasons for your choice.

(a) H_2S or PH_3 (d) CH_4 or SiH_4

(b) H_2O or HCl   (e) CH_3OH or CH_3NH_2

(c) H_2O or H_2S (f) HI or H_2S

Inductive and Steric Effects

Acidity is also influenced by the presence of polar functional groups, which induce a shift of electron density within the molecule. We can see the effect of electron-donating and -releasing groups on acidity by comparing a series of similarly constructed acids. For example, we can see from the pK_a values of butanoic acid, 4-chlorobutanoic acid, 3-chlorobutanoic acid, and 2-chlorobutanoic acid (Table 6-2) that the electronegative chlorine atom enhances acidity and that its effect is greater the closer it is to the acidic —CO_2H site. The electronegative atom (chlorine) withdraws electron density through the series of σ bonds. This **inductive effect**—a charge polarization through a series of σ bonds—causes a shift of electron density from the

Table 6-1 ▸ Approximate pK_a Values of Organic and Inorganic Acids

Compound	pK_a	Compound	pK_a	Compound	pK_a
$HOSO_2O$—H **Sulfuric acid**	−10	HS—H **Hydrogen sulfide**	7	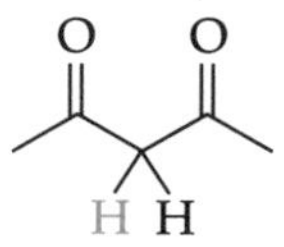 2,4-Pentadienone	9
I—H **Hydroiodic acid**	−10	ArS—H **Thiophenol**	7		
Br—H **Hydrobromic acid**	−9	H_3N^+—H **Ammonium ion**	9	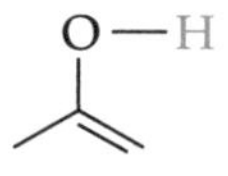 Acetone enol	11
Cl—H **Hydrochloric acid**	−7	HCN **Hydrogen cyanide**	9		
$ArSO_2O$—H **An arylsulfonic acid**	−6.5	ArO—H **A phenol**	10	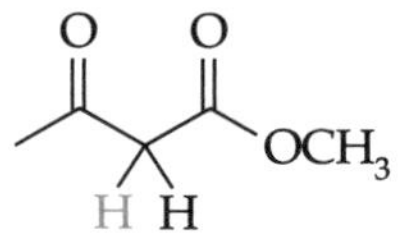 Methyl acetoacetate	11
H_2O^+—H **Hydronium ion**	−1.7	CH_3O—H **Methanol**	15.5		
O_2NO—H **Nitric acid**	−1.5	HO—H **Water**	15.7	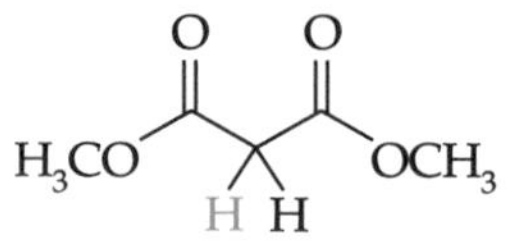 Dimethyl malonate	13
F—H **Hydrofluoric acid**	3	CH_3CH_2O—H **Ethanol**	16		
CH_3COO—H **Acetic acid**	5	$(CH_3)_3CO$—H **t-Butanol**	18	 Cyclopentadiene	16
HCO_3—H **Carbonic acid**	5				

acid site and, hence, stabilizes the anion formed by deprotonation. An even stronger electron-withdrawing substituent, such as a nitro (—NO_2) group, would enhance the acidity of the carboxylic acid even more. The pK_a of nitroacetic acid ($O_2NCH_2CO_2H$) is 1.68, whereas that of chloroacetic acid ($ClCH_2CO_2H$) is 2.68.

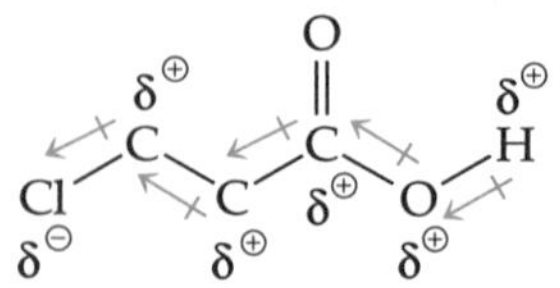

Because an inductive effect is transmitted *through bonds,* the effect is greater when transmission is through fewer bonds; that is, when the electronegative element is closer to the acidic site. Furthermore, the greater the number of electronegative atoms, the greater is the stabilization by electron withdrawal. Thus, acidity increases in the series: acetic acid; mono-; di-; and trichloroacetic acid (Table 6-2). Compared with changes in the element X in an H–X bond, these inductive effects are small, but nonetheless important. For example, trichloroacetic acid is five times as acidic as dichloroacetic acid.

Table 6-1 ▸ Approximate pK_a Values of Organic and Inorganic Acids (*continued*)

Compound	pK_a	Compound	pK_a	Compound	pK_a
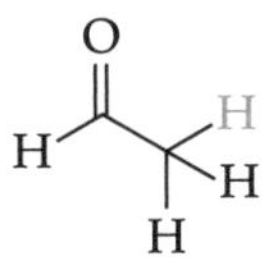 Acetaldehyde	17	N≡C—CH$_2$—H Acetonitrile	25	H—N(iPr)$_2$ Diisopropylamine	40
Acetamide	17	CH$_3$C≡C—H Propyne	25	PhCH$_2$—H Toluene	41
Acetone	19	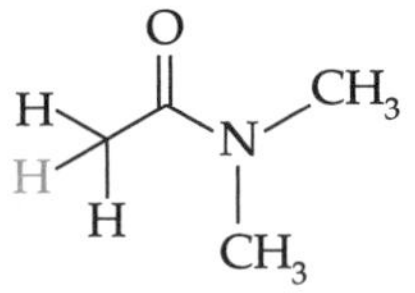 N,N-Dimethylacetamide	30	H$_2$C=CHCH$_2$—H Propene	43
Methyl acetate	25	Ph$_3$C—H Triphenylmethane	31.5	Ph—H Benzene	43
		 1,3-Pentadiene	33	H$_2$C=C—H Ethene	44
		H—H Dihydrogen	35	CH$_3$—H Methane	48
		H$_2$N—H Ammonia	38	CH$_3$CH$_2$—H Ethane	50
				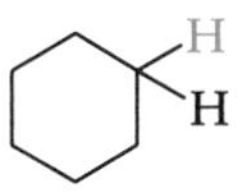 Cyclohexane	51

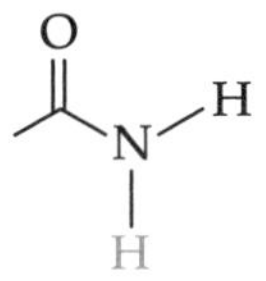

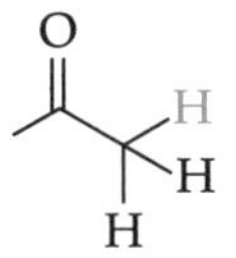

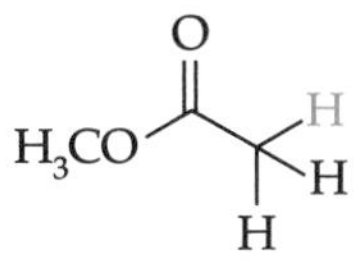

Table 6-2 ▸ Inductive Effects on Acidity

Compound	pK_a	Compound	pK_a
Cl$_3$CCO$_2$H Trichloroacetic acid	0.6	CH$_3$CH$_2$CHCO$_2$H Cl 2-Chlorobutanoic acid	2.9
Cl$_2$CHCO$_2$H Dichloroacetic acid	1.3	CH$_3$CHCH$_2$CO$_2$H Cl 3-Chlorobutanoic acid	4.1
ClCH$_2$CO$_2$H Chloroacetic acid	2.9	CH$_2$CH$_2$CH$_2$CO$_2$H Cl 4-Chlorobutanoic acid	4.5
CH$_3$CO$_2$H Acetic acid	4.9	CH$_3$CH$_2$CH$_2$CO$_2$H Butanoic acid	4.9

EXERCISE 6-O

Predict which compound in each of the following pairs is more acidic and give reasons for your choice:

(a)

Cl—CH₂—C(=O)—OH   or   F—CH₂—C(=O)—OH

(b)

CH₃—CHCl—CH₂—C(=O)—OH   or   CH₃—CH₂—CHCl—C(=O)—OH

(c) F_3CCH_2—OH   or H_3CCH_2—OH

The replacement of a hydrogen atom by an alkyl substituent decreases the acidity of dissolved alcohols. Thus, the solution-phase acidities of methanol, ethanol, 2-propanol, and *t*-butanol decrease by more than two pK_a units as the carbon bearing the OH group is more fully alkylated (Table 6-1). The order of acidity in solution (MeOH > EtOH > *i*-PrOH > *t*-BuOH), where solvation and intermolecular association are important, is reversed in the gas phase, where effects are observed in isolated molecules. The observed solution-phase pK_as must therefore be sensitive to intermolecular effects. The replacement of a hydrogen atom by a group of comparable electronegativity but much larger size stabilizes an ion in the gas phase but induces a pronounced destabilizing **steric effect** on the solvation of the anion (conjugate base) in solution. Because an alkyl group is nonpolar and less polarizable than most solvents, its presence inhibits the stabilizing interaction of a polar solvent molecule with the anion (Figure 6-12) for much the same reason that "like dissolves like" (see Chapter 1).

Hybridization Effects

From the pK_a values in Table 6-1, we see that the hybridization of the carbon atom attached to the acidic hydrogen greatly influences the acidity of simple carbon acids: with more *p* character, the acidity of a C–H bond *decreases* appreciably. Thus, the pK_a of ethane (with an sp^3-hybridized C–H bond) is about 50, that of ethene (with an sp^2-hybridized C–H bond) is about 44, and that of ethyne (with an sp-hybridized C–H bond) is about 25.

The explanation for this order is that, in each case, the acid dissociation equilibrium generates an anion whose lone pair of electrons is held in a different kind of hybridized orbital. We have seen in Chapter 2 that sp-hybridized atoms are more electronegative than sp^2- or sp^3-hybridized atoms. Thus, a lone pair in an sp^3-hybridized orbital (25% *s* character) is held farther from the nucleus than one in an sp^2-hybridized orbital (with 33% *s* character), which is farther from the nucleus than one in an sp-hybridized orbital (50% *s* character). Because it is more favorable for the negative charge of an anion to be in an orbital that is closer to the positively charged nucleus, an sp-hybridized anion is more stable than an sp^2-hybridized anion, which is more stable than an sp^3-hybridized anion.

Resonance Effects

We have seen in Chapter 3 that allyl cations and radicals are stabilized by resonance. The allyl anion, formed by the deprotonation of propene, is similarly stabilized if the electron pair released by deprotonation is accommo-

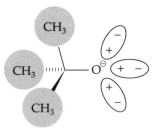

versus

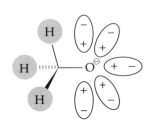

▲ **FIGURE 6-12**
A comparison of the different solvation of the sterically hindered *t*-butoxide anion and the methoxide anion. Solvent molecules more easily approach the negatively charged oxygen atom in the smaller methoxide anion.

dated in a *p* orbital formed as the original sp^3-hybridized center becomes sp^2-hybridized.

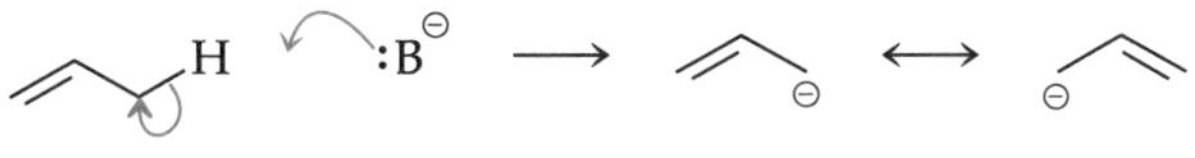

Delocalization of negative charge along the three-atom system, together with equal resonance contributions from the two structures, appreciably enhances the stability of this anion. The stabilization accompanying this resonance delocalization causes both the allyl and enolate ions to be planar because, in the flat geometry, orbital interaction is strongest. Thus, the sp^3-hybridized C–H bond in propene (pK_a 43) is more acidic than that in propane (pK_a 50). When the conjugation of the allyl anion is extended further, the additional resonance contributors stabilize the anion and make the protonated form more acidic. Thus, deprotonation of 1,3-pentadiene ($pK_a \approx$ 33) is easier than that of propene because of the additional resonance stabilization of the more highly conjugated anion derived from the diene.

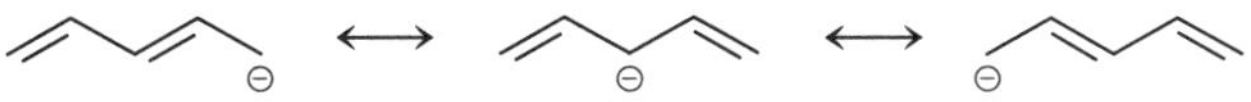

There is also resonance stabilization of the anion that results from deprotonation of a C–H group adjacent to a carbonyl group. The resulting anion is stabilized not only because of delocalization like that encountered in the allyl anion, but also because one of the resonance contributors has negative charge on a more-electronegative oxygen atom.

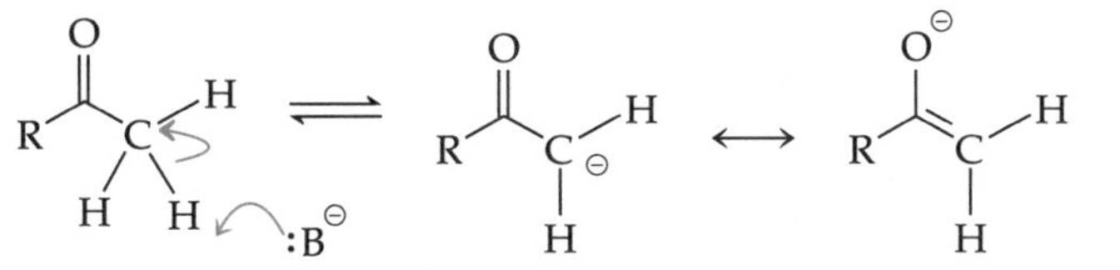

This anion, referred to as an enolate, is one of the most-important anions of organic chemistry. Despite the fact that deprotonation adjacent to the carbonyl group of an aldehyde or a ketone requires the breaking of a C–H rather than an O–H bond, the α C–H acidity is sufficiently enhanced that it is only about five pK_a units less acidic than an alcohol O–H group.

EXERCISE 6-P

Draw the significant resonance contributors for the enolate anion generated from each of the following compounds:

(a)

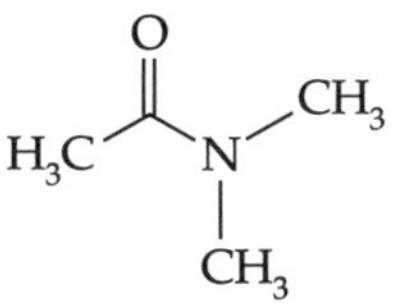

(d)

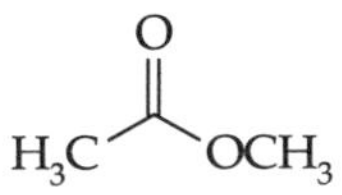

(b)

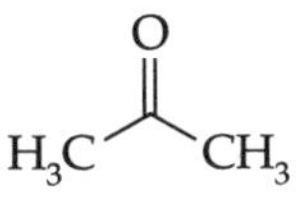

(e)

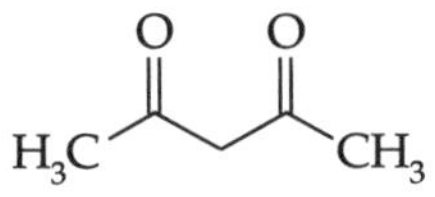

(c)

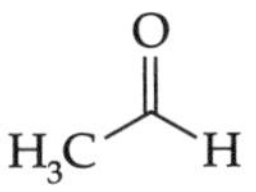

(f)

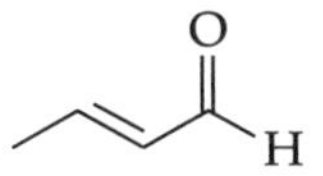

Predict which hydrogen is most acidic in each of the following compounds:

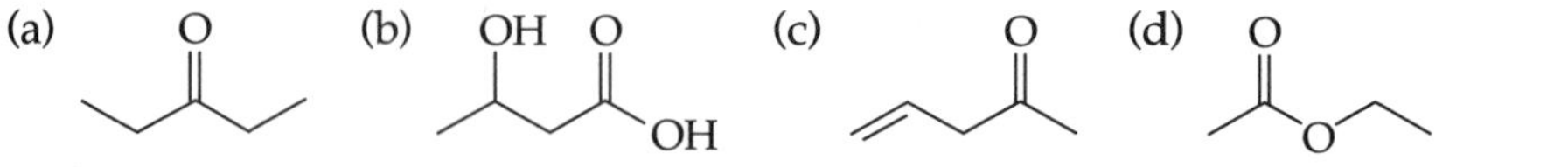

Enolate Anion Stability

Let us consider in more detail how the structure of the carbonyl group influences the acidity of an α C–H. Typical values for the pK_a for an aldehyde (17), a ketone (19), an ester (25), and a 1,3-diketone (9) are listed in Table 6-1. The difference between values observed for aldehydes and ketones is the result of compensating factors; that is, an alkyl group attached to a carbonyl carbon of a ketone behaves as if it were electron releasing, stabilizing the ketone relative to the aldehyde. (Indeed, the two bonds of the carbonyl group in a ketone are together worth 179 kcal/mole, whereas those in an aldehyde are worth only 176 kcal/mole. With the same trend, the σ and π bonds of the carbonyl group of formaldehyde are worth only 173 kcal/mole.) Although alkyl substituents also stabilize alkenes (Chapter 2), they do so by electron release, which is very much less important in the negatively charged enolate anion.

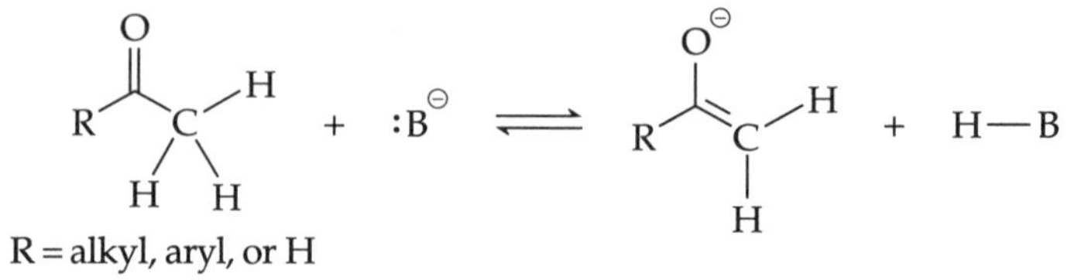

To compare the acidity of an aldehyde or a ketone with that of a typical ester or amide, we must keep in mind that an acid derivative bears a heteroatom bound to the carbonyl group. Oxygen or nitrogen can influence acidity by an inductive effect, but we must also consider the resonance interaction of the adjacent heteroatom with the carbonyl π bond. Resonance contributors such as those shown at the far right below for the ester or amide groups diminish the ability of the carbonyl oxygen to accommodate further charge, as is needed for the stabilization of the enolate anion.

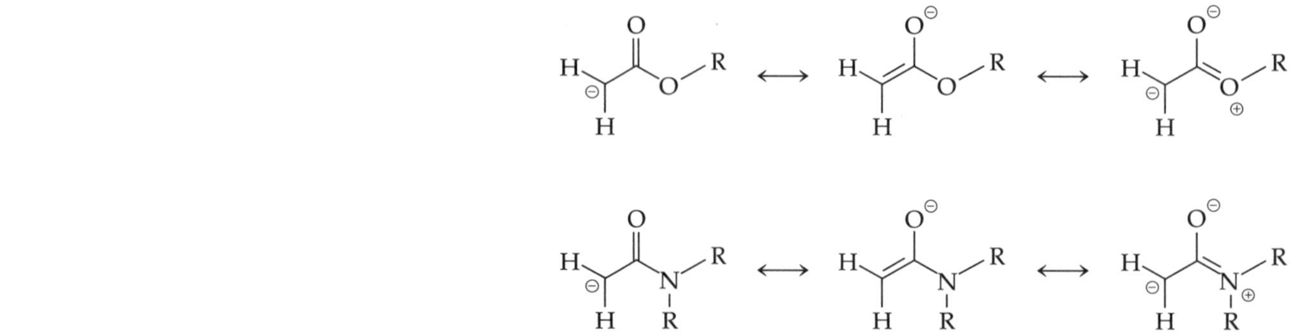

Thus, the α-C–H acidity of esters and amides is somewhat lower than that of aldehydes or ketones. Nitrogen (in an amide) is better able than oxygen (in an ester) to act as a π electron donor toward a carbonyl group; that is, electron donation from the less-electronegative nitrogen atom induces greater partial double bond character in the C–N bond of an amide. The stronger contribution of amide resonance in the structure at the lower right than of ester resonance in that at the upper right against the enolate stabilization in the structures at the left is responsible for the lower α-C–H acidity in an amide than in an ester. (This argument applies only to tertiary amides because, in primary and secondary amides, the hydrogen on nitrogen is more acidic than the α-C–H.)

The placement of two carbonyl groups in a 1,3 relation induces further delocalization (and stability) in the resulting anion, in the same way as already seen for the allyl and pentadienyl anions. The acidity of a 1,3-diketone (and other 1,3-dicarbonyl groups) is thus quite high.

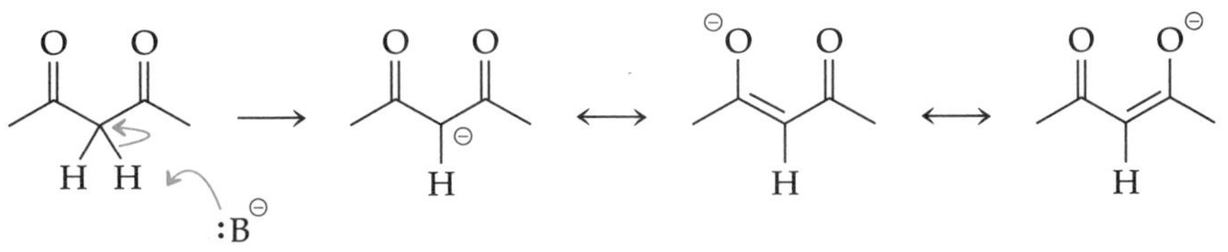

Aromaticity

The effects of aromaticity contribute significantly to the stability of an anion. For example, deprotonation of cyclopentadiene generates the cyclopentadienide anion (Figure 6-13), which contains six (that is, $4n + 2$) electrons in a Hückel aromatic system: the pK_a of cyclopentadiene (16) is very different from that of 1,3-pentadiene (33) and is, indeed, very close to that of water, despite the cleavage of a C–H rather than an O–H bond. Deprotonation of cycloheptatriene results in an anion for which we can draw seven identical resonance contributors. Nonetheless, cycloheptatriene has a pK_a of about 40 and is thus significantly less acidic than cyclopentadiene. (As an exercise, draw the remaining six resonance structures for the cycloheptatrienyl anion.) This decreased acidity is the direct result of the difference in the number of electrons in the two anions: six in cyclopentadienyl anion, corresponding to a Hückel aromatic system ($4n + 2$, $n = 1$) and eight in cycloheptatrienyl anion ($4n$, $n = 2$), a system that, though delocalized, is not a Hückel aromatic. This large difference in acidity (10^{24}) can thus be attributed directly to the effects of aromaticity on the stabilization of cyclopentadienide as a cyclic, conjugated, planar, delocalized anion.

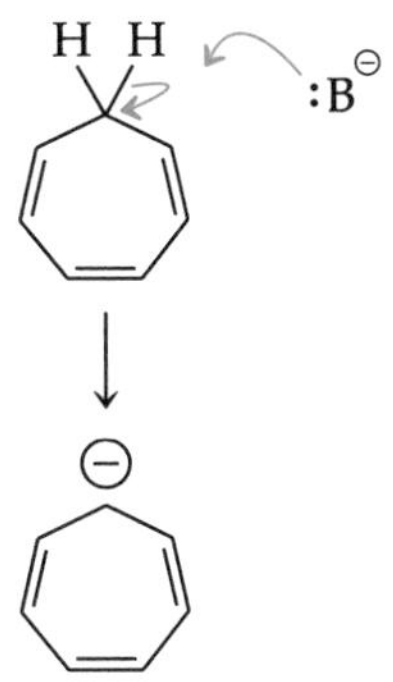

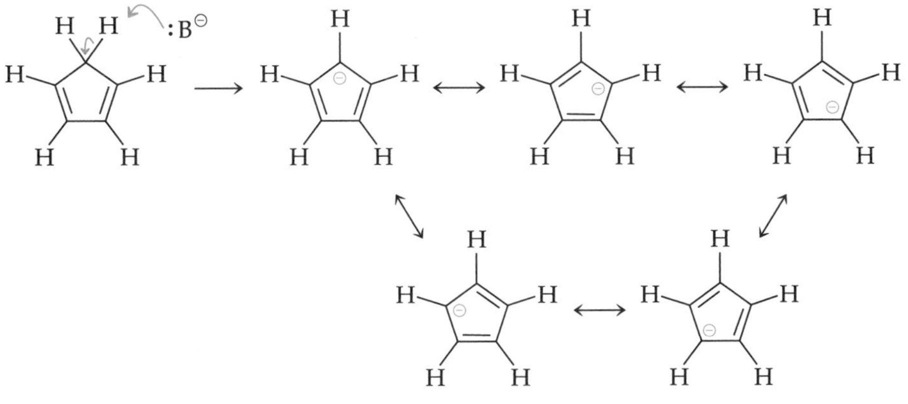

▲ FIGURE 6-13
Resonance contributors of the cyclopentadienide anion.

EXERCISE 6-R

Which member of each of the following pairs of compounds is more readily deprotonated?

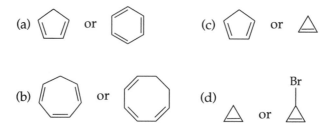

These factors govern the acidity of a given functional group and, hence, the position of its chemical equilibrium with its conjugate base. Acidity is important not only as a concept used to illustrate thermodynamic equilibria, but also as a means of ranking the relative stabilities of anions and the thermodynamic feasibility of various reactions under acidic and basic conditions.

In summary, the following effects control relative acidity. Everything else being equal, HX is a stronger acid than HY if:

1. X is a more electronegative atom than Y ($HF > H_2O > NH_3 > CH_4$);

2. the H–X bond is weaker than the H–Y bond ($HI > HBr > HCl > HF$);

3. X is a group bearing more electronegative atoms closer to the site of negative charge in the conjugate base X^- than in Y^- ($Cl_3CCO_2H > Cl_2CHCO_2H > ClCH_2CO_2H > CH_3CO_2H$);

4. X^- is less sterically blocked from solvation than Y^- ($MeOH > EtOH > i\text{-}PrOH > t\text{-}BuOH$);

5. X^- has a greater fractional s character than Y^- ($RC{\equiv}CH > R_2C{=}CRH > R_3C{-}CR_2H$);

6. the negative charge in X^- can be delocalized over a larger number of atoms than in Y^- ($CH_2{=}CH{-}CH{=}CH_2{-}CH_3 > CH_2{=}CH{-}CH_3 > CH_3CH_2CH_3$ or $CH_3COCH_2COCH_3 > CH_3COCH_3 > CH_3CH_2CH_3$);

7. the negative charge in X^- can be delocalized onto a more-electronegative atom than in Y^- ($CH_3COCH_3 > CH_3CH{=}CHCH_3 > CH_3CH_2CH_3$ or $CH_3CHO > CH_3CO_2CH_3 > CH_3CON(CH_3)_2$); or

8. the negative charge in X^- is more stabilized by aromaticity than in Y^- (cyclopentadiene > 1,3-pentadiene > cycloheptatriene).

Look carefully at Table 6-1 to see specific examples that illustrate each of these effects.

6-8 ▸ Kinetic and Thermodynamic Control

Unfortunately, a study of the changes in bond energy in the conversion of a starting material into a product provides information about $\Delta H°$ but not about the energy of the relevant transition state or any possible reactive intermediates. In most cases, it is not possible to predict reaction kinetics solely on the basis of the thermodynamics of a reaction. Nonetheless, as we study various reactions, we will learn empirically (that is, from experimental observations) which reactions are rapid and which ones are slow. With care, we can extrapolate these observations (within limits) to predict rates for similar reactions.

Reactions are generally considered to take place under either kinetic or thermodynamic control. This distinction is based on the extent to which the reverse reaction (from product to reactant) takes place under the reaction conditions employed. If the reverse reaction is rapid, equilibrium is quickly established. When the position of the equilibrium is fixed by $\Delta G°$, the reaction is said to be under **thermodynamic control.** If the reverse reaction cannot occur (or does so very slowly) under the reaction conditions, the reaction is said to be under **kinetic control.**

The distinction between kinetic and thermodynamic control is most relevant when a reaction can partition between two products of different stabilities. In general terms, a reaction proceeds under thermodynamic control when the difference in activation energies for two competing forward and reverse reactions is small enough that the energy difference between products governs the partitioning. The equilibrium thus established is therefore governed by $\Delta G°$ (the difference in free energy between the two products) rather than the difference in the activation energies.

In contrast, a reaction proceeds under kinetic control when a large difference in activation energies allows one transition state to be reached more readily than the other. No equilibrium is established between products and the preference for one product over another is determined by the relative heights of the activation barriers for the two processes. As stated in Section 6-1, in an exothermic reaction, the activation energy barrier for the forward reaction is less than that for the reverse reaction (by $\Delta H°$). For $\Delta H°$ values larger than a few kilocalories per mole, the rates of the forward and backward reactions become so different that equilibrium is generally difficult to establish within a reasonable period of time. Reactions with large values of $\Delta H°$ are thus generally considered to be under kinetic control. Such reactions are finished, for practical purposes, once the energy barrier from the reactant has been surmounted.

Let us construct a reaction profile to illustrate these two types of reaction control. Suppose that a reactant, R, can be converted into either of two products of different stabilities: P_1 or P_2 (Figure 6-14, on page 218). A specific example of this general type is the protonation of an enolate anion (R) to form either a ketone (P_1) or an enol (P_2), a reaction whose thermodynamics we considered earlier in this chapter.

The ketone is more stable than the enol, but the conversion of the enolate into the ketone requires a higher activation energy than that needed to convert the enolate into the enol. If we supply sufficient energy for a reactant to overcome both of the barriers, $\Delta H^{\ddagger}_1$ and $\Delta H^{\ddagger}_2$, sufficient energy is available to interconvert the ketone, the enol, and the enolate so that equilibrium is established. Under these **reversible** conditions, the more stable product (the

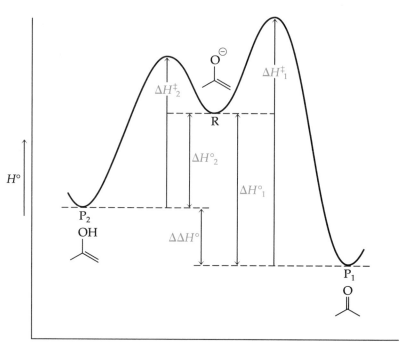

▲ FIGURE 6-14

Kinetic (R → P_2) and thermodynamic (R → P_1) control in the protonation of the acetone enolate anion.

ketone) is ultimately formed, with the distribution between the two products being governed by the enthalpy difference $\Delta\Delta H°$, assuming $\Delta\Delta S° \approx 0$. Because this difference is greater than a few kilocalories per mole, the product composition is completely dominated by the more-stable product. This reaction is then considered to proceed under thermodynamic control. Here, thermodynamics favors the weaker acid (pK_a of ketone $\approx$ 19) over the stronger (pK_a of enol $\approx$ 10).

However, when the difference in activation energies is relatively large, it is sometimes possible to achieve reaction conditions in which the rate of formation of the ketone is so small as to be unobservable, whereas the reaction to produce enol proceeds at a convenient rate. Because the activation energies for producing the enolate anion from both ketone and enol are larger than that leading from the enolate to the enol, both reverse (deprotonation) reactions proceed at such slow rates as to have no effective consequence. Under these conditions, the enol is observed because it is formed faster than the ketone, despite its lower thermodynamic stability. Under these conditions, the reaction would be said to proceed under kinetic control.

EXERCISE 6-S

Suggest a chemical reason why protonation of the enolate to give the enol may be faster than that to form ketone. (Hint: Think about charge density in the resonance-stabilized enolate anion.)

Suppose that we choose to conduct the reaction in Figure 6-14 so that only a small fraction of the reactant molecules have sufficient energy to overcome $\Delta H^{\ddagger}_1$ or $\Delta H^{\ddagger}_2$. Under such conditions, irreversible reaction to P_2 proceeds smoothly at a reasonable rate. If, on the other hand, we had chosen a higher temperature at which the reaction proceeds rapidly in either direction, equilibrium is established, favoring the more-stable species, P_1. Thus, adjusting the reaction temperature is often the easiest way of shifting from kinetic to thermodynamic control. Conducting a reaction at low temperature generally favors the kinetic products, but the specific temperature at which thermodynamics dominates depends on the specific reaction being considered.

EXERCISE 6-T

For each of the following energy diagrams, indicate whether kinetic or thermodynamic control is more likely as $R \rightarrow P_1$ or P_2 or both.

(a)

(b)

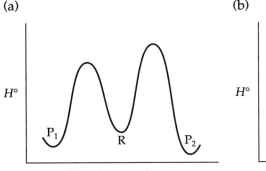

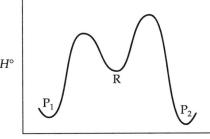

(c)

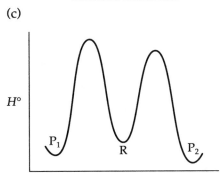

The distinction between kinetic and thermodynamic control is sometimes a qualitative one because few reactions are so exothermic that equilibrium cannot be achieved under some conditions. We use the term kinetic control to refer to those reactions in which the conversion from reactant into one product that is less stable than others can be driven to essential completion under chosen laboratory conditions before significant reverse reaction takes place.

6-9 ▸ Reaction Rates: Understanding Kinetics

Let us now consider other methods for controlling reactions. The rate at which any reaction proceeds is governed by the energy of the highest-lying transition state: the best method for controlling reactivity depends on whether the rate-determining step consists of a single species or whether it requires collision of two or more reagents. A reaction having only a single species in the rate-determining step is referred to as a **unimolecular reaction.**

THREE-BODY COLLISIONS

Termolecular reactions are extremely rare simply because the probability of three bodies colliding with the correct orientations is very low. The Battle of Gettysburg during the American Civil War was extremely intense, with many millions of bullets fired by both sides in very short periods of time. (More Americans were killed in the Civil War than in all wars in the twentieth century combined.) Nonetheless, historians have uncovered only examples in which two bullets met and fused in midair.

One that requires a collision between two reactants in the rate-determining step is referred to as **bimolecular**. Reactions requiring collision between more than two species are rare.

Unimolecular Reactions

Typically, a unimolecular reaction consists either of homolytic cleavage to radical fragments or of heterolytic cleavage to ionic fragments, followed by fast conversion of these reactive intermediates into products. In either case, the rate is governed by the number of reactant molecules per unit time that have sufficient energy to overcome the activation barrier that separates reactants and products. Estimation of the energy required for this transformation is derived from **transition state theory.** This theory recognizes that a given reaction proceeds efficiently only if the energy necessary for a reactant to approach the transition state is available.

In general, the facility with which a unimolecular reaction takes place can be enhanced either by increasing the fraction of molecules that can pass over an energy barrier or by decreasing the barrier. The former is accomplished by changing the temperature, whereas the latter is effected by altering the structure of the substrate undergoing the reaction or the way in which the reaction is conducted (for example, by increasing the polarity of a solvent in a reaction that includes the formation of ions).

Activation energy barriers govern rates in the same way that entropy and enthalpy govern thermodynamics. When entropy is negligible, the Arrhenius equation (equation 12) describes the relation between the observed rate constant k_{obs}, and the activation energy, $\Delta H^{\ddagger}$, in which A is a fitting factor characteristic of the reaction, R is the ideal gas constant, and T is the temperature.

$$k_{obs} = Ae^{-\Delta H^{\ddagger}/RT} \tag{12}$$

and

$$\ln k_{obs} = \ln A - \frac{\Delta H^{\ddagger}}{RT}$$

This equation shows how activation energies can be determined in the laboratory: a plot of the logarithm of the observed rate constant (at various temperatures) against $1/T$ will have a slope equal to $-\Delta H^{\ddagger}/R$.

When the Arrhenius equation is rewritten as in equation 13, the form of the equation becomes parallel to equation 2, in Section 6-6.

$$\Delta H^{\ddagger} = -RT \ln\left(\frac{k_{obs}}{A}\right) \tag{13}$$

Thus, activation energies ($\Delta H^{\ddagger}$) relate to reaction rate constants (k_{obs}) much as free energy changes (ΔG°) relate to the equilibrium constant (K).

EXERCISE 6-U

Calculate the relative rate of (k_1/k_2) of two reactions (A → B) and (C → D) occurring at room temperature with identical *A* values if the difference in activation energies is:

(a) 0   (b) 1 kcal/mole   (c) 2 kcal/mole   (d) 5 kcal/mole

At a given temperature, an array of many molecules has a fixed distribution of individual energies. A typical distribution, as first described by Ludwig Boltzmann and shown in Figure 6-15, describes the probability that an individual molecule has a specific energy (greater or lesser than average) at a fixed temperature T_1. The resulting curve is called a **Boltzmann distribution.** It tells us the fraction of molecules that have energies greater than some fixed value (for example, the activation energy, $\Delta H^{\ddagger}$) at a defined temperature T_1. At a higher temperature, T_2, the curve describing the energy distribution would be broadened and shifted to the right, indicating both a higher mean energy of the array of molecules and a greater fraction of molecules with an energy greater than a fixed value.

For a reaction that proceeds with the activation energy shown by the vertical line in Figure 6-16 (on page 222), only those molecules with an energy greater than the activation energy barrier can overcome that barrier and proceed to product. At the higher temperature, T_2, a significantly larger fraction of the molecules have sufficient energy to overcome the barrier and form product. The rate of a unimolecular reaction therefore increases with increasing temperature.

EXERCISE 6-V

How would the shape of a typical Boltzmann curve change if:

(a) temperature is increased?

(b) temperature is decreased?

(c) activation energy is increased at room temperature?

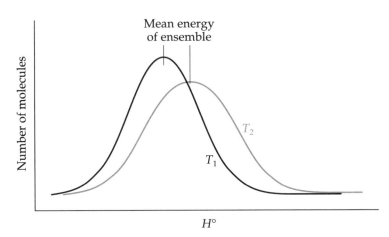

▲ FIGURE 6-15

Distribution of energies of an array of molecules at two temperatures ($T_1 < T_2$).

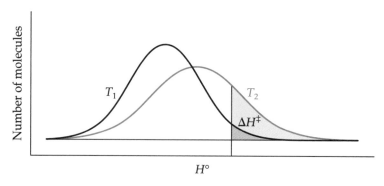

▲ **FIGURE 6-16**

The fraction of molecules with sufficient free energy for reaction ($\Delta G° > \Delta H^{\ddagger}$) is shown by the area under the curve in color.

Bimolecular Reactions

In a bimolecular (or higher-order) reaction, the rate is governed by three factors: the number of collisions between the reacting partners within the time period of interest, the energy of the colliding molecules, and the orientation at the moment of collision. The number of collisions per unit of time is a function of the number of the reacting species per unit of volume. If we increase the number of molecules per volume—that is, if we increase the **concentration**—we proportionally increase the number of collisions and hence the number of effective collisions. Thus, the rate of a bimolecular reaction is proportional to the concentration of each of the reactants.

As in a unimolecular reaction, the complex formed by the collision of two reagents must have sufficient energy to overcome the activation energy barrier. Increasing the temperature increases the average kinetic energy of both reactants, and hence the probability increases that a collision between the two more-activated reagents has sufficient energy to overcome the barrier and be converted into product. Thus, an increase in temperature has a similar effect on unimolecular and bimolecular reactions (as long as entropy effects remain minor).

For a collision to induce a chemical change, two molecules must approach one another in the correct orientation that leads to the transition state. Because most organic molecules are not spherically symmetrical, the fraction of accessible geometries that can lead to product, possibly aided by electrostatic attractions or repulsions, influences the fraction of productive energy-sufficient collisions. Because the number of *productive* collisions increases in parallel with the number of *all* collisions, the reaction rates of bimolecular (or more-complex termolecular reactions) can be controlled by increasing not only temperature, but also the relative concentration of each of the reactants required to form the transition state.

Conclusions

We can use a reaction energy profile to easily distinguish endothermic, exothermic, and thermoneutral reactions. A reaction in which a reactive intermediate is formed is said to proceed through more than one stage—that is, it is a multistep reaction. One that proceeds without intermediates is referred to as concerted. Energy profiles graphically illustrate the formation of reactive intermediates in multistep reactions and allow easy determination of the rate-determining step. The important organic intermediates are carbocations, free radicals, carbanions, carbenes, radical cations, and radical anions.

Reaction profiles illustrate enthalpy changes in the conversion of a reactant into a product and describe activation barriers in the reaction sequence. In most organic reactions, entropy changes contribute only slightly, if at all, to free energy, although the importance of entropy depends on the reaction temperature and on the stoichiometry of a reaction.

Activation barriers describe the energy required to reach the transition state, the structure and energy of which can be related to reactants or products according to the Hammond Postulate. According to the principle of microscopic reversibility, the forward and reverse reactions of a reversible transformation occur through the same transition state.

Chemical equilibria can be used to measure the thermodynamics of a given conversion. Under strictly reversible conditions, a reaction proceeds to the more-stable product and is under thermodynamic control. If insufficient activation energy is available to allow a reaction to be fully reversible, a reaction is said to proceed under kinetic control, and the product distribution obtained is controlled by the difference in activation barriers (rates) rather than in $\Delta G°$ (product stability).

Acid-base equilibria constitute one of the most-important classes of chemical equilibria. The position of such an equilibrium is described by pK_a, and the acidities of organic compounds are sensitively influenced by differences in electronegativity, hybridization, bond energies, resonance stabilization, aromaticity, and inductive effects in the anion formed by the deprotonation of a neutral organic molecule. A large positive pK_a is indicative of a weak organic acid.

Rates of reaction are governed by activation barriers; and at a given time, only a fraction of the individual molecules possess sufficient energy to overcome a fixed reaction barrier. Activation barriers are large in transition states in which bond breaking has been substantial, whereas they are smaller in transition states that have high degrees of bond making. Biomolecular reactions are described by collision theory such that increasing the concentration of either reactant enhances the probability of a productive collision.

Summary of New Reactions

Diels–Alder Reaction

Acid–Base Reactions: Deprotonation

$$HX + B: \rightleftharpoons X^{\ominus} + {}^{\oplus}BH$$

Deprotonation α to a Carbonyl Group: Formation of an Enolate Anion

Keto-Enol Tautomerization

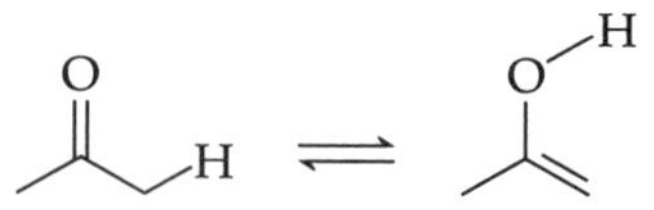

Review Problems

6-1 Draw a reaction profile that corresponds to each of the following descriptions:

(a) an exothermic concerted reaction

(b) an endothermic reaction taking place in two steps through a reactive intermediate

(c) an exothermic reaction occurring in three steps in which the second step is rate determining

6-2 Rank the following intermediates according to stability (most stable first). Explain your choices.

(a) $CH_3CH_2CH_2CH_2^{\oplus}$, $CH_3CH^{\oplus}CH_2CH_3$, $(CH_3)_2C^{\oplus}CH_2CH_3$, $(CH_3)_3C^{\oplus}$

(b) $CH_3CH_2CH_2CH_2\cdot$, $CH_3CH\cdot CH_2CH_3$, $(CH_3)_2C\cdot CH_2CH_3$, $(CH_3)_3C\cdot$

(c) $CH_3CH_2CH_2CH_2^{\ominus}$, $CH_3CH^{\ominus}CH_2CH_3$, $(CH_3)(C_6H_5)C^{\ominus}CH_2CH_3$, $(CH_3)_3C^{\ominus}$

(d) $CH_2\colon$, $CH_3CH\colon$, $C_6H_5CH\colon$, $(C_6H_5)_2C\colon$

(e) $H_2CO^{\ominus}$, $(C_6H_5)_2CO^{\ominus}$, $H_2C{=}CH_2^{\ominus}$, $H_2C{=}C(C_6H_5)_2^{\ominus}$

(f) $(C_6H_6)^{\oplus}$, $p\text{-}NO_2(C_6H_5)^{\oplus}$, $p\text{-}CH_3(C_6H_5)^{\oplus}$, $p\text{-}Cl(C_6H_5)^{\oplus}$

6-3 Draw resonance structures to indicate possible sites of formal charge or of the unpaired electron in:

(a) naphthalene cation radical

(b) benzophenone anion radical

6-4 At low temperature, *cis*-1,2-diphenylcyclopropane forms an anion at the benzylic position upon treatment with strong base. When quenched with D_2O, a mixture of 1-deutero-*cis*-1,2-diphenylcyclopropane and 1-deutero-*trans*-1,2,-diphenylcyclopropane is formed. Draw the structures of both the intermediate anion and the product in three dimensions. Does the structure of the product allow you to say anything about whether the carbanionic carbon in the intermediate anion is a center of chirality?

6-5 For each of the following reaction types, is the entropy change positive (favorable), negative (unfavorable), or near zero (negligible)?

(a) a large molecule fragments into three smaller ones

(b) a small molecule condenses with another small molecule to make a large one

(c) an ester is hydrolyzed by water to produce a carboxylic acid and an alcohol

6-6 Which diene and dienophile would you choose to synthesize each of the following products by a Diels-Alder reaction?

(a)

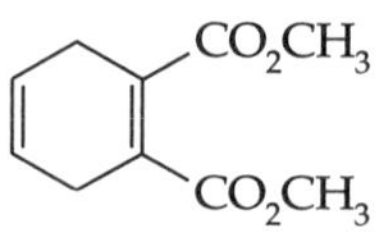

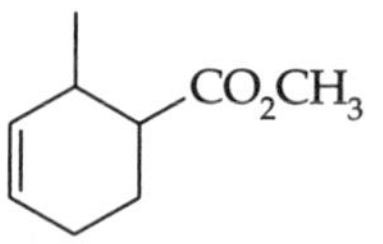

(b)

(c)

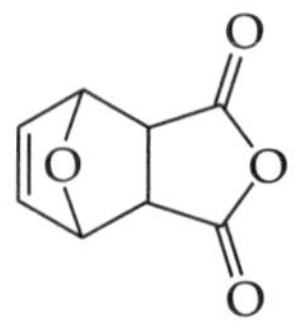

(d)

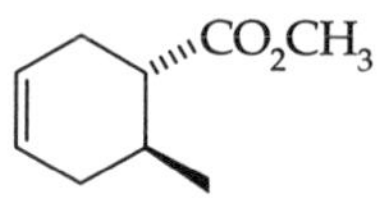

6-7 In the following proposed Diels-Alder reactions, reaction 1 proceeds but reaction 2 fails. Explain why.

Reaction 1:

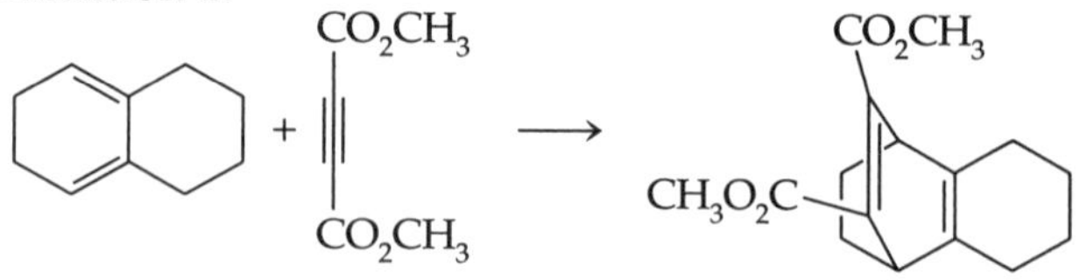

Reaction 2:

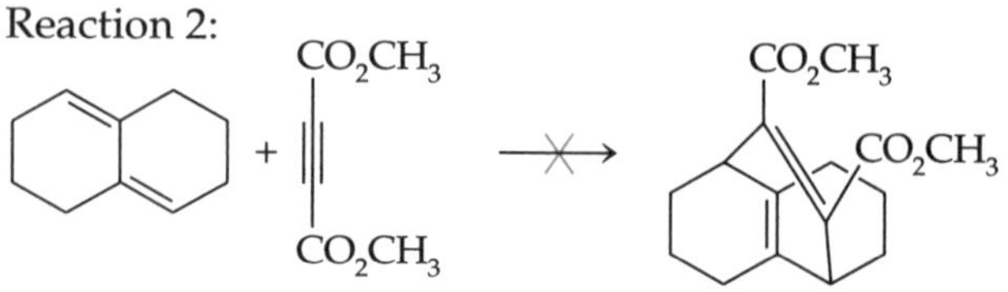

6-8 Reorder the following sets of compounds according to increasing pK_a:

(a) cyclohexanol, phenol, cyclohexanecarboxylic acid

(b) 1-butyne, 1-butene, butane

(c) propanoic acid, 3-bromopropanoic acid, 2-nitropropanoic acid

(d) phenol, toluene, benzene

(e) dimethyl ether, ethanol, methyl acetate ($CH_3CO_2CH_3$)

(f) hexylamine, aniline, hexanoamide

(g) benzoic acid, *p*-chlorobenzoic acid, 2,4,6-trichlorobenzoic acid

(h) ethanoic acid (acetic acid), 1,2-ethanedioic acid (oxalic acid), 1,3-propanedioic acid (malonic acid) (Hint: A carboxylic acid group acts as an effective electron-withdrawing group to the adjacent σ-bond system.)

(i) protonated forms of pyrrole, pyridine, *N*-methylpyrrole

6-9 Crotonaldehyde ($CH_3CH\!=\!CHCHO$) has a pK_a of 20, despite the fact that it lacks enolizable hydrogens α to the carbonyl group.

(a) Determine which hydrogen is removed upon interaction with base.

(b) Write one or more resonance structures to account for the stability of the anion (conjugate base).

6-10 Octylamine is insoluble in water, but it dissolves in dilute sulfuric acid. Octanoamide ($C_7H_{15}CONH_2$) does not dissolve either in water or in dilute sulfuric acid. Rather, octanoamide dissolves in dilute aqueous base. Propose an explanation for the contrasting solubilities of the amine and amide in both acid and base.

6-11 For each of the following molecules, determine which proton is most acidic; that is, which proton would be removed by treatment with one equivalent of base?

(a) OH

(c) N NH₂

(b) NH

(d) H N O

6-12 Compound A is converted into a more-stable product B upon heating without any additional reagent. The reaction profile shows the formation of a reactive intermediate as A reacts.

(a) Draw an energy diagram that illustrates the key features of this reaction.

(b) Suppose that the reaction is conducted at a higher temperature. Does this change the shape of the energy diagram?

(c) Does the rate of reaction depend on the concentration of A?

6-13 In the following molecules, determine which proton is most acidic; that is, which proton would be removed by treatment with one equivalent of base?

(a)

(b)

(c)

(d)

6-14 Consider a reaction in which reactants C and D combine in the rate-determining step. Determine which of the following statements applies to this reaction, and explain your reasoning.

(a) Doubling the concentration of C doubles the rate of the reaction.

(b) Doubling the concentration of D cuts the rate of reaction in half.

(c) Doubling the concentration of both C and D doubles the rate of the reaction.

(d) Increasing the temperature increases the rate of reaction.

Chapter 7

Mechanisms of Organic Reactions

In Chapter 6, we learned that the first step in predicting chemical reactivity is to estimate whether the conversion of interest is thermodynamically feasible. In this chapter, we will learn to group various chemical reactions according to reaction type and to describe how typical reactions take place. The specific sequence in which bonds are made and broken as a reactant is converted into a product is known as the **reaction mechanism**. If we are to completely understand a chemical reaction, we must explicitly follow the flow of electrons in each step as electrons move to or from a bonding orbital. To do so, we will use curved arrows to indicate the electron motion in each stage of several common organic reactions and, thus, to describe their mechanisms.

We will analyze the mechanisms of several specific reactions that are representative of quite different reaction types: (1) a concerted nucleophilic substitution; (2) a multistep nucleophilic substitution that proceeds through an intermediate carbocation; (3) a multistep nucleophilic substitution in which two cationic intermediates are formed sequentially; and (4) a homolytic substitution proceeding through a free radical intermediate. But, before we examine the mechanistic details of any of these reactions, we must understand what is taking place as the reaction proceeds. We will learn how to classify each reaction; how to determine its energetic feasibility; and how to represent the bonding changes that take place. This introductory chapter focuses on how reaction mechanisms can be clearly defined by following electron flow. In later chapters, we will learn how these transformations are used in synthesis and will see more specific examples.

7-1 ▸ Classification of Reactions

In considering a new reaction, we first determine what is accomplished—whether the number of atoms in the product differs from that in the reactant, whether any atoms in the product are different from those in the reactant, and whether the positions of any atoms in the product differ from their original positions in the reactant. Depending on the answers to these questions, we then classify a given chemical conversion as one of seven major organic reaction types: additions, eliminations, substitutions, condensations, rearrangements, isomerizations, and oxidation-reduction reactions.

Addition Reactions

In an **addition reaction,** two reactant molecules combine to form a product containing the atoms of both reactants. Two examples of addition reactions are shown in Figure 7-1. In the first, water and cyclohexene combine to produce cyclohexanol; and, in the second, hydrogen is added to cyclohexene (in the presence of a metal catalyst) to form cyclohexane in a reaction called a *catalytic hydrogenation* (recall Chapter 2). In each case, the product incorporates all of the atoms of both reactant molecules.

Some addition reactions require the presence of a catalyst, which does not appear in the product. As stated in Chapter 2, a *catalyst* can be defined as a reagent that facilitates a reaction without itself ultimately forming chemical bonds in the product or appearing in the stoichiometric equation describing the reaction. For example, the addition of water to an alkene proceeds at a reasonable rate only in the presence of a strong acid, and the addition of hydrogen to a carbon–carbon double bond occurs only when a metal surface is present. However, in neither case does the catalyst appear in the product; that is, it is unchanged. After the reaction, the catalyst is free to participate in another reaction cycle. That a catalyst is needed to accelerate the rate of an addition reaction does not influence the classification of the reaction.

Elimination Reactions

An **elimination reaction** is the inverse of an addition. In an elimination, a single complex molecule splits into two simpler products, as shown in Figure 7-2. For example, cyclohexyl bromide can be induced to undergo elimination (the formal loss of HBr) by treatment with a base. Under these conditions, HBr is not observed directly because, in the presence of a base, it undergoes an acid-base reaction to form a salt. However, we recognize the formal loss of HBr from cyclohexyl bromide in forming cyclohexene, irrespective of its final form (here, as ethanol and bromide ion). It is because the HBr formed in the elimination is immediately converted into water and

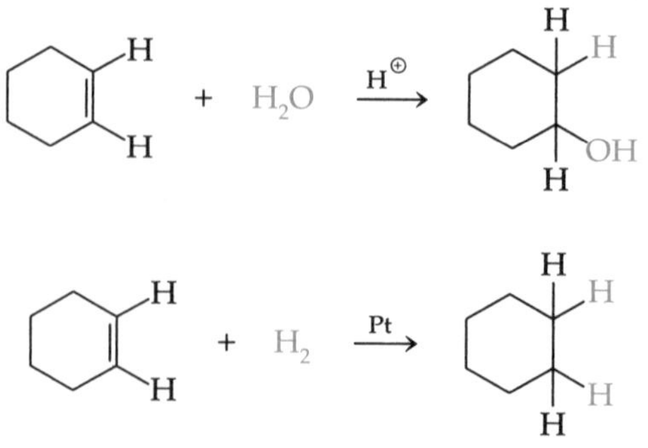

▲ FIGURE 7-1

Addition reactions combine in one product molecule all the atoms present in two reactant molecules. Shown here are two typical addition reactions: (upper reaction) hydration, in which water is added to the C═C double bond; and (lower reaction) catalytic hydrogenation, in which two hydrogen atoms are added across the C═C double bond. (These reactions will be treated more thoroughly in Chapters 10 and 12.)

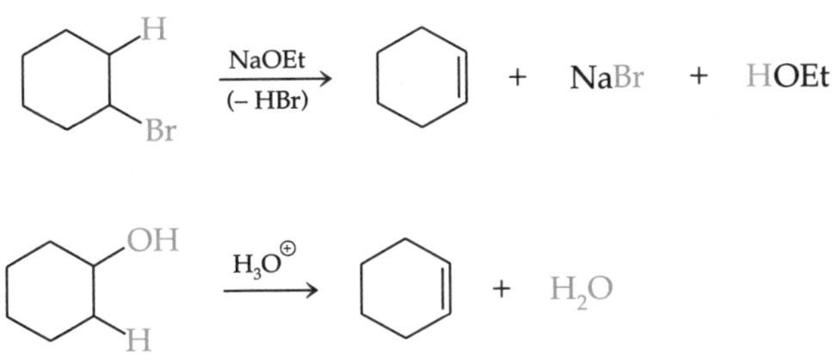

▲ FIGURE 7-2

An elimination reaction is the opposite of an addition reaction: one reactant molecule contains all the atoms present in two product molecules. Shown here are two typical elimination reactions: (upper reaction) dehydrobromination, in which the formal loss of HBr results in an alkene group; and (lower reaction) dehydration, in which the loss of water also results in a C=C double bond. (These reactions will be treated more thoroughly in Chapter 9.)

bromide under the reaction conditions that it is shown in parentheses in Figure 7-2. In the second reaction, the treatment of cyclohexanol with acid produces cyclohexene and water upon heating, accomplishing a reversal of the addition reaction considered earlier in Figure 7-1.

Substitution Reactions

In a **substitution reaction,** one atom or group of atoms in a molecule is re-placed by another. For example, a hydrogen atom in cyclohexane is re-placed by a bromine atom when the alkane is exposed to Br_2 in the presence of light or heat (Figure 7-3). In the products (cyclohexyl bromide and HBr), the positions of hydrogen and bromine are switched from their original po-sitions in the reactants. Another example of a substitution can be found in the treatment of cyclohexyl iodide with sodium bromide (Figure 7-3). Again, the positions of iodine and bromine in the reactant have been inter-changed in the products.

Condensation Reactions

A **condensation reaction** consists of the interaction of two molecules of in-termediate complexity to form a more-complex product, usually with the loss of a small molecule. For example, the combination of an alcohol with a carboxylic acid in the presence of an acid catalyst produces an ester, a more-complex molecule, while water is formed simultaneously (Figure 7-4). A

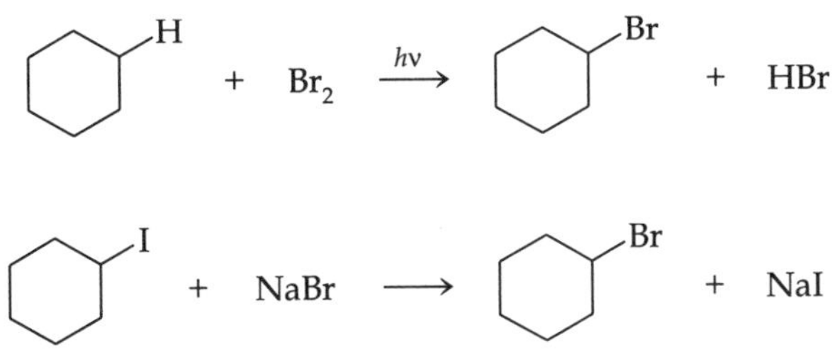

▲ FIGURE 7-3

In a substitution reaction, some atom or group in the reactant is replaced by a different atom or group in the product. In the upper reaction, a hydrogen atom present in cyclohexane is replaced by a bromine in the product, cyclohexyl bromide. In the lower reaction, an iodine present in the reactant is replaced by a bromine in the product. (The upper reaction is treated later in this chapter, and the lower one more thoroughly in Chapter 8.)

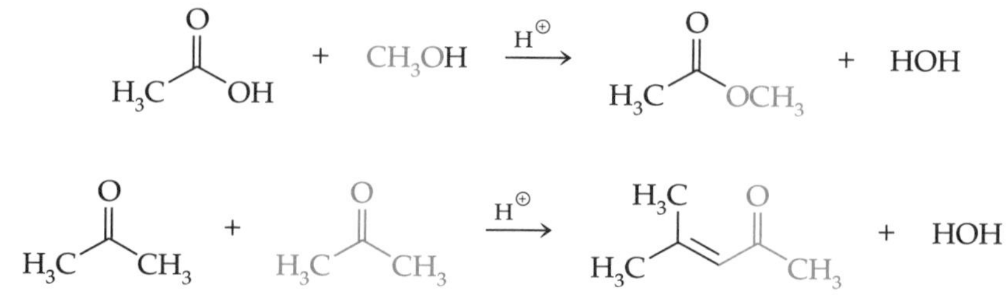

▲ FIGURE 7-4

In a condensation reaction, a more-complex organic product is formed from simpler organic reactants, along with a simple molecule. In the upper reaction, two different organic reactants (an acid and alcohol) combine to form an ester. The product ester has fewer atoms than the sum of those present in the two reactants because water is formed as a by-product. In the lower reaction, two molecules of a single reactant (a ketone) combine to form a ketone of higher molecular weight, again with water formed as a second product. (These reactions will be treated more thoroughly in Chapters 12 and 13.)

second example is an aldol condensation, in which two molecules of acetone combine to form the product, an enone, plus water. Condensation reactions are often reversible, but the position of the equilibrium can often be controlled (that is, shifted toward product) if the small molecule is removed (for example, by distillation) as it is formed. This, in turn, shifts the equilibrium in accord with le Chatelier's Principle as the reacting system attempts to replenish the "missing" product. **Le Chatelier's Principle** asserts that an equilibrium between A and B producing C and D can be shifted toward C and D by increasing the concentrations of A or B or both (pushing from the left) or decreasing those of C or D or both (pulling from the right). It can be shifted toward A and B by increasing the concentrations of C or D or both or by decreasing those of A or B or both.

$$A \ + \ B \ \rightleftharpoons \ C \ + \ D$$

Rearrangement Reactions

In a **rearrangement reaction,** the molecular skeleton is altered; that is, the sequence of carbon atoms attached along a chain is changed. These reactions may also include other changes in the molecule: for example, one functional group may be converted into another.

Typically, rearrangement reactions consist of several steps, which makes them both scientifically interesting and mechanistically complex. Two examples are given in Figure 7-5. In the Beckmann rearrangement, an

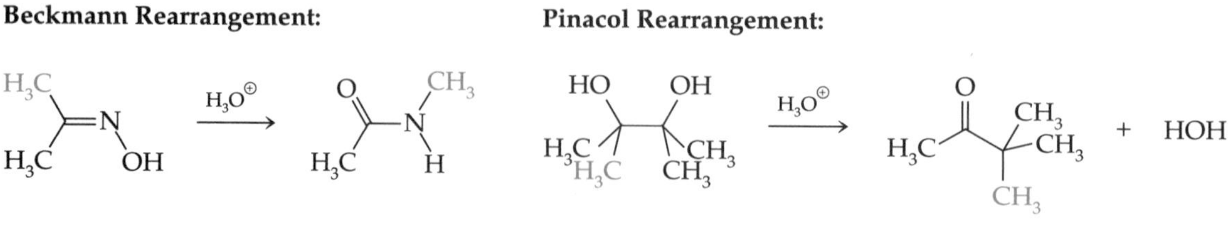

Beckmann Rearrangement: **Pinacol Rearrangement:**

▲ FIGURE 7-5

In a rearrangement reaction, the atoms or groups present in the reactant appear in the product connected in a different fashion. The reactant and product can have the same empirical formula, as in the Beckmann rearrangement, or they may have different numbers and types of atoms, as in the pinacol rearrangement, in which water is formed as a by-product. (These reactions are treated more thoroughly in Chapter 14.)

▲ FIGURE 7-6
The interconversion of the three isomers of butene (C_4H_8) is an example of an isomerization reaction. Geometric isomerization relates compounds differing only in the position of atoms or groups in space; positional isomers differ in the position of a functional group in the molecule.

alkyl group originally attached to carbon becomes attached to nitrogen, and an N—O bond in the reactant is broken, with a C=O bond appearing in the product. In the pinacol rearrangement, a methyl group migrates from one carbon to the adjacent carbon as a molecule of water is lost.

The pinacol rearrangement could also be classified as an elimination. In general, reactions that fall into more than one category are classified as the more-complex process.

Isomerization Reactions

An **isomerization** is a reaction in which compounds with the same molecular formula, but different structures, are interconverted. An isomerization differs from a rearrangement in that the carbon skeleton remains intact, but the disposition in space of substituents or functional groups is changed. In an isomerization, the molecular formulas of the reactant and product are always the same; whereas, in a rearrangement, they can be the same (as in the Beckmann rearrangement) or different (as in the pinacol rearrangement). There are two types of isomerization reactions: geometric and positional. In a **geometric isomerization,** all atoms in the product remain attached to the same atoms as in the reactant, but the disposition in space of the bonds connecting them is changed. In a **positional isomerization,** the position (or positions) of one or more substituents or functional groups in the product differs from the original position in the reactant. For example, the conversion of *cis*-2-butene into *trans*-2-butene is a geometric isomerization and that of 2-butene into 1-butene (Figure 7-6) or of *n*-butyl bromide into *s*-butyl bromide is a positional isomerization.

Oxidation and Reduction Reactions

The remaining class of reactions consists of those in which there is a formal change in oxidation level of one or more carbon atoms in a molecule. Because such reactions were discussed in Chapters 2 and 3, we will not repeat the examples here. We note only that these reactions can often also be classified as substitutions (when the number of heteroatoms at a given carbon is changed), additions (when hydrogen is added across a multiple bond), or eliminations (when the elements of molecular hydrogen have formally been removed from adjacent atoms). Often such reactions are referred to as redox reactions so as to emphasize the need for oxidizing or reducing reagents for their accomplishment.

EXERCISE 7-A

For each of the following conversions, define the relevant reaction type:

(a)

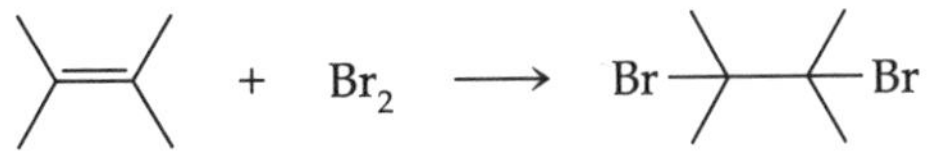

(b) 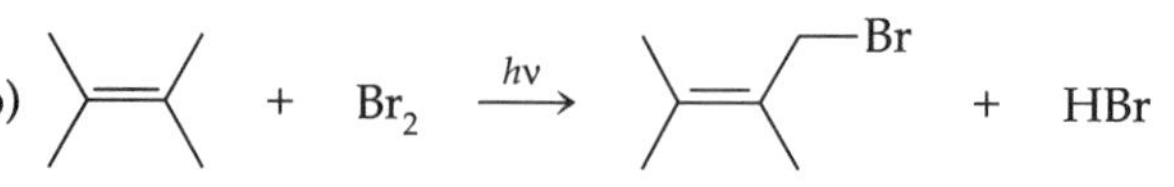 $\quad$![structure] + Br$_2$ $\xrightarrow{h\nu}$![structure with Br] + HBr

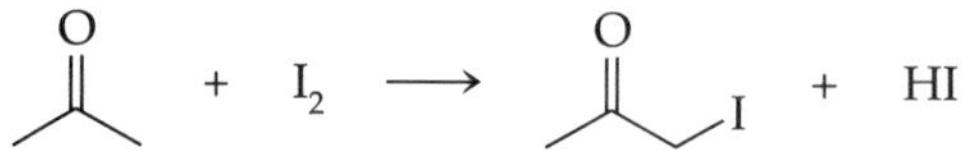

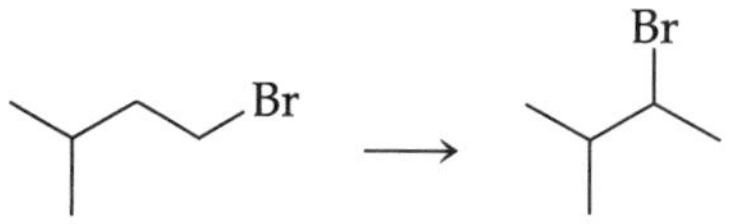

(c) CH$_3$OH + HCl $\longrightarrow$ CH$_3$Cl + H$_2$O

(d) 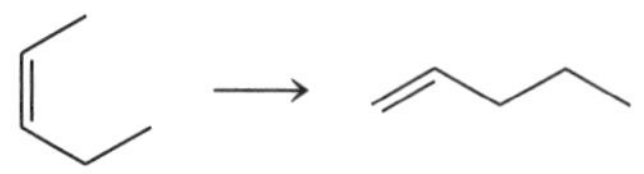 ![cyclobutene] $\longrightarrow$![pentadiene structure]

(e)  $\quad$ H$_3$C$-$C(=O)$-$OH + CH$_3$OH $\longrightarrow$ H$_3$C$-$C(=O)$-$OCH$_3$ + H$_2$O

(f) $\quad$![acetone] + I$_2$ $\longrightarrow$![iodoacetone] I + HI

(g) $\quad$![branched bromide] $\longrightarrow$![branched bromide product]

(h) H$_2$C$=$CH$_2$ + ![butadiene] $\longrightarrow$![cyclohexene]

Correctly classifying a specific reaction type does *not* imply that we know *how* the reaction has taken place. For example, substitution could, in principle, occur either by a direct replacement of groups or by a sequence of addition and elimination reactions. To illustrate this difference, we can conceive of a reaction sequence in which 1,1,2-triphenylethene interacts with molecular bromine to yield an addition product, 1,2-dibromo-1,1,2-triphenylethene.

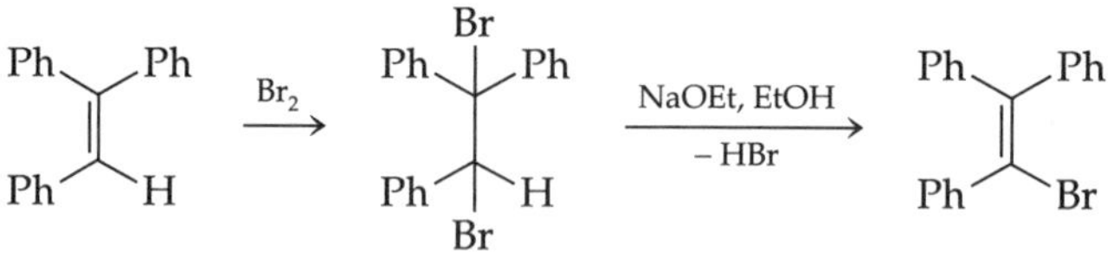

![reaction scheme: Ph2C=CHPh + Br2 → addition product (Br groups) → NaOEt, EtOH, –HBr → substitution product]

By the elimination of HBr, we can obtain a formal substitution product from the original reactant (replacement of H in 1,1,2-triphenylethene by Br) through a sequence of addition and elimination reactions. It is insufficient, therefore, to specify the kind of reaction without describing how the reaction proceeds. A detailed description of electron flow, including the identity of any intermediate formed, is referred to as a reaction mechanism; in subsequent chapters, we will emphasize reaction mechanisms as an important method for intellectually organizing the reactions of organic chemistry.

7-2 ▸ Bond Making and Bond Breaking: Thermodynamic Feasibility

All chemical reactions entail bond making or bond breaking or both. From Chapter 3, we know that a σ bond between atoms A and B can sometimes be cleaved so that the two shared electrons are distributed equally (one electron to A, one to B), producing neutral species called radicals. As men-

Homolysis:

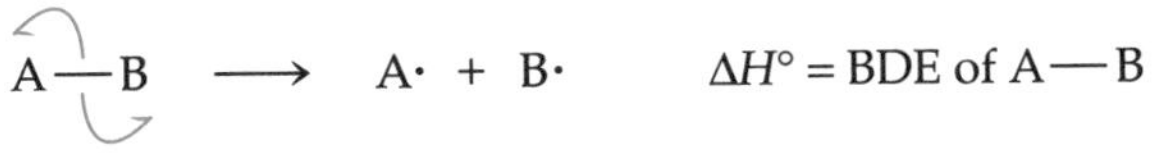

$$A \overset{\frown}{-} B \longrightarrow A\cdot + B\cdot \qquad \Delta H° = \text{BDE of } A-B$$

▲ FIGURE 7-7

In a homolytic cleavage, the two electrons of the covalent bond initially connecting two atoms are partitioned so that one electron is associated with each atom. A bond is broken and two radicals are formed. (BDE = bond-dissociation energy.)

tioned in Chapter 3, this process is referred to as *homolytic cleavage*, or *homolysis* (Figure 7-7).

In the alternative mode of bond breaking, the two electrons of a σ bond move as a pair to one of the initially bonded atoms. This process, called *heterolytic cleavage*, or *heterolysis* (Figure 7-8) produces a positive and a negative ion. The direction of the electron flow (to A or B) is governed by the relative electronegativity of the two atoms participating in the bond, with the more-electronegative atom becoming negatively charged.

The convention employed to describe these two modes of electron movement is to use a half-headed arrow for a single electron (in a homolytic cleavage) and a full-headed arrow for an electron pair (in a heterolytic cleavage). To understand the relatively small number of reaction types that will be encountered in the remainder of this book, it is very important that we recognize the precise meaning of this **arrow notation.** The use of half- or full-headed curved arrows to indicate motion of electrons is the best way to clearly indicate how a reaction occurs. *Bear in mind that the curved arrows indicate motion of electrons, not atoms.* When electrons move, atoms follow. The tail of the curved arrow marks the origin of the electron(s), and the head marks its terminus.

Energy Changes in Homolytic Reactions

Homolysis and heterolysis consist solely of bond breaking. In many chemical reactions, however, bond breaking is often accompanied by bond making, in which another reagent assists the cleavage. For example, when a radical center is already available, it can assist in homolytic cleavage, as shown in Figure 7-9. Thus, when a radical, R· , with one unpaired electron interacts with an A–B σ bond, the electron that becomes accessible to A in the homolytic cleavage of the A–B bond enables A to form a new bond with R· , the second electron of the newly formed R–A bond being contributed by R· . Meanwhile, the other electron in the original A–B covalent bond becomes localized on a new radical, B· .

The coupling of bond making and bond breaking is very common. Many chemical bonds are very strong, and breaking them costs a lot of

Heterolysis:

$$A \overset{\frown}{-} B \longrightarrow A^{\oplus} + B^{\ominus}$$

▲ FIGURE 7-8

In a heterolytic cleavage, the two electrons of the covalent bond initially connecting two atoms move as a pair onto one of the atoms. A bond is broken and two ions are formed. The atom that takes up the two electrons becomes an anion and that left with no electrons from the bond becomes a cation. The enthalpy change for this bond cleavage is influenced by the bond strength and the solvation energy for the ions formed.

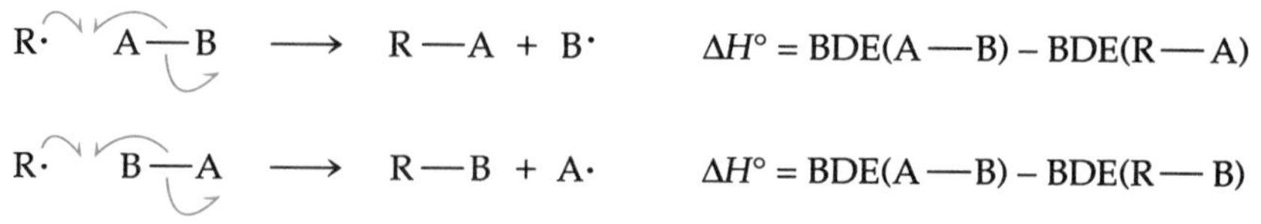

$$R \cdot \quad A \text{—} B \longrightarrow R \text{—} A + B \cdot \qquad \Delta H^\circ = BDE(A \text{—} B) - BDE(R \text{—} A)$$

$$R \cdot \quad B \text{—} A \longrightarrow R \text{—} B + A \cdot \qquad \Delta H^\circ = BDE(A \text{—} B) - BDE(R \text{—} B)$$

▲ FIGURE 7-9

In an assisted homolysis, some or all of the energy required to cleave the A–B bond is offset by the energy gained in simultaneously forming the R–A (upper reaction) or R–B (lower reaction) bond. (BDE = bond-dissociation energy.)

energy. When a part of this lost energy is regained in the formation of a new bond, the reaction proceeds much more easily.

Because half-headed arrows are used to indicate the motion of *one* electron, we need three half-headed arrows to describe this reaction. In this assisted bond homolysis, the enthalpy difference includes not only the bond dissociation of A–B, as was required in simple homolysis (Figure 7-7), but also the energy gained by the formation of a bond between R and A (Figure 7-9).

Had the radical, R· , attacked the other end of the molecule, the same kind of electron flow would have produced a bond between R and B and a localized electron on atom A, forming a radical. The relevant bond energies then would be those of A–B and R–B.

The enthalpy change of these reactions represents the balance between the bond-dissociation energy of A–B and the bond energy of the newly formed R–A or R–B bond. The bond energies of R–A and R–B need not be equal, and, as a result, the ΔH° values for these two reactions can vary. Because the entropy change in such a reaction is small, the reaction proceeds along the more energetically favorable route when steric hindrance is not a factor; that is, R· will form a bond with A· or with B· according to which ΔH° value is more negative (or less positive).

EXERCISE 7-B

From the bond energies listed in Table 3-6, calculate ΔH° for each of the following reactions and predict which C–H bond would be preferentially cleaved by interaction with a bromine atom:

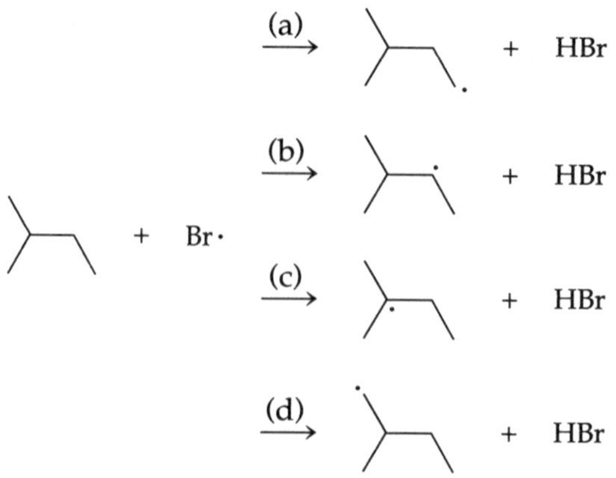

Energy Changes in Heterolytic Reactions

Like homolytic reactions, heterolytic reactions can be assisted by external reagents. Heterolytic cleavages are usually most likely to occur with polar σ bonds. An *electrophile* interacts more favorably with the electron-rich end of

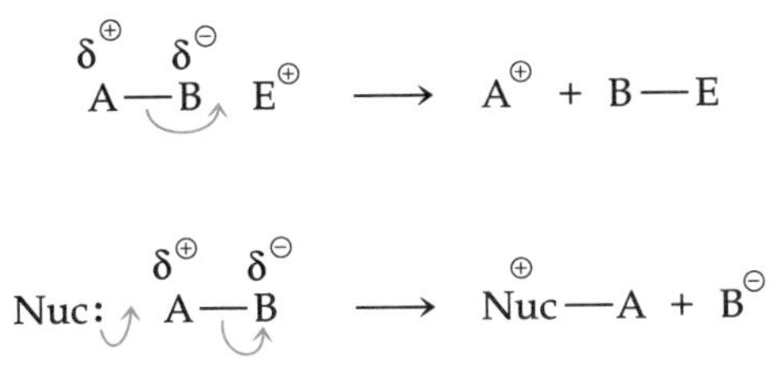

$$\overset{\delta^{\oplus}}{A}\!-\!\overset{\delta^{\ominus}}{B}\quad \overset{\oplus}{E} \quad \longrightarrow \quad \overset{\oplus}{A} \;+\; B\!-\!E$$

$$Nuc\!:\;\overset{\delta^{\oplus}}{A}\!-\!\overset{\delta^{\ominus}}{B} \quad \longrightarrow \quad \overset{\oplus}{Nuc}\!-\!A \;+\; \overset{\ominus}{B}$$

▲ FIGURE 7-10

In a polar A–B σ bond in which B is the more-electronegative atom, A is partially positively charged and B is partially negatively charged. Therefore, an electrophile (E⁺) prefers to associate with B and a nucleophile (Nuc:) prefers A. In an assisted heterolysis, the energy required to cleave the A–B bond is partly offset by the energy gained in simultaneously forming the B–E bond (upper reaction) or the Nuc–A bond (lower reaction). Because ions are formed in these heterolytic reactions, we cannot use simple bond-dissociation energies to calculate the enthalpy change but must also consider the energy needed to form and solvate the polar reactant and product ions.

an A–B polar bond. The polarization of a covalent bond between A and B such that electrons are shifted toward the more-electronegative atom B facilitates the formation of a bond between B and an electrophile (Figure 7-10). Because this reaction entails the simultaneous movement of an electron pair (that is, of *two* electrons), we use a full-headed arrow to describe this electron motion.

Because of these same electrostatic factors, a *nucleophile* is attracted to the positive end of the A–B bond and can assist heterolytic cleavage by donating electrons to the developing positive charge at A, forming a Nuc–A bond as the A–B bond is cleaved. As this takes place, the two electrons of the A–B bond shift onto B. Here, two electron pairs move, as indicated by two full-headed curved arrows.

The heterolytic cleavage shown in Figure 7-8 and the assisted heterolytic reactions shown in Figure 7-10 differ from homolytic cleavages in that they include the formation of ions. How easily these ions are formed depends on the polarity of the solvent. In contrast, homolytic cleavages produce neutral radicals and are not greatly affected by the nature of the solvent. Solvation of the ions formed in a heterolytic cleavage provides a significant amount of energy, but the precise values are difficult to measure and depend on specific reaction conditions. As a result, it is much more difficult to describe enthalpy changes in heterolytic cleavages than those in homolytic cleavages.

An indirect means is usually used to predict the relative heterolytic bond energies in a series of similar compounds in which parallel bond breaking occurs. For example, because we know that a tertiary cation is more stable than a secondary one, we know that reaction 1 costs less energetically than reaction 2.

$$(CH_3)_3C\!-\!Br \longrightarrow (CH_3)_3\overset{\oplus}{C} \;+\; \overset{\ominus}{Br} \tag{1}$$

$$(CH_3)_2CHBr \longrightarrow (CH_3)_2\overset{\oplus}{CH} \;+\; \overset{\ominus}{Br} \tag{2}$$

$$\Delta H°(1) < \Delta H°(2)$$

Likewise, even without knowing the specific pK_as (recall Chapter 6), we know that reaction 3 is less costly energetically than reaction 4 because an enolate anion is more stable than an alkyl anion.

$$CH_3COCH_2\!-\!M \longrightarrow CH_3CO\overset{\ominus}{CH_2} \;+\; \overset{\oplus}{M} \tag{3}$$

$$CH_3CH_2CH_2\!-\!M \longrightarrow CH_3CH_2\overset{\ominus}{CH_2} \;+\; \overset{\oplus}{M} \tag{4}$$

$$\Delta H°(3) < \Delta H°(4)$$

CRACKING TOWERS: BREAKING BONDS IN CRUDE OIL

Petroleum, or crude oil, is a complex mixture of many different compounds but most are aliphatic and aromatic hydrocarbons. As it comes from the well, petroleum contains many more compounds of high molecular weight than are needed. Various methods have been developed to degrade larger hydrocarbons into smaller ones, processes that are referred to as "cracking." In thermal cracking, hydrocarbons are heated to as high as 760° C—a temperature high enough to yield sufficient energy when molecules collide so as to break carbon–carbon bonds in a homolytic fashion, producing carbon free radicals. Alternatively, in the presence of an acidic, inorganic catalyst, bonds are broken by heterolytic cleavage, with the generation of carbocations. In addition to producing hydrocarbons that are more suitable for use as fuels in internal combustion engines, cracking also yields propene and butene, which are used to make many different products, including plastics.

Thus, the trends that we have learned about the relative stabilities of intermediates (here, of carbocations and carbanions) serve us well in ordering reactivity in heterolytic reactions.

If we are concerned only with the net reaction (rather than with the intermediate ion-forming steps), we can still use the table of bond-dissociation energies to calculate $\Delta H°$ of a reaction, even if it proceeds through heterolytic steps. For example, even though reaction 5 proceeds through ions, we can nonetheless calculate the reaction enthalpy by subtracting the relevant bond energies of the reactants from those of the products. This approach assumes similar solvation energies for the reactants and products, a quite reasonable assumption for pairs of similar, neutral reagents.

$$(CH_3)_3C\text{---}Br + HCl \longrightarrow (CH_3)_3C\text{---}Cl + HBr \qquad (5)$$

In summary, reactions in which bonds are concurrently broken and formed (assisted homolytic and heterolytic reactions) proceed much more readily and with different energetic requirements than do those in which only bond cleavage takes place. Because of difficulties in measuring heterolytic bond strengths, it is often difficult to predict quantitatively the thermodynamics of individual steps in such reactions, although trends in a series of similar compounds are predictable. In contrast, thermochemical calculations are often easy for homolytic cleavages.

7-3 ▸ How to Study a New Organic Reaction

Once we have recognized its reaction type and have established that a proposed reaction is sensible thermodynamically, we can next determine what conditions are necessary for the proposed conversion. There are several ways by which we can fully describe an organic reaction. To have mastered a particular reaction, we should be able to answer all of the following four questions:

1. Given the reactant and reagents, together with a set of reaction conditions, what is the expected product(s)?

$$\text{Reactant} \xrightarrow{\text{Reagents}} \text{?}$$

2. Given a reactant and a product, what reagents and conditions favor this transformation?

$$\text{Reactant} \xrightarrow{?} \text{Product}$$

3. Given both a product and a set of reagents, what is a reasonable starting material(s)?

$$\text{?} \xrightarrow{\text{Reagents}} \text{Product}$$

4. What are the intermediates in the conversion and what electron flow accomplishes their formation and reaction?

EXERCISE 7-C

Review the reactions that you have learned so far in order to supply the missing information for the following reactions:

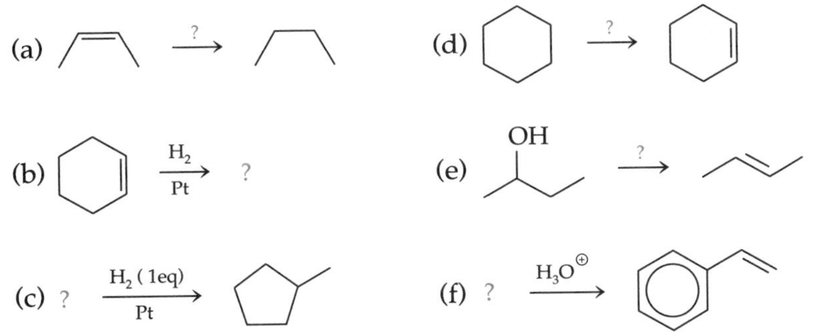

(a)

(b)

(c)

(d)

(e)

(f)

In addressing the first two questions, we must become familiar with a range of reagents and reaction conditions. In answering the third question, we must sometimes propose a series of reactions that would achieve in several steps what would be difficult to accomplish in one. To do this, we use an approach known as **retrosynthetic analysis,** in which we work backward. Having chosen a target product, we choose a reasonable precursor, which in turn has a logical precursor, and so forth. Thus, when finished, we have a plan for building molecules of increasing complexity from readily available starting materials through a series of reactions. This approach enables us to logically plan the construction of interesting new molecules or propose new synthetic routes to complex existing molecules (for example, natural products). This area, called **organic synthesis,** is a very important subfield of organic chemistry. It will be covered in more detail in Chapter 15.

The fourth question focuses on exactly how electrons (and hence atoms) move when a reactant is converted into a product. This detailed description, or *reaction mechanism,* is the underpinning of our understanding of organic chemistry and constitutes much of the critical information that allows us to predict new reactions with confidence. A study of how reactions occur is called **mechanistic organic chemistry,** and the subfield of organic chemistry that relates structure to reactivity in explaining reaction mechanisms is called **physical organic chemistry.**

A major strength of organic chemistry is that, if we learn the answers to these four questions *really well* for a small number of reactions, we can generalize to other reactions of the same classification. Thus, it is not necessary to learn thousands of reactions; instead, we will learn a few and apply what we have learned to similar cases.

(Read the preceding paragraph to yourself three times or however many it takes until it sinks in and you really believe it.)

We cannot extrapolate what we know about one organic reaction to another until we know its reaction mechanism. Typically, a chemical reaction consists of bond making and bond breaking as the reactant is converted into a product: a *reaction mechanism* is nothing more complicated than the sequence of elementary steps by which this event occurs. We follow these steps by showing the flow of electrons (using curved arrows) as some bonds are broken and others formed. Reaction mechanisms, which describe how electrons move to make and break chemical bonds along a reaction coordinate, are therefore of *very great importance* to organic chemistry.

To understand a reaction mechanism, we must establish the identities of all intermediates formed en route from reactant to product. If we know something about the energies of these intermediates (even if only roughly), we can approximate the structure and energy of the transition states leading to the formation of these intermediates and can predict relative reactivity for closely related reactions. In the following sections, we will specifically consider the mechanisms of three representative kinds of reactions: those having no reactive intermediates—that is, *concerted reactions*; those involving *heterolytic cleavage*, and therefore ionic intermediates; and those involving *homolytic cleavage*, and hence radical intermediates. Our goal is to show how **arrow pushing** (in which curved arrows describe the movement of electrons as a reaction proceeds) helps us to define a reaction mechanism.

Although our focus in this chapter is on describing several types of reaction mechanisms, the examples used to illustrate these concepts do accomplish chemical transformations that are useful in synthesis. You should begin now to assemble study aids, consistent with the individual learning method most effective for you, that can organize these reactions according to what they accomplish and how they proceed. Many students find it useful to prepare "flash cards" that summarize important features of each reaction.

EXERCISE 7-D

For each of the following bond cleavages, use curved arrows to describe the electron flow:

(a) $CH_3O{-}OCH_3 \xrightarrow{\Delta} CH_3O^{\cdot} + {}^{\cdot}OCH_3$

(b)

(c)

(d)

7-4 ▸ Mechanism of a Concerted Reaction: Concerted Nucleophilic Substitution (S_N2)

As defined in Chapter 6, a concerted reaction proceeds directly from reactant to product without forming any detectable intermediates, whether ionic or neutral. The absence of charged intermediates is usually established by observing that rates for concerted reactions are far less affected by solvent polarity than are those for reactions in which ions are formed. Because radicals often react with oxygen on almost every collision, the absence of radical intermediates is often established by observing an insensitivity of the reaction to the presence or absence of oxygen. We have seen one concerted mechanism in Chapter 6; namely, the Diels-Alder reaction.

Figure 7-11 depicts another example, a concerted **self-exchange reaction** in which an incoming bromide ion interacts with methyl bromide, causing the C–Br bond to break (displacing bromide) while forming a new carbon–bromine bond. Because the reactants and products in this reaction are identical, this reaction is thermoneutral and produces no chemical change (unless the bromine atoms are isotopically different). We use a full-headed curved arrow (a) to indicate that the two electrons of one of the lone pairs of the bromide ion move toward carbon to form what ultimately becomes a carbon–bromine σ bond. The new bond cannot form, however, without the breaking of the original C–Br bond; otherwise, carbon would have to accommodate ten electrons in its valence shell. As a result, a second full-headed curved arrow (b) indicates that the two electrons originally in the carbon–bromine σ bond of the starting material must move from a bonding orbital between these atoms to become a nonbonded lone pair on bromide as the bond is broken.

These two bond-making (a) and bond-breaking (b) steps are simultaneous; no intermediates are formed. In the transition state for this reaction, carbon is partially bonded to the incoming and departing bromines. This partial bonding is sometimes shown by dotted lines, with the structure enclosed in brackets to indicate that, as a transition state, it has no intrinsic stability and cannot be isolated. This dotted line notation is not nearly as precise as the curved-arrow notation and is *not* recommended as the primary way to think about mechanism. Do not use it if it does not help you see the electron flow. (And be very clear not to imply that you think the structures shown with dotted lines have more than a transient existence.)

Because the attacking reagent bears a nonbonded electron pair and is therefore nucleophilic and because the original C–Br bond has been replaced by a new one, this reaction is called a **nucleophilic substitution.** (Recall that nucleophilic means "nucleus loving," so that nucleophiles are

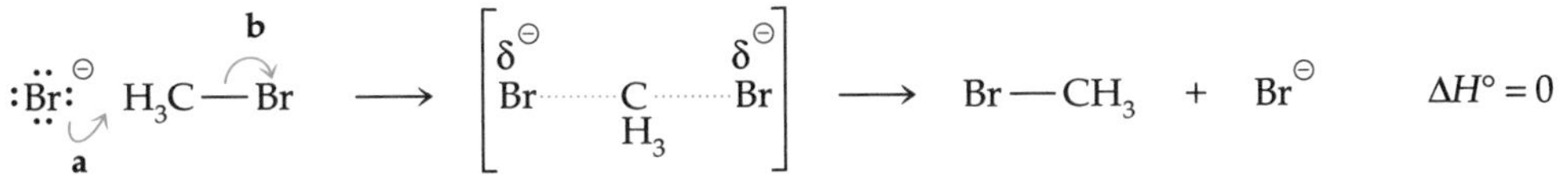

$$:\overset{..}{\underset{..}{Br}}:^{\ominus} \quad H_3C\overset{b}{\underset{a}{\frown}}Br \quad \longrightarrow \quad \left[\begin{array}{ccc} \overset{\delta\ominus}{Br} & \cdots & \underset{H_3}{C} & \cdots & \overset{\delta\ominus}{Br} \end{array}\right] \quad \longrightarrow \quad Br\!-\!CH_3 \ + \ Br^{\ominus} \qquad \Delta H^\circ = 0$$

▲ FIGURE 7-11

In this bimolecular nucleophilic displacement (an S_N2 reaction), the incoming nucleophile (bromide ion) forms a bond to carbon as the carbon–bromine bond in the alkyl bromide reactant is broken. Because the bond making and bond breaking occur together in a single step without the formation of an intermediate, this reaction is concerted.

attracted to a positive charge. Thus, anions are nucleophilic.) An effective nucleophile has high electron density, so that an anion is more reactive as a nucleophile than the corresponding neutral reagent. A less-electronegative atom bearing a nonbonded electron pair is a better nucleophile than a more-electronegative one because it can more easily donate its electron pair in approaching the transition state. Two reagents (the starting material and the nucleophile) participate in the transition state of the rate-determining step, which makes this reaction type a **bimolecular nucleophilic substitution,** abbreviated **S$_N$2.** Because two reagents participate in bond making and bond breaking in the rate-determining step, the rate of this reaction is affected by the concentrations of both the substrate and the nucleophile.

The same kind of reaction can produce a net chemical change if the incoming and outgoing halide ions are different. For example, in the upper reaction in Figure 7-12, a bromide ion displaces iodide from methyl iodide, producing a carbon–bromine bond and an iodide ion. Because a carbon–bromine bond is stronger than a carbon–iodine bond, this reaction is energetically favorable and thus exothermic ($\Delta H°$ is negative, assuming that other factors are equal). Here, iodide is the **leaving group,** pulling the electrons originally in the covalent C–I bond toward itself, whereas bromide is the nucleophile, donating an electron pair toward carbon.

The lower reaction in Figure 7-12 is the reverse of this reaction, in which iodide displaces bromide from methyl bromide to produce methyl iodide plus bromide. For this reaction, $\Delta H°$ is positive and the reaction is endothermic. In this unfavorable reaction, bromide is the leaving group and iodide is the nucleophile.

EXERCISE 7-E

Which reagent in each of the following pairs is the more-reactive nucleophile?

(a) $^\ominus$NH$_2$ or NH$_3$ (c) Cl$^\ominus$ or I$^\ominus$

(b) OH$_2$ or NH$_3$ (d) HS$^\ominus$ or HO$^\ominus$

Let us compare the transition states for the reactions of bromide ion with methyl bromide (Figure 7-11) and methyl iodide (upper reaction in Figure 7-12). In the self-exchange reaction, the carbon–bromine bond of methyl bromide must stretch in the transition state as a new carbon–bromine bond is being formed. (Otherwise, as mentioned earlier, carbon would have to accommodate more than eight valence electrons.) Breaking an intact covalent bond is energetically costly, but forming a new bond is energetically favorable. Because the entering group and the leaving group are identical, there is no energetic preference for bonding either to

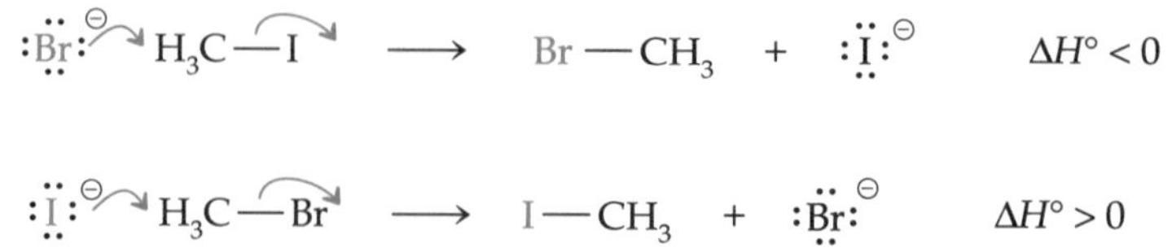

$$:\!\ddot{B}r\!:^\ominus \curvearrowright H_3C\!-\!I \longrightarrow Br\!-\!CH_3 + :\!\ddot{I}\!:^\ominus \qquad \Delta H° < 0$$

$$:\!\ddot{I}\!:^\ominus \curvearrowright H_3C\!-\!Br \longrightarrow I\!-\!CH_3 + :\!\ddot{B}r\!:^\ominus \qquad \Delta H° > 0$$

▲ **FIGURE 7-12**
The thermodynamics of a substitution reaction can be estimated from the bond energies of the reactant and products. A nucleophilic substitution in which bromide displaces iodide from an alkyl halide is exothermic, whereas one in which iodide displaces bromide is endothermic.

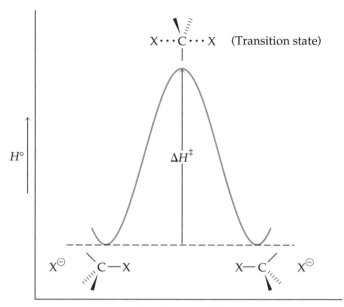

▲ FIGURE 7-13
A reaction profile for a self-exchange reaction is completely symmetrical. At the transition state, the strength of the bond being made to the incoming nucleophile is exactly equivalent to the strength of the bond being broken as the leaving group departs.

the incoming bromide or to the outgoing one, and the transition state is symmetrical.

In general, for an S_N2 reaction with $\Delta H° = 0$, the transition state geometrically resembles the reactant and product equally. The degree of bond making and bond breaking is exactly the same, and a smooth conversion from reactant into product takes place (Figure 7-13). To attain this transition state, the incoming nucleophile (Br^-) in our self-exchange reaction must have attacked carbon from the carbon end of the C–Br bond; that is, from the side opposite that from which the leaving-group Br departs. This **backside attack** (Figure 7-14, top) is required to avoid both the placement of

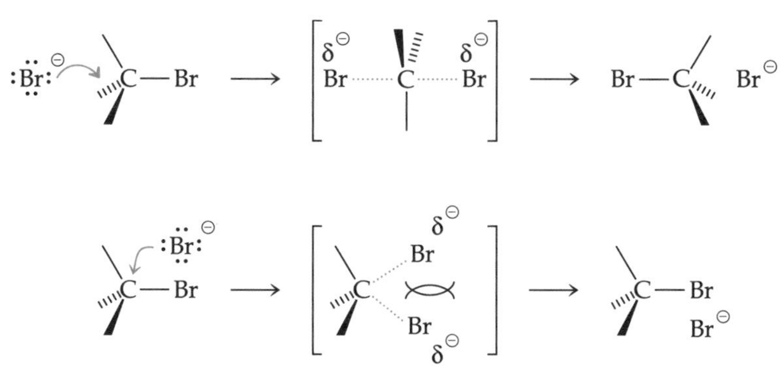

▲ FIGURE 7-14
In principle, there are two possible transition states for an S_N2 reaction: that resulting from backside attack (top), in which the nucleophile attacks from the side opposite that of the leaving group; and that from frontal attack (bottom), in which the nucleophile attacks on the same side as the leaving group. The upper transition state, attained by backside attack with inversion of configuration at carbon, is strongly favored over the lower transition state.

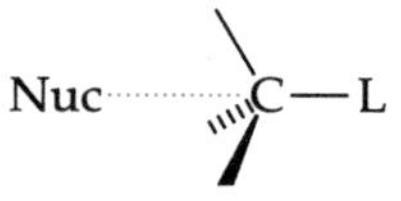

An early transition state

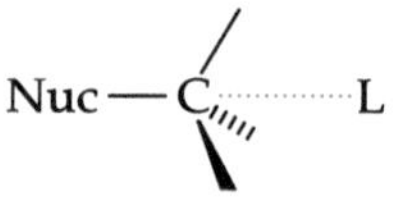

A late transition state

▲ FIGURE 7-15

A transition state with very little bond making and bond breaking resembles the organic reactant and is called early. A transition state with substantial amounts of bond making and bond breaking resembles the product and is called late.

groups of like charge (anions) in the same region of space and the three-center, four-electron arrangement (anti-Hückel, somewhat like the cyclopropenyl anion) that would exist in a frontal attack (Figure 7-14, bottom). Backside displacement causes an **inversion of configuration** at a center of asymmetry. Thus, when this displacement reaction occurs at a chiral carbon, a bromide of opposite configuration is formed.

If, instead of the symmetrical transition states in Figures 7-13 and 7-14, the bonds to the incoming nucleophile (Nuc) and the leaving group (L) are not equally strong, as in Figure 7-12, the amounts of bond making (C–Nuc) and bond breaking (C–L) in the transition state would differ. As stated in Chapter 6, a transition state in which only a small degree of breaking of the C–L bond has occurred is *early*; one with substantial C–L stretching and appreciable bonding to the incoming nucleophile is *late* (Figure 7-15). Depending on the identities of the nucleophile and leaving group, the range of S_N2 reactions spans a continuum from early to symmetrical to late transition states.

EXERCISE 7-F

Using the Cahn-Ingold-Prelog rules that you learned in Chapter 5, assign absolute configuration at each chiral center in the reactants and products of each of these S_N2 reactions:

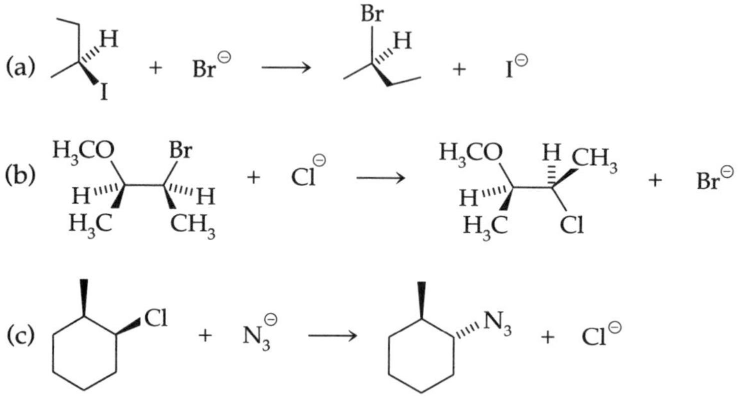

Backside displacement in an S_N2 reaction requires the nucleophile's approach to carbon to be close enough to permit partial bonding. Thus, the approach of the incoming nucleophile is strongly affected by the bulk of the

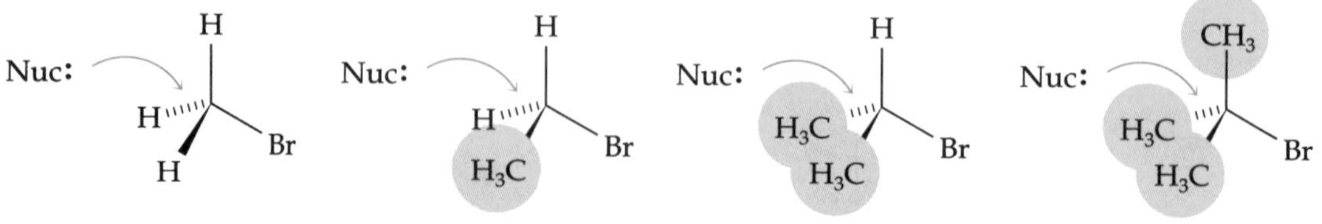

▲ FIGURE 7-16

Backside attack in an S_N2 reaction becomes more difficult when the carbon bears a larger number of bulky alkyl substituents. Thus, a concerted bimolecular nucleophilic substitution is easy for the methyl halide at the left and for the primary halide in the second structure, but it becomes more difficult for the secondary halide shown as the third structure and virtually impossible for the tertiary halide at the far right.

substituent groups present on the carbon bearing the leaving group. The ease with which displacement occurs is greatest at primary C–L bonds; displacement occurs less readily at secondary C–L bonds and even less so at tertiary ones.

Figure 7-16 illustrates a nucleophile's attack to break a carbon–bromine bond at a methyl, ethyl, isopropyl, and *t*-butyl center. The van der Waals radii of the alkyl groups are drawn to roughly show the larger steric demand of an alkyl group over that of hydrogen. Clearly, the ease of nucleophilic displacement within this series follows the order: methyl > 1° > 2° >> 3°. In fact, concerted displacements are so difficult at tertiary centers that other reactions occur instead.

EXERCISE 7-G

Which compound in each of the following pairs is more active toward a nucleophile under S_N2 conditions?

(a) 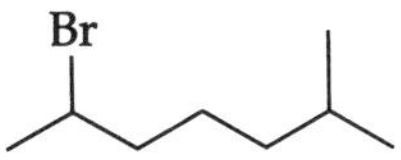 [structure with Br] or [structure with Br]

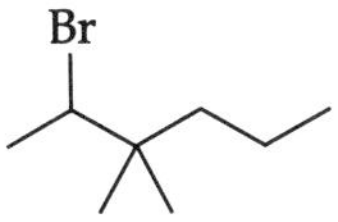

(b) 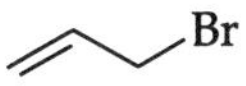 [structure with Br] or [structure with Br]

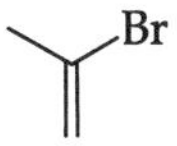

(c) 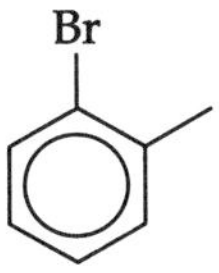 [structure with Br] or [structure with Br]

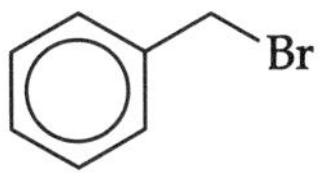

EXERCISE 7-H

Arrange the following isomeric bromides in order of decreasing relative rates in an S_N2 displacement.

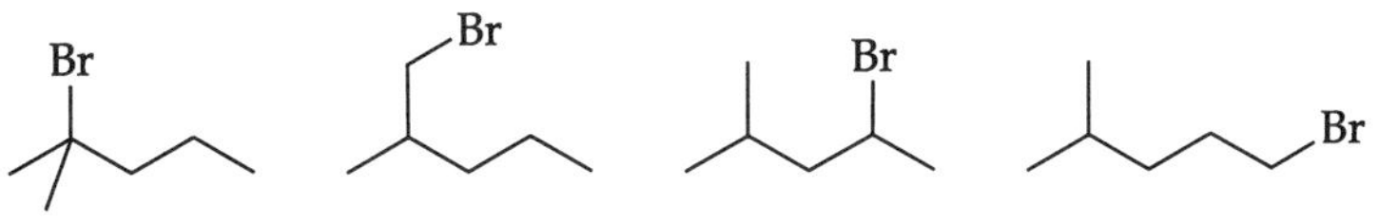

[structures with Br]

With the S_N2 reaction, it is possible to convert an alkyl halide into any of several different functional groups (Table 7-1, on page 244). Different nucleophiles produce amines, alcohols, ethers, thioethers, exchanged halides, or azides. By learning the mechanism shown in Figure 7-11, we have also learned how each of these related transformations takes place. Because these reactions constitute useful methods for preparing each of these products, they are of both synthetic and mechanistic interest.

In summary, a concerted nucleophilic displacement (S_N2) occurs when an organic substrate bearing a covalently bound leaving group is attacked from the backside by an incoming nucleophile, causing inversion of configuration at carbon. A primary or secondary alkyl halide undergoes substitution by a nucleophile capable of forming a stronger C–Nuc bond than the C–L (leaving group) bond in the starting material, a general requirement for an exothermic reaction.

Table 7-1 ▸ Preparation of Some Typical Functional Groups by S_N2 Displacement

$$R-Br + Nuc \longrightarrow R-Nuc$$

in which R = methyl,
primary alkyl,
or secondary alkyl

Source of Nucleophile	Product
NH_3	Amine (RNH_2)
NaOH	Alcohol (ROH)
$NaOCH_3$	Ether ($ROCH_3$)
$NaSCH_3$	Thioether ($RSCH_3$)
NaCl	Alkyl chloride (RCl)
KI	Alkyl iodide (RI)
NaN_3	Alkyl azide (RN_3)

EXERCISE 7-1

Which of the following substrates are good candidates for reaction with NaN_3 in acetone (typical S_N2 reaction conditions)?

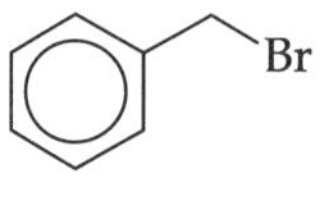

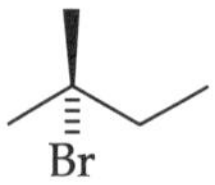

7-5 ▸ Mechanism of Two Multistep Heterolytic Reactions: Electrophilic Addition and Nucleophilic Substitution (S_N1)

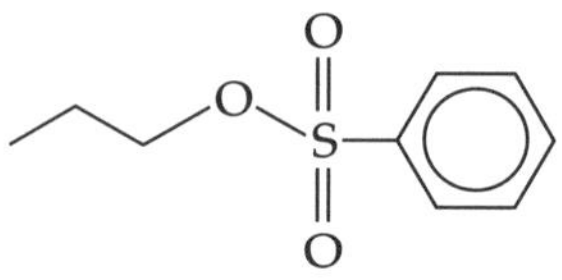

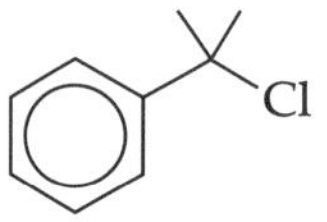

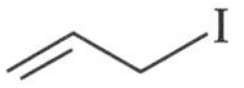

Reactions that include the formation of intermediates (Chapter 6) take place in distinct steps. Here, we consider the mechanisms of two reactions in which bond cleavage is heterolytic and the intermediates formed are cations. The first reaction includes the formation of a carbocation; the second, the formation of a carbocation followed by an oxonium ion intermediate. As we will learn, the formation of cations is inferred from two observations about the rates of reaction: (1) a correlation of relative reactivity with the stabilities of the intermediate carbocations; and (2) enhanced rates in polar solvents that stabilize a high-energy, charged intermediate.

Electrophilic Addition of HCl to an Alkene

The addition of HCl to cyclohexene takes place in two steps.

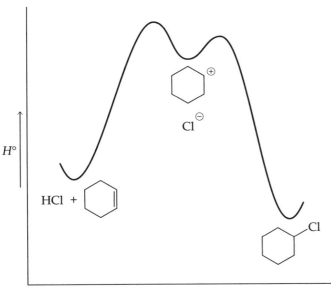

In the first step, the π electrons of the double bond in cyclohexene are donated toward the electrophile (HCl) to form a carbon–hydrogen σ bond. The full-headed arrow indicates that the two electrons of the π bond move in this step to form a new C–H bond as the two electrons in the H–Cl bond shift to Cl. The proton thus acts as an electrophile. Because the overall transformation is an addition, we call this reaction an **electrophilic addition.**

The protonation of one carbon converts the other carbon originally participating in the double bond into a cation. (As an exercise, calculate the formal charge of this carbon.) This first step is difficult because more bonds are broken than are formed: both a C=C π bond and a H—Cl σ bond are broken, whereas only a C—H σ bond is formed, resulting in a carbocation and a chloride ion. This reaction takes place only with great difficulty in the gas phase with no solvent molecules to stabilize the ions generated. However, a polar solvent greatly stabilizes these ionic intermediates and, as a result, the reaction is accelerated by solvents such as water. Nonetheless, this solvation energy cannot compensate completely for the substantial bond breaking in this first endothermic step.

In the second step, the chloride ion formed in the first step reacts with the carbocation to form a carbon–chlorine σ bond. This step is very easy because it consists only of bond making. Therefore, the endothermic first step is the slow step and is rate determining. Its transition state closely resembles the intermediate carbocation. A reaction profile is shown in Figure 7-17. The factors that stabilize the intermediate cation and chloride ion also stabilize the transition state. In turn, the transition-state energy is reduced,

◀ FIGURE 7-17
A reaction profile for a multistep electrophilic addition shows that the reaction takes place by formation of a carbocation in the rate-determining step.

Reaction coordinate

as is the required activation energy, resulting in a faster reaction. This reaction is accelerated by polar solvents because the critical intermediates are ions: the cation is stabilized by solvents containing heteroatoms involved in polar bonds and the anion is stabilized by solvents having hydrogen atoms bound to heteroatoms. Water contains both features and is particularly good at stabilizing charged intermediates, both cations and anions.

We know that tertiary cations are more stable than secondary ones, which are, in turn, more stable than primary cations. When we considered the relative ordering of reactivity of alcohols in Chapter 3, we saw that the stability of the resulting carbocation determines the relative facility for cleavage of the carbon–oxygen bond. Thus, t-butyl alcohol is cleaved by acid (ionized) more readily than is s-butyl alcohol, which is more reactive than n-butyl alcohol. This order of reactivities follows from, and is therefore the same as, the order of carbocation stabilities: tertiary > secondary > primary.

Similarly, a reaction such as that shown in Figure 7-17 is faster when the intermediate cation is tertiary. Therefore, in the addition of HCl to methylcyclohexene, the reaction proceeds through a more-stable, tertiary cation intermediate by a reaction pathway with a lower activation energy barrier (Figure 7-18). Note that the two carbon atoms of the double bond in 1-methylcyclohexene are not equivalent. Protonation at C-2 gives the tertiary ion in the rate-determining step in the upper reaction. Protonation at C-1 forms a secondary carbocation in the lower reaction. Because secondary cations are much less stable than tertiary ones, the formation of the secondary cation does not compete effectively with formation of the tertiary cation and the isomeric product shown at the lower right is not formed.

These energy considerations are illustrated in Figure 7-19, an energy profile very similar to that for the addition of HCl to cyclohexene. The reaction that forms the more-stable, tertiary cation proceeds from the reactants in the center to the products at the left, whereas the slower process (forming the secondary cation) proceeds to the products at the right. Under either kinetic or thermodynamic control, 1-chloro-1-methylcyclohexane is formed: this product is more stable than the isomeric product at the far right, and it is formed through a pathway with a lower activation energy.

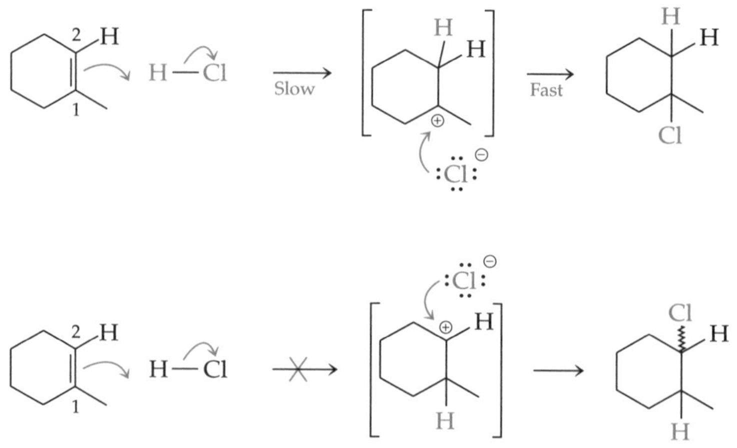

▲ FIGURE 7-18
Protonation of 1-methylcyclohexane at C-2 forms a tertiary cation, whereas protonation at C-1 produces a secondary one. Because a tertiary carbocation is more stable than a secondary one, the upper reaction is thermodynamically easier than the lower one.

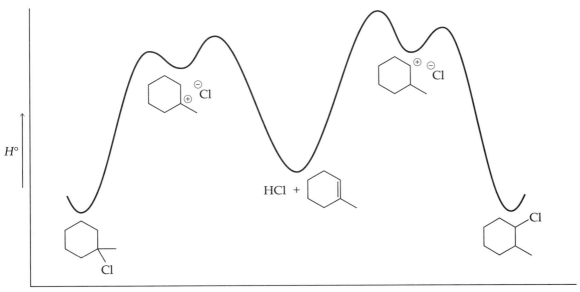

Reaction coordinate

▲ **FIGURE 7-19**
A comparison of the energy diagrams for the two possible regiochemistries
for electrophilic addition of HCl to 1-methylcyclohexene.

Although 1-methylcyclohexene is more stable than cyclohexene because of the methyl group on the double bond, the effect of this substituent
on the relative stabilities of the possible cations is even greater. Thus, the
addition of a proton to methylcyclohexene has a lower activation energy
than that for cyclohexene itself, and the former reaction occurs more readily.
The addition of HX to a carbon–carbon double bond takes place in a
step-by-step fashion, with the positions of H and X in the product being
governed by the stability of the intermediate cation. Electrophilic attack at
the less highly substituted carbon gives the more-stable carbocation and,
thus, protonation by acid occurs at the less highly substituted carbon atom
of the double bond. This preferred orientation defines the **regiochemistry,**
or positional isomerism, of the reaction. We can predict the regiochemistry
of the addition of HX to unsymmetrical alkenes to be that resulting from the
addition of the proton to the less-substituted carbon. The Russian chemist
Vladimir Markovnikov was the first to make this observation; thus, the
preference for this regiochemistry is referred to as **Markovnikov's Rule.** Although Markovnikov did not fully understand the chemical basis for his
rule, which was uncovered many years later, he was an excellent scientist
who generalized from his experimental observations so as to predict the
course of new reactions.

EXERCISE 7-J

Predict the preferred regiochemistry for the addition of HCl to each of the
following compounds on the basis of carbocation stability:

(a)   (b)   (c)   (d)

Multistep Nucleophilic Substitution (S$_N$1): Hydrolysis of Alkyl Bromides

Another reaction that includes the formation of an intermediate cation is the conversion of an alkyl halide into an alcohol, with the replacement of the halogen by an OH group from water. Because this reaction proceeds through cationic intermediates, it is a multistep **heterolytic substitution.** In this **hydrolysis** reaction, the C–X bond is broken *before* a bond is formed with the nucleophile. Because this first step involves only the starting material, this nucleophilic substitution is unimolecular, and the reaction is therefore referred to as an **S$_N$1 reaction.** In this terminology, S$_N$ describes the overall reaction (a substitution in which the substituent is a nucleophile), and the number 1 relates to the molecularity of the rate-determining step (unimolecular); that is, the rate-determining step consists only of *bond breaking* in the substrate and does not involve the nucleophile. The same terminology was used earlier to describe a concerted nucleophilic substitution as an S$_N$2 reaction: namely, as a bimolecular nucleophilic substitution in which two species (the substrate and nucleophile) participate in the rate-determining step.

Consider, for example, the hydrolysis of *t*-butyl bromide and isopropyl bromide (Figure 7-20, in which R = *t*-Bu or *i*-Pr). These nucleophilic substitutions are mechanistically different from those considered in Section 7-4 because they take place in several steps, the first step being the heterolytic cleavage of the carbon–bromine bond to generate a carbocation and a bromide ion. (In contrast, the S$_N$2 substitutions discussed earlier are concerted and do not proceed through any reaction intermediate.) After the carbocation is formed in the first, difficult step, water attacks the cation in a second step, forming a new carbon–oxygen bond. In this step, electron density flows from the lone pair on water's oxygen to the carbocationic carbon, forming a bond between carbon and oxygen. As a result, the oxygen now bears a formal positive charge in this intermediate, which is referred to as an oxonium ion. Loss of a proton restores neutrality to this oxygen in the last step of the overall process.

The cation formed in the first step of the heterolytic cleavage of the carbon–bromine bond of *t*-butyl bromide is tertiary, whereas that derived from isopropyl bromide is secondary. Thus, the upper reaction in Figure 7-21 is less endothermic than the lower reaction. Because this step of the S$_N$1 reaction consists only of bond breaking, it is endothermic and the transition state for the C–Br cleavage resembles a carbocation.

To the extent that a tertiary cation is more stable than a secondary one, the transition state leading to the former is favored and a lower activation energy barrier is encountered. With a lower energy barrier, a greater fraction of the reactant molecules are able to reach the transition state at a given

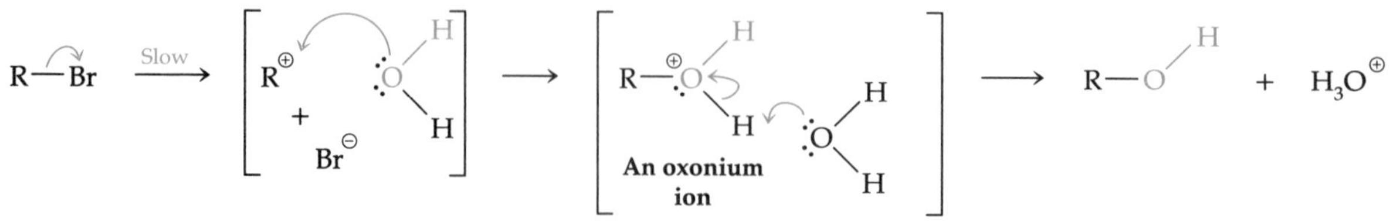

▲ **FIGURE 7-20**

In the first step of an S$_N$1 reaction, bond breaking results in the formation of a carbocation; this step is rate determining. Trapping of this cation by water in the second step is rapid because only bond making is required. The oxonium ion formed is then deprotonated to produce the observed alcohol product.

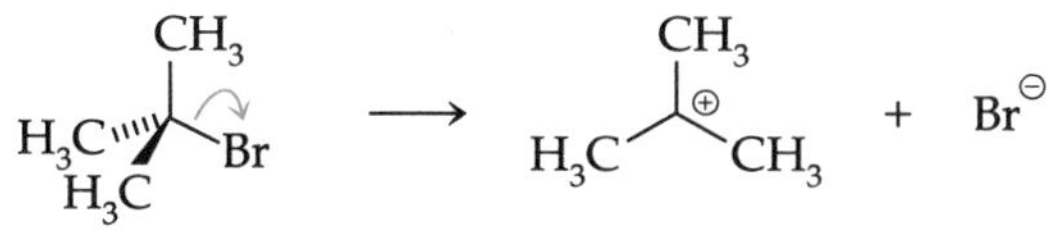

t-Butyl bromide

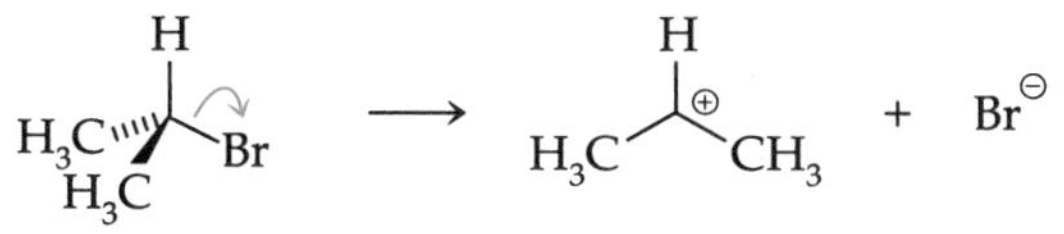

i-Propyl bromide

▲ FIGURE 7-21

An S_N1 reaction proceeds more readily when a more-stable cation is formed. Because the C–X bond is cleaved in the first (rate-determining) step, a weaker C–X bond is cleaved more readily than a stronger one. The order of reactivity toward S_N1 reaction is: RI > RBr > RCl > RF.

temperature (Chapter 6) and the reaction is faster. The order of reactivity for an S_N1 reaction is thus $3° > 2° > 1°$, opposite that for an S_N2 reaction.

EXERCISE 7-K

Which of the following pairs of compounds is likely to react more rapidly under S_N1 conditions?

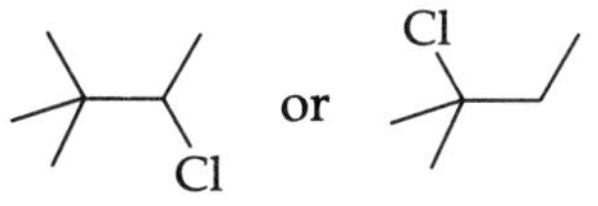

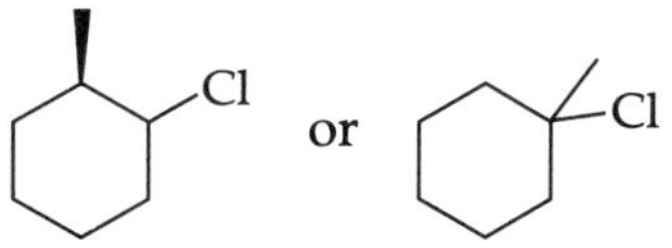

The carbocation is not the final product, however, and hydrolysis requires that this intermediate react with water to form an oxonium ion, deprotonation of which gives the final product, tertiary butyl alcohol (Figure 7-20). Of the two cations (the carbocation and oxonium ion), the carbocation is the less stable because carbon lacks a full complement of valence electrons in a carbocation, whereas each second-row atom in the oxonium ion has access to eight valence electrons (Figure 7-22).

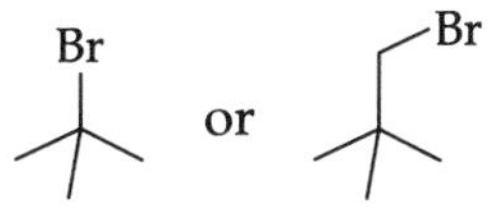

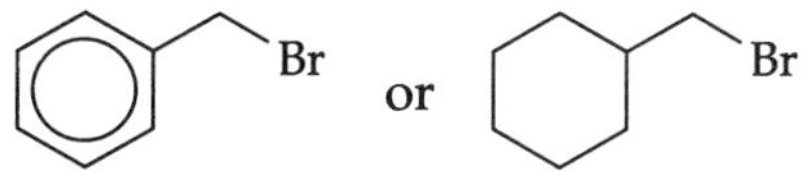

Formation of the carbocation, entailing only bond breaking, is thermodynamically unfavorable. Because this step has a large activation energy, it is undoubtedly rate determining, and typically reasonable reaction rates are obtained only upon heating. Because the rate-determining step is unimolecular, the concentration of the nucleophile (water) does not appear in the reaction rate expression because it does not enter into the reaction until after the rate-determining step. The reaction rate therefore depends *only* on the concentration of the reactant. Trapping of the carbocation by water in the second step consists only of bond making and is exothermic. In the final step, a proton on the oxonium ion is transferred to a base—in this case, water. The deprotonation is very fast, as is generally true for reactions in which a proton is transferred from one heteroatom to another and the identity of the charged atom does not change (here, from an oxygen in the

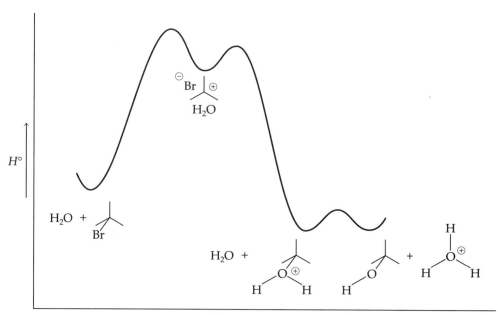

▲ FIGURE 7-22

Complete reaction profile for the S_N1 hydrolysis of *t*-butyl bromide. The rate-determining step (carbocation formation) is followed by fast nucleophilic capture of the cation by water, forming an oxonium ion, which then transfers a proton to a neutral water molecule.

oxonium ion to an oxygen in the hydronium ion). Because the solvent acts as the nucleophile and traps the cationic intermediate, these S_N1 reactions are called **solvolysis** reactions. When water is the solvent, a solvolysis is called a **hydrolysis.**

We can also approach the carbocationic intermediate from the alcohol, by the reversed sequence shown in Figure 7-23. Protonation of the alcohol results in the formation of an oxonium ion, dehydration of which provides the cation. From the principle of microscopic reversibility, this carbocation is the same as that formed in the solvolysis of the corresponding alkyl bromide. We conclude that the rate-determining step of both the forward and the reverse reactions is cation formation. This example clearly shows why it is very important that we understand the structure and reactivity of intermediates in organic chemistry: when we really understand a reaction in one direction, we also understand its reverse. For example, the same carbocation can be trapped by halide to form alkyl halide or by water to form alcohol: the intermediate is the same whether the reaction is proceeding from halide to alcohol or from alcohol to halide. (As you continue your study of

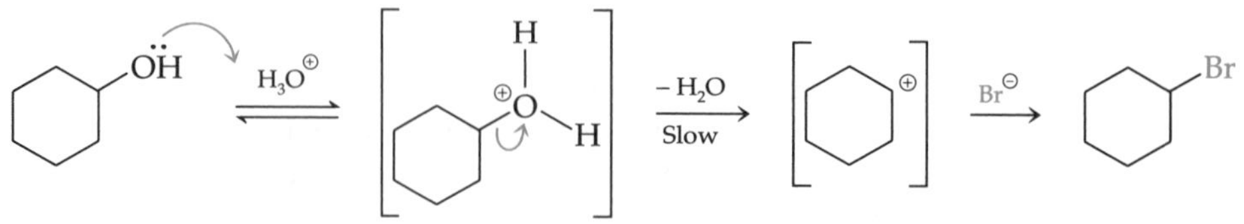

▲ FIGURE 7-23

An alcohol can be converted into an alkyl halide by an S_N1 reaction that is the reverse of the hydrolysis of an alkyl halide. In both reactions, formation of the carbocationic intermediate occurs in the rate-determining step.

organic chemistry, you will find with increasing frequency that your understanding of reaction mechanisms begins to dove-tail and that you will already know a great deal about the reactions of new functional groups.)

The order of carbocation stability (3° ~ benzylic > 2° ~ allylic > 1° > CH_3) also provides a driving force for rearrangement reactions. Whenever a more-stable carbocation can be produced by shifting a substituent from an adjacent position to a cationic carbon, this shift takes place rapidly. For example, under typical S_N1 conditions, the hydrolysis of 2-bromo-3-methylbutane produces 2-methyl-2-butanol through a rearranged tertiary cation produced when hydrogen shifts with its electrons from C-3 of the reactant to the carbocation at C-2.

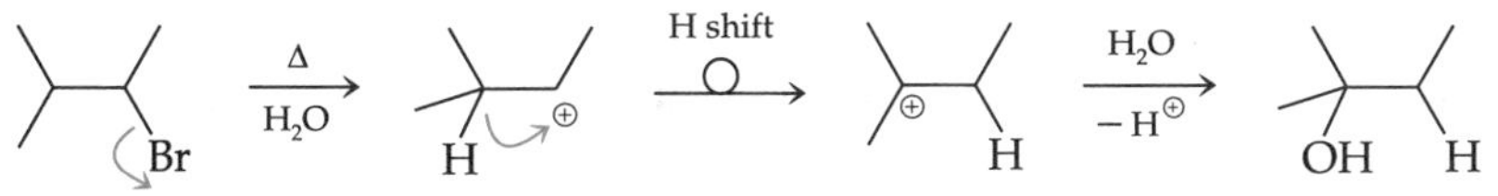

Either hydrogen or an alkyl group can shift in this way to yield a more-stable carbocation.

In summary, hydrolysis occurs through a series of steps if the intermediate cation is particularly stable; for example, a tertiary carbocation. Thus, a tertiary alkyl bromide is hydrolyzed in an S_N1 reaction to the corresponding alcohol more readily than a secondary or primary bromide. Trapping of a stable carbocation by water generates an oxonium ion, from which loss of a proton gives the product alcohol. In practice, secondary and tertiary alcohols are formed by heating alkyl halides in water.

EXERCISE 7-L

Predict, on the basis of carbocation stability, which member of each of the following pairs hydrolyzes at the faster rate:

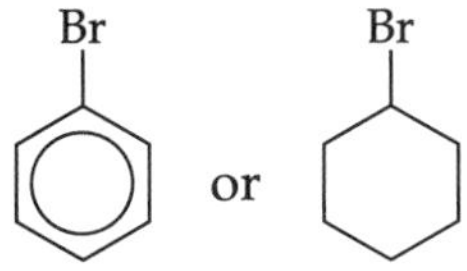

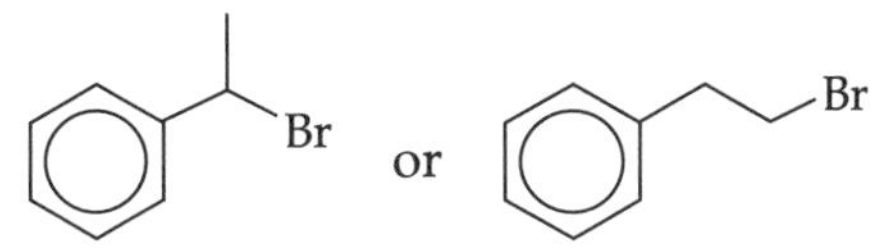

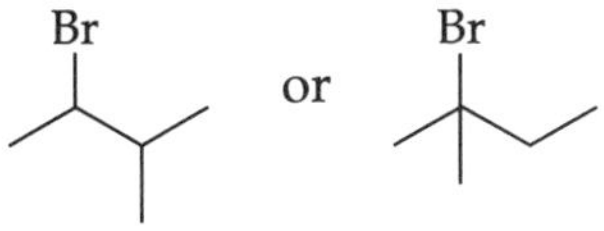

7-6 ▸ Mechanism of a Multistep Homolytic Cleavage: Free Radical Halogenation of Alkanes

One of the principal means of introducing functional groups into alkanes is homolytic substitution. In a homolytic reaction, radical intermediates are formed and the electrons of the σ bonds undergoing cleavage do not remain paired as they do in reactions in which ionic intermediates are formed. To describe these homolytic substitution reactions, we use two half-headed arrows to show the movement of single electrons.

$$CH_3CH_3 + Cl_2$$

$$\downarrow \text{\textit{hv} or } \Delta$$

$$CH_3CH_2Cl + HCl$$

▲ FIGURE 7-24
Alkanes are converted into alkyl halides by free radical (homolytic) substitution initiated by exposing chlorine or bromine to heat (Δ) or light (*hv*) in the presence of the alkane.

OXIDATIVE DESTRUCTION OF BACTERIA BY RADICALS

The bond between two oxygen atoms in an organic molecule is generally very weak (~36 kcal/mole). Homolytic cleavage of this bond generates two radicals that can readily oxidize organic materials and, by doing so, can destroy bacteria. Hydrogen peroxide (HO-OH), the simplest example of a compound with an oxygen–oxygen bond, has been used for centuries as a dilute solution in water as a cleansing and antiseptic agent. When applied to an open wound or indeed any tissue, hydrogen peroxide forms bubbles and foams as it is decomposed into molecular oxygen and water by the iron present in hemoglobin. Although this gas evolution mainly helps to clean the wound, it contributes significantly to the use of hydrogen peroxide as an antiseptic because many people feel that something should visibly happen—as, for example, the dark staining that occurs when iodine is used as a disinfectant.

Energetics of Homolytic Substitution in the Chlorination of Ethane

In the **free radical halogenation** of ethane, a C–H bond is replaced by a C–X bond, with X being one of the halogens. This reaction proceeds through free radical intermediates and is thus a **homolytic substitution** reaction. The stoichiometry of the reaction is defined in Figure 7-24, and we can estimate the $\Delta H°$ of this reaction by comparing the energy of the bonds consumed in the reaction (C–H and Cl–Cl) with those of the bonds formed (C–Cl and H–Cl). By using the energies from Table 3-6, we calculate $\Delta H°$ as shown in Figure 7-25 and find that this homolytic substitution is exothermic.

EXERCISE 7-M

Calculate $\Delta H°$ for each of the following possible conversions:

(a) 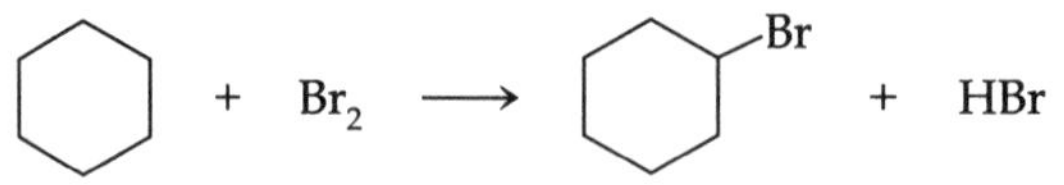 (hexane ring) + Br$_2$ ⟶ (cyclohexane with Br) + HBr

(b) 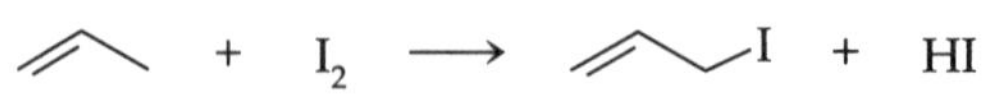 (structure) + I$_2$ ⟶ (structure with I) + HI

(c) 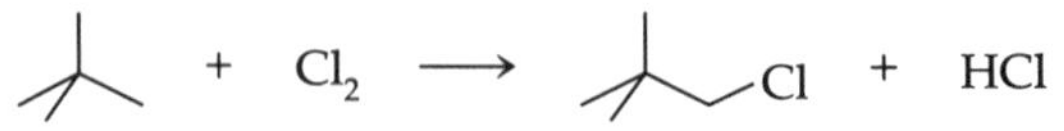 (structure) + Cl$_2$ ⟶ (structure with Cl) + HCl

▶ FIGURE 7-25
To calculate the enthalpy change of a chemical reaction, we subtract the bond energies of bonds formed in the products from those broken in the reactants.

$$CH_3CH_2-H + Cl-Cl \xrightarrow{\textit{hv}} CH_3CH_2-Cl + H-Cl$$

$$\downarrow \qquad\qquad \downarrow \qquad\qquad\qquad \downarrow \qquad\qquad \downarrow$$

$$CH_3CH_2 \cdot \;\; \cdot H \qquad Cl \cdot \;\; \cdot Cl \qquad CH_3CH_2 \cdot \;\; \cdot Cl \qquad H \cdot \;\; \cdot Cl$$

$$\Delta H° = 100 \qquad \Delta H° = 58 \qquad \Delta H° = 80 \qquad \Delta H° = 103$$

$$\Delta H° = (100 + 58) - (80 + 103) = -25 \text{ kcal/mole}$$

Steps in a Radical Chain Reaction

A radical reaction takes place in steps and, as before, we must evaluate all of the steps in order to identify the one that is rate determining. The reaction requires light or heat for initiation. In this **initiation step,** two free radicals are produced from a stable starting material (Figure 7-26). The energy required for this homolytic fission is the bond-dissociation energy of a weak σ bond: here, the chlorine–chlorine bond. Light or heat is used to induce fission of this bond, generating two reactive radical fragments. This endothermic initiation step is critical in beginning the reaction; but it is not a part of the overall stoichiometry. Because they require that a bond be broken without the simultaneous formation of another one, initiation steps consume a substantial amount of energy. Fortunately, initiation steps need not be stoichiometric, because even a few radicals formed in an initiation step can begin a radical chain that can form a large number of product molecules.

The radicals produced in the initiation step then interact with the alkane in a process in which the C–H bond is homolytically cleaved as the H–X bond is formed (Figure 7-27). Again, referring to the table of bond-dissociation energies (Table 3-6), we find that this step is exothermic by 100 − 103 = −3 kcal/mole. Notice that this step does not generate more free radicals than are present as reactants: it simply converts one reactant radical into a different product radical. One bond is broken as another is formed. Such a step is therefore referred to as a **propagation step.** In a propagation step, the number of product radicals is equal to the number of reactant radicals.

The resulting alkyl radical then interacts with Cl_2 (second step of Figure 7-27). The chlorine–chlorine bond is broken at the same time that a carbon–chlorine bond is formed in an exothermic step ($\Delta H° = 58 − 80 = −22$ kcal/mole). Like the preceding step, this propagation reaction does not change the number of reactive intermediates: an alkyl radical is consumed as a chlorine atom is formed. Notice that the reactive intermediate that initiates the first propagation step is formed as a product in the second. This chlorine atom can then serve as a reactant to repeat the first step. The reactions in Figure 7-27 alternate in a **radical chain reaction** until the reactant alkane or chlorine or both are consumed.

When we add the two propagation reactions in Figure 7-27 (and cancel species appearing on both sides), we obtain the net reaction shown in Figure 7-25. The sum of all propagation steps in a homolytic free radical reaction must add to the net stoichiometric equation: otherwise, one of the reactions is not balanced.

When the initiation reaction is continued (by supplying continuous heat or light), the number of radicals increases until they begin to encounter

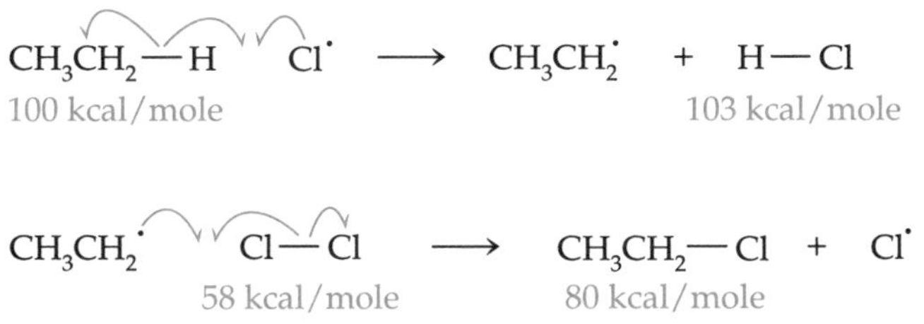

Cl—Cl

$\downarrow$ *hv* or Δ

Cl· + Cl·

$\Delta H° = +58$ kcal/mole

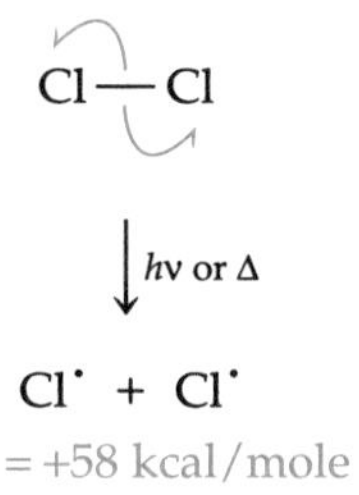

▲ FIGURE 7-26
In an initiation step, a covalent bond is homolytically cleaved to produce two radicals. The energy needed to break the bond can be supplied either as light or as heat.

CH_3CH_2—H   Cl· $\longrightarrow$ CH_3CH_2· + H—Cl
100 kcal/mole   103 kcal/mole

CH_3CH_2·   Cl—Cl $\longrightarrow$ CH_3CH_2—Cl + Cl·
58 kcal/mole   80 kcal/mole

▲ FIGURE 7-27
The propagation steps in a free radical chlorination consume one radical while producing another. Both steps are exothermic and proceed efficiently. The product radical in one step is the reactant radical in the next, and so these reactions repeatedly cycle until the alkane or chlorine is almost completely consumed.

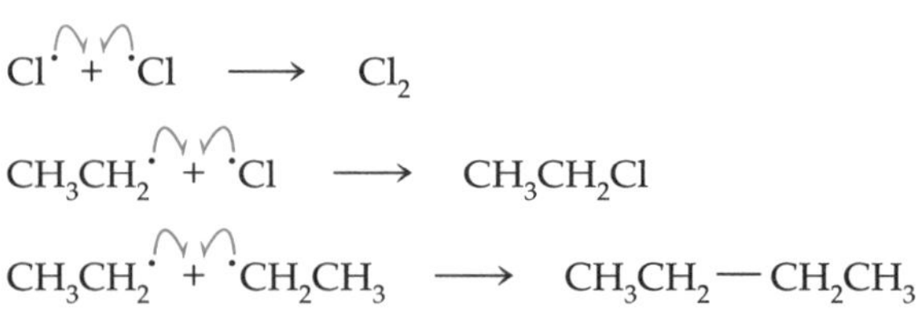

$$\text{Cl} \cdot + \cdot \text{Cl} \longrightarrow \text{Cl}_2$$

$$\text{CH}_3\text{CH}_2 \cdot + \cdot \text{Cl} \longrightarrow \text{CH}_3\text{CH}_2\text{Cl}$$

$$\text{CH}_3\text{CH}_2 \cdot + \cdot \text{CH}_2\text{CH}_3 \longrightarrow \text{CH}_3\text{CH}_2 - \text{CH}_2\text{CH}_3$$

▲ **FIGURE 7-28**

In a termination step, a covalent bond is formed as each of two radicals donate its unpaired electron to form a σ bond. This bond formation releases energy and blocks further propagation steps by consuming the reactive free radical.

each other, at least occasionally. Two radicals can then combine exothermically to form a σ bond (Figure 7-28) in a process that is exactly opposite that of the initiation step. The reactions shown convert two reactive radicals into one stable product. They are therefore referred to as **termination reactions,** because they stop the radical chain reaction by consuming a reactive intermediate without producing another.

The two different radicals available from each of the propagation steps can combine with themselves or with each other. One would therefore expect the formation in a termination step of not only small amounts of molecular chlorine (by the combination of two chlorine atoms), but also small amounts of alkyl chloride (R· + Cl·) and carbon–carbon (R· + R·) coupling products. Because termination steps begin to become important only when radical concentrations increase, which is usually when one of the reactant molecules is consumed, termination steps do not contribute significantly to the observed product distribution.

EXERCISE 7-N

Calculate $\Delta H°$ for each of the following proposed propagation steps:

(a)  $\diagup\!\!\diagdown$ + Br· $\longrightarrow$ $\diagup\!\!\diagup\!\!\diagdown$ + HBr

(b) 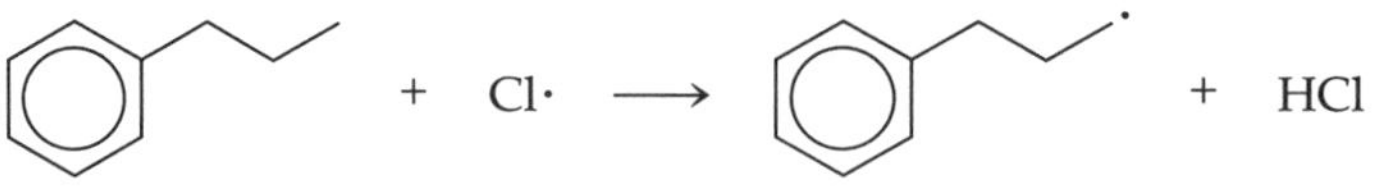 [benzene ring]$\diagup\!\!\diagdown$ + Cl· $\longrightarrow$ [benzene ring]$\diagup\!\!\diagup\!\!\diagdown\cdot$ + HCl

(c) 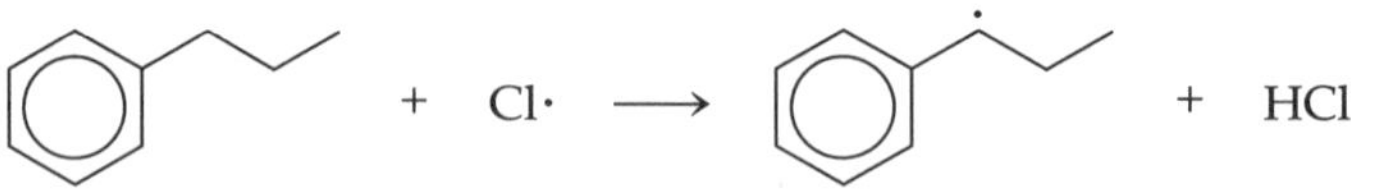 [benzene ring]$\diagup\!\!\diagdown$ + Cl· $\longrightarrow$ [benzene ring]$\overset{\cdot}{\diagup}\!\!\diagdown$ + HCl

---

NATURAL DEFENSE SYSTEMS

The bombardier beetle (genus *Brachinus*) uses a radical chain reaction to produce the propellant enabling it to shoot an irritating mixture of steam, water, and HCN at an attacker. The chain decomposition of hydrogen peroxide is sufficiently exothermic to warm water above its boiling point, building pressure which is released when the beetle targets its enemy.

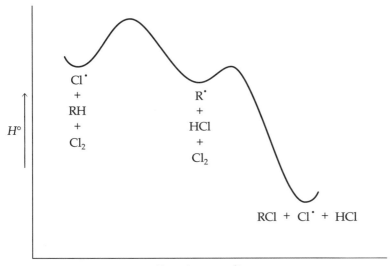

▲ FIGURE 7-29
This reaction profile shows that two propagation steps are needed in a free
radical substitution reaction.

Although a homolytic substitution takes place through several steps,
the propagation steps account for the bulk of product formation. The net
stoichiometry and thermodynamics for the reaction derive therefore from a
consideration of the two propagation steps alone and are represented by the
reaction profile shown in Figure 7-29.

Relative Reactivity of Halogens

Free radical halogenation is not limited to reactions with chlorine. Ethane
also reacts with bromine by a homolytic pathway essentially identical with
that for chlorination. The energies of the various bonds broken and formed
in chlorination and bromination are quite different, however, and these differences can significantly affect the usefulness of free radical chlorination
and bromination.

$$CH_3CH_2 \!-\! H + Br\cdot \longrightarrow CH_3CH_2\cdot + H\!-\!Br \qquad \Delta H° = 100 - 87 = +13 \text{ kcal/mole}$$
$$CH_3CH_2\cdot + Br\!-\!Br \longrightarrow CH_3CH_2\!-\!Br + Br\cdot \qquad \Delta H° = 46 - 68 = -22 \text{ kcal/mole}$$

Abstraction of hydrogen by chlorine is slightly exothermic, whereas
that by bromine is endothermic, as a result of the stronger bond in HCl than
in HBr. Because of this difference, the transition state in chlorination is earlier (more reactantlike) than that in bromination. The later transition state in
bromination is therefore more radicallike (Figure 7-30). As we will see in the
next subsection, a late transition state, resembling the radical intermediate
more than the starting material, affords higher selectivity and better regiochemical control.

We can also use thermodynamics to explain why free radical chains are
not used for fluorination or iodination (Figure 7-31). Abstraction of a hydrogen atom from cyclohexane by fluorine would be exothermic by approximately 39 kcal/mole, a value much too large to allow for safe and effective
control in a self-propagating reaction without special precautions. Free

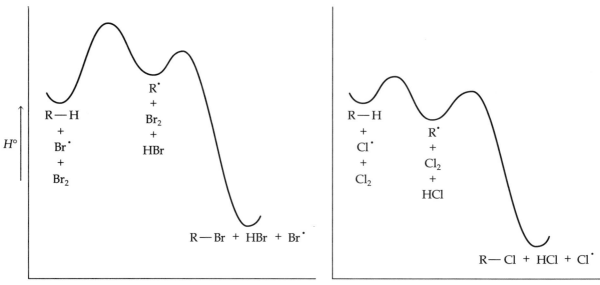

▲ FIGURE 7-30

A comparison of the transition states for hydrogen atom abstraction by bromine (left) and chlorination (right) is described by their reaction profiles. Because the first step is endothermic for bromination and exothermic for chlorination, the transition state for bromination is late and has substantial radical character, whereas that for chlorination is early and resembles the reactant alkyl halide rather than the intermediate radical.

radical fluorination would be so exothermic that one would need to worry about extensive local heating and explosion, although reasonable yields of fluorination products have been obtained under very carefully controlled conditions. Iodination, on the other hand, is very unfavorable thermodynamically and, even if it were driven by the removal of HI by reaction with base, its propagation steps would be intolerably slow.

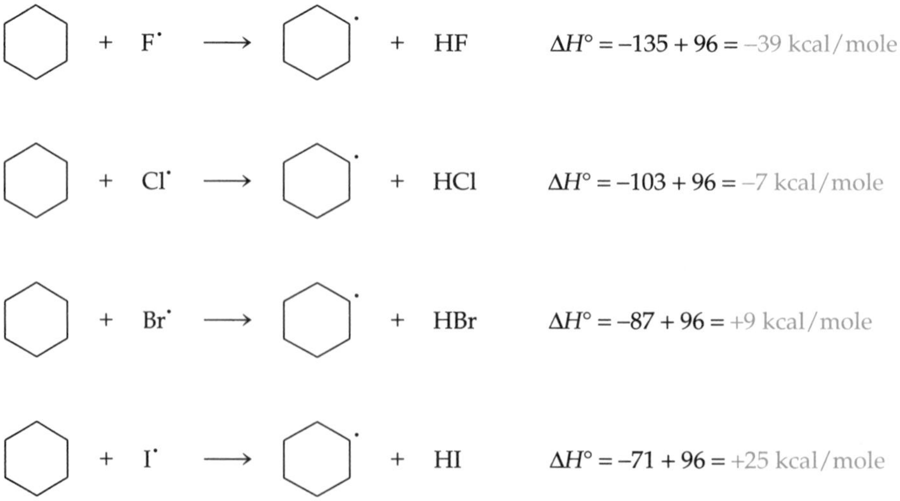

$$\text{(hexane)} \quad + \quad F^{\cdot} \quad \longrightarrow \quad \text{(hexyl radical)}^{\cdot} \quad + \quad HF \qquad \Delta H^{\circ} = -135 + 96 = -39 \text{ kcal/mole}$$

$$\text{(hexane)} \quad + \quad Cl^{\cdot} \quad \longrightarrow \quad \text{(hexyl radical)}^{\cdot} \quad + \quad HCl \qquad \Delta H^{\circ} = -103 + 96 = -7 \text{ kcal/mole}$$

$$\text{(hexane)} \quad + \quad Br^{\cdot} \quad \longrightarrow \quad \text{(hexyl radical)}^{\cdot} \quad + \quad HBr \qquad \Delta H^{\circ} = -87 + 96 = +9 \text{ kcal/mole}$$

$$\text{(hexane)} \quad + \quad I^{\cdot} \quad \longrightarrow \quad \text{(hexyl radical)}^{\cdot} \quad + \quad HI \qquad \Delta H^{\circ} = -71 + 96 = +25 \text{ kcal/mole}$$

▲ FIGURE 7-31

The thermodynamic feasibility of hydrogen atom abstraction by a halogen atom varies with the identity of the halogen, ranging from a reaction so exothermic that it is difficult to control with fluorine to one so endothermic as to be impossible under normal laboratory conditions with iodine.

Chlorofluorocarbons (CFCs) are important industrial chemicals that have a variety of uses, including as refrigerants and degreasing solvents. They are named by a special system referred to as the rule of 90. Thus, to determine the composition of CFC-12, the numerical suffix, 12, is added to 90 (12 + 90 = 102). The result is read as one carbon (1 02), no hydrogens (10 2), and 2 fluorines (10 2); the remaining atoms necessary to fill the valence(s) of the carbon(s) are chlorines (2). Thus, CFC-12 is CCl_2F_2.

Unfortunately, CFCs decompose on exposure to ultraviolet light in the upper atmosphere, generating (among other products) chlorine atoms. In turn, these chlorine atoms serve as catalysts for the decomposition of ozone into molecular oxygen:

$$CCl_2F_2 \xrightarrow{\text{Sunlight}} \cdot Cl$$

$$2\,O_3 \xrightarrow{\cdot Cl} 3\,O_2$$

Ozone in the upper atmosphere absorbs harmful ultraviolet light and thus plays a very important, protective role (for example, against uv-induced skin cancer) for life on earth. A hole in the ozone layer near the South Pole has permitted higher-than-normal ultraviolet irradiation in the far southern regions of South America, and consequent blinding of entire flocks of sheep and of the Indians tending them has been reported to result from retinal damage induced by unfiltered UV light. Replacements for the CFCs, primarily hydrofluorocarbons that will not damage the ozone layer, are beginning to be produced on a large scale.

Regiocontrol in Homolytic Substitution

Isomeric alkyl halides can be produced when different types of hydrogen–carbon bonds are present in an alkane. We can write a reaction sequence for the halogenation of propane in parallel to that for ethane, recognizing that hydrogen can be abstracted from either of two types of C–H bonds, at the two equivalent end carbons (C-1 and C-3) or at the middle carbon (C-2).

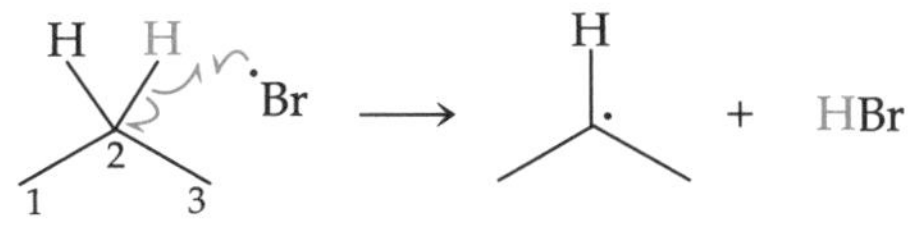

or

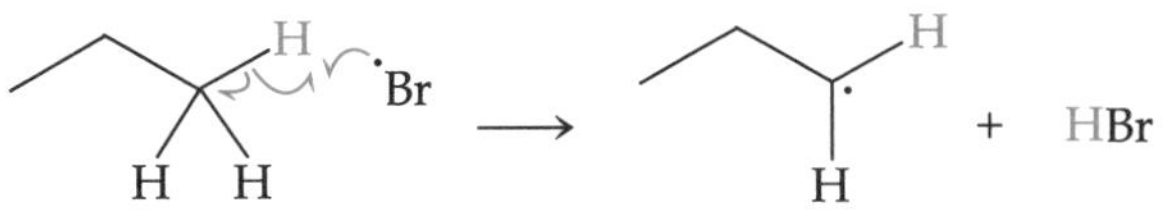

Statistics favor the six primary hydrogens at the C-1 and C-3 positions over the two secondary hydrogens at C-2. If the end and middle C–H bonds were equally reactive, we would expect three times as much 1-bromopropane as 2-bromopropane.

Experimentally, however, the ratio of 1-bromopropane to 2-bromo-propane is 8:92, far from statistical, whereas only a small deviation from the statistical prediction is observed in chlorination. This "preference" (revealed by a different ratio from that predicted statistically) for one positional isomer over another is called **regiocontrol,** and these reactions are said to be **regioselective.** The higher-than-expected formation of 2-bromopropane is consistent with the greater stability of the secondary radical. Reactions differ greatly in the degree of regioselectivity: here, bromination is highly selective, whereas chlorination is hardly selective at all.

EXERCISE 7-O

Predict which hydrogen would be preferentially substituted in the free radical bromination of each of the following compounds by drawing the product that would be expected from treatment with Br_2 and light:

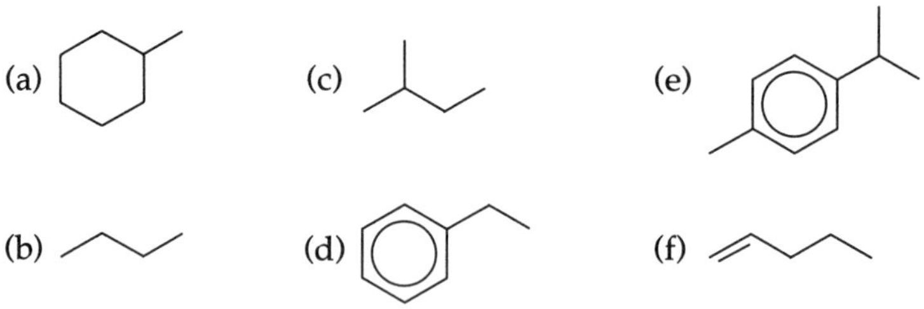

(a)   (c)   (e)

(b)   (d)   (f)

The greater predominance of the 2-halogenated product in bromination than in chlorination can be explained by the use of the Hammond Postulate. The late transition state in an endothermic bromination resembles the radical intermediate, and the difference in stabilities of the 1-propyl and 2-propyl radicals is more important than it is in the early transition states of chlorination. Because the secondary radical is more stable than the primary radical, hydrogen abstraction by bromine from the secondary site (leading to a secondary radical and then ultimately to the secondary halide) occurs to a much larger extent than the same abstraction by chlorine. As a result, the reaction's regioselectivity [preference for halogenation at the secondary (more-stable) radical site] is greater in bromination than in chlorination.

We conclude that the less-reactive bromine atom abstracts hydrogen through a later transition state, leading to higher selectivity, than the more-reactive (and less-selective) chlorine atom. For preparative purposes, bromination gives useful yields of 2-bromopropane, but chlorination gives a difficult-to-separate mixture of 1-chloropropane and 2-chloropropane.

EXERCISE 7-P

Predict whether each of the following alkyl halides can be synthesized by direct free radical halogenation of the indicated hydrocarbon. If not, indicate what product would be formed.

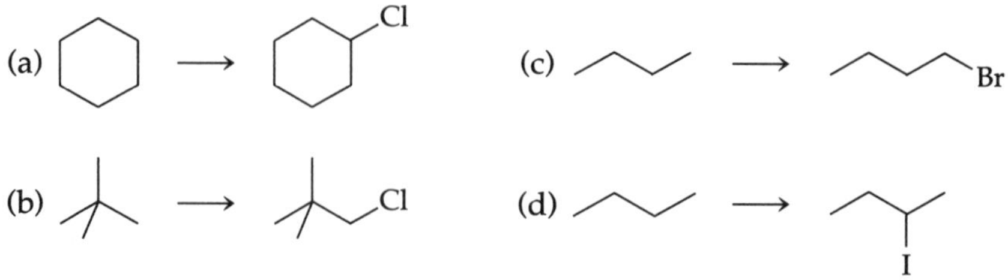

(a)   (c)

(b)   (d)

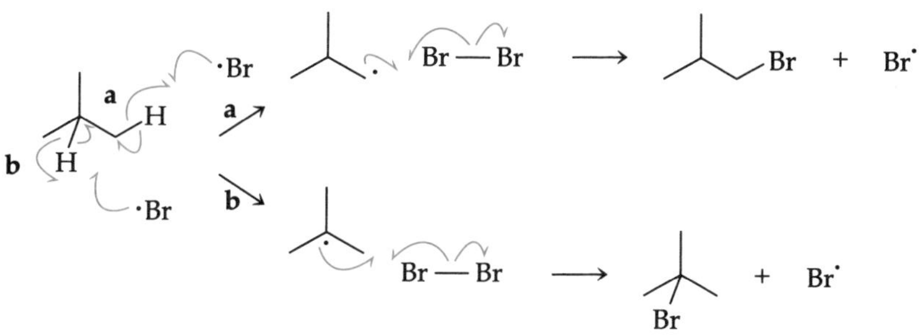

▲ FIGURE 7-32

Abstraction of a primary hydrogen (path *a*) produces a primary radical, whereas abstraction of a tertiary hydrogen (path *b*) produces a tertiary radical. Because bromination is highly selective, path *b* is taken to the virtual exclusion of path *a*.

Analogous arguments would also pertain to the free radical chlorination and bromination of 2-methylpropane.

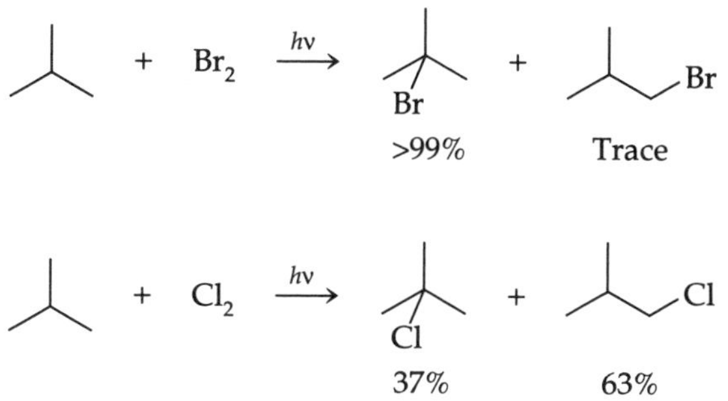

The product yields of the primary and tertiary bromination and chlorination products differ even more significantly from a purely statistical distribution than in the halogenation of propane. In the mechanism for the bromination (Figure 7-32), regioselective abstraction of the tertiary hydrogen by Br· (path *b*) is favored over that of the primary hydrogen (path *a*).

The high selectivity of bromination makes it particularly attractive as a means of introducing a halogen into hydrocarbons. Even higher selectivity for bromination can be attained when *N*-bromosuccinimide is used as the bromine atom source. *N*-Bromosuccinimide has a weak nitrogen–bromine bond that can be fragmented into radicals upon warming or exposure to visible light. It causes the replacement of allylic or benzylic C–H σ bonds by C–Br bonds (Figure 7-33).

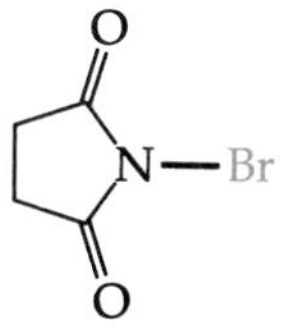

N-Bromosuccinimide

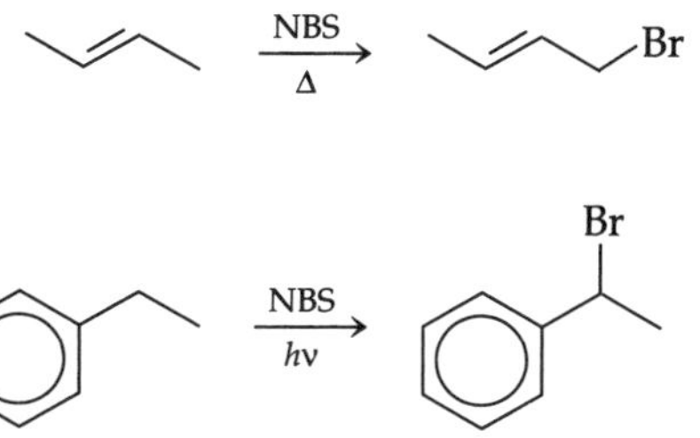

▲ FIGURE 7-33

Free radical substitution of an allylic (upper reaction) and benzylic (lower reaction) hydrogen by bromine through the use of the reagent *N*-bromosuccinimide (NBS).

Table 7-2 ▸ How to Use the Reactions in This Chapter to Make Various Functional Groups

Functional Group	Reaction
Alcohol	Acid-catalyzed hydration of alkenes; or hydrolysis of alkyl halides
Aldehyde	Aldol condensation of aldehydes
Alkene	Acid-catalyzed dehydration of alcohols; or dehydrohalogenation of alkyl halides; or olefin isomerization
Alkyl halide	Free radical halogenation of hydrocarbons; or hydrochlorination of alkenes; or allylic or benzylic bromination by N-bromosuccinimide; or halide exchange of alkyl halide
Amide	Beckmann rearrangement of an oxime
Amine	S_N2 amination of alkyl halide
Ester	Acid-catalyzed esterification of carboxylic acids
Ketone	Pinacol rearrangement of a diol

7-7 ▸ Synthetic Applications

Alkyl halides are important synthetic intermediates because halogen can be easily replaced by other groups. We have seen in this chapter that alkyl halides can be produced by three possible routes: addition of HX to an alkene, heterolytic substitution of an alcohol, or homolytic substitution of alkane. The resulting alkyl halides can then be converted to other functional groups by the displacement reactions listed earlier in Table 7-1.

As you learn new reactions, it is important to compare different routes that achieve a common objective. If you think about the unique characteristic of each reaction, you will learn how to use it discriminately. Table 7-2 is a way of compiling the reactions considered in this chapter. Refer to the section in the chapter that deals with each reaction to be sure that you understand how it is best employed.

Conclusions

Most organic reactions can be classified as one of seven major types: additions, eliminations, substitutions, condensations, rearrangements, isomerizations, or oxidation/reduction reactions. Some of these reactions take place through reactive intermediates, whereas others proceed directly from starting material to product in a single step. A reaction that includes the formation of one or more reactive intermediates is called a multistep reaction. Reactions in which there are no intermediates are referred to as concerted.

There are two distinctly different mechanisms for bond cleavages: homolytic and heterolytic. In homolytic cleavages, which are governed by bond-dissociation energies, single electrons move separately to form radical intermediates. In heterolytic cleavages, two electrons move as a pair, resulting in the formation of ions. Heterolytic reactions are much more affected by solvent polarity than are homolytic reactions.

Reaction mechanisms such as those presented in this chapter are best described (and understood) by the use of curved-arrow notation in which a full-headed arrow indicates the motion of two electrons and a half-headed arrow indicates the motion of one electron. The curved arrows indicate movement of electrons, not atoms. When electrons move, atoms follow.

Concerted nucleophilic substitution occurs through backside attack by a nucleophile on the face remote from a leaving group. This reaction, called an S_N2 reaction, occurs with inversion of configuration and is sensitive to the concentrations of both the substrate and the nucleophile. The observed order of reactivity in an S_N2 reaction ($1° > 2° \gg 3°$) depends on steric access of the nucleophile to the reactive carbon atom. The S_N2 reaction effects several kinds of functional group interconversions.

Electrophilic addition takes place in several steps and proceeds through a carbocation intermediate. A hydrohalogenation reaction begins by protonation at one carbon of a $C=C$ double bond, generating a carbocation at the other sp^2-hybridized carbon. The rate-determining step is that leading to the cation, and the character of its transition state is closely related to that of the cation. The product is formed in a second, fast step in which the cation is captured by the halide anion. The regioselectivity is governed by cation stability, which is consistent with Markovnikov's Rule.

Multistep nucleophilic substitution through an S_N1 mechanism also takes place through the formation of an intermediate carbocation. In the hydrolysis of an alkyl halide, this cation is captured by water to form a second intermediate, an oxonium ion, which loses a proton to form the final alcohol product. Thus, there can be more than one reactive intermediate along the reaction coordinate. The rate-determining step of an S_N1 reaction is that leading to the carbocation, and the activation barrier in this reaction is affected by the energy needed for C–X bond heterolysis and by the stability of the carbocationic intermediate. The reactivity order for S_N1 reactions ($3° > 2° \gg 1°$) is governed by cation stability and the reaction rate depends only on substrate concentration (*not* on the concentration of the nucleophile).

Free radical halogenation also occurs by a multistep mechanism and is accomplished by homolytic cleavage. Free radical chain reactions consist of three steps: initiation, propagation, and termination partial reactions. The net stoichiometry of such a reaction is controlled by propagation. Propagation steps usually include the formation of a carbon radical by hydrogen abstraction from an alkane. Regiocontrol in homolytic substitution is governed by radical stability ($3° > 2° > 1°$) and by the earliness or lateness of the transition state. Bromine is more regioselective than chlorine because bromine is less reactive and more selective in abstracting hydrogen through a later (more radicallike) transition state.

Summary of New Reactions

Acid–catalyzed Hydration of Alkenes

Acid–catalyzed Dehydration of Alcohols

Dehydrohalogenation

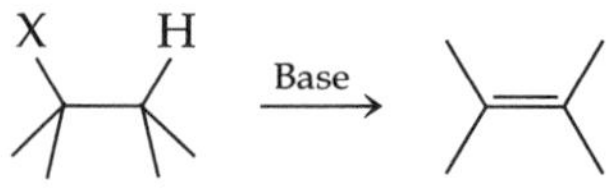

Free Radical Halogenation

$$R—H + X_2 \xrightarrow{h\nu} R—X + H—X \qquad X = Br, Cl$$

Nucleophilic Substitution (Bimolecular): S_N2

$$R—X + {}^{\ominus}Nuc \longrightarrow R—Nuc + {}^{\ominus}X \qquad R: 1° > 2° > 3°$$

(See Table 7-1 for ${}^{\ominus}Nuc$)

Acid-catalyzed Esterification

$$RCO_2H + R'OH \xrightarrow{H_3O^{\oplus}} RCO_2R' + H_2O$$

Aldol Condensation

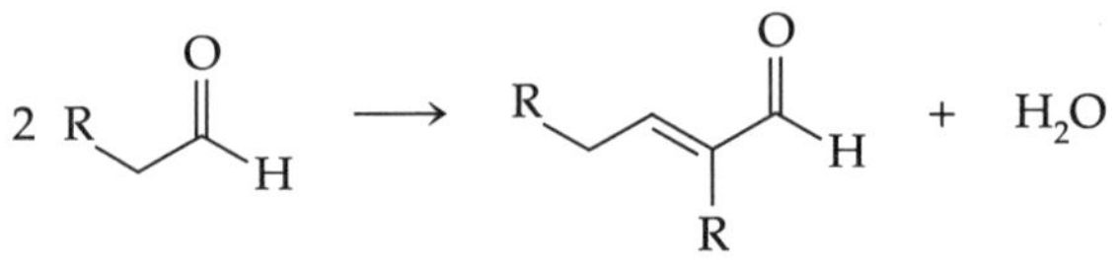

Pinacol Rearrangement

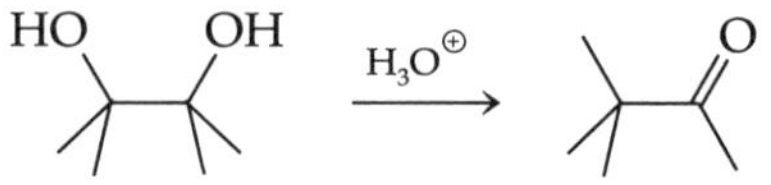

Beckmann Rearrangement

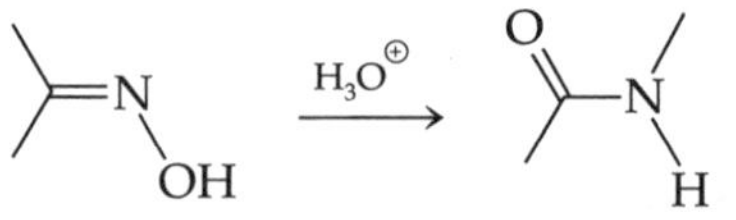

Olefin Isomerization

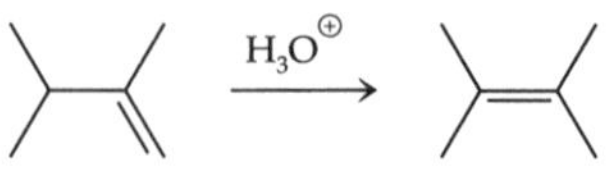

Hydrohalogenation of Alkenes

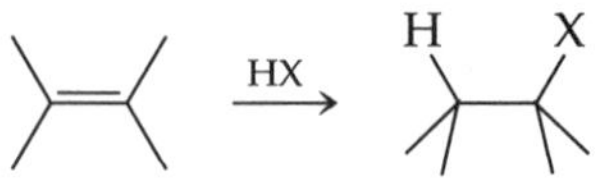

Hydrolysis of Alkyl Halides

$$RX + H_2O \longrightarrow ROH + HX$$

Allylic and Benzylic Bromination by *N*-Bromosuccinimide

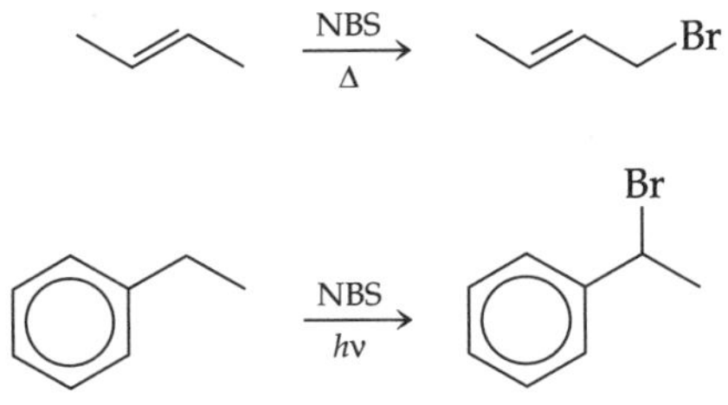

Review Problems

....

7-1 Classify each of the following reactions as an addition, elimination, substitution, condensation, rearrangement, geometric isomerization, or oxidation/reduction.

(a)

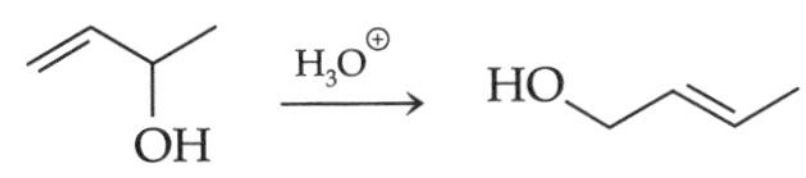

(b)

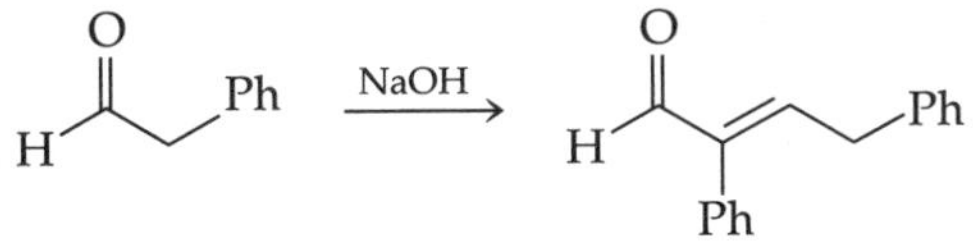

(c)

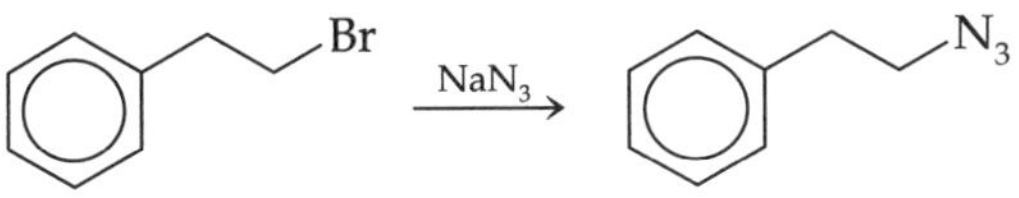

(d)

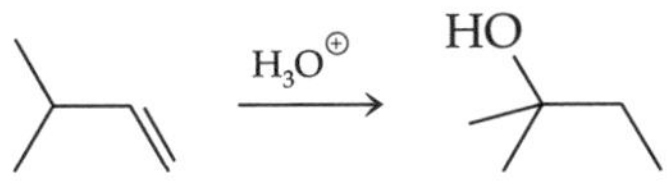

(e)

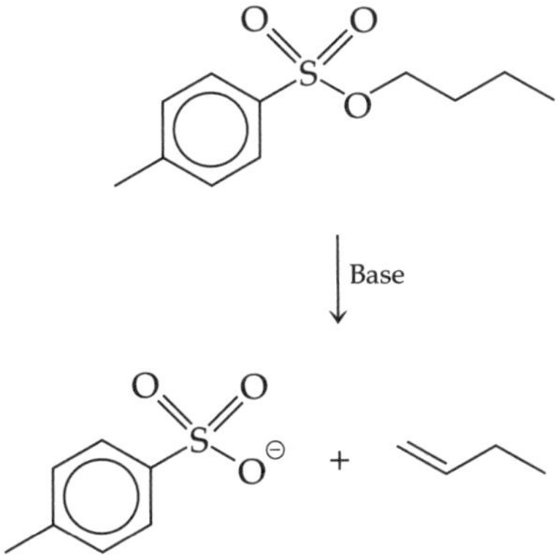

(f)

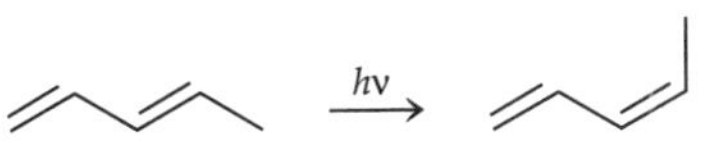

(g)

7-2 The catalytic hydrogenation of an alkene is sometimes called an addition reaction. (Recall that it was also called a reduction in Chapter 2.)

(a) Explain why both classifications are reasonable.

(b) The conversion of 2-propanol into 2-propanone can be considered either an elimination or an oxidation. Explain how this reaction can be viewed as either reaction type.

7-3 Suppose that you wished to make each of the following compounds by an S_N2 reaction. Identify the alkyl halide that you would need, together with the necessary nucleophile.

(a) $CH_3CH_2CH_2CH_2N_3$

(b) $CH_3CH_2CH_2CH_2CN$

(c) CH_3OCH_3

(d) tetrahydrofuran

(e) $CH_3CH_2CH_2CH_2SH$

(f) $CH_3CH_2CH_2SCH_2CH_3$

(g) CH_3OSO_2Ph

(h) $CH_3CH_2P(C_6H_5)_3^{\oplus}Br^{\ominus}$

7-4 Epoxides can be formed through an intramolecular S_N2 reaction. Given what you know about pK_a values, write a mechanism for the following reaction:

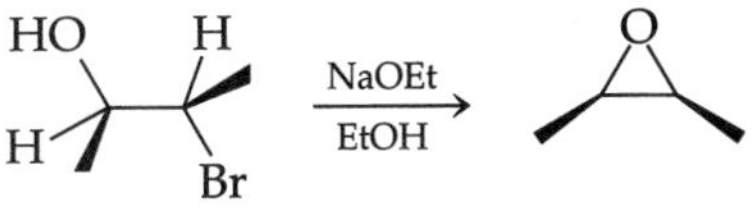

7-5 Azide is known to react by an S_N2 pathway thousands of times more rapidly with 2-bromopentane than with its isomer neopentyl bromide (1-bromo-2,2-dimethylpropane) despite the fact that the leaving group is at a secondary site in the former compound and at a primary site in the latter. Explain.

7-6 To reach the conclusion that the reaction with 2-bromopentane cited in Problem 7-5 did indeed occur through an S$_N$2 reaction, the chemists studying the reaction did several additional experiments in which they:

(a)   used optically active (*R*)-2-bromopentane.

(b)   doubled the concentration of alkyl bromide.

(c)   doubled the concentration of azide ion.

Predict what they would have seen in each experiment if the reaction really took place through an S$_N$2 pathway.

7-7 Choose the member of the following pairs of unsaturated hydrocarbons that is more reactive toward acid-catalyzed hydration and predict the regiochemistry of the alcohols formed from that compound.

(a)   or

(b)   or

(c)   or

7-8 When allowed to stand in dilute aqueous acid, (*R*)-2-butanol slowly loses its optical activity. Write a mechanism that can account for this racemization.

7-9 When the acid-catalyzed hydration of 3-methyl-1-butene is carried out in D$_2$O, the D and OD groups in the alcohol product do not appear on adjacent carbons. Write a detailed mechanism, using curved arrows to show electron flow, that identifies the hydration product formed and explains this observation. (Hint: Carbocations can rearrange by shifting a hydrogen atom or alkyl group from an adjacent position if a driving force permits the formation of a more-stable cation.)

7-10 Rank the following alcohols according to their rates of reactivity toward treatment with HBr. Explain your ranking.

(a)   *t*-butyl alcohol, *s*-butyl alcohol, *n*-butyl alcohol

(b)   *p*-methoxybenzyl alcohol, *p*-nitrobenzyl alcohol, benzyl alcohol

(c)   benzyl alcohol, *p*-methylphenol, α,α-dimethylbenzyl alcohol

7-11 Under forcing conditions (such as hot concentrated sulfuric acid), ethyl ether can eliminate ethanol by a mechanism similar to the acid-catalyzed dehydration discussed in this chapter.

(a)   Using curved arrows, write a mechanism by which ethyl ether is converted into ethylene and ethanol.

(b)   Predict the product that would be obtained if tetrahydrofuran or dioxane (common organic solvents) were so treated.

Tetrahydrofuran   Dioxane

7-12 Given what you now know about the mechanism for acid-catalyzed dehydration, propose a detailed mechanism for the pinacol rearrangement mentioned (without detail) on pages 230–231.

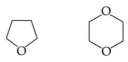

7-13 Heating many alkyl chlorides or bromides in water effects their conversion into alcohols through an S$_N$1 reaction. Order each of the following sets of compounds with respect to solvolytic reactivity:

(a)

(b)

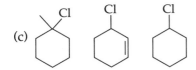

(c)

7-14 When alkyl halides are treated with aqueous silver nitrate, silver halide precipitates and an alcohol is formed. From what you know about the S_N1 reaction, propose a mechanism for the following conversion. (Hint: Consider a possible rearrangement to produce a more-stable cationic intermediate.)

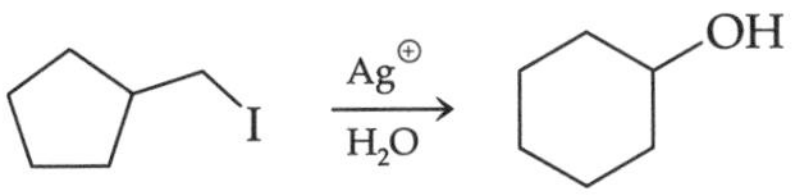

7-15 Suppose that the following reactions were proposed as routes for making the indicated product. Determine whether they are likely to proceed as written. If not, write the expected product and explain why the indicated reaction would not occur.

(a)

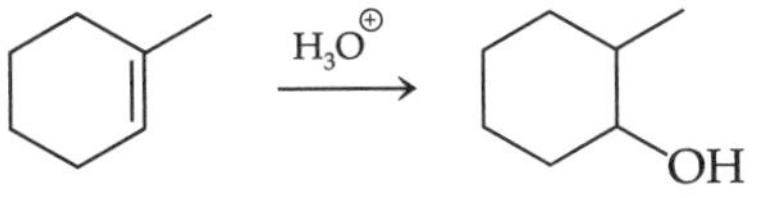

(b)

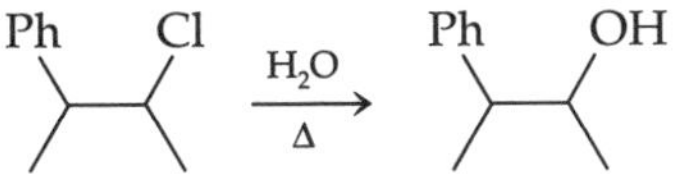

(c)

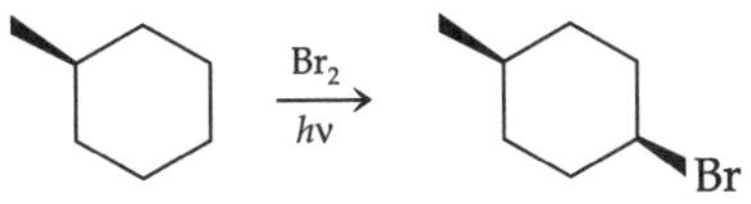

(d)

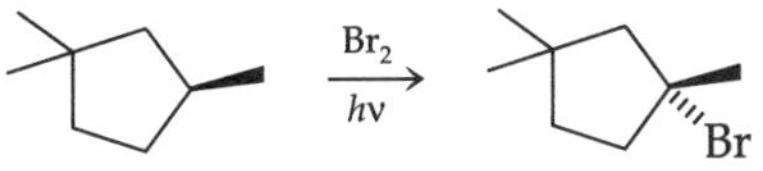

(e)

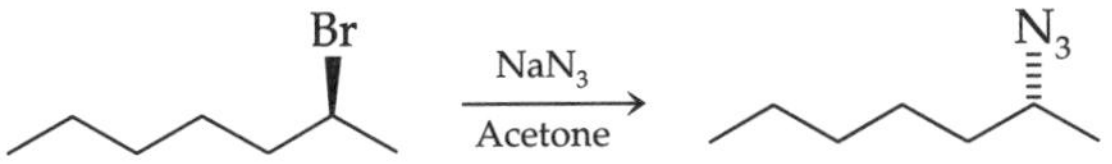

7-16 Homolytic chlorination and bromination are effective means for producing alkyl halides from alkanes. Chlorine is somewhat less expensive than bromine and, if you were running a chemical plant where it was important to keep the costs of reagents needed for large-scale (many tons) conversions as low as possible, it would be useful to use chlorine when possible. The following reactions are known free radical brominations. Decide whether chlorine could be used instead of bromine to prepare the analogous alkyl chloride in good yield. Explain your reasoning.

(a)

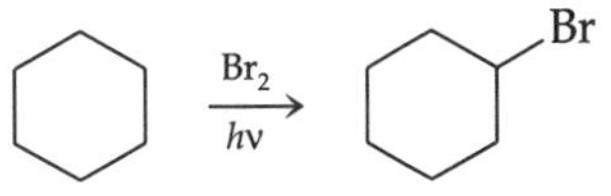

(b)

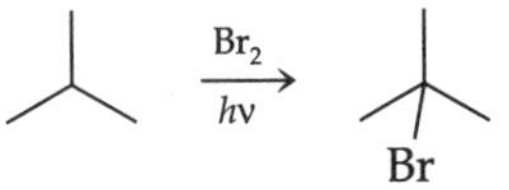

(c)

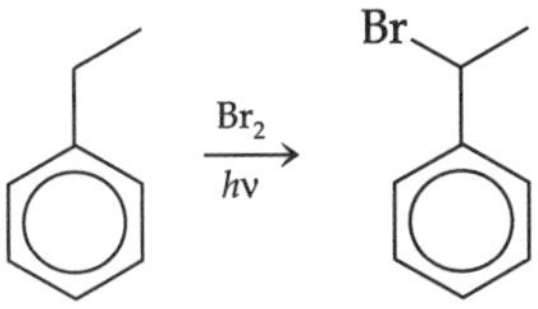

(d)

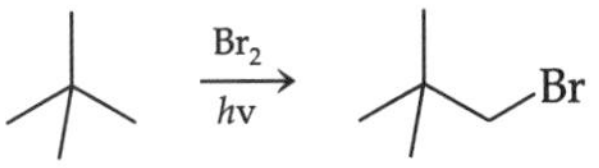

7-17 As an alternative to free radical halogenation, consider the following proposed reaction of cyclohexane with chlorine:

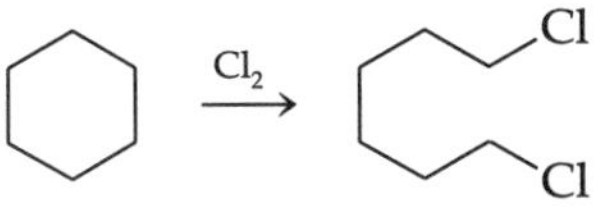

(a) Assuming that such a reaction would be initiated by the same route as in homolytic substitution (Cl—Cl → 2 Cl·), propose a mechanism by which the indicated reaction could proceed through a radical chain.

(b) For each of the propagation steps that you wrote for part *a*, use the table of bond-dissociation energies (Table 3-6) to calculate the expected enthalpy change. (Assume that a C–C bond between secondary carbons is worth about 84 kcal/mole.) Does the calculation explain why this proposed reaction is not observed in the laboratory (that is, why cyclohexyl chloride is observed instead)?

7-18 In seeking a source for gasoline and other low-weight hydrocarbons (for example, butadiene, which is used in large quantities as a component of plastics and rubber, as described in Chapter 16), the petroleum industry runs large "cracking towers" in which complex mixtures of higher-molecular-weight alkanes are heated to very high temperatures. Under these conditions, the alkanes "crack" through the homolysis of C–C and C–H bonds.

Consider propane to be a model for hydrocarbon cracking. Refer to the table of bond-dissociation energies (Table 3-6) to decide whether "cracking" would be more efficiently initiated by cleavage of a C–C, a primary C–H, or a secondary C–H bond.

Chapter 8

Nucleophilic Substitution at sp^3-Hybridized Carbon

Homolytic substitution and nucleophilic substitution (two of the reactions considered in Chapter 7) are frequently used to alter the functional groups present in a molecule. We can use homolytic substitution to introduce either chlorine or bromine into a hydrocarbon and, in a subsequent step, employ the S_N2 reaction to replace the halogen by another group. In this chapter, we look in more detail at S_N2 reactions, transformations that occur by bimolecular backside nucleophilic displacement.

To use S_N2 reactions at sp^3-hybridized carbon to form new carbon–carbon and carbon–heteroatom bonds, we must become familiar with sets of reagents that can act as nucleophiles. First, we review how heteroatomic nucleophiles are used to interconvert functional groups and learn some new examples not considered in Chapter 7. Then, we will consider the reactions of nucleophiles in which the electron-rich atom is carbon. We will see how carbon nucleophiles are generated and how they can attack electron-deficient sp^3-hybridized carbon atoms to form carbon–carbon bonds. Later (in Chapter 13), we will consider how these same reagents can participate in nucleophilic additions and substitutions at sp^2-hybridized atoms.

The availability of methods for the formation of new carbon–carbon bonds is very important in building new carbon skeletons; that is, in the synthesis of both naturally occurring molecules and new materials. These reactions are used routinely by chemists seeking to make new compounds, to model (and hence understand) the reactions that nature uses to assemble natural products in biosynthesis, and to prepare new drugs to chemically target specific diseases.

8-1 ▸ Review of Mechanisms of Nucleophilic Substitution

Let us begin by reviewing what we learned about nucleophilic substitution reactions in Chapter 7. We know that concerted (S_N2) and two-step (S_N1) nucleophilic substitution reactions at tetrahedral carbon are the extreme

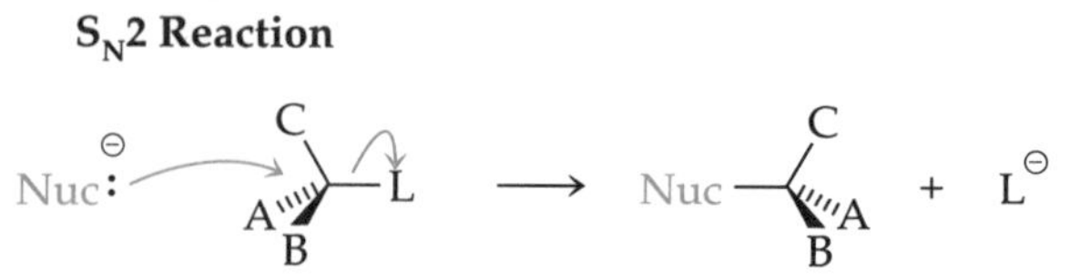

S_N2 Reaction

(L = X, OTs)

▲ **FIGURE 8-1**

A concerted nucleophilic substitution takes place by backside attack in which an electron-rich nucleophile (Nuc:) approaches the side opposite that from which a leaving group (L) departs. A group can function as an effective leaving group if it can take up the electrons originally constituting the carbon–L bond. Leaving groups are often bound to carbon through an electronegative atom, as in halides (—X) or tosylates (—OTs). The course of the S_N2 reaction causes an inversion of the configuration of the carbon atom to which L was originally attached.

mechanisms for the replacement of an electronegative leaving group. As shown in Figure 8-1, the concerted pathway, an S_N2 reaction, consists of backside attack by a nucleophile and results in the displacement of a leaving group, which departs with the electrons initially constituting its σ bond to carbon. Backside attack means that the nucleophile must approach from the side opposite that to which the leaving group is attached. Because both the nucleophile and the substrate are partially bonded in the transition state, the reaction is bimolecular. (The "2" in S_N2 indicates the number of molecules taking part in the rate-determining step.) The rate of a bimolecular reaction depends on the concentration of both of these species, and second-order kinetics are observed.

$$\text{Rate} \propto [\text{R–L}][\text{Nuc}]$$

Backside attack results in inversion of configuration at the carbon atom undergoing substitution.

At the other extreme is the S_N1 reaction, in which the leaving group first departs with the electrons from the C–L bond, forming a trigonal, sp^2-hybridized carbocation (Figure 8-2). This step consists only of bond breaking and undoubtedly is rate determining. The loss of the leaving group is followed by a much faster bond-making step in which an external nucleophile (:Nuc) uses a pair of electrons to form a new bond to the carbocation. The transition state of the first (rate-determining) step has appreciable carbocationic character, and the facility of an S_N1 substitution relates directly to the stability of the cation formed. Because the rate-determining step of an

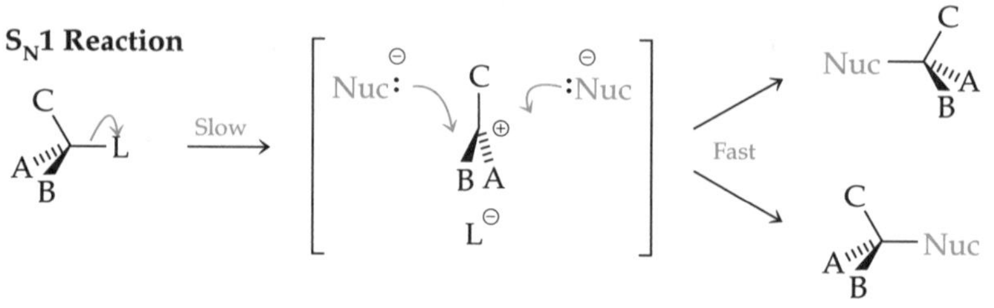

▲ **FIGURE 8-2**

In a two-step nucleophilic substitution (S_N1 reaction), the leaving group, L, departs with the electrons of the C–L bond in the first (slow) step. The planar carbocation is then trapped by fast attack by the substituting nucleophile. Because this fast nucleophilic attack is equally easy on either face of the planar carbocation, both possible stereoisomers are formed. The S_N1 reaction therefore leads to a racemic product when L is attached to a center of chirality.

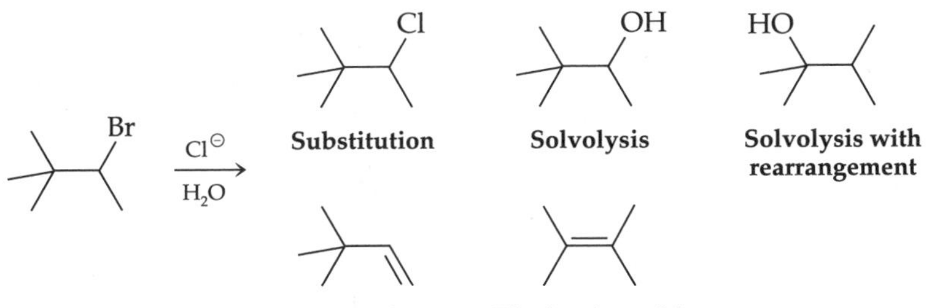

▲ FIGURE 8-3
Because they proceed through carboactions, two-step (S_N1) nucleophilic substitutions are complicated by side reactions so that the expected substitution is only one of several possible products.

S_N1 reaction involves only the substrate, this reaction is unimolecular. (As before, the "1" in S_N1 indicates the number of molecules taking part in the rate-determining step.) The rate of an S_N1 reaction is not influenced by the concentration of the nucleophile, and first-order kinetics are observed. Thus, the reaction rate depends only on the concentration of the substrate.

$$\text{Rate} \propto [\text{R–L}]$$

Those reactions that follow the S_N2 pathway are more valuable as tools for forming new compounds in a stereochemically controlled way. S_N2 reactions proceed with stereocontrol (inversion of configuration), whereas S_N1 reactions proceed with racemization because they involve a planar carbocation. Other factors that complicate the S_N1 reaction include rearrangements, eliminations, and competing solvolysis reactions (Figure 8-3). All of these competing pathways occur because these reactions proceed through cationic intermediates. Except in special cases in which these other reactions can be turned to advantage, S_N1 reactions generally produce more-complex mixtures of products than do their S_N2 counterparts.

A HIGHLY TOXIC ALKYL HALIDE

Mustard gas was used extensively in World War I and was stockpiled by many nations as a chemical weapon deterrent until several international treaties in the 1980s banned its use and called for its destruction. When this gas contacts the skin or lung, the water present rapidly displaces HCl, producing high local concentrations of acid that cause extensive blistering, tissue destruction, and, in severe exposure, death.

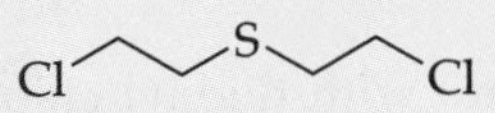

"Mustard gas"

EXERCISE 8-A

For each of the following substitution reactions, predict the product, including the absolute stereochemistry of any centers of chirality. Use R,S nomenclature to specify absolute configuration of any centers of chirality in the reactant and product.

(a) 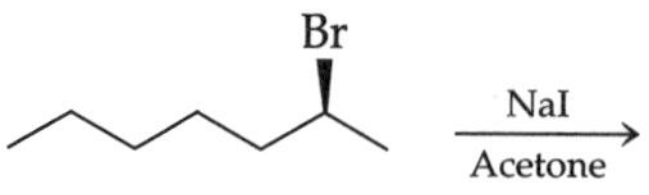 $\xrightarrow[\text{Acetone}]{\text{NaI}}$

(b) 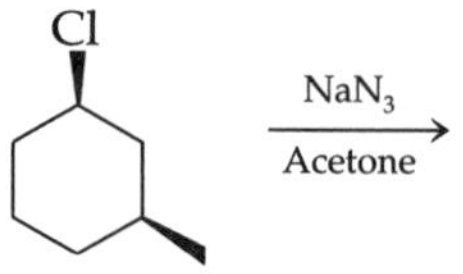 $\xrightarrow[\text{Acetone}]{\text{NaN}_3}$

(c) 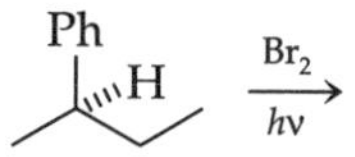 $\xrightarrow[h\nu]{\text{Br}_2}$

(d) 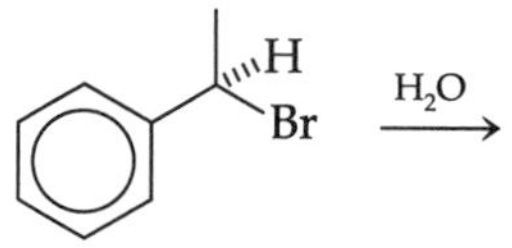 $\xrightarrow{\text{H}_2\text{O}}$

EXERCISE 8-B

In principle, secondary alkyl halides can react by either an S_N1 or an S_N2 pathway.

(a) Predict the stereochemical consequence of each of those pathways in the hydrolysis of (2R,4R)-2-bromo-4-pentanol to a diol.

(b) If the starting material were optically active, would the product also be optically active?

(c) Would your answer to part *b* be different if the starting material had been (2S,4R)-2-bromo-4-pentanol?

8-2 ▸ Functional-Group Transformations through S_N2 Reactions

Because of the complications inherent in S_N1 reactions, S_N2 reactions are often preferred for the transformation of one functional group into another, several of which were listed in Tables 7-1 and 7-2. We will deal first with nucleophiles other than carbon. In virtually all of these reactions, the organic substrate is a primary or secondary alkyl halide or tosylate (Figure 8-4).

Alkyl halides can be synthesized by free radical halogenation of alkanes (Chapter 7)

$$\text{RH} \xrightarrow[h\nu]{X_2} \text{RX} \qquad (X = \text{Br, Cl})$$

or by treatment of an alcohol with thionyl chloride (SOCl₂), a phosphorus halide (PX₃ or PX₅), or phosphorus oxyhalide (POX₃).

$$\text{ROH} \xrightarrow{\text{SOCl}_2} \text{RCl}$$

$$\text{ROH} \xrightarrow[\text{(or PX}_5)]{\text{PX}_3} \text{RX} \qquad (X = \text{Br, Cl})$$

$$\text{ROH} \xrightarrow{\text{POX}_3} \text{RX} \qquad (X = \text{Br, Cl})$$

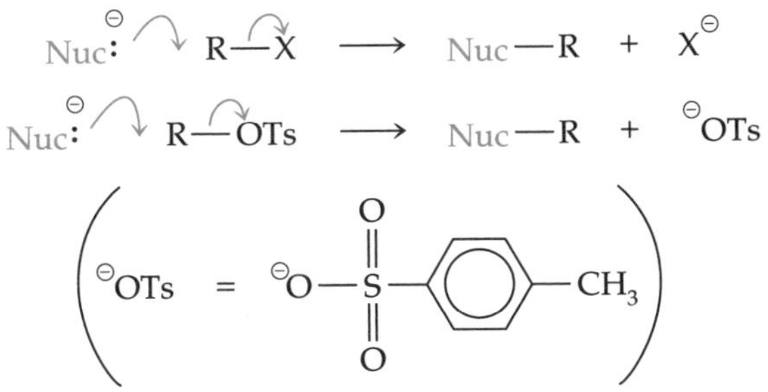

$$\overset{\ominus}{Nuc:} \curvearrowright R \overset{\frown}{-} X \longrightarrow Nuc-R + X^{\ominus}$$

$$\overset{\ominus}{Nuc:} \curvearrowright R \overset{\frown}{-} OTs \longrightarrow Nuc-R + \overset{\ominus}{OTs}$$

▲ FIGURE 8-4
Alkyl halides (RX) and alkyl tosylate (ROTs) are frequently used as substrates in S_N2 reactions.

Tosylates, *p*-toluenesulfonate esters, are prepared by the reaction of an alcohol with *p*-toluenesulfonyl (tosyl) chloride. Because of the high stability of a tosylate anion (ArSO$_3^-$), the —OTs group often behaves like a halide as a leaving group in an S_N2 displacement.

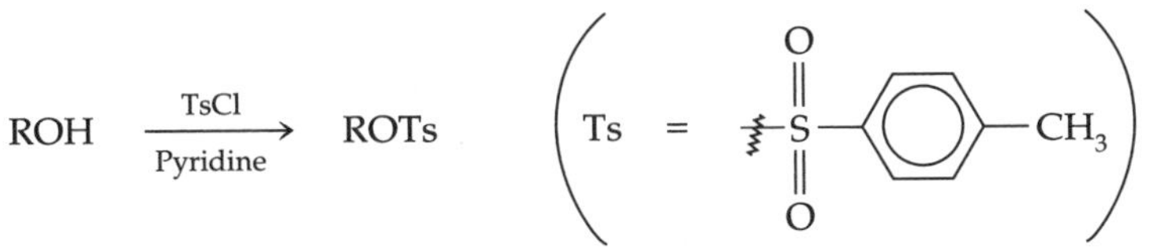

$$ROH \xrightarrow[\text{Pyridine}]{\text{TsCl}} ROTs$$

Williamson Ether Synthesis

As discussed in Chapter 3, alkoxides can be produced from alcohols by treatment with either a base or an alkali metal.

$$ROH + Na \longrightarrow RO^{\ominus} + Na^{\oplus} + \tfrac{1}{2}H_2 \uparrow$$

Alkoxide ions are good nucleophiles and displace halide ions from alkyl halides (or tosylate ions from alkyl tosylates), resulting in the formation of a new carbon–oxygen bond.

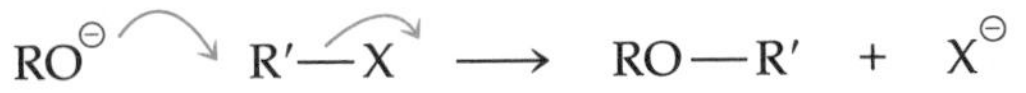

$$RO^{\ominus} \curvearrowright R' \overset{\frown}{-} X \longrightarrow RO-R' + X^{\ominus}$$

The ether produced in this way has a larger number of carbon atoms than either of the starting materials and is thus a more-complex structure. This reaction, called the **Williamson ether synthesis,** is a straightforward application of an S_N2 reaction for the construction of a complex organic molecule from simpler starting materials.

Because an S_N2 reaction is required for the Williamson ether synthesis, it is useful for the synthesis of ethers only when the alkyl halide (or tosylate) is primary or secondary. Thus, for the synthesis of tertiary butyl ethyl ether by this route, we must use the tertiary butoxide ion as the nucleophile and ethyl bromide as the electrophile.

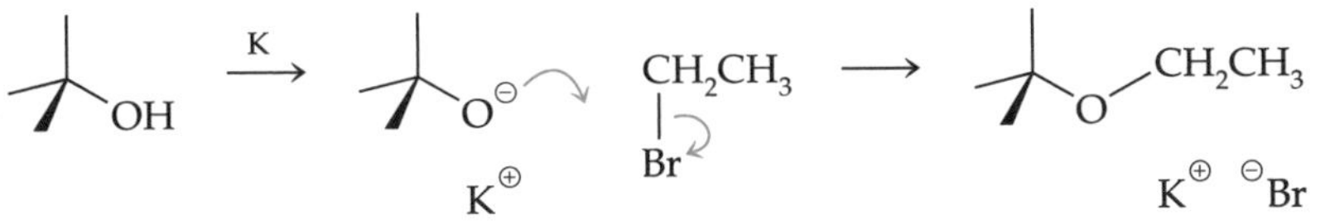

The alternate combination of ethoxide ion with *t*-butyl bromide would be unsuccessful because of steric hindrance to reaching the transition state for

ETHERS AS ANESTHETICS

Chemistry has revolutionized medicine and, unlike the medical doctors of earlier centuries, today's physician is much more likely to treat and cure disease with chemicals than with a knife. The use of nitrous oxide and ethyl ether as anesthetics is one of the earliest examples of the profound effect that chemistry has had on the practice of medicine. Nonetheless, the practical utilization of early observations on the effects of these two compounds on the perception of pain did not come easily or quickly. Joseph Priestley was the first chemist to investigate the chemical properties of nitrous oxide (in 1772, two years before his discovery of oxygen). Some twenty-six years later, another English chemist, Humphry Davy, began exploring the medical uses of gases and tested nitrous oxide on himself by inhaling large quantities in a very short period of time. Apparently, he quite enjoyed the experience, but his suggestion that nitrous oxide be used in surgery was not followed.

It was not until well into the nineteenth century that dentists in Boston began demonstrating the effectiveness of nitrous oxide. In the first public demonstration in 1844, however, a tooth was pulled before the gas had taken effect; and the dentist, Horace Wells, was booed and hissed from the amphitheater. Two of his followers, William Morton and Charles Jackson, took up the use of nitrous oxide but also began experimenting with ethyl ether. These three, Wells, Morton, and Jackson, fought among themselves for the honor of being recognized for the discovery of anesthesia and even took their dispute to Congress for resolution (which, in typical fashion, failed to act). Meanwhile, a physician in the state of Georgia had been using ethyl ether in his practice since 1842. (The AMA and the ADA acknowledge Wells as being the discoverer of anesthesia.)

the S_N2 reaction at the tertiary halide site. Instead, ethoxide ion would act as a base and effect an elimination of HBr. This reaction is dominant because of its lower activation energy. In fact, the reaction fails completely with tertiary halides and often leads to poor yields with secondary halides.

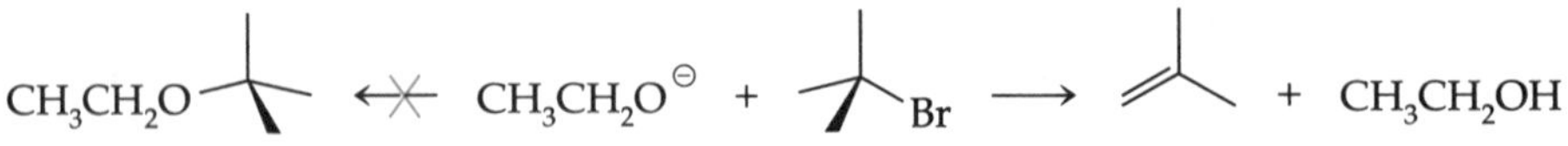

CH_3CH_2O
$CH_3CH_2O^{\ominus}$
+
Br
+ CH_3CH_2OH

Many variants of the Williamson ether synthesis are possible: for example, we can make ethers from the corresponding alcohols. Figure 8-5 shows the synthesis of *n*-propyl ether from *n*-propanol. Reaction of this alcohol with sodium produces *n*-propoxide ion, and *n*-propyl tosylate can be prepared by a separate reaction with tosyl chloride. Combining these two reagents results in an S_N2 displacement, giving *n*-propyl ether.

Thioethers can be synthesized by a pathway similar to the Williamson ether synthesis through the interaction of an alkyl or aryl thiolate with an alkyl bromide or tosylate.

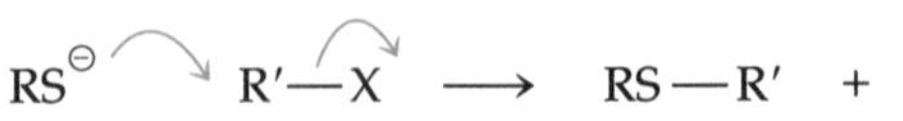

$RS^{\ominus}$
R′—X
RS—R′ + $X^{\ominus}$

A thioether

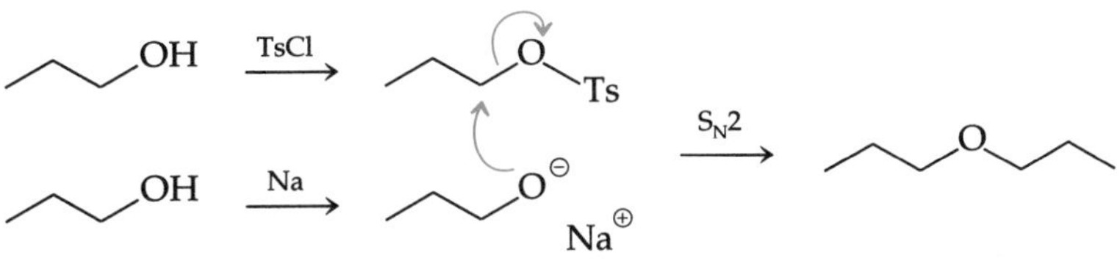

▲ FIGURE 8-5

An alcohol can be the source of both the substrate and the nucleophile in a Williamson ether synthesis. Here, the nucleophile (alkoxide) is formed by treatment of the alcohol with metallic sodium; the substrate tosylate is formed by treating the alcohol with tosyl chloride. The ether is formed in an S_N2 reaction between these two reagents.

An important use of ethers is as solvents for organic reactions. Because of the lone pairs of electrons on oxygen, ethers can coordinate with cations (for example, protons or metal ions) and can solvate, and hence stabilize, cationic intermediates. However, ethers are relatively nonpolar and lack the hydroxyl groups of polar protic solvents such as water that are effective in solvating anions. Thus, ethers constitute a class of solvents in which the solubility of organic and some inorganic reagents is high, yet heterolytic cleavage of the relatively nonpolar bonds of an ether itself is more difficult than in most polar reactions. As a result, ethers are themselves inert to most reaction conditions and are useful solvents.

EXERCISE 8-C

Propose the preferred reagents for synthesizing each of the following ethers. If more than one choice is possible, list both.

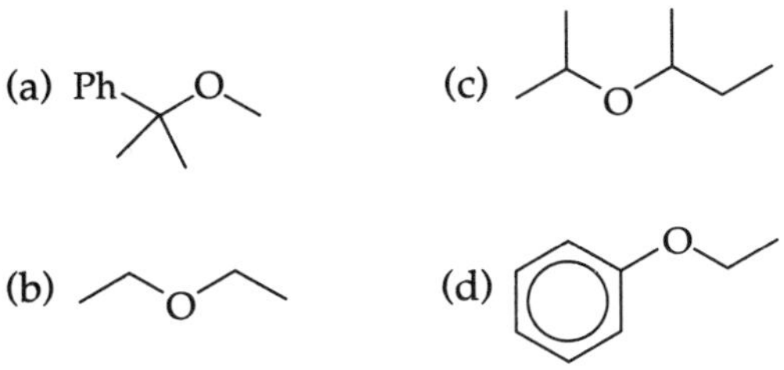

(a)   (c)

(b)   (d)

Reaction of Alkyl Halides with Nitrogen Nucleophiles

Nitrogen lies at the left of oxygen in the periodic table. Accordingly, ammonia is more nucleophilic than is water, and alkylamines are more nucleophilic than are alcohols. Indeed, ammonia and alkylamines react readily as neutral molecules with alkyl halides and tosylates in an S_N2 displacement reaction (Figure 8-6), unlike oxygen nucleophiles, which must be converted first into alkoxide anions for use in the Williamson ether synthesis. The initial product of the reaction of ammonia with an alkyl halide is an ammonium salt. This ion lacks a lone pair on nitrogen and is not nucleophilic. However, the ammonium salt has several acidic hydrogens on nitrogen and is in equilibrium with the free alkylamine through a simple acid-base reaction. Removal of a proton restores the lone pair to nitrogen. The resulting alkylamine has reactivity similar to that of ammonia except that, because of the replacement of hydrogen by a carbon substituent, it is somewhat bulkier and less reactive than ammonia itself. The alkylamine also reacts with the alkyl halide, resulting in a dialkylammonium salt that can be

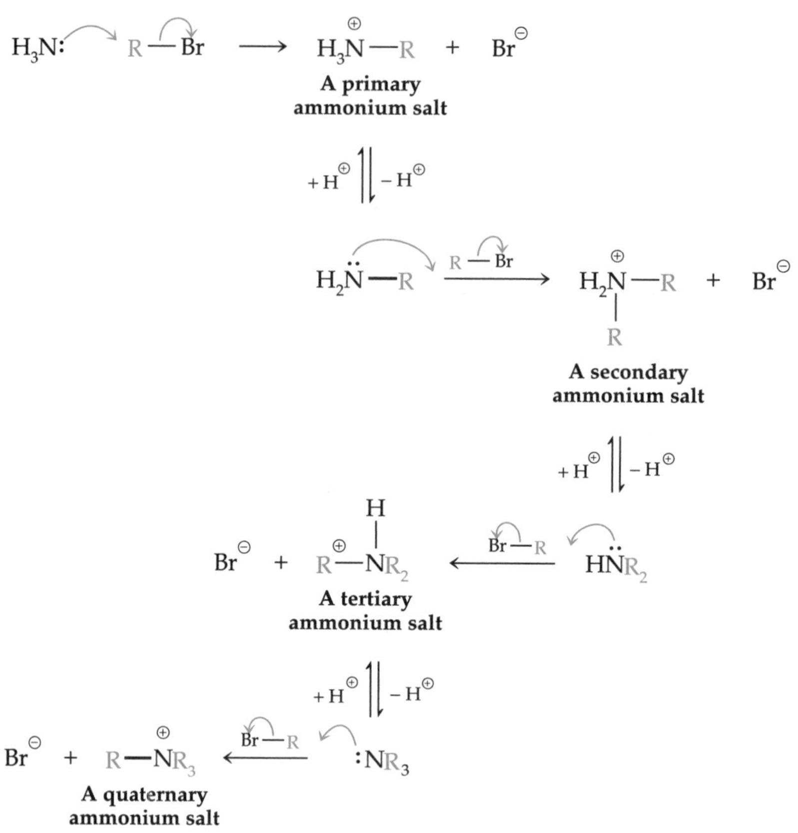

$$H_3N: \quad R \!-\! Br \longrightarrow H_3\overset{\oplus}{N}\!-\!R \;+\; Br^{\ominus}$$

**A primary
ammonium salt**

$$+H^{\oplus} \updownarrow -H^{\oplus}$$

$$H_2\ddot{N}\!-\!R \xrightarrow{\; R \!-\! Br \;} H_2\overset{\oplus}{\underset{|}{N}}\!-\!R \;+\; Br^{\ominus}$$
$$\qquad\qquad\qquad\qquad\qquad R$$

**A secondary
ammonium salt**

$$+H^{\oplus} \updownarrow -H^{\oplus}$$

$$\overset{\textstyle H}{\underset{\textstyle }{}}$$
$$Br^{\ominus} \;+\; R\!-\!\overset{\oplus}{\underset{|}{N}}R_2 \xleftarrow{\; Br\!-\!R \;} H\ddot{N}R_2$$

**A tertiary
ammonium salt**

$$+H^{\oplus} \updownarrow -H^{\oplus}$$

$$Br^{\ominus} \;+\; R\!-\!\overset{\oplus}{N}R_3 \xleftarrow{\; Br\!-\!R \;} :NR_3$$

**A quaternary
ammonium salt**

▲ FIGURE 8-6

Ammonia and neutral amines are sufficiently nucleophilic to attack an alkyl halide without further activation. Although an ammonium salt is formed directly in the S_N2 reaction, acid-base equilibration provides some of the neutral amine at equilibrium. The product is therefore also nucleophilic and can participate in a second attack. When this sequence is repeated, a complex mixture of primary, secondary, and tertiary amines and a quaternary ammonium salt is formed. The S_N2 reaction of ammonia with an alkyl halide or an alkyl tosylate is therefore not a synthetically useful way to form a primary alkyl amine unless ammonia is used in large excess.

deprotonated to form the dialkylamine. This amine is also nucleophilic and the alkylation on nitrogen is repeated, forming the trialkylammonium ion, the trialkylamine, and ultimately the quaternary ammonium salt in which the positively charged nitrogen is bound to four carbons. There are no protons on nitrogen in this ammonium salt and no further reaction is possible.

The rates of these alkylation reactions decrease progressively as the degree of substitution at nitrogen increases. However, the differences in these rates of reaction is not large, and it is difficult to control alkylation of nitrogen so as to obtain only (or mainly) the primary, secondary, or tertiary amine. Only if a large excess of ammonia is used is the reaction limited to the formation of mainly the primary amine. For example, mixing ammonia with methyl bromide in equal amounts gives a complex mixture in which all four products are present. **Selectivity** (the formation of one product to the exclusion of other possible products) is difficult to obtain in this sequence of transformations because the initially formed product can participate in the same type of reaction by which it was formed, and the rate of this subsequent step is comparable to the first. There are similar questions of selectivity for many organic reactions. In most cases, reactions are selective only if the rate of the first reaction is significantly faster than subsequent steps.

For the alkylation of nitrogen to be synthetically useful without an excess of the nucleophile, the product must be so constituted that the product nitrogen is not nucleophilic. In the **Gabriel synthesis,** this is accomplished by replacing two of the hydrogens of ammonia with amide bonds in phthalimide, as shown in Figure 8-7. The N–H group of phthalimide is highly acidic (pK_a 9) and can be deprotonated with potassium hydroxide, generating the anion shown. This anion is stabilized by delocalization over three heteroatoms (nitrogen and the two carbonyl group oxygens), yet it is still nucleophilic. There are two lone pairs on nitrogen in the anion so that the localized pair acts as a nucleophile to alkylate an alkyl halide or tosylate (RX or ROTs), forming a new nitrogen–carbon bond. The product of this

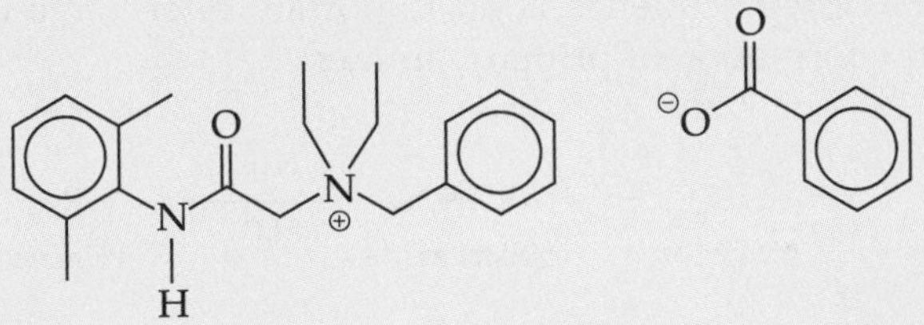

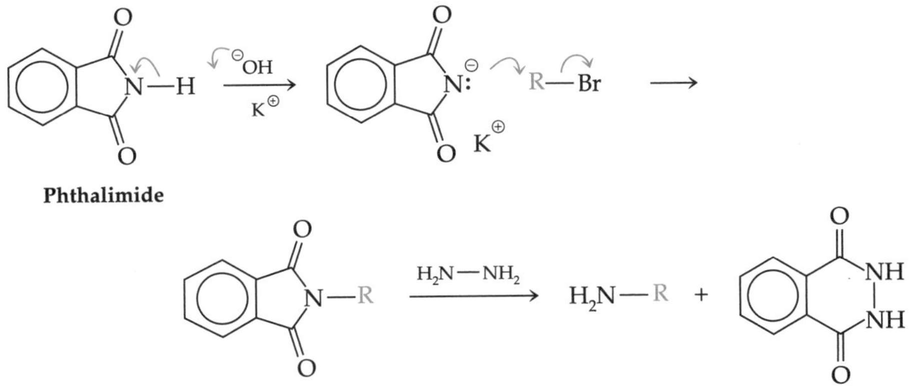

Phthalimide

▲ FIGURE 8-7

In the Gabriel synthesis, the anion formed by deprotonation of phthalimide acts as a nucleophile in an S$_N$2 displacement. The *N*-alkylated product is no longer nucleophilic, and secondary reactions, which hampered the direct displacement in Figure 8-6, are avoided. Phthalimide thus acts as a blocker for secondary reactivity; it is removed by treatment with hydrazine in a last step.

alkylation has no protons on nitrogen; thus, an anion of the product cannot be formed. Furthermore, the nitrogen atom bears three sterically blocking carbon substituents and the lone pair of electrons on nitrogen in the alkylated phthalimide is delocalized into the π systems of both carbonyl groups. As a result, the nitrogen atom in the alkylated product is essentially nonnucleophilic.

Reaction of the *N*-alkylphthalimide with hydrazine, a particularly good nitrogen nucleophile, releases the primary amine. The difficulty in controlling the alkylation of a simple amine is circumvented in the Gabriel synthesis by the construction of a system in which the product does not readily undergo the reaction by which it was formed.

The azide ion ($^-N_3$) is a particularly reactive nucleophile. Reaction of primary and secondary alkyl halides with this nitrogen nucleophile results in the formation of alkyl azides by an S_N2 mechanism. In turn, alkyl azides undergo reduction with reagents such as $NaBH_4$ to form primary alkyl amines. This two-step sequence is an important alternative to the Gabriel synthesis for the formation of primary amines.

$$R-X \; + \; ^{\ominus}N_3 \; \longrightarrow \; R-N_3 \; \xrightarrow[\text{2. H}_2\text{O}]{\text{1. NaBH}_4} \; R-NH_2$$

Primary   **Azide ion**   **Alkyl azide**   **Primary amine**
or secondary
alkyl halide

EXERCISE 8-D

(a) Write resonance structures for the conjugate base of phthalimide. Do these structures explain phthalimide's high acidity?

(b) Write resonance structures for phthalimide itself. Compare the structures in parts *a* and *b* to explain why it is necessary to deprotonate phthalimide before it is sufficiently active to enter into an S_N2 displacement. (In contrast, ammonia displaces bromide from methyl bromide without further activation.)

Phosphines as Nucleophiles

Phosphorus is in the same column of the periodic table as nitrogen, and thus it is not surprising that **phosphines** (PR_3) and amines undergo similar reactions as nucleophiles. Indeed, phosphorous compounds generally have higher reactivity as nucleophiles than their nitrogen counterparts because of the higher polarizability of phosphorus. Thus, an S_N2 displacement reaction at an alkyl halide by a trisubstituted phosphine produces a **phosphonium salt**, which is similar to the ammonium salts discussed in the preceding section.

$$Ph_3P\!: \quad H_3C-X \; \longrightarrow \; Ph_3\overset{\oplus}{P}-\underset{\underset{H}{|}}{C}H_2 \overset{\ominus}{Br} \; \longrightarrow \; Ph_3\overset{\oplus}{P}-\overset{\ominus}{C}H_2$$

A phosphonium
ylide

Li—Bu

A phosphonium
salt

Because of the positive charge in the phosphonium salt, the acidity of the protons on the adjacent carbon is enhanced; and treatment of a phosphonium salt with a strong base such as an alkyllithium effects deprotonation, forming a neutral compound that can be formulated as a zwitterion, a species with both a positively charged atom and a negatively charged one. Zwitterions that bear opposite charges on adjacent atoms are called **ylides** and, because of the negative charge density on carbon, **phosphonium ylides** have appreciable nucleophilic character.

8-3 ▸ Preparation of Carbon Nucleophiles

A carbon nucleophile has a nonbonded lone pair of electrons on carbon in at least one significant resonance contributor. Many carbon nucleophiles (except cyanide) are too reactive to be stored conveniently as reagents and are usually produced *in situ* as needed. We must therefore interrupt our discussion of bond-forming reactions through S_N2 reaction pathways to consider how sp^2- and sp^3-hybridized carbon nucleophiles are formed. In this section, we describe the formation of carbon–metal bonds at sp^3- and sp^2-hybridized carbons, and in the next section, we consider the use of these organometallic (or carbanionic) reagents as nucleophiles in S_N2 reactions.

The formation of a carbon–carbon bond by an S_N2 reaction requires that one terminus of the new bond be derived from an electron-deficient reagent and the other from an electron-rich partner. Thus, in one substrate of a nucleophilic substitution reaction, an electrophilic carbon is attached to a more-electronegative atom, for which a large number of leaving group substituents are possible (HOR^+, X, SR, tosylate, etc.). However, the group attached to a nucleophilic carbon must be less electronegative than carbon: this list is much shorter, usually being restricted to either a proton or a metal. If a C–H bond is sufficiently acidic, a strong base can remove the proton, producing a carbanion that is highly active as a nucleophile. With less acidic C–H bonds, the anion formed by deprotonation is sometimes too unstable and the carbon atom is better activated as a nucleophile by binding to a metal. In comparison with a C–H bond, a carbon–metal bond is strongly polarized, placing substantial negative charge on carbon. As a result, when a carbon atom is bound to a metal, the resulting reagent, called an **organometallic compound,** has a carbon center that is highly nucleophilic. We will study the reactions of carbanions and organometallic reagents with electron-deficient sp^3-hybridized atoms in the remainder of this chapter and those with electron-deficient sp^2-hybridized atoms in Chapter 13.

$\overset{\delta\ominus}{\diagdown}\overset{\delta\oplus}{\rule{0pt}{0pt}}$—L

Nucleophilic carbon
(L = H, metal)

$\overset{\delta\oplus}{\diagdown}\overset{\delta\ominus}{\rule{0pt}{0pt}}$—L

Electrophilic carbon
(L = X, OTs, etc.)

sp-Hybridized Carbon Nucleophiles: Cyanide and Acetylide Anions

The high acidity of *sp*-hybridized carbon–hydrogen bonds (Chapter 6) makes deprotonation relatively easy. These carbanions, bearing a free electron pair on carbon, are negatively charged and are even more nucleophilic than amines. For example, the acidity of H—CN and of a terminal alkyne (RC≡C—H) permits formation of the related carbanions, cyanide ($^-$CN) and **alkynide anions** (RC≡C$^-$).

$$N\equiv C-H \xrightarrow{\overset{\ominus}{:}Base} N\equiv C\overset{\ominus}{:}$$

$$R-C\equiv C-H \xrightarrow{\overset{\ominus}{:}Base} R-C\equiv C\overset{\ominus}{:}$$

(Alkynide anions are also called **acetylide anions**.) A terminal alkyne (1-alkyne) has an acidic proton (hydrogen bound to the terminal sp-hybridized carbon) that can be removed with strong base to form a nucleophilic anion ($RC{\equiv}C^-$).

$$R{-}C{\equiv}C{-}H \xrightarrow{\text{NaNH}_2} R{-}C{\equiv}C{:}^{\ominus} + Na^{\oplus} + NH_3$$

The resulting acetylide anion resembles cyanide ion in its electronic configuration; but, because of the difference in electronegativity between carbon and nitrogen, terminal alkynes are less acidic than HCN (pK_a 25 versus 9) and their anions are less stable. Deprotonation of a terminal alkyne to generate an acetylide anion requires a strong base such as sodium amide (NaNH$_2$, pK_a of NH$_3 \sim 38$).

The acidity of terminal alkynes forms the basis of a simple chemical test for the presence of this functional group. Reaction of a terminal alkyne with an aqueous solution of $Ag^+(NH_3)_2{}^-OH$ results in deprotonation and the formation of an insoluble silver acetylide that precipitates as a thin, silvery layer on the surface of the reaction vessel.

$$R{-}C{\equiv}C{-}H \xrightarrow{\overset{\oplus}{Ag}(NH_3)_2 \ {}^{\ominus}OH} R{-}C{\equiv}C{:}^{\ominus} Ag^{\oplus} \downarrow$$

Precipitate

EXERCISE 8-E

Draw a Lewis dot structure for cyanide ion (^-CN), and calculate the formal charge on carbon and on nitrogen. Does the formal charge calculated predict higher nucleophilic activity at carbon or at nitrogen?

sp^2- and sp^3-Hybridized Carbon Nucleophiles: Organometallic Reagents

The bond between a metal and carbon is polarized toward carbon because all metals are less electronegative than carbon. As a result, carbon in organometallic reagents is nucleophilic and forms bonds with electrophilic reagents.

Insertion of a Metal into a Carbon–Halogen Bond. The formation of an organometallic reagent can be accomplished by treatment of an alkyl halide with a zero-valent metal, usually lithium or magnesium. These metals have a high affinity both for nucleophilic carbon and for halide ions, and the reaction is therefore exothermic. With divalent metals such as magnesium, the organometallic reagent has the metal between the carbon and the halogen that were bound to each other in the alkyl halide. For this reason, the formation of an organometallic reagent in this way is often called an **insertion** reaction.

$$H_3C{-}Br + Mg \longrightarrow H_3C{-}Mg{-}Br$$

1. Organolithium reagents. Let us first consider the structures of organolithium compounds, a class of reagents in which carbon is directly bound to lithium, to illustrate general concepts of bonding in organometallics. **Organolithium reagents** can be prepared by treatment of an alkyl halide with lithium metal. For example, the reaction of *n*-butyl bromide with lithium metal generates *n*-butyllithium and lithium bromide (Figure 8-8).

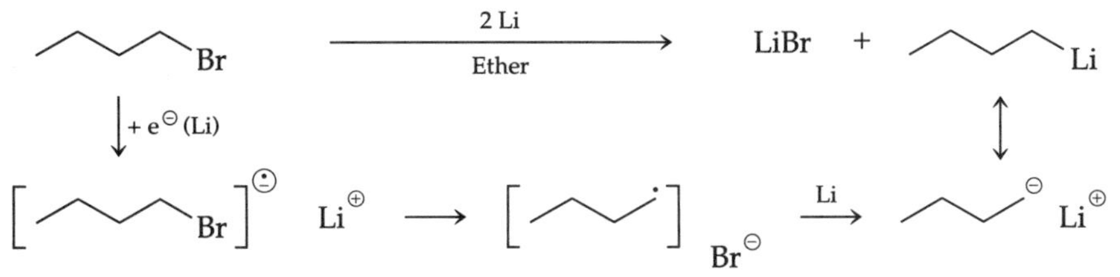

▲ **FIGURE 8-8**

Organolithium reagents are often prepared by replacing a carbon–halogen bond with a carbon–lithium bond. Lithium metal can donate an electron to the σ antibonding orbital of the C–X bond, causing the σ bond to weaken in the radical anion and fragment into a carbon free radical and a halide ion (here, bromide). This radical is then further reduced by a second equivalent of lithium metal to produce a carbanion that is tightly paired with the lithium cation concurrently formed in the reduction.

Although specific details of how this reaction occurs are not known, we have a general idea of how the reaction proceeds from the known properties of lithium metal and of alkyl halides. Lithium is a monovalent alkali metal, possessing one more electron than is required for a filled valence shell. As a result, one electron is readily removed and lithium metal acts as a strong reducing agent. Lithium metal can transfer this electron to an alkyl bromide, where the extra electron in the resulting radical anion must be located in a C–Br σ antibonding orbital, resulting in a greatly weakened carbon–bromine bond. Because bromine is more electronegative than carbon, two of the three electrons associate with bromine upon cleavage of the C–Br bond, leaving one electron on carbon. The resulting free radical is easily reduced by a second equivalent of lithium metal, generating a carbanion that ion pairs with a lithium cation.

In the upper, stoichiometric reaction in Figure 8-8, butyllithium is drawn as having a covalent bond to lithium, whereas the same species is represented as an ion pair at the lower right. Because of tight ion pairing between these two oppositely charged species, they are very strongly associated, and many people consider the distinction between a highly polarized carbon–lithium bond and a tight ion pair to be semantic.

Differences in size and electronegativity between carbon and the metal increase in the progression through the alkali metal series (M = Li$^+$, Na$^+$, K$^+$, Cs$^+$). The fraction of ionic character (that is, more negative charge on carbon) is thus larger with metals such as sodium and potassium than in organolithiums.

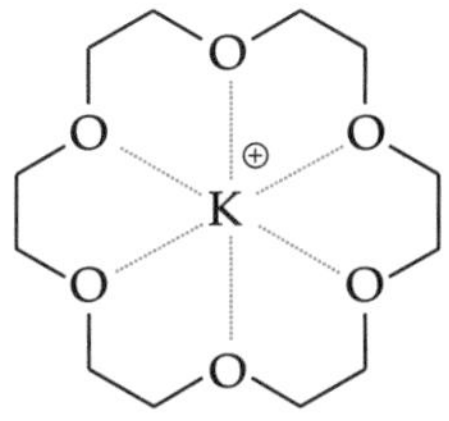

The degree of ion pairing can also be influenced by solvation. In particular, dipolar, aprotic solvents (Lewis basic solvents lacking acidic OH groups) have a high affinity for complexation with alkali metal ions. For example, ethers such as tetrahydrofuran (THF) or cyclic polyethers (for example, 18-crown-6 ether) and dimethylsulfoxide, shown in the margin, have a strong affinity for binding to alkali metals and can compete with the carbanion for ion pairing. In the presence of such solvents, the alkyllithium ion pair exhibits more carbanionic character.

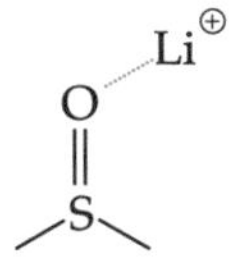

2. Grignard reagents. Metals nearer the center of the periodic table are less electropositive than the alkali metals, and a greater degree of covalent bonding exists between carbon and the metal. An example of such bonding is found in reagents with carbon–magnesium bonds, referred to as Grignard

Tetrahydrofuran (THF)

18-Crown-6 ether

Dimethyl sulfoxide (DMSO)

reagents to acknowledge their discoverer, Victor Grignard, a French chemist to whom the Nobel Prize was awarded in 1912 for this work. **Grignard reagents** (RMgX) are produced by electron transfer from magnesium metal to alkyl halides in the same way as described earlier for organolithiums.

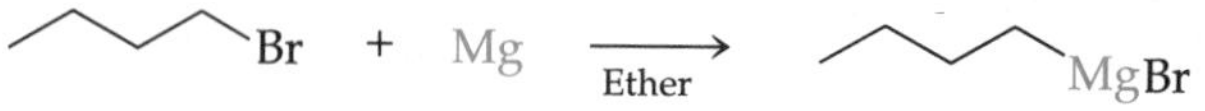

n-Butylmagnesium bromide

The reaction of magnesium with organic halides to form Grignard reagents is not restricted to alkyl halides. The replacement of a carbon–halogen bond by a carbon–metal bond also occurs at *sp²*-hybridized carbon atoms. Thus, aryllithium and aryl Grignard reagents also can be generated by insertion into aryl halide or vinyl halide bonds.

Other organometallics can be prepared by similar insertions into a C–X bond. For example, diethylzinc is formed when ethyl bromide is treated with zinc metal. An equilibrium between R—M—Br and R_2M + RBr favors the Grignard reagent (RMgBr) when M = Mg and dialkylzinc when M = Zn.

$$Et—Br \ + \ Zn \ \longrightarrow \ Et—Zn—Br \ \rightleftharpoons \ Et—Zn—Et \ + \ Br—Zn—Br$$

3. Organometallic reagents from dihalides. Side reactions complicate the formation of an organometallic if the substrate bears two halogen atoms. For example, zinc can be inserted into a carbon–iodine bond of a geminal diiodide. (A **geminal dihalide** is one in which two halogen atoms are attached to the same atom.)

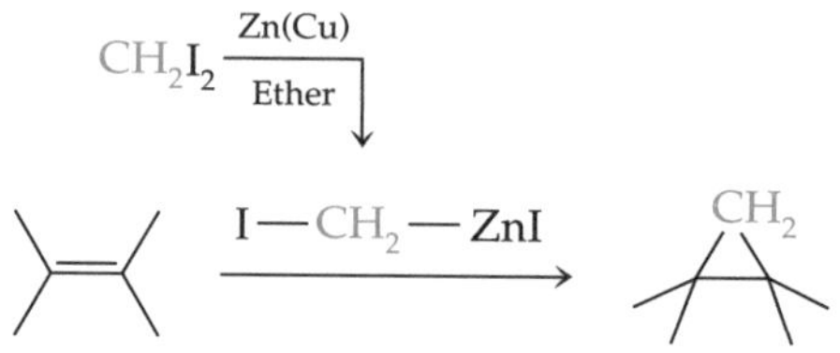

The resulting species, known as the **Simmons-Smith reagent,** has carbene character by virtue of opposite polarizations of the carbon–zinc and carbon–iodine bonds. This organometallic reagent adds to the carbons of a double bond as if it were a free singlet carbene (Chapter 6), producing a cyclopropane.

Vicinal dihalides, those that have two halogen atoms bound to adjacent carbon atoms, react with zinc in much the same way as geminal dihalides. However, the first-formed, intermediate organozinc species can undergo elimination of the second halide, forming an alkene.

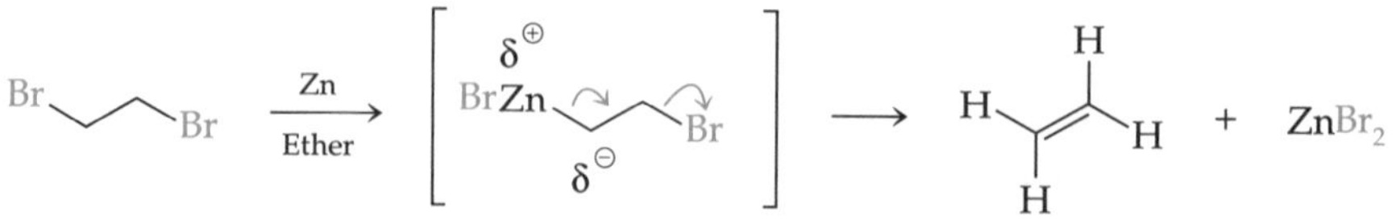

The organozinc reagent formed from a vicinal dihalide is therefore not sufficiently stable to act as a nucleophile in attacking another alkyl halide.

Transmetallation. Alkyllithiums and Grignard reagents are often employed as synthetic precursors to other organometallic reagents. For example, treatment of alkyllithiums with copper halide generates lithium dialkylcuprates in which two alkyl groups are bound to copper.

$$\text{RLi} \;+\; \text{CuI} \;\xrightarrow[\text{Ether}]{} \; \underset{\substack{\textbf{Alkylcopper} \\ \textbf{(polymer)}}}{(\text{RCu})_n} \;\xrightarrow[\text{Ether}]{\text{RLi}} \; \underset{\textbf{Lithium dialkylcuprate}}{\text{R}_2\text{CuLi}} \;\longleftrightarrow\; \text{R}_2\text{Cu}^{\ominus} \;\, ^{\oplus}\text{Li}$$

This overall reaction takes place in two stages: the first, **transmetallation** (an exchange of metals), is followed by nucleophilic attack by a second molecule of alkyllithium on the alkylcopper reagent.

Similar metal exchange reactions can also produce **organozinc, organocadmium,** and **organomercury** compounds.

$$2\,\text{RMgBr} \;+\; \text{MBr}_2 \;\rightleftharpoons\; \text{R}_2\text{M} \;+\; \text{MgBr}_2$$

$$(\text{M} = \text{Cd, Hg, Zn})$$

As in the organocuprate reagents, the transmetallation equilibrium favors the organometallic reagent with the less-electropositive metal—here, the organozinc, organocadmium, or organomercury. These organometallic compounds are often very toxic and, unfortunately, represent the environmental fate of improperly disposed cadmium and mercury in many lake bottoms.

Deprotonation Alpha to a Carbonyl Group: Enolates and Their Equivalents. An enolate anion produced by the deprotonation of a C–H bond adjacent to a carbonyl group is the most generally useful nucleophilic carbon species. Enolates can be produced by the treatment of carbonyl compounds bearing α hydrogens with alkoxide (Figure 8-9) or with stronger bases such as a lithium dialkylamide (LiNR₂). Enolates can be formed from aldehydes, ketones, or esters. In all cases, they have two significant resonance contributors, in one of which the negative charge is located on oxygen and in the other on carbon. Delocalization of charge thus stabilizes the enolate anion while retaining nucleophilic character at the α carbon.

Reprotonation of an enolate on oxygen results in an enol, a species with appreciable nucleophilic character on carbon. (Recall from Chapter 6 that the process by which a proton is moved from the α carbon to the oxygen of the carbonyl group is referred to as *tautomerization*.) A significant resonance contributor of the enol places one of the oxygen lone pairs in a double bond to carbon, shifting the π electrons to the α carbon and placing formal negative charge at that position. As a result, the α carbon in either an enol or an enolate anion can act as a nucleophile.

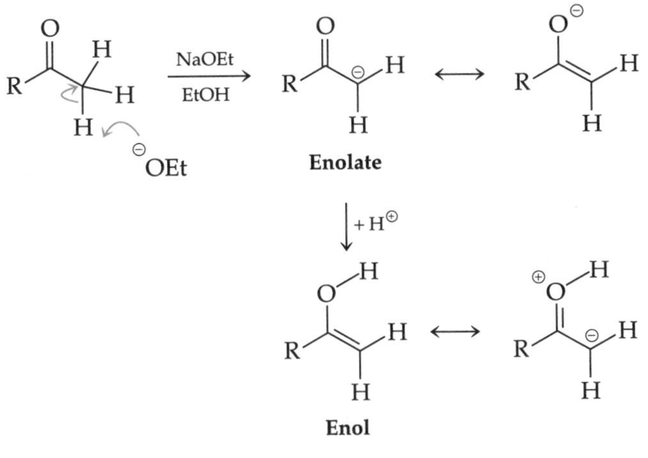

▲ **FIGURE 8-9**

The acidity of C–H bonds adjacent to a carbonyl group allows for the formation of enolate anions upon treatment with base (alkoxide or lithium dialkylamide). Reprotonation of the enolate on oxygen produces an enol.

An **enamine** is the nitrogen analog of an enol. Because it is isoelectronic with an enol, an enamine also is active as a nucleophile. As we will see in Chapter 12, enamines are usually prepared by treatment of aldehydes or ketones with amines.

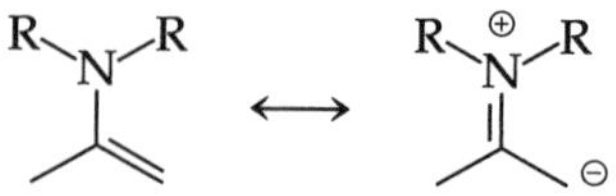

An enolate anion is greatly stabilized by the dispersal of the negative charge over both carbon and oxygen. Anions generated from β-dicarbonyl compounds are even further stabilized by charge delocalization over two adjacent carbonyl groups and are thus particularly stable (recall the discussion of acidity in Chapter 6). Similarly, anions formed from β-ketoesters, from β-diesters, and from compounds in which a nitrile group is attached to the α carbon of a carbonyl group also are particularly stable (Figure 8-10). Thus, carbon nucleophiles can be generated by the formation of organometallics, by transmetallation, or by acid-base reactions to generate alkynide ions, enolate anions, enols, or enamines.

EXERCISE 8-F

Write resonance structures for the enol of acetone and the dimethylamine enamine of acetone, and explain why each of these species has nucleophilic reactivity at the α carbon.

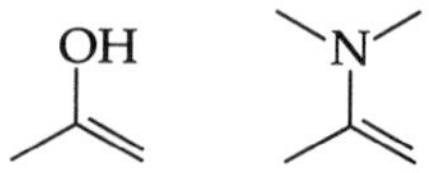

Enol   **Enamine**

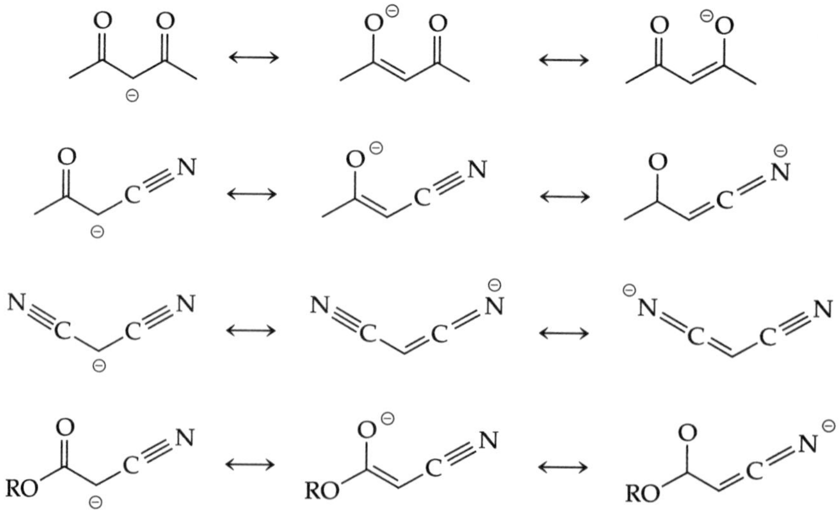

▲ FIGURE 8-10

Enolate anions can be produced from many carbonyl-containing functional groups, and a nitrile (—C≡N) group can similarly stabilize adjacent negative charge. Shown here are resonance-stabilized anions formed by the deprotonation of a β-diketone, a β-ketonitrile, a β-dinitrile (malononitrile), and an α-cyanoester.

Protonation as a Limitation

The partial negative charge that makes carbanions and organometallic reagents active as nucleophiles also makes them susceptible to a complicating side reaction. Carbanions and organometallic compounds are strong bases. As a result, treatment of an organometallic with water causes the protonation of carbon and the formation of the metal hydroxide. Because enolate anions are formed by equilibrium deprotonation by alkoxide in protic media, they are not as susceptible to protonation as are the organometallic reagents. Thus, their reactions are not restricted to aprotic solvents, as are the organoalkali or Grignard reagents.

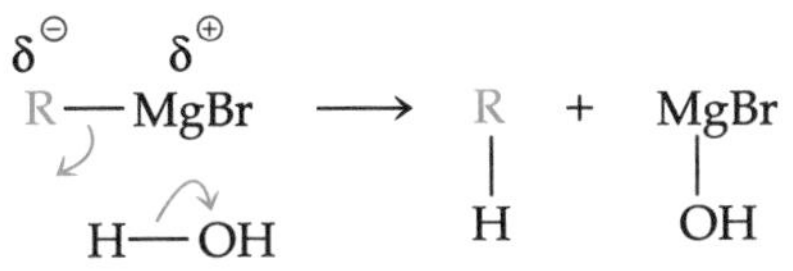

$$\overset{\delta\ominus}{R} - \overset{\delta\oplus}{MgBr} \longrightarrow \underset{H}{R} + \underset{OH}{MgBr}$$

$$H - OH$$

This reaction limits the utility of an organometallic reagent, making it impossible to use these reagents in protic solvents such as water or alcohol or even to generate the reagent if an OH, NH_2, or other acidic group is present in the precursor. Thus, it is critical that all reactions in which organolithiums or Grignard reagents are used be conducted in dry, nonprotic solvents such as ethyl ether or tetrahydrofuran.

The facility with which organometallic reagents are protonated can be turned to advantage. As shown in Figure 8-11, the conversion of benzyl bromide into toluene can take place by protonation of an initially formed Grignard reagent. This route has another desirable feature because it makes possible the introduction of deuterium in specific sites when D_2O is used in place of H_2O. If water is replaced by D_2O, a carbon–halogen bond in the precursor to the Grignard reagent is replaced by a carbon–deuterium bond, resulting in a molecule with deuterium labeling at a defined position.

EXERCISE 8-G

From which of the following organic halides can a stable Grignard reagent be made by treatment with magnesium metal and dry ethyl ether?

(a) 2-bromo-1-butanol

(b) 2-chloro-1-phenylpropane

(c) *p*-bromotoluene

(d) *p*-bromophenol

(e) 3-bromopropanoamide

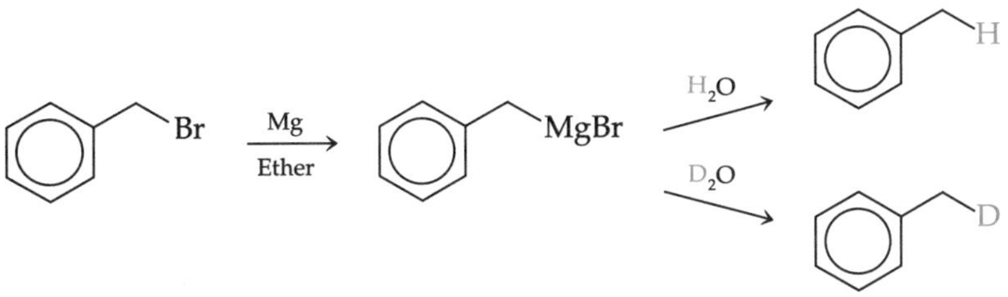

▲ FIGURE 8-11

Because Grignard reagents are protonated by water, a C–Br bond of an alkyl, aryl, or vinyl halide can be converted into the corresponding C–H or C–D bond. Insertion of magnesium into the C–X bond forms a Grignard reagent that is protonated to a hydrocarbon labeled by hydrogen if normal water is used for the quench or by deuterium if the quenching water is isotopically labeled as D_2O.

Labeling of compounds is very important in the pharmaceutical industry. It is often desirable to determine the fate of a drug as it is degraded by the body. By feeding a compound labeled with deuterium (or tritium) at a specific site to test animals and then analyzing the excretion products, pharmaceutical chemists can determine how large molecules are cleaved biologically into smaller ones that are soluble and can be excreted.

Because organozinc compounds are also protonated by acid, zinc also can be used to effect the reduction of simple alkyl halides; that is to convert C–X into a C–H bond.

$$\text{R}-\text{Br} + \text{Zn} \longrightarrow \overset{\delta^{\ominus}}{\text{R}}-\overset{\delta^{\oplus}}{\text{ZnBr}} \xrightarrow{\text{HOAc}} \text{R}-\text{H}$$

In this case, the reaction can be carried out in an ether solvent containing a small amount of water or in an acid, such as acetic acid (HOAc). As in the protonation of Grignard reagents, protonation of the carbon end of the polar carbon–zinc bond results in the formation of an alkane through a zinc-mediated reduction of the carbon–bromine bond.

8-4 ▸ Reactivity of Carbon Nucleophiles in S$_N$2 Reactions

Now that we know how to prepare carbanions and organometallic reagents, we are ready to consider their reactivity as nucleophiles in S$_N$2 reactions. Like carbanions, each organometallic reagent has a highly polarized carbon–metal bond that localizes appreciable negative charge on carbon. It is this high electron density that makes these reagents active as nucleophiles. In this section, we consider the possibilities and limitations in the use of carbon nucleophiles to form carbon–carbon bonds.

S$_N$2 Reactions by *sp*-Hybridized Carbon Nucleophiles

Cyanide Ion. Cyanide (⁻CN) is a particularly stable anion with a negative charge on carbon. It is an effective nucleophile in part because of its small size.

$$\text{N}\equiv\text{C:}^{\ominus} \quad \text{H}_3\text{C}-\text{Br} \xrightarrow{-\text{Br}^{\ominus}} \text{N}\equiv\text{C}-\text{CH}_3 \xrightarrow{\text{H}_3\text{O}^{\oplus}} \underset{\text{HO}}{\overset{\text{O}}{\|}}{\text{CH}_3}$$

An organonitrile

The lone pair on carbon can be used in a nucleophilic displacement reaction with alkyl halides, following an S$_N$2 pathway. This is an important reaction because a new carbon–carbon bond is formed, and the size and complexity of the carbon skeleton of the product is greater than that of the starting material. Furthermore, organonitriles can be hydrolyzed to carboxylic acids under acidic or basic reaction conditions. (Notice that the oxidation level of carbon in a nitrile is identical with that of a carboxylic acid. Therefore, conversion of the nitrile into more-common organic functional groups is a relatively easy reaction.) This S$_N$2 displacement, followed by hydrolysis, makes possible the synthesis of a carboxylic acid containing one more carbon than the reactant alkyl halide.

EXERCISE 8-H

For each of the following reactions, propose a sequence of reagents and conditions that could be used to effect the conversion.

(a)

(b)

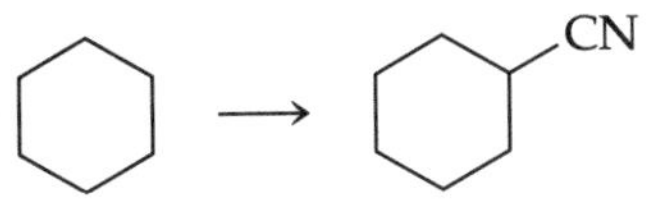

(c)

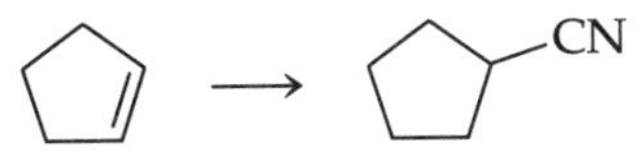

Acetylide Anions. Acetylide anions can be used as effective nucleophiles in carbon–carbon bond-forming reactions. Because the reaction is an S$_N$2 displacement, it takes place most efficiently with primary alkyl halides.

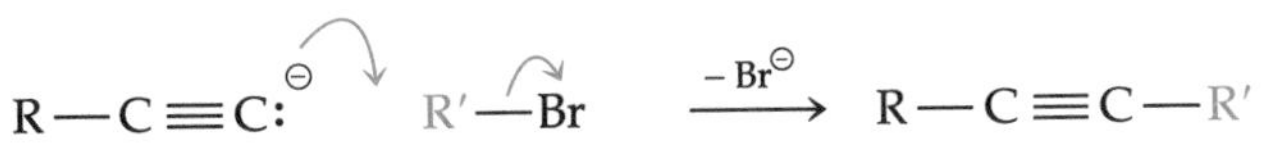

$$R-C\equiv C:^{\ominus} \qquad R'-Br \qquad \xrightarrow{-Br^{\ominus}} \qquad R-C\equiv C-R'$$

Because this transformation extends the carbon chain, it is a useful method for forming alkynes of higher molecular weight.

Alkylation of Other Organometallics

Similar reactivity is observed for organocopper reagents. For example, an S$_N$2 displacement by lithium dialkylcopper on an alkyl halide also forms a new carbon–carbon bond.

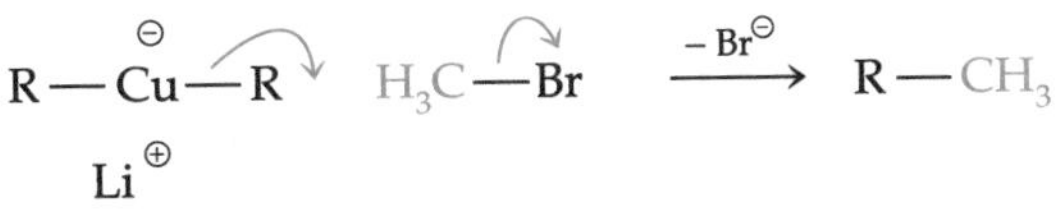

$$R-\overset{\ominus}{Cu}-R \quad H_3C-Br \quad \xrightarrow{-Br^{\ominus}} \quad R-CH_3$$
$$Li^{\oplus}$$

S$_N$2 Opening of Ethylene Oxide by Grignard Reagents

The negative charge density on carbon in an organometallic reagent makes this family of reagents active nucleophiles. A strained epoxide (for example, ethylene oxide) can be opened by nucleophilic attack by a Grignard reagent.

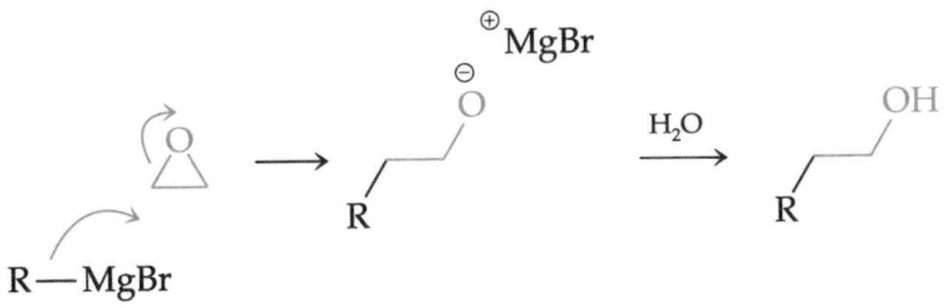

After hydrolysis, an alcohol bearing two more carbons than were present in the Grignard reagent is produced. This displacement does not take place with simple unstrained ethers.

Alpha–Halogenation of Enolate Anions

Haloform Reaction. Enolate anions generated in aqueous or alcoholic solution can displace halide in S_N2 reaction with iodine or bromine, effecting the halogenation of the enolate at the α position. This S_N2 reaction is the first step of a sequence called the iodoform reaction in which a ketone with three α-hydrogens is converted into a carboxylic acid and iodoform (CHI_3).

Let us specifically consider the reaction of acetophenone in basic aqueous solution with iodine, in which α-carbon–halogen bonds are formed by S_N2 displacement of iodide from I_2. Because acetophenone is a considerably weaker acid than water, only a very small concentration of the enolate is formed; but, because the enolate is so reactive, it is removed from the equilibrium by reaction and quickly reformed in an acid-base pre-equilibrium. As iodine approaches the negatively charged enolate, the I–I bond becomes polarized so as to allow nucleophilic displacement (Figure 8-12). The α-hydrogens in the resulting α-iodoacetophenone are even more acidic than in acetophenone itself by virtue of inductive electron withdrawal by iodine. Thus, in base, deprotonation occurs rapidly to form an iodoenolate. Nucleophilic displacement of I^- from I_2 with the iodoenolate as nucleophile (again, on I_2) results in a diiodoacetophenone, in which the remaining carbon–hydrogen bond is even more acidic. Repetition of the deprotonation-displacement reactions produces triiodoacetophenone. With no additional α-hydrogens available for deprotonation, hydroxide can then assume a secondary role as a nucleophile, attacking the carbon end of the C–O dipole, which has been activated by inductive withdrawal by the triiodomethyl group attached. The resulting tetrahedral intermediate reforms the C=O double bond by expelling $^-CI_3$, a particularly stable carbanion by virtue of the three attached polarizable halogens. (How this reaction takes place will be considered in more detail in Chapter 12.) After protonation, a bright yellow precipitate of iodoform (CHI_3) results. The formation of this precipitate is considered to be a positive iodoform test.

This reaction, which also works with Br_2 or Cl_2, is called a **haloform reaction,** although no precipitate is observed with the other halogens. The key feature of the haloform reaction is the increase in the acidity of the α-C–H with each successive halogenation. The **iodoform test** is used as a diagnostic for the presence of a CH_3CO (acetyl) group in a molecule, because CHI_3 cannot be formed unless there are three α-hydrogens that can be replaced.

EXERCISE 8-I

Which of the following molecules will give positive iodoform tests?

(a) 3-pentanone

(b) 2-pentanone

(c) pentanal

(d) acetophenone

(e) acetic acid

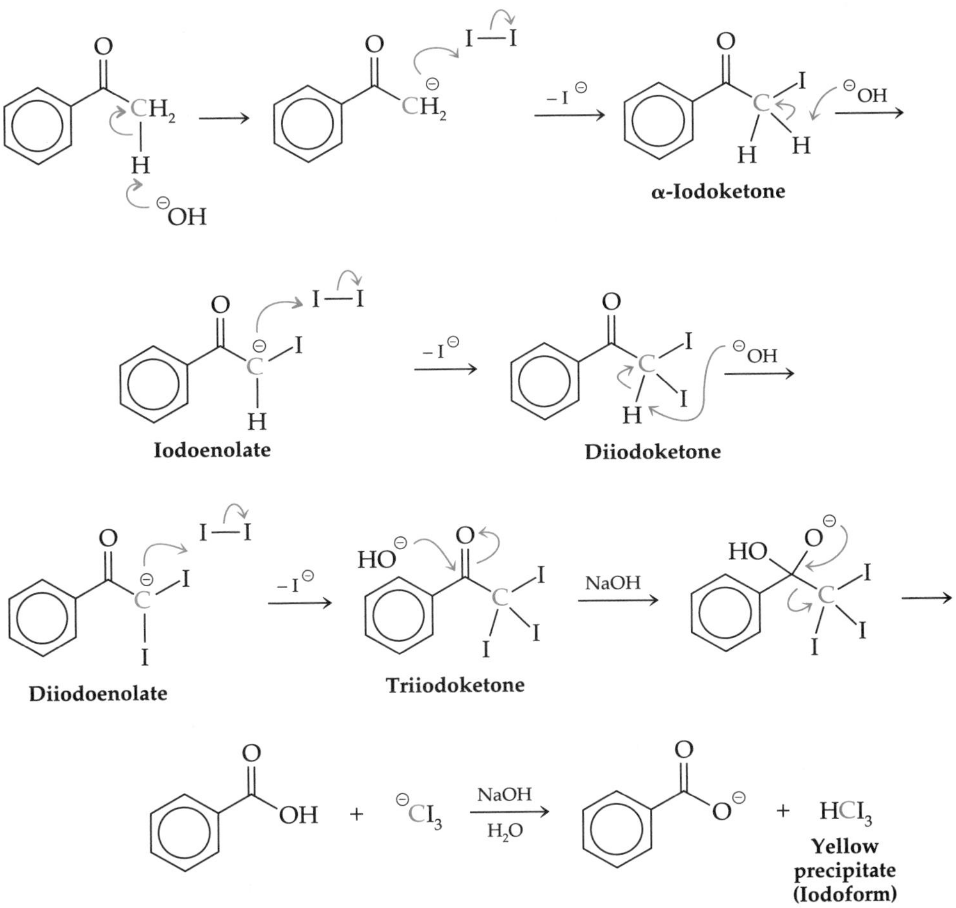

▲ FIGURE 8-12

The iodoform reaction begins with the formation of an enolate anion from a ketone of the general type $RCOCH_3$. The enolate acts as a nucleophile to displace I^- from I_2 by an S_N2 route. The α-iodoketone produced has two acidic hydrogens, and enolization occurs again. The resulting iodoenolate again reacts by an S_N2 pathway, producing a diiodoketone. Upon repetition of this sequence, a triiodoketone is formed. Largely because of the stability of the triiodomethyl anion, the triiodoketone is converted by base into a carboxylic acid and iodoform, which precipitates as a yellow solid. The iodoform reaction can thus be used as a chemical test for the initial presence of the acetyl functional group ($—COCH_3$).

Hell-Volhard-Zelinski Reaction. Monohalogenation α to a carboxylic acid can be obtained by treatment of a carboxylic acid bearing α hydrogens with bromine in the presence of phosphorus tribromide (Figure 8-13), a reaction referred to as the **Hell-Volhard-Zelinski reaction.** In phenylacetic acid, for example, α-bromination occurs in high yield. The mechanism of this reaction is less important than its synthetic utility—for example, in the synthesis of α amino acids.

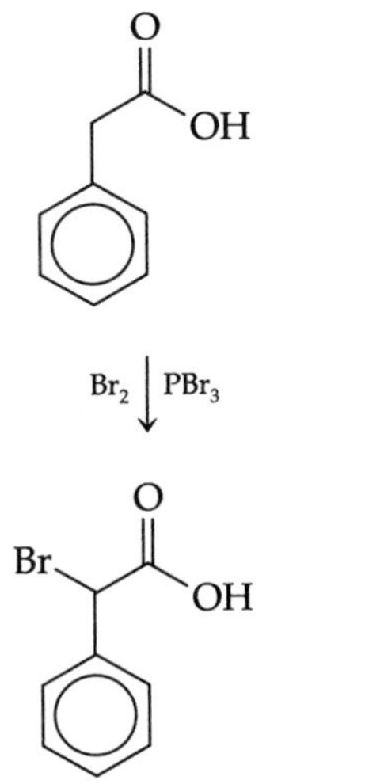

▲ FIGURE 8-13

The Hell-Volhard-Zelinski reaction causes monobromination at the α position of a carboxylic acid. The α-bromination occurs through an S_N2-like pathway similar to that in the iodoform reaction shown in detail in Figure 8-12.

EXERCISE 8-J

It is known that, in the first step of the Hell-Volhard-Zelinski reaction, the carboxylic acid is converted into an acid bromide (that is, an acyl bromide exactly analogous to an acid chloride). This acid bromide then undergoes enolization, with the enol displacing bromide from Br_2.

(a) Draw the structure of the active enol that would be obtained from the acid bromide of acetic acid and show, with curved arrows, the flow of electrons that accomplishes α-bromination.

(b) Explain why a second α-bromination does not occur as readily as the first.

Enolate Anion and Enamine Alkylation

An important S_N2 reaction is the formation of a carbon–carbon bond at a position α to a carbonyl group. Reaction of an enolate anion with an alkyl halide or an alkyl tosylate results in the formation of a new carbon–carbon bond at the α position. As before, this reaction is the result of a nucleophilic displacement by an enolate anion in an S_N2 backside attack (Figure 8-14).

Similar carbon–carbon bond formation also occurs with enamines. Although enamines are not negatively charged, they still have partial negative charge at the α-carbon and react with electrophiles such as methyl bromide. Although the rate of this reaction is lower than that of enolates, enamines accomplish the same transformation; that is, the formation of a carbon–carbon bond (Figure 8-15). The resulting **iminium ion** is hydrolyzed with water after the alkylation reaction is complete, forming the ketone. Because enamines are formed from ketones, the overall sequence forms a carbon–carbon bond α to a carbonyl group and is an alternative to direct alkylation of the ketone (as the enolate anion). It is particularly useful in those cases in which the carbonyl compound contains another functional group that also would react with the strong base necessary to form the enolate. α-Alkylations of enolate anions or enamines, being S_N2 reactions, work best with alkyl halides that are not sterically encumbered; that is, with methyl and primary alkyl halides.

In contrast, neutral enols are not sufficiently reactive to serve as nucleophiles with common electrophilic carbon reagents. They do, however, react with chlorine, bromine, and iodine to form carbon–halogen bonds.

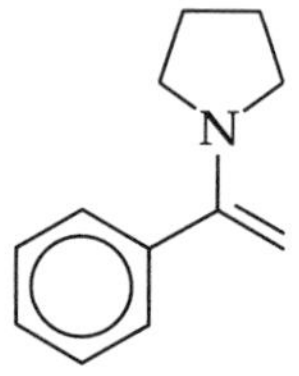

**Pyrrolidine enamine
of acetophenone**

EXERCISE 8-K

Write a complete mechanism for the alkylation of the pyrrolidine enamine of acetophenone (shown at the left) by an alkyl bromide.

Two serious limitations of enolate and enamine alkylation are: (1) enolization can occur on either side of a dialkyl ketone; and (2) the enolate can attack the starting carbonyl compound in an important competing side reaction—a condensation to be discussed in Chapter 13. Furthermore, the

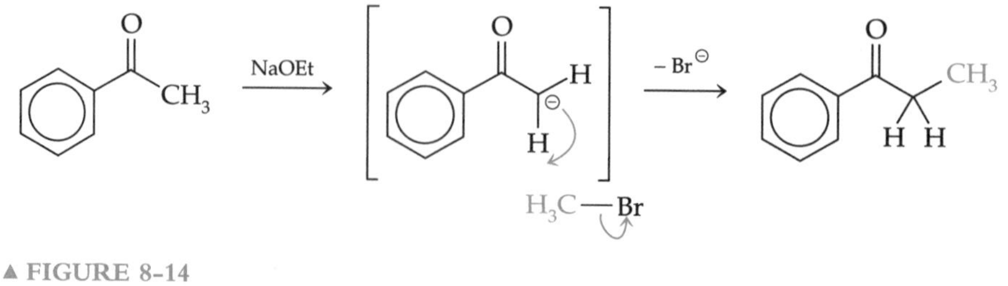

▲ **FIGURE 8-14**

When an enolate anion displaces halide from an alkyl halide or tosylate from an alkyl tosylate, a new carbon–carbon bond is formed.

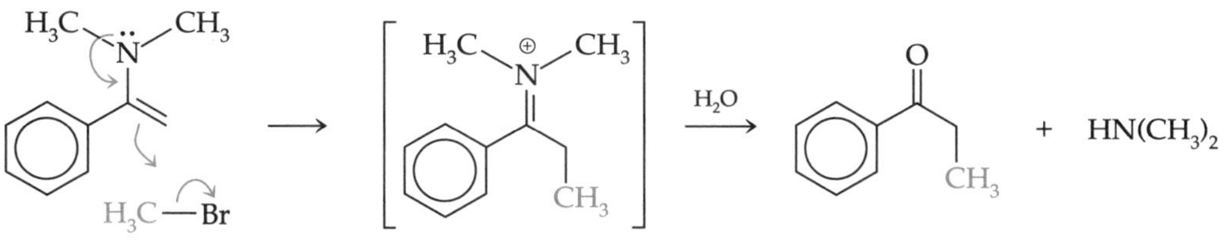

▲ FIGURE 8-15
Because an enamine is isoelectronic with an enolate anion, the electron flow is
the same when a carbon-carbon bond is formed at the α position. Thus, the
first step is an S$_N$2 reaction. The resulting iminium ion is converted by water
into a ketone. Notice that the final product is the same as that produced by the
enolate alkylation in Figure 8-14.

product itself is a ketone that may have acidic hydrogens at the α position.
Acid-base reaction of the product with the original enolate can result in proton transfer and the formation of the enolate anion derived from the product. The further alkylation that can then result often complicates the α-alkylation of simple carbonyl compounds. As discussed in the next sections,
these problems can be easily solved by the use of an ester enolate, a β-di-
ketoenolate, or a β-ketoester enolate as the nucleophilic enolate anion.

Ester Enolate Anion Alkylations

Esters with α-C–H bonds also form enolate ions that are reactive as nucleophiles in S$_N$2 reactions. Consider the reaction of an ester enolate anion generated in the presence of an alkyl halide (Figure 8-16). Direct displacement
of a halide through the now familiar backside S$_N$2 transition state is rapid,
forming a new carbon–carbon bond. Consistent with what we know about
S$_N$2 reactivity, this displacement occurs most readily if the alkyl halide is
not sterically encumbered; that is, the facility of the reaction with respect to
the alkyl halide is methyl halide > primary halide > secondary halide. Tertiary halides do not undergo substitution; rather they undergo elimination
of HX to form an alkene in an undesired side reaction. (Elimination is often
a significant side reaction with secondary halides as well.)

   As in the alkylation of aldehyde and ketone enolate anions, similar side
reactions limit the utility of this method. The synthesis of α-branched acids
and esters is more commonly addressed by the use of more-stable enolate
anions, as discussed next.

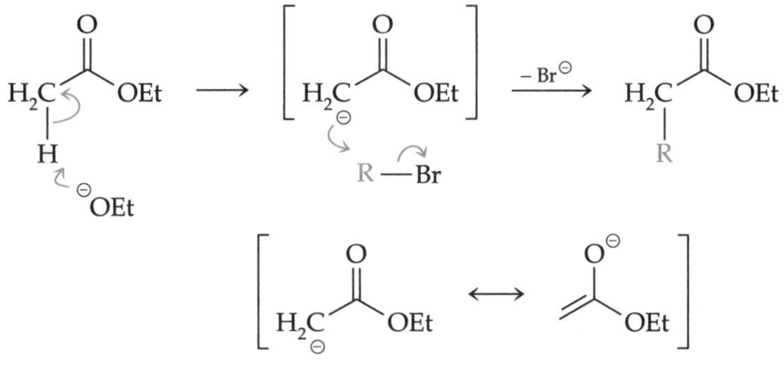

▲ FIGURE 8-16
An ester enolate anion can be alkylated by an S$_N$2 pathway, forming a new
carbon–carbon bond at the position α to the carbonyl group.

Alkylation of Beta-Dicarbonyl Compounds

Acetoacetic Ester Synthesis. As we learned in Chapter 6, a proton attached to a carbon between two carbonyl groups is highly acidic. For example, in **acetoacetic ester** (a β-ketoester), the pK_a at the α hydrogen (pK_a 11) is such that the enolate anion is readily formed upon treatment with relatively mild bases such as sodium alkoxides. There are three important resonance contributors for the derived anion, two with negative charge on oxygen and a third with formal charge on carbon (Figures 8-10 and 8-17).

As with simple enolate anions, reprotonation can occur on either carbon or oxygen. Depending on the structure, β-dicarbonyl compounds—especially β-diketones—often have significant concentrations of the enol at equilibrium. The enol derives significant additional stabilization by delocalization of a lone pair from the enolic oxygen through the carbon–carbon π bond to the remaining carbonyl group.

EXERCISE 8-L

Draw the structure of the three tautomers of a neutral acetoacetic ester and, on the basis of bond-dissociation energies and conjugative effects, predict the relative order of stability of these three tautomers.

As we have already seen in the Gabriel synthesis, an anion stabilized by two adjacent carbonyl groups can be used as a nucleophile in an S_N2 reaction (Figure 8-17). The β-ketoester enolate anion is more stable (and less

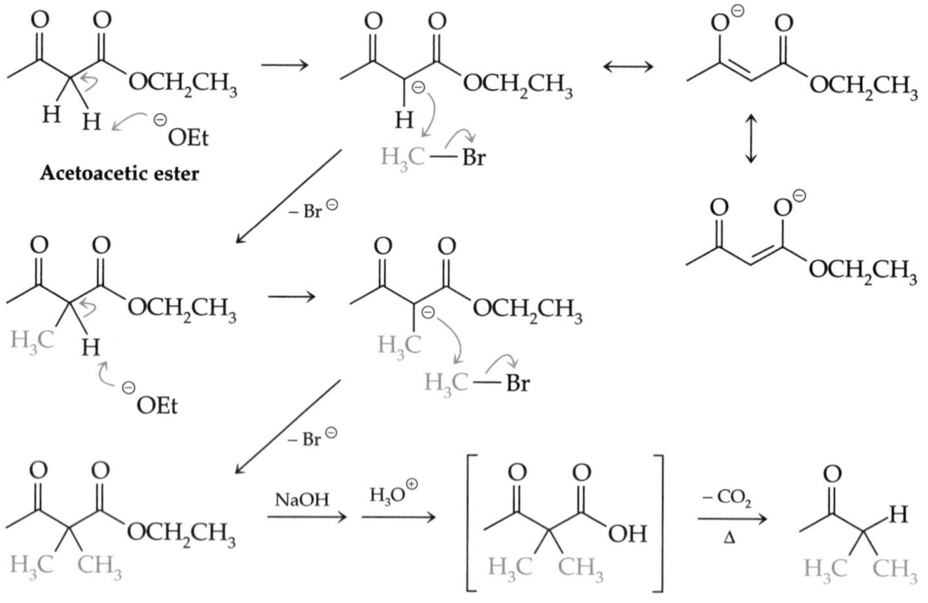

▲ FIGURE 8-17

An enolate anion is easily formed by deprotonation of a β-ketoester or β-diketone (the reaction of ethyl acetoacetate is shown here). This stabilized anion participates in an S_N2 alkylation on carbon. The product still has an acidic α hydrogen, and enolate anion formation can be repeated with the first alkylated product. A second S_N2 alkylation forms a second carbon–carbon bond at the α carbon. This second alkyl group can be the same as the first (as shown here) or it can be completely different. The mono- or dialkylated β-ketoester is then hydrolyzed by treatment with aqueous base. Neutralization with acid produces a β-ketoacid, which loses CO_2 upon heating. The final product is a ketone alkylated once or twice at the α position.

reactive) than a simple enolate anion. More importantly, the basicity of the β-ketoenolate anion is reduced to an even greater extent than is its nucleophilicity so that reaction with an alkyl halide produces significantly higher yields of the product of S_N2 displacement. The desired S_N2 reaction forms a carbon–carbon bond, with lower amounts of side reactions.

This alkylation accomplishes the net replacement of a hydrogen on the carbon between the two carbonyl groups by an alkyl group. In a substituted acetoacetic ester, there is still another α hydrogen: the process of deprotonation-alkylation can therefore be repeated, forming a doubly alkylated product (in which both of the α hydrogens of acetoacetic ester are replaced by alkyl groups). The rate of the second alkylation is significantly slower than the first, however, because of additional steric hindrance in the transition state of the second alkylation. Thus, it is possible to stop at the monoalkylation stage, with one substituent bound to the α carbon, if desired, or to continue in a second step to attach a second alkyl group (either the same or a different one).

Decarboxylation of Beta-Ketoacids. The acetoacetic ester synthesis is particularly useful not only because alkylation (once or twice) takes place efficiently at the α position, but also because the ester group, having done its job, is easily lost. Upon heating in acid, the ester group is replaced by a proton as CO_2 and ROH are lost.

The final product described in Figure 8-17 requires not only alkylation of the β-ketoester, but also the loss of the ester group. Because an ester is at the same oxidation level as a carboxylic acid, it can be hydrolyzed to the corresponding acid in acid or base. In one conformer of the resulting acid (formed by rotation about the carbon–carbon bond to the carboxylic acid carbon), the carboxylic acid proton can hydrogen bond to the carbonyl group of the β-ketone.

Here, a lone pair of electrons on the ketone oxygen acts as Lewis base to the acidic proton. However, this spatial arrangement then allows a special reaction possibility by which the proton can be transferred from the carboxylic acid oxygen to the ketone group with migration of the additional σ and π electron density. The result of this process, which proceeds through a particularly stable six-membered transition state, is the production of an enol and carbon dioxide. The π bonds between carbon and oxygen in carbon dioxide are exceptionally strong and the production of this molecule serves as a driving force for this reaction. The enol produced rapidly tautomerizes to the corresponding ketone. This process results in the loss of CO_2 and is called **decarboxylation.** It occurs at or slightly above room temperature for most β-ketoacids.

EXERCISE 8-M

Unlike β-ketoacids, α-ketoacids do not undergo decarboxylation upon heating. Explain why.

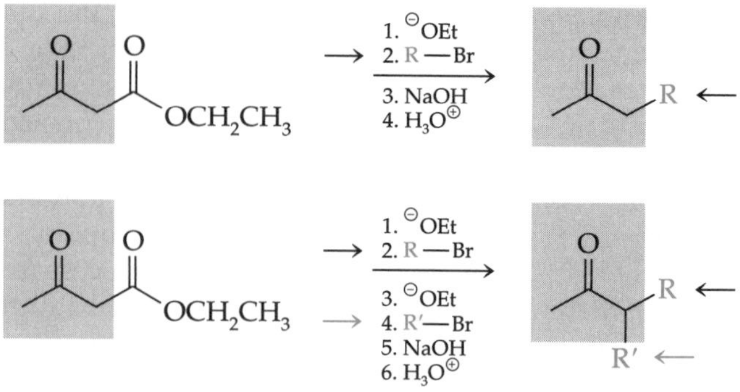

▲ FIGURE 8-18

The sequence of steps shown in the upper reaction produces a monoalkylated ketone, whereas that in the lower reaction produces a dialkylated ketone. In both cases, base forms the enolate anion, which acts as a nucleophile to displace halide from an alkyl halide. This sequence is repeated in the lower reaction, introducing a second alkyl group. Hydrolysis with base produces a carboxylate ion that loses CO_2 when treated with acid.

We can view the overall process of alkylation of acetoacetic ester, followed by hydrolysis and decarboxylation, as a method for preparing methyl ketones in which one or two additional alkyl groups are attached to the other α carbon (Figure 8-18).

Because the final product is formed after disposing of the ester group, you may ask why the ester needs to be present initially for this reaction. Although in principle it is possible to form these same products from acetone itself by sequential alkylation, there are significant advantages to carrying out the process with the ester group present. First, without the additional stabilization afforded by the second carbonyl group, a much stronger base is required to effect the deprotonation. Furthermore, once one alkyl group has been added to acetone, the ketone becomes unsymmetrical. Because there are two different α carbons of comparable acidity, mixtures of regioisomers result from the second alkylation. As we will see in Chapter 13, monocarbonyl compounds can also undergo condensation as a significant side reaction to the simple alkylation. Generally, β-ketoesters are too unreactive to undergo these competing condensation reactions. Thus, we can view the ester group as a temporary functional group that facilitates the reaction by providing special, enhanced acidity to a single carbon α to the ketone carbonyl group and can then be removed after this role has been completed.

Malonic Ester Synthesis. Closely related to the acetoacetic ester alkylation is the malonic ester synthesis. **Malonic ester** differs from acetoacetic ester (Figure 8-19) in having two ester groups (instead of an ester and a ketone as in acetoacetic ester) bound to a single carbon atom. Following the same sequence as that for the acetoacetic ester synthesis—namely, enolate anion formation, alkylation once or twice, followed by hydrolysis of the ester groups to the corresponding acids and decarboxylation—gives rise, after tautomerization, to a substituted acetic acid. Decarboxylation of the β-diacid, the last step in Figure 8-19, occurs through a six-electron transition state, with the loss of carbon dioxide providing the driving force. However, it is not as rapid as the loss of carbon dioxide from a β-ketoacid. Decarboxylation of β-diacids requires elevated temperatures, sometimes as high as 100 °C.

We can view this process in exactly the same way as the acetoacetic ester synthesis; that is, as a route to a substituted acetic acid bearing one or two alkyl groups at the α carbon (Figure 8-20). The initial presence of the

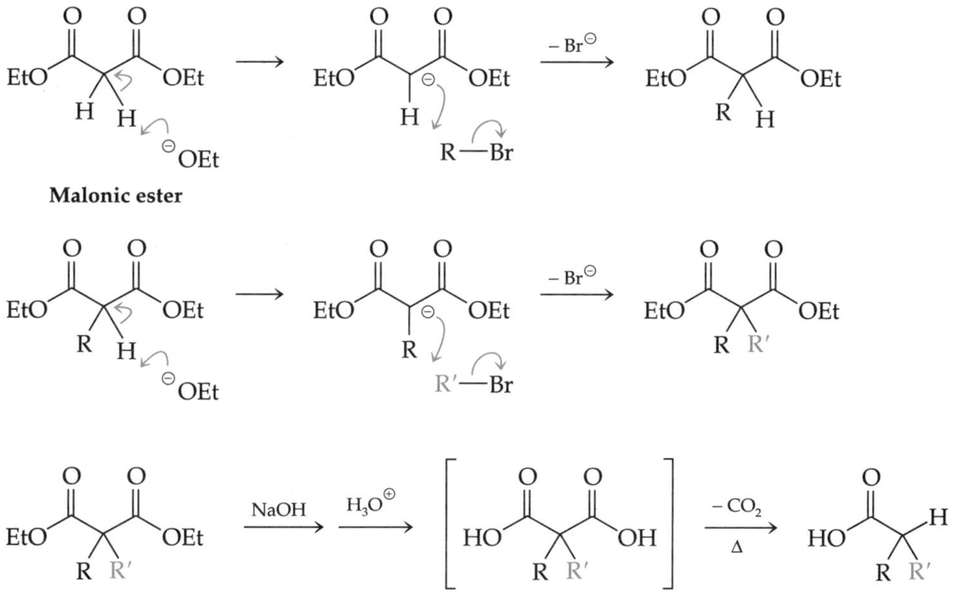

Malonic ester

▲ FIGURE 8-19

Malonic ester synthesis (shown here with dimethyl malonate) follows the same sequence of enolate anion formation, S$_N$2 alkylation, hydrolysis, and decarboxylation as that in the acetoacetic ester synthesis shown in Figure 8-17. The difference is that the product obtained after the hydrolysis step is a β-diacid in the malonic ester synthesis rather than the β-ketoacid in the aceto-acetic ester synthesis. The loss of CO_2 produces an acid instead of a ketone.

additional ester group facilitates the reaction sequence by enhancing the acidity of the α-carbon that acts as a nucleophile.

We now know two routes for carbon–carbon bond formation adjacent to a carbonyl group: direct alkylation by an S$_N$2 route of a carbonyl enolate anion or an enamine (followed by hydrolysis); and alkylation of a stabilized ester enolate (malonic or acetoacetic ester), again by an S$_N$2 route to gener-ate an alkylated ester, the hydrolysis and pyrolysis of which leads, respec-tively, to a carboxylic acid or a ketone. In the acetoacetic ester and malonic ester syntheses, stabilized enolate anions act as nucleophiles, and so these routes afford more practical methods.

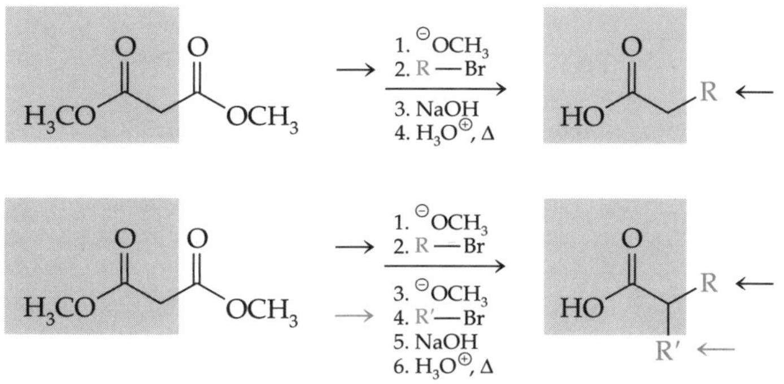

▲ FIGURE 8-20

The sequence of steps shown in the upper reaction produces monoalkylated acid, whereas that in the lower reaction produces dialkylated acid. In both cases, base forms the enolate anion, which acts as a nucleophile to displace halide from an alkyl halide. This sequence is repeated in the lower reaction, introducing a second alkyl group. Hydrolysis with base produces a diacid that loses CO_2 when treated with acid and heat.

BARBITURATES FROM DIALKYLMALONATES

Thiobarbiturates are anesthetics that are used both by themselves and in conjunction with other anesthetics. They are prepared by the reaction of thiourea with substituted malonic esters, as shown below for sodium pentothal. This compound has also been used as a "truth serum," causing the interrogated person to whom the drug has been administered to relax his defenses in defending false information.

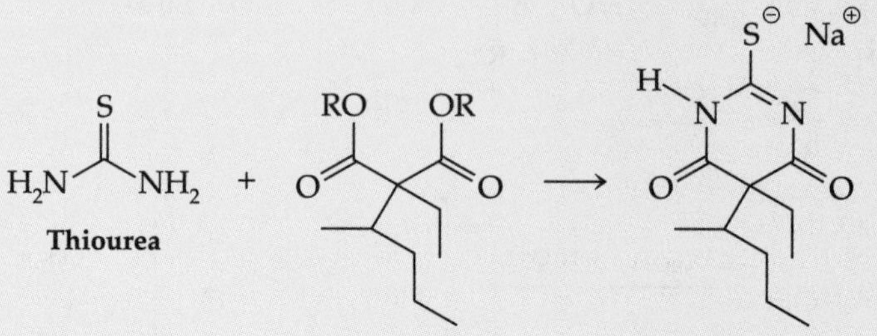

Thiourea   Sodium pentothal

EXERCISE 8-N

Write a detailed mechanism showing specific electron flow for the decarboxylation of a monoalkylated malonic acid.

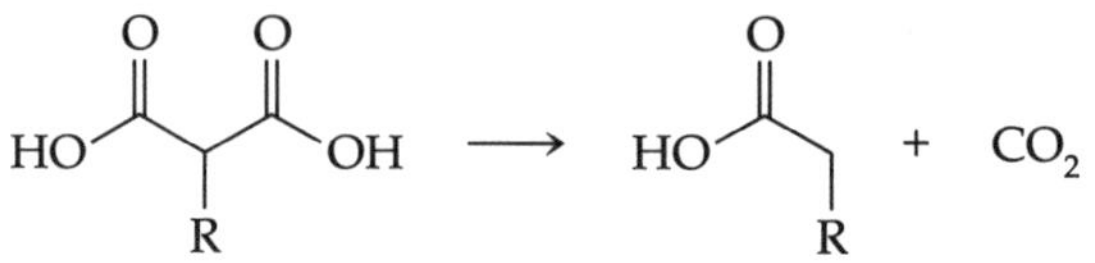

EXERCISE 8-O

Choosing any reagents, show how 2-methylbutyric acid and 3-methyl-2-pentanone could be prepared by the use of the acetoacetic ester or malonic ester synthesis.

2-Methylbutyric acid   3-Methyl-2-pentanone

8-5 ▸ Synthetic Methods: Functional-Group Conversion

As in Chapter 7, it is useful to keep track of new reactions that allow us to alter functional groups. Table 8-1 is a way of compiling the reactions presented in this chapter. The question posed is: How does one attach the carbon skeleton of an alkyl halide (or alkyl tosylate) to various functional groups? (This reaction thus also converts the alkyl halide functional group

Table 8-1 ► How to Use S$_N$2 Reactions to Make Various Functional Groups

Functional group	Reaction
Alcohol	Opening of ethylene oxide by a Grignard reagent
Alkane	Grignard hydrolysis; or protonation of organozincs; or organocopper alkylation
Alkyne	Acetylide alkylation
Amine	Gabriel synthesis; or amine alkylation
Azide	Alkylation of N$_3^-$
α-Bromoester	Hell-Volhard-Zelinski reaction
Carboxylic acid	Cyanation of an alkyl halide, followed by hydrolysis (one-carbon extension); or iodoform reaction of a methylketone; or bromination of an acid (Hell-Volhard-Zelinski reaction); or decarboxylation of a substituted malonic acid
Ester	Ester enolate alkylation; or acetoacetic ester alkylation
Ether	Williamson ether synthesis
Grignard reagent	Magnesium metal insertion
Ketone	Ketone enolate alkylation; or ketone enamine alkylation; or acetoacetic ester synthesis; or decarboxylation of a β-ketoacid
Organocopper	Transmetallation of an organolithium
Organolithium	Lithium insertion
Organozinc	Zinc insertion
Phosphonium salt	Phosphine alkylation

into another one.) In conceptually grouping reactions, you should take special note of those that form new carbon–carbon bonds.

As an exercise to complement those in the problem set at the end of the chapter, quiz yourself on each of the reactions listed in Table 8-1 and in the Summary of New Reactions (at the end of this chapter), according to the criteria set out in Section 7-3:

1. Can I predict the product (including stereochemistry and regiochemistry), given the reactants?

2. Do I know the reagent(s) and conditions necessary to convert the organic group into the product?

3. For a desired product formed by a given reaction path, can I correctly choose an appropriate starting material?

4. Given the reactants, conditions, and products, can I write a detailed reaction mechanism showing specific electron flow that describes how the reaction proceeds?

For synthetic utility, the third question is of utmost importance. Although we have emphasized reaction mechanisms in the textual discussion, practical organic chemistry requires the use of sequences of known reactions to make new molecules. This can be achieved only if we know reactions literally "backward" and "forward." You can be sure of knowing these reactions well if you can drill yourself on both the mechanism and the functional-group interconversion accomplished in each transformation.

Only by understanding how various products can be obtained by several alternate pathways will it be possible to plan syntheses intelligently. Having a variety of functional-group interconversions at hand makes it easier to integrate new reactions with those that we already know.

Conclusions

Nucleophilic substitutions are of two major types: S_N1 reactions consisting only of cleavage of the bond between carbon and a leaving group in a rate-determining unimolecular step; and S_N2 reactions, in which partial bond formation with the incoming nucleophile occurs simultaneously with cleavage of the bond to the leaving group. The S_N1 reaction takes place in a sequence of steps, with formation of a carbocation intermediate as the rate-determining step. The S_N2 reaction, in contrast, occurs in a single concerted step, with no intermediates. There is inversion of stereochemistry in an S_N2 reaction because of the required backside attack by the incoming nucleophile; whereas, in an S_N1 reaction, a planar, achiral cationic intermediate is formed and racemization results.

Because S_N1 reactions occur through intermediate carbocations, they are complicated by competing addition, elimination, and rearrangement reactions. S_N1 reactions are therefore generally less useful in synthesis than S_N2 reactions.

The S_N2 reaction proceeds through a pentavalent transition state and is therefore sensitive to steric effects, which dictate a reactivity order (primary > secondary > tertiary) opposite that observed for an S_N1 reaction. The facility of cleavage of the bond between carbon and the leaving group in an S_N2 reaction is influenced by the C–L bond strength and the ability of the leaving group to accommodate negative charge. Conversely, a more-active nucleophile more rapidly donates an electron pair to the electron-deficient center, with anionic nucleophiles (formed by deprotonation) being more active than their neutral precursors. A good nucleophile is therefore often a good base.

Carbon nucleophiles are generated in one of three ways:

1. treatment of alkyl halides with zero-valent metals; for example, lithium or magnesium, producing organolithiums or Grignard reagents;

2. transmetallation; that is, exchange of other metals for the alkali metal or magnesium in organolithiums or Grignard reagents; or

3. treatment of compounds bearing acidic hydrogens with base. This type of C–H cleavage is especially easy when the acidic C–H bond is α to the carbonyl group of an aldehyde, ketone, or ester. Terminal alkynes also possess an acidic C–H bond and deposit a characteristic precipitate upon treatment with basic $Ag(NH_3)^+$.

The negatively charged carbon in alkali or alkaline earth organometallics is associated with a counterion in order to preserve electroneutrality. The strength of the interaction between the positive and negative centers determines the type of bond between the metal and carbon, with free ions and pure covalent bonds presenting the extremes. Most alkali metal salts exist as ion pairs, whereas most Grignard reagents (RMgBr) are essentially covalently bound—that is, there is a σ bond between carbon and the metal.

Nucleophilic substitutions are employed as the critical step in several functional-group transformations, including carbon–carbon bond formation. In such a reaction, an electron-rich reagent (often an anion) reacts with an organic molecule bearing a leaving group at a center of partial positive charge. The carbon–metal bond in an organometallic compound is polarized because the metal is electropositive and the carbon atom is electronegative. As a result, the carbon end of an organometallic σ bond bears partial negative charge and can be protonated, halogenated, or alkylated.

The sequence by which a methyl ketone is iodinated three times at the α position, before being converted into a carboxylic acid and iodoform, constitutes a chemical test for the acetyl ($-COCH_3$) functional group. The Hell-Volhard-Zelinski reaction allows α-monobromination of a carboxylic acid. An enolate anion bears significant negative charge both on the α carbon and on the carbonyl oxygen, and reaction of an enolate with an electrophilic species (for example, X_2 or RX) affords net substitution at the α position. Analogous α-alkylation occurs in the isoelectronic enamine. Alkylation of an enolate or alkynyl anion forms a new carbon–carbon bond that allows for the construction of more-complicated structures.

The specific S_N2 substitutions considered in this chapter include: the Williamson ether synthesis; amination of alkyl halides; the Gabriel synthesis of primary amines; α-halogenation; carbon–carbon bond formation by cyanide displacement; alkylation of alkynyl anions; epoxide opening by Grignard reagents; and α-alkylation of enolate anions of simple ketones, acetoacetic esters, and malonic esters. Although the mechanisms of these reactions are similar, some have special secondary twists, as in the decarboxylation of alkylated acetoacetic acids or malonic acids to form respectively α-alkylated ketones or acids.

Summary of New Reactions

Conversion of Alcohols into Alkyl Bromides

$$ROH + PBr_3 \text{ (or } PBr_5) \longrightarrow RBr$$

Conversion of Alcohols into Alkyl Chlorides

$$ROH + SOCl_2 \text{ (or } POCl_3) \longrightarrow RCl$$

Conversion of Alcohols into Tosylates

$$ROH + p\text{-Me}(C_6H_4)SO_2Cl \longrightarrow p\text{-Me}(C_6H_4)SO_2OR$$

Halide Exchange

$$CH_3CH_2-Br \xrightarrow{NaCl} CH_3CH_2Cl$$

Hydrolysis of Alkyl Halides

$$CH_3CH_2—Br \xrightarrow{NaOH} CH_3CH_2OH$$

Williamson Ether Synthesis

$$RX + {}^{\ominus}OR' \longrightarrow ROR'$$

Synthesis of Thioethers

$$RX + {}^{\ominus}SR' \longrightarrow RSR'$$

Amination of Alkyl Halides

$$RX + NH_3 \longrightarrow RNH_2 + R_2NH + R_3N + R_4\overset{\oplus}{N}\overset{\ominus}{X}$$

$$CH_3CH_2—Br \xrightarrow{\text{Excess } NH_3} CH_3CH_2NH_2$$

Gabriel Synthesis

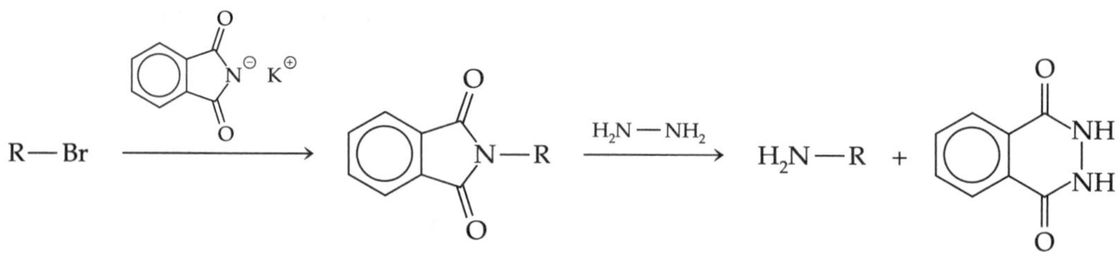

Formation of Alkyl Azides

$$RX + {}^{\ominus}N_3 \longrightarrow RN_3$$

Phosphonium Ylide Formation

$$Ph_3P: + H_3C—Br \longrightarrow Ph_3\overset{\oplus}{P}—CH_2—H \;\; {}^{\ominus}Br \xrightarrow{BuLi} Ph_3\overset{\oplus}{P}—\overset{\ominus}{C}H_2$$

A phosphonium salt

A phosphonium ylide

Nitrile Formation (with One-Carbon Chain Extension)

$$R—X + {}^{\ominus}C{\equiv}N \longrightarrow R—C{\equiv}N$$

Nitrile Hydrolysis

$$R—C{\equiv}N \xrightarrow{H_3O^{\oplus}} RCO_2H$$

Alkynyl Anion Alkylation

$$R—C{\equiv}C^{\ominus} + R'X \longrightarrow R—C{\equiv}C—R'$$

Generation of Organolithiums

$$RX + Li \longrightarrow RLi + LiX$$

Generation and Reaction of Grignard Reagents (Protonation and Two-Carbon Extension)

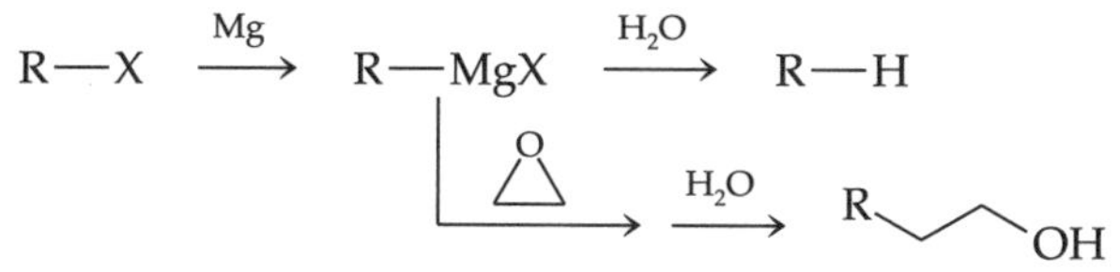

Organozincs: Reduction of Alkyl Halides

$$R—X \xrightarrow{\text{Zn}} \left[R—ZnX \right] \xrightarrow{\text{H}_2\text{O}} R—H$$

Transmetallation

$$R—Li + MX_n \longrightarrow R_nM + n\,LiX$$

$$RMgX + MX_n \longrightarrow R_nM + n\,MgX_2$$

Deuteration of Organometallics

$$R_nM + D_2O \longrightarrow R—D$$

Iodoform Reaction: α-Halogenation with Oxidation

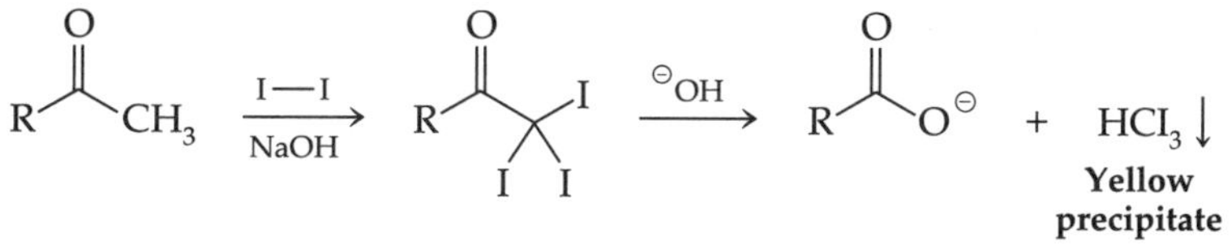

Hell-Volhard-Zelinski Reaction: α-Bromination of Acids

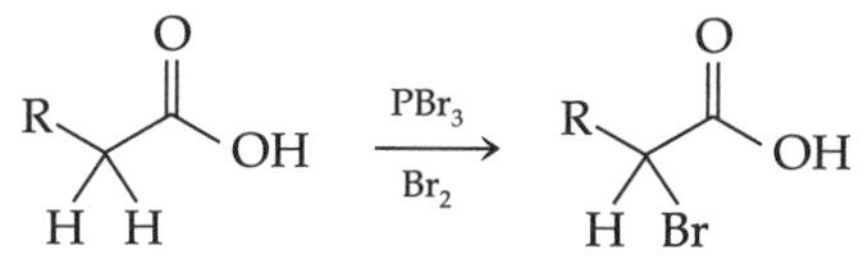

Enolate Alkylation

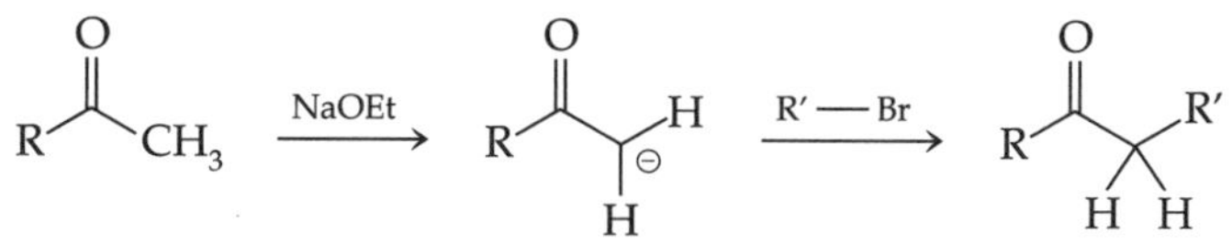

Enamine Alkylation

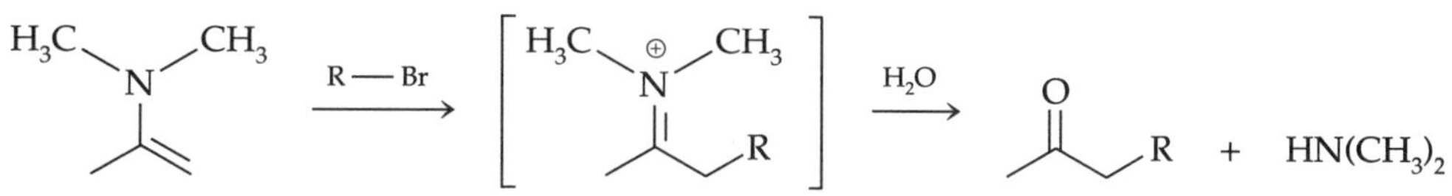

Ester Enolate Alkylation

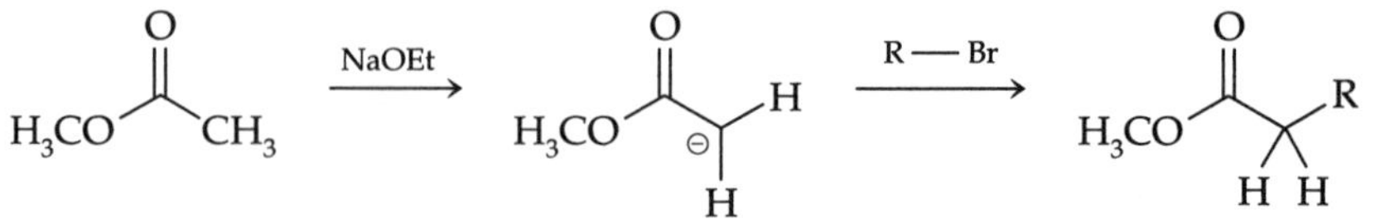

Malonic Ester Synthesis

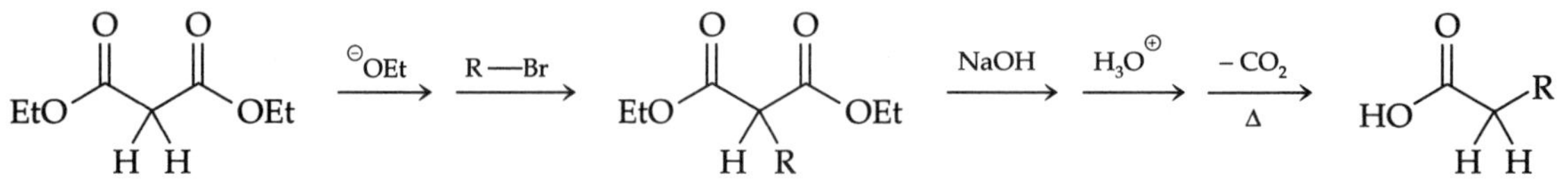

Hydrolysis and Decarboxylation of β-Ketoesters

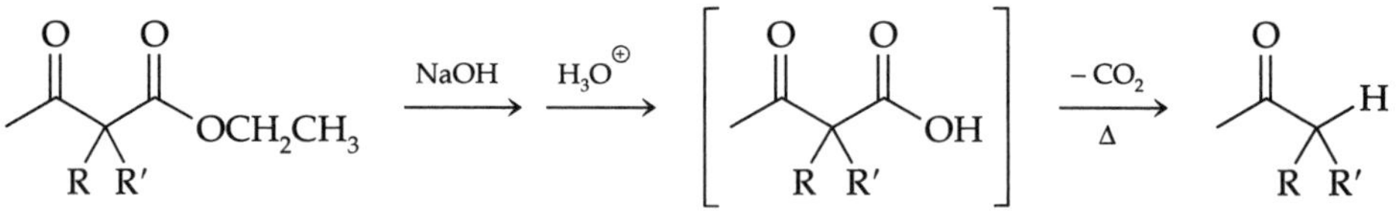

Acetoacetic Ester Synthesis

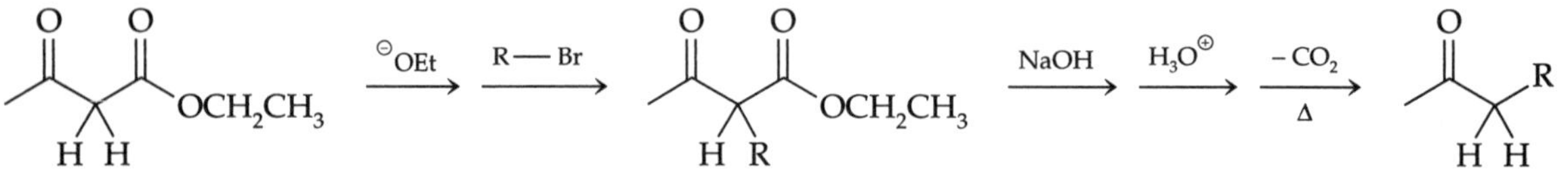

Hydrolysis and Decarboxylation of β-Diesters

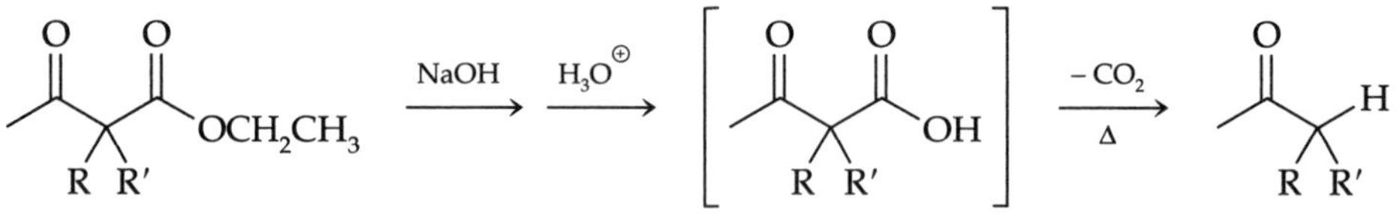

Review Problems

8-1 Assuming that the following reactions take place by an S_N2 displacement, choose the faster reaction of the following pairs and explain your reasoning:

(a) reaction of cyanide with *n*-iodoheptane or *n*-chloroheptane

(b) reaction of ethanol or sodium ethoxide with *n*-butyl bromide

(c) reaction of azide with *n*-butyltosylate or *s*-butyltosylate

(d) reaction of isopropoxide with ethyl bromide or of ethoxide with 2-bromopropane

8-2 S_N1 reactions proceed through planar carbocations, and only under special circumstances does an S_N1 reaction give a nonracemic product. For example, although the reaction rate depends only on the concentration of the reactant, the isolated product obtained when (S)-2-bromo-*n*-propylmethyl ether is hydrolyzed is (S)-2-hydroxy-*n*-propyl ether. In addition, the solvolysis of this reactant is much faster than that of 2-bromopropane.

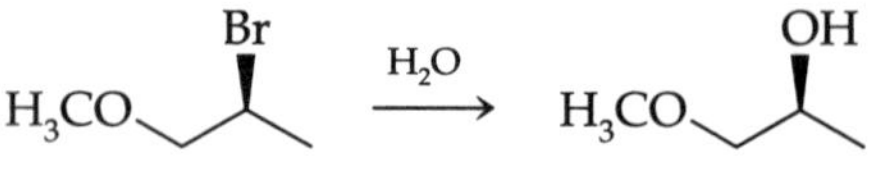

(a) Suggest an explanation for these observations.

(b) Predict the stereochemical course that would be expected if sodium cyanide in acetone had been used as the nucleophile instead of water.

8-3 For each of the following reactions, predict the expected product. Where a center of chirality is created or destroyed, indicate the expected absolute configuration.

(a) 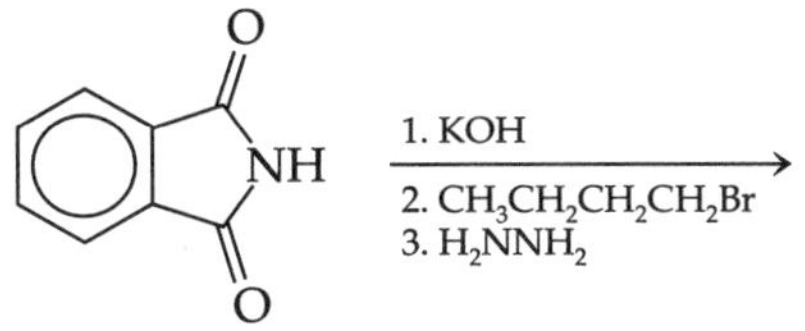 [phthalimide structure] $\xrightarrow{\begin{array}{l}\text{1. KOH}\\\text{2. CH}_3\text{CH}_2\text{CH}_2\text{CH}_2\text{Br}\\\text{3. H}_2\text{NNH}_2\end{array}}$

(b) 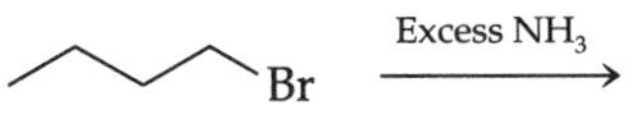 [structure] Br $\xrightarrow{\text{Excess NH}_3}$

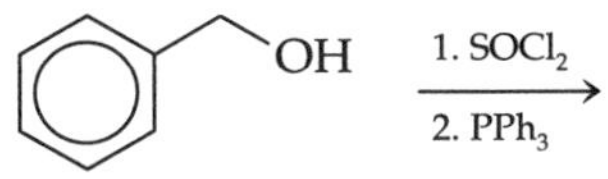

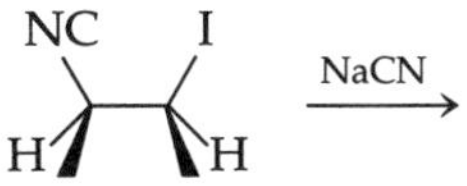

(c) 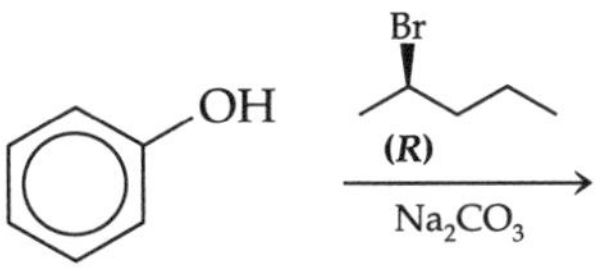 [phenol with (R)-2-bromopentane] $\xrightarrow{\begin{array}{l}(R)\\\text{Na}_2\text{CO}_3\end{array}}$

(d) [benzyl alcohol] $\xrightarrow{\begin{array}{l}\text{1. SOCl}_2\\\text{2. PPh}_3\end{array}}$

(e)  [allyl OTs] $\xrightarrow{\text{NaCN}}$

(f) 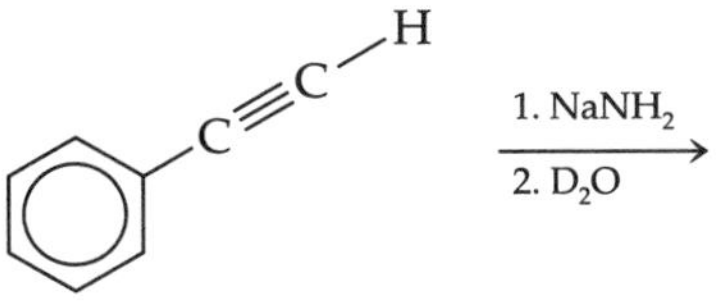 [phenylacetylene] $\xrightarrow{\begin{array}{l}\text{1. NaNH}_2\\\text{2. D}_2\text{O}\end{array}}$

(g) [NC-CHI structure] $\xrightarrow{\text{NaCN}}$

(h) 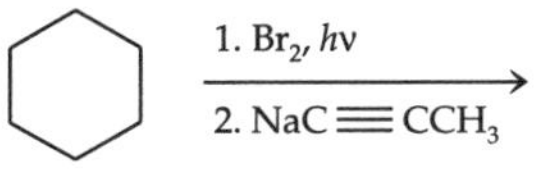 [cyclohexane] $\xrightarrow{\begin{array}{l}\text{1. Br}_2, h\nu\\\text{2. NaC}\equiv\text{CCH}_3\end{array}}$

(i) 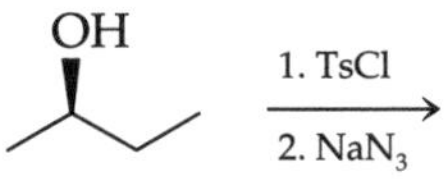 [2-butanol] $\xrightarrow{\begin{array}{l}\text{1. TsCl}\\\text{2. NaN}_3\end{array}}$

(j) 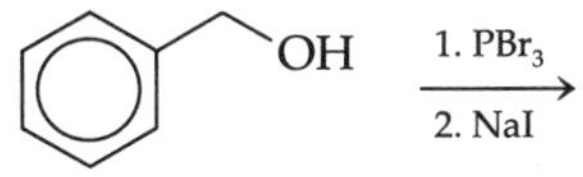 [benzyl alcohol] $\xrightarrow{\begin{array}{l}\text{1. PBr}_3\\\text{2. NaI}\end{array}}$

(k) 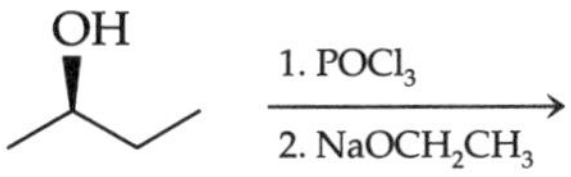 [2-butanol] $\xrightarrow{\begin{array}{l}\text{1. POCl}_3\\\text{2. NaOCH}_2\text{CH}_3\end{array}}$

(l) [phenethyl thiol] $\xrightarrow{\begin{array}{l}\text{1. Na}\\\text{2. CH}_3\text{Br}\end{array}}$

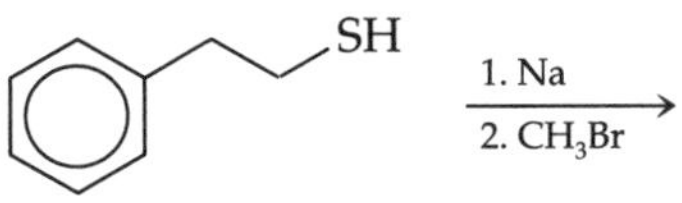

(m) 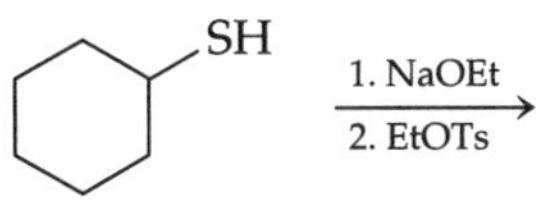 [cyclohexanethiol] $\xrightarrow{\begin{array}{l}\text{1. NaOEt}\\\text{2. EtOTs}\end{array}}$

(n) CH_3OTs $\xrightarrow{\begin{array}{l}\text{1. PPh}_3\\\text{2. BuLi}\end{array}}$

8-4 Using the reactions that you have learned in this chapter and the preceding ones, suggest a route by which the carbon skeletons of the following compounds can be synthesized from a two- or three-carbon alkyne and any alkyl halide containing three carbons or fewer. (You may use any other reagents needed; more than one step may be required.)

(a) 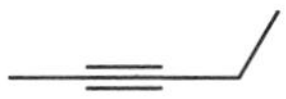 [alkyne structure]

(b)  [structure]

(c) [structure]

(d) 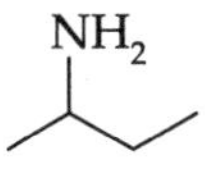 [amine structure with NH$_2$]

8-5 Propose a sequence of reactions and the appropriate reagents that can be used to effect the following conversions. If special conditions or solvents are needed, specify them.

(a) benzyl chloride into benzyllithium
(b) toluene into benzylmagnesium bromide
(c) benzyl alcohol into lithium dibenzylcuprate
(d) diphenyldiiodomethane into diphenylcarbene
(e) benzyl bromide into dibenzylmercury

8-6 2-Octanone has two sites for possible alkylation.

(a) Draw the structures of the two enolate anions that could be formed by treatment with sodium ethoxide, and decide which is favored thermodynamically.
(b) Upon treatment of 2-octanone with base, followed by methylbromide, the major product isolated is 3-nonanone. Explain.

8-7 The replacement of specific protons in a molecule by deuterium often simplifies the interpretation of an ^{1}H NMR spectrum. Suggest a way that the deuterated compounds in parts *a* through *h* could be prepared from any nondeuterated organic precursor, using D$_2$O as the source of deuterium. Describe the changes in the ^{1}H NMR spectrum that would be expected after the exchange.

(a) 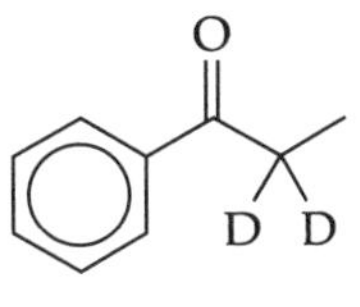 [ketone with D D structure]

(b) 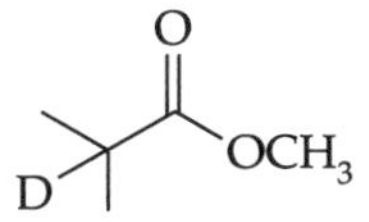 [ester with D, OCH$_3$ structure]

(c) [alkyne with D structure]

(d) [CH$_2$D structure]

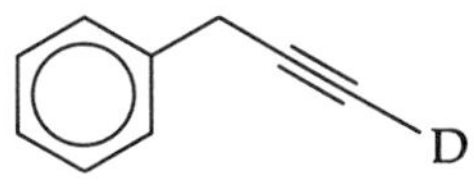

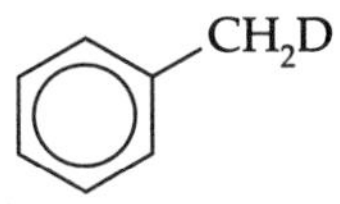

(e) 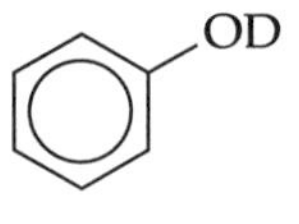 [structure: phenol with OD group]

(g) 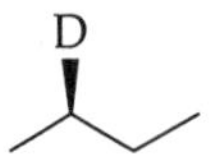 [structure: sec-butyl with D at stereocenter]

(f) 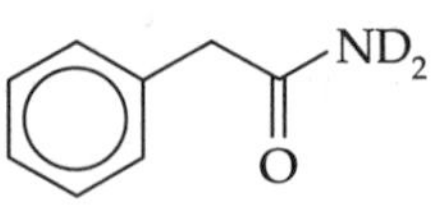 [structure: phenylacetamide with ND₂ group]

(h) 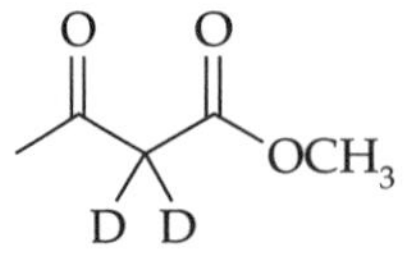 [structure: methyl β-ketoester with two D atoms, OCH₃]

8-8 Identify the organic halide that, after conversion into a Grignard reagent and treatment with ethylene oxide, would produce each of the following alcohols:

(a) 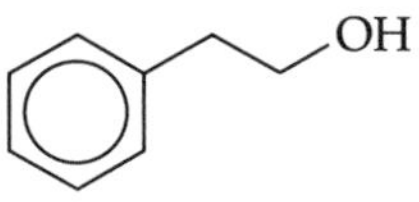 [structure: 2-phenylethanol, OH]

(b) 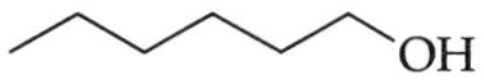 [structure: straight-chain alcohol, OH]

(c)  [structure: cyclohexyl-substituted alcohol, OH]

8-9 Suggest a method by which α functionality could be introduced into pentanoic acid to prepare the following compounds:

(a) α-bromopentanoic acid

(b) α-aminopentanoic acid

(c) α-cyanopentanoic acid

8-10 Using curved arrows, write a complete reaction mechanism for each of the following conversions. (You may omit, for now, details of the hydrolysis of the ester to the acid.)

(a) 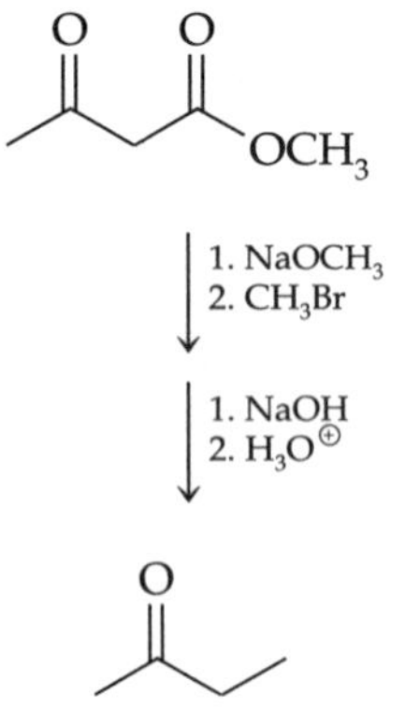 [structure: methyl acetoacetate, OCH₃]

1. NaOCH₃
2. CH₃Br

↓

1. NaOH
2. H₃O⊕

[structure: methyl ketone product]

(b) 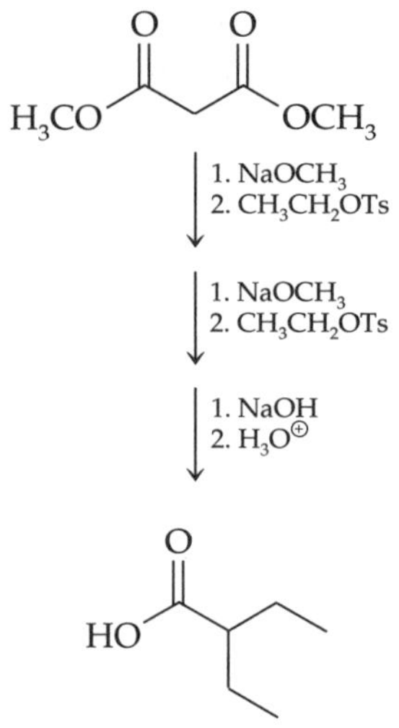 [structure: dimethyl malonate, H₃CO ... OCH₃]

1. NaOCH₃
2. CH₃CH₂OTs

↓

1. NaOCH₃
2. CH₃CH₂OTs

↓

1. NaOH
2. H₃O⊕

↓

[structure: 2-ethylbutanoic acid, HO]

8-11 Propose a reasonable synthetic route for each of the following molecules from acetoacetic ester or malonic ester. You may use any alkyl halide or tosylate.

(a) 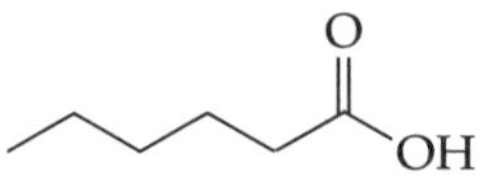 [structure: carboxylic acid, OH]

(b) 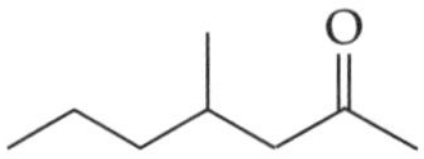 [structure: branched ketone]

(c) 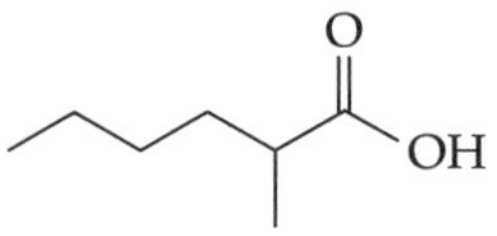 [structure: branched carboxylic acid, OH]

(d) 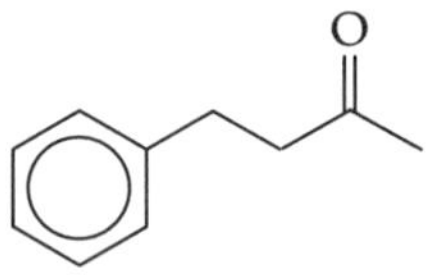 [structure: phenyl-substituted ketone]

Chapter 9

Elimination
Reactions

Most compounds that can undergo nucleophilic substitution (that is, those that bear a leaving group at an sp^3-hybridized atom) can also undergo elimination. In most elimination reactions, two groups on adjacent atoms are lost as a double bond is formed.

In a typical elimination, a substrate bearing two groups, A and B, on adjacent atoms undergoes cleavage of the two bonds connecting the carbon skeleton to A and B.

$$A \diagdown\diagup B \longrightarrow \diagdown\diagup \quad + \quad A^\oplus \quad + \quad B^\ominus$$
$$(\text{or } A \!-\! B)$$

Two of the four electrons from these σ bonds appear in the product as a π bond, and the remaining two electrons appear either as a covalent bond between the two fragments A—B or as an electron pair localized on A or B, producing a cation and an anion. At the same time, the carbon atoms to which the groups A and B were attached rehybridize from sp^3 to sp^2.

In most organic elimination reactions, one of the eliminated groups is hydrogen and the other is a leaving group like those that we have already encountered in substitution reactions.

$$H \diagdown\diagup L \longrightarrow \diagdown\diagup \quad + \quad H^\oplus \quad + \quad L^\ominus$$

Because hydrogen is more electropositive than most leaving groups (which usually have an electronegative atom at the point of attachment), an ionic cleavage of these two substituents produces the equivalent of H^+ and a negatively charged leaving group, L^-.

In this chapter, we will learn the mechanisms for several kinds of eliminations. We will find that certain structural features of the starting materials for these reactions lead to control of regiochemistry and stereochemistry in the product alkenes. In these cases, elimination reactions are valuable methods for the preparation of alkenes.

9-1 ▸ Mechanistic Options for Eliminations

We can conceive of three mechanisms for an elimination reaction that differ in the timing of cleavage of the two σ bonds: (1) first C–L, then C–H; (2) first C–H, then C–L; or (3) C–L and C–H simultaneously.

In the first option (Figure 9-1), the C–L bond is broken heterolytically to form a carbocation. (This is the same step that initiates the S_N1 reaction discussed in Chapters 7 and 8.) In a second step, this carbocation loses a proton from an adjacent carbon atom to form a π bond. The first step consists only of bond breaking, with no concomitant bond formation; whereas, in the second step, a C–H σ bond is cleaved at the same time that a C=C π bond is formed. Furthermore, deprotonation is usually assisted by the transfer of a proton to a base with the formation of a second covalent bond. As a result, the transition-state energy for the second step is lower than that of the first, and the loss of L^- is the rate-determining step. (Recall from Chapter 7 that formation of a carbocation by loss of the leaving group is the rate-determining step in S_N1 reactions as well.)

Whether an elimination or a substitution ultimately occurs depends on the relative rates of deprotonation and of nucleophilic attack on the carbocation. As a result, we should consider how reaction conditions influence the reactivity of the carbocationic intermediate so as to favor elimination or substitution. For example, elimination is favored by entropy because there are more chemical species in the product (two or three) than in the reactant (one), whereas entropy effects are usually negligible in substitution reactions, in which the same number of chemical species is usually present in the reactants and in the products. Entropy effects become increasingly important as the temperature is raised, making elimination competitive with substitution.

Notice that the rate-determining step of this pathway is unimolecular and endothermic, with a transition state closely resembling a carbocation. Therefore, using terminology similar to that for describing nucleophilic substitutions, we call the reaction shown in Figure 9-1 an **E1 reaction,** indicating that a unimolecular heterolytic breaking of the carbon–L σ bond (with the formation of a carbocation) is the slow step of the elimination. Because carbocations are formed, E1 reactions are favored in compounds in which the leaving group is at a tertiary or secondary position.

In the second option for the timing of cleavage in an ionic elimination, the first step consists of the removal of a proton, H^+, by a base, generating a

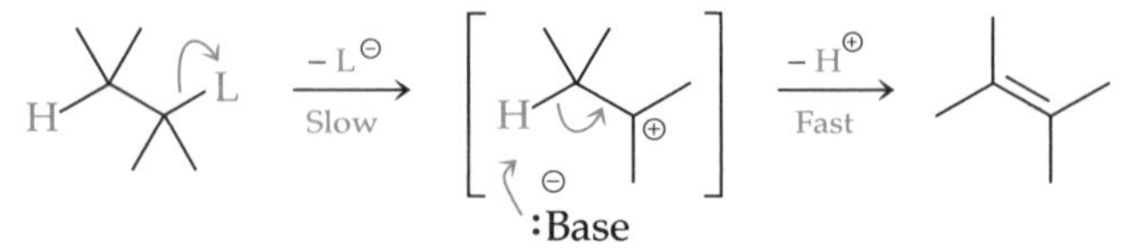

▲ FIGURE 9-1

In an E1 elimination, the rate-determining step is unimolecular loss of the leaving group, L^-, to form a carbocation. The elimination is completed by a second fast step in which a proton is lost. If a nucleophile attacks the cation faster than the proton is lost, nucleophilic substitution (by an S_N1 mechanism) is observed instead of elimination.

RECORDING INFORMATION THROUGH A COLOR-PRODUCING ELIMINATION REACTION

When you sign a credit card receipt on carbonless paper, you are inducing an acid-catalyzed elimination reaction. The carbonless paper contains small microcapsules (3–8 micrometers in diameter) that contain a colorless compound. When you apply pressure (by writing), you break the capsule and release the solution to the acid-treated paper. As in the acid-catalyzed eliminations discussed here, a bond between carbon and oxygen is broken, producing a highly colored (blue, purple, or black) cation. One such dye is shown here.

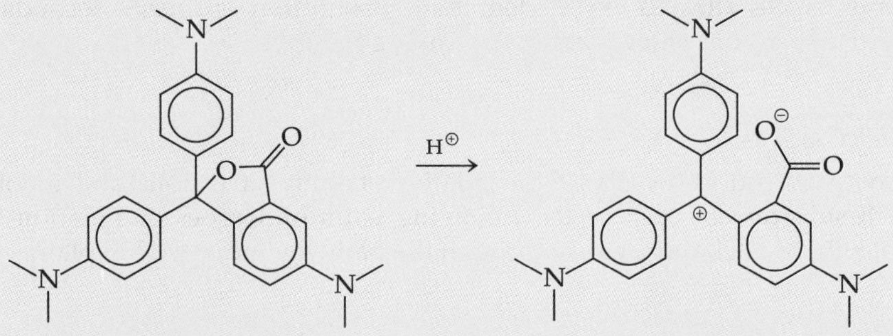

carbanion from which the leaving group is lost in a second step (Figure 9-2). Because deprotonation is fast and reversible, the reaction rate is controlled by how fast the leaving group is lost from this anionic intermediate. The loss of L^- from the anion (that is, the conjugate base) in the second, rate-determining step is unimolecular. This elimination reaction is called an **E1-CB reaction.** In the E1 mechanism, L^- is lost from the neutral substrate in the rate-determining step; whereas, in the E1-CB mechanism, L^- is lost from the anionic conjugate base of the neutral substrate in the rate-determining step. Because simple alkyl carbanions are very unstable, alkyl halides and alcohols that lack another functional group that would stabilize the anionic intermediate do not undergo elimination by this pathway.

In the third mechanism, two σ bonds are broken and a π bond is formed simultaneously. In this concerted pathway, deprotonation by base occurs at the same time that the C–L bond is broken (Figure 9-3). This elimination forms the conjugate acid of the base, a π bond between the carbon atoms, and a leaving group, L^-, within a single transition state. It is bimolecular because both the base and the organic reactant are participants in the transition state of the rate-determining step. This concerted elimination

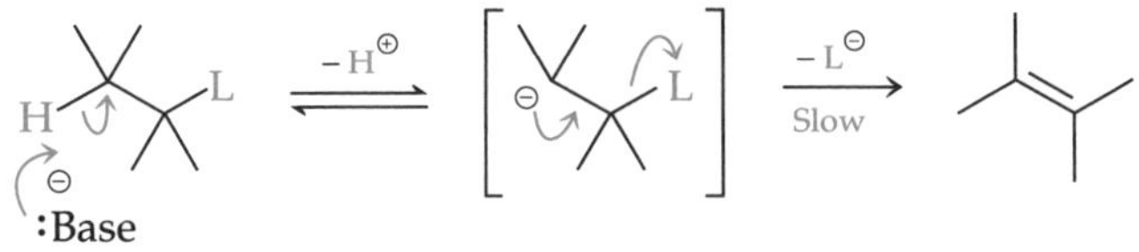

▲ FIGURE 9-2
In an E1-CB elimination, an acid-base pre-equilibrium deprotonates the neutral starting material to form its conjugate base. The loss of the leaving group, L^-, from the conjugate base in the second unimolecular step is rate determining.

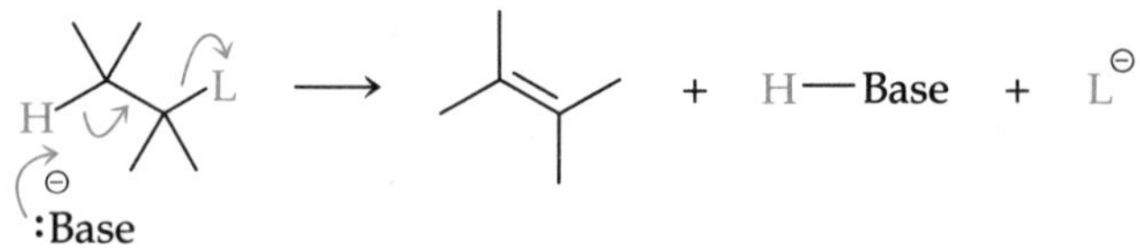

▲ FIGURE 9-3
In an E2 elimination, H^+ and L^- are lost at the same time through a concerted, bimolecular rate-determining step.

is therefore called an **E2 reaction.** The transition state of lowest energy in an E2 reaction has the orbitals that form the π bond in the product alkene aligned for maximal overlap. Unlike the E1 reaction, the E2 process does not proceed through an intermediate carbocation, and the rate of this reaction is less affected by the degree of substitution—primary, secondary, or tertiary—at the center bearing the leaving group, L.

EXERCISE 9-A

From what you know about the relative stability of cations and anions, which substrate in each of the following pairs undergoes elimination (of H_2O, HBr, or HCl) more easily through the pathway indicated? Explain.

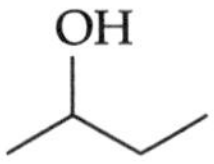

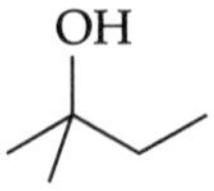

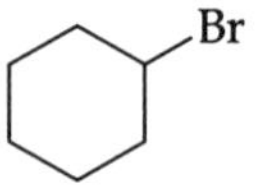

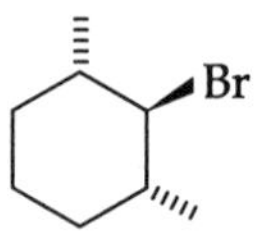

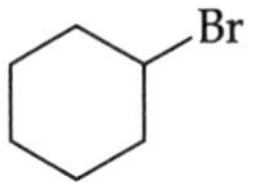

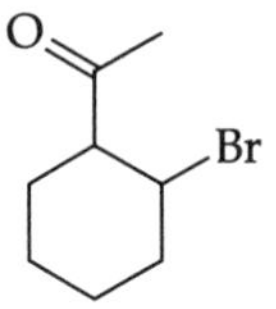

9-2 ▸ E1 versus E2 Elimination Reactions: Dehydrohalogenation of Alkyl Halides

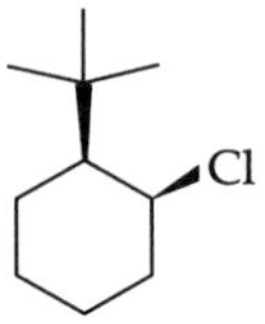

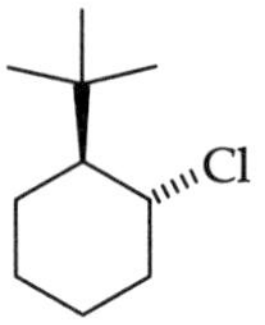

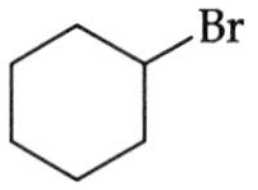

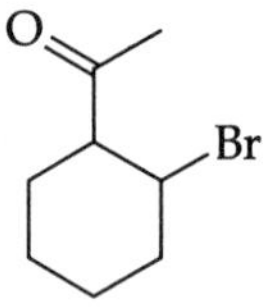

First, we will consider the elimination of HX from an alkyl halide (called **dehydrohalogenation**) by a step-by-step (E1) and a concerted (E2) pathway. We will see that the regiochemistry and stereochemistry of a dehydrohalogenation can be predicted by simple rules.

Regiochemistry

What factors facilitate E1 over E2 elimination and which isomeric alkene is formed when more than one is possible? That is, how does *regiochemistry* differ in these two reactions? First, we examine the E1 reaction, which takes place with the formation of a carbocation in the rate-determining step. The stability of this intermediate therefore influences the rate of elimination. Consider the elimination of HBr (called **dehydrobromination**) from 1-

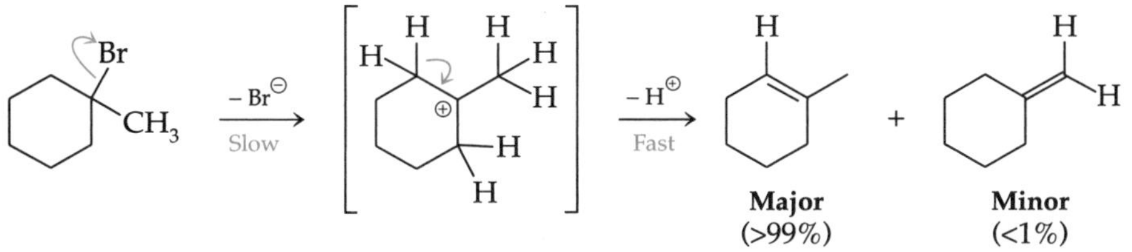

▲ FIGURE 9-4

A tertiary carbocation is formed in the rate-determining step in the E1 dehydrobromination of 1-bromo-1-methylcyclohexane. Fast deprotonation from either of the adjacent positions completes the reaction.

bromo-1-methylcyclohexane. Here, the loss of bromide produces a tertiary cation (Figure 9-4). This species is symmetrical, and loss of a proton from either of the adjacent methylene groups leads to the same product, 1-methyl-1-cyclohexene, in which the double bond is in the ring **(endocyclic).** On the other hand, loss of a proton from the methyl group produces methylenecyclohexane, in which the double bond is outside the ring **(exocyclic).** As stated in Chapter 2, the more highly substituted an alkene, the more stable it is. Thermodynamics favors the formation of the more-stable alkene—in this case, the trisubstituted alkene formed when the double bond is within the ring (endocyclic). Thus, the regiochemical preference for the formation of 1-methyl-1-cyclohexene in Figure 9-4 is dictated by product stability.

Another example is the bromide in Figure 9-5 in which the hydrogens on the ring carbon atoms adjacent to the carbocation formed in the first step are not equivalent. There are three possible products, one exocyclic and two endocyclic isomers. Again, the major product is the most stable: here, the most highly substituted alkene is 1,2-dimethylcyclohexene.

Similar considerations apply to acyclic compounds. For example, consider the case in which a leaving group such as Br$^-$ is attached to a secondary carbon and in which elimination can give two possible products, as in 2-bromobutane. In an E1 elimination, ionization gives rise to the 2-butyl cation (Figure 9-6, on page 308), a species with nonequivalent hydrogens on adjacent carbons. Loss of a proton from C-3 in this carbocation (path *a*) results in the formation of 2-butene as a mixture of *cis* and *trans* isomers. On the other hand, if a hydrogen on C-1 is lost (path *b*), 1-butene is the product. The three alkenes formed in paths *a* and *b* are not equally stable. The double bond in each of the 2-butene isomers bears two alkyl substituents, whereas that in 1-butene has only one. Therefore, path *a* should dominate over path *b* because the three transition states leading to these products have the same order of stability as the products. When the more-stable alkene is formed, the regioselectivity is said to be governed by **Zaitsev's Rule,** which predicts preferential formation of the thermodynamically more stable, more highly substituted regioisomer. The three examples shown in Figures 9-4 through 9-6 follow this rule.

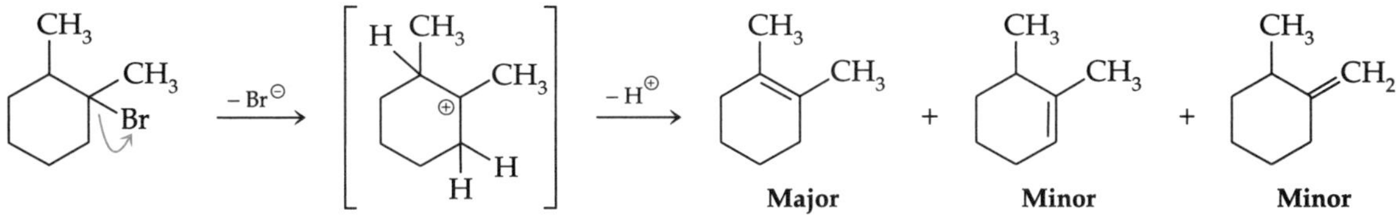

▲ FIGURE 9-5

The product distribution obtained in an E1 elimination is determined by the relative stability of the possible alkenes. The more highly substituted double bond is favored by thermodynamics.

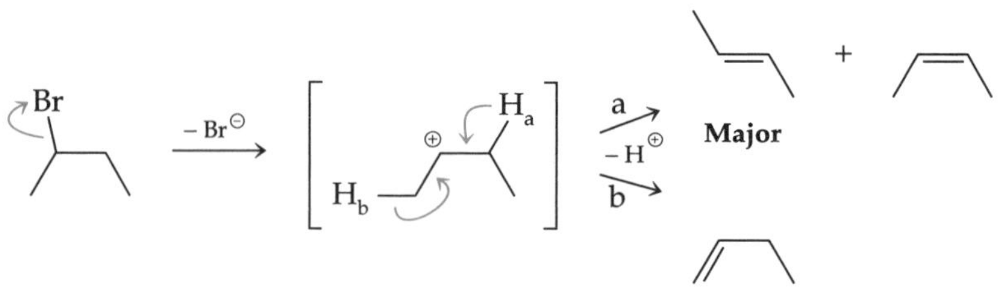

▲ FIGURE 9-6
The loss of H_a from the carbocation formed in the rate-determining step of the E1 dehydrobromination of 2-bromobutane gives 2-butene, as a mixture of *cis* and *trans* isomers. The loss of H_b from this same carbocation produces 1-butene. The most-stable product, *trans*-2-butene, is formed in highest chemical yield.

E_1 reactions proceed through cationic intermediates, and we learned in Chapter 3 that tertiary carbocations are more stable than secondary ones, which are in turn more stable than primary ones. When a more-stable cation can be produced by shifting a hydrogen or carbon atom from an adjacent atom, a rearrangement takes place rapidly. This property of carbocations affects the regiochemistry of E_1 eliminations. When bromide is lost, for example, from 2-bromo-3,4-dimethylpentane, a secondary carbocation is formed.

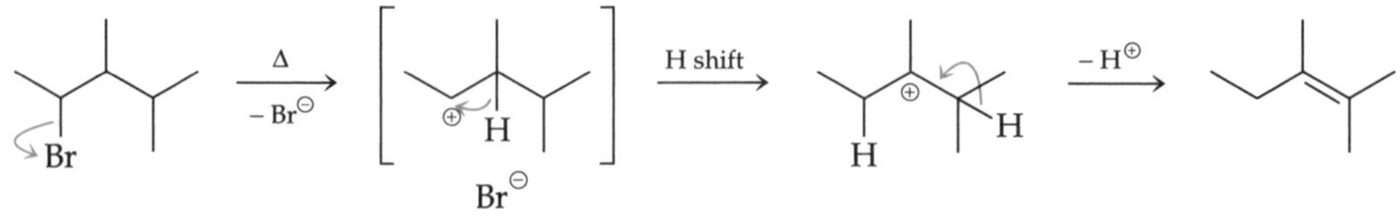

If a C-3 hydrogen migrates to C-2 with the electrons in the C-3–H bond, a more-stable tertiary carbocation is produced. Deprotonation of this cation takes place from C-4 to form the more-stable tetrasubstituted double bond. Because of this rearrangement, hydrogen is lost from an atom not initially adjacent to the carbon to which the leaving group was bound.

Regiochemistry can also be altered by shifting reaction conditions to favor a different mechanism. In the E2 mechanism (Figure 9-7), covalent bonds to the proton and leaving groups are broken simultaneously; so the approach of the base is a part of the rate-determining step. The stability of the product of an E2 reaction is an important factor in determining the rate, but it is not always the main determinant of the product regiochemistry,

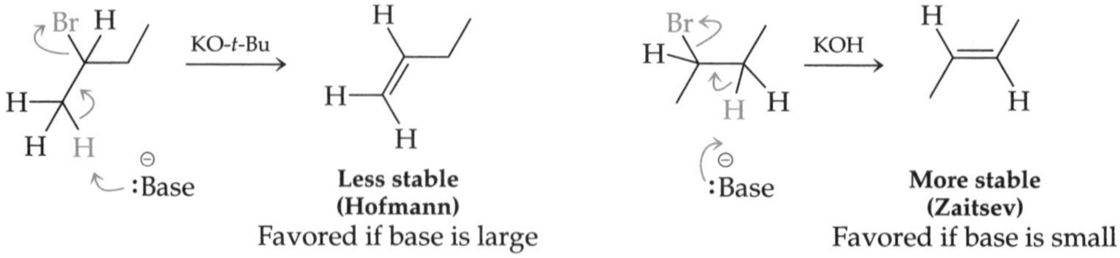

▲ FIGURE 9-7
The reaction at the left is an E2 elimination that produces mainly the Hofmann product (thermodynamically less stable but formed more rapidly), whereas the reaction at the right produces predominantly the Zaitsev (thermodynamically more stable) product. In the transition state at the left, the approach of the base is much less sterically encumbered at the primary site than it is at the secondary site in the transition state at the right.

and a less-stable product is sometimes formed more rapidly. The E2 reaction is not as exothermic as the second step of the E1 reaction, in which the unstable, cationic intermediate is converted into product. Therefore, there is a greater degree of interaction between the base and the hydrogen in the transition state of the E2 reaction, as well as greater sensitivity to the size of the base. For example, in the E2 reaction of 2-bromobutane, the base can approach a proton at either the C-1 or the C-3 position, with simultaneous C–H bond cleavage, π-bond formation, and elimination of bromide ion.

The formation of the less-substituted alkene in an elimination reaction is referred to as a **Hofmann elimination.** Notice, for example, that there are three hydrogens on C-1 in 2-bromobutane, but only two on C-3 (Figure 9-7). Statistically, therefore, it is more favorable to remove a proton from C-1 (to form 1-butene) than from C-3 (to form 2-butene). Furthermore, the steric congestion around the hydrogens on C-1 is lower than that for the hydrogens on C-3, and thus a base encounters lower steric repulsion in the transition state leading to 1-butene. This factor becomes important and can even dominate when very large bases are used. In such cases, the less-substituted (and less-stable) alkene dominates the product mixture, and the reaction is kinetically controlled. A reaction that gives the less-stable isomer as the major product is referred to as one that gives a Hofmann orientation. Hofmann elimination is encountered most frequently in E2 reactions; E1 eliminations usually give Zaitsev products.

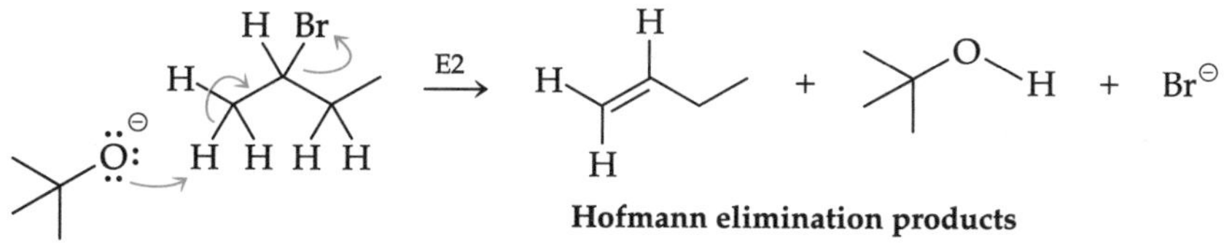

Hofmann elimination products

At the other extreme, a small base such as sodium hydroxide is less sensitive to steric interactions than a larger one such as potassium *t*-butoxide. With a small base, the thermodynamic stability of the product prevails in the transition state, and the two isomeric 2-butenes are formed preferentially. Thus, larger bases favor the Hofmann elimination product (1-butene), whereas the use of smaller bases results in the Zaitsev orientation (*cis*- and *trans*-2-butene).

EXERCISE 9–B

For each of the following elimination reactions, indicate whether the regiochemistry of the indicated elimination is of the Hofmann or Zaitsev type.

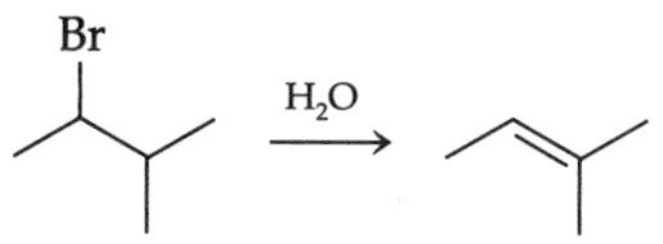

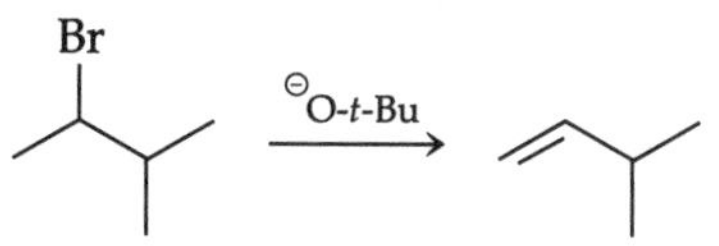

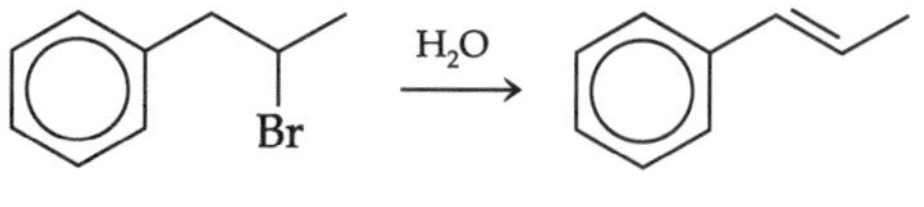

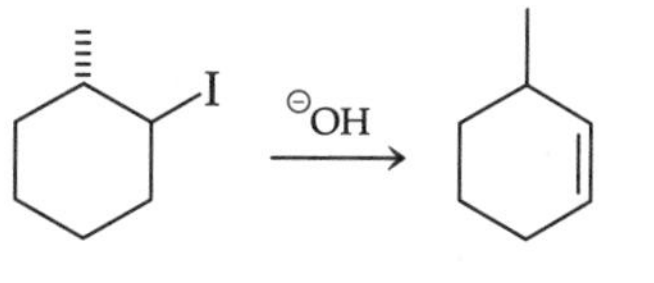

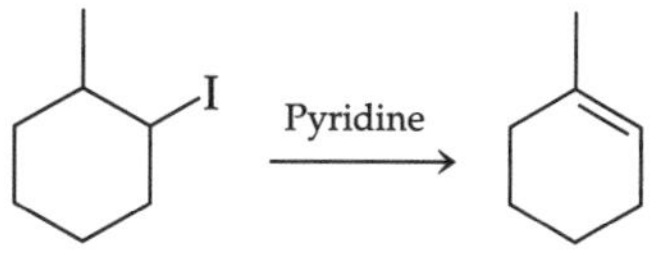

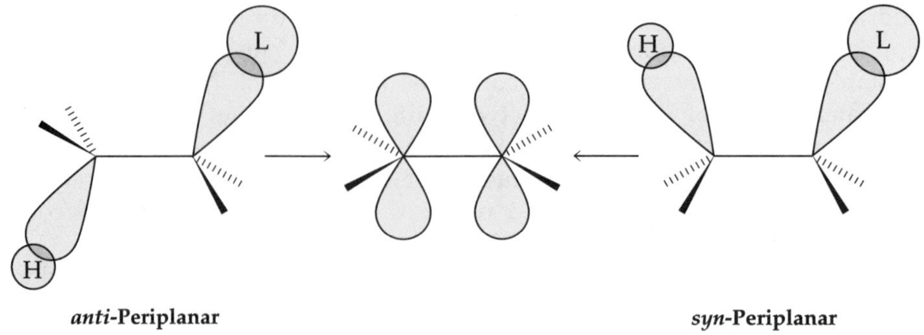

anti-**Periplanar**   syn-**Periplanar**

▲ **FIGURE 9-8**

In the *anti*-periplanar arrangement, the C–H and C–L σ bonds are in an *anti*, staggered conformation; whereas, in the *syn*-periplanar arrangement, these two bonds are eclipsed. In the vast majority of cases, eliminations take place through an *anti* periplanar transition state, in part because the *syn*-periplanar conformation is the less-stable (eclipsed) arrangement.

Another, even more significant factor also influences the regiochemical outcome of E2 reactions: namely, the π bond can be formed in the transition state of a concerted reaction only to the extent that the C–H and C–L σ bonds are coplanar in the reactant. This is because the π bond is formed by the overlap of the *p* orbitals formed as the C–H and C–L bonds are broken. This spatial relation can be achieved in only two ways, referred to as **anti-periplanar** and **syn-periplanar,** as shown in Figure 9-8.

Consider the dehydrobromination of *trans*-2-methylbromocyclohexane, an unsymmetrically substituted secondary alkyl halide.

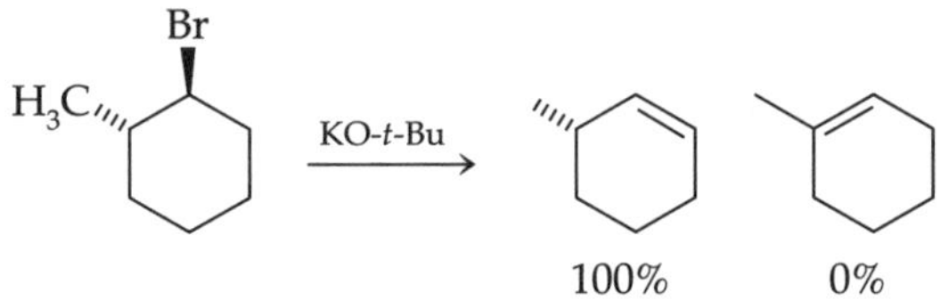

Experimentally, E2 elimination through an *anti*-periplanar transition state gives 3-methylcyclohexene, not 1-methylcyclohexene, as the only product. To define the factors responsible for this regiocontrol, we must consider the relative orientation of the groups, H and L, taking part in the reaction. Let us first consider the starting material in a chair conformation with the bromine placed at an axial position, as shown in the three-dimensional representations in Figure 9-9. From an examination of a Newman projection viewed along the bond from C-1 to C-2, we see that no elimination is possible because the C–Br bond is coplanar with the C–CH₃ bond—not with the C–H bond, as is required for the E2 elimination transition state. Therefore, elimination cannot take place so as to produce a π bond between C-1 and C-2.

Let us next consider the Newman projection viewed along the bond from C-6 to C-1 when bromine is in the axial position. The C–Br bond is now coplanar with the axial hydrogen on C-6. Elimination is possible in this *anti*-periplanar arrangement, giving rise to the formation of 3-methylcyclohexene.

Now consider the situation with bromine in an equatorial position (recall from Chapter 5 that axial and equatorial conformers can be interconverted by a ring flip). We find (Figure 9-10) that there are no C–H bonds in proper orbital alignment for concerted elimination (coplanar with the C–Br

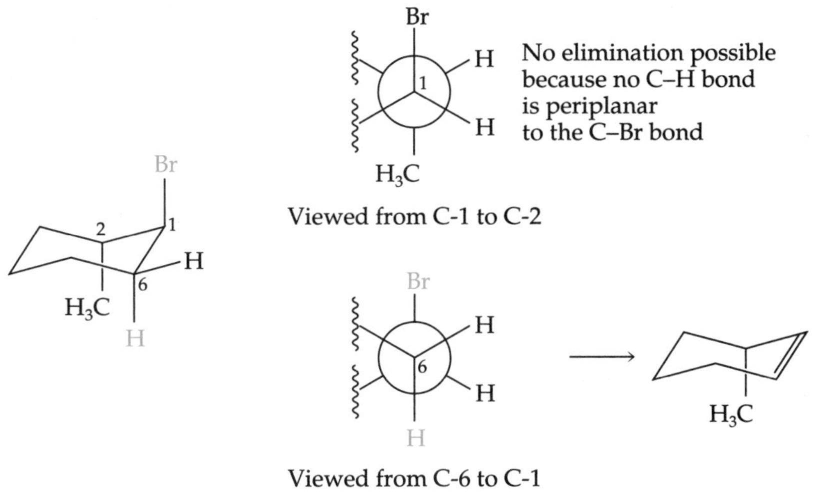

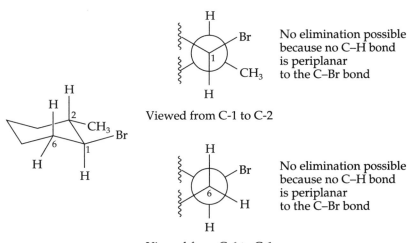

▲ FIGURE 9-9

In the chair conformer shown at the left, bromine is *anti*-periplanar with a hydrogen at C-6 but is *gauche* to a hydrogen at C-2. Concerted elimination can only produce a double bond between C-1 and C-6, yielding 3-methylcyclohexene.

bond) on either C-2 or C-6, and elimination cannot occur from this conformation because E2 elimination specifically requires a periplanar transition state. This requirement for orbital alignment in a periplanar transition state is an example of **stereoelectronic control** of regiochemistry. Even though the conformer at the bottom of Figure 9-9 constitutes only a small fraction of the equilibrium mixture, it is the only conformer from which elimination can occur because it is the only conformer in which the alignment of the C–H and C–Br bonds is *anti*-periplanar.

▲ FIGURE 9-10

A ring flip of the chair conformation shown in Figure 9-9 produces the conformation shown at the left. When bromine is in an equatorial position, it is *anti*-periplanar to the C–C ring bonds rather than to any C–H bond. Elimination is therefore impossible from this ring-flipped conformer, even though it is the more-stable conformer and is present in highest abundance at equilibrium.

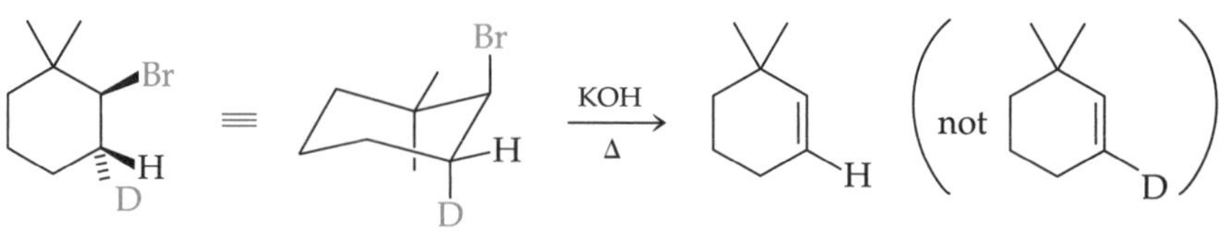

▲ FIGURE 9-11
DBr (rather than HBr) is lost because deuterium is *anti*-periplanar to bromine
in the ring-flipped conformer that places bromine in an axial position.

The requirement for a periplanar alignment of bonds in the transition
state for an E2 elimination reaction can be demonstrated in a more-subtle
way by looking for a preference for *cis* or *trans* elimination in an isotopically
labeled substrate. In 2,2-dimethyl-*trans*-6-deuterocyclohexyl bromide, the
elimination must take place through the formation of a π bond between C-1
and C-6 because of the presence of two methyl groups at C-2. Thus, we can
look for elimination of either HBr or DBr (Figure 9-11). The isolated product
obtained upon treatment with strong base is that resulting from the loss of
DBr, not HBr. The deuterium located *trans* to bromine in the starting material can assume an *anti*-periplanar alignment suitable for the E2 elimination,
whereas this is not possible for the *cis* hydrogen. That is, the requirement
for a periplanar arrangement in the transition state means that the groups
lost must be *trans*, not *cis*, to each other.

In summary, elimination reactions that take place through an E2 mechanism proceed through a transition state in which the carbon–hydrogen and
carbon–leaving-group bonds are in a periplanar arrangement. For cyclohexane derivatives, only the *anti*-periplanar arrangement can be achieved
readily. For the bonds to hydrogen and the leaving group to have this
arrangement, both groups must be axial.

EXERCISE 9-C

Predict the product expected from an E2 elimination through an *anti*-
periplanar transition state from each of the following substrates.

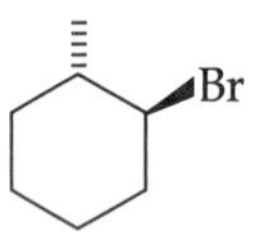

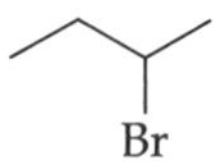

Leaving Groups

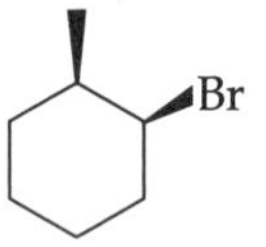

How is the efficiency of heterolytic fission in an E1 or an E2 elimination influenced by the identity of the leaving group? In an E1 mechanism, the C–L
bond is broken in the rate-determining step. Clearly, the weaker the C–L
bond, the easier the bond cleavage, because less energy is needed to reach
the transition state in which this bond is substantially broken. A comparison of the facility for ionization of a series of *t*-butyl halides (Figure 9-12) reveals that the rate decreases with increasing strength of the carbon–halogen
bond in the series: R—I (51 kcal/mole) > R—Br (68 kcal/mole) > R—Cl
(82 kcal/mole). Because the same carbocation is generated in each case, the
order of reactivity must result from differences in the halide. Of significance
here, in addition to the bond strength, is the stability of the halide anion
generated in the reaction.

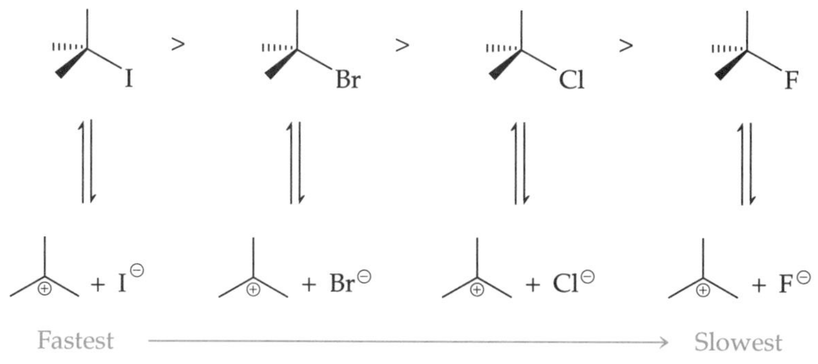

Fastest ————————————————————————→ Slowest

▲ FIGURE 9-12

The rate of cleavage of a C–X bond decreases in the progression from the bottom to the top of the column of halogens in the periodic table. This trend is parallel to the progression of bond-dissociation energies for these bonds.

Recall from Chapter 6 that HF is the weakest halogen acid (HX) and HI the strongest and that the H–F bond is the strongest and the H–I bond is the weakest. The effect of the stabilities of the anions on the rates of cleavage of alkyl halides is analogous to their effect on the acidity of the acid: HI is the most acidic of the halogen acids, and alkyl iodides are the most reactive of the alkyl halides. The tendency of a given halogen to act as an effective leaving group (departing with the two electrons that initially constitute the σ bond) is thus roughly inversely related to the basicity of the ion formed: the weaker the base (and the stronger the conjugate acid HX), the better the leaving group.

In an E2 elimination, both the C–H and the C–L bonds are broken in the rate-determining step. Because the C–L bond is broken in the key transition state of both the E1 and the E2 reactions, C–L bond strength affects the rates of both cleavages in roughly the same way.

Nucleophiles

Recall from Chapter 6 that the affinity of an anion (or a nucleophile) for a carbocation (or other electrophile) is referred to as nucleophilicity. Because basicity is a measure of the affinity of an anion or a nucleophile for a proton, it is not surprising that these two properties, nucleophilicity and basicity, are often parallel. In general, with the same nucleophilic atom, the stronger the base, the better the nucleophile and the poorer the leaving group.

There is one important distinction between nucleophilicity and basicity: the way in which the terms are usually employed. In general, nucleophilicity refers to relative rates of reactions, whereas basicity (and acidity) refers to relative thermodynamic stabilities. For example, although *t*-butoxide is a stronger ($>10^3$ times) base than methoxide, the latter is generally a better nucleophile because it is less sterically hindered. Because a proton is very small, basicity is relatively unaffected by steric interactions. On the other hand, as we have seen in Chapters 7 and 8, steric hindrance to reaching a transition state greatly affects the rate of a bimolecular nucleophilic substitution (S_N2) reaction.

Charge-intensive reagents that have electron density or negative charge localized on a single atom generally have a higher basicity-to-nucleophilicity ratio than do charge-dispersed reagents. Charge-intensive reagents are sometimes called **hard;** charge-delocalized reagents are sometimes called **soft.** Hard bases are often stronger bases than soft ones.

Choose the member of each of the following pairs of compounds that is likely to be the stronger base. Which would be the better nucleophile?

(a) $^{\ominus}NH_2$ or NH_3

(b) $^{\ominus}OH$ or H_2O

(c) $^{\ominus}OH$ or $^{\ominus}SH$

(d) 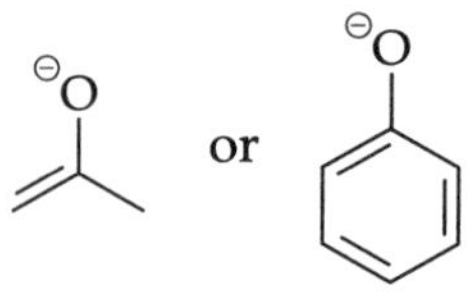 [structure] or [structure]

(e) 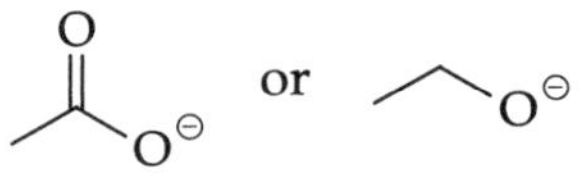 [structure] or [structure]

Stereochemistry

The preference for one geometric isomer over another is influenced by the mechanism of elimination. Again, let us first consider the E1 reaction. Figure 9-13 illustrates Newman projections of the carbocation formed by loss of bromide ion from 2-bromobutane. These representations are obtained by viewing down the bond from C-2 to C-3. As a carbocation, C-2 is in the foreground with its σ bonds orthogonal to the vacant p orbital. We hold C-2 in this fixed position while we rotate the substituents on the sp^3-hybridized C-3 at the other end of the σ bond between C-2 and C-3. To form a π bond, the C–H bond that is to be broken must be aligned with the vacant p orbital of the carbocation, because only in this way can electron density flow into the developing π bond as C-3 rehybridizes from sp^3 to sp^2. In the structure at the upper left in Figure 9-13, elimination by loss of H_a puts the methyl groups on the same side of the double bond, producing *cis*-2-butene.

We can bring a different hydrogen atom into coplanarity with the empty p orbital by rotating the rear carbon counterclockwise by 60°, bringing the methyl group originally at four o'clock to the two o'clock position. Cleavage of the C–H_b bond also gives *cis*-2-butene, with the two methyl

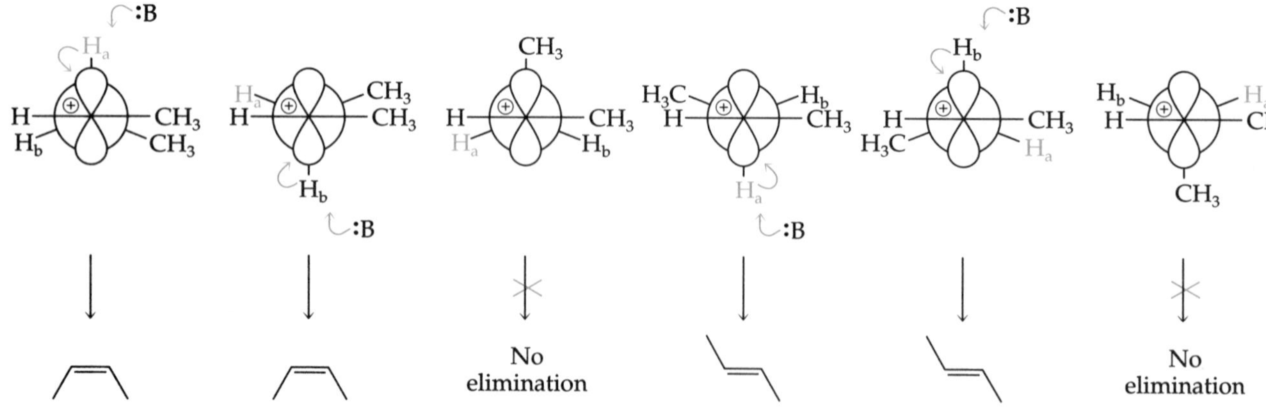

▲ FIGURE 9-13

Shown here are Newman projections of the 2-butyl cation viewed down the C-2–C-3 bond. With C-2 held steady, C-3 is rotated counterclockwise by 60° to obtain the conformers shown. Only in the first, second, fourth, and fifth conformers is a C–H bond aligned with the p orbital of the carbocation. Only in these conformers can elimination take place.

groups again on the same side of the double bond. If we continue our counterclockwise rotation by another 60°, the C–CH$_3$ bond becomes coplanar with the vacant p orbital and elimination is not possible because no C–H bond is properly aligned with the p orbital. Rotation by another 60°, however, again affords a conformer in which the C–H$_a$ bond is aligned with the vacant p orbital. In this arrangement, the methyl groups are located on opposite sides of the incipient π bond, and deprotonation by an attacking base produces *trans*-2-butene. We have now rotated through 180°. If we were to continue the rotation by another 180°, returning to the starting arrangement, we would find an analogous set of isomers.

Let us now compare those arrangements that lead respectively to *cis*-2-butene and to *trans*-2-butene (Figure 9-13). These two sets of conformers are *not* of equal energy. Those at the left, leading to *cis*-2-butene, have a *gauche* interaction between C-1 and C-4 that is absent from the conformers at the right that lead to *trans*-2-butene. These latter conformers therefore dominate the conformational equilibrium. The same steric interactions that influence the conformational distribution of the cation also affect the transition states for elimination: therefore, the major product (*trans*-2-butene) is that derived from the more-stable cation conformers.

EXERCISE 9-E

In each structure below, indicate which hydrogens are (or can be) *anti*-periplanar to the bromine leaving group.

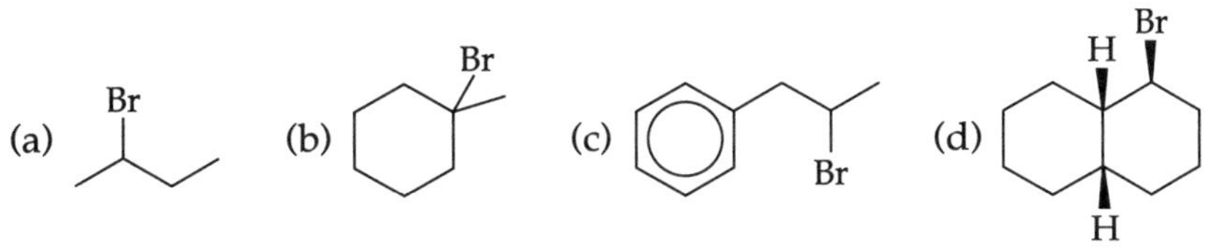

Stereochemistry is also a factor in the E2 reaction. Only two staggered arrangements of 2-bromobutane meet the requirement discussed earlier for an *anti*-periplanar arrangement in the transition state of the concerted elimination. These conformers are shown in Figure 9-14; again, a difference in conformer stability is revealed in the transition-state energies and accordingly in the observed product distribution. Because of the destabilizing *gauche* interaction in the upper conformer, the lower conformer is favored: this conformer leads to the formation of *trans*-2-butene.

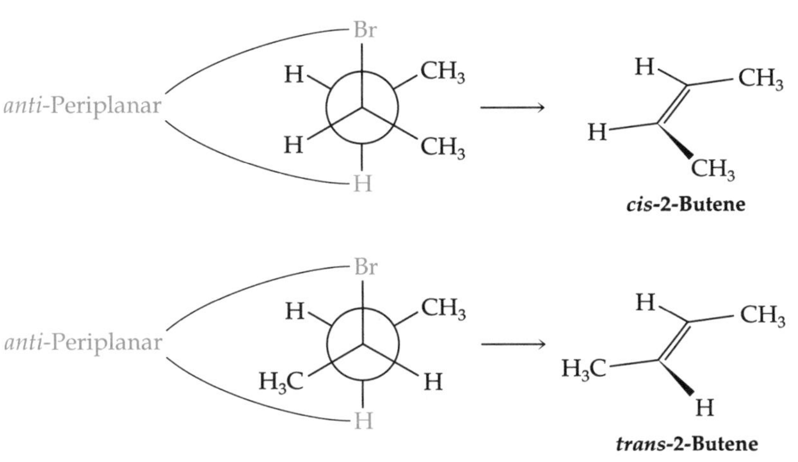

◄ FIGURE 9-14

In the upper, *gauche* conformer, elimination of HBr through an *anti*-periplanar transition state produces *cis*-2-butene, whereas in the lower, *anti* conformer, this same mode of elimination leads to *trans*-2-butene.

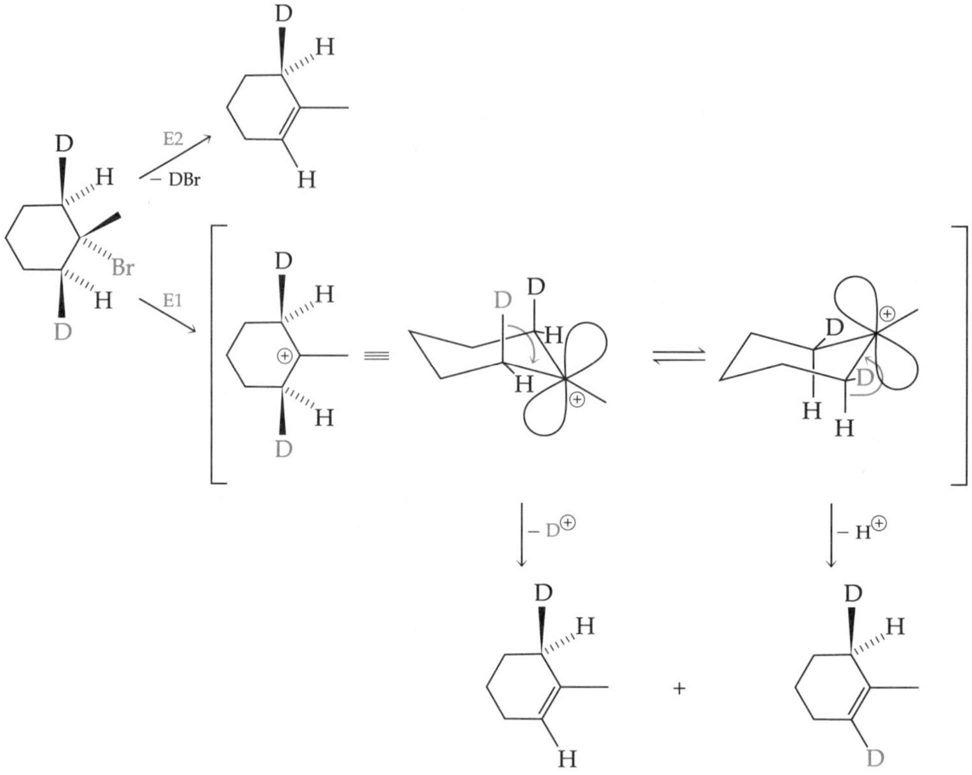

▲ FIGURE 9-15

The E2 elimination is stereospecific, leading to exclusive loss of DBr through an *anti*-periplanar transition state. The E1 elimination is not stereospecific, and two products are obtained: one by loss of HBr and the other by loss of DBr. The vacant *p* orbital of the carbocation formed as an intermediate in the E1 elimination can align with either the C–H or the C–D σ bond.

The requirement for an *anti*-periplanar transition state makes possible a stereoselectivity in E2 eliminations that cannot be achieved readily under E1 conditions. Suppose that 1-methylcyclohexyl bromide were labeled stereospecifically with deuterium in positions 2 and 6 (Figure 9-15). With the stereochemistry for the methyl and bromine groups shown, an E2 elimination gives *trans* product exclusively, through an *anti*-periplanar transition state, with the loss of D–Br to give 1-methylcyclohexane. In the E1 pathway, however, a tertiary cation is formed. Because the cation is sp^2-hybridized and planar, each of the two possible chair conformers provides for coplanar alignment of the vacant *p* orbital with either the C–D or C–H bonds so that loss of either hydrogen or deuterium occurs with nearly equal ease. (One would not expect exactly equal amounts of these two products because of a small isotope effect that makes the C–D bond slightly stronger.)

9-3 ▸ Elimination of HX from Vinyl Halides

The elimination of HX from a vinyl halide produces an alkyne. As we did for eliminations at sp^3-hybridized atoms (Section 9-1), we can write the three pathways shown in Figure 9-16 for eliminations at sp^2-hybridized atoms. In the E1 route, a halide ion is lost and a vinyl cation is formed. If this divalent cation is sp-hybridized, it is particularly unstable because it has access to only six valence electrons, of which four are accommodated in C–R and C–C sp-hybridized σ bonds and the remaining two are held in a *p*

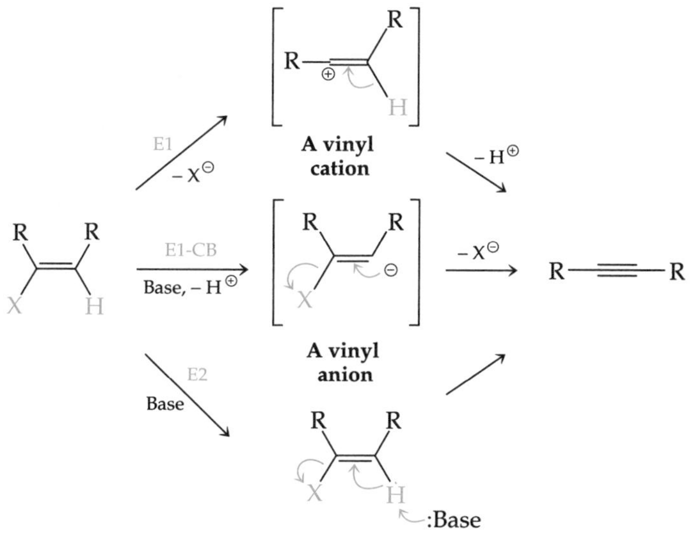

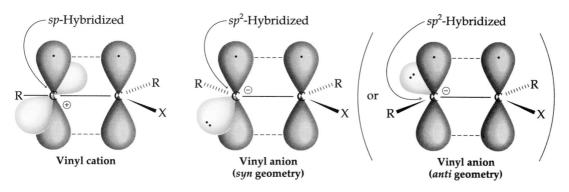

▲ FIGURE 9-16

Eliminations at the sp^2-hybridized carbon atom of a vinyl halide can proceed through three possible mechanisms: (upper) an E1 mechanism through an intermediate vinyl cation; (middle), an E1-CB mechanism through a vinyl anion, or (lower) a concerted loss of HX in an E2 mechanism.

orbital, far from the nucleus (Figure 9-17). An E1 elimination is therefore more difficult in a vinyl halide than in an alkyl halide and is only rarely observed.

In the E1-CB route (Figure 9-16), deprotonation occurs first, generating a vinyl anion. In the vinyl anion, shown in its two geometric isomers in Figure 9-17, the anionic carbon is sp^2-hybridized, bearing two hybrid σ bonds, a lone pair in a hybrid orbital, and a π bond. This atom has eight valence electrons and a lone pair in an orbital having more *s* character than in an sp^3-hybridized anion. The vinyl anion is therefore more stable than a tetrahedral sp^3-hybridized carbanion. The unshared electron pair in the vinyl anion is held in an sp^2-hybridized orbital and is therefore closer to the nucleus than in the sp^3-hybridized anions considered earlier in this chapter. As a result, vinyl anions are unusually stable (Chapter 6), as is indicated by the acidity of the vinyl C–H bond (pK_a 44). Thus, initial deprotonation is more

sp-Hybridized	sp^2-Hybridized	sp^2-Hybridized
Vinyl cation	**Vinyl anion** (*syn* geometry)	**Vinyl anion** (*anti* geometry)

▲ FIGURE 9-17

In the vinyl cation shown at the left, the cationic carbon is sp-hybridized, bearing two hybrid σ bonds, a π bond, and a vacant *p* orbital, and the electrons are accordingly held at a position further from the nucleus. The vinyl cation is therefore less stable than a trigonal sp^2-hybridized carbocation. For these same reasons, an sp^2-hybridized vinyl anion (shown with the lone pair *syn* to the leaving group at the center and *anti* at the right) is more stable than the sp^3-hybridized anion produced in an E1-CB reaction of an alkyl halide.

favorable in this elimination than at an sp^3-hybridized atom (Figure 9-2). The lone pair in the hybrid orbital is orthogonal to the π system and coplanar with the C–X σ bond. This is true both for the isomer in which the lone pair is *cis* to the halogen leaving group (as shown at the center of Figure 9-17) and for the *trans* arrangement (at the right).

The lone pair of electrons in the anion is coplanar with the leaving group, X, and the next step (completing the elimination) is often rate determining. At the extreme, the loss of X is concerted with deprotonation, giving rise to an E2 elimination (Figure 9-16). Very strong bases, such as sodium amide (NaNH$_2$), are generally employed for this purpose.

The product alkyne has no stereochemical features, and so it is difficult to find experimental details that can help us decide which mechanism is the correct one. Nonetheless, because the vinyl cation is so unstable, the E1-CB and concerted E2 eliminations are much more likely mechanisms for these eliminations than is the E1 route.

Vinyl halides themselves can be prepared by elimination of HX from geminal or vicinal dihalides, bearing two halogens either on the same carbon atom or on adjacent ones (Figure 9-18). When alkoxide bases are used, it is possible to stop with the loss of one equivalent of HX. With a stronger base such as sodium amide (NaNH$_2$), a second equivalent is lost and an alkyne is produced.

Vicinal dibromides are prepared by treating alkenes with Br$_2$ in CCl$_4$ in a reaction that will be discussed in more detail in Chapter 10. The loss of two equivalents of HBr from the resulting vicinal dibromide constitutes a method by which an alkene can be converted into an alkyne.

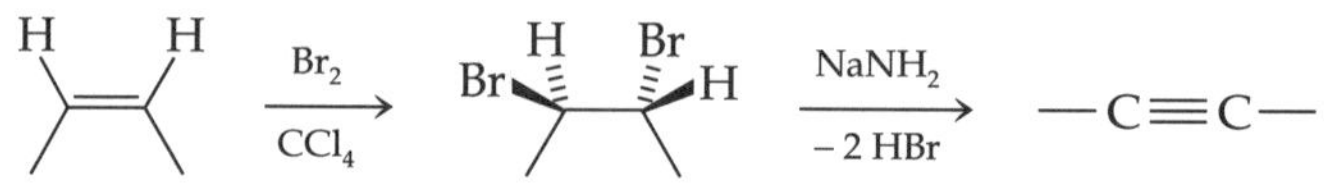

EXERCISE 9-F

In the base-catalyzed elimination induced by treating 2-bromo-(Z)-2-butene with NaNH$_2$, 1-butyne is formed. Suggest a mechanism for this transformation, and give reasons for the observed regiochemistry.

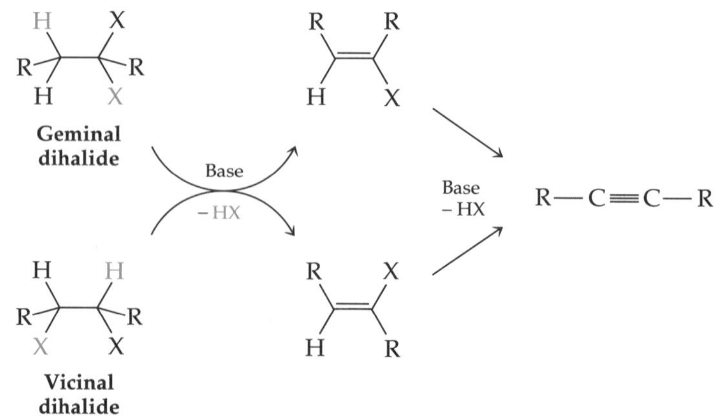

▲ FIGURE 9-18

Upon treatment with strong base, dehydrohalogenation can occur twice to produce an alkyne from either a geminal dihalide or a vicinal dihalide. If a vinyl halide is desired, a weaker base such as an alkoxide is usually employed.

FIGURE 9-19

Of the three possible mechanisms for the dehydrohalogenation of an aryl halide, the E1 route proceeding through a phenyl cation is unlikely because of the great instability of this intermediate. As with vinyl halides, elimination through a phenyl anion (in an E1-CB pathway) or through a concerted E2 pathway is much more likely.

9-4 ▸ Elimination of HX from Aryl Halides

Let us focus on the elimination of HBr from bromobenzene (Figure 9-19). An E1 mechanism would proceed with the loss of bromide ion to form a phenyl cation, a species destabilized by the hybridization effects that destabilize the vinyl cation shown in Figure 9-17. However, the carbon of a phenyl cation is constrained within a six-membered ring, and so this intermediate would be even further destabilized by angle strain. The phenyl cation is therefore even less stable than a vinyl cation, and is an even less likely intermediate. Indeed, no direct evidence has been obtained for the formation of a phenyl cation in the course of an elimination.

The phenyl anion formed by deprotonation of bromobenzene has the same hybridization as the vinyl anion. The sp^2-hybridized orbital containing the lone pair of the phenyl anion formed from bromobenzene is coplanar with the carbon–bromine σ bond, and elimination in the next step occurs easily. Whether this elimination takes place in two steps or in one concerted step, as in the concerted E2 reaction shown in Figure 9-19, is unclear, but the same product, **benzyne,** is formed by either route.

The structure of benzyne is interesting because it contains a formal triple bond within a six-membered ring. The orbitals that form the second π bond generated in the elimination reaction must be perpendicular to the π system of the aromatic ring, as in all alkynes (Chapter 2). In benzyne, however, the preferred angle for *sp*-hybridization (180°) is not possible because it would require four of the six atoms of the ring to be arranged in a colinear fashion. Angle distortion therefore makes the triple bond in benzyne much less stable, and thus more reactive, than one in a simple acyclic alkyne. As a result, benzyne undergoes addition reactions very readily. For example, the addition of water or ammonia to form phenol or aniline, respectively, occurs very rapidly.

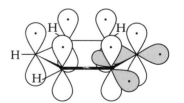

Benzyne

The *p* orbitals in color overlap to form an additional π bond

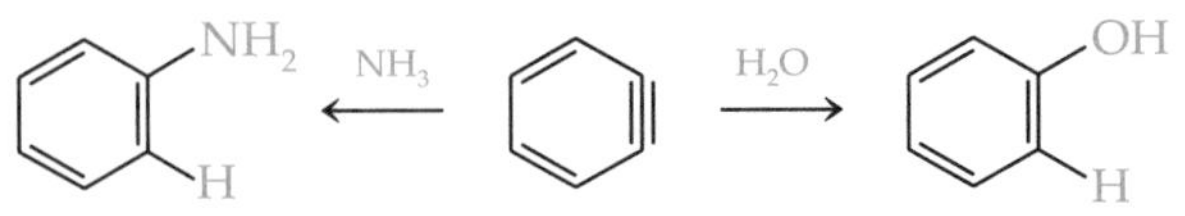

The mechanisms of addition reactions will be treated in greater detail in Chapter 10. For now, simply recognize that the elimination reaction to form benzyne followed by an addition reaction constitutes a route by which an aryl halide bond can be replaced by an aryl–OH or aryl–NR$_2$ bond in a net substitution.

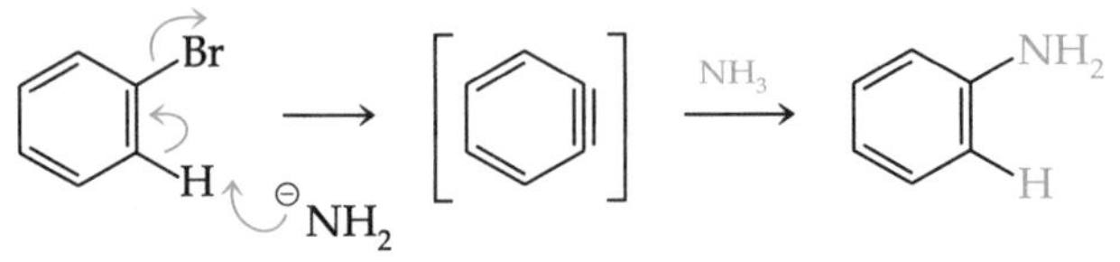

Because this substitution takes place through an elimination-addition sequence, the simple classification of the reaction as a substitution is mechanistically inadequate.

EXERCISE 9-G

Write a mechanism for the formation of phenol from iodobenzene upon treatment with concentrated KOH.

9-5 ▸ Dehydration of Alcohols

The elimination of water (**dehydration**) from alcohols takes place under acidic conditions and thus differs from the elimination of hydrogen halides. As we learned in Chapters 3 and 7, dehydration is facilitated by acid because protonation of the hydroxyl group effectively converts this leaving group from hydroxide ion into water. Because H$_3$O$^+$ is a stronger acid than H$_2$O, its conjugate base (H$_2$O) is a better leaving group than $^-$OH ion. Whenever a stable carbocation is produced by dehydration of a protonated alcohol, an E1 elimination can occur.

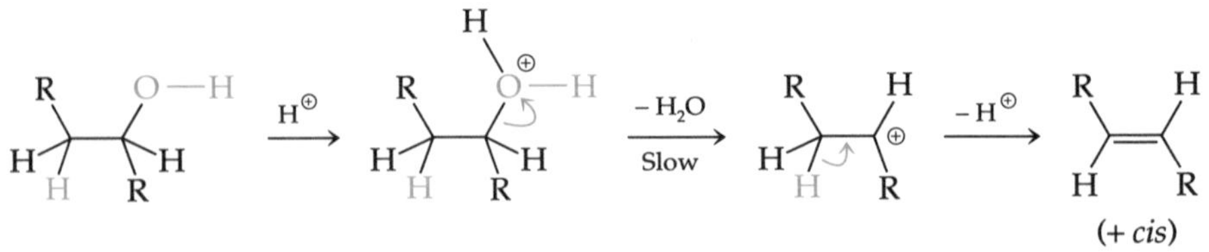

Because an unstable primary carbocation would be formed in the E1 dehydration of a primary alcohol, the acid-catalyzed E1 elimination through a carbocation is difficult. Instead, a proton is lost from an adjacent carbon at the same time as water is lost. This allows for the formation of an alkene without the intermediate formation of an unstable carbocation.

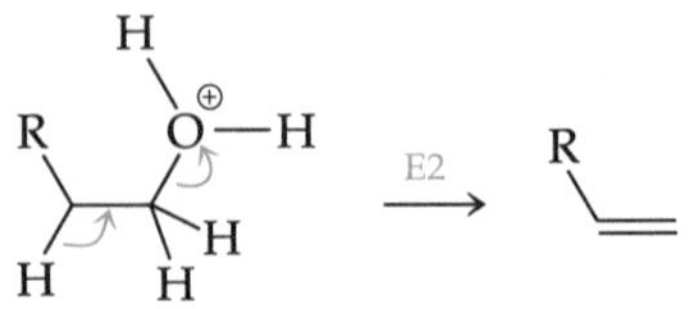

Dehydration is particularly easy when a conjugated double bond is formed. For example, an alcohol that bears a carbonyl group two carbons away (a β-hydroxyaldehyde or -ketone) readily undergoes dehydration with formation of an α,β-unsaturated carbonyl compound.

**An E$_2$ Dehydration
in Base: Not Observed**

A direct E2 elimination of hydroxide ion under basic conditions, as shown in the margin, would require a significantly higher activation energy and has not been observed with simple alcohols because hydroxide ion (or its deprotonated form, O^{2-}) is such a poor leaving group. Only acid-catalyzed dehydrations take place with reasonable chemical efficiency.

EXERCISE 9-H

For the acid-catalyzed dehydration of cyclohexanol, use curved arrows to indicate electron movement in the E1 and E2 mechanisms. Show the sequence of any intermediates. What would the rate-limiting step be in each sequence? Why doesn't cyclohexanol undergo elimination following an E2 pathway under basic conditions?

In conclusion, acid-catalyzed dehydration of secondary and tertiary alcohols is usually accomplished by an E1 pathway and proceeds through a carbocation intermediate. This cation has two reactivity options (S_N1 substitution and rearrangement) in competition with simple elimination. Dehydration of primary alcohols takes place by E2 elimination from the protonated alcohol. Simple alcohols do not undergo base-catalyzed dehydration because OH^- is a much poorer leaving group than is H_2O.

9-6 ▸ Elimination of X_2

Elimination reactions are not restricted to loss of HX or H—OH. Elimination of two electronegative atoms on adjacent atoms (for example, a vicinal dihalide) can also be accomplished by treatment with an active metal. The formation of an alkene is often accompanied by the formation of a halide salt of the metal, as illustrated in the margin. Here, the net conversion includes the oxidation of zinc and the reduction of the two carbons of the organic dihalide. Because the metal is converted from its zero valent (metallic) state into a +2 ion, the metals most useful for this conversion are those that can easily support a +2 change in oxidation level. Indeed, zinc and tin are most often used to effect these conversions. This net formation of an alkene from a vicinal dibromide frequently occurs through an organozinc intermediate (Chapter 8) in which the metal has been inserted into the carbon–bromine bond.

9-7 ▸ Oxidation of Alcohols

The oxidations of primary alcohols to aldehydes and of secondary alcohols to ketones may be formally considered to be eliminations of dihydrogen.

Often these reactions are accomplished by treatment with metal ions that are readily reduced. For example, oxidations with chromate ion (Cr^{6+}) are commonly used to prepare carbonyl compounds from the corresponding alcohols. Often, metals in relatively high oxidation states exist in a number of equilibrating ionic forms in solution. For example, chromium trioxide exists in equilibrium with chromic acid and alkali dichromates, and the same red orange mixture is obtained when any of these reagents is placed into aqueous acid. Regardless of the specific reagent employed, these reactions are called **chromate** or **chromic acid oxidations.**

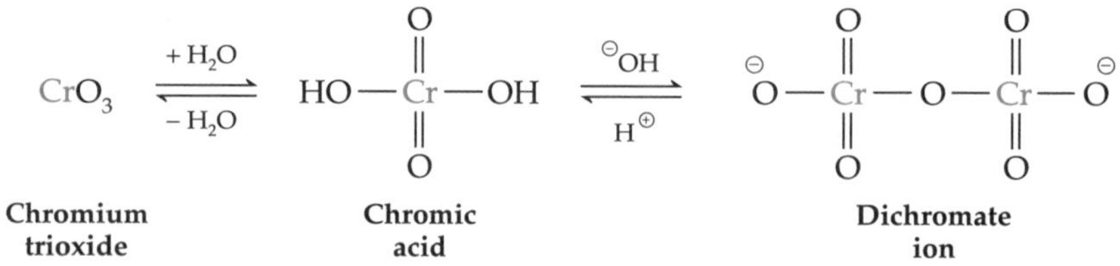

Chromium trioxide	Chromic acid	Dichromate ion

Treatment of an alcohol with one of these reagents produces an alkyl chromate ester by a route that will become clear when we consider nucleophilic additions in Chapter 12.

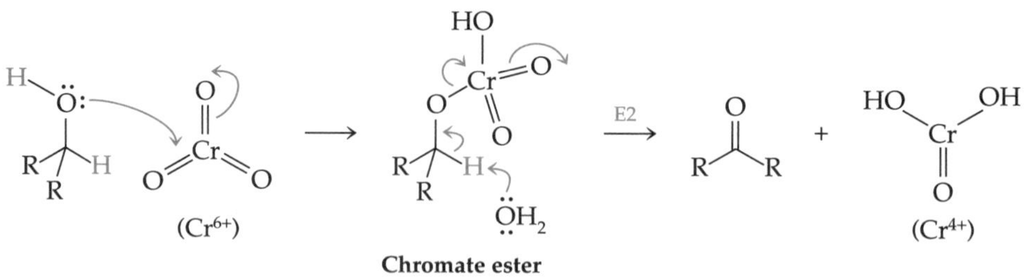

Chromate ester

Because the chromium atom is highly electron deficient and can change its oxidation state, the oxygen–chromium bond can be broken in what is effectively an E2 elimination. In this reaction, water acts as a base to remove a proton from the carbon to which the OH group was originally attached. The electrons originally in the C—H bond form a C=O double bond as the electrons in the oxygen–chromium bond are shifted to chromium, altering its oxidation state from chromium(VI) to chromium(IV). A number of secondary reactions then ensue, because chromium ultimately appears as Cr(III), but for now the significant detail is that the chromate ester intermediate is converted in an E2-like step from a chromium(VI) into a chromium(IV) species, a reduction that accompanies the two-electron oxidation of the alcohol to the ketone.

These oxidations convert secondary alcohols into ketones and primary alcohols into aldehydes and then into carboxylic acids. It is difficult to stop the oxidation of a primary alcohol at the aldehyde oxidation level in aque-

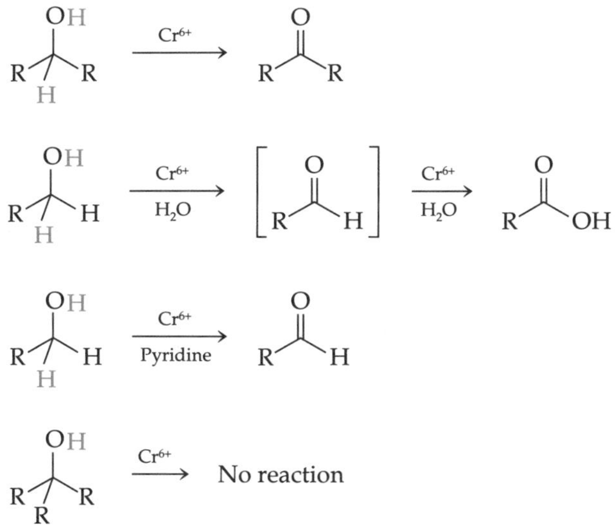

▲ FIGURE 9-20
Chromic acid oxidations produce a ketone from a secondary alcohol (top reaction). A carboxylic acid is produced from a primary alcohol when aqueous chromic acid is employed (second reaction), but an aldehyde can be obtained if the reaction is conducted in pyridine (third reaction). No oxidation is observed with tertiary alcohols for a chromic acid oxidation attempted at room temperature (bottom reaction).

ous solution; however, if the reaction is conducted in pyridine, the oxidation stops at the aldehyde stage. A tertiary alcohol is resistant to chromate oxidation because it lacks a hydrogen on the carbinol carbon (Figure 9-20).

Chromate oxidation constitutes a useful chemical test for the presence of an oxidizable substrate (alcohol or aldehyde) because the orange-red Cr(VI) reagent is converted into deep green Cr(III) as the oxidation proceeds. Other reagents that can be used to accomplish such conversions are listed in Table 3-5. Like CrO_3, these reagents have in common a stable, lower oxidation level of the metal ion.

In the chromate oxidation of alcohols, the carbinol hydrogen is removed as a proton from the chromate ester by a weak base (water). Such metal-centered oxidations differ significantly from biological oxidations in which this C–H bond is broken with the opposite polarity; that is, by the

---

HIGHWAY SAFETY THROUGH CHEMISTRY

The change in color from red orange to green on reduction of Cr(VI) to Cr(III) is the basis of the "breathalyzer" test. Scientific studies have shown that the concentration of alcohol in the blood correlates well with that in exhaled air from the lungs. Passing a defined volume of air through a tube containing chromate ion results in oxidation of the alcohol to acetic acid and reduction of the chromium to the green +3 oxidation state. The greater the concentration of alcohol in the breath, the farther the green color will have progressed down the tube. (The green color of jade also is due to the presence of Cr^{3+} salts.)

$$CH_3CH_2OH \; + \; Cr^{6+} \; \longrightarrow \; CH_3CO_2H \; + \; Cr^{3+}$$
$$\text{Orange} \qquad\qquad\qquad\qquad\qquad \text{Green}$$

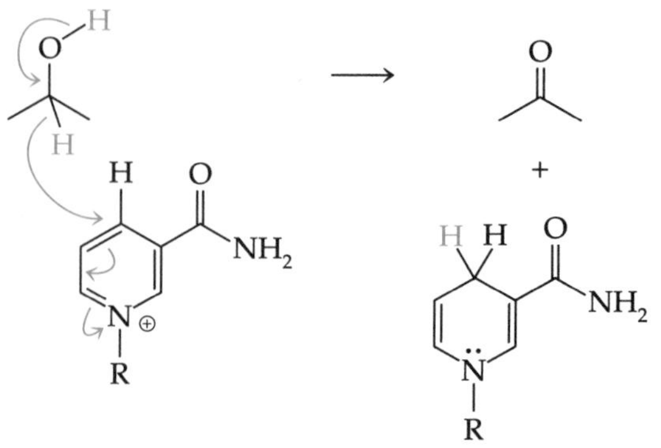

▲ FIGURE 9-21

In biological oxidations, the electron flow in the alcohol is the reverse of that in a chromic acid oxidation. The electrons from the O–H σ bond form the C=O double bond, releasing the electrons from the C–H σ bond. Hydrogen is therefore transferred as hydride to an electron-deficient site, such as that found in nicotinamide.

loss of hydrogen as hydride, H$^-$ (Figure 9-21). Hydride is a nucleophilic rather than electrophilic species, and the inverse of these oxidation reactions (namely, the addition of hydride as a nucleophile to a carbonyl group) also is an important reaction.

EXERCISE 9-I

For each of the following substrates, write the product expected (if any) in a chromate oxidation.

(a)

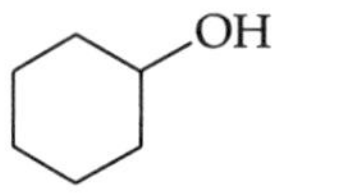

(b)

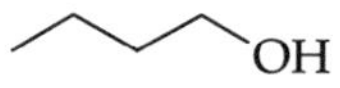

(c)

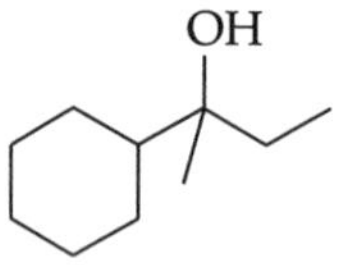

(d)

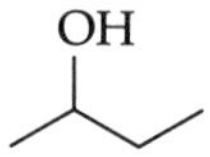

(e) $CH_3CH_2CH_2CO_2H$

(f)

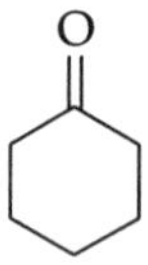

EXERCISE 9-J

The oxidation of an alcohol to an aldehyde and then to a carboxylic acid in the presence of water probably includes oxidation of the hydrate of the aldehyde (a geminal diol). This hydrate is produced by the addition of water to the aldehyde in a step that does not involve chromium. Write a mechanism by which propanol can be converted into propanoic acid using chromic acid (H_2CrO_4) in water. (You need not concern yourself with the details of formation of chromium–oxygen bonds or how the hydrate is formed.)

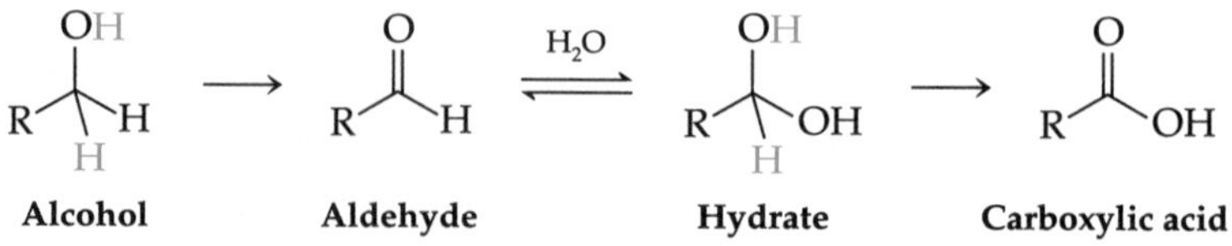

Alcohol   Aldehyde   Hydrate   Carboxylic acid

PHENOL OXIDATION: FIXING IMAGES IN PHOTOGRAPHY
AND PRESERVING FOODS

Hydroquinone is a dihydroxybenzene, or bis-phenol, that readily undergoes oxidation to form quinone. Perhaps the largest volume use of hydroquinone is in the photographic industry, where it is used to reduce silver halides (mainly AgCl and AgBr) to metallic silver. Small particles of silver formed upon the absorption of light in a film emulsion are the basis for images in most photographs. Other phenols also are readily oxidized, and compounds such as butylated hydroxytoluene (BHT) are used as preservatives for foods, as well as in other products such as rubbers and plastics.

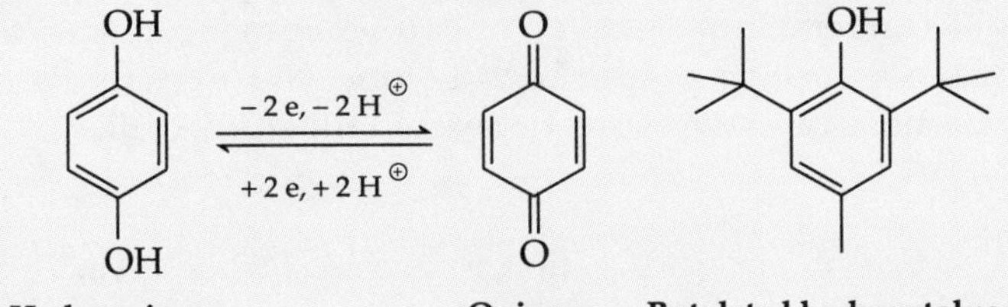

Hydroquinone   **Quinone**   **Butylated hydroxytoluene**
                                                    (BHT)

9-8 ▸ Oxidation of Hydrocarbons

The conversions of alkanes into alkenes (and of alkenes into alkynes) can similarly be considered oxidations in which dihydrogen is eliminated. Although these reactions are difficult, they can sometimes be accomplished under extreme conditions—for example, in a cracking tower for the processing of petroleum. In a refinery, a gaseous stream of hydrocarbon is heated to some temperature significantly above 300 °C as it passes over alumina, causing cleavage of C–H and C–C bonds and yielding a product mixture rich in alkenes. However, these direct dehydrogenations are not feasible under usual laboratory conditions. Instead, such oxidation reactions are accomplished by sequential steps, which are more easily controlled in the laboratory.

For example, the conversion of cyclohexane into cyclohexene can be achieved in two steps (free radical halogenation followed by dehydrohalogenation) in good synthetic yield under normal laboratory conditions, whereas direct dehydrogenation is difficult indeed.

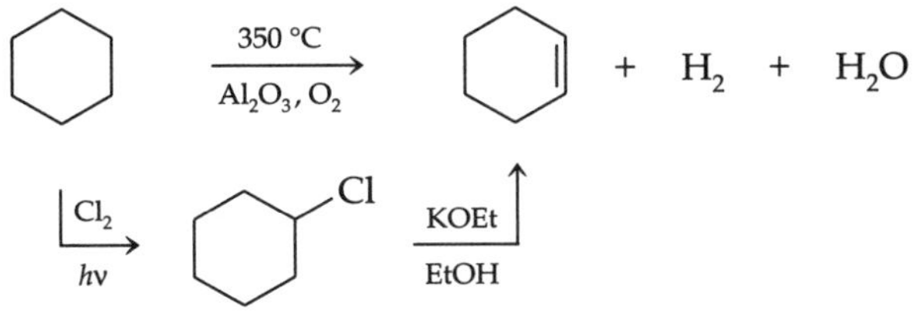

EXERCISE 9-K

Suggest a sequence of reagents that could be employed to accomplish the following transformations (parts *b* and *c* are on the next page):

(a)

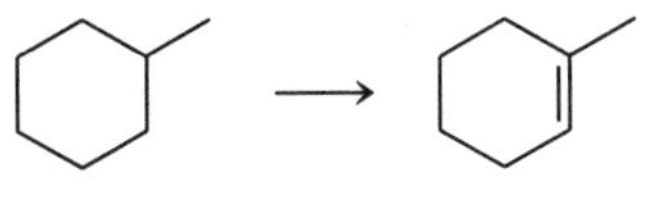

(b) 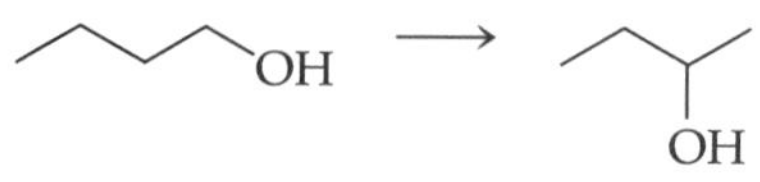 $\diagup\!\!\!\diagdown\!\!\!\diagup$OH $\longrightarrow$ $\diagup\!\!\!\diagdown\!\!\!\diagup$ (c) 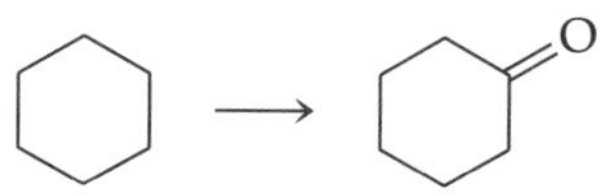 ⬡ $\longrightarrow$ ⬡=O
OH

9-9 ▸ Synthetic Methods

As in the preceding chapters, these elimination reactions can be grouped according to the functional-group conversion accomplished. Table 9-1 provides such a summary. This table is intended to draw your attention to the synthetic applicability of each reaction. It is useful to have several possible ways to make a given functional group in planning syntheses. Furthermore, you thoroughly understand a reaction only if you can recognize the reactant needed to make a specific product. As before, you may find it useful to write these reactions on index cards to be used for study drill.

Table 9-1 ▸ How to Use Elimination Reactions to Make Various Functional Groups

Functional Group	Reaction
Aldehyde	Chromate oxidation of a primary alcohol in pyridine
Alkene	Dehydrohalogenation of an alkyl halide; or dehydration of an alcohol; or debromination of a vicinal dibromide; or radical halogenation of an alkane, followed by dehydrohalogenation
Alkyne	Dehydrohalogenation of a vinyl halide; or dehydrohalogenation of a geminal or vicinal dihalide; or bromination of an alkene, followed by dehydrohalogenation
Aniline	Ammoniation of benzyne
Benzyne	Dehydrohalogenation of an aryl halide
Carboxylic acid	Chromate oxidation of a primary alcohol in water; or chromate oxidation of an aldehyde in water
Ketone	Chromate oxidation of a secondary alcohol
Phenol	Hydration of benzyne

Conclusions

In an elimination reaction, two σ bonds are cleaved from adjacent positions, forming a π bond. When the groups that are eliminated are hydrogen and a halide ion, two mechanisms are commonly encountered: (1) the rate-determining loss of the leaving group to generate a carbocation (E1 elimination); or (2) a concerted reaction in which a base abstracts a proton while a double bond is being formed as the leaving group leaves with its electron pair (E2 elimination). A third mechanism, the E1-CB route, is normally encountered in a dehydrohalogenation only when another group is present to stabilize the intermediate anion formed in an acid-base pre-equilibrium.

Because the E1 elimination proceeds through intermediate formation of a carbocation, it is complicated by side reactions characteristic of that intermediate; that is, substitutions, rearrangements, and polymerizations. The regiochemistry of an E1 elimination is often controlled by product stability (Zaitsev elimination), whereas E2 reactions are controlled either by product stability or by the ease with which a base can approach (a Hofmann elimination). The ease of an E1 reaction is controlled by the facility with which the intermediate cation is accessed: thus, E1 eliminations commonly take place at tertiary halide centers, less commonly at secondary centers, and almost never at primary centers.

Concerted elimination sequences (E2 reactions) sometimes shift product distributions toward Hofmann elimination products (less highly substituted alkenes). The major feature governing the facility of a Hofmann elimination is the relative size of the base and the degree of steric hindrance at the site of deprotonation. The E2 elimination requires precise alignment of hydrogen and the leaving group in a periplanar, preferably *anti*-periplanar, transition state. Regiochemistry for E2 eliminations is dictated by access to this coplanar alignment by the required bonds. Stereochemical preferences in either the E1 or the E2 eliminations are governed by the relative stabilities of the conformers through which elimination takes place.

Dehydrohalogenation of vinyl bromides is unlikely to proceed through an E1 mechanism because of the instability of a vinyl cation. Instead, alkynes are formed either in a concerted pathway or through a vinyl anion. Alkyne formation occurs upon treatment with strong base.

The elimination of HX from an aromatic halide produces benzyne, an interesting compound in which ring aromaticity is maintained while an orthogonal π bond is formed between two of the atoms of the ring. The geometric distortion from the ideal 180° angle for *sp*-hybridized atoms makes benzyne exceedingly reactive, and an elimination-addition sequence provides a route for aromatic substitutions that convert aryl bromides into phenols and anilines.

Dehydration of alcohols through an E1 mechanism occurs readily under acid-catalyzed conditions. Dehydration of β-hydroxycarbonyl compounds is especially easy because of the formation of a conjugated double bond.

The elimination of X_2 from vicinal dihalides can be accomplished by treatment with an easily oxidizable zero-valent metal such as zinc or tin.

Although oxidation reactions constitute formal eliminations of H_2, the mechanisms by which they occur are often complex. Metal-centered oxidations of alcohols take place through a simultaneous change in the oxidation level of the metal ion or complex.

The direct elimination of two hydrogen atoms from alkanes is not a practical laboratory method for the oxidations of hydrocarbons, which are instead oxidized in the laboratory by sequences of functional-group manipulations.

Summary of New Reactions

Hofmann Dehydrohalogenation of Alkyl Halides

Zaitsev Dehydrohalogenation of Alkyl Halides

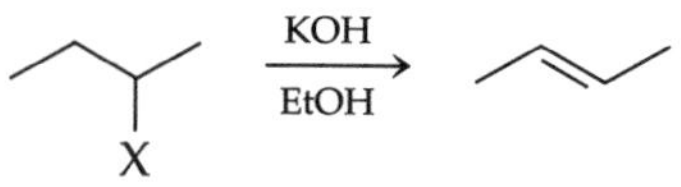

Dehydrohalogenation of Vinyl Halides

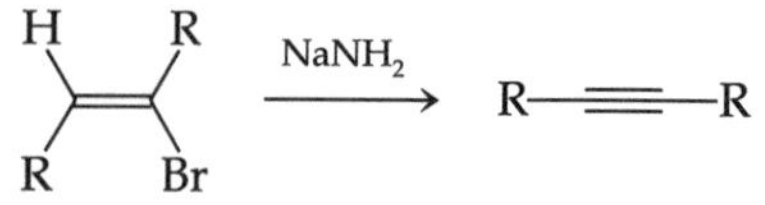

Halogenation–Dehydrohalogenation of Alkenes

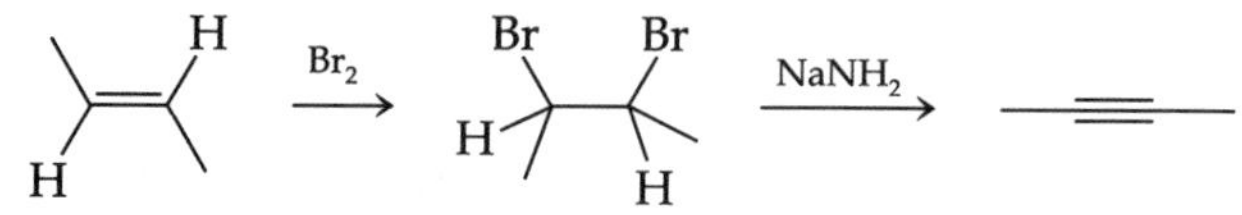

Dehydrohalogenation of Aryl Halides

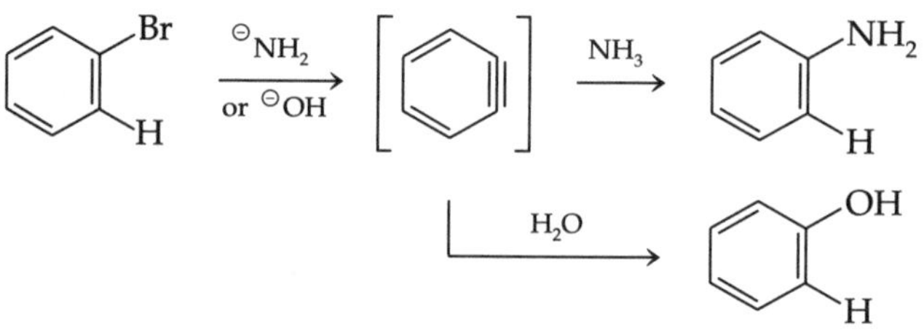

Alcohol Dehydration

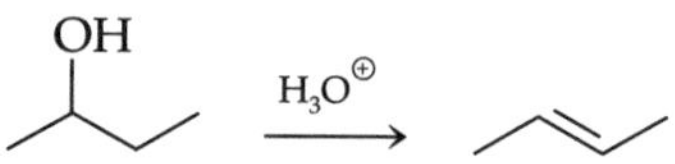

Dehalogenation of Vicinal and Geminal Dihalides

Alcohol Oxidation

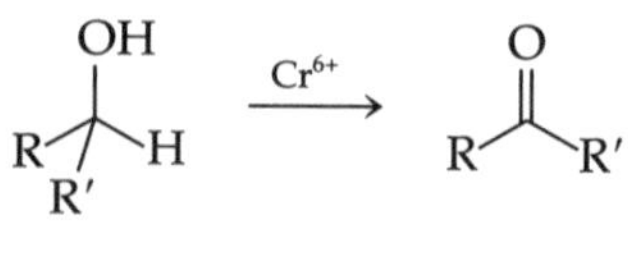

Review Problems

9-1 Use curved arrows to show the electron flow in the preferred reaction mechanism for each of the following eliminations. Briefly discuss the reasoning that led you to choose an E1, E1-CB, or E2 mechanism.

(a)

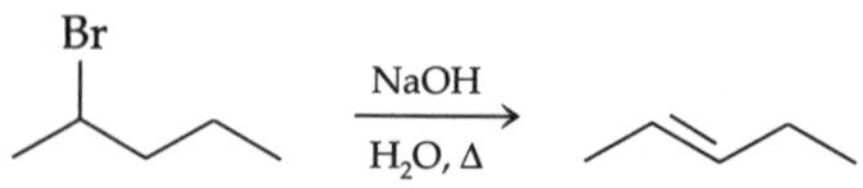

(b) 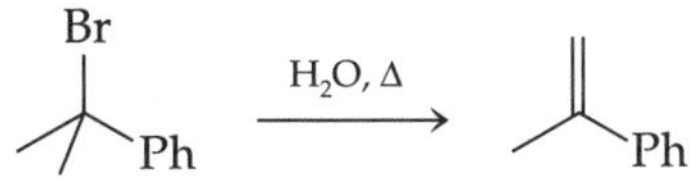 $\xrightarrow{H_2O, \Delta}$

(c) 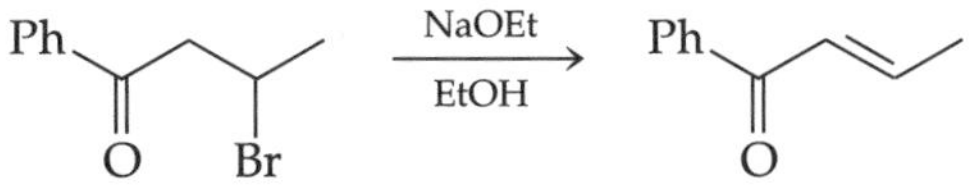 $\xrightarrow[\text{EtOH}]{\text{NaOEt}}$

9-2 Of the following pairs of structures, choose the compound that better fits the description:

(a) Gives more Zaitsev product in an E2 reaction

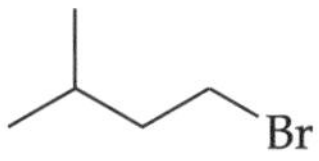

or

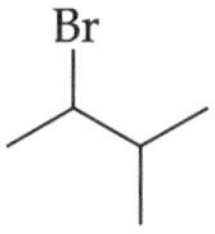

(b) Reacts more rapidly with cold aqueous HBr

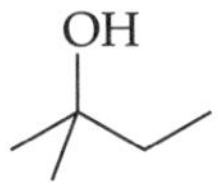

or

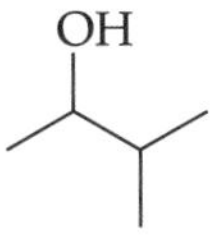

(c) Undergoes elimination with less competing nucleophilic substitution upon treatment with HBr

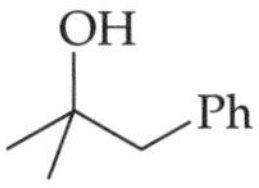

or

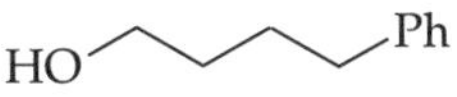

(d) Gives a mixture of two alkenes by E1 reaction induced by sulfuric acid

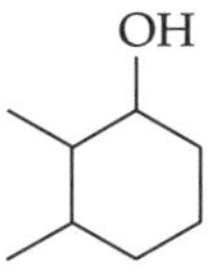

or

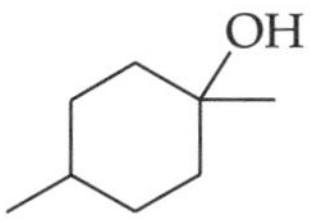

(e) Gives exactly three alkenes by E2 reaction

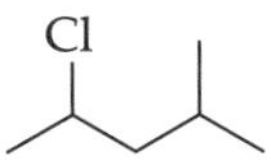

or

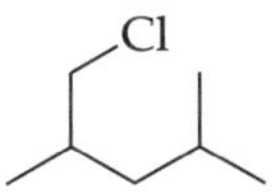

9-3 Predict the major product expected in each of the following reactions:

(a) 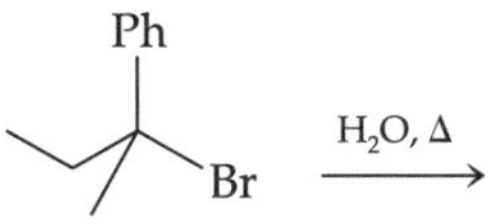 $\xrightarrow{H_2O, \Delta}$

(b) 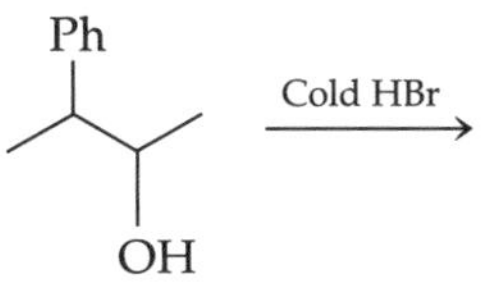 $\xrightarrow{\text{Cold HBr}}$

(c) 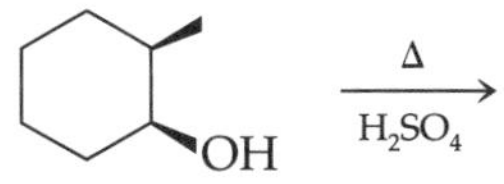 $\xrightarrow[\text{H}_2\text{SO}_4]{\Delta}$

(d)  $\xrightarrow[\text{EtOH}]{\text{NaOEt}}$

(e) 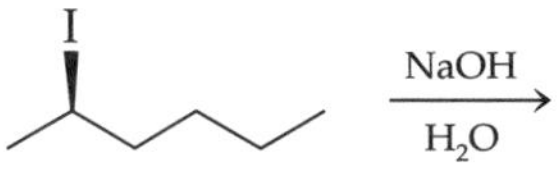 $\xrightarrow[\text{H}_2\text{O}]{\text{NaOH}}$

(f) 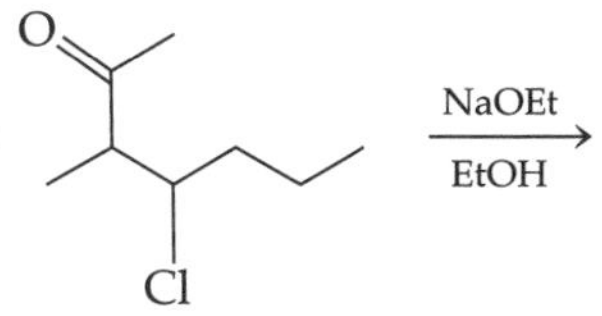 $\xrightarrow[\text{EtOH}]{\text{NaOEt}}$

9-4 Predict whether the amount of Hofmann elimination observed in each of the following reactions is larger, smaller, or unchanged from that observed in the treatment of 2-bromobutane with sodium ethoxide in ethanol:

(a) 2-chlorobutane with sodium ethoxide in ethanol

(b) 2-iodobutane with sodium ethoxide in ethanol

(c) 2-methyl-2-bromobutane with hot water

(d) 2-bromobutane with sodium *t*-butoxide

(e) 2-bromobutane with sodium hydroxide

9-5 Assume that treatment with base of the following cyclohexyl halides effects an elimination through an E2 mechanism. Draw Newman projections to represent the transition states for all possible products and predict the preferred geometry of the product.

(a) 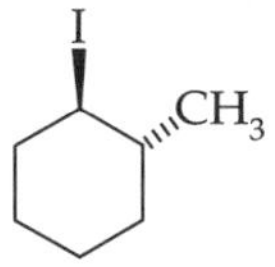 (b)

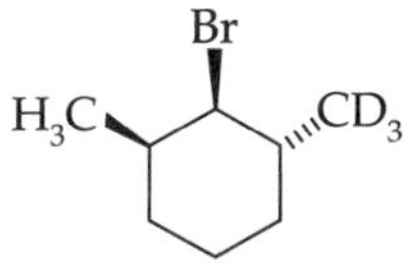

9-6 If the starting material is labeled with deuterium as indicated, predict how many deuteria will be present in the major elimination product and where they will appear. (Continued on the next page.)

(a) 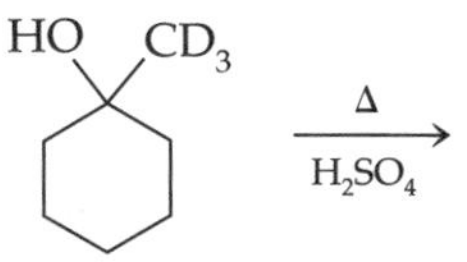 $\xrightarrow[\text{H}_2\text{SO}_4]{\Delta}$

(b) 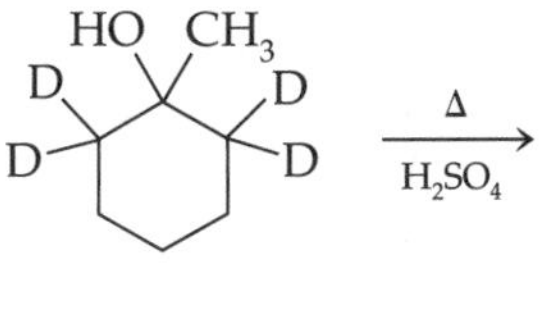 $\xrightarrow[\text{H}_2\text{SO}_4]{\Delta}$

(c) 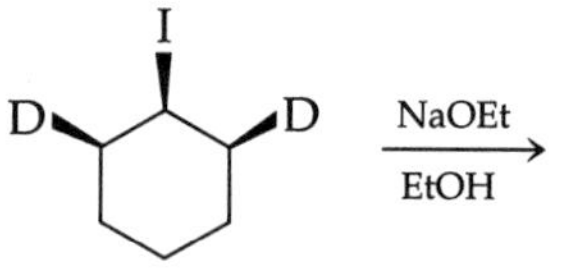 $\xrightarrow[\text{EtOH}]{\text{NaOEt}}$

(d) 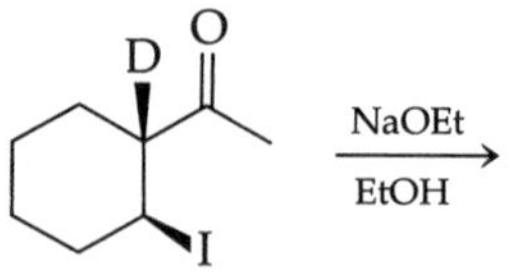 $\xrightarrow[\text{EtOH}]{\text{NaOEt}}$

(e) 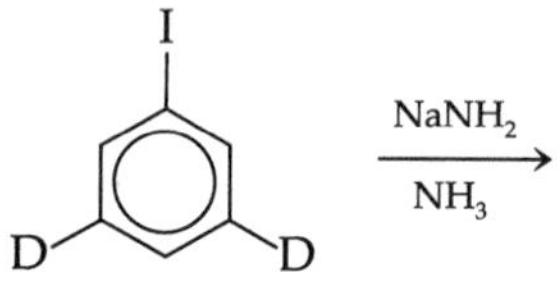 $\xrightarrow[\text{NH}_3]{\text{NaNH}_2}$

(f) 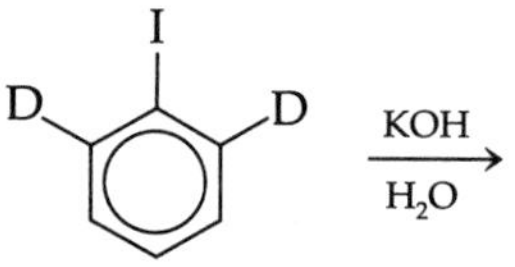 $\xrightarrow[\text{H}_2\text{O}]{\text{KOH}}$

(g) 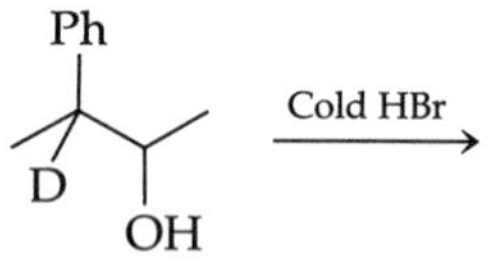 $\xrightarrow{\text{Cold HBr}}$

9-7 In the following reactions, a rearranged skeleton is observed in the principal product. Write a mechanism for each reaction, using curved arrows to indicate electron flow, that leads to the observed product.

(a)  $\xrightarrow{\text{Cold HBr}}$

(b) 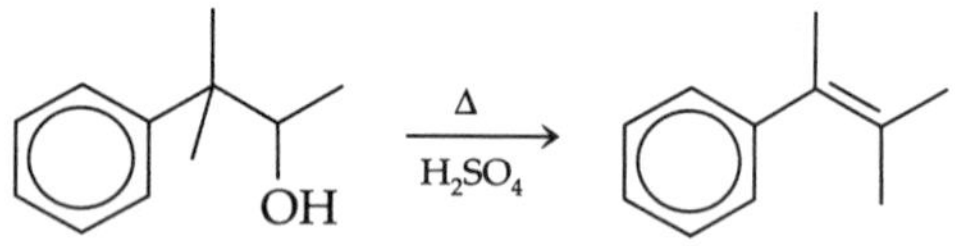 $\xrightarrow[\text{H}_2\text{SO}_4]{\Delta}$

9-8 Draw the structure of the major product expected when 2-butanol is treated with each of the following reagents in the indicated sequence:

(a) 1. PBr$_3$; 2. NaOEt, EtOH
(b) 1. hot H$_2$SO$_4$; 2. Br$_2$, CCl$_4$; 3. NaNH$_2$, NH$_3$
(c) 1. K$_2$Cr$_2$O$_7$, H$_2$SO$_4$; 2. I$_2$, NaOH
(d) 1. cold HBr; 2. KO-t-Bu, t-BuOH
(e) 1. SOCl$_2$ in pyridine; 2. NaOH in EtOH; 3. Br$_2$ in CCl$_4$; 4. Zn in Et$_2$O

9-9 For each of the following alcohols, describe what, if anything, you would see upon treatment with aqueous chromic acid. What would you see with the Lucas reagent (Chapter 3)?

(a) 2-octanol   (e) 3-methyl-1-octanol
(b) 3-octanol   (f) 3-methyl-3-octanol
(c) cycloheptanol   (g) 3-octanone
(d) 1-methylcyclo-   (h) octanal
    heptanol

9-10 Propose three routes employing different organic starting materials by which you could synthesize 2-butyne.

9-11 In each of the following reactions, two or more regioisomers are found as products. Draw the structures of these isomers.

(a) 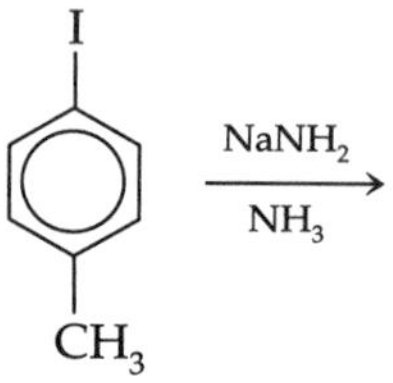 $\xrightarrow[\text{NH}_3]{\text{NaNH}_2}$

(b) 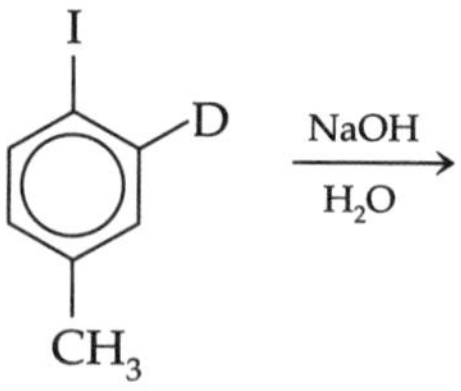 $\xrightarrow[\text{H}_2\text{O}]{\text{NaOH}}$

(c) 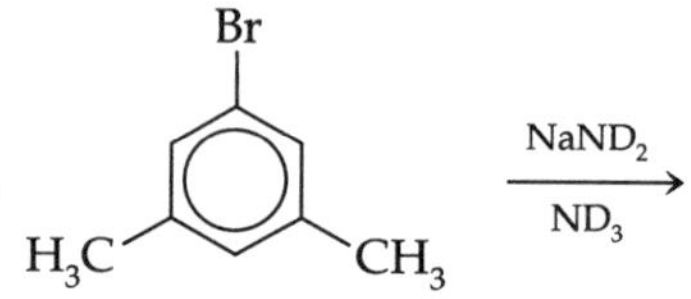 $\xrightarrow[\text{ND}_3]{\text{NaND}_2}$

(d) 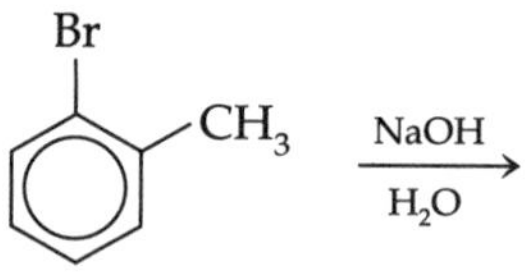 $\xrightarrow[\text{H}_2\text{O}]{\text{NaOH}}$

9-12 One of the hardest predictions to make before a reaction is run is how important elimination will be in an attempt to do a substitution. Shown here are S$_N$1 and S$_N$2 reactions. Draw the structures of the elimination products that might be competitively formed, and suggest how you would use spectroscopy to determine whether the major product that you have isolated is a substitution or elimination product. Write a reaction mechanism that shows how each of the elimination products might be formed.

(a) 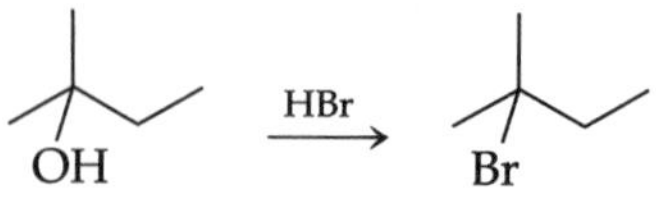 $\xrightarrow{\text{HBr}}$

(b) 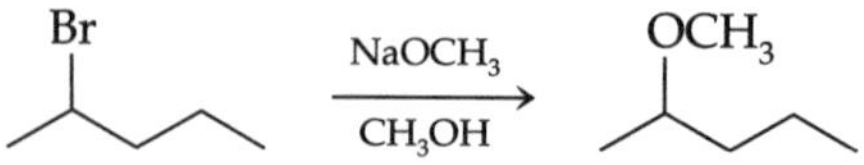 $\xrightarrow[\text{CH}_3\text{OH}]{\text{NaOCH}_3}$

Chapter 10

Electrophilic Addition to Carbon–Carbon Multiple Bonds

The elimination of HL to form a multiple bond was the subject of Chapter 9. We will now consider the inverse reaction: namely, the addition of HL to an alkene. The principle of microscopic reversibility dictates that the pathway of lowest energy must be the same in the forward and the reverse directions if reaction conditions are comparable. Thus, some of the intermediates discussed in Chapter 9 are also intermediates in the addition reactions discussed in this chapter.

Many of the important functional groups of organic chemistry contain multiple bonds in addition to the sp^3-hybridized atoms whose reactions are discussed in Chapters 7 through 9. In alkenes, the characteristic C=C double bond constitutes a site of high electron density that is readily attacked by electron-deficient reagents called *electrophiles* rather than by the nucleophiles or bases discussed in earlier chapters. In this chapter, we consider the interaction of electrophiles with unsaturated hydrocarbons and focus in particular on addition reactions initiated by electrophilic attack on alkenes, alkynes, and dienes.

10-1 ▸ Electrophilic Addition to Alkenes

The characteristic reaction of this family is *electrophilic addition*, a process in which a C=C π bond is replaced by two σ bonds.

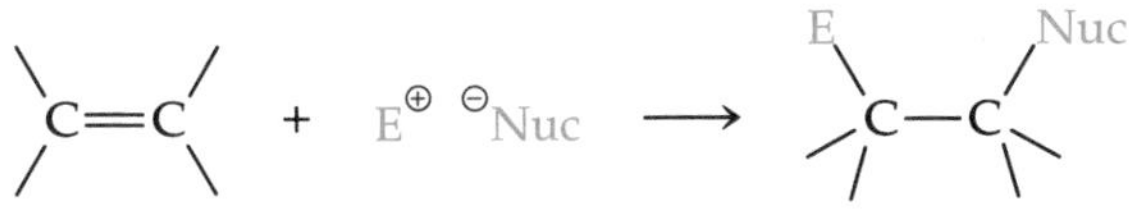

$$\diagdown C = C \diagup \quad + \quad E^{\oplus} \; {}^{\ominus}Nuc \quad \longrightarrow \quad \diagup C - C \diagdown$$

The electrons in the π bond of an alkene are less tightly bound than those in σ bonds. They are therefore much more polarizable and can interact strongly with positively charged reagents. This interaction constitutes the first step of an electrophilic addition.

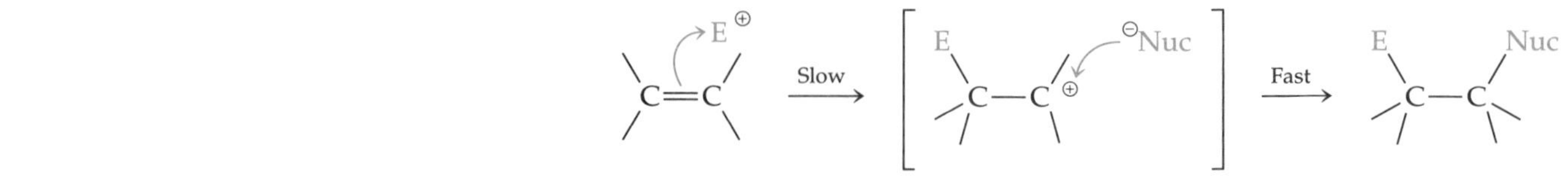

▲ FIGURE 10-1

In an electrophilic addition, an electron-deficient reagent, E^+ (often a proton, H^+), approaches the π cloud of a C=C double bond. The electrons of the π bond flow toward the electrophile, forming a C–E σ bond and a carbocation. This cation is rapidly captured by a nucleophile, which donates its electrons to form a C–Nuc σ bond. Because the second step consists only of bond making, it is faster than the first. The step in which the carbocation is formed is therefore the rate-determining step.

A general route for electrophilic addition is shown in Figure 10-1. This reaction is initiated by the addition of an electrophile (often a proton) to a C=C double bond, as in the electrophilic addition of HCl to cyclohexene briefly discussed in Chapter 7. In this step, the two electrons originally in the π bond shift toward the electrophile, generating a new σ bond. Localization of these two electrons in a σ bond leaves the other carbon of the original alkene positively charged. The resulting carbocation is easily attacked in a second step by a reagent that acts as a nucleophile. When the nucleophile attacks this carbocation, lone-pair electron density is donated to form a new σ bond between the nucleophile and carbon, thus completing the addition. Overall, in an electrophilic addition, two new σ bonds are formed, one to an electrophile and one to a nucleophile, at the two carbon atoms originally participating in a carbon–carbon π bond. This process is called an *addition* because the product incorporates all atoms of both reactants; it is called *electrophilic* because it is initiated by interaction of the π bond with an electrophile.

Is an electrophilic addition to an alkene thermodynamically favorable? Overall, the π bond of the alkene and a σ bond between the electrophile and the nucleophile are broken. In place of these bonds, two new σ bonds are formed, one between the electrophile and one carbon of the alkene and the other between the nucleophile and the other carbon. A calculation of the enthalpy change accompanying these transformations requires information about the specific bonds being made and broken, but we can nonetheless draw some qualitative conclusions about the thermodynamics of such reactions. Most σ bonds to carbon are stronger than either a typical carbon–carbon π bond (Table 3-6) or a bond between common electrophile–nucleophile pairs. As a result, the combined strength of the two bonds formed in this process (C—E and C—Nuc) usually exceeds that of the bonds consumed (C=C and E—Nuc), and most electrophilic additions are exothermic.

The sequence shown in Figure 10-1 represents a two-step reaction mechanism for this transformation. To understand the controlling features of the reaction, we must establish which of the two steps is rate determining. The rate-determining step is that in which the transition state of highest energy is formed, and those factors that stabilize (or destabilize) this transition state directly affect relative reactivity.

In the first step of electrophilic addition, the π bond and the E–Nuc bond are broken as a σ bond (between a carbon and the electrophile) is formed. As a result, this first step is endothermic in most cases. In contrast, the second step consists of only bond making and is highly exothermic. Thus, the first step, producing a carbocation, is likely to be rate determining.

EXERCISE 10-A

For each of the following reagents, consider normal charge polarization to identify that part of the reagent that can act as an electrophile.

(a) H—Br

(b) 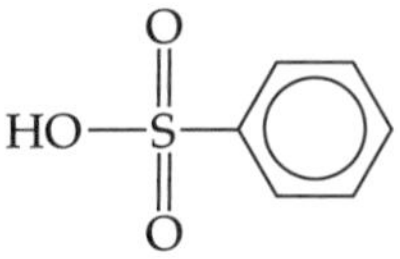 HO—S(=O)(=O)—C₆H₅

(c) 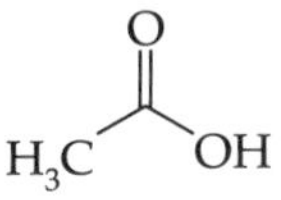 H₃C—C(=O)—OH

(d) Cl_2

(e) Br_2

(f) HOCl

(g) 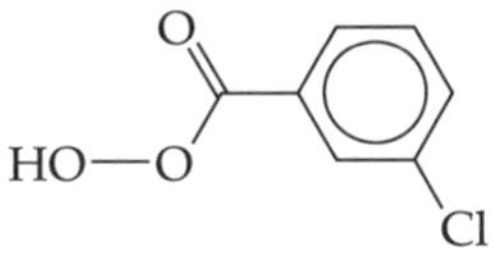 HO—O—C(=O)—C₆H₄Cl

(h) 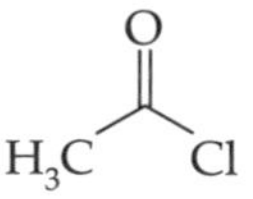 H₃C—C(=O)—Cl   + $AlCl_3$

(i) H_2O

Because the stability of the transition state for an endothermic reaction is substantially influenced by the product, the transition state of the first step in this sequence resembles the cationic intermediate. We can therefore expect the rates of electrophilic addition reactions to increase with increasing stability of the intermediate cation.

10-2 ▸ Addition of HX

Let us consider a specific example: the addition of hydrogen chloride to *cis*-2-butene to produce 2-chlorobutane, first in the gas phase and then in solution.

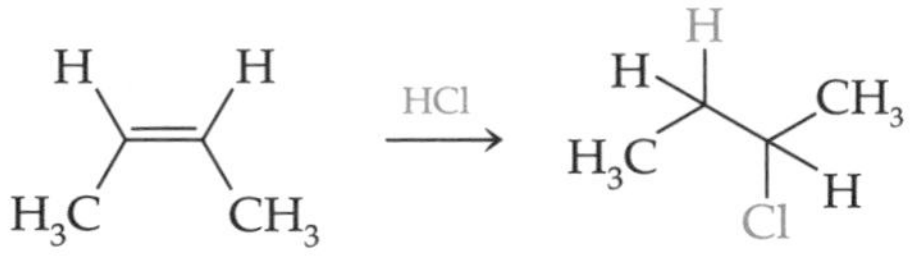

Gas-Phase Reactivity

Because chlorine is more electronegative than hydrogen, the σ bond connecting these atoms in HCl is strongly polarized toward chlorine, leaving hydrogen electron deficient and hence electrophilic. Thus, the proton assumes the role of electrophile in initiating the addition (Figure 10-2, on page 334). The partially positively charged end of HCl approaches the π density of the multiple bond (either above or below the carbon plane) because this is the region where the electron density is greatest. The proximity of the electrophile causes electrons to flow from the π bond to form a σ bond, as chlorine is converted into chloride ion as it accepts the two electrons originally present in the polar covalent H–Cl bond. As a σ bond is formed between carbon and the proton, the other carbon initially participating in the π bond takes on carbocationic character. This ion pair is relatively unstable in the gas phase because of the absence of stabilizing interactions between a solvent and both the cation and the anion. In a rapid second step, chloride ion collapses with the carbocation to form the second σ bond. (Notice that

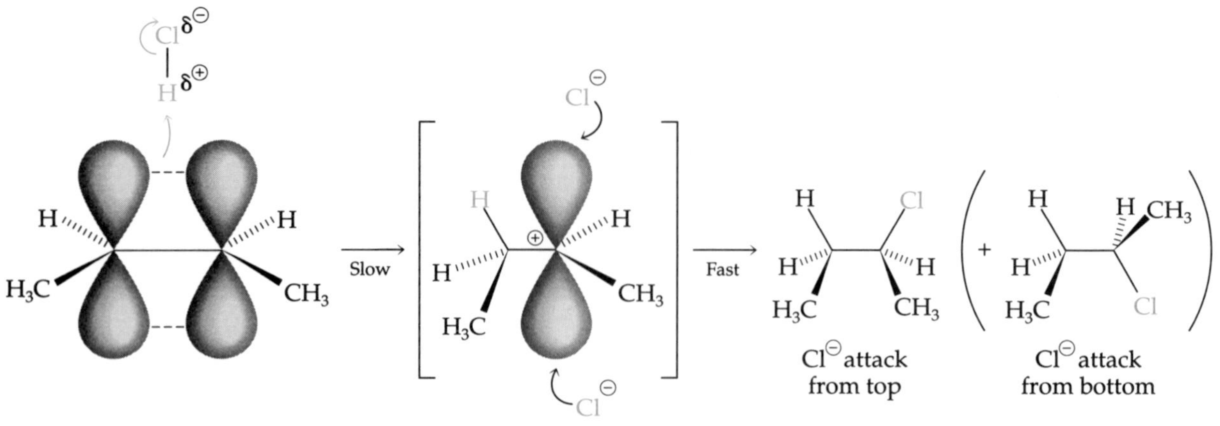

▲ FIGURE 10-2

The electrostatic attraction between the partial positive charge on hydrogen in the polar H–Cl σ bond and the electron cloud of the alkene π bond brings these reagents into a geometry in which electrons can flow from the π orbital to form a σ bond between carbon and hydrogen. Chloride, formed by taking up the electrons of the H–Cl σ bond, then approaches the vacant *p* orbital of the carbocation. Because the cation is sp^2-hybridized and planar, this approach is equally easy on the top and bottom faces. Equal amounts of each enantiomer are therefore formed at the new center of chirality.

bond formation between carbon and chlorine from above and below the carbon plane results in the two enantiomers of the product.)

Solution-Phase Reactivity: The Effect of Water on HX Addition

Electrophilic additions of H–X are much faster when conducted in water than in the gas phase because the polar solvent stabilizes the intermediate ions. The reaction of *cis*-2-butene with HCl in water, as shown in Figure 10-3, has many of the same features as the reaction in the gas phase. However, HCl is completely dissociated in water; thus, in aqueous solution, the *hydronium ion* is the electrophile.

$$HCl + H_2O \rightleftharpoons H_3O^+ + Cl^{\ominus}$$

Hydronium ion, H_3O^+, can be thought of as a proton solvated by water; chloride ion also is highly solvated. (As a short-cut, organic chemists sometimes write a proton, H^+, to represent the solvated electrophile (H_3O^+ in water) participating in the initial attack. Although this shorthand is convenient, one should clearly recognize that an unsolvated proton is a quite unstable species that is not found in solution.) Solvation thus plays a major role in the generation of electrophiles from acids in aqueous solution.

As a hydronium ion approaches an alkene, water is released as one of the O–H σ bonds of the hydronium ion is broken and a σ bond to carbon is formed. The $H–OH_2^+$ bond is weaker than an H–Cl bond, and no additional charge separation develops. Both of these factors make this step much more facile thermodynamically than with gaseous HCl. The resulting carbocation, though identical with that formed in the gas phase, is somewhat solvated and therefore stabilized in water.

The reaction medium now contains two potential nucleophiles, chloride ion and water. When chloride ion is the nucleophile, the same product is formed as in the gas-phase reaction.

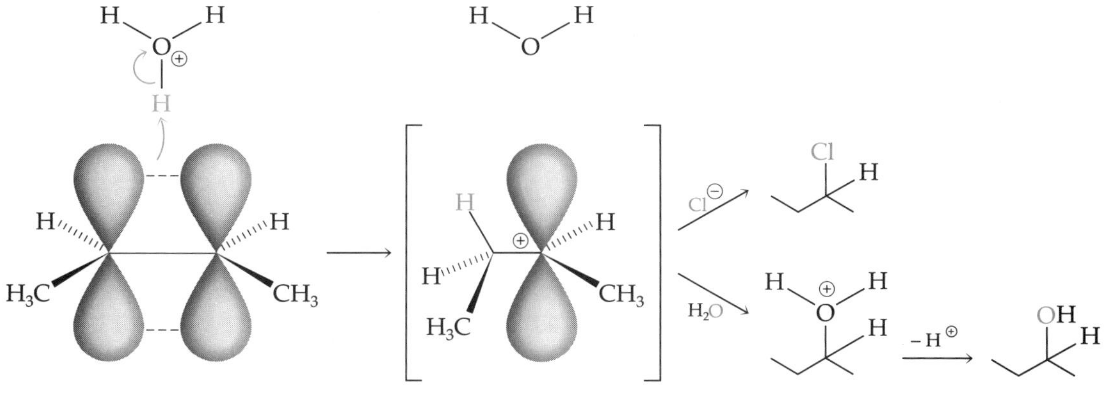

▲ FIGURE 10-3

In solution, H–Cl is dissociated, and so hydronium (H_3O^+) acts as the attacking electrophile. Electron flow from the alkene π bond to H_3O^+ forms a C–H σ bond, freeing neutral water. (The initial attack by H_3O^+ also takes place with equal ease from the bottom face of the π bond.) The planar, sp^2-hybridized carbocation formed in this rate-determining step is then captured by nucleophilic attack either by chloride or by water. Although chloride, an anion, is a more-reactive nucleophile, it is present in much lower concentration than the solvent water. Both the chloride and alcohol are racemic because attack is equally easy on the top and bottom faces of the carbocation.

A HYPERVALENT CARBOCATION

The strengths of acids, as measured by pK_a values, vary from very weak ones such as hydrocarbons to the other extreme—that at which the acids are much stronger than sulfuric acid. Acids with acidities equal to or greater than sulfuric acid are called "super acids." Some of these acids are sufficiently acidic that they donate a proton to alkanes, even to methane to form the CH_5^+ ion, first prepared by the American chemist George Olah. Such cations are called "hypervalent" because they formally possess "extra" bonds beyond those normally found at second-row atoms. As such, they differ in character from the trivalent carbocations encountered as intermediates in this chapter. There can be no "classical" formalism for a hypervalent ion because it does not conform to the rules of valency (C = 4, H = 1). One possible "nonclassical" structure is shown below; here, the added proton is associated side-on with the electron density of one of the C–H bonds of methane.

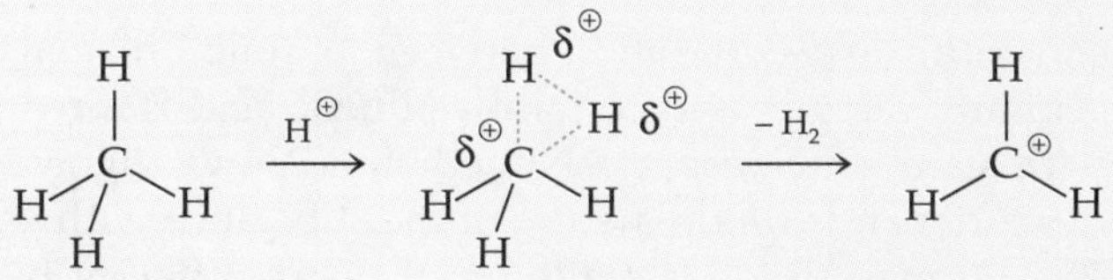

These nonclassical cations are unstable and even undergo loss of H_2.

EXERCISE 10-B

Assuming that cation stability governs the barrier for protonation in H–X additions, predict which compound in each of the pairs in parts *a* through *e* (on page 336) will be more rapidly hydrochlorinated in a polar solvent.

(a) $H_2C=CH_2$ or [structure]

(d) [structure] or [structure]

(b) [structures] or [structure]

(e) [structure] or $H_2C=CH_2$

(c) [structure] or [structure]

Hydration

When water (rather than chloride ion) acts as the nucleophile, an *oxonium ion* is formed. The positive charge in this ion can be relieved by **deprotonation,** a process in which the proton is transferred to solvent—in this case, water. This cleavage of an oxygen–hydrogen σ bond completes the formation of an alcohol, and the overall process is called **hydration.** Thus, unlike the gas-phase reaction, there are two potential products formed in an electrophilic addition of HCl in water.

The relative amounts of alkyl chloride and alcohol produced in this reaction (Figure 10-3) are determined by the relative nucleophilicity of chloride ion and water. We know from Chapter 6 that chloride ion, being negatively charged and polarizable, is a much better nucleophile than is neutral water. Therefore, alkyl chloride formation dominates. On the other hand, with a dilute solution of an acid with a less nucleophilic anion, such as the bisulfate ion (HSO_4^-) from sulfuric acid, the major product will be that resulting from reaction with water.

In the last step of hydration, a proton is transferred from the oxonium ion to water, replacing the hydronium ion consumed in the first step. Thus, the proton is not consumed, and the formation of alcohol by hydration is **acid catalyzed.** An acid is required to initiate the reaction but is reformed as the product is generated. The hydration of an alkene thus requires acid, but only water and the alkene are consumed in forming the product.

In general, acids of the type H-X, in which X is a relatively electronegative atom, are effective reagents for electrophilic addition, which can be generalized as shown in Figure 10-4. When H and X add across a C=C

PEOPLE VS. COCKROACHES

People tend to draw a correlation between "strong" reagents such as sulfuric acid and danger, and certainly sulfuric acid is a reagent to be treated with care and respect. Nonetheless, how vigorously such reagents react depends on the nature of the substance with which they are reacting. Human tissue is made up of many different hydrophilic molecules, containing virtually all common functional groups; human skin is relatively rapidly destroyed by concentrated sulfuric acid. On the other hand, some insects, such as cockroaches, are coated with a mostly hydrocarbon layer that is therefore relatively inert to sulfuric acid. Indeed, a cockroach will swim about, apparently merrily, on the surface of concentrated sulfuric acid. (Please wear safety glasses if you try this at home—roaches splash terribly when doing the backstroke.)

◀ FIGURE 10-4
Both hydrohalogenation and
hydration proceed as
electrophilic additions through
an intermediate carbocation.

Hydrohalogenation if X = halogen (Cl, Br, or I)
Hydration if X = OH

double bond, the result is *hydrohalogenation* if X is a halide (X = Cl, Br, I) or
hydration if X = OH. Both reactions are initiated by protonation (by a hydronium ion) in the initial electrophilic attack, which results in the breaking of
the carbon–carbon double bond. Because this step is the slower one, the reaction rate is determined by the stability of the resulting carbocation.

Regiochemistry

Let us now consider what happens when 1-butene rather than 2-butene undergoes electrophilic addition of HCl (Figure 10-5). Because the alkene is no
longer symmetrical, protonation can be at either C-1 or C-2, producing either a secondary or a primary cation. We know from Chapters 3 and 6 that a
secondary cation is more stable than a primary one. Because carbocation
formation is the rate-determining step of an electrophilic addition, the
greater stability of the secondary cation formed dictates that protonation at
C-1 is dominant. Capture of this cation by chloride ion generates 2-
chlorobutane, whereas capture of the less-stable primary cation leads to 1-
chlorobutane. Preferential formation of the more-stable secondary cation
thus leads to the secondary chloride as the major adduct.

Cation stability also controls the direction of addition to other substrates. For example, the addition of HCl to 2-methylpropene leads to predominantly 2-chloro-2-methylpropane through the more-stable tertiary
cation (upper reaction in Figure 10-6, on page 338). Similarly, the addition
of HCl to styrene (lower reaction in Figure 10-6) produces 1-chloro-1-
phenylethane through a highly stabilized benzylic cation.

Recall from Chapter 7 that, according to Markovnikov's Rule, the less
highly substituted position in an unsymmetrically substituted alkene is attacked by the electrophile in an electrophilic addition reaction. The physical
basis for this principle is that the rate-determining step in electrophilic addition is formation of the carbocation and that the order of carbocationic
stability is tertiary ~ benzylic > allylic ~ secondary > primary ~ vinyl >
phenyl.

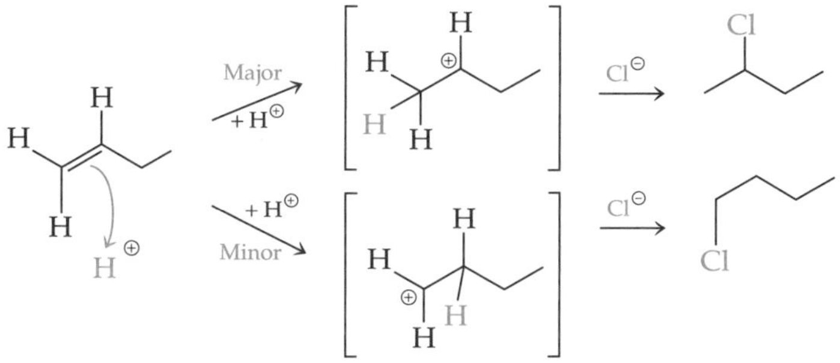

▲ FIGURE 10-5
Electrophilic protonation of 1-butene at C-1 produces a secondary carbocation
at C-2, whereas protonation at C-2 leads to a primary cation at C-1. Because a
secondary cation is more stable than a primary one, the upper route is the
major reaction pathway.

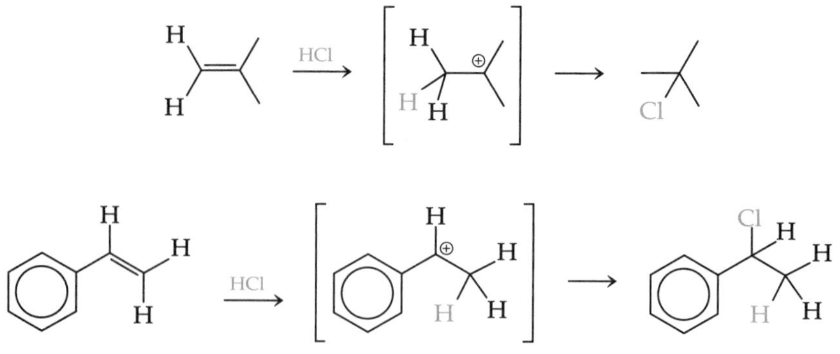

▲ FIGURE 10-6

Carbocation stability governs the regiochemistry of the protonation in an electrophilic hydrohalogenation. In the upper reaction, protonation at C-1 produces a tertiary cation, rather than the primary cation that would have been formed by protonation at C-2. In the lower reaction, protonation at C-2 forms a benzylic cation, which is substantially more stable than the primary cation that would have been produced by protonation at C-1.

EXERCISE 10-C

What is the identity of the carbocation that is formed as an intermediate in the acid-catalyzed hydration of each of the following compounds?

Addition to Dienes

Because the stability of the allylic cation is comparable to that of a secondary cation, HX addition to a conjugated diene, through an allylic cation intermediate, is quite favorable. As shown in Figure 10-7, Markovnikov's Rule predicts that the electrophile will attack at the end of the conjugated system to form an allylic cation. Resonance delocalization of the allylic cation places positive charge at C-2 and C-4. Trapping of the allylic cation by chloride at each of these positions leads to 3-chloro-1-butene and

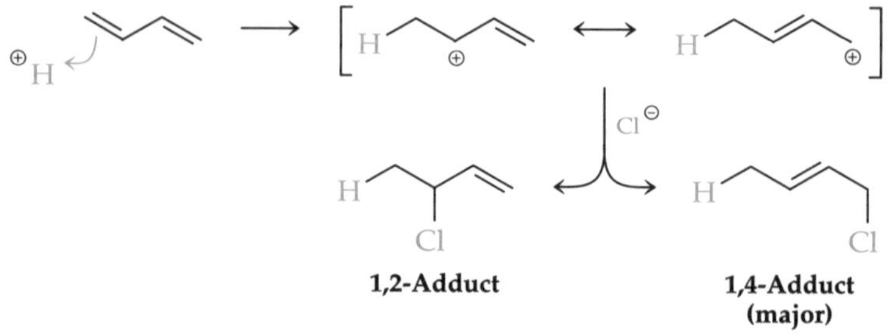

1,2-Adduct **1,4-Adduct**
 (major)

▲ FIGURE 10-7

Protonation of a diene proceeds to form the more-stable cation, in accordance with Markovnikov's Rule. With butadiene, protonation at C-1 produces a resonance-stabilized allylic cation. Nucleophilic capture of the cation at the two sites that bear formal positive charge in these resonance contributors leads to two products—namely, the 1,2- and 1,4-adducts.

1-chloro-2-butene, respectively, the 1,2 and 1,4 adducts. This terminology refers to the positions of the added H and the X group along the carbon skeleton.

The stabilities of these two products are not equal: the double bond in 1-chloro-2-butene is disubstituted and that in 3-chloro-1-butene is only monosubstituted. When the reaction takes place at room temperature (a condition that leads to thermodynamic control), more of the more-stable alkene, shown at the right in Figure 10-7, is formed than that shown at the left. The addition of H–X across a four-carbon system (as in the 1,4 adduct), is an example of **conjugate addition.**

EXERCISE 10-D

Predict the regiochemistry of both the 1,2 and the 1,4 adducts formed by treatment of each of the following dienes with HBr.

(a) 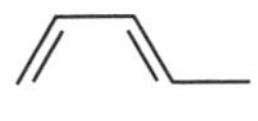   (b) 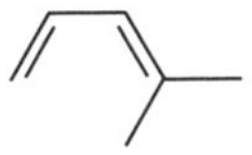   (c)

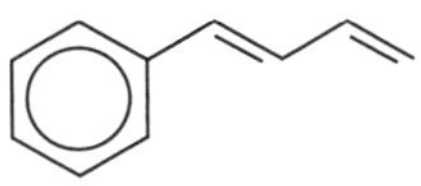

Stereochemistry

The intermediacy of a carbocation dictates that the stereochemistry of HX electrophilic additions be **stereorandom**—that is, without a stereochemical preference at the cationic site. For the hydrochlorination of 2-phenyl-1-butene (Figure 10-8), for example, we can use Markovnikov's Rule to explain why chlorine is bound at C-2 in the product. This addition makes C-2 a center of chirality, despite the fact that this carbon was not chiral in the reactant. As a result, the carbocation is **prochiral,** meaning that an achiral carbon becomes a center of chirality as a reaction proceeds. Because the carbocation is planar, however, chloride attacks the top and bottom lobes of the vacant *p* orbital with equal facility in the second step of the reaction. Equal amounts of the two possible enantiomers are generated, therefore, and the product isolated is racemic and, hence, optically inactive. Thus, stereorandom addition is obtained in an electrophilic addition whenever the fully equilibrated planar carbocationic intermediate is attacked with equal ease from its top and bottom faces.

Sometimes this random stereochemistry (deriving from a planar carbocation) can produce geometric isomers. For example, in the addition of deuterium chloride to cyclohexene (Figure 10-9), the initial deuteration (addition of D⁺) generates a planar cyclohexyl cation. Attack occurs equally on the front and the back faces of this cation (that is, on the same side to which the deuterium was added and on the opposite side), so that the product isolated contains equal amounts of *cis* and *trans* isomers.

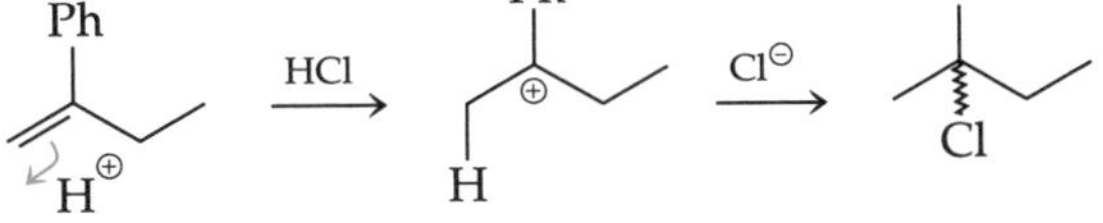

▲ FIGURE 10-8

Protonation of 2-phenyl-1-butene takes place at C-1 to produce a benzylic cation. This cation is attacked equally easily on both faces, leading to racemic chloride. The attack by chloride is stereorandom, as indicated by the wavy line.

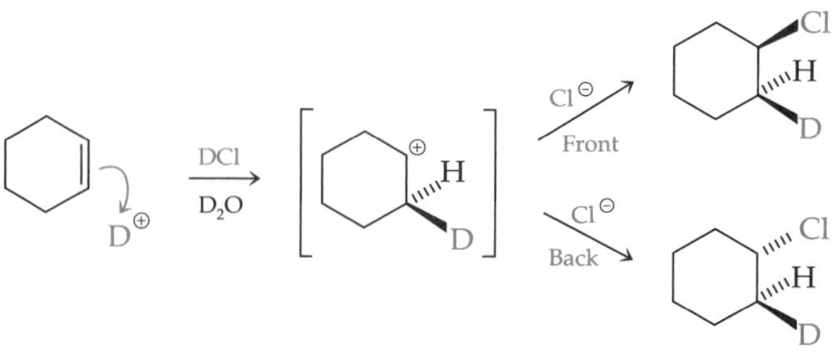

▲ FIGURE 10-9

Stereorandom σ bond formation in the second step of the electrophilic addition of DCl to cyclohexene produces equal amounts of the product in which chloride is *cis* and *trans* to deuterium. These nonmirror image isomers are diastereomers and have slightly different physical and chemical properties.

The formation of *cis* product by delivery of an electrophile and nucleophile on the same face of a multiple bond is referred to as a ***syn* addition.** That in which the addition is on opposite faces of the double bond is called an ***anti* addition.** In the reaction shown, equal amounts of the *syn* and *anti* adducts are formed, making the observed reaction stereorandom.

A *syn* addition was discussed in Chapter 2 in relation to the catalytic hydrogenation of alkenes in which two hydrogen atoms are delivered to the double bond in *syn* fashion from the surface of the metal catalyst. As we shall see later in this chapter, the opposite stereochemical possibility (*anti* addition) occurs in the addition of Br_2 or Cl_2 to alkenes. Because reactions that proceed through free carbocations show stereocontrol (equal amounts of *syn* and *anti* addition products), neither catalytic hydrogenation nor halogenation (addition of X_2) can proceed through a free carbocation.

Rearrangements

The Markovnikov orientation observed in electrophilic addition is a direct consequence of carbocation stability. Another consequence of the stability of a carbocationic intermediate in electrophilic addition is the possibility for rearrangement. A group at an adjacent carbon rapidly migrates to the carbocationic center whenever a thermodynamic driving force exists for such a migration—resulting, for example, in the conversion of a secondary cation into a tertiary one.

By using these ideas, we can predict the course of the hydrobromination of 3-phenyl-1-propene (Figure 10-10). Protonation by electrophilic at-

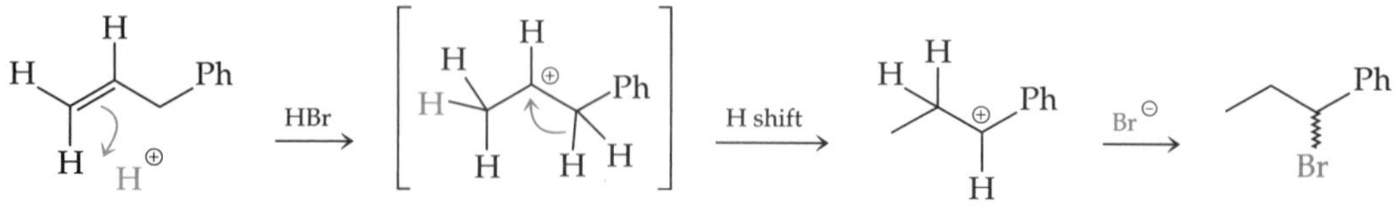

▲ FIGURE 10-10

Protonation of 3-phenyl-1-propene takes place at C-1 to produce the more-stable, secondary carbocation. A hydrogen shift from C-3 produces an even more stable benzylic carbocation. Capture of this planar carbocation by bromide takes place with equal ease on each face, leading to equal amounts of the two enantiomeric bromides. The lack of stereochemical control in the formation of the C–Br bond is indicated by the wavy line representing the covalent bond to bromine.

tack at C-1 generates a secondary (rather than primary) cation. Whenever a more-stable carbocation is produced by a shift of a group from an adjacent atom, a rearrangement occurs rapidly. Here, a hydrogen originally bound to C-3 migrates with its electrons to this secondary center. This bond shift produces a rearranged carbocation that is benzylic as well as secondary. Capture of this cation by bromide gives rise to the observed racemic product.

Because of this rearrangement, bromine is found in the product at a carbon atom that did not originally participate in π bonding. Thus, we have two lines of evidence for a carbocationic rearrangement: (1) isomerization of the carbon skeleton and (2) the presence of the nucleophile in the product at a position not taking part in the reactant's double bond.

EXERCISE 10-E

Write a full reaction mechanism for each of the following transformations and explain the driving force for the formation of the rearranged skeleton.

(a) $\xrightarrow{\text{H}_3\text{O}^\oplus}$ (b) $\xrightarrow{\text{HBr}}$

HX Addition to Alkynes

Because the π system of an alkyne closely resembles that of an alkene, we might expect similar reactivity to result in the addition of H—X to triple bonds. Using our knowledge of the mechanism of electrophilic addition, let us work through the intermediates formed to understand the relative reactivity of alkynes and alkenes. The electrophilic addition of HBr to an alkyne (Figure 10-11, on page 342) results in protonation, employing the π electrons of one of the bonds in the triple bond. The resulting vinyl cation is digonal, bearing σ bonds to two, rather than three, substituents. As stated in Chapter 9, the vinyl cation is appreciably less stable than a comparably substituted trigonal sp^2-hybridized carbocation.

The vinyl cation has a stability somewhat like that of a primary sp^2-hybridized cation. Thus, its formation by protonation of an alkyne is more difficult than the corresponding protonation of an alkene to form a secondary, or more-stable, cation, and the rate of protonation is somewhat slower. In fact, this reaction probably would not take place at all if it were not for the higher electron density of a triple bond than that of a double bond. As in electrophilic additions to alkenes, however, the second step (capture of the vinyl cation by bromide) should be particularly facile, generating a vinyl bromide. Because the vinyl cation is linear, it is easily attacked on either side of the remaining alkenyl double bond, generating a mixture of both *cis*- and *trans*-vinyl halides.

The vinyl halide formed in this step is an alkene derivative, and also is subject to electrophilic attack by HBr. Electron withdrawal by the electronegative bromine atom leaves less electron density in the π bond of a vinyl bromide than in that of a simple alkene, and further electrophilic attack is slower. There are two regiochemical possibilities for the protonation: at the carbon bearing hydrogen or at that bearing bromide. Both cations are secondary, but that formed by the latter protonation bears an adjacent bromine atom. An additional resonance contributor can be written by employing

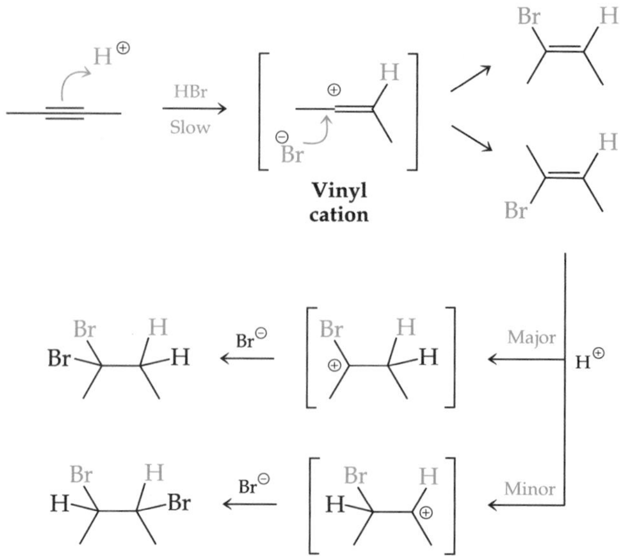

▲ FIGURE 10-11

Protonation of an alkyne is somewhat slower than that of an alkene because this reaction produces a vinyl cation, a particularly unstable cationic intermediate (Chapter 9). It is rapidly captured by a nucleophilic halide anion to form a vinyl halide. Further protonation is preferred at the carbon atom of the alkene *not* bound to bromine: this allows the lone pairs of the covalently bound bromine atom to stabilize the resulting cation. When this cation is trapped by a second bromide ion, geminal rather than vicinal dihalide is formed.

one of bromine's lone pairs to interact with the vacant *p* orbital, stabilizing this cation and making it more stable than the alternate one.

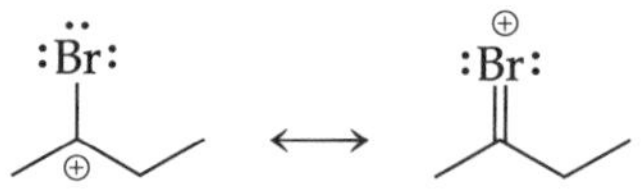

As in all electrophilic additions, the more-stable cation is formed preferentially, dictating the observed regiochemistry. Capture of this cation by bromide gives rise to a geminal dihalide (Figure 10-11); that is, a dihalide in which the two bromine atoms are bound to the same carbon atom.

The observed regiochemistry for the addition (to form geminal halides) follows from the preferential formation of the more-stable cation; that is, in accord with Markovnikov's Rule. With a terminal alkyne, protonation takes place so as to produce the more-stable secondary vinyl cation, leading to Markovnikov addition (Figure 10-12). The second addition again takes place so as to produce a geminal halide.

Although alkynes are attacked by electrophiles somewhat more slowly than are simple alkenes, it is nonetheless possible to stop (with judicious control of stoichiometry) at the alkenyl bromide stage, in which one molecule of HBr has been added across the triple bond. This control is possible because of the deactivation of the alkenyl bromide toward electrophilic attack by the electronegative bromine atom introduced in the first addition.

The hydration of alkynes also can be achieved by electrophilic addition, usually by treatment with a mercuric salt in an aqueous acidic solution. The role of the mercuric ion has not been established unambiguously, but it probably acts as a Lewis acid in the first stage of the reaction, facilitating the

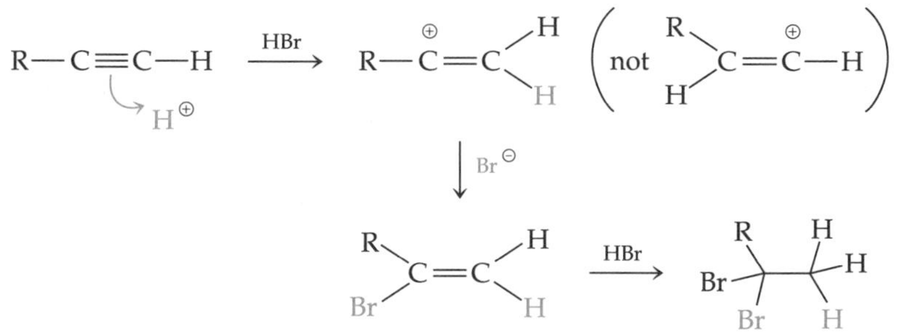

▲ FIGURE 10-12
Markovnikov addition to a terminal alkyne is observed because a secondary
vinyl cation (rather than the less stable primary vinyl cation) is formed upon
protonation.

rather difficult formation of what would formally be a vinyl cation. The
availability of *d* electrons on mercury makes it possible to further stabilize
the cation as a bridged structure called a **mercuronium ion** (Figure 10-13).
(As we will see in Section 10-4, bridged cations are also intermediates in halo-
genations and other addition reactions of third-, fourth-, and fifth-row elements.) Attack by water on the mercuronium ion takes place in accord with
Markovnikov's Rule; that is, at the more highly substituted carbon. The

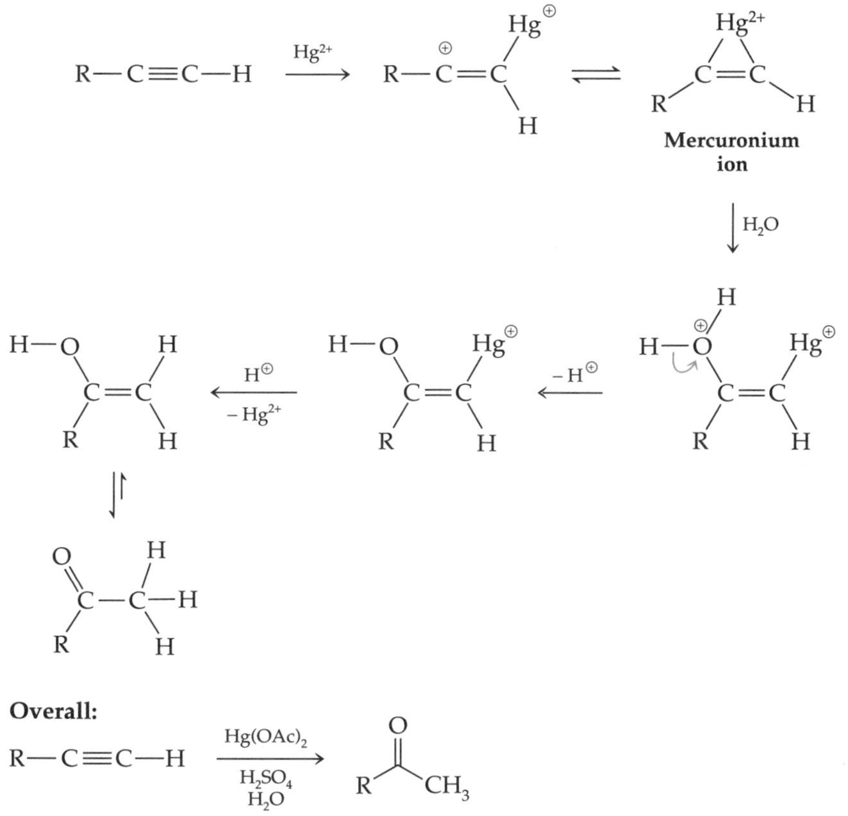

▲ FIGURE 10-13
In the hydration of a 1-alkyne, mercuric ion binds to C-1 to form the more
highly substituted vinyl cation, which likely exists as a mercuronium ion.
Nucleophilic attack by water, followed by deprotonation of the resulting
oxonium ion, produces an enol. Acid-catalyzed tautomerization leads to the
observed methyl ketone.

resulting oxonium ion then loses a proton to produce an enol that also contains a carbon–mercury bond. The C–Hg bond is then replaced by a C–H bond upon treatment with acid. Because tautomerization of the resulting enol to the more-stable keto form is rapid in aqueous acid, the enol is converted into the corresponding ketone. Starting with a 1-alkyne, a methyl ketone is formed. With a nonterminal alkyne, mixtures of the two possible ketones are usually obtained because there is only a small difference in energy between the two possible vinyl cations formed in the initial electrophilic attack.

EXERCISE 10-F

For each of the following HX additions, predict the regiochemistry of each adduct and determine whether the product mixture is optically active. If not, determine whether the inactivity results from the absence of chiral centers, the formation of equal amounts of enantiomers, or the formation of a *meso* compound.

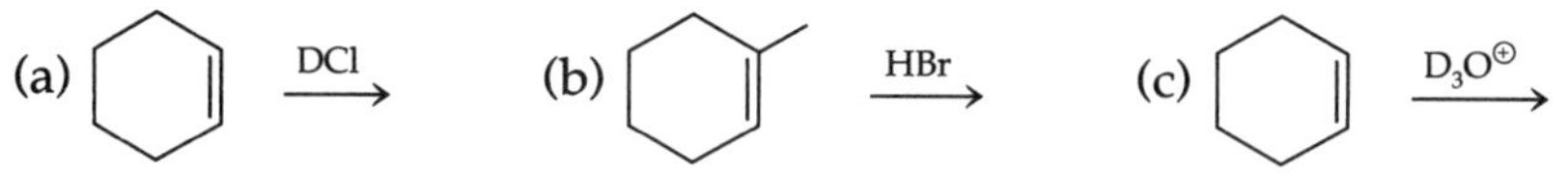

(a) $\xrightarrow{\text{DCl}}$ (b) $\xrightarrow{\text{HBr}}$ (c) $\xrightarrow{\text{D}_3\text{O}^\oplus}$

10-3 ▸ Reversing Markovnikov Regiochemistry

If we are to use electrophilic additions as synthetic methods for making alkyl halides and alcohols, it is necessary to have a technique complementary to simple H–X addition in which normal (Markovnikov) regiochemical preference is reversed. Free radical hydrobromination and hydroboration-oxidation are shown in this section to be useful methods for preparing alkyl halides and alcohols by anti-Markovnikov addition.

Radical Addition of HBr

Although Markovnikov's Rule applies to the electrophilic addition of HBr to carbon–carbon double bonds in most organic solvents, the reverse regiochemistry is sometimes observed when the reaction is conducted in ether.

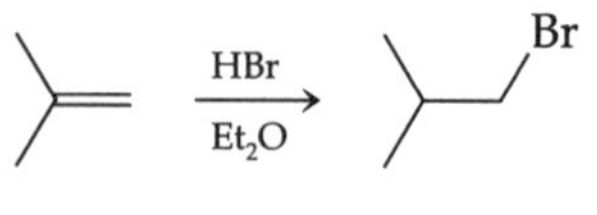

$$\xrightarrow[\text{Et}_2\text{O}]{\text{HBr}}$$

Because the reaction mechanism determines the regiochemistry of a reaction, this switch in regiochemistry means that the anti-Markovnikov hydrobromination product was formed by a different mechanism than that discussed in Section 10-2.

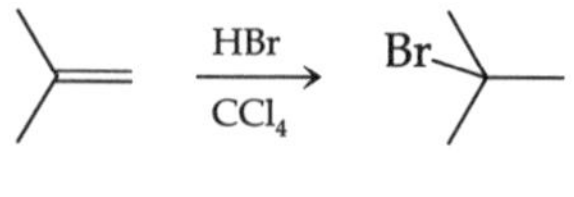

$$\xrightarrow[\text{CCl}_4]{\text{HBr}}$$

Ether solvents are notorious for containing small amounts of **peroxide** (ROOR) or hydroperoxides (ROOH) formed by partial decomposition of the ether upon standing. Peroxides and hydroperoxides are characterized by a

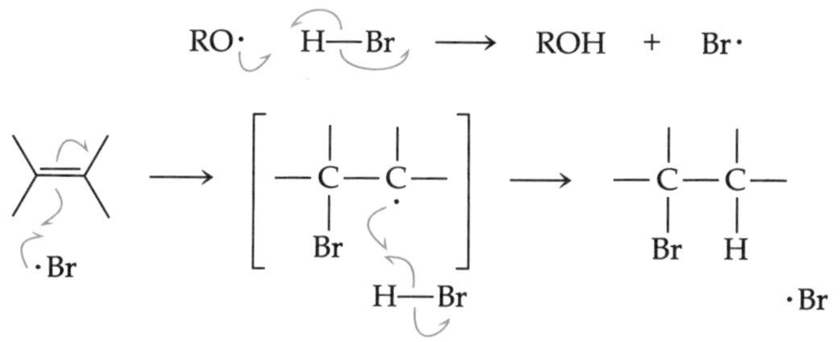

$$RO\cdot \quad H\!-\!Br \quad \longrightarrow \quad ROH \quad + \quad Br\cdot$$

$$\underset{\cdot Br}{\diagdown\!\!=\!\!\diagup} \quad \longrightarrow \quad \left[\begin{array}{c} | \quad | \\ -C\!-\!\overset{\cdot}{C}- \\ | \quad | \\ Br \end{array} \right. \underset{H-Br}{} \quad \longrightarrow \quad \begin{array}{c} | \quad | \\ -C\!-\!C- \\ | \quad | \\ Br \quad H \end{array}$$
$$\cdot Br$$

▲ FIGURE 10-14

When peroxides are present, alkoxy and hydroxy radicals can initiate a radical chain addition. The alkoxy radical abstracts hydrogen from H–Br, producing a bromine atom. This electron-deficient radical attacks a double bond, forming a C–Br bond and the more-stable carbon free radical. This radical, in turn, abstracts hydrogen from H–Br, thus leading to an H-Br adduct. Because the bromine radical attacks the double bond (rather than a proton as in an electrophilic addition), the opposite regiochemistry is observed for this radical addition. The two steps in the lower reaction constitute propagation steps, which are repeated in a cycle to yield product, even when only a few initiating radicals are available.

weak oxygen–oxygen bond that can be cleaved homolytically either by gentle warming or by light to initiate a radical chain like that described in Chapter 7.

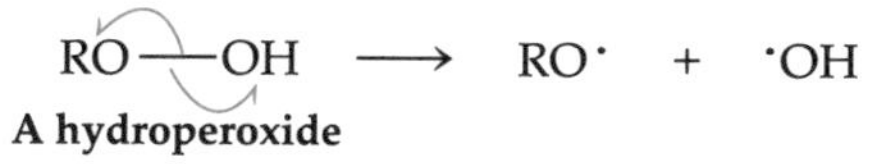

$$\underset{\textbf{A hydroperoxide}}{RO\!-\!OH} \quad \longrightarrow \quad RO\cdot \quad + \quad \cdot OH$$

Homolytic cleavage of the O–O peroxide bond generates alkoxy and hydroxy radicals that can serve as initiators for a radical chain reaction, making it possible to form products in ether whose regiochemistry is opposite that observed in solvents when a mechanism that proceeds through cationic intermediates is followed.

Interaction of an alkoxy radical with HBr results in the rapid abstraction of a hydrogen atom, generating a reactive bromine atom (Figure 10-14) that then attacks the alkene to begin a radical chain propagation. Thus, bromine adds to the double bond, forming a σ bond in which one electron is contributed by the bromine radical and the other by the π system of the reactant alkene. The second electron of the π bond remains in a *p* orbital on carbon.

The bromine atom, although neutral, is electron deficient, lacking one electron from a filled-shell configuration. It therefore interacts with double bonds in much the same way as a positively charged electrophile. Because a radical requires only a single electron from the double bond to form a σ bond, however, this radical attack generates a free radical rather than a carbocation. (Recall that half-headed arrows are used to indicate the motion of a single electron.) This sequence constitutes chain propagation in that a reactant free radical is converted into a product that also is a free radical: the number of reactive radicals on the left and right sides of the equation for that step are equal. When this alkyl free radical attacks another molecule of HBr, abstracting hydrogen, a formal hydrohalogenation of the C=C double bond is accomplished as a bromine atom is regenerated. This bromine atom can then attack a second alkene, and the cycle can be repeated again and again.

These two propagation steps consume, respectively, one mole of alkene and one mole of HBr to generate one mole of the adduct. The first step, which generates a carbon free radical, is thermodynamically more difficult than the second. The second step, which consumes the alkyl free radical, generating a C–H σ bond, is faster. The intermediate formation of the radical determines the regiochemistry of this reaction. Because radicals and cations follow the same order of reactivity (3° > 2° > 1°), and because the identity of the attacking reagent is reversed (H⁺ in electrophilic hydrobromination and Br in radical hydrobromination), the regiochemistry of this free radical addition also is reversed from that observed in an electrophilic addition (from that predicted from Markovnikov's Rule). This reversed regiochemistry in free radical additions is therefore said to have occurred in an **anti-Markovnikov sense.**

Let us consider, for example, the free radical addition of HBr to 1-methylcyclohexene. Following the same reaction mechanism as in Figure 10-14, the alkoxy radical, formed by the decomposition of trace amounts of peroxide, abstracts hydrogen from HBr to produce a bromine atom, which adds at C-2 to generate the more-stable tertiary radical. The alternative secondary radical is not formed to an appreciable extent because it is considerably less stable. (Notice that the regiochemistry of the addition is fixed at this point.)

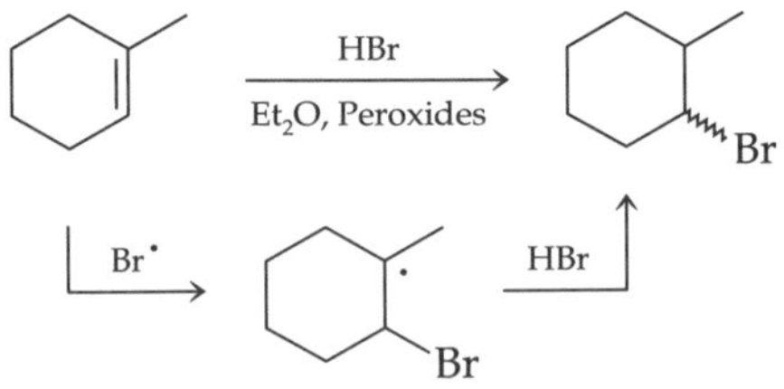

Although the order of radical stability is the same as that observed for carbocations, the identity of the attacking species has changed from being an electrophile (a proton, H⁺) to a bromine free radical. Thus, the anti-Markovnikov product is formed.

The reaction is completed as this planar tertiary radical abstracts hydrogen from a second molecule of HBr. Hydrogen abstraction takes place to an equal extent on the two faces of this planar intermediate so that equal amounts of *cis* and *trans* addition of HBr result. Thus, conducting HBr additions in the presence of peroxides allows us to circumvent the usual regiochemical preference in HBr additions, but it does not lead to controlled stereochemistry. Radical addition is restricted to hydrobromination and fails with other hydrogen halides.

EXERCISE 10-G

To obtain each of the following products, would you use a solvent (such as ethanol and water) that favors polar intermediates or one (such as ether) that favors radical reactions?

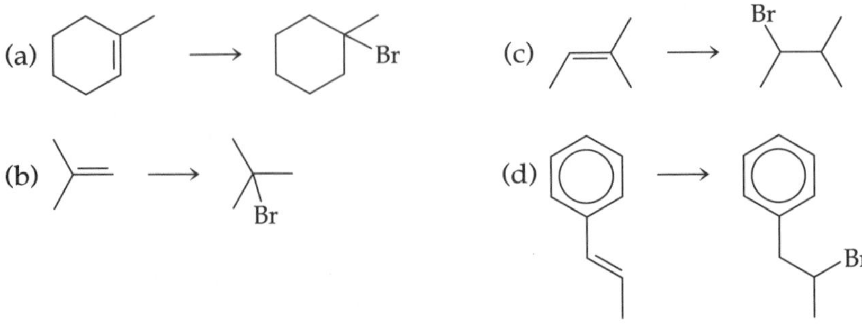

Hydroboration–Oxidation

Anti-Markovnikov hydration can be achieved by employing a sequence of two reactions (hydroboration followed by oxidation) that also avoid carbocation intermediates. Thus, products with opposite regiochemistry are formed in the hydration of 1-methylcyclohexene in dilute acid and in the **hydroboration-oxidation** sequence.

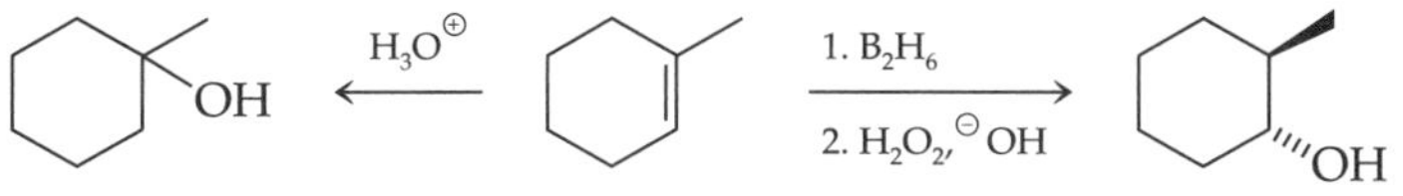

Let us consider the latter reaction more closely. The agent employed in this reaction is diborane, B_2H_6. Diborane exists in equilibrium with small amounts of borane, BH_3.

$$B_2H_6 \;\rightleftharpoons\; 2\,BH_3$$

Because boron has only three valence electrons, it can form only three covalent bonds and remain neutral. Thus, trivalent borane is highly electron deficient and acts effectively as a Lewis acid to coordinate with virtually any source of electron density. (For this reason, it exists predominantly as a dimer rather than as a monomer.) When brought into contact with a double bond, however, a boron–hydrogen bond can interact directly with the alkene π orbital to form the four-membered transition state shown in Figure 10-15. Thus, a carbon–boron and a carbon–hydrogen bond are formed on the same face of the double bond. This first step, referred to as **hydroboration,** is therefore a *syn* addition. Hydroboration has proved to be a very important reaction in both organic and inorganic chemistry. For his extensive and pioneering studies on organoboron chemistry, Herbert C. Brown of Purdue University was awarded a Nobel Prize in chemistry in 1979.

Although we will not consider how the next step, a peroxide-induced oxidation, proceeds mechanistically, carbon–boron bonds can commonly be replaced by carbon–oxygen bonds upon treatment with alkaline hydroperoxide. When this occurs, the substrate shown is converted into an alcohol. The orientation of the C–O bond formed in the oxidation step is identical with that of the original C–B bond, so that configuration is retained in this conversion. The *syn* addition achieved through hydroboration thus leads to *syn* addition in the formation of the final alcohol product.

To probe regiocontrol in hydroboration, let us consider the reaction of methylcyclohexene (Figure 10-16). There are two potential transition states for the hydroboration of methylcyclohexene: one in which boron is placed so as to bond to the more-substituted carbon and one in which it bonds to the less-substituted carbon. It is much more favorable sterically to place the

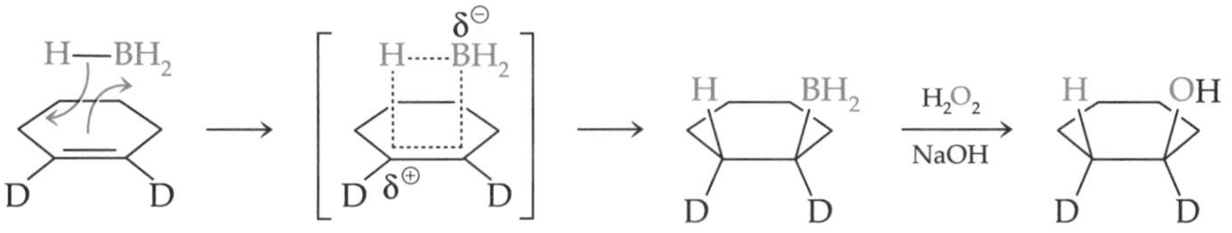

▲ **FIGURE 10-15**

Because hydroboration is a concerted reaction in which hydrogen and BH_2 are delivered simultaneously to one face of the alkene, *syn* addition is observed. When BH_2 is replaced by an OH group in the oxidation step, the configuration at the center of chirality is retained and the same stereochemical relation (*syn*) is maintained between the H and OH groups.

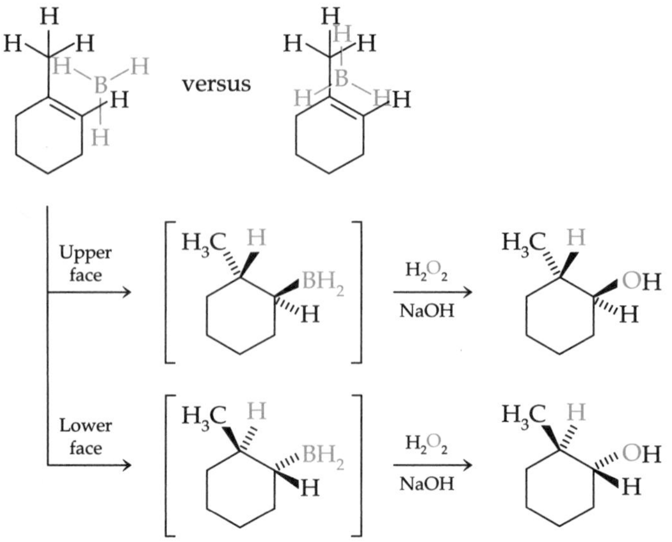

▲ FIGURE 10-16

Hydroboration of an unsymmetrically substituted double bond delivers the small hydrogen atom to the more-crowded carbon of the original C=C double bond and the larger BH$_2$ group to the less sterically congested carbon atom. This favors the regiochemistry shown at the upper left over that shown at the upper right. The regiochemistry having been determined, the concerted attack can occur with equal facility from the upper or lower faces of the alkene π bond.

larger BH$_2$ group away from the large substituent than to have the large groups occupying the same region in space. Furthermore, BH$_3$ is an electrophilic species, and this regiochemistry is consistent with Markovnikov's Rule.

The structure at the upper left of Figure 10-16 leads to *syn* addition of BH$_2$ and H with the boron at C-2, and the alcohol formed by oxidation of this alkylborane (by replacement of C—BH$_2$ by C—OH with retention of configuration) also has the final OH group located *syn* to H and hence *trans* to the methyl group at C-2. Although two centers of chirality are generated in this hydroboration-oxidation, equal amounts of the two enantiomers are formed because borane attacks the top and bottom faces of the alkene with equal facility. Thus, *syn* addition leads specifically to a racemic modification of these chiral compounds.

EXERCISE 10-H

For each of the following reactions, predict whether the desired stereochemistry and regiochemistry can be attained with acid-catalyzed hydration, hydroboration-oxidation, or not at all.

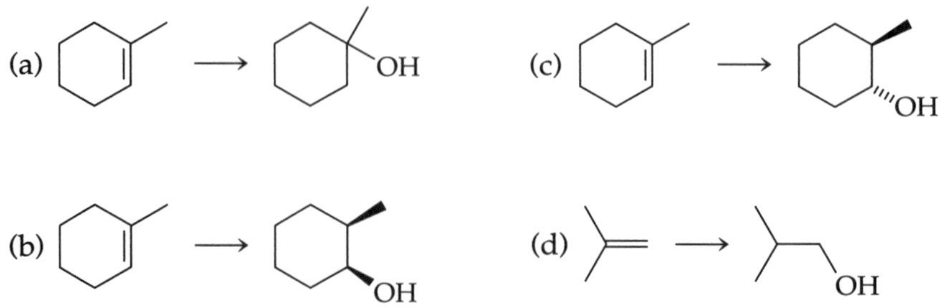

(e)

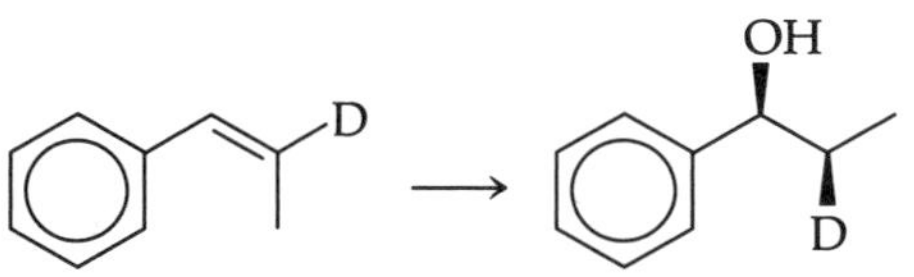

(f)

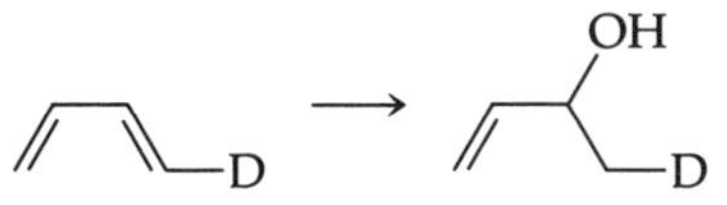

10-4 ▸ Other Electrophiles

In the electrophilic additions discussed so far, protons (from HX) have been the most-common electrophilic species. Other positively charged or partially positively charged reagents also can fill this role.

Oxymercuration–Demercuration

A reaction in which mercuric ion (Hg^{2+}) reacts as the electrophile in the first step of addition is shown in Figure 10-17. Mercuric acetate in acetic acid delivers ^+HgOAc as a positively charged metal ion to the π electron density of the double bond. Sigma-bond formation between carbon and mercury results in a carbocation that is then attacked either by acetate ion (a weak nucleophile) or by water. In dilute aqueous solution, water wins this competition, and the resulting oxonium ion, after deprotonation, leads to an alcohol with a carbon–mercury bond. This carbon–mercury bond can be converted into a carbon–hydrogen bond by reduction with sodium borohydride. Again, we will not consider the details of the mechanism of this reduction, although it is known that it does not include charged intermediates. The overall process, referred to as **oxymercuration-demercuration,** effects hydration of the alkene.

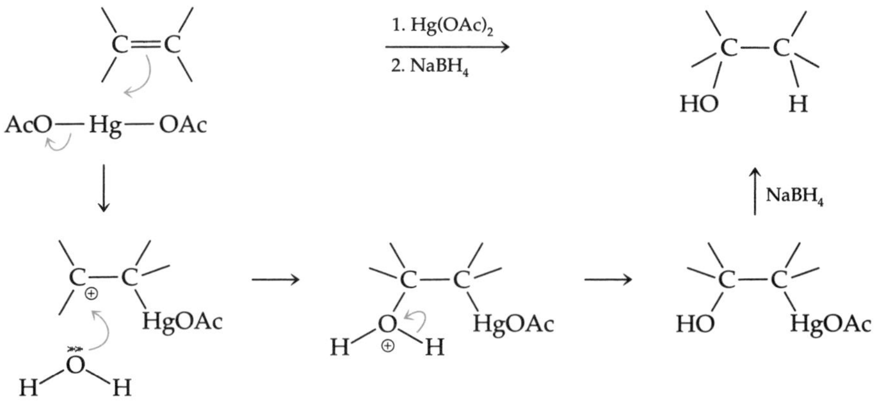

▲ FIGURE 10-17
When mercuric ion acts as the electrophile, the carbocation formed is very rapidly trapped by water. This nucleophilic attack precludes cationic rearrangement, implying that the carbocation is not "free." In a final step, the C–Hg bond is reduced with sodium borohydride, resulting in a C–H bond. Oxymercuration-demercuration gives hydrated product with the regiochemistry predicted from Markovnikov's Rule but without cationic rearrangement.

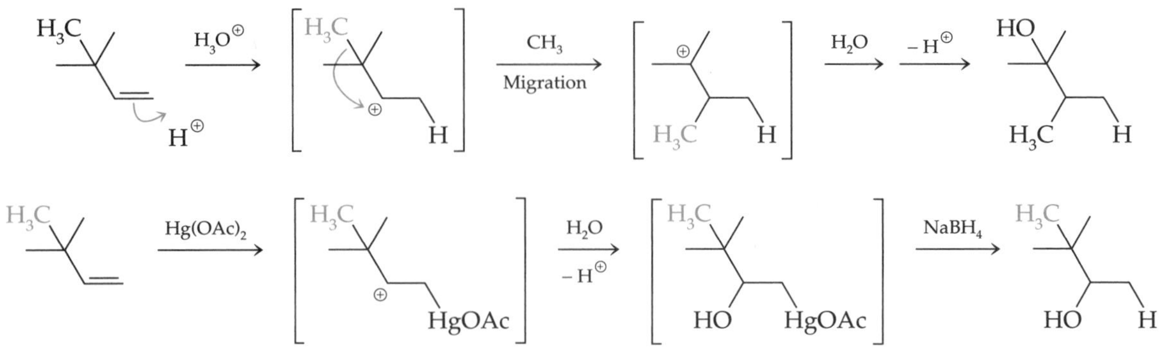

▲ FIGURE 10-18
Rearrangements that occur in an acid-catalyzed hydration (upper reaction) are avoided in an oxymercuration-demercuration pathway (lower reaction).

The cation formed by electrophilic attack by mercuric acetate is apparently not free to react by normal carbocation pathways because rearrangement does not accompany this reaction. The cationic character in the transition state of the rate-determining step does, however, control the regiochemistry of this addition, and oxymercuration-demercuration proceeds with a regiochemistry in accord with Markovnikov's Rule.

The utility of this reaction compared with direct acid-catalyzed hydration can be seen by comparing the products formed from *t*-butylethylene under each set of reaction conditions. As shown in the upper reaction in Figure 10-18, acid-catalyzed hydration proceeds in a Markovnikov sense to give the more-stable secondary cation. This cation, however, can rearrange to an even more stable tertiary cation by migration of a methyl group from the *t*-butyl group (C-3 to C-2). Such carbocation rearrangements are fast when there is a thermodynamic driving force (here, the conversion of a secondary cation into a tertiary cation). This intermediate is then attacked by water to form the tertiary alcohol after deprotonation of the intermediate oxonium ion.

In contrast, in oxymercuration the secondary cation is trapped before it can rearrange, leading ultimately to a different alcohol. This lack of rearrangement is believed to be the result of stabilization of the secondary cation by electron donation from mercury (Figure 10-19), resulting in a blocked carbocation—possibly a three-membered-ring species. Because of the internal stabilization in this intermediate, the driving force for carbocationic rearrangement is substantially reduced. (We will see similar three-membered intermediates formed from bromine and chlorine shortly.) The secondary cation can thus be trapped by water without skeletal rearrangement.

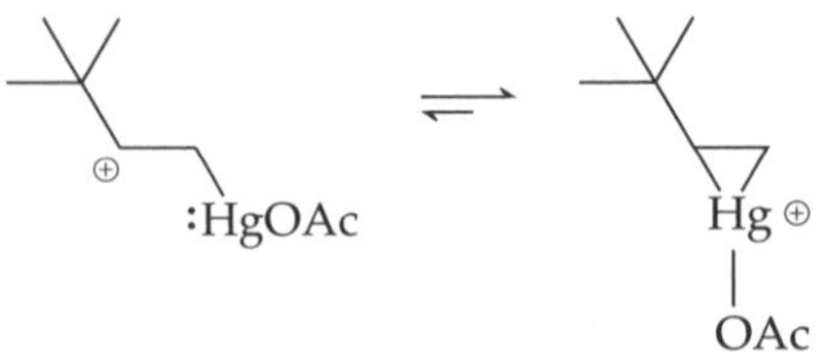

▲ FIGURE 10-19
Because mercuric ion has available nonbonding electron pairs, it can stabilize an adjacent carbocation by back donation. The resulting cation is no longer a free carbocation, and normal carbocationic rearrangements are blocked.

For each of the following reactants, predict the product that would be formed by:

1. oxymercuration-demercuration
2. acid-catalyzed hydration
3. hydroboration-oxidation

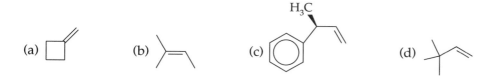

(a)   (b)   (c)   (d)

Addition of X_2

Dihalogens also function as electrophiles in addition reactions, producing vicinal dihalides in high chemical yield. The addition of bromine to double (or triple) bonds is very rapid and, because Br_2 in CH_2Cl_2 is deep red, its consumption in forming the colorless dibromide can be used as a color indicator for the presence of double bonds in a molecule. The product of bromination of an alkyne is a tetrahalide.

Although the X–X bond in a dihalogen (Cl_2, Br_2, or I_2) is nonpolar, it becomes polarized as it approaches a center of electron density such as the π cloud of a carbon–carbon double bond (Figure 10-20). The electron density in the π bond causes the electrons in the bromine–bromine σ bond to shift toward the atom remote from the π orbital. As the reactants get closer

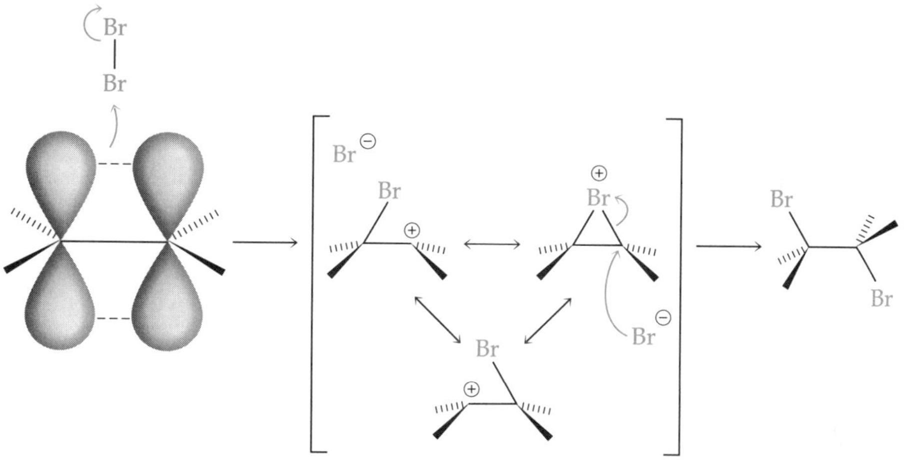

▲ **FIGURE 10-20**
Bromine becomes polarized as it approaches a π cloud, so that the nearer bromine atom acts as an electrophile. A σ bond is formed between carbon and bromine as the electrons of the π bond are donated toward the partially positively charged bromine atom. The nonbonding electron pair on the attached bromine backbonds to the cationic carbon, forming a cyclic bromonium ion in which bromine bears formal positive charge. Backside nucleophilic attack by bromide ion (an S_N2 pathway) opens the ring, producing the dibromide in which the two bromines have been added on opposite sides of the π bond—that is, in an *anti* addition.

together, electrons flow from the carbon–carbon π bond toward the nearer bromine as the remote bromine leaves as a bromide ion, taking with it the electron pair in the original σ bond that linked the halogen atoms.

At first glance, this process appears quite similar to that in which a proton is delivered to the π bond from a strong acid (Figure 10-2). However, Br has lone pairs that can bridge, and therefore stabilize, the incipient carbocation. Bonding of a lone pair from bromine to the carbocation forms a cyclic intermediate referred to as a **bromonium ion** in which bromine, rather than carbon, bears formal positive charge. (Notice that the bromonium ion is very similar in structure to the bridged mercury ion in Figure 10-19.) This cyclic ion is symmetrical, and the bridging bromine atom effectively blocks further attack on that face of the double bond. The nucleophile required in the second step is therefore forced to attack from the opposite face, remote from the bridging bromine, to donate its electrons to form a second σ bond. This transition state resembles that of the backside attack of the S_N2 reactions discussed in Chapters 7 and 8. As bromide ion attacks, one of the carbon–bromine bonds of the cyclic bromonium ion is broken, and the electrons are returned to bromine as a lone pair. The resulting adduct is a *vicinal dibromide*, bearing two bromines on adjacent atoms.

The stereochemical outcome of this *anti* addition can be seen in the addition of bromine to a cyclic alkene in which the two bromines are oriented on opposite sides of the ring. For example, with cyclohexene, bromine addition proceeds to give *trans*-1,2-dibromide.

In the past, carbon tetrachloride was often used as a solvent for the addition of bromine to an alkene. Unfortunately, this solvent has been shown to have detrimental effects on the health of those exposed for a long period of time. Dichloromethane (CH_2Cl_2) can be used in place of CCl_4, although it, too, has recently come under suspicion of causing cancer because of the effect of long-term, high-level exposure on laboratory animals bred to be susceptible to carcinogenic compounds.

Through the *anti* addition of bromine, two centers of chirality are produced in the product. Nonetheless, the product is an optically inactive mixture. This result is the same as that for the other electrophilic additions so far considered, although its origin is somewhat different. Both of the carbons bound to bromine in the bromonium ion derived from cyclohexene are stereocenters (Figure 10-21). One can imagine a mirror plane in the molecule (through the bromine atom), and indeed the two stereocenters present are of opposite configuration (one *R* and one *S*).

Attack by bromide ion resulting in opening of this bromonium ion is equally likely at each of these two centers (Figure 10-21). Because the process is an S_N2 reaction, inversion of configuration occurs. Thus, reaction at the carbon of *S* configuration in the bromonium ion leads to the *R,R* dibromide, whereas attack at the *R* carbon leads to the *S,S* stereoisomer. These two isomers constitute the mirror-image components of the racemic mixture produced. [Note that generation of the *R,S* diastereomeric dibromide (a *meso* compound) by this pathway is not possible; indeed, none is observed.]

An alternate way of rationalizing the attack by bromide ion from the opposite face of a bromonium ion is to recognize that, although the bridging bromine bears a formal positive charge in this cyclic cation, it is still surrounded by electron pairs. Approach of the negatively charged bromide ion

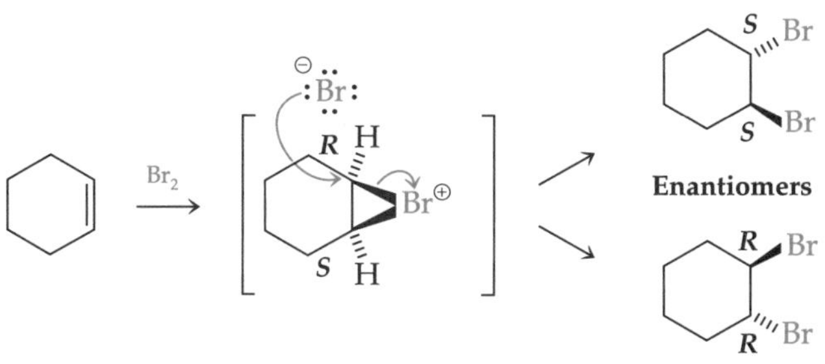

▲ FIGURE 10-21

Formation of a cyclic bromonium ion in the first step of electrophilic bromination produces two centers of chirality. Backside nucleophilic opening of the *R* center inverts it to the *S* configuration in the dibromide; alternatively, the same nucleophilic attack at the *S* center (not shown) inverts it to the *R* configuration. Had the initial attack been from the bottom face of the π bond (not shown), the opposite configuration at each of the stereocenters would have been formed in the cyclic bromonium ion; but, with backside nucleophilic attack by bromide (from the top), the same pair of enantiomers [(*R*,*R*) and (*S*,*S*)] would have been produced.

to the same face as the bridging bromine would result in a region of high electron density. This is avoided when bromide ion approaches from the back face, as in all S$_N$2 reactions.

This *anti* addition, in which two substituents assume a *trans* relation in cyclic systems, also has significance for acyclic compounds. *Anti*-addition to *cis*-2-butene produces the isomers shown in Figure 10-22. This is the *R*,*R* / *S*,*S* pair rather than *R*,*S* (*meso*) compound, as is seen more easily with the Fischer projections shown at the right in Figure 10-22.

EXERCISE 10-J

Write a detailed mechanism for the bromination of *trans*-2-butene. Identify whether the product mixture is a *meso* compound or *d*,*l* pair. (Be careful in assigning the stereochemistry of the bromonium ion: the situation here is different from that of the bromination of *cis*-2-butene.)

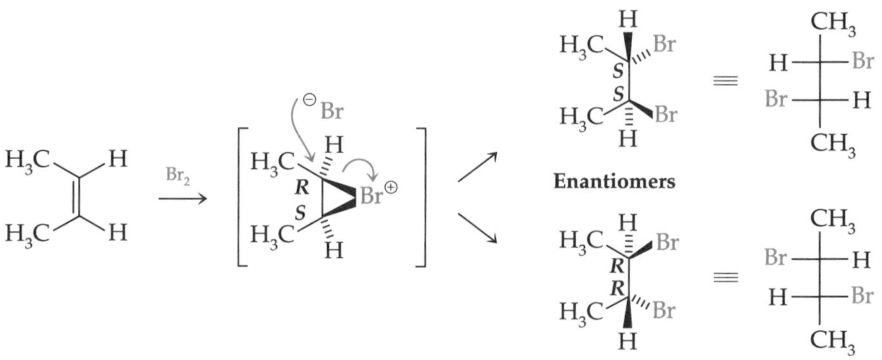

▲ FIGURE 10-22

Delivery of an electrophilic bromine from the top face of the π bond produces the cyclic bromonium ion shown. Nucleophilic attack at the *R* center leads to the *S*,*S* enantiomer; that at the *S* center (not shown) gives the *R*,*R* enantiomer. Because these two compounds are formed in equal amounts, an optically inactive racemic mixture is obtained.

We can imagine that similar, **cyclic halonium ions** could also be formed with chlorine and iodine. Apart from the higher electronegativity of chlorine, the **chloronium ion** is quite analogous to the bromonium ion. Iodine is both less electronegative and more polarizable than bromine. However, molecular iodine does not react with simple alkenes to form diiodoalkane products. This lack of reactivity is not due to kinetics or the inability to form a bridged ion; rather, it is because the strengths of the two carbon–iodine bonds of the product are less than the sum of those of the carbon–carbon π bond and I–I bond, making the reaction endothermic. Fluorine is much more reactive than the other halogens and reacts even with C–H bonds, resulting in complex mixtures. Thus, the addition of a dihalogen through cyclic halonium ions is important only for Br_2 and Cl_2.

Sulfur, which is adjacent to chlorine in the periodic table, also is known to form bridged cyclic structures analogous to the halonium ions.

EXERCISE 10-K

Predict the structure of the intermediate cation and the stereochemistry of the final product in the reaction of each of the following reagents with cyclohexene.

(a) ICl   (b) PhSBr   (c) BrCl   (d) HOCl

The halogenation of an alkene generates a vicinal dihalide, which, as discussed in Chapter 9, can lose the two vicinal halogen atoms by treatment with zinc. The high yields obtained in such elimination reactions make bromination-debromination a means of reversibly protecting a double bond.

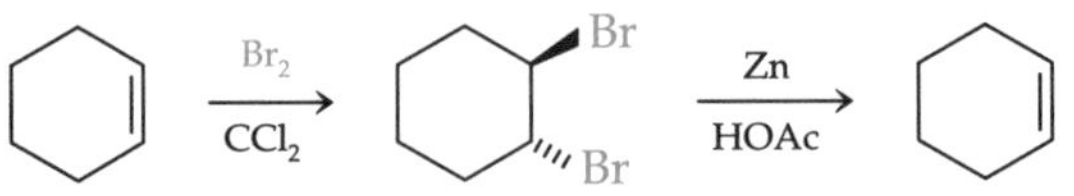

For example, cyclohexene can be brominated by treatment with Br_2 in CH_2Cl_2. In the dibromide, further reaction at the carbons of the original double bond is blocked, making it possible to use reagents (those that would ordinarily attack a double bond) to alter functionality in other parts of the molecule. The double bond can then be regenerated by removal of the two bromides by treatment with Zn. This reverse reaction thus lets us use the vicinal dibromide as a **protected alkene,** and the bromines are referred to as a **protecting group.** Here, a reactive functional group (the double bond) is "protected" from other reagents used to achieve transformations in other parts of the molecule. The concept of the protection of functional groups, as well as its applications, will be described in more detail in Chapter 15.

Because vicinal dihalides can be twice dehydrohalogenated by treatment with base (Chapter 9), halogenation also provides an intermediate through which an alkene can be converted into an alkyne, as shown here for *cis*-2-butene.

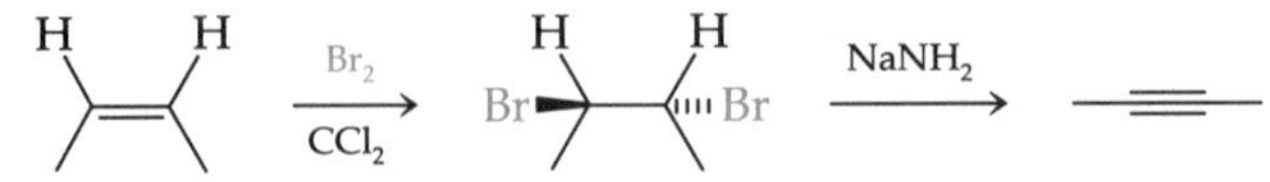

Carbenes

As described in Chapter 6, a singlet carbene has a vacant p orbital that is perpendicular to the plane containing the two σ bonds and a lone pair. This vacant p orbital makes the carbene electrophilic, allowing it to attack double bonds in the same way as other electrophiles considered in this chapter. The flow of electrons from a carbon–carbon π bond toward the vacant carbene p orbital would result in the formation of a σ bond and the development of cationic character at the other carbon of the alkene. As in the formation of the cyclic bromonium ion, however, the carbon atom to which this new σ bond is formed bears a lone pair that can backbond with the incipient carbocation, simultaneously forming another carbon–carbon bond. As in halonium ions, both new σ bonds are formed simultaneously; that is, the formation of a cyclopropane derivative is concerted, and there are no intermediates.

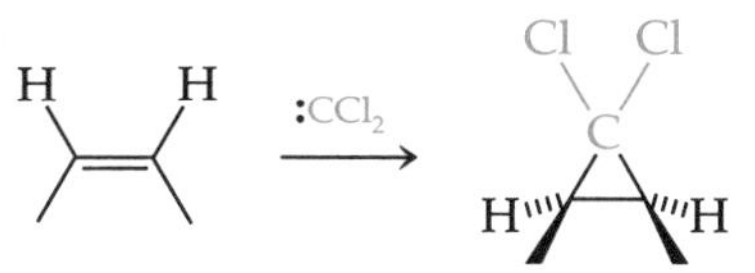

Because there is no free carbocation, the relative stereochemistry of the substituents on the starting alkene is maintained. Thus, when dichlorocarbene adds to *cis*-2-butene, the *cis* relation between the methyl groups is found in the product. This retention of stereochemistry in a carbene addition is excellent evidence for the concerted nature of its addition to the double bond.

Note that a concerted addition (parallel to the formation of a cyclic bromonium ion) is possible only from a singlet carbene. Because a triplet has two singly occupied orbitals, it is more likely to act as a radical than as an electrophile. With a triplet carbene (Figure 10-23), one expects the formation of a *biradical* intermediate because there is a single electron on each of the two developing centers of reactivity. (Recall from Chapter 2 that a biradical is a reactive intermediate in which two noninteracting radical sites are

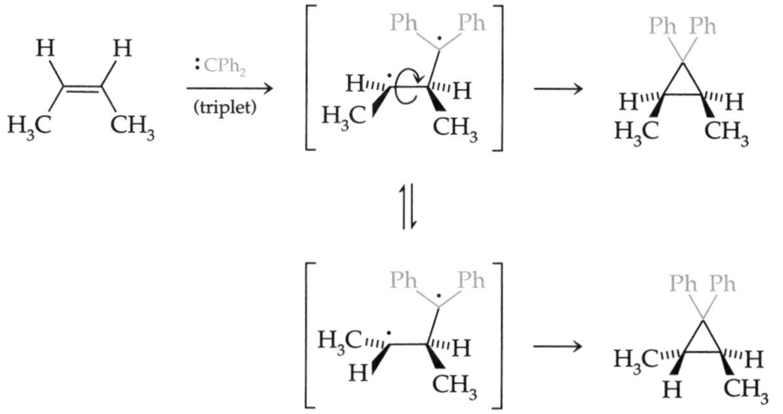

▲ FIGURE 10-23

The addition of a triplet carbene to an alkene proceeds in two steps through a biradical intermediate. After one σ bond is formed, free rotation about the C–C bond of what was the alkene scrambles the stereochemistry and both *cis*- and *trans*-cyclopropanes are obtained. This stereorandom result is unlike that obtained with singlet carbenes, in which retention of the original stereochemistry about the alkenyl bond is observed.

present within a single molecule.) This biradical can exist long enough for rotation to occur about the σ bond that was part of the double bond originally, which means that stereochemistry is lost. In a slower second step, the biradical can then collapse to a mixture of cyclopropanes: both *cis* and *trans* products are obtained.

Thus, to understand the stereochemistry of carbene addition, we must know whether the carbene is a singlet or a triplet. The multiplicity of a carbene is determined by the way in which it is formed. One method for forming carbenes is treatment of a geminal diiodide with a zinc-copper couple in the **Simmons-Smith reaction.**

$$\text{Ph}_2\text{CI}_2 \quad \xrightarrow{\text{Zn(Cu)}} \quad \text{Ph}_2\text{C:}$$

The carbenoid formed in the Simmons-Smith reaction is a singlet and adds stereospecifically. Triplet carbenes are usually formed through photochemical routes. The details of photochemical reactions are usually dealt with in more advanced courses.

EXERCISE 10-L

Predict the structures obtained by the addition of singlet dichlorocarbene (:CCl₂) to each of the following compounds and indicate whether the products are *meso* compounds or *d,l* pairs.

(a) ⬡ (b) /═\ (c) ∨∿∧

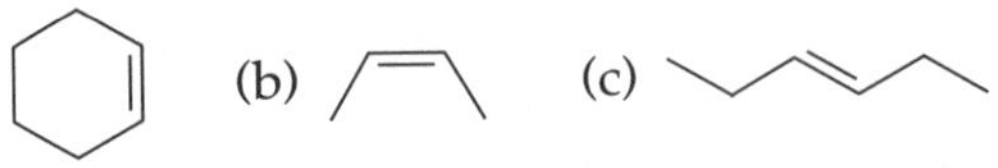

Epoxidation

Because oxygen ordinarily bears two nonbonding lone pairs, an electrophilic oxygen atom can form a bridged structure analogous to those that we have seen for sulfur, chlorine, and bromine. In analogy with the reaction of a dihalogen, an oxygen–oxygen σ bond in hydrogen peroxide (HO—OH) should become polarized as it approaches a double bond (Figure 10-24). An electrophilic attack by polarized hydrogen peroxide, however, would require hydroxide ion to act as a leaving group, which, as discussed in Chapters 8 and 9, is difficult. Because hydroxide ion is highly basic, heterolytic cleavage of the oxygen–oxygen bond is difficult. As a result, this reaction of hydrogen peroxide fails. However, peracids do react with alkenes to form three-membered-ring compounds containing oxygen. **Peracids** are oxygenated relatives of carboxylic acids: RCO₃H. Because of their stability and storability as laboratory reagents, *meta*-chloroperbenzoic acid and peracetic acid are often used for this purpose. Both of these reagents convert alkenes into **epoxides**, also known as **oxiranes,** in high yields.

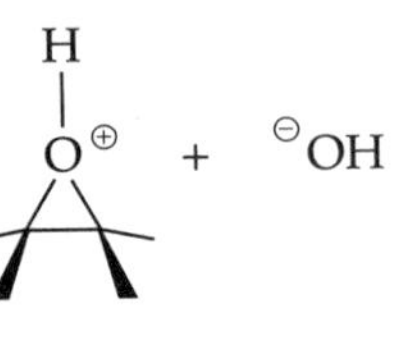

▲ FIGURE 10-24
Electrophilic attack by hydrogen peroxide on an alkene fails because hydroxide ion is a poor leaving group.

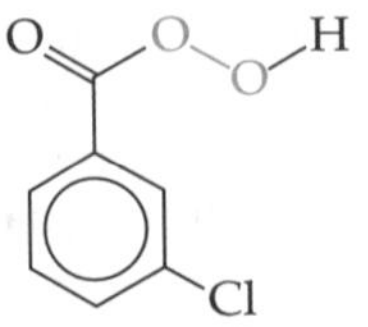

m-Chloroperbenzoic acid (MCPBA)

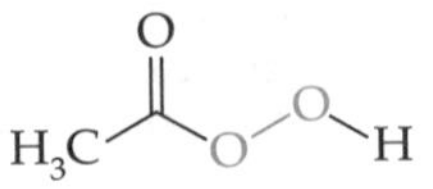

Peracetic acid

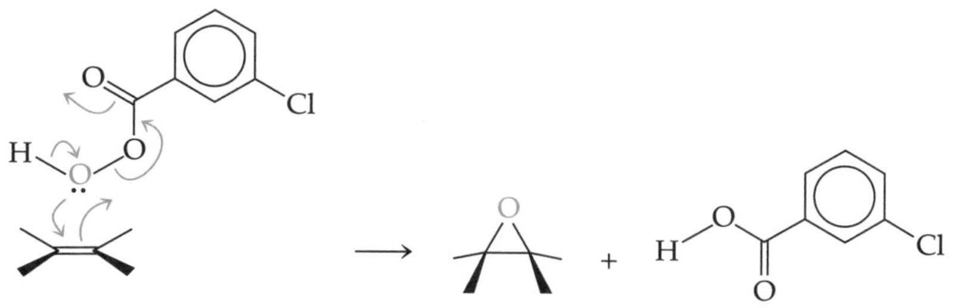

▲ FIGURE 10-25
The electrophilic (—OH) oxygen in a peracid initiates electrophilic attack on an alkene. As π electrons flow from the C=C double bond, the electrophilic oxygen backbonds by a pathway parallel to that for the formation of a halonium ion. The leaving group is a carboxylic acid, which can readily accommodate the O—O electron pair from the original σ bond.

The mechanism of peracid epoxidation is believed to include a cyclic, concerted transition state, as shown in Figure 10-25, in which oxygen is transferred to the alkene at the same time that the carbon–carbon bond is broken and the proton is transferred to the carbonyl oxygen.

EXERCISE 10-M

Write a reaction mechanism for the formation of epoxide by treating *cis*-2-butene with peracetic acid, and predict whether the product would be optically active.

Epoxides, like cyclic halonium ions, can undergo ring opening through backside attack by nucleophiles.

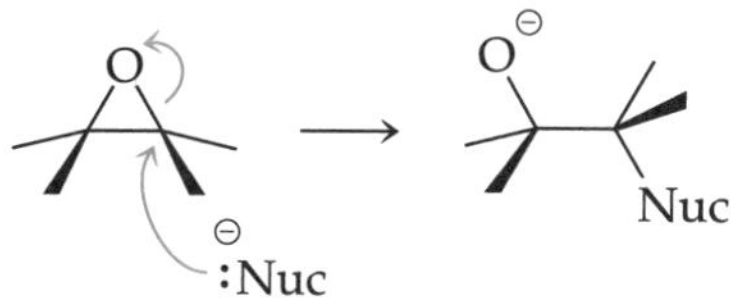

Ring opening of epoxides is quite slow unless a Lewis acid (or a proton) is complexed with oxygen. For example, the opening of epoxides with NaOH

ETHYLENE OXIDE: THE SIMPLEST EPOXIDE

Ethylene oxide, the simplest epoxide, is produced in very large quantities for incorporation into a number of products, including the antifreeze ethylene glycol and a variety of plastics. In addition, ethylene oxide is an effective sterilizing agent for medical equipment and a fumigant for food and clothing. Because of its low boiling point (11°C), it is readily removed after use.

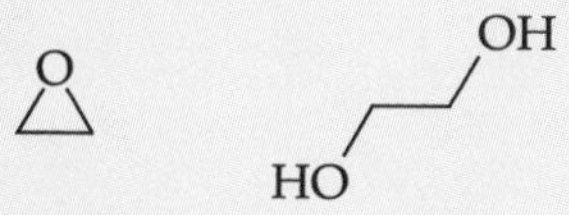

Ethylene oxide   **Ethylene glycol**

in H$_2$O requires elevated temperatures, as is done commercially for the synthesis of ethylene glycol from ethylene oxide.

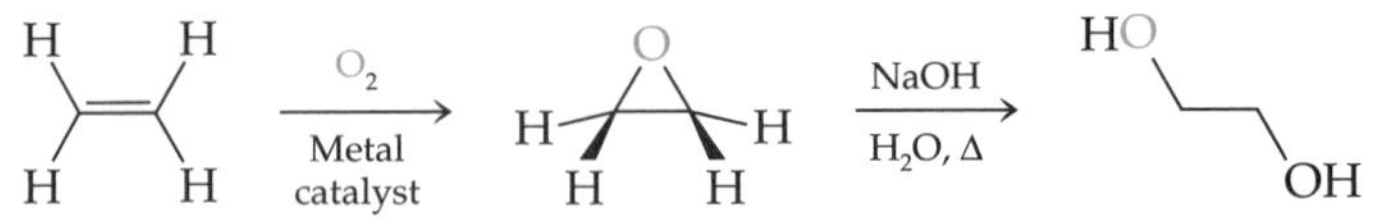

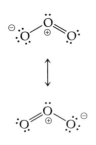

(Reagents such as peracetic acid are far too expensive for use in the production of bulk chemicals. In this case, ethylene oxide is produced from ethylene by the use of molecular oxygen and a metal catalyst.) In contrast, Grignard reagents react with epoxides at room temperature and below, forming a new carbon–carbon bond in the product alcohol (Chapter 8).

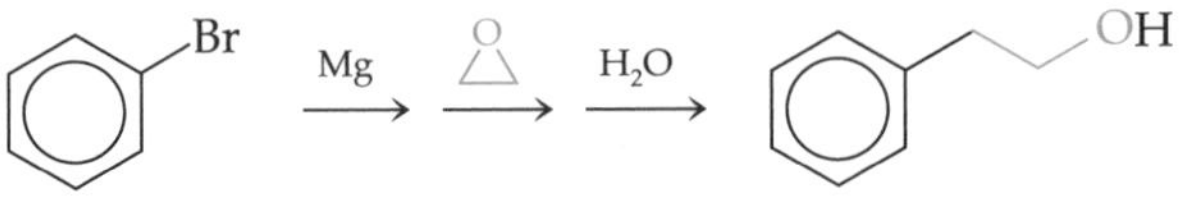

Epoxides are relatively reactive under biological conditions, and some molecules containing more than one of these functional groups have been demonstrated to induce cancer in laboratory test animals. Although the mode of biological action is not absolutely clear, carcinogenicity is thought to result from sequential reactions with both strands of double-stranded DNA.

EXERCISE 10-N

Suggest reaction mechanisms for each of the steps in the conversion of bromobenzene into 2-phenylethanol (the opening of ethylene oxide by phenyl Grignard reagent) as shown above.

Ozonolysis

Ozone (O$_3$) also acts as an electrophilic agent. It exists in a zwitterionic form in which the central oxygen formally bears positive charge. (Recall from Chapter 2 that a zwitterion is a neutral molecule that bears two oppositely charged centers in at least one significant resonance structure.) Two resonance contributors make the terminal oxygens partially negatively charged, but the formal oxygen–oxygen multiple bond confers part of the electrophilicity of the central positively charged oxygen to the terminal oxygens. This zwitterion can thus initiate electrophilic attack in which electrons from the carbon–carbon π bond of an alkene are polarized toward the terminal oxygen, allowing the shift of electrons shown in Figure 10-26. The resulting structure is referred to as a **molozonide,** a compound

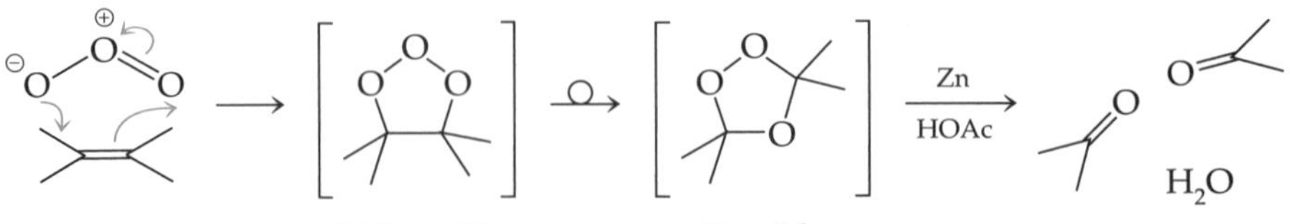

Molozonide   Ozonide

▲ FIGURE 10-26

Electrophilic attack by ozone proceeds through a six-electron transition state to form a molozonide that rearranges to an ozonide. Reductive workup with zinc in acetic acid results in the cleavage of what was a C=C double bond into carbonyl fragments.

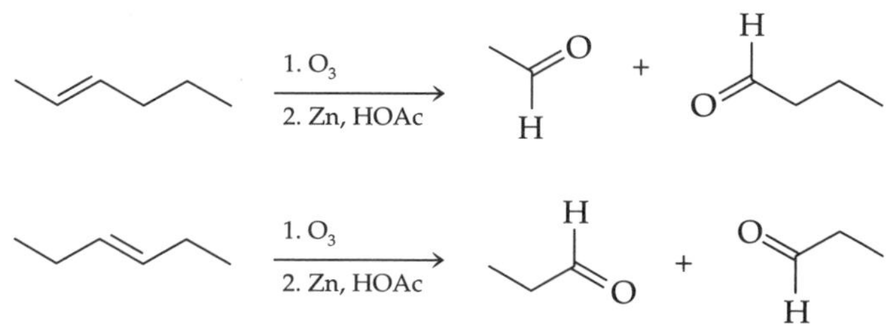

Ozonolysis is an oxidative degradation method by which C=C double bonds are converted into carbonyl fragments.

containing two weak oxygen–oxygen bonds. This compound is relatively unstable and rearranges to an **ozonide** by routes that need not be considered in a chapter on addition reactions. This rearrangement accomplishes the breaking of both the σ and the π bonds between atoms originally joined by a double bond, replacing them with two double bonds to oxygen. Upon treatment of an ozonide with zinc in acetic acid, the ozonide is reduced to the corresponding carbonyl compounds. The ozonolysis of alkynes, when followed by treatment with zinc and acetic acid, leads to carboxylic acids at each of the originally *sp*-hybridized atoms.

Ozonolysis thus effects the net conversion of a double bond into a pair of carbonyl compounds. This occurs by cleavage of the carbon skeleton in a process known as an **oxidative degradation.** Ozonolysis is a technique commonly used to simplify complex structures by replacing a C=C double bond with two carbonyl groups. This method has proved to be very effective in determining structures of complex naturally occurring compounds. Ozonolysis degrades a complex molecule to simpler fragments that can be more easily analyzed. In these simpler fragment molecules, the termini of what was a C=C double bond appear as aldehydes or ketones.

A simple example of this degradation is shown in Figure 10-27. One can distinguish *trans*-2-hexene from *trans*-3-hexene, despite their close chemical structures and reactivities, on the basis of the products formed upon ozonolysis. *Trans*-2-hexene gives rise to two products (acetaldehyde plus butanal), whereas *trans*-3-hexene affords only one (two equivalents of propanal). Ozonolysis is particularly important in the degradation of functionally complicated molecules of nature whose structures can be more readily deduced through analysis of the simpler corresponding carbonyl compounds than of the alkene itself.

Oxidative degradation can also be accomplished by treatment with hot aqueous potassium permanganate (Figure 10-28). This treatment is more vigorous than ozonolysis and, when a hydrogen is bound to the alkene carbon, the aldehyde that is formed is further oxidized to an acid by perman-

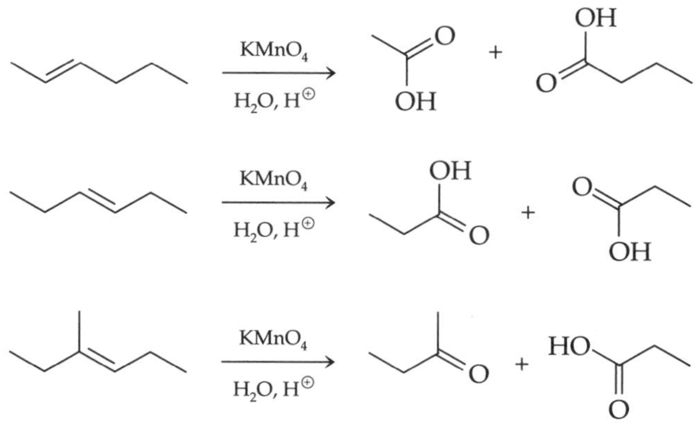

▲ FIGURE 10-28

Oxidative degradation with hot aqueous $KMnO_4$ converts a =CHR terminus of a C=C double bond into a carboxylic acid (RCO_2H). A =CR_1R_2 terminus (bottom reacton) is converted into a ketone (O=CR_1R_2) by this method.

ganate. For example, the oxidation of *trans*-2-hexene and *trans*-3-hexene with aqueous permanganate results in the acids shown in Figure 10-28. Permanganate is purple and its reduction product, MnO_2, is brown. The fading of the purple color of the permanganate ion is indicative of oxidative degradation and is another color test that can be used to quickly indicate the presence of oxidizable functional groups in an organic molecule. Alkenes, alkynes, alcohols, and aldehydes give positive permanganate oxidation tests.

EXERCISE 10-O

Predict the product(s) expected from ozonolysis of each of the following compounds:

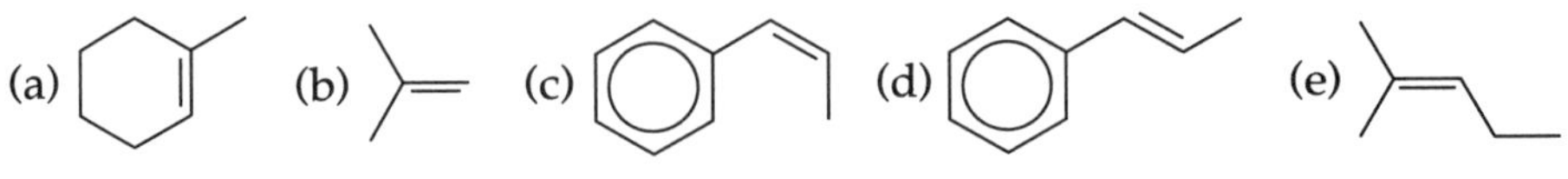

Carbocations as Electrophiles

The carbocation formed by the protonation of an alkene is itself an electrophilic species and can attack the starting alkene if its original concentration is high. For example, the protonation of styrene generates the carbocation shown in Figure 10-29 (with a regiochemistry consistent with Markovnikov's Rule). This species itself can be attacked by another molecule of styrene, forming a second carbon–carbon σ bond. Several fates are possible for this more-complex structure: as in the simple addition of H—X to an alkene (Section 10-2), the cation can be trapped by an anion to form a simple adduct (route *a*). This product has the formal elements of H—X

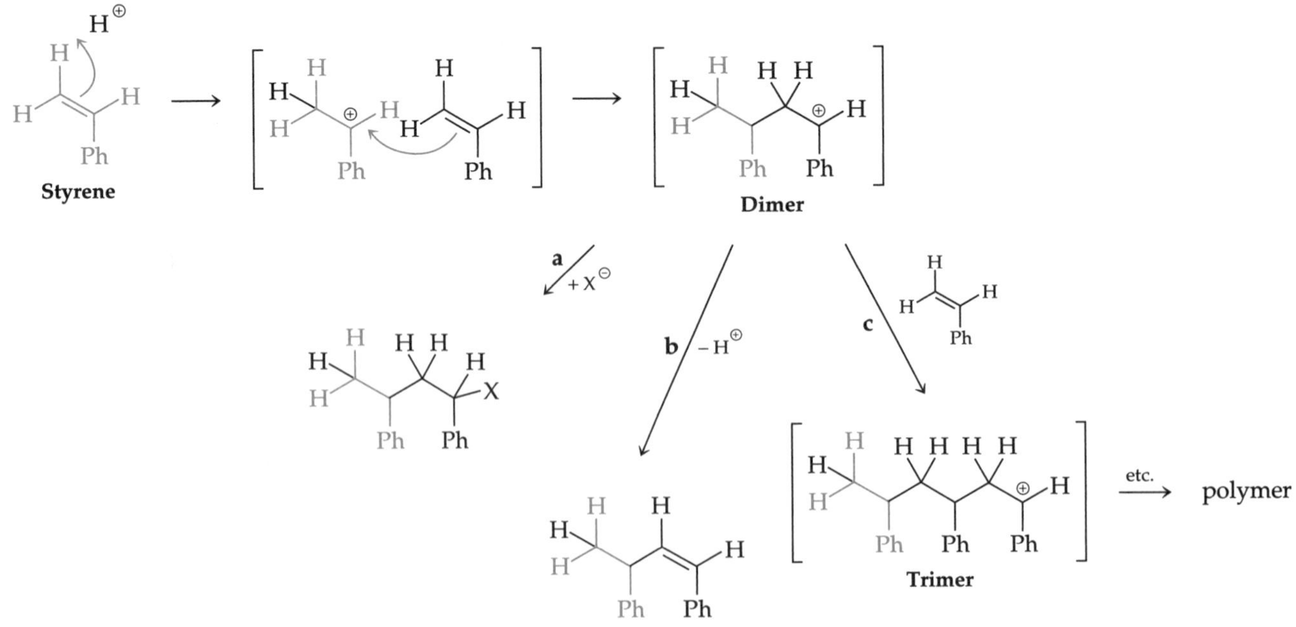

▲ FIGURE 10-29

Carbocations generated by protonation of an alkene can themselves act as electrophiles to attack another molecule of alkene. The cation formed by the combination of two alkenes can be captured by a nucleophile, X^- (path *a*); lose a proton to form a neutral dimer (path *b*); or attack a third molecule of alkene (path *c*).

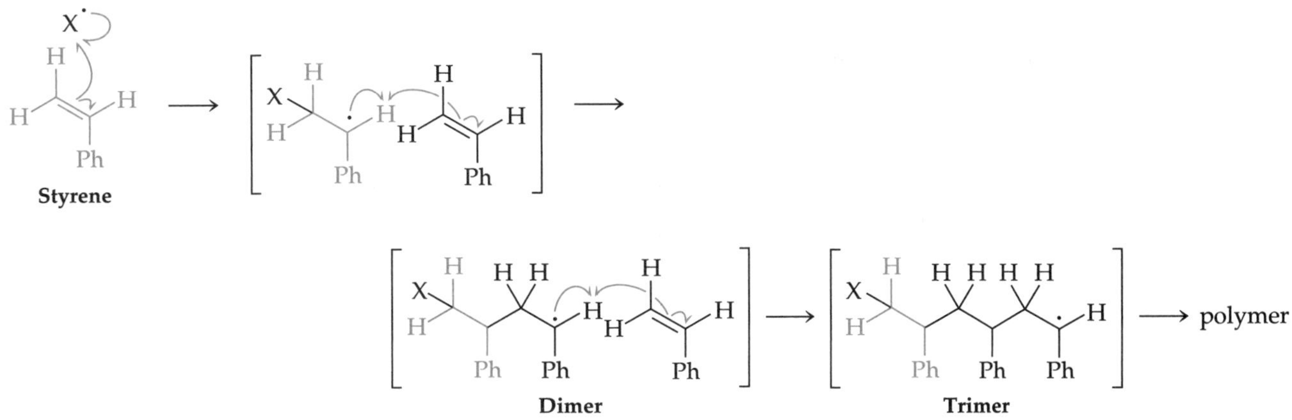

▲ FIGURE 10-30
A carbon free radical produced by σ bonding with a radical initiator can itself add to another molecule of alkene. The dimer radical formed by the combination of two alkenes can be terminated by combination with another radical; lose a hydrogen atom to form a neutral dimer; or attack a third molecule of alkene. Because termination and loss of hydrogen are less important with radicals than with cations, polymerization is often more efficient through radical routes.

added across a **dimer** of the starting alkene. Alternately, this more-complex cation can lose a proton (route *b*), as in the second step of an E1 elimination discussed in Chapter 9. However, it also acts as an electrophile attacking another molecule of starting material (route *c*) to make an even longer chain. In the **trimer** shown at the lower right in Figure 10-29, three molecules of styrene are bound together. This reaction could continue again and again, ultimately forming a **polymer**—in this case, **polystyrene.** This important process is discussed in much more detail in Chapter 16.

10-5 ▸ Radical Additions

Recall from Chapters 2 and 6 that carbon radicals, as well as carbocations, are electron deficient. For this reason, radicals and cations have the same order of stability: tertiary > secondary > primary. Accordingly, the reactivity of radicals and cations toward double bonds is similar.

The electrophilic attack by a carbocation described in the preceding subsection is very similar to the attack by a carbon free radical on a double bond: the only significant difference is that radical intermediates rather than cations are formed. These reactions are also similar to those initiated by bromine and shown in Figure 10-14. When the free radical formed by interaction with an initiator attacks another molecule of alkene, the reactions shown in Figure 10-30 are observed. Thus, polystyrene can be prepared by either radical or cationic polymerization pathways.

It is also possible for a carbon free radical to attack a remote alkene group present within the same molecule. This intramolecular carbon-carbon bond formation is becoming increasingly important as a synthetic tool for forming rings.

Table 10-1 ▶ How to Use Addition Reactions to Make Various Functional Groups

Functional Group	Addition Reaction
Alcohol	Acid-catalyzed hydration of alkenes; or oxymercuration-demercuration; or hydroboration-oxidation
Aldehyde	Ozonation-reduction (ozonolysis) of alkenes
Alkyl halide	Electrophilic hydrohalogenation; or peroxide-initiated radical hydrohalogenation
Borane	Hydroboration
Carboxylic acid	Hot $KMnO_4$ oxidation of alkenes containing $=CHR$; or ozonation of alkynes
Cyclopropane	Carbene addition to alkenes
Dihalide (vicinal)	Electrophilic bromination or chlorination
Diol	Opening of epoxides by aqueous base
Epoxide	Peracid oxidation of alkenes
Ketone	Acid-catalyzed alkyne hydration; or ozonation-reduction (ozonolysis) of alkenes; or hot $KMnO_4$ oxidation of alkenes containing $=CR_1R_2$
Tetrahalide	Electrophilic halogenation of an alkyne

10-6 ▶ Synthetic Methods

As in preceding chapters, addition reactions can be grouped according to the functional-group conversion accomplished in each type. Table 10-1 provides such a summary. This table summarizes the methods by which an alkene can be converted into any of eleven other types of compounds. When combined with the methods presented in Chapter 9 for preparing alkenes, these reactions are powerful techniques for interconverting a number of functional groups.

Conclusions

Electrophiles approach regions of electron density, particularly π orbitals, in alkenes, dienes, and alkynes, so as to form the more-stable of two possible cations. In the rate-determining step, the electrons of the π bond move toward the electrophile, forming a new bond between carbon and the electrophile. In a second step, the other carbon of the original double bond becomes bound to a nucleophile.

Because of the intermediacy of carbocations in electrophilic addition reactions, regiochemistry is controlled by the formation of the more-stable cation. Consistent with Markovnikov's Rule, the order of carbocation stability (tertiary > secondary > primary) dictates that the more highly substituted carbon becomes a carbocation in electrophilic addition. Because carbocations are planar, the addition of a nucleophile in the second step occurs equally easily on both faces, and both stereoisomers are present in the product mixture in those cases in which stereocenters are formed. Carbocations also undergo skeletal rearrangements when a driving force exists: when secondary carbocations are formed in electrophilic addition reactions, they often rearrange to more-stable tertiary cations.

Reversal of the normal (Markovnikov) sense of addition is accomplished by changing reagents and, hence, the mechanism of addition. For example, hydrobromination of alkenes occurs in polar media according to Markovnikov's Rule, whereas in ether in the presence of a free radical initiator, the reverse regiochemistry is obtained. Anti-Markovnikov hydration is achieved by hydroboration-oxidation proceeding through a *syn* addition of borane in the first stage: regiochemistry of the hydroboration is controlled by steric bulk and the electrophilicity of BH_3.

Addition to 1,3-dienes occurs in a regiochemical sense consistent with the most-stable cationic intermediate (an allylic cation). This intermediate cation adds nucleophiles so as to form both 1,2 and 1,4 (conjugate) adducts.

Stereochemical control is afforded in some electrophilic additions to alkenes because of the intermediacy of bridged cationic intermediates; for example, in the cyclic bromonium ion formed as an intermediate in the halogenation of an alkene. Stereochemical control is also present in singlet carbene additions to form cyclopropanes and in peracid oxidations to form epoxides, both of which are concerted reactions that proceed without the formation of intermediates. The first step of ozonolysis also is concerted, forming a molozonide, but rapid rearrangement followed by reduction ultimately produces carbonyl compounds, with the cleavage of both the σ and π bonds of the original alkene.

The addition of bromine to an alkene and oxidative cleavage of an alkene with permanganate are used as characteristic color tests for the presence of C–C double or triple bonds.

Summary of New Reactions

Hydrohalogenation and Hydration

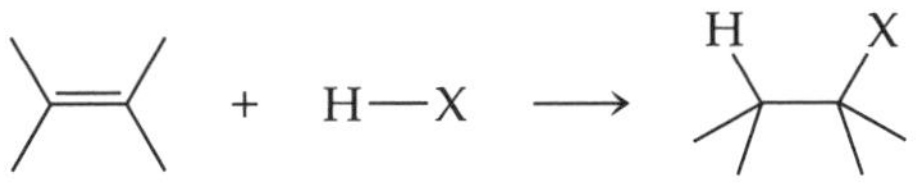

X = halide or OH
(Markovnikov)

Conjugate Hydrohalogenation

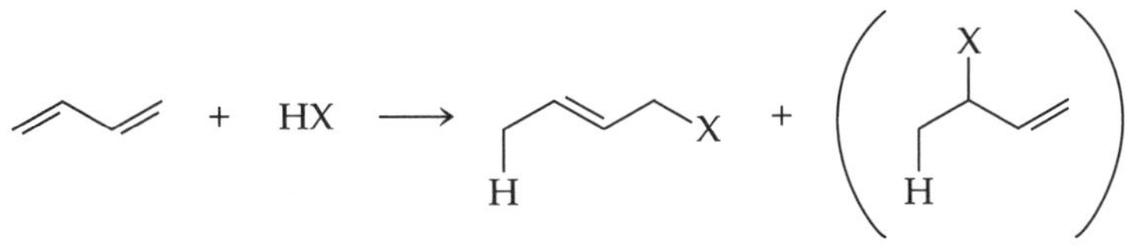

1,4-Adduct   **1,2-Adduct**

Hydrohalogenation of Alkynes

Hydration of Alkynes

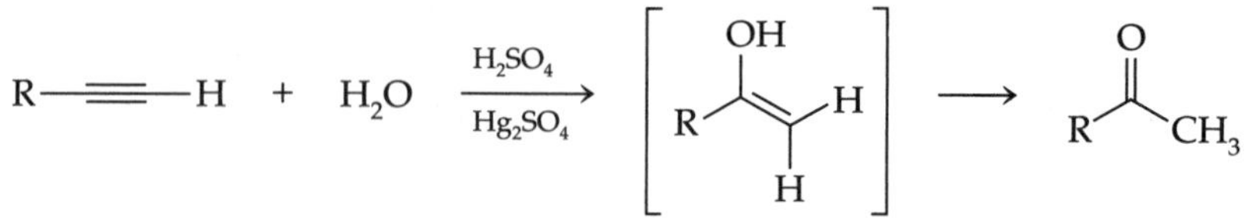

R—≡—H + H_2O $\xrightarrow{\begin{array}{c} H_2SO_4 \\ \hline Hg_2SO_4 \end{array}}$ $\left[\begin{array}{c} OH \\ R\diagup{=}\diagdown H \\ H \end{array} \right]$ $\longrightarrow$ $R\overset{O}{\underset{}{\diagdown}} CH_3$

Free Radical Hydrohalogenation

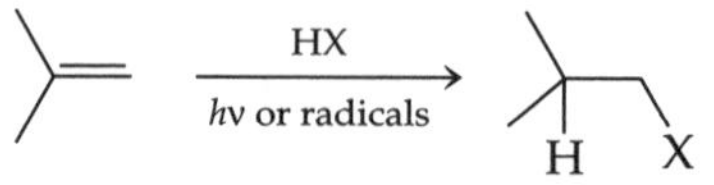

$\xrightarrow[\text{hv or radicals}]{HX}$ (anti-Markovnikov)

Hydroboration–Oxidation

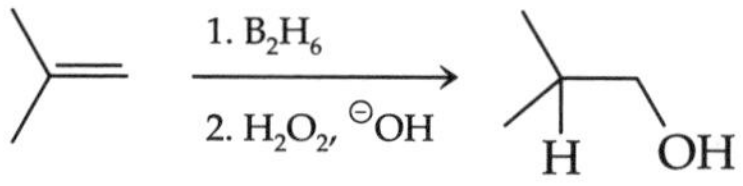

$\xrightarrow[\text{2. } H_2O_2,\ ^{\ominus}OH]{\text{1. } B_2H_6}$ (anti-Markovnikov, *syn* addition)

Oxymercuration–Demercuration

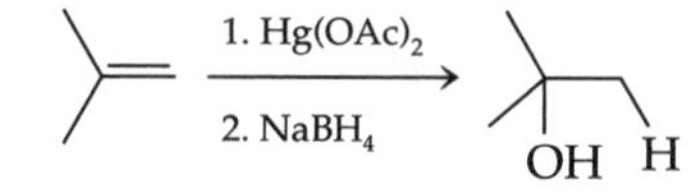

$\xrightarrow[\text{2. } NaBH_4]{\text{1. } Hg(OAc)_2}$ (Markovnikov addition, without rearrangement)

Halogenation of Alkenes and Alkynes

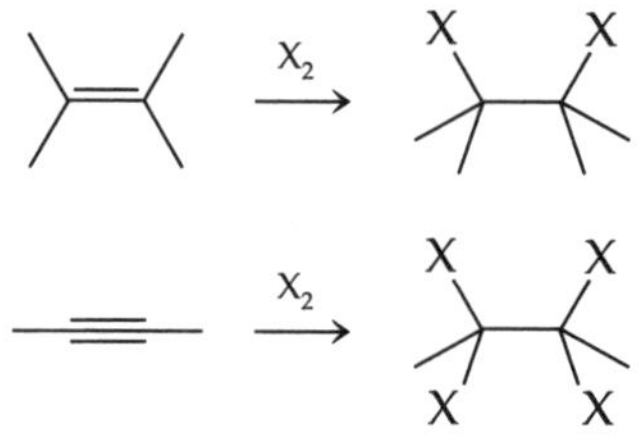

$\xrightarrow{X_2}$

$\xrightarrow{X_2}$

Carbene Cyclopropanation

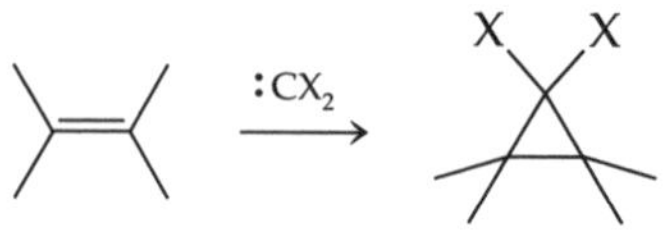

$\xrightarrow{:CX_2}$

Peracid Epoxidation

$\xrightarrow{RCO_3H}$

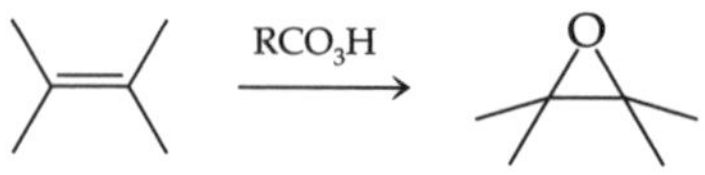

Ozonolysis

$\xrightarrow[\text{2. Zn}]{\text{1. } O_3}$

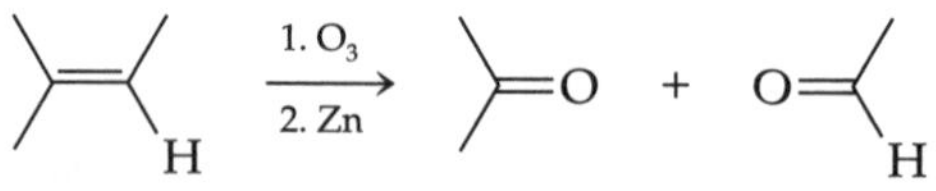

Oxidative Cleavage by Potassium Permanganate

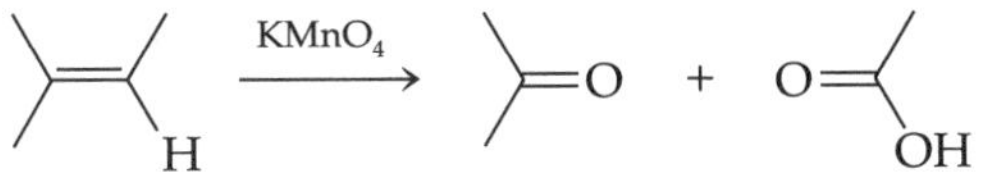

Cationic Polymerization

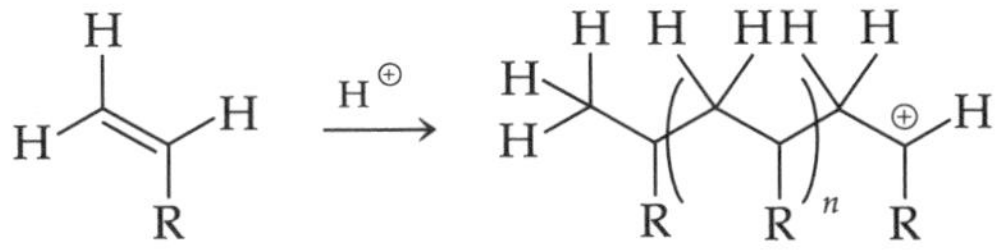

Radical Polymerization

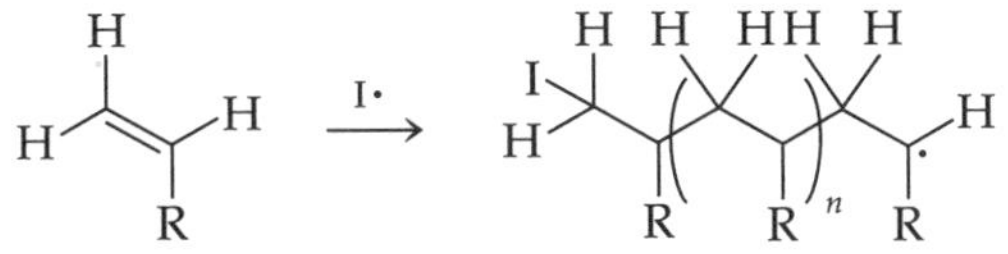

(I· = initiator)

Review Problems

10-1 What product(s) would you expect when 1-methylcyclohexene is treated with each of the following reagents?

(a) HBr in MeOH

(b) HI in H_2O

(c) HBr in ether

(d) 1. B_2H_6; 2. H_2O_2, NaOH

(e) 1. $Hg(OAc)_2$; 2. $NaBH_4$

(f) Br_2 in CH_2Cl_2

(g) :CH_2(singlet)

(h) *m*-chloroperbenzoic acid

(i) 1. O_3; 2. Zn, HOAc

(j) Cl_2 in CH_2Cl_2

(k) hot $KMnO_4$

(l) dilute aqueous H_2SO_4

10-2 What products (if any) do you expect from the reaction of 2-butyne with the following reagents?

(a) Pt, excess H_2

(b) one mole HBr

(c) two moles HBr

(d) one mole Br_2 in CH_2Cl_2

(e) two moles Br_2 in CH_2Cl_2

(f) aqueous $HgSO_4$

(g) 1. O_3; 2. Zn, HOAc

(h) hot $KMnO_4$

10-3 What major product(s), if any, would you expect from the reaction of 1,3-butadiene with the following reagents?

(a) excess H_2, Pt

(b) one equivalent of Br_2

(c) dilute aqueous H_2SO_4

(d) HCl

10-4 What reagent (or series of reagents) can transform 1-butene into each of the following compounds?

(a) 1-bromobutane

(b) 2-bromobutane

(c) 1-butanol

(d) 2-butanol

(e) butane

(f) 1-butyne

(g) 2-butyne

(h) *s*-butyllithium

(i) propanoic acid

(j) 2-aminobutane

(k) 2-heptanone

(l) hexanoic acid

(m) 1-hexanol

10-5 Suggest a sequence of reagents that can convert 1-pentanol into each of the following products.

(a) 1-chloropentane

(b) 1-pentene

(c) 2-pentanol

(d) 2-pentene

(e) 2-pentanone

(f) pentanoic acid

(g) 1-pentanal

(h) 2-bromopentane

10-6 Determine the correct starting material required to prepare each of the following products by using the indicated reagents:

(a) Br_2 in CCl_4 to prepare *meso*-2,3-dibromobutane

(b) D_2/Pt to prepare (*d,l*)-2,3-dideuterobutane

(c) Cl_2 in CCl_4 to prepare (*d,l*)-2,3-dichlorobutane

(d) HBr in CH_2Cl_2 to prepare racemic 2-bromobutane

(e) methylene (:CH_2) to prepare (*d,l*)-*trans*-dimethylcyclopropane

(f) *m*-chloroperbenzoic acid to prepare the *meso*-epoxide

10-7 Arrange the following alkenes in order of their reactivity toward acid-catalyzed hydration and explain your reasoning.

(a) 1-hexene, 2-methyl-1-pentene, 2-hexene

(b) 2-methylpropene, *cis*-2-butene, *trans*-2-butene

(c) 1-phenyl-1-butene, 1-phenyl-2-butene, 2-phenyl-1-butene

10-8 The conditions commonly used for acid-catalyzed hydration of alkenes can sometimes lead to competing positional isomerization. Propose a mechanism by which 1-pentene can be converted into *trans*-2-pentene upon treatment with acid, justifying at each step any relevant stereochemistry or regiochemistry.

10-9 When HBr is added to a simple alkene in the presence of ether solvents containing peroxides, the regiochemistry obtained is the reverse of that observed in polar solvents. From your knowledge of the radical addition mechanism, what product would you expect upon treatment of 1,3-butadiene with HBr in peroxide-containing ethers?

10-10 Chlorine adds to *trans*-2-butene in methylene chloride to give *meso*-2,3-dichlorobutane; but, if the reaction is conducted in water, a chlorohydrin (a compound bearing OH and Cl on adjacent carbons) is obtained. Using curved arrows to indicate

electron flow, suggest a mechanism for this reaction that also predicts its stereochemical course.

10-11 Explain why 2-butyne is less reactive than *trans*-2-butene toward most electrophiles such as bromine and why it is nonetheless possible to stop after a single equivalent of bromine has been added to the alkyne.

10-12 Draw in three dimensions the structures expected for all products of addition of singlet diphenylcarbene to *trans*-2-butene. Are there centers of asymmetry in the product(s)? Compare your result with that obtained by treatment of *trans*-2-butene with *m*-chlorobenzoic acid.

10-13 Write detailed reaction mechanisms for each of the following reactions. Explain any relevant regio- or stereochemistry.

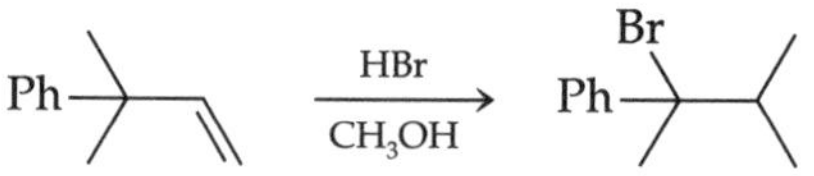

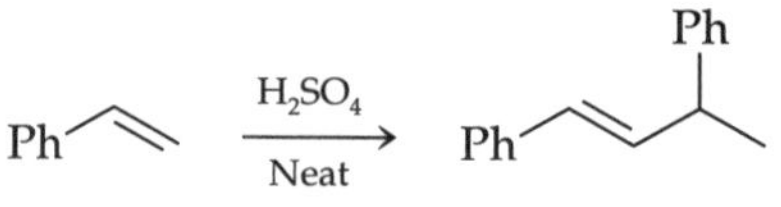

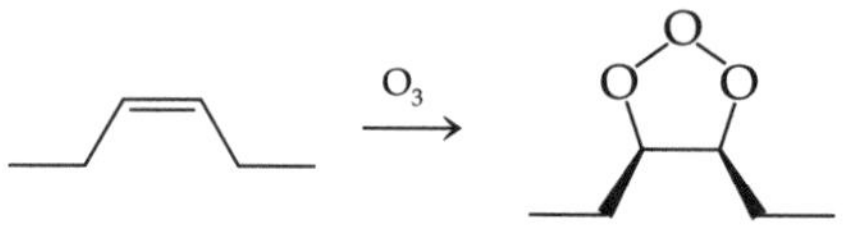

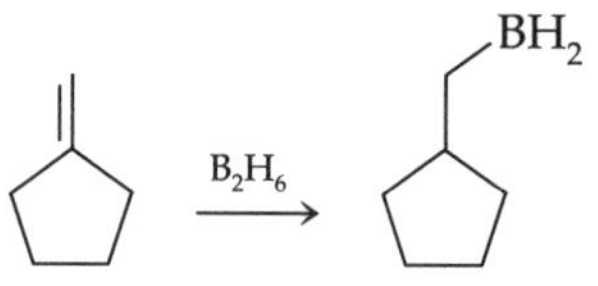

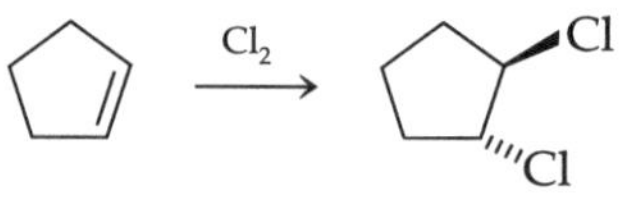

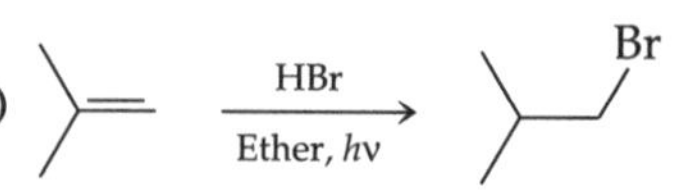

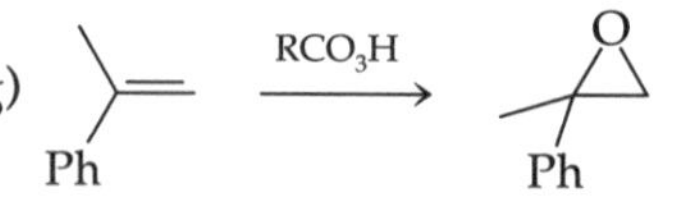

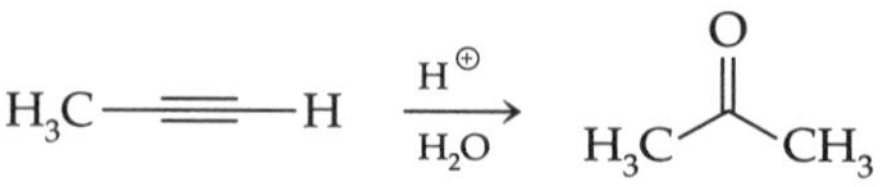

10-14 Choose a chemical and a spectroscopic method that can be used to distinguish the following pairs of compounds. Describe clearly what you would see with each compound with each technique.

(a) cyclohexane and cyclohexene

(b) 1-hexene and 1-hexyne

(c) 2-hexene and 1-hexanol

(d) 2-hexene and 2-bromohexane

(e) 2-hexene-1-ol and 1-hexanol

(f) 2-hexanol and 2-bromohexane

(g) 1-hexanal and hexane

10-15 Osmium tetroxide can add to an alkene, giving an intermediate osmate ester that is hydrolyzed stereospecifically to a 1,2-diol. For example, when cyclohexene is treated with OsO_4, *meso*-1,2-cyclohexanediol is ultimately formed. Assuming that the hydrolysis occurs with retention of configuration, determine whether the stereochemistry for the formation of the osmate ester is *syn* or *anti*, and write a mechanism for its formation that is consistent with your answer.

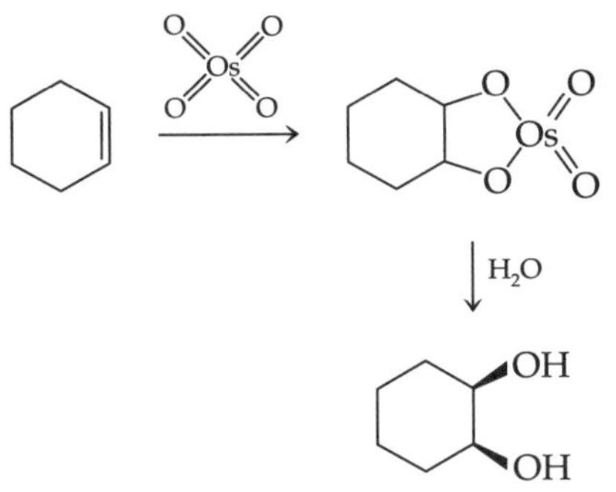

10-16 Limonene is a naturally occurring $C_{10}H_{16}$ hydrocarbon that causes the odor of lemons (a "citrus" smell). When treated with excess hydrogen in the presence of a platinum catalyst, it takes up two equivalents of hydrogen. When treated with ozone, followed by zinc in acetic acid, it forms 1 mole of formaldehyde and 1 mole of the tricarbonyl compound shown below.

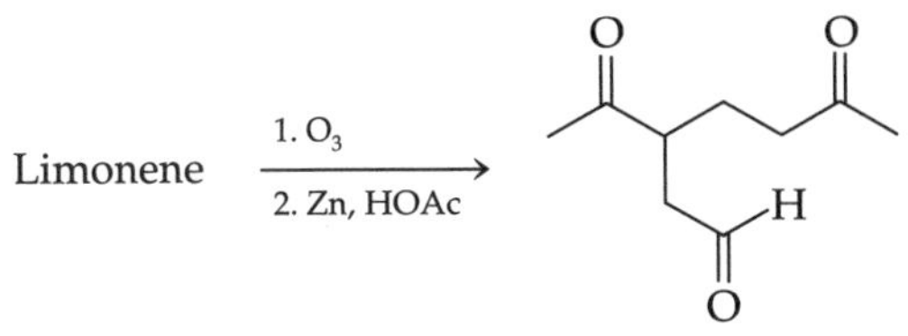

Propose one or more structures that are consistent with this data. If you cannot distinguish between two or more structures on the basis of the data given, propose a chemical or a spectroscopic method or both by which the structures that you have identified could be distinguished. How many unique peaks would appear in the ^{13}C NMR spectrum of each structure that you have found?

10-17 Hydroboration-oxidation of an alkyne proceeds through the same syn addition mechanism as that for an alkene. The C–B bond of the resulting vinyl borane also is replaced with retention of configuration in the oxidation step. From what you know about the mechanism of hydroboration-oxidation of alkenes, what products would be formed by hydroboration-oxidation of each of the following alkynes?

(a) 1-hexyne

(b) 3-hexyne

(c) phenylacetylene

Chapter 11

Electrophilic Substitution of Aromatic Molecules

The π cloud of an aromatic ring, like that of a C=C double bond, makes it a potential chemical target for electrophiles. With aromatic molecules, however, electrophilic attack leads to substitution products; whereas, with alkenes, electrophilic attack produces addition products. This difference in reactivity follows from the energetic importance of aromaticity in planar, cyclic, conjugated compounds that contain $4n + 2$ electrons. In this chapter, we compare the electrophilic attack of a C=C double bond on an aromatic ring with that on a simple alkene. The mechanisms of these reactions make it easy to see why electrophilic attack on aromatic rings leads to substitution, whereas that on alkenes gives addition products. We will see how the active electrophiles needed for aromatic substitution are produced and will learn the identity of the products formed when the electrophiles attack aromatic rings. These products incorporate substituent groups that can be chemically modified by various secondary reactions: some of these reactions were presented in earlier chapters; others will be considered here. For example, we will see how an electron-withdrawing group such as —NO$_2$ can be converted into an electron-donating group such as —NH$_2$. We will also learn how the relative reactivity and regiochemistry of an electrophilic aromatic substitution are affected by substituents initially present on the aromatic ring and how to plan syntheses in which several substituents are placed on the ring. Finally, we will learn how the stability of the intermediate carbocation influences the preferred position for electrophilic substitution on fused aromatic rings.

11-1 ▸ Mechanism of Electrophilic Aromatic Substitution

The π electrons in an aromatic compound such as benzene are highly delocalized, and an electrophile can approach this region of high electron density just as it would in an electrophilic attack on an alkene, which was discussed in Chapter 10. In the electrophilic attack on an alkene (Figure 11-1, lower reaction), the loss of a π bond is compensated by the energy gained in forming the σ bond to the electrophile (E⁺). The second step, the formation of a bond to the nucleophile (Nuc:), is exothermic. However, because aromaticity contributes significantly to the stability of aromatic compounds, cations formed by electrophilic attack on an aromatic ring (upper reaction) and on an alkene (lower reaction) follow different reaction pathways—namely, substitution and addition.

If we write a resonance structure for benzene in which the three π bonds are localized between pairs of carbon atoms, we can see how the flow of electrons would exactly parallel that in an electrophilic attack on an alkene (Figure 11-1). Like the reaction with alkenes, electrophilic attack on benzene breaks a π bond while forming a cation. However, the formation of the σ bond to an aromatic ring converts one carbon atom into an sp^3-hybridized center. This atom then lacks a p orbital and cannot participate in delocalized π bonding. Sigma-bond formation thus disrupts ring aromaticity. The formation of a cationic intermediate is therefore more costly energetically, and the attack by an electrophile on an aromatic ring is much more difficult than attack by the same electrophile on a simple carbon–carbon double bond.

Nonetheless, the cation formed in an electrophilic attack on a benzene ring is resonance stabilized: the positive charge in this cation is found on three sp^2-hybridized carbon atoms as the remaining four π electrons are delocalized over five atoms. The extra stabilization afforded by delocalization in this cation is what makes electrophilic attack on aromatic rings possible at all, given the high cost of disrupted aromaticity. Without this resonance stabilization, such attacks would not be thermodynamically feasible.

Unlike a carbocationic intermediate formed in electrophilic addition to a double bond, those formed by electrophilic attack on aromatic rings are

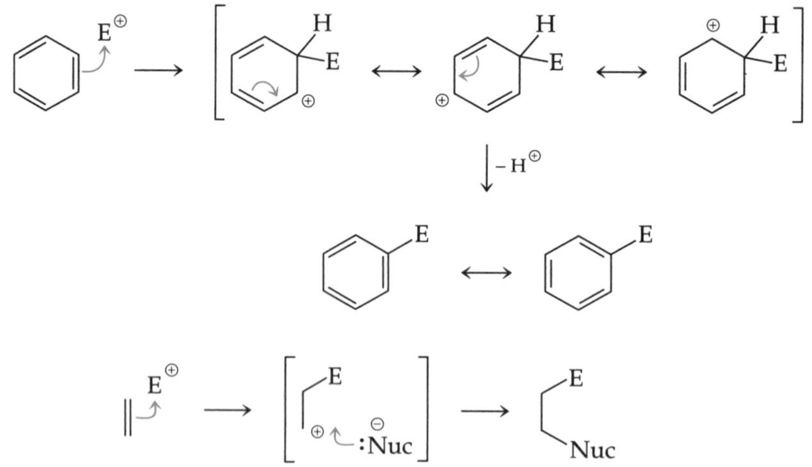

▲ FIGURE 11-1

The first step of an electrophilic aromatic substitution (upper reaction) is exactly parallel to that of electrophilic addition (lower reaction): a pair of π electrons shifts toward the electrophile (E⁺), forming a C–E σ bond and a carbocation. In the upper reaction, cation formation is very costly energetically, with the loss of aromaticity that occurs when the σ bond is formed being only partly compensated by resonance delocalization in the intermediate cation. However, because three significant resonance contributors exist for this cation, it is much more stable than a simple trigonal cation. Aromaticity is restored by loss of a proton, producing a net substitution product; aromaticity would have been lost permanently if the cation had been trapped by a nucleophile in an addition like that shown in the lower reaction.

not trapped by nucleophiles in a second fast step. Nucleophilic capture at any one of three carbons bearing positive charge in the resonance-stabilized cationic intermediate would lead to a product lacking aromaticity.

Instead, the cation formed by electrophilic attack on the aromatic ring has another fate, similar to that seen earlier for oxonium ions; that is, it is deprotonated by a weak base—in this case, leading to a neutral, aromatic product. Thus, when the electrons initially present in the C–H bond at the site of electrophilic attack are donated back to the aromatic π system, aromaticity is restored and a substitution product is formed.

The reaction accomplished in this series of steps is the replacement of a C–H bond on the ring by a C–E bond.

The mechanism outlined in Figure 11-1 is called **electrophilic aromatic substitution** because electrophilic attack on the π system initiates a reaction in which one of the ring hydrogens is replaced by the electrophile. In this mechanism, a σ bond is formed between a ring carbon and the attacking electrophile in the first step, forming a delocalized cation that is deprotonated in a second, fast step. The driving force for this deprotonation is the restoration of aromaticity in the substituted product.

As in an electrophilic addition to an alkene, the initial electrophilic attack (to break a π bond and form a carbocation) is the slow step in an aromatic electrophilic substitution. Thus, attack by the electrophile is rate determining, and the reaction is facilitated by those features that either stabilize the carbocation after electrophilic attack or enhance the electron demand (and, hence, the reactivity) of the electrophile.

11-2 ▸ The Introduction of Groups by Electrophilic Aromatic Substitution: Activated Electrophiles

Because cation formation on an aromatic ring is accompanied by the loss of aromatic stabilization, the electrophiles necessary for electrophilic aromatic substitution must be more active than those needed for electrophilic addition to alkenes. Some common electrophilic reagents that can effect electrophilic aromatic substitution are shown in Figure 11-2. Because of the high reactivity needed to disrupt aromaticity in the rate-determining first step in Figure 11-1, each of these reagents must be converted into a more-active electrophile than is usually required for an electrophilic addition to an alkene. Thus, chlorination or bromination of benzene (replacement of H by Cl or Br) generally occurs only in the presence of a Lewis acid; **nitration** (replacement of H by —NO$_2$) of an aromatic ring occurs in the presence of nitric acid with sulfuric acid; **sulfonation** (replacement of H by —SO$_3$H)

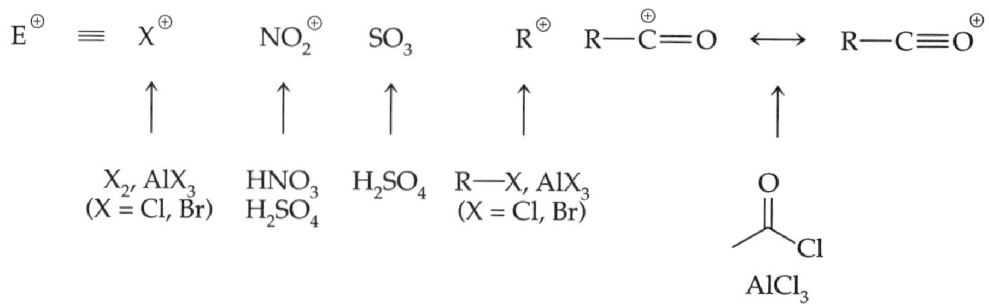

$$E^{\oplus} \equiv X^{\oplus} \quad NO_2^{\oplus} \quad SO_3 \quad R^{\oplus} \quad R\!-\!\overset{\oplus}{C}\!=\!O \;\longleftrightarrow\; R\!-\!C\!\equiv\!\overset{\oplus}{O}$$

$$\uparrow \qquad \uparrow \qquad \uparrow \qquad \uparrow \qquad\qquad\qquad \uparrow$$

$$X_2,\, AlX_3 \quad HNO_3 \quad H_2SO_4 \quad R\!-\!X,\, AlX_3$$
$$(X = Cl,\, Br) \quad H_2SO_4 \qquad\qquad (X = Cl,\, Br)$$

▲ FIGURE 11-2
The more-rigorous conditions for electrophilic aromatic substitution (compared with electrophilic addition) require more-reactive electrophiles. A greater fraction of positive charge is developed on an electrophilic atom when its precursor interacts with either a Brønsted or a Lewis acid.

occurs with fuming sulfuric acid; **alkylation** (replacement by —R) occurs when an alkyl halide interacts with a Lewis acid; and **acylation** [replacement by —C(O)R] occurs when an acid chloride interacts with a Lewis acid. An apt generalization about all electrophilic aromatic substitutions is that the electrophilic atom must bear a larger fraction of positive charge to induce σ-bond formation in the initial step than is needed for electrophilic addition to an alkene. Let us consider each reaction in greater detail.

Halogenation

The mechanism for electrophilic aromatic chlorination is shown in Figure 11-3. When benzene and chlorine are mixed, no reaction ensues because the induced polarization of the chlorine–chlorine bond is insufficient to cause C–Cl σ-bond formation and its resulting disruption of the π bonding of the aromatic ring. However, in the presence of a Lewis acid, such as $AlCl_3$ or $FeCl_3$ (acting as an electron acceptor), more positive charge develops on one of the chlorines, permitting electrophilic attack.

$$:\!\overset{..}{\underset{..}{Cl}}\!-\!\overset{..}{\underset{..}{Cl}}\!: \quad AlCl_3 \;\longrightarrow\; \overset{\delta\oplus}{Cl}\cdots\cdots Cl\cdots\cdots\overset{\delta\ominus}{AlCl_3}$$

After this rate-determining attack by electrophilic chlorine (formally Cl^+, a species that may have no real existence), deprotonation can be assisted by the Lewis base ($AlCl_4^-$) formed by coordination of chloride to aluminum chloride. Parallel reactions also ensue with bromine in the presence of Lewis acids (for example, $Br_2/AlBr_3$ or $Br_2/FeBr_3$) but fail with fluorine and iodine.

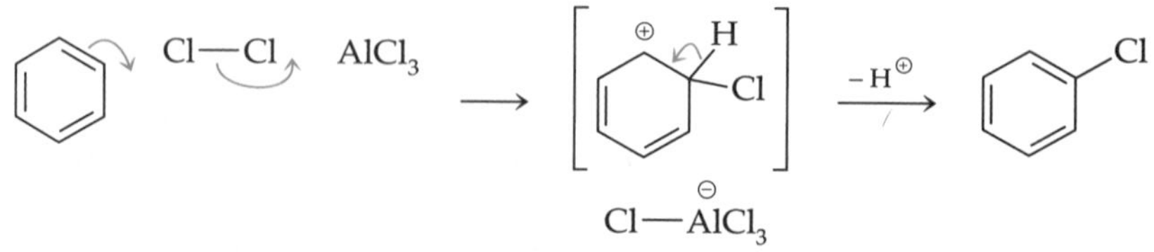

▲ FIGURE 11-3
Lewis acid, when complexed with Cl_2, polarizes the Cl–Cl bond to place appreciable positive charge on one of the chlorines (that nearest the aromatic ring). Electrons are then donated from the aromatic π system to form a C–Cl σ bond. Upon deprotonation of the intermediate cation, two electrons from the C–H σ bond are released to restore ring aromaticity in the chlorinated substitution product.

TOXICITY OF CHLORINATED AROMATICS

Dioxins are a group of chlorinated aromatic hydrocarbons that are formed in trace amounts in the production of many chlorinated compounds. For example, 2,4,5-trichlorophenoxyacetic acid (2,4,5-T, also known as Agent Orange) is an effective herbicide that kills many different kinds of plants and was widely used as a defoliant during the war in Vietnam. In the manufacture of 2,4,5-T, small amounts of the dioxin 2,3,7,8-TCDD also are generated. This compound is extremely toxic in very small doses and has been shown to cause mutations in laboratory animals. Some of the health problems that later developed in soldiers who were exposed to Agent Orange during the war in Vietnam also are related to contact with this toxin.

Hexachlorophene, an effective topical antibacterial agent, has a structure similar to that of 2,3,7,8-TCDD and has been shown to cause neurotoxic symptoms in laboratory animals given very large doses. Because of the possible risks associated with long-term use, hexachlorophene has been banned for most uses.

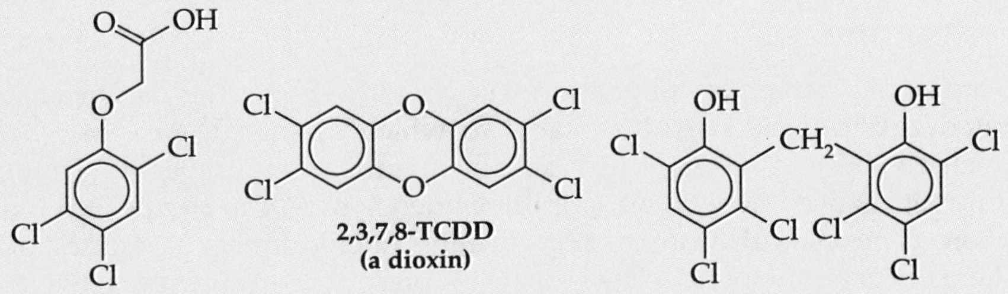

2,4,5-Trichlorophenoxy-
acetic acid
(Agent Orange)

2,3,7,8-TCDD
(a dioxin)

Hexachlorophene

Nitration

A Lewis dot structure of nitric acid reveals that all three oxygen atoms bear unshared electron pairs and are potentially basic. Any of these sites can be protonated by sulfuric acid (Figure 11-4). In one of these isomeric protonated nitric acids, there is an effective leaving group. From the third structure in Figure 11-4, loss of water generates the electrophilic nitronium ion (NO_2^+). This highly reactive ion can interact with an aromatic hydrocarbon as shown in Figure 11-5 to generate a resonance-stabilized carbocation, deprotonation of which gives nitrobenzene.

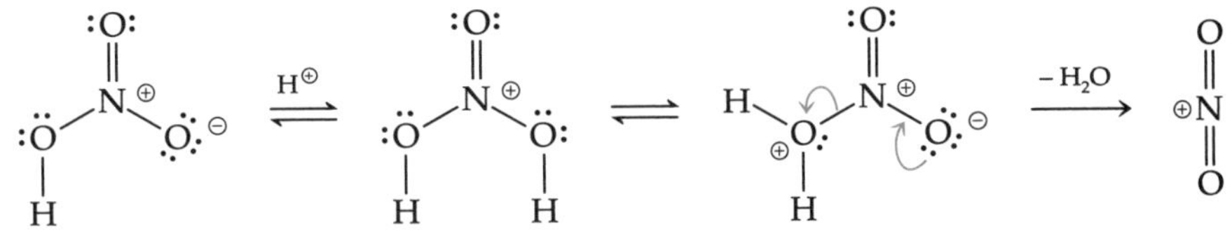

▲ FIGURE 11-4
The three oxygen atoms of nitric acid are potential sites for protonation by a strong mineral acid. Although several protonated species coexist in an acid-base equilibrium, only the structure that is doubly protonated on a single oxygen can conveniently lose water to form the nitronium ion, NO_2^+.

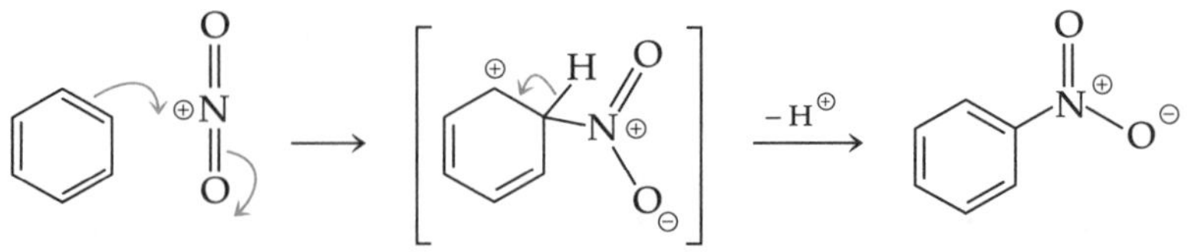

▲ FIGURE 11-5
The nitronium ion formed by treatment of nitric acid with sulfuric acid is a highly active electrophile. It can attack an aromatic ring, forming a σ bond to carbon in the rate-determining step of the aromatic electrophilic substitution. Deprotonation of the resonance-stabilized cation completes the substitution, as aromaticity is restored to the ring.

EXERCISE 11-A

Draw a Lewis dot structure for NO_2^+ and compare it with that for CO_2. From what you know about hybridization, do you expect NO_2^+ to be linear or bent?

Sulfonation

Electrophilic aromatic sulfonation can be induced by the use of concentrated sulfuric acid or sulfuric acid to which dissolved, gaseous SO_3 (as much as 30%) has been added, so-called fuming sulfuric acid. (The term "fuming" is used here because of the tendency of SO_3 to escape into the air where it mixes with water vapor, forming microdroplets of sulfuric acid.) There are actually many different species present in sulfuric acid that act as electrophiles. The most reactive among them is SO_3, and the reaction of this electrophile with benzene is shown in Figure 11-6.

Friedel–Crafts Alkylation

Reaction of an alkyl halide with an aromatic compound in the presence of a Lewis acid results in replacement of a hydrogen by an alkyl substituent. This net carbon–carbon bond-forming reaction is referred to as a **Friedel-Crafts alkylation,** in acknowledgment of the contributions of the two chemists, Charles Friedel, a Frenchman, and James Crafts, an American, who first discovered this reaction in 1877. Shown in Figure 11-7, this reaction is quite similar to that shown in Figure 11-3 for the chlorination of benzene.

The Lewis acid polarizes the carbon–chlorine bond of the alkyl chloride in the same way as it does a dihalogen, forming an intermediate carbocation.

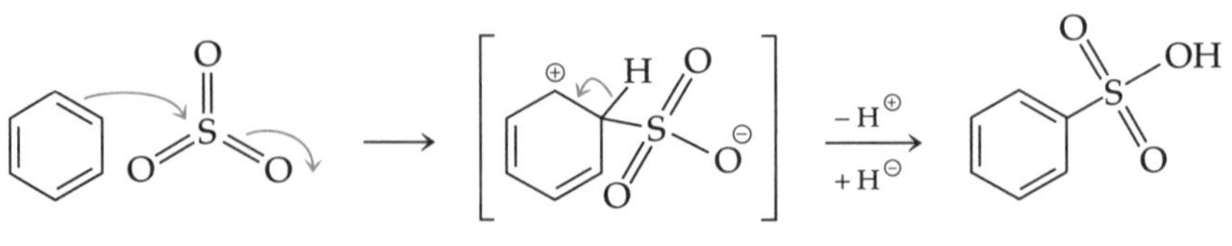

▲ FIGURE 11-6
Reaction of SO_3 as an electrophile with benzene results in an intermediate zwitterion. Loss of a proton regenerates the aromatic system, and protonation of the sulfonate group forms the product, benzenesulfonic acid.

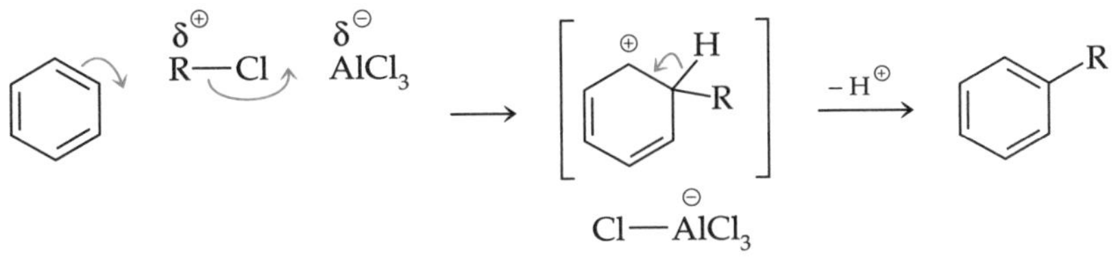

▲ **FIGURE 11-7**
In the Friedel-Crafts alkylation, a carbocation is formed when a Lewis acid
such as aluminum chloride coordinates with an alkyl chloride. As we learned
in Chapter 10, a carbocation is an active electrophile and attacks the aromatic
ring, forming a C–C σ bond. The resulting resonance-stabilized cation is
deprotonated to form the product, an alkylbenzene.

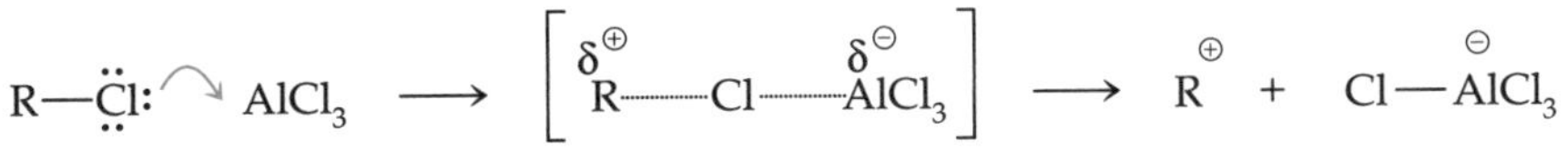

However, the formation of a carbocation as the active electrophile allows rearrangement reactions to occur. When an alkyl chloride is activated by a
Lewis acid to generate what would be a primary cation, rearrangement to a
secondary cation takes place as chloride ion is lost. The resulting secondary
cation then attacks the aromatic ring.

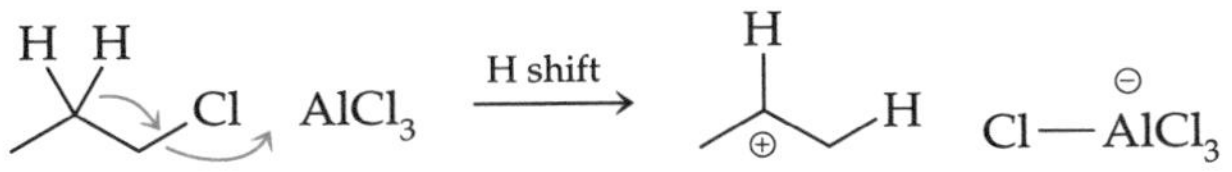

(The rearrangement is unimolecular and the electrophilic attack is bimolecular: this distinction contributes significantly to the difference in rate.) Thus,
Friedel-Crafts alkylation with a primary alkyl halide produces a product in
which the alkyl group attached to the ring corresponds to the more-stable,
rearranged carbocation. For example, the reaction of *n*-propyl chloride
under Friedel-Crafts conditions forms isopropylbenzene as the isolated
product (Figure 11-8). Thus, the initially formed primary cation rearranged
to the secondary ion more rapidly (as shown above) than it could attack the
aromatic ring (as shown in Figure 11-8).

The Friedel-Crafts alkylation reaction requires that a relatively stable
cation can be formed from the alkyl halide by interaction with the Lewis
acid. Neither aryl halides nor vinyl halides can be activated by Lewis acid

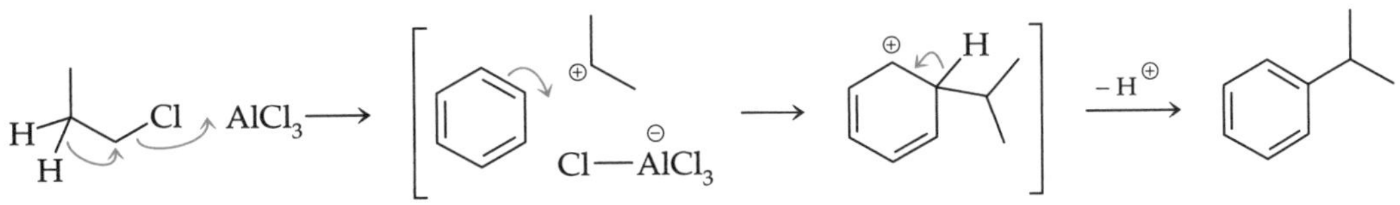

▲ **FIGURE 11-8**
Cationic rearrangements are very fast when a more-stable cation results. Here,
AlCl$_3$ acts as a Lewis acid to abstract chloride ion from *n*-propyl chloride. The
carbocation, formed as chloride is lost, rearranges by a hydride shift from C-2,
possibly even as it is formed, producing the isopropyl cation. This cation
reacts as an electrophile, adding to the aromatic ring and forming a
carbon–carbon σ bond. Deprotonation of the resonance-stabilized cation leads
to the substitution product in which the carbon in the alkyl group attached to
the ring is different from that bearing the chlorine atom in the starting alkyl
halide.

because the resulting cations are very unstable. Thus, the group that becomes attached to the ring in a Friedel-Crafts alkylation must be derived from an sp^3-hybridized alkyl halide.

EXERCISE 11-B

What product would be formed in each of the following Friedel-Crafts alkylations?

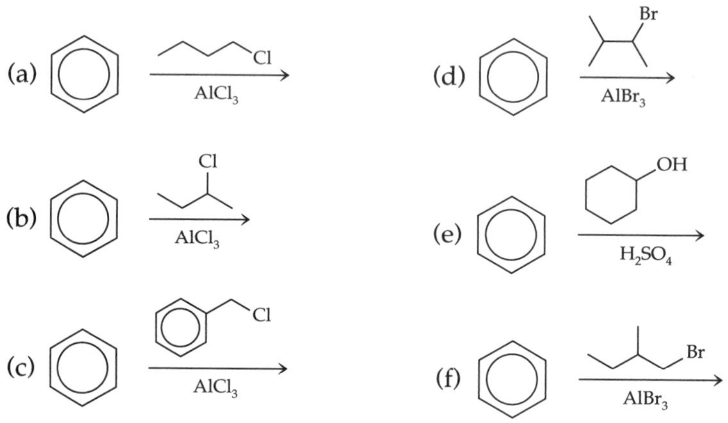

(a)   benzene $\xrightarrow[\text{AlCl}_3]{}$ ... Cl

(d)   benzene $\xrightarrow[\text{AlBr}_3]{}$ ... Br

(b)   benzene $\xrightarrow[\text{AlCl}_3]{}$ ... Cl

(e)   benzene $\xrightarrow[\text{H}_2\text{SO}_4]{}$ ... OH

(c)   benzene $\xrightarrow[\text{AlCl}_3]{}$ ... Cl

(f)   benzene $\xrightarrow[\text{AlBr}_3]{}$ ... Br

Friedel–Crafts Acylation

The —C(O)R group is referred to as an acyl group and the attachment of an acyl group to another carbon skeleton is called **acylation. Friedel-Crafts acylation** (also discovered by Friedel and Crafts) results in acylation of an aromatic ring by treatment of a carboxylic acid chloride with a Lewis acid. The Friedel-Crafts acylation is initiated by polarization of the carbon–chlorine bond of an acid chloride by treatment with $AlCl_3$. The cation thus formed is an **acylium ion** (Figure 11-9). The acylium ion bears an oxygen atom (with nonbonded lone pairs) at a position adjacent to the positive charge; this allows for a significant resonance contributor with the positive charge on oxygen. As a result, acylium ions are relatively stable and do not undergo rearrangement reactions.

Although the acylium ion bears significant positive charge on both carbon and oxygen, it acts as an electrophile solely at carbon. Reaction at oxygen

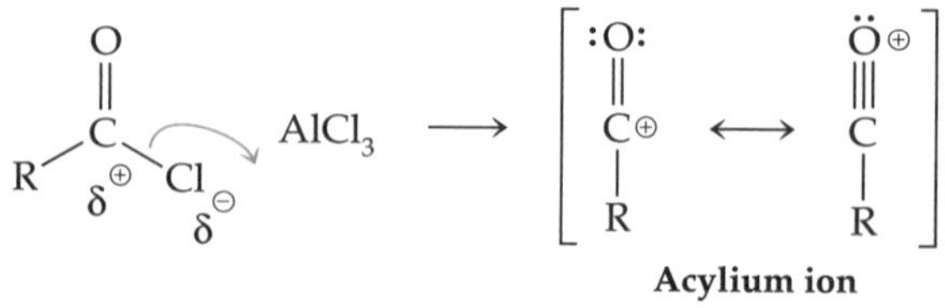

Acylium ion

▲ FIGURE 11-9

When chloride is removed from a carboxylic acid chloride by interaction with a Lewis acid, an acylium ion is formed. This ion bears formal positive charge on carbon and is a highly reactive electrophile. By using one of the lone pairs of oxygen to "backbond" to form a triple bond between carbon and oxygen, a resonance contributor for the acylium ion delocalizes the positive charge to oxygen. The existence of this second resonance contributor stabilizes the acylium ion, easing its formation by interaction with $AlCl_3$.

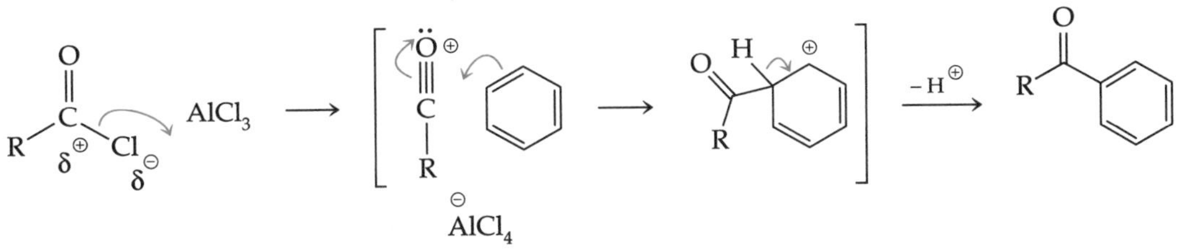

▲ FIGURE 11-10
Lewis acid complexation with a carboxylic acid chloride produces an acylium
ion. Friedel-Crafts acylation consists of electrophilic attack by an acylium ion
on an aromatic compound, forming a C–C σ bond. Deprotonation restores
aromaticity and produces the observed ketone product.

would result in a species still bearing positive charge on oxygen and having a
trivalent carbon. Thus, a new carbon–carbon σ bond forms between a ring
carbon and the carbon of the acylium ion. The resonance-stabilized cation
formed in this electrophilic attack is deprotonated, producing a ketone product (Figure 11-10).

   Friedel-Crafts acylation is important because it results in the attachment of a straight-chain (unbranched) carbon fragment to an aromatic ring.
(This is not possible in some Friedel-Crafts alkylations because of fast skeletal rearrangements.) The Friedel-Crafts acylation can attach an unbranched
acyl group to the ring, producing a ketone. Because it is relatively easy to
reduce aryl ketones under either acidic or basic conditions to the corresponding hydrocarbon (that is, to convert a C=O double bond into a CH_2
group), it is possible to convert the straight-chain ketone substituent attached in a Friedel-Crafts acylation into the corresponding unbranched hydrocarbon chain. This reduction can be accomplished under acidic conditions by treatment with zinc in HCl (called a **Clemmensen reduction**) or
under basic conditions by treatment with basic hydrazine (NH_2NH_2), called
a **Wolff-Kishner reduction,** as shown in Figure 11-11. We will defer a consideration of the details of these reactions until Chapter 12, but this need
not stop us from using them in planning chemical transformations.

   The sequence of Friedel-Crafts acylation followed by reduction makes
it possible to attach long-chain hydrocarbons to a ring without the rearrangements that accompany Friedel-Crafts alkylation. For example,
Friedel-Crafts alkylation with 1-propyl chloride proceeds through a rearranged secondary cation to produce isopropylbenzene (Figure 11-8),
whereas Friedel-Crafts acylation by propanoyl chloride, followed by reduction, gives rise to *n*-propylbenzene (Figure 11-12).

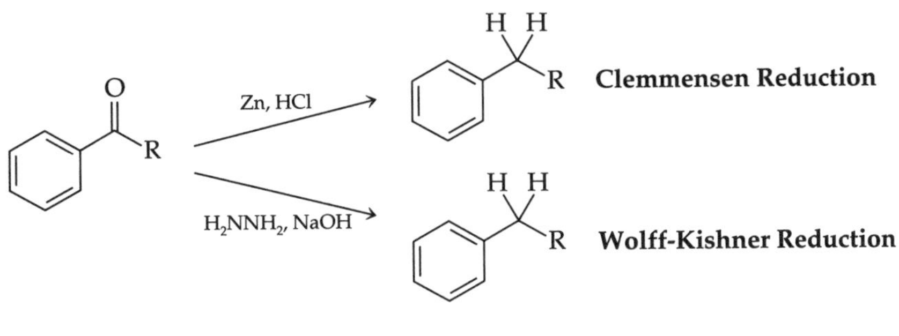

▲ FIGURE 11-11
A C=O group of a ketone can be converted into a —CH_2— group by
reduction under acidic (Clemmensen reduction) or basic (Wolff-Kishner
reduction) conditions.

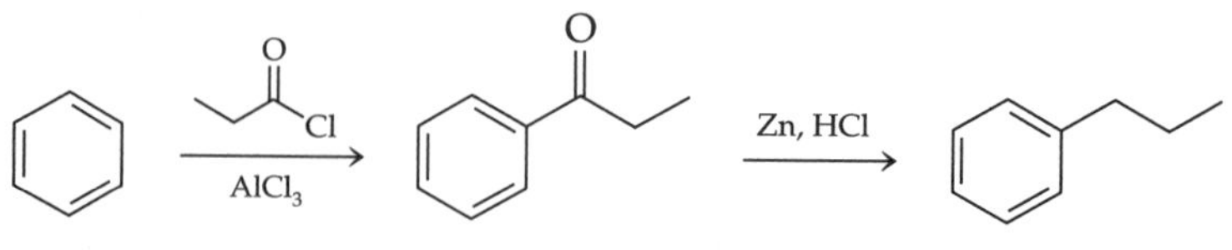

▲ FIGURE 11-12
Friedel-Crafts acylation permits a linear three-carbon skeleton to be attached to an aromatic ring. Clemmensen reduction of the acylated product converts the side chain into an unbranched alkyl group. By the use of this two-step sequence, it is possible to avoid the carbocationic rearrangements that lead to branched product in the Friedel-Crafts alkylation.

11-3 ▸ Reactions of Substituents and Side Chains

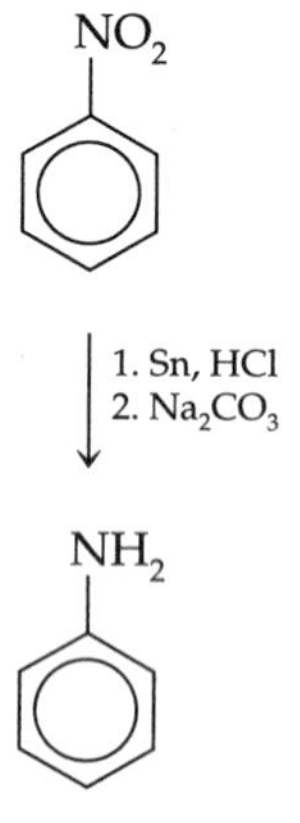

▲ FIGURE 11-13
A nitro (—NO_2) group can be converted into an amino (—NH_2) group by treatment with a reducing metal such as tin or zinc in the presence of acid. Sodium bicarbonate is used to neutralize the acid from the first step.

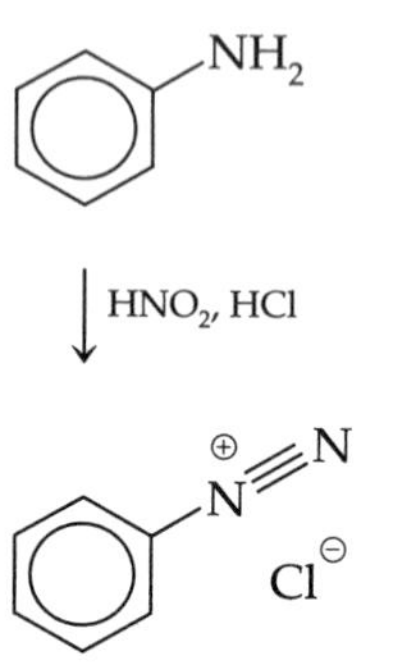

Other substituents introduced by electrophilic aromatic substitution also can be chemically altered by standard methods. For example, nitro groups in either aromatic or aliphatic compounds can be reduced to the corresponding NH_2 group by treatment with either Sn or Zn metal in HCl (Figure 11-13). Because the —NO_2 group can be introduced by electrophilic nitration, this sequence is an alternative to the benzyne ammonolysis described in Chapter 9 as a method for preparing aniline (aminobenzene). The aniline functional group is significant in a number of naturally occurring compounds that are biologically active.

Anilines are also important because they constitute the starting material for the production of azo dyes. The treatment of a primary aniline with nitrous acid, HNO_2, produces a stable **diazonium salt,** [Ar—N≡N]$^+$ X$^-$, by a reaction called **diazotization** (Figure 11-14). Because diazonium salts are cations, they are active electrophiles and can attack other, electron-rich aromatic rings. For example, benzenediazonium chloride reacts with *N,N*-

▲ FIGURE 11-14
Diazotization converts the —NH_2 group of aniline into an —$^+$N≡N substituent.

A WARNING FOR THOSE WHO ENJOY COCKTAILS WITH FUNNY UMBRELLA HATS

Red No. 2 is a commercial dye now used only for dyeing wool and silk. Before its use was restricted in 1976 by the FDA, this dye was used for coloring food, especially maraschino cherries. However, Red No. 2 was shown to be a mutagen in one especially sensitive test. The battle between those who argued that there was no rationale for using, even at very low levels, any food color that might be dangerous and those who contended that no one consumed enough cherries to be at significant risk raged for years. The "compromise" that resulted banned Red No. 2 but permitted existing stocks of colored cherries to be sold.

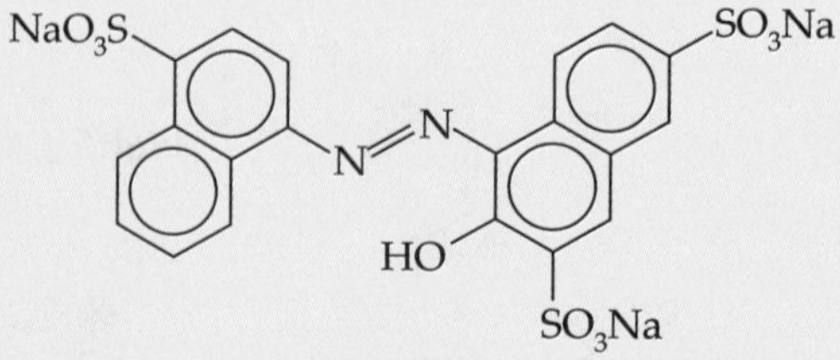

Red No. 2

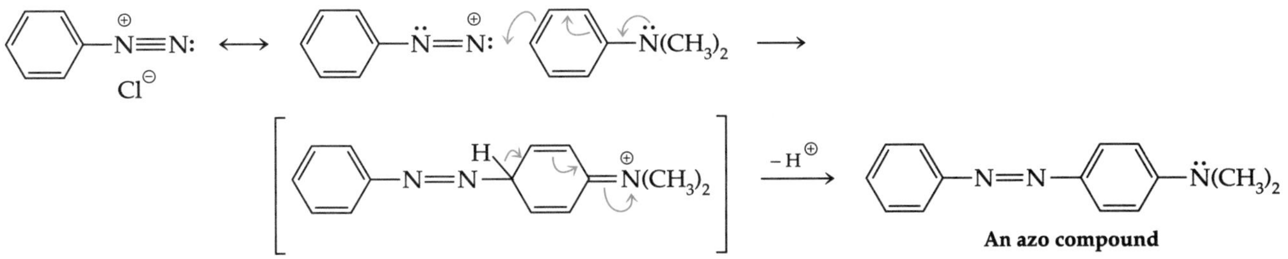

▲ FIGURE 11-15
Electrophilic attack by the terminal nitrogen atom of a diazonium salt forms a
C–N σ bond to an electron-rich aromatic ring. Deprotonation of the resonance-
stabilized cation produces the highly colored azo compound.

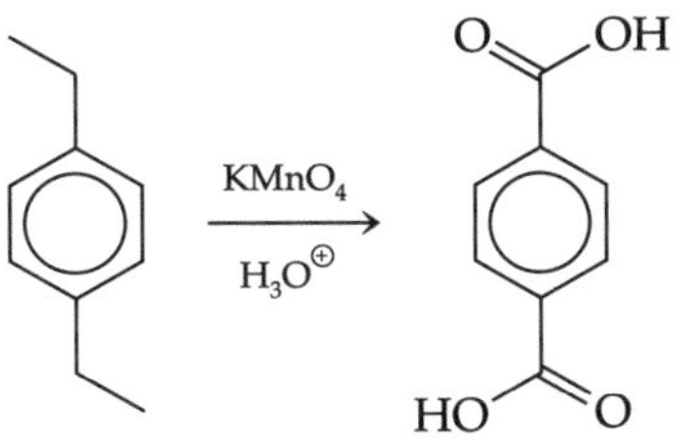

Terephthalic acid

▲ FIGURE 11-16
Heating a compound in aqueous $KMnO_4$ is a vigorous oxidation method.
Both branched and straight-chain alkyl groups of any length, when attached
to an aromatic ring, are oxidatively degraded to $-CO_2H$ groups. Thus,
benzoic acids can be prepared by Friedel-Crafts alkylation or acylation
followed by oxidative degradation of the attached chains.

dimethylaniline in the *para* position to produce *p*-(dimethylamino)azo-
benzene, a bright yellow solid that was, at one time, used as a colorant in
margarine (Figure 11-15). Azo compounds are highly colored because the
$-N{=}N-$ linkage extends the π system between aromatic rings, resulting in strong absorption in the visible region. With various substituents on
the two aromatic rings, azo compounds of nearly every color have been
prepared. Azo dyes were among the first synthetic colorfast agents that
could be used for dyeing wool and cotton.

The oxidation of side chains on aromatic rings is another means of imparting functionality to an aromatic ring. The oxidation of alkyl-substituted
aromatic rings with hot aqueous $KMnO_4$ results in oxidative cleavage of the
side chain, forming a $-CO_2H$ group irrespective of the length of the side
chain. For example, both alkyl groups in *p*-diethylbenzene are cleaved by
hot $KMnO_4$ to form the diacid terephthalic acid (Figure 11-16).

Attachment to an aromatic ring does not interfere with the characteristic reactivity of substituents present on a side chain. For example, a chain
introduced by electrophilic substitution can undergo free radical halogenation at the benzylic position, as discussed in Chapter 7 and shown in Figure
11-17. This reaction is regiospecific for CH_2-group adjacent to the aromatic
ring, a result of enhanced radical stability at that position.

EXERCISE 11-C

Identify the reagents (or sequence of reagents) needed to effect each of the
transformations in parts *a* through *g* on the next page.

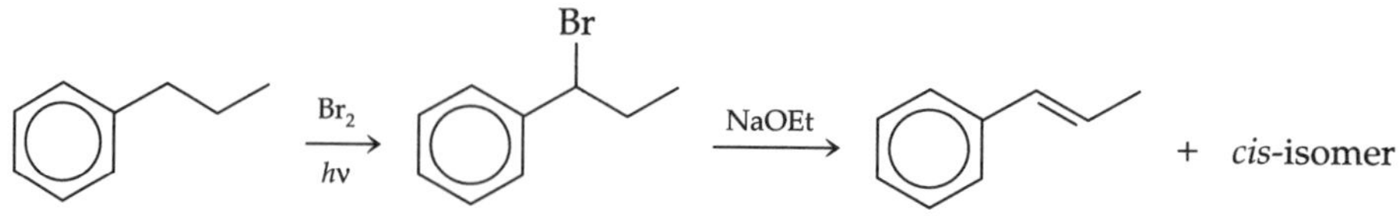

▲ FIGURE 11-17

The product of electrophilic substitution retains the expected reactivity of any group present. Shown here is a free radical bromination at the benzylic position of *n*-propylbenzene. (The starting material can be prepared from benzene by Friedel-Crafts acylation, followed by reduction, as shown in Figure 11-12.) The bromide can be dehydrobrominated by treatment with strong base by the routes described in Chapter 9. Treatment of this benzylic bromide with base leads to elimination, as shown here; treatment with a nucleophile would induce substitution, as described in Chapter 8.

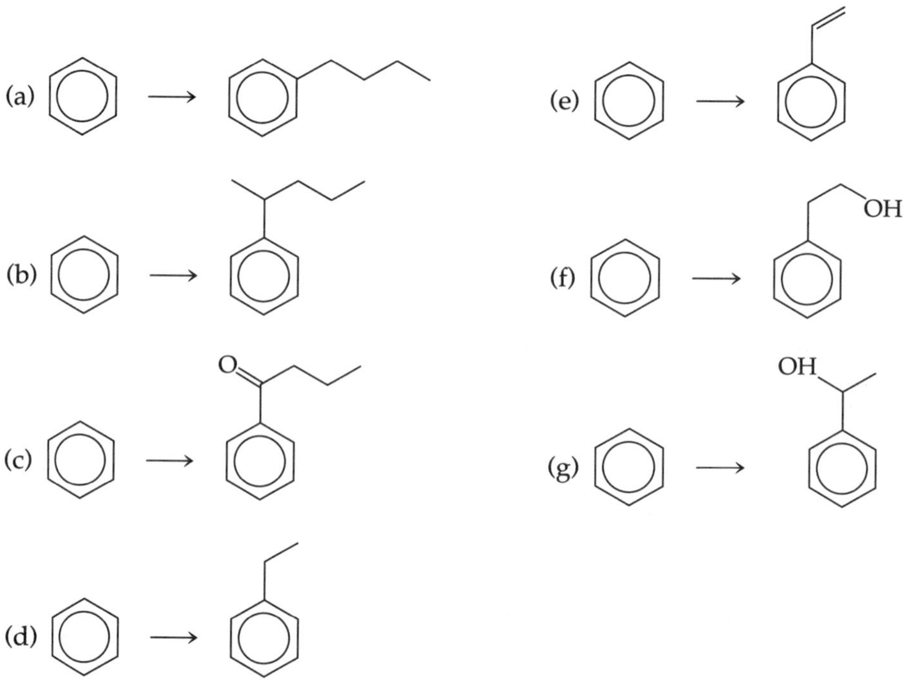

11-4 ▸ Substituent Effects

Let us consider the effect of substituents already bound to the benzene ring on the regiochemistry and kinetics of further electrophilic substitutions. We will see that the electron density of the atom attached to the aromatic ring significantly affects both the rate of the electrophilic substitution and the preference for the position at which further substitution takes place.

The electrophilic bromination of toluene (Figure 11-18), for example, does not give a statistical mixture of the three possible bromine substitution products. A purely statistical distribution of products would give a 2:2:1 mixture of *ortho, meta,* and *para* substitution in the absence of other factors, reflecting the number of hydrogens available for substitution. In the electrophilic bromination of toluene, however, substantially greater than statistical ratios of *ortho* and *para* products are observed.

This reaction takes place by a route parallel to the chlorination shown in Figure 11-3: electrophilic bromine is generated by treatment of molecular bromine with a Lewis acid. Attack by this electrophilic reagent can occur at

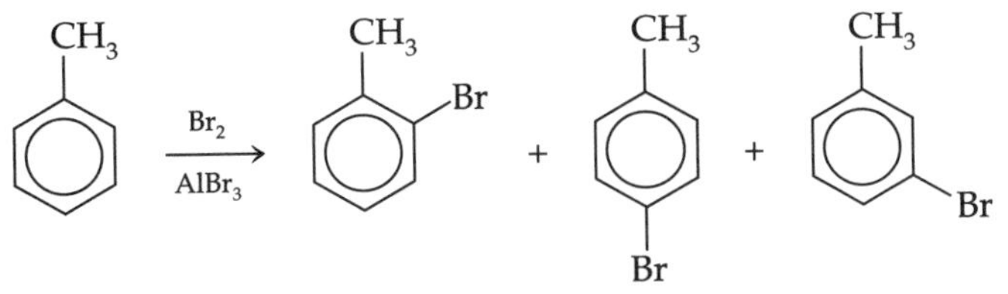

▲ FIGURE 11-18
Electrophilic bromination of toluene produces a distribution of
monobrominated product having far less of the *meta*-substituted isomer than
would be expected from the statistical abundance of hydrogens that could be
replaced.

the *ortho, meta,* or *para* position. An isomeric resonance-stabilized cation results from attack at each position, and three resonance contributors can be
written for each cation (Figure 11-19). For the cation produced by attack at
either the *ortho* or the *para* position, one of the contributing resonance structures bears positive charge at the site substituted by the methyl group. To
the extent that an alkyl substituent stabilizes a cation (recall that tertiary

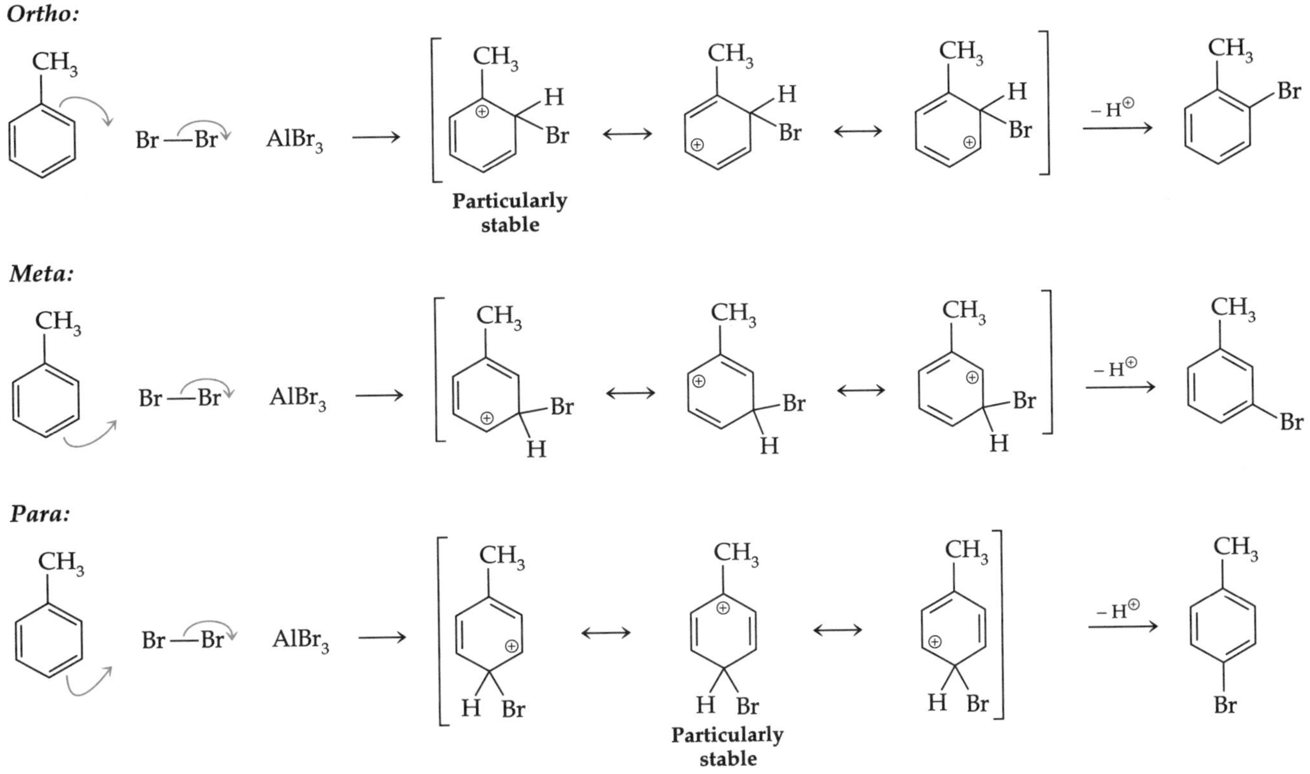

Ortho:

Meta:

Para:

▲ FIGURE 11-19
The resonance-stabilized cations formed by attack at the *ortho* or *para* positions
bear positive charge at a site to which the methyl group of toluene is attached.
When attack is at the *meta* position, the positive charge in the resonance-
delocalized cation is not at the atom to which the methyl group is attached.
Because more highly substituted cations are more stable, the cations formed
by attack at the *ortho* and *para* positions are favored over that produced by
meta attack.

cations are more stable than secondary or primary ones), these structures are particularly stable. The existence of a resonance contributor with unusual stability stabilizes each of the delocalized cations formed from *ortho* or *para* attack, and facilitates their formation in the rate-determining step. Deprotonation, as described earlier for benzene, then gives rise to substitution products. Notice that attack at the *meta* position clearly does not result in a particularly stabilized cationic intermediate.

This preference for the *ortho* and *para* products can be simply explained by the **directive effect** of the electron-releasing (cation-stabilizing) methyl group: the methyl group directs electrophilic attack to the *ortho* and *para* positions so as to form the more-stable intermediates. If this electronic effect were the only important one, however, we would expect two-thirds of the product to be *ortho* substituted and one-third to be *para* substituted. Finding a larger-than-expected fraction of *para*-substitution product (Figure 11-18) means that the *ortho*-substituted product is disfavored by some additional factor. The proximity of a methyl group makes the transition state for *ortho* attack more crowded than that leading to *para* substitution. For example, the electrophilic bromination of *t*-butylbenzene produces almost exclusively *para* product because the bulky alkyl group blocks *ortho* attack. Thus, electronic factors favor *ortho* and *para* attack, with steric hindrance further favoring *para* substitution.

Not only does the methyl group of toluene control regiochemistry, by directing the attack of the electrophile to the *ortho* and *para* positions, but it also accelerates the reaction. The bromination of toluene is much faster than the bromination of benzene because the methyl group behaves as an electron-releasing group, thus increasing the π-electron density in the ring that is attacked by the electrophile. It also stabilizes the intermediate cation and the transition state of the rate-determining step that leads to its formation.

Electron Donors and Acceptors

To the extent that a group attached to a benzene ring is electron releasing, it stabilizes the carbocations formed by attack at the *ortho* and *para* positions (Figure 11-20). This electron release also accelerates the reaction by stabilizing the transition state leading to the intermediate cation. The higher electron density in the π system is the electron source for the σ bond to the electrophile.

Still more effective is the activation of a ring by a strong electron donor that has an unshared electron pair at the atom bound to the ring (for exam-

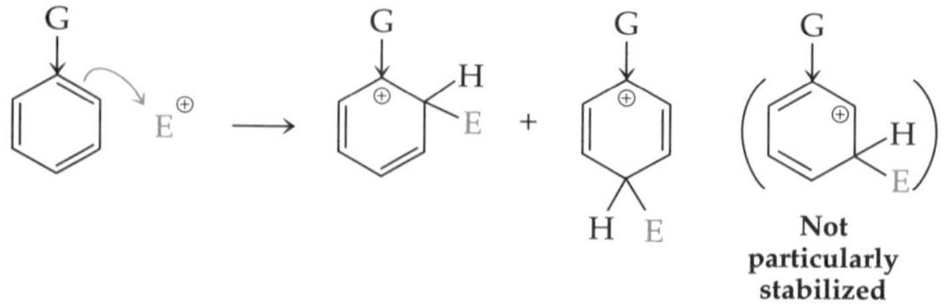

▲ FIGURE 11-20
When an attached group, G, is electron releasing (compared with hydrogen), it provides special stabilization to a cation formed by electrophilic attack at a position *ortho* or *para* to its position. The group interacts directly with a site bearing positive charge in these intermediates, but it is incorrectly positioned to stabilize the cation formed by *meta* attack. This stabilization of the intermediate formed in the rate-determining step also accelerates the rate of attack at these positions because the transition states leading to the intermediate also are stabilized (Hammond Postulate).

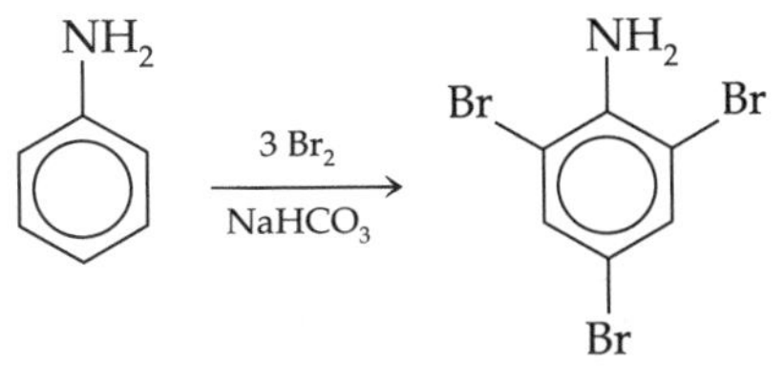

▲ FIGURE 11-21

A strongly electron releasing group such as —NH_2 activates the ring sufficiently that halogenation can occur without activation by a Lewis acid. The amino group is so effective in activating the ring toward electrophilic attack that tribromination occurs—all three specifically activated *ortho* and *para* positions are brominated. (Sodium bicarbonate is used here to neutralize HBr produced as the reaction by-product.)

ple, an —NH_2 group). Bromination of aniline takes place at all three *ortho* and *para* positions even without the activation of a Lewis acid (Figure 11-21) because of the significant activation by the strong electron-releasing capacity of the —NH_2 group.

The strong activation by groups bearing lone pairs at the atom directly attached to the benzene ring derives from the release of electrons from the substituent, stabilizing an additional resonance contributor in which positive charge located on carbon in the cationic intermediate can be shifted to the heteroatom (upper structures in Figure 11-22).

When the lone pair on the atom attached to the ring also takes part in a resonance interaction with another group, the substituent is less able to release electrons to the ring. Groups such as —NHC(O)R or —OC(O)R still

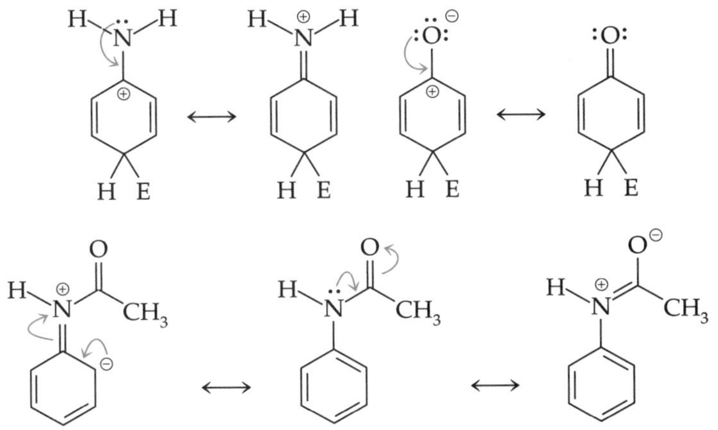

▲ FIGURE 11-22

When a substituent is attached to an aromatic ring through an atom that bears a nonbonded lone pair, the electron pair can be donated to the ring directly. This electron donation provides for an additional highly stabilized resonance contributor for the intermediate cation, as shown here for the cations formed by *para* attack on aniline (upper left-hand structures) and on phenoxide ion (upper right-hand structures). When the electron density in the ring is increased by donation from the substituent, the reaction rate is enhanced. Amide (—NHC(O)R) or ester (—OC(O)R) substituents are less effective than a simple amino or hydroxy group in releasing electrons to an aromatic ring because of contributions from structures like that at the right. Nonetheless, because of the existence of resonance contributors like that at the lower left, they activate the ring toward electrophilic attack and direct to the *ortho* and *para* positions.

act as electron donors but are somewhat less effective in activating electrophilic substitution than is a simple —NH$_2$ or —OH group. In N-phenylacetamide, for example, the existence of the resonance structure at the lower right in Figure 11-22 interferes with donation of electron density to the ring, making the —NHC(O)Ph group less effective than a simpler —NH$_2$ group. The resonance structure at the lower left in Figure 11-22, however, still permits effective electron release to the ring and induces an enhanced rate of electrophilic attack. This group, however, limits reactivity with bromine so that monosubstitution is observed (Figure 11-23). Because an ester or amide is easily hydrolyzed, this indirect route can be used to prepare monosubstituted (mainly *para*) anilines and phenols.

Alkyl groups, lacking a delocalizable lone pair on the attached atom, stabilize the carbocationic intermediates to a somewhat smaller degree than do the amide or ester groups. Alkyl groups cannot stabilize the transition state through π-electron release in the way that these strong or moderate electron donors can.

By inference, electron-withdrawing groups should have the opposite effect. Thus, electron-withdrawing groups particularly destabilize the cationic intermediates formed by attack at the *ortho* and *para* positions (Figure 11-24). Because the resonance-stabilized cation substituted by an electron-withdrawing group is destabilized compared with that formed from benzene itself, its formation also occurs more slowly: an electron-withdrawing group reduces the electron density of the π system and makes electrophilic attack more difficult. The particular instability of a cationic intermediate formed by electrophilic attack at the *ortho* and *para* positions is avoided by attack at the *meta* position (Figure 11-25). Hence, *meta* substitution dominates.

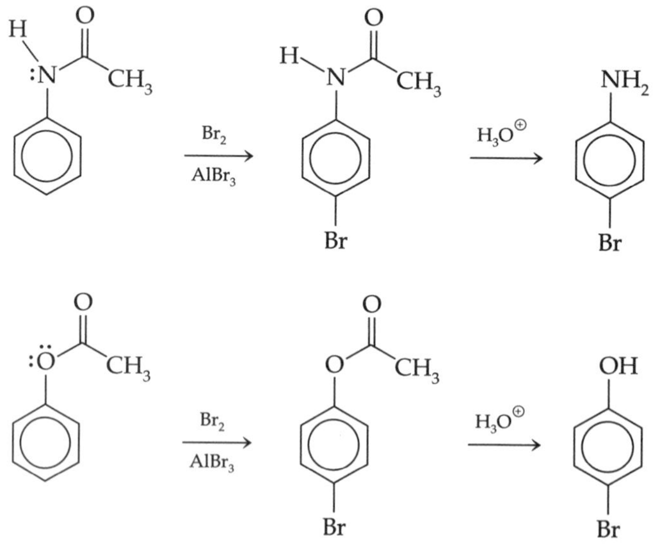

▲ FIGURE 11-23
Because the —NHC(O)R group in N-phenylacetamide (upper reaction) and the —OC(O)R group in phenyl acetate (lower reaction) bear a nonbonded electron pair on the atom attached to the ring, they are activating substituents toward electrophilic attack. Resonance with the carbonyl group (see Figure 11-22), however, makes them less activating than —NH$_2$ or —OH groups. Bromination can therefore be stopped at the monosubstitution stage, unlike the bromination of aniline in Figure 11-21. The —NHC(O)R and —OC(O)R substituents are also sufficiently bulky that steric interactions disfavor *ortho* substitution, leading to monobrominated *para* product as the major one. These groups can be readily hydrolyzed in aqueous acid, and this sequence thus provides a route for the preparation of *p*-substituted anilines and phenols.

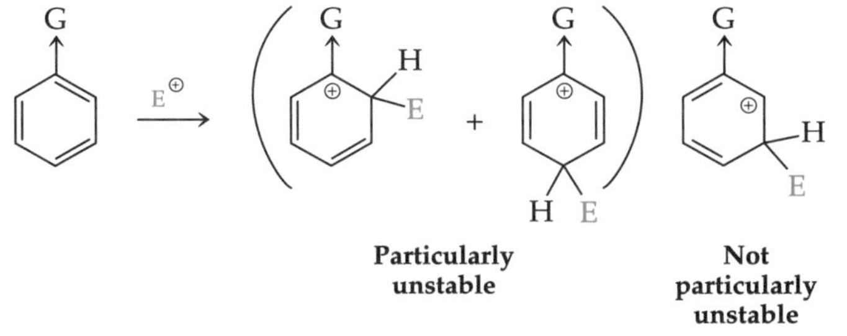

Particularly
unstable

Not
particularly
unstable

▲ **FIGURE 11-24**

Like that of an electron-releasing group, the effect of an electron-withdrawing group is greatest on the intermediate cations formed by *ortho* or *para* attack. However, because an electron-withdrawing group *destabilizes* the cation, its effect is to retard, rather than enhance, the attack at these positions so that *meta*-substitution product is formed.

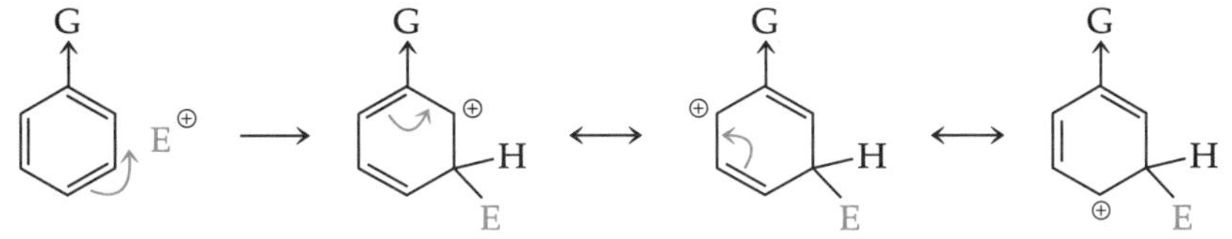

▲ **FIGURE 11-25**

The cation formed by *meta* attack bears positive charge at atoms *not* directly bound to the electron-withdrawing group, G. Hence, this cation is less destabilized by the electron-withdrawing group than that formed by *ortho* or *para* electrophilic attack.

Electron withdrawal through the σ skeleton also deactivates toward electrophilic substitution by reducing electron density in the ring. Because this effect is attenuated when felt through a greater number of σ bonds, this inductive withdrawal has a larger effect on the *ortho* than the *para* position, for the same reasons given in Chapter 6 that 4-chlorobutanoic acid is less acidic than 3-chlorobutanoic acid, which is less acidic than 2-chlorobutanoic acid, which is less acidic than butanoic acid itself.

Two typical electron-withdrawing groups, $-NO_2$ and $-CO_2Me$, are shown in Figure 11-26. In both cases, the atom attached to the benzene ring

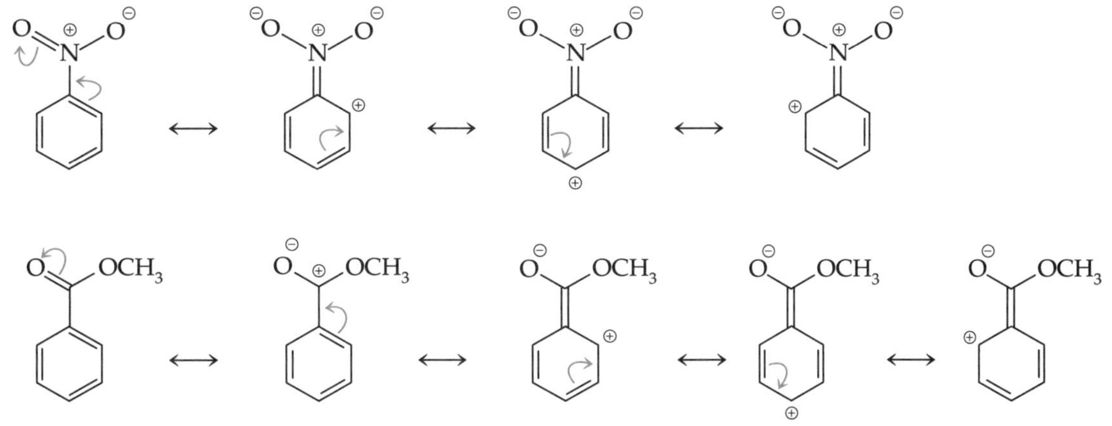

▲ **FIGURE 11-26**

In at least one resonance contributor, an electron-withdrawing group bears formal positive charge at a ring atom. (In some cases, such as the $-NO_2$ group shown in the upper resonance structures, all resonance structures bear formal charge at the attached atom.) Electron density then flows from the ring, making it more difficult for electrophilic attack to occur.

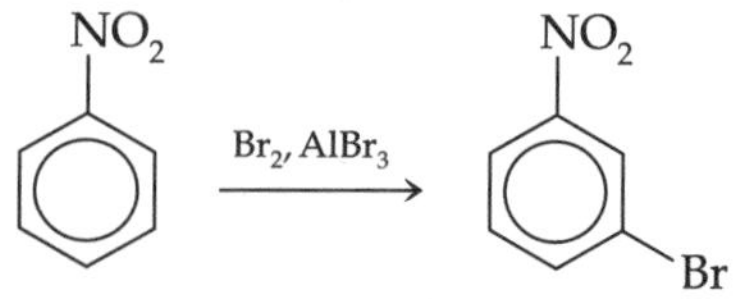

▶ **FIGURE 11-27**
The nitro group (—NO$_2$) is
strongly electron withdrawing.
It therefore slows electrophilic
bromination and directs further
substitution to the *meta*
position.

is either formally positively charged or partially positively charged by
virtue of significant resonance contributors. This electron deficiency makes
electrophilic attack much more difficult, particularly at the *ortho* and *para*
positions: even *meta*-substitution product is obtained at a slow rate. For example, the bromination of nitrobenzene is slower than that of benzene itself
and gives predominantly *m*-bromonitrobenzene (Figure 11-27).

An amino substituent can be changed from a strong donor (—NR$_2$) to
an acceptor (—$^\oplus$NHR$_2$) by protonation. In the protonated form, the lone
pair on nitrogen is not available for π-release to the ring, and the substituent
instead functions as an electron-withdrawing group. The effects of an electron-withdrawing group are sometimes so significant that the reaction
fails—for example, it is impossible to conduct a Friedel-Crafts alkylation or
acylation on a ring deactivated by the presence of a strong electron-withdrawing group such as —NO$_2$, —C≡N, or —CO$_2$R.

Table 11-1 contains a brief summary of the electron-releasing capacity
of several common substituent groups. Those groups that bear a lone pair
on the atom directly attached to the ring most effectively release electron
density. They activate the ring most and exert the strongest directive effect
toward the *ortho* and *para* positions. Those that bear positive charge (or partial positive charge) at the atom adjacent to the ring deactivate the ring toward electrophilic attack and direct substitution toward the *meta* position.

An Exception: Electrophilic Substitution
of Halogen-substituted Aromatics

For most substituents of aromatic rings, those that are electron donating direct *ortho*, *para* and activate the ring kinetically toward electrophilic attack,
and those that are electron withdrawing direct *meta* while deactivating the
ring. However, halogen substitution is unique in that it has the effect of directing *ortho*, *para* but *slowing* the rate of electrophilic attack. Chlorobenzene,
for example, is nitrated about fifty times more slowly than benzene itself
(Figure 11-28) but almost exclusively at the *ortho* and *para* positions. Similar
effects are observed with the other halobenzenes (fluorobenzene, bromobenzene, and iodobenzene). The rationale for this apparent contradiction
is that the halogen substituents are highly electronegative (Chapter 3) and
thus withdraw electron density from the aromatic ring through the σ system, an "inductive effect." This σ withdrawal, shown in Figure 11-29, is
caused by the high electronegativity of halogen in a C–X σ bond. The chlorine atom is partially negatively charged and the ring carbon to which it is
attached is partially positively charged. The approach of the electrophile to
the partially positively charged ring is thus significantly retarded, accounting for the slower rate at which the halides undergo electrophilic substitution.

On the other hand, a covalently attached halogen atom possesses three
nonbonded lone pairs on the atom (the halide) directly attached to the benzene ring. Some of this electron density can be released through the ring π
system to the *ortho* and *para* positions as the active electrophile approaches
(Figure 11-30), a "resonance effect." Thus, as an electrophile attacks the ring,
producing a resonance-stabilized cation, positive charge density is highest

Table 11-1 ▸ Electron-releasing and Electron-withdrawing Groups

Electron-releasing Groups (Donors): *o, p* Directors

Strong   —N̈H₂   —ÖH   —Ö:⊖   —ÖR   —N̈R₂

Moderate

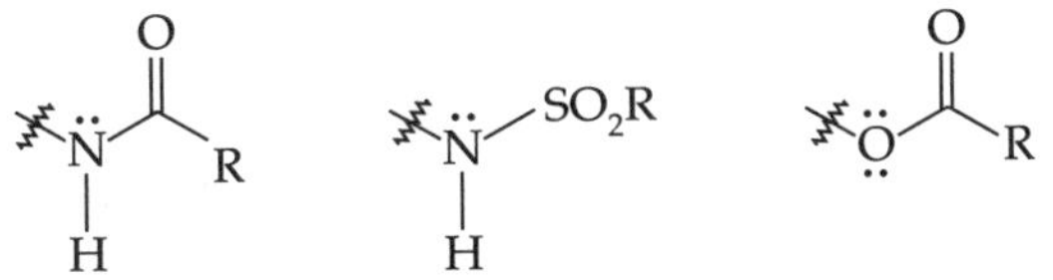

Weak   —R   (R = alkyl or aryl)

Electron-withdrawing Groups (Acceptors): *m* Directors

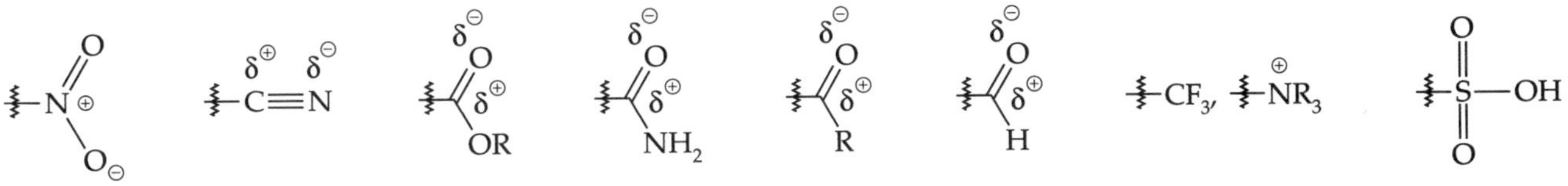

Halogens (σ Withdrawers, π Donors): *o, p* Directors

—F   —Cl   —Br   —I

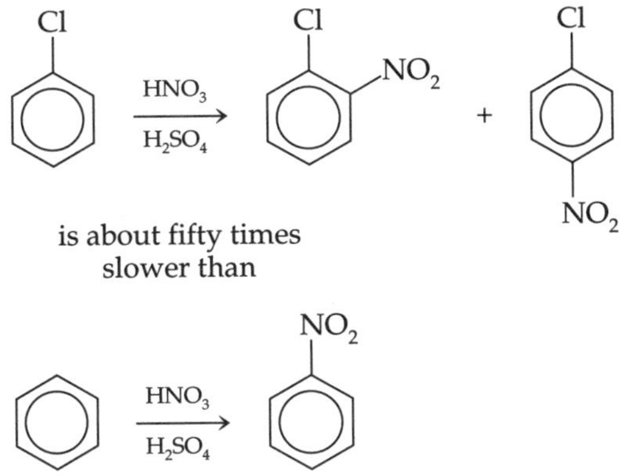

is about fifty times slower than

▲ FIGURE 11-28
The halogens are anomalous among substituents in directing *ortho, para* while slowing the rate of electrophilic attack.

Sigma Withdrawal

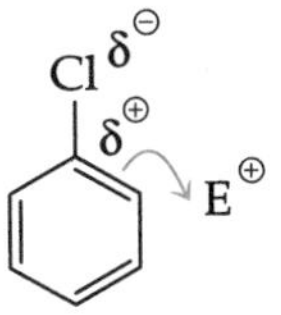

▲ FIGURE 11-29
The high electronegativity of a halogen polarizes a C–X bond so that carbon bears partial positive charge. This leaves the ring electron deficient and makes electrophilic attack difficult. The inductive deactivation is greater at the *ortho* than the *para* position because its influence decreases with distance from the electronegative chlorine atom.

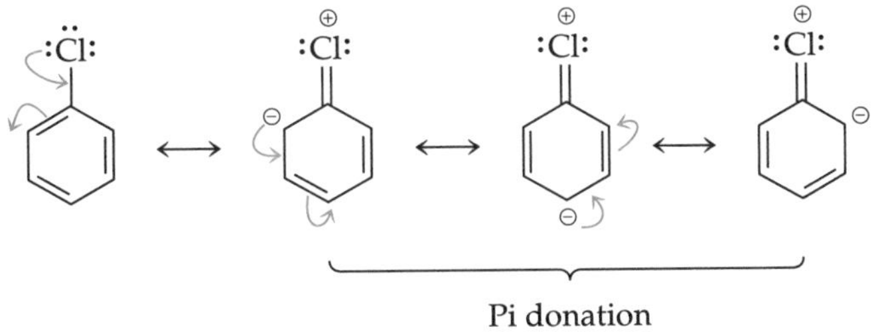

Pi donation

▲ FIGURE 11-30
Back donation of one of the lone pairs of the halogen shifts electron density in the π system toward the *ortho* and *para* positions, facilitating electrophilic attack at those sites.

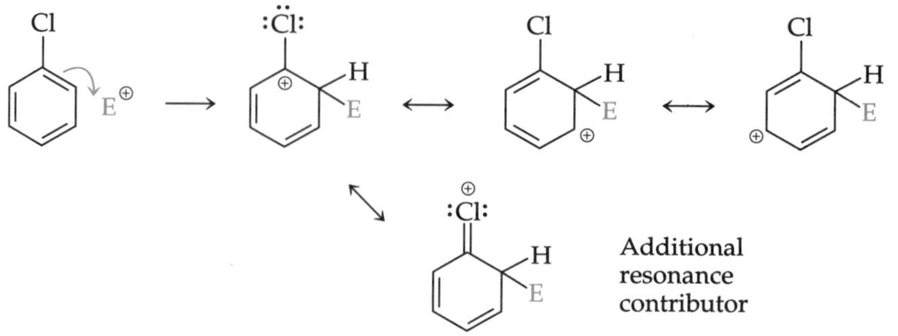

Additional
resonance
contributor

▲ FIGURE 11-31

The cation formed when an electrophile attacks at a position *ortho* to a halogen substituent bears formal positive charge on the carbon bearing the halogen in one resonance contributor. Backbonding of a nonbonded lone pair of electrons on the halogen stabilizes this form and facilitates the formation of the cation in the rate-determining step.

at positions *ortho* and *para* to the site to which the halogen is bound. This allows electron release from the halogen to provide an additional resonance form, and thus enhanced stability, to the cation (Figure 11-31).

EXERCISE 11-D

Draw the structures for *para* attack that are analogous to those shown in Figure 11-31. Do the same stabilizing effects on the intermediate cation apply?

Such stabilizing resonance contributors are absent in the intermediates formed from electrophilic attack at the *meta* position (Figure 11-32). The lone pairs on the halogen therefore offer no additional stability beyond that available in the unsubstituted compound. Because the regiochemistry of electrophilic aromatic substitution is determined by the relative stability of the possible isomeric cations, *ortho* and *para* substitutions are favored over the analogous process for *meta* attack. Thus, halobenzenes are beautiful examples for demonstrating σ-electron withdrawal with simultaneous π-electron release.

Each of the halides exerts analogous effects, although the magnitude of the influence on rate and on positional selectivity varies from F to Cl to Br to I. The ability of a given halogen to contribute additional stabilization to the intermediate cation depends on its electronegativity and how closely it

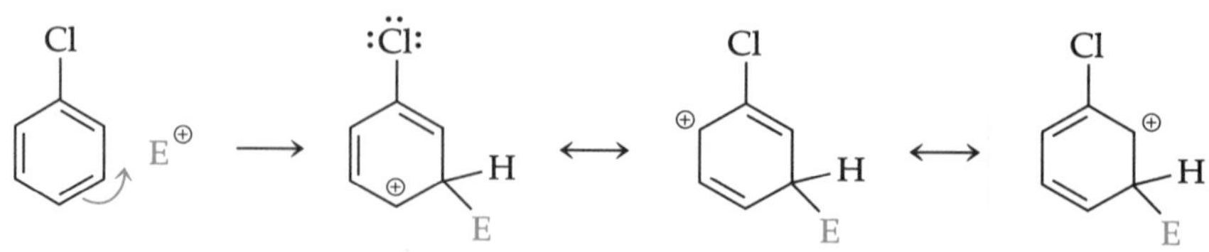

▲ FIGURE 11-32

The cation formed when an electrophile attacks at a position *meta* to a halogen substituent bears formal positive charge only at sites other than that to which the halogen is bound. Backbonding by π lone pairs from the halogen cannot stabilize the *meta*-substituted cation. The *meta*-substitution product is therefore formed only in minor yield.

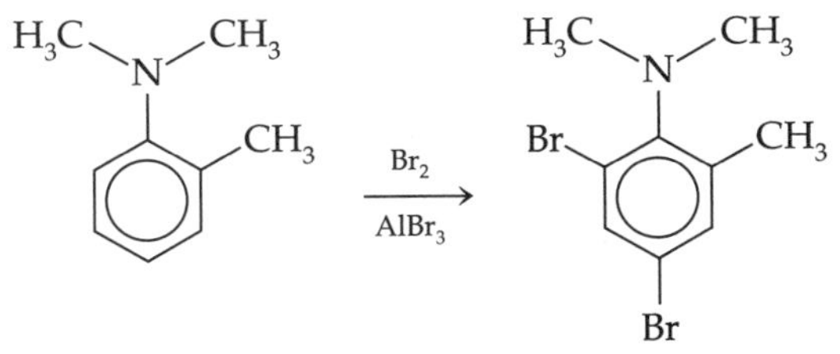

▲ FIGURE 11-33
The *o,p* directive effect of the stronger electron donor ($-$NMe$_2$) dominates over that of the weaker donor ($-$Me).

matches carbon in orbital size. Only with similarly sized *p* orbitals can a halogen atom effectively donate its electrons in a resonance sense. Recall that electronegativity decreases in the progression down a column of the periodic table as size mismatching with carbon increases. Again, we find offsetting inductive and resonance effects.

EXERCISE 11-E

Write the resonance structures for an electrophilic attack at the *para* position of fluorobenzene, justifying why fluorobenzene undergoes electrophilic substitution at the *ortho* and *para* positions.

Multiple Substituents

When more than one group is attached to a ring, the effect of the stronger donor prevails. In *o*-methyl-*N,N*-dimethylaniline, for example, substitution is directed by the strongly activating $-$NMe$_2$ group (Figure 11-33) to its *ortho* and *para* positions, rather than by the more weakly activating methyl group.

Thus, electrophilic aromatic substitution takes place when a sufficiently activated electrophile approaches an aromatic π system, with substitution rather than addition being the outcome in order to restore ring aromaticity. Rate acceleration and *ortho, para* direction result when electron-releasing groups are present on a ring. A slower reaction rate and *meta* direction occur when electron-withdrawing groups are present.

EXERCISE 11-F

Predict the regiochemistry of the monosubstitution product expected in each of the following conversions. (Continued on the next page.)

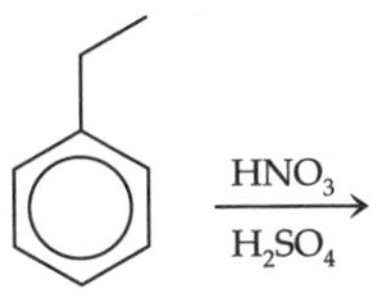

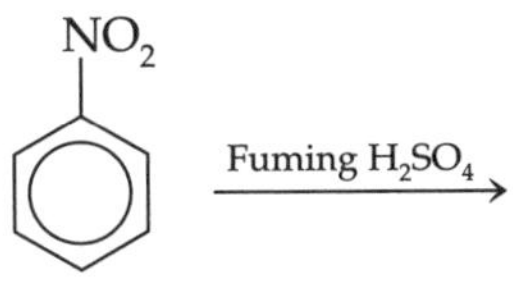

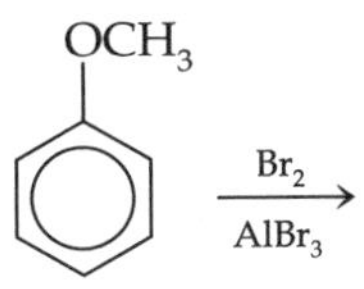

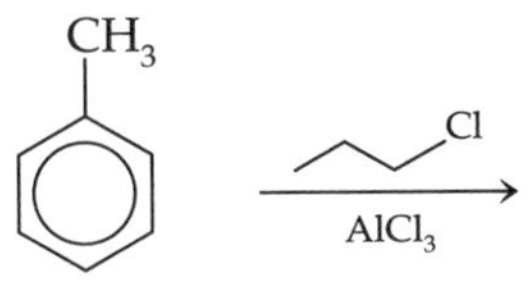

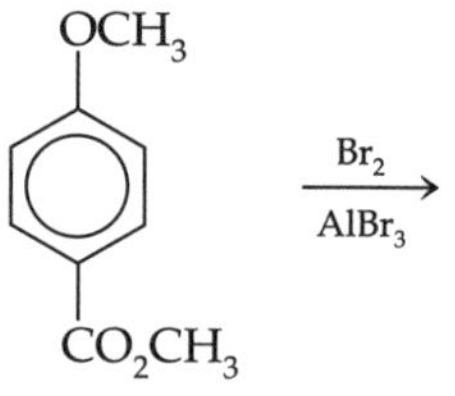

Using Substituent Effects in Synthesis

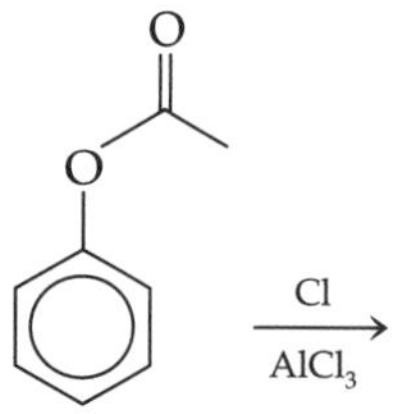

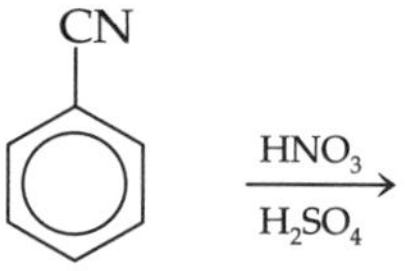

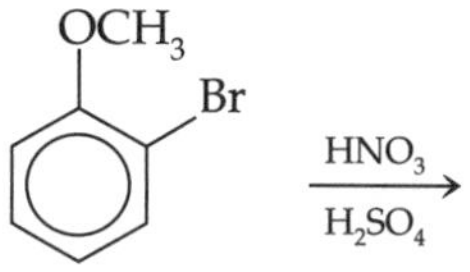

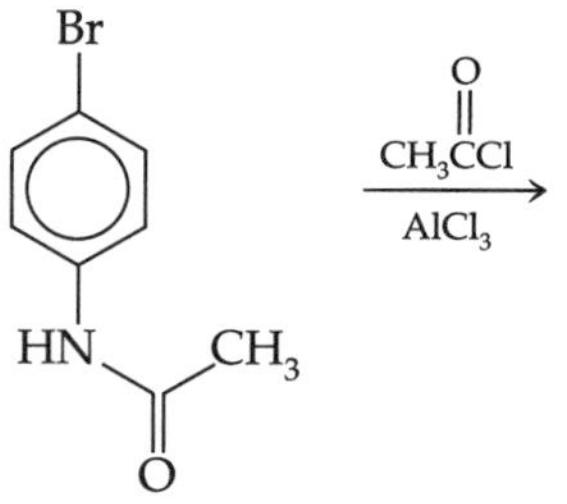

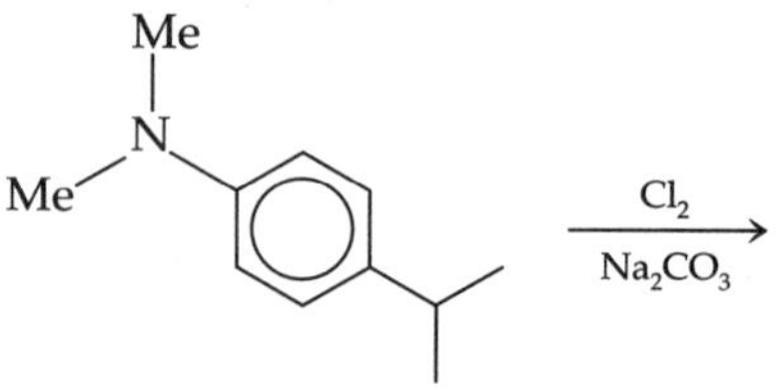

As explained in Section 11-3, many substituents possess characteristic chemical reactivity that enables them to be converted from electron-withdrawing groups into electron-releasing ones, or vice versa. For example, an electron-withdrawing —NO₂ group can be reduced to an electron-releasing —NH₂ group and an electron-releasing alkyl group can be oxidized to an electron-withdrawing —CO₂H group. The order in which substituents are introduced to an aromatic ring can control the regiochemistry of the isomeric products, and the ability to alter substituents further enhances the chemist's control over regiochemistry.

As an example, let us consider how we might prepare *m*-chlorobenzoic acid and *p*-bromobenzoic acid from benzene. We recognize first that, for both products, we must replace (1) a ring hydrogen atom by a halogen and (2) a second hydrogen by a —CO₂H group. This can be accomplished by the oxidation (with hot KMnO₄) of an alkyl chain introduced by a Friedel-Crafts alkylation. If we chlorinate benzene first, the chlorine substituent will direct further substitution to the *ortho* and *para* positions (Figure 11-34), as described in the preceding subsection. Thus, Friedel-Crafts alkylation places the alkyl group at the wrong position for the *meta* product. If we reverse the order, first alkylating to yield toluene and then chlorinating (Figure 11-35), we obtain the same mixture, because —CH₃ also is *o,p* directing. However, if we oxidize the methyl group introduced by a Friedel-Crafts alkylation in the first step in Figure 11-35, we obtain a *meta*-directing —CO₂H group. Chlorination of benzoic acid then gives the desired product (Figure 11-36, on page 392).

By using routes parallel to those in Figures 11-34 and 11-35, we can prepare our second synthetic target, *p*-bromobenzoic acid (Figure 11-37, page 392). The choice between the routes shown in Figure 11-37 is then made on secondary grounds: perhaps the separation of the mixture of solid brominated benzoic acids is easier than the separation of the liquid brominated

ASPIRIN: A SIMPLE AROMATIC PHARMACEUTICAL

Aspirin, acetylsalicylic acid, is the world's most widely used remedy to reduce pain and to lower fever. Although its first description in the medical literature in 1899 was for the treatment of rheumatic fever, its phenolic precursor, salicylic acid, was described earlier (1876) as being effective in controlling fever and in treating gout and arthritis. Still earlier (1763), homeopathic medical practitioners reported that willow (*salix*) bark, when chewed, was effective in treating malaria. Only later was it determined that extraction of willow bark yields salicin, a compound that could be hydrolyzed and oxidized to salicylic acid.

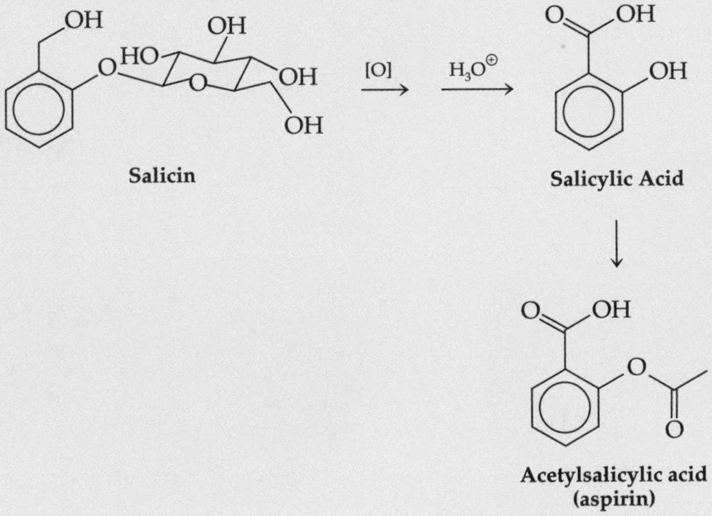

Salicin   **Salicylic Acid**

**Acetylsalicylic acid
(aspirin)**

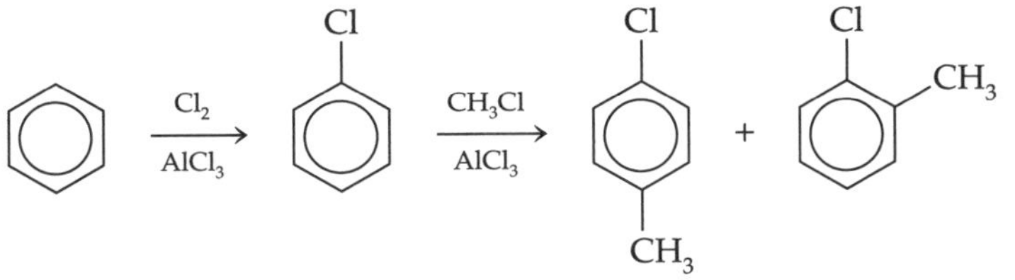

▲ FIGURE 11-34

Chlorination of benzene produces chlorobenzene. The chlorine substituent is deactivating and *o,p* directing. A nonstatistical mixture of *o*- and *p*-chlorotoluene is therefore formed because of greater deactivation of the *ortho* positions than the *para* position by both greater σ inductive withdrawal and more-significant steric hindrance by the bulky chlorine substituent.

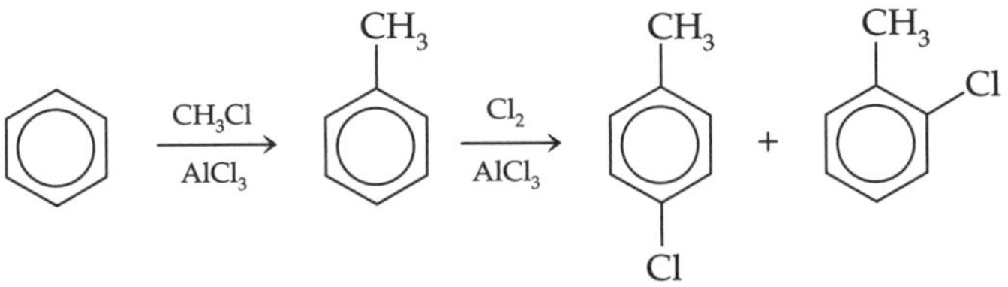

▲ FIGURE 11-35

Friedel-Crafts alkylation of benzene produces toluene. Because the —CH₃ group is activating and *o,p* directing, a mixture of *o*- and *p*-chlorotoluene is again formed.

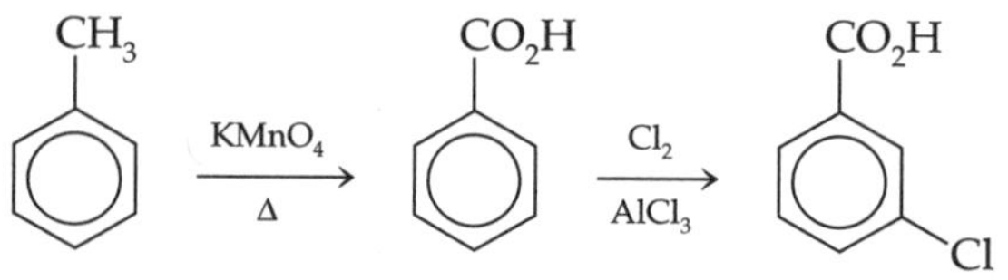

▲ FIGURE 11-36
When toluene, formed by Friedel-Crafts alkylation of benzene, is oxidized by hot potassium permanganate, benzoic acid is formed. Because the —CO₂H group is electron withdrawing, it directs chlorination to the *meta* position, producing the desired *m*-chlorobenzoic acid.

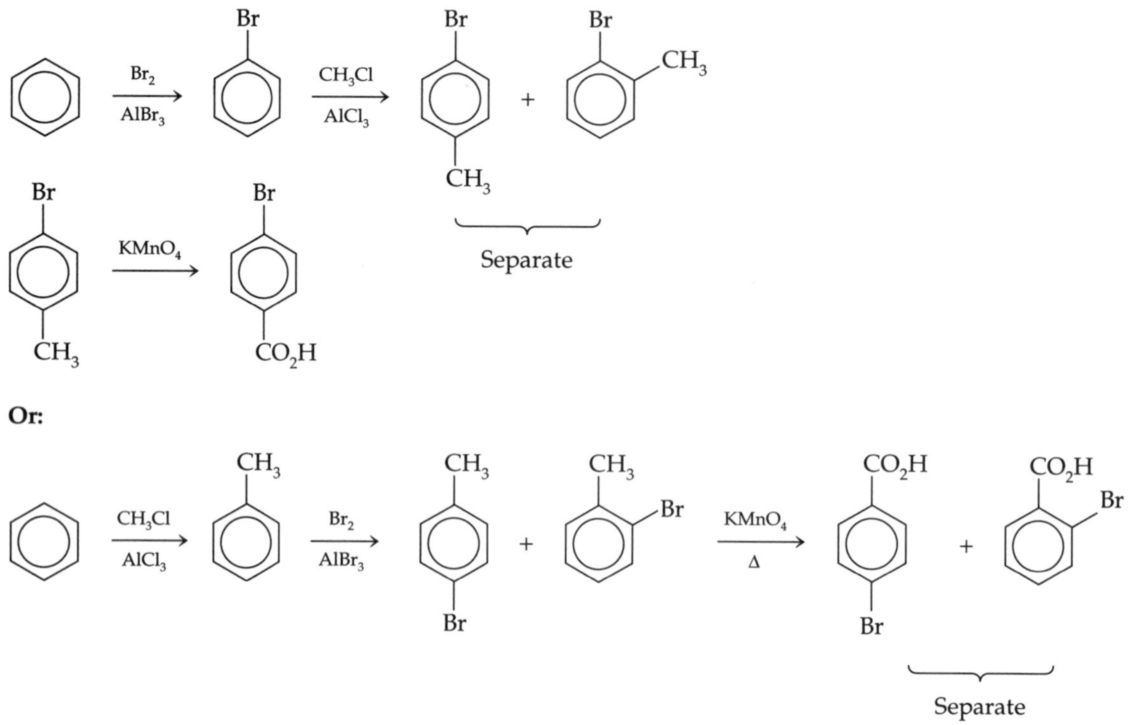

▲ FIGURE 11-37
Methylation of bromobenzene (formed by electrophilic bromination of benzene) by Friedel-Crafts alkylation affords a mixture of *o*- and *p*-bromotoluene (upper sequence). The same mixture is formed by bromination of toluene (lower sequence). The *ortho* and *para* isomers of bromotoluene can be separated, and *p*-bromotoluene can then be oxidized to *p*-bromobenzoic acid (upper sequence). Alternatively, the mixture of *o*- and *p*-bromotoluene can be oxidized with hot potassium permanganate to a mixture of *o*- and *p*-bromobenzoic acids, which would then be separated (lower sequence).

toluenes. Or, perhaps toluene or bromobenzene is available in addition to benzene, so that a step in one of the sequences could be omitted. For our purposes, it is important to note that several routes are possible for this synthesis. An important element of design for chemical synthesis is not only knowing what reactions are needed, but also the order in which they can be used.

EXERCISE 11-G

Identify the reagents required and give the order in which they should be used to generate each of the following products from benzene.

(a) [structure: toluene with CH₃ and NO₂ ortho]

(b) [structure: aniline NH₂ with Br meta]

(c) [structure: aniline NH₂ with Br para]

(d) [structure: NO₂ and COCH₃ meta]

(e) [structure: NO₂ and ethyl meta]

(f) [structure: NH₂ and ethyl meta]

(g) [structure: NO₂ and Br ortho]

(h) [structure: Cl and NO₂ meta]

11-5 ▸ Electrophilic Attack on Polycyclic Aromatic Compounds

The same rationale that was used in explaining substituent directive effects (resonance stabilization of transition states leading to the most-stable cationic intermediate) can also explain regiochemical preference in electrophilic attack on polycyclic aromatics. This electrophilic reactivity is important because it is likely to be related to the carcinogenicity associated with these compounds. (Recall from Chapter 2 that benzopyrene is a known chemical carcinogen.)

The reaction of naphthalene with electrophiles is illustrative of these π resonance effects. Figure 11-38 shows the cationic intermediates formed by

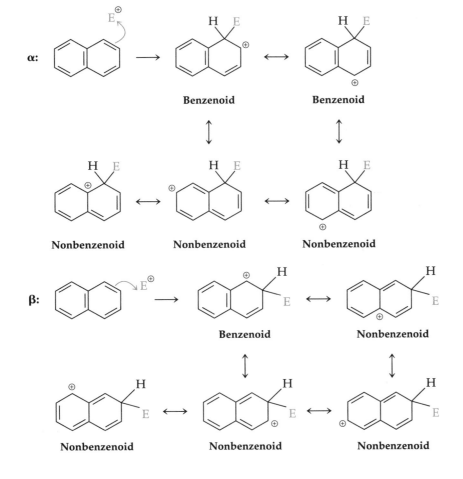

◀ FIGURE 11-38
Electrophilic attack at both the α and β positions produces a cation with five significant resonance contributors. In the cation formed by α attack, two of these structures are benzenoid, whereas only one is benzenoid in the cation formed by β attack.

ROYAL PURPLE: THE DYE THAT NATURE FORGOT

William Perkin was just eighteen when he decided to spend his vacation from London's Royal College of Chemistry in 1856 working in his home laboratory, somewhat naively trying to make quinine. In one attempt, he tried to attach an allyl group to a mixture of aniline and toludine (methylaniline), producing a mixture that gave a black sludge after being oxidized with chromic acid. The residue, which contained the first synthetic dye, dissolved in alcohol and dyed cloth purple. It was fortunate that this dye was purple: since the Middle Ages, purple was the color of royalty because only kings could afford purple or lavender cloth that did not fade, the only source of a color-fast purple dye having been a small shellfish that grew in the Tyrian Sea.

With a method in hand to make a mauve dye, Perkin interrupted his education, obtained a patent on the process, and persuaded his father to lend him money to build a factory to manufacture the several dyes he quickly learned to synthesize. Eighteen years later, Perkin sold his factory, leaving him sufficiently wealthy at age 36 that he could be free to do research without financial concern. In doing so, he discovered a number of important organic reactions. The highest award of the American Chemical Society is named in his honor.

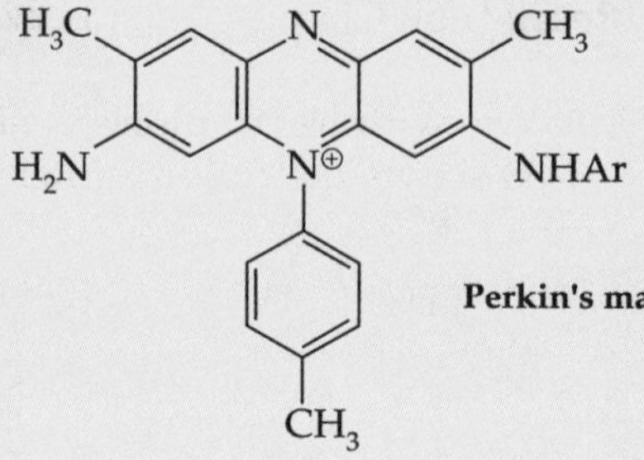

Perkin's mauve

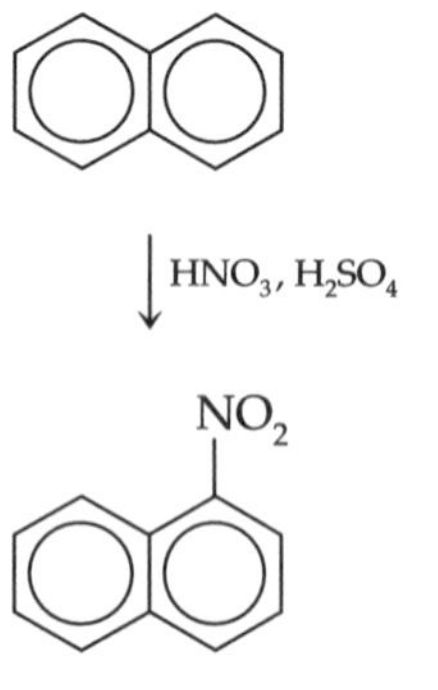

electrophilic attack at the α and β positions of naphthalene. In both cases, five resonance contributors can be drawn in which the positive charge is delocalized throughout the fused rings. With α attack, however, two of these structures retain three formal double bonds within a single six-membered ring. This structural element is called a **benzenoid ring**, which, in accord with Hückel's Rule (Chapter 2), exhibits particular stability and contributes significantly to the resonance stabilization of the cation. The transition state leading to the cation with a larger number of benzenoid contributors is more stable. Thus, the cation formed by α attack is more stable than the one formed by β attack, and electrophilic attack is easier at the α position.

In accordance with this view, nitration occurs exclusively at the α position (Figure 11-39). This extra stabilization of the transition state leading to the σ-bound cation intermediate also accelerates the reaction, relative to the nitration of benzene.

▲ FIGURE 11-39
The enhanced stability of the cation formed by electrophilic attack at the α position accounts for the formation of nitration product in which the —NO_2 group is found exclusively at that site.

EXERCISE 11-H

Predict the monosubstitution product formed in each of the following reactions:

(a) [structure: naphthalene] $\xrightarrow[\text{AlCl}_3]{\text{Cl}_2}$

(c) [structure: naphthalene with NO$_2$ substituent] $\xrightarrow[\text{AlBr}_3]{\text{Br}_2}$

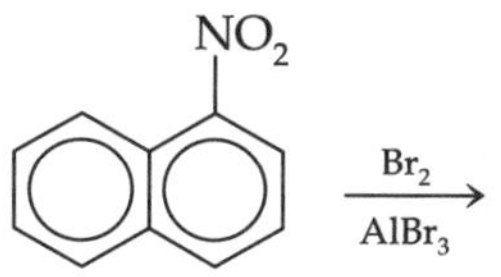

(b) [structure: anthracene] $\xrightarrow[\text{AlCl}_3]{\text{CH}_3\text{COCl}}$

(d) [structure: naphthalene with CH$_3$ substituent] $\xrightarrow[\text{H}_2\text{SO}_4]{\text{HNO}_3}$

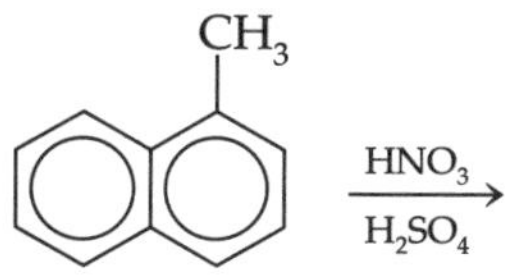

11-6 ▸ Synthetic Applications

Table 11-2 is a summary of the reactions discussed in this chapter, grouped according to their synthetic utility. These reactions can be used in sequence to attain conversions not specifically listed. An important intellectual challenge is to integrate these new reactions with those presented in earlier chapters.

Table 11-2 ▸ How to Use Electrophilic Aromatic Substitution to Make Various Functional Groups

Functional Group	Reaction
Alkyl aromatic	Friedel-Crafts alkylation (RX, AlX$_3$); or Wolff-Kishner reduction of aryl ketones (basic NH$_2$NH$_2$); or Clemmensen reduction of aryl ketones (Zn, HCl)
Aniline	Reduction of aryl nitro compounds (Zn, HCl)
Aryl carboxylic acid	Oxidation of alkylbenzenes with hot KMnO$_4$
Aryl halide	Halogenation of aromatic hydrocarbons (X$_2$, AlX$_3$), X = Cl, Br
Aryl ketone	Friedel-Crafts acylation (RCOCl, AlCl$_3$)
Aryl nitro compound	Nitration of aromatic hydrocarbons (HNO$_3$, H$_2$SO$_4$)
Aryl sulfonic acid	Sulfonation of aromatic hydrocarbons (fuming H$_2$SO$_4$)
Azo compound	Treatment of electron-rich aromatics with diazonium salts
Diazonium salt	Diazotization of anilines (HNO$_2$)

Conclusions

Despite the attendant loss of aromaticity, electrophilic attack on an aromatic ring proceeds to form a new σ bond and a resonance-stabilized cation. Electron density in the π system of the aromatic ring and cationic stabilization control both the rate and the positional selectivity of electrophilic attack. The resonance stabilization available in a re-aromatized ring, however, dictates that deprotonation of the cationic intermediate ensues, restoring aromaticity and producing net substitution rather than addition.

Electron-donating substituents stabilize the conjugated intermediates and accelerate the rate of reaction. Such substituents also direct further substitution to the *ortho* and *para* positions in substituted benzenes by virtue of stronger interaction with the positive charge immediately adjacent to the substituent. Electron-withdrawing substituents, in contrast, inhibit the reaction and slow the rate of electrophilic attack. This inhibition is most intense at the *ortho* and *para* positions, allowing *meta* substitution to become dominant in compounds bearing electron-withdrawing groups. Halogenated aromatic compounds combine σ withdrawal with π donation, leading to slower electrophilic substitution but with *ortho, para* direction. Similar effects influence the regiochemical choices for electrophilic substitution in fused aromatic rings.

Substituents introduced onto an aromatic ring retain their characteristic reactivity. For example, side chains or simple substituents can be oxidized, reduced, halogenated, and so forth. The oxidation-reduction reactions often invert the electron-donating or electron-withdrawing character of a substituent and therefore alter its directive effect on further substitution. One of the most-important lessons to be learned from aromatic substitution is the nature of electronic and steric effects of substituents on chemical reactivity, a deeper understanding of which will be helpful in future courses in chemistry and biochemistry.

Summary of New Reactions

Aromatic Halogenation

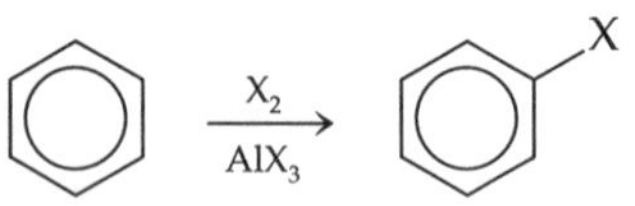

$(X = Cl \text{ or } Br, \text{ not } F \text{ or } I)$

Aromatic Nitration

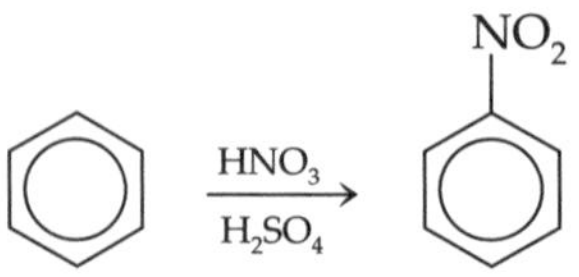

Aromatic Sulfonation

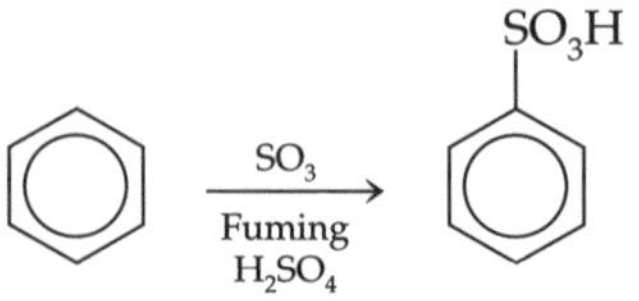

Friedel-Crafts Alkylation

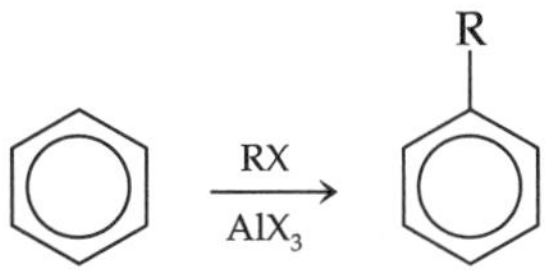

$$\text{benzene} \xrightarrow[\text{AlX}_3]{\text{RX}} \text{R-substituted benzene}$$

Friedel-Crafts Acylation

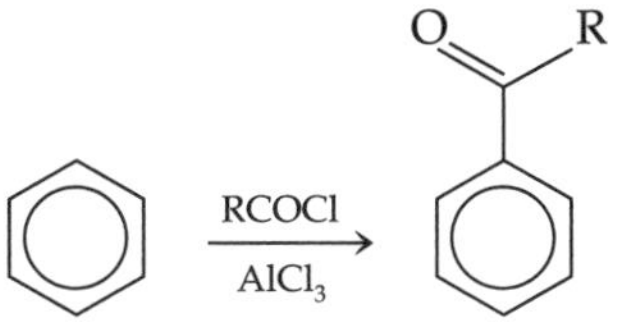

$$\text{benzene} \xrightarrow[\text{AlCl}_3]{\text{RCOCl}} \text{aryl ketone}$$

Clemmensen Reduction

$$\underset{R}{\overset{O}{\|}}\underset{R'}{C} \xrightarrow[\text{HCl}]{\text{Zn}} \text{RCH}_2\text{R}'$$

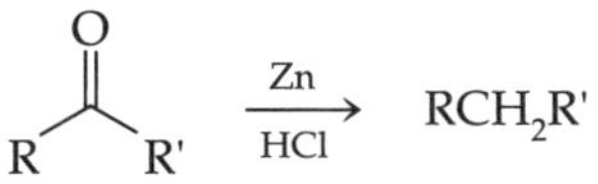

Wolff-Kishner Reduction

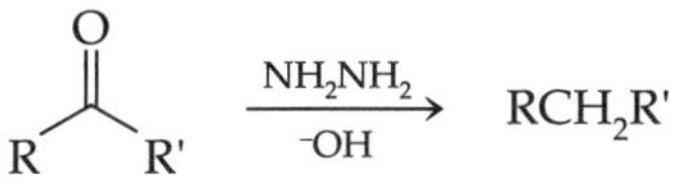

$$\underset{R}{\overset{O}{\|}}\underset{R'}{C} \xrightarrow[^-\text{OH}]{\text{NH}_2\text{NH}_2} \text{RCH}_2\text{R}'$$

Reduction of Nitro Compounds

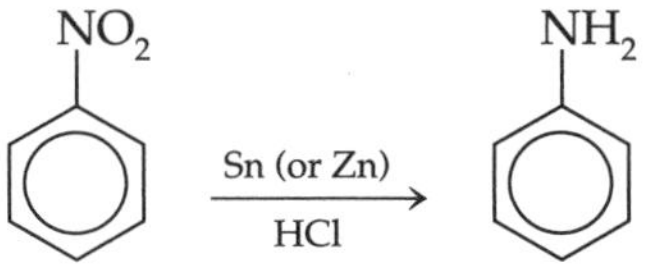

$$\text{C}_6\text{H}_5\text{NO}_2 \xrightarrow[\text{HCl}]{\text{Sn (or Zn)}} \text{C}_6\text{H}_5\text{NH}_2$$

Diazotization of Anilines

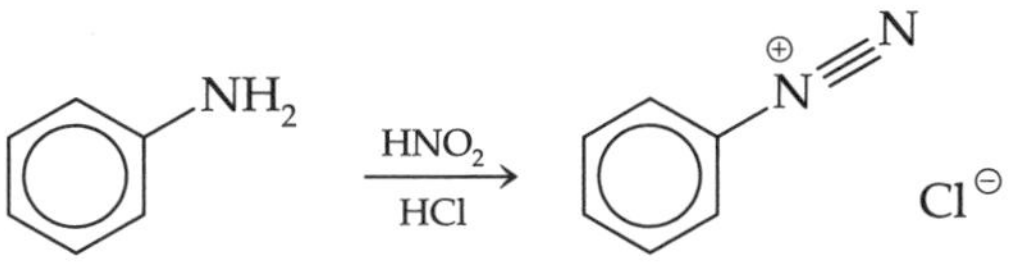

$$\text{C}_6\text{H}_5\text{NH}_2 \xrightarrow[\text{HCl}]{\text{HNO}_2} \text{C}_6\text{H}_5\overset{\oplus}{\text{N}}\equiv\text{N} \quad \text{Cl}^{\ominus}$$

Diazo Coupling

$$\text{C}_6\text{H}_5\overset{\oplus}{\text{N}}\equiv\text{N} \; \text{Cl}^{\ominus} + \text{C}_6\text{H}_5-\text{N(CH}_3)_2 \longrightarrow \text{C}_6\text{H}_5-\text{N}=\text{N}-\text{C}_6\text{H}_4-\text{N(CH}_3)_2$$

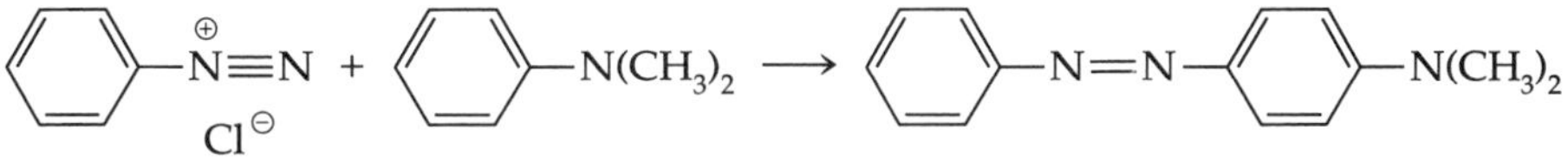

Oxidation of Alkyl Side Chains

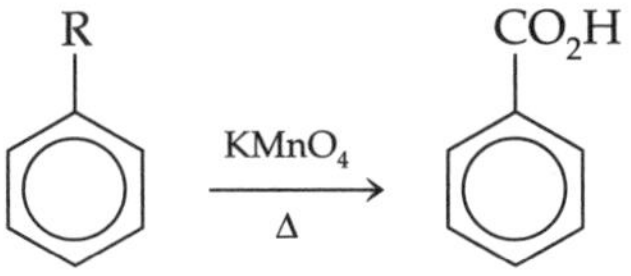

$$\text{R-C}_6\text{H}_5 \xrightarrow[\Delta]{\text{KMnO}_4} \text{C}_6\text{H}_5\text{CO}_2\text{H}$$

Hydrolysis of Aromatic Amides

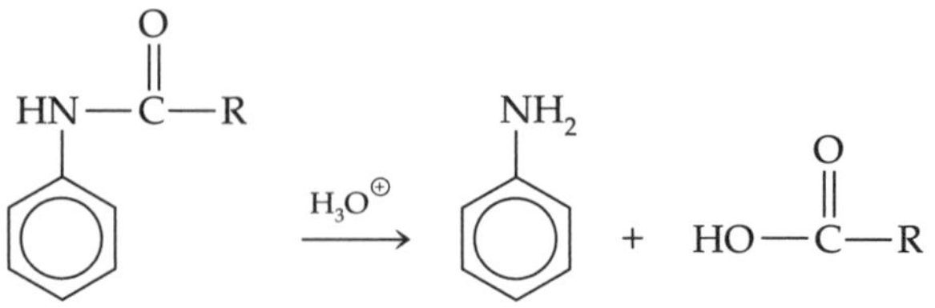

Hydrolysis of Aromatic Esters

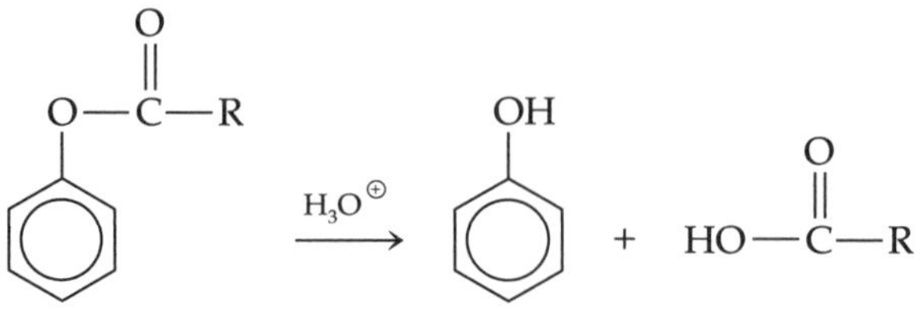

Review Problems

11-1 Predict the preferred regiochemistry for the product(s) that you would expect to obtain upon treating each of the following compounds with chlorine and AlCl$_3$:

(a) toluene

(b) nitrobenzene

(c) acetanilide

(d) anisole

(e) *p*-chlorophenol

(f) acetophenone

(g) *o*-chlorotoluene

11-2 Draw the structure of the major product obtained from each of the following sequences:

(a)

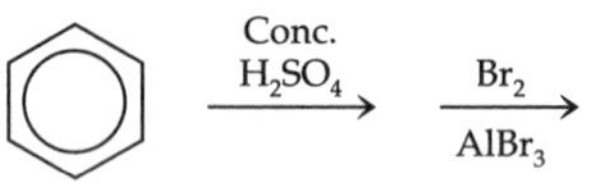

(b)

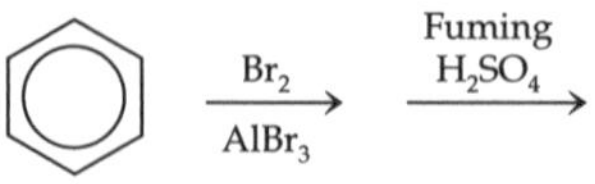

(c)

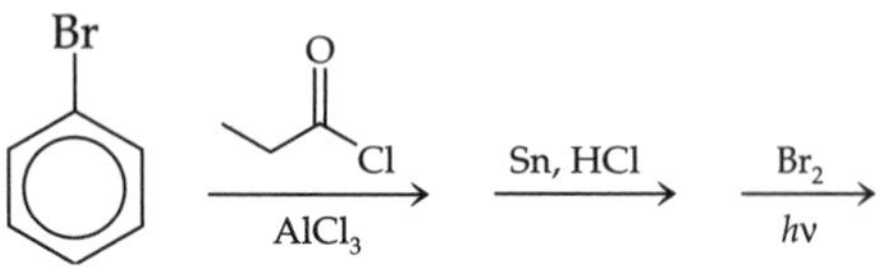

(d)

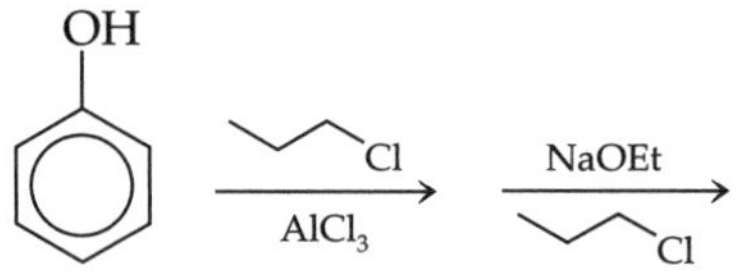

(e)

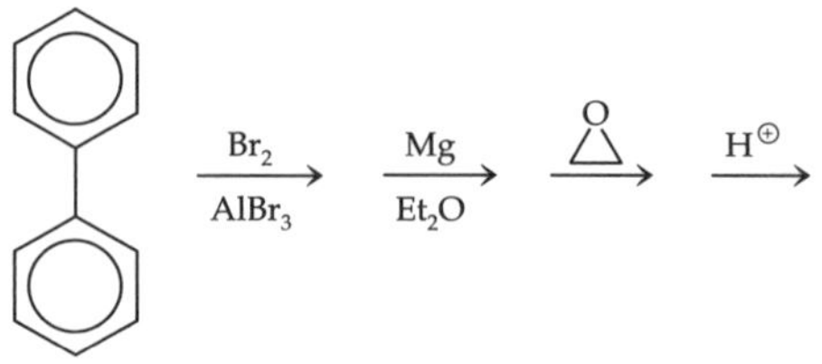

(f)

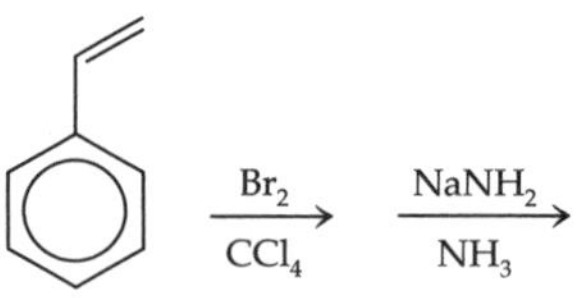

11-3 Predict the major product expected from the reaction of Br$_2$ and AlBr$_3$ with phenylbenzoate, being sure to explain why one ring is more active than the other.

11-4 Suggest a reasonable sequence of reagents that could convert toluene into each of the following compounds.

(a) *p*-nitrobenzoic acid

(b) *m*-nitrobenzoic acid

(c) *p*-nitrobenzyl alcohol

(d) *p*-toluenesulfonic acid (HOTs)

11-5 Write reaction mechanisms that can account for the formation of each of the following products, being sure to explain any important regiochemical and stereochemical control elements:

(a)

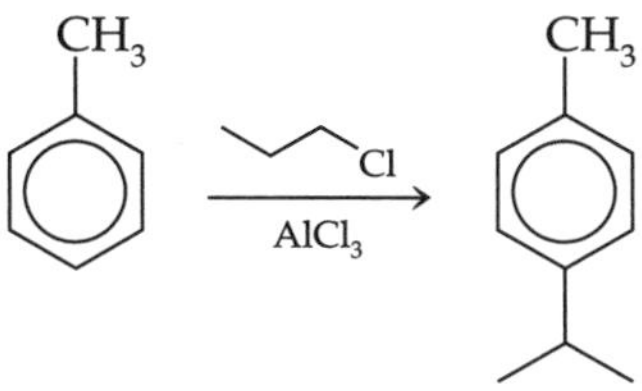

(b)

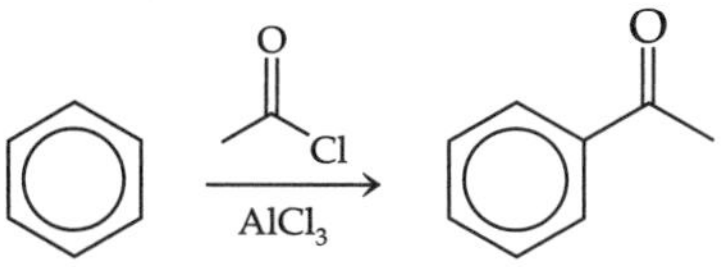

(c)

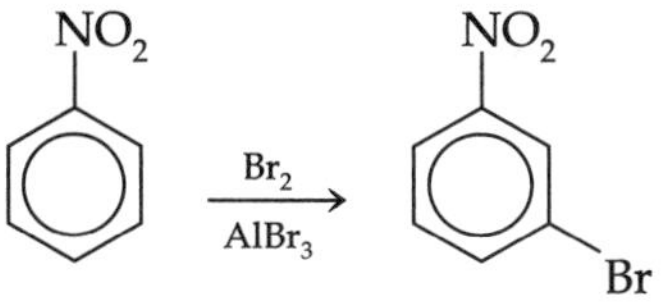

(d)

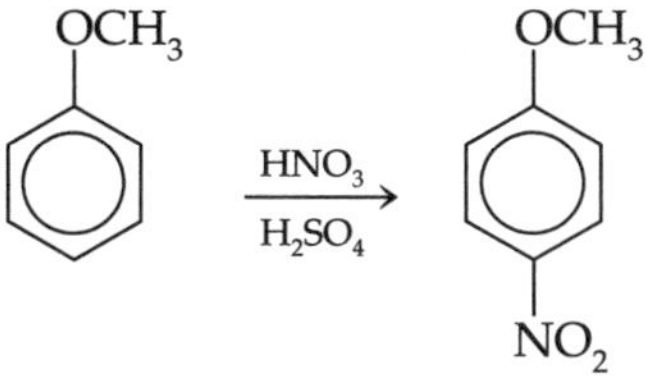

(e)

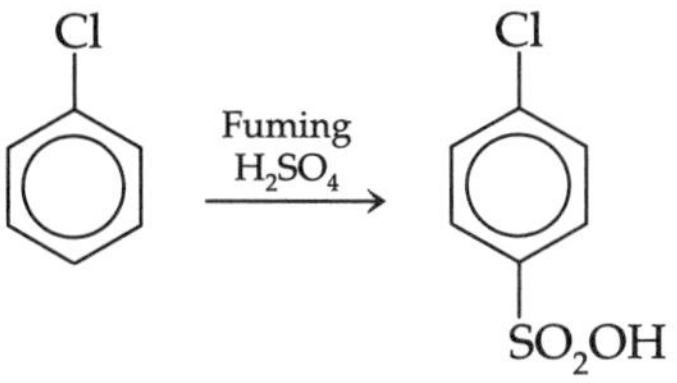

(c)

(d)

(e)

(f)

(g)

(h)

(i)

11-6 Predict the major product (or products) that would be obtained when each of the following compounds is nitrated.

(a) *N*-phenylacetamide
(b) methyl benzoate
(c) fluorobenzene
(d) *n*-propylphenylether
(e) 1-methylnaphthalene

11-7 Propose a synthesis for each of the following compounds from either benzene or toluene. You may use any inorganic reagent that you wish and any organic compound containing four carbons or less.

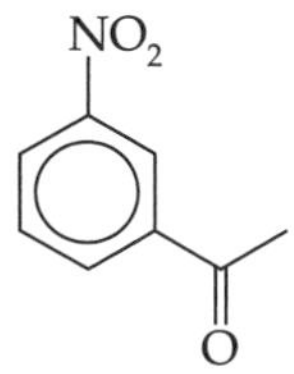

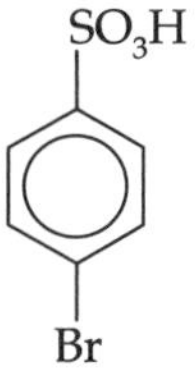

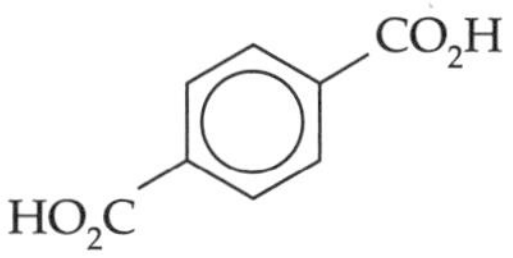

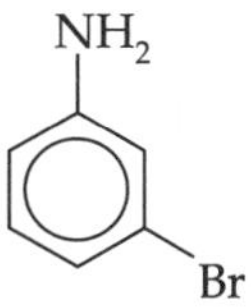

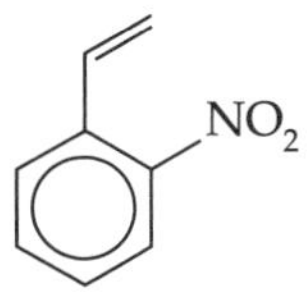

(a)

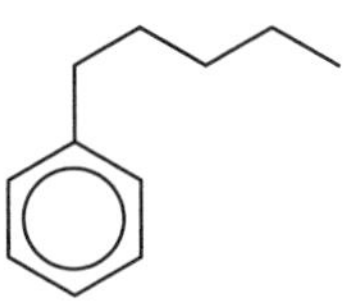

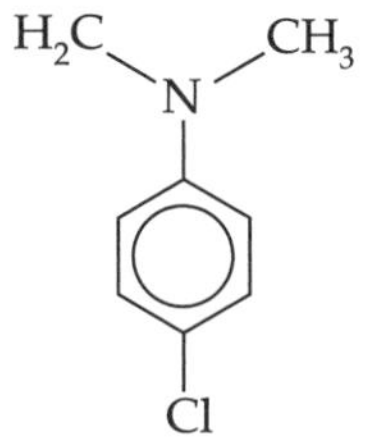

(b)

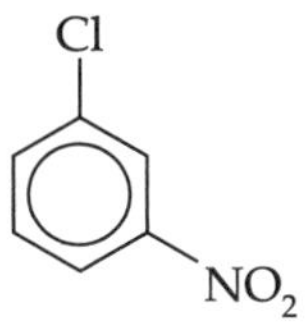

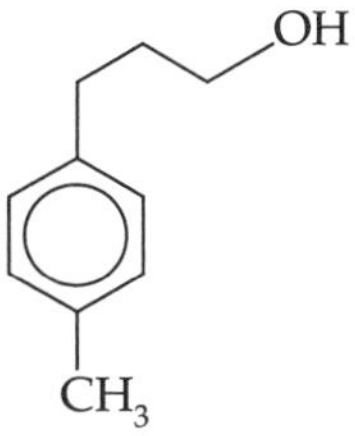

(j)

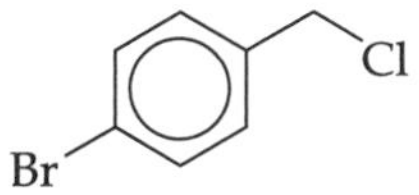

11-8 Identify the starting material needed to prepare the indicated compounds from the reagents given in parts *a* through *e* on the next page.

(a) X $\xrightarrow{\quad}$

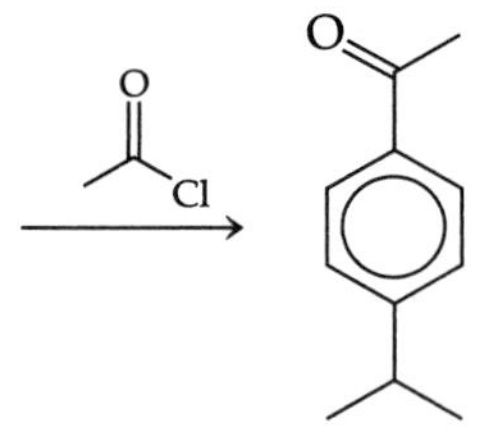

(b) X $\xrightarrow[\text{NaOH}]{\text{H}_2\text{NNH}_2}$ $\xrightarrow[\text{AlBr}_3]{\text{Br}_2}$

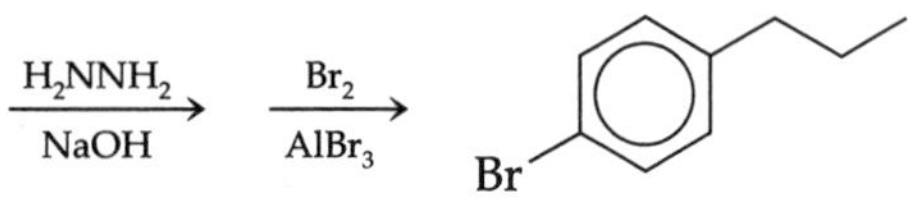

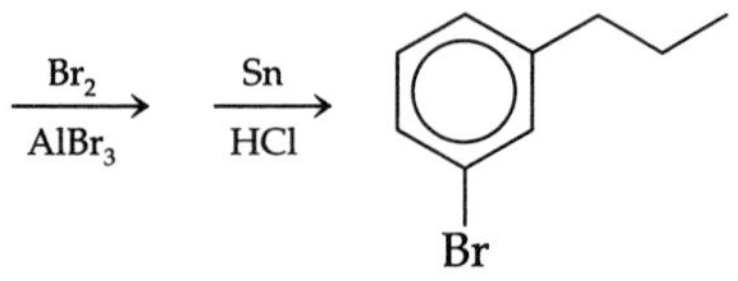

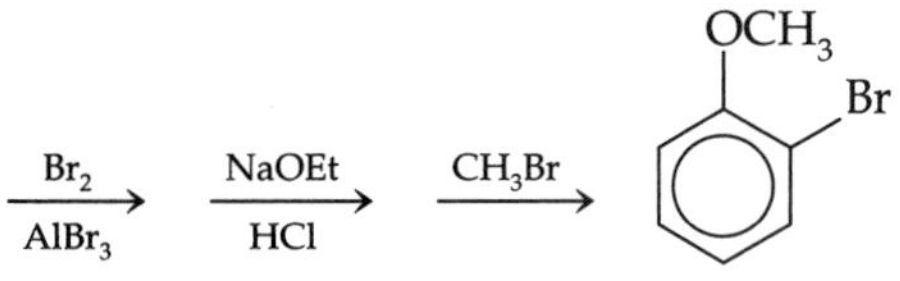

(c) X $\xrightarrow[\text{AlBr}_3]{\text{Br}_2}$ $\xrightarrow[\text{HCl}]{\text{Sn}}$

(d) X $\xrightarrow[\text{AlBr}_3]{\text{Br}_2}$ $\xrightarrow[\text{HCl}]{\text{NaOEt}}$ $\xrightarrow{\text{CH}_3\text{Br}}$

(e) X $\xrightarrow[\text{HCl}]{\text{Zn}}$ $\xrightarrow[\text{H}_2\text{SO}_4]{\text{HNO}_3}$ $\xrightarrow[\text{HCl}]{\text{Sn}}$

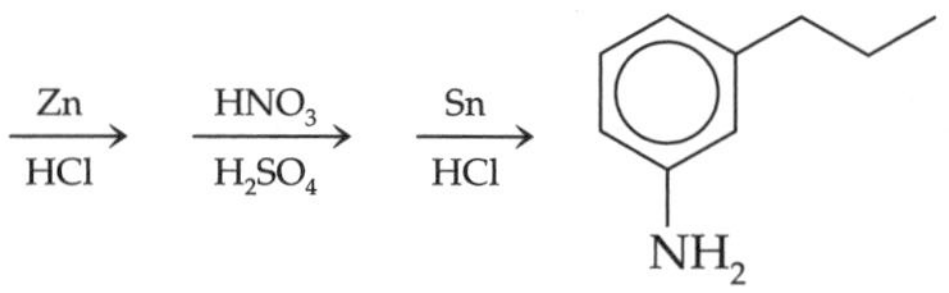

11-9 Write a mechanism, using curved arrows to indicate electron flow, for the Friedel-Crafts acylation of naphthalene at C-1 and at C-2. Explain why the major product is that formed by substitution at C-1.

11-10 BHT, the major antioxidant used as a food preservative in the United States, is a mixture of positional and structural isomers of butylated hydroxytoluenes. The major component in BHT, 2,6-di-*t*-butyl-4-methylphenol, is made industrially from *p*-methylphenol and isobutene (2-methylpropene). Explain the preferred regiochemistry in the major product, and propose a mechanism for the formation of 2,6-di-*t*-butyl-4-methylphenol from these reagents in acidic methanol.

Chapter 12

Addition and Substitution by Heteroatomic Nucleophiles at sp^2-Hybridized Carbon

Many of the reactions described in Chapters 10 and 11 proceed through the intermediate formation of carbocations derived from organic compounds containing multiple bonds. In Chapter 10, we saw how an electrophile interacts with an electron-rich carbon–carbon multiple bond, initiating electrophilic addition. In contrast, as described in Chapter 11, electrophilic attack on an aromatic π system produces net substitution. In this chapter and the next, we consider reactions with a reverse electron demand—namely, those in which addition or substitution is initiated by the attack of a nucleophile on the C=O bond of a carbonyl group. As we might predict because of the polarization of the C=O π bond, nucleophiles (electron-rich reagents) covalently bond to the electron-deficient carbonyl carbon, producing an anionic intermediate in which negative charge is shifted to oxygen. Nucleophilic attack on aldehydes or ketones leads to net addition, whereas that on carboxylic acid derivatives (acids, acid chlorides, anhydrides, esters, thiol esters, or amides) produces net substitution.

The reactions of nucleophiles with carbonyl compounds is a sufficiently broad and important class that two chapters are required for its effective coverage: in this chapter, the reactions of noncarbon nucleophiles are discussed; and, in Chapter 13, those of carbon nucleophiles are covered. (Refer to Chapter 8 to review how these reagents are generated.) The mechanisms for nucleophilic substitutions and additions are similar, whether the nucleophilic atom is carbon or a heteroatom. Those in which carbon is the nucleophile proceed through the formation of carbon–carbon bonds and result in the formation of products with larger and often more-complex carbon skeletons. The additions or substitutions of heteroatomic nucleophiles covered in this chapter result in the conversion of one functional group into another. Nonetheless, common characteristic mechanisms pertain to both the reactions to be discussed here and those covered in Chapter 13 because both consist of attack by a nucleophile on a carbonyl carbon to form a tetrahedral intermediate, which is converted ultimately into either an addition or a substitution product.

12-1 ▸ Nucleophilic Addition to Carbonyl Groups

Because of the high electron density in the π system of a carbon–carbon multiple bond, alkenes and alkynes are unlikely candidates for attack by electron-rich nucleophilic reagents. However, the π bond between carbon and a heteroatom is polarized, resulting in significant partial positive charge on carbon and significant partial negative charge on the heteroatom. We can view this polarization in a carbonyl group as resulting from significant contributions from two resonance structures, one in which the C=O π bond is intact and a second in which the electrons originally in the π bond are shifted to oxygen, producing formal positive charge on carbon and negative charge on oxygen. Because the hybrid has significant contributions from each of these resonance structures, the carbon end of the C=O functional group is partially positively charged and, hence, a particularly active site for nucleophilic attack.

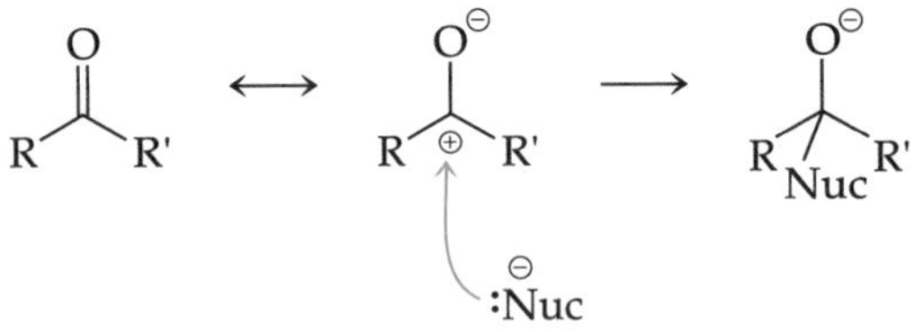

Furthermore, we can understand why nucleophiles add more readily to C=O π bonds than to C=C π bonds by considering the products that result from each of these proposed nucleophilic attacks. Addition to a carbonyl group produces a negatively charged oxygen, a species of much greater stability than the carbanion formed upon the addition of a nucleophile to an alkene.

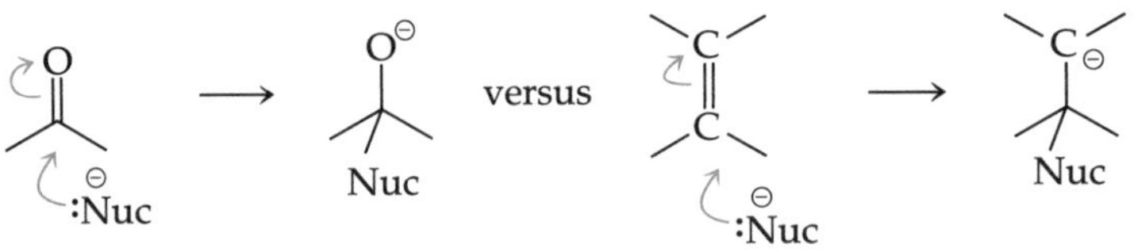

The interaction of a nucleophile with a carbonyl carbon results in the formation of a carbon–nucleophile σ bond that provides energetic compensation for the accompanying rupture of the carbonyl π bond. The two electrons originally in the C=O π bond are shifted to the more-electronegative oxygen atom, thus placing surplus electron density on the atom best able to accommodate negative charge. As shown in Figure 12-1, the polarization of

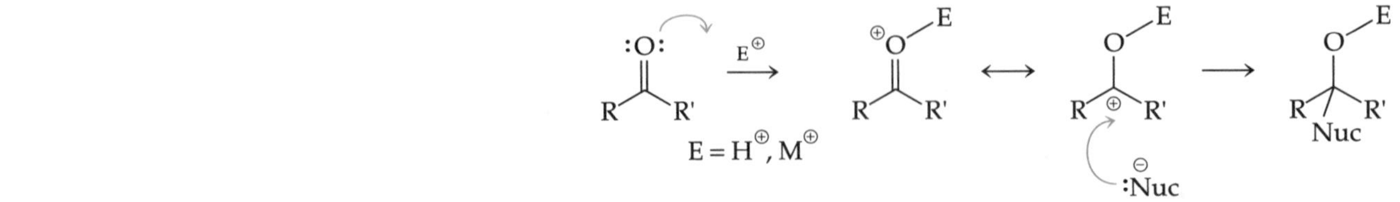

▲ FIGURE 12-1

Complexation with an electrophile activates a carbonyl group toward nucleophilic attack by the formation of a cation that is easily attacked by an electron-rich reagent. The product incorporates a new C–Nuc σ bond.

SIMPLE CARBONYL COMPOUNDS IN CHEMICAL COMMUNICATION
BETWEEN INSECTS

Pheromones are chemicals used for communication between individual members of a species. Pheromones are used as sex attractants, as
trail markers, and as alarms. When you see ants following in line
across your kitchen, know that they are following a chemical trail to
food or water mapped by a successful explorer who excretes a specific
compound to help his coworkers. The placement in traps of molecules
that have been identified as sex attractants to inhibit breeding made it
possible to control both the voracious gypsy moth (which threatened
New England's pine forests) and the screwworm (which created serious problems for Texas cattlemen) without resorting to widespread
spraying of insecticides.

To be effective, a pheromone must be both narrowly specific, so
that only one species will respond, and highly potent. Most known sex
attractants contain between ten and seventeen carbon atoms—a range
that permits sufficient molecular complexity for the creation of a molecule that is unique to a given species but simple enough for easy
biosynthesis as demanded. Many pheromones are simple carbonyl
compounds or acid derivatives. For example, 9-ketodecenoic acid is
the sex attractant used by a honeybee queen in her nuptial flight, but it
also serves to develop ovaries when ingested by worker bees. One
species of ant *(Iridomyrmex priunosus)* uses 2-heptanone to warn other
members of the hill of danger.

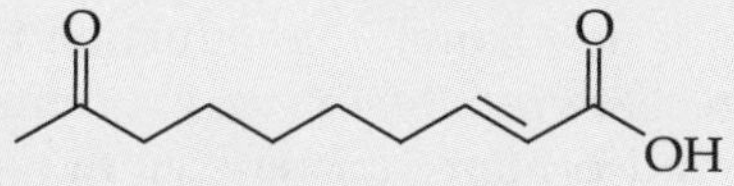

**9-Ketodecenoic acid
(honeybee queen substance)**

Do human pheromones exist? Studies indicate that mammals do
not give an automatic, standardized response to chemicals in the way
that insects do. Two steroids, however, have been used commercially
to induce the mating stance in sows, thus facilitating the artificial in-
semination of pigs. Folk medicine in Africa and Asia abounds with
supposed examples of human aphrodisiacs, none of which have been
proved objectively to be effective. One of them, xylomollin, was first
synthesized in the laboratory by one of the authors of this text.

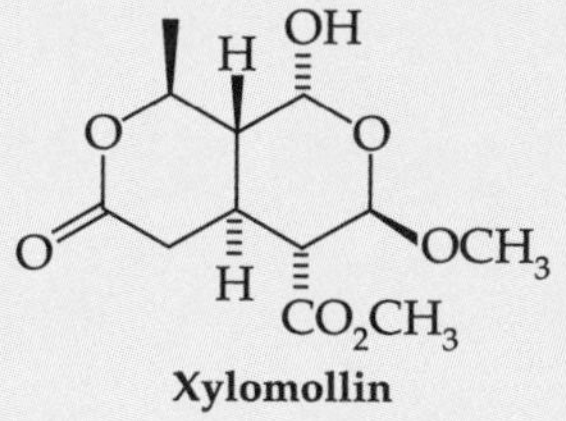

Xylomollin

the carbonyl group is further increased by prior coordination of the carbonyl oxygen with a proton or metal ion. Nucleophilic addition is facilitated by this complexation because the nucleophile interacts with an intermediate bearing a full positive charge. As a result, the product formed is
neutral if the attacking nucleophile is anionic.

12-2 ▸ Complex Metal Hydride Reductions

A new C–H bond is formed at a carbonyl carbon when a complex metal hydride delivers the equivalent of H^- to the $C{=}O$ double bond. An aldehyde or ketone is thus converted into an alcohol; an ester is converted into a primary alcohol; and an amide is converted into an amine.

Aldehydes and Ketones

Among the simplest possible nucleophilic reagents is the hydride ion (H^-). However, because of the low solubility of alkali metal hydride salts (LiH, NaH, and KH) in organic solvents, these reagents cannot be used as sources of nucleophilic H^-. On the other hand, complex reagents in which hydride is bound to boron or aluminum are soluble in organic solvents and do provide the equivalent of the hydride ion as a nucleophile; the addition of a hydride ion to a carbonyl group effects net reduction. The two most-common sources of hydride are sodium borohydride, $NaBH_4$, and lithium aluminum hydride, $LiAlH_4$. Indeed, 44 g of $LiAlH_4$ will dissolve in 100 ml of ethyl ether.

Let us consider the reaction of acetone with sodium borohydride (Figure 12-2). Sodium borohydride, $NaBH_4$, is sufficiently stable that protic solvents such as alcohols can be used for this reaction. In alcohol, ionization occurs to form Na^+ and $(BH_4)^-$. As the negatively charged complex hydride $(BH_4)^-$ approaches the dipolar $C{=}O$ double bond, a bond between carbon and hydrogen forms at the same time that the π-bond electron pair shifts onto oxygen and the bond between the hydrogen and boron breaks. When the reaction is carried out in ethanol, the reduction is facilitated by hydrogen-bonding interactions between the carbonyl group and solvent hydroxyl groups (shown as a second proton-transfer step in Figure 12-2). The resulting alkoxide ion reacts in a Lewis acid-base fashion with BH_3, forming a bo-

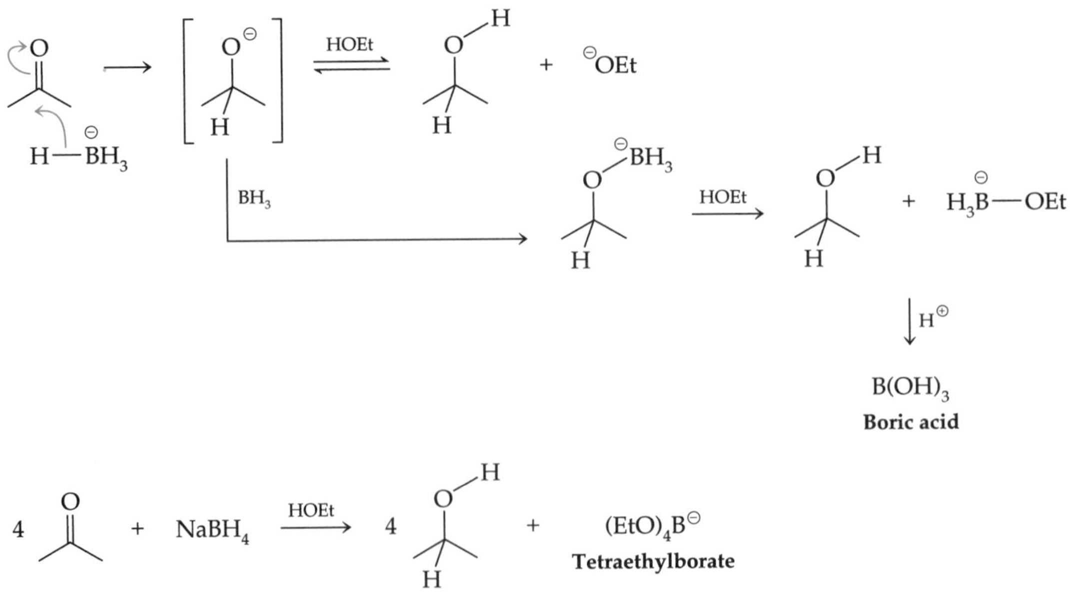

▲ FIGURE 12-2

In a sodium borohydride reduction of an aldehyde or ketone, an alcohol is produced. The reduction of acetone with $NaBH_4$ is initiated by the delivery of a hydride to the carbonyl carbon as an electron pair shifts onto oxygen. Protonation of the borate produces the alcohol product.

SYNTHETIC PERFUMES: A SIMPLE ESTER

The first commercial synthetic perfume was marketed in 1882 by a Parisian glove maker named Houbigant: it was an ethanolic solution of an aromatic lactone, coumarin, that had been prepared in the laboratory for the first time in 1868. It was called "Fougère Royale," or Royal Fern, owing to the fact that coumarin has a fernlike odor reminiscent of cut grass.

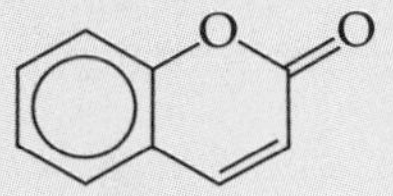

Coumarin

---

rate with an O–B σ bond. This borate has three remaining B–H bonds, each of which in turn is a hydride equivalent for the reduction of another molecule of the starting ketone. Upon treatment with acid, the borate decomposes, forming boric acid. Overall, one mole of $NaBH_4$ reduces four moles of ketone. (In practice, there is also some side reaction between the reagent and the solvent, producing hydrogen gas. As a result, a small excess of $NaBH_4$ is typically used.) The product alcohol—in this case, 2-propanol—is formed by a reduction of the $C\!=\!O$ double bond; the reducing agent $NaBH_4$ thus effects a nucleophilic addition of hydride to the $C\!=\!O$ double bond.

Lithium aluminum hydride also effects nucleophilic addition of hydride to carbonyl groups. For example, reduction of acetone with $LiAlH_4$ results in the delivery of a hydride equivalent (Figure 12-3) through a pathway parallel to the reaction shown in Figure 12-2. However, $LiAlH_4$ is more reactive than $NaBH_4$ and must be used in aprotic solvents such as ether. In the absence of a proton source from the solvent, the neutral AlH_3 complexes with the negatively charged oxygen, producing an **aluminate** containing an O–Al bond. This species still contains three aluminum–hydrogen bonds and provides another hydride equivalent to another carbonyl compound. By further reaction, all of the hydrogens of AlH_4^- are successfully delivered to the $C\!=\!O$ double bond of additional molecules of the ketone being reduced, ultimately producing the tetraalkylaluminate shown. The aluminum–oxygen bond has, as does the boron–oxygen bond, appreciable polar character, much like the ionic bond in a sodium alkoxide. Therefore, it is easily hydrolyzed, forming the alcohol product by protonation of the oxygen. (Any unreacted or partially reacted aluminum hydride is rapidly

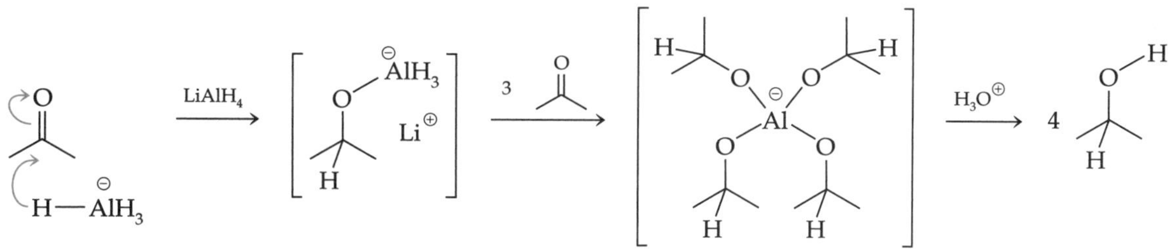

▲ FIGURE 12-3

Reduction of acetone with $LiAlH_4$ proceeds through a route parallel to that with $NaBH_4$ (Figure 12-2). The aluminate anion formed after multiple reduction is decomposed upon later treatment with aqueous acid to form four equivalents of 2-propanol.

$$\text{LiAlH}_4 \xrightarrow{\text{H}_2\text{O}} \text{Al(OH)}_3 \;+\; \text{LiOH} \;+\; 4\,\text{H}_2\uparrow$$

▲ **FIGURE 12-4**
Lithium aluminum hydride is easily protonated by water or alcohol, causing the release of hydrogen gas. To prevent this complication, nonprotic ether solvents are usually employed for LiAlH$_4$ reductions.

protonated by water to form hydrogen gas.) Overall, the stoichiometry is the same for NaBH$_4$ and LiAlH$_4$: in the absence of side reactions, one mole of reagent provides four hydride equivalents that can reduce four moles of ketone.

EXERCISE 12-A

(a) Write a complete mechanism for the LiAlH$_4$ reduction of 2-butanone, showing all electron flow.

(b) The product formed in part *a* contains a center of chirality. Will the alcohol formed by this route be optically active? Why or why not?

The reduction with lithium aluminum hydride and the second step, in which the aluminate salt is decomposed by acid, must be conducted as two separate steps. Otherwise, the acid-base reaction shown in Figure 12-4 intervenes, decomposing the hydride reagent and generating hydrogen faster than hydride can be added to the carbonyl carbon.

Reduction of Derivatives of Carboxylic Acids

All ketones, aldehydes, and carboxylic acid derivatives have carbonyl groups, but the reactions of carboxylic acid derivatives with hydride reagents differ from those of aldehydes and ketones in a very significant way. Once hydride has been added to the carbonyl group of a carboxylic acid derivative, a reaction occurs that is not possible in the reduction of ketones and aldehydes. In the reaction of lithium aluminum hydride with an ester such as ethyl acetate (Figure 12-5), an aluminate is initially formed, in a fashion quite similar to that shown in Figure 12-3 for the reduction of aldehydes and ketones. In the ester, however, there is a good leaving group ($^-$OR) bound to the carbonyl carbon. Electron density flows from the O–Al bond to regenerate a π bond as the leaving group takes up the electrons originally present in the σ bond between the OR group and the carbonyl carbon. The aldehyde produced in this step is more reactive toward hydride reducing agents than is the ester from which it was derived, and it is reduced more rapidly by aluminum hydride by the same route as that shown in Figure 12-3. An alcohol is produced when the resulting intermediate is treated with acid. Thus, reduction of an ester with lithium aluminum hydride results in the formation of a primary alcohol, with the OR group of the original ester lost as the corresponding alcohol.

The reduction of a tertiary amide by lithium aluminum hydride, illustrated in Figure 12-6 with *N,N*-dimethylacetamide, begins in the same way as the reduction of an ester. The initial adduct has one oxygen and one nitrogen substituent rather than the two oxygen groups present in the tetrahedral intermediate formed in ester reduction. Oxygen is more electronegative than nitrogen, and the carbon–oxygen σ bond is more easily broken as

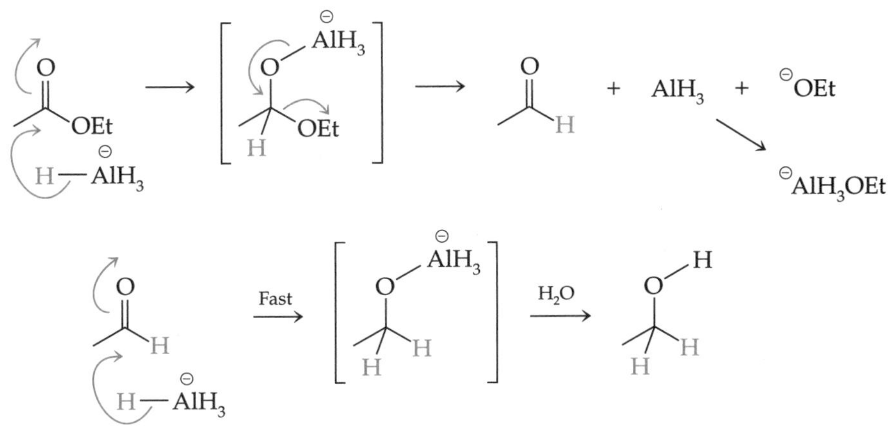

▲ FIGURE 12-5
Lithium aluminum hydride reduction of an ester produces a primary alcohol.
$LiAlH_4$ delivers hydride to the carbonyl carbon of an acid derivative in the
same way as to an aldehyde or ketone.

the lone pair on nitrogen forms a carbon–nitrogen double bond. The resulting salt is an *iminium ion*. The iminium ion is highly electrophilic and
rapidly undergoes hydride reduction by a second equivalent of $LiAlH_4$,
producing an amine with all C–C and C–N bonds originally present still intact.

The pathway for the reduction of primary and secondary amides (Figure 12-7) is similar to that for tertiary amides. However, the iminium ion
that is formed has a proton on nitrogen that is rapidly lost, forming a neutral imine. The C=N double bond in an imine is polarized in the same direction as a C=O double bond in a carbonyl compound, with carbon bearing partial positive charge and the heteroatom bearing partial negative
charge. As a result, hydride attack at the carbon end of the C=N double
bond is rapid, effecting reduction and the formation of a new C—H σ bond
with a shift of the π electron density onto nitrogen. The resulting negatively
charged nitrogen atom coordinates with aluminum, producing a complex
that, after treatment with mild acid, leads to the free amine. Thus, metal hydride reductions of all three classes of amides generate the corresponding
amine as the initial C=O group is converted into a —CH_2— unit.

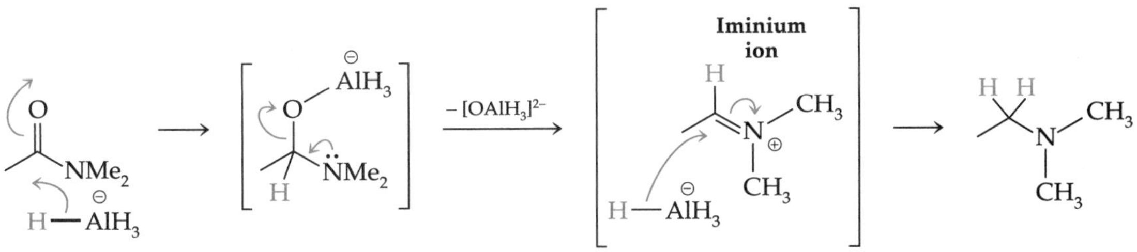

▲ FIGURE 12-6
Reduction of a tertiary amide by $LiAlH_4$ converts the carbonyl group into a
—CH_2— unit, thereby converting an amide into an amine. The complex
metal hydride delivers a hydride to the carbonyl group, producing a C—H σ
bond. The lone pair on nitrogen acts to displace the aluminate anion while
forming a C=N bond in an iminium ion. A second hydride is then delivered
to carbon, leading to the observed amine product.

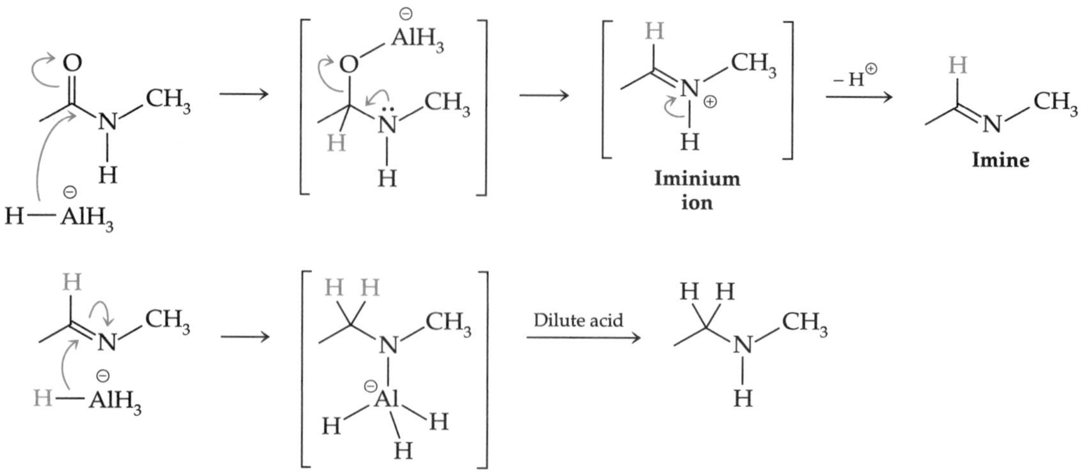

▲ FIGURE 12-7
Reduction of a secondary amide by LiAlH₄ produces a secondary amine by a
pathway parallel to that for a tertiary amide (Figure 12-6), except that the
iminium ion loses a proton to produce a neutral imine.

EXERCISE 12-B

Write a detailed mechanism for the LiAlH₄ reduction of acetamide.

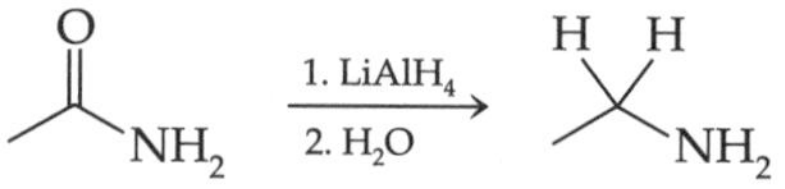

Relative Reactivity of Carbonyl Compounds toward Hydride Reducing Agents

The relative reactivity of various carbonyl functional groups toward nucleophilic attack roughly correlates with the magnitude of the positive charge
density on the carbonyl carbon. In a nucleophilic hydride transfer reaction,
the π bond of the carbonyl group is destroyed; the rate of hydride reduction
is affected by factors that increase or decrease the stability of the π bond of
the carbonyl group in the starting material and that affect the stability of the
intermediate tetrahedral anion. For example, as stated in Chapter 3, carboxylic acid esters react more slowly than ketones and aldehydes because
resonance donation of electrons from the oxygen of the —OR group stabilizes the ester and shifts electron density toward the carbonyl group. Resonance stabilization is even greater in amides because nitrogen is less electronegative than is oxygen, but it is weaker in thiol esters than in simple
esters. This stabilization can be viewed as a contribution from the zwitterionic amide and ester resonance structures shown at the right in Figure 12-8.
Because this resonance stabilization is lost as the π bond is broken and the
nucleophile is added to the carbonyl carbon, resonance-stabilized reactants
have higher activation energies and, consequently, are slower to react with
complex metal hydrides than are aldehydes or ketones.

The order of reactivity toward nucleophiles of several functional
groups containing C═O double bonds is:

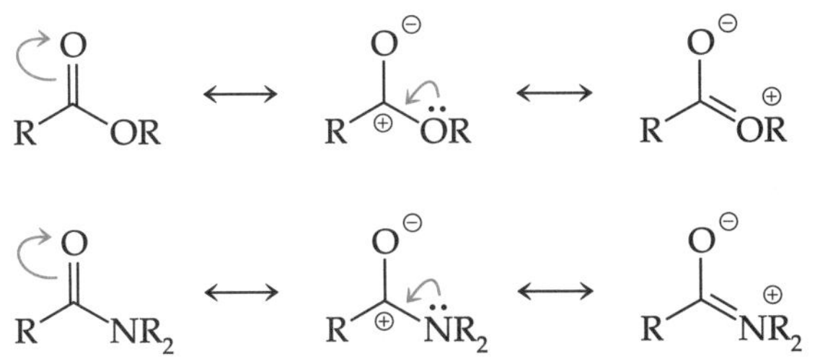

▲ FIGURE 12-8

Resonance donation by the nonbonded lone pair of electrons on the —OR oxygen of an ester (or on the —NR$_2$ group of an amide) leads to the zwitterionic resonance structures shown at the right. These additional structures stabilize the ester or amide relative to an aldehyde or ketone, making esters and amides less reactive toward nucleophiles.

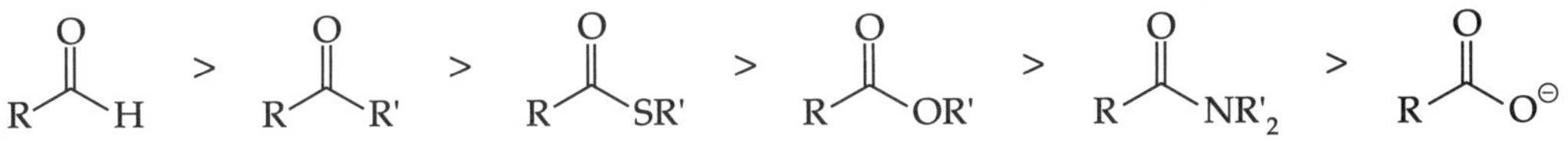

Aldehydes and ketones are more reactive than the carboxylic acid derivatives: the relative reactivity of these derivatives is governed by how effectively the heteroatom bound to carbon can release electrons to the carbonyl π system. Because the hydrogen atom attached to the C=O double bond in an aldehyde is smaller than the alkyl group of a ketone and because hyperconjugative electron release from the additional alkyl group of a ketone stabilizes the C=O double bond, activation energy barriers for nucleophilic attack on ketones are somewhat larger than those on aldehydes. Therefore, aldehydes are more reactive than ketones.

The relative reactivity of the various carbonyl functional groups has important consequences. Both sodium borohydride and lithium aluminum hydride reduce aldehydes and ketones, but the more-reactive reagent, lithium aluminum hydride, is required for the reduction of esters, thiol esters, and amides.

EXERCISE 12-C

Recall that complex metal hydrides react with acids as weak as water to generate hydrogen in an acid-base reaction. Why are carboxylic acids much more difficult to reduce with LiAlH$_4$ than are esters?

EXERCISE 12-D

For each of the following pairs of compounds, consider whether it would be possible to reduce the first compound in a mixture of the first and second. If so, what reducing agent could be used to effect the selective reduction?

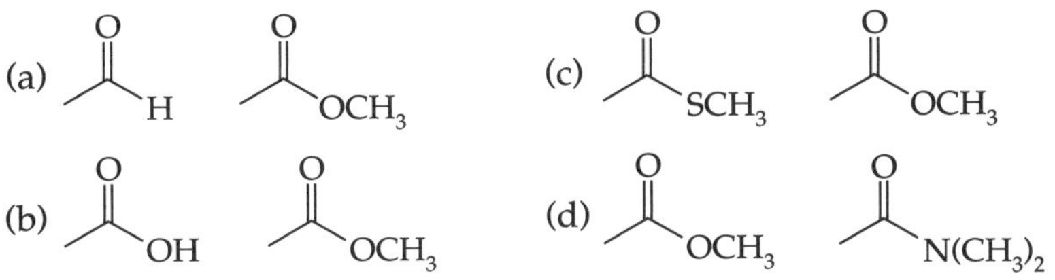

12-3 ▸ Biological Hydride Reductions

The biological equivalent of a complex metal hydride reducing agent is **nicotinamide adenine dinucleotide (NADH).**

Like $NaBH_4$ and $LiAlH_4$, NADH transfers a hydride equivalent to a reducible substrate. The complex metal hydride reductions bear close resemblance to NADH reductions, which will be treated in Chapter 17. Like $NaBH_4$ and $LiAlH_4$, NADH delivers a hydride to the carbon end of a $C{=}O$ double bond in the presence of a catalyst (the enzyme alcohol dehydrogenase), restoring aromaticity in NAD^+ and forming a reduced alcohol. The reverse transfer of a hydride from an alcohol to NAD^+ accomplishes alcohol oxidation by a process quite different from that discussed in Chapter 8 for metal-centered redox reactions.

12-4 ▸ Nonhydride Chemical Reductions

The addition of hydrogen across multiple bonds can be accomplished by methods other than hydride reductions. Two of these methods, catalytic hydrogenation and electron transfer (dissolving metal) reduction-protonation, are reasonably general methods for achieving the reduction of unsaturated organic compounds. Neither method can be described mechanistically with full-headed, curved arrows because each takes place through steps in which only one electron is transferred.

Catalytic Hydrogenation

Catalytic hydrogenation proceeds by a mechanism that is entirely different from that of the reduction of a carbonyl group by a complex hydride reagent, in which negative charge is developed on the carbonyl oxygen as hydride is added. Instead, as mentioned in Chapter 2, *catalytic hydrogenation* requires the activation of molecular hydrogen (H_2) by interaction with the surface of a noble metal. A **noble metal** is one that is very stable in the zero oxidation state. Among the most-common noble metals used in catalytic hydrogenations are platinum and palladium, but finely divided nickel and other metals also can be used. Typically, noble metal catalysts are used as highly dispersed powders on a large-surface-area support (carbon or alumina). Sometimes, the zero-valent metal is generated *in situ* by reduction of the corresponding oxide. For example, platinum oxide (PtO_2), when treated with hydrogen, generates finely divided platinum (and water), which is highly active in catalytic hydrogenations.

The interaction of molecular hydrogen with the surface of platinum results in the rupture of the H–H bond and the formation of two metal–hydrogen bonds (Figure 12-9). A multiple bond (for example, the $C{=}C$ bond of an alkene) interacts with this activated form of hydrogen, which is then transferred, resulting in the net addition of H_2 to the $C{=}C$ π

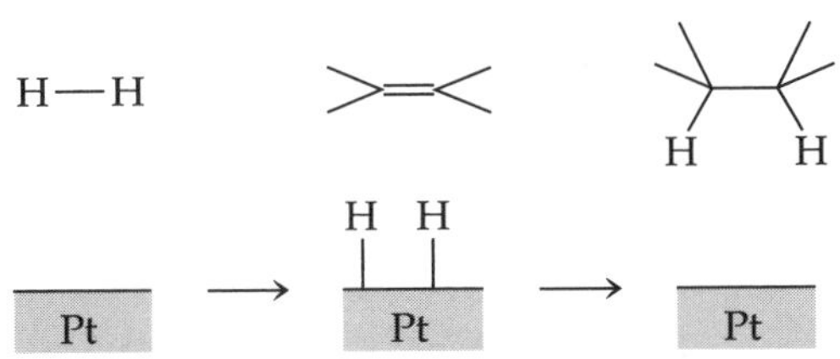

▲ FIGURE 12-9
In a catalytic hydrogenation, a metal surface (usually Pt, Pd, or Ni) activates gaseous hydrogen by chemically binding the two hydrogen atoms. Hydrogen is then transferred, in a *syn* addition, to a double bond.

bond, to form two C–H covalent bonds. This conversion effects reduction of the double bond, while the metal catalyst is regenerated in its initial form, ready for interaction with another molecule of hydrogen. Carbon–oxygen and carbon–nitrogen double bonds also are reduced with hydrogen and a metal catalyst, although the rates of reduction of these stronger and more-polarized bonds are much slower than those of carbon–carbon double bonds.

The metal surface is absolutely critical for these reactions, but it is not consumed in the net chemical transformation. It is said to act therefore as a *catalyst*; that is, as a species that accelerates the rate of a reaction without it-self being consumed. For this reason, these reactions are called catalytic hy-drogenations.

In the mechanism shown in Figure 12-9, both hydrogens are delivered to the same face of the molecule. Catalytic hydrogenation is thus a stereo-specific *syn addition*; that is, the hydrogens add to give *cis* products. This can be observed in the addition of one equivalent of H_2 to an alkyne.

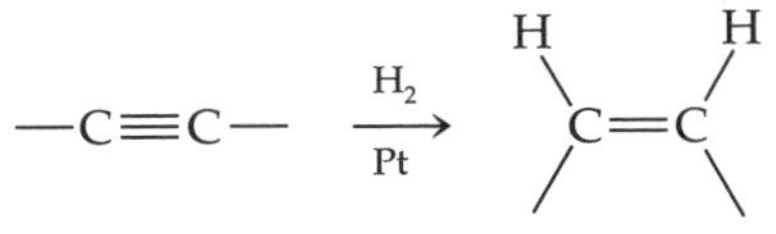

However, *syn* addition cannot be demonstrated in a carbonyl or imine hy-drogenation because the alcohol or amine formed does not have a geomet-ric isomer at the C–O or C–N bond. Only when an alkene is appropriately substituted can the consequence of the stereospecific reduction be observed. However, the rate of reduction drops with increased alkyl substitution on the double bond, and alkenes can also undergo positional isomerization in the presence of noble metals. For example, the reduction of the bicyclic alkene in Figure 12-10 affords a 9:1 mixture of the *trans* and *cis* reduced products. Thus, the observation that catalytic reductions proceed by the *cis* addition of hydrogen is usually significant only for the reduction of disub-stituted alkynes.

The ease with which a multiple bond is reduced under catalytic hydro-genation conditions is related to the strength of the π bond. For example, because of electron-electron repulsion, the two π bonds of an alkyne are in-dividually weaker than the π bond that remains in the product alkene after the addition of two hydrogen atoms.

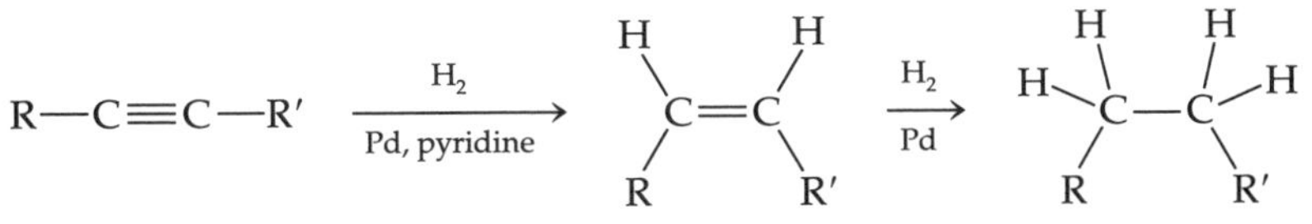

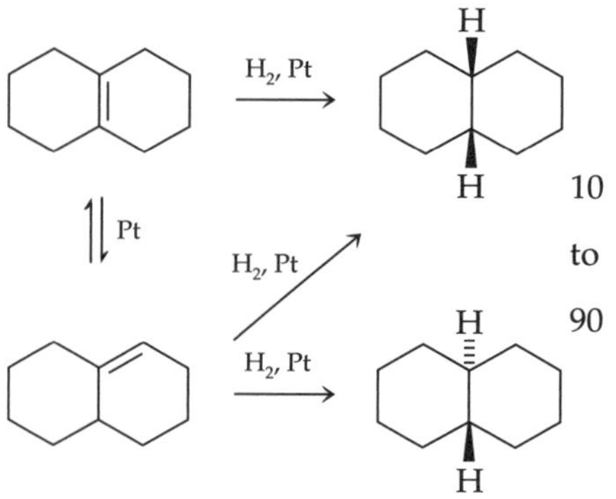

▲ **FIGURE 12-10**
Reduction of tri- and tetrasubstituted alkenes is often accompanied by double-bond migration followed by reduction.

Thus, the rate of catalytic hydrogenation of an alkyne is faster than that of the resulting alkene and, with care, the process can be limited to the addition of one equivalent of hydrogen. It is often useful, however, to deactivate the catalyst somewhat—for example, by adding small amounts of an amine (such as pyridine or quinoline) that can bind to the surface of the metal. (These species are referred to as **catalyst poisons**.) Nonetheless, further reduction can be accomplished, and whether the alkene or the further-reduced alkane is the product is often determined by the reaction time. The progress of such catalytic hydrogenations can be monitored by the uptake of hydrogen from the gas phase as the reduction proceeds.

Catalytic reductions of aromatic compounds also are possible, but such reactions are quite slow. The addition of one equivalent of hydrogen to an aromatic compound such as a benzene requires disruption of the aromatic system and is considerably slower than the addition of hydrogen to the remaining π bonds.

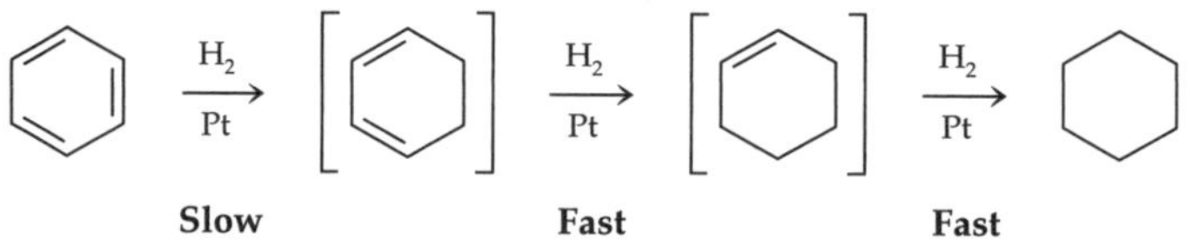

Thus, in contrast with the reduction of alkynes, in which it is possible to produce an alkene, it is not possible to isolate any partially reduced products at an intermediate oxidation level; for example, a cyclohexadiene.

The π bond of a carbonyl group is considerably stronger than that of an alkene (93 versus 63 kcal/mole); and catalytic hydrogenation of aldehydes, ketones, and (especially) esters requires the use of high temperature and pressure to increase the concentration of hydrogen in the solution and the reaction rate. Thus, molecules containing both an alkene and a carbonyl group undergo catalytic hydrogenation at room temperature and atmospheric pressure only at the carbon–carbon π bond, as illustrated for both unconjugated and conjugated enones in Figure 12-11.

Why then are hydride reductions of carbonyl π bonds possible, whereas neither alkynes nor alkenes react with either $LiAlH_4$ or $NaBH_4$? This difference is the result of the relative stability of the negatively charged intermediates formed in hydride reductions. The addition of hydride to a carbonyl group places negative charge on oxygen. In contrast, the addition of hydride to a carbon–carbon double or triple bond would produce a much

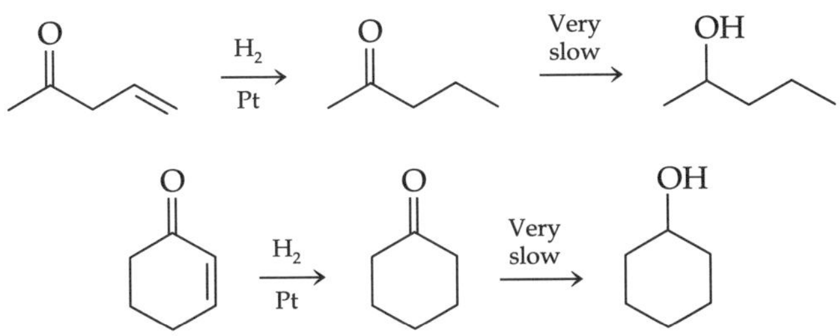

▲ FIGURE 12-11
Alkene groups are reduced more readily by catalytic hydrogenation than are carbonyl groups. Carbon–carbon double bonds can therefore be selectively converted by catalytic hydrogenation into saturated hydrocarbons in the presence of C=O multiple bonds.

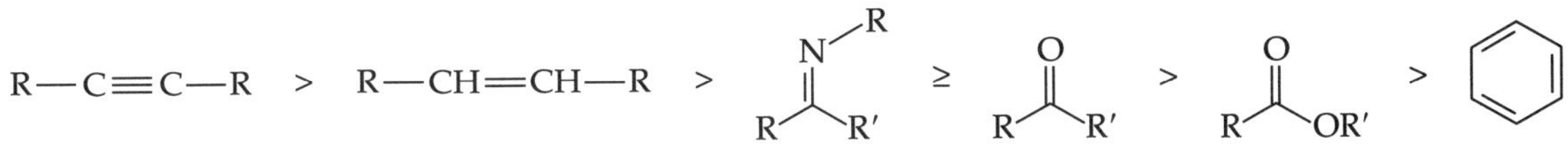

▲ FIGURE 12-12
The ease of catalytic hydrogenation (with H_2 and a metal catalyst) depends on the functional group being reduced.

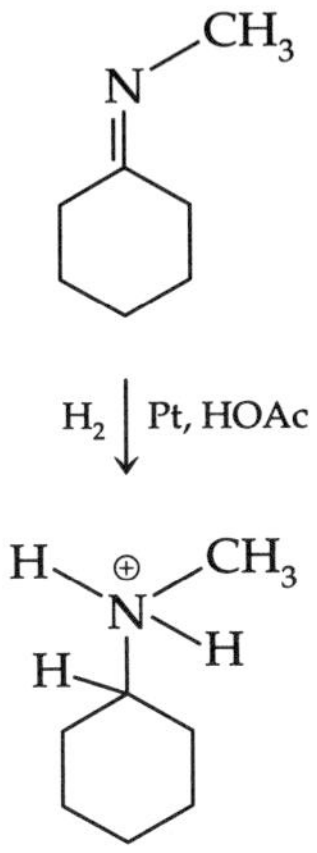

less-stable carbanionic intermediate with negative charge on a much less electronegative carbon atom.

Other carbon–heteroatom multiple bonds are reduced through catalytic hydrogenation as well. For example, imines undergo reduction to form amines. Because the carbon–nitrogen π bond in an imine is weaker than the carbon–oxygen π bond in a carbonyl group but stronger than the carbon–carbon π bond in a simple alkene, imines are reduced at rates somewhere between those of the other two functionalities. However, amines complex with noble metals and greatly reduce their effectiveness as catalysts. (Quinoline, for example, is a catalyst poison.) Thus, the reduction of an imine is best carried out in the presence of acid (for example, by using acetic acid as a solvent) so that the product amine is protonated, thus blocking surface complexation.

The relative ease with which catalytic hydrogenation reduces the functional groups discussed so far is shown in Figure 12-12. Alkynes are hydrogenated most easily, and esters are the most difficult to reduce. Even though the catalytic hydrogenation of esters and aromatics requires high temperature and pressure and special catalysts, it is often the preferred method for accomplishing industrial-scale reductions. Those factors that seem undesirable for small-scale reactions in the laboratory are often advantageous when applied to large quantities. For example, excess reagent (hydrogen) is easily removed, as is the insoluble metal catalyst, and there are no waste by-products formed from the reagent, as is the case in reductions with complex metal hydrides.

Dissolving–Metal Reductions

We have now considered two different reduction schemes: one consisting of the addition of a hydride ion and then a proton; and the other the addition of two hydrogen atoms (catalytic hydrogenation). A third method for effecting reduction provides an electron source (often as a zero-valent alkali metal) for generating a radical anion or dianion intermediate, which is then

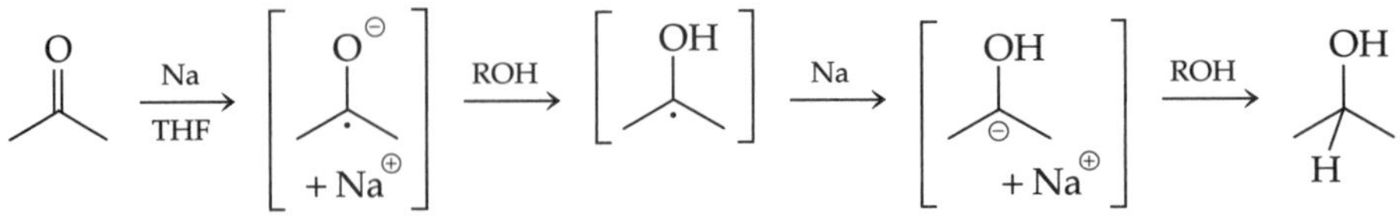

▲ FIGURE 12-13
Alkali metals can transfer one electron to a carbonyl group, producing a ketyl radical anion. When protonated, the radical ion is converted into a radical that is again reduced by a second equivalent of alkali metal. The resulting carbanion is then protonated to produce an alcohol. This reduction mechanism thus transfers an electron, a proton, a second electron, and a second proton. An alternate sequence in which two electrons and then two protons are transferred is also possible.

protonated *in situ*. For example, as discussed in Chapter 6, the addition of an electron to a carbonyl group generates a radical anion (a ketyl), which is ion-paired with the alkali metal cation formed when the electron is transferred from the neutral metal, as illustrated in Figure 12-13 with acetone. In the presence of a proton source, the radical anion is protonated, producing a carbon (ketyl) radical that is, in turn, further reduced by the addition of an electron from a second atom of sodium. Protonation of the resulting carbanion leads to the fully reduced product. In some cases, the ketyl radical anion formed on the surface of sodium is further reduced by the addition of another electron before it is protonated. The resulting dianion then adds two protons, resulting in the same product shown in Figure 12-13. Although the sequence may be different, both routes take place by adding two electrons and two protons, giving rise to a net reduction.

Because sodium can also react directly with an alcohol (generating hydrogen gas), these reactions are usually conducted in an ether solvent or in liquid ammonia containing only enough alcohol to serve as a proton source.

$$ROH \; + \; Na \; \longrightarrow \; RO^{\ominus} \; + \; Na^{\oplus} \; + \; \tfrac{1}{2} H_2 \uparrow$$

In the course of these reactions, the sodium metal disappears as it reacts and is converted into soluble sodium salts. Such reactions are therefore called **dissolving-metal reductions.**

In principle, these reactions require not a metal but rather a source of electrons. Indeed, they can also be accomplished by the addition of electrons in an electrochemical cell.

The facility of these reactions depends on the ability of the organic compound to accommodate an extra electron to form a radical anion. Dissolving-metal reductions work well for carbonyl groups and alkynes, but alkali metals do not transfer electrons efficiently to simple alkenes, which are therefore not reduced by these reagents. Thus, it is possible to reduce an alkyne to an alkene without further reduction to the alkane. When a dianion intermediate is formed, the negative charges are disposed *trans* to each other to minimize electron repulsion as much as possible. Protonation of this dianion leads to *trans* addition (Figure 12-14). This stereochemical course is opposite that of the *cis* addition of hydrogen to an alkyne by catalytic hydrogenation.

EXERCISE 12-E

Write the structures of the products, if any, expected from each of the chemical reductions in parts *a* through *g*.

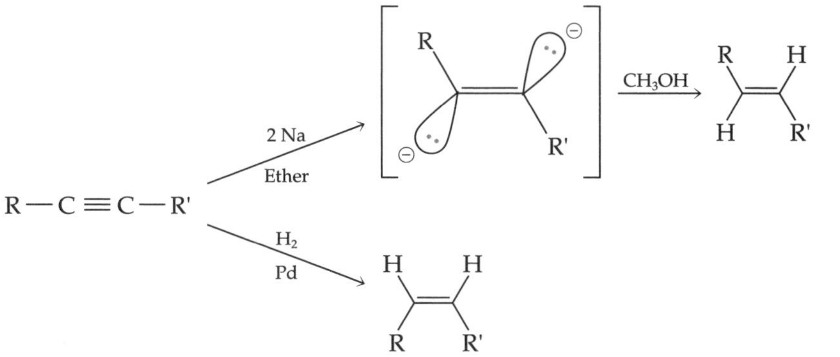

▲ FIGURE 12-14
Dissolving metal reduction of an alkyne produces a *trans*-alkene, whereas catalytic hydrogenation leads to a *cis*-alkene.

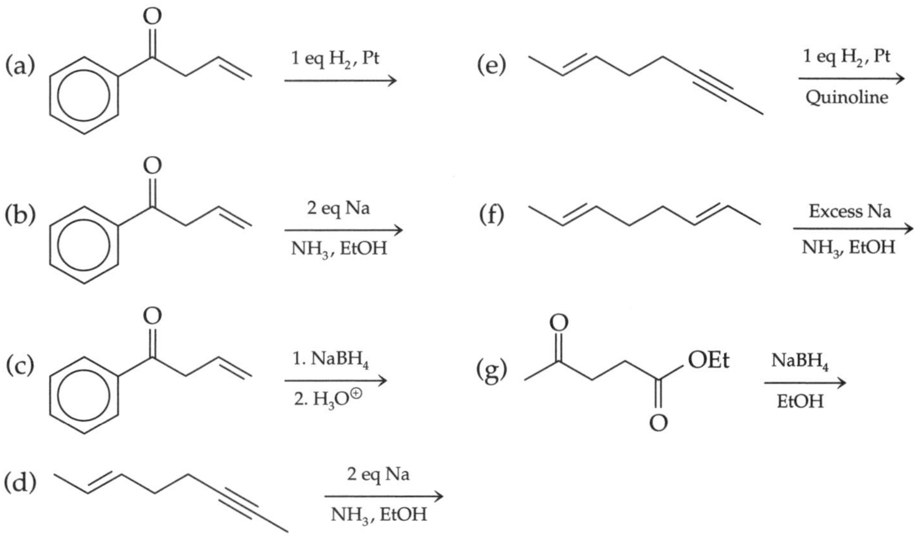

Metal hydride reductions, catalytic hydrogenations, and dissolving-metal reductions are therefore complementary methods for the reduction of different organic functional groups. The direct electron-mediated reductions have particular biological relevance, although in living systems less-rigorous reducing agents than alkali metals are used to form the radical anion intermediates.

12-5 ▸ Anions as Nucleophiles

Other negatively charged reagents also act as nucleophiles and attack the carbon end of the C=O group. Let us now consider how reactivity for a given carbonyl compound depends on the activity of the nucleophile.

As we know from the trends discussed in Chapter 6, the acidity of an acid, HX, increases as the position of X in the periodic table progresses from the left to the right: in acidity, $CH_4 < NH_3 < H_2O < HF$. This also means that **basicity,** the affinity of an atom or anion for a proton, decreases in the same left-to-right progression (NH_3 is a stronger base in water than is HF). A strong base has a high affinity for a proton, and the conjugate acid of a strong base is thus a weak acid.

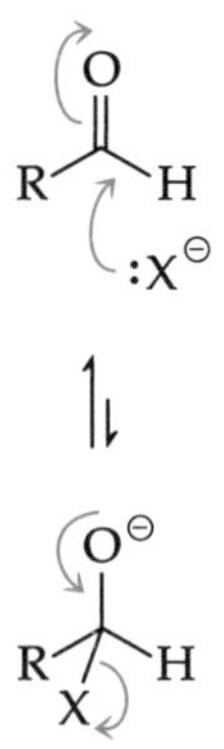

▲ **FIGURE 12-15**

Although a halide anion is a nucleophilic species, nucleophilic attack by a halide at a carbonyl group is reversible because the C—X bond formed is weaker than a C=O π bond. No net addition product can be isolated.

Nucleophilicity refers to the affinity of an atom or an anion for an electrophilic carbon atom. *Basicity,* which refers to the affinity of an atom or an anion for a proton, sometimes correlates well with nucleophilicity. The elements that lie farther to the left in the periodic table are less electronegative and, as a result, more nucleophilic. Thus, because HCl is a strong acid, Cl⁻ is a weak base and is often a weaker nucleophile than either oxygen or nitrogen anions.

This relation also holds in the progression from the top to the bottom of the periodic table. The larger, less-electronegative halide ions (for example, iodide) are more nucleophilic than the smaller, more-electronegative fluoride ion; and sulfur anions are more nucleophilic than oxygen anions. Differences in polarizability (the larger atoms are more polarizable), as well as in electronegativity, contribute to differences in nucleophilicity within a column of the periodic table. Because two different factors affect nucleophilicity, we must be careful to restrict this analysis to comparisons *within* the same row or *within* the same column of the periodic table.

In Chapter 8, we saw that halide ions are effective nucleophiles for S_N1 and S_N2 reactions at sp^3-hybridized carbon atoms. Nonetheless, halide ions do not effect nucleophilic addition or substitution at carbonyl groups. It is mechanistically useful to understand why not. Figure 12-15 proposes how a halide ion might attack an aldehyde. As in the reaction with hydride, shown in Figure 12-2, a tetrahedral carbon would be formed as the nucleophile approached the positive end of the C=O dipole. The resulting alkoxide, however, would very closely resemble the intermediate formed upon hydride attack on an ester; that is, the electrons on the negatively charged oxygen atom could be used to reform the C=O double bond and expel a leaving group. Halide ion is a much better leaving group than alkoxide because it is a very weak base and is thus a quite stable anion. [This is the same argument used in Figure 12-6 to explain why hydride reduction of amides proceeds to give amines rather than alcohols; that is, why ⁻OH (or [OAlH₃]²⁻) is the leaving group rather than ⁻NR₂.] If the halide were to be lost from the tetrahedral intermediate, as in Figure 12-15, the starting material would be regenerated and no net reaction would be observed. For this reason, halides are not effective nucleophiles for either addition or substitution reactions. A better nucleophile, however, would be a poorer leaving group, and the analogous tetrahedral intermediate would be protonated to form addition products.

EXERCISE 12-F

Use curved arrows to represent the flow of electrons in a reaction mechanism that explains how an acid chloride can be converted into a carboxylate anion upon treatment with aqueous base. Explain why the reverse reaction, the conversion of a carboxylic acid into an acid chloride by treatment with NaCl does not occur.

Whether or not the product of nucleophilic addition is more stable than the starting material is determined by the relative strengths of bonds made and broken in the nucleophilic attack. Generally, there are two important criteria that must be considered if net reaction is to be observed. First, the strength of the bond between the nucleophile and the carbonyl carbon must be stronger than that of the carbonyl π bond. The fact that carbon–chlorine bonds (average C—Cl, 81 kcal/mole) are weaker than a typical C=O π bond (93 kcal/mole) is one reason that nucleophilic attacks by halide ions fail. Second, the anion formed by attack by the nucleophile (an alkoxide) must be more stable than the nucleophilic anion itself. Again, halide ions are more stable (weaker bases) than an alkoxide ion. Thus, the addition of a halide ion to a carbonyl group is not favorable energetically and can be easily discounted.

Simply looking at the organic part of a reaction, however, cannot lead to an accurate energetic analysis. For example, as mentioned earlier, simple hydride reagents such as KH and NaH are ineffective hydride donors not because of weak bonds but because of their low solubility in organic solvents. Conversely, reagents such as $NaBH_4$ and $LiAlH_4$ are more soluble and are thus effective in transferring a hydride to a carbonyl group. A complete thermodynamic analysis of a given reaction would require that we take into consideration the strengths of all bonds formed and broken, as well as the solvation and lattice energies of all intermediates and products. Such an analysis is beyond the scope of this book and in many cases is not yet even experimentally possible.

12-6 ▸ Addition of Oxygen Nucleophiles

Unlike halide ions, oxygen nucleophiles productively attack carbonyl compounds. Common oxygen nucleophiles that effect nucleophilic addition include water, hydroxide ion, and alcohols. In this section, we consider how these nucleophiles react with aldehydes and ketones to produce, respectively, hydrates, disproportionation products through the Cannizzaro reaction, and acetals and ketals.

Addition of Water: Hydrate Formation

For the reasons given in Section 12-5, halides do not effect nucleophilic addition to carbonyl groups. However, elements at the left of the halogens in the periodic table form stronger bonds and are less stable as free anions than are halides. Figure 12-16 illustrates the attack by hydroxide ion (an oxygen nucleophile) on an aldehyde. When hydroxide ion attacks the carbonyl group, the carbonyl oxygen becomes negatively charged, and an alkoxide is formed. Thus, there is little difference in the stability of the two negatively charged species; that is, between the tetrahedral alkoxide and the nucleophilic hydroxide ion. Although the π bond of the aldehyde is still somewhat stronger (93 kcal/mole) than the σ carbon–oxygen bond that replaces it (86 kcal/mole), the energy difference is smaller than it would be if a halide were the attacking nucleophile. Protonation of this intermediate alkoxide by water produces a **geminal diol** (a **hydrate**) and hydroxide. The overall reaction can thus be viewed as the addition of water across the π

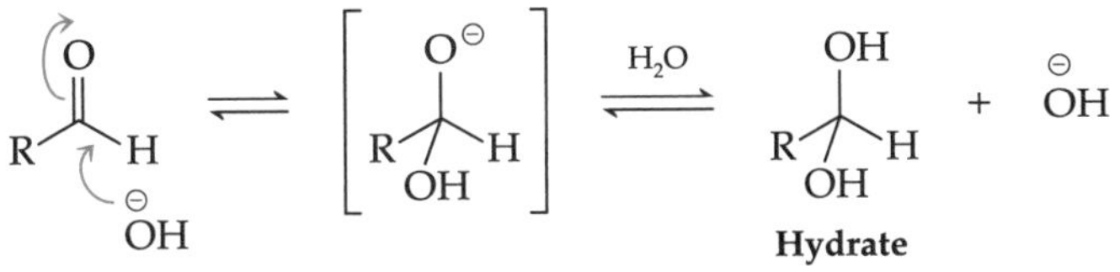

▲ FIGURE 12-16
When hydroxide ion acts as a nucleophile to attack a carbonyl carbon, the resulting alkoxide bears two comparably strong C—O σ bonds. Because a C—O σ bond is only slightly weaker than a C═O π bond, the addition is only partly reversible. Protonation on oxygen produces a hydrate.

bond, catalyzed by hydroxide. In fact, this hydration of an aldehyde (as well as that of a ketone) is very rapid in alkaline solution. Because the product hydrate is less stable than the starting carbonyl compound, however, this hydration is easily reversed. In most aldehydes, there is no net chemical consequence of this reversible addition. However, it can be detected and its rate measured by the use of water enriched in ^{17}O or ^{18}O, with the heavy isotope of oxygen finding its way into the carbonyl group.

EXERCISE 12-G

Propose a mechanism by which normal acetone can be labeled with ^{18}O.

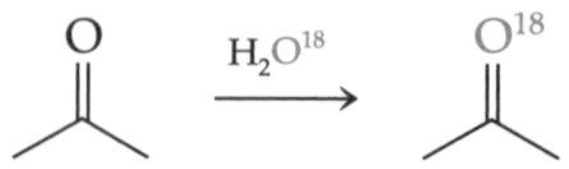

In acidic water (at low pH), the concentration of hydronium ion increases, whereas that of hydroxide ion is correspondingly reduced (remember that $[H^+][^-OH] = 10^{-14}$).

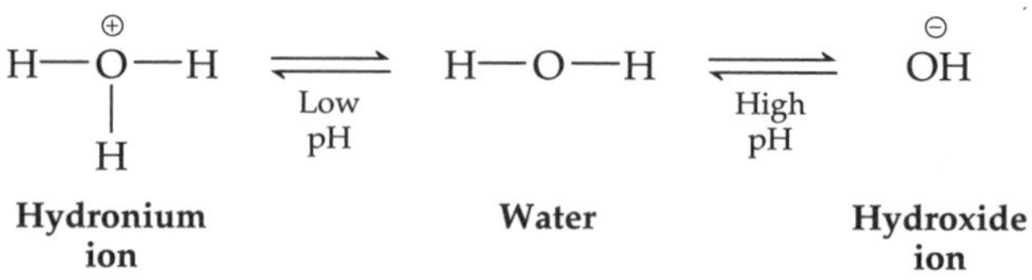

Hydronium   **Water**   **Hydroxide**
ion   **ion**

The nucleophilicity of a neutral water molecule is much lower than that of a hydroxide ion, and one might expect a correspondingly slower attack on a carbonyl group by water. (The hydronium ion is not nucleophilic because the oxygen bears a formal positive charge.) However, in acidic solution, water does add to the $C=O$ double bond. One possible explanation is shown in Figure 12-17. Here, the carbonyl group is rapidly and reversibly converted by protonation into a significantly better electrophile, one that is sufficiently reactive to be attacked by a less-nucleophilic, neutral water molecule. Thus, nucleophilic attack occurs in acidic solution even with a weak nucleophile such as water, generating an oxonium ion. Deprotonation of this species affords the hydrate. Here, again, we have the addition of water across the π bond but now in a process that is catalyzed by acid.

An important point to be inferred from a comparison of the reactions described in Figures 12-16 and 12-17 is that nucleophilic attack can be accel-

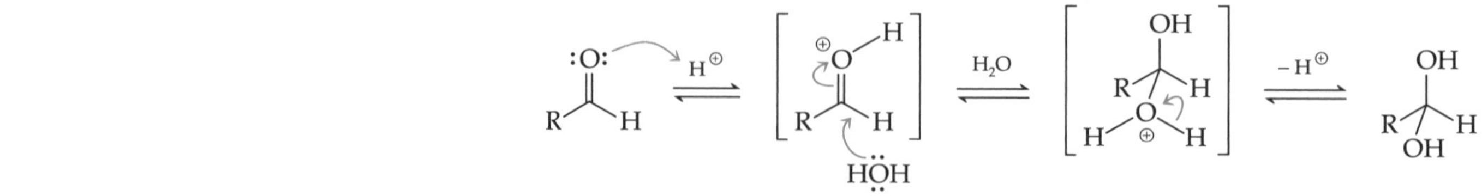

▲ FIGURE 12-17
In acid, the carbonyl oxygen is protonated, producing a cation that can be attacked by a weaker nucleophile, such as water. The oxonium ion formed in this nucleophilic attack on carbon is deprotonated to a neutral hydrate. Each of these steps is completely reversible: high concentrations of water shift the equilibrium toward the hydrate; removal of water shifts the equilibrium toward the aldehyde.

Table 12-1 ▸ Equilibrium Constants for Hydrate Formation

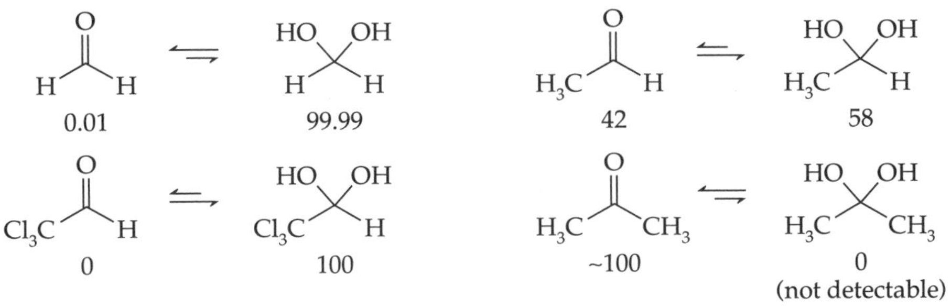

erated under both basic and acidic conditions. In the presence of base, the nucleophile is deprotonated, resulting in an anion with enhanced nucleophilicity. In the presence of acid, the nucleophile is neutral (and hence less reactive than the anion), but the carbonyl compound can be activated toward reaction with the nucleophile by protonation on oxygen.

Each of the steps in both Figure 12-16 and Figure 12-17 is reversible so that an equilibrium is established between the carbonyl compound and its hydrate. By definition, a catalyst is not consumed in a reaction and can have no effect on the overall energetics of the reaction. The equilibrium position of hydration is therefore the same regardless of whether an acid or a base is used as catalyst.

The position of this equilibrium is governed by the stability of the hydrate adduct relative to the starting carbonyl compound. Structural features in the starting material that stabilize the carbonyl group include the presence of electron-donating groups or, conversely, the absence of electron-withdrawing substituents. The relative amounts of four carbonyl compounds and their hydrates are given in Table 12-1.

EXERCISE 12-H

For each of the following carbonyl compounds, predict whether the amount of hydrate present at equilibrium in aqueous acid is larger or smaller than that present when acetone is dissolved in the same acidic medium. Explain your reasoning clearly.

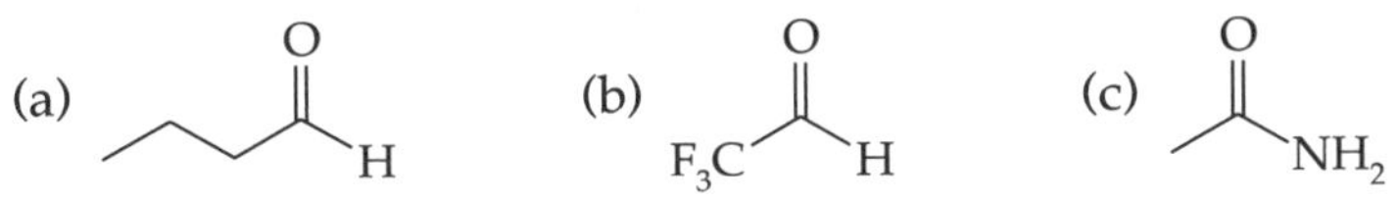

(a)   (b)   (c)

Hydroxide Ion as a Nucleophile:
The Cannizzaro Reaction

The **Cannizzaro reaction** converts some aldehydes into equal amounts of the corresponding carboxylic acid and alcohol upon treatment with sodium or potassium hydroxide (Figure 12-18). This reaction works only with aldehydes (such as benzaldehyde) that lack α hydrogens. For such aldehydes, α-deprotonation by hydroxide cannot lead to an enolate anion; instead, hydroxide adds to the carbonyl carbon. Reversal of this addition is quite rapid, but the reverse reaction simply reforms the starting material. In an alternate pathway, the carbonyl group can be reformed from the tetrahedral

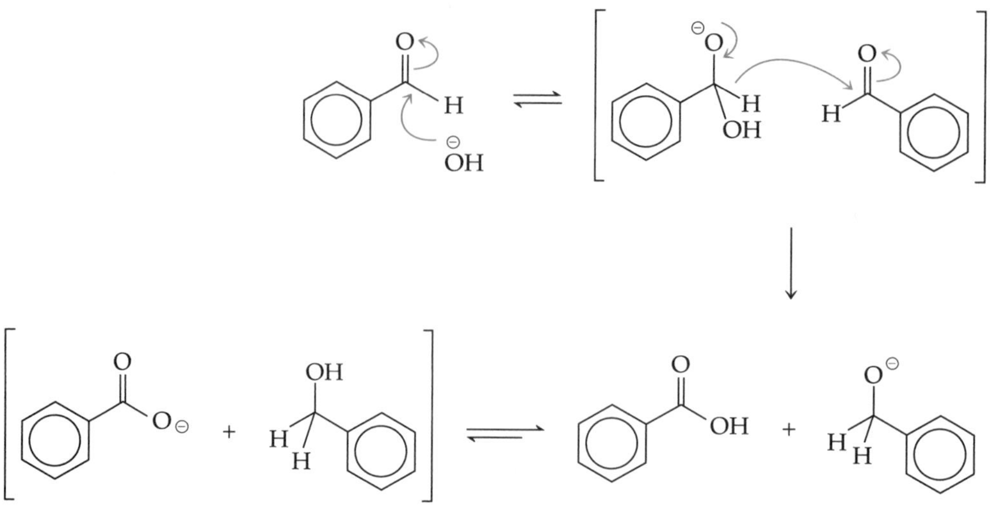

▲ FIGURE 12-18

In the Cannizzaro reaction, nucleophilic attack by hydroxide ion leads to an intermediate like that produced in base-catalyzed hydrate formation. Although this C—O σ-bond formation is reversible, it is also possible that, as the electrons on oxygen reform the C═O double bond, a hydride can be transferred to a second molecule of aldehyde. The carboxylic acid and alkoxide ion produced in this step then equilibrate in an acid-base exchange. When the reaction mixture is neutralized, a carboxylic acid and an alcohol are isolated. This reaction fails for aldehydes that have α-hydrogens because hydroxide ion instead attacks the C–H bond to form an enolate anion.

intermediate if hydride ion (instead of hydroxide ion) is lost. Simple loss of hydride ion is not possible because this ion is very unstable owing to its concentrated charge. However, hydride can be transferred simultaneously to an electrophile—in this case, benzaldehyde—by a route very similar to that of the complex metal hydride reductions presented in Section 12-2. Thus, as the carbonyl group is reformed from the tetrahedral intermediate derived from one molecule, hydride is transferred to a second molecule of starting material, so that a carboxylic acid and an alkoxide ion are formed. Proton transfer between these products generates a carboxylate anion and a neutral alcohol. This hydride transfer mechanism results in the reduction of the aldehyde group and is similar to those discussed in Section 12-3 for biological reductions.

The hydroxide-induced conversion of two molecules of benzaldehyde into one molecule of carboxylic acid and one of alcohol effects a **disproportionation,** a reaction in which a species of intermediate oxidation level (an aldehyde) is both oxidized (to an acid) and reduced (to an alcohol). The Cannizzaro reaction is understandable from what we know about nucleophilic attack and the stability of tetrahedral intermediates because this reaction illustrates how knowledge of the mechanisms of a small number of reactions (here, hydrate formation and hydride reduction) is useful in understanding new reactions.

Addition of Alcohols

Hemiacetals and Acetals. The addition of alcohol across the carbonyl π bond of an aldehyde takes place by a reaction pathway that is essentially identical with that for the addition of water. Let us first consider the reaction of an alcohol and an aldehyde in the presence of base (Figure 12-19). As we have seen, alcohols are acidic, and treatment with an alkali metal results

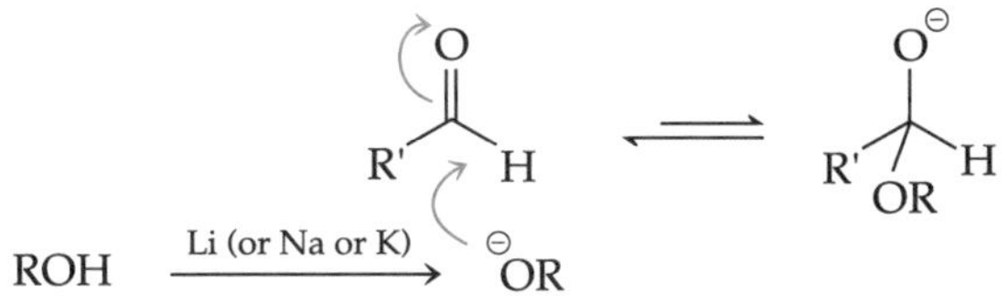

▲ **FIGURE 12-19**
Treatment of an alcohol with an alkali metal produces the alkali metal alkoxide salt. The alkoxide is sufficiently nucleophilic to attack a carbonyl carbon, but the equilibrium with the tetrahedral intermediate favors the aldehyde and alkoxide ion.

in the formation of an alkoxide anion. (The reaction with potassium metal is so exothermic that dangerous conditions can result when potassium salts are made from primary or secondary alcohols in this way. The hydrogen generated can combine explosively with oxygen to form water in a highly exothermic reaction.) Alternatively, alkoxides can be generated *in situ* by treatment of the neat alcohol with strong base. Alkoxide ions are effective nucleophiles and rapidly add to the carbonyl carbon. However, as in the reaction of an aldehyde with hydroxide, this first step is readily reversible; the addition product is less stable than the starting carbonyl compound and the addition is disfavored entropically. Consequently, the result of this addition cannot be observed.

Catalysis by acid, however, does produce an observable result, as illustrated in Figure 12-20 by the reaction of methanol with acetaldehyde by a path parallel to an acid-catalyzed hydration reaction. The species at the right in the first line, $CH_3CH(OH)(OCH_3)$, is called a **hemiacetal** and, like the hydrate, is thermodynamically less stable than the starting material. However, under acidic conditions, further transformations ultimately convert the aldehyde and two equivalents of alcohol into an **acetal**

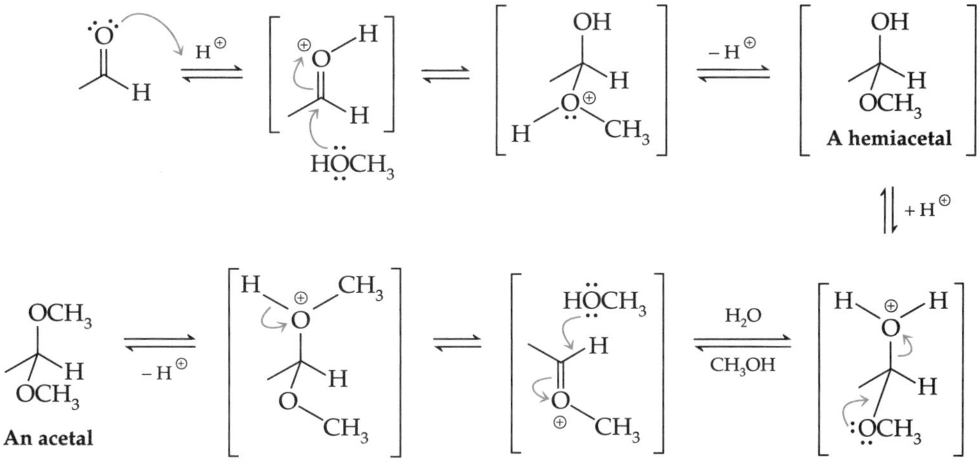

▲ **FIGURE 12-20**
In acid, protonation of the carbonyl oxygen activates the C=O group toward nucleophilic attack. An alcohol can therefore attack, producing a hemiacetal after deprotonation. The hemiacetal can be dehydrated if the —OH group is protonated, with the loss of water being assisted by a nonbonded electron pair from the —OCH₃ group. The resulting cation is electronically analogous to a protonated carbonyl group and is again attacked by a second alcohol molecule. After deprotonation, an acetal is produced. As with the acid-catalyzed hydration, acetal formation is completely reversible: high concentrations of alcohol shift the equilibrium toward the acetal, whereas high concentrations of water shift it toward the aldehyde.

[CH$_3$CH(OCH$_3$)$_2$] and water. Note that a hemiacetal has a general formula RCH(OH)(OR), whereas an acetal has the formula RCH(OR)$_2$.

The sequence takes place by the conversion of an —OH group of the hemiacetal into a good leaving group by protonation. This requires the presence of acid. The loss of water from the protonated hemiacetal is assisted by a lone-pair donation from the alkoxy oxygen, and an intermediate cation analogous to a protonated carbonyl group is formed. Addition of a second molecule of alcohol to the carbonyl carbon results in the formation of an oxonium ion intermediate. The sequence is finished by deprotonation, producing an acetal—a tetrahedral carbon with two geminal alkoxy groups. As we will see in Chapters 16 and 17, hemiacetals, acetals, and analogous functional groups are of great importance in the chemistry of sugars and nucleic acids.

By comparing the bonds present in the reactants (the carbonyl compound and two molecules of alcohol) with those present in the product (acetal and water), we find the same change in bonding as in the hydration of an aldehyde: the carbon–oxygen π bond is replaced by a new carbon–oxygen σ bond. Because the π bond is stronger, the formation of both the hydrate and the acetal is endothermic. However, the formation of the acetal differs in one very important way: a second product, water, is formed. The addition of a species that either removes (for example, a molecular sieve) or reacts exothermically with water can make the overall process exothermic. Under such conditions, aldehydes can be converted into acetals in good yield. For example, as it is formed, the water can be removed from the reaction by azeotropic distillation so that the reaction is "pulled" toward the acetal (recall LeChatelier's Principle). The formation of the acetal can also be favored by the use of the alcohol as solvent. Under these conditions, the high concentration of starting material "pushes" the reaction toward the products.

The idea that reactions can be "pushed" and "pulled" toward the desired product is very important in biochemical transformations. Active living systems cannot be at thermodynamic equilibrium: an influx of starting materials (food) and expulsion of waste products is required for activity. As new starting materials are ingested, they are pushed toward product by a temporary increase in their concentration. In turn, these products are starting materials for other reactions with their own products. Each of these reactions is balanced near its own equilibrium, but the constant output of material drives the chemical reactions that translate into energy of metabolism of the living system.

Hemiketals and Ketals. The reactions just described for aldehydes that produce hemiacetals and acetals also take place with ketones, forming the structurally analogous **hemiketal** and **ketal** functional groups shown in Figure 12-21.

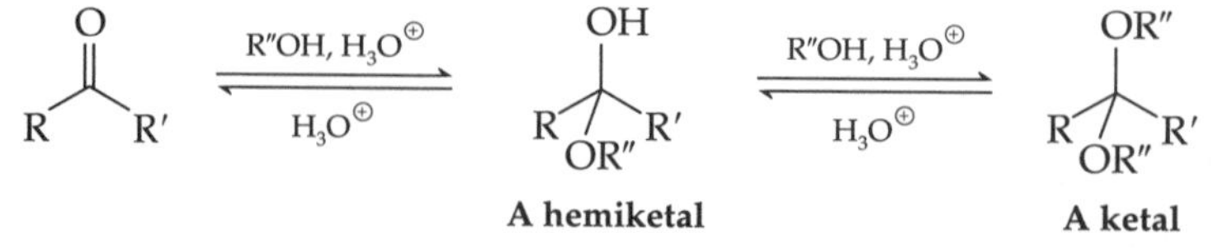

A hemiketal   **A ketal**

▲ **FIGURE 12-21**

A hemiketal [RR'C(OH)(OR'')] and ketal [RR'C(OR'')$_2$] are produced from a ketone by a mechanism parallel to that given in Figure 12-20 for the alcoholysis of an aldehyde. Acidic alcohol favors ketal formation; aqueous acid favors production of the carbonyl compound.

Write a complete reaction mechanism for the formation of a ketal between acetophenone and ethylene glycol. Then write a mechanism for the hydrolysis of the ketal to reform the starting materials.

The formation of both acetals and ketals is reversible in the presence of aqueous acid. The mechanisms for the conversions of ketal into hemiketal and of hemiketal into ketone are identical with those for the conversions of acetal into hemiacetal and of hemiacetal into aldehyde. As in the reversible brominations discussed in Chapter 10, reversible acetal and ketal formation are methods for protecting carbonyl groups, permitting the reversible masking of the reactivity of the $C=O$ π bond.

12-7 ▸ Addition of Nitrogen Nucleophiles

Nitrogen-containing compounds that bear a lone pair of electrons on nitrogen are active nucleophiles. Common nitrogen nucleophiles that attack carbonyl groups to effect nucleophilic addition include ammonia, primary and secondary amines, hydrazine derivatives, and hydroxyl amine. In this section, we consider how these nucleophiles react with aldehydes and ketones to produce, respectively, imines and Schiff bases, enamines, hydrazones, and oximes.

Amines

Imine Formation. Nitrogen functional groups are more basic and more nucleophilic than their oxygen counterparts in comparable structures. Thus, the reactions of nitrogen nucleophiles take place under less-stringent conditions. The reaction of ammonia with an aldehyde or ketone, as illustrated in Figure 12-22 with acetone, begins by nucleophilic attack on the carbonyl carbon by a pathway similar to that of the addition of water or alcohol to a carbonyl π bond. However, the higher nucleophilicity of ammonia makes it unnecessary to employ either acid or base catalysis to initiate this reaction. The nitrogen lone pair attacks the carbonyl carbon to generate a tetrahedral, zwitterionic intermediate. Rapid deprotonation at nitrogen and reprotonation at oxygen form a neutral species; but, because this intermediate is in a protic medium, it can be protonated again, either on oxygen or nitrogen. Protonation on nitrogen is not productive, but protonation on oxygen sets the stage for loss of water with nitrogen's lone pair of electrons forming a $C=N$ double bond. The resulting *iminium ion* loses a proton to form a neutral *imine*. Imines of aldehydes and those derived from ammonia are particularly unstable; not easily isolated, they are readily converted into the starting ketone and amine in the presence of water.

Imines can be reduced to amines, however, by hydride reagents or by catalytic hydrogenation. An electron-deficient derivative of $NaBH_4$—namely, sodium cyanoborohydride, $NaBH_3CN$—is frequently used as the complex metal hydride in this reaction. The conversion of a carbonyl group

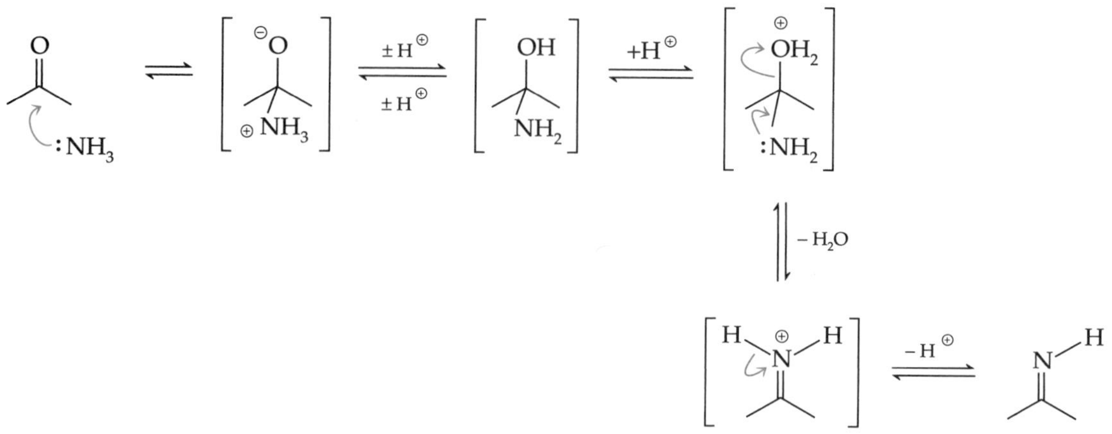

▲ FIGURE 12-22

An imine is formed by a mechanism exactly parallel to that for hydrate or acetal formation (Figures 12-17 and 12-20), except that ammonia replaces water as the active nucleophile. Because nitrogen nucleophiles are more reactive than oxygen nucleophiles, neither acid nor base catalysis is needed.

into an amine through an intermediate imine is called **reductive amination.** Although a reducing agent is required only for the second step, the amine and metal hydride are added together so that the imine can be reduced as it is formed.

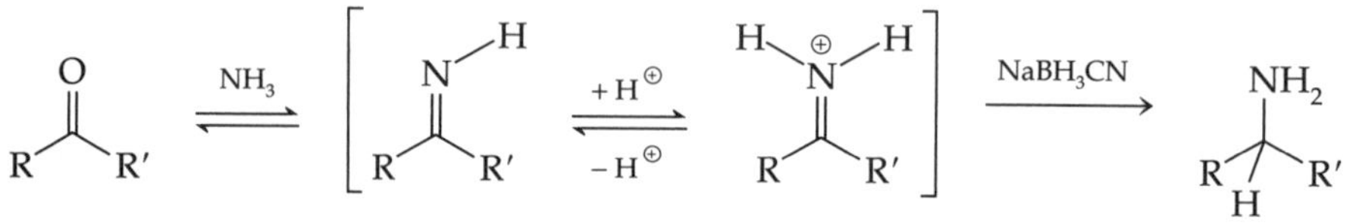

Because of the presence of the electron-withdrawing cyano group on boron in sodium cyanoborohydride, the reagent is less reactive than sodium borohydride and does not reduce aldehydes and ketones. Indeed, it is the protonated imine that undergoes reduction in these reductive amination reactions.

Tautomerism to Enamines. Imines bearing a hydrogen at the carbon α to the C=N double bond (such as that derived from acetone) can tautomerize to generate an **enamine,** by protonation-deprotonation.

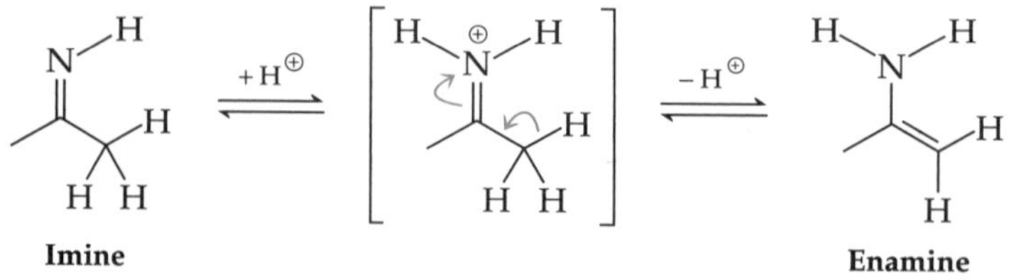

This imine-enamine tautomerization is similar to the keto-enol tautomerization discussed in Chapter 3 and is catalyzed by either acid or base. The enamine is an important intermediate in that the α-carbon bears significant negative charge density, much like an enolate anion. Thus, enamines undergo reactions as carbon nucleophiles that closely resemble those already described for enolate anions in Chapter 8.

EXERCISE 12-J

Suggest a mechanism for the reduction of acetone imine to 2-aminopropane by sodium cyanoborohydride.

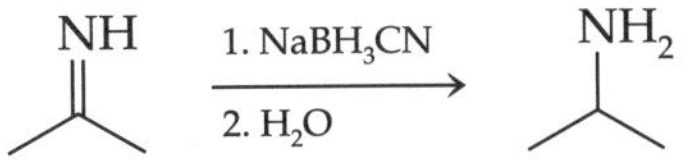

_____

Imine (Schiff-Base) Formation. Ketones also react with primary and secondary amines such as methylamine and dimethylamine to form iminium ions. With a secondary amine, a proton is lost from the α-carbon and an enamine is formed, as illustrated for cyclohexanone in Figure 12-23. Reaction of a primary amine with a ketone forms both an imine (also known as a **Schiff base,** $R_2C=NR'$) and an enamine, bearing a proton on nitrogen. These two species are simply proton tautomers and are in rapid equilibrium. Except with very large alkyl groups on nitrogen, the imine form is favored.

_____
EXERCISE 12-K

Write a full mechanism, using curved arrows, to illustrate how an imine is formed when acetone is treated with *n*-butylamine.

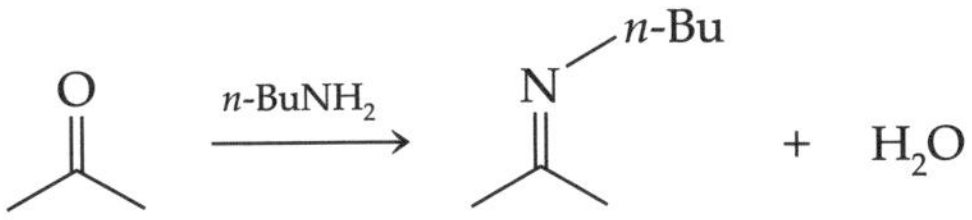

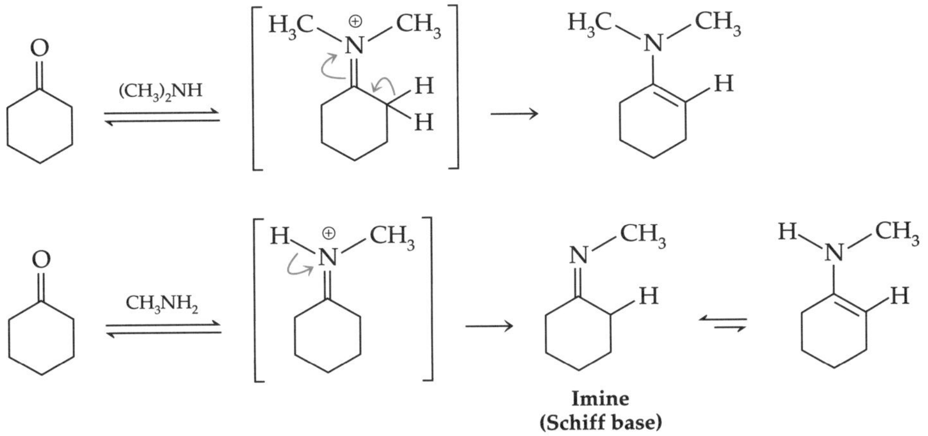

_____

**Imine
(Schiff base)**

▲ **FIGURE 12-23**

Iminium ions are formed by nucleophilic attack by either a secondary amine (upper reaction) or a primary amine (lower reaction), followed by dehydration. The positive charge of both iminium ions is relieved by deprotonation. In the iminium ion formed by attack by a secondary amine, the proton lost must be an α-hydrogen, producing a neutral enamine. In the iminium ion formed from a primary amine, the proton lost can be either an α-hydrogen (leading to an enamine) or an N—H hydrogen (leading to an imine). An imine bearing a carbon substituent on nitrogen is called a Schiff base.

Other Nitrogen Nucleophiles

A number of other nitrogen-containing nucleophiles form derivatives of ketones. Hydrazine (H_2N—NH_2) reacts with a ketone such as cyclohexanone to form a **hydrazone** by a mechanism that is identical with that for the formation of imines.

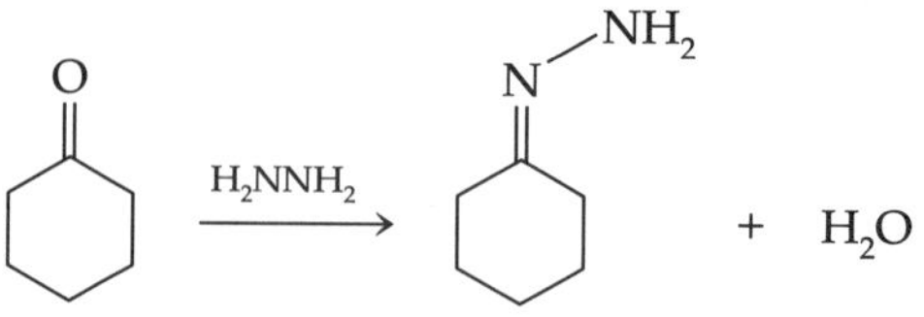

Hydrazone

Hydrazones are often solids, and so the reaction with hydrazine is a useful method for converting a liquid carbonyl compound into a solid derivative that can be more easily characterized—for example, by its melting point.

Phenylhydrazines are used to make phenylhydrazones of ketones and aldehydes (Figure 12-24). The product of the reaction of a ketone with 2,4-dinitrophenylhydrazine, a 2,4-dinitrophenylhydrazone (DNP), is a brightly colored (often red orange) solid, whose formation is frequently used as a qualitative test for the presence of aldehydes and ketones.

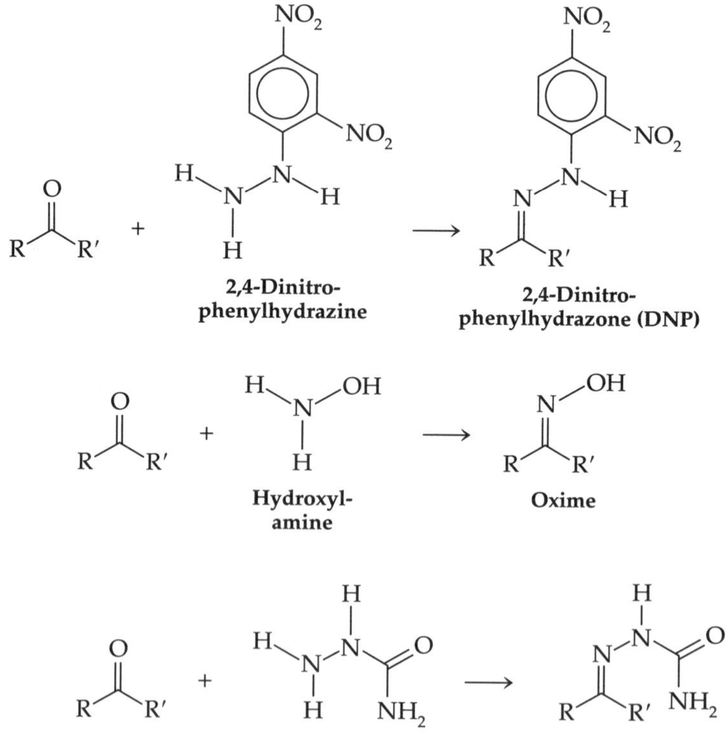

▲ FIGURE 12-24
Derivatives are prepared when the lone pair of electrons on an —NH_2 group attached to another heteroatom attacks the carbonyl carbon of an aldehyde or ketone. With hydrazines (H_2N—NHR), hydrazones are produced; with hydroxylamine (H_2N—OH), oximes are produced; and with semicarbazides (H_2N—$NHCONH_2$), semicarbazones are produced. Formation of these solid derivatives is often used as a chemical test for the presence of an aldehyde or ketone.

Oximes and semicarbazones are derivatives (often solids) formed by the treatment of an aldehyde or ketone with hydroxylamine or semicarbazide (Figure 12-24). Such a reagent (for example, a hydrazine or a semicarbazide) often has more than one possible nucleophilic site available for reaction with a carbonyl group. In 2,4-dinitrophenylhydrazine (shown in the margin), the terminal nitrogen (in color) is more nucleophilic than the other amino nitrogen. The amino nitrogen attached to the aromatic ring has greatly diminished nucleophilicity because of delocalization of the lone pair of electrons into the aromatic π system, as is evident in the lower resonance contributor in which the NH group adjacent to the ring bears formal positive charge. (The nitrogen atom in an —NO_2 group is formally positively charged, as described in Chapter 11, and is not nucleophilic.) The electron density at the terminal NH_2 group is not affected by this interaction and, for this reason, can more effectively serve as an active nucleophile. Similarly, the two nitrogens of semicarbazide directly attached to the carbonyl carbon are resonance donors to the carbonyl oxygen. Therefore, the remaining nitrogen (shown in color) is the more-nucleophilic atom and is the atom that is bound to carbon in a **semicarbazone.**

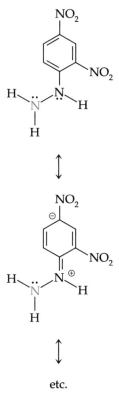

2,4-Dinitrophenyl-
hydrazine

Semicarbazide

In principle, either the oxygen or the nitrogen of hydroxylamine (H_2N—OH) could react as a nucleophile. However, the less-electronegative atom within a row of the periodic table is generally more basic and more nucleophilic. Thus, nitrogen of hydroxylamine reacts as a nucleophile, combining with an aldehyde or a ketone to form an **oxime** (Figure 12-24).

EXERCISE 12-L

Write a full mechanism, in parallel with that for imine formation and using curved arrows, to illustrate how a hydrazone is formed when acetone is treated with hydrazine.

12-8 ▸ Nucleophilic Acyl Substitution of Carboxylic Acids and Derivatives

There are several families of carboxylic acid derivatives in which a heteroatom is bound to a C=O group. Examples are carboxylic acid chlorides, anhydrides, esters, thiol esters, and amides. Let us consider a general scheme in which a heteroatomic nucleophile (X^-) attacks the carbonyl carbon of a carboxylic acid derivative whose heteroatom is represented by Y (Figure 12-25). In this reaction, one carboxylic acid derivative, RCOY, is converted into another, RCOX, by a process called **nucleophilic acyl substitution.** (Recall that —C(O)R is an acyl group.)

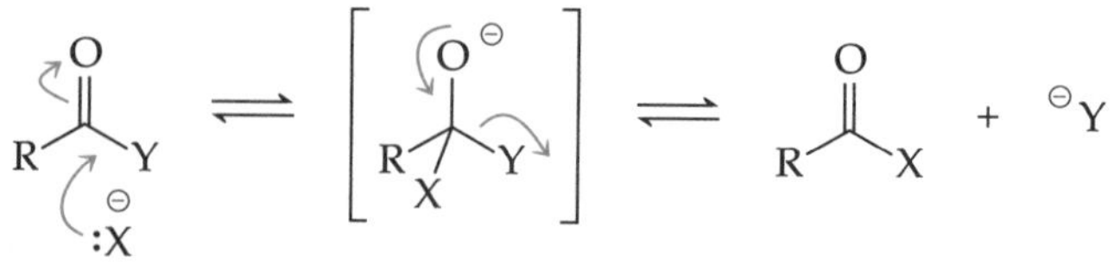

▲ FIGURE 12-25

In a nucleophilic acyl substitution, one heteroatomic group, Y, is replaced by another, X. Here, Y can be —OH in an acid, —Cl in an acid chloride, —OCOR in an acid anhydride, —OR in an ester, —SR in a thiol ester, or —NR$_2$ in an amide.

These conversions can be accomplished either by the use of a good nucleophile, as is implied by the negatively charged X$^-$ in Figure 12-25, or by the enhancement of the electrophilicity of the starting material by protonation (Figure 12-26). Each reaction step in nucleophilic acyl substitution is potentially reversible, and the position of the overall equilibrium is determined mainly by the relative stabilities of the reactant and product carboxylic acid derivatives. Other important factors are the relative stabilities of X$^-$ and Y$^-$ (Figure 12-25) and the relative strengths of the H–X and H–Y bonds formed when the reaction takes place in an acidic medium.

It is useful to review the order of stability of various carboxylic acid derivatives (Figure 12-27). Toward nucleophilic attack, the most reactive derivative is the acid chloride, whereas the least reactive is the carboxylate ion. In all cases, reactivity is determined by the degree of electron delocalization from the heteroatom (for example, Cl or N) into the carbonyl π system. For the acid chloride, this delocalization is of minor importance both because chlorine is relatively electronegative and because it is larger than carbon so that, consequently, the relevant orbitals are mismatched in size. Sulfur is about the same size as chlorine, although less electronegative; thiol esters, therefore, are much more stable than carboxylic acid chlorides. The central oxygen of a carboxylic acid anhydride must interact equally with each of the adjacent carbonyl groups; therefore, the degree of electron donation of its lone pairs is lower than that of an ester. Carboxylic acid esters have a slightly greater resonance stabilization than do neutral carboxylic acids because the alkyl group of an ester releases electron density to the ester oxygen by hyperconjugation. Amides are quite stable because nitrogen is comparable in size to carbon and is less electronegative than oxygen.

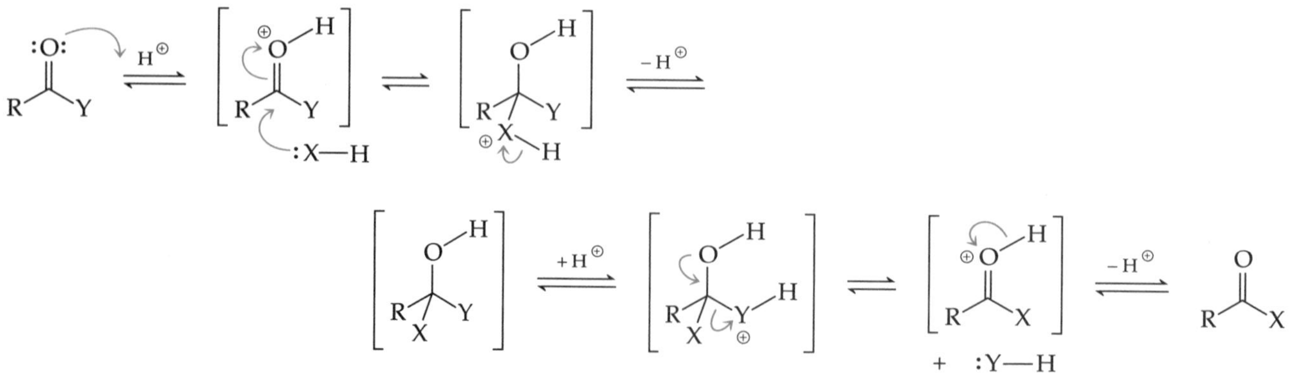

▲ FIGURE 12-26

Protonation of the carbonyl oxygen of a carboxylic acid derivative converts it into a cation that is more easily attacked by nucleophiles. (Compare this higher reactivity with that in the acid-catalyzed acetal formation shown in Figure 12-20.)

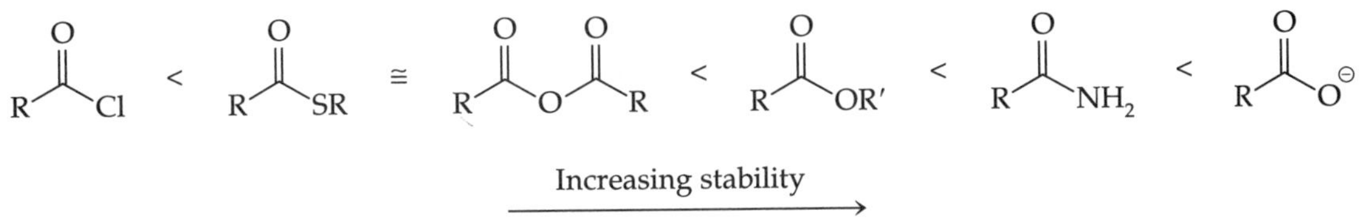

Increasing stability

▲ **FIGURE 12-27**
The more easily the attached heteroatom Y in a carboxylic acid derivative
RCOY can release electrons, the more stable is its C=O group and the less
reactive it is toward nucleophilic attack.

The carboxylate ion is the most stable of all carboxylic acid derivatives because its two resonance forms have identical energy.

It is possible to convert a more-reactive derivative into a less-reactive
one by simple nucleophilic acyl substitution under equilibrium conditions,
but the reverse conversion is not possible under the same conditions. For
example, an amide can be easily prepared from an acid chloride, but the
conversion of an amide into the corresponding acid chloride cannot be accomplished directly. Similarly, an ester can be converted into an amide, but
the reverse conversion cannot take place directly. Each of these acid derivatives, however, can be hydrolyzed under alkaline conditions, producing a
carboxylate ion. We will see shortly how a carboxylate ion can be converted
into an acid and then into a carboxylic acid chloride, allowing us to cycle
through the entire range of acid derivatives.

Hydrolysis of Carboxylic Acid Derivatives

All carboxylic acid derivatives can be hydrolyzed under acidic conditions to
carboxylic acids and under basic conditions to carboxylate ions. In most
cases, the equilibrium is driven toward the carboxylic acid (or carboxylate)
by the greater resonance stability of the product formed. For example, let us
consider the hydrolysis of a thiol ester under both basic and acidic conditions. In base, as illustrated in Figure 12-28, hydroxide ion attacks the carbonyl carbon of the thiol ester, forming a negatively charged, tetrahedral intermediate. This species can either revert to starting material by rupture of
the just-formed C–O bond or proceed to product by cleavage of the C–S
linkage. The reaction is not complete at this point, however, because the
leaving group, the thiolate anion, is a sufficiently strong base to essentially
quantitatively convert the carboxylic acid into its carboxylate.

An energy diagram for this process is shown in Figure 12-29. The carboxylic acid lies lower in energy than the thiol ester, and deprotonation

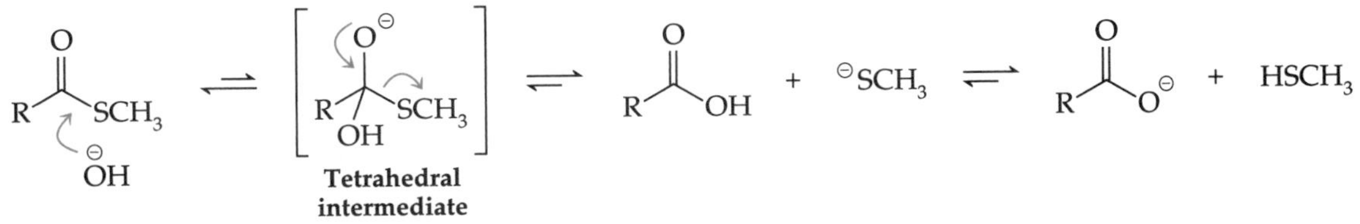

▲ **FIGURE 12-28**
Nucleophilic attack by hydroxide ion on carbon in a thiol ester leads to an
anionic tetrahedral intermediate. When the C=O bond is reformed as the
electrons on the negatively charged oxygen atom move toward carbon, the
thiomethyl anion ($^-SCH_3$) is lost. Acid-base equilibration produces the
isolated carboxylate anion and methanethiol.

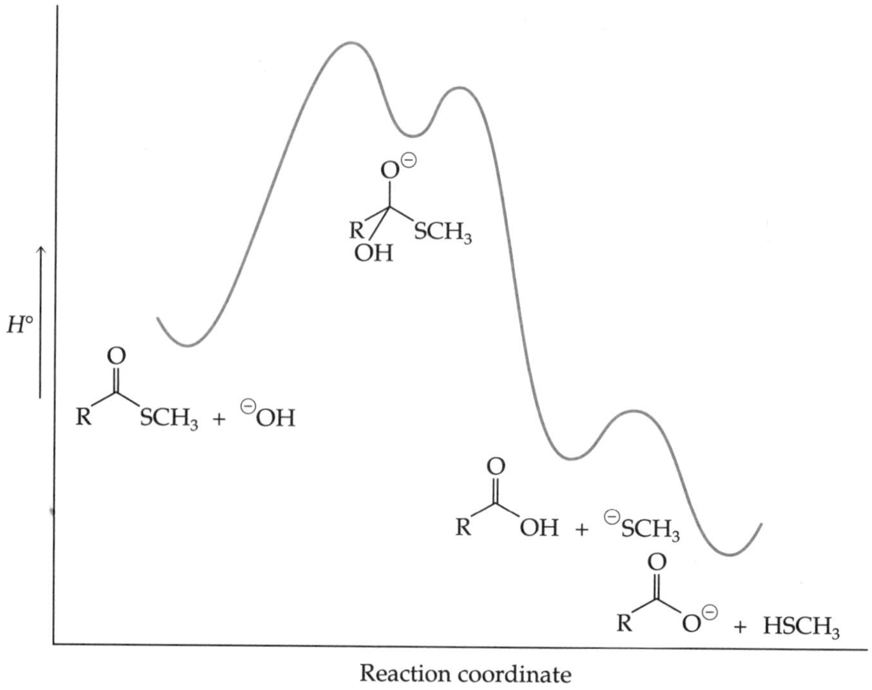

▲ FIGURE 12-29
The hydrolysis of a thiol ester is highly exothermic because of the high
stability of the carboxylate anion.

forms the carboxylate ion, which is further stabilized by resonance delocalization. Thus, under basic conditions, thiol esters are converted essentially
quantitatively into carboxylate ions. This is not a base-*catalyzed* reaction because the original nucleophilic species, the hydroxide ion, is consumed: for
each equivalent of thiol ester produced, one equivalent of hydroxide must
be used. Such reactions are referred to as **base-induced reactions.**

Under acidic conditions, the major features of the reaction are the
same, although the details differ (Figure 12-30). Here, the reaction is initiated by the transfer of a proton from the acidic medium to the carbonyl
oxygen of the thiol ester. Reaction with neutral water as a nucleophile then

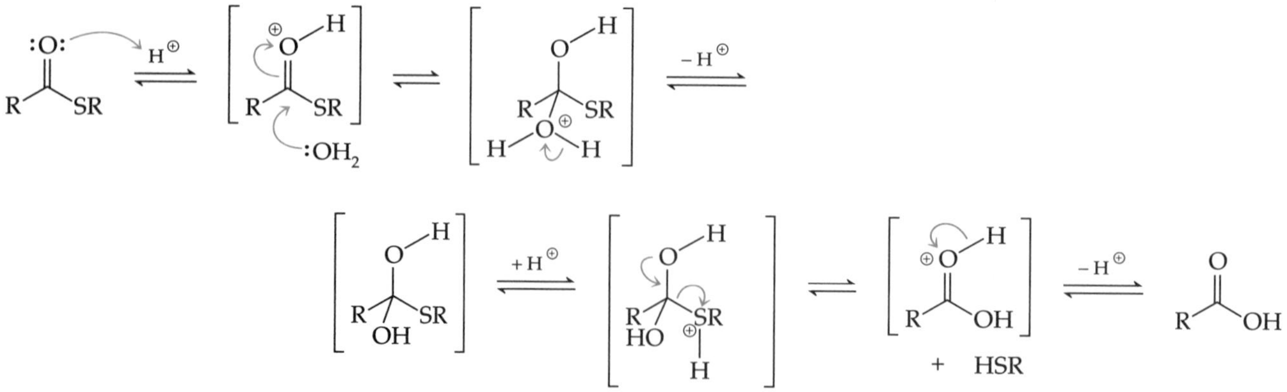

▲ FIGURE 12-30
In the acid-catalyzed hydrolysis of a thiol ester, the carbonyl group is
protonated, making attack by the weakly nucleophilic water molecule easier.
Proton exchange in the tetrahedral intermediate facilitates loss of neutral RSH,
producing a carboxylic acid after deprotonation.

leads to a positively charged tetrahedral intermediate, which rapidly loses a proton to form a neutral species. Reprotonation on sulfur provides an opportunity for reformation of the carbonyl group with expulsion of a thiol. Loss of a proton from the initially formed species provides the carboxylic acid. Although this sequence includes a number of protonation and deprotonation steps, they balance overall so that there is no net consumption of acid. Thus, the process is acid-*catalyzed*, in contrast with the base-*induced* reaction just discussed.

Interconversion of Carboxylic Acids and Esters

The thiol ester hydrolyses illustrated in Figures 12-28 and 12-30 are examples in which the product (acid) is more stable than the reactant (thiol ester), the equilibrium generally favoring the more-stable carboxylic acid. Analogous interconversions between other acid derivatives in which the reactant and product are more similar in energy (for example, between a carboxylic acid ester and the corresponding carboxylic acid) are reactions in which the position of the equilibrium can be controlled by the use of an excess of one reagent (the acid is favored in water) or by the removal of one of the products (the ester is favored when water is present at low concentrations). Under different conditions, it is possible to hydrolyze an ester or to esterify a carboxylic acid.

The acid-catalyzed hydrolysis of an ester to a carboxylic acid (for example, methyl acetate and water to acetic acid and methanol) is illustrated in Figure 12-31. Because the difference in resonance stabilization between the reactant and product is small (recall Figure 12-25), the equilibrium can be shifted by the application of LeChatelier's Principle. As discussed in Chapter 6, an equilibrium describes a state in which the rate of conversion of starting material into product exactly equals that from product back into starting material. These rates are influenced not only by the relative activation energies, but also by the concentrations of the species required for the forward and backward reactions. Thus, the equilibrium in Figure 12-31 can be shifted toward the carboxylic acid by the use of water as solvent, whereas the reverse reaction of the carboxylic acid and an alcohol to form the ester can be carried out effectively and efficiently by the use of the alcohol as the reaction solvent. The same acid-catalyzed reaction path can therefore be used for ester hydrolysis or for acid esterification.

From our general scheme (Figure 12-25) for the interconversion of carboxylic acid derivatives with anionic nucleophiles, we see that nucleophilic attack by hydroxide on an ester produces a carboxylic acid under alkaline conditions. However, the acid thus produced is rapidly converted into a carboxylate anion by reaction with a second hydroxide ion (Figure 12-32). From the order of stability of the carboxylic acid derivatives given in Figure 12-27, we see that the carboxylate ion is very stable; thus, the equilibrium greatly favors the carboxylate ion and methanol. Indeed, the energy difference between the ester and the carboxylate anion is sufficiently large, and the rate of the reverse reaction sufficiently low, that this reaction is often described as irreversible. The difference in stability between a carboxylate ion

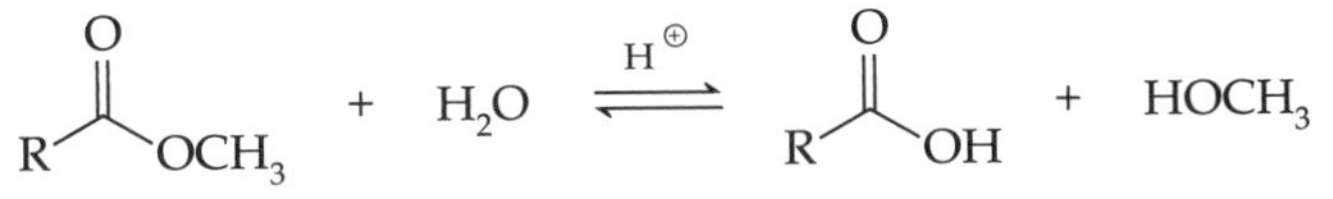

$$ R \overset{\displaystyle O}{\underset{}{\big\|}} OCH_3 \ + \ H_2O \ \underset{}{\overset{H^{\oplus}}{\rightleftharpoons}} \ R \overset{\displaystyle O}{\underset{}{\big\|}} OH \ + \ HOCH_3 $$

▲ FIGURE 12-31
Ester hydrolysis produces a carboxylic acid and an alcohol.

$$R\text{—}\overset{O}{\underset{\|}{C}}\text{—}OCH_3 + {}^{\ominus}OH \rightleftharpoons \left[R\text{—}\overset{O}{\underset{\|}{C}}\text{—}OH + {}^{\ominus}OCH_3 \right] \rightleftharpoons R\text{—}\overset{O}{\underset{\|}{C}}\text{—}O^{\ominus} + HOCH_3$$

▲ FIGURE 12-32

The equilibrium between a carboxylic acid and ester can be shifted toward the acid in aqueous base because hydroxide (or alkoxide) can deprotonate the carboxylic acid as it is produced, thus removing it from the equilibrium at the left in this figure.

and the other carboxylic acid derivatives is sufficiently large that under most standard conditions it is not possible to proceed from the carboxylate to an ester or any other carboxylic acid derivative. Basic conditions can therefore be used to hydrolyze an ester but not to esterify an acid.

EXERCISE 12-M

One method for driving an acid-catalyzed esterification reaction is through the removal of water by the use of a Dean-Stark trap. In this apparatus, the reaction mixture is heated under reflux, but the condensing vapors do not return directly to the flask. Instead, they are diverted to a side arm in which the condensed solvent is collected before it is returned to the distillation pot. When a water-immiscible solvent, such as benzene, that forms a low-boiling azeotrope with water is used, water is removed from the flask and transferred to the bottom of the side arm (recall that benzene is less dense than water). Explain how LeChatelier's Principle applies in the use of this apparatus.

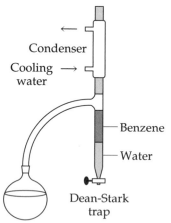

Condenser

Cooling → water

— Benzene

— Water

Dean-Stark trap

Reaction mixture being heated to reflux

WHAT AIRPORT BEAGLES KNOW ABOUT ESTERIFICATION

The structural differences between morphine, codeine, and heroin are relatively minor. Morphine has a phenolic —OH group and a secondary allylic —OH group: in codeine, the phenolic —OH is converted into —OCH₃; in heroin, both —OH groups are acetylated (—OCOCH₃). Morphine occurs naturally in the opium poppy and accounts for as much as 40% of the dried weight of sap collected from the seed pods. (The common poppy flower has none of this alkaloid.) Illicit drug laboratories therefore prepare heroin by the treatment of morphine with acetic anhydride. Because acetic acid is produced in the final hydrolysis, drug enforcement agents seek out these covert laboratories with dogs specially trained to recognize the characteristic pungent odor of acetic acid.

Morphine   **Codeine**   **Heroin**

EXERCISE 12-N

In contrast with the hydrolysis of carboxylic acid esters under acidic conditions, which is an equilibrium process, the hydrolysis of carboxylic acid esters in the presence of base can proceed in only one direction. Explain.

Transesterification

The interconversion of one carboxylic acid ester into another is called **transesterification.** For example, a methyl ester can be converted into an ethyl ester under acidic conditions (Figure 12-33). Because the two esters are of comparable energy, the principal factor affecting the equilibrium position is the relative concentrations of the corresponding alcohols. When one of the alcohols is more volatile than the other, the reaction can be driven to that side by removal of the product alcohol. In the present case, because methanol boils at a lower temperature than ethanol, methyl esters can be efficiently converted into ethyl esters by carrying out the reaction in ethanol at a temperature at which methanol boils off as it is produced.

Transesterification can also be accomplished, in principle, by base catalysis. However, acid conditions are preferred for two practical reasons. First, under alkaline conditions, esters undergo not only nucleophilic addition reactions, but also α-deprotonation and complicating side reactions. Second, any water present as a contaminant reacts with the base to form hydroxide ion, which in turn acts as a nucleophile, converting ester into carboxylic acid and then into carboxylate ion. Because this process is not reversible, some of the material is lost as a carboxylate anion by-product.

Amide Hydrolysis.

Amides can be hydrolyzed under either acidic or basic conditions; but, because of the stability of the amide functional group, basic hydrolysis is quite slow and it is frequently much easier to employ acidic conditions. (Primary and secondary amides also participate in acid-base equilibration with ⁻OH. As anions, they strongly resist nucleophilic attack.)

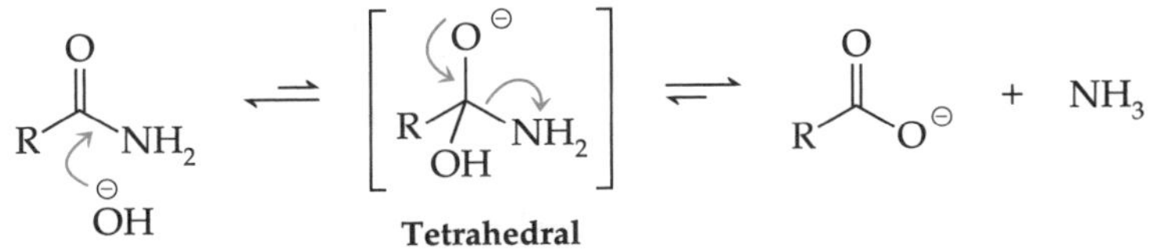

Hydrolysis of amides in aqueous acid follows a reaction pathway parallel to that for ester hydrolysis, but the reaction is slower than with esters because amines are poorer leaving groups than alcohols.

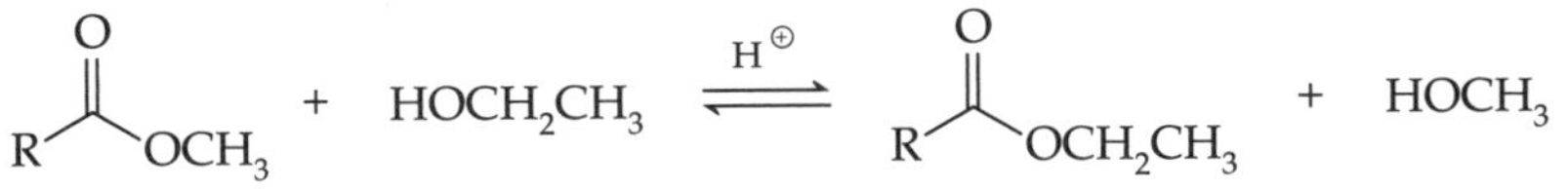

▲ **FIGURE 12-33**

In a transesterification, the —OR group of an ester is replaced by a different one, —OR'. Because of the similarity of the reactant and the product, this equilibrium is shifted toward one ester by conducting the reaction in the alcohol to be attached—that is, the ethyl ester is favored if the reaction is conducted in ethanol, whereas the methyl ester is favored in methanol.

The resistance of carboxylic acid amides to hydrolysis is a very important characteristic of the amide functional group. Proteins and peptides are large molecules made up of smaller ones joined by amide linkages, and the stability of the C(O)—NH bond is of great biochemical significance.

Substitution Reactions of Acid Chlorides

Because carboxylic acid chlorides are the most reactive (and least stable) of the carboxylic acid derivatives, they can be converted readily into any of the other derivatives. For example, carboxylic acid chlorides react with water at neutral pH to form carboxylic acids and HCl. Unlike other carboxylate derivatives, the acid chloride is sufficiently reactive that neither protonation of the carbonyl oxygen atom nor a highly nucleophilic species such as hydroxide ion is required for its hydrolysis by nucleophilic acyl substitution. Similarly, acid chlorides are easily converted into esters upon treatment with alcohol or into amides by treatment with ammonia or a primary or secondary amine.

$$\underset{R}{\overset{O}{\parallel}}{\overset{}{C}}\text{—Cl} \ + \ H_2O \ \longrightarrow \ \underset{R}{\overset{O}{\parallel}}{\overset{}{C}}\text{—OH} \ + \ HCl$$

Because of the ease with which they can be converted into other carboxylic acid derivatives, acid chlorides play a very important role in carboxylic acid chemistry. In those cases in which it is desirable to avoid strongly acidic conditions, a weak base such as pyridine can be incorporated into the reaction medium, as illustrated in Figure 12-34 for the formation of a carboxylic acid amide from an acid chloride and an amine.

Carboxylic acid chlorides are the least stable of the carboxylic acid derivatives; therefore, they cannot be formed directly from other acid derivatives by simple nucleophilic acyl substitution. Instead, a highly activated sulfurous acid derivative is often employed. One of the most-common techniques for preparing carboxylic acid chlorides is the treatment of a carboxylic acid with thionyl chloride, $SOCl_2$.

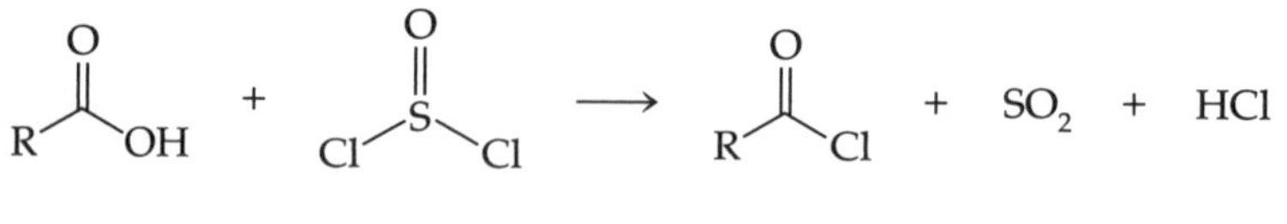

$$\underset{R}{\overset{O}{\parallel}}{\overset{}{C}}\text{—OH} \ + \ \underset{Cl}{\overset{O}{\parallel}}{\overset{}{S}}\text{—Cl} \ \longrightarrow \ \underset{R}{\overset{O}{\parallel}}{\overset{}{C}}\text{—Cl} \ + \ SO_2 \ + \ HCl$$

Let us consider the reaction of a carboxylic acid with thionyl chloride. In the first step, the carbonyl oxygen acts as a nucleophile, attacking the sulfur of thionyl chloride and forming a tetravalent sulfur intermediate (Figure 12-35). This species is quite similar to the tetrahedral intermediate in the reactions of carboxylic acid derivatives. In an analogous fashion, chloride ion is lost from this intermediate, reforming the sulfur–oxygen π bond and pro-

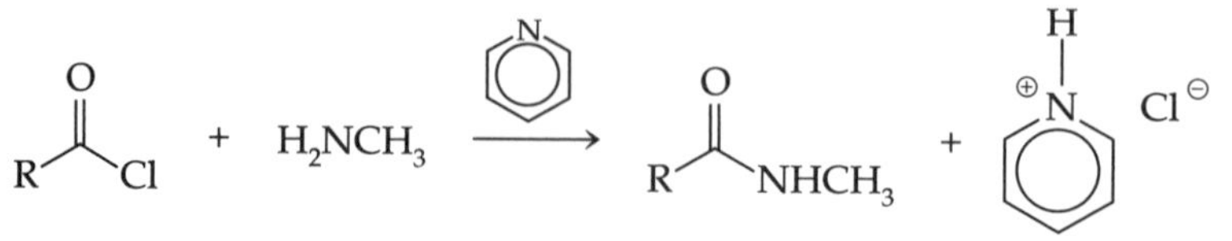

▲ **FIGURE 12-34**
The reaction of an acid chloride with a primary amine produces a secondary amide and HCl. When pyridine takes part in this amidation reaction, the HCl is converted into a pyridinium salt.

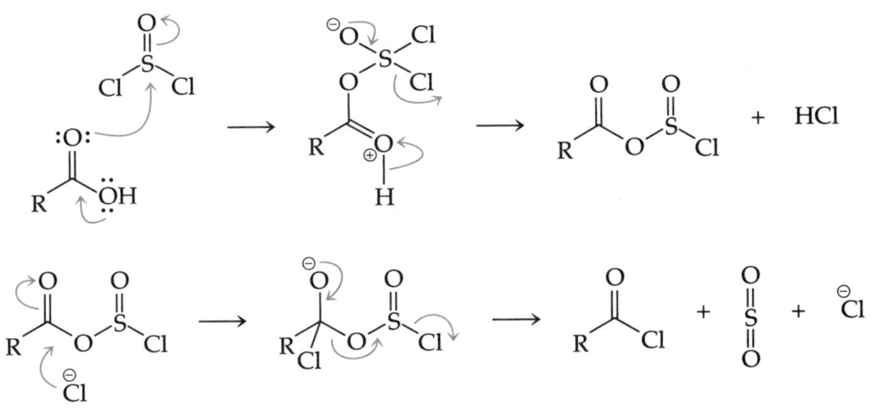

▲ FIGURE 12-35

The conversion of a carboxylic acid into an acid chloride is initiated when a lone pair on oxygen in the carboxylic acid attacks the sulfur atom of thionyl chloride. A tetrahedral sulfur atom is formed. Deprotonation of the oxonium ion site in this intermediate and back donation of an electron pair from oxygen produce chloride and a mixed acid anhydride (upper reaction). The C$=$O group of the mixed anhydride is strongly activated toward nucleophilic attack (lower reaction) and a chloride ion is sufficiently nucleophilic to form a C–Cl σ bond. Reformation of the C$=$O bond of the acid chloride is accompanied by loss of SO_2 and chloride ion.

ducing a second intermediate, an acyl chlorosulfite anhydride. This attack achieves the overall replacement of a chlorine atom in thionyl chloride by the acyloxy unit of the carboxylic acid. In the next step, the freed chloride ion acts as a nucleophile, attacking the activated carbonyl group. Unlike the attack by a halide ion on an aldehyde or ketone, which simply reverses, this halide attack forms a tetrahedral intermediate in which an even better leaving group (—SO_2Cl) is present. Collapse of this tetravalent intermediate forms the acid chloride, sulfur dioxide, and another chloride ion. Overall, this transformation takes a relatively stable carboxylic acid species (the acid itself) to the most-reactive derivative (the acid chloride). Thus, when used in conjunction with the other carboxylic acid derivative interconversions, this reaction can be used to convert any carboxylic acid derivative into another, even if the product is less stable. For example, by a sequence of reactions, a more-stable carboxylic acid amide can be converted into a less-stable thiol ester: first, the amide can be hydrolyzed to a carboxylic acid, which can be converted into the carboxylic acid chloride by treatment with thionyl chloride. Finally, the reaction of the acid chloride with an alkyl thiol produces a thiol ester.

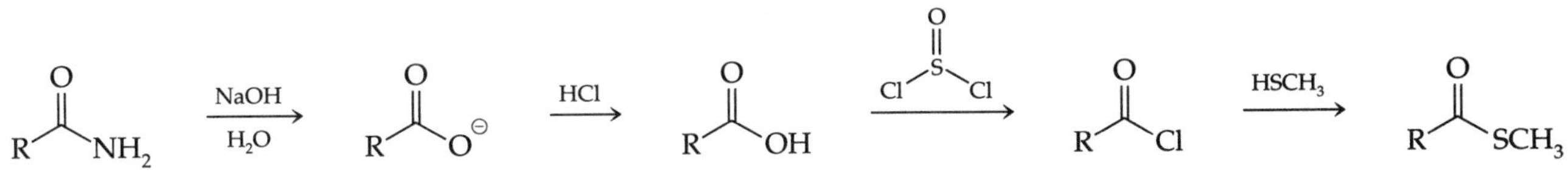

Overall, the conversion of an amide into the corresponding thiol ester is an uphill process. The driving force for this conversion comes from the quite exothermic conversion of thionyl chloride into sulfur dioxide and HCl as the acid chloride is produced.

Thionyl chloride is structurally similar to a carboxylic acid chloride and is highly reactive. This conversion is accompanied by the formation of two very stable, small molecules, SO_2 and HCl, both of which are gases at

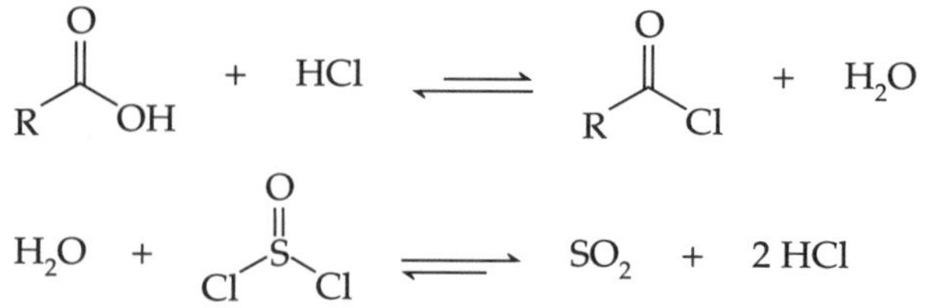

$$\underset{R}{\overset{O}{\underset{}{\parallel}}}\underset{OH}{\longleftarrow} + \ HCl \ \rightleftharpoons \ \underset{R}{\overset{O}{\underset{}{\parallel}}}\underset{Cl}{\longleftarrow} + \ H_2O$$

$$H_2O \ + \ \underset{Cl}{\overset{O}{\underset{}{\parallel}}}\underset{Cl}{\overset{S}{\longleftarrow}} \ \rightleftharpoons \ SO_2 \ + \ 2 \ HCl$$

▲ FIGURE 12-36

The direct conversion of an acid into an acid chloride (upper reaction) is unfavorable because of the higher stability of a carboxylic acid than an acid chloride, and the equilibrium lies to the left as written. The simple hydrolysis of thionyl chloride (lower reaction) is highly favored because of the high stability of HCl and SO_2 and because entropy favors the formation of three molecules from two reactants; thus, the lower equilibrium lies far to the right. When coupled, the net reaction (the conversion of a carboxylic acid and thionyl chloride into an acid chloride, SO_2, and HCl by the mechanism in Figure 12-35) is favored.

room temperature. Both entropy and enthalpy provide a strong energetic driving force for the otherwise thermodynamically unfavorable direct conversion of the carboxylic acid into its chloride. Furthermore, because SO_2 rapidly escapes from the reaction as a gas, the reverse reaction is impossible. Thus, we have one unfavorable reaction (the conversion of the carboxylic acid into the acid chloride) being driven by the simultaneous conversion of thionyl chloride into SO_2 and HCl. These two reactions are not independent and one cannot occur without the other: chloride ion is required for the first, and water is needed for the second. However, we can write two partial, or "half," reactions (Figure 12-36). In this case, the upper reaction is quite endothermic, whereas the lower reaction is exothermic. We can view these two reactions as **coupled reactions** for the formation of the acid chloride, the lower reaction providing sufficient energy to overcome the endothermicity of the upper one. Coupled reactions are often depicted with intersecting, curved reaction arrows (not to be confused with electron-flow arrows), as shown in Figure 12-37 (on page 437). This way of looking at reactions is especially attractive if both half reactions contain molecules of particular interest. Later in this book, we will use the concept of coupled reactions when we consider energetics of biochemical transformations.

EXERCISE 12-O

For each of the following pairs, choose the substrate that is more readily hydrolyzed and give reasons for your choice.

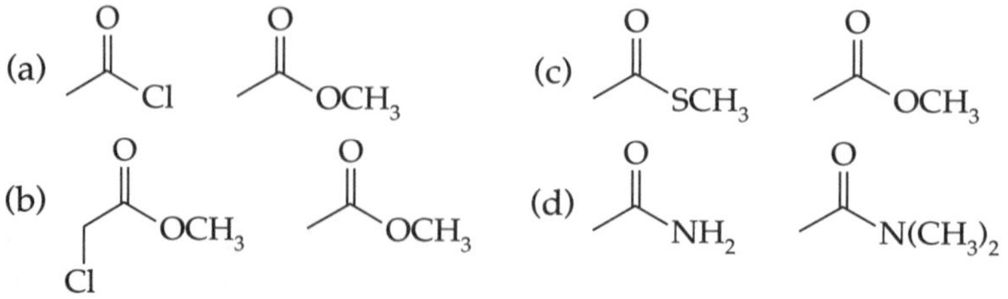

(a) $\underset{Cl}{\overset{O}{\underset{}{\parallel}}}\quad\underset{OCH_3}{\overset{O}{\underset{}{\parallel}}}$ (c) $\underset{SCH_3}{\overset{O}{\underset{}{\parallel}}}\quad\underset{OCH_3}{\overset{O}{\underset{}{\parallel}}}$

(b) $\underset{Cl}{\overset{O}{\underset{}{\parallel}}}\underset{OCH_3}{\quad}\quad\underset{OCH_3}{\overset{O}{\underset{}{\parallel}}}$ (d) $\underset{NH_2}{\overset{O}{\underset{}{\parallel}}}\quad\underset{N(CH_3)_2}{\overset{O}{\underset{}{\parallel}}}$

Reactions of Acid Anhydrides

An acid anhydride is similar in reactivity to an acid chloride and can be formally derived from the dehydration of two moles of a carboxylic acid.

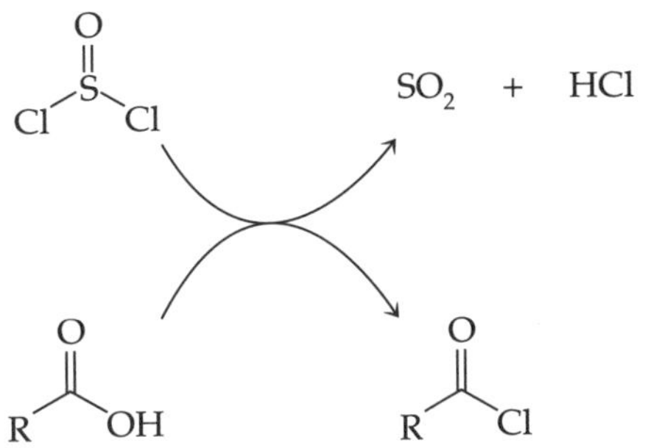

▲ FIGURE 12-37
Coupling the reactions of thionyl chloride with water and of a carboxylic acid with HCl makes the formation of a carboxylic acid chloride energetically possible.

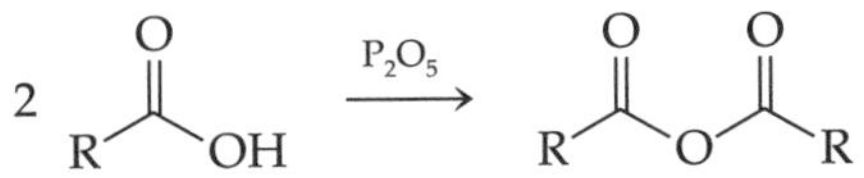

Although P_2O_5 is generally an effective desiccant and is sometimes used for this purpose, acyclic anhydrides are more difficult to prepare than the correspondingly activated acid chlorides. (For example, acetic anhydride is one of only a handful of acyclic anhydrides that are commercially available.) The difficulty in preparing anhydrides is caused by the high exothermicity of the hydrolysis of an anhydride linkage.

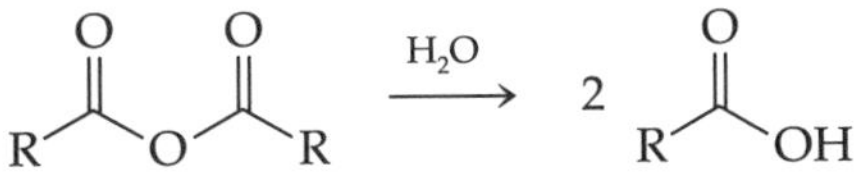

Similar hydrolyses of the structurally analogous phosphoric acid anhydrides are very important biologically and constitute a major method by which energy is stored in living organisms.

Formation of Carboxylic Acids from Nitriles

An organic nitrile contains a carbon atom at the same oxidation state as that in a carboxylic acid (count the number of bonds to heteroatoms). These compounds are formed either by the use of cyanide as a nucleophile in an S_N2 displacement from an alkyl halide (Chapter 8) or by treatment of a primary amide with a strong dehydrating reagent such as $POCl_3$.

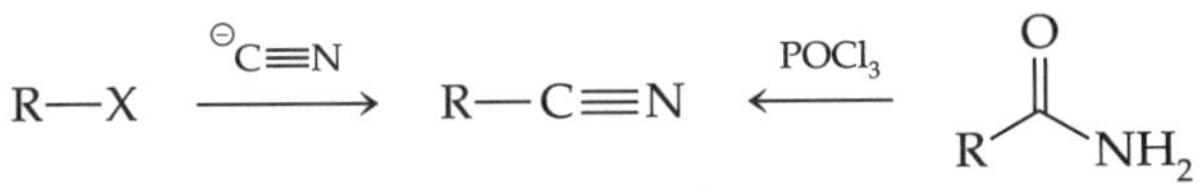

Acid-catalyzed hydrolysis of nitriles converts these species first into carboxylic acid amides (Figure 12-38) and then into the corresponding carboxylic acids. In turn, these can be converted into other acid derivatives, as described earlier. Thus, nitriles are versatile intermediates that can be converted into a range of other functional groups.

EXERCISE 12-P

Suggest a mechanism for the acid-catalyzed hydrolysis of acetonitrile ($CH_3C\equiv N$) to acetic acid and ammonia.

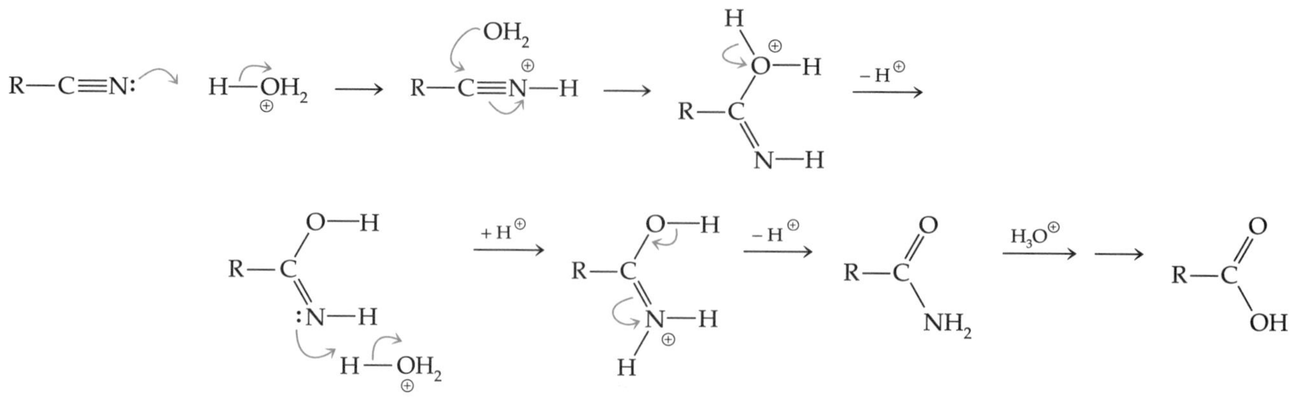

▲ FIGURE 12-38

Acid-catalyzed hydration of a nitrile produces an amide. This reaction begins by protonation of the nitrile nitrogen, thus activating the nitrile carbon to nucleophilic attack by water. Proton exchange and tautomerization lead to the amide. The amide can then be further hydrolyzed under these conditions to the carboxylic acid, as described in Section 12-8.

12-9 ▸ Sulfonic Acid Derivatives

Derivatives of sulfonic acids are similar to those of carboxylic acids except that they contain hexavalent sulfur rather than tetravalent carbon. As with carboxylic acids, there are several derivatives of sulfonic acids: among the most important are sulfonyl chlorides, sulfonic esters, and sulfonamides. Like carboxylic acids, sulfonic acids can be converted into sulfonyl chlorides by treatment with thionyl chloride. Sulfonic acid derivatives undergo nucleophilic substitution by reaction pathways that parallel those of carboxylic acid derivatives. For example, the sulfonyl group of benzenesulfonyl chloride is polarized in the same way as the carbonyl group of a carboxylic acid chloride (Figure 12-39) and is easily attacked by oxygen and nitrogen nucleophiles. Reaction of a sulfonyl chloride with an alcohol follows a sequence parallel to that for the formation of a carboxylic acid ester from an acid chloride. Thus, upon nucleophilic attack by the oxygen atom of an alcohol, a zwitterionic intermediate is produced from which chloride ion is lost. In the resulting sulfonic ester, there is appreciable polarization in the carbon–oxygen (O–R) bond as a result of strong electron withdrawal by the two S=O groups of the sulfonic ester. Furthermore, this same electron withdrawal is

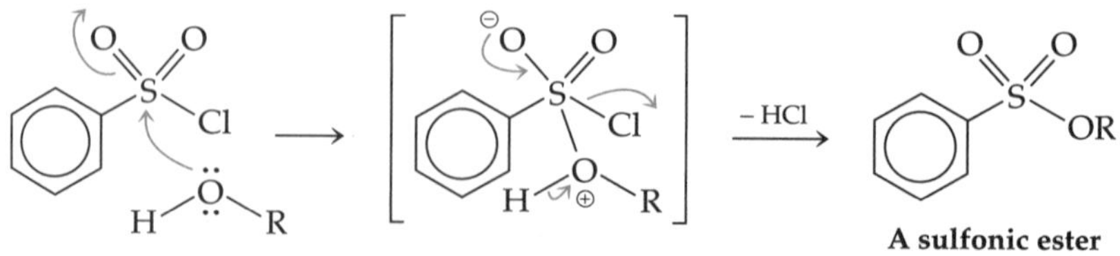

A sulfonic ester

▲ FIGURE 12-39

As a third-row element, sulfur can expand its valence shell and exist as a hexavalent atom in stable compounds. The S=O bond in a sulfonyl chloride or sulfonic acid is polarized, like a C=O bond, and can be similarly attacked by nucleophiles, forming an intermediate analogous to the tetravalent intermediates formed upon nucleophilic attack on a carbonyl group. Upon reformation of the S=O double bond and expulsion of chloride ion, a sulfonyl chloride is converted into a sulfonic ester.

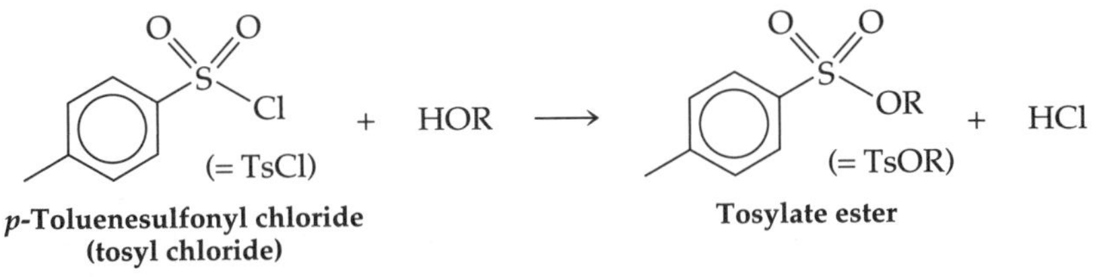

▲ FIGURE 12-40
Tosylate esters are formed when tosyl chloride reacts with an alcohol, in parallel to
the formation of an ester when a carboxylic acid chloride reacts with an alcohol.

effective in stabilizing the negative charge in the $ArSO_3^-$ ion; as a result, this
anion is a very effective leaving group. For example, the $p\text{-MeC}_6H_4SO_3^-$
group, called tosylate and often abbreviated as TsO^-, has a leaving-group
ability similar to that of a halide ion, as discussed in Chapter 8.

When p-toluenesulfonyl chloride (tosyl chloride, Figure 12-40) is
treated with an alcohol, a **tosylate ester** is formed. In this product, the R–O
bond is highly polar and exhibits reactivity similar to that of a C–Br bond.
Therefore, functionalization of alcohols for substitution by an S_N2 pathway
(bimolecular backside displacement) can be accomplished by converting
the alcohol into a tosylate ester, which is much more reactive in both S_N1
and S_N2 substitutions than is the alcohol from which it was derived. This
route is particularly useful for primary and secondary alcohols for which
acid-catalyzed dehydration (which takes place through the formation of a
carbocation) is unfavorable or accompanied by significant side reactions or
rearrangements.

EXERCISE 12-Q

Suggest a reagent (or a series of reagents) that can be used to accomplish
each of the following conversions.

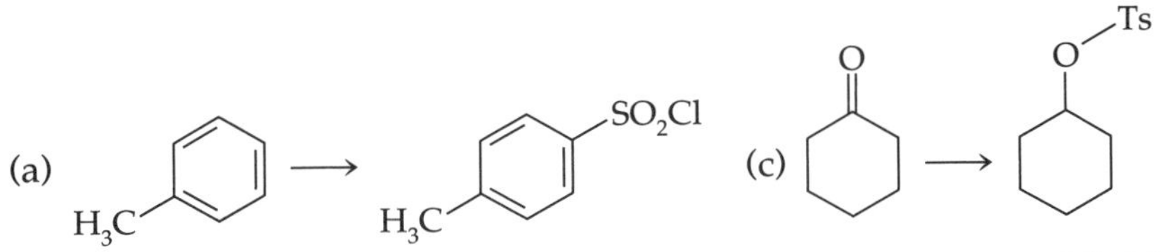

Amine nucleophiles react with sulfonyl chlorides, resulting in the substitution of chlorine by nitrogen and the formation of sulfonamides. For example, treatment of benzenesulfonyl chloride with dimethylamine produces a sulfonamide.

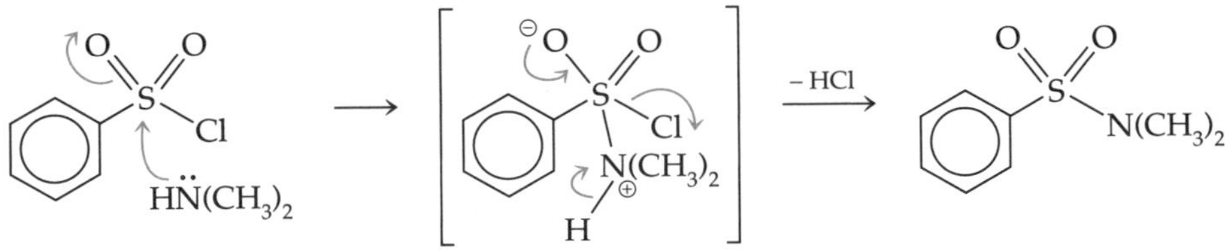

SULFONAMIDES: NOT ONLY ANTIBIOTICS

The biological activity of sulfonamides is not limited to human use: Glean (chlorsulfuron) is used as a herbicide to control broad-leaved weeds that plague cereal crops. After killing the competing weed by inhibiting an enzymatic transformation, chlorsulfuron is hydrolyzed to a soluble chlorosulfonic acid, urea (a fertilizer), and a cyanuric acid derivative (a triazole that is also a fungicide).

Glean

The mechanism of this reaction is essentially identical with that which produces an amide from an amine and a carboxylic acid chloride and a sulfonic acid ester from a sulfonic acid chloride. Sulfonamides have significant practical value because they are potent antibiotic agents. We will explore their chemistry further in Chapter 23, where we will see how they kill bacteria chemically without simultaneously doing damage to the cells of the mammalian host.

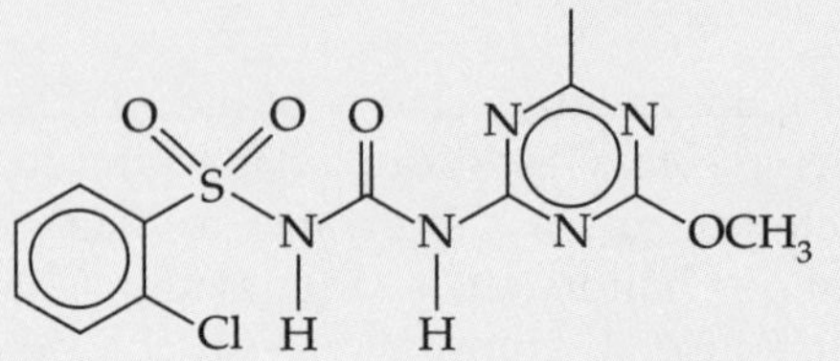

12-10 ▸ Phosphoric Acid Derivatives

Reaction of an alcohol with halogenated derivatives of phosphoric acid is an alternate method for converting an alcohol into the corresponding alkyl bromide or chloride. This reaction takes place by nucleophilic attack by the oxygen of the alcohol at the phosphorus atom of a phosphoryl halide ($POCl_3$ or $POBr_3$).

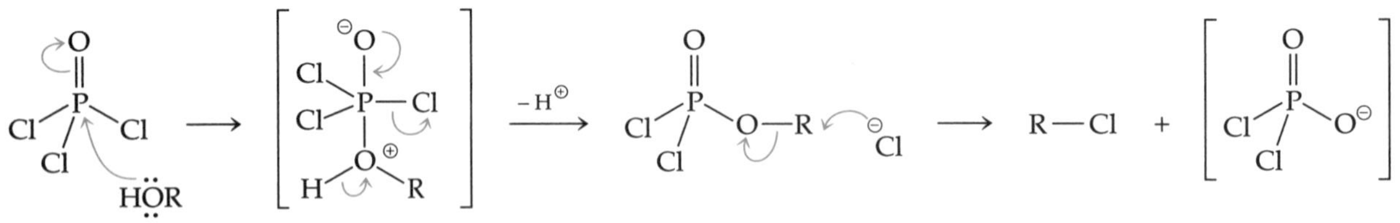

Like the S═O double bond, the P═O double bond is highly polarized, with phosphorus bearing appreciable partial positive charge, thus facilitating nucleophilic addition and consequent rapid loss of chloride ion. The oxygen–carbon bond in the resulting intermediate is even more highly polarized than in a tosylate ester, and nucleophilic substitution by chloride ion follows rapidly. For primary and secondary alcohols, the reaction follows an S_N2 pathway; for a tertiary alcohol, a multistep S_N1 reaction is required. Although the carbon–chlorine bond in the product is weaker than the carbon–oxygen bond in the starting material, this difference is more than offset by the high strength of the phosphorus–oxygen bond that is formed in the inorganic anionic by-product. Once again, an apparently energetically unfavorable conversion of one organic material into another is driven by the exothermicity of the coupled transformation.

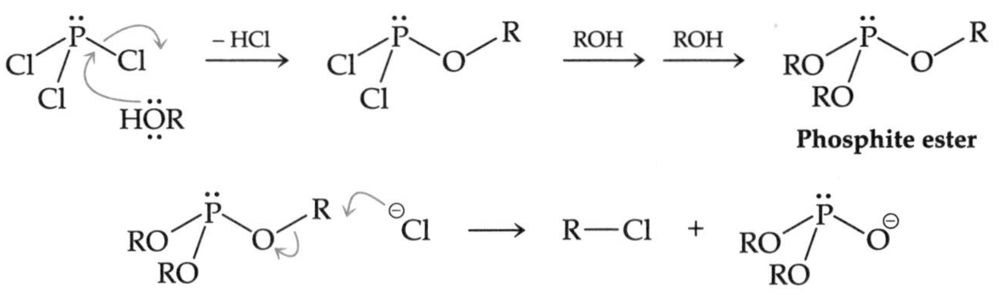

▲ FIGURE 12-41

Backside concerted S_N2 reaction by the oxygen atom of an alcohol on a P–Cl bond leads ultimately to a phosphite ester. The C–O bond in a phosphite ester is polarized so that Cl⁻ can effect an S_N2 displacement, producing an alkyl chloride.

Alternatively, an alcohol can be converted into an alkyl bromide or chloride by treatment with phosphorus trichloride (PCl_3), phosphorus pentachloride (PCl_5), or phosphorus tribromide (PBr_3). In each case (illustrated with PCl_3 in Figure 12-41), a simple S_N2 reaction of three equivalents of the alcohol with the reagent leads to the formation of phosphorus–oxygen bonds in a phosphite ester. The high strength of the phosphorus–oxygen bond provides the driving force for this step. Chloride ion then effects S_N2 displacement, yielding the alkyl chloride. (Each of the three alkyl groups of the phosphite ester undergoes this reaction.) Reaction of an alcohol with phosphorus tribromide produces the same phosphite ester, which then undergoes displacement with bromide ion to produce the alkyl bromide.

EXERCISE 12-R

An alkyl chloride can be prepared from the corresponding alcohol and thionyl chloride by a reaction that resembles that of the alcohol with $POCl_3$. Write a detailed mechanism for the conversion of ethanol into ethyl chloride by the use of $SOCl_2$.

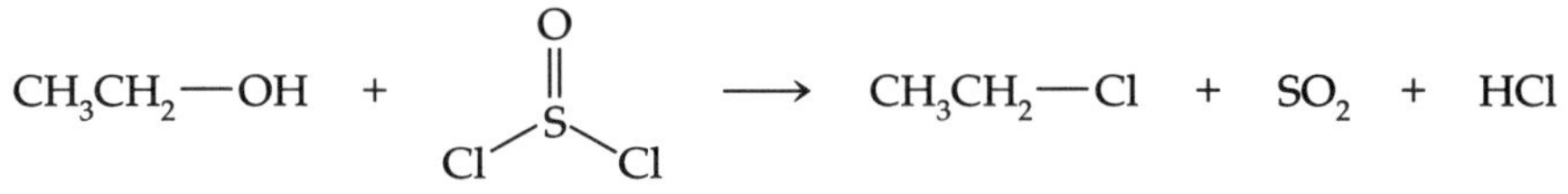

12-11 ▸ Synthetic Applications

As in preceding chapters, the new reactions considered in this chapter are summarized in Table 12-2 (on page 442) according to their usefulness for interconverting various functional groups.

Conclusions

Nucleophilic addition to an sp^2-hybridized carbon usually occurs only if a heteroatom is part of the π system; that is, at a C=O or C=N double bond. With such compounds, negative charge shifts onto the more-electronegative oxygen or nitrogen atom upon nucleophilic attack. The reactivity of various carbonyl compounds is determined by carbonyl group stability, which is affected by such factors as resonance stabilization and electron

Table 12-2 ▸ How to Use Nucleophilic Additions and Substitutions (and Related Reactions) to Make Various Functional Groups

Functional Group	Reaction
Acetal	Treatment of aldehydes with acidic alcohol
Acid chloride	Treatment of acids with thionyl chloride
Acid	Cannizzaro disproportionation of aldehydes; or amide hydrolysis; or acid chloride hydrolysis; or ester hydrolysis; or thiol ester hydrolysis; or acid anhydride hydrolysis
Alcohol	Complex metal hydride reduction of aldehydes and ketones; or lithium aluminum hydride reduction of esters; or Cannizzaro disproportionation of aldehydes
Alkane	Catalytic hydrogenation of alkynes or alkenes
Alkene	Catalytic hydrogenation of alkynes (*cis*); or dissolving metal reduction of alkynes (*trans*)
Alkyl chloride	Alcoholysis of $POCl_3$ or $SOCl_2$
Amide	Amidation of acid chlorides; or amidation of esters; or hydrolysis of nitriles
Amine	Lithium aluminum hydride reduction of amides; or reductive amination of aldehydes or ketones; or complex metal hydride reduction of imines
Ester	Alcoholysis of acid chlorides; or transesterification
Hydrate	Hydration of aldehydes or ketones
Hydrazone	Amination of aldehydes or ketones by hydrazines
Imine	Amination of aldehydes or ketones; or tautomerization of enamines
Ketal	Treatment of ketones with acidic alcohol
Nitrile	Dehydration of amides
Oxime	Amination of aldehydes or ketones by hydroxylamine
Phosphite ester	Alcoholysis of PCl_3
Semicarbazone	Amination of aldehydes or ketones by semicarbazide
Sulfonic ester	Alcoholysis of sulfonyl chlorides
Sulfonamide	Amination of sulfonyl chlorides

release from substituent groups. The order of reactivity toward nucleophiles is: aldehydes > ketones > esters > amides > carboxylate ions.

Nucleophilic attack on a carbonyl group can result either in addition or in substitution. Addition products are formed if a poor leaving group is bound to the carbonyl carbon, as in aldehydes and ketones. Substitution takes place if a good leaving group is present, as in acid chlorides, anhydrides, esters, amides, and nitriles. In both cases, nucleophilic attack at the carbonyl carbon leads to the formation of a tetrahedral intermediate. In nucleophilic addition, this tetrahedral intermediate is trapped, usually by protonation at oxygen; whereas, in nucleophilic acyl substitution, one of the σ bonds at carbon is fragmented as the electrons on the carbonyl oxygen in the tetrahedral intermediate reform the C=O double bond.

Nucleophilic addition takes place if complex metal hydrides are used as reducing agents. Primary alcohols are produced from aldehydes, and secondary alcohols from ketones. With esters or amides, nucleophilic attack by hydrides produces an intermediate carbonyl compound or imine, which is then attacked by a second hydride equivalent. The net reduction takes an ester to a primary alcohol and an amide to an amine.

There are other reducing reagents besides complex metal hydrides. For example, biological reductions are accomplished with the cofactor NADH, which delivers the equivalent of a hydride to a reducible substrate. Catalytic hydrogenation also effects a nonhydridic reduction, which allows for stereospecific *syn* (*cis*) addition across a triple bond. The opposite stereochemical result (*anti*, or *trans*, addition across an alkyne triple bond) results from a dissolving metal reduction.

Nucleophilicity, the affinity of an atom or a group of atoms for a partially positively charged carbon, often parallels basicity, the affinity for a proton. For example, oxygen nucleophiles are less reactive than are nitrogen nucleophiles, which in turn are less reactive than carbon nucleophiles, following the order of basicity. Because nucleophilicity is inversely related to leaving group ability, the opposite order is followed if the group is being displaced. Care must be taken in using this analogy, however, because basicity is generally measured as an equilibrium value, whereas nucleophilicity always refers to relative reactivity (or the rate of reaction).

The nucleophilic addition of water to a ketone or aldehyde occurs rapidly with mild acid or base catalysis to reversibly form a hydrate, whereas the addition of alcohols yields a ketal or acetal (or a hemiketal or hemiacetal). The positions of these equilibria are controlled by the concentration of water or alcohol and the inherent stability of the carbonyl compound.

Hydroxide ion attack on an aldehyde lacking an α-hydrogen results in disproportionation through the Cannizzaro reaction. In this sequence, the tetrahedral intermediate acts as a hydride delivery reagent, transferring H^- to a second molecule, thereby simultaneously forming an oxidized (carboxylic acid) and reduced (alcohol) derivative of the starting aldehyde.

Nitrogen nucleophiles also add to carbonyl compounds. Depending upon the substitution at nitrogen, either imines (from primary amines) or enamines (from secondary amines) are formed. When imines are formed in the presence of a reducing agent, amines are produced through reductive amination. Nucleophilic attack by hydrazine (or its derivatives) gives rise to hydrazone (or its derivatives). Because such derivatives are often intensely colored solids, their formation can be used as a qualitative indicator of the presence of an aldehyde or ketone functional group.

Carboxylic acid derivatives are interconverted by nucleophilic acyl substitution. When an oxygen nucleophile attacks an ester, transesterification or hydrolysis to a carboxylic acid results. Thiol esters are more reactive than their oxygen counterparts because of the mismatch in orbital size between sulfur and the carbonyl π system. Acid anhydrides and acid chlorides are very easily attacked by oxygen nucleophiles (water or alcohols), producing the corresponding acids or esters. The high reactivity of these substrates derives from σ electron withdrawal by Cl in an acid chloride and by the carboxyl substituent in an anhydride.

The nucleophilic attack by water on nitriles effects hydrolysis to carboxylic acids. Nucleophilic substitution reactions also occur with sulfonic acid esters of alcohols. Tosylates are alcohol derivatives that are reactive toward nucleophilic substitution. Sulfonamides are amides of sulfonic acid, analogous to carboxylic acid amides, and some are biologically active as antibiotics. Certain phosphorus and sulfur reagents ($POCl_3$, PI_3, PBr_3, PCl_5, and $SOCl_2$) are useful for the conversion of alcohols into alkyl halides.

Summary of New Reactions

Complex Metal Hydride Reduction of Aldehydes and Ketones

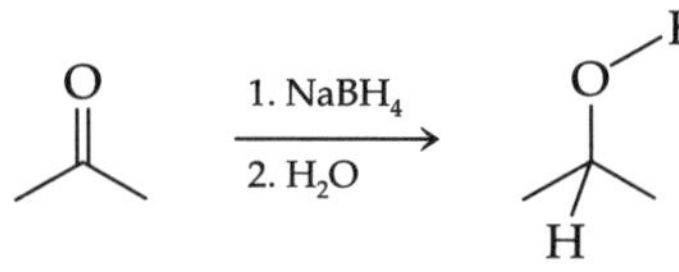

Complex Metal Hydride Reduction of Esters

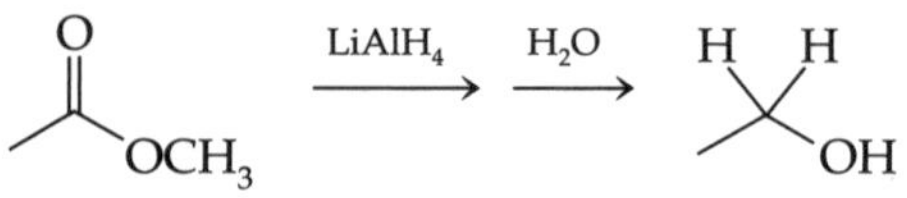

Complex Metal Hydride Reduction of Amides

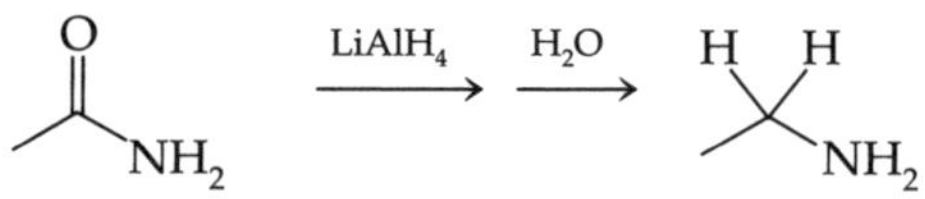

Catalytic Hydrogenation of Alkynes

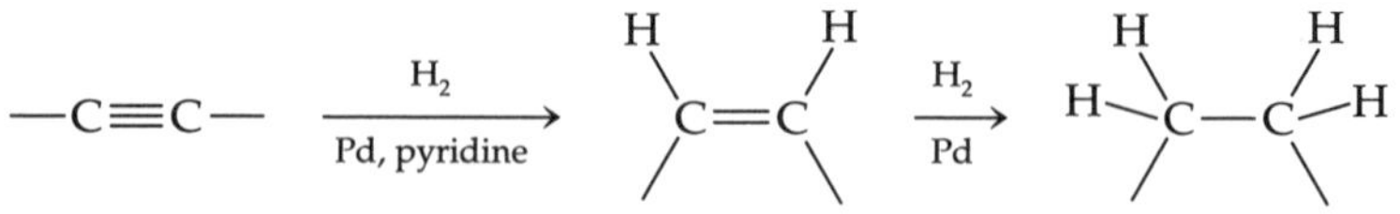

Dissolving-Metal Reduction of Alkynes

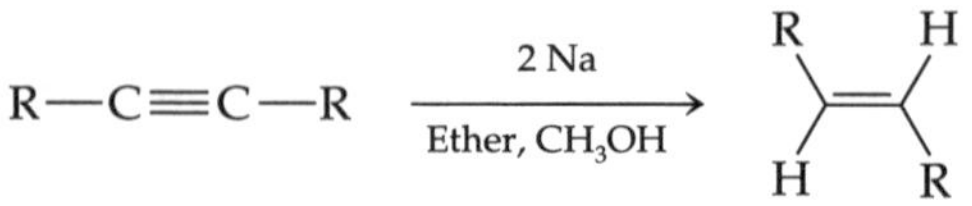

Hydrate Formation

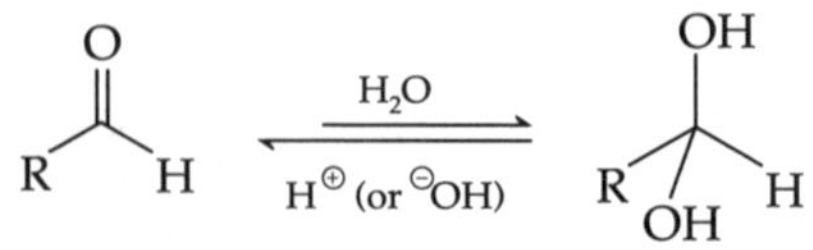

Acetal-Ketal Formation

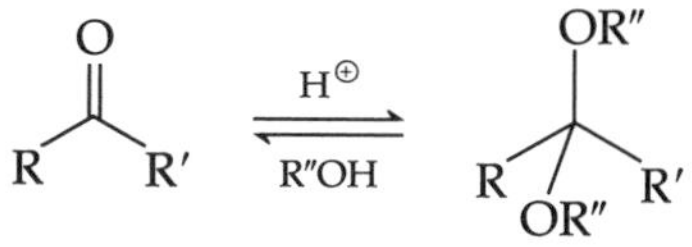

Cannizzaro Reaction

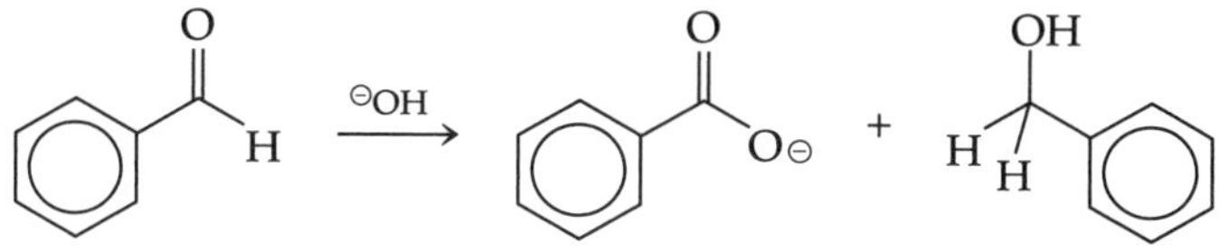

Imine (Schiff-Base) Formation

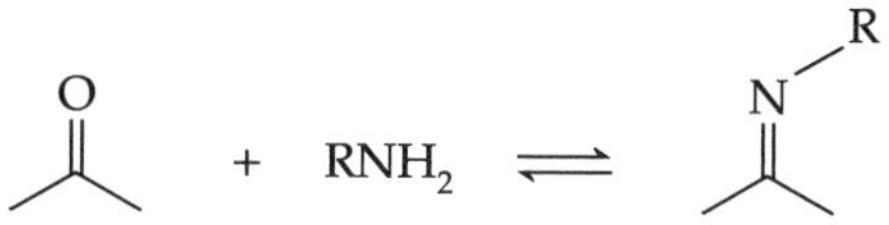

Enamine Formation

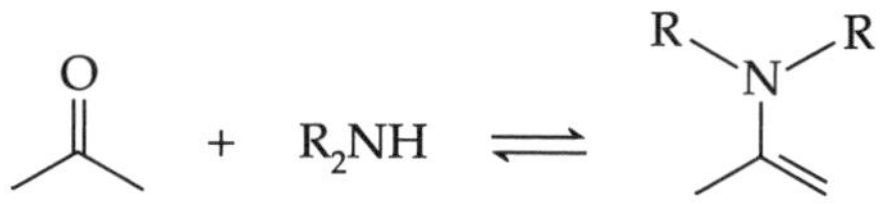

Reductive Amination

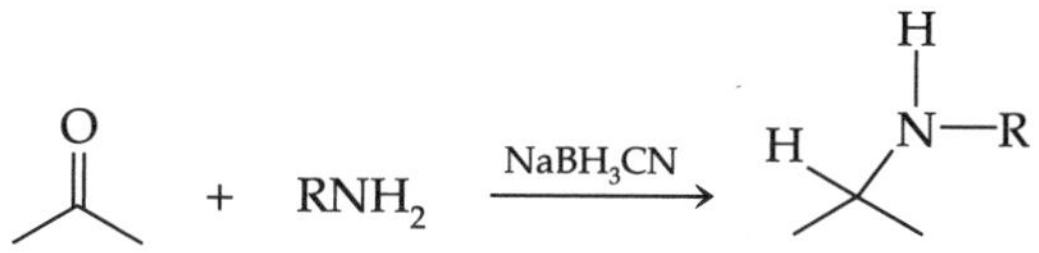

Solid Derivatives of Aldehydes and Ketones

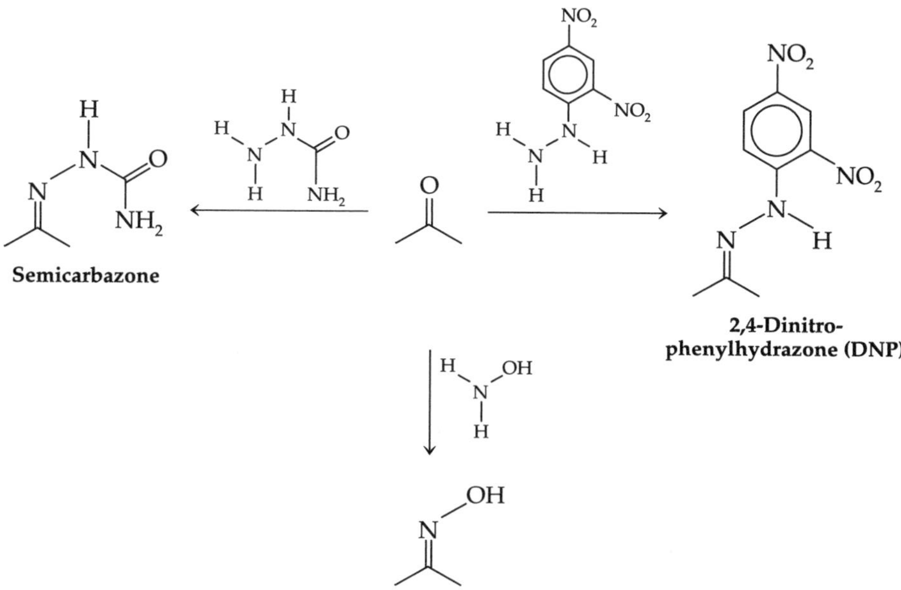

Semicarbazone

2,4-Dinitro-
phenylhydrazone (DNP)

Oxime

Interconversion of Acid Derivatives by Nucleophilic Acyl Substitution

Acid chlorides → esters, amides, acids

$$RCOCl + R'OH \longrightarrow RCO_2R'$$
$$RCOCl + RNH_2 \longrightarrow RCONHR$$
$$RCOCl + H_2O \longrightarrow RCO_2H$$

Anhydrides → esters, amides, acids

$$RCO_2COR + R'OH \longrightarrow RCO_2R'$$
$$RCO_2COR + R'NH_2 \longrightarrow RCONHR'$$
$$RCO_2COR + H_2O \longrightarrow RCO_2H$$

Transesterification

$$RCO_2R' + R''OH \longrightarrow RCO_2R'' + R'OH$$

Ester and thiol ester hydrolysis

$$RCO_2R' + H_2O \longrightarrow RCO_2H + R'OH$$
$$RCOSR' + H_2O \longrightarrow RCO_2H + R'SH$$

Amide hydrolysis

$$RCONH_2 + H_2O \longrightarrow RCO_2H + NH_3$$

Nitrile hydrolysis

$$RCN + H_2O \longrightarrow RCO_2H + NH_3$$

Amide Dehydration

$$RCONH_2 + POCl_3 \longrightarrow RC{\equiv}N$$

Preparation of Acid Chlorides

$$RCO_2H + SOCl_2 \longrightarrow RCOCl$$

Preparation of Sulfonyl Esters and Amides

$$RSO_2Cl + R'OH \longrightarrow RSO_3R'$$
$$RSO_2Cl + R'NH_2 \longrightarrow RSO_3NHR'$$

Phosphoryl Chlorides in Conversion of Alcohols into Chlorides

$$ROH + POCl_3 \longrightarrow RCl$$

Review Problems

12-1 Determine the structure of the product(s) formed, if any, when pentanal is treated with each of the following reagents.

(a) 1. $NaBH_4$; 2. H_3O^+

(b) 1. $LiAlH_4$; 2. H_3O^+

(c) H_2, Pt

(d) H_2CrO_4

(e) CrO_3, pyridine

(f) *n*-propylamine

(g) phenylhydrazine

(h) CH_3CONH_2

(i) $HOCH_2CH_2OH, H_3O^+$

(j) aqueous NaCl

(k) $SOCl_2$

(l) NH_2OH

(m) cold $KMnO_4$

12-2 Determine the structure of the product(s)

formed, if any, when methyl pentanoate is treated with each of the following reagents.

(a) 1. NaBH$_4$; 2. H$_3$O$^+$

(b) 1. LiAlH$_4$; 2. H$_3$O$^+$

(c) H$_2$, Pt

(d) H$_2$CrO$_4$

(e) *n*-propylamine

(f) aqueous NaCl

12-3 Determine the structure of the product(s) formed, if any, when pentanoamide is treated with each of the following reagents.

(a) 1. NaBH$_4$; 2. H$_3$O$^+$

(b) 1. LiAlH$_4$; 2. H$_3$O$^+$

(c) H$_2$, Pt

(d) aqueous ethanol, H$_3$O$^+$

(e) *n*-propylamine

(f) aqueous NaCl

12-4 Determine the structure of the product(s) formed, if any, when acetyl chloride is treated with each of the following reagents.

(a) H$_2$O

(b) *n*-propanol, acid

(c) (CH$_3$)$_2$NH

(d) NH$_3$

(e) C$_6$H$_6$ and AlCl$_3$

(f) CH$_3$CH$_2$SH, pyridine

(g) CH$_3$COO$^-$Na$^+$

(h) C$_6$H$_5$OH, pyridine

(i) H$_2$, Pt

12-5 Determine the reagent that would be needed to effect each of the following conversions:

(a) CH$_3$CH$_2$COCl into CH$_3$CH$_2$COOH

(b) CH$_3$CH$_2$COOH into CH$_3$CH$_2$COCl

(c) CH$_3$CH$_2$COOCH$_3$ into CH$_3$CH$_2$CONH$_2$

(d) CH$_3$CH$_2$COOCH$_3$ into CH$_3$CH$_2$COOCH$_2$CH$_3$

(e) CH$_3$CH$_2$COOH into CH$_3$CH$_2$CONH$_2$

(f) CH$_3$CH$_2$CN into CH$_3$CH$_2$COOH

12-6 Using curved arrows to indicate electron flow, propose a detailed reaction mechanism for each of the following conversions.

(a)

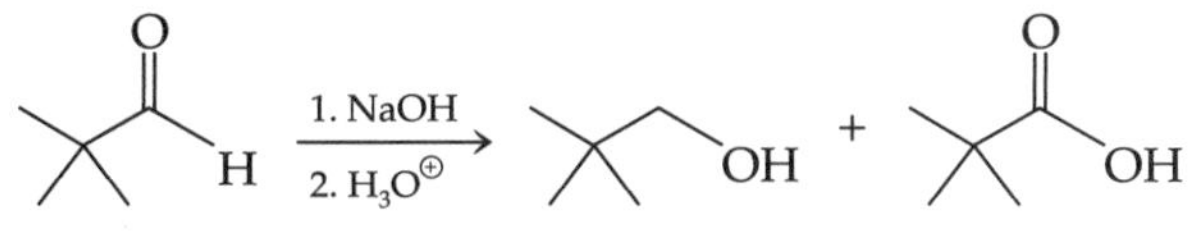

(b)

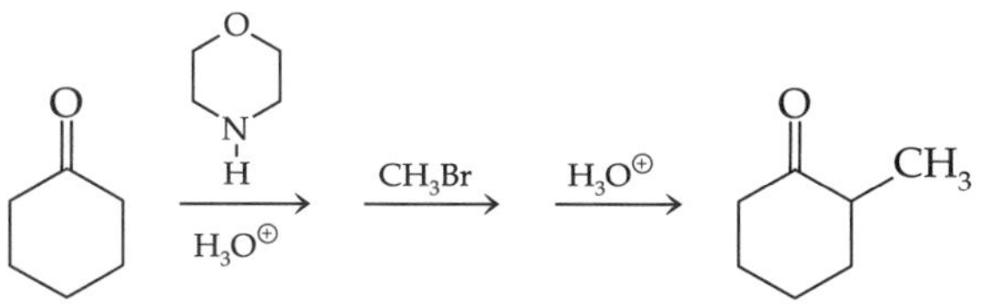

(c)

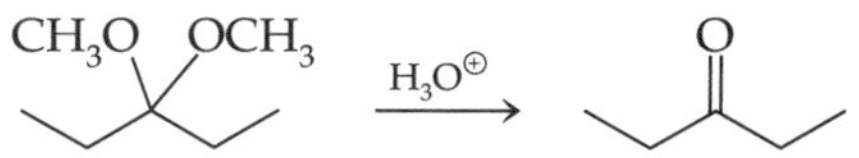

(d)

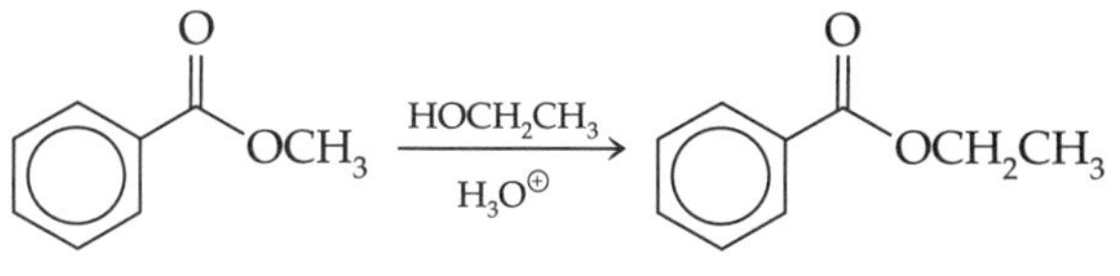

(e)

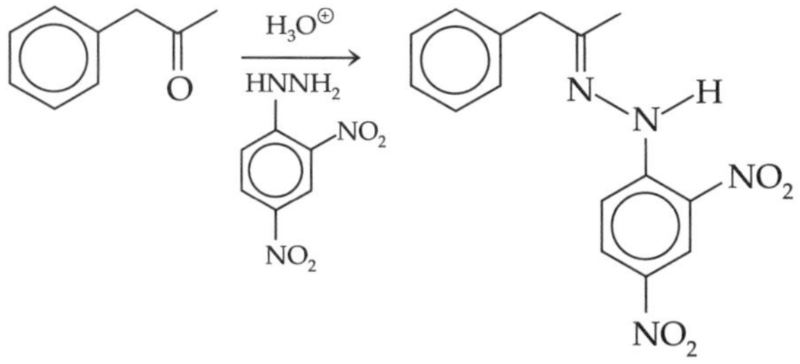

(f)

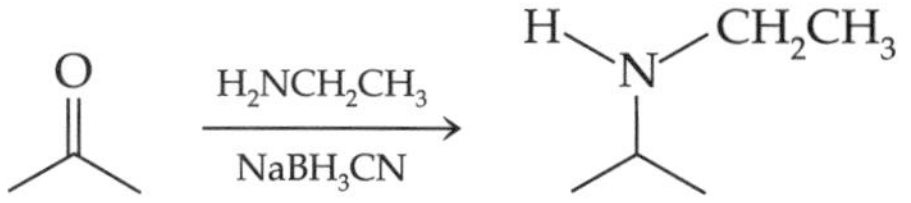

(g)

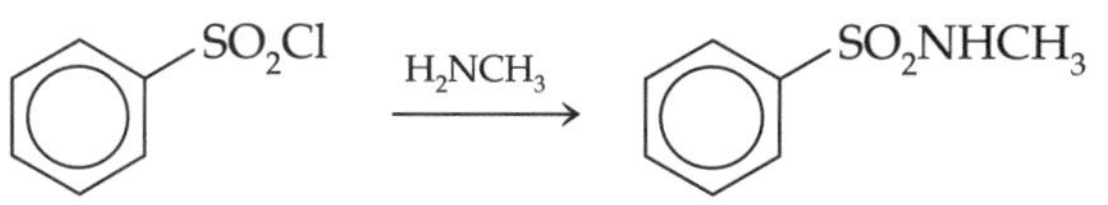

12-7 Predict the product expected from each of the following reactions (continued on page 448):

(a)

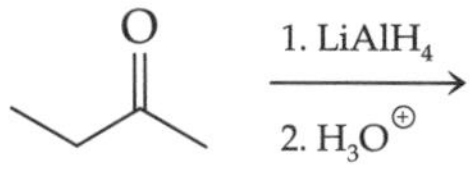

(b)

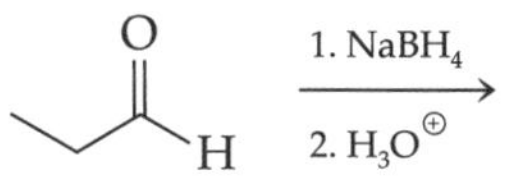

(c)

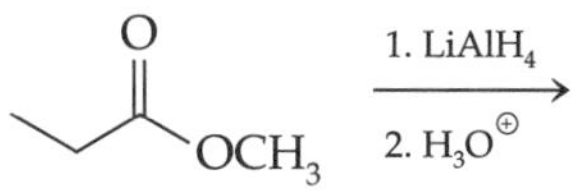

(d)

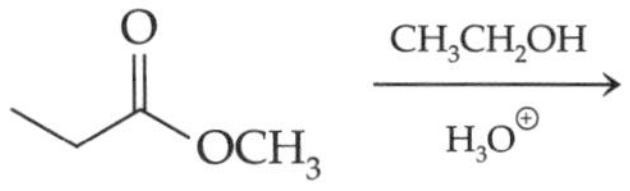

(e)

(f) 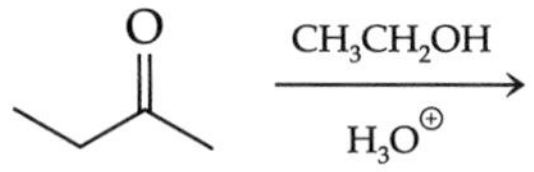 $\xrightarrow[\text{H}_3\text{O}^{\oplus}]{\text{CH}_3\text{CH}_2\text{OH}}$

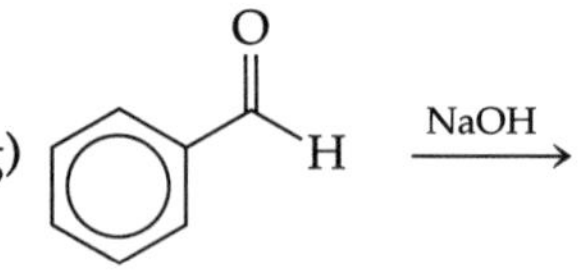

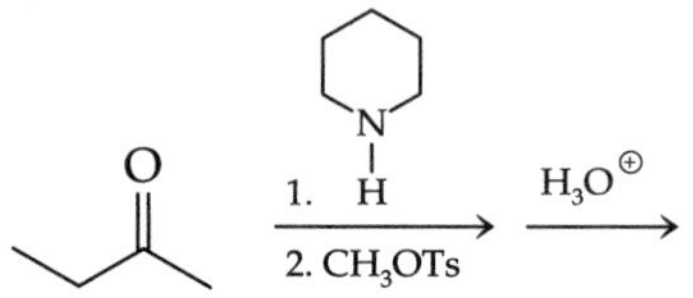

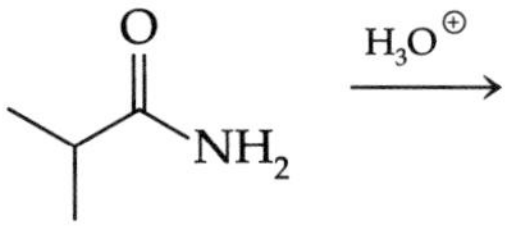

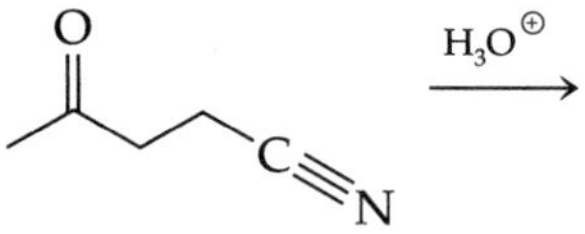

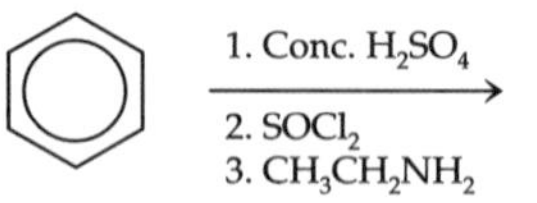

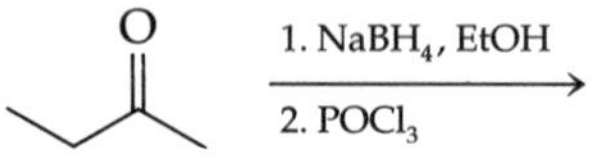

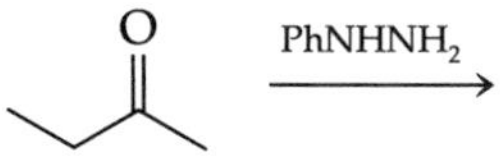

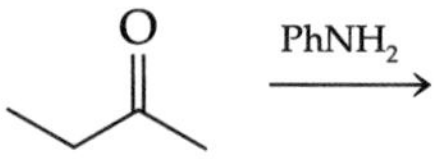

(g) $\xrightarrow{\text{NaOH}}$

(h) $\xrightarrow[\text{2. CH}_3\text{OTs}]{\text{1. } \quad}$ $\xrightarrow{\text{H}_3\text{O}^{\oplus}}$

(i) $\xrightarrow{\text{H}_3\text{O}^{\oplus}}$

(j) $\xrightarrow{\text{H}_3\text{O}^{\oplus}}$

(k) $\xrightarrow[\substack{\text{2. SOCl}_2 \\ \text{3. CH}_3\text{CH}_2\text{NH}_2}]{\text{1. Conc. H}_2\text{SO}_4}$

(l) $\xrightarrow[\text{2. POCl}_3]{\text{1. NaBH}_4\text{, EtOH}}$

(m) $\xrightarrow{\text{PhNHNH}_2}$

(n) $\xrightarrow{\text{PhNH}_2}$

(o) $\xrightarrow[\text{2. CH}_3\text{CH}_2\text{NH}_2]{\text{1. SOCl}_2}$

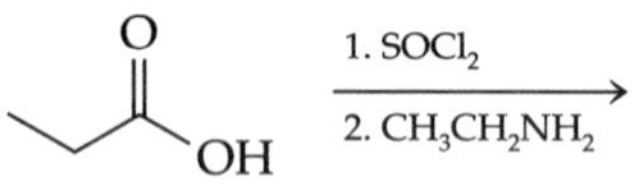

(p) 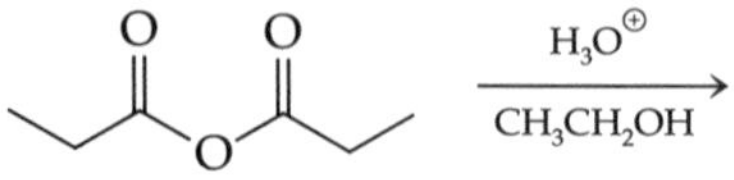 $\xrightarrow[\text{CH}_3\text{CH}_2\text{OH}]{\text{H}_3\text{O}^{\oplus}}$

(q) 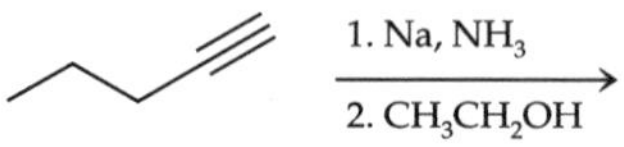 $\xrightarrow[\text{2. CH}_3\text{CH}_2\text{OH}]{\text{1. Na, NH}_3}$

(r) 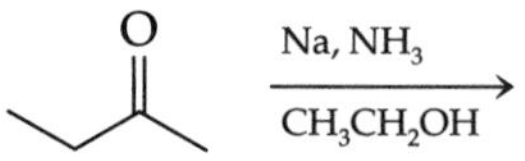 $\xrightarrow[\text{CH}_3\text{CH}_2\text{OH}]{\text{Na, NH}_3}$

(s) 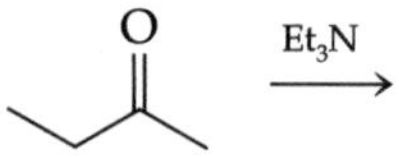 $\xrightarrow{\text{Et}_3\text{N}}$

(t) 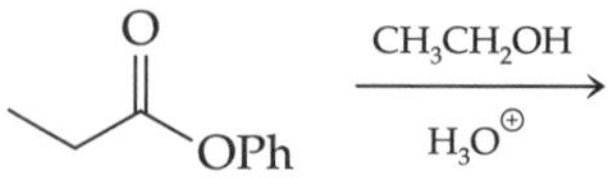 $\xrightarrow[\text{H}_3\text{O}^{\oplus}]{\text{CH}_3\text{CH}_2\text{OH}}$

(u) 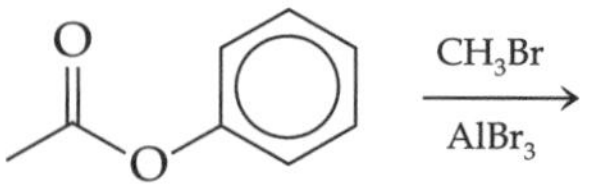 $\xrightarrow[\text{AlBr}_3]{\text{CH}_3\text{Br}}$

12-8 The following sequence has been successfully as a route for the conversion of aldehydes into ketones.

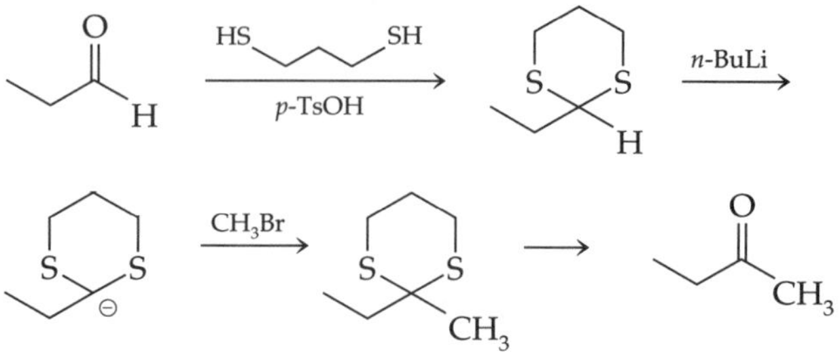

(a) Write a reaction mechanism for the formation of the thioacetal in the first step. Is this reaction easier or harder than the formation of the corresponding acetal with ethylene glycol ($\text{HOCH}_2\text{CH}_2\text{OH}$)?

(b) A key step in this sequence is the deprotonation in the second step. Explain why this deprotonation occurs with the thioacetal shown here but fails to occur with a normal acetal.

(c) Given your knowledge of acetal chemistry, propose a reagent that could be used to induce the last step of this sequence.

12-9 Propose a chemical test and a spectroscopic method that could be used to distinguish the compounds in each of the following pairs:

(a) 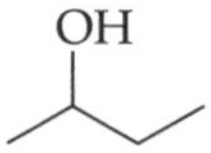 and

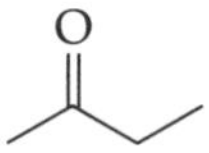

(b) 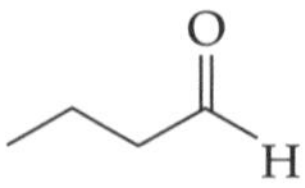 and

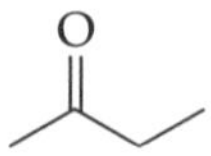

(c) 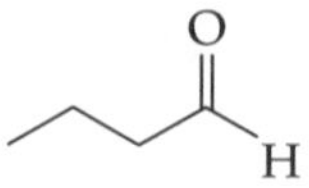 and

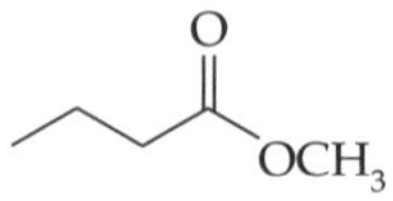

(d) 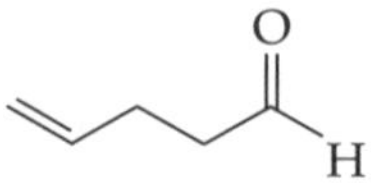 and

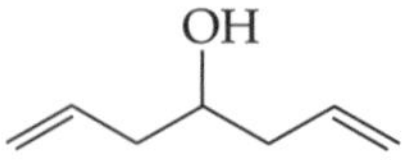

(e) 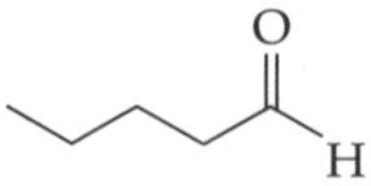 and

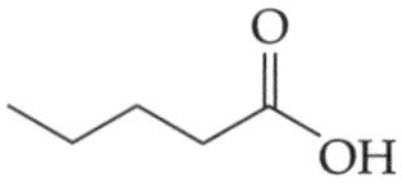

(f) 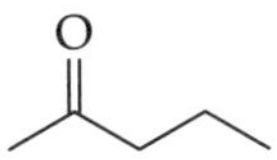 and

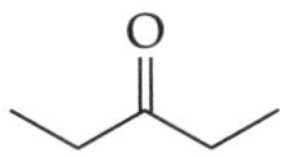

(g) 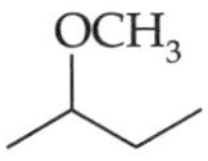 OCH₃ and OH

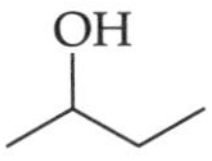

(h) 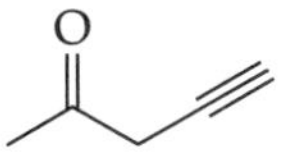 and

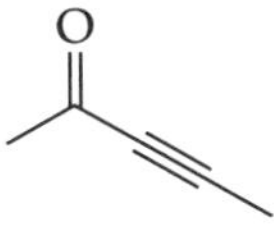

(i) 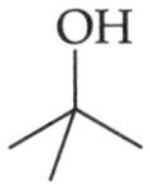 OH and

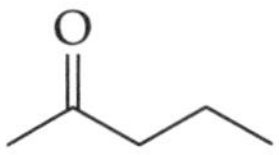

(j) 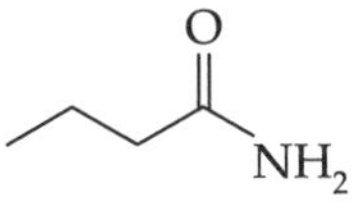 and

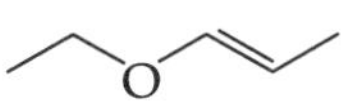

(k) 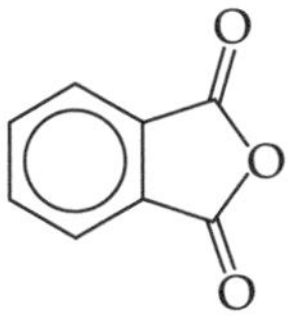 and

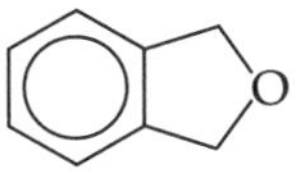

12-10 In many biological redox reactions, nicotinamide is an important oxidant because its alkylated form is very easily reduced; that is, it easily accepts electrons to generate an anion that is then protonated. Consider the following reaction a laboratory model for the biological reduction of an alkylated nicotinamide.

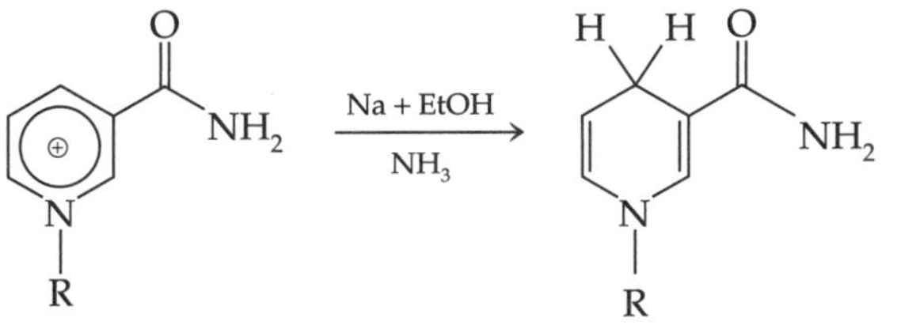

Is the reduction of this nicotinamide easier or harder than the reduction of the related alkylated pyridinium salt?

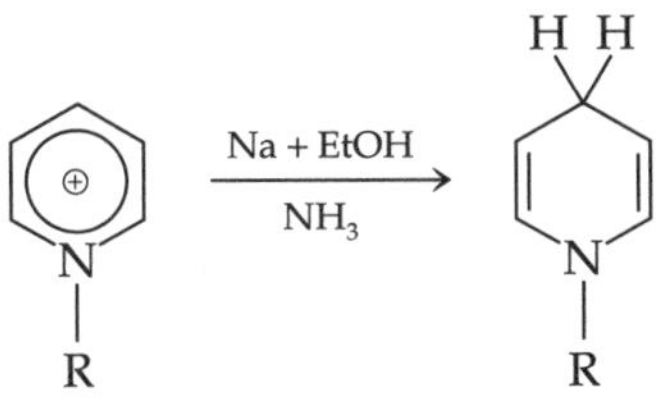

(Hint: Write resonance structures for the products obtained from each of these cations.)

12-11 Propose an efficient synthetic route that could convert each of the following reactants into the indicated products.

(a)

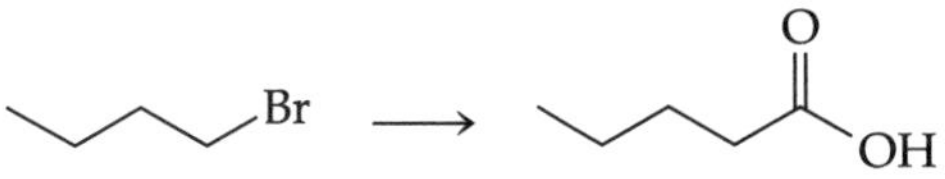

(b)

(c)

(d)

(e)

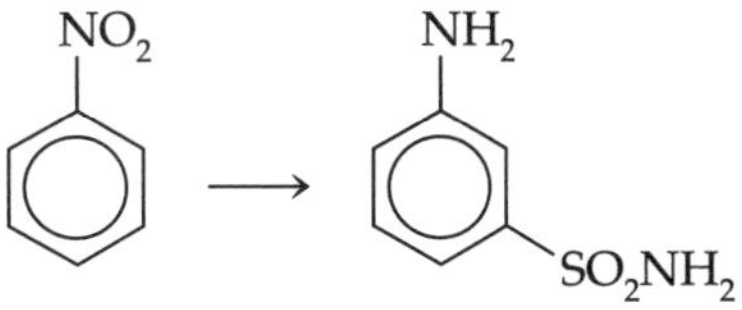

(f)

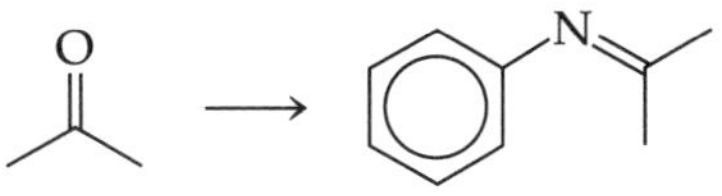

(g)

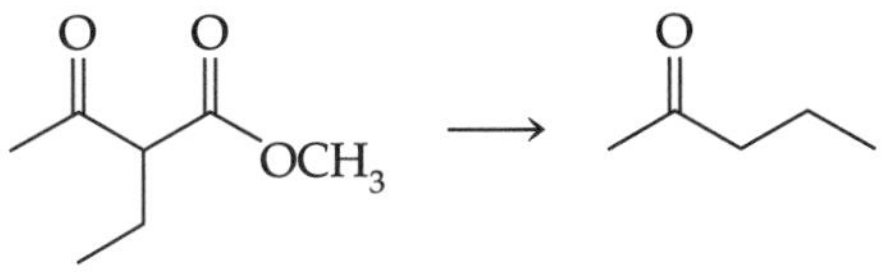

(h)

(i)

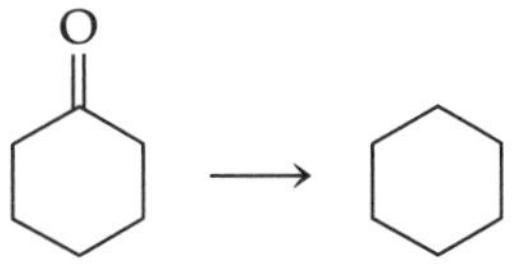

(j)

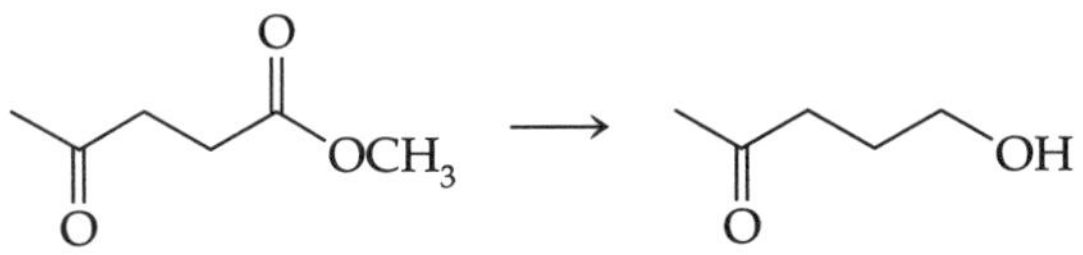

12-12 When an imine (Schiff base) is formed by the reaction of acetone with aniline, the rate of reaction is found to depend on pH. The reaction rate is faster at both low pH and high pH ranges than at intermediate ranges. Suggest an explanation.

Chapter 13

Addition and Substitution by Carbon Nucleophiles at sp^2-Hybridized Carbon

A principal goal of organic chemistry is to make new molecules that have interesting properties or that duplicate the structural features of naturally occurring molecules having significant biological activity. To understand how such molecules can be constructed both in the laboratory and in nature, we must know how to manipulate functional groups and how to build molecules of greater structural complexity from simple precursors.

We have seen, in Chapter 8, how substitution by carbon nucleophiles at sp^3-hybridized centers can lead to new carbon skeletons and new functionality. In this chapter, we will learn how these same carbon nucleophiles are used in addition and substitution reactions at sp^2-hybridized carbon atoms for carbon–carbon bond construction, combining two organic reagents to form a more-complex product. We will see how carbonyl compounds are used both as precursors of the nucleophile and as the electrophile in these reactions. The mechanisms of carbon–carbon bond formation by nucleophilic addition and substitution reactions of carbon nucleophiles are exactly parallel to those presented in Chapter 12 for nitrogen, oxygen, and halogen nucleophiles. Like the heteroatomic nucleophiles discussed in Chapter 12, carbon nucleophiles form new σ bonds by attacking a carbonyl carbon, converting a planar carbonyl group into a tetrahedral intermediate.

Reactions that form new carbon–carbon bonds constitute the tools needed for increasing the complexity of molecules by extending their carbon skeletons. In this chapter, we will learn the details of the aldol and Claisen condensations that account for their high efficiency and wide utility. As we shall see in later chapters, these reactions are frequently encountered in natural pathways for the construction of biologically important molecules **(biosynthesis)**.

13-1 ▸ Reaction of Carbon Nucleophiles with Carbonyl Groups

A number of carbon nucleophiles whose syntheses were described in Chapter 8 add to the carbonyl groups of aldehydes and ketones. In this section, we consider a number of nucleophilic addition reactions: of cyanide ion to make cyanohydrins, of Grignard reagents to make a variety of oxygen-containing functional groups, of organolithiums and organocuprates to effect 1,2 and 1,4 additions to enones, of resonance-stabilized carbanions to produce new carbon–carbon bonds at the α,β position of an enone in a Michael reaction, and of phosphonium ylides to prepare alkenes in the Wittig reaction.

Cyanide

One of the simplest carbon nucleophiles is cyanide ion (⁻CN). The addition of ⁻CN ion to a carbonyl group results in the formation of a new carbon–carbon bond in the product α-cyanoalcohol, called a **cyanohydrin.** Cyanide is both a good leaving group and a good nucleophile; thus, cyanohydrin formation is usually reversible, as shown here for the reaction with acetone.

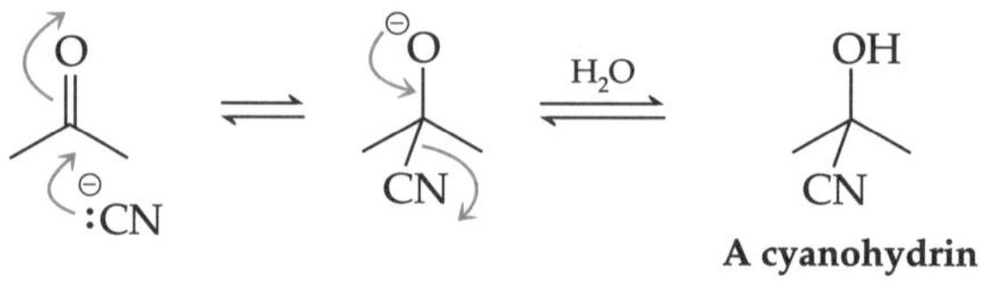

A cyanohydrin

TOXICITY OF A NATURALLY OCCURRING CYANOHYDRIN DERIVATIVE

Amygdalin (also called laetrile) is a naturally occurring cyanohydrin that can be isolated from bitter almond seeds and peach and apricot pits. For some years, it was touted as an anticancer drug and, although not approved for use in the United States, was administered to many cancer patients who went abroad for chemotherapy. Unfortunately, it has been shown to be highly toxic and ineffective as a cancer treatment.

Its high toxicity derives from its nonselective release of HCN under physiological conditions. In the laboratory, cyanohydrin formation is reversed upon treatment of amygdalin with acid, producing HCN, benzaldehyde, and two equivalents of glucose.

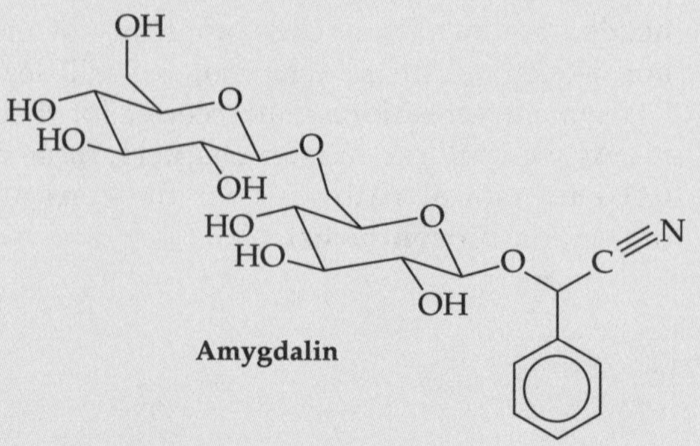

Amygdalin

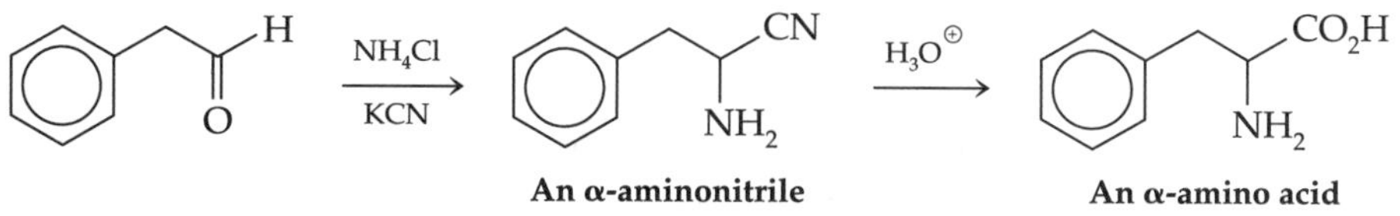

▲ FIGURE 13-1

In the first step of a Strecker synthesis, ammonia (present in equilibrium with
NH_4Cl) converts an aldehyde into an imine. Nucleophilic attack by cyanide
on the imine produces an α-aminonitrile, hydrolysis of which gives an α-
amino acid.

In a variation of this reaction known as the **Strecker synthesis,** an aldehyde is treated with ammonium chloride in the presence of potassium
cyanide to form an α-aminonitrile (Figure 13-1). Hydrolysis of the nitrile
leads to an α-amino acid. Here, phenylacetaldehyde is converted into
phenylalanine. This reaction begins in the same way as the aminations presented in Chapter 12; that is, by the formation of an imine from ammonia
(present in equilibrium with its conjugate acid, the ammonium ion) and the
aldehyde. Cyanide ion then adds to this imine, forming the α-aminonitrile.
Hydrolysis of the nitrile to a carboxylic acid provides a racemic amino acid.
These important compounds are the "repeat" units in proteins and peptides.

EXERCISE 13-A

Using your knowledge of the mechanisms of nucleophilic additions and
substitutions from Chapter 12, provide a rational mechanism for the reactions shown in Figure 13-1.

Grignard Reagents

The electron density in the carbon–magnesium bond of a Grignard reagent
(R–MgX) is highly polarized toward carbon because of the large difference
in electronegativity between these elements. As a result, the carbon is quite
nucleophilic and adds readily to aldehydes and ketones (Figure 13-2). After
nucleophilic addition is complete, neutralization with dilute acid results in
protonation of the newly formed alkoxide salt, producing an alcohol. The
addition of Grignard reagents to carbonyl groups is quite general: Grignard

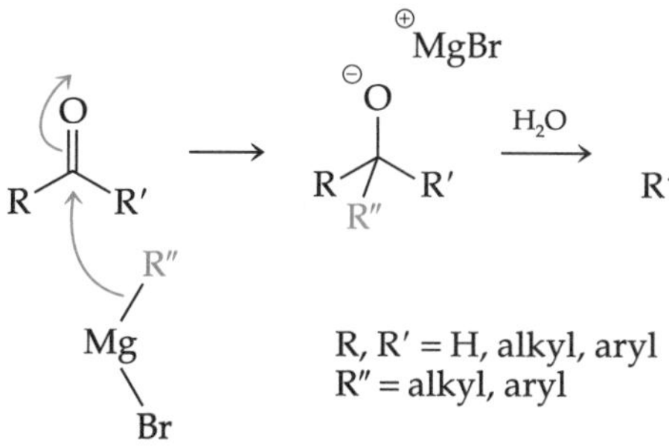

R, R′ = H, alkyl, aryl
R″ = alkyl, aryl

▲ FIGURE 13-2

The C–Mg bond of a Grignard reagent is highly polar, with appreciable
electron density on carbon. The Grignard reagent therefore acts as a
nucleophile, forming a σ bond between its carbon and a carbonyl carbon
while electrons from the C=O π bond are shifted onto oxygen. The resulting
magnesium alkoxide salt is converted into an alcohol by treatment with water.

reagents can attack formaldehyde, aldehydes, and ketones, providing synthetically useful routes, respectively, for primary, secondary, and tertiary alcohols.

Nucleophilic attack by a Grignard reagent on an ester forms a tetrahedral intermediate that is readily transformed into a ketone by loss of ⁻OR. Because ketones are more reactive toward nucleophilic attack than are esters, the initially formed ketone then reacts rapidly with a second equivalent of Grignard reagent. Protonation of the alkoxide salt with dilute acid affords a tertiary alcohol (Figure 13-3). Note that the tertiary alcohol formed by the reaction of a Grignard reagent with an ester must have two identical groups attached to the carbinol carbon, whereas all three alkyl groups can be different when a tertiary alcohol is produced from a ketone.

Grignard reagents also react through the same mechanism with the carbon atom in carbon dioxide, producing a resonance-stabilized carboxylate anion that, after acidification, gives a carboxylic acid.

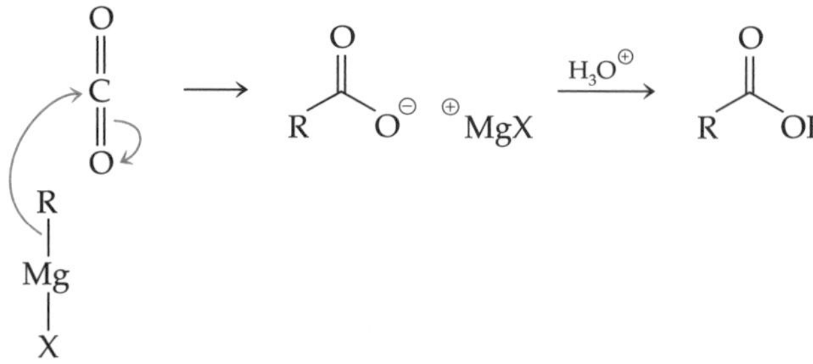

Recall from Chapter 8 that Grignard reagents also react with ethylene oxide to form a new carbon–carbon σ bond in the product, a primary alcohol. In this reaction, two carbons are added to those present in the Grignard reagent (Figure 13-4). In contrast, the reaction of a Grignard reagent with formaldehyde produces a primary alcohol that has just one more carbon than did the alkyl halide from which the Grignard reagent was prepared.

The synthetic utility of Grignard reagents is summarized in Table 13-1, which tells us that the reaction of Grignard reagents with the appropriate electrophile leads to primary, secondary, and tertiary alcohols, as well as to carboxylic acids and alkanes. A reaction that leads to a product that has been assembled by C–C bond formation and that retains an alcohol func-

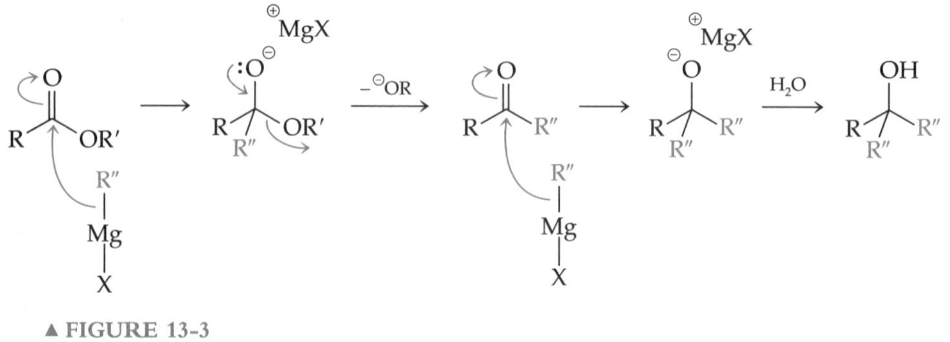

▲ FIGURE 13-3

Nucleophilic acyl substitution ensues when a Grignard reagent attacks an ester. Loss of alkoxide from the initially formed tetrahedral intermediate produces a ketone, which is attacked by a second molecule of Grignard reagent by the pathway shown in Figure 13-2. The tertiary alcohol ultimately produced from an ester bears two identical groups on the carbinol carbon. In the nucleophilic addition of a Grignard reagent to a ketone (Figure 13-2), a tertiary alcohol also is produced, but three different groups may be present on the carbinol carbon.

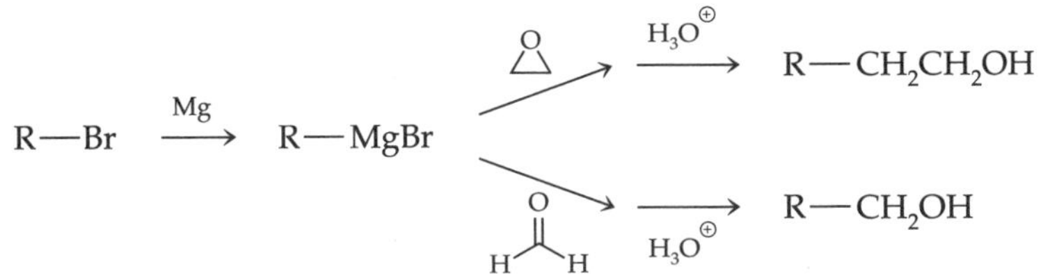

▲ FIGURE 13-4

Primary alcohols can be prepared by the reaction of a Grignard reagent with either ethylene oxide (upper reaction) or formaldehyde (lower reaction). The carbon skeleton of the alcohol prepared from ethylene oxide has two more carbon atoms than the Grignard reagent, whereas that prepared from formaldehyde has one more carbon atom.

tional group for subsequent manipulation is a valuable synthetic tool for the construction of complex molecules.

EXERCISE 13-B

For each of the following targets, several different synthetic routes are possible that employ a Grignard reagent. For each product, provide at least two different sets of Grignard reagent and coreactant that will lead to the observed product. Describe reaction conditions that might be used to prepare the target molecule.

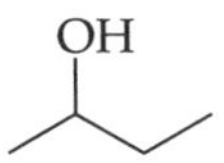

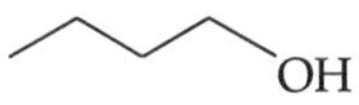

Organolithium Reagents

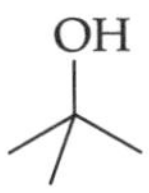

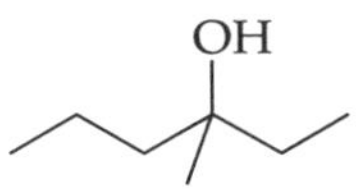

Organolithium reagents react with carbonyl-containing compounds in virtually the same way as Grignard reagents. Typical examples are the reaction of benzaldehyde with butyllithium or with a lithium acetylide, as shown in Figure 13-5.

We should not forget, however, that the very feature that imparts nucleophilic character (an unshared electron pair) to these reagents also makes them basic. Thus, a significant side reaction that can take place with carbonyl compounds bearing an acidic α hydrogen is deprotonation. We will return to this alternate mode of reaction later in the chapter. Nucleophilic

Table 13-1 ► Synthetic Utility of Grignard Reactions		
Reactants		**Product**
$RMgX + H_2CO$	$\longrightarrow$	RCH_2OH
$RMgX + R'CHO$	$\longrightarrow$	$RR'CHOH$
$RMgX + R'COR''$	$\longrightarrow$	$RR'R''COH$
$RMgX + R'CO_2R''$	$\longrightarrow$	$R_2R'COH$
$RMgX + CO_2$	$\longrightarrow$	RCO_2H
$RMgX +$ 	$\longrightarrow$	RCH_2CH_2OH
$RMgX + H_2O$	$\longrightarrow$	RH

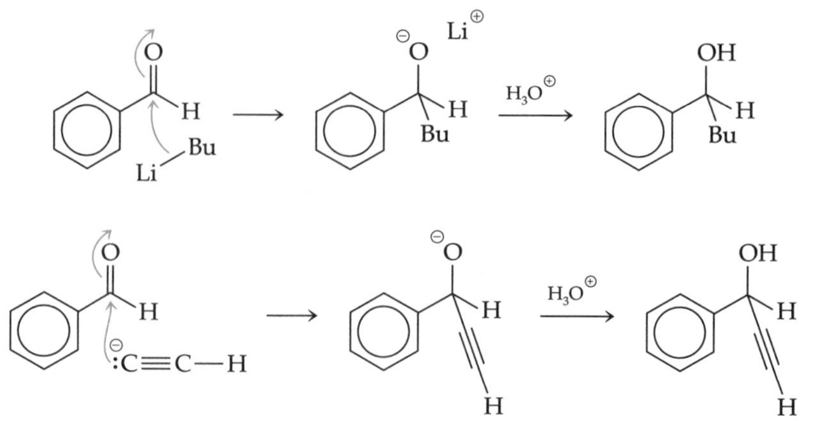

▲ FIGURE 13-5
Organolithiums and other organoalkali metals act as nucleophiles to attack carbonyl groups, forming C–C σ bonds. Because benzaldehyde has no α protons, high yields of adduct are produced. Carbonyl compounds that do have α protons also react by enolization.

addition and substitution are observed with organometallic reagents only when nucleophilic attack is faster than competing acid-base chemistry. This is frequently the case with both organolithium and organomagnesium (Grignard) reagents but, when enolization takes place at a comparable rate, mixtures are produced and chemical yields of addition products are lower.

Conjugate Addition

With α,β-unsaturated carbonyl compounds, an alternate mode of carbon–carbon bond formation takes place at the β carbon rather than at the carbonyl carbon. We can use resonance structures to explain the electrophilic character of the β carbon.

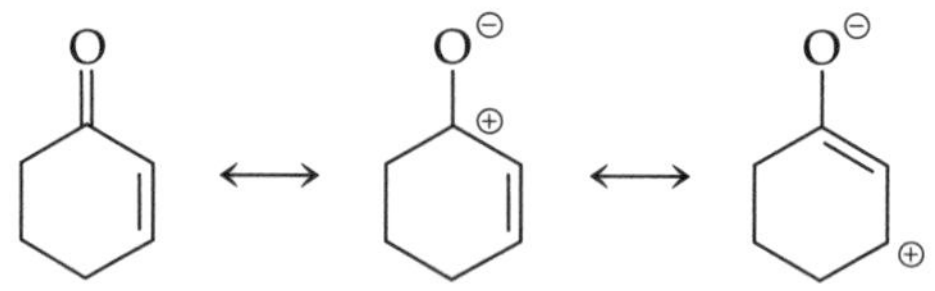

To the extent that the right-hand structure contributes to the hybrid, the β carbon has electrophilic character. The partial positive charge on this carbon enhances reactivity toward a nucleophile. Thus, an α,β-unsaturated carbonyl compound can be attacked by nucleophiles either at the carbonyl carbon (as in simple aldehydes, ketones, and esters) or at the β carbon.

When a nucleophile attacks the β position of an α,β-unsaturated ketone, a σ bond is formed at that position as the electron pair of the carbon–carbon π bond shifts toward oxygen, ultimately forming an enolate anion (Figure 13-6).

In this conjugate addition (a nucleophilic analog of the electrophilic 1,4 addition to dienes discussed in Chapter 10), the initially formed anion is protonated on oxygen to produce an enol. Here, the proton and the nucleophile have added in a 1,4 (conjugative) sense. Tautomerization of the enol to the more-stable carbonyl group completes the sequence. The final product results from formal addition of H and Nuc across the carbon–carbon double bond, but we must not forget that this conjugate addition requires a carbonyl group: a nucleophile would never attack an isolated C=C double

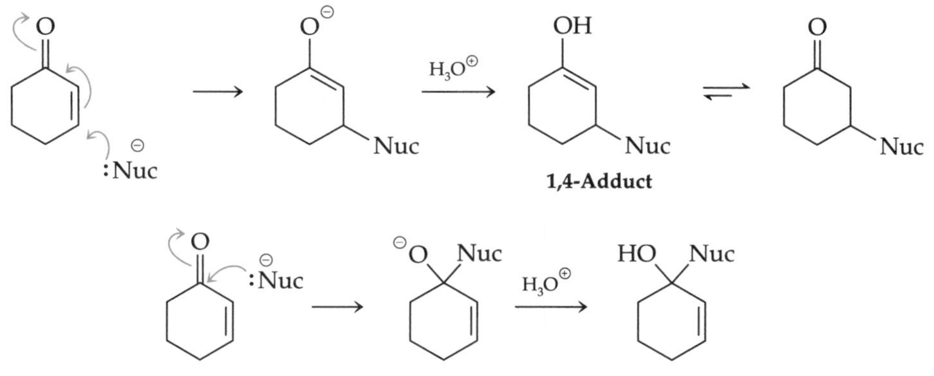

▲ FIGURE 13-6
Both the β position and the carbonyl carbon in an α,β-unsaturated ketone bear significant partial positive charge and are attacked by nucleophiles. Nucleophilic attack at the β position forms a new C–C σ bond and an enolate anion (upper reaction). After protonation and tautomerization, a ketone is produced by this conjugate (1,4) addition. Nucleophilic attack at the carbonyl carbon instead produces an allylic alcohol after protonation, in a 1,2 addition (lower reaction).

bond. The critical role of the carbonyl group in the enone is to provide stabilization (through resonance delocalization) of the negatively charged intermediate formed upon attack by the nucleophile at the β position.

Alternatively, a nucleophile can add to the carbonyl carbon of an α,β-unsaturated ketone. In this 1,2 addition, a new carbon–carbon bond is formed between the nucleophile and the carbonyl carbon. Protonation of the negatively charged oxygen in this 1,2 adduct produces an allylic alcohol rather than the ketone ultimately produced by 1,4 addition. This 1,2 mode of nucleophilic attack thus closely resembles a Grignard addition to a simple aldehyde or ketone, and the additional π bond between the α and β carbons plays no critical role.

Whether the 1,2 or 1,4 mode of addition is followed in the reaction of a nucleophile with an α,β-unsaturated carbonyl group depends on the identity of both the nucleophile and the electrophile. In general, charge-intensive (hard) nucleophiles such as Grignard reagents and alkyllithiums add to the carbonyl carbon because the addition product also is charge intensive (with negative charge concentrated on oxygen). The degree of charge concentration in the nucleophile is comparable to that in the alkoxide product. In contrast, the negative charge in the enolate anion formed in 1,4 addition is delocalized by resonance. Thus, charge-diffuse (soft) nucleophiles tend to add in the 1,4 mode. An example of each of these modes is shown in Figure 13-7. Dialkylcuprate reagents (such as lithium dimethylcuprate) add in a conjugate 1,4 sense, giving cyclohexanone alkylated at the β position, whereas alkyllithiums (such as methyllithium) preferentially add in a 1,2 fashion, producing a tertiary allyl alcohol.

When resonance-stabilized anions are used as nucleophiles in a conjugate addition, the reaction is called a **Michael addition.** In such nucleophiles, charge dispersal is extensive (spread over two or more atoms), and attack at the β position is highly preferred. The Michael addition of the anion of malononitrile to methyl vinyl ketone is shown in Figure 13-8. Here, the charge-diffuse nucleophile approaches the β carbon, causing a shift of charge to oxygen, to form an enolate anion. Protonation on oxygen, followed by keto-enol tautomerization, gives the observed adduct.

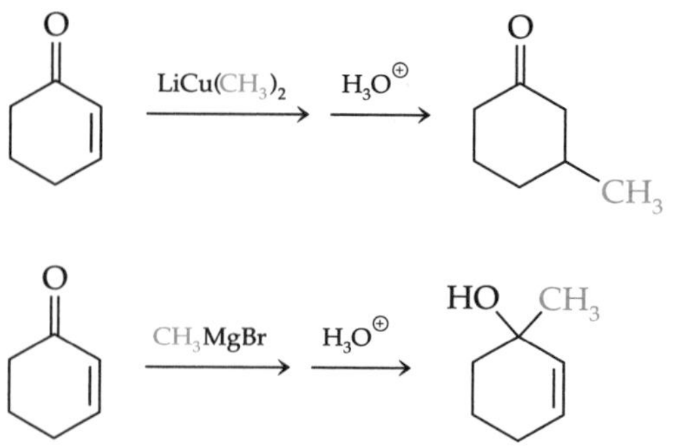

▲ FIGURE 13-7

Charge-diffuse nucleophiles such as dialkyl coppers attack cyclohexenone at the β position to form conjugate addition products, whereas charge-intensive nucleophiles such as Grignard reagents or organolithiums attack the carbonyl carbon to produce 1,2 adducts.

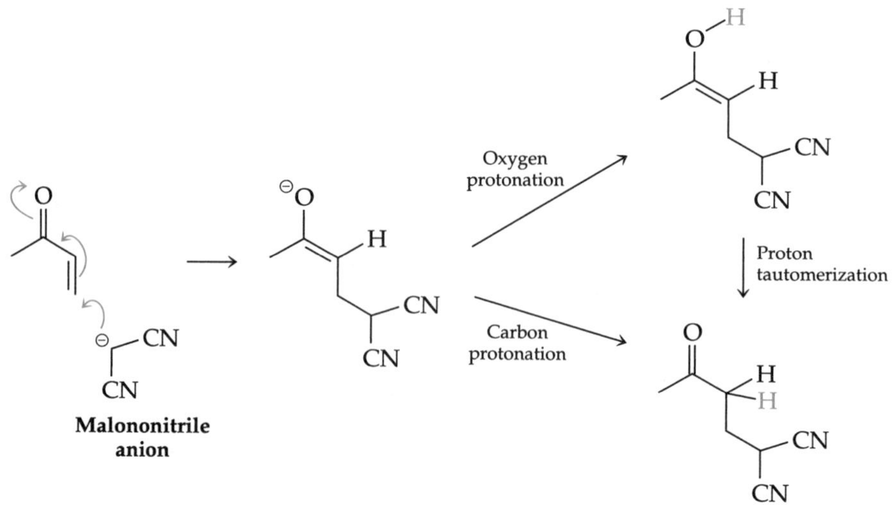

▲ FIGURE 13-8

Conjugate addition takes places by nucleophilic attack of the resonance-stabilized malononitrile anion at the β position of methyl vinyl ketone. The resulting enolate anion is protonated either on oxygen to form an enol or on carbon to produce a ketone. The enol tautomerizes to the ketone under basic reaction conditions by deprotonation-reprotonation.

In some cases, nucleophiles add at comparable rates in both the simple and the conjugate senses. Under conditions in which the additions are reversible, the 1,4-addition product dominates because the stronger C=O bond of the carbonyl group is preserved. For example, amines attack the β position of an α,β-unsaturated ketone (Chapter 12) rather than at the carbonyl carbon.

EXERCISE 13-C

Predict whether 1,2 or 1,4 addition is more likely for each of the following reagents. Draw the product expected from the reaction of each reagent with cyclohex-2-enone.

(a) CH_3CH_2MgBr (b) $(CH_3CH_2)_2CuLi$ (c) CH_3Li

The Wittig Reaction

Carbon nucleophilicity is not confined to simple organometallic reagents; carbon nucleophiles are also formed by deprotonation adjacent to an electron-withdrawing group. An example was given in Chapter 8, regarding the production of phosphonium ylides.

Consider the ylide formed by deprotonation of the phosphonium salt formed by nucleophilic (S_N2) substitution of ethyl bromide by triphenylphosphine. A strong base such as a lithium dialkylamide or *n*-butyllithium is required. Because the carbon center in the resulting ylide bears significant negative charge, it is nucleophilic and adds to the carbonyl carbon of a ketone such as acetone, as shown in Figure 13-9. The resulting intermediate, called a **betaine,** is zwitterionic, with a negatively charged oxygen atom separated by two atoms from a positively charged phosphorus atom. In the next stage of reaction, these charged atoms bond together, forming a four-membered ring. This ring then opens to generate a carbon–carbon double bond and the exceptionally strong phosphorus–oxygen double bond.

This transformation is called the **Wittig reaction** in recognition of its discovery by Georg Wittig, for which he received the Nobel Prize in chemistry in 1979. Overall, this reaction converts an aldehyde or ketone into an alkene in which the double bond of the product joins two fragments that were starting materials (Figure 13-10). (Notice that the conversion accomplished in the Wittig reaction is almost the inverse of the cleavage of an alkene by ozonolysis.) In Figure 13-10, the conversion of an alkyl bromide and ketone into an alkene is shown separately from the conversion of triphenylphosphine into triphenylphosphine oxide and of butyllithium into butane. The use of coupled arrows summarizes the principal organic conversion and implies that these processes are *not* independent. Furthermore, these reactions consist of several steps and cannot take place by simultaneously mixing all required reagents. This method of summarizing transformations as *coupled reactions* is convenient for summarizing a net conversion,

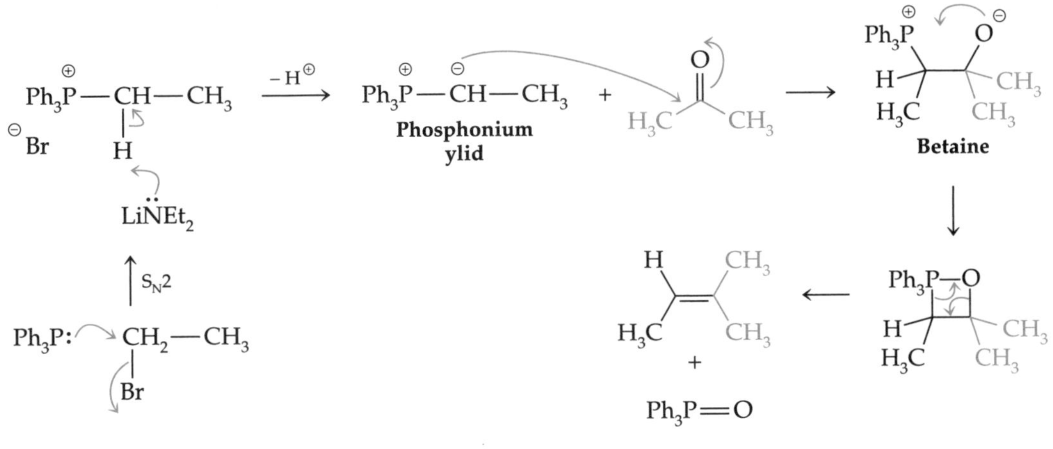

▲ FIGURE 13-9
A phosphonium salt can be made nucleophilic by the removal of a hydrogen from the carbon adjacent to the positively charged phosphonium ion. The resulting ylide acts as a carbon nucleophile to attack an aldehyde or ketone in the Wittig reaction. As a new C–C bond is formed between the ylide carbon and the carbonyl carbon, electron density shifts to oxygen, producing a zwitterion called a betaine. The betaine can form a highly strained four-membered ring that quickly reopens to form an alkene and a phosphine oxide. The formation of the very strong P=O bond provides the driving force for the reaction.

Overall:

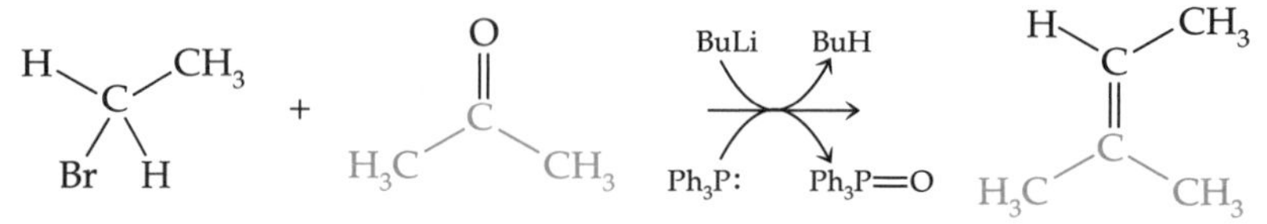

▲ FIGURE 13-10

In a Wittig reaction, a C=O bond is replaced by a C=C bond to what was the α carbon of a phosphonium salt, which in turn was derived from a C—X bond of an alkyl halide.

especially when the molecules are complicated. We will encounter coupled reactions more frequently in later chapters where biochemical transformations are described.

EXERCISE 13-D

Identify two possible combinations of organic reagents from which each of the following alkenes could be prepared by a Wittig reaction.

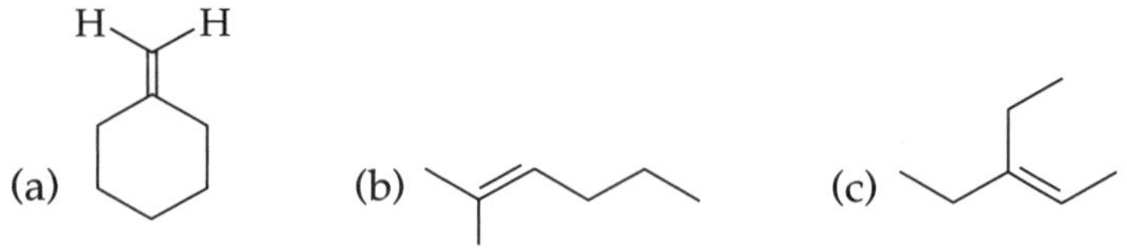

(a)   (b)   (c)

13-2 ▸ Enolates and Enols as Nucleophiles: The Aldol Condensation

The reactions in Section 13-1 are quite useful for building large molecules from small ones in the laboratory. However, they differ in one significant characteristic from biosynthetic transformations: all require strongly basic and highly reactive nucleophiles that are rapidly protonated in water, the solvent for most biologically relevant transformations. The remainder of this chapter deals with reactions that take place through more-stable and less-reactive carbon nucleophiles—molecules that more closely resemble those taking part in nature's construction of complex biomolecules.

So far, we have seen carbonyl compounds such as ketones and esters only as the electrophilic partners in carbon–carbon bond-forming reactions. Nonetheless, they are potential nucleophiles as well. This reactivity can be revealed by the removal of a proton originally bound to a carbon α to the carbonyl group. Two nucleophilic species can be derived from carbonyl compounds: enolate anions and enols. The more-nucleophilic enolate anion is generated by deprotonation with base, whereas the enol is formed by proton tautomerization (proton addition followed by proton loss) under acidic conditions, as illustrated with acetone in Figure 13-11.

The enol or enolate anion generated from a carbonyl compound is a nucleophile that can attack the aldehyde, ketone, or ester from which it was formed because the highly polar carbonyl group makes the carbon end of a C=O double bond highly electrophilic. Thus, two carbonyl compounds are joined by a new carbon–carbon bond upon treatment with acid or base.

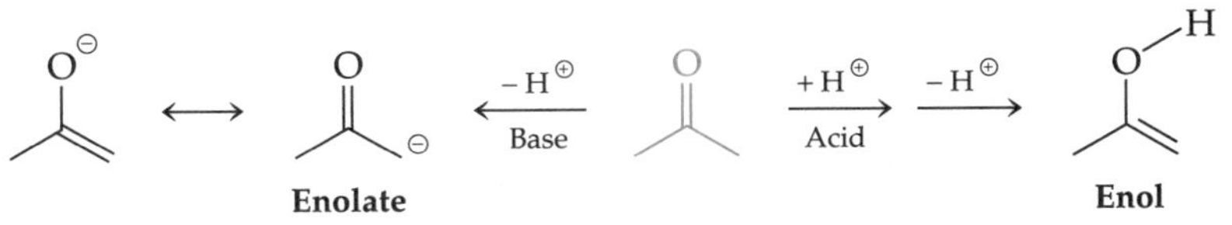

▲ FIGURE 13-11

A ketone (such as acetone) can be made nucleophilic under acidic or basic conditions. In base, the ketone is deprotonated to an enolate anion. In acid, protonation on the carbonyl oxygen, when followed by α-deprotonation (loss of the α hydrogen), produces an enol.

Base-catalyzed Condensation of Aldehydes

Enolate anions are stable carbon nucleophiles because of charge delocalization over the three-atom system. There are two significant resonance contributors for enolate anions, one with negative charge on carbon and the other with negative charge on oxygen. Enolate anions can be formed by deprotonation of the α carbon of an aldehyde, ketone, or ester (Figure 13-12). With negatively charged oxygen and nitrogen bases such as alkoxides and amide anions, this deprotonation is quite rapid, but the equilibrium position depends on the relative acidity of the carbonyl compound and the conjugate acid formed upon protonation of the base. Aldehydes, ketones, and esters have comparable acidities, ranging (in the order given) from 17 to 25. Because alcohols (the conjugate acids of alkoxide bases) are more acidic (pK_a 16–18) than are simple carbonyl compounds, only a low concentration of enolate anion is formed upon treatment with alkoxide. In contrast, dialkylamides such as lithium diisopropylamide, LDA [$LiN(CH(CH_3)_2)_2$], are much stronger bases (the pK_a values of their conjugate acids are typically about 36). Deprotonation with these bases is therefore essentially complete.

Let us now consider a specific simple example of the reaction of a carbonyl compound (acetaldehyde) with itself (Figure 13-13). Acetaldehyde can be deprotonated by base; in our example, hydroxide ion is used to form the enolate anion. This enolate anion is formed only in small amounts and is thus in the presence of a relatively large amount of the original carbonyl compound with which it reacts through nucleophilic addition. A new carbon–carbon bond is formed between the enolate carbon and the carbonyl

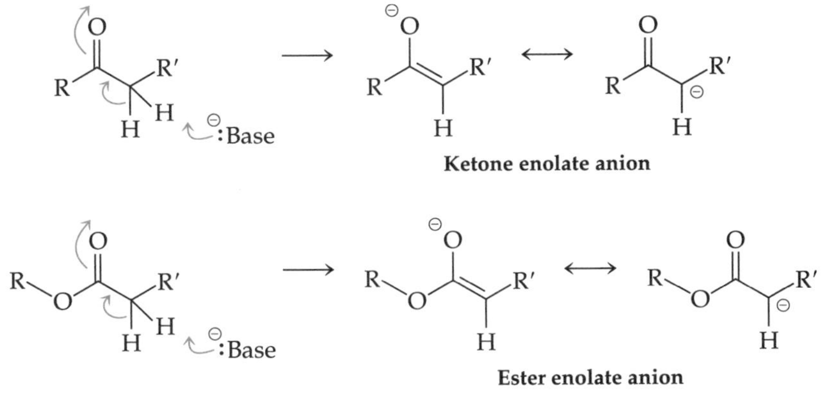

Ketone enolate anion

Ester enolate anion

▲ FIGURE 13-12

Aldehydes or ketones (upper reaction) or esters (lower reaction) can be converted into enolate anions by treatment with alkoxide or lithium dialkylamide.

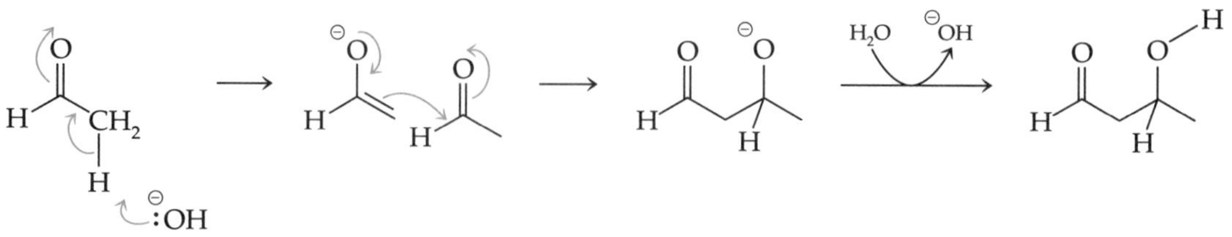

▲ FIGURE 13-13

The enolate anion of acetaldehyde (produced here in the first step) can attack neutral acetaldehyde, forming a new C–C bond between the α carbon of the enolate anion and the carbonyl carbon of the neutral aldehyde. Protonation of the resulting alkoxide produces a β-hydroxyaldehyde. Because base is regenerated when the neutral product is formed, this is a base-catalyzed reaction.

carbon of a second neutral molecule of acetaldehyde (Figure 13-13). The addition of a proton from water completes the reaction. Because the product contains both an aldehyde and an alcohol functional group, it is known as an **aldol** and its formation from two molecules of acetaldehyde is referred to as an **aldol reaction.** Notice that hydroxide ion is regenerated in the last step so that base is not consumed. The aldol reaction is therefore base catalyzed.

Upon heating, the reaction of an aldehyde (or a ketone) with itself under basic conditions does not stop at the aldol. The base (hydroxide) next removes a proton from the α carbon in the aldol product. Hydroxide is then lost, forming an α,β-unsaturated aldehyde (Figure 13-14). The overall reaction from aldehyde to a higher-molecular-weight unsaturated aldehyde is called an **aldol condensation.** Recall that condensation reactions are those that bring together two molecules to make a larger one with the loss of some small molecule (in this case, water).

Several features of the aldol condensation can sometimes lead to significant complications. The product of an aldol condensation of acetaldehyde is itself an aldehyde and can react as an electrophile with acetaldehyde enolate in two different ways, by 1,2 and 1,4 addition (Figure 13-15); these products also are aldehydes. Thus, for this reaction to be of synthetic value, the relative rate of the initial reaction leading to the aldol condensation product must be appreciably higher than those of subsequent reactions of this product. Selectivity for the reaction of the starting material rather than that of the product is important for all reactions used to prepare products in quantity. The higher electrophilicity of a carbonyl group in a simple aldehyde or ketone than that of an α,β-unsaturated carbonyl compound satisfies this requirement for most aldol condensations.

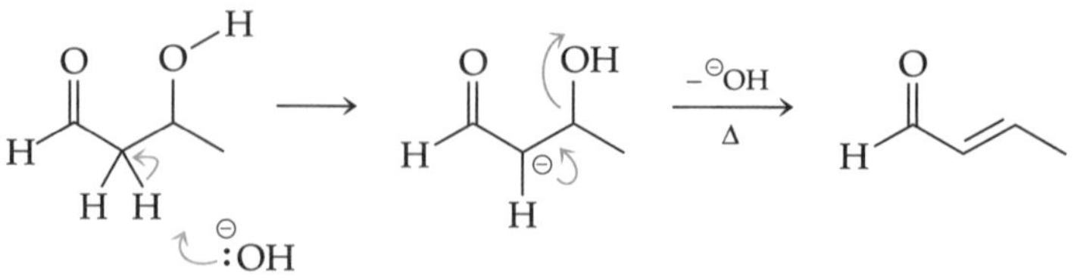

▲ FIGURE 13-14

An α hydrogen in an aldol is acidic because it is adjacent both to a carbonyl group and to a potential leaving group (⁻OH). Dehydration of the β-hydroxyalcohol produced in the aldol reaction is therefore possible in base. An α,β-unsaturated aldehyde is then produced.

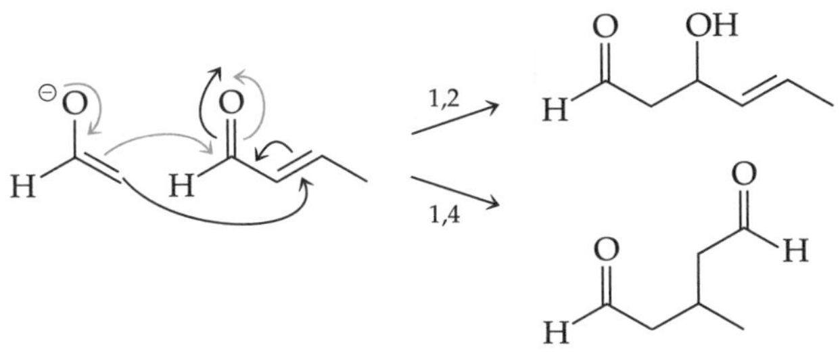

▲ FIGURE 13-15
The enolate anion of acetaldehyde can attack the α,β-unsaturated aldehyde formed in the aldol condensation in either a 1,2 (upper reaction: orange arrows) or 1,4 (lower reaction: black arrows) addition. This reaction is generally slower than the simple aldol reaction.

EXERCISE 13-E

Draw the structure of the aldol condensation product expected from each of the following compounds upon treatment with KOH/EtOH.

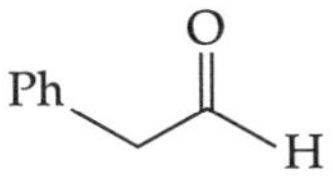

Acid-catalyzed Condensation of Aldehydes

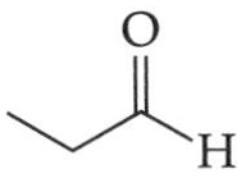

The aldol reaction can also be conducted under acid-catalyzed conditions, in which the enol rather than the enolate is the reactive nucleophile (Figure 13-16). Protonation of the carbonyl group of acetaldehyde is the first step needed to initiate the tautomerization to the enol. Deprotonation at the α carbon to relieve the positive charge in the protonated carbonyl produces the enol. In the enol, electron donation from oxygen confers nucleophilic character to the α carbon atom. The enol is not as reactive a nucleophile as the enolate anion, but it is sufficiently so to attack the protonated carbonyl

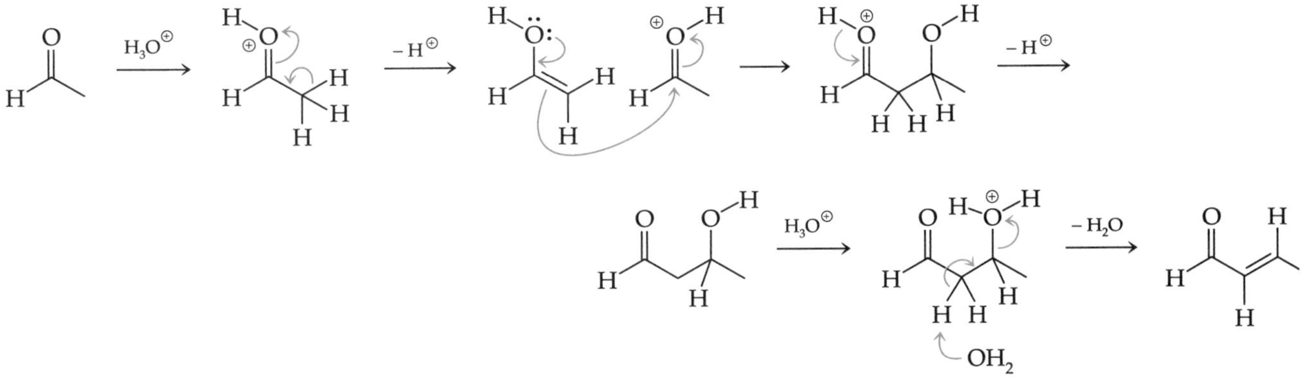

▲ FIGURE 13-16
In an acid-catalyzed aldol condensation, the aldehyde tautomerizes to its enolic form, making the α carbon nucleophilic. The enol, although weakly nucleophilic, can attack an activated carbonyl group, such as that present when another molecule of aldehyde is protonated on oxygen. Deprotonation of the resulting cationic intermediate produces a β-hydroxyalcohol, and acid-catalyzed dehydration leads to the α,β-unsaturated aldehyde—namely, the aldol condensation product.

group of a second molecule of acetaldehyde. Loss of a proton from this intermediate produces the same aldol product as in the base-catalyzed aldol reaction. Dehydration of the aldol is accomplished by the routes discussed in Chapter 9 for the acid-catalyzed dehydration of simple alcohols. This acid-catalyzed dehydration is so efficient that it is often difficult to stop at the aldol stage under acid-catalyzed conditions.

EXERCISE 13-F

Would the structures of the aldol products obtained under acid-catalyzed conditions for each of the aldehydes shown in Exercise 13-E differ from those obtained in base?

Aldol Condensations of Ketones

Ketones also undergo aldol condensations when treated with acids and bases, as shown for acetone in Figure 13-17. Treatment of acetone with base generates the enolate anion, which acts as a nucleophile to attack a second molecule of acetone. After protonation of the ionic intermediate, the aldol-like product is formed. Again, dehydration of the β-hydroxycarbonyl compound occurs readily, forming the α,β-unsaturated enone. With unsymmetrical ketones, two different enolates can typically be formed, and several different aldol condensation products are produced, differing in both regio- and stereochemistry (Figure 13-18).

EXERCISE 13-G

By using curved arrows, suggest a mechanism by which the α,β-unsaturated ketone at the lower left in Figure 13-18 is formed.

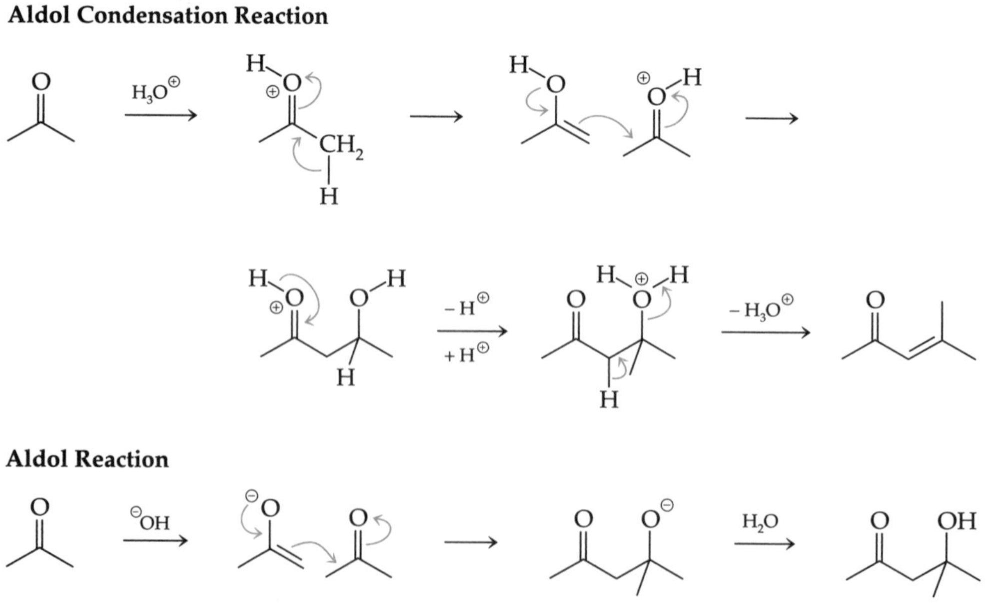

Aldol Condensation Reaction

Aldol Reaction

▲ FIGURE 13-17
The mechanisms of the aldol reaction and the aldol condensation of a ketone are exactly parallel to those of an aldehyde. Shown here are an acid-catalyzed aldol condensation (upper reaction) and a base-catalyzed aldol reaction (lower reaction) of acetone.

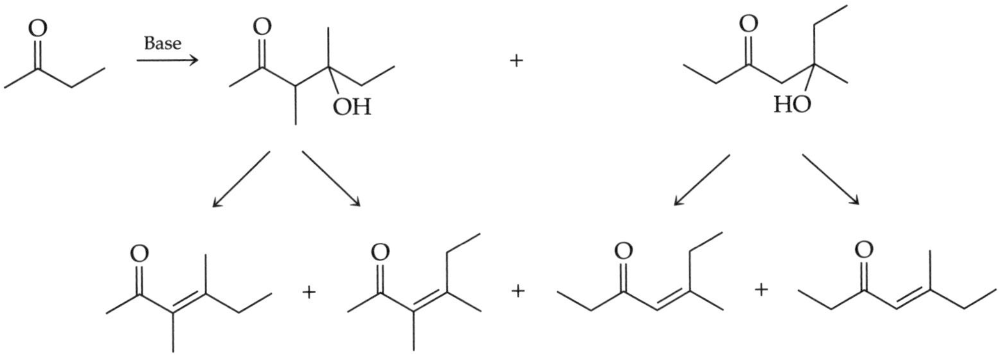

▲ FIGURE 13-18

Treatment of 2-butanone with base produces two possible enolate anions, with negative charge either at C-1 or at C-3. These two enolate anions lead to two different β-hydroxyketones. When water is lost, each of these β-hydroxyketones gives two geometrically isomeric α,β-unsaturated ketones as isolated aldol condensation products.

Crossed Aldol Condensations

The synthetic utility of the aldol reaction would be greatly enhanced if the enolate of one substrate could be used to attack the carbonyl group of another, forming a crossed condensation product. Unfortunately, complex product mixtures are obtained when this is attempted in the laboratory. Even if each carbonyl compound is symmetrical and can generate only a single enolate anion, four possible cross products can be formed because each component can serve as either a nucleophile or an electrophile. The situation becomes even more complicated when one or both of the starting materials can form two different enolate ions. When acetone and 2-butanone are heated in base, for example, all of the products shown in Figure 13-19 are formed. The complexity of the mixture, together with the difficulty of separating such chemically similar products, makes this reaction of little value.

A crossed aldol reaction is usually practical only when one of the carbonyl components lacks α hydrogens. In this case, it is possible to form the enolate anion of only one substrate, limiting the number of different products (Figure 13-20). In this example, only the enolate anion of acetone can be formed because benzaldehyde lacks α hydrogens. Because the reactivity of an aldehyde toward nucleophilic attack is higher than that of a ketone (Chapter 12), the reaction of acetone with itself is slower than the crossed aldol reaction, and the only product observed is that obtained by acetone enolate's attack on benzaldehyde. The loss of water from the resulting aldol product is especially easy because of extended conjugation in

AN α,β-ENONE JUST LIKE GRANDMA USED TO MAKE

Maltol is an α-hydroxy, α,β-unsaturated cyclic ketone used to impart a "freshly baked" aroma and flavor to breads and cakes.

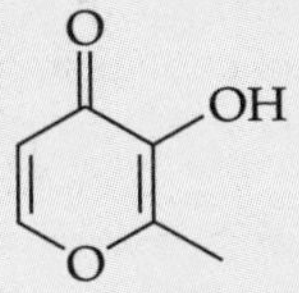

Maltol

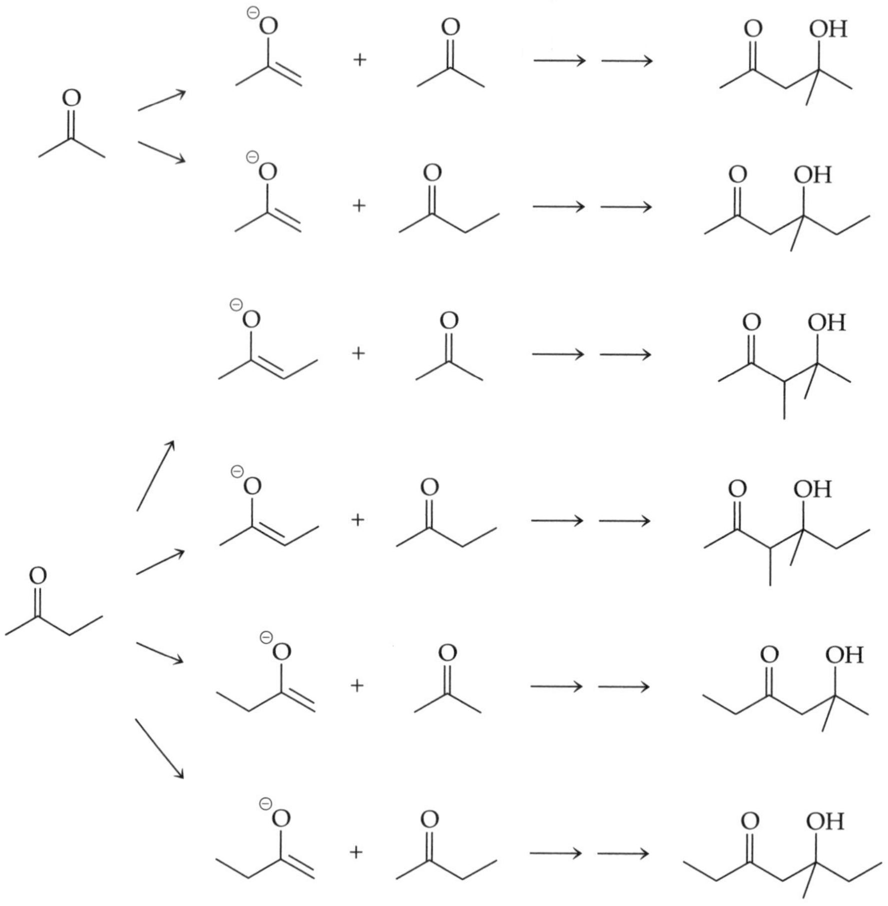

▲ FIGURE 13-19

A very complex array of products is formed when a crossed aldol reaction is attempted with a mixture of two ketones. With acetone and 2-butanone, three enolate anions are formed in the first deprotonation step. Each enolate anion can attack either neutral ketone to produce six possible β-hydroxyalcohols.

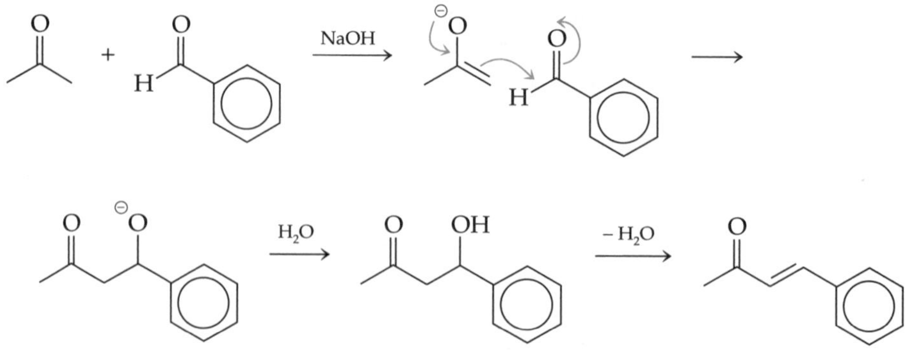

▲ FIGURE 13-20

The complexity of a crossed aldol reaction can be made manageable if one of the carbonyl reactants cannot form an enolate anion. Because benzaldehyde has no α hydrogens, it can act only as an electrophile. Here, acetone enolate anion attacks benzaldehyde, forming a C–C bond and ultimately producing aldol condensation product. Although attack by acetone enolate anion on acetone also is possible, it is slower than the attack on the more-reactive aldehyde.

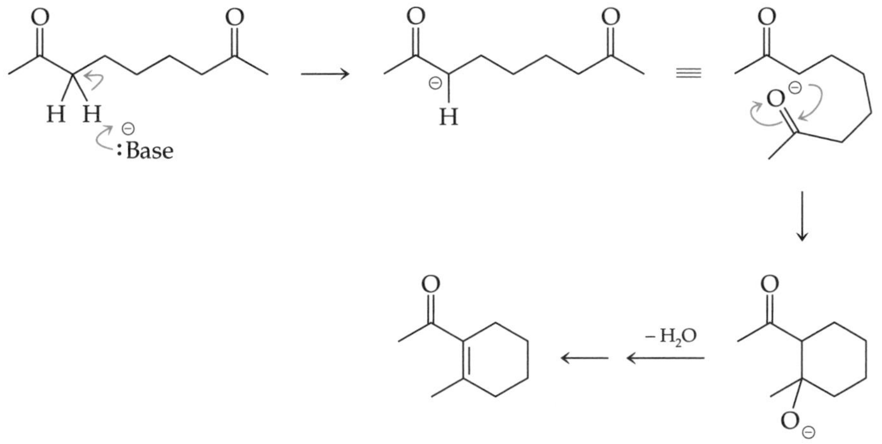

▲ FIGURE 13-21
Intramolecular aldol condensation can occur in a compound bearing two carbonyl groups. In nona-2,8-dione, enolization occurs at C-3 so that cyclization to form a six-membered ring can take place when the enolate anion attacks the C-8 carbonyl group. (Enolization at C-1 could lead to condensation only through a more-strained eight-membered-ring transition state.) The mechanism of this intramolecular attack is the same as that discussed earlier for intermolecular C–C bond formation in the aldol condensation.

the condensation product. In later chapters, we will see how the problem of multiple products in a crossed aldol reaction can be solved so that only one product is formed; namely, through catalysis by enzymes in biological systems. Enzymes permit a chemical selectivity that is not usually possible in a simple chemical reaction in the laboratory.

Intramolecular Aldol Condensation

The aldol reaction of a single molecule containing two carbonyl groups results in the formation of a cyclic product, as shown in Figure 13-21. Dehydration is especially easy from a cyclic aldol and, except under certain reaction conditions, the isolated product is the aldol condensation product, an α,β-unsaturated ketone.

In the **Robinson ring annulation,** the intramolecular aldol reaction is used to add a ring to a cyclic starting material, as shown in Figure 13-22.

EXERCISE 13-H

Using curved arrows, provide a detailed reaction mechanism for the cyclization shown in Figure 13-22.

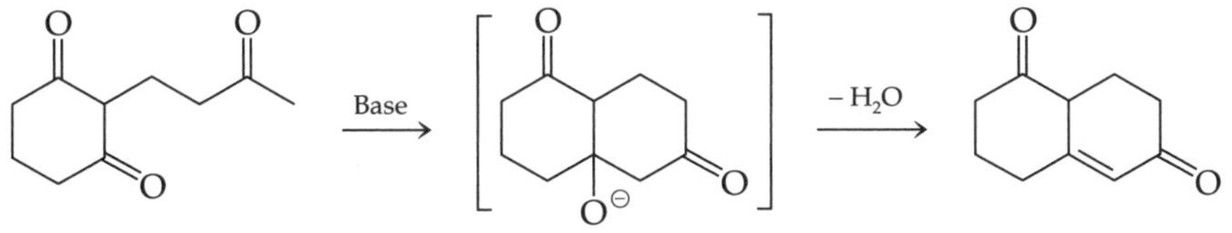

▲ FIGURE 13-22
A six-membered ring is formed when enolization takes place at C-1. (Because a highly strained four-membered ring would be formed in an aldol reaction of the enolate anion formed by deprotonation at C-3, the alternate possible product is not observed.)

Ring annulation (the attachment of new rings to existing cyclic structures) results in the formation of fused-ring systems. Fused rings are found in a variety of biologically active materials such as the steroid sex hormones: the male hormone testosterone and the female hormone estradiol, the principal active component of birth control pills.

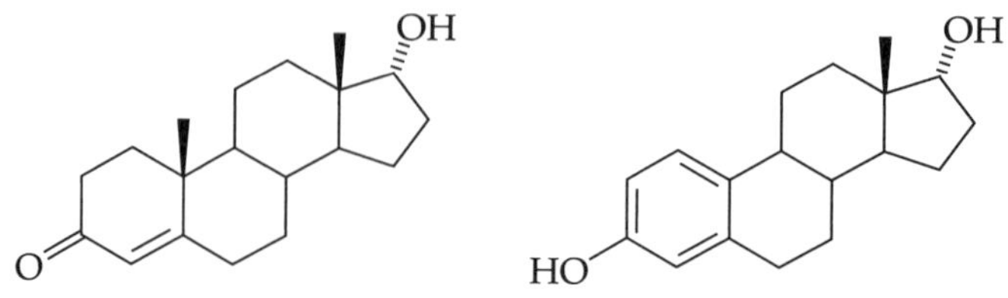

Testosterone   Estradiol

13-3 ▸ The Claisen Condensation

As in the aldol reaction, a new carbon–carbon bond is formed between an enolate anion and a carbonyl carbon when an ester is treated with base. The enolate anion generated upon deprotonation of an ester is sufficiently nucleophilic to react with the carbonyl group of another equivalent of ester in a nucleophilic acyl substitution. This reaction is known as the **Claisen condensation** and results in the formation of a β-ketoester. Although this reaction formally resembles the aldol reaction, there are significant differences. For example, the Claisen condensation is almost always carried out in the presence of base.

Base-induced Claisen Condensation

The base-induced Claisen condensation begins with the generation of an ester enolate anion (Figure 13-23). As in the aldol reaction, an alkoxide is usually used as base to deprotonate at the α position. Deprotonation α to the ester carbonyl forms an ester enolate anion, a carbon nucleophile that then attacks the carbonyl carbon of a neutral ester molecule. As in the nu-

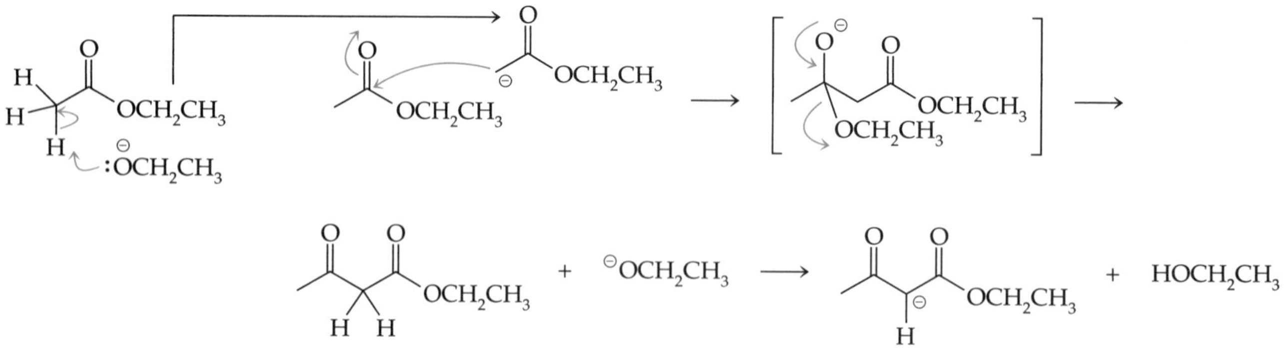

▲ FIGURE 13-23

An ester enolate anion is formed when an ester with α hydrogens is treated with alkoxide. As with an aldehyde enolate anion in the aldol reaction, the resulting ester enolate anion attacks a second neutral ester carbonyl carbon. A tetrahedral intermediate is produced and, when the C=O double bond is reformed as the electron pair on oxygen reforms a π bond, the —OR group originally present on the ester is lost as alkoxide. The product of this base-induced Claisen condensation is a β-ketoester. The high acidity of the α hydrogens assures fast deprotonation in base, producing a resonance-stabilized anion.

cleophilic acyl substitutions discussed in Chapter 12, the negatively charged, tetrahedral intermediate loses alkoxide to form the neutral β-ketoester. As we learned in Chapter 6 (Table 6-1), the C–H bond between the two carbonyl groups of a β-ketoester is especially acidic ($pK_a = 11$). Thus, deprotonation by alkoxide of the β-ketoester formed in the Claisen condensation is thermodynamically favorable. Indeed, this acid-base reaction, which consumes base, constitutes a principal driving force for the Claisen condensation. As a result, the Claisen condensation is best carried out with a full equivalent of base.

EXERCISE 13-I

Draw the structures of the Claisen condensation products expected from each of the following esters upon treatment with KOEt in EtOH.

(a) $CH_3CO_2CH_2CH_3$ (c) $CH_3CH_2CO_2CH_3$

(b) $CH_3CO_2CH_3$ (d) $PhCH_2CO_2Ph$

EXERCISE 13-J

Recall from the acetoacetic ester synthesis discussed in Chapter 8 that the conversion of β-ketoesters into ketones can be accomplished by hydrolyzing the ester while heating to effect decarboxylation. Predict the product that would be formed if the product of a Claisen condensation of the following esters were heated in aqueous acid:

(a) 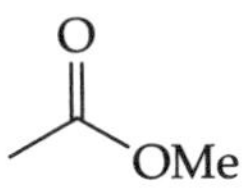   (b)

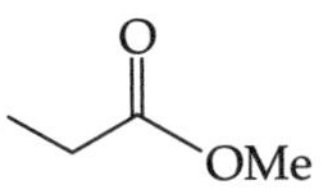

Crossed Claisen Condensations

The reaction of one ester enolate anion with a different ester presents the same problem that arises in crossed aldol condensations. Only when one of the ester components lacks α hydrogens (and therefore cannot form an enolate anion) does a crossed Claisen condensation proceed cleanly. For example, Figure 13-24 shows the condensation of the enolate anion of methyl acetate with methyl formate. Methyl formate lacks an α hydrogen and the condensation reaction proceeds in the fashion shown. Formate esters also are especially reactive toward nucleophiles for both electronic and steric reasons (recall the differences between aldehydes and ketones). Crossed Claisen condensations are also effective with carbonate, oxalate, and benzoate esters.

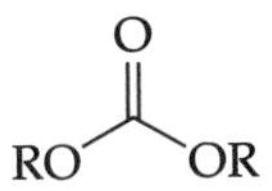

| Carbonate ester | 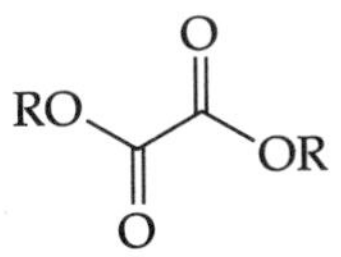 Oxalate ester | 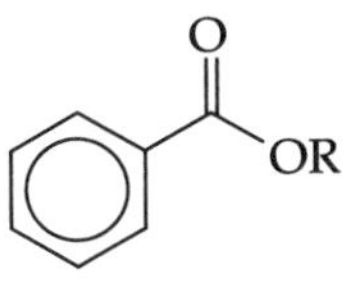 Benzoate ester | 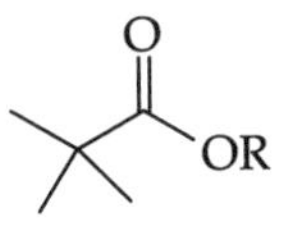 Pivalate ester |

Aliphatic esters completely substituted at the α position (like the pivalate ester shown) also have no α hydrogens, but they are relatively poor electrophiles because of steric hindrance to the formation of a tetrahedral intermediate. As a result, they are rarely useful in Claisen condensations.

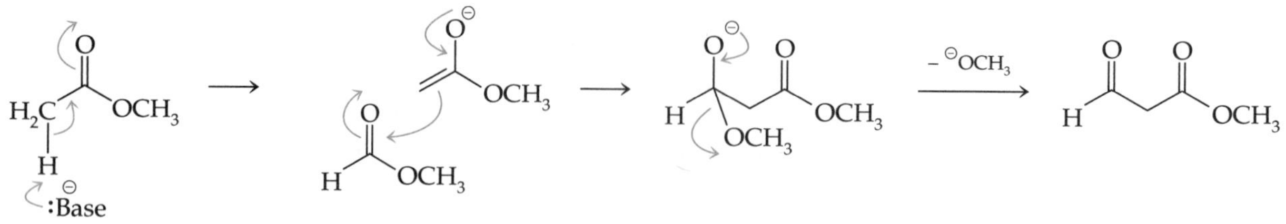

▲ FIGURE 13-24
As in the crossed aldol reaction, a crossed Claisen condensation gives a synthetically useful product only if one of the esters lacks α hydrogens and is a more-active electrophile than the other. Here, the ester enolate anion of methyl acetate attacks methyl formate to initiate the Claisen condensation. No enolate anion is possible from a formate ester, and the formate ester is more reactive than an acetate ester toward nucleophilic attack for the same reasons that an aldehyde is more reactive than a ketone.

EXERCISE 13-K

Predict the product of Claisen condensation of the enolate anion of methyl acetate ($CH_3CO_2CH_3$) with carbonate, oxalate, benzoate, and pivalate esters.

EXERCISE 13-L

Propose appropriate starting materials for the synthesis of each of the following by a route employing a Claisen condensation:

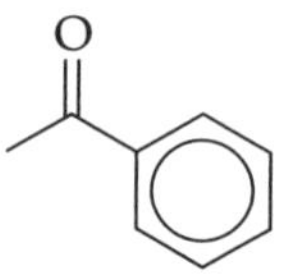

Reformatsky Reaction

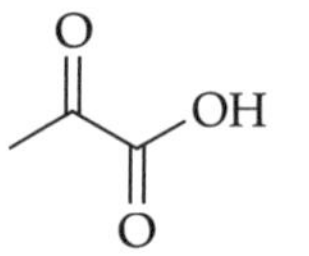

Unless one reactant lacks α hydrogens, the reaction of two ketones (or aldehydes) in the aldol condensation reaction or the reaction of two esters in the Claisen condensation forms a complex mixture of products. The **Reformatsky reaction** addresses this problem by having a preformed ester enolate act as a nucleophile to attack a ketone or an aldehyde. Simply to treat a mixture of ketone and ester with base does not lead to a synthetically useful product because the ketone is both more acidic and more electrophilic than the ester. This combination therefore results only in the aldol condensation of the ketone. However, the ester enolate can be formed first in the absence of the ketone by reduction of an α-bromoester with zinc (Figure 13-25). (The Hell-Volhard-Zelinski reaction discussed in Chapter 8 can be used to prepare an α-bromoacid; this acid can then be esterified to form the starting material for the Reformatsky reaction.) In the Reformatsky reaction, a ketone is then added to this preformed **zinc enolate,** affording the crossed condensation product in good yield.

Dieckmann Condensation

An intramolecular variant of the Claisen condensation is known as the **Dieckmann condensation.** With the diester shown in Figure 13-26, for example, deprotonation at the position α to either ester group in this symmet-

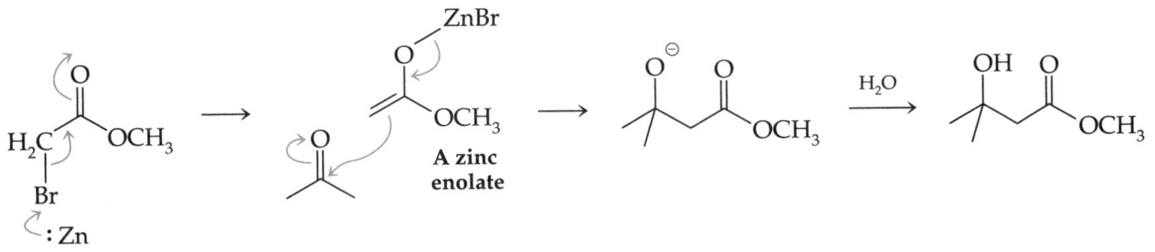

▲ FIGURE 13-25
A zinc enolate is formed by insertion of zinc metal into an α-C–Br bond. The resulting enolate then attacks an aldehyde or ketone more rapidly than it can attack its ester precursor. A β-hydroxyester is thus produced.

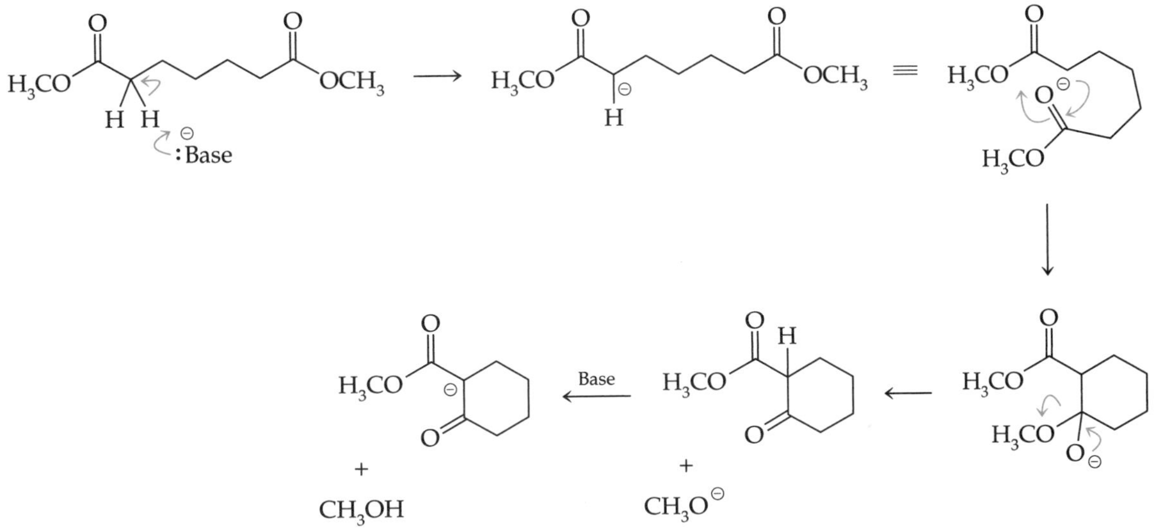

▲ FIGURE 13-26
As in the intramolecular aldol reaction shown in Figure 13-21, an intramolecular Claisen condensation can occur in a compound bearing two ester groups. In the diester shown here, the two possible sites for enolate formation are identical. The ester enolate anion then attacks the other ester group through a six-membered transition state. The resulting tetrahedral intermediate loses methoxide to produce the β-ketoester expected in a Claisen condensation.

rical molecule produces an ester enolate anion, which then attacks the remaining ester. A β-ketoester is formed upon loss of ⁻OR from the cyclized tetrahedral intermediate. As with the Claisen condensation that joins two molecules, a primary driving force of the Dieckmann cyclization is the formation of the anion of the product β-ketoester. As a result, the Dieckmann condensation of an unsymmetrical diester (Figure 13-27) proceeds only in the direction that results in a β-ketoester with an acidic hydrogen between the two carbonyl groups.

EXERCISE 13-M

Draw the structure of the ester enolate needed to form each of the two possible Dieckmann condensation products shown in Figure 13-27.

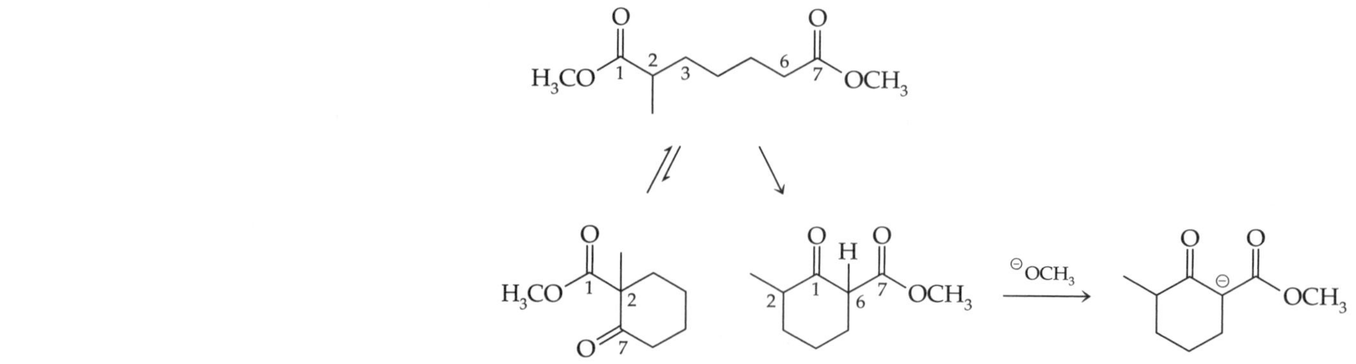

▲ FIGURE 13-27

Although an ester enolate anion can be formed at either C-2 or at C-6, only the C-6 enolate anion leads to product, because the β-ketoester formed in the cyclization of the C-6 enolate anion (at the right) bears an acidic proton at the α position, whereas that formed from the C-2 enolate anion (at the left) does not. The β-ketoester at the right is therefore deprotonated to a resonance-stabilized anion under the basic conditions of the reaction.

The pharmaceutical industry relies significantly on pathways like those discussed in this chapter as methods for making new compounds for the treatment of disease states. We will see the application of some of these reactions for the synthesis of modern pharmaceutical agents in Chapter 15.

13-4 ▶ Synthetic Applications

The reactions discussed in this chapter are particularly important as synthetic methods because they produce new carbon–carbon bonds. Applications of these reactions in making structurally more complex carbon skeletons with various functional groups are listed in Table 13-2.

Conclusions

The reaction of a carbon nucleophile with a carbonyl carbon results in the formation of a carbon–carbon bond. When cyanide ion is used as the nucleophile, a cyanohydrin is formed. Both alkyllithium and Grignard reagents form carbon–carbon bonds by nucleophilic attack on a carbonyl group. Addition of these nucleophiles to formaldehyde, aldehydes, and ketones results in primary, secondary, and tertiary alcohols, respectively. The reaction of these reagents with esters results in tertiary alcohols in which two of the carbon substituents of the carbinol carbon are the same. A carboxylic acid is formed in the reaction of a Grignard reagent with carbon dioxide.

Specific reactions considered in this chapter include: nucleophilic additions (cyanohydrin formation, the Strecker synthesis, Grignard reactions, and Michael additions) and nucleophilic substitution (Grignard reaction with an ester). The Wittig reaction constitutes a method for the formation of alkenes from aldehydes or ketones. The Michael reaction leads to the conjugate addition of a nucleophile to the β carbon of an α,β-unsaturated ketone.

Enolate anions and enols act as nucleophiles to attack aldehydes and ketones, effecting an aldol reaction. The aldol can be dehydrated to an α,β-unsaturated carbonyl compound, the aldol condensation product. When an ester enolate anion attacks an ester, the initial adduct reacts further by loss of an alkoxy group, yielding a β-ketoester by nucleophilic substitution.

Table 13-2 ▸ How to Use Nucleophilic Additions and Substitutions by Carbon Nucleophiles (and Related Reactions) to Make Various Functional Groups

Functional Group	Reaction
Acid	Carbonation of Grignard reagents; or Strecker synthesis of α-amino acids
Alcohol	Treatment of ethylene oxide with Grignard reagents (primary); or treatment of formaldehyde with Grignard reagents (primary); or treatment of aldehydes with Grignard reagents (secondary); or treatment of ketones with Grignard reagents (tertiary); or treatment of esters with Grignard reagents (tertiary); or treatment of aldehydes or ketones with organolithiums; or treatment of aldehydes or ketones with acetylides; or aldol reaction: treatment of aldehydes or ketones with acid or base (β-hydroxycarbonyl compound)
Aldehyde	Aldol reaction: treatment of aldehydes or ketones with acid or base (β-hydroxyaldehyde or α,β-enal)
Alkene	Wittig reaction of aldehydes or ketones
Cyanohydrin	Cyanation of aldehydes or ketones
Ester (β-Ketoester)	Claisen condensation: treatment of esters with base; or Reformatsky reaction of β-bromoester
Ketone	Treatment of an α,β-enone with dialkyl copper; or Michael addition of a stabilized anion to an enone; or aldol reaction: treatment of ketones with acid or base (β-hydroxyketone or α,β-enone); or Claisen condensation: treatment of an ester with base (β-ketoester); or decarboxylation of a β-ketoacid from hydrolysis of a Claisen condensation product

Aldol and Claisen condensations (and Dieckmann cyclizations) are valuable tools for the construction of carbon–carbon bonds between two carbonyl groups. Crossed aldol and Claisen condensations are synthetically useful only if one of the carbonyl groups lacks α hydrogens and, hence, the ability to form an activated enol or enolate anion. An intramolecular version of the aldol condensation (Robinson ring annulation) is an effective method for the formation of fused rings. The Reformatsky reaction provides a method for condensing an ester enolate equivalent with a ketone.

Summary of New Reactions

Cyanohydrin Formation

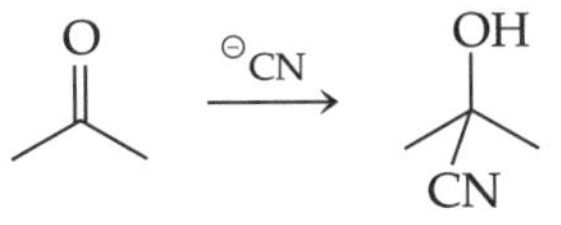

Strecker Synthesis

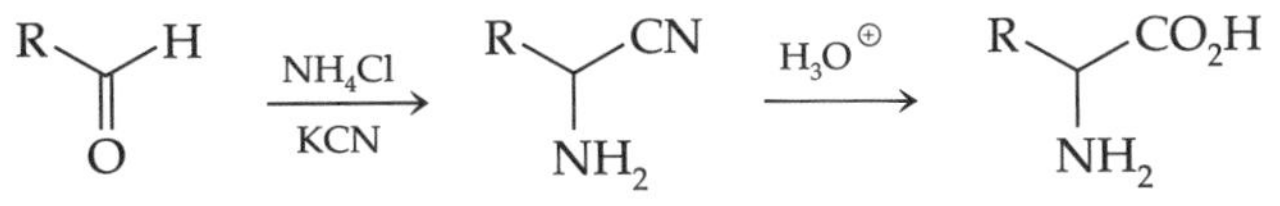

Grignard Additions (See Table 13-1)

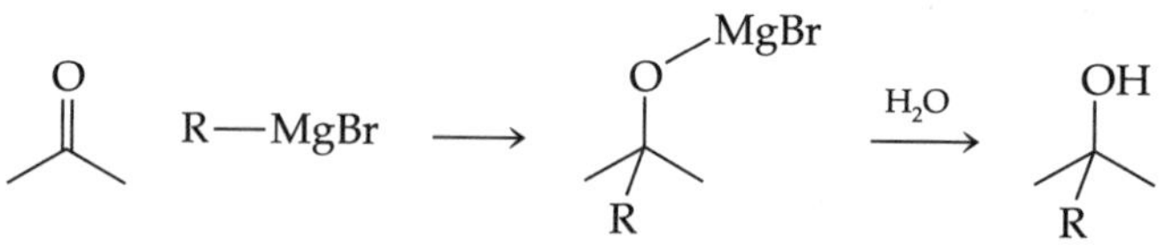

Addition of Organometallics

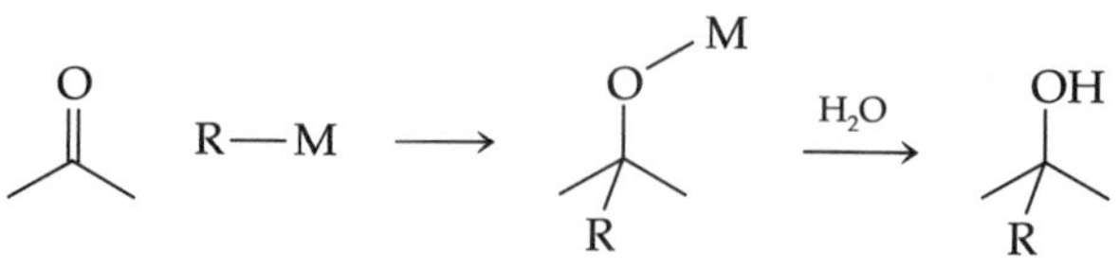

Michael Addition

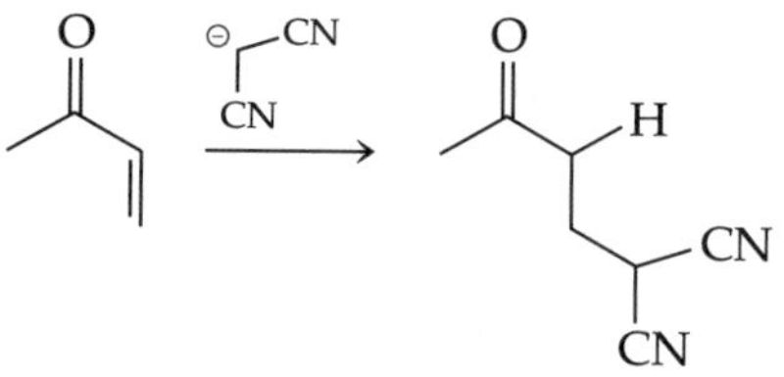

Wittig Reaction

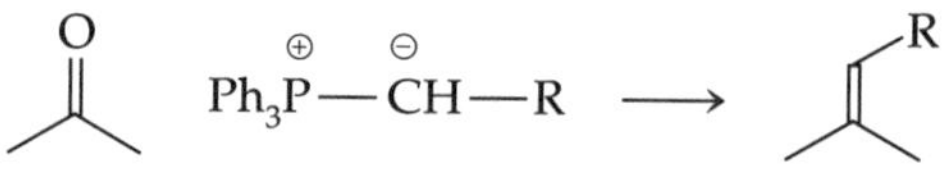

Aldol Condensation

(The Robinson annulation includes an intramolecular aldol condensation.)

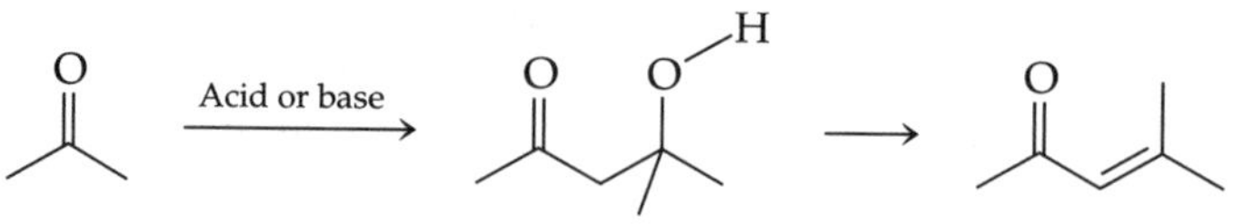

Aldol **Aldol condensation**

Claisen Condensation

(The Dieckmann condensation is an intramolecular variant of the Claisen condensation.)

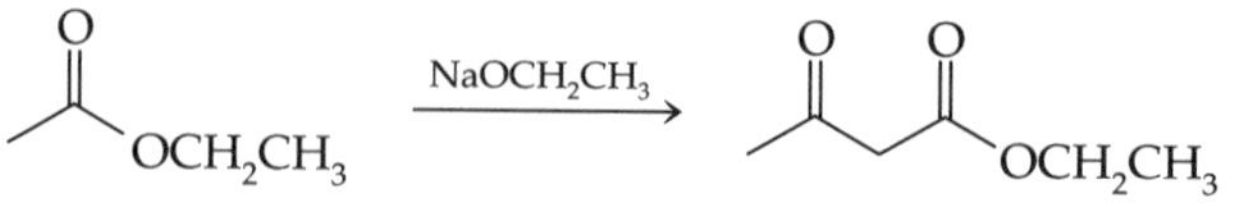

Reformatsky Reaction

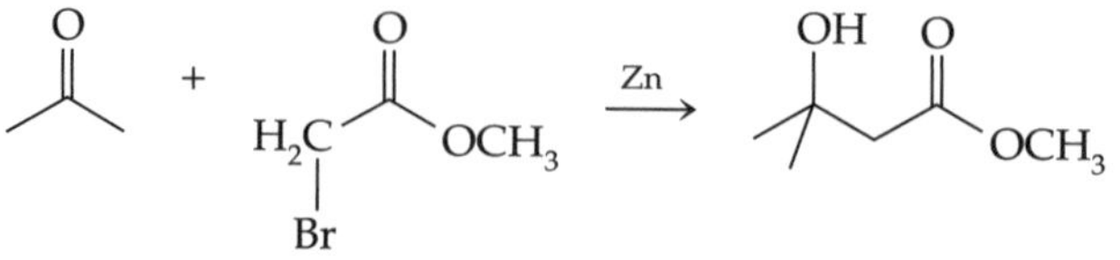

Review Problems

13-1 Predict the major product expected when 1-pentanal is treated with each of the following reagents:

(a) MeMgBr

(b) PhCH$_2$PPh$_3$$^+$, *n*-butyllithium

(c) NaOH

(d) LiCuPh$_2$

(e) H$_3$O$^+$, Δ

(f) NaCN

13-2 Predict the major product expected when methyl pentanoate is treated with each of the following reagents:

(a) MeMgBr   (b) NaOCH$_3$ (c) H$_3$O$^+$, H$_2$O

13-3 Predict the major product expected when methyl 2-pentenoate is treated with each of the following reagents:

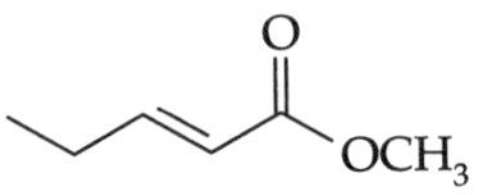

Methyl 2-pentenoate

(a) MeMgBr

(b) 1. NaOH; 2. H$_3$O$^+$, H$_2$O

(c) LiCuPh$_2$

(d) H$_3$O$^+$, H$_2$O

13-4 Indicate the reagent needed to convert *n*-butylbromide into each of the following products through a Grignard synthesis:

(a)

(b)

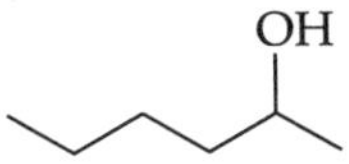

(c)

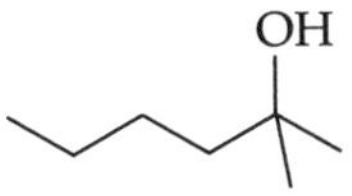

(d)

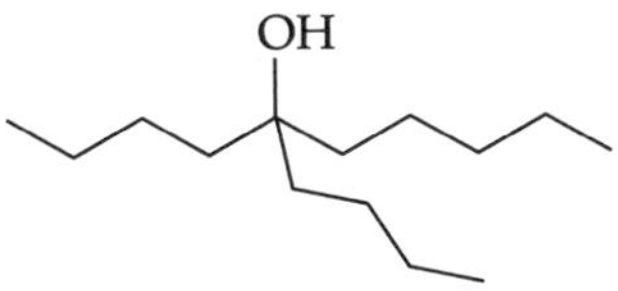

(e)

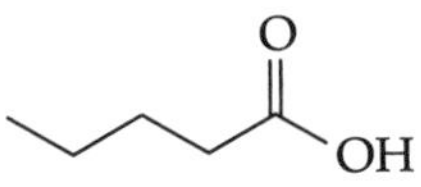

(f)

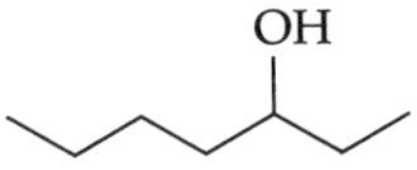

(g)

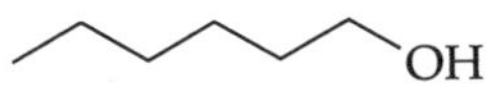

13-5 Indicate the reagent or sequence of reagents needed to effect each of the transformations in parts *a* through *k*.

(a)

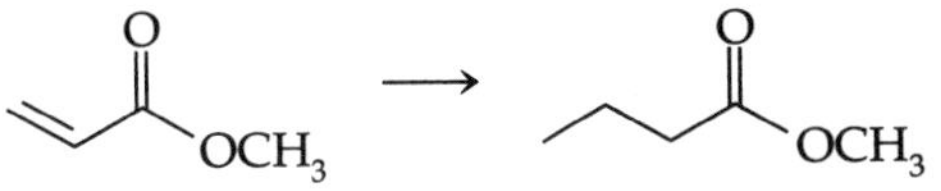

(b)

(c)

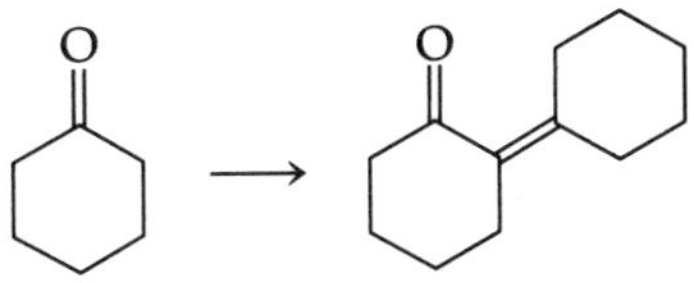

(d)

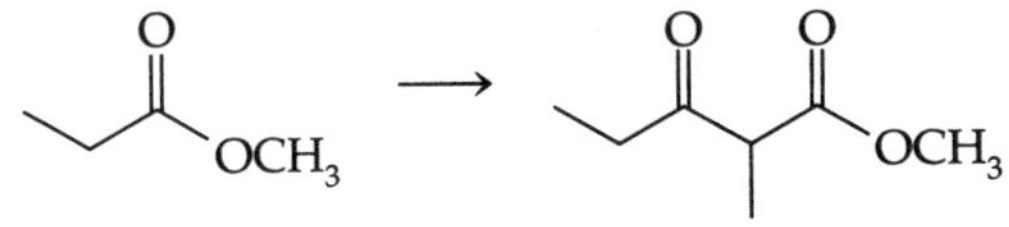

(e)

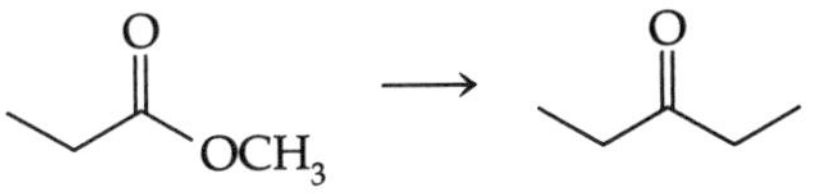

(f)

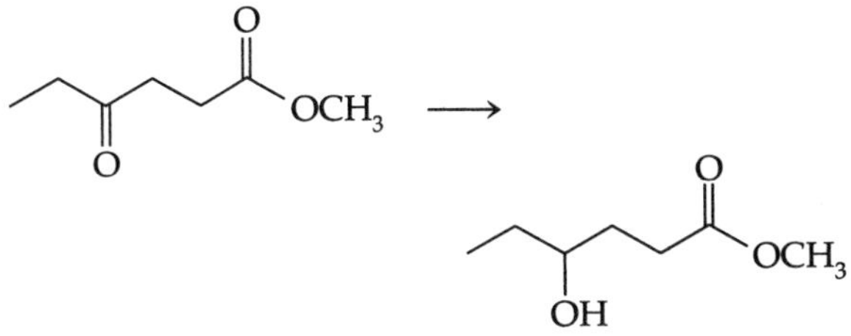

(g)

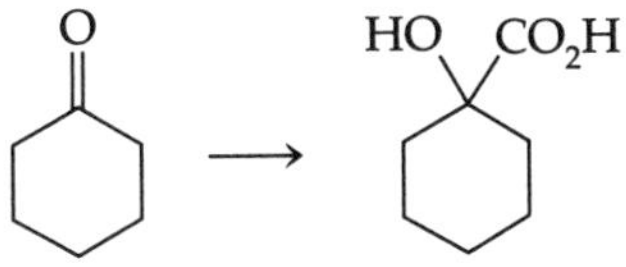

(h)

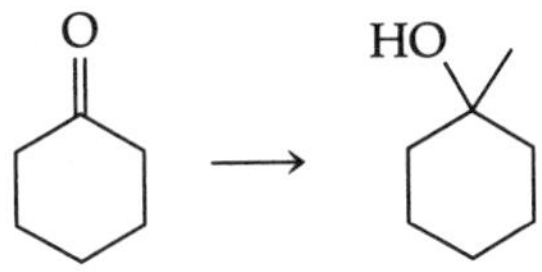

(i)

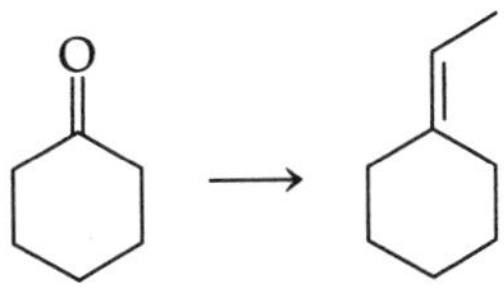

(j)

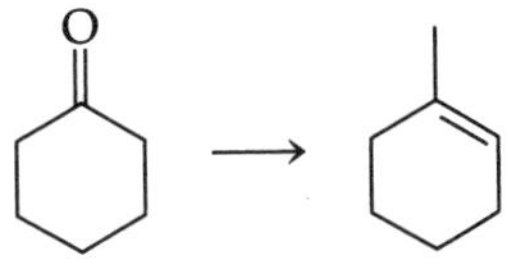

(k)

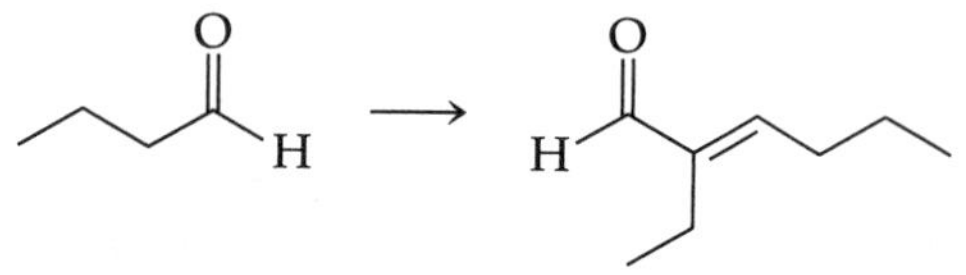

13-6 In the acid-catalyzed aldol reaction of compound A, an α,β-unsaturated aldehyde is obtained; but, in that of the isomeric aldehyde B, a β-hydroxy alcohol is formed. Explain.

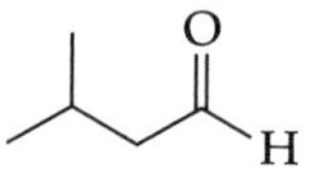

     A 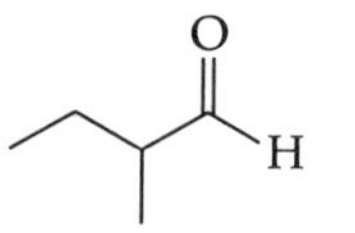   **B**

13-7 Suggest an efficient route for the synthesis of each of the following compounds from any starting material containing four or fewer carbons, acetoacetic ester, malonic ester, and any inorganic reagents. (More than one step may be needed.)

(a)

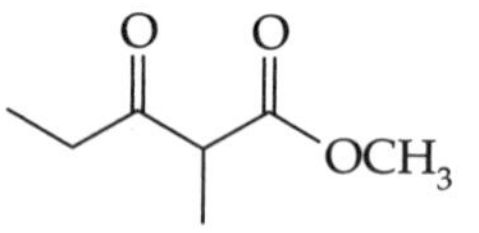

(b)

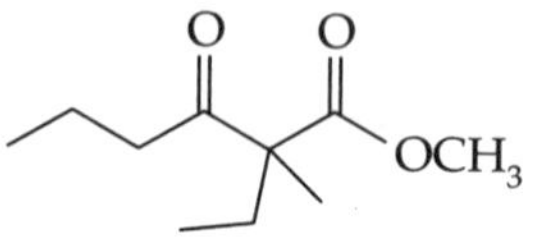

(c)

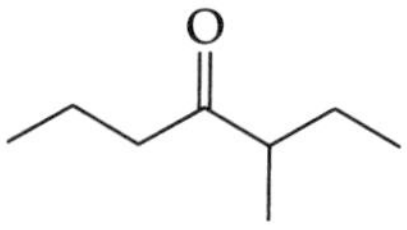

(d)

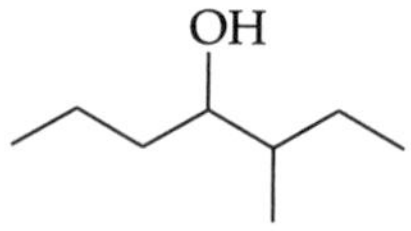

(e)

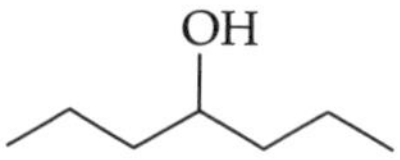

(f)

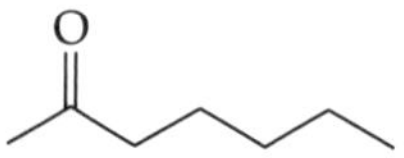

(g)

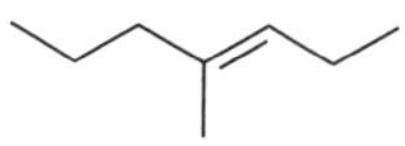

(h)

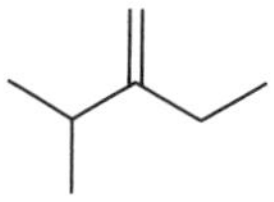

13-8 Using curved arrows to indicate electron flow, write a reaction mechanism for each of the following transformations:

(a)

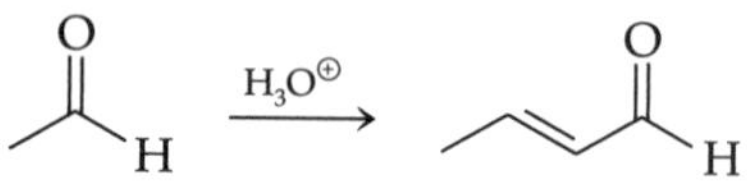

(b)

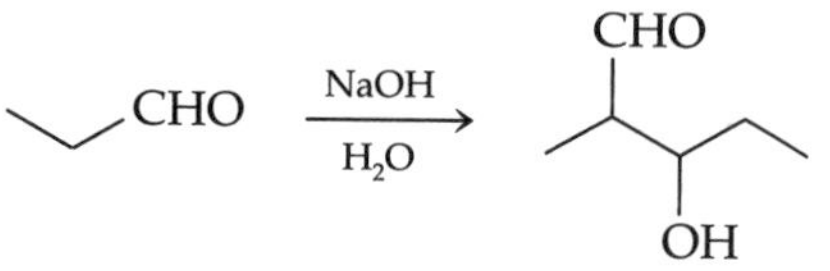

(c)

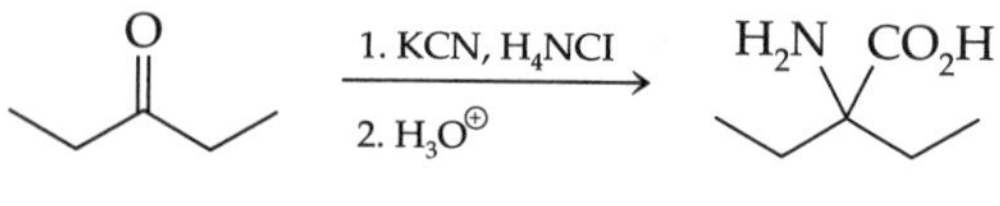

(d)

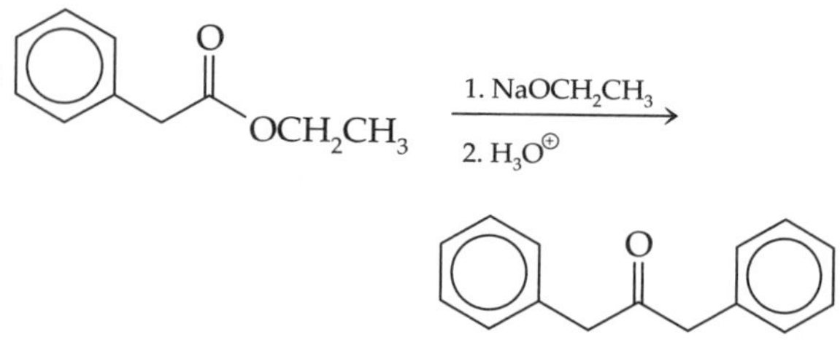

(e)

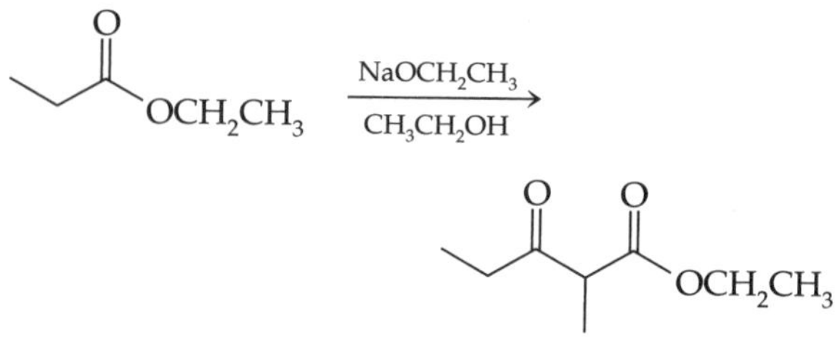

(f)

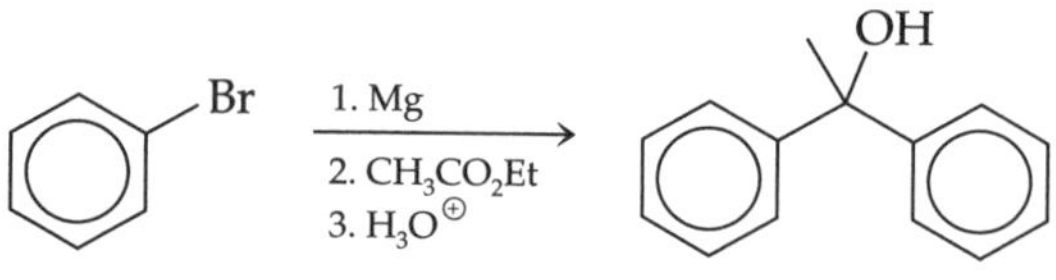

(g)

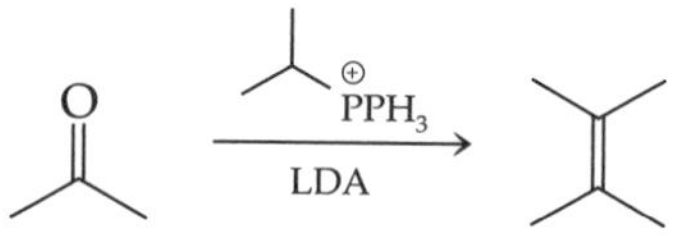

(h)

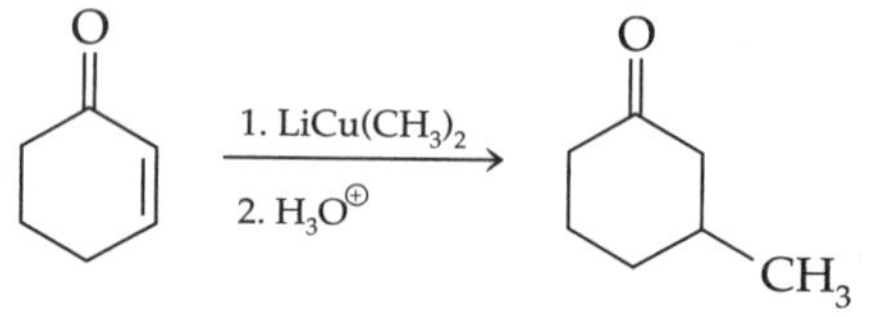

13-9 Explain why the base-induced aldol condensation in the following reaction proceeds in good yield

$$C_6H_5CH=CHCHO + CH_3CH=CHCHO \longrightarrow$$

$$C_6H_5(CH=CH)_3CHO \ (87\%)$$

but that in the second reaction gives the indicated product in less than 20% yield.

$$C_6H_5CH_2CH=CHCHO + CH_3CH=CHCHO \longrightarrow$$

$$C_6H_5CH_2(CH=CH)_3CHO$$

13-10 Which compound in each of the following pairs can more readily undergo a Claisen condensation? Explain.

(a) $CH_3CO_2CH_3$ or CH_3COSCH_3

(b) $C_6H_5CO_2CH_3$ or $C_6H_5CH_2CO_2CH_3$

(c) $(CH_3)_3CCO_2CH_3$ or $CH_3CH_2CH_2CH_2CO_2CH_3$

Chapter 14

Skeletal–Rearrangement Reactions

In the preceding chapter, we learned methods for constructing large molecules by making carbon–carbon bonds. Here, we consider how we can change the connectivity of an existing organic backbone by using reactions that result in skeletal rearrangements.

From Chapters 3, 7, and 10, we know that cations can rearrange, resulting in a change in the sequence of attachment of atoms. In this way, an atom or group of atoms becomes attached to a different carbon atom in the product from that to which it was attached in the starting material. We will see in this chapter how some similar transformations and some quite different rearrangement reactions lead to the formation of new carbon–carbon, carbon–nitrogen, and carbon–oxygen bonds.

14-1 ▸ Carbon–Carbon Rearrangements

A chain of bound carbon atoms constitutes the backbone of every organic compound and the structure of a given molecule is critically dependent on the sequence in which these atoms are attached. Any reaction that shifts the position of a carbon atom and its substituents within a molecule effects an isomerization that greatly affects both the physical and the chemical properties of the resulting isomer. In this section, we will examine those reactions in which a carbon–carbon bond is broken in one part of a molecule and is reformed at another position.

Cation Rearrangements

Shifts of hydrogen from one atom to the next are often quite rapid in carbocations when a more-stable carbocation can be formed from a less-stable one. For example, when 2-methyl-1-propanol is treated with aqueous acid, water is lost and a tertiary cation is formed when a hydrogen shifts from C-2 to C-1. The readdition of water to the tertiary cation results in the formation of 2-methyl-2-propanol. Because the position of the OH group in the product alcohol has changed from its original position in the starting alcohol, this reaction is classified as an isomerization because the carbon skeleton is unchanged.

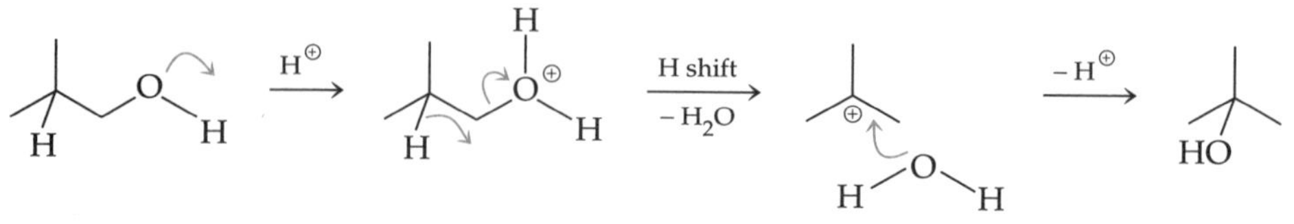

In this isomerization, the identity of the atom to which the functional group is bound has changed, but the sequence of attachment of carbon atoms along the backbone has not. The skeleton would have been altered, however, if a carbon atom with its substituents, rather than hydrogen, had migrated to the developing carbocationic center.

Wagner-Meerwein Rearrangement. In most of the cationic shifts described so far, a hydrogen atom migrates, but alkyl groups also can shift (along with the electrons from the σ bond connecting the group to the adjacent atom). In those cases in which an alkyl group (that is, a carbon substituent) migrates, there are changes in the carbon skeleton, and the reaction is referred to as a **Wagner-Meerwein rearrangement.**

Wagner-Meerwein rearrangements are very similar in detail to those in which hydrogen atoms migrate. Let us consider as an example the solvolysis of 1-bromo-2,2-dimethylpropane, as shown in Figure 14-1. If a simple heterolytic cleavage of the carbon–bromine bond were to occur, a highly unstable primary cation would be formed; this would require a large activation energy. However, if the adjacent methyl group migrates to C-1 at *the same time* that the bromide ion leaves, a much more stable tertiary cation is formed. The simultaneous migration of the alkyl group and departure of the leaving group to form a tertiary cation is faster than the simple loss of the leaving group to form a primary cation; and the products observed are those that result from further reaction of the rearranged cation. In our example, two elimination products and an alcohol are observed. The alcohol

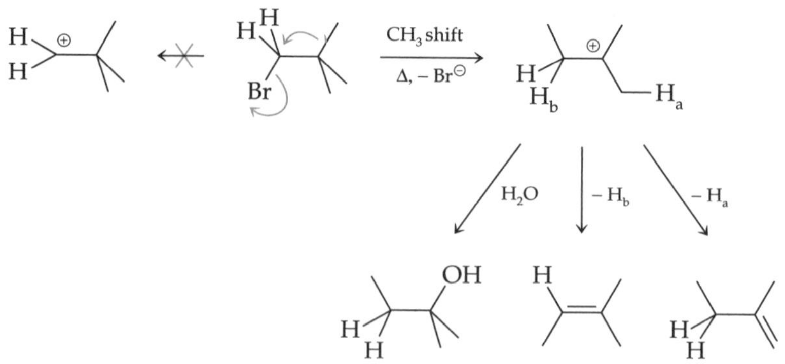

▲ FIGURE 14-1

Heterolytic cleavage of the C–Br bond of 2,2-dimethyl-1-bromopropane does not lead to the primary cation shown at the upper left because the more-stable tertiary cation shown at the upper right is formed if a methyl group migrates from C-2 to C-1 as the C–Br bond is broken. As discussed in Chapter 7, this cation can be trapped by water to produce an alcohol or, as discussed in Chapter 9, it can be deprotonated to give either Zaitsev or Hofmann elimination product.

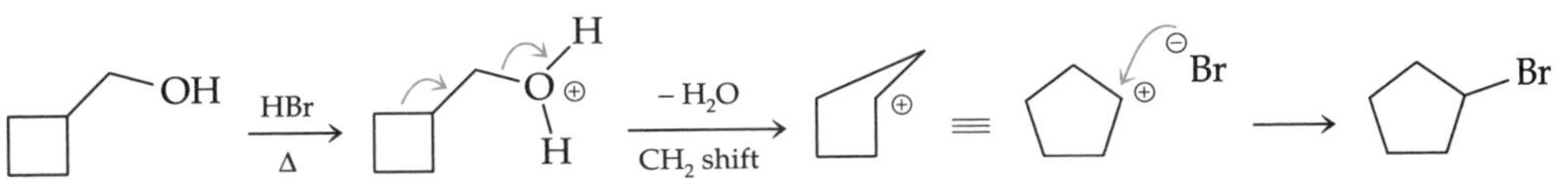

▲ FIGURE 14-2
Treatment of cyclobutylmethanol with HBr produces an oxonium ion from which water can be lost. Migration of a ring carbon to the carbinol carbon with the simultaneous loss of water produces a secondary cyclopentyl cation, instead of the primary cyclobutyl methyl cation that would be formed if the dehydration took place without skeletal rearrangement. This cation is then trapped by bromide, producing the rearranged alkyl halide.

results from reaction of the tertiary cation with water, forming an oxonium ion from which loss of a proton generates the product alcohol with a rearranged carbon skeleton. Alternatively, the cation can lose a proton from either of two different adjacent sites (H_b or H_a). All three observed products derive from the rearranged cation in which the carbon skeleton differs from that of the starting material by the migration of a methyl group from C-2 to C-1 in a Wagner-Meerwein rearrangement.

When a more-stable intermediate can be formed as a group migrates, rearrangement nearly always occurs, no matter how the cation is formed. For example, the same products shown in Figure 14-1 are expected from hydrogen migration in 3-methyl-2-bromobutane. When a driving force for rearrangement exists (with cations, 3° > 2° > 1°), alkyl group migrations almost always take place faster than a less-stable cation can be trapped by solvent or another nucleophile.

Cation rearrangements can also be thermodynamically driven by factors other than the degree of substitution of the cation: ring strain is also important. As an example, consider the treatment of cyclobutylmethanol with strong acid (Figure 14-2). Protonation of the alcohol oxygen forms an oxonium ion, from which simple loss of water would generate a primary cation. However, the carbinol carbon is next to a strained four-membered ring, and an adjacent methylene group (CH_2) migrates to this center at the same time that water is lost. Both a reduction in strain (a four-membered ring becoming a five-membered ring) and an increase in the degree of substitution (from a primary to a secondary cation) are accomplished by this migration. The resulting cyclopentyl cation is then captured in a slower step by an external nucleophile. When treated with aqueous HBr, cyclobutylmethanol is converted into cyclopentyl bromide.

Carbon-skeletal rearrangements with ring expansion also occur with cations formed by the protonation of alkenes. For example, Figure 14-3

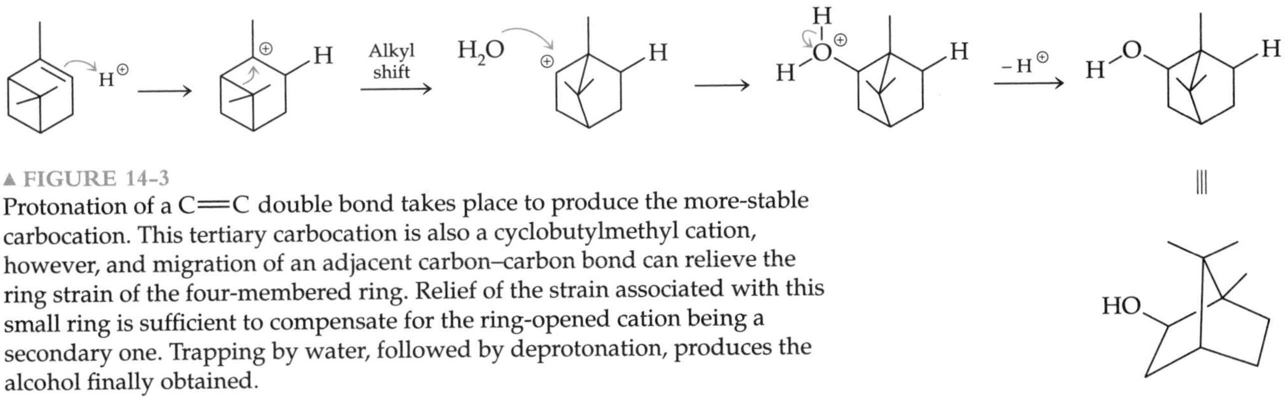

▲ FIGURE 14-3
Protonation of a C═C double bond takes place to produce the more-stable carbocation. This tertiary carbocation is also a cyclobutylmethyl cation, however, and migration of an adjacent carbon–carbon bond can relieve the ring strain of the four-membered ring. Relief of the strain associated with this small ring is sufficient to compensate for the ring-opened cation being a secondary one. Trapping by water, followed by deprotonation, produces the alcohol finally obtained.

shows the protonation of α-pinene to form the tertiary carbocation, which is highly favored over the alternate regiochemistry that would generate a secondary cation (recall Markovnikov's Rule). Migration of a carbon group from the adjacent position produces a secondary cation that is nonetheless more stable than the initial tertiary carbocation because ring strain in the four-membered ring of the starting material is relieved. The secondary cation is then trapped by water, ultimately producing a product alcohol with a different carbon skeleton from that of the starting material.

EXERCISE 14-A

Treatment of the alcohol shown below with acid entails the migration of a carbon substituent to form a stable cation. Predict the structure of the product and suggest a mechanism for its formation.

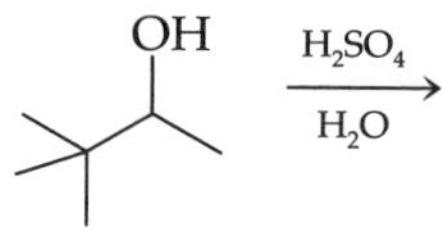

Stabilized by Oxygen

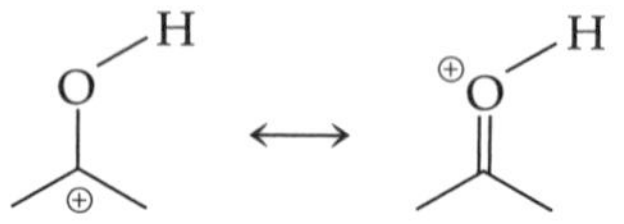

Destabilized by Oxygen

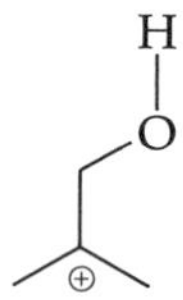

Pinacol Rearrangement. Rearrangements through cationic intermediates also take place in molecules containing more than one functional group, as in the **pinacol rearrangement** (Figure 14-4). The pinacol rearrangement (of a 1,2-diol) begins in the same way as alcohol dehydration: by protonation of one of the hydroxyl groups to form an oxonium ion. Pinacol itself (2,3-dimethyl-2,3-butanediol) is symmetrical and, because the two hydroxyl groups are identical, it makes no difference which group is protonated. The cation that would be formed by simple loss of water would be less stable than a rearranged cation because of the presence of the electron-releasing OH group on the adjacent carbon. Therefore, migration of a methyl group is simultaneous with the loss of water, leading to a carbocation directly substituted by an oxygen substituent. This cation is greatly stabilized by donation of lone-pair electron density from oxygen to carbon, as shown in the margin. (In contrast, a cation in which the oxygen atom is bound to the adjacent atom has no such resonance stabilization and is, in fact, destabilized by inductive electron withdrawal by the highly electronegative oxygen atom.) Loss of a proton from this cationic intermediate

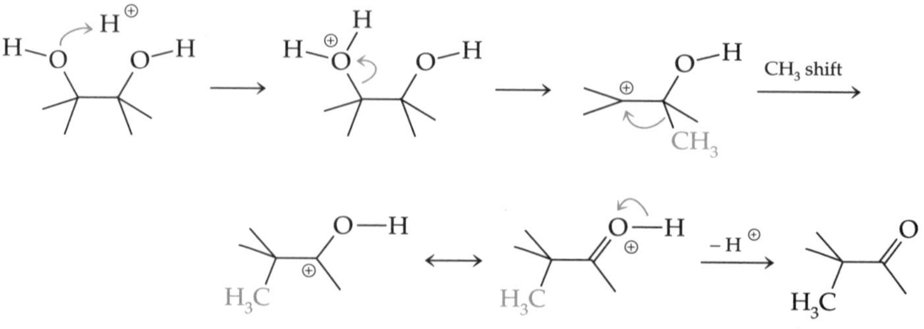

▲ FIGURE 14-4

In a pinacol rearrangement, the driving force for the migration of an alkyl group is the strong π-electron release of a lone pair of electrons from the —OH group when bound to the carbocation.

produces the ketone product. The net difference in bonding between the starting material and the product (diol versus ketone and water) is due to the replacement of a carbon–oxygen σ bond to the second OH group by a carbon–oxygen π bond. Because the latter is stronger by about 7 kcal/mole, the pinacol rearrangement is thermodynamically favorable.

EXERCISE 14-B

Both of the diols shown below undergo the pinacol rearrangement. Predict the product in each case and suggest a mechanism, using curved arrows to indicate electron flow, by which the conversion takes place.

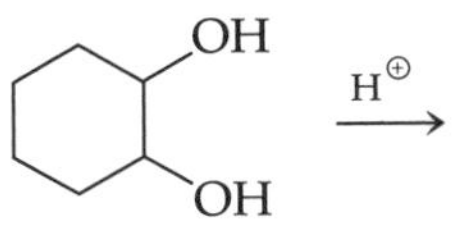

An Anionic Rearrangement

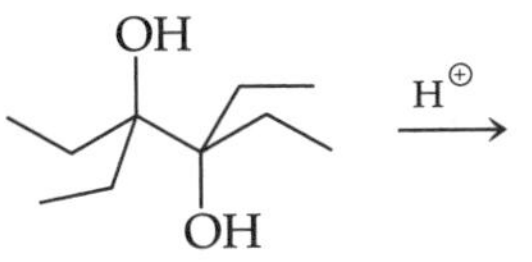

Rearrangements of carbanions are much less common than those of cations. However, such reactions do occur when a more-stable anion is produced. An example of this reaction is the **benzilic acid rearrangement** (Figure 14-5) in which treatment of an α-diketone with hydroxide ion leads to a product acid with a rearranged carbon skeleton.

This rearrangement begins by nucleophilic attack of hydroxide on one of the carbonyl carbons of benzil, in the same fashion as in the nucleophilic acyl substitution reactions in Chapter 12. This first step results in the conversion of one of the carbonyl groups into a tetrahedral intermediate bearing a negatively charged oxygen. This negative charge serves as an electronic "push" for migration of the C–C σ bond to the other carbonyl carbon.

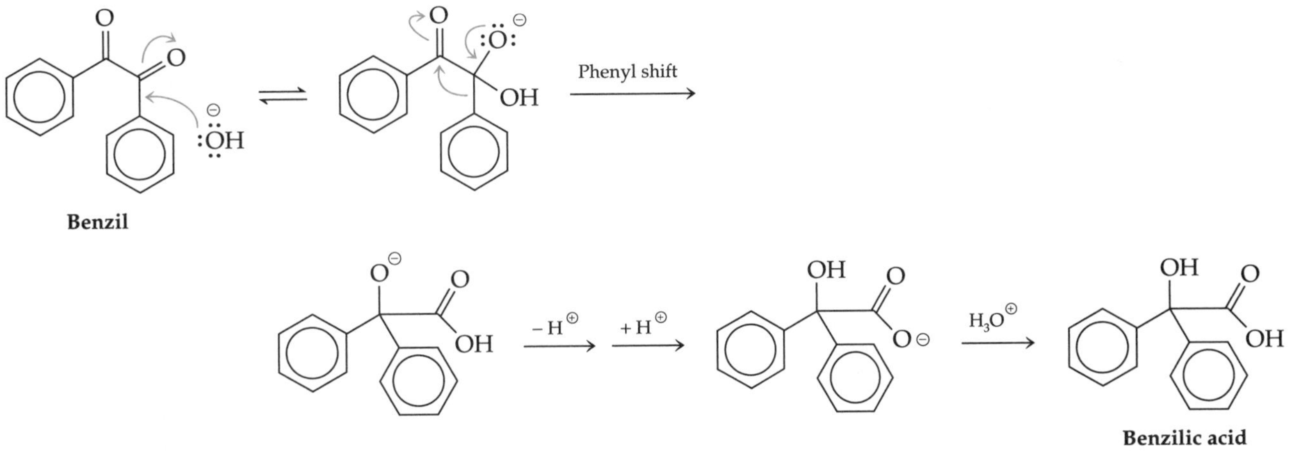

▲ FIGURE 14-5

The first step in a benzilic acid rearrangement is the attack of hydroxide on the carbonyl carbon, parallel to the first step of the Cannizzaro reaction in Chapter 12. Although this step is reversible, no productive reaction is observed until the phenyl ring migrates to the adjacent carbonyl as the electrons on oxygen reform the C=O π bond. Acid-base equilibration and neutralization produce the observed benzilic acid.

In this way, the first carbonyl group is reformed while the π bond of the second is broken. Rapid proton exchange follows, generating the carboxylate anion of the product. The decrease in basicity from the original reagent (hydroxide) to the product (carboxylate) is an important component of the thermodynamic driving force of the benzilic acid rearrangement. Furthermore, α-diketones such as benzil are destabilized by the proximity of the two partially positively charged carbonyl carbons. Protonation upon addition of acid produces benzilic acid, in which two phenyl groups are attached to one carbon. In contrast, these groups were attached in the starting material to adjacent carbonyl groups. Notice that this reaction causes both carbonyl carbons to change oxidation level, one being oxidized and the other being reduced.

Diketones with α hydrogens are preferentially deprotonated by hydroxide, generating an enolate anion that participates in aldol chemistry rather than the benzilic acid rearrangement. Thus, this rearrangement is restricted to α-diketones in which no α hydrogens are present.

EXERCISE 14-C

Using your knowledge of nucleophilic addition to carbonyl compounds, predict whether benzil (PhCOCOPh) or benzophenone (PhCOPh) would be more reactive toward a nucleophile. Explain your reasoning.

EXERCISE 14-D

When 1,1,1-triphenyl-2-bromoethane is treated with lithium metal in THF, 1,1,2-triphenylethane is isolated after neutralization. Propose a mechanism for the formation of this product. (Hint: Remember from Chapter 8 that the treatment of alkyl halides with metals results in one-electron transfer followed by carbon–halogen bond cleavage to generate alkyl radicals, which are reduced in the presence of excess dissolving metal to the corresponding carbanion–alkali-metal cation pair.)

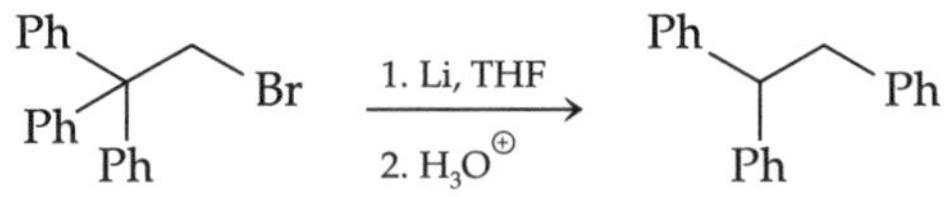

$$\begin{array}{c}\text{Ph} \\ \text{Ph} \end{array}\!\!\!\!\!\!\!\!\!\!\underset{\text{Ph}}{\diagup}\!\!\!\!\!\!\!\!\!\diagdown \text{Br} \quad\xrightarrow[\text{2. H}_3\text{O}^{\oplus}]{\text{1. Li, THF}}\quad \begin{array}{c}\text{Ph}\\ \end{array}\!\!\!\!\!\underset{\text{Ph}}{\diagdown}\!\!\!\!\!\!\!\!\diagup \text{Ph}$$

A Pericyclic Rearrangement: The Cope Rearrangement

As explained in Chapter 7, the Diels-Alder reaction proceeds with high efficiency because it takes place through a transition state comprising six electrons (a Hückel number characteristic of aromaticity), as shown in Figure 14-6. Such a reaction is called **pericyclic** to indicate that the product is formed in **concerted** fashion (without intermediates) through a transition state that can be described as a cyclic array of interacting orbitals. Because this pericyclic reaction results from the combination of two starting materials, it is also referred to as a **cycloaddition reaction.**

In the Diels-Alder reaction, a six-electron delocalized transition state results from the interaction of two molecules, one contributing two and the other four π electrons. The Diels-Alder reaction converts the three π bonds in the reactant into two σ bonds and one π bond in the product. Thus, this cycloaddition requires the *intermolecular* interaction of the π systems of two reactants to form a single cyclic product. Because there is a net conversion

▲ FIGURE 14-6

The Diels-Alder reaction is concerted, taking place through a six-electron transition state.

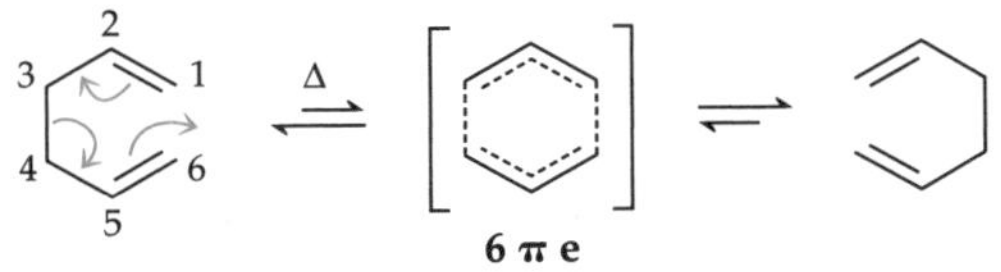

▲ FIGURE 14-7

In a six-electron cyclic transition state, the σ bond between C-3 and C-4 in the 1,5-hexadiene reactant is broken as a new one forms between C-1 and C-6 in the product. At the same time, the π bonds shift to form a degenerate 1,5-hexadiene.

of π into σ bonds, a cycloaddition reaction is generally thermodynamically favorable. The reverse process, called a **cycloreversion,** fragments a cyclic molecule into two or more smaller π systems.

Other transition states having cyclic arrays of six electrons can result in skeletal rearrangements. Electron delocalization similar to that in the transition state for the Diels-Alder reaction is found in the rearrangement of 1,5-hexadiene, as shown in Figure 14-7. In this process, a new carbon–carbon σ bond is formed between C-1 and C-6 at the same time that the bond between C-3 and C-4 is broken. Simultaneously, both π bonds shift and take up new positions between different carbons. This reaction is called a **pericyclic rearrangement** to draw attention to the cyclic nature of its transition state, which connects one end of the π system to the other in this unimolecular rearrangement. It is also called a **Cope rearrangement** in honor of its discoverer, Arthur Cope of the Massachusetts Institute of Technology.

Like that of the Diels-Alder reaction, the transition state for this pericyclic reaction has six π electrons, two from the σ bond and a total of four from the two π bonds. The product of this reaction has the *same* number of σ and π bonds as the reactant, but their positions have shifted within the molecule. This type of reaction is therefore called a **sigmatropic rearrangement,** and the bond migration is called a **sigmatropic shift.** In 1,5-hexadiene, the atoms joined by the σ bond are shifted three carbons in one direction (from C-4 to C-6) and three carbons in the other (from C-3 to C-1); this arrangement is therefore called a [3,3] sigmatropic shift. In addition, in the example shown in Figure 14-7, the product is chemically identical with the starting material (except with respect to the specific identity of the individual carbon atoms). Such processes are called degenerate; this is an example of a **degenerate rearrangement.**

Because the product is the same as the reactant in a degenerate rearrangement, how can we determine that the bond changes shown in Figure 14-7 have actually taken place? One way is to replace specific hydrogen atoms in the starting material with their isotope, deuterium, as shown in Figure 14-8. In the reactant, the deuteriums are bound to vinyl carbons; in the product, at allylic positions. The change is observable because vinylic hydrogens and hydrogens attached to sp^3-hybridized atoms absorb in different regions of the ^{1}H NMR spectrum.

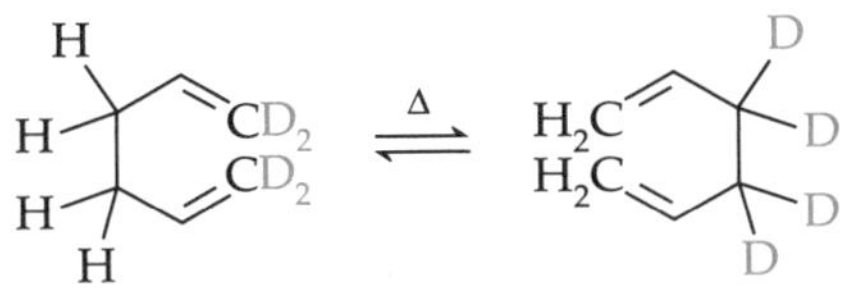

▲ FIGURE 14-8

Isotopic labeling of some positions with deuterium makes it possible to demonstrate that a degenerate rearrangement has taken place.

THE WOODWARD–HOFFMANN RULES

The discovery that pericyclic reactions (like those described in this chapter) were concerted and that their stereochemical course could be predicted from rules derived from theory was made in 1965 by Robert B. Woodward and Roald Hoffmann, both then at Harvard University. The rules that they formulated, called the Woodward-Hoffmann rules, explained many puzzles about such reactions that had for years stumped mechanistic chemists, who had described them as "no mechanism reactions." Based on the simple counting of electrons in interacting π systems, these rules are one of the few examples in chemistry about which one could say: "Exceptions: there are none." This work was acknowledged in the 1981 Nobel Prize to Roald Hoffmann, who was then at Cornell University. (Woodward died in 1979; he had already received a Nobel Prize in chemistry in 1965 for his many contributions to the art of organic synthesis.)

▲ FIGURE 14-9
The Cope rearrangement of a substituted 1,5-hexadiene is *not* degenerate. The product incorporates disubstituted double bonds, whereas the reactant has monosubstituted double bonds.

There is no energetic driving force for the sigmatropic rearrangement shown in Figure 14-8, nor indeed for any degenerate rearrangement. However, in the example in Figure 14-9, the change from the monosubstituted π bonds in the starting material to disubstituted π bonds in the product makes this reaction exothermic so that the product is energetically favored.

Other substituents also can make an otherwise degenerate Cope rearrangement exothermic. For example, a Cope rearrangement of the allylic alcohol in Figure 14-10 initially produces an enol. Tautomerization by proton transfer of this intermediate leads to the ultimate product, an aldehyde. The net change in bonding in this rearrangement converts a carbon–carbon π bond into a carbon–oxygen π bond (63 versus 93 kcal/mole) and an oxygen–hydrogen σ bond into a carbon–hydrogen σ bond (111 versus 99 kcal/mole). Thus, this reaction is sufficiently exothermic (approximately 18 kcal/mole) to be essentially irreversible.

EXERCISE 14-E

Is the conversion of allyl alcohol into the corresponding aldehyde a redox reaction? If so, which atoms undergo a change in oxidation level?

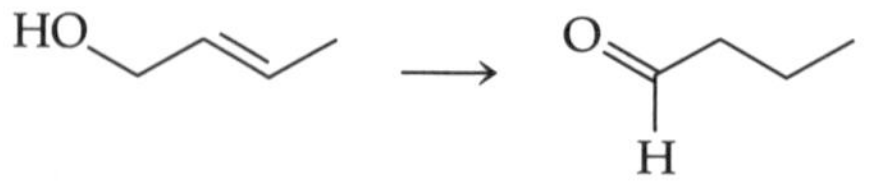

Later in this chapter, we will see that sigmatropic rearrangements also take place in systems in which heteroatoms are present within the hexadienyl skeleton itself.

▶ FIGURE 14-10
When an enol is formed in a Cope rearrangement, rapid tautomerization converts the enol into its carbonyl form, producing an aldehyde.

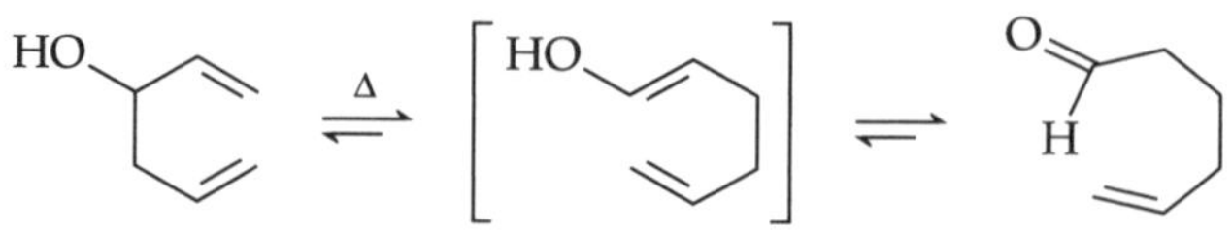

For a molecule to be able to undergo the Cope rearrangement, or, indeed, any pericyclic reaction, the molecule must be capable of achieving a geometry in which the two terminal atoms (here, C-1 and C-6) can interact and bond. When this condition is met, these rearrangements occur simply upon heating the substrate.

There is another way to assemble a six-π-electron transition state starting with a single reactant molecule with three π bonds, as shown in Figure 14-11. This reaction is called an **electrocyclic reaction** and consists of an *intramolecular* cyclization: the three π bonds of the starting triene are converted into one σ bond and two π bonds in the product. Unlike the Cope rearrangement, which is essentially thermoneutral, a six-electron electrocyclization is exothermic by approximately 20 kcal/mole, the difference in energy between a carbon–carbon π and σ bond. An electrocyclic reaction thus consists of the formation of a ring from an acyclic precursor π system; its reverse process (an electrocyclic ring opening) takes a cyclic reactant to a product that has one ring fewer.

▲ FIGURE 14-11
An electrocyclic reaction is an intramolecular interaction of the ends of a π system. Like the Diels-Alder reaction in Figure 14-6 and the Cope rearrangement in Figure 14-7, the electrocyclic reaction of 1,3,5-hexatriene proceeds through a six-electron transition state.

EXERCISE 14-F

Classify each of the following transformations as a cycloaddition, an electrocyclic reaction, a sigmatropic rearrangement, or a nonpericyclic reaction.

(a) 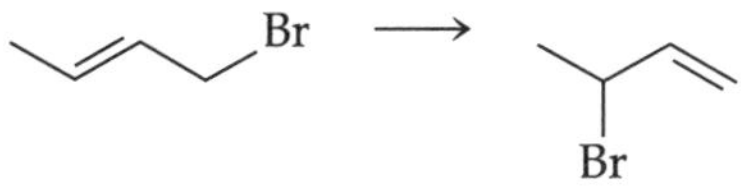 [structure: allyl bromide → branched allyl bromide]

(b) 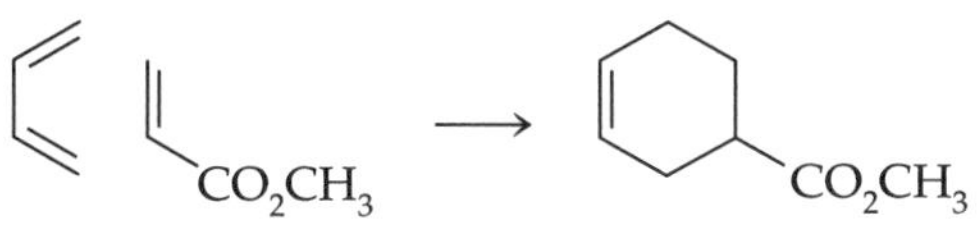 [structure: diene + CO_2CH_3 alkyne → cyclohexene with CO_2CH_3]

(c) 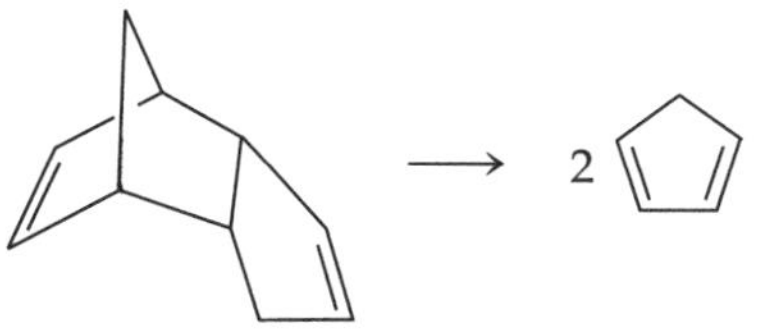 [structure: tricyclic → 2 cyclopentadiene]

(d) 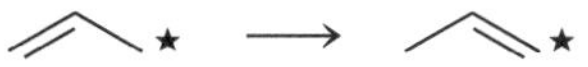 [structure] ⭑ ⟶ [structure] ⭑

⭑indicates ^{13}C

(e) 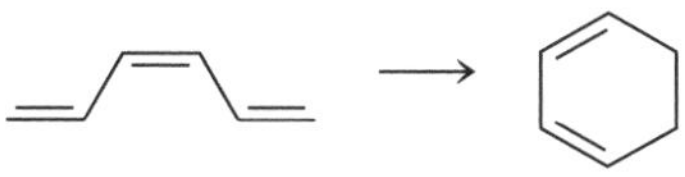 [structure: triene → cyclohexadiene]

(f) 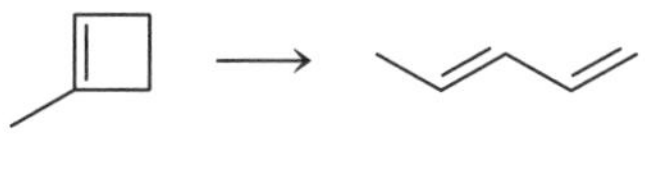 [structure: bicyclobutene → diene]

(g) 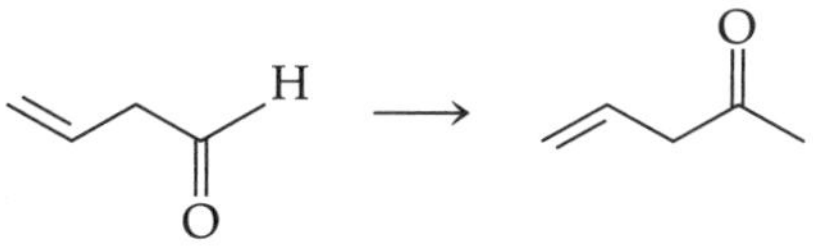 [structure: aldehyde with H → ketone]

(h) [structure: cyclobutene → isobutylene diene]

PERICYCLIC REACTIONS IN THE PHARMACEUTICAL INDUSTRY

Pericyclic reactions are often valuable tools for the synthetic chemist. For example, the industrial synthesis of vitamin D_2 starts by photochemical conversion of the steroid ergosterol into the hexatriene precalciferol by a retroelectrocyclic reaction. Precalciferol undergoes a thermal 1,7-hydrogen shift (the hydrogen is shown in color) to form a different hexatriene, vitamin D_2. This vitamin is also known as calciferol, in recognition of its key role in calcium uptake.

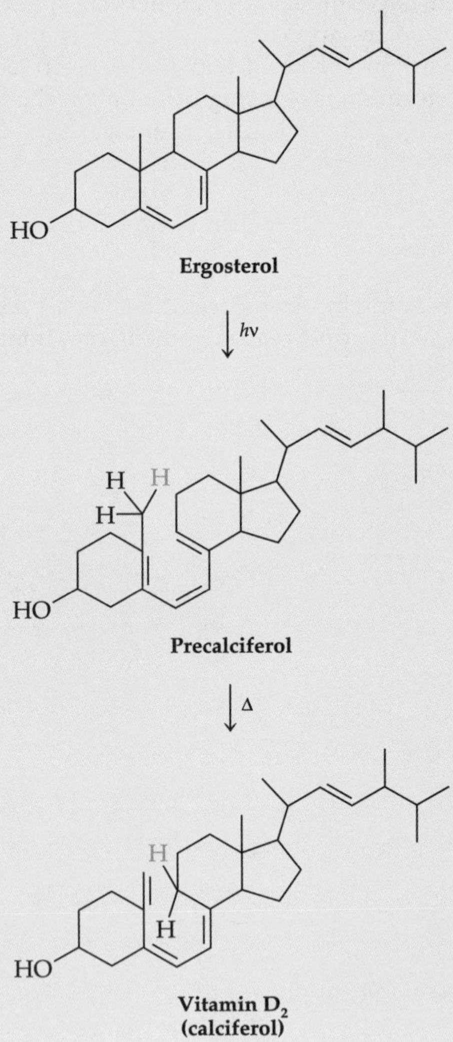

Ergosterol

↓ *hv*

Precalciferol

↓ Δ

Vitamin D_2
(calciferol)

EXERCISE 14–G

The 1,3,5-hexatriene shown in Figure 14-11 is the *cis* geometric isomer. Would an electrocyclic reaction be possible for the *trans* isomer? Explain your reasoning clearly.

14-2 ▸ Carbon–Nitrogen Rearrangements

We have seen several examples of carbon-skeleton rearrangements in which a carbon substituent migrates to another carbon atom. There are several other rearrangement reactions in which alkyl groups migrate to heteroatoms. In this section, we will consider two such transformations, the Beckmann and Hofmann rearrangements, in which migration is to nitrogen. These arrangements, as well as the Baeyer-Villiger oxidation discussed in the next section, have in common: (1) a good leaving group, L, attached to a heteroatom, X; (2) a free lone pair of electrons on the heteroatom; and (3) a migrating group, R (alkyl or aryl), on the adjacent carbon atom.

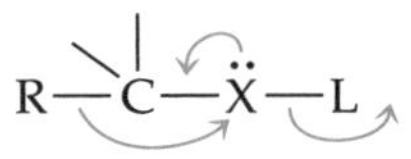

The Beckmann Rearrangement

In the **Beckmann rearrangement,** an oxime is converted into an amide. Recall from Chapter 12 that an oxime is easily obtained by treatment of an aldehyde or ketone with hydroxylamine. A comparison of the ketone from which the oxime is formed with the rearranged amide shows that oxime formation, followed by the Beckmann rearrangement, effectively inserts an —NH— unit between the carbonyl carbon and the α carbon of a ketone.

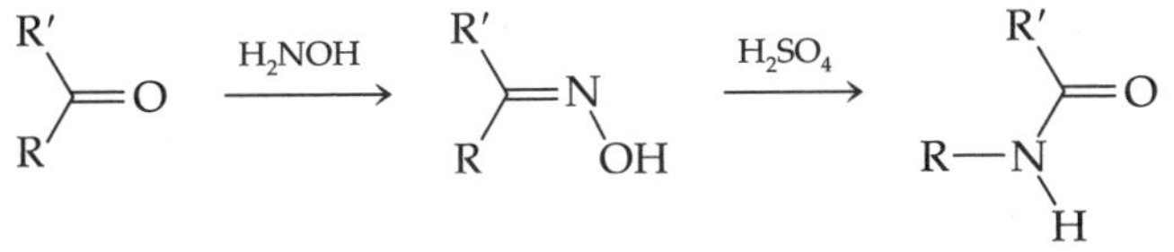

The mechanism of the Beckmann rearrangement begins by the conversion of the OH group of the oxime into a good leaving group. As shown in Figure 14-12, this is usually accomplished by protonation with a strong acid

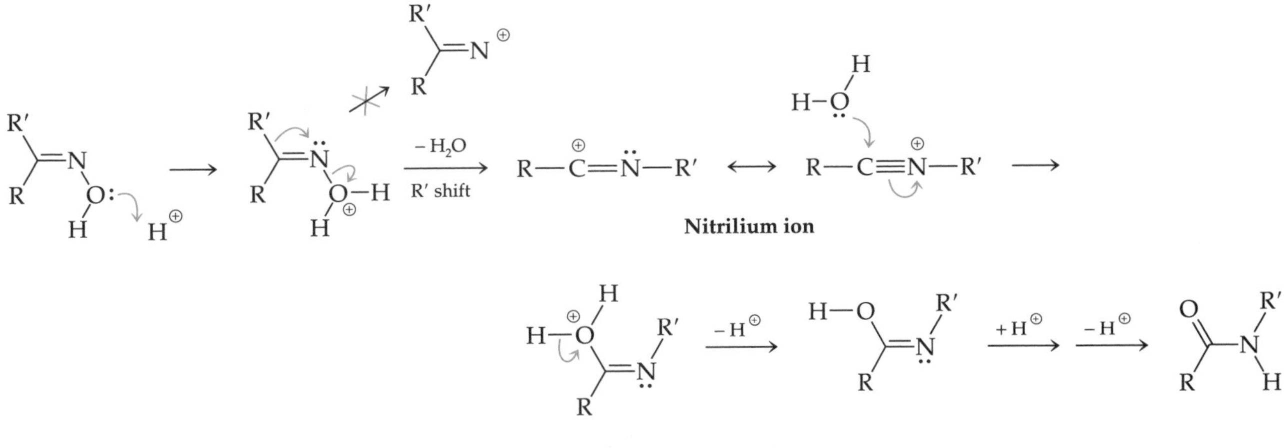

▲ FIGURE 14-12

Protonation of the —OH group of an oxime converts it into a much better leaving group. Water is lost at the same time that an alkyl group attached to the imino carbon migrates to nitrogen, and an electron pair on nitrogen is donated back toward carbon to produce a second π bond. The nitrilium ion thus produced is highly activated toward attack by even a weak nucleophile such as water. Deprotonation and tautomerization lead to the amide product.

such as H_2SO_4. Simple loss of water from this oxonium ion by heterolytic cleavage of the nitrogen–oxygen bond would form a very unstable cation with positive charge on a nitrogen atom lacking an octet of electrons. However, as in reactions such as the pinacol rearrangement, the simultaneous migration of an alkyl group (R′) results in the formation of a resonance-stabilized nitrilium cation.

$$R—C{\equiv}\overset{\oplus}{N}—R' \quad\longleftrightarrow\quad R—\overset{\oplus}{C}{=}\overset{..}{N}—R'$$

A nitrilium ion

This ion is similar electronically to a protonated nitrile and reacts readily with water. Deprotonation and tautomerization of the resulting intermediate gives rise to the product amide. The last steps are essentially identical with those of nitrile hydrolysis, the mechanism of which was given in Chapter 12. However, the Beckmann rearrangement can be accomplished under conditions that are milder than those required for the acid-catalyzed hydrolysis of an amide to a carboxylic acid.

As specific examples, benzophenone oxime is converted into *N*-phenyl-benzamide, and *N*-cyclohexylacetamide is formed from cyclohexyl-methylketone, as shown in Figure 14-13. A comparison of the structure of the starting ketones with that of the products reveals that the Beckmann rearrangement accomplishes the insertion of an —NH— group between the carbonyl carbon and the α carbon. In the second example, two different alkyl groups are attached to the starting ketone. The migration of the larger substituent, as shown, is the usual outcome for the Beckmann rearrangement of unsymmetrical ketones.

EXERCISE 14-H

What is the starting material needed for the synthesis of each of the following amides by a Beckmann rearrangement?

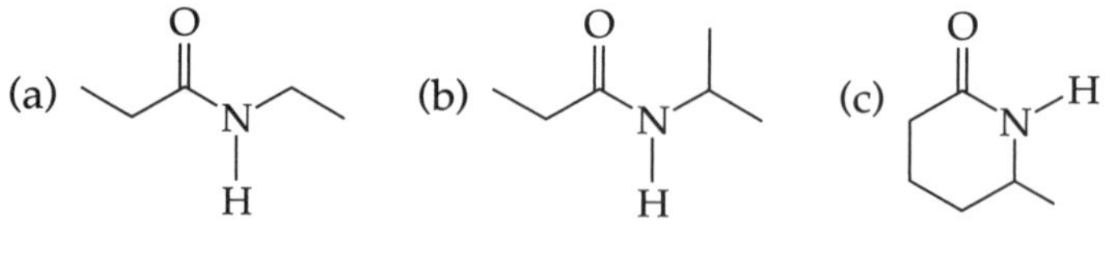

A LARGE–SCALE COMMERCIALLY SIGNIFICANT REARRANGEMENT

The Beckmann rearrangement of the oxime of cyclohexanone is carried out on a very large scale industrially because the product, caprolactam, is the direct precursor of Nylon 6, a versatile polymer that has many applications—among them, as fibers in the manufacture of carpeting. Concentrated sulfuric acid is used both as the acid catalyst and as the solvent for the reaction. However, caprolactam is soluble in sulfuric acid, which must be neutralized in order to isolate the organic product. Ammonia is used for this purpose, and the large quantity of ammonium sulfate produced as by-product is sold as fertilizer.

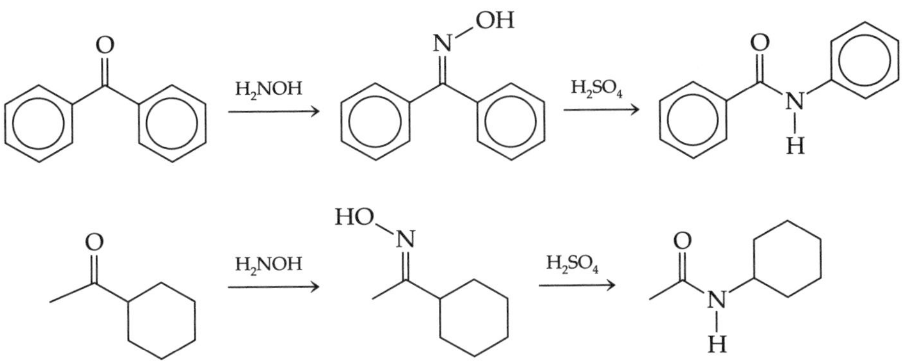

▲ FIGURE 14-13

The combination of oxime formation and the Beckmann rearrangement converts a ketone into an amide by a formal —NH— insertion. Here, *N*-phenylacetamide is produced from benzophenone and *N*-cyclohexylacetamide is produced from cyclohexylmethylketone.

EXERCISE 14-I

Write a detailed reaction mechanism showing a complete electron flow for the reaction sequence required to prepare *N*-cyclohexylacetamide from methylcyclohexyl ketone.

EXERCISE 14-J

The Beckmann rearrangement of cyclopentanone oxime is slower than that of cyclohexanone oxime, which is much slower than that of the oxime of an acyclic ketone. Why is the reaction rate affected by the presence of the ring? (Hint: Consider the geometry of each of the intermediates formed along the rearrangement path for the oxime of cyclopentanone.)

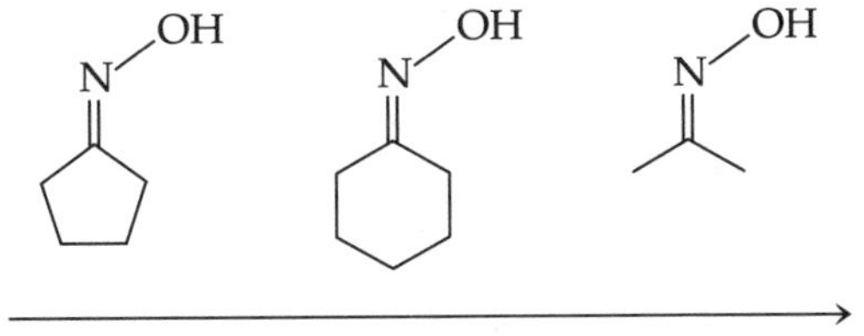

Increasing rate of Beckmann rearrangement

The Hofmann Rearrangement

The **Hofmann rearrangement** results from the treatment of a primary amide with bromine and hydroxide in water, ultimately forming an amine in which the carbonyl group of the starting amide has been lost.

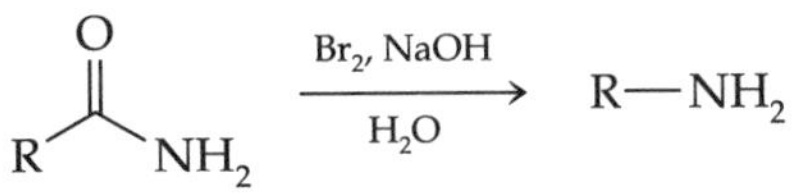

The Hofmann rearrangement occurs through a pathway similar to that of the Beckmann rearrangement. The combination of base and Br_2 converts the amide into an *N*-bromoamide by the same reaction pathway as that for

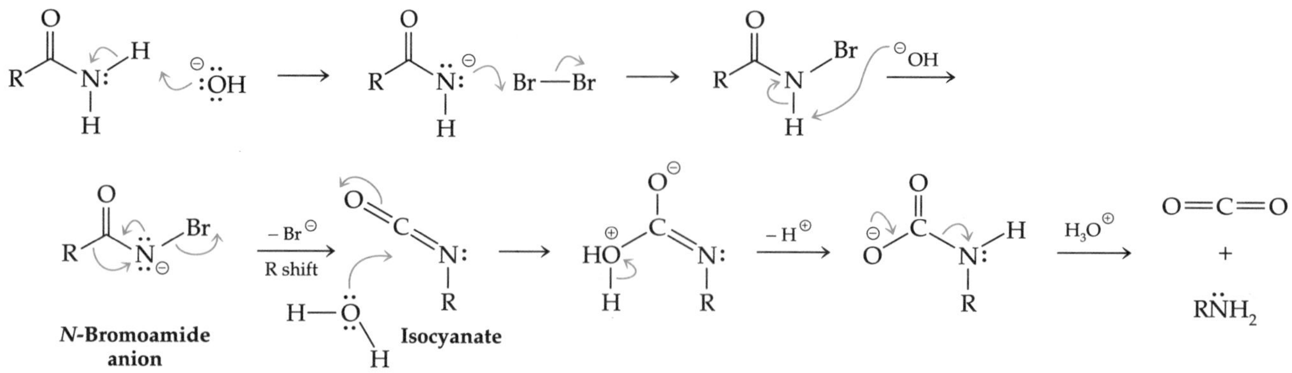

N-Bromoamide anion **Isocyanate**

▲ FIGURE 14-14

The formation of the *N*-bromoamide in the first step of the Hofmann rearrangement is exactly parallel to the formation of α-iodoketone in the first step of the iodoform reaction (Chapter 8). Deprotonation produces an anion that is highly activated for rearrangement. As in the mechanism of the Beckmann rearrangement, bromide ion is lost at the same time that an alkyl group attached to the carbonyl migrates to the amide nitrogen and an electron pair on nitrogen backbonds to produce a π bond. The isocyanate thus produced is highly activated toward attack by water. Deprotonation and tautomerization lead to an unstable carbamic acid that decomposes to CO_2 and an amine.

the conversion of a ketone into an α-bromoketone (Chapter 8). First, an acidic proton is removed from nitrogen by hydroxide ion. The resulting anion then reacts rapidly with Br_2, a very reactive electrophile (Figure 14-14).

A comparison of the structure of this *N*-bromoamide with the protonated oxime in the Beckmann rearrangement reveals a leaving group (L = Br) attached to an atom (N) that bears a lone pair and is adjacent to a carbon atom that bears a potential migrating group (R). Thus, as in the Beckmann rearrangement, the weak bond between nitrogen and the leaving group is cleaved heterolytically with the loss of bromide ion—in this case, as the alkyl group migrates to nitrogen and the lone pair on nitrogen forms a π bond to carbon. This species is called an **isocyanate.** Because of the presence of a carbon doubly bound to two heteroatoms, an isocyanate is even more reactive toward nucleophilic attack by water than are the aldehydes, ketones, and esters discussed in Chapter 12. It is therefore rapidly attacked by water. Acidification of the solution effects protonation of the carboxylate to form an unstable *N*-carboxylic acid (a carbamic acid) that readily loses carbon dioxide.

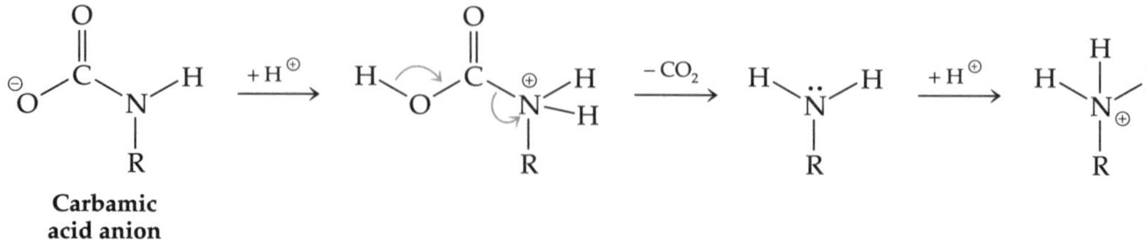

Carbamic acid anion

─────────────

EXERCISE 14-K

Methyl isocyanate is the reagent whose inadvertent release as a gas from a chemical plant in India caused thousands of deaths. Consider the reaction of CH_3NCO with water, and speculate about why it might be so toxic in a human being.

The Hofmann rearrangement provides a method for the synthesis of amines from carboxylic acids, as illustrated for the conversion of benzoic acid into aniline.

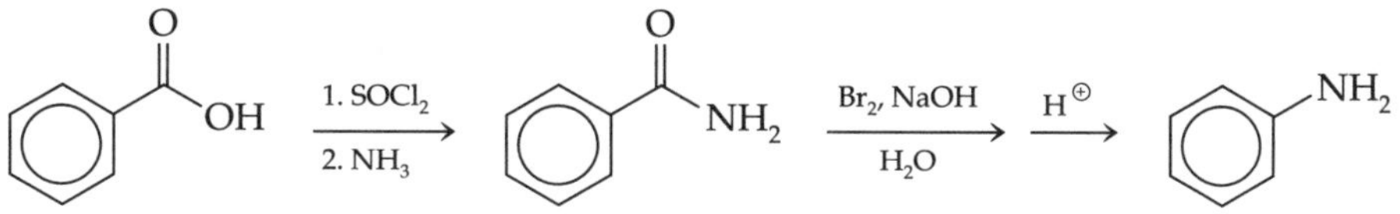

EXERCISE 14-L

Write a detailed mechanism for each of the steps in a Hofmann rearrangement of benzamide to aniline.

EXERCISE 14-M

Suggest a route by which each of the following conversions can be accomplished through a method that employs a Hofmann rearrangement. In some cases, additional steps will be necessary to prepare the starting amide for this rearrangement.

(a)

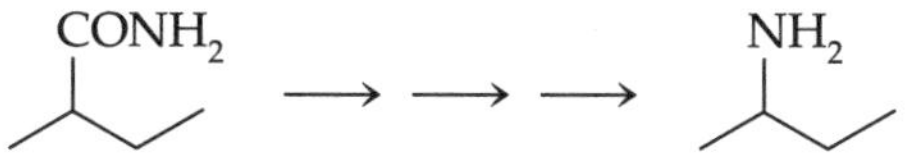

(b)

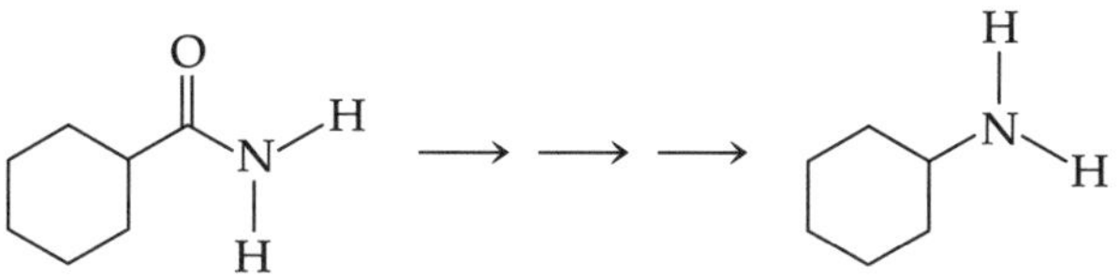

(c)

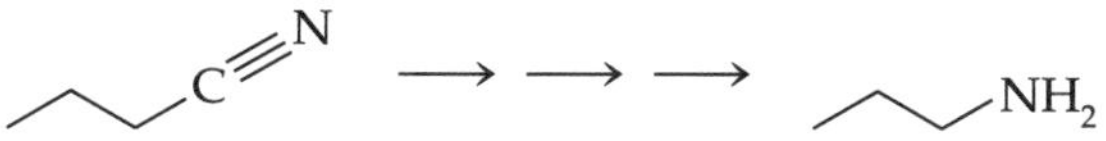

(d)

14-3 ▶ Carbon–Oxygen Rearrangements

In the Sections 14-1 and 14-2, we learned about migrations of alkyl and aryl groups to a different atom, either carbon or nitrogen. In this section, we turn our attention to the migration of an alkyl or aryl group to an oxygen atom. We consider two such rearrangements: first, a multistep reaction (the Baeyer-Villiger oxidation) that converts a ketone into an ester by the insertion of an —O— atom between the carbonyl carbon and an α carbon atom by a mechanism similar to that of the Hofmann rearrangement; second, a concerted, pericyclic reaction (the Claisen rearrangement) that converts an allyl vinyl ether into a γ,δ-unsaturated carbonyl compound by a mechanism similar to that of the Cope rearrangement.

The Baeyer–Villiger Oxidation

The overall effect of the Beckmann rearrangement is the insertion of a nitrogen atom between the carbonyl carbon and the α carbon of a ketone, forming an amide (through the oxime). The **Baeyer-Villiger oxidation** accomplishes a very similar transformation of a ketone, with the insertion of an oxygen atom into the carbonyl–α-carbon bond to form an ester.

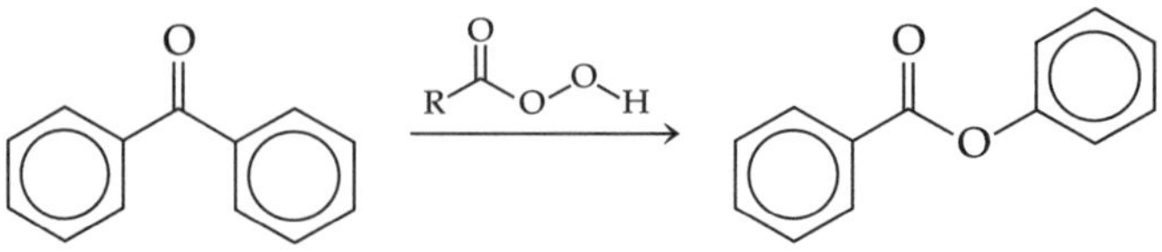

The reaction takes place when a ketone is treated with a peracid, a carboxylic acid that has one additional oxygen. Peracids are powerful oxidizing agents, and this reaction is called an oxidation even though, as we will see, it is quite similar mechanistically to the rearrangements already discussed.

The most-common peracids employed for Baeyer-Villiger oxidations are *m*-chloroperbenzoic acid (MCPBA) and peracetic acid. The first is crystalline and relatively stable when pure. However, it is somewhat more expensive than peracetic acid, which can be prepared as a solution in acetic acid simply by adding hydrogen peroxide and sulfuric acid. All peracids are very unstable in the presence of metals and metal ions, and even atmospheric dust contains a sufficient concentration of metal ions (such as iron oxides) to catalyze the decomposition of peracids to form the acid and molecular oxygen.

The Baeyer-Villiger oxidation begins with the formation of an intermediate resulting from addition of the OH group of the peracid across the π bond of the ketone.

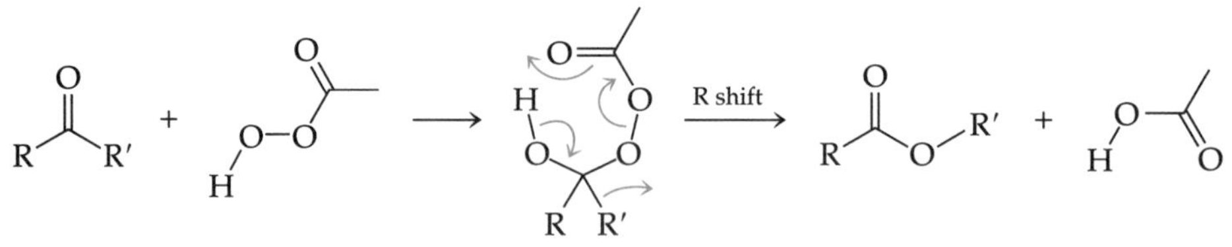

The mechanism of this acid-catalyzed reaction is similar to the steps in the formation of a hemiketal and ketal from a ketone and an alcohol, which were discussed in Chapter 12. Thus, protonation of the carbonyl group activates it toward nucleophilic attack by the terminal oxygen of the peracid. In a cyclic transition state, the C=O bond is reformed as the alkyl group migrates to oxygen and a carboxylic acid leaves. This step is similar to a Beckmann or Hofmann rearrangement, except that the leaving group is a carboxylic acid and the heteroatom to which the group migrates is oxygen. The result of this unimolecular rearrangement, as shown, is the product ester and the acid derived from the peracid.

The Baeyer-Villiger oxidation can be used with either acyclic or cyclic ketones. For example, the Baeyer-Villiger oxidation of cyclohexanone generates a **lactone** (a cyclic ester).

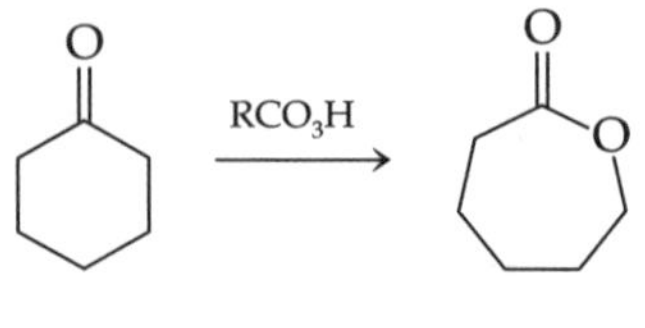

With unsymmetrical ketones, the more highly substituted carbon migrates preferentially.

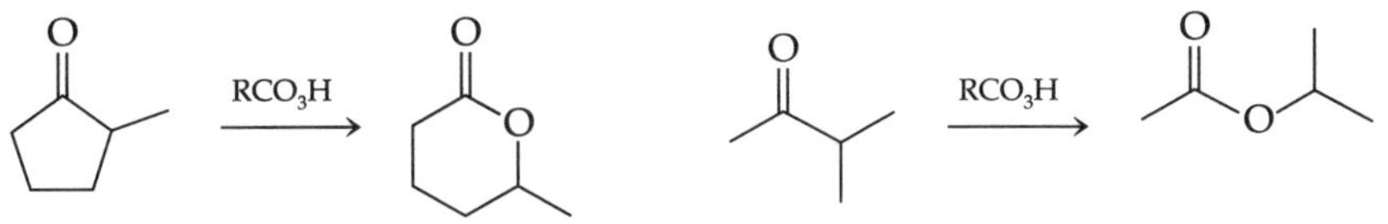

EXERCISE 14-N

For each of the following conversions, suggest a sequence of reagents, employing a Baeyer-Villiger oxidation as the last step.

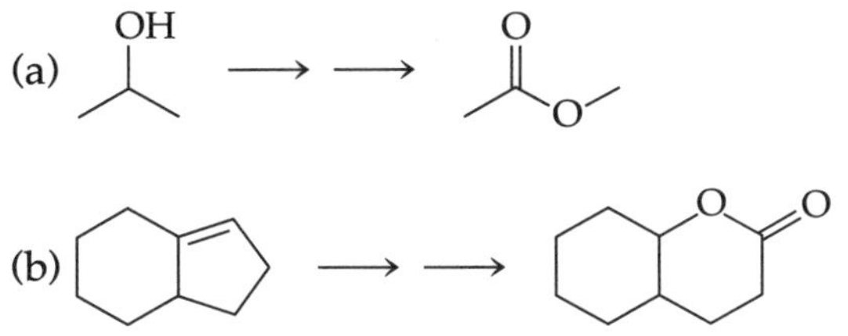

(a)

(b)

A Pericyclic Rearrangement: The Claisen Rearrangement

The **Claisen rearrangement** is a pericyclic reaction very similar to the Cope rearrangement; it, too, takes place through a six-membered transition state having two π bonds and a σ bond. Indeed, the Claisen rearrangement is often referred to as an **oxa-Cope rearrangement** because these two processes differ only by the presence of an oxygen atom in the skeleton. The reactant is an allyl vinyl ether and the product is a γ,δ-enone. The formation of a carbon–oxygen π bond in the product of the Claisen rearrangement makes this process quite exothermic and contributes to a lower transition-state energy from that observed in an analogous all-carbon skeleton (in the Cope rearrangement). The temperature required is therefore lower as well. As a result, Claisen rearrangements are quite useful tools for the formation of ketoalkenes.

An analogous transformation takes place with allyl phenyl ether. Tautomerization of an α hydrogen to oxygen is driven by aromatization of the ring.

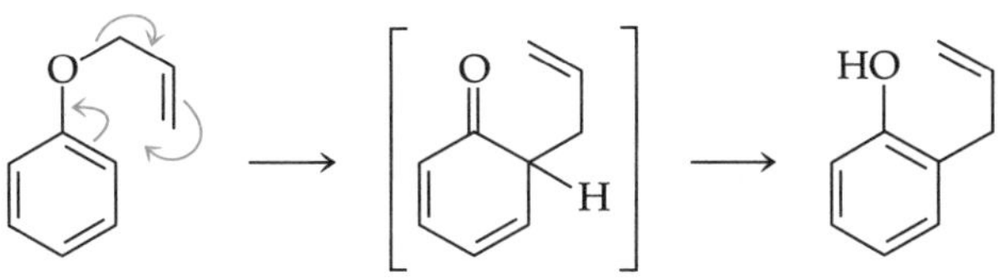

In the phenolic product, the allyl group that was originally bound to oxygen is attached to carbon. Thus, this rearrangement affords an alternative route to Friedel-Crafts alkylation as a method for introducing alkyl substituents onto the *ortho* position of a phenolic ring.

Two other examples of a substrate in which heteroatoms are included in the chain of atoms undergoing concerted, pericyclic rearrangements are shown in Figure 14-15. In the first example, a divinyl hydrazine rearranges to form a diimine, the hydrolysis of which gives rise to the corresponding dialdehyde (Chapter 12). The second example is the rearrangement of an

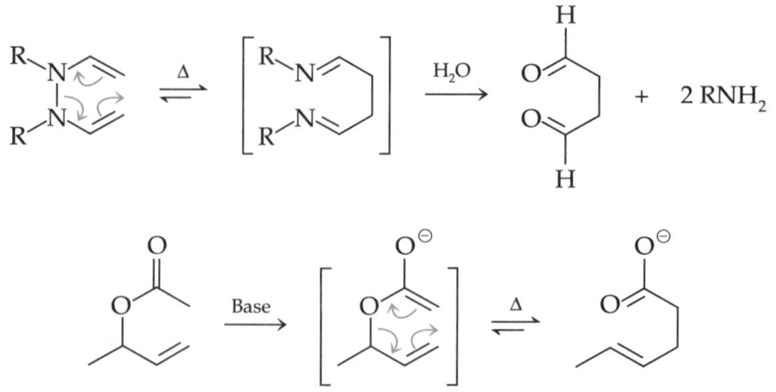

▲ FIGURE 14-15

The concerted rearrangements of 1,5-hexadienes bearing heteroatoms are accomplished through six-electron transition states, as in the Cope and Claisen rearrangements.

allylic ester enolate ion by a pathway very similar to the Claisen rearrangement, except that an additional oxygen atom is present as part of an ester enolate anion.

EXERCISE 14-O

Show the starting materials that would be converted into the following products by a Claisen rearrangement. (Hint: Re-examine the Claisen rearrangement in the reverse direction; that is, from product to starting material.)

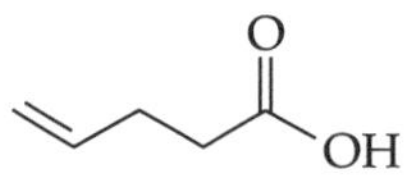

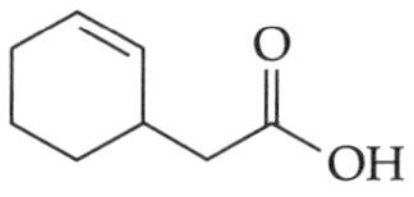

14-4 ▸ Synthetic Applications

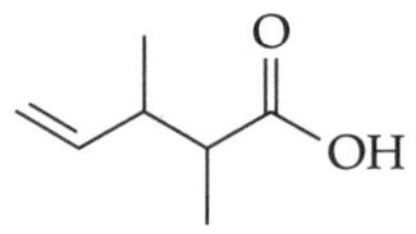

Several of the rearrangements considered in this chapter alter not only the sequence of the attachments of backbone atoms, but also the identity of the functional group present. Table 14-1 regroups the reactions presented herein

Table 14-1 ▸ How to Use Rearrangements to Prepare Various Functional Groups

Functional Group	Reaction
Acid	Benzilic acid rearrangement
Alkene	Cope rearrangement
Amide	Beckmann rearrangement
Amine	Hofmann rearrangement
Ester	Baeyer-Villiger oxidation
Ketone	Pinacol rearrangement; or Claisen rearrangement

according to the functional-group transformation accomplished in each re-action.

The reactions considered in Chapters 6 through 14 constitute the major reaction types that you will need for the rest of this course. The remaining chapters will show how these reactions are incorporated into what practicing chemists do, for example, with applications in synthesis, polymer chemistry, and bioorganic chemistry. It is timely therefore to put these conversions into context with the other reactions that you have learned. Table 14-2 tabulates the reactions studied so far for forming various types of bonds. This table also appears as a reference appendix at the end of this book.

Table 14-2 ▸ Summary of Methods for the Preparation of Functional Groups

C—H

1. Catalytic hydrogenation of alkenes and alkynes:

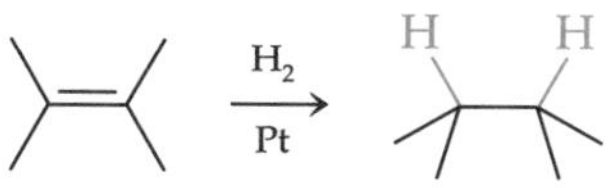

2. Hydrolysis of Grignard reagents:

$$R{-}MgBr \xrightarrow{H_2O} R{-}H$$

3. Clemmensen reduction of an aldehyde or ketone:

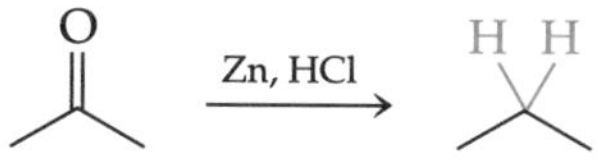

4. Wolff-Kishner reduction of an aldehyde or ketone:

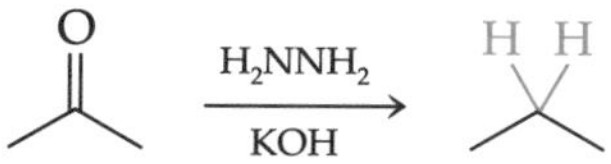

5. Decarboxylation of β-ketoacids:

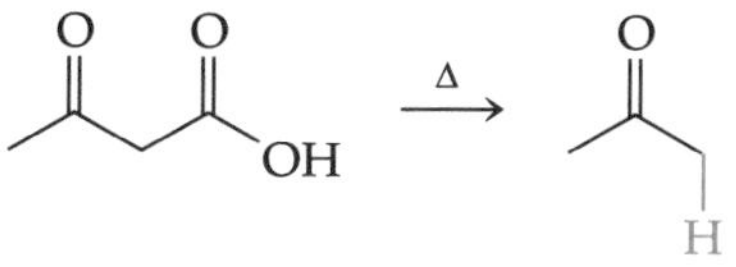

C—C

1. S_N2 displacement by cyanide:

$$R{-}Br \xrightarrow{\ominus C\equiv N} R{-}C\equiv N$$

2. S_N2 displacement by acetylide anions:

$$R{-}Br \ + \ \ominus{\equiv\!\equiv}{-}R' \ \longrightarrow \ R{-}{\equiv\!\equiv}{-}R'$$

3. Grignard addition:

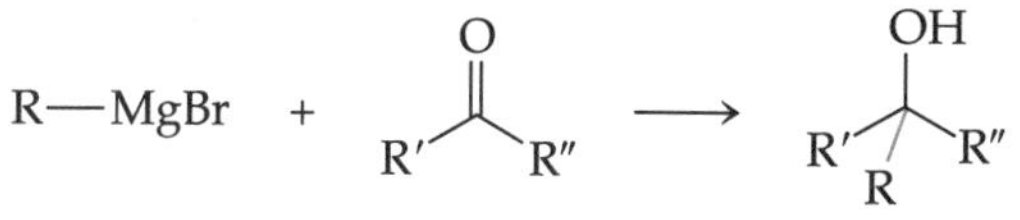

R' and R" = H, alkyl, aryl
R = alkyl, aryl

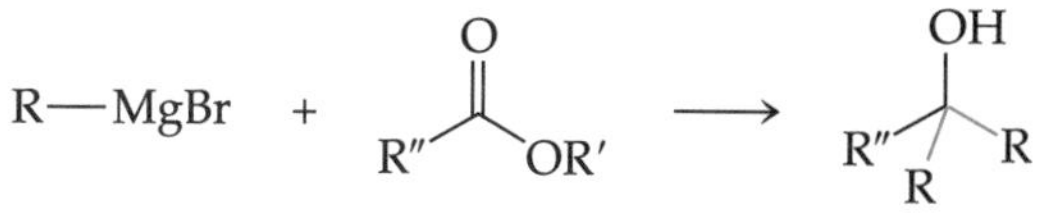

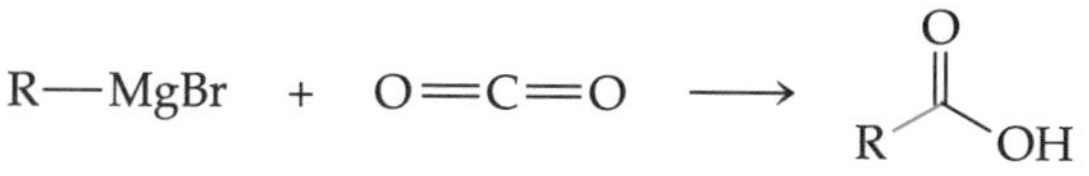

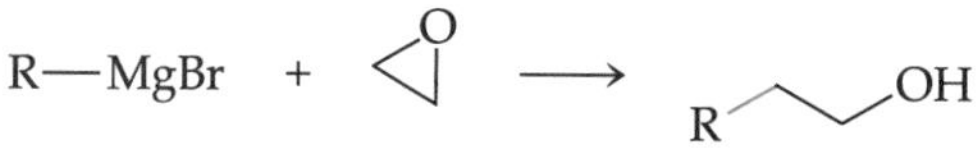

4. Friedel-Crafts acylation:

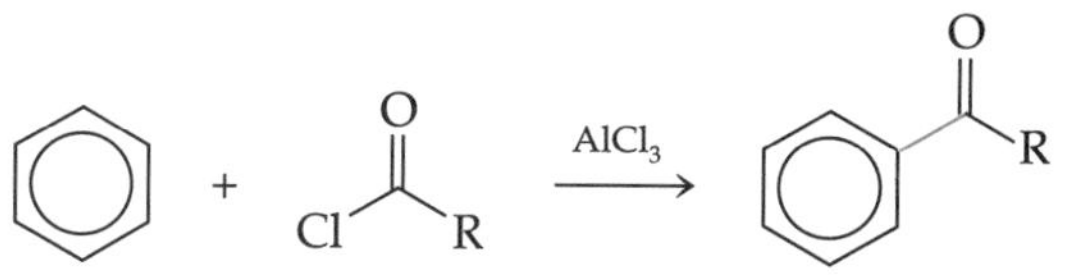

5. Friedel-Crafts alkylation:

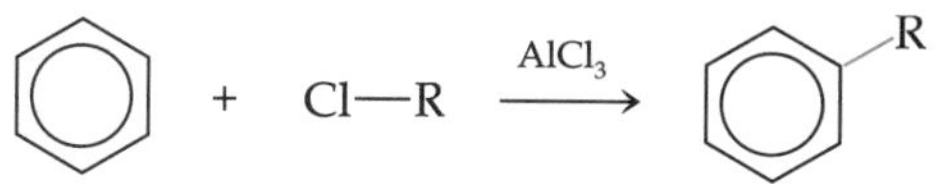

6. Diels-Alder reaction:

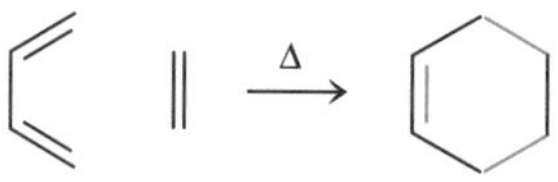

7. Conjugate addition to α,β-unsaturated carbonyl groups:

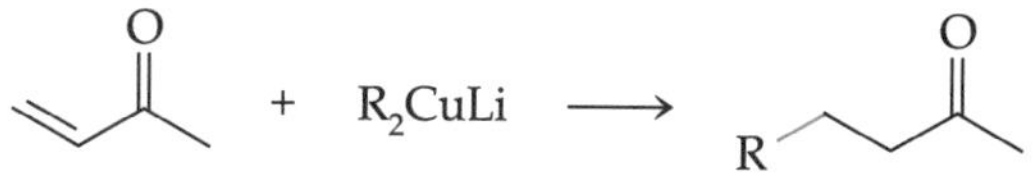

Table 14-2 ▸ Summary of Methods for the Preparation of Functional Groups (*continued*)

8. Michael reaction:

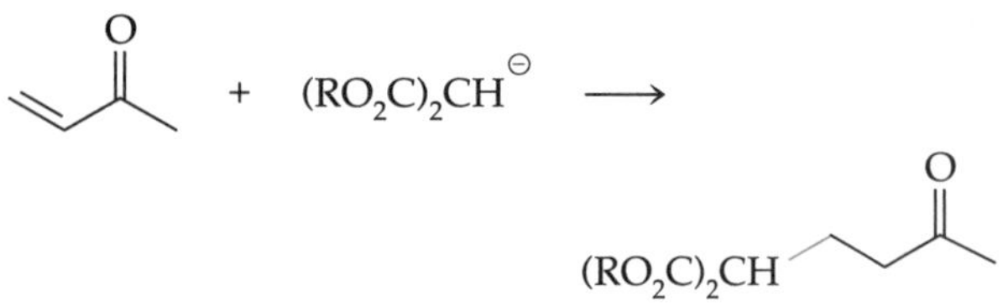

$$\text{(from CH}_2\text{=CH–CO–CH}_3) + (RO_2C)_2CH^\ominus \longrightarrow (RO_2C)_2CH\text{–CH}_2\text{CH}_2\text{–CO–CH}_3$$

9. Aldol reaction:

(acetone) $\xrightarrow{\ ^\ominus \text{Base}\ }$ (4-hydroxy-4-methyl-2-pentanone)

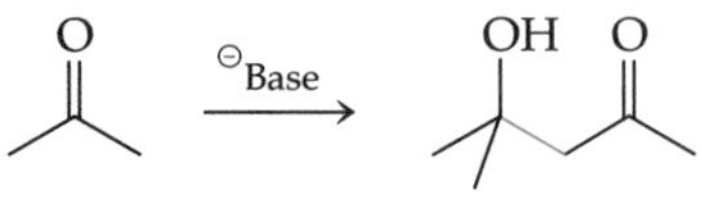

10. Acetoacetic ester synthesis:

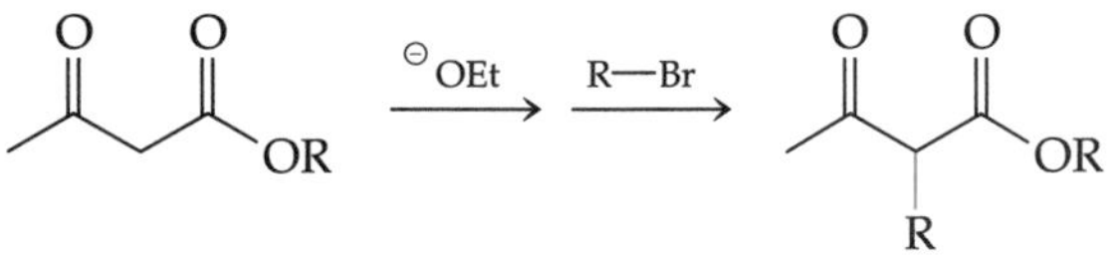

(ethyl acetoacetate) $\xrightarrow{\ ^\ominus \text{OEt}\ \ \ R\text{—Br}\ }$ (alkylated acetoacetate, R at α-carbon)

11. Malonic ester synthesis:

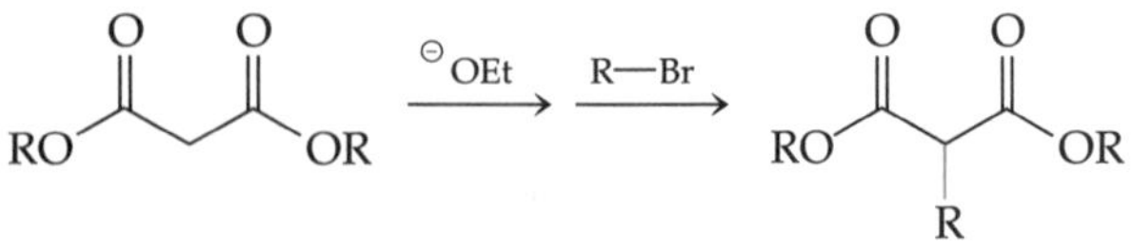

(diethyl malonate) $\xrightarrow{\ ^\ominus \text{OEt}\ \ \ R\text{—Br}\ }$ (alkylated malonate, R at central carbon)

12. Claisen condensation:

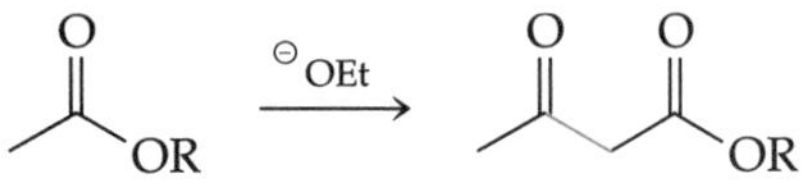

(ethyl acetate) $\xrightarrow{\ ^\ominus \text{OEt}\ }$ (ethyl acetoacetate)

13. Cope rearrangement:

(1,5-hexadiene) $\longrightarrow$ (cyclohexene derivative)

14. Claisen rearrangement:

(allyl vinyl ether) $\longrightarrow$ (γ,δ-unsaturated carbonyl)

C=C

1. Dehydrohalogenation:

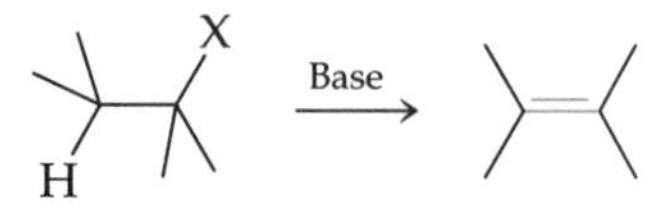

(alkyl halide) $\xrightarrow{\ \text{Base}\ }$ (alkene)

2. Dehydration:

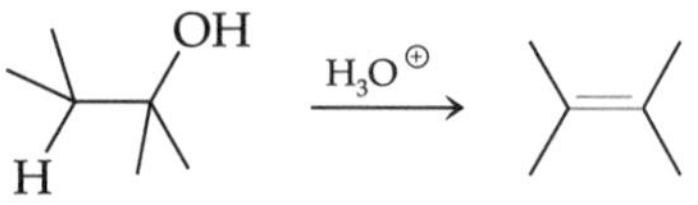

(alcohol) $\xrightarrow{\ H_3O^\oplus\ }$ (alkene)

3. Hofmann elimination:

($^\oplus NR_3$ substrate) $\xrightarrow{\ ^\ominus \text{OH}\ }$ (alkene)

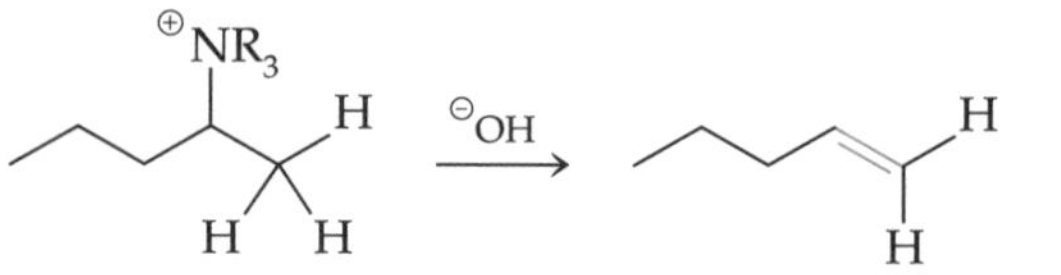

4. Catalytic hydrogenation of alkynes:

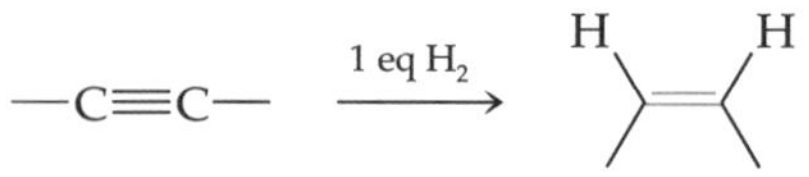

$$-C{\equiv}C- \xrightarrow{\ 1\ eq\ H_2\ } \text{(cis-alkene)}$$

5. Dissolving metal reduction of alkynes:

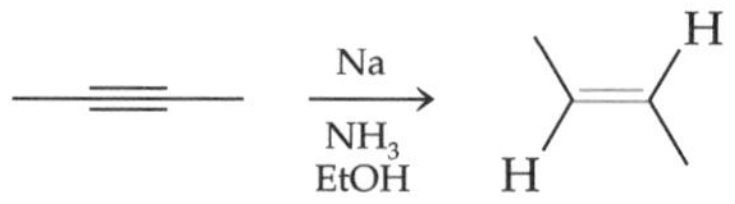

(alkyne) $\xrightarrow[\substack{NH_3 \\ EtOH}]{\ Na\ }$ (trans-alkene)

6. Reductive elimination:

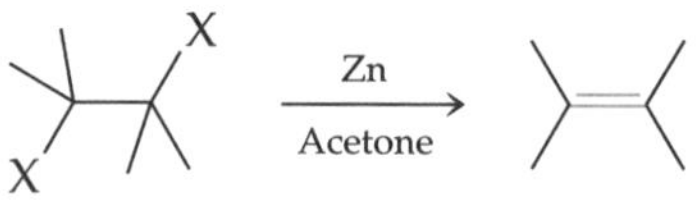

(vicinal dihalide) $\xrightarrow[\text{Acetone}]{\ Zn\ }$ (alkene)

7. Wittig reaction:

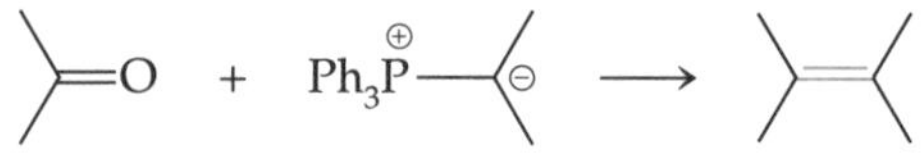

(ketone) =O $+$ $Ph_3\overset{\oplus}{P}$—$^\ominus$ $\longrightarrow$ (alkene)

8. Aldol condensation:

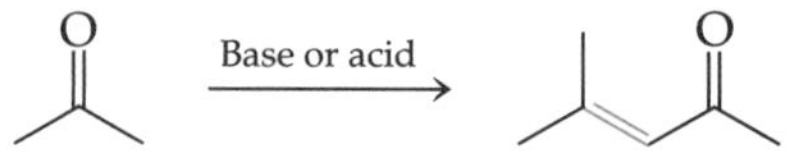

(acetone) $\xrightarrow{\ \text{Base or acid}\ }$ (4-methyl-3-penten-2-one)

C≡C

1. Dehydrohalogenation:

(vicinal dibromide) $\xrightarrow{\ ^\ominus \text{Base}\ }$ $-C{\equiv}C-$

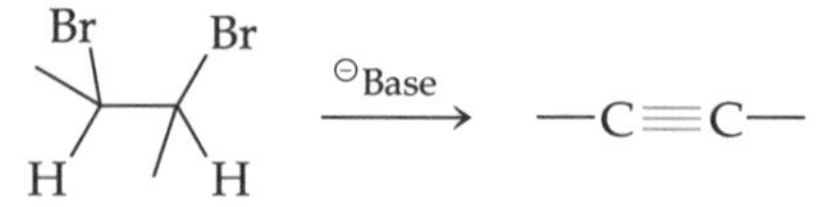

2. S$_N$2 displacement by acetylide anions:

$$-C{\equiv}C^\ominus + R\text{—Br} \longrightarrow -C{\equiv}C\text{—}R$$

C—X

1. Free radical halogenation:

$$R\text{—H} \xrightarrow[h\nu]{\ X_2\ } R\text{—X}$$

2. Addition of HX:

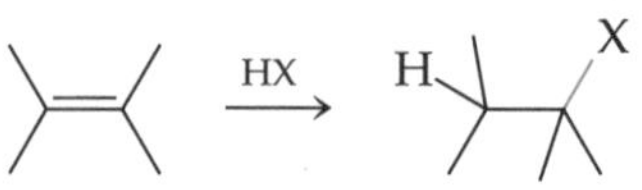

(alkene) $\xrightarrow{\ HX\ }$ (alkyl halide)

Table 14-2 ▸ Summary of Methods for the Preparation of Functional Groups (*continued*)

3. Conversion of alcohols into alkyl halides:

$$R—OH \xrightarrow{PX_3, POX_3, HX, \text{ or } SOX_2} R—X$$

4. Electrophilic aromatic substitution:

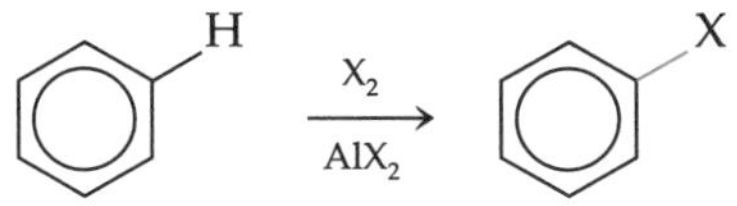

5. α-Halogenation of ketones:

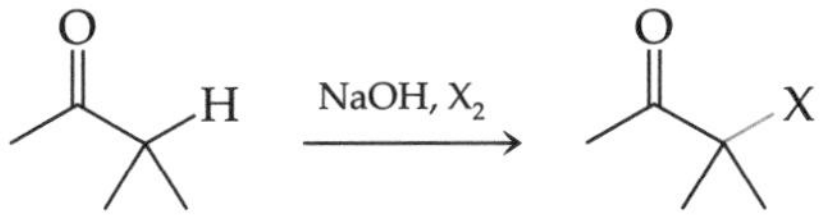

6. Hell-Volhard-Zelinski reaction:

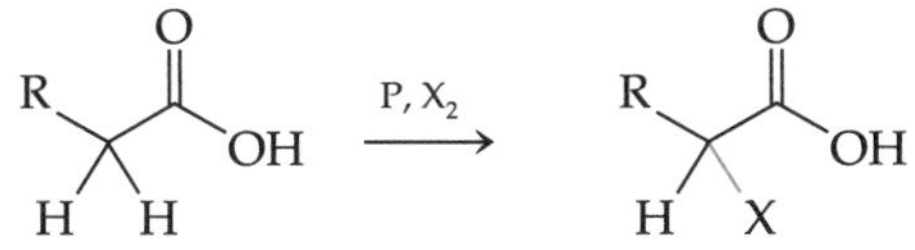

C—OH

1. Hydrolysis of alkyl halides:

$$R—X \xrightarrow{^{\ominus}OH} R—OH$$

2. Hydration of alkenes:

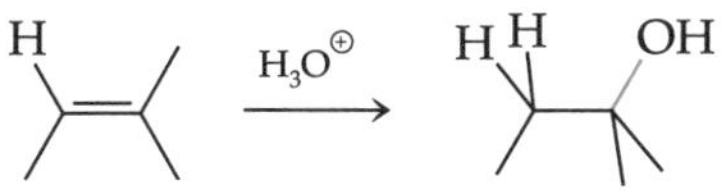

3. Oxymercuration-demercuration:

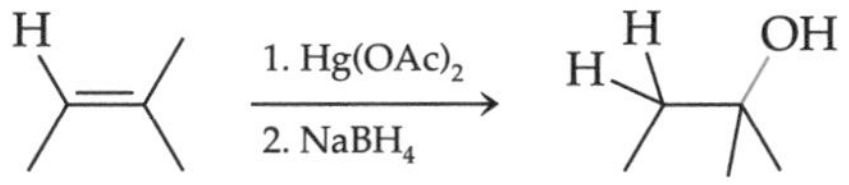

4. Hydroboration-oxidation:

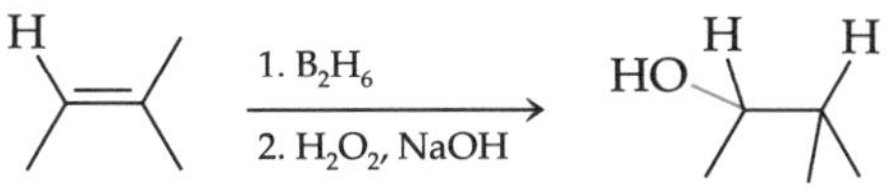

5. Hydration of benzyne:

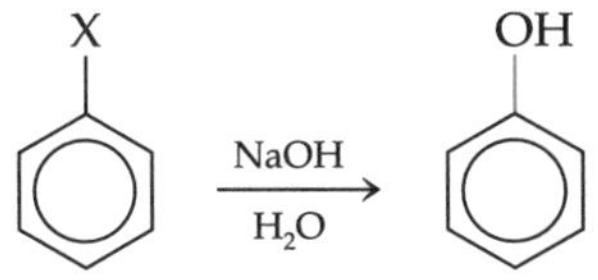

6. Nucleophilic opening of epoxides:

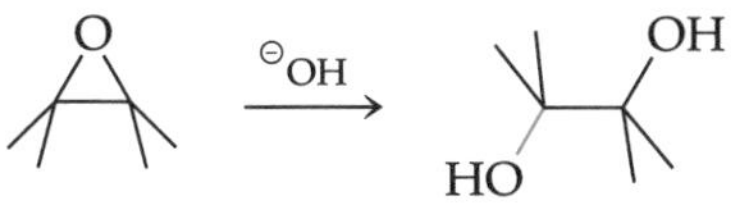

7. Grignard reaction of aldehydes and ketones:

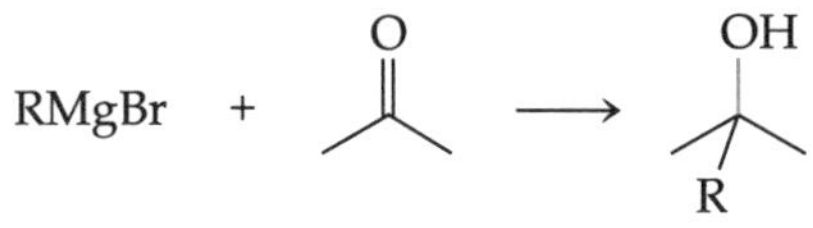

8. Grignard reaction of esters:

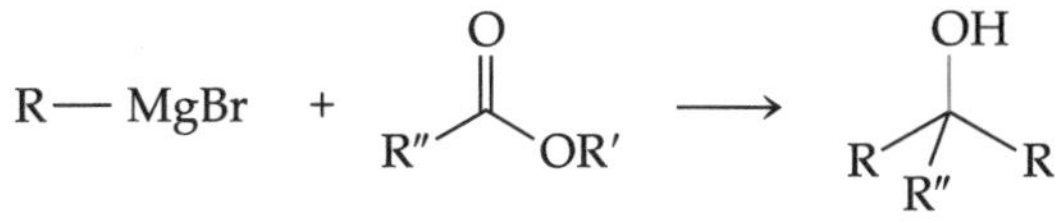

9. Metal hydride reductions of aldehydes and ketones:

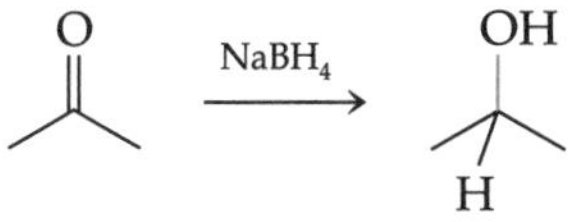

10. Metal hydride reduction of esters:

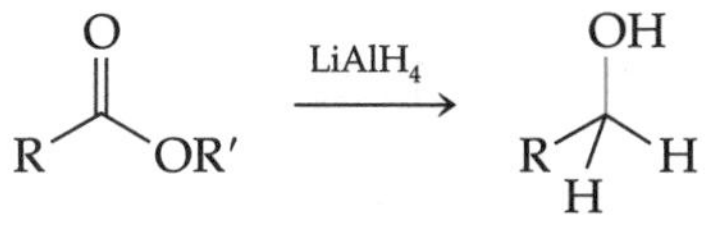

11. Aldol reaction:

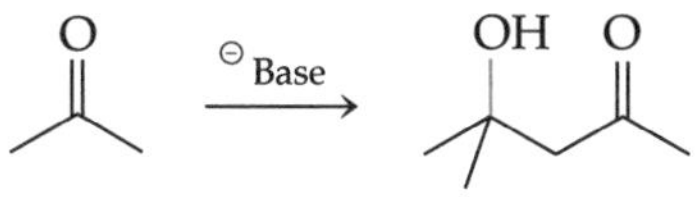

12. Cannizzaro reaction:

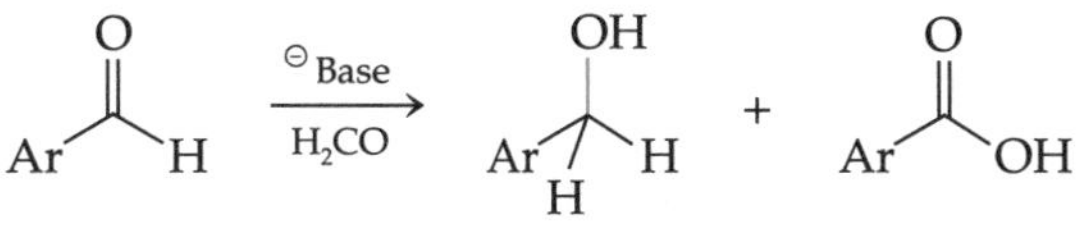

R—C≡N

1. S_N2 displacement by cyanide:

$$R—Br + {}^{\ominus}C≡N \longrightarrow R—C≡N$$

2. Cyanohydrin formation:

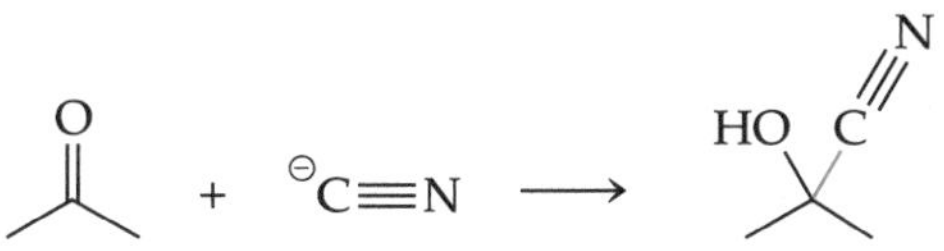

3. Dehydration of amides:

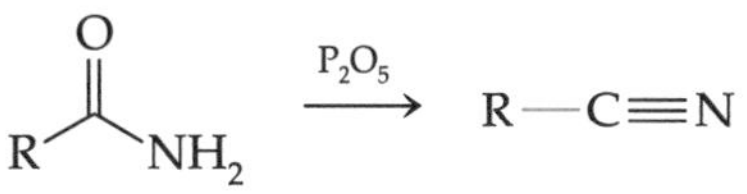

Table 14-2 ▸ Summary of Methods for the Preparation of Functional Groups (*continued*)

R—NH₂

1. Aminolysis of alkyl halides:

$$R—X \xrightarrow{NH_3} R—NH_2$$

2. Reduction of aromatic nitro compounds:

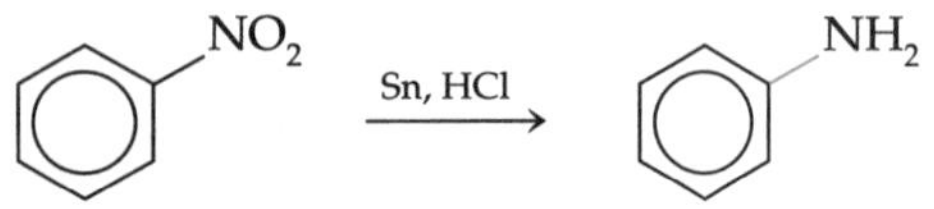

Ar—NO₂ $\xrightarrow{\text{Sn, HCl}}$ Ar—NH₂

3. Amination of benzyne:

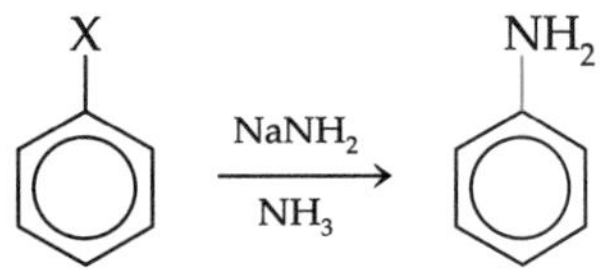

Ar—X $\xrightarrow[\text{NH}_3]{\text{NaNH}_2}$ Ar—NH₂

4. Reductive amination of ketones:

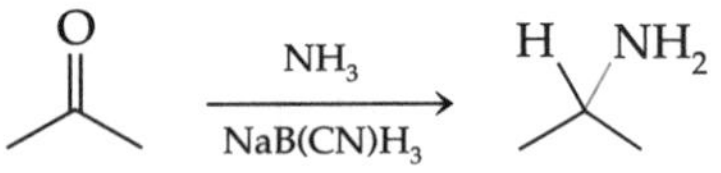

ketone $\xrightarrow[\text{NaB(CN)H}_3]{\text{NH}_3}$ amine

5. Lithium aluminum hydride reduction of amides:

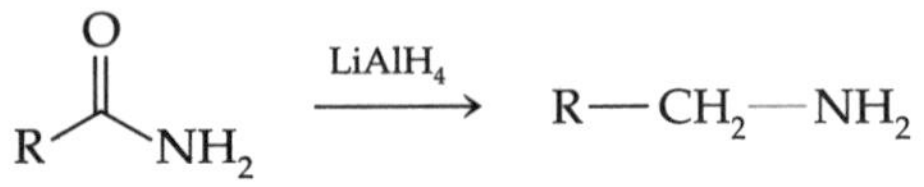

$$R—C(=O)—NH_2 \xrightarrow{\text{LiAlH}_4} R—CH_2—NH_2$$

6. Hofmann rearrangement:

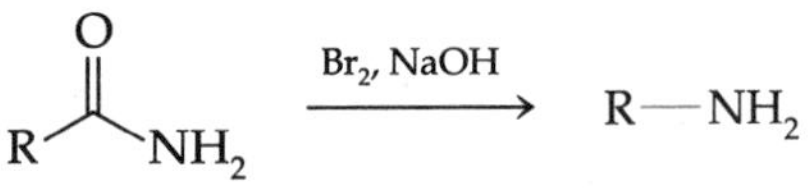

$$R—C(=O)—NH_2 \xrightarrow{\text{Br}_2,\ \text{NaOH}} R—NH_2$$

7. Catalytic hydrogenation of nitriles:

$$R—C≡N \xrightarrow{\text{H}_2,\ \text{Ni(Ra)}} R—CH_2—NH_2$$

8. Gabriel synthesis:

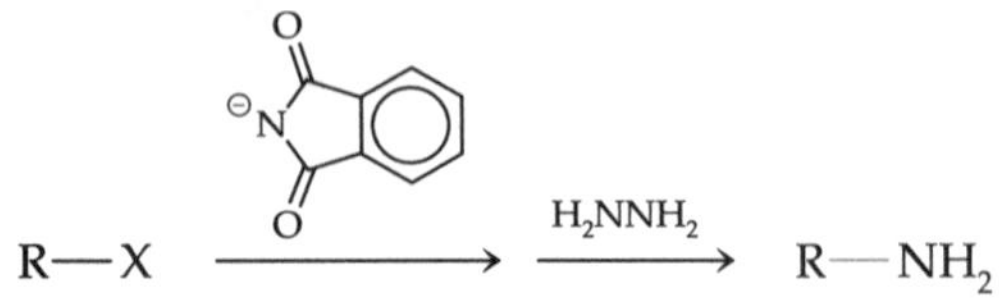

$$R—X \xrightarrow{\text{H}_2\text{NNH}_2} R—NH_2$$

9. Catalytic hydrogenation of azides:

$$R—X \xrightarrow{\text{NaN}_3} R—N_3 \xrightarrow[\text{Pt}]{\text{H}_2} R—NH_2$$

R—O—R′

1. Williamson ether synthesis:

$$R—O^{\ominus} + Br—R' \longrightarrow R—O—R'$$

2. Peracid oxidation of alkenes:

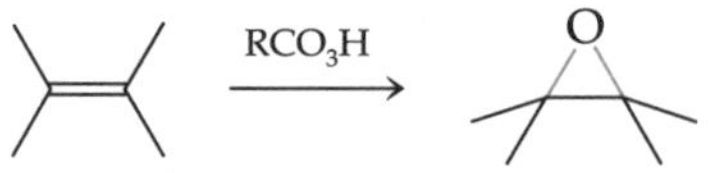

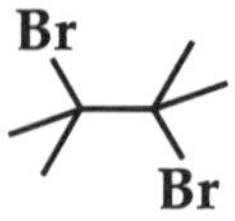

alkene $\xrightarrow{\text{RCO}_3\text{H}}$ epoxide

Br / Br

1. Bromination of alkenes:

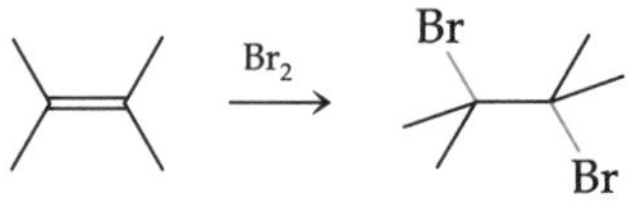

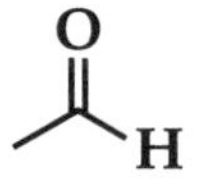

alkene $\xrightarrow{\text{Br}_2}$ dibromide

(aldehyde) O / H

1. Oxidation of primary alcohols:

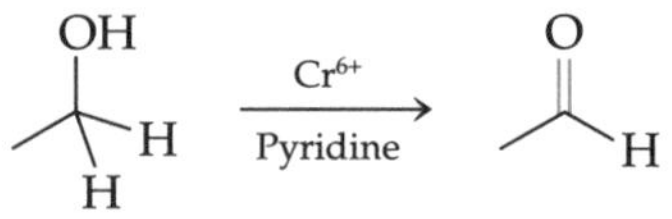

$$R—CH_2OH \xrightarrow[\text{Pyridine}]{\text{Cr}^{6+}} R—CHO$$

2. Aldol reaction:

acetaldehyde $\xrightarrow{^{\ominus}\text{Base}}$ aldol

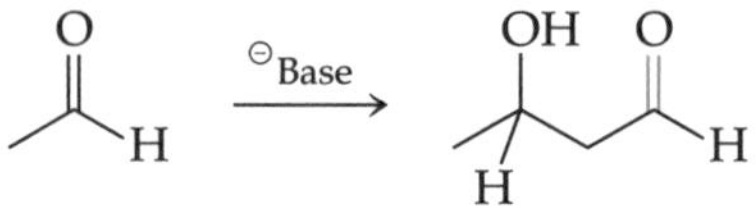

3. Ozonolysis of alkenes:

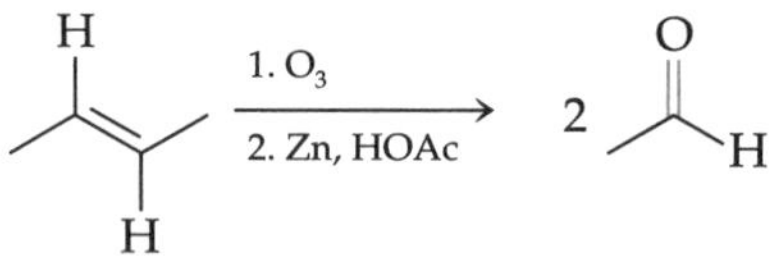

alkene $\xrightarrow[\text{2. Zn, HOAc}]{\text{1. O}_3}$ 2 aldehyde

4. Acetal hydrolysis:

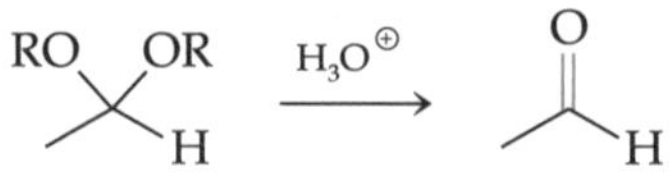

acetal $\xrightarrow{\text{H}_3\text{O}^{\oplus}}$ aldehyde

(ketone) O

1. Ketal hydrolysis:

ketal $\xrightarrow{\text{H}^+,\ \text{H}_2\text{O}}$ ketone

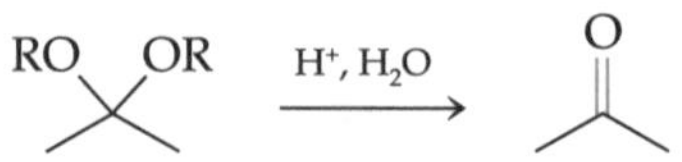

2. Hydrolysis of terminal alkynes:

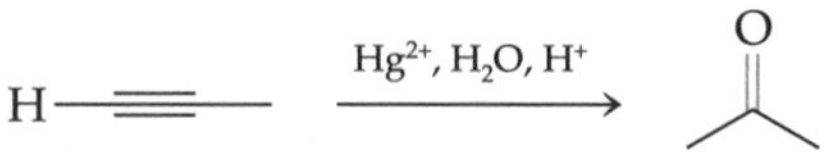

H—C≡C—R $\xrightarrow{\text{Hg}^{2+},\ \text{H}_2\text{O},\ \text{H}^+}$ ketone

Table 14-2 ▸ Summary of Methods for the Preparation of Functional Groups (*continued*)

3. Chromate oxidation of secondary alcohols:

4. Oxidation of aldehydes:

4. Friedel-Crafts acylation:

5. Permanganate oxidation of alkyl side chains of arenes:

5. Claisen condensation:

6. Iodoform reaction:

6. Decarboxylation of β-ketoacids:

7. Carboxylation of Grignard reagents:

$$R-Br \xrightarrow[\text{2. } CO_2]{\text{1. Mg}} \xrightarrow{H^+}$$

7. Pinacol rearrangement:

8. Decarboxylation of β-diacids:

8. Claisen rearrangement:

9. Benzilic acid rearrangement:

1. Hydrolysis of carboxylic acid derivatives:

$$\xrightarrow[\text{(or NaOH)}]{H^+, H_2O}$$

X = Cl, OR, OAc, NR₂, SR

2. Nitrile hydrolysis:

$$-C{\equiv}N \xrightarrow[\text{(or NaOH)}]{H^+, H_2O}$$

3. Oxidation of primary alcohols:

$$-CH_2-OH \xrightarrow[H_2O]{Cr^{6+}}$$

1. Treatment of acid chloride with thionyl chloride:

$$\xrightarrow[\text{(or PCl}_3 \text{ or PCl}_5)]{SOCl_2}$$

1. Acid dehydration:

$$\xrightarrow{\Delta \text{ or } P_2O_5}$$

Table 14-2 ▸ Summary of Methods for the Preparation of Functional Groups (*continued*)

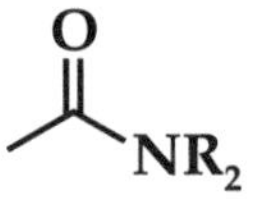

$\underset{\text{NR}_2}{\overset{\text{O}}{\|}}$

1. Amidation of carboxylic acid derivatives:

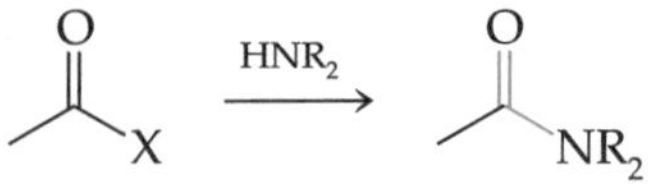

$$\underset{X}{\overset{O}{\|}} \xrightarrow{HNR_2} \underset{NR_2}{\overset{O}{\|}}$$

X = Cl, OR, OAc

2. Beckmann rearrangement:

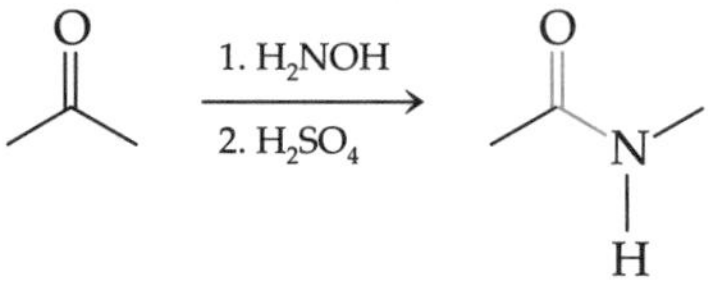

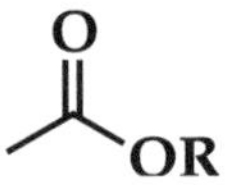

$$\overset{O}{\|} \xrightarrow[\text{2. } H_2SO_4]{\text{1. } H_2NOH} \underset{\underset{H}{N}}{\overset{O}{\|}}$$

$\underset{\text{OR}}{\overset{\text{O}}{\|}}$

1. Esterification of carboxylic acids:

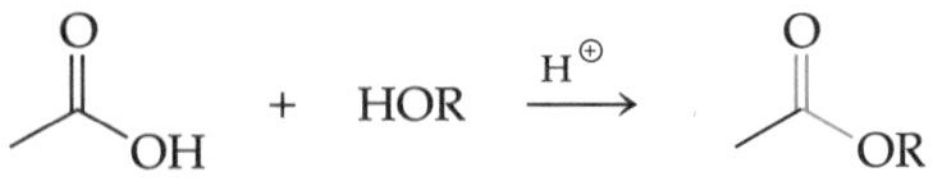

$$\underset{OH}{\overset{O}{\|}} + \text{HOR} \xrightarrow{H^\oplus} \underset{OR}{\overset{O}{\|}}$$

2. Transesterification:

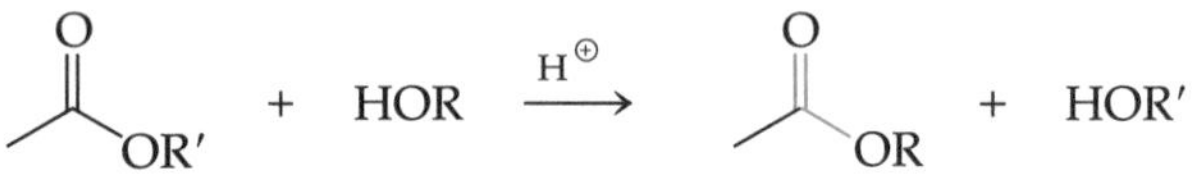

$$\underset{OR'}{\overset{O}{\|}} + \text{HOR} \xrightarrow{H^\oplus} \underset{OR}{\overset{O}{\|}} + \text{HOR}'$$

3. Baeyer-Villiger oxidation:

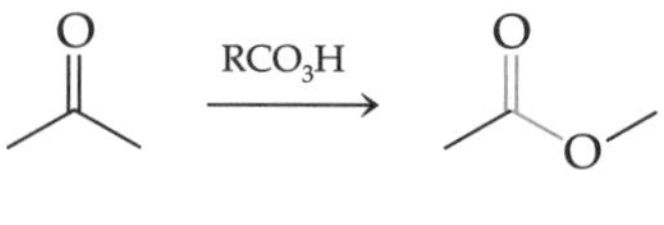

$$\overset{O}{\|} \xrightarrow{RCO_3H} \underset{O}{\overset{O}{\|}}$$

$\underset{}{\overset{\text{RO}\quad\text{OR}}{\diagdown\!/}}$

1. Acetal (ketal) formation:

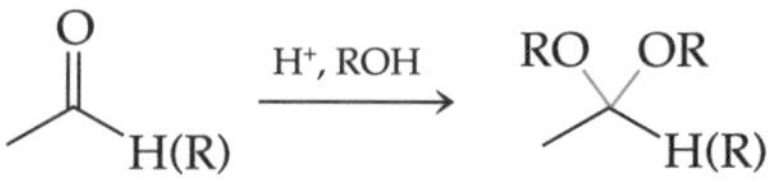

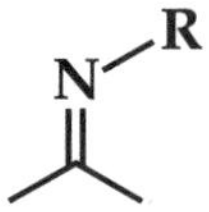

$$\underset{H(R)}{\overset{O}{\|}} \xrightarrow{H^+, ROH} \underset{H(R)}{\overset{RO\quad OR}{\diagdown\!/}}$$

$\underset{}{\overset{\qquad R}{N\diagdown}}$

1. Imine formation:

$$\overset{O}{\|} + H_2N\!-\!R \longrightarrow \overset{\qquad R}{N\diagdown}$$

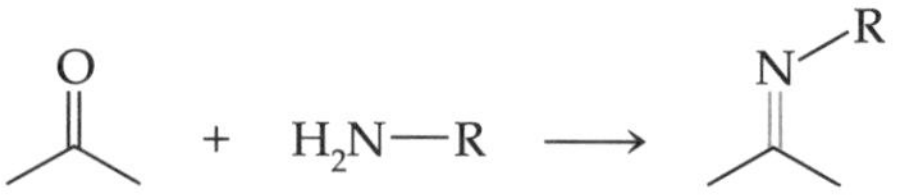

Conclusions

We have seen in this chapter several rearrangement reactions that result in changes in connectivity in a carbon skeleton. These rearrangement reactions take place by the migration of alkyl or aryl groups from one site to an adjacent atom. In the Wagner-Meerwein rearrangement, an alkyl group migrates to an adjacent carbocation (or incipient carbocation). These migrations are controlled by cation stability so as to form the more-stable intermediate (3° > 2° > 1° or with release of appreciable ring strain). The driving force for the key step in a pinacol rearrangement is also the formation of a more-stable carbocation (a protonated carbonyl).

Anionic rearrangements are rarer than their cationic counterparts, although the benzilic acid rearrangement constitutes one method by which such carbon migrations occur within α-diketones. Again, the driving force is the formation of a more-stable intermediate—in this case, a carboxylate anion.

The pericyclic reactions considered in this chapter are of three general types: cycloadditions, sigmatropic shifts, and electrocyclic reactions. A cycloaddition requires the interaction of the π systems from two molecules, whereas an electrocyclic reaction takes place by the interaction of the *p* orbitals at the ends of a single π system within one molecule. A sigmatropic shift describes the migration of a σ bond across a π system. Two classes of sigmatropic shifts that achieve specific skeletal rearrangements are discussed: in the Claisen and the Cope rearrangements, both ends of a σ bond shift by three carbons, forming a new σ bond at those positions and producing a rearranged backbone. A heteroatom can be incorporated at the σ bond, at a multiple bond, or as a substituent on the carbon framework.

In a number of rearrangements, a group bound to carbon (or an equivalent) migrates to an attached heteroatom (at the α position) that bears a leaving group and a lone pair. Examples of such migrations are the Beckmann rearrangement (converting a ketone through an oxime into an amide) the Hofmann rearrangement (converting an amide into the corresponding amine), and the Baeyer-Villiger oxidation (converting a ketone into an ester).

Summary of New Reactions

Pinacol Rearrangement

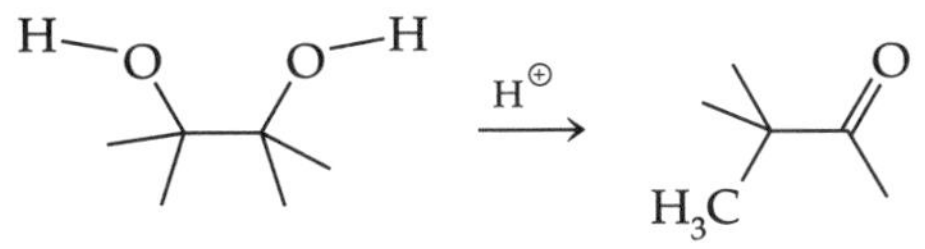

Benzilic Acid Rearrangement

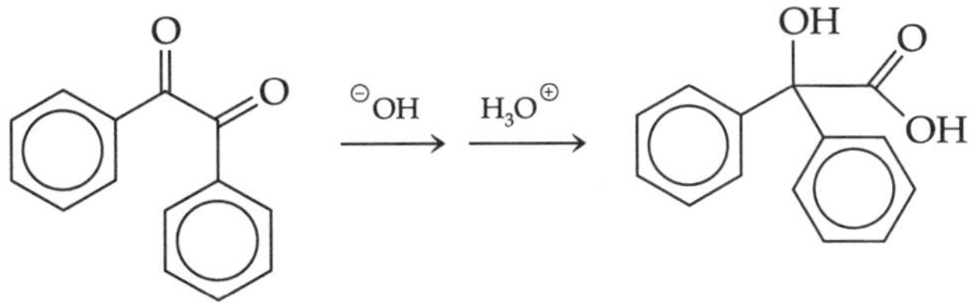

Cope Rearrangement

Beckmann Rearrangement

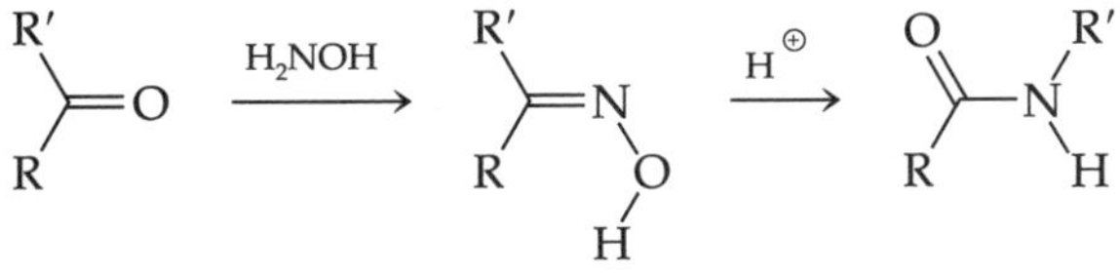

Hofmann Rearrangement

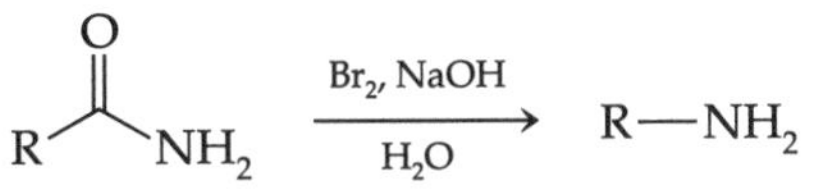

Baeyer–Villiger Oxidation

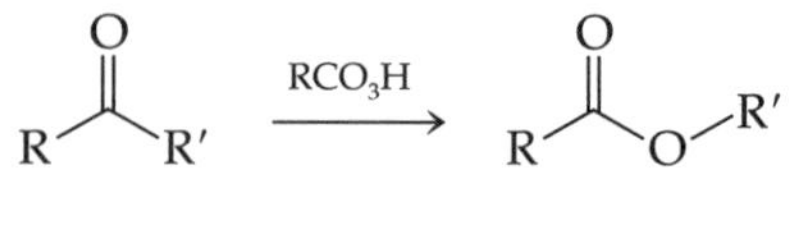

$$R \overset{O}{\underset{}{\|}} R' \xrightarrow{RCO_3H} R \overset{O}{\underset{}{\|}} O{-}R'$$

Pericyclic Reactions

Cycloaddition (for example, a Diels-Alder reaction)

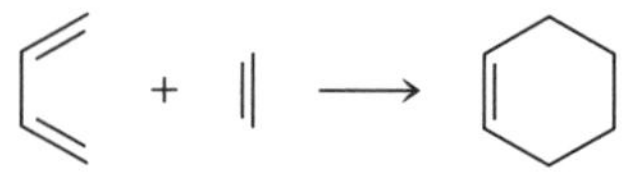

Sigmatropic shift (for example, a Cope rearrangement)

(or, for example, a Claisen rearrangement)

Review Problems

14-1 For each of the following reactions, predict the major product expected when the reactant is treated with the listed sequence of reagents.

(a) 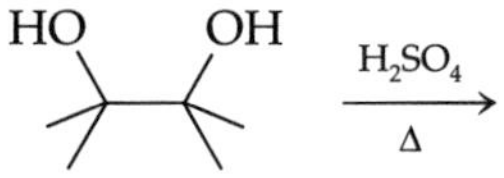 $\underset{\Delta}{\xrightarrow{H_2SO_4}}$

(b) 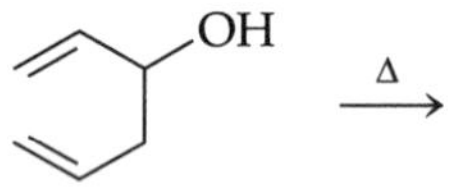 $\xrightarrow{\Delta}$

(c)  $\xrightarrow{SOCl_2} \xrightarrow{NH_3} \xrightarrow{Br_2, NaOH}$

(d) 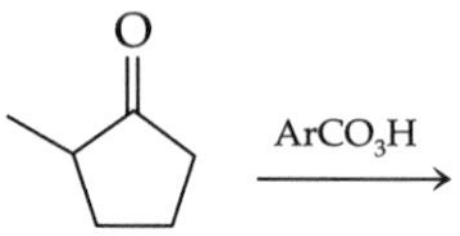 $\xrightarrow{ArCO_3H}$

(e) 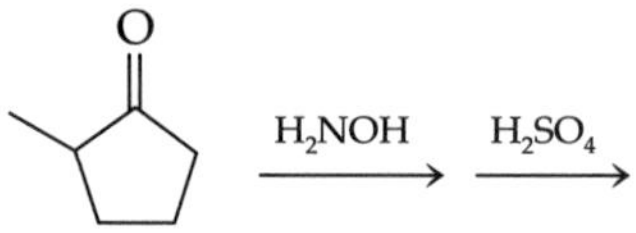 $\xrightarrow{H_2NOH} \xrightarrow{H_2SO_4}$

(f) 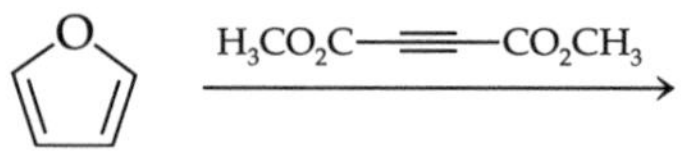 $\xrightarrow{H_3CO_2C-\!\!\!\equiv\!\!\!-CO_2CH_3}$

(g) 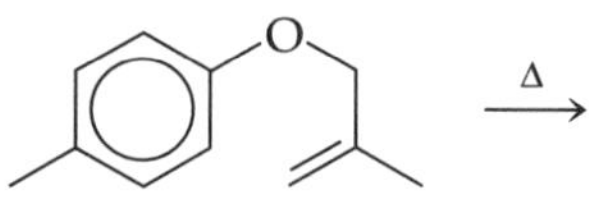 $\xrightarrow{\Delta}$

14-2 Specify the reagents and conditions needed to convert cyclohexanone into each of the following products:

(a)

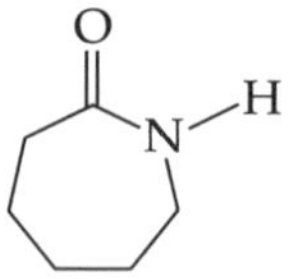

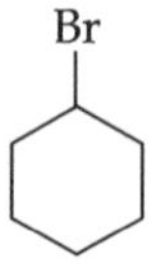

(e) 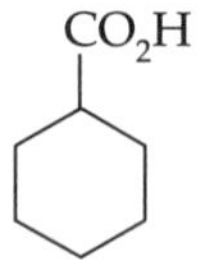 CO_2H

(b)

(f) 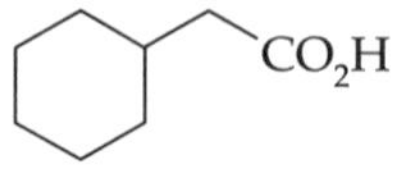 CO_2H

(c) OH

(g) 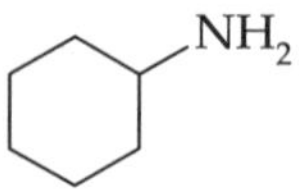 NH_2

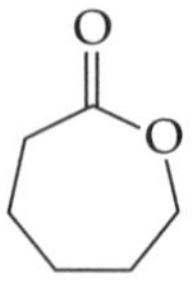

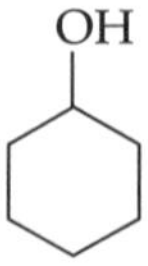

(d) Br

14-3 Define the starting materials and reagents that would be required to prepare 1-butylamine by the use of each of the following reactions:

(a) a Gabriel synthesis

(b) a Hofmann rearrangement

(c) a Beckmann rearrangement

14-4 Write detailed mechanisms, using curved arrows to indicate electron flow, for each of the following reactions.

(a) 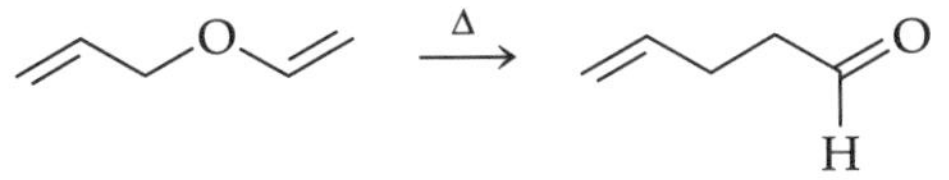 [structure] $\xrightarrow{\Delta}$ [structure with CHO]

(b) 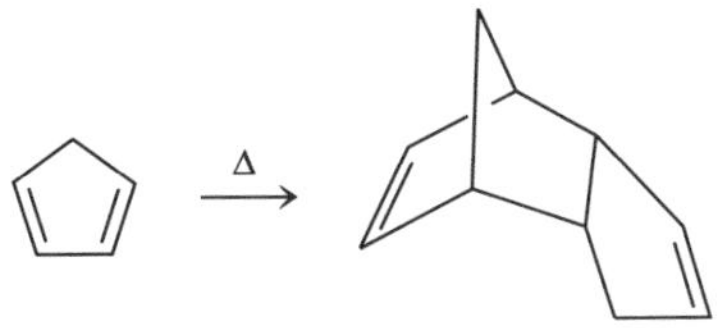 [cyclopentadiene structure] $\xrightarrow{\Delta}$ [bicyclic structure]

(c) 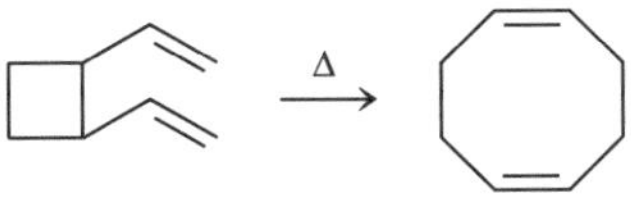 [structure] $\xrightarrow{\Delta}$ [cyclooctatetraene structure]

(d) 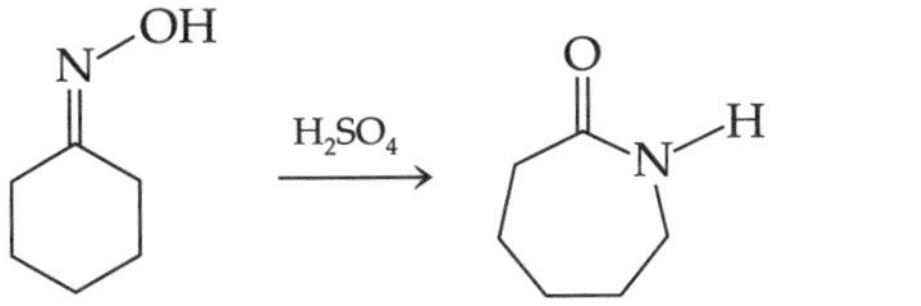 [cyclohexanone oxime, N–OH] $\xrightarrow{H_2SO_4}$ [caprolactam structure N–H]

(e) 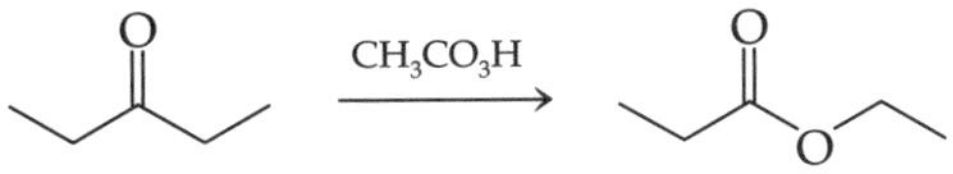 [ketone] $\xrightarrow{CH_3CO_3H}$ [ester structure]

(f) 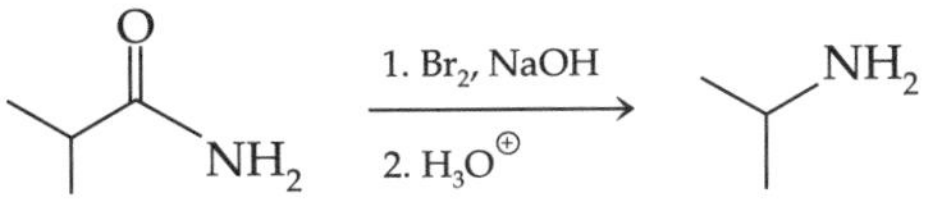 [structure with NH_2] $\xrightarrow[\text{2. } H_3O^{\oplus}]{\text{1. } Br_2, NaOH}$ [amine NH_2]

(g) 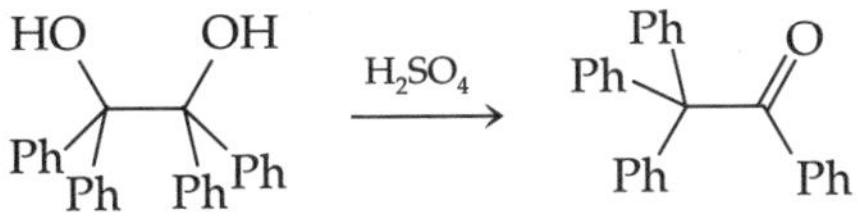 HO OH [structure Ph, Ph, Ph, Ph] $\xrightarrow{H_2SO_4}$ Ph [structure Ph, Ph, Ph]

14-5 The usual method for the attachment of alkyl chains to aromatic rings is electrophilic aromatic substitution. (Recall the Friedel-Crafts acylation and alkylation from Chapter 11). An allyl group can be attached to the aromatic ring of a phenol, however, by a sequence in which a Williamson ether synthesis is followed by a Claisen rearrangement. With this sequence in mind, write a mechanism by which the following reaction takes place.

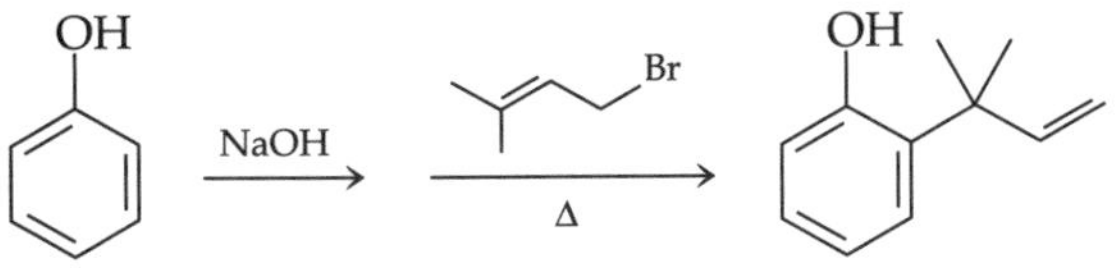

[phenol structure] $\xrightarrow{NaOH}$ [allyl bromide structure, Br] $\xrightarrow{\Delta}$ [product structure with OH]

14-6 In each of the following rearrangements, the functional group(s) present in the molecule is altered. Describe the changes expected as the reaction proceeds in (1) the infrared spectrum, (2) either the carbon or ^1H NMR spectrum, and (3) the mass spectrum of the reaction mixture that could be used to follow the course of the rearrangement.

(a) 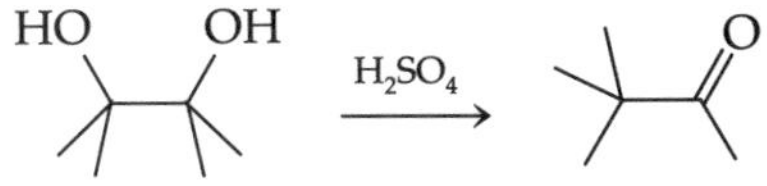 HO OH [structure] $\xrightarrow{H_2SO_4}$ [ketone structure]

(b) 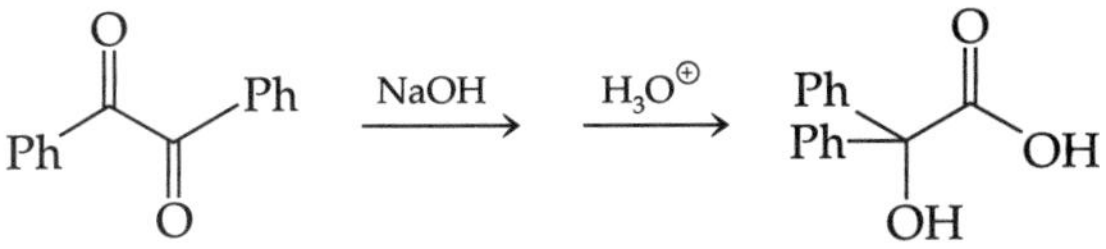 [diketone Ph, Ph] $\xrightarrow{NaOH}$ $\xrightarrow{H_3O^{\oplus}}$ Ph Ph [structure with OH, OH]

(c) 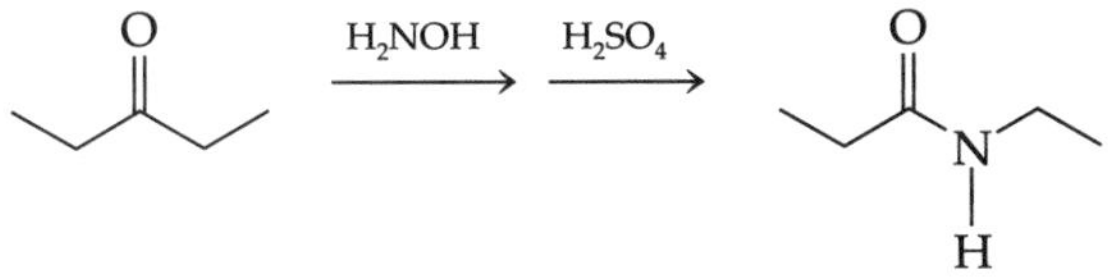 [ketone] $\xrightarrow{H_2NOH}$ $\xrightarrow{H_2SO_4}$ [amide structure N–H]

(d) 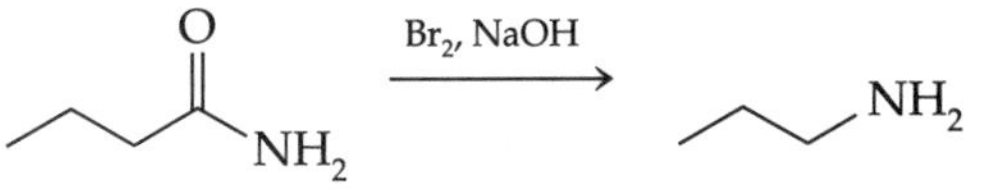 [amide structure with NH_2] $\xrightarrow{Br_2, NaOH}$ [amine structure NH_2]

(e) 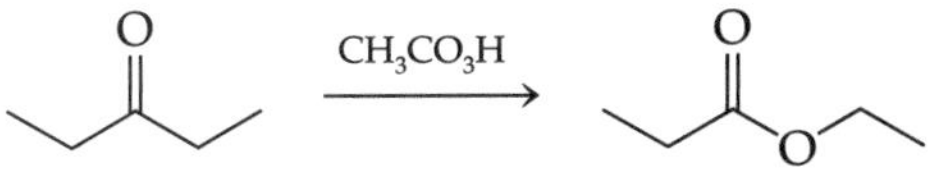 [ketone] $\xrightarrow{CH_3CO_3H}$ [ester structure]

(f) 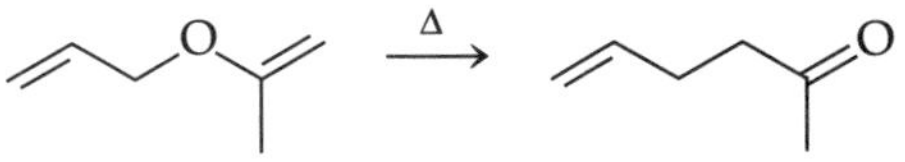 [allyl vinyl ether structure] $\xrightarrow{\Delta}$ [ketone structure]

14-7 Consider the various methods by which 2-butanol can be converted into each of the compounds in parts *a* through *t*. In particular, suggest the reagents and conditions necessary to accomplish each of these functional-group conversions.

(a) 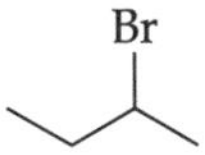 [structure with Br]

(b) 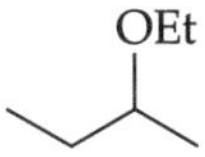 [structure with OEt]

(c) 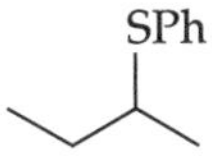 [structure with SPh]

(d)  [alkene structure]

(e)  [alkene structure]

(f)  [alkane structure]

(g)  [alkyne structure]

(h)  [structure with OH]

(i) 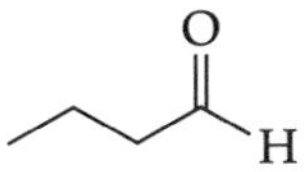 [aldehyde structure with H]

(j) 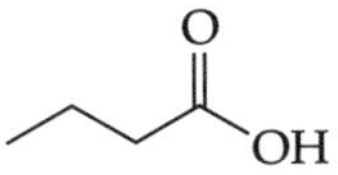 [carboxylic acid structure with OH]

(k) 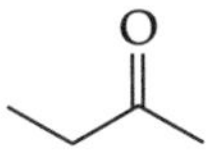 [ketone structure]

(l)

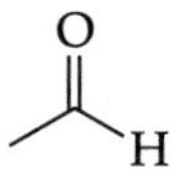

(m)

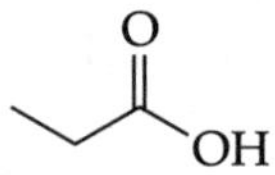

(n)

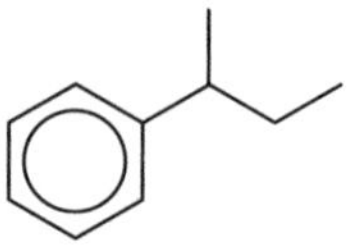

(o)

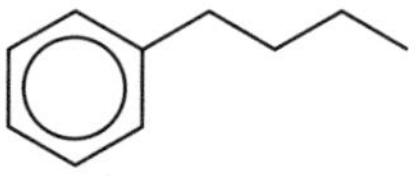

(p)

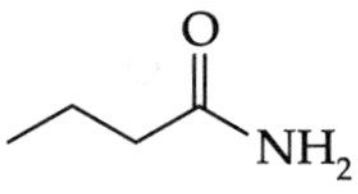

(q)

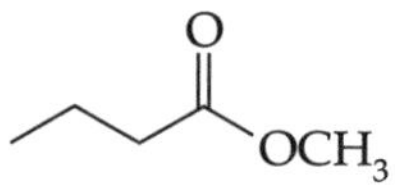

(r)

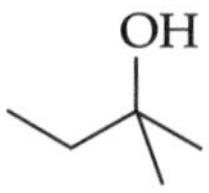

(s)

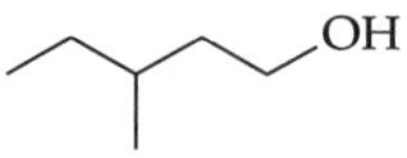

(t)

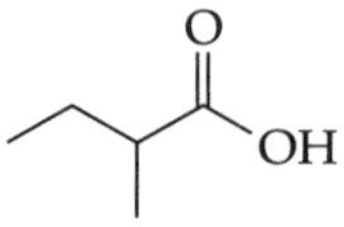

Chapter 15

Multistep Syntheses

We have now seen a wide range of reactions and studied their mechanisms in detail. With this information in hand, we are in a position to view these reactions as processes that transform one species into another and, thus, as tools for chemical synthesis. Although we are familiar with the functional-group transformations that our set of reactions can accomplish, each transformation by itself may not be an impressive change. On the other hand, when a number of these transformations are carried out in sequence, the structural resemblance of the ultimate product to the initial starting material can be far from obvious.

As an example, consider the transformation of 2-propanol into 2-butanone as illustrated in Figure 15-1. Even though we have not covered a single reaction that can induce this specific conversion, we will shortly see that, by using a series of reactions that we do know, we will be able to achieve such a new conversion. In this chapter, we will learn to recognize clues provided by the starting material and the product that make the choice of these reactions more immediately apparent.

Why might we be interested in combining reactions together? The field of organic chemistry owes its diversity to the almost unlimited number of structures that are possible based upon carbon. Because each chemical reaction generally makes only a relatively minor change in structure, we must use several of them to prepare complex molecules such as those found in nature (as we will explore in later chapters) from simple and readily available ones such as acetone, ethanol, and ethyl acetate.

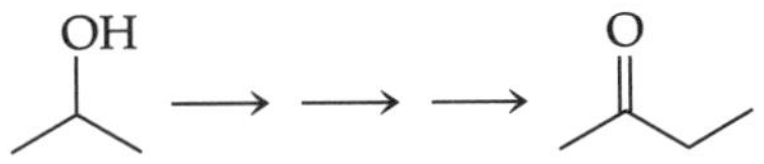

▲ FIGURE 15-1
The transformation of 2-propanol into 2-butanone requires the formation of a new carbon–carbon bond and the adjustment of the oxidation level of the carbinol carbon. We have learned no single reaction that can accomplish this conversion but can use several known reactions in sequence to accomplish the task.

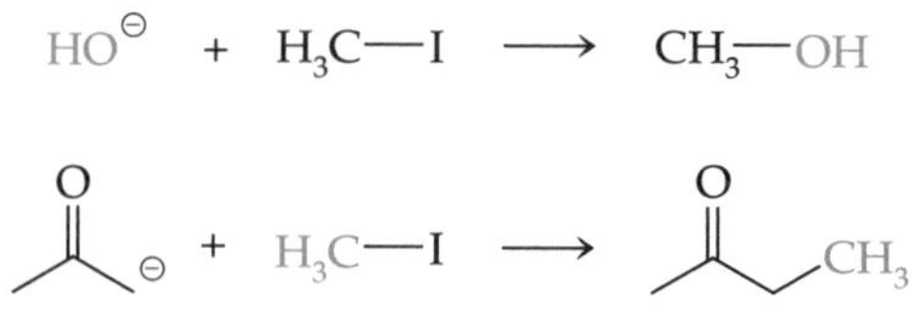

▶ FIGURE 15-2
S_N2 reactions can either alter the functional group present (upper reaction) or extend a carbon chain by carbon–carbon bond formation.

$$HO^{\ominus} \ + \ H_3C{-}I \ \longrightarrow \ CH_3{-}OH$$

$$\overset{O}{\underset{\ominus}{\big|\big|}} \ + \ H_3C{-}I \ \longrightarrow \ \overset{O}{\underset{CH_3}{\big|\big|}}$$

15-1 ▶ Grouping Chemical Reactions

We have considered many different kinds of chemical transformations in the preceding chapters, grouping them according to their reaction mechanisms. For example, both of the reactions shown in Figure 15-2 take place by S_N2 reaction pathways. Yet, when we view them from the perspective of what they accomplish rather than how they occur, we might choose to put these reactions into entirely different categories. For the purpose of combining reactions into sequences to construct complex molecules from simple ones, it is more convenient to separate them into the following three categories:

1. carbon–carbon bond-forming processes;
2. oxidation-reduction reactions; and
3. functional-group transformations.

This classification scheme naturally fits our needs for remembering chemical transformations for synthetic purposes. Because all sequences that result in the transformation of a small organic molecule into a larger one require carbon–carbon bond formation, these reactions are particularly important to synthesis. However, because many carbon–carbon bond-forming reactions require carbonyl functional groups, it is also important that we know both how to make these functional groups and how to remove them, because we will often find that they are not present in the product.

Let us return briefly, then, to the example in Figure 15-1 to suggest one possible solution, as provided in Figure 15-3. Here, we find one possible sequence (of three steps) that effects the transformation: oxidation to acetone; formation of the enolate anion; and alkylation with methyl iodide to form 2-butanone. With regard to the overall transformation, the most-important process is the carbon–carbon bond-forming reaction because that reaction, in effect, elaborates the smaller molecule into the larger one. This is not to say that the other reactions are unimportant, because, without the oxidation, we would not have the carbonyl functional group necessary for the carbon–carbon bond-forming reaction. Yet, the oxidation step serves only to provide us with the requisite functional group for the key reaction.

Before we consider other methods for creating sequences like that illustrated in Figure 15-3, it is useful to review some of the reactions from the preceding chapters, placing them in the three categories listed above. Table 15-1 gives the major carbon–carbon bond-forming reactions; Table 15-2, the

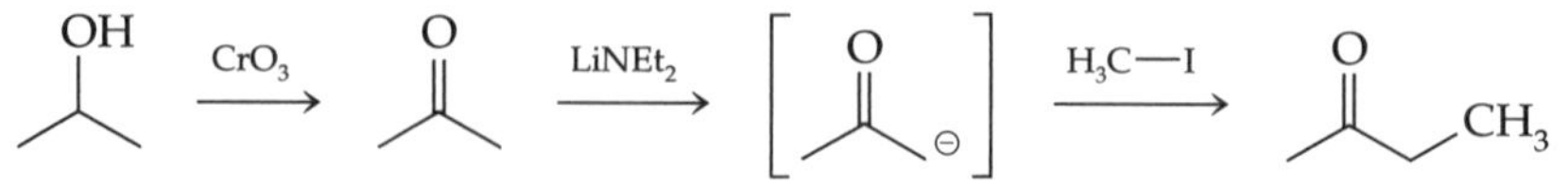

▲ FIGURE 15-3
The transformation proposed in Figure 15-1 can be accomplished in several steps. Chromate oxidation of a secondary alcohol converts it into a ketone. The treatment of this ketone with strong base (for example, lithium diethylamide) produces an enolate anion that attacks an alkyl halide by an S_N2 reaction, yielding the desired product.

Table 15-1 ▸ Carbon–Carbon Bond-forming Reactions

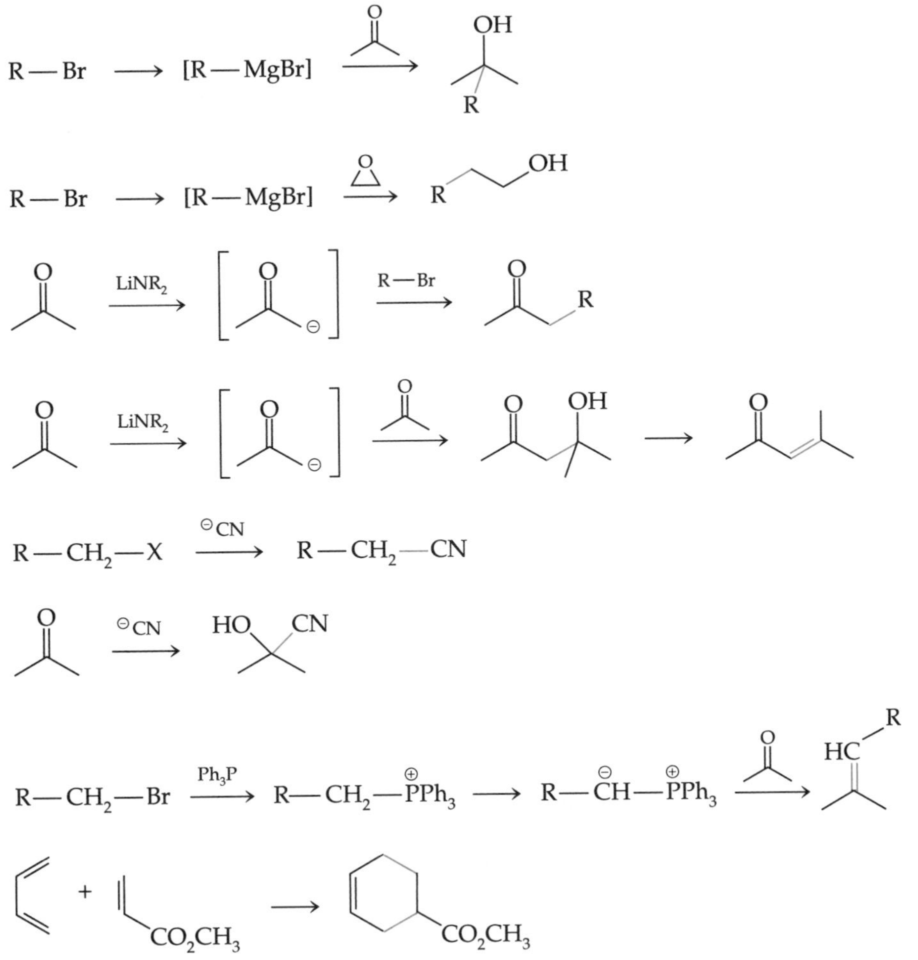

Table 15-2 ▸ Oxidation-Reduction Reactions

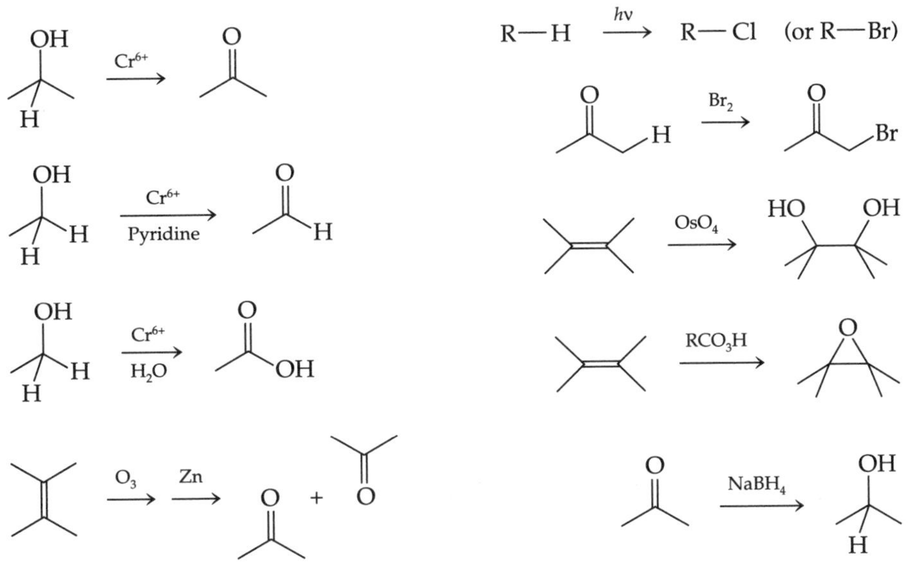

Table 15-2 ▸ Oxidation–Reduction Reactions (continued)

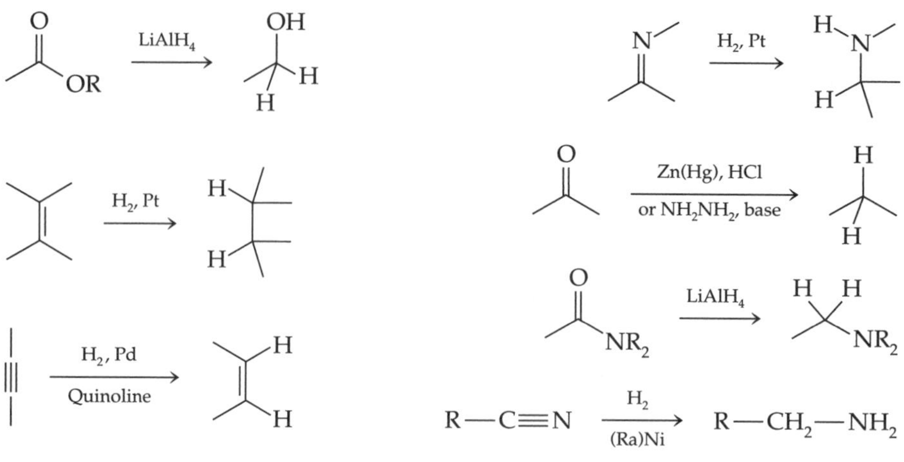

Reactions with Simultaneous Oxidation of One Carbon and Reduction of Another:

$$X = Cl, Br, I, OR, NR_2$$

oxidation-reduction transformations; and Table 15-3, several important functional-group transformations. Included in Table 15-2 as a separate subsection are reactions such as the addition of water to an alkene in which

Table 15-3 ▸ Functional-Group Transformations

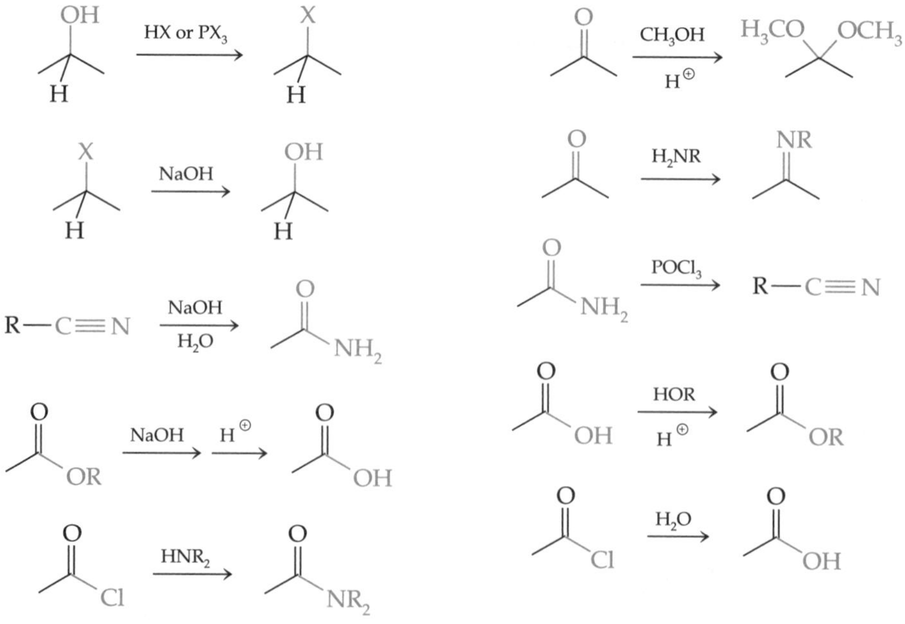

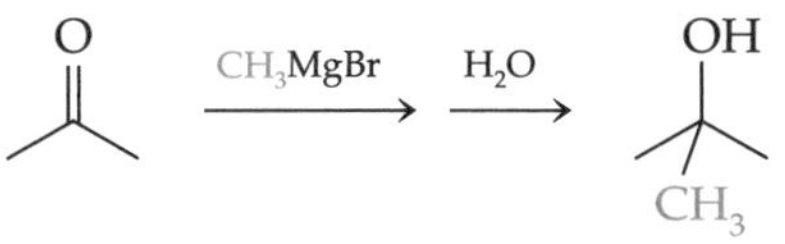

▲ FIGURE 15-4

A ketone is converted into a tertiary alcohol upon hydrolysis of the salt formed by treatment with a Grignard reagent.

there is no net change in oxidation, but in which one carbon undergoes a reduction that is balanced by the oxidation of another. Although there is no net redox change, the oxidation and reduction levels of some atoms within the molecule change in the course of these reactions.

It is certainly possible (and indeed is often the case) that a specific reaction fits more than one category. In Figure 15-3, the oxidation of an alcohol to a ketone is both an oxidation and a functional-group transformation. As a second example, illustrated in Figure 15-4, the addition of a Grignard reagent to a ketone (followed by protonation) fits all three categories: it is a carbon–carbon bond-forming reaction, a reduction of the carbonyl carbon, and a functional-group transformation of a ketone into an alcohol. In such cases, we always regard the reaction as belonging to that category that is higher on the list (that is, a reaction that is both a functional-group transformation *and* an oxidation-reduction reaction is classified as the latter). Thus, the reaction in Figure 15-4 is considered a carbon–carbon bond-forming process.

EXERCISE 15-A

Classify the following reactions as carbon–carbon bond-forming reactions, oxidation-reductions, or functional-group transformations.

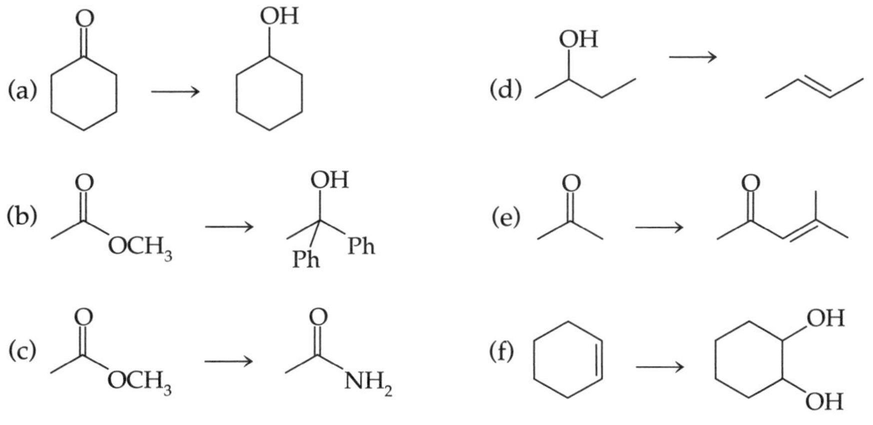

15-2 ▸ Retrosynthetic Analysis: Working Backward

How shall we proceed to develop a solution to problems such as that presented in Figure 15-1? We might start by "trying" (as a thought experiment) various reactions on the starting material, 2-propanol. Although this appears to be a major task, we can narrow our search by noting that at some point we must form a carbon–carbon bond because there are four carbon atoms in the product, one more than in the starting material. Thus, we systematically examine the carbon–carbon bond-forming reactions in Table 15-1 to identify whether any could apply to the starting material. In this case, no reaction is directly applicable.

Rather than immediately giving up, we look at the product to see whether it might be the product of one of our chemical transformations. Again, if possible, we select a reaction that forms a carbon–carbon bond because we know that this must occur at some point in the sequence. From Table 15-1, we see that the alkylation of a simple ketone does indeed accomplish the needed transformation. Furthermore, we see that the necessary starting materials would be acetone (as its enolate) and an alkylating agent such as methyl iodide. (Recall from Chapter 7 that the third way of analyzing a reaction asks us to identify the correct starting materials for forming a desired product under defined conditions.)

At this point, we realize that the number of carbons in acetone (the immediate precursor for 2-butanone by this analysis) and the given starting material, 2-propanol, is the same. Thus, we have no further need of carbon–carbon bond-forming reactions and can restrict our analysis to those in categories 2 and 3 (oxidation-reduction and functional-group transformation). We find that an oxidation of 2-propanol to acetone connects the starting material (2-propanol), through the intermediate formation of acetone, ultimately with 2-butanone. The total sequence is illustrated in Figure 15-3. Rather than following each of the possible reactions of 2-propanol through many steps, we have greatly simplified the analysis by starting with the last step and thinking about how to make the most logical precursor for that step. Because this approach begins with the product and goes back, step-by-step to the reactant, it is called **retrosynthetic analysis.**

Why should we *work backward*? Although it may seem unnatural because what we want to accomplish in fact is the transformation in the forward direction, there is a very simple reason. In any sequence that progresses from smaller to larger molecules, the number of options rises dramatically in the forward direction but diminishes in the backward retrosynthetic analysis. Diagrammatically, this approach, shown in Figure 15-5, looks much like a Christmas tree: any one of a number of reactions can apply to the starting

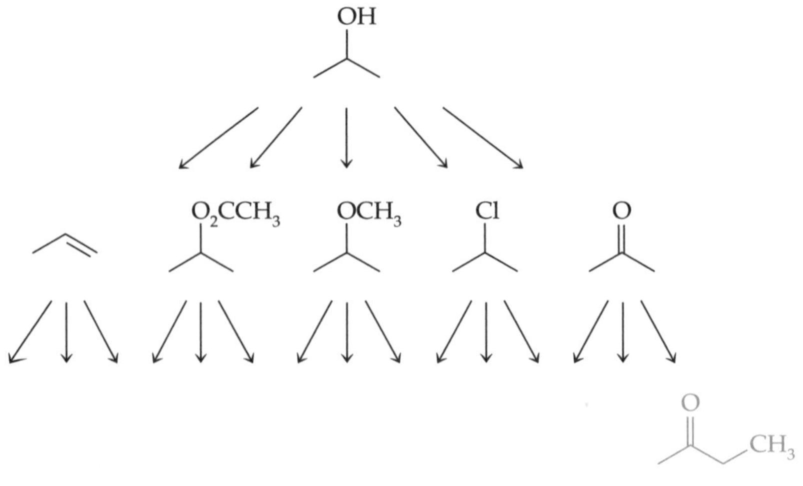

▲ FIGURE 15-5

Although a number of reactions of a secondary alcohol are possible, the oxidation to a ketone at the far right produces a compound (propanone) that is specifically activated for forming the additional carbon–carbon σ bond needed in the product.

SYNTHESIS OF VERY LARGE AND COMPLEX MOLECULES

One of the most-complex molecules ever synthesized in the laboratory is vitamin B_{12}, prepared by Robert B. Woodward of Harvard University and Albert Eschenmoser of the Eidgenössische Technische Hochschule in Zurich. Its synthesis required the effort of more than one hundred chemists working together over a period of eleven years. Woodward's group was also the first to synthesize chlorophyll *a*, the macrocycle responsible for the green color of plants. The experimental description of the synthesis of chlorophyll *a* included the research of seventeen coworkers and required an entire issue of a journal. The synthesis was considered so special that the journal's editors changed the cover of the journal to green (from its customary blue) to match the color of chlorophyll. [Check it at your library: *Tetrahedron* **46**, 7599 (1990)]. However, Woodward's favorite color was blue—he owned more than two hundred ties, all the same shade of blue.

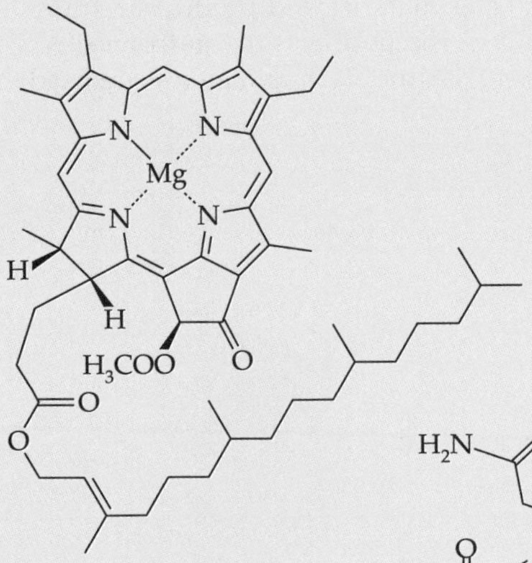

Chlorophyll *a*

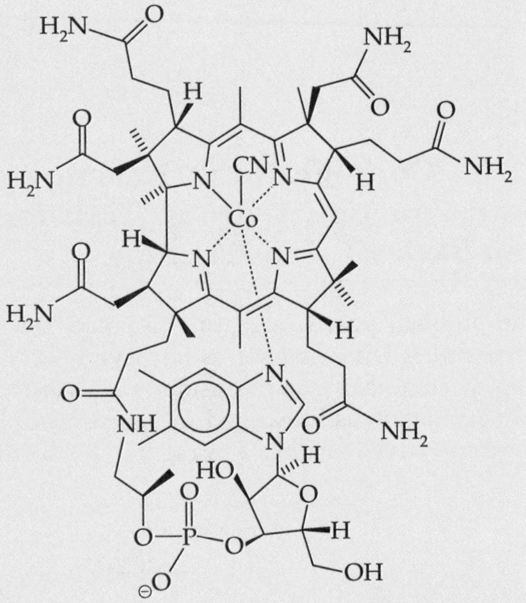

Vitamin B_{12}

material and, furthermore, many additional reactions can apply to each of the initial products. Although it is certainly possible that more than one of the branches will ultimately lead to the product desired, some pathways may never lead to this goal. Conversely, by applying our thinking in the backward direction and utilizing the knowledge that the product has more carbon–carbon bonds than does the starting material, we follow a path that, in a sense, "funnels" along a useful sequence.

This way of thinking of transformations is often confusing at first because reactions are taught and learned in the forward rather than in this backward direction. It is critical that you "relearn" reactions in this sense: product as originating by one of the three classes of transformations from a given starting material.

EXERCISE 15-B

The following transformations take place in multireaction sequences. Examine the starting materials and products listed below and determine which carbon–carbon bonds in each product would likely have been made in a sequence that required the smallest number of carbon–carbon bond-forming reactions. Do not concern yourself with the actual reactions. Consider *only* how many carbons are in each starting material and product and how the carbons of the starting material and of the product are interconnected. Assume that no sources of carbon other than the starting material are used.

(a) 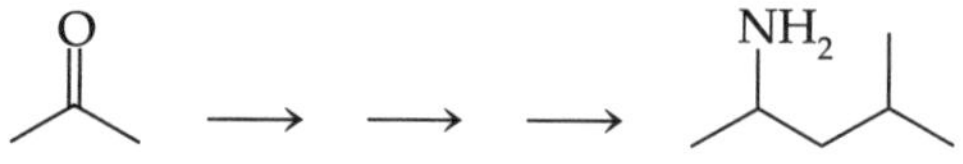   → → →

(b) 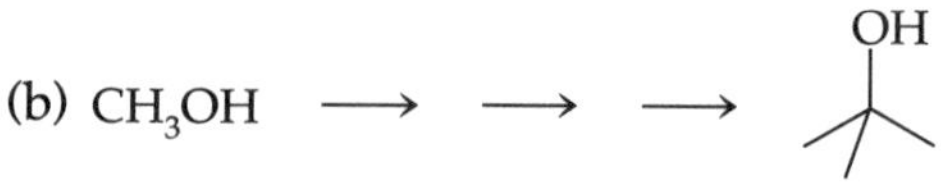   CH$_3$OH → → →

(c) 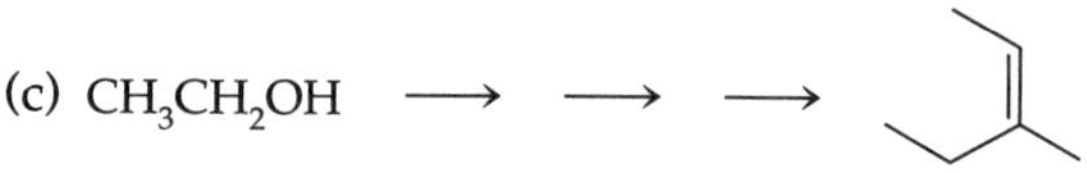   CH$_3$CH$_2$OH → → →

15-3 ▸ Complications: Reactions Requiring both Functional-Group Transformation and Skeletal Construction

The problem posed in Figure 15-1 and solved in Figure 15-3 might now seem rather trivial after this retrosynthetic analysis. Let us add one more step of complexity and consider the overall transformation of 2-propanol into 2-butanol, as illustrated in Figure 15-6. It is certainly tempting at this point to use the synthesis in Figure 15-3 for the preparation of 2-butanone,

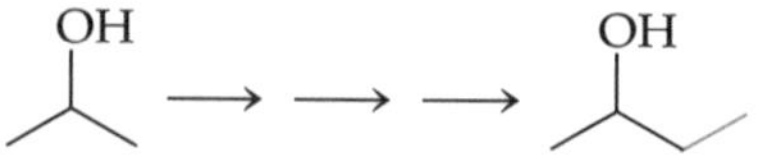

▲ FIGURE 15-6

The transformation of 2-propanol into 2-butanol requires the addition of a carbon atom (by the formation of a new carbon–carbon bond). We have learned no single reaction by which an alkyl chain can be attached to a carbon atom adjacent to a carbinol carbon. We must therefore conduct a retrosynthetic analysis to plan a sequence of reactions to accomplish the synthesis.

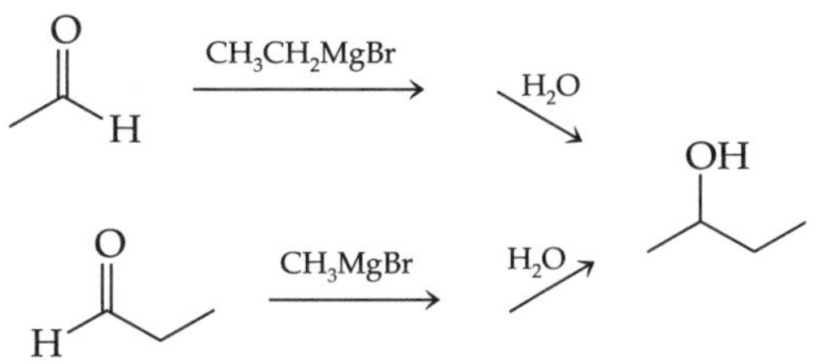

▲ FIGURE 15-7

2-Butanol can be prepared by a Grignard synthesis through the reaction either of the ethylmagnesium bromide with acetaldehyde (upper reaction) or of methylmagnesium bromide with propanal (lower reaction).

realizing that we need only to reduce 2-butanone to arrive at our new objective, 2-butanol.

Let us see if we can arrive logically at this same conclusion by retrosynthetic analysis. Thus, we compare the ultimate product, 2-butanol, with the starting material to once again ascertain that somewhere in the sequence a carbon–carbon bond must be formed. Returning to our compilation of carbon–carbon bond-forming reactions (Table 15-1), we find a transformation, the reaction of a Grignard reagent with a carbonyl group, that directly forms an alcohol. Furthermore, this is the only carbon–carbon bond-forming reaction that we have seen so far that directly produces this functional group.

Next, we apply this chemistry to imagine what starting materials we might use for a Grignard synthesis of 2-butanol. The Grignard reaction forms a carbon–carbon bond to a carbinol carbon in the product, and so only two sequences are possible because there are only two such carbon–carbon bonds in 2-butanol. These sequences are illustrated in Figure 15-7. Of these two reactions, the upper reaction is less convenient because it breaks 2-butanol into two, two-carbon fragments, neither of which could be immediately derivable from the three-carbon starting material 2-propanol, whereas the lower one uses a three-carbon starting material. But, if we are to use the lower transformation, we must derive a connection between 2-propanol and propanal. (Rest assured that this does not look trivial even to practiced chemists.) Perhaps we are stuck, having traveled backward ("up our Christmas tree"), following a lead that took us down a blind alley.

We then return to our retrosynthetic analysis to see whether there is some other carbon–carbon bond-forming reaction that might directly produce 2-butanol. There is none, but the formation of the required carbon–carbon bond need not be the last step in the sequence. Perhaps some other transformation (an oxidation, a reduction, or a functional-group transformation) can be used as the ultimate reaction.

The next logical step is to consider that the final reaction might be an oxidation (or reduction). When we examine the oxidation-reduction reactions in Table 15-2, we see that the reduction of a ketone produces a secondary alcohol (Figure 15-8). Indeed, this transformation connects the simpler problem originally depicted in Figure 15-1 to the more-complex one presented in Figure 15-6.

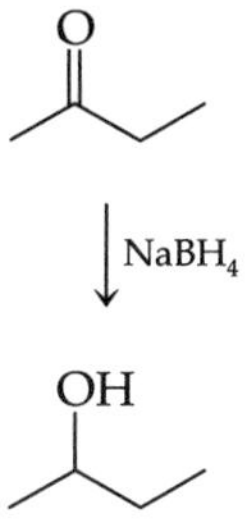

In summary, then, a retrosynthetic analysis does not always lead automatically to the shortest sequence of transformations connecting starting materials and products, but it does allow us to choose alternate pathways. With some practice, we can recognize fairly quickly when to abandon those sequences that would require an extraordinary number of steps or an unrealistically difficult step.

▲ FIGURE 15-8

The reduction of a ketone to a secondary alcohol can be conveniently accomplished by treatment with a complex metal hydride reducing agent.

EXERCISE 15-C

Develop a sequence to transform propanal into 2-butanol by the reactions that have been covered so far. (You may use any reagents, including organic ones, as long as the three carbons of propanal are incorporated into the product.)

15-4 ▸ A Multistep Example

Often, more than one short sequence connects a given starting material to a desired product. As an example, consider the transformation of 2-propanol into 3-methyl-2-butanone (Figure 15-9).

We begin by recognizing that there are two more carbon atoms in the product than in the starting material. At some point, at least one carbon–carbon bond must be formed, and it is reasonable to determine whether this might be accomplished as the last step in the sequence. Indeed, there is such a reaction, exactly the same alkylation of a ketone that we used in the preceding problem (the last step of Figure 15-3). As illustrated in Figure 15-10, 2-butanone (from Figure 15-3) can be converted into an enolate that is alkylated by methyl iodide to form the desired product, 3-methyl-2-butanone. Because we now intersect with the product of the preceding problem, we have an overall sequence that can be used to accomplish the desired transformation. There is one difficulty, however: the formation of an enolate from an unsymmetrical ketone generally leads

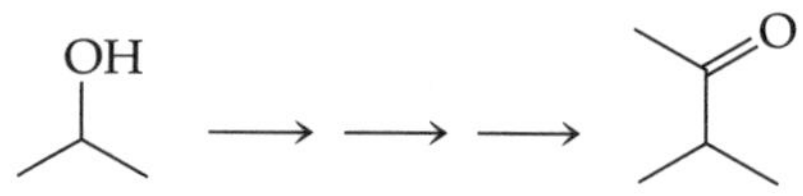

▲ FIGURE 15-9
The conversion of 2-propanol into 3-methyl-2-butanone requires a carbon–carbon bond-forming step because there are two more carbon atoms in the product than in the reactant.

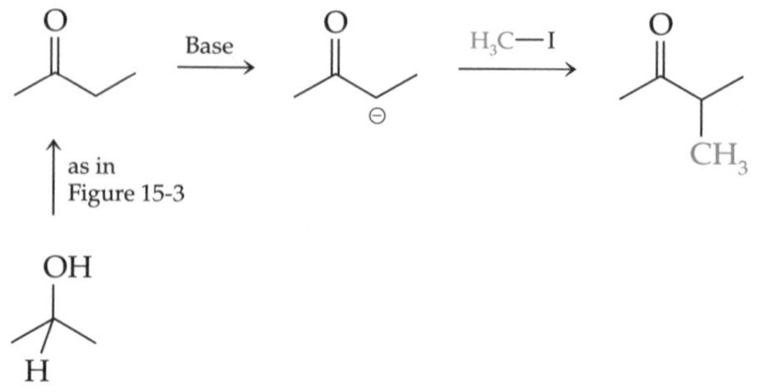

▲ FIGURE 15-10
The desired product can be prepared by α methylation of 2-butanone. Because we have already devised a route for the conversion of 2-propanol into 2-butanone, as in Figure 15-3, we have a retrosynthetic route from 2-propanol to 3-methyl-2-butanone.

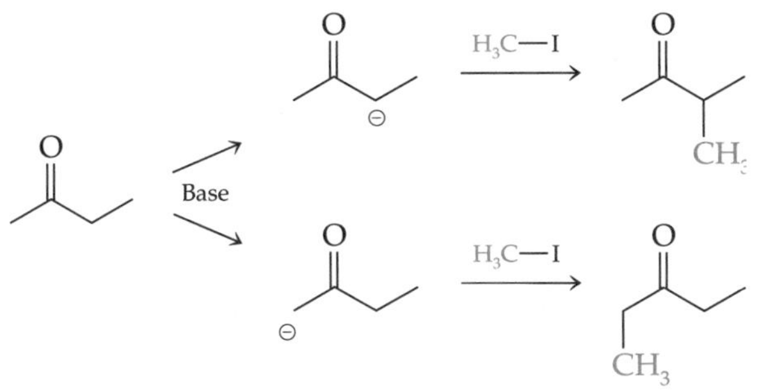

▲ FIGURE 15-11
Because there are two acidic carbons in 2-butanone, two isomeric enolate anions are formed. Methylation can therefore take place both at C-3 (upper reaction) and at C-1 (lower reaction). A reaction that affords a mixture of isomeric ketones is not desirable in a synthesis, if the reaction can be avoided.

to a mixture of regioisomers. Thus, this synthesis would produce not only 3-methyl-2-butanone, as desired, but also 3-pentanone (Figure 15-11).

Let us return, then, to the original objective (Figure 15-9) and search for a sequence that does not have this problem. The list of carbon–carbon bond-forming reactions that afford products that have ketone groups contains only the enolate alkylation reaction that we have already examined. We move on, then, to consider other possible last steps in which an oxidation or reduction would provide a route to the desired ketone. Indeed, the oxidation of a secondary alcohol, as illustrated in Figure 15-12, could be used to produce the desired 3-methyl-2-butanone.

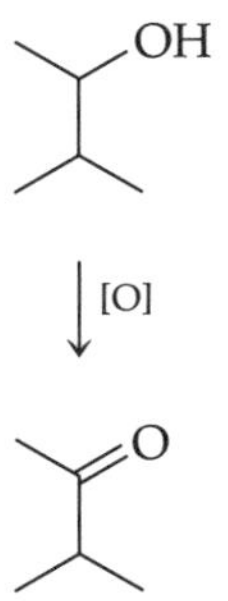

We should then consider how we might arrive at this alcohol, 3-methyl-2-butanol. As before, we look first for a reaction that forms a carbon–carbon bond. From Table 15-1, we choose the reaction of a Grignard reagent (derived from the appropriate alkyl halide) with a carbonyl group to produce a secondary alcohol. The Grignard synthesis of alcohols always results from the formation of a carbon–carbon bond to the carbinol carbon. In this case, there are two different bonds of this type and thus two possible combinations of Grignard reagent and carbonyl compound that lead to this alcohol. These options are illustrated in Figure 15-13 in which route A starts with 2-bromopropane, which, after conversion into the Grignard reagent, reacts with acetaldehyde. The alternative sequence, B, is the reaction of 2-methylpropanal with the Grignard reagent derived from methyl bromide.

One of the two sequences, A or B, might be more efficient or shorter than the other, but how are we to choose between them? The original objective, described in Figure 15-9, concerned the conversion of 2-propanol into

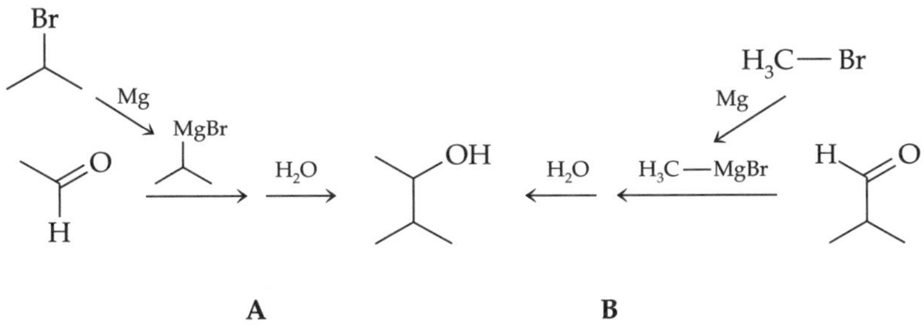

▲ FIGURE 15-12
Chromate oxidation of a secondary alcohol produces a ketone in good yield.

A   **B**

▲ FIGURE 15-13
Two possible routes for a Grignard synthesis of 3-methyl-2-butanol are: (A) treatment of acetaldehyde with isopropylmagnesium bromide; and (B) treatment of 2-methylpropanal with methylmagnesium bromide.

3-methyl-2-butanone. In our first analysis, we determined that at some point we would need to form a carbon–carbon bond because the product has five carbons, whereas the starting material has only three. We can see, then, that a sequence that includes carbon–carbon bond formation between a three-carbon unit derived from isopropanol and another unspecified fragment with two carbons requires only one carbon–carbon bond-forming step. Conversely, two carbon–carbon bond-forming steps would be needed if fragments containing only one carbon each were used.

Of the two alternative sequences in Figure 15-13, route A combines a three-carbon unit (2-bromopropane) with a two-carbon unit (acetaldehyde) to produce a five-carbon alcohol, whereas sequence B results in the formation of the same product from a four-carbon (and a one-carbon) starting material. Thus, we deduce that an additional carbon–carbon bond-forming step would be required to prepare the four-carbon aldehyde from 2-propanol if pathway B were followed. The first approach then is to investigate pathway A.

Combining the steps in Figures 15-12 and route A in Figure 15-13 results in the sequence illustrated in Figure 15-14. Our synthesis is not yet complete, however, because we still need to prepare 2-bromopropane from 2-propanol. Clearly, this process does not require carbon–carbon bond formation and, by counting bonds to heteroatoms, we see that the oxidation levels of all carbons in the two species are the same. Because neither a carbon–carbon bond formation nor an oxidation-reduction is needed, we turn to Table 15-3, listing functional-group transformations, to see which one might be appropriate. Indeed, halides (both chlorides and bromides) can be conveniently prepared (using any of a variety of reagents) from the corresponding alcohols. Thus, it remains only to fill in the blank depicted by the dashed arrow with one such reagent—for example, HBr—to complete the sequence.

EXERCISE 15-D

Devise at least one multistep sequence for the transformation of ethanol (A) into alkene (B).

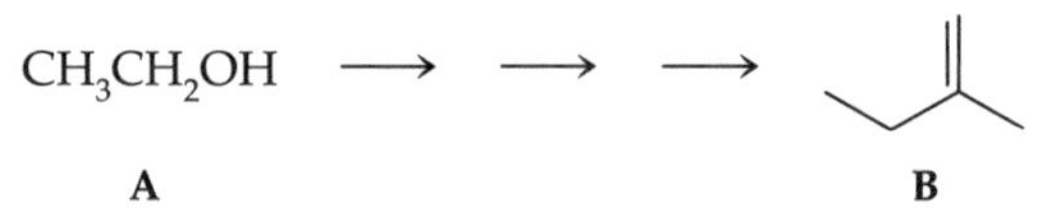

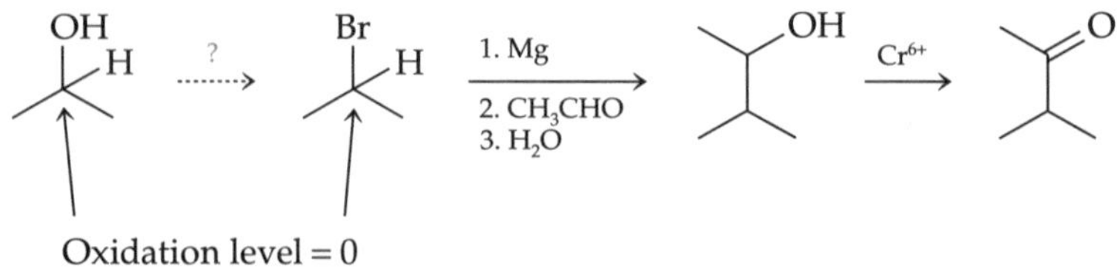

▲ FIGURE 15-14
3-Methyl-2-butanone can be prepared by oxidation of 3-methyl-2-butanol (Figure 15-12). This intermediate alcohol can be prepared by the reaction of isopropylmagnesium bromide with acetaldehyde (route A of Figure 15-13). The desired synthesis can be completed, therefore, if the given starting material (2-propanol) can be converted into a precursor for this three-carbon Grignard reagent.

15-5 ▸ Selecting the Best Synthetic Route

At this point, we might suspect that the "Christmas tree" representation shown in Figure 15-6 is an oversimplification because it implies that there is only one preferred route connecting our starting material with our ultimate product. Certainly, when all options are considered, it is often possible to end up with more than one workable route.

Let us return to pathway B in Figure 15-13, which does lead to 3-methyl-2-butanone, although by a sequence judged to be less efficient overall than the alternative pathway, A. To be able to use this sequence as part of the conversion shown in Figure 15-10, we must find reactions that can be used to efficiently convert 2-propanol into the intermediate 2-methylpropanal.

We can, and should, view this simply as another problem in retrosynthetic analysis. Thus, we first consider the carbon content of the two species to see whether we need to effect carbon–carbon bond formation at some point. Indeed, because 2-propanol has three carbons and methylpropanal has four, we need to form a carbon–carbon bond. From our list of reactions that make such bonds (Table 15-1), we choose the enolate alkylation. In this case, the starting material is propanal, and the formation of its enolate (with a strong base such as lithium dialkylamide), followed by reaction of the anion with an alkyl halide, results in the desired 2-methylpropanal (Figure 15-15). We have now simplified the overall problem, because the starting material for this step, propanal, has the same number of carbons as does 2-propanol.

Once again, we can consider the conversion of 2-propanol into propanal to be a separate synthesis and conduct an initial investigation of possible routes for its synthesis as part of our retrosynthetic analysis (Figure 15-16). Although the number of carbons in these two species is the same, the oxidation levels of the carbons are different. Furthermore, oxygen is bound to a different carbon atom in the reactant and the product. Overall, we must reduce C-2 of 2-propanol and oxidize C-1 so as to afford the aldehyde functional group. To find routes to accomplish this conversion, we turn to the table of oxidation-reduction reactions (Table 15-2). Because we

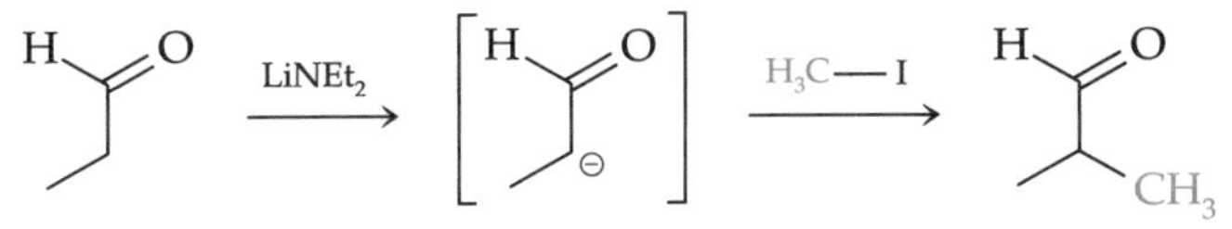

▲ FIGURE 15-15
The enolate anion of propanal is alkylated by methyl iodide to produce 2-methylpropanal.

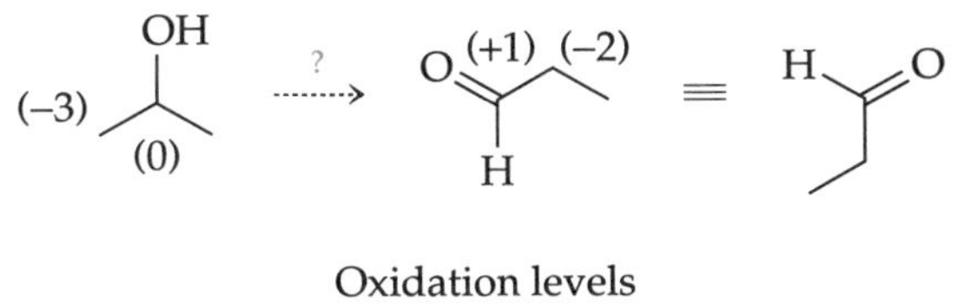

Oxidation levels

▲ FIGURE 15-16
2-Propanol must be oxidized in order to convert it into propanal, but this is not a simple oxidation reaction because the atom to which oxygen is bound in the reactant (C-2) differs from that to which it is bound in the product (C-1).

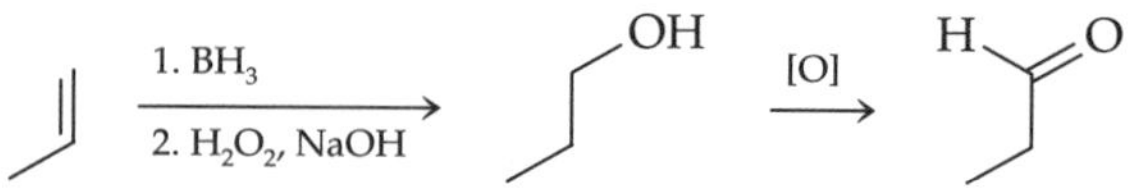

$$\text{propene} \xrightarrow[\text{2. H}_2\text{O}_2,\text{ NaOH}]{\text{1. BH}_3} \text{1-propanol} \xrightarrow{[\text{O}]} \text{propanal}$$

▲ FIGURE 15-17

Hydroboration-oxidation of propene leads to 1-propanol by an anti-Markovnikov hydration. Oxidation of the alcohol with CrO$_3$ in pyridine produces propanal.

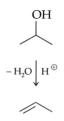

▲ FIGURE 15-18
Acid-catalyzed dehydration of 2-propanol leads to propene.

need to accomplish both an oxidation and a reduction, we must pay special attention to those reactions that simultaneously effect both transformations. These are the dehydration of an alcohol and the reverse process, the hydration of an alkene to form an alcohol.

The latter process can be accomplished in two significantly different ways that result in different regiochemical outcomes. Thus, simple hydration of an alkene follows a Markovnikov orientation and results in the hydroxyl group of the alcohol being on the more-substituted carbon of the original alkene. Conversely, hydroboration-oxidation places the hydroxyl group on the less-substituted carbon.

Can we use one or more of these processes to produce the desired product from acetaldehyde (Figure 15-16)? Because none of them produce a carbonyl group, the answer is no; and we are thus forced to examine other possible transformations that produce acetaldehyde, keeping in mind that the major tasks to be accomplished are the oxidation of C-1 and the reduction of C-2. This aldehyde can in turn be prepared by oxidation of the corresponding primary alcohol, 1-propanol. This compound could be formed by hydroboration-oxidation of propene (Figure 15-17).

Once again we have a new problem to analyze, but one that should be relatively straightforward by this time. Indeed, as shown in Figure 15-18, we can prepare propene from 2-propanol by dehydration.

By combining the reactions shown in Figures 15-18, 15-17, 15-15, and route B of Figure 15-13, we have a second route for the conversion of 2-propanol into 3-methyl-2-butanone as had been proposed in Figure 15-9.

EXERCISE 15-E

Suggest reagents that could be used to accomplish each of the transformations *a* through *g*.

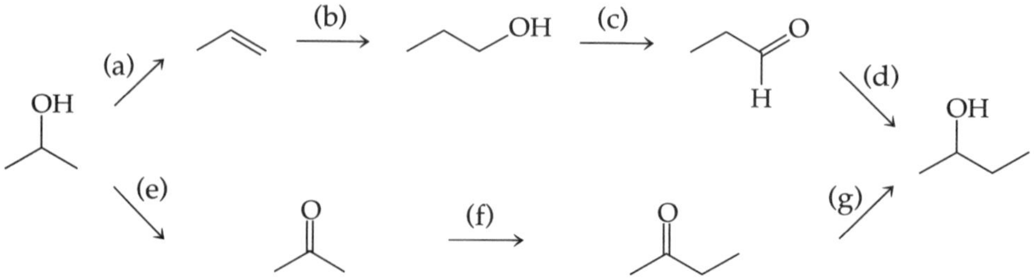

EXERCISE 15-F

Compare and contrast the positive and negative features of the two routes proposed in Exercise 15-E for the conversion of 2-propanol into 2-butanol.

SYNTHESIS OF VERY LARGE QUANTITIES OF MATERIALS

Industrial chemists must often take into account factors relating specifically to the scale of reactions when considering various options. Although laboratory chemists most often perform reactions in flasks smaller than 20 liters, this size is much too small for the manufacture of bulk chemicals. To help in making the transition to the scale of industrial production, consider that in 1991 9 *billion* pounds of styrene was produced, almost all of which was used in the manufacture of the plastic polystyrene (the material in polystyrene foam). If all of a year's production was amassed in one place, it would require a cubical container 150 meters on a side.

15-6 ▸ Criteria for Evaluating Synthetic Efficiency

Retrosynthetic analysis has now led us to two separate routes from 2-propanol to 3-methyl-2-butanone (Figure 15-19). Certainly, we should feel a real sense of accomplishment in having achieved two different syntheses, but how can we determine which one is "best"? There is no simple answer to such a question because there are many factors to be considered. Some of the more-important criteria for an efficient synthesis are:

1. number of steps;
2. yield of each step;
3. reaction conditions;
4. ease of purification of intermediates; and
5. cost of starting materials, reagents, and personnel time.

Often, these factors are not independent, but let us first consider each of them in turn. To simplify this analysis, we will assume that other criteria are fixed. For example, when analyzing the effect of the number of steps, we assume that all other factors (in this case, criteria 2 through 5) are the same.

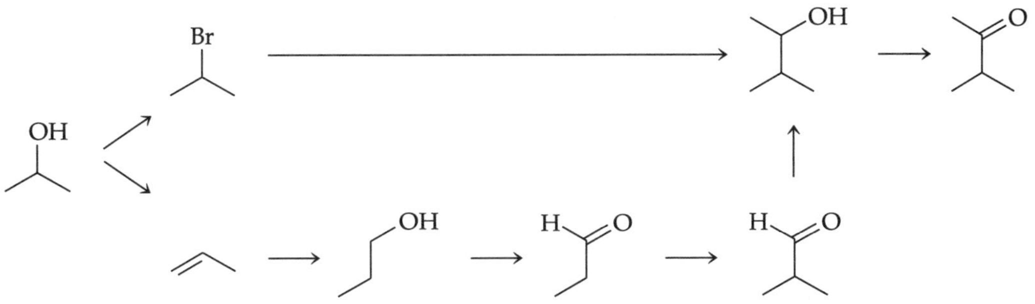

▲ FIGURE 15-19
Proposals for two separate routes for converting 2-propanol into 3-methyl-2-butanone. The upper reaction adds a two-carbon fragment in one step, whereas the lower reaction adds two separate one-carbon units. The upper reaction is a three-step synthesis; the lower reaction requires six steps.

Table 15-4 ▸ Overall Yield versus Number of Steps and Yield per Step

Number of Steps	Yield/Step (%)		
	90	75	50
1	90	75	50
2	81	56	25
3	73	42	12
4	66	32	6
5	53	18	3
6	48	13	1.5

Let us assume that the chemical yield in each step of a multistep synthesis is 90%, 75%, or 50%. We can then easily calculate the effect of added steps on overall yield (Table 15-4). After two steps, the yield drops from that observed in a single step: this decrease is more dramatic if the yield per step is lower. With a 50% yield per step, the overall yield after five transformations is only 3%.

Thus, when the yields in the individual steps are the same, the sequence consisting of the lowest number of steps is preferable. Furthermore, a sequence of three steps, even if each step proceeds with 90% yield, results in an overall yield that is slightly lower than that attained in a single step with a 75% yield. We also see how quickly the effect of modest yields (such as 50% per step) can reduce the amount of material available from a synthesis.

Because generally the objective of a synthesis is to prepare usable quantities of a product, we can also look at the effect of overall yield in a slightly different fashion. Assume that the objective is to prepare 10 grams of a product from a starting material whose molecular weight, for the sake of simplicity, is the same. The five-step sequence with 50% yield per step requires that we begin our synthesis with 333 grams of starting material (Figure 15-20), whereas the five-step sequence with 90% yield per step requires only about 17 grams.

The number of steps, the overall yield, and the yield per step are clearly important, but another factor must also be considered. The type of synthesis that we have dealt with so far is referred to as a **linear synthesis:** that is, one that effects sequential transformations, as schematically represented in Figure 15-20 by a five-step sequence proceeding with 50% yield per step. Let us consider an alternative possible route, called a **convergent synthesis,** as illustrated in Figure 15-21 in which two separate starting materials, A and D, are taken along separate routes to form intermediates C and F, which are combined to form the ultimate product, G. Although we again have a sequence with a total of five steps and have assumed that each of the steps proceeds in 50% yield, it is not possible for us to derive a simple overall yield for the sequence because there are two branches. Nonetheless, we can examine the effect of this branching if we consider the amounts of starting material required, rather than the overall yield.

Assume for simplicity that half of the mass of the ultimate product, G, is moved along each of the sequences A → C and D → F. To produce 10 grams of the ultimate product, G, requires 10 grams of C and 10 grams of F because the yield in the step that combines these two is only 50%. Furthermore, 40 grams of A is required to produce C and 40 grams of D is required to produce F. Thus, overall, a total of 80 grams of starting materials is required to produce 10 grams of the ultimate product. This contrasts mark-

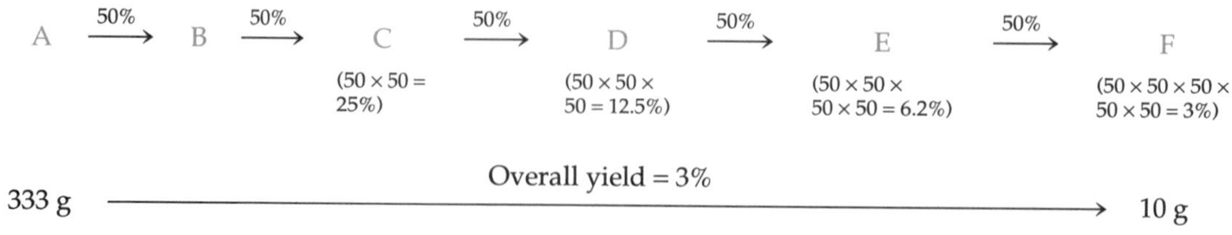

$$A \xrightarrow{50\%} B \xrightarrow{50\%} C \xrightarrow{50\%} D \xrightarrow{50\%} E \xrightarrow{50\%} F$$

C: $(50 \times 50 = 25\%)$
D: $(50 \times 50 \times 50 = 12.5\%)$
E: $(50 \times 50 \times 50 \times 50 = 6.2\%)$
F: $(50 \times 50 \times 50 \times 50 \times 50 = 3\%)$

333 g ⟶ Overall yield = 3% ⟶ 10 g

▲ **FIGURE 15-20**
Compounded losses in each step make it necessary to use large quantities of starting material in a multistep synthesis.

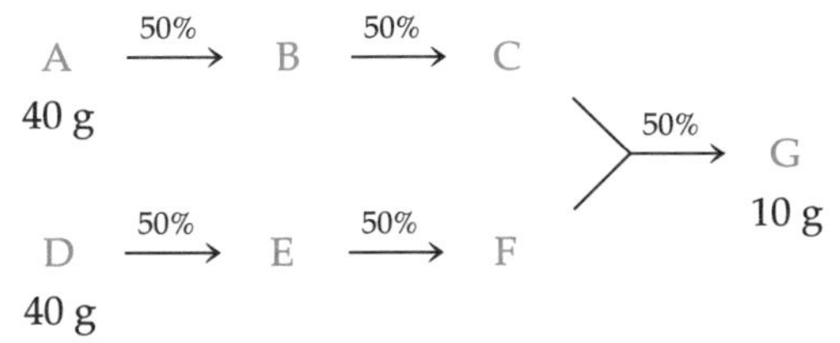

$$A \xrightarrow{50\%} B \xrightarrow{50\%} C$$

40 g

$$\qquad\qquad\qquad\qquad \xrightarrow{50\%} G$$

10 g

$$D \xrightarrow{50\%} E \xrightarrow{50\%} F$$

40 g

▲ FIGURE 15-21

Although this convergent synthesis consists of the same number of steps as the linear synthesis shown in Figure 15-20, much smaller quantities of starting materials are needed here to attain the same amount of product.

edly with the 333 grams of starting material required by the linear synthesis in Figure 15-20, even though the number of steps and the yield per step for the linear and convergent syntheses are the same. Although it is not always possible to develop a convergent synthesis, it is clearly advantageous to do so whenever appropriate.

The overall yield, which determines the amount of starting material required for a synthesis, is not the only consideration of importance. Reactions that use simple, inexpensive reagents and solvents and that do not require elaborate experimental precautions make a synthesis easier to carry out. Conversely, reactions that require very low or high temperatures, inert atmospheres (to prevent contamination by water or oxygen or both), or unusual solvents should be avoided when simpler alternatives are available, even with some sacrifice in yield.

Another important criterion is the number and the nature of by-products. In a simple sense, this consideration is but one facet of chemical yield, but there are other ramifications of the production of by-products besides the simple reduction in the amount of the desired product. For example, reconsider the alkylation of 2-butanone (Figure 15-11), which can yield two products (resulting from alkylation at C-3 and C-1). Before proceeding to the next step, these products must be separated and the desired compound purified. In this example, the similarity between these two isomeric ketones makes this separation difficult. Separations like this often consume more time and effort than the chemical transformations themselves, and some of the desired material is frequently lost in the process.

How, then, can we select between various possible synthetic routes? In some cases, the choice is simplified because most or all of the factors just described favor only one of several possible sequences. That, indeed, is the case for the choices shown in Figure 15-19. The upper reaction requires only three transformations, whereas the other requires six. Furthermore, the longer route includes a hydroboration-oxidation sequence and, thus, requires use of the toxic and highly flammable reagent diborane. Finally, the hydroboration-oxidation sequence does not form the primary alcohol exclusively: rather, it gives a mixture of the dominant primary alcohol with significant amounts of the secondary alcohol (approximately 20%). We therefore have no difficulty in deciding to use the short sequence to accomplish our transformation of 2-propanol into 3-methyl-2-butanone. This example is rather unusual, however; in many cases, the alternate routes are not so markedly different. Indeed, even for practicing synthetic chemists, the selection process is often difficult.

EXERCISE 15-G

Calculate the net yield in the following synthesis, given the individual yield shown for each step. Calculate how much starting material would be needed to prepare 10 g of product by this route. Categorize each of the steps *a* through *f* as a carbon–carbon bond-forming, an oxidation-reduction, or a functional-group transformation.

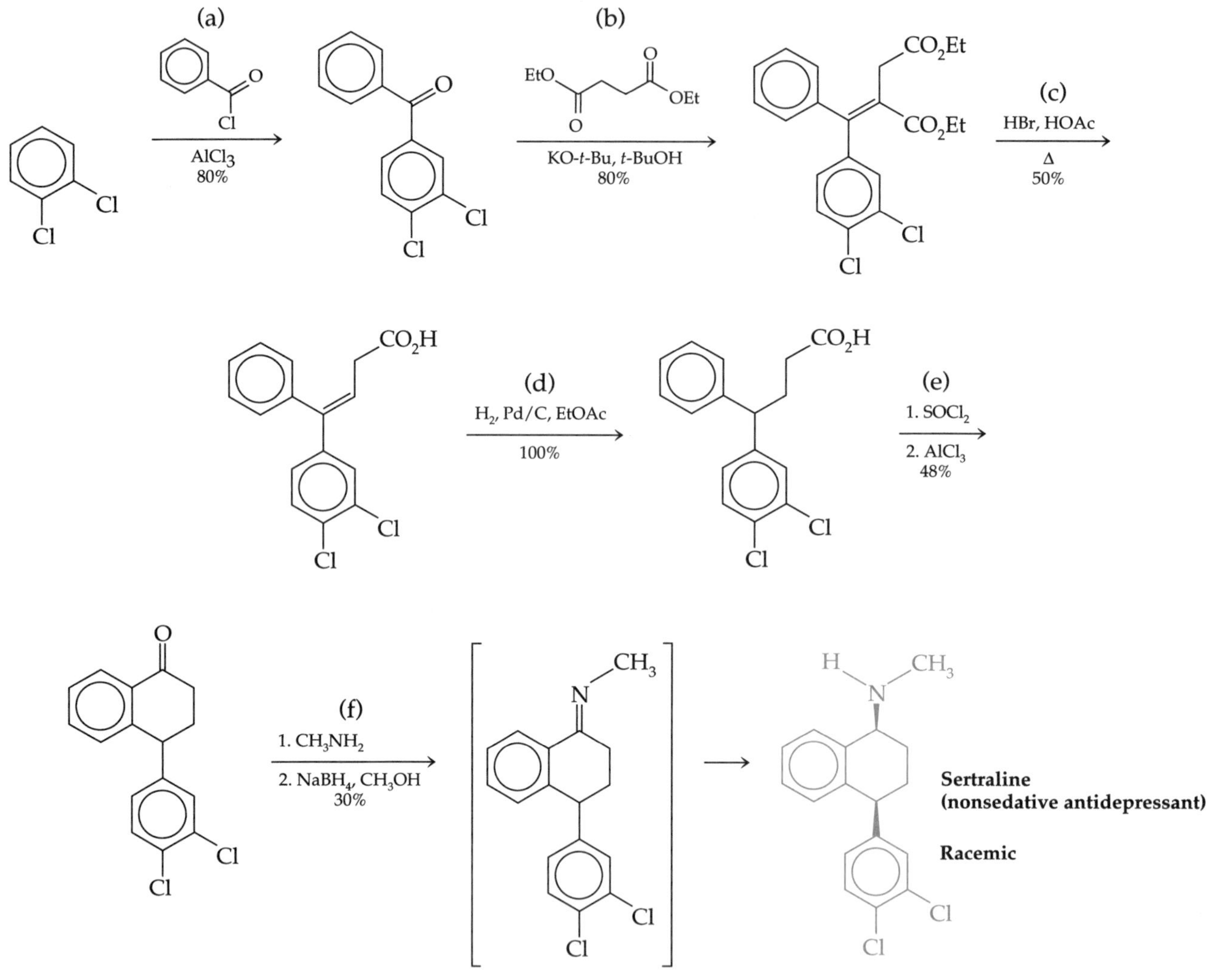

15-7 ▸ "Real World" Examples: Functional-Group Compatibility

The examples considered so far have been simple molecules, generally having only a single functional group. However, the types of molecules of interest to the practicing synthetic chemist are often complicated structures containing many different functional groups. For example, the ten molecules shown in Figure 15-22 are the leading prescription pharmaceuticals sold in 1990. Their generic descriptors are listed with their trade names in the figure. Some are naturally occurring materials: for example, **digoxin**

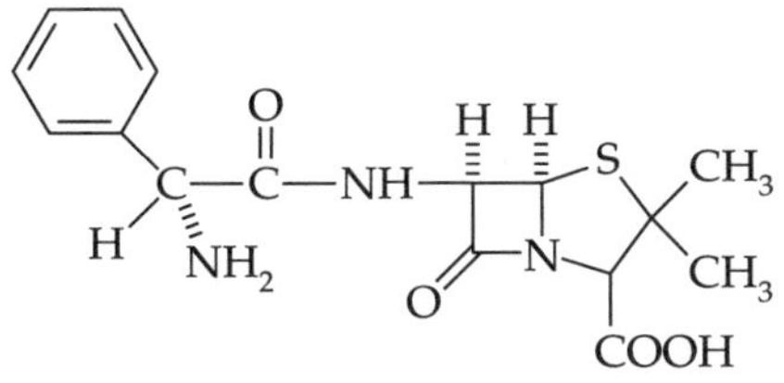

Amoxicillin (Amoxil, SmithKline Beecham)
Antibiotic

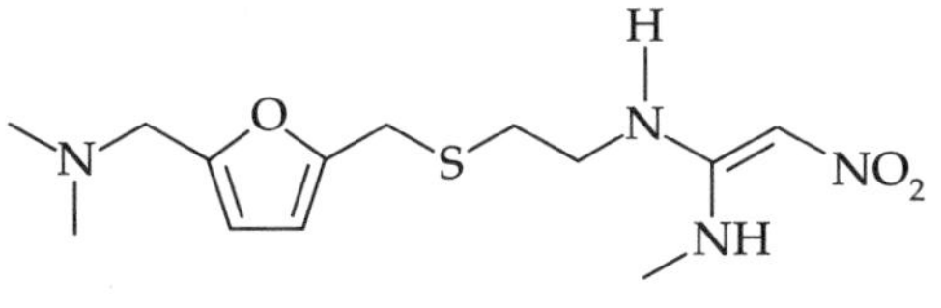

Rantidine (Zantac, Glaxo)
Antiulcerative

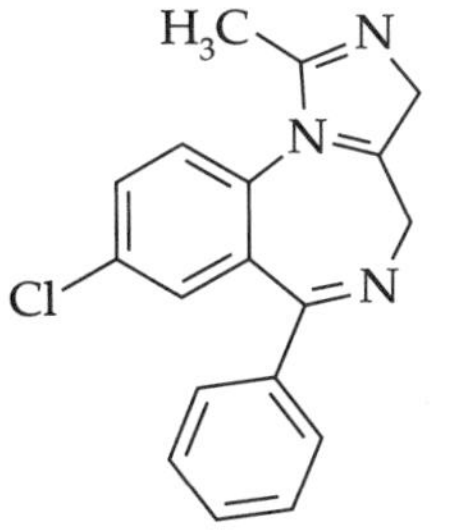

Alprazolam (Xanax, Upjohn)
Antianxiety

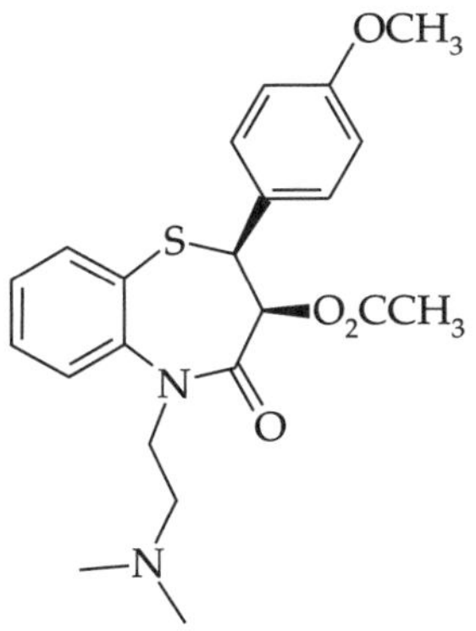

Diltiazem (Cardizem, Marion Merrell Dow)
Calcium channel blocker, coronary vasodilator

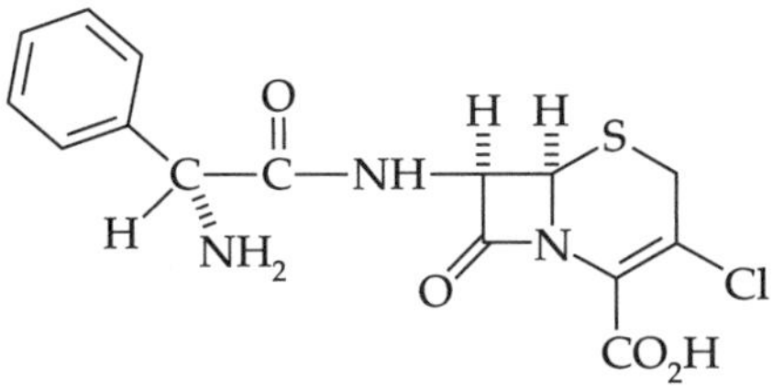

Cefaclor (Ceclor, Eli Lilly)
Antibiotic

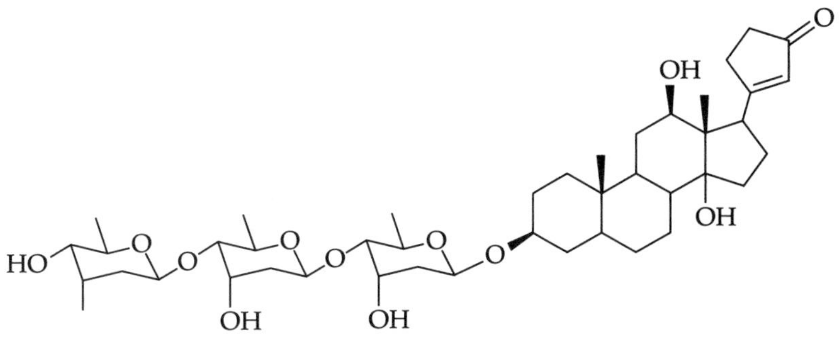

Digoxin (Lanoxin, Burroughs Wellcome)
Heartbeat regulator

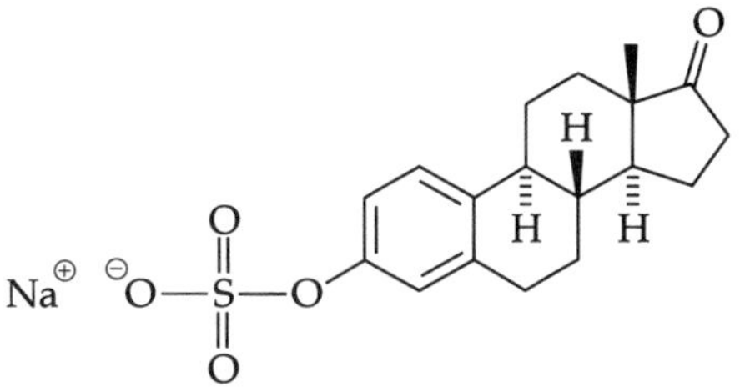

Conjugated estrogen (Premarin, Wyeth-Ayerst)
Antimenopausal, estrogen replacement

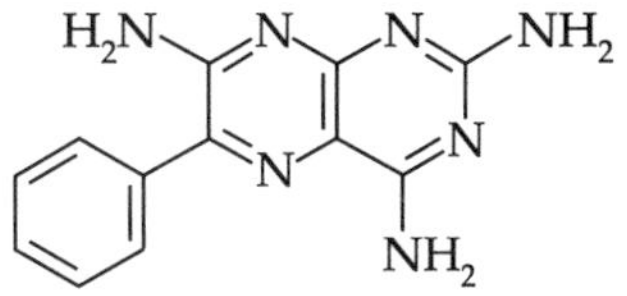

Triamterene (Dyazide, SmithKline Beecham)
Diuretic

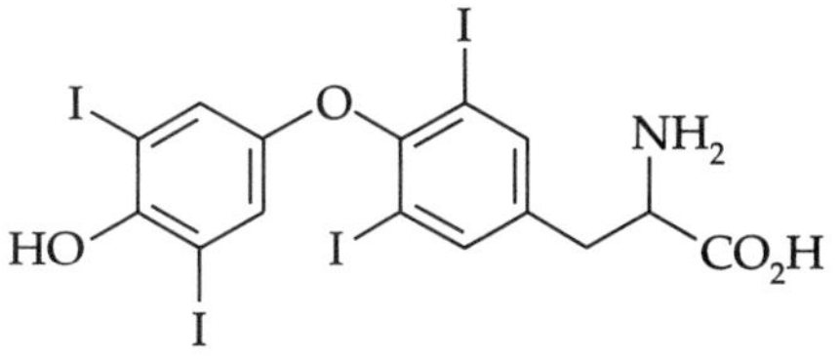

Levothyroxine (Synthoid, Boots)
For treatment of hypothyroidism

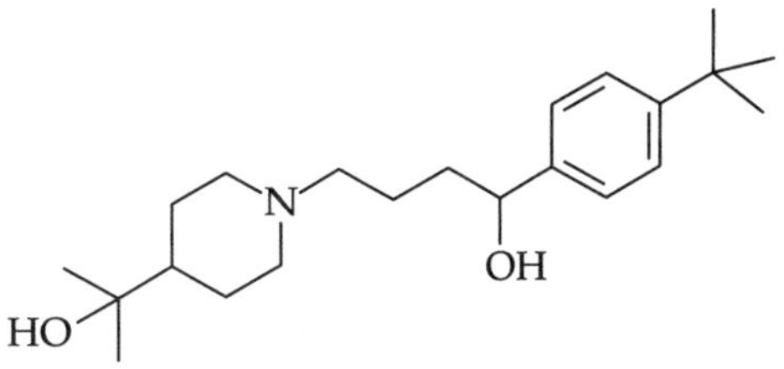

Trefenadine (Seldane, Marion Merrell Dow)
Antihistamine

▲ **FIGURE 15-22**
The current ten most widely used prescription drugs are multifunctional
organic molecules. Their generic names (as well as their trade names and
manufacturers) are listed.

(**Lanoxin**) and **levothyroxine** (**Synthoid**) are obtained for commercial use by isolation from their natural sources. It is often more economical to do so than to prepare complex organic materials synthetically in the laboratory. Furthermore, naturally occurring materials can often be modified in relatively minor ways to provide highly active pharmaceutical agents, as is the case for **amoxicillin** (**Amoxil**) and **cefaclor** (**Ceclor**), which are highly effective **antibiotics** belonging to the **penicillin** and **cefalosporin** classes, respectively, as well as for **conjugated estrogen** (**Premarin**). The remaining five compounds, **ranitidine** (**Zantac**), **alprazolam** (**Xanax**), **triamterene** (**Dyazide**), **diltiazem** (**Cardizem**), and **trefenadine** (**Seldane**) are prepared by multistep syntheses.

In devising synthetic sequences for the preparation of such complex molecules, the chemist must consider whether more than one functional group in an intermediate might react with a reagent whose primary target is in another part of the molecule. Generally, it is desired that only one of the functional groups interacts with the reagent. Thus, the question of **functional-group compatibility** must be considered.

NAMING PHARMACEUTICAL AGENTS

The success in the market place of Tagamet for the treatment of ulcers stimulated a search by many pharmaceutical companies for new β-blockers that would be more effective and have fewer side effects. At Glaxo, this research resulted in the structurally quite similar compound marketed as Zantac (ranitidine).

Cimetidine (Tagamet)

Ranitidine (Zantac)

Sales of Zantac account for more than half of Glaxo's gross revenues of $4.9 billion. However, when patent protection expires in 1995, many other manufacturers will be able to produce and sell Zantac but will have to use the generic name, ranitidine. How are names for pharmaceutical agents selected? The company that first prepares a successful drug is free to choose any proprietary name that it wishes, and marketing specialists use their expertise in picking a name that is easy to pronounce and, perhaps, even "catchy." An independent group helps in deciding on a generic name by providing a list of ten possibilities from which the discovering company may select. At this point, the name that is least easy to pronounce and, especially, to remember is chosen in the hope that, even after patent protection has expired, physicians and consumers will still opt for the trade name under which the drug was first released, preserving market share for the discovering company.

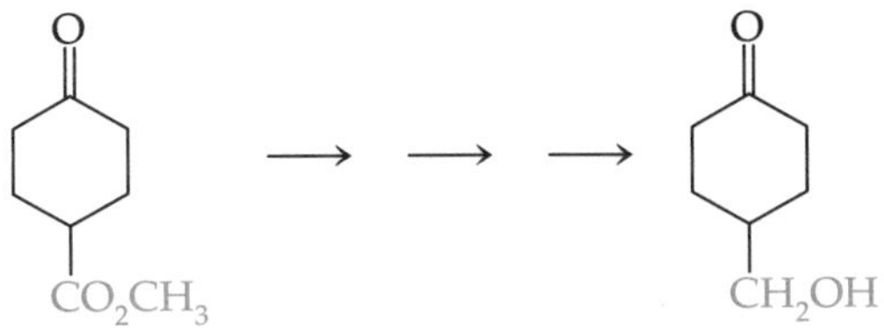

▲ FIGURE 15-23
In the desired conversion, an ester is reduced to an alcohol in the presence of a
ketone group. Because a ketone is usually more easily reduced than an ester,
this reaction is nontrivial.

For example, let us consider the possible conversion of the ketoester in
Figure 15-23 into the product ketoalcohol. Retrosynthetic analysis of the
problem reveals that the starting material and the product have the same
number of carbons; therefore, the conversion does not require a carbon–
carbon bond-forming reaction. Indeed, all that is needed is the reduction of
the ester functional group in the starting material to a primary alcohol
group in the product, a reaction that we know can be accomplished with
$LiAlH_4$. However, the starting material has two functional groups, a ketone
and an ester; unfortunately, both of them are reduced by $LiAlH_4$. In fact, the
resonance stabilization of an ester functional group makes it somewhat
more resistant to reduction than is a ketone. Reaction of the ketoester with
$LiAlH_4$ results in the reduction of both functional groups, producing the
diol, as shown in Figure 15-24.

Had the problem been differently constituted such that the reduction of
the ketone carbonyl group was required (as illustrated in Figure 15-25), then
a milder reducing reagent, $NaBH_4$, could have been used to *selectively* reduce only the ketone group.

Analogous concerns apply to many other functional groups as well. In
planning a synthesis, it is necessary to be aware of possible competing reactions that can be induced by a chosen reagent on other parts of a molecule.
As we will see, one solution to effecting a transformation on only the less-
reactive functional group is to first convert the more-reactive one into a different functional group that does not react with the chosen reagent.

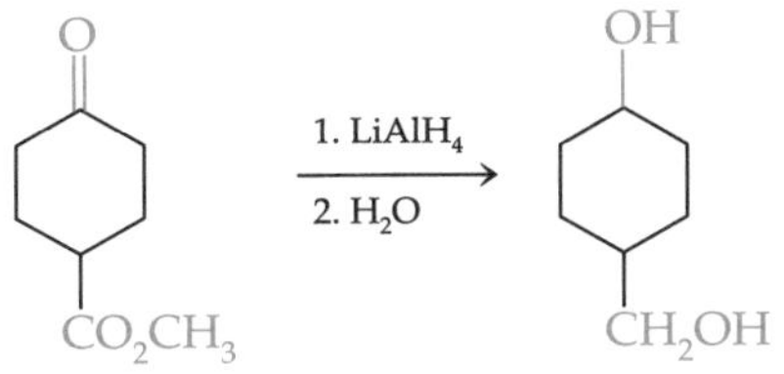

▲ FIGURE 15-24
Because both ketone and ester are reduced by $LiAlH_4$, a diol is produced.

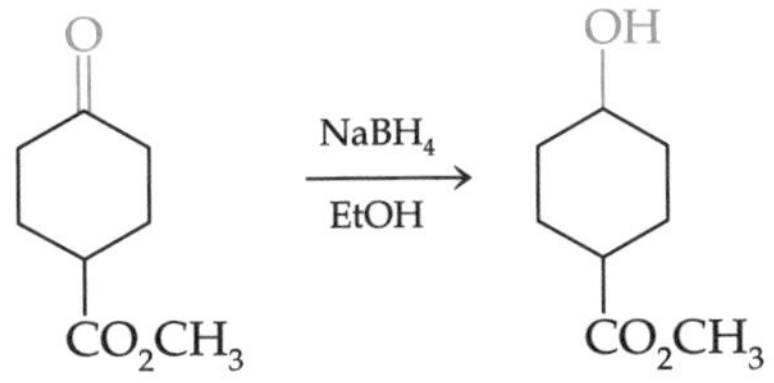

▲ FIGURE 15-25
The higher activity of a ketone than that of an ester toward nucleophilic attack
makes it possible to selectively reduce a ketone with $NaBH_4$ in the presence of
an ester.

SEMISYNTHETIC PHARMACEUTICAL AGENTS

It is common in the pharmaceutical industry to make use of natural compounds for the production of valuable chemicals. For example, some antibiotics are produced by microbes from which they are simply isolated and purified. However, some microbes do not produce products that have all of the necessary and desirable features (activity, stability, and few side effects, for example). In this case, it is often possible to "help" a microbe produce a desired product that is similar to one naturally produced. Feeding microbes the necessary structural subunits often results in the incorporation of these pieces into the final product, as, for example, in the industrial production of the antibiotic penicillin V. This antibiotic is prepared by feeding *Penicillium* molds 2-phenoxyethanol, which the mold oxidizes and uses for the amide side chain of penicillin V.

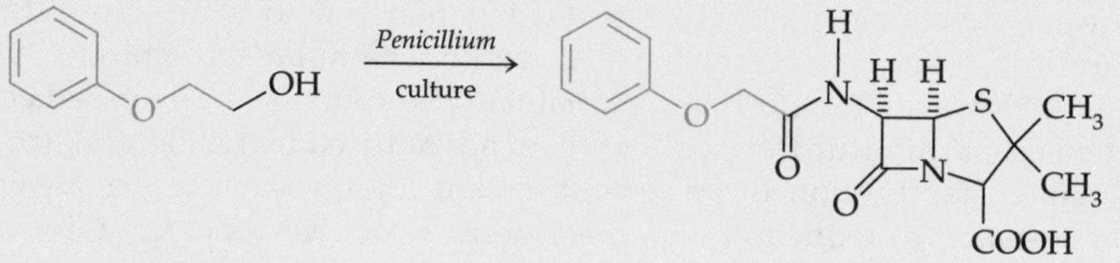

2-Phenoxyethanol   Penicillin V

EXERCISE 15-H

For each of the following transformations, determine whether the cited reagent is compatible with the other functional groups in the molecule. If not, predict the nature of the problem that would be encountered if the indicated transformation were attempted.

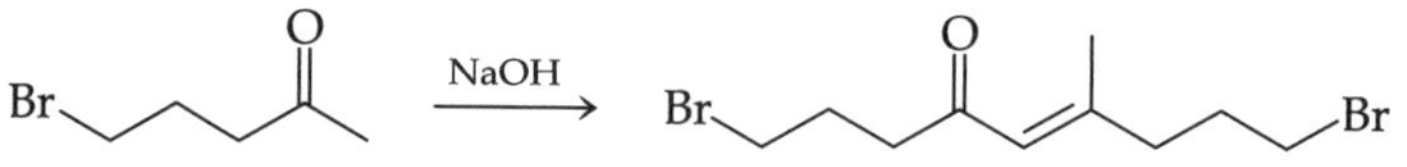

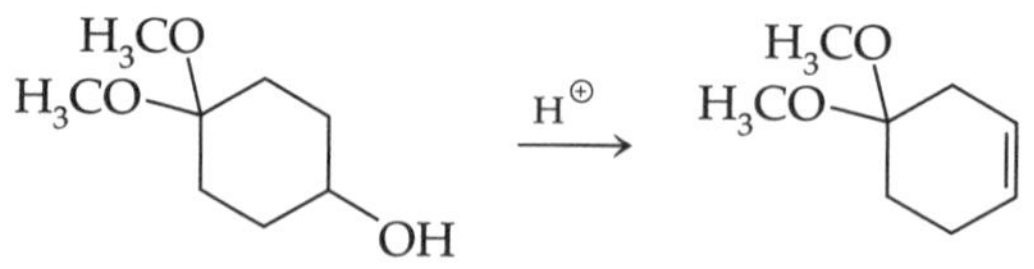

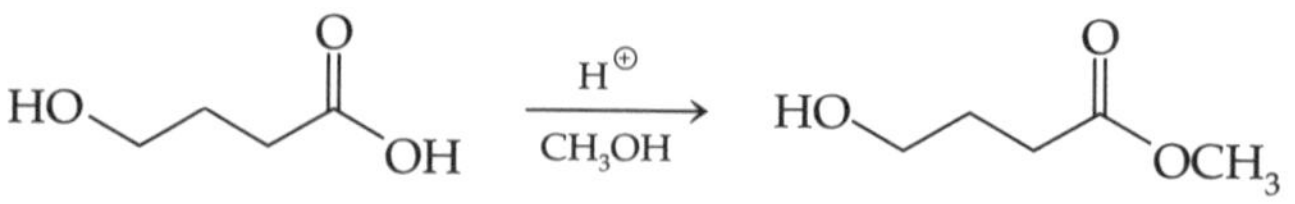

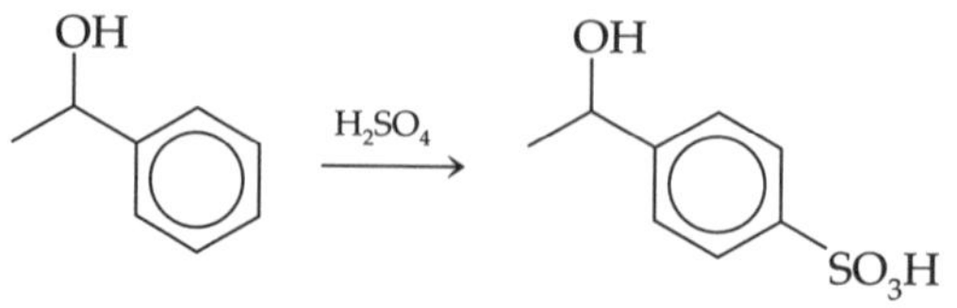

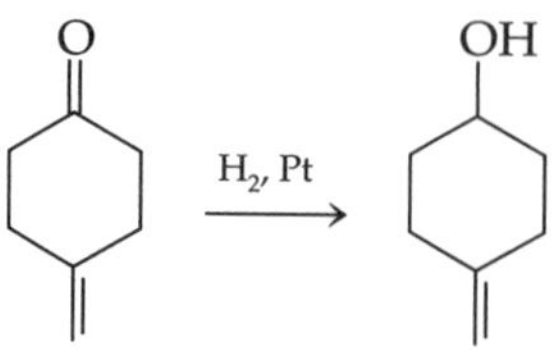

(f)

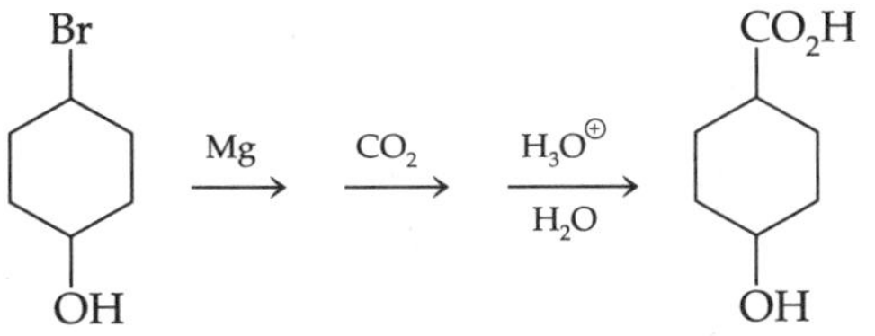

(g)

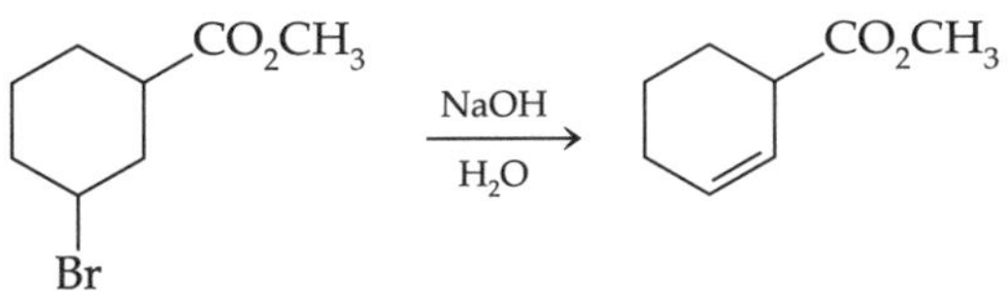

15-8 ▸ Protecting Groups

The problem posed in Figure 15-23 presents a difficult situation because we do not have a reduction reaction that directly affords the desired ketoalcohol. Thus, we must turn to the next group of reactions (functional-group transformations) to see if one of those might instead serve as the last step in the sequence. In Table 15-2, we find that ketals can be converted into ketones by hydrolysis with water in the presence of acid catalysts.

Protecting Groups for Aldehydes and Ketones

Assuming that a ketal hydrolysis (shown in Figure 15-26) can be used as the last step in the synthesis, we must then define what should be the next-to-last step, keeping in mind that we have not yet accomplished the required reduction. We ask the question: What reduction (or oxidation) reaction might be used to form the ketal alcohol? The ketal ester can indeed be reduced with LiAlH$_4$ to provide the desired intermediate (Figure 15-27). With

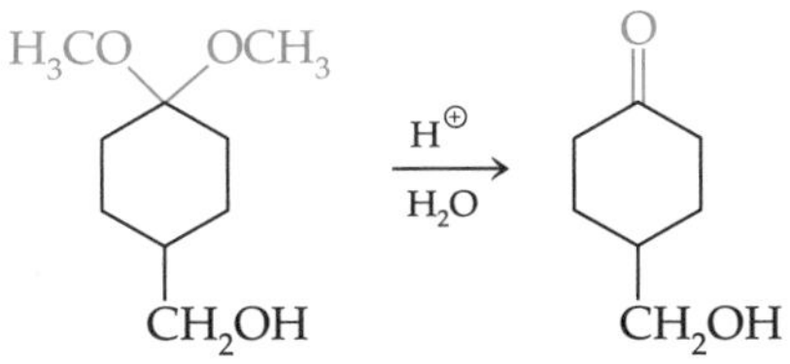

▲ FIGURE 15-26
A ketone is formed when a ketal functional group is hydrolyzed in aqueous acid.

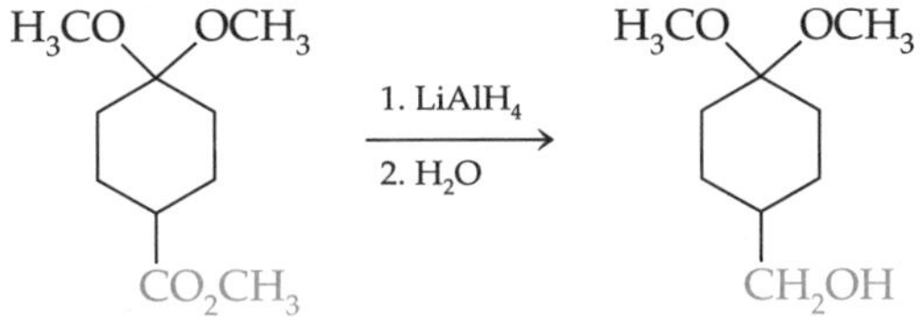

▲ FIGURE 15-27
A ketal functional group is relatively inert to LiAlH$_4$; so selective reduction of the ester takes place.

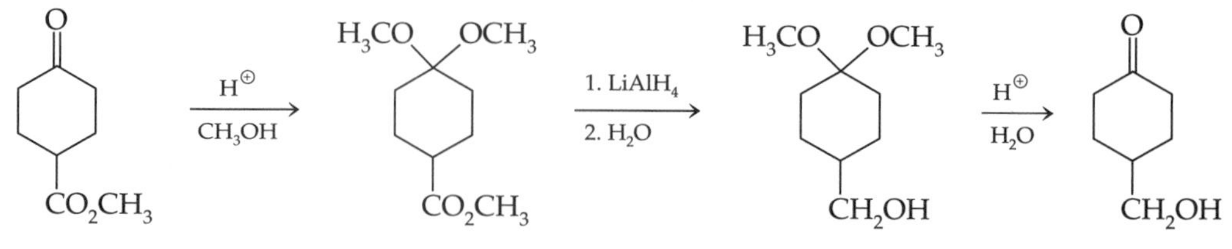

▲ FIGURE 15-28
When the ketone group is protected as a ketal, selective reduction of the ester to a primary alcohol can be accomplished. Deprotection of the ketal by acid-catalyzed hydrolysis then affords the desired ketoalcohol.

the required reduction having been accomplished, whatever reactions remain to connect our original starting material with the ketal ester must come from the functional-group transformations listed in Table 4-3. Because ketones can be converted into ketals in the presence of the alcohol group (as in the first step in the sequence shown in Figure 15-28), we can circumvent the original problem of having two reactive functional groups present in our starting material, both of which would undergo reaction with a common reagent. By converting the more-reactive ketone functional group into a ketal, we have *protected* it from reduction. Once that critical step is accomplished, the ketone functional group can be regenerated from the ketal.

Because of the role that the ketal plays in temporarily protecting the ketone functional group, ketals are often referred to as **protecting groups.** There are rather stringent requirements placed on the reactions that chemists employ to protect functionality. First, the use of a protecting group requires two additional steps (protection and deprotection), and the yield of both reactions must be high. Second, the reactions must produce few by-products, which can pose significant separation problems. Finally, reactions that protect and deprotect a functional group should preferably require the use of only simple reagents and reaction conditions.

The formation of a ketal from a ketone and its subsequent hydrolysis is a protection scheme for the ketone that fits all of the foregoing requirements (Figure 15-28). Both reactions generally proceed with high chemical yields, no major by-products are formed, and the reaction conditions are relatively mild and affect few other functional groups.

A wide variety of protecting groups have been developed for the most-common functional groups, but we will limit our consideration to one example for each of three other functional groups: alcohols, amines, and carboxylic acids.

EXERCISE 15-I

Useful protecting groups must have three features:

1. The reactions used to form them must be readily reversible.
2. Both the protection and deprotection steps must proceed in high yield.
3. A characteristic reactivity of the starting functional group must not be present in the protected form.

Evaluate whether each of the transformations listed in Table 15-3 meets the two criteria listed above, selecting those that you think might serve as methods for protecting functional groups.

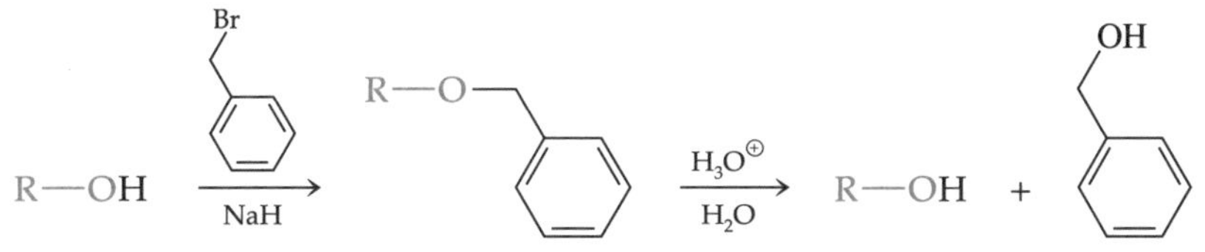

▲ FIGURE 15-29
Under basic conditions, an alcohol is converted into an alkoxide, making it sufficiently active as a nucleophile to displace bromide from benzyl bromide (the product is called a benzyl ether and the —H₂C Ph group is often abbreviated Bn). The alcohol can be reformed by acid-catalyzed hydrolysis of the resulting benzyl ether.

Protecting Groups for Alcohols

There are many protecting groups for alcohols, possibly more than for any other functional group. Benzyl ethers are simple and quite useful examples. Benzyl ethers can be formed readily by the reaction of an alcohol with benzyl bromide under basic conditions, as shown in Figure 15-29 (a Williamson ether synthesis). The normal acidity of an alcohol is masked in the ether, which is inert to almost all basic and mildly acidic reaction conditions. However, benzyl ethers are readily cleaved in strong acid by an S_N1 substitution at the benzylic carbon, as well as by reduction with H_2 and a metal catalyst such as Pd.

EXERCISE 15-J

Write a detailed mechanism for the alcohol deprotection (benzyl ether hydrolysis) shown as the second step of Figure 15-29 for methanol (R = CH₃). Consider carefully which carbon–oxygen bond is more likely to be cleaved under acidic conditions. Are there situations with other R groups in which it would be less clear which carbon–oxygen bond would be cleaved more rapidly?

Protecting Groups for Carboxylates

The acidity of a carboxylic acid can be masked by its conversion into a tertiary butyl ester (Figure 15-30). In contrast with normal esterification by nucleophilic acyl substitution, the formation of tertiary butyl esters is an acid-catalyzed, Markovnikov addition to the π bond of isobutylene (2-methylpropene).

EXERCISE 15-K

Write a detailed mechanism, showing electron flow, for the first conversion shown in Figure 15-30.

Hydrolysis of the tertiary butyl ester under basic conditions (by nucleophilic acyl substitution) is difficult because of the additional steric interference presented by the three methyl groups of the tertiary butyl group. However, the acid-catalyzed esterification can be reversed in the presence of water, and the treatment of tertiary butyl esters with relatively mild acid

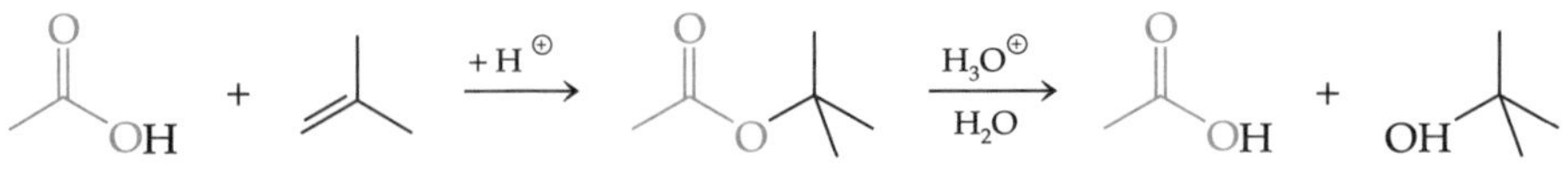

▲ FIGURE 15-30

A *t*-butyl ester is formed in the acid-catalyzed reaction of a carboxylic acid with isobutylene (2-methylpropene). The acid can be regenerated by acid-catalyzed hydrolysis of the ester.

under aqueous conditions reforms the carboxylic acid and tertiary butyl alcohol. Both the formation of tertiary butyl esters and their subsequent cleavage to reform the carboxylic acids take place under conditions that are significantly milder than would be necessary to form and cleave esters by a nucleophilic acyl substitution pathway (for example, methyl esters).

EXERCISE 15-L

The cleavage of *t*-butyl esters can also be achieved with neat trifluoroacetic acid, a carboxylic acid that is considerably stronger than acetic acid because of stabilization of the carboxylate by the three electron-withdrawing fluorine atoms. Under these conditions, however, isobutylene, not *t*-butyl alcohol, is formed. Explain this difference.

Protecting Groups for Amines

Finally, we consider the protection of amines, a functional group that is considerably less acidic than either carboxylic acids or alcohols but has more significant nucleophilic character as a neutral species. The reaction of a primary amine with tertiary butyl chloroformate (Figure 15-31) in the presence of a weak base (such as pyridine) forms a **urethane** (alternatively referred to as a **carbamate**). The urethane functional group [$R_2NCO(OR)$] resembles an ester on one side and a carboxylic acid amide on the other. Indeed, the lone pair of electrons on nitrogen in a urethane is delocalized into the carbonyl group in the same way as in an amide. The interaction of an electrophile with this lone pair requires disruption of resonance stabilization. As a result, urethanes are not sufficiently nucleophilic to react with most electrophiles. In deprotection, however, the ester part of the urethane resembles the tertiary butyl ester considered in Figure 15-30. Thus, treatment with

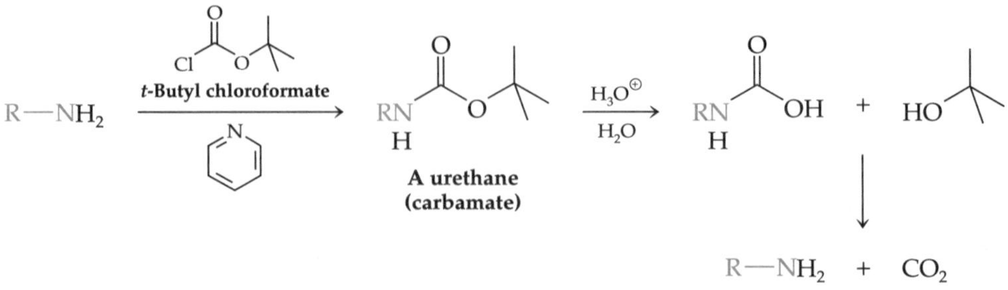

A urethane (carbamate)

▲ FIGURE 15-31

The formation of a urethane by the reaction of a primary amine with a chloroformate takes place by the same mechanism by which an amide is formed in the reaction of an amine with an acid chloride. The acid-catalyzed hydrolysis of the urethane produces the same intermediate that is formed in the Hofmann rearrangement (by hydrolysis of an isocyanate). The loss of CO_2 from this acid derivative is rapid, regenerating the starting amine.

mild aqueous acid results in the formation of tertiary butyl alcohol and a nitrogen-substituted carboxylic acid (a carbamic acid) that undergoes rapid decarboxylation to form the amine.

EXERCISE 15-M

Write significant resonance contributors for a urethane, for an amide, and for an ester. Which of these functional groups has the most charge on the carbonyl oxygen? Which the least? Does the amide or the urethane have more charge on nitrogen? Does the ester or the urethane have more charge on the ether oxygen?

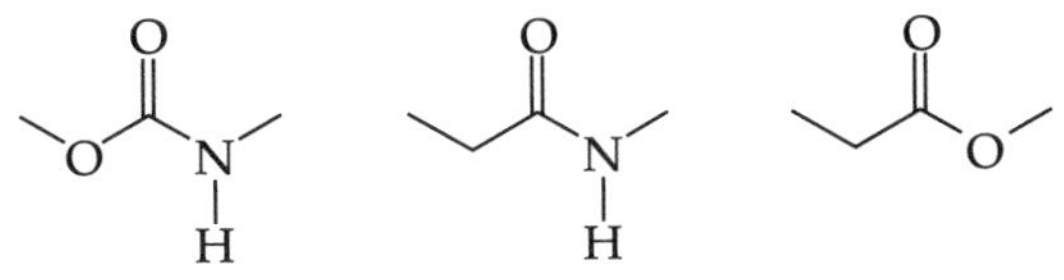

| A urethane | An amide | An ester |

EXERCISE 15-N

Write a detailed reaction mechanism for the last step of Figure 15-31. Is this conversion faster under acidic or under basic conditions? Explain why this decarboxylation has a lower activation energy than that for a simple carboxylic acid.

Let us now consider how we might use protecting groups as part of a process to convert the ketoalcohol prepared in Figure 15-28 into a ketone bearing a methyl group at the α carbon of the ketone (Figure 15-32). By comparing the starting material and product, we see that we need to form a carbon–carbon bond. From Table 15-1, we select the conversion of a ketone into its enolate, followed by alkylation with an electrophile (in this case, methyl iodide). However, although the two functional groups present in the starting material are quite different (a ketone and a primary alcohol), both are alkylated when treated with base and an alkyl halide: respectively, by enolate alkylation and Williamson ether synthesis (Figure 15-33). Furthermore, because the alcohol is more acidic than is the ketone, deprotonation and alkylation of the latter group is not possible without the conversion of the alcohol into its methyl ether.

Thus, the desired carbon–carbon bond-forming reaction can be effectively accomplished only if we temporarily mask the reactivity of the alcohol group. A satisfactory approach is illustrated in Figure 15-34. First, the alcohol is converted into its benzyl ether. The alkylation is then accomplished, and the benzyl ether protecting group for the alcohol is removed.

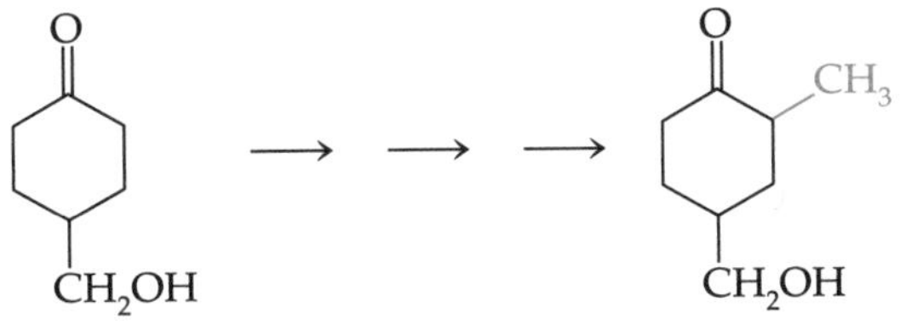

▲ FIGURE 15-32
This proposed conversion is a synthetic challenge because it requires the α methylation of a ketone in the presence of an alcohol.

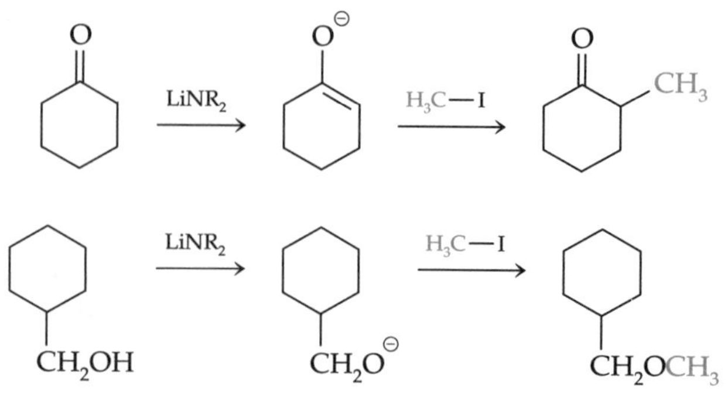

▲ FIGURE 15-33

Sequential treatment with strong base and methyl iodide effects alkylation of both a ketone (upper reaction) and an alcohol (lower reaction). These compounds serve as models for the reactivity expected for the ketoalcohol in Figure 15-32.

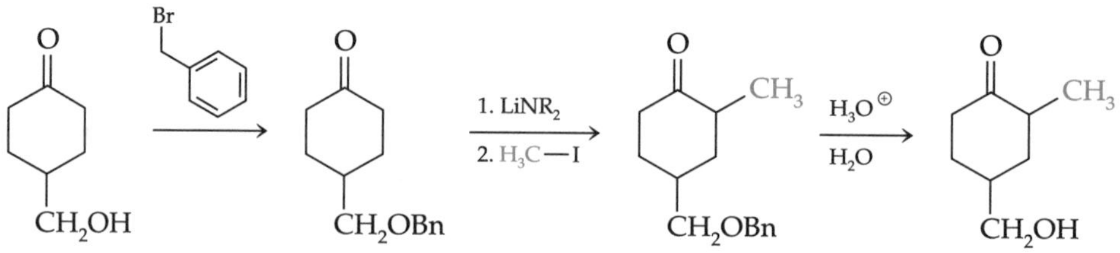

▲ FIGURE 15-34

Conversion of the primary alcohol into a benzyl ether blocks its further alkylation. The resulting compound can be alkylated at the α carbon of the ketone, by treating the ketone enolate (produced by the reaction with lithium dialkylamide) with methyl iodide. The benzyl ether can then be removed by acid-catalyzed hydrolysis.

15-9 ▸ Practical Examples

Having explored the thought process required for retrosynthesis, we will briefly analyze several commercially important, relatively short synthetic sequences. We will not attempt to develop a retrosynthetic analysis; rather, we will concentrate on how each of the transformations used fits into the overall scheme and accomplishes the needed changes.

Ibuprofen and Ketoprofen

Ibuprofen and **ketoprofen** are two pharmaceutical agents that have a variety of uses. Originally sold by prescription only, ibuprofen is now available as an over-the-counter general pain reliever.

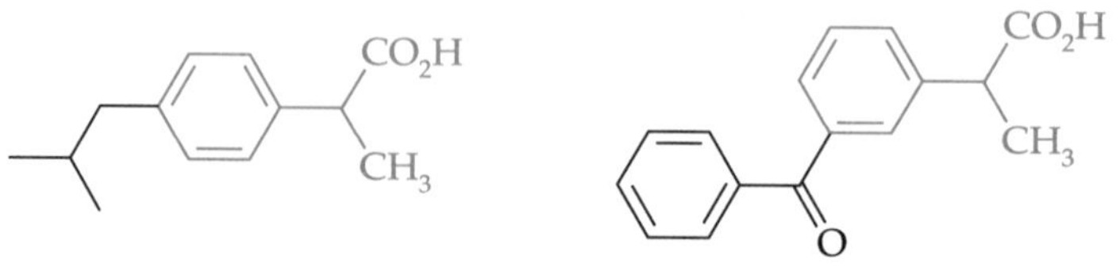

Ibuprofen

Ketoprofen

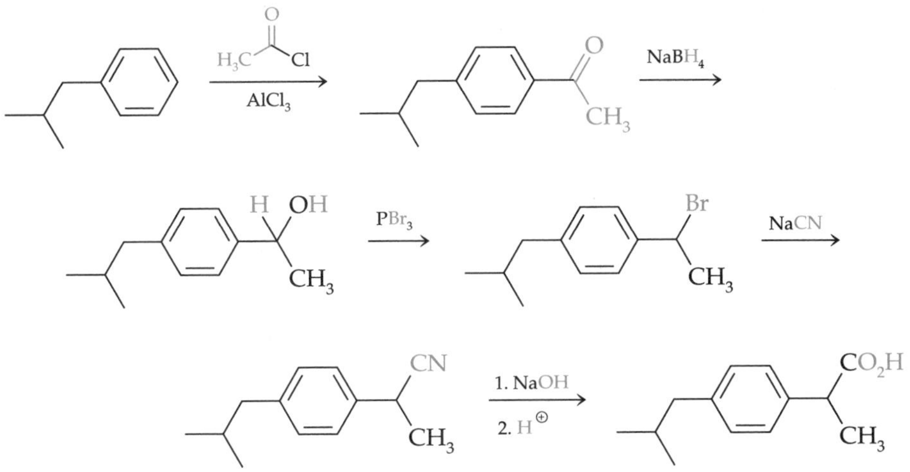

▲ FIGURE 15-35
The synthesis of ibuprofen begins with a Friedel-Crafts acylation of isobutyl benzene. Reduction of the ketone to an alcohol and conversion into a bromide permits the introduction of an additional carbon atom as a nitrile in an S_N2 displacement. Hydrolysis of the nitrile gives the carboxylic acid product.

If we concentrate on the right-hand part of these molecules, we see that each has an aromatic ring as a substituent on propanoic acid. Indeed, many pharmaceutical agents have this same structural feature, bearing various aromatic rings with different functionality. Such "second generation" drugs are often developed through a trial-and-error approach by pharmaceutical chemists who are trying to develop new agents that require lower dosages and have fewer side effects than existing products.

The syntheses of ibuprofen (Figure 15-35) and ketoprofen (Figure 15-36) address the construction of the phenylpropanoic acid unit in distinctly different ways.

We can view the problem of forming this unit as the elaboration of a benzylic carbon with a carboxylic acid and a methyl group. Both syntheses depend on the displacement of a benzylic bromide by cyanide to incorporate the required carboxylic acid functional group (the fourth step in Figure 15-35 and the second step of Figure 15-36). In ibuprofen, the benzylic carbon is secondary and already bears the necessary methyl group. In contrast, the ketoprofen synthesis requires displacement by cyanide at a primary benzylic bromide site. Recall that an S_N2 substitution is subject to steric interference and that often E_2 elimination pathways are also followed by secondary alkyl halides. From this aspect, the ketoprofen synthesis is superior.

The ibuprofen synthesis (Figure 15-35) begins by Friedel-Crafts acylation of isobutylbenzene. Reduction of the ketone and conversion of the resulting alcohol into the bromide precedes the critical cyanide displacement. Hydrolysis of the nitrile then installs the required acid in the observed product.

The ketoprofen synthesis (Figure 15-36) begins by benzylic bromination of the methyl group of 3-methylbenzophenone. After the S_N2 displacement by cyanide ion, the introduction of a methyl group is needed at a position α to the nitrile. We have seen that nitriles can be deprotonated at the α position to form an anion that can act as a nucleophile (recall the Michael addition). There are then two acidic hydrogens that might be replaced by

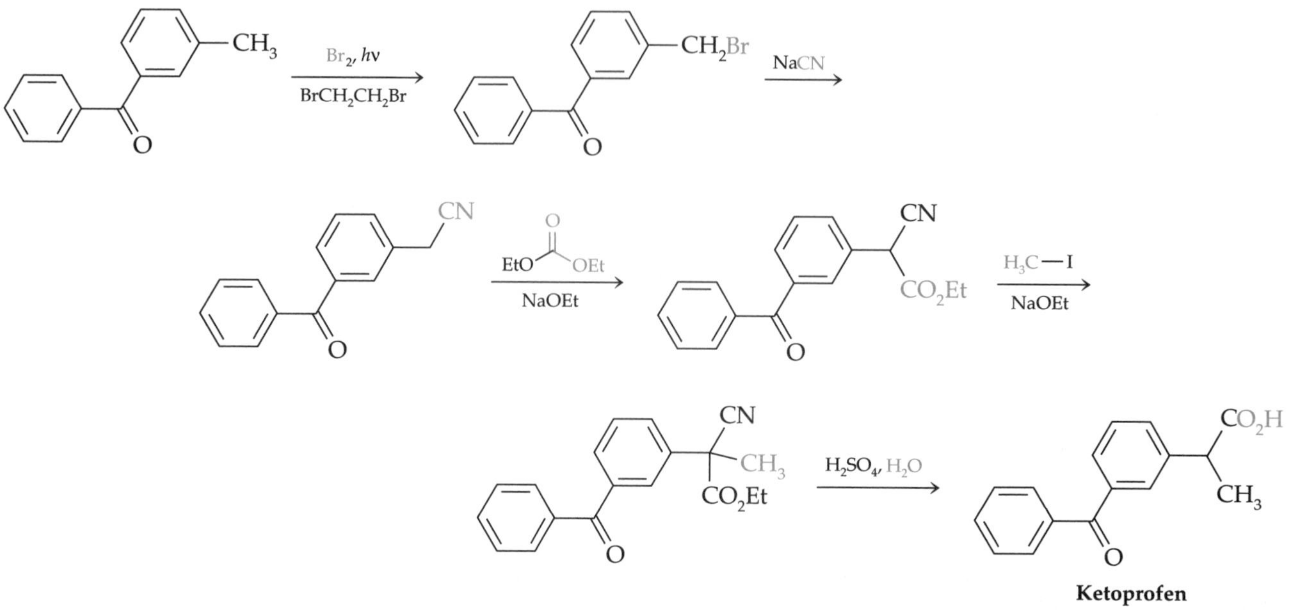

▲ FIGURE 15-36

The synthesis of ketoprofen begins with a benzylic bromination, introducing a nitrile by an S$_N$2 replacement. The remaining benzylic hydrogens are then sufficiently acidic that an anion can be formed by deprotonation with alkoxide. Nucleophilic acyl substitution of diethyl carbonate introduces a —CO$_2$Et group, increasing the acidity of the remaining hydrogen. Deprotonation produces an anion that is alkylated by methyl iodide. The nitrile and ester are then hydrolyzed, producing the same intermediate as derived from the alkylated diester in a malonic ester synthesis—namely, a β-diacid from which CO$_2$ is lost at room temperature.

sequential reaction with an electrophile (for example, methyl iodide). Even if this double alkylation were not a problem, the deprotonation would require the use of a strong base (such as lithium dialkyl amide), which would add substantially to the cost of the process.

Alternatively, the nitrile can be converted into a nitrile ester by treatment with a weak base and diethyl carbonate. Although there is still an acidic hydrogen between these two functional groups, the anion that is formed by deprotonation is stabilized by resonance delocalization with the nitrile, the benzophenone group, and an ester functional group, making it insufficiently reactive to undergo further reaction with diethyl carbonate. On the other hand, this anion reacts (but only once) with methyl iodide as shown in Figure 15-36. Because this alkylation results in a quaternary center, no further alkylation is possible. The synthesis is finished by hydrolysis of both the nitrile and the ester groups to carboxylic acids. The resulting dicarboxylic acid is a malonic acid derivative and undergoes spontaneous loss of carbon dioxide to form ketoprofen.

EXERCISE 15-O

Write a mechanism for the reaction of a nitrile with base and diethyl carbonate.

Valium

The synthesis of **diazepam** (**Valium**) is shown in Figure 15-37. Valium was the first of a long series of **psychoactive pharmaceutical agents** that are used as **sedatives.** A comparison of the structure of Valium in Figure 15-37 with that of Xanax (Figure 15-22), the currently most popular antianxiety agent, shows the structural similarity between these two drugs.

Let us examine the structure of Valium to see if there are some features that can be readily addressed by a retrosynthetic analysis. We find a tertiary amide and an imine, as well as two aromatic rings. Because all are common functional groups, we can well imagine that they might be formed by one of the functional-group transformations listed in Table 15-3. Thus, the combination of an aminocarboxylic acid and an aminoketone, as illustrated in

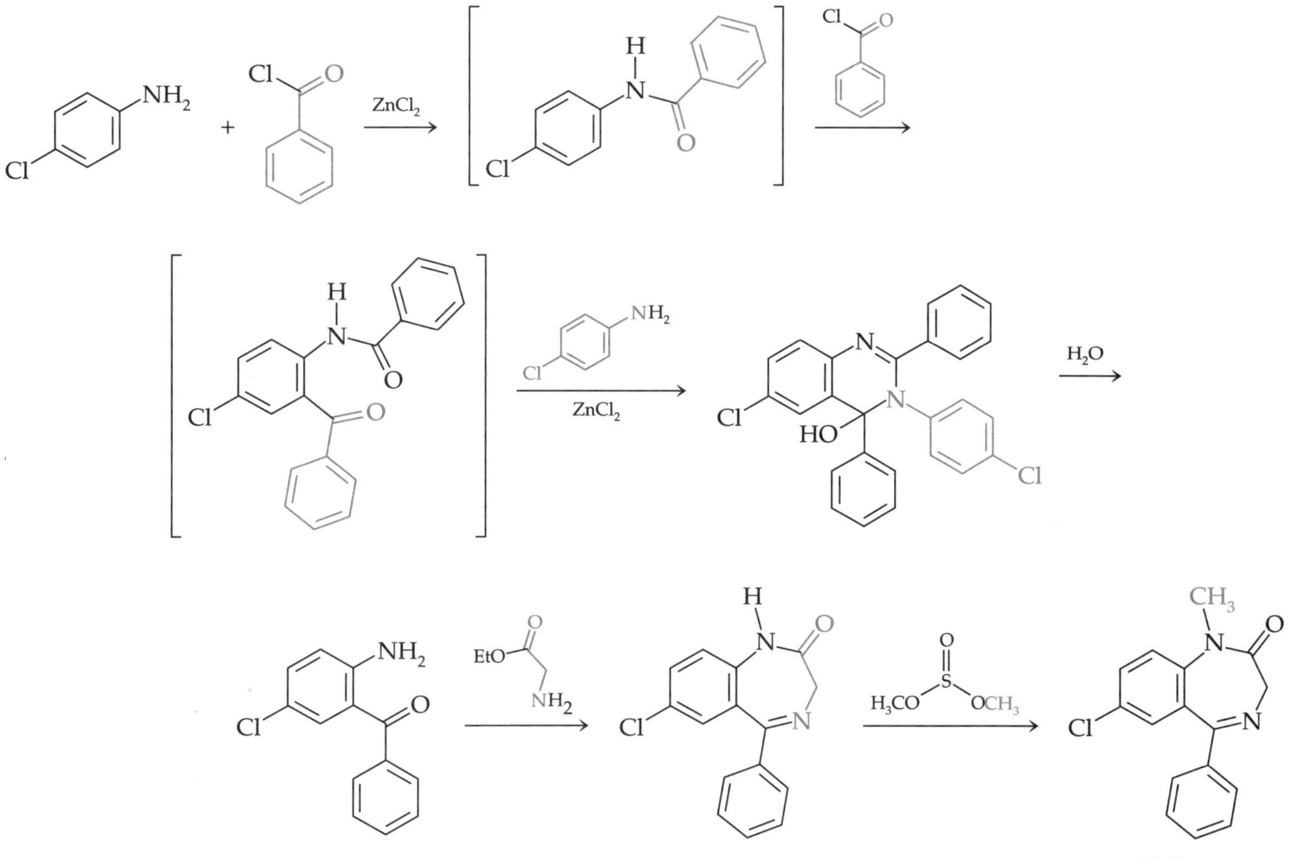

▲ FIGURE 15-37

The first step in the synthesis of Valium is a double acylation of an aromatic amine; first, by nucleophilic acyl substitution of an acid chloride by the —NH₂ group, and then by an electrophilic aromatic substitution on the ring. The resulting intermediate is trapped by a second equivalent of aniline, leading to a six-membered ring with two nitrogens. Hydrolysis reverses this cyclization and frees the simple electrophilic substitution product. A seven-membered ring is then produced by the intramolecular interaction with an aminoester. This cyclization takes place by amide formation between the ring —NH₂ group and the ester group of the coreactant and by imidation of the ketone group by the primary amine. Methylation of the amide N—H by dimethyl sulfate completes the synthesis.

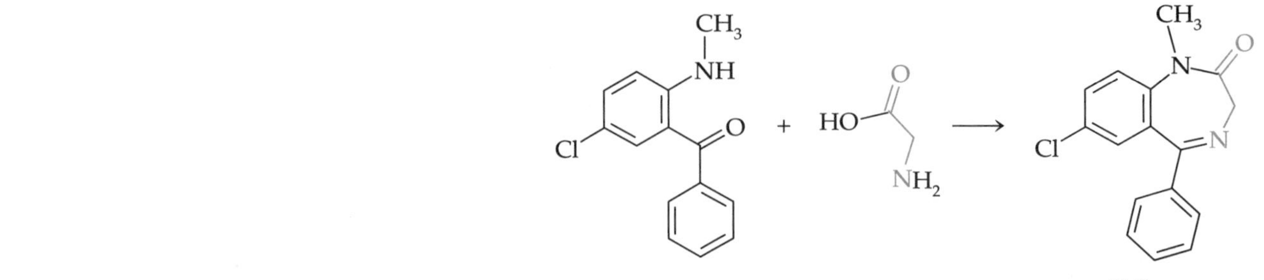

▲ FIGURE 15-38

A bifunctional molecule such as an amino acid can undergo two chemical reactions (formation of an amide at the acid group and formation of an imide at the amino group) when it is allowed to react with the complementary groups positioned correctly for ring formation.

Figure 15-38, should result in the formation of Valium. Indeed, this reaction is very close (bearing an extra *N*-methyl group) to the next-to-last transformation in the commercial preparation shown in Figure 15-37. In that sequence, an aminoester reacts with a ketoamine to form the seven-membered ring in Valium. The final reaction converts the secondary amide by alkylation of nitrogen into the tertiary amide.

EXERCISE 15-P

The reactions of dimethyl sulfate ($CH_3OSO_2OCH_3$) and dimethyl carbonate ($CH_3OCOOCH_3$) with a nucleophile lead to different products. Compare the mechanisms of these two reactions and explain why the reactions of these two reagents follow different pathways.

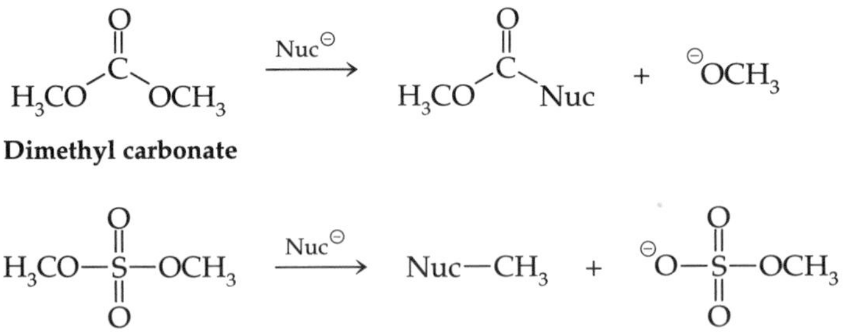

EXERCISE 15-Q

From the discussion of the ibuprofen and ketoprofen syntheses, explain why it might be more desirable to carry out the alkylation of nitrogen with a methyl group after the amide is formed (as in the commercial procedure in Figure 15-37) rather than before (as in Figure 15-38).

If a cyclization similar to that in Figure 15-38 is to be the last step of the synthesis, we must have a route to prepare the acylated aniline shown at the left of that figure. The first step of the Valium synthesis (Figure 15-37) is the reaction of benzoyl chloride with *p*-chloroaniline. At first, this might seem to be a straightforward Friedel-Crafts acylation in which the required ketoamine could be prepared as in Figure 15-39. However, recall that

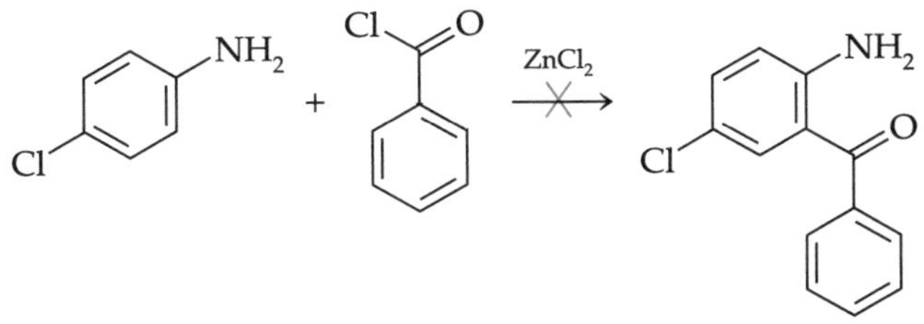

▲ FIGURE 15-39

This Friedel-Crafts acylation does not lead directly to the indicated product because the —NH₂ group is a more-active electrophile than is the π cloud of the aromatic ring.

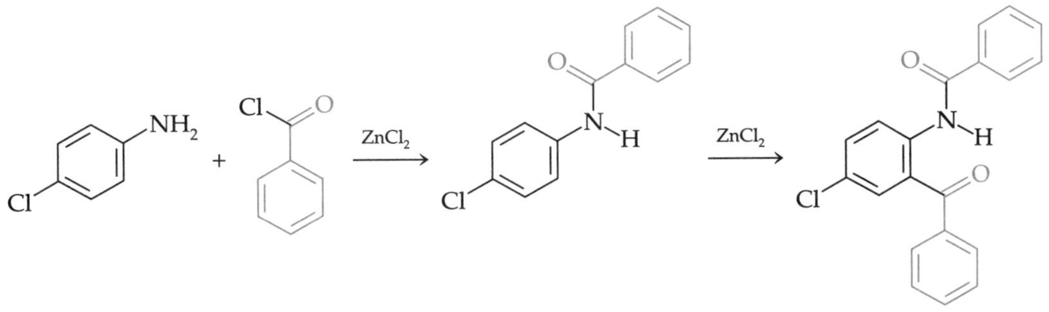

▲ FIGURE 15-40

Friedel-Crafts acylation of the amide formed by nucleophilic acyl substitution in the first step is directed to the *ortho* position because the *para* position is substituted by chlorine.

amides are formed from the reaction of carboxylic acid chlorides with amines. This is precisely what happens, rather than a Friedel-Crafts acylation of the ring (Figure 15-40). Once this occurs, the nitrogen is no longer nucleophilic because its lone pair of electrons is delocalized into the carbonyl π system. The aromatic ring then undergoes the desired Friedel-Crafts acylation and, because the amide is more activating than chlorine, the reaction takes place at a position *ortho* to nitrogen.

The reaction does not stop here: in the presence of zinc chloride, the substituted aniline adds to this intermediate at the keto group, producing an intermediate that attacks the carbonyl group of the amide. Although this sequence may seem complicated, it is in fact nothing more than a clever protection of the reactive nitrogen by the very reagent that is used to accomplish the desired reaction.

From this point on, you should begin to consider new reactions not only in regard to how they occur, but also in regard to what they accomplish. It will also be useful to begin to look at new molecules from the point of view of how they might have been constructed from smaller ones.

Conclusions

Synthetically useful reactions can be classified into three groups: carbon–carbon bond-forming reactions; oxidation-reduction reactions; and functional-group transformations. An analysis that dissects the steps needed to transform a simple molecule into a relatively more complex one is best accomplished in a reverse direction, proceeding from the ultimate product back to starting material (retrosynthetic analysis). A proposed pathway can be evaluated for synthetic efficiency by considering the number of steps, the yield of each step, the required reaction conditions, the ease of purification of intermediates, and the cost of reagents and personnel time. A convergent synthesis, in which short separate routes combine to form a desired product, is generally preferred to a linear synthesis, which takes place by a series of sequential transformations.

A protecting group is a functional group that is readily interconverted with another group but has significantly different reactivity toward common reagents. Ketals and acetals are used as protecting groups for ketones and aldehydes, benzyl ethers as protecting groups for alcohols, *t*-butyl esters as protecting groups for carboxylic acids, and urethanes as protecting groups for amines. Such protecting groups are important in the construction of complex molecules that may contain many functional groups. Protecting groups are used to control reactivity in synthetic intermediates in which a desired conversion at one site may be incompatible with the presence of another, different group at another position.

These concepts can be used to analyze syntheses (such as those used in industry) that incorporate many of the reactions presented in Chapters 7 through 14.

Summary of New Reactions

Benzyl Ether Hydrolysis

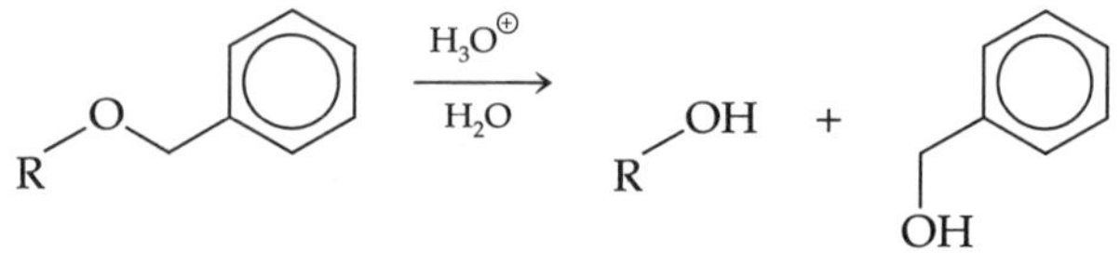

t-Butylester Formation

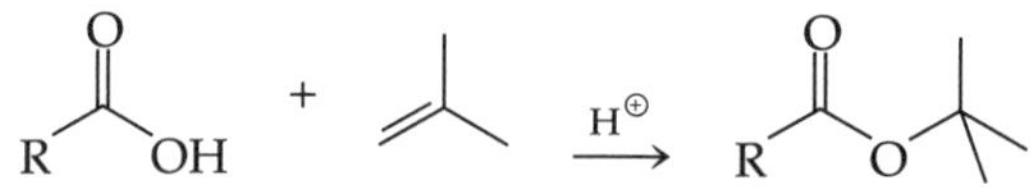

t-Butylester Cleavage

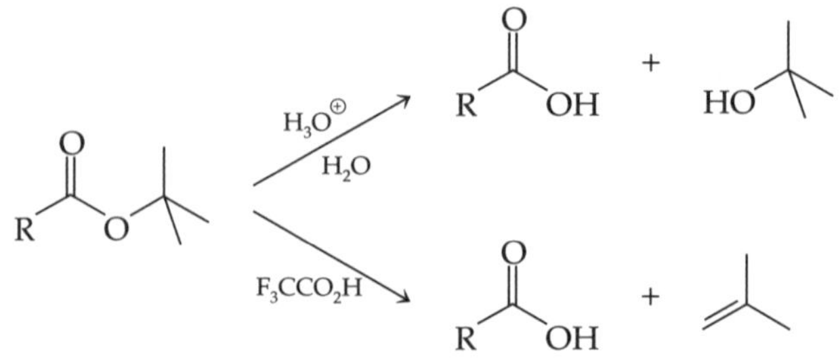

Urethane Formation

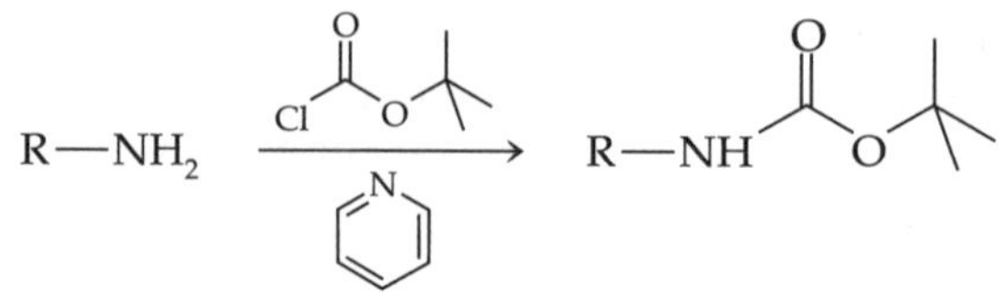

Urethane Hydrolysis

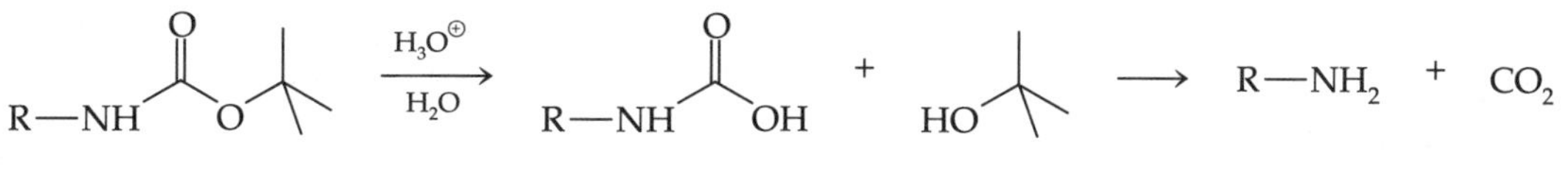

Review Problems

··

15-1 For each of the reactions shown schematically here, classify the transformation as a carbon–carbon bond formation, an oxidation-reduction, or a functional-group transformation.

(a)

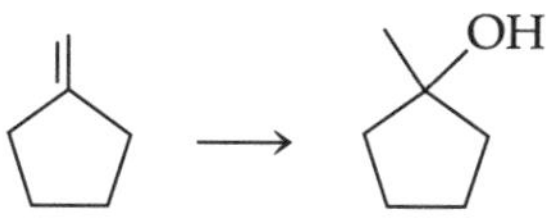

(b)

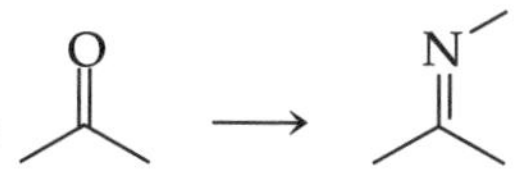

(c)

(d)

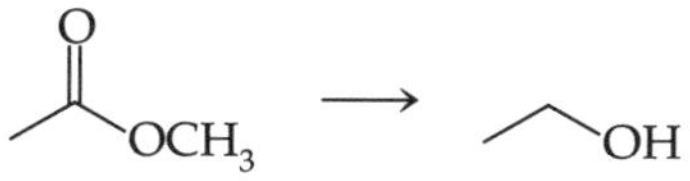

(e)

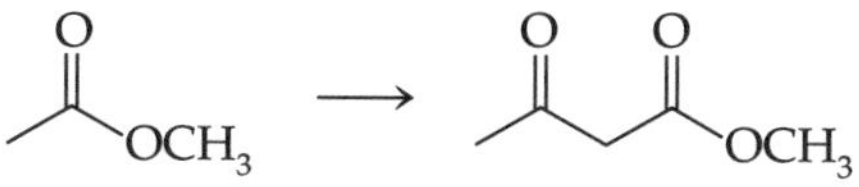

(f)

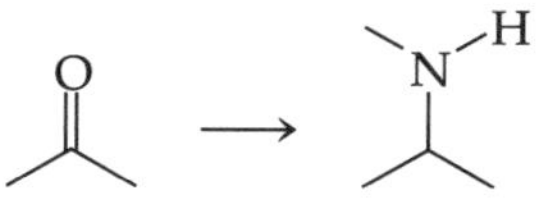

(g) $CH_3Br \longrightarrow$

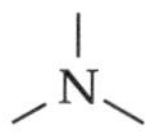

15-2 What reagents can be used to accomplish each of the transformations (parts *a* through *g*) in Problem 15-1.

15-3 Provide an efficient route for the conversion of methanol and ethanol into *t*-butyl alcohol (you may use any inorganic reagents needed).

$CH_3OH + CH_3CH_2OH \longrightarrow \longrightarrow \longrightarrow$

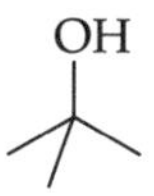

15-4 Provide a short and efficient route for the conversion of methanol and ethanol into 2-butanol.

$CH_3OH + CH_3CH_2OH \longrightarrow \longrightarrow \longrightarrow$

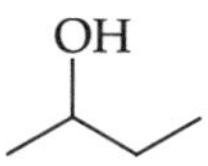

15-5 Each of the following syntheses requires only one reaction to accomplish the transformation. However, each starting material is bifunctional in that one group is to undergo the reaction while the other remains unchanged. This may require the use of a protecting group. Decide if a protecting group is needed and, if so, which one can be used in each case? Provide the three steps necessary (protection, transformation, and deprotection) required for each synthesis.

(a)

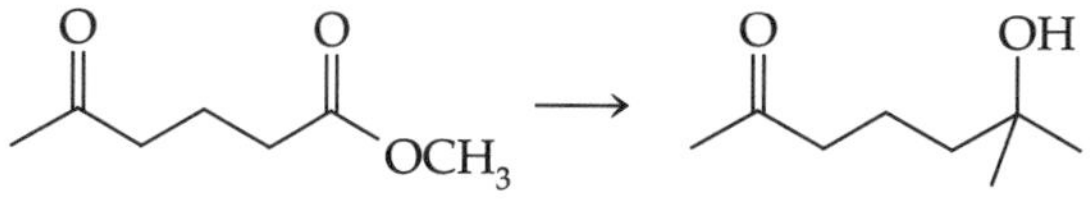

(b)

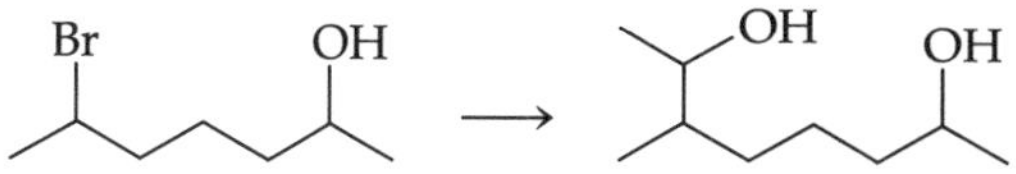

(c)

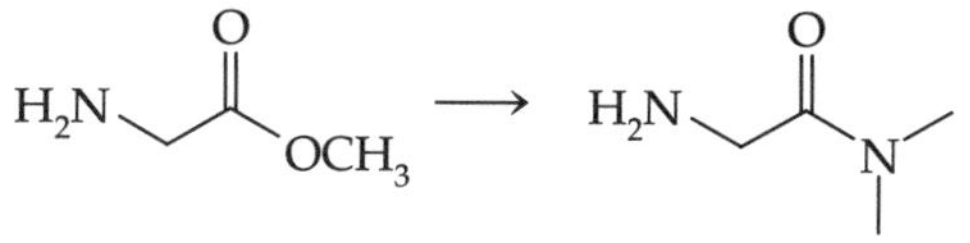

15-6 The transformation of a ketone into a ketal is the most-common way of protecting this functional group. Review your understanding of the mechanisms of carbonyl group transformations by providing mechanisms for both the formation and the hydrolysis of the dimethyl ketal of acetone.

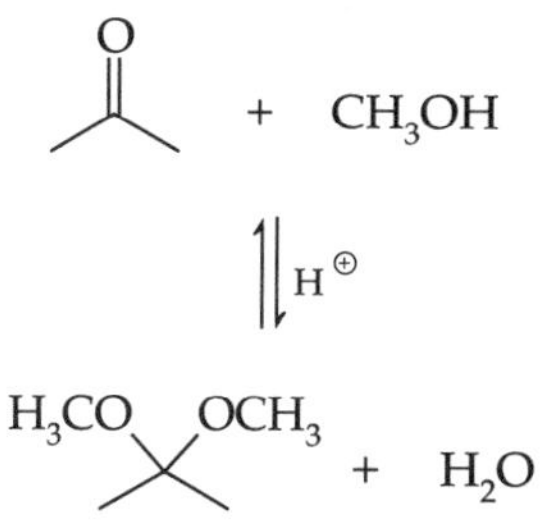

15-7 Develop as short a synthesis as possible for the following *cis*-alkene (*cis*-2-pentene), starting from methanol and ethanol as the only sources of carbon. (You may use any inorganic reagents needed.)

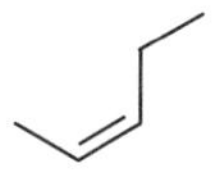

15-8 What properties of the ultimate product in Problem 15-7 might lead to serious practical difficulties in the final step of its synthesis?

15-9 How might the synthesis that you developed in Problem 15-7 be modified in a simple way to prepare 2-methyl-2-pentene?

15-10 Develop a short synthesis of the following alcohol from starting materials that do not have rings. (Hint: Because a cyclic product is ultimately desired, examine Table 15-3 for reactions that make rings.)

Chapter 16

Polymeric Materials

To this point, we have studied the properties and transformations of relatively small molecules, typically with molecular weights less than 1000 and containing no more than twenty or thirty carbon atoms. However, the world around us is full of molecules that are tens and even hundreds of times as large. Examples range from high-molecular-weight hydrocarbons found as complex mixtures in such materials as coal and crude petroleum to highly specialized molecules such as enzymes and DNA that are important to living systems. When these large molecules are composed of many repeating subunits, they are called *polymers,* a term derived from the Greek *polumeres* (having many parts). Johns Jakob Berzelius introduced this term in 1830, twenty-two years before the birth of Jacobus Henricus van't Hoff, who first described a three-dimensional tetrahedral carbon, providing a good indication of the importance of polymeric materials in organic chemistry. It is also of interest that Berzelius was the first to use the terms catalyst, isomer, and protein as well, all of which we will encounter in this chapter.

We will first examine, using a simple example, how a large molecule can be conceptually derived from repeating subunits and then deal in some detail with the chemical and physical properties of polymeric materials classified by the different functional groups present. Figure 16-1 illustrates a part of a large molecule in which the subunit set off in the box can be seen to repeat along the long, chainlike molecule. Most of the functional groups present are carboxylic acid esters; thus, quite logically, this macromolecule belongs to the class of polymers known as **polyesters.**

As we know from Chapter 12, esters can be prepared from alcohols and carboxylic acids. We can thus conceive of making the polyester illustrated in Figure 16-1 from the simple hydroxy acid known as lactic acid. In this case, lactic acid is referred to as the **monomer** from which polylactic acid is formed by a process called **polymerization.** We can consider this process to begin with two lactic acid molecules, with the carboxylic acid functional group of one reacting with the alcohol unit of the other. The product, a *dimer,* not only contains the ester functional group formed in this reaction, but also retains one alcohol and one carboxylic acid group (Figure 16-2).

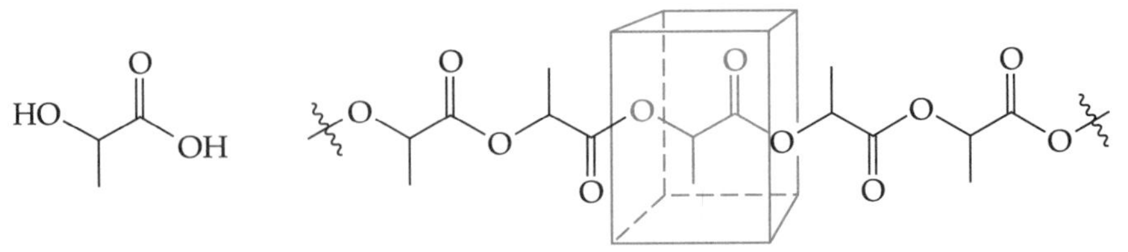

Lactic acid   **Polylactic acid**

▲ FIGURE 16-1

In polylactic acid, a large number of lactic acid molecules, $CH_3CH(OH)CO_2H$, are covalently bound through ester linkages to form a long chain. The box contains the "repeat unit" in polylactic acid. This unit includes the bond that connects the OH group of one lactic acid molecule to the carbonyl carbon of the next.

Thus, the two functional groups present in this dimer of lactic acid are the very same ones present in the monomer. Reaction of either the alcohol or the acid group of the dimer with a third molecule of lactic acid leads to a *trimer* containing three **repeat units** and still retaining one alcohol and one acid functional group. In theory, this process could continue indefinitely, with the chain lengthening at either end or at both ends. This example illustrates a feature that all monomers have in common: functionality such that a *minimum* of two bonds can be established, thereby linking each monomer to two others.

EXERCISE 16-A

The dimer of lactic acid illustrated in Figure 16-2 is capable of forming a cyclic structure by forming a second ester linkage through the reaction of the carboxylic acid group with the alcohol at the other end of the molecule, making a bis-lactone. Draw a structure for this bis-lactone (also known as a lactide). Write a reaction mechanism, assuming acid catalysis, for the conversion of the dimer into the lactide. (Hint: Recall Chapter 12.)

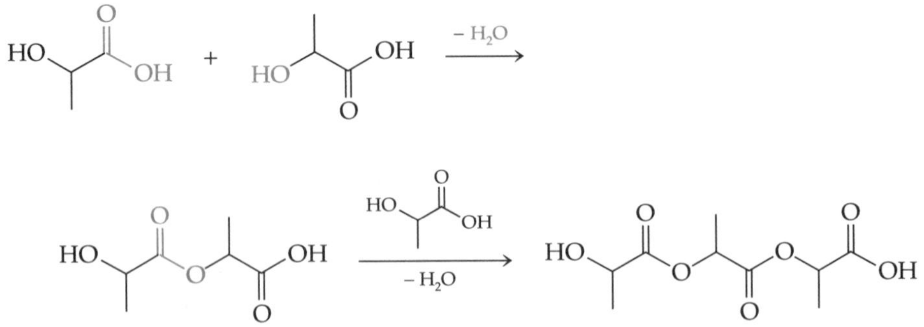

Dimer   **Trimer**

▲ FIGURE 16-2

Reaction of the OH group of one molecule of lactic acid with the carboxylic acid group of another results in the formation of an ester linkage by nucleophilic acyl substitution. The resulting ester is called a dimer because it is the condensation product derived from two equivalents of the starting acid. This dimer has both an OH group and a CO_2H group, and ester formation can be repeated again and again at both ends of the molecule with additional molecules of lactic acid. The formation of a trimer (derived from condensation of three equivalents of lactic acid) is the next step of this polymerization.

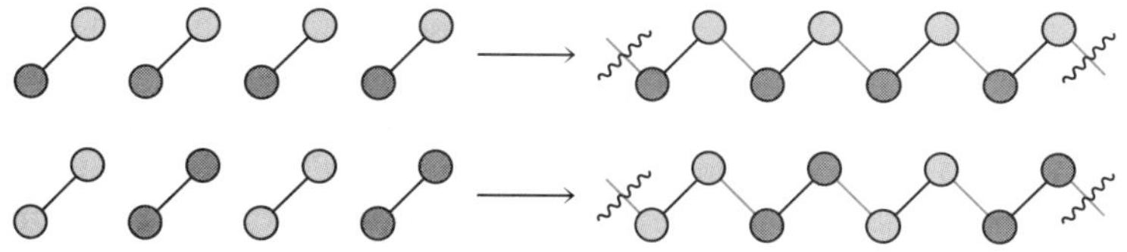

▲ FIGURE 16-3

A regular polymer is formed when a covalent bond is formed between two different functional groups of a monomer. This is illustrated in the upper reaction in which the linkages are between gray and orange balls. A polymer can also be formed between two different molecules, each of which has two functional groups that are the same (lower reaction). Although this polymer has the same number and type of linking bonds as the polymer formed in the upper reaction, the chain sequence is different.

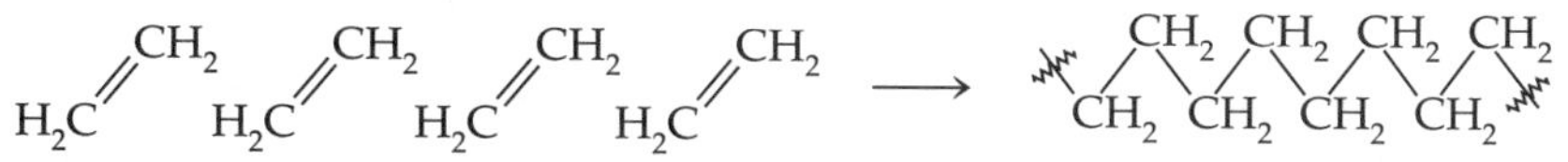

▲ FIGURE 16-4

In polyethylene, the σ bonds formed in the polymerization of ethylene are indistinguishable from the σ bonds present within the original monomer.

This concept can be illustrated graphically in a general sense, with the monomer shown as a stick with two balls representing the functional groups that link the monomers together. In lactic acid, the two functional groups are different and are represented by orange and gray balls in Figure 16-3. However, a monomer unit need not contain two different functional groups. For example, a polyester could have been derived from the combination of a dicarboxylic acid and a diol. Here, the monomer units have the same functional group at each end, but a 1:1 pairing of the monomers in the polymer is required.

It is easy to visualize the monomer units necessary to form a polyester. The functional group in the polymer, an ester, is directly related to the two functional groups from which it is made. As we will see, many polymers are formed by linking monomer units together with carbon–carbon bonds; it is often difficult, if not impossible, to distinguish between the carbon–carbon bonds present in the original monomer and those formed as a result of linking these units together. For example, ethylene undergoes polymerization to form long hydrocarbon chains in which the carbon–carbon σ bonds present in the monomer are indistinguishable from those formed in the polymerization. Nonetheless, a knowledge of the chemistry used to form such polymers facilitates the identification of the individual monomer units (Figure 16-4).

16-1 ▸ Linear and Branched Polymers

In the foregoing examples, each of the monomer units is capable of attachment to another at two sites. Of necessity, the linkage of these monomer units leads to a **linear polymer** in which the monomer units are attached end-to-end. Although each of the possible functional groups that can link the monomer units imparts different characteristics to the resulting polymer, linear polymers have common physical properties. For example, individual polymer chains associate with each other by electrostatic and van der Waals attractions. Because of the high molecular weight of a polymer, the number of such attractions on a molar basis is quite large and polymers often exist as solids or highly viscous liquids. The viscosity (or rigidity) of these polymers decreases at higher temperatures. The decrease in viscosity results from progressively greater disruption of the attractive intermolecular interactions between the polymer molecules that, although large, result from the sum of many relatively weak van der Waals and dipole–dipole forces.

In contrast, it is possible to build extremely large, multidimensional molecules that have distinctly different properties from those of linear polymers. These polymers are said to be **branched** or **cross-linked.** In such a polymer, chemical bonds interconnect chains so that a complex network results. The most-extreme example is diamond, in which the smallest repeat unit is a single carbon that, in the polymer, is connected to four others by carbon–carbon bonds. In fact, the diamond arrangement is made under such drastic conditions of high temperature and pressure that we are unable to define how it is formed in detail. Nonetheless, the bonding of each carbon to four partners in the diamond lattice results in a material that is connected in three dimensions by very strong, carbon–carbon bonds. As a consequence, diamond does not melt or soften as it is heated and is totally insoluble in all known solvents.

The structure of graphite is somewhat analogous, except that the carbons are sp^2-hybridized and thus linked to only three neighboring carbons. Because of the planar arrangement of bonding to such a carbon, graphite is composed of planar sheets of carbon resembling fused polyaromatic arrays. Because each sheet is planar, there is a fairly strong van der Waals attractive interaction between the sheets, and graphite exhibits the same insolubility as does diamond. On the other hand, the attraction of the sheets to each other, although strong, depends upon van der Waals interactions that change little as one sheet slides over the other. Thus, the sheets can be easily moved with virtually no resistance, and graphite therefore serves as an excellent lubricant. Diamond and graphite are examples of three-dimensional and two-dimensional polymers, respectively, in which all of the "monomer" units are identical.

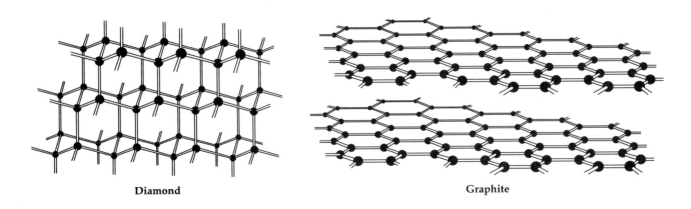

Diamond   Graphite

So far we have seen two distinctly different types of polymers, the linear arrays derived from monomer units that have only two possible attachment points and branched polymers, which are two- or three-dimensional arrays formed when each of the monomers can be attached to another by either three bonds, as in graphite, or four bonds, as in diamond. The three-dimensional network of chemical bonds in the polymer generally leads to a

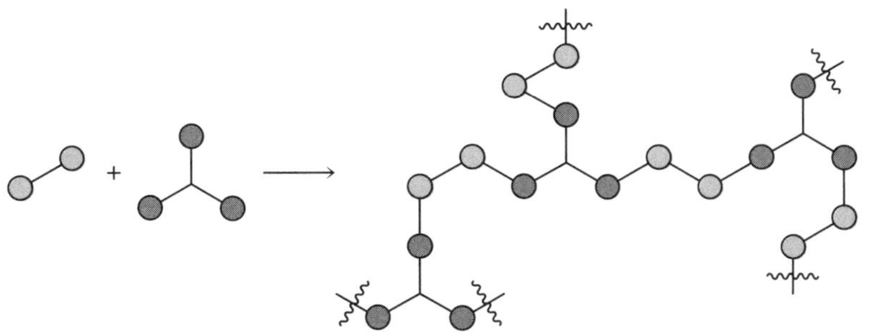

▲ FIGURE 16-5
If an individual monomer unit has three active functional groups, a branched arrangement of chains can emanate at each site. Such a polymer cannot be described by a sequential attachment of atoms, as was done in Figure 16-3.

material that is harder and less flexible than the corresponding linear polymers with similar functional groups. Polymers can also be made from mixtures of different monomers, with one unit having three bonding sites and the other only two. Such a possibility is represented schematically in Figure 16-5, in which the orange balls represent alcohol functional groups and the gray balls represent carboxylic acids.

16-2 ▸ Types of Polymerization

The chemical transformations that result in polymers can be divided into two major classes. The polyesters exemplify the type of material referred to as a **condensation polymer.** The reaction used to form such a polymer—that is, the reaction of an acid with an alcohol to produce a polyester—produces a small molecular by-product—in this case, water. In contrast, in the conversion of ethylene into polyethylene, all atoms present in the monomer are retained in the polymeric product. Because the latter process consists of the addition of one ethylene molecule to the next (recall electrophilic addition in Chapter 10), the resulting polymers are known as **addition polymers.**

These two types of polymerization often produce polymers whose structures and properties differ; and it is for this reason that they are usually treated as separate categories. However, it is sometimes possible to make quite similar polymers, the nylons being examples, by either addition or condensation polymerization.

16-3 ▸ Addition Polymerization

As discussed in Chapter 11, carbon–carbon π bonds are susceptible to electrophilic attack in a process that results in the breaking of the π bond and the simultaneous formation of a new σ bond between the electron-deficient reagent (the electrophile) and a carbon of the original π bond. The other carbon of the π bond becomes a cation and thus activated as an electrophile for reaction with a second equivalent of alkene. This interaction forms yet another carbon–carbon σ bond and repetition of the bond formation results in a carbon chain that continues to grow until the alkene is consumed. Recall as well that both radicals and cations are electron deficient and therefore electrophilic: both can initiate addition polymerization.

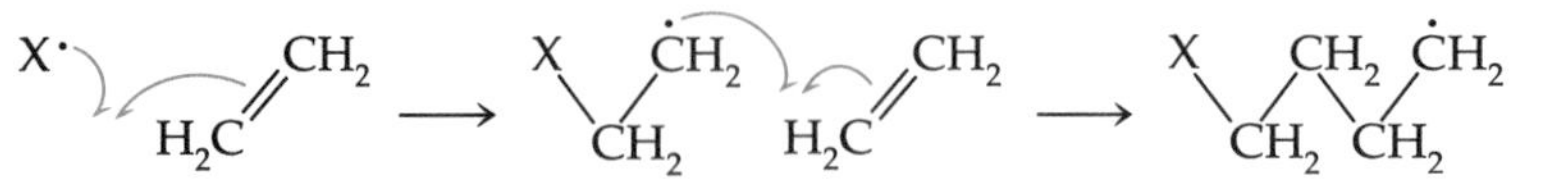

$$X\cdot \underset{H_2C}{\overset{CH_2}{\diagup}} \longrightarrow X \underset{CH_2}{\overset{\dot{C}H_2}{\diagdown}} \underset{H_2C}{\overset{CH_2}{\diagup}} \longrightarrow X \underset{CH_2 \ CH_2}{\overset{CH_2 \ \dot{C}H_2}{\diagdown}} \qquad \Delta H° = -20 \text{ kcal/mole}$$

▲ FIGURE 16-6

Radical polymerization takes place as a C–C π bond is broken, forming a σ
bond to the attacking radical and a carbon radical. This alkyl radical then
attacks another C–C π bond, forming a new C–C σ bond in a process that is
exothermic by 20 kcal/mole. The resulting radical can repeat the sequence
again and again.

Radical Polymerization

Figure 16-6 illustrates the reaction of a radical with ethylene, resulting in
the formation of a new σ C–X bond at the expense of the carbon–carbon π
bond. Because the product of this initial reaction is itself a radical, it is also
capable of adding to yet another molecule of the alkene in a process that regenerates a carbon-centered radical with a net change in bonding from one
π to one σ carbon–carbon linkage. Thus, each step in this polymerization
process is exothermic by approximately 20 kcal/mole, the difference in energy between a carbon–carbon π and σ bond. Each addition of a radical to
ethylene lengthens the growing polymer chain by two carbons. The overall
process, called **radical polymerization,** is depicted in Figure 16-7, in which
the bonds represented in color in the product are those formed in the polymerization.

  The polyethylene made by radical polymerization can have a molecular weight ranging from 14,000 to 1,400,000 (corresponding to between 500
and 50,000 ethylene monomer units). However, because the attractive van
der Waals interactions holding these chains loosely together are much
weaker than covalent linkages, the chains (like the sheets of graphite considered earlier) can move relative to each other (Figure 16-8). Thus, polyeth-

$$X\cdot \text{⌇⌇⌇⌇⌇⌇⌇} \longrightarrow X \text{⌁⌁⌁⌁⌁}\cdot$$

Polyethylene

▲ FIGURE 16-7

The σ bonds formed in the radical polymerization of ethylene (shown here in
orange) are identical with those originally present in the alkene.

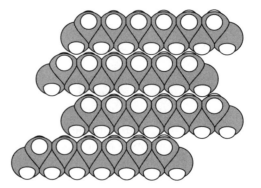

▲ FIGURE 16-8

Because van der Waals attractions are individually weak, polymer chains can
slide along each other, producing a flexible material. The representation used
in this figure is called space filling because the atoms (carbon is gray,
hydrogen is white) are drawn to scale based on van der Waals radii.

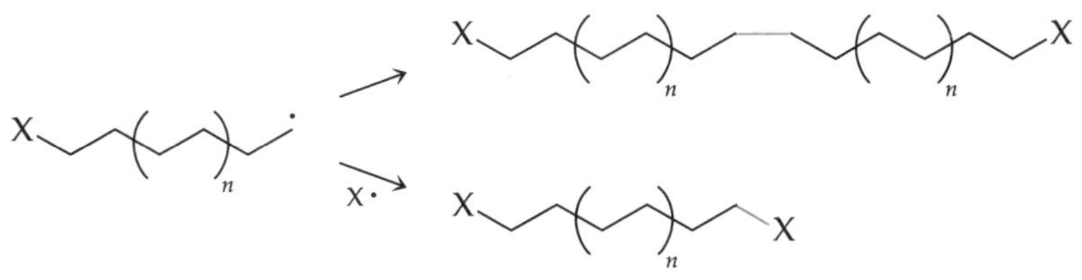

▲ FIGURE 16-9

In a termination step, two radicals combine to form a stable molecule. A growing polymeric radical can be terminated either by combination with another radical chain (upper reaction) or with the radical initiator (lower reaction).

ylene and most other linear polymers are somewhat flexible. (Some linear polymers, such as that used to make rubber bands, are especially flexible or "stretchy.") Flexibility increases with temperature because thermal motion becomes easier and disrupts intermolecular attractive interactions. It is thus possible to form polymers such as polyethylene into various shapes by heating and then molding them while they are relatively soft. Such materials are called **plastics,** a term derived from the Greek *plastikos* meaning fit to be molded.

Radical polymerization of ethylene need not go on indefinitely. As we have already seen in the free-radical halogenation of alkanes in Chapter 7, processes that simultaneously consume two radicals terminate a radical chain reaction. This is also true of radical polymerization, which can be terminated by the reaction between the radical at the growing end of the polymer chain and another such radical or the radical initiator (Figure 16-9).

EXERCISE 16-B

How might the relative probability of the two termination reactions shown in Figure 16-9 change as monomer is converted into ethylene?

EXERCISE 16-C

Abstraction of a hydrogen atom from a position adjacent to the growing end in ethylene polymerization by another carbon radical also results in termination, with the formation of an alkene. Write a reaction mechanism for this process. How many chains are terminated by this abstraction? Would you expect this process to be exothermic or endothermic?

Ionic Polymerization

Ethylene and other alkenes also undergo polymerization by processes in which the growing end is either a carbocation (**cationic polymerization**) or a carbanion (**anionic polymerization**), as illustrated in Figure 16-10. Cationic and radical polymerizations are quite similar in that both take place through the interaction of an electron-deficient (and therefore electrophilic) species (either the radical initiator or a carbocation) with the π electron density of an alkene. At first inspection, the reaction of a simple carbon–carbon π bond with an anion, especially a carbanion, appears to be an unusual reaction, because it requires the interaction of an electron-rich species with a π bond, also of high electron density. (Recall from Chapter 11 that the characteristic addition reaction of alkenes is electrophilic addition.) Nonetheless, keep in mind that, regardless of how ethylene is converted

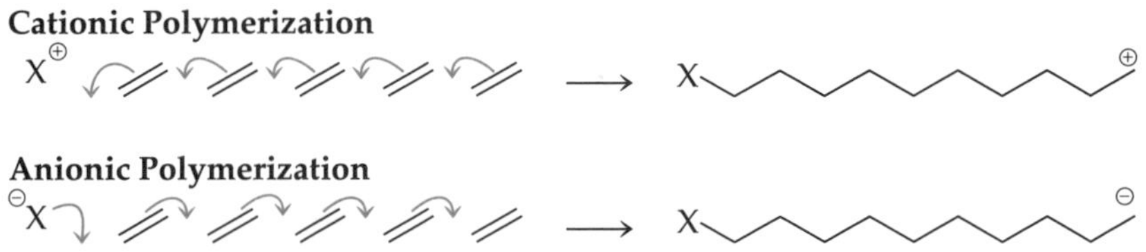

Cationic Polymerization

Anionic Polymerization

▲ **FIGURE 16-10**

In an ionic polymerization, a new covalent bond is formed between an alkene and either a cation (upper reaction) or an anion (lower reaction). In the cationic polymerization, the π electrons flow toward the cationic initiator, forming a C–C σ bond and a carbocation. In the anionic polymerization, an electron pair flows from the initiator to form a new C–C σ bond as the π electrons of the alkene move to the other carbon, producing a carbanion. A sequence of such steps affords the polymer, terminated by either a cation or an anion.

into a polymer, the thermodynamics of its formation is the same. Thus, all radical, cationic, and anionic polymerizations are exothermic by approximately 20 kcal/mole for *each* ethylene incorporated into the polymer.

Anionic polymerization of ethylene is initiated by the addition of an alkyllithium reagent (such as butyllithium) to ethylene. Unlike the free-radical process, the anionic polymerization of ethylene does not have termination steps because one carbanion does not react with another. As a result, the growing anionic chain is sometimes called a **living polymer,** meaning that the end of the polymer chain is chemically active until a quenching terminator is added. Polyethylene formed by a living, anionic polymerization has much longer polymer chains and a correspondingly higher molecular weight than can be attained by radical polymerization. These longer chains intertwine to a greater extent than those of lower-molecular-weight polyethylene formed by radical polymerization; the resulting polymer is therefore more rigid and less flexible and has a higher density than a polymer formed by radical polymerization. Because the interactions between the various polymer chains in polyethylene are stronger than those of each chain with solvent, polyethylene is a very insoluble plastic.

A variety of plastics are made by polymerization of substituted ethylenes. For example, **chloroethylene** forms the polymer **poly(vinyl chloride), PVC,** and **styrene** is converted into **polystyrene** (Figure 16-11). Poly(vinyl chloride) is used in many applications, from plastic bags to water pipes, and polystyrene is the plastic in **Styrofoam.**

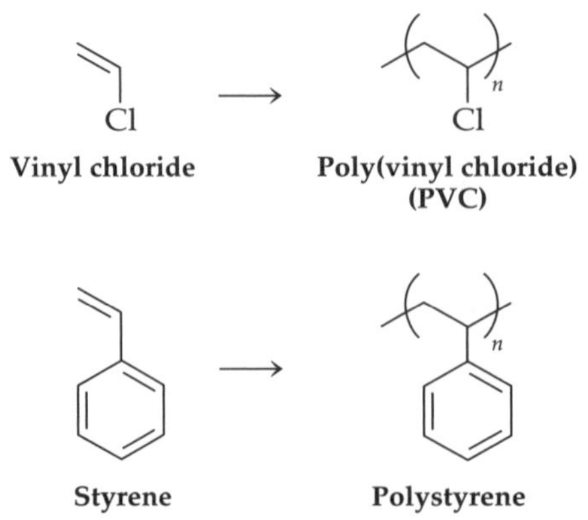

Vinyl chloride   Poly(vinyl chloride) (PVC)

Styrene   Polystyrene

▲ **FIGURE 16-11**

PVC and polystyrene are addition polymers prepared from vinyl monomers (substituted alkenes).

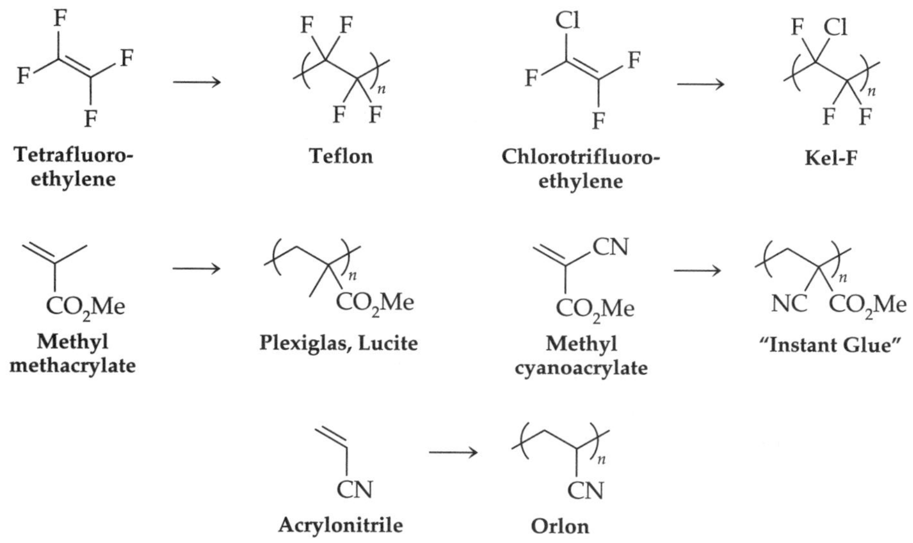

▲ FIGURE 16-12

Many common polymers and plastics are formed from vinyl monomers.

EXERCISE 16-D

Explain clearly why polystyrene polymerizes in a head-to-tail fashion instead of head-to-head. (Hint: Recall from Chapter 6 the factors that affect radical, carbocation, and carbanion stability.)

Some other substituted ethylene monomers used for preparing linear polymers are illustrated in Figure 16-12. Each of these polymers, or plastics, has unique properties that depend on the functionality present. For example, **Teflon** is an inert plastic because of the absence of carbon–hydrogen bonds and the high strength of the carbon–fluorine bond (110 kcal/mole) and is unreactive with all reagents except molten lithium, sodium, and potassium. Teflon is also a very "slippery" material because the fluorine atoms do not participate in significant attractive interactions with other groups. The presence of the carbomethoxy group in **Plexiglas** imparts a high index of refraction, as well as optical clarity, to this material; hence, this plastic is used as a lightweight replacement for glass. Indeed, **glass** itself is a polymer based on a three-dimensional network of tetrahedrally

SERENDIPITY IN SCIENCE

Chance discoveries have played major roles in the development of both science and technology. For example, R.J. Plunkett, a chemist at DuPont, was working with tetrafluoroethylene, taking samples for experiments from a metal cylinder in which the gas was stored. One day, to his surprise, no gas came from the cylinder when the valve was opened even though he was sure that he had not used all of it. Indeed, upon (carefully) cutting open the cylinder with a metal saw, he discovered a white powder, the first sample of the polymer known as Teflon. The best scientists are indeed those with inquiring minds who are constantly questioning unexpected results.

arranged silicon atoms linked by oxygen. The carbomethoxy group in Plexiglas also contributes to the high solubility of this material in organic solvents, especially those with carbonyl groups such as acetone and ethyl acetate. As a result of this high solubility, the smooth, transparent surface characteristic of Plexiglas can be easily destroyed if it is splashed with these solvents. The surface layers of the plastic dissolve but then redeposit unevenly as the solvent evaporates.

EXERCISE 16-E

For one of the polymers in Figure 16-12, the monomer unit is not the shortest repeating structural subunit. Identify this polymer. Is it possible to determine in which direction the polymerization occurred by examining a middle section of a polymer? Why or why not?

Butadiene and other conjugated dienes also form polymers. For example, the polymer derived from **isoprene (2-methylbutadiene)** is synthetic rubber. The rubber tree and several other plants produce a similar polymer known as **latex rubber,** which differs from synthetic rubber in that essentially all of the double bonds in the natural material have the *cis* configuration. **Chloroprene (2-Chlorobutadiene)** is a structurally similar monomer that leads to a synthetic rubber called **neoprene.** Replacement of the methyl group of isoprene with a chlorine atom reduces the ability of neoprene to associate with hydrocarbons, and thus neoprene is more resistant to gasoline and oils than is latex rubber.

Isoprene   Synthetic rubber

Chloroprene   Neoprene

Although synthetic and naturally occurring rubbers are similar in many ways, some of their properties differ because the synthetic polymers are quite complex mixtures. Consider the polymer formed by end-to-end polymerization of butadiene upon radical, cationic, or anionic initiation. The cationic polymerization begins by formation of a bond between a cationic initiator and one of the terminal atoms, resulting in a stabilized, allylic cation (Figure 16-13). Addition of this species to another monomer unit results in a linear polymer. However, two different carbons bear positive charge density, and, thus, there are two different sites for the addition of the next monomer unit. The diene polymer can therefore be either linear, through bond formation at either terminal carbon of the diene, or branched. Even if all the chains are linear, isomerism is possible because not all of the double bonds need to be formed in a *trans* geometric configuration (Figure 16-14). In contrast, latex rubber is produced in a living plant through catalysis by enzymes. These natural catalysts produce a polymer in which essentially all of the double bonds have the *cis* geometry.

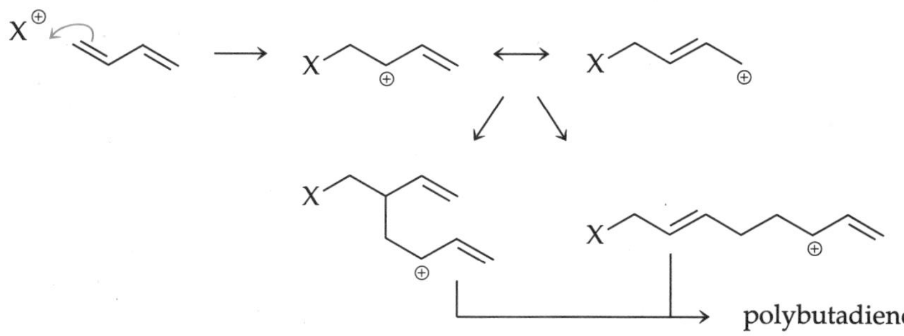

▲ FIGURE 16-13

Both linear and branched polymers can be formed in the cationic polymerization of butadiene. Analogous structures are also formed with radical and anionic initiators.

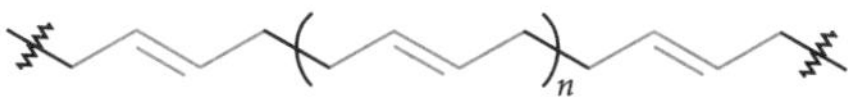

▲ FIGURE 16-14

The repeat unit in polybutadiene is $-CH_2CH=CHCH_2-$, in which the double bonds need not be all *trans* as shown here. The presence of a mixture of geometric isomers in synthetic polybutadienes differentiates them from naturally occurring compounds that are enzymatically formed and are geometrically regular.

NATURAL POLYENES

Because of the ease with which conjugated dienes undergo polymerization, they are rare in plants. Notable exceptions of highly conjugated polyenes are vitamin A, essential for vision, and enanthotoxin from the hemlock water dropwort, the most-poisonous plant in England.

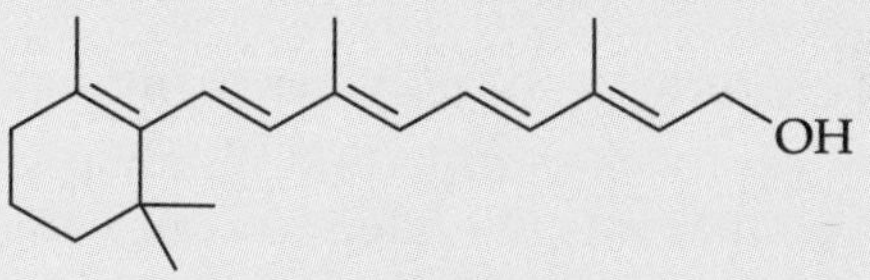

Vitamin A

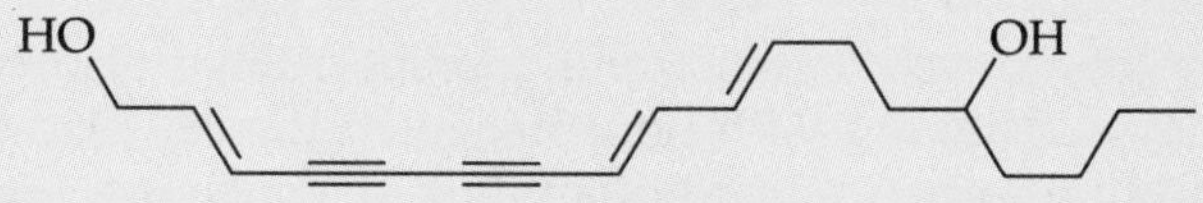

Enanthotoxin

Cross-Linking

Most of the polymers described so far are derived from monomers with possible bonding sites at only two positions. As a result, the polymer chains are held together only by relatively weak van der Waals attractive interactions. In contrast, glass is a very rigid material in which the tetravalent silicon atoms provide for a highly interconnected, three-dimensional covalent network.

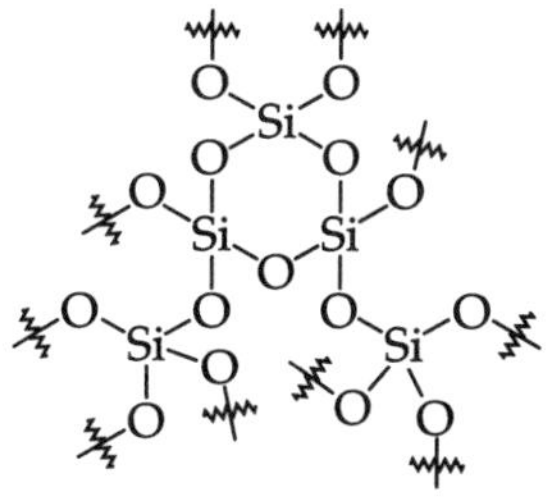

Glass

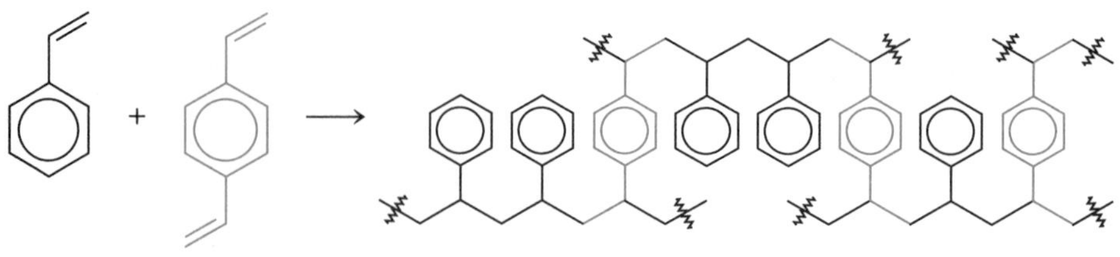

▲ FIGURE 16-15

Each time that a *p*-divinylbenzene is included in a growing polystyrene chain, a cross-link can be formed with another chain. The degree of cross-linking can be controlled therefore by how much of the cross-linking monomer (*p*-divinylbenzene) is present in the monomer mixture that is fed into the reaction vessel.

Cross-linking is a process in which a bifunctional molecule (such as a diene) participates in polymerization and is incorporated into two separate polymer chains. Polymerization of a mixture of a simple alkene (such as styrene) with a diene (such as **p-divinylbenzene**) affords an opportunity for each alkene unit in the diene to be incorporated into a separate chain, thus linking the chains together. The divinylbenzene can be viewed as a monomer capable of forming four bonds, thus forming cross-links between growing polymer chains (Figure 16-15). The extent of cross-linking attained depends on the relative concentrations of styrene and divinylbenzene. But, even if divinylbenzene comprises only a very small percentage of the mixture, the resulting plastic is much more rigid than polystyrene itself because of additional covalent bonds between the growing chains.

EXERCISE 16-F

Draw a short section of polyisoprene consisting of three isoprene units (fifteen carbons) in which:

(a) all of the double bonds have the polymer chains in a *trans* configuration;

(b) all of the double bonds have the polymer chains in a *cis* configuration;

How would you expect the properties of only one of the types (part *a* or *b*) to differ from one that was a random mixture of geometric and regiochemical isomers?

Diene polymers such as latex or neoprene rubber also can be made more rigid by cross-linking with nonconjugated dienes. One such linking agent is **5-vinylnorbornene,** large quantities of which are produced each year by the Diels-Alder reaction of butadiene with cyclopentadiene.

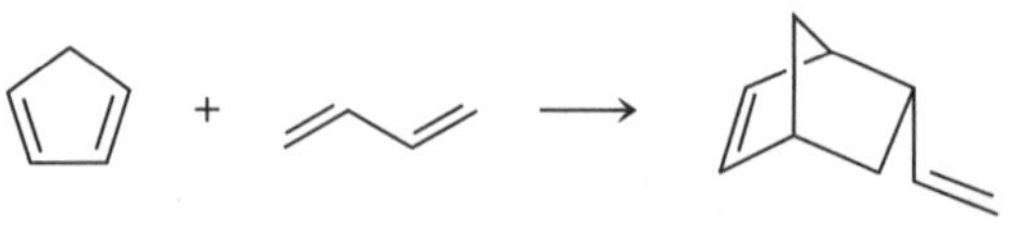

5-Vinylnorbornene

The inclusion of 5-vinylnorbornene in a monomer mixture results in a more-rigid polymer by enhancing the degree of cross-linking between chains.

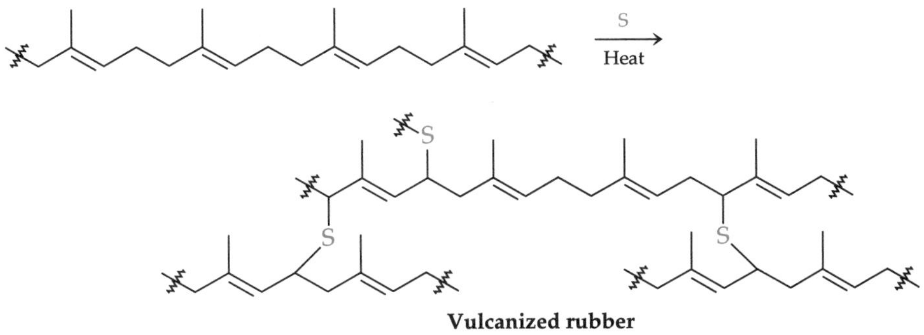

Vulcanized rubber

▲ FIGURE 16-16
Vulcanization strengthens and rigidifies the synthetic polymer derived from isoprene by cross-linking, through C–S covalent bonds, probably at allylic positions.

Alternatively, individual polymer chains can be linked by sulfur bridges. Transformation of the rather gummy and soft latex rubber into the much more rigid material that is used, for example, in automobile tires is accomplished by heating with sulfur in a process known as **vulcanization,** a process discovered by accident by Charles Firestone. (The term refers to Vulcan, the Roman god of fire, who was thought to be very strong.) The reactions almost certainly involve radicals, but details about the reaction mechanism are not well understood. In Figure 16-16, the bridges are depicted as being the result of simple allylic substitution of sulfur for hydrogen.

Interestingly, neither the original polymer nor the cross-linked rubbers derived from isoprene have functionality that leads to color. The characteristic black color of the rubber is caused by the presence of **carbon black,** a material similar to graphite, which acts as a lubricant and imparts a greater lifetime to the rubber under conditions of repeated flexing. Unfortunately, the production of large quantities of carbon black in eastern Europe, using outdated technology, results in the emission of substantial amounts of this material into the atmosphere and causes significant industrial pollution.

Heteroatom–containing Addition Polymers

The polymers discussed so far have very low solubility in water because they lack functionality that can form hydrogen bonds. In this section, we will learn about polymers that contain oxygen—either as alcohol or ether functional groups. As discussed in Chapter 3, the presence of heteroatoms along a carbon chain increases solvent-solute interactions, enhancing the solubility of the heteroatom-containing compound.

Polyols. The polyol known as **poly(vinyl alcohol), PVA,** is very water soluble. From its name, it might be inferred that this polymer is made from vinyl alcohol, but this enol is not present in significant amounts in equilibrium with its much more stable keto form, acetaldehyde (Figure 16-17). Rather, radical polymerization of vinyl acetate is used to form poly(vinyl acetate), a polymer with a hydrocarbon chain substituted with acetate esters. The esters are cleaved by acid- or base-catalyzed transesterification with methanol, forming methyl acetate and PVA (Figure 16-17). The resulting polymer is much more soluble in water than are hydrocarbon polymers such as polyethylene and polystyrene. Aqueous polymer solutions have a higher viscosity but lower surface tension than pure water. Poly(vinyl

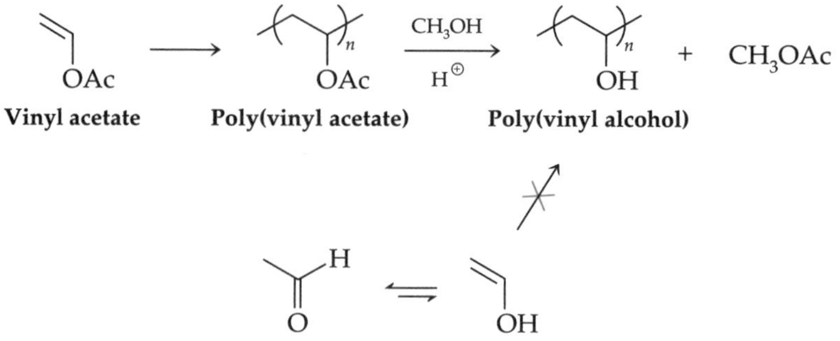

▲ FIGURE 16-17

The equilibrium between acetaldehyde and vinyl alcohol is strongly dominated by the more-stable aldehyde. It is not feasible therefore to polymerize it directly to form poly(vinyl alcohol). Instead, vinyl acetate is polymerized to poly(vinyl acetate), which is hydrolyzed, usually with acid catalysis, to PVA.

alcohol) is therefore included in formulations for specific applications, ranging from hair sprays and styling gels to lubricants for molding rubber.

Polyethers. The addition-polymerization polymers discussed so far have carbon–carbon bonds linking the monomers. There are also many heteroatom-linked polymers formed by condensation polymerization, several of which are discussed in Section 16-4. However, one very important class of addition polymers is formed by the polymerization of simple epoxides. For example, the addition of a nucleophile such as hydroxide ion to ethylene oxide results in ring-opening and affords the monoanion of ethylene glycol. (We have seen a similar reaction of ethylene oxide with Grignard reagents.) This ion can itself serve as a nucleophile, reacting with yet another molecule of ethylene oxide (Figure 16-18). Polymerization in this fashion is quite similar to the anionic polymerization of ethylene. Each step in this **ring-opening polymerization** results in the release of the ring strain of the three-membered epoxide ring and is thus exothermic by approximately 25 kcal/mole.

The resulting polyether polymer is called **poly(ethylene glycol)**, shortened to **PEG.** (As we shall see in Chapter 19 in dealing with molecular recognition, naturally occurring cyclic polyethers are important in the biological transport of cations across membranes.) Synthetic PEGs are used in cosmetic creams, lotions, and deodorants, as well as antistatic agents. These plastics are marketed commercially as **Carbowaxes,** as well as under

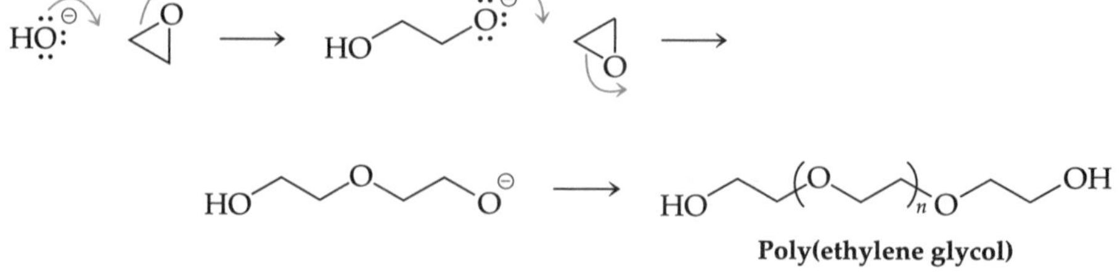

▲ FIGURE 16-18

Poly(ethylene glycol) is prepared by an S_N2 ring-opening of ethylene oxide by an alkoxide in a pathway parallel to the Grignard synthesis of two-carbon extended primary alcohols.

other trade names. They have high water solubility because of hydrogen bonding to the ether oxygens and have many of the same applications as poly(vinyl alcohol).

EXERCISE 16-G

Poly(vinyl alcohol) and poly(ethylene glycol) are similar in that both have oxygen functional groups. Assuming nearly equal molecular weights, which of these polymers would have the greater solubility in water? in hydrocarbon solvents? In a comparison of polymers of the same molecular weight, which would be more viscous? Explain your reasoning.

Polyacetals. Another oxygen-containing polymer is **paraformaldehyde,** a polyacetal formed by the addition polymerization of formaldehyde, which takes place as an aqueous solution of formaldehyde is concentrated.

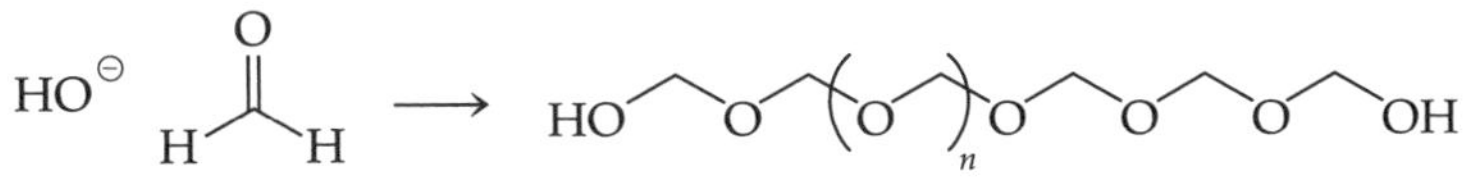

Notice that both ends of the polymer are hemiacetals rather than acetals, the functional group that constitutes the polymer backbone, and are active sites for further polymer growth. Also notice that all atoms of the monomer are retained within the polymer, making this polyacetal an addition polymer.

The polymerization is only slightly exothermic; as a result, paraformaldehyde undergoes depolymerization in water, reforming formaldehyde. Because formaldehyde is a strong antibacterial agent, aqueous solutions of paraformaldehyde are used for disinfection, and the polymer is the active ingredient in some contraceptive creams.

Polyacetals (and polyketals) made from formaldehyde, as well as other aldehydes and ketones, are strong plastics that are resistant to fatigue and have good electrical properties, making them quite useful as components for computer hardware and automobile parts. To stabilize these materials toward hydrolysis, the hemiacetals (or hemiketals) at the ends are "capped" by reaction with either acetic anhydride (to form an ester) or ethylene oxide. These polyacetals (or polyketals) require more strongly acidic or basic conditions for their hydrolysis than does paraformaldehyde.

EXERCISE 16-H

If you buy "formaldehyde" from a chemical supplier, it is really supplied as a trioxane, a cyclic trimer. Write a clear, detailed reaction mechanism for all steps in the conversion of formaldehyde into this trimer, catalyzed by a protic acid such as HCl. Can the reaction also be catalyzed by base? Why or why not?

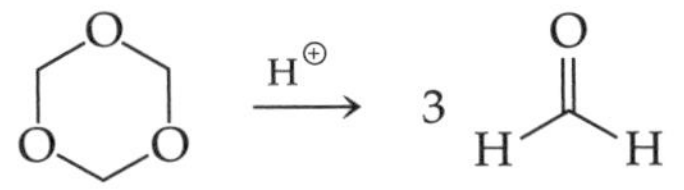

Trioxane

16-4 ▸ Condensation Polymers

Unlike an addition reaction in which all atoms of the reactants are incorporated within the product, a condensation reaction forms a more-complex organic molecule from two less-complex ones, with the expulsion of a small molecule. Some examples of condensation reactions include the acid-catalyzed esterification of a carboxylic acid discussed in Chapter 12 and the aldol and Claisen condensations discussed in Chapter 13, in all of which water is formed as a by-product. When such reactions are repeated many times with an appropriately functionalized monomer, a condensation polymer is formed.

Polyesters

Because the interaction of dicarboxylic acids with diols to form polyesters produces one equivalent of water for each of the links formed in the polymer chain, polyester formation is a condensation polymerization. Esters also can be used, in which case the reaction is a transesterification. **Dacron,** a commercially important example of such a polymer used as a fiber, is formed by the reaction of dimethyl terephthalate with ethylene glycol, a condensation reaction in which methanol is produced as a by-product (Figure 16-19).

Polycarbonates are another class of condensation polymers that result from transesterification. A diol often employed in this reaction is **bisphenol A,** produced by reaction between phenol and acetone in the presence of a Lewis acid (the "A" in the name comes from *a*cetone), as shown in Figure 16-20.

Because phenols are better leaving groups than are aliphatic alcohols, diphenyl carbonate is more reactive in transesterification than is, for example, dimethyl carbonate. The resulting polymer has many of the same properties as those of both Plexiglas and polystyrene. All of these materials have high optical clarity, but, in addition, polycarbonate is much stronger and more rigid with great impact resistance.

EXERCISE 16-I

Write a complete, detailed reaction mechanism for the conversion of phenol and acetone into bisphenol A in the presence of a protic acid. (Can you guess what bisphenol B is?)

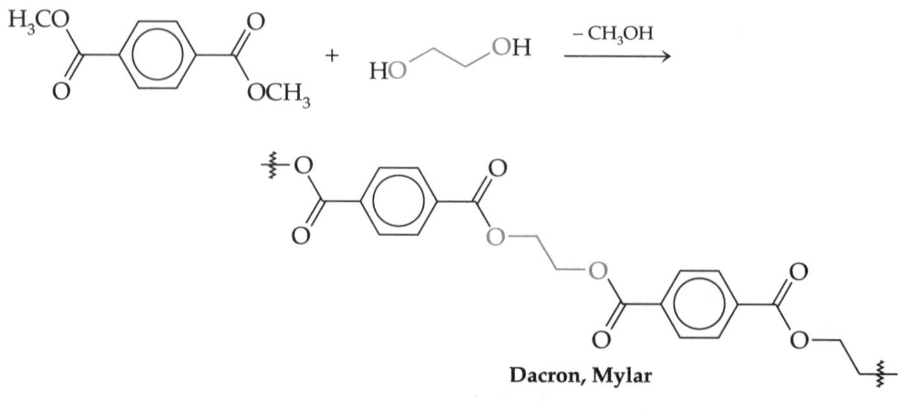

▸ FIGURE 16-19
The interaction of a diacid derivative with a diol produces a polyester.

Dacron, Mylar

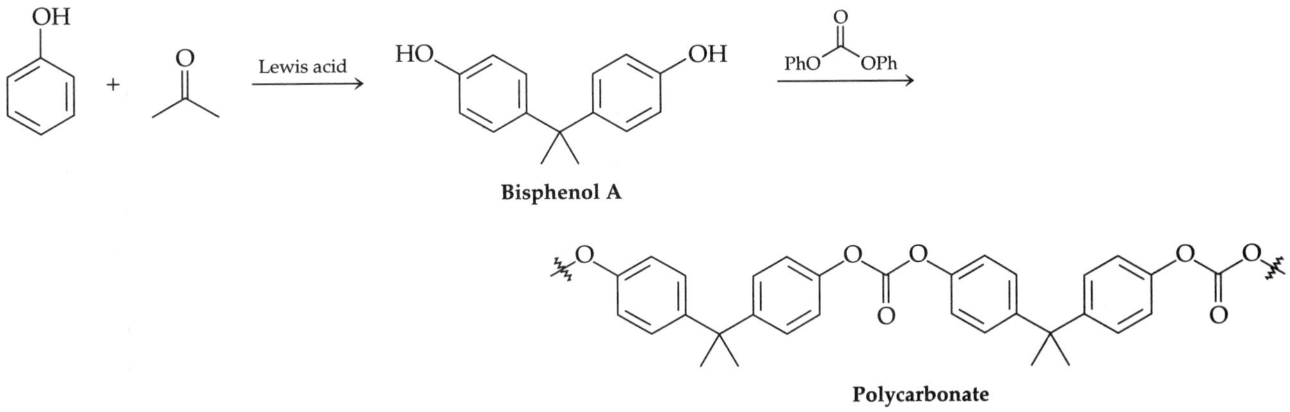

▲ FIGURE 16-20
Reaction of bisphenol A with a diaryl carbonate to form a polycarbonate.

EXERCISE 16-J

Write a mechanism for polymer formation under alkaline conditions with hydroxide ion as base from bisphenol A and (a) diphenyl carbonate and (b) dimethyl carbonate. Explain why the reaction with diphenyl carbonate is more efficient.

Polysaccharides

Like polyesters, polysaccharides contain oxygen atoms along the polymer backbone. Naturally occurring **polysaccharides,** however, are polyacetals formed by condensation of a hemiacetal group of one monomer with an alcohol group of another, taking place with the loss of water. The monomer units of polysaccharides, **saccharides,** are often called **sugars** or **carbohydrates** in recognition of their molecular formulas as $C_m(H_2O)_n$, suggesting structures that are hydrates of carbon. These **biopolymers** are derived from simple sugars such as glucose and have a variety of essential biological functions. For example, as shown is Figure 16-21, **starch** is formed mainly

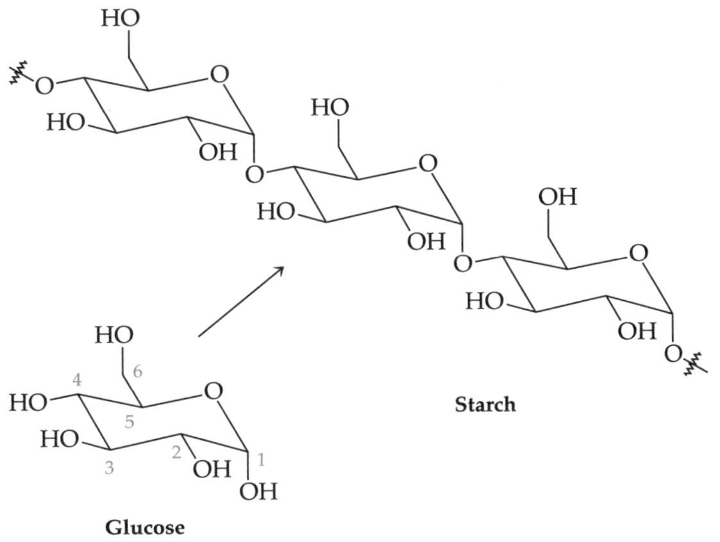

◄ FIGURE 16-21
Starch is a polymeric form of glucose, joined through the C-1 and C-4 hydroxyl groups. Because one equivalent of water has been lost in the formation of each acetal linkage, starch is a condensation polymer.

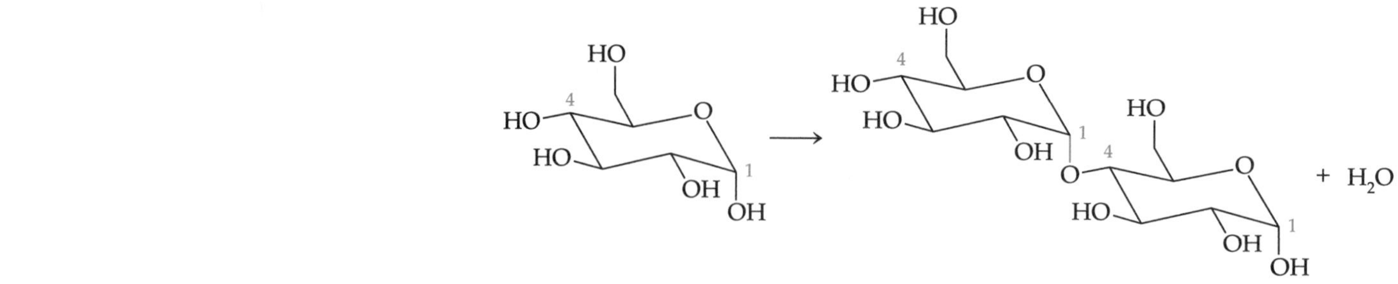

▲ FIGURE 16–22

A hemiacetal at C-1 of one glucose molecule condenses with a hydroxy group at C-4 to form the acetal linkage connecting two glucose molecules. The dimeric product still has a free hemiacetal group at one end (the far right) of the molecule and a C-4 hydroxyl group at the other end (the far left). The dimer is thus able to repeat the first reaction to continue chain growth.

from the reaction of the hydroxyl group at C-4 of one glucose molecule with C-1 of another. Notice that glucose is itself a hemiacetal: as we will see in Chapter 17, this cyclic form is more stable than the corresponding open-chain hydroxyaldehyde.

As was discussed in Chapter 12, the reaction of a hemiacetal with an alcohol produces an acetal and a molecule of water. In a similar reaction, two glucose molecules form a disaccharide (Figure 16-22). Hence, starch and related polysaccharides are examples of condensation polymers.

EXERCISE 16-K

There are two possible reaction pathways for the conversion of an acyclic hemiacetal such as glucose into an acetal by reaction with an alcohol and acid. Using the simple cyclic hemiacetal shown below, write a complete detailed reaction mechanism for the formation of the cyclic acetal by a process that:

(a) cleaves the carbon–oxygen bond within the six-membered ring;

(b) cleaves the carbon–oxygen bond of the hydroxyl group.

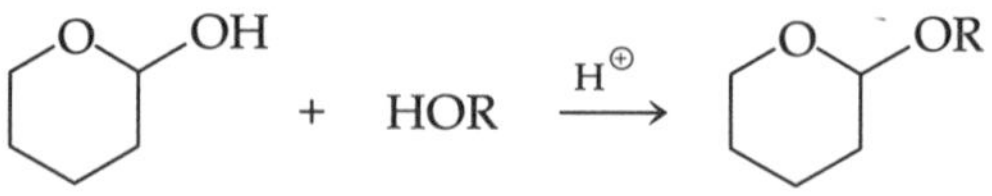

The presence of many oxygen atoms, both as acetal and as hydroxyl groups, affords water solubility to starch (also known as **amylose**) despite its high molecular weight (it can comprise as many as 4000 glucose units). In contrast, **amylopectin** is a water-insoluble starch consisting of as many as a million glucose units in which occasional links with the C-6 hydroxyl groups result in a highly branched structure (Figure 16-23). **Cellulose** is a very similar biological polymer that differs in structure from starch mainly in the stereochemistry of the linkage between glucose units (Figure 16-24). Cellulose generally has between 3000 and 5000 glucose units and is thus

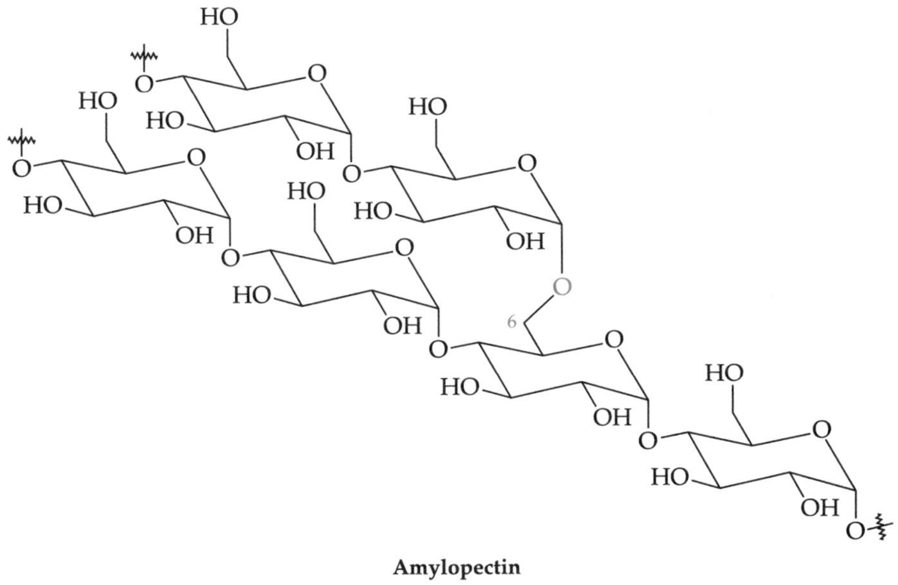

Amylopectin

▲ FIGURE 16-23
The highly branched structure of amylopectin makes it much less water soluble than glucose.

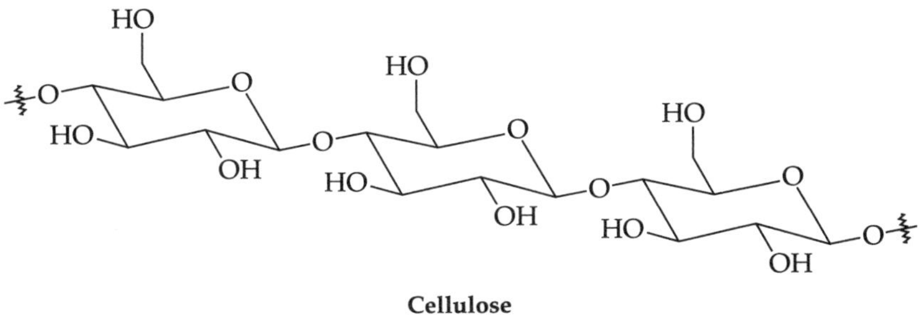

Cellulose

▲ FIGURE 16-24
In cellulose, the acetal linkage at C-1 is equatorial; whereas, in glucose (Figure 16-21), it is axial.

similar in size to amylose. However, unlike amylose, cellulose is water insoluble. As a direct result of the stereochemistry at the linkage of one glucose to another, the polymer chains of cellulose fit together much better than do those of starch. Starch is used by plants as a storage medium for glucose and needs to be accessible to individual cells for the metabolic requirements of the plant, whereas cellulose is mainly a structural material that must withstand rain, humidity, and so forth. This structural variance allows polymers with very different biological functions to be produced from a common monomer.

The reaction of cellulose with acetic anhydride converts many of the hydroxyl groups into acetate esters. Because of this change in functionality, the properties of **cellulose acetate** are quite different from those of cellulose, the most important property being optical transparency. This modified natural polymer is more soluble in organic solvents than is cellulose itself and can be processed into thin sheets. Photographic film is one example of a commercial application of cellulose acetate.

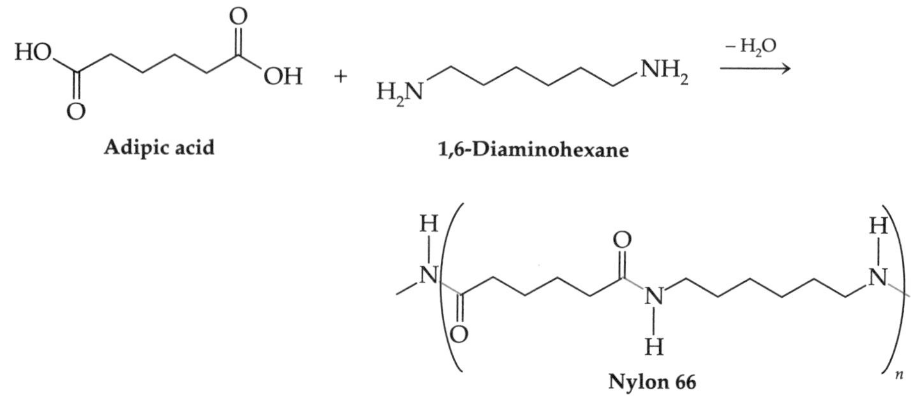

▲ FIGURE 16-25

Because one equivalent of water is lost for each amide linkage formed, a
polyamide like nylon 66 is a condensation polymer.

Polyamides

A condensation product formed between a diacid and a diamine is called a
polyamide. In a route parallel to polyester formation, adipic acid and **1,6-
diaminohexane** react (at high temperature) to form a polyamide known as
nylon 66 (Figure 16-25).

EXERCISE 16-L

Consider the possibility of the reaction of acetic acid with methyl amine to
form methylacetamide. Could the reaction be either acid or base catalyzed?

This designation is used to distinguish this polymer from **nylon 6,** produced by the polymeric ring opening induced by the catalytic interaction of
a nucleophile with caprolactam. This seven-membered cyclic amide (a **lactam**) is prepared industrially by the Beckmann rearrangement of cyclohexanone oxime in sulfuric acid (Figure 16-26). To isolate the product, the acid
(which is used as solvent) must be neutralized, often with ammonia. The
by-product, ammonium sulfate, is "disposed" of by selling it as fertilizer.

EXERCISE 16-M

Write a mechanism for the Beckmann rearrangement illustrated in Figure
16-26 and for the ring-opening polymerization that leads to nylon 6.

The terms condensation and addition refer to the methods by which
the polymers are produced and therefore do not necessarily provide clues
to the character of the polymer formed. Regarding the polyamides under
consideration here, the numbers refer to the number of carbon atoms present in the monomeric units. Whereas nylon 66 is produced by the condensation polymerization of two different six-carbon monomers, one a diacid and
the other a diamine, in which water is lost, nylon 6 is the result of addition
polymerization of a single six-carbon monomer (the cyclic amide). Both
nylon 6 and nylon 66 are excellent plastics used in making very long lasting
fibers that are quite flexible, in part because of the conformational freedom
of the chains.

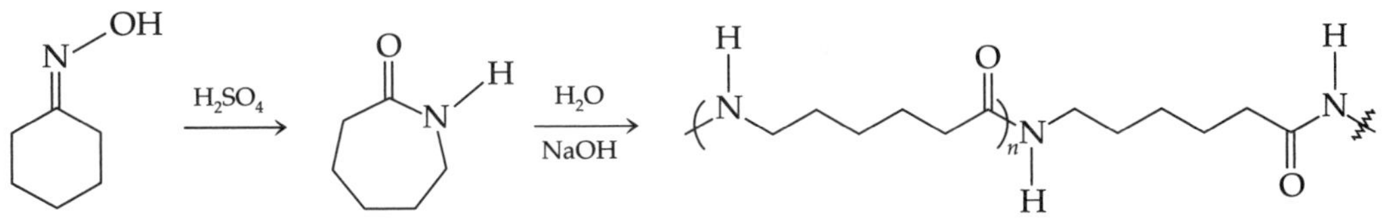

Caprolactam

Nylon 6

▲ FIGURE 16-26
In the polymerization of caprolactam to produce nylon 6, a cyclic amide (a lactam) reacts with hydroxide ion to form a free amine and a carboxylate. The amino group reacts with another equivalent of caprolactam by nucleophilic acyl substitution, effecting transamination (similar to the transesterification discussed in Chapter 11) and the formation of a dimer.

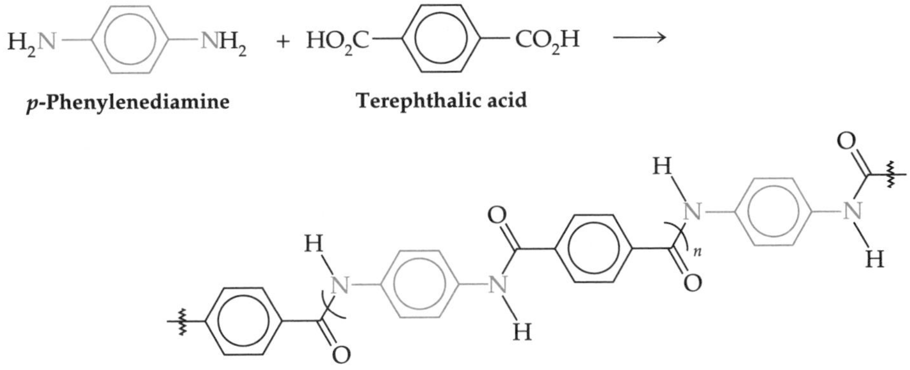

p-Phenylenediamine

Terephthalic acid

PPTA

▲ FIGURE 16-27
Rotation about most of the bonds in PPTA is restricted, making it a strong and rigid plastic.

Condensation polymerization of terephthalic acid with *p*-phenylenediamine results in a polymer, known as **PPTA** (*p*-*p*henylene*t*erephthal*a*mide), that has very unusual properties (Figure 16-27). In contrast with nylon 6 and nylon 66, which have considerable conformational freedom, PPTA is quite rigid. As a result, this polymer can be formed into fibers that have great tensile strength and resist both compression and elongation. PPTA is considerably stronger than steel on a per weight basis.

Polypeptides

The amide linkage present in a polyamide resembles those found in **polypeptides,** which are natural polymeric materials derived from **α-amino acids** that are linked by amide groups formed by a condensation reaction between the amino group of one α-amino acid and the carboxylic acid group of another (Figure 16-28). Peptides and proteins are naturally occurring polymers that have many different forms and a variety of functions. The distinction between the terms peptide and protein is one that relates to molecular size. The term peptide refers to polymers containing no more than 100 amino acid subunits, whereas the term protein refers to a larger molecule. The synthetic (manufactured) polyamides, in fact, were developed to mimic the properties found in silk and in animal hairs such as wool, both of which are composed mainly of polypeptides.

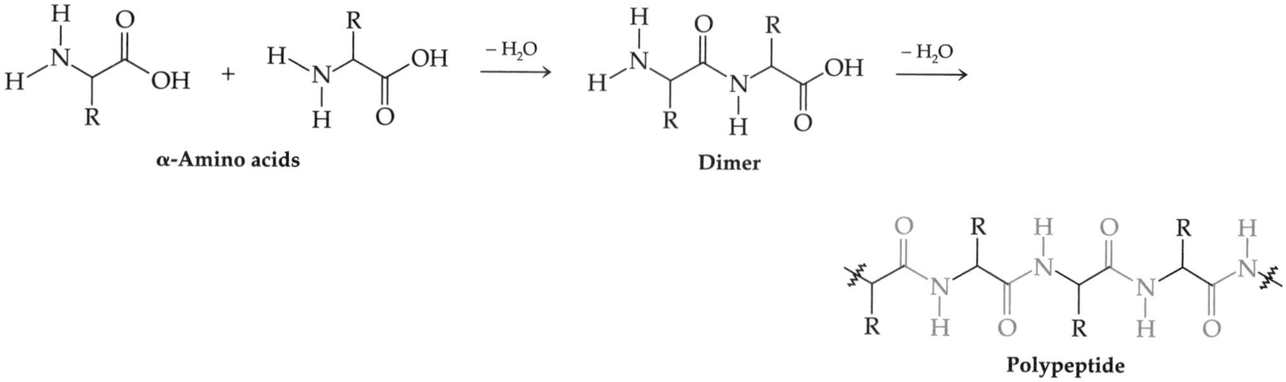

▲ FIGURE 16-28

A polypeptide has the general structure of the product in this polymerization. The peptide linkage is an amide (shown in color) formed by condensation of a carboxylic acid group of one α-amino acid with the amino group of another. The naturally occurring α-amino acids differ in the identity of the R group.

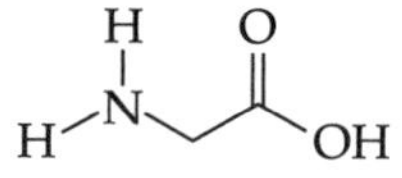

Silk is a polymer of the amino acids **glycine** and **alanine** and can be represented by the general structure in Figure 16-28 in which the R groups are a random mixture of methyl groups and hydrogens. **Wool** (Figure 16-29) is structurally more complex, having sulfur–sulfur bonds that link individual chains one to another and form a matrix somewhat like that of vulcanized rubber. These bonds are the result of significant amounts of the sulfur-containing amino acid **cysteine** in wool. Although a bond between two sulfurs may seem unusual, it is nonetheless very easily formed upon exposure of thiol functional groups to oxidants, and even molecular oxygen will effect this transformation.

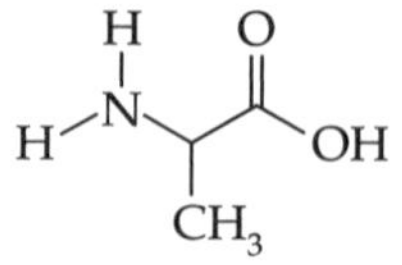

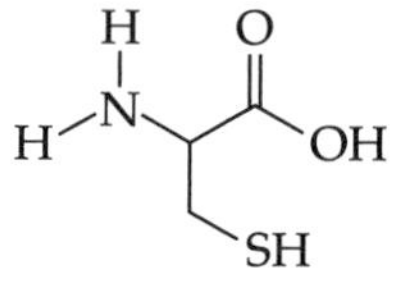

All amino acids except glycine (H_2N—CH_2—CO_2H) contain a center of chirality and are generally found only in one enantiomeric form in nature. (We will see an exception to this in later discussions dealing specifically with the function of peptides in biological systems.) The regularity of the handedness of the amino acids that compose the polypeptide chain determines the three-dimensional structure of the peptide, as well as how the polar amide functional groups interact both within and between chains.

Similar in structure to the polyamides are the **polyurethanes,** which have as one of the components in the chain the urethane (or carbamate) group. Urethanes are formed by the reaction of an isocyanate with an alco-

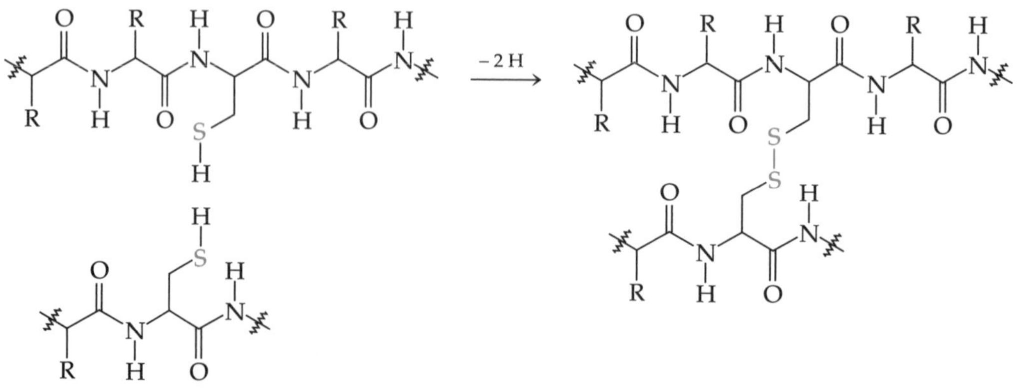

▲ FIGURE 16-29

Wool is cross-linked by dithiol bonds between cysteine α-amino acid units.

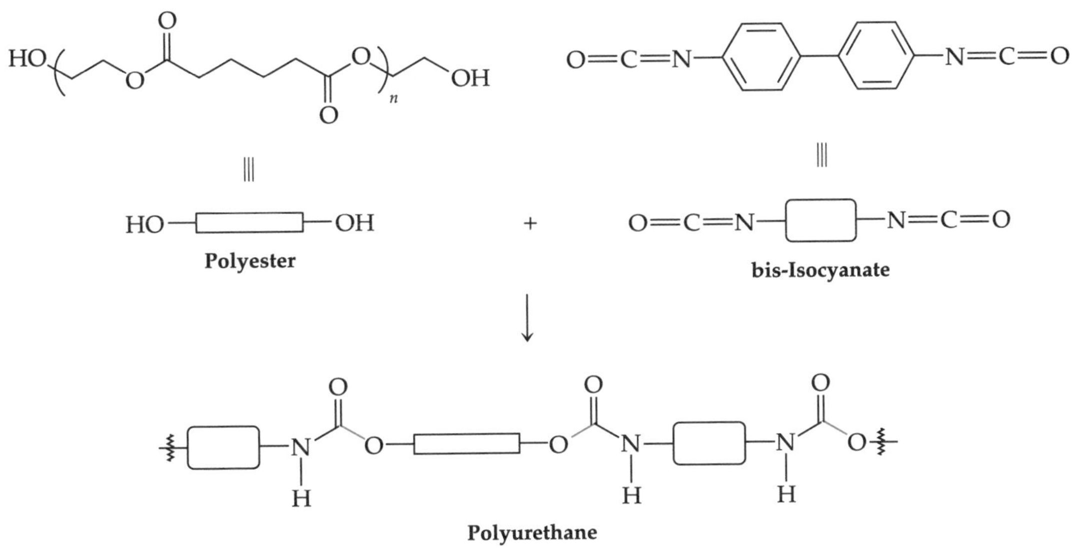

▲ FIGURE 16-30

A polyester oligomer terminated with two alcohol functional groups condenses with a bis-isocyanate to form a polyurethane. The condensation step is mechanistically similar to the isocyanate hydrolysis in the Hofmann rearrangement discussed in Chapter 14.

hol. (Recall from Chapter 14 that isocyanates are critical intermediates in rearrangements.) Polyurethanes are typically derived from low molecular weight polyesters (with terminal hydroxyl groups) and bis-isocyanates, which combine to form much longer chains linked by both ester and urethane groups. Block diagrams represent each of the basic units, the polyester and the bis-urethane, in Figure 16-30 so that the functional groups participating in the polymerization can be seen more clearly.

MEDICAL APPLICATIONS OF POLYMERS

Biologically compatible polymers are an increasingly important type of specialty polymer. The polymer used in making contact lenses must be quite hydrophilic to permit easy lubrication of the eye. A hydrogel in which free alcohol groups are attached to a poly(methyl methacrylate) polymer is therefore used. Polymers used for making dental impressions, however, cannot form local hydrous pockets but must effectively wet the oral tissue. Cross-linkable polyethers capped with groups to be opened by a ring-opening polymerization (like that discussed in this chapter for the polymerization of epoxides) are used.

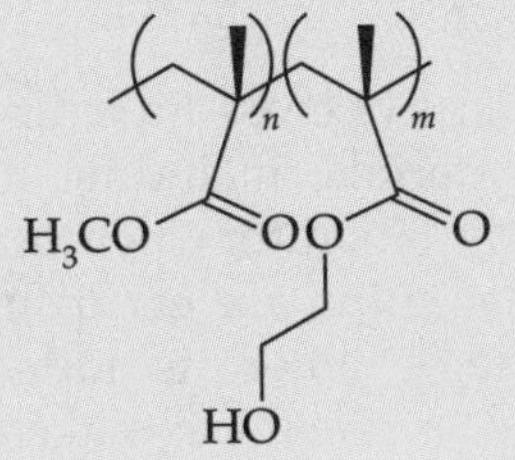

Hydrogel 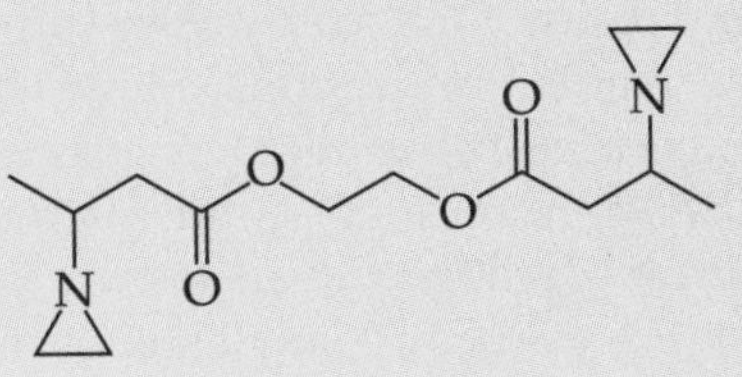   **Monomer for dental impressions**

When a material with a low boiling point (for example, CO_2 or a volatile hydrocarbon such as methane or ethane) is dissolved under pressure into one of the starting materials and is then allowed to expand and vaporize as the polymerization proceeds, a material is obtained with tiny "void" spaces sealed by the surrounding polymer. The resulting polyurethane foam is a valuable, lightweight material for building insulation and padding.

EXERCISE 16-N

An alternative way of producing a gas that forms bubbles in polyurethane foam is to include a small amount of water along with the other components in the polymerization mixture. The water reacts with the isocyanate functional group to produce, ultimately, an amine and carbon dioxide, which is volatile. Write a complete, detailed mechanism for this conversion, assuming that a protic acid serves to catalyze the reaction. Do the same for the reaction in which hydroxide ion serves as the initiating species.

16-5 ▸ Extensively Cross Linked Polymers

Even in vulcanized rubber and copolymers of polystyrene and divinylbenzene, the number of cross-linking bonds is usually small in comparison with the number of monomer units present. On the other hand, monomers with three (or more) points of connection produce polymers that extend in two and three dimensions rather than in the linear arrangement found in simpler polymers such as polyethylene. The resulting extensively cross-linked polymers are often very hard. **Bakelite** is one example of a **resin** (a highly viscous polymeric glass) derived from the reaction of phenol with formaldehyde (Figure 16-31).

Each reaction in the formation of Bakelite can be viewed as an electrophilic aromatic substitution, taking place at both the *ortho* and the *para* positions of phenol. These carbon–carbon bond-forming reactions are indeed very similar to those in the preparation of bisphenol A (Figure 16-20). The resulting polymer is very highly branched: it is therefore quite rigid and does not soften significantly at elevated temperatures, which makes it a useful material for cooking utensils, dishes, and bowling balls.

EXERCISE 16-O

Write a mechanism for the reaction of formaldehyde with phenol in the presence of a Lewis acid. Why does the reaction take place *ortho* and *para*, but not *meta*, to the hydroxyl group?

Epoxy resins constitute a very important class of cross-linked polymers that have many applications as structural materials, including the commonly known **epoxy glues.** For example, many laboratory bench tops are now made of epoxy resin, and microelectronic chips are encapsulated in this material. Because the resins are expensive, a "filler" is added in the same way that sand and small stones are added to cement to make concrete (for micro chips, the filler is silicon dust). The chemistry of epoxy resins is straightforward but the structures are complex; for this reason, block diagrams, like that shown on the next page for bisphenol A, represent structures in the discussion that follows.

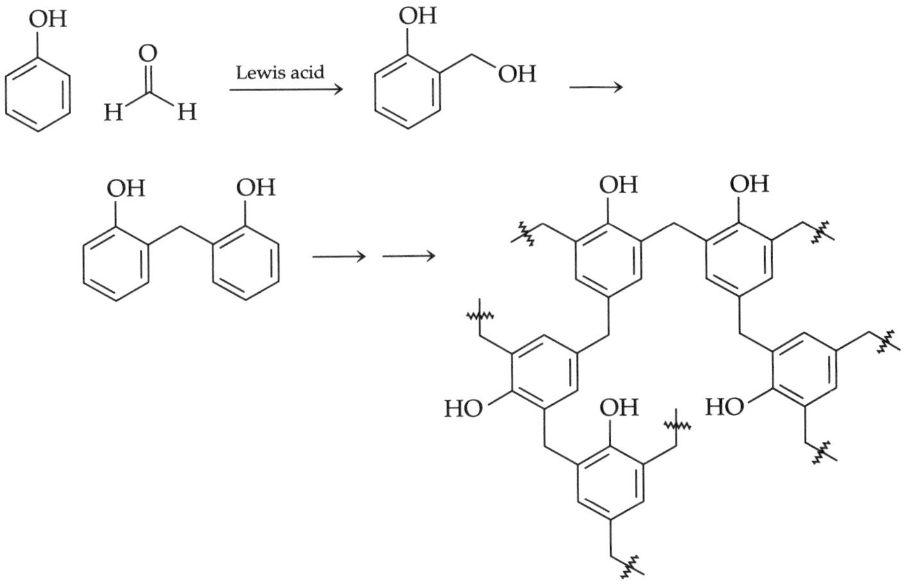

Bakelite

▲ FIGURE 16-31

The activating effect of the phenolic —OH group directs alkylation by phenol to all three *ortho* and *para* positions. The benzylic alcohol alkylates another phenolic ring, producing a highly cross linked, very rigid polymer.

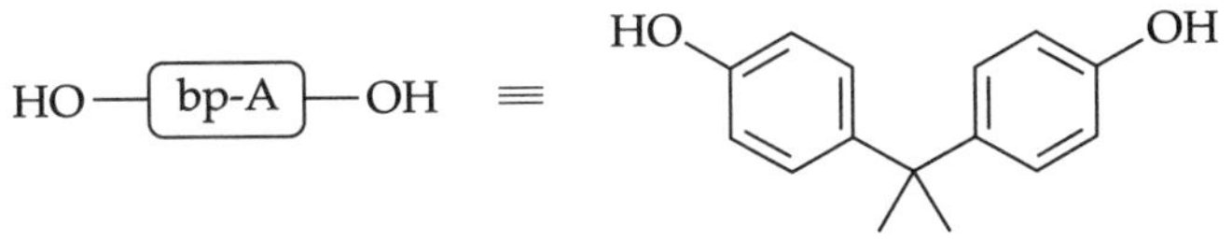

Bisphenol A

   Reaction of bisphenol A with **epichlorohydrin** results in the formation of a bis-epoxide (Figure 16-32). This, in turn, reacts with additional bis-phenol A in a one-to-one ratio, forming first a simple one-to-one adduct. This epoxy alcohol then reacts either with more bisphenol A, forming a diol, or with more bis-epoxide to form a larger bis-epoxide. These reactions continue, ultimately forming a mixture of short polymers composed of diols, epoxy alcohols, and bis-epoxides. This mixture is still fluid and is the clear, nearly odorless component of the two-part epoxy glues. The reaction of this complex mixture of linear polymers with a triamine, the fishy-smelling component (Figure 16-33), leads to the opening of the terminal epoxides and the formation of a three-dimensionally linked network that is much more rigid than the original resin.

EXERCISE 16-P

The reaction of epichlorohydrin with an alcohol (or other nucleophile) can be viewed as a simple S_N2 displacement at the carbon bearing the chlorine atom. Labeling studies, however, indicate that the nucleophile is sometimes bound to the carbon at the other end of the molecule. Suggest a mechanism that can account for these observations. In your answer, use methoxide as the nucleophile.

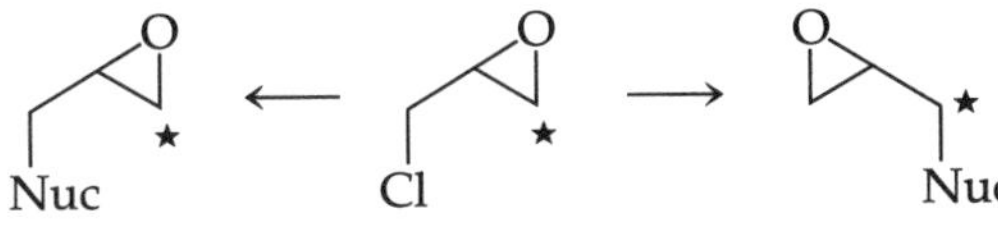

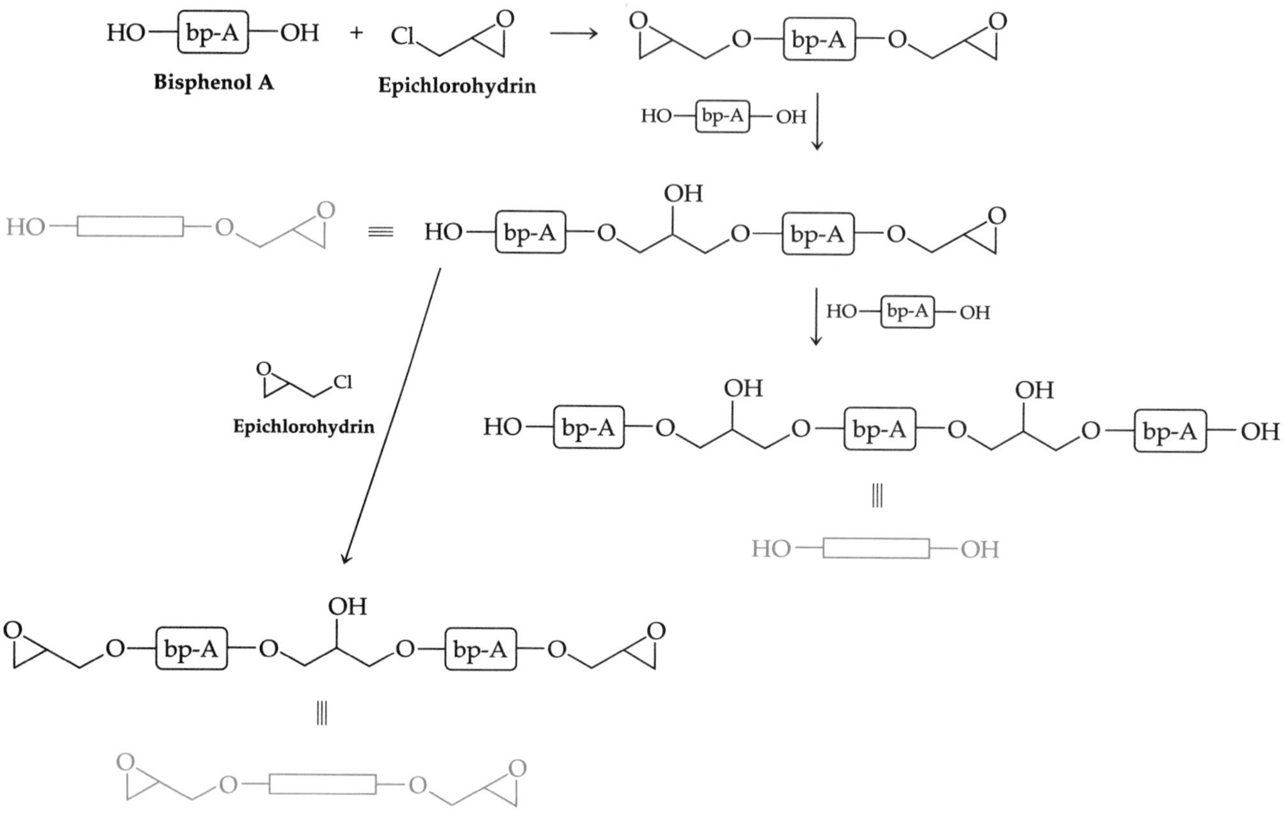

▲ FIGURE 16-32

Reaction of bisphenol A with two equivalents of epichlorohydrin by two S_N2 displacements of chloride ion produces a diepoxide. Reaction of this diepoxide with additional bisphenol A at one of the epoxides produces an epoxyphenol; further reaction at the remaining epoxide group produces a longer molecule containing three bisphenol A subunits and two parts derived from epichlorohydrin. At the same time, some of the epoxyphenol reacts with additional epichlorohydrin, producing a bis-epoxide. In all, there are three types of products formed: bis-phenol; monophenol-monoepoxide; and bis-epoxide. Each product is shown in a simplified schematic representation, emphasizing the functional groups that are present.

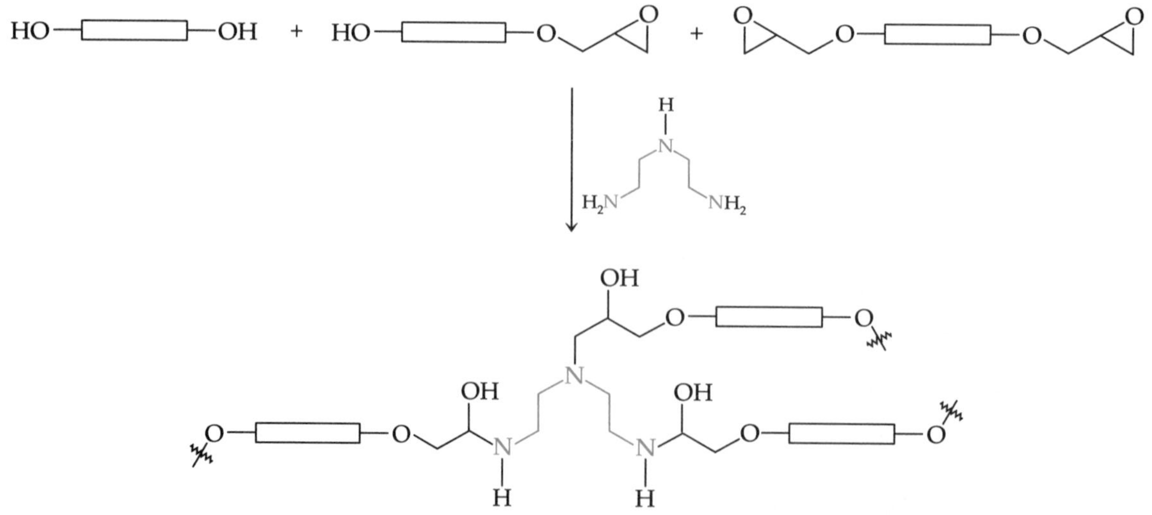

▲ FIGURE 16-33

When the mixture of diols, epoxyalcohols, and diepoxides shown in Figure 16-32 is treated with a triamine, a highly cross linked polymer is formed.

16-6 ▸ Three-Dimensional Structure

The functional groups present in a polymer can dramatically affect its bulk properties. The contrast between the water insolubility of polyethylene and other hydrocarbon polymers and the water solubility of materials such as the poly(vinyl alcohols) and polysaccharides is dramatic. Indeed, these differences are expected, based on our knowledge of the properties that various functional groups impart to small molecules. Recall, however, that there is large difference in water solubility between amylose and cellulose, even though both of these biopolymers are approximately the same size and have the same functionality. However, they differ in the stereochemistry of the linkage between the glucose subunits. This comparatively small stereochemical difference in the monomer subunits translates into large differences in the three-dimensional shapes of the polymers. In this section, we examine some of the aspects of three-dimensional structure and how this structure affects the properties of the polymer.

Polypropylene

Polypropylene is the simplest polymer for which bulk properties are influenced significantly by stereochemistry. When propylene is polymerized, the methyl groups along the polymer backbone can be oriented relative to each other in one of three ways: random; alternating; or all on one side. Simple radical, anionic, and cationic polymerizations afford a random orientation of the methyl groups, producing a stereochemistry referred to as **atactic.** In contrast, a complex, metal-based initiator known as the **Ziegler-Natta catalyst** produces a polymer in which almost all of the methyl groups are on the same side of a regularly oriented polymer backbone, referred to as **isotactic** (Figure 16-34). A polymer chain with groups alternating between front and back is known as **syndiotactic.** Methods are not yet available to make syndiotactic polymers in sufficient quantities to be made into commercial products.

The Ziegler-Natta polymerization uses an organometallic Lewis acid catalyst prepared by the treatment of a trialkylaluminum with titanium trichloride. In addition to the stereoregular polymers shown in Figure 16-34, a high-density ethylene with greater strength and heat resistance than was previously possible can be synthesized with the Ziegler-Natta catalyst. For this work, Karl Ziegler of the Max Planck Institute for Coal Research in Mülheim-Ruhr, Germany, and Giulio Natta of the Milan Polytechnic Institute received the Nobel Prize in chemistry in 1963, only ten years after the introduction of their catalyst. Van der Waals interactions between chains with a regular arrangement are stronger than those between chains with randomly oriented groups. As a result, isotactic polypropylene is harder and more rigid than the atactic polymer.

Let us first look at the chirality at each of the carbons in isotactic polypropylene. If the isotactic polypropylene formed by Ziegler-Natta

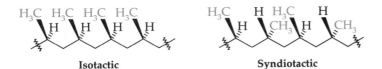

Isotactic   **Syndiotactic**

▲ FIGURE 16-34

In an isotactic polymer, identical groups point in the same direction along a stereoregular, conformationally fixed backbone, whereas their direction alternates in a syndiotactic polymer. Ziegler-Natta polymerization leads to the isotactic arrangement.

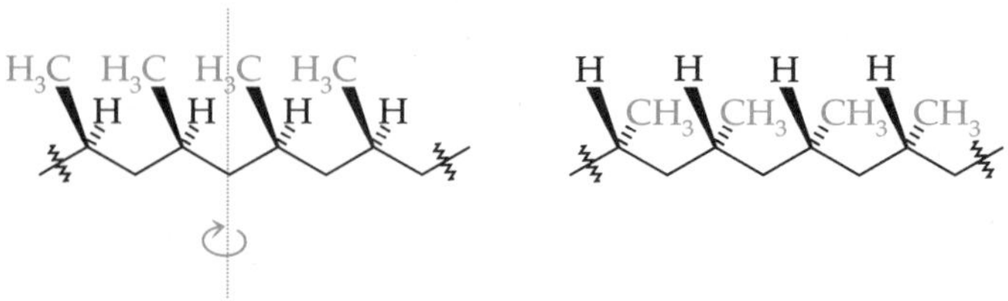

▲ FIGURE 16-35

Isotactic polypropylene can be viewed either from the front or back of the polymer chain. These nearly identical molecules (differing only at the end carbons) are interconverted by rotation about the central methylene group.

polymerization (Figure 16-34) were rotated end-to-end, an arrangement in which all of the methyl groups point to the back would be obtained (Figure 16-35). We could also arrive at almost the same arrangement by interchanging the hydrogen and methyl group (and thus the configuration) on each of the carbons. We say *almost* the same because the two ends of the polymer chain are not identical: one bears the initiator as a substituent. For a large polymer chain, this difference (though real) is insignificant because the vast majority of the methyl-bearing carbons are far from either end of the chain. Thus, although each of the methyl-bearing carbons in isotactic polypropylene is a chiral center, its absolute stereochemistry is not important. Equal amounts of polymer of each absolute configuration are formed under normal conditions, but chains of opposite handedness are almost indistinguishable.

In contrast, we will see in the next section that some biopolymers contain chiral centers at which the differences between the four substituents are large. In this case, the absolute stereochemical relation between individual polymer chains is indeed important and has a significant consequence on the properties of the resulting polymers.

EXERCISE 16-Q

Re-examine the polymers illustrated in Figures 16-11 and 16-12 from the point of view of stereochemistry. Which ones can be formed in one of the three different forms (atactic, isotactic, and syndiotactic)? Draw a structure that illustrates the possible stereochemistries.

Polypeptides

Naturally occurring polypeptides are universally derived from α-amino acids. More than twenty amino acids (differing in the alkyl substituent, R) are commonly found in natural polypeptides (Figure 16-36). (These monomers are tabulated in Chapter 18, where amino acid chemistry is

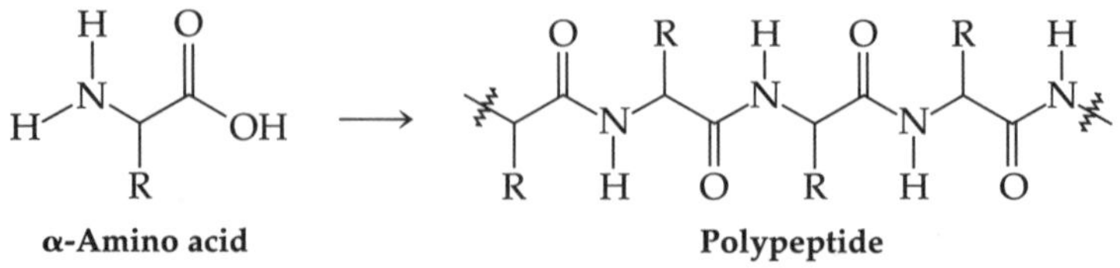

▲ FIGURE 16-36

More than twenty different chiral α-amino acids are the monomers of naturally occurring polypeptides.

discussed in more detail.) This diversity of building blocks allows for a virtually unlimited number of peptides that differ in the sequence of the constituent amino acids. Nonetheless, these polypeptides have certain features in common that derive from the presence of regularly repeating amide groups. This linkage is referred to as a **peptide bond,** and hence the term *polypeptide* explicitly refers to the repeating occurrence of the amide linkage.

To understand the three-dimensional properties of polypeptides, we must first look at the unique properties of the amide functionality. As already noted in Chapter 3, there is significant electron delocalization in amides, with donation of the nitrogen lone pair of electrons to the carbonyl group π system (Figure 16-37). The additional stability imparted by this delocalization has important implications in the chemistry of amides in general and specifically in the structure that this group imparts to polypeptides. One consequence of this overlap is the planarity of the amide bond; that is, an important contribution from the right-hand resonance structure in Figure 16-37 requires all of the atoms shown in the hybrid at the bottom of Figure 16-37 to be in the same plane. Experimental evidence is consistent with an energetic contribution from this π overlap of about 18 kcal/mole. Indeed, the six atoms discretely taking part in the amide bond are essentially coplanar in virtually all amides.

There are two geometric isomers for the monosubstituted amides found in polypeptides. The arrangement in Figure 16-37 with the carbon *syn* and the hydrogen *anti* to the carbonyl oxygen (known as the *syn* or Z isomer) is considerably more stable than the alternate *anti* (or E) arrangement. (The origin of this energetic difference has not yet been well defined.) As a consequence of lone-pair–π overlap and the defined geometry, the amide group is a much more rigid linker for polypeptides than would be implied by the resonance structure illustrated at the left in Figure 16-37.

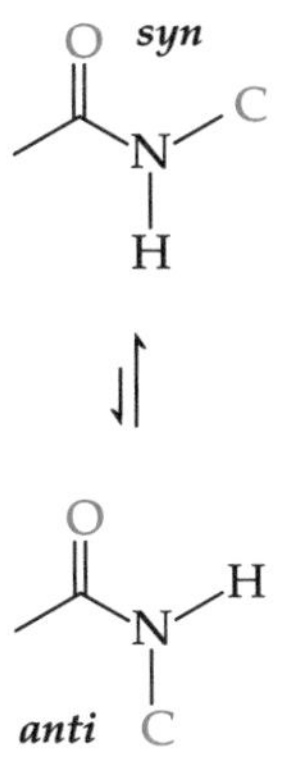

A second important feature imparted to amides by electron delocalization is the very significant partial negative charge on oxygen, positive charge on nitrogen, and an increase in the polarization of the nitrogen–hydrogen bond, which leaves the hydrogen highly electron deficient. The presence of both an electron-rich oxygen and an electron-deficient hydrogen in the amide functional group provides the opportunity for strong hydrogen bonding *between* amides (Figure 16-38).

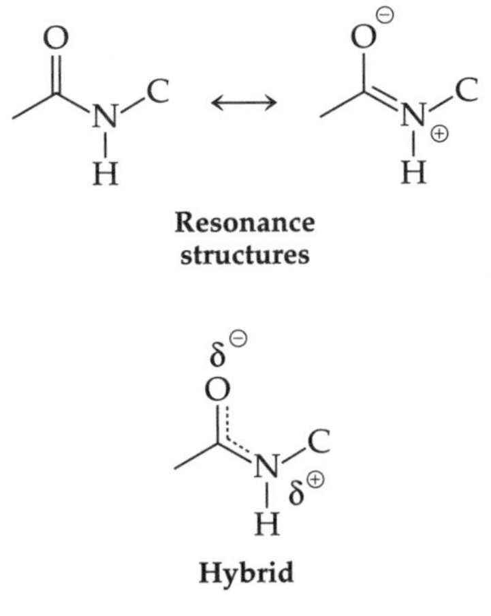

▲ FIGURE 16-37
Delocalization of the lone pair of electrons on the nitrogen atom of an amide results in partial double bond character in the C–N bond. Because of this π-bond character, there is restricted rotation about the C–N bond and all of the atoms shown lie in the same plane.

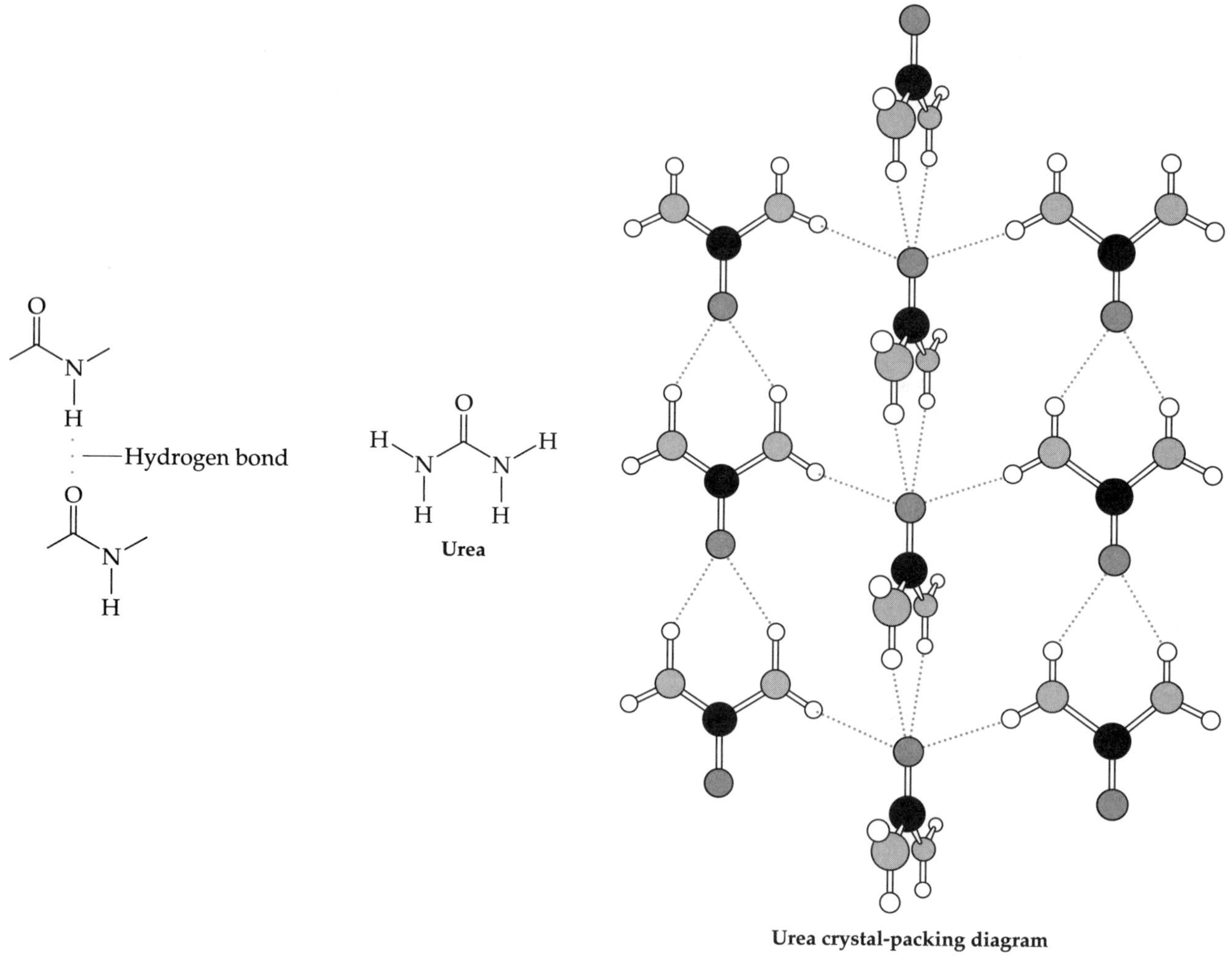

Urea crystal-packing diagram

▲ FIGURE 16-38

Intermolecular hydrogen bonding is responsible for the regular order between amide groups in separate urea molecules packed as a solid crystal.

The crystal structure of urea illustrated in Figure 16-38 shows how hydrogen bonds can hold individual molecules together. If we compare the melting point of urea (133° C) with that of acetone (–95° C), a molecule of very similar molecular weight and functionality but without a hydrogen-bond donor, we can understand the importance of these hydrogen-bonding interactions.

Estimates for the magnitude of the energy of hydrogen bonds in various environments vary from less than 1 to about 5 kcal per hydrogen bond. It is quite likely that the magnitude of the energy of this interaction differs quite markedly with relatively subtle changes in distance and bond angle.

Note that the hydrogen bond illustrated in Figure 16-38 is an arrangement in which the nitrogen–hydrogen and oxygen–carbon bond systems are colinear. This arrangement is most commonly found for hydrogen bonds in amides although not in the crystal structure of urea.

EXERCISE 16-R

One way to view the hydrogen bond is as a proton partially bound to two different electron-rich groups. These two bonds are generally quite different, one linkage being very strong and the other very weak, as can be deduced from the different bond lengths.

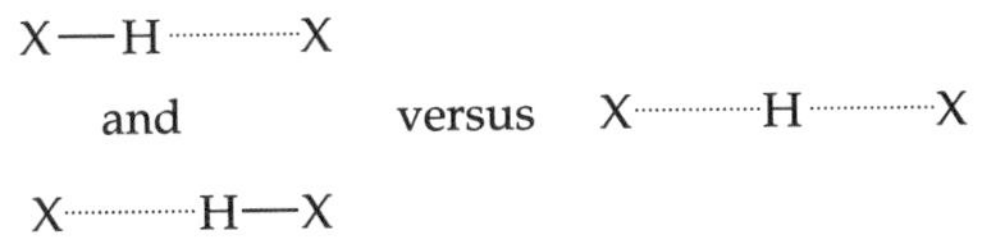

$$X—H\cdots\cdots X$$

and   versus $X\cdots\cdots H\cdots\cdots X$

$$X\cdots\cdots H—X$$

Explain why this should be the case rather than there being two equal half-bonds between the hydrogen and its neighbors. Furthermore, explain the linear arrangement of atoms described for urea in view of the simple concepts of bonding arrangements that were developed in Chapters 1 and 2.

The amide groups in a polypeptide can interact through hydrogen bonding in either an *intra*molecular or an *inter*molecular sense. We consider the latter arrangement first because it is a bit easier to see. Figure 16-39 shows a polypeptide in which the amino hydrogen atom and the carbonyl oxygen atom of one α-amino acid are on the same side of the chain. We might then imagine that this pair can interact by hydrogen bonding with another pair in an adjacent polypeptide molecule. This is illustrated in Figure 16-40 (on page 572) for the interaction of three polypeptide chains. Both the upper and lower peptide chains have pairs of amino and carbonyl groups that can participate in further hydrogen bonding with other peptide chains. This structure, then, can be extended in a virtually unlimited fashion to form an essentially flat sheet of peptides. Because *unlike* groups interact by hydrogen bonding, the interaction is intermolecular and the chains must alternate direction (from carbonyl to amino group along the chain) to achieve multiple hydrogen-bonding between chains. The arrows in Figure 16-40 indicate the direction of chain growth—from the amino (—NH) group to the carboxyl (—CO₂H) group of each amino acid.

Recall that the stereochemistry of naturally occurring amino acids is almost universally *S*. As a consequence, the substituent groups (the R groups) nearest to each other on adjacent chains in this sheetlike arrangement of polypeptides are on the same side. If one of these substituents is a small alkyl group such as methyl and the other is hydrogen (or if both are hydrogen), there is essentially no steric interaction between the groups on adjacent chains. On the other hand, there is a substantial repulsive interaction if both are large alkyl groups. This interaction results in a twisting of the peptide chains such that the alkyl groups rotate away from each other, reducing the steric interaction between these groups (Figure 16-41). This deviation from a totally planar arrangement results in what is referred to as a *pleating* of the peptide sheets and the lower arrangement in Figure 16-41 is referred

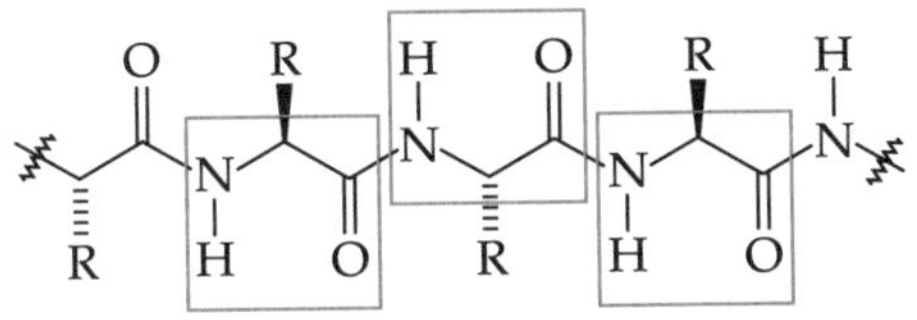

▲ **FIGURE 16-39**
This polypeptide is oriented so that the amide N—H bond points in the same direction as the C=O bond of the carbonyl group of the α-amino acid of which it is a part (shown in box). This places the N—C bond *syn* to the carbonyl group of the amino acid to which it is linked.

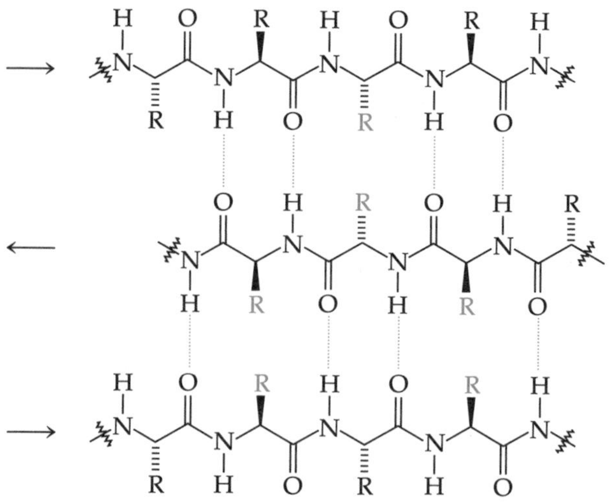

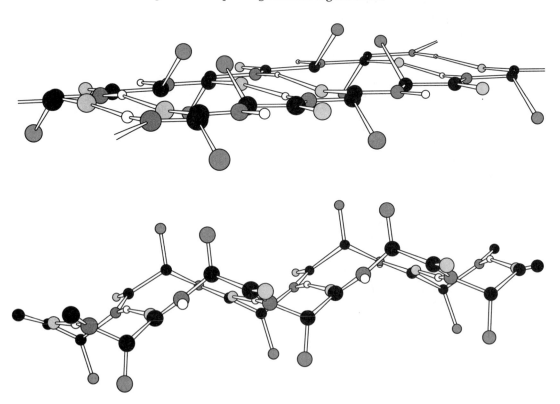

▲ **FIGURE 16-40**

When adjacent peptide chains are oriented in opposite directions, they are properly positioned for intermolecular hydrogen bonding with another planar peptide backbone located on either side. (The arrows at the left indicate the direction from the amino to the carboxyl terminus of the polypeptide chain.) An extended sheetlike structure resulting from this interaction is similar to that leading to the urea packing forces in Figure 16-38.

▲ FIGURE 16-41

When the polymer backbone of the peptide is arranged in the sheetlike structure of Figure 16-40, the side-chain substituents (shown in color) on the α carbons are directed toward the same region of space, as shown in the upper structure. To prevent the steric interaction that results when the substituents are large, the α carbons rotate from the plane occupied by the amide bonds, as shown in the lower structure. The resulting structure is called a β-pleated sheet.

to as a **β-pleated sheet.** This pleating is not without other consequences: in reducing the steric interaction between the alkyl groups, it brings the hydrogens on the centers of chirality closer together. Furthermore, the twisting distorts the hydrogen bonds, further destabilizing the pleated-sheet arrangement. If all or most of the R groups are not hydrogen, then the pleated-sheet arrangement is less stable than an alternate arrangement in which the —NH and C=O groups are *intramolecularly* hydrogen bound.

Consider, for example, the geometry needed for intramolecular hydrogen bonding between peptide units in the same chain. This arrangement for hydrogen bonding would clearly require the polypeptide chain to be three-dimensional, rather than planar. In fact, intramolecular hydrogen bonding can be accommodated without major bond-angle distortion if the hydrogen bonding is between the carbonyl of one amino acid and the NH group of the third amino acid down the chain. This results in a coiled arrangement in which each carbonyl group can hydrogen bond with the N—H of the amide of a different amino acid unit.

This helical arrangement is referred to as an **α-helix,** so called because, as viewed from one end, the chain coils in a clockwise direction as the chain proceeds away from the viewer. (This is an occasion to use your molecular models again.)

At first, one might expect that the chain could coil equally well in either a left- or a right-handed fashion. However, the twisting of the chain required to achieve this intramolecular hydrogen bonding places the alkyl and hydrogen substituents of the amino acid units in distinctly different positions (Figure 16-42). In the α-helix, the alkyl substituents of (S)-amino acids are oriented more or less directly away from the helical structure, whereas the hydrogens are pointed toward its interior. If the helix were

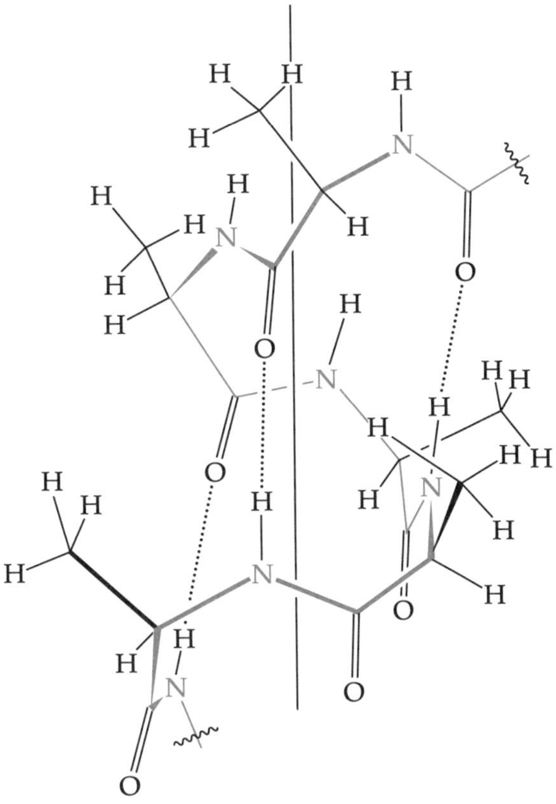

◀ **FIGURE 16-42**
Coiling in an α-helix takes place so as to direct the large groups (in this case, the methyl groups of alanine in polyalanine) bound to the α carbon toward the exterior of the helical coil.

α-Helix

coiled in the opposite (counterclockwise) direction, producing a left-handed helix (a **β-helix**), these positions would be interchanged and a very substantial steric interaction between the chain and the alkyl groups would result.

EXERCISE 16-S

Draw three strands of peptides oriented in a parallel (rather than antiparallel) fashion to each other. Which arrangement, parallel or antiparallel, has more hydrogen bonds?

Many different molecules found in nature are based on the peptide bond. Some, whose molecular weights exceed 100,000, are composed of many hundreds of individual amino acid units. As mentioned earlier, those composed of one hundred amino acids or fewer are known as polypeptides, whereas the larger species are called **proteins.** Each polypeptide or protein has a unique sequence of amino acids, which is referred to as its **primary structure.** The three-dimensional structure of large polypeptides is often quite complex, consisting of a combination of both helical and β-pleated-sheet arrangements located in different parts of the chain. These arrangements, fixed in local regions, constitute the **secondary structure** of the polypeptide, as illustrated in Figure 16-43 for a rather small protein (in this case, one that plays a critical role in the immune system). This schematic

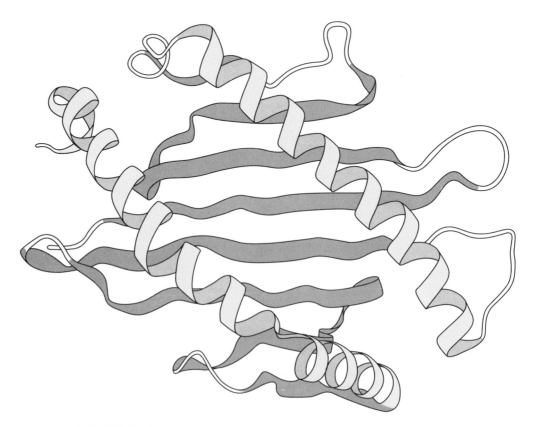

▲ FIGURE 16-43

The tertiary structure of the binding domain of the major histocompatibility complex class II protein (a critical component of the immune system) shows regions in which both α-helices (the coils in color) and β-pleated sheets (the ribbons in gray) are folded into a complex, three-dimensional arrangement.

representation also illustrates what is referred to as the **tertiary structure** of a protein; that is, how the β-pleated sheets and α-helices are spatially dispersed. The protein shown has seven distinct regions that are β-pleated-sheet structures, twisted and folded around each other so that adjacent chains are antiparallel, and two regions that are helical: the combination of these regions makes a complex three-dimensional arrangement. The precise tertiary structure of the protein results directly from the type and location of its individual constituent amino acids. Often, several large polypeptide or protein molecules join together to form a discrete complex, called a **quaternary structure,** which also ultimately derives its precise arrangement from the sequence of individual amino acids present.

Major research is currently underway to discover rules that may permit the prediction of a protein's tertiary structure from its primary sequence of amino acids. Such analyses are complicated by the observation that solvents also are important in determining three-dimensional structure. Solvents (such as water) that can form relatively strong hydrogen bonds to the peptide groups can produce solvated polypeptide structures that are similar in energy to the α-helix, in which all hydrogen bonds are internal. For example, in dioxane, a solvent that can serve only as a hydrogen-bond acceptor, polyalanine is nearly completely coiled in a helical arrangement. Conversely, in water, the degree of coiling is only about 50%. We must not forget that, although much smaller by comparison, solvent molecules constitute the environment in which large polymers exist. The shape of macromolecules thus varies with the solvent in a usually quite understandable fashion.

Cellulose and Starch

Earlier in this chapter, it was pointed out that cellulose is much less soluble in water than is starch of comparable molecular weight. We are now in a position to understand this difference from a molecular point of view. Recall that the structures of these naturally occurring polyacetals are quite similar, differing mainly in the stereochemistry at C-1 (Figures 16-22 and 16-24). This difference has a dramatic effect on the ability of each of these materials to participate in *intramolecular* hydrogen bonding between the hydroxyl groups. Because the linkage in starch has an axial oxygen substituent on one of the six-membered rings, the individual rings can coil, forming a helix. In this arrangement, hydrogen bonding occurs between the hydroxyls on adjacent rings of the same polymer chain. (This is shown schematically in Figure 16-44 with rectangles representing the individual glucose subunits.) In contrast, the oxygen linking the units in cellulose is an equatorial substituent on both rings and, as a result, it is not possible to twist the chain so that adjacent rings can participate in intramolecular hydrogen bonding (Figure 16-44). Hydrogen bonding is thus highly favored between glucose units on adjacent chains, and a sheetlike arrangement similar to the β-pleated-sheet network in proteins is highly favored. These sheets form layers like those in graphite, with hydrogen bonds that lock the individual chains into a rigid, three-dimensional matrix, as shown in the two views in Figure 16-45.

Thus, relatively subtle differences in molecular structure result in quite dramatic differences in three-dimensional structure. For example, the substitution of an alkyl group for a hydrogen atom in an amino acid residue of a peptide leads to a change in arrangement from a sheet to an α-helix. In turn, these structural changes lead to marked chemical and physical differences that affect the function of biopolymers. Certainly, cellulose would not be very useful as a structural material if it were readily soluble in water.

Cellulose

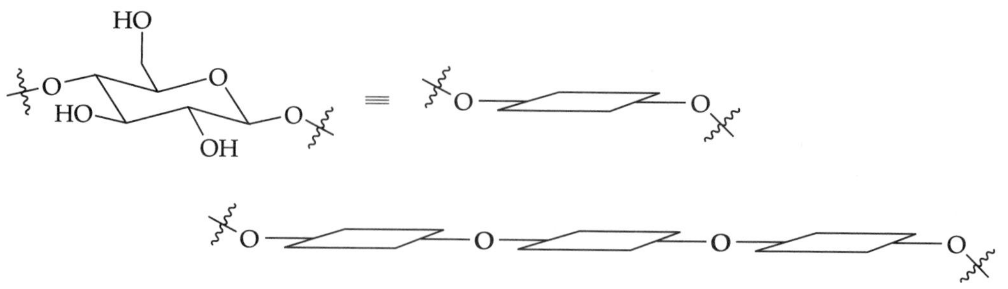

Starch

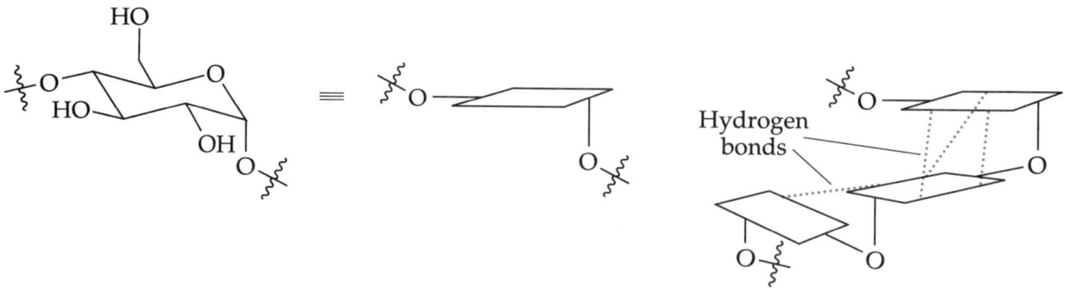

▲ **FIGURE 16-44**
The linear structure of cellulose stands in striking contrast with the coiled arrangement of starch when the carbohydrate rings are represented by rectangles.

Conclusions

The molecules described in this chapter are large polymers. The properties of a polymer are uniquely determined by the characteristics of its component functional groups and its three-dimensional structure. Although the chemical characteristics of the functional group(s) present can be modeled by those of the monomer, the structure and physical properties are unique to the polymer. The difference in structure between linear and branched polymers contributes to the macroscopic properties of hardness and strength that must be matched to applications for these materials.

Monomers can have a variety of functional groups, including simple alkyl groups, aromatic substituents, esters, alcohols, amines, and halocarbons. These substituents either are appendages on the polymer skeleton or directly participate in the chemical bonding that links the monomer units together to form polymers. These linkages include carbon–carbon bonds in vinyl polymers, carbon–oxygen bonds in polyesters, polyacetals, and polyethers, and carbon–nitrogen bonds in polyamides and peptides. Each of these functionalities, along with other functional groups present as substituents, contributes to the bulk properties of the polymer. Thus, for example, polyethylene is totally insoluble in most solvents, poly(methyl methacrylate) dissolves readily in polar organic solvents, and poly(vinyl alcohol) dissolves in water.

There are two broad classifications of polymers, addition and condensation, based on the type of reaction required for polymerization. Addition polymerization takes place through an intermediate cation, radical, or anion. The polymers formed with ionic intermediates are referred to as "living" polymers because the reactions are not self-terminating and are complete only when all of the starting materials have been consumed.

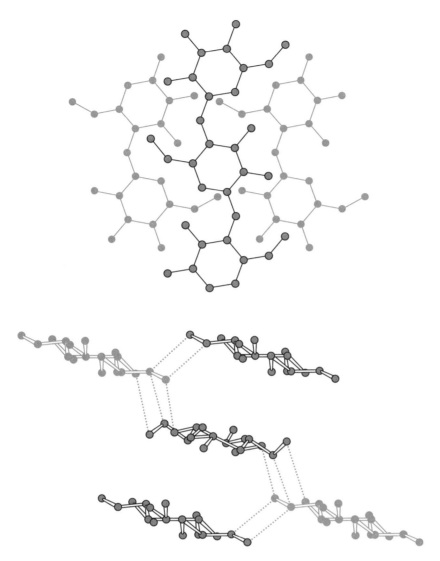

▲ FIGURE 16–45

Two views of the relative arrangement of cellulose chains as found in, for example, wood fibers. In the upper view (a "top" view), the six-membered rings of the glucose units are clearly visible. In the lower picture, in which the chains are viewed end on, the dotted lines represent the hydrogen bonding between glucose hydroxyl groups on adjacent chains that contribute to the strong, intermolecular attractive interactions that account for the strength and rigidity of cellulose. (Redrawn with permission from *Chemistry in Britain*, October 1989, p. 107.)

The properties of a polymer can vary significantly with stereochemistry as a result of variations in the strength of intermolecular attractive interactions. Covalent cross-linking can make a polymer very hard and rigid. Hydrogen bonding plays an important role in the three-dimensional structure of nitrogen- and oxygen-containing polymers. Intermolecular hydrogen bonding produces a β-pleated-sheet secondary structure in peptides, whereas intramolecular hydrogen bonding leads to a coiled, α-helical secondary structure. Hydrogen bonds contribute substantially to the secondary, tertiary, and quaternary structures of proteins. Intramolecular hydrogen bonding in starch reduces the number of intermolecular hydrogen bonds and, as a result, starch is water soluble. In contrast, cellulose has little opportunity to participate in intramolecular hydrogen bonding; thus, the polymer chains are held quite tightly together by multiple contacts.

Summary of New Reactions

Radical Polymerization

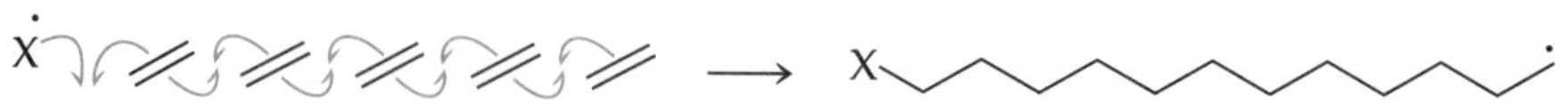

Anionic Polymerization

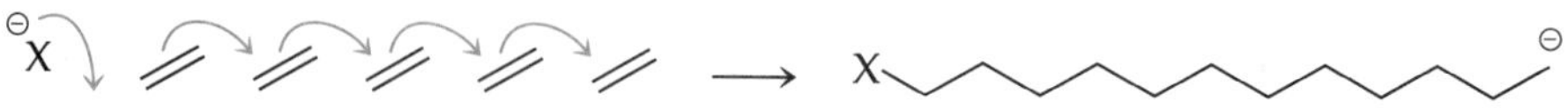

Cationic Polymerization

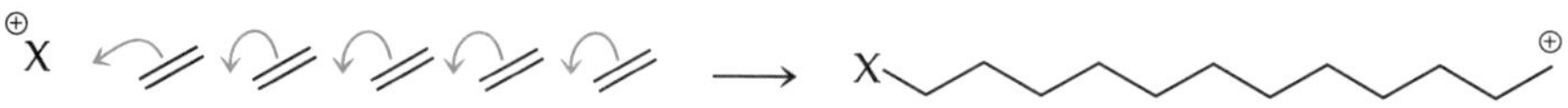

Polyacetal Formation

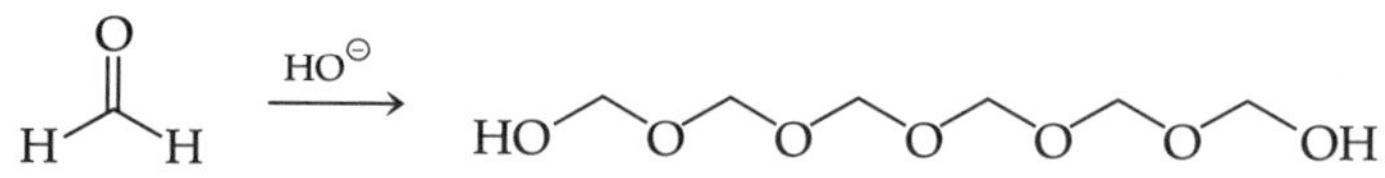

Review Problems

16-1 Identify the minimum structural repeat unit in each of the following polymers.

(a) 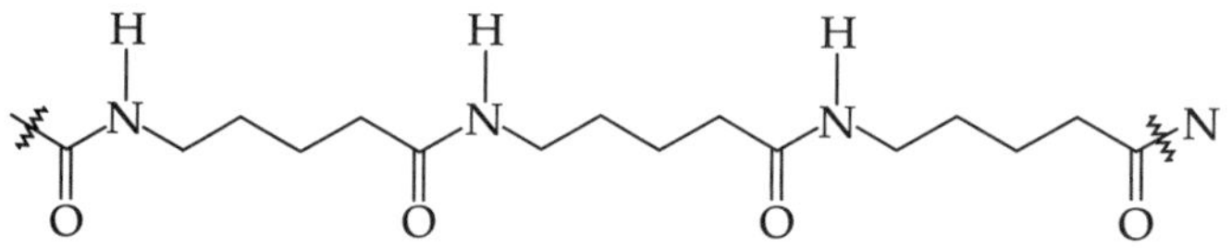

(b) 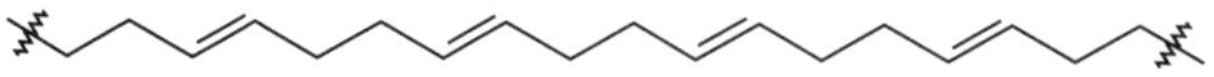

(c) 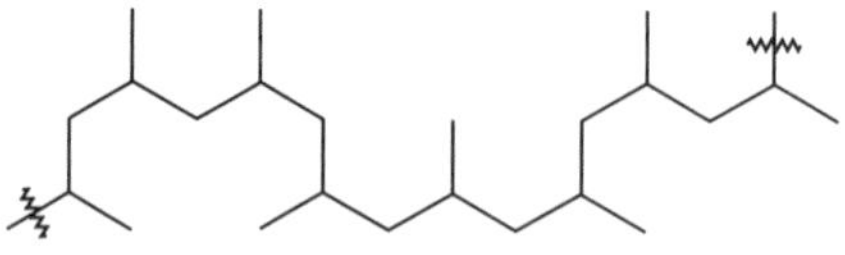

(d) 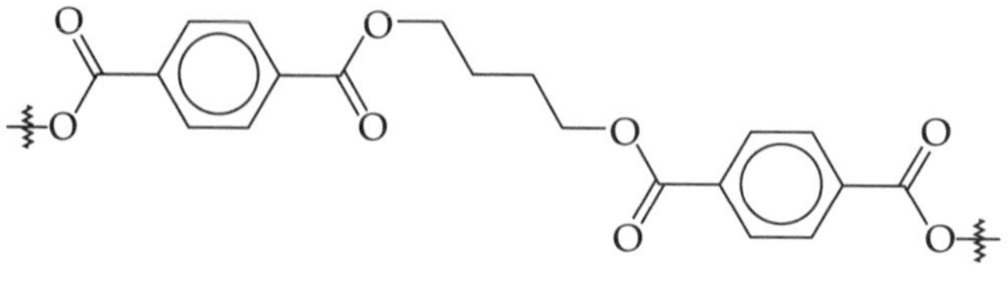

(e) 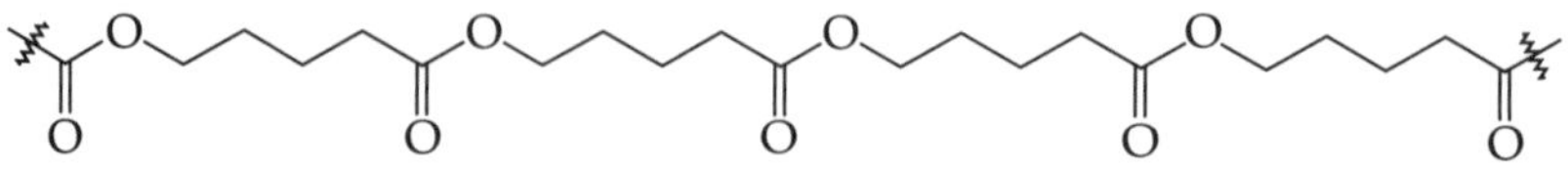

(f) 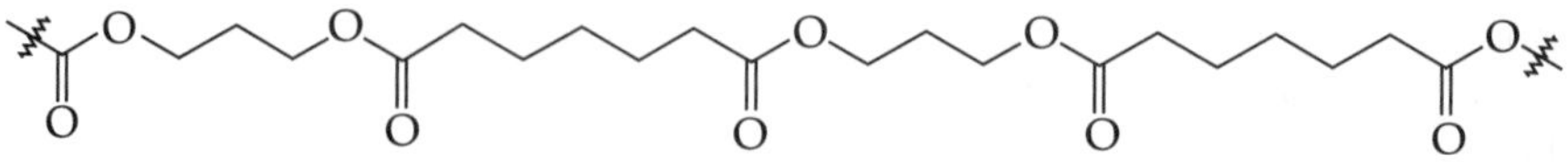

16-2 Identify the functional group that might have been formed in the synthesis of each of the polymers in Problem 16-1. Write structures for the monomer unit(s) in each case.

16-3 The polymer formed from formaldehyde known as paraformaldehyde, or polyacetal, is not very stable under both acidic and basic conditions. However, conversion of the hemiacetal end groups of this polymer into either —CH$_2$CH$_2$—OH groups by reaction with ethylene oxide or into acetate esters by reaction with acetic anhydride produces polymers (known as Celcon and Delrin) that are much more stable.

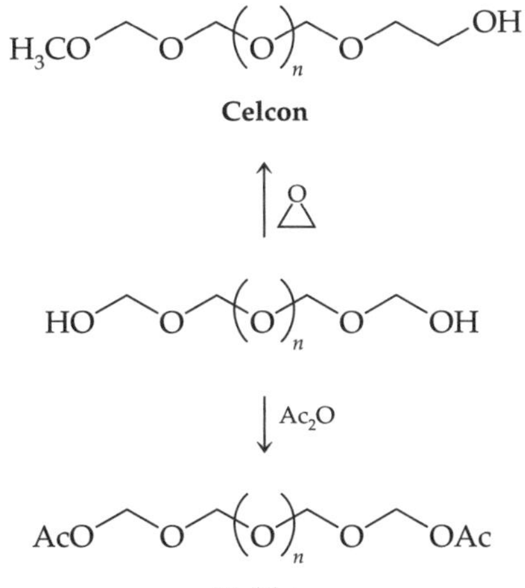

Celcon

Deldrin

Suggest a reason for this enhanced stability. One of these polymers is much more stable to strong base than the other. Which is it and why?

16-4 Large quantities of ethylene are converted by oxidation with molecular oxygen into ethylene oxide, which is then treated with aqueous base to form ethylene glycol for use in making polyesters such as Dacron. Increasing the amount of water in the reaction causes a decrease in the amount of di- and tri(ethylene glycol) compared with ethylene glycol. Suggest a reason for this relation. When the reaction is conducted as a very dilute solution in water, almost all of the ethylene oxide is converted into ethylene glycol. Why might this not be a practical method for making ethylene glycol on a very large scale?

16-5 The reaction of phenol with formaldehyde in the presence of acid produces a material called bisphenol F. Through analogy with bisphenol A, write a structure for this material and suggest a mechanism for its formation.

16-6 Terephthalic acid (the *para* isomer of benzene dicarboxylic acid) is prepared industrially from *p*-dimethylbenzene (*p*-xylene). Suggest a reagent or sequence of reagents that you could use in the laboratory to carry out this transformation.

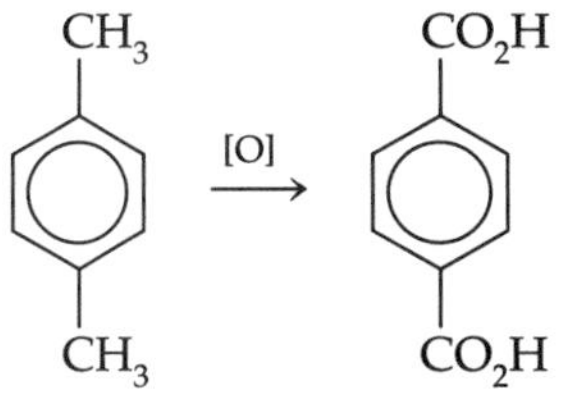

p-**Xylene**   **Terephthalic acid**

16-7 Polyamides can be made from β-amino acids in much the same way as they are made from α-amino acids. Assuming a completely extended structure for these polymers as shown below, examine possible hydrogen-bonding interactions between chains. Would you expect these hydrogen-bonding interactions to be stronger (on a per weight basis) or weaker than those for α-amino acids? How do the relations between the chains differ for polymers derived from α- and β-amino acids?

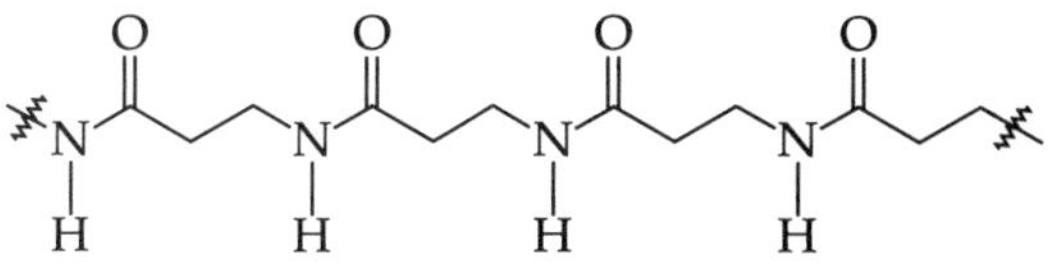

Chapter 17

Structures and Reactions of Naturally Occurring Compounds Containing Oxygen Functional Groups

Except in Chapter 16, small molecules have served so far to illustrate the concepts relevant to a particular organic reaction. However, in this chapter and subsequent ones, we will deal with biologically important reactions of larger organic molecules. To do so, we must first become familiar with their structures and chemical properties so that we can then examine how the interrelation between the various classes of compounds forms the chemical basis for living systems. We will learn how the general structural features of these molecules relate to their important biological functions. The chemical transformations that these molecules undergo will be treated in detail in subsequent chapters.

17-1 ▸ Lipids

The **lipids** are a broad group of molecules whose common characteristic is their solubility in nonpolar solvents. The lipids contain only simple functional groups. Because they are composed mainly of carbon and hydrogen, they dissolve readily in hydrocarbon solvents.

Fats and Waxes

Fats and **waxes** are the simplest of the lipids. They are esters of long, straight-chain acids (known as **fatty acids**) and either the triol **glycerol** in fats, referred to as **triglycerides,** or long, straight-chain alcohols in waxes.

Triglycerides

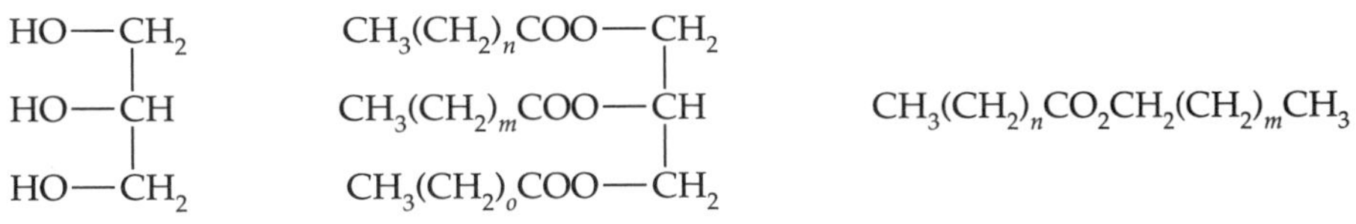

$$HO-CH_2$$
$$HO-CH$$
$$HO-CH_2$$
Glycerol

$$CH_3(CH_2)_nCOO-CH_2$$
$$CH_3(CH_2)_mCOO-CH$$
$$CH_3(CH_2)_oCOO-CH_2$$
Fats

$$CH_3(CH_2)_nCO_2CH_2(CH_2)_mCH_3$$
Waxes

VARIED USES OF ETHANOL

Ethyl alcohol was probably the first organic chemical to be made routinely by civilized man. Fermentation of sugars and starches still accounts for most of the alcohol produced for human consumption. There are many industrial applications of ethanol, both as the alcohol itself and as a part of larger molecules. Most industrial ethanol is made by hydration of ethylene produced from the cracking of petroleum feedstock. However, the quantity of ethanol produced in this way has been dropping steadily; from 1.3 billion pounds in 1981 to only 546 million pounds in 1990. In part, this decline in the synthesis of ethanol from ethylene is due to the vast overproduction of wine grapes, especially in Italy. As a result, excess wine is distilled to produce ethanol for industrial use, as well as for mixing with gasoline to make gasohol.

Several common fatty acids are depicted in Table 17-1, and the alcohols that are their ester partners in typical waxes are listed in Table 17-2. Both fats and waxes are found in nature as mixtures of many components. Thus, a sample of a fat contains many different triglycerides in which the individual components are constituted from several of the fatty acids. Furthermore, even with only two or three fatty acids present as ester units, there can be several different combinations. For example, an individual triglyceride in a fat sample from a biological source composed primarily of esters of stearic and palmitic acid may have some molecules with three stearyl ester groups, some with two stearyl and one palmityl, and so forth. Because glycerol has one secondary and two primary alcohol groups, several structural isomers with subtly different properties are possible when all of the attached acids are not identical. Waxes isolated from natural sources, such as the moisture-barrier coating of many insects and plants, also consist of mixtures of esters in which the chain lengths of both the alcohol and the acid components vary.

EXERCISE 17-A

Consider a triglyceride in which all three of the acid units are unique. In this case, carbon-2 of the glycerol unit is a chiral center (use molecular models if needed to understand this). Determine the total number of possible isomers, including both structural and stereochemical detail, and provide a schematic representation for each isomer such as that shown below.

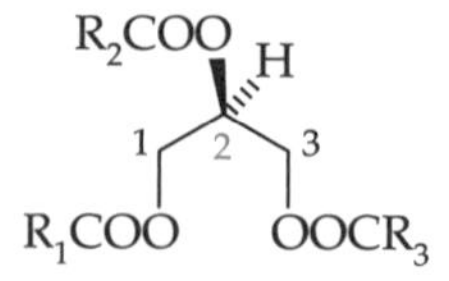

$$\text{R}_2\text{COO} \quad \text{H}$$
$$\underset{\underset{\text{R}_1\text{COO} \qquad \text{OOCR}_3}{1 \quad 2 \quad 3}}{}$$

You might at first be surprised to see that all of these common, naturally occurring acids and alcohols have an even number of carbon atoms. As we will see in detail in Chapter 22, this follows from the biological synthesis of long alkane chains in which acetic acid, a two-carbon precursor, undergoes a Claisen condensation. In long-chain fatty acids (and alcohols),

Table 17-1 ► Common Fatty Acids

Structure		Number of Carbons	mp (°C)
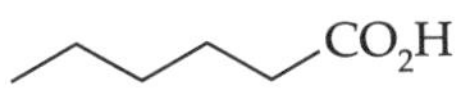 (CH₃…CO₂H)	Caproic acid	6	−3
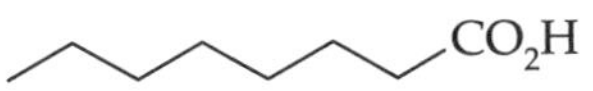	Caprylic acid	8	17
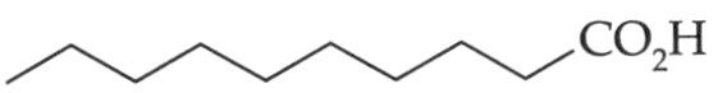	Capric acid	10	31
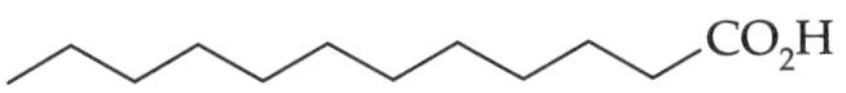	Lauric acid	12	44
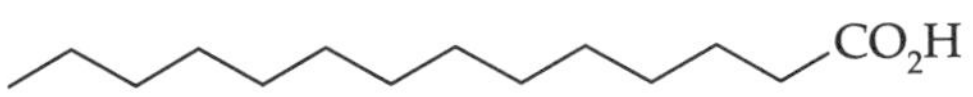	Myristic acid	14	58
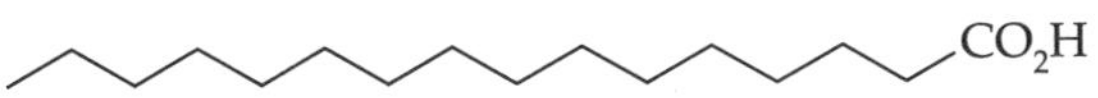	Palmitic acid	16	63
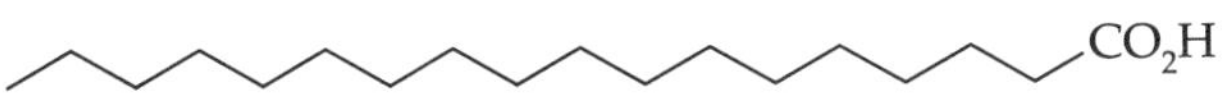	Stearic acid	18	70
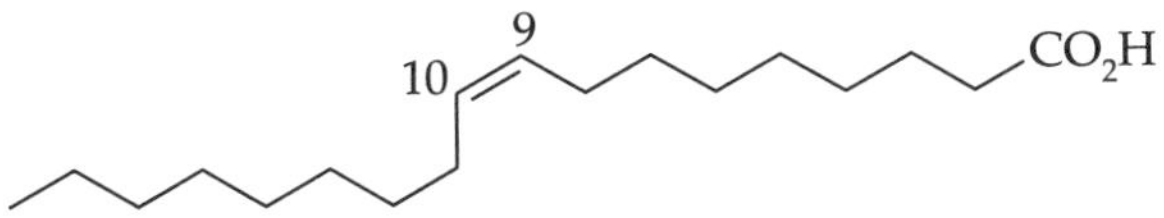	Oleic acid	18	4
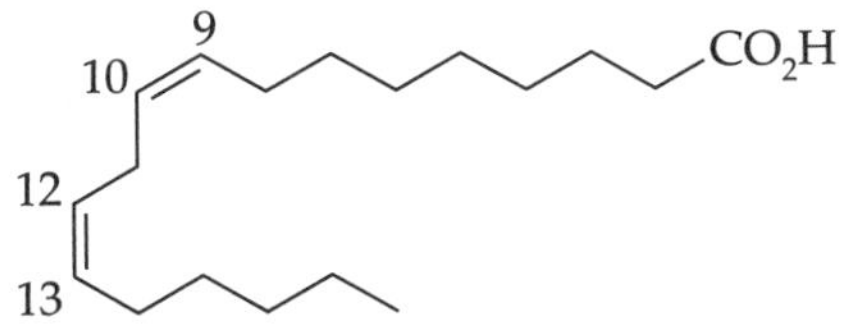	Linoleic acid	18	−5
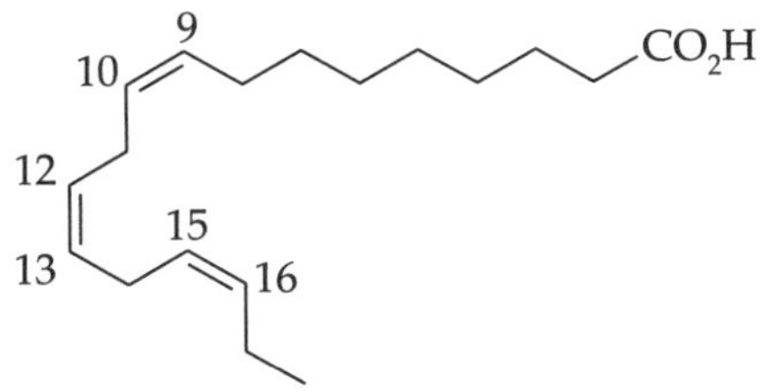	Linolenic acid	18	−12

the hydrocarbon portion of the molecule clearly dominates the structure. The waxes, then, can be viewed as structurally similar to short chains of polyethylene, described in Chapter 16. Not surprisingly, the macroscopic properties of waxes (as well as fats) also are reminiscent of polyethylene.

Table 17-2 ► Common Long-Chain Alcohols

Formula		Number of Carbons
$CH_3(CH_2)_{14}CH_2OH$	Cetyl alcohol	16
$CH_3(CH_2)_{22}CH_2OH$	Carnaubyl alcohol	24
$CH_3(CH_2)_{24}CH_2OH$	Ceryl alcohol	26
$CH_3(CH_2)_{28}CH_2OH$	Myricyl alcohol	30

They are nearly completely insoluble in water; but, as their classification as lipids would imply, they are quite soluble in hydrocarbon solvents. That they are mixtures prevents them from being crystalline. Indeed, it is a difficult task to separate any single pure component from naturally occurring fats and waxes.

The melting points for the fatty acids given in Table 17-1 increase gradually as the number of carbons in the chain grows. Although fats and waxes derived from these acids are not crystalline because they exist as mixtures, their **viscosity,** or "stiffness," nonetheless increases as the fatty acids are lengthened. Exceptions to the trend in melting points are the three **unsaturated fatty acids** listed in Table 17-1. Here, the term unsaturated has the same meaning as for hydrocarbons in Chapter 2: unsaturated fatty acids are differentiated from their saturated counterparts by the presence of double bonds. Notice that each additional double bond results in an increasingly lower melting point for the three acids: oleic, linoleic, and linolenic acid.

All of the double bonds in these three acids have the *cis* configuration, which interrupts the regular zigzag arrangement found in the extended conformation of the saturated fatty acids. This "bend" in the chain inhibits close contact between the molecules, resulting in a lower degree of crystallinity. These three fatty acids impart a lower viscosity to the fats that contain them. The geometry and position of the double bond(s) are indicated as, for example, *cis*-Δ^9-octadecenoic acid for oleic, in which Δ^9 refers to the carbon of the chain at which the double bond starts.

Saponification

Fatty acid esters can be readily hydrolyzed with aqueous hydroxide by a process known as **saponification,** so called because the carboxylic acid salts formed in this way are the constituents of soap. This reaction has been known for thousands of years, consisting at first of the treatment of animal fats with an alkaline solution obtained by soaking ashes in water.

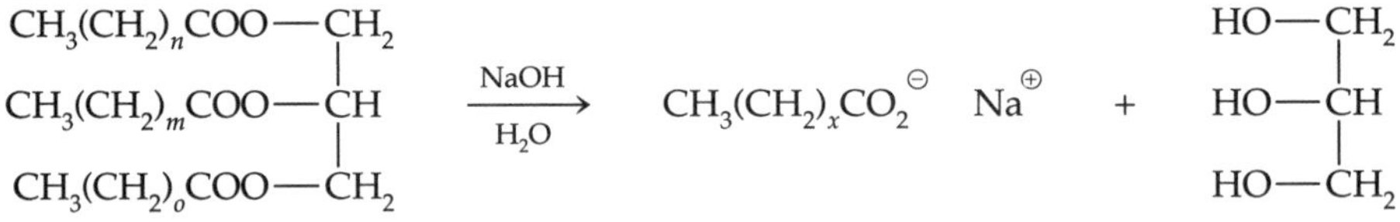

$$CH_3(CH_2)_nCOO-CH_2$$
$$CH_3(CH_2)_mCOO-CH \xrightarrow[H_2O]{NaOH} CH_3(CH_2)_xCO_2^{\ominus}\ Na^{\oplus}\ +\ HO-CH$$
$$CH_3(CH_2)_oCOO-CH_2 \qquad\qquad\qquad\qquad HO-CH_2$$

Fats   **Salts of fatty acids**   **Glycerol**

EXERCISE 17-B

Hydrolysis of carboxylic acid derivatives takes place by nucleophilic acyl substitution by water under both acidic and basic conditions (recall Chapter 12). Write a complete, detailed reaction mechanism for the hydrolysis of a carboxylic acid ester in the presence of sodium hydroxide. Now do the same for the reaction that would take place in water in the presence of hydrochloric acid. Only one of these processes is catalytic. Which is it? Explain why the other is not.

EXERCISE 17-C

The monoesters of 1,2- (and 1,3-) diols rapidly equilibrate, as illustrated at the top of the next page for the two isomeric structures shown. Write a detailed mechanism for this interconversion. Would you expect this transformation to be faster or slower than the hydrolysis of a carboxylic acid ester in the presence of base? Explain.

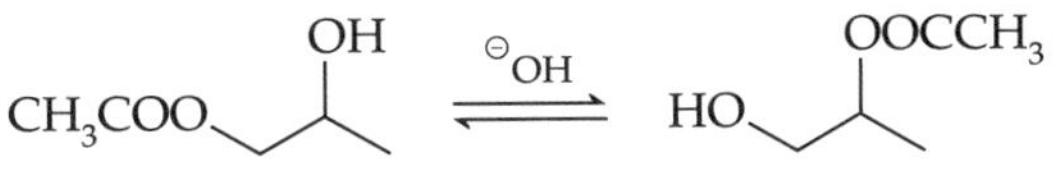

$$\text{CH}_3\text{COO} \overset{\text{OH}}{\diagdown} \underset{\ominus \text{OH}}{\overset{\ominus \text{OH}}{\rightleftharpoons}} \text{HO} \overset{\text{OOCCH}_3}{\diagdown}$$

Micelles

Salts of long-chain fatty acid carboxylic acids (carboxylate anions) have an unusual property that results from the presence in a single molecule of both a **hydrophobic** component (the hydrocarbon unit) and a **hydrophilic** one (the carboxylate). Because of the hydrophilic groups, the salts of long-chain fatty acids dissolve readily in water, but in so doing they form structures known as **micelles,** which are globular collections of many molecules in which most of the carboxylates are on the surface, as in Figure 17-1. The carboxylate "heads" are located on the surface of the micelle where they can interact with the surrounding water, whereas the hydrocarbon "tails" are intermixed in the center. This central region is very hydrocarbonlike and is held together by van der Waals attractive interactions, like those discussed in relation to polymers in Chapter 16. Micelles are roughly spherical simply because it is this shape that has the maximum volume-to-surface ratio: in contrast, the hydrocarbon chains are quite disordered in the interior because, although the chains have the same "diameter" along their length, the volume available to accommodate them decreases toward the center of the sphere. Because the central region of a micelle is structurally ill defined,

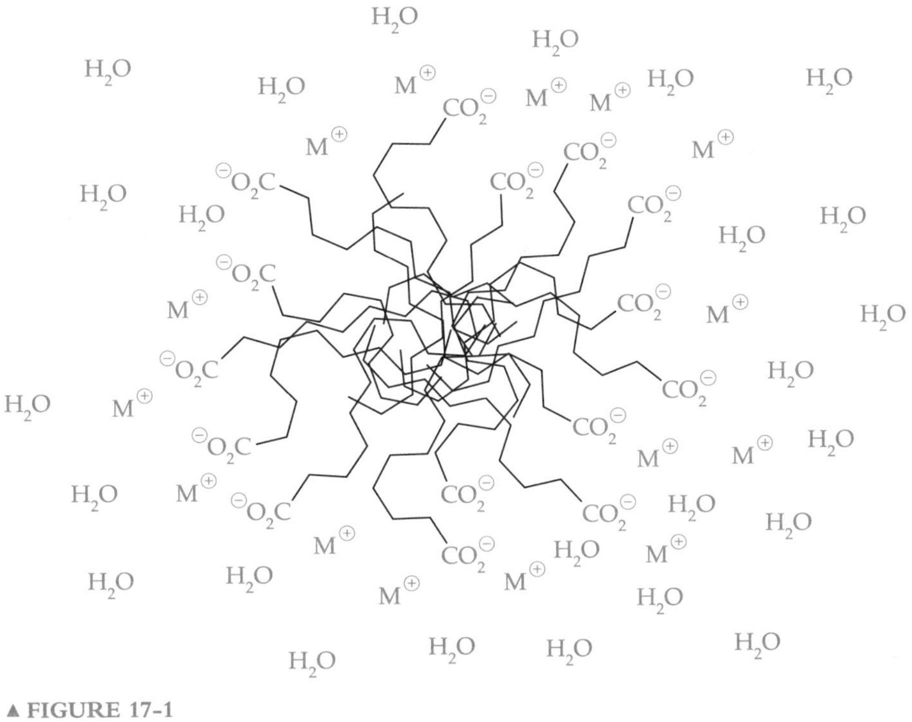

▲ FIGURE 17-1

At concentrations above a critical value, long-chain fatty acids associate to form a micelle. The hydrocarbon tails of these molecules are directed toward the core of the micelle, making the core hydrocarbonlike and, hence, hydrophobic. In contrast, the polar carboxylate ion ends of the acids are directed toward the external water phase, producing a highly polar interface between the micelle and the aqueous solution. This interface contains both the carboxylate anion and its counterion, M^+, as well as substantial quantities of solvating water molecules. This polar head region is therefore highly hydrophilic.

ORGANIC CHEMISTRY IN THE COMPUTER INDUSTRY

As scientists learn more about our complex environment and ecosystem, it sometimes becomes necessary to re-evaluate the use of some chemicals. For example, chlorocarbons such as trichloroethane have many industrial uses because they have a good balance of solvent properties for both hydrocarbonlike and polar materials, are relatively nontoxic, and do not readily burn. They have been used as cleaning agents both in the dry-cleaning business and in the computer industry, where the removal of all contaminants is essential for the production of microcomputer chips containing billions of circuits.

Unfortunately, some of these solvents inevitably escape into the atmosphere where they ultimately migrate to the upper layers and contribute to the destruction of the ozone layer. Recently, AT&T has experimented with the replacement of tricloroethane with other solvents as cleaning agents for chip manufacture and discovered that *n*-butyl butyrate is a satisfactory substitute. Although the ester used by AT&T is synthetic, *n*-butyl butyrate is found naturally in many fruits, including cantaloupes, peaches, plums, and pineapples, and has a smell that is characteristic of quite ripe bananas.

other hydrocarbonlike molecules can readily dissolve in this region. For example, soaps solubilize nonsaponified fats and other water-insoluble materials in the interiors of micelles and thus act as cleaning agents, removing the oily materials that bind dirt to clothing and to our bodies.

The calcium salts of long-chain carboxylates are not water soluble and form a scummy film as they precipitate. Because many ground sources of water pass through limestone deposits and thus have high concentrations of calcium ions, soaps have been replaced for most cleaning purposes by detergents that are similar in structure but are based on sulfonic acids. The calcium (and other metal) salts of sulfonic acids are more soluble in water than are carboxylic acid salts. In the past, phosphates also were added to

$SO_3^{\ominus}$ $Na^{\oplus}$

$CH_3(CH_2)_n$

**Sodium alkyl-
benzene sulfonate**

$CH_3(CH_2)_nSO_3^{\ominus}$ $Na^{\oplus}$

**Sodium alkyl
sulfonate**

BUBBLES

The foaming that we associate with soaps and detergents is not connected with micelle formation; thus, foaming cleaners are not necessarily more effective in removing dirt. Nonetheless, people expect soaps to foam and thus chemists working in the detergent industry often formulate their products to foam in order to obtain better acceptance in the market place. In some circumstances, foaming can be harmful, as, for example, in dishwashers and clothes washers, which lack the space to hold the foam. Foaming can also denature protein molecules, which unfold from their native structures in order to stretch from one side of a bubble to the other. For this reason, vials of insulin suspensions should be rolled between the hands for mixing rather than shaken. Perhaps James Bond, in insisting that his vodka martinis be "shaken, not stirred," wanted to be sure that any enzymes present were denatured.

detergents to solubilize calcium salts. This practice is currently banned for most cleaning agents (dishwashing detergents being one exception) because phosphate serves as a source of phosphorus, resulting in the overgrowth of algae and other undesirable plants in streams, rivers, and lakes.

Bilayer Membranes

The fundamental structure of a micelle is the result of the formation of distinctly different hydrophobic and hydrophilic regions. Comparable arrangements are also found at many of the boundary regions, known as **membranes,** present in living cells. Membranes are found on the outer surface of a cell, separating it from the external aqueous medium. In addition, membranes form the boundaries of internal structures of the cell such as the cell nucleus, mitochondria, and endoplasmic reticula.

The fundamental building blocks of membrane are **phospholipids,** which are quite similar in structure to the carboxylates of soaps. Phospholipids are dicarboxylate, monophosphate esters of glycerol (Figure 17-2). The phosphoric acid unit is itself a diester, monoesterified on one side by one of the hydroxyl groups of glycerol and on the other by one of a number of alcohols that also contain amino groups. In aqueous solution near pH 7, the phosphoric acid is ionized and the amine protonated. The resulting zwitterion is a very polar species that is quite hydrophilic. In contrast, the other end of the molecule bears the long-chain hydrocarbon tails of the saturated and unsaturated fatty acids and is quite hydrophobic. The constituent molecules are free to twist, rotate, and generally reorganize within the membrane. However, the tight contact of the aminophosphate groups with the external, aqueous environment prevents these groups from penetrating into the hydrocarbon interior of the membrane.

In a cell membrane, these phospholipids form a structure known as a **fluid mosaic** arrangement, which is basically a **bilayer.** This structure is formed by two layers that point their hydrocarbon tails toward each other. Figure 17-3 is a schematic rendering of a lipid bilayer structure formed from phospholipids; the polar head groups are shown as spheres. This lipid bilayer structure was first proposed in 1925 by two Dutch chemists, E. Gorter and F. Grendel. They based their structural suggestion on the observation that twice as much lipid is present in the total mass of a red blood cell as would be required to cover it with a single layer. The bilayer is held together both by van der Waals interactions between the hydrocarbon tails of the phospholipids and by the association of the polar head groups with water.

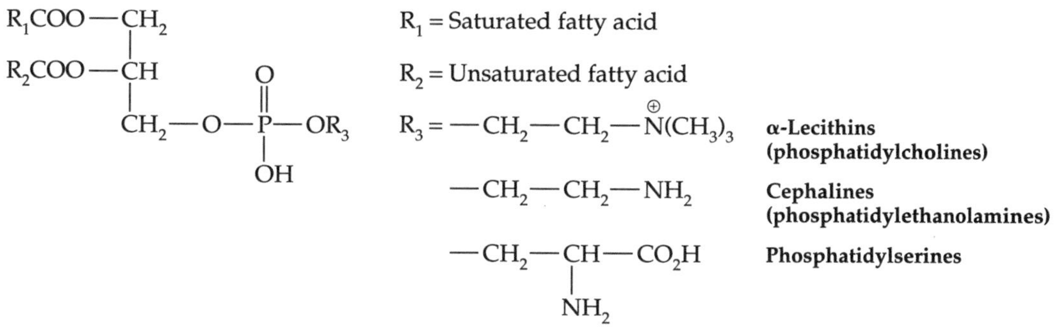

$R_1COO—CH_2$

$R_2COO—CH$

$CH_2—O—\overset{\overset{O}{\|}}{\underset{\underset{OH}{|}}{P}}—OR_3$

$R_1 =$ Saturated fatty acid

$R_2 =$ Unsaturated fatty acid

$R_3 = —CH_2—CH_2—\overset{\oplus}{N}(CH_3)_3$ **α-Lecithins (phosphatidylcholines)**

$—CH_2—CH_2—NH_2$ **Cephalines (phosphatidylethanolamines)**

$—CH_2—\underset{\underset{NH_2}{|}}{CH}—CO_2H$ **Phosphatidylserines**

▲ FIGURE 17-2

Phospholipids are esters of glycerol, $HOCH_2CH(OH)CH_2OH$, in which two of the ester bonds are to fatty acids and the third is to a phosphoric acid derivative.

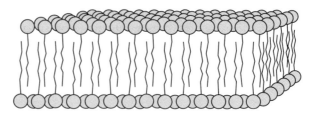

▲ FIGURE 17-3

A lipid bilayer is formed when the hydrocarbon tails of a fatty acid or phospholipid are directed together, away from bulk water, whereas the polar head groups are exposed to the aqueous solution. The sandwich structure in a bilayer is a three-dimensional array for self-organization that is a structural alternative to the roughly spherical structures of micelles.

Although the sum of these interactions provides sufficient energy to stabilize the bilayer structure, each one independently is of small energetic value. Hence, the three-dimensional structure of the bilayer is flexible and can be readily distorted and reshaped in much the same way as polyethylene, discussed in Chapter 16. The name *fluid mosaic* derives from this lack of rigidity in the bilayer structure.

The bilayer structure of the membrane of cells is complicated by the presence of a number of other molecules that serve very important auxiliary functions (Figure 17-4). For example, there are protein molecules present that have hydrophobic exteriors and internal regions containing several charged amino acid residues. Because of their hydrophobic exterior, these proteins "dissolve" readily in the hydrophobic interior of the phospholipid bilayers and function as "channels" through which polar or charged mole-

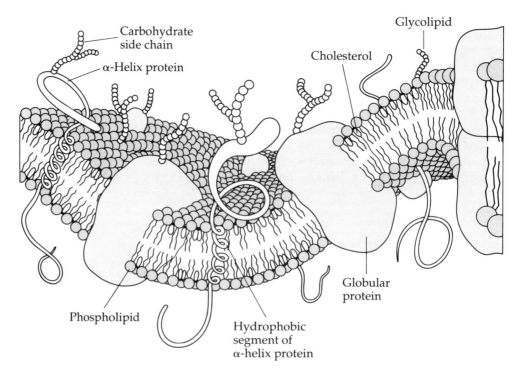

Carbohydrate side chain

α-Helix protein

Glycolipid

Cholesterol

Phospholipid

Hydrophobic segment of α-helix protein

Globular protein

▲ FIGURE 17-4

Large molecules that span the thickness of the lipid bilayer are active in the transport of neutral molecules or ions from one side of the membrane to the other. Often these channels are composed of helical peptides, sometimes containing a large number of ionized (charged) side chains. Such peptides are particularly effective at transporting ions and small drug molecules.

cules can move from one side of the membrane to the other. Collections of these protein molecules form complex tubes connecting one side of the membrane with the other. Without these channels, it would not be possible for polar biological molecules and metal ions to pass through the membrane and into the interior of the cell. Because of their high polarity, these small biomolecules are coated with a very tightly held water shell, which in turn is further hydrogen bound to the surrounding water. These water molecules would have to be shed and the hydrogen bonds broken for these polar molecules to be accepted into the hydrocarbon inner region of the bilayer.

When a bilayer is extended to form a closed surface, the resulting structure is called a **vesicle,** a cutaway view of which is shown in Figure 17-5. Vesicles and micelles hold promise as potential vehicles for the delivery of pharmaceutical agents. Many possible therapeutic agents, though quite effective when delivered to a specific site of action, either are destroyed by enzymes or are toxic elsewhere in the body. For example, peptide hormones are known to control essential functions such as the reproductive cycle, and other peptides such as insulin play important regulatory roles as catalysts, controlling key biochemical pathways. However, it is not possible to correct a deficiency of these peptides by administering them orally because enzymes in the stomach, called **peptidases,** rapidly cleave peptide bonds as part of the natural process of digestion. A peptide hormone "encapsulated" inside a micelle or vesicle may be protected by this coating until the micelle, with its contents, has passed through the stomach wall.

EXERCISE 17-D

It is important that we understand how little material is required to form a bilayer membrane. Assume that a molecule is approximately cubic and measures 20 Å on a side. How many of these molecules would it take to build a cube that is 1/1000 of a millimeter on a side (approximately the size of a human red blood cell)? How many molecules are on each of the surfaces of the cube? What fraction of the total volume occupied by the component molecules is present on the surface? (1 Å = 10^{-7} mm.)

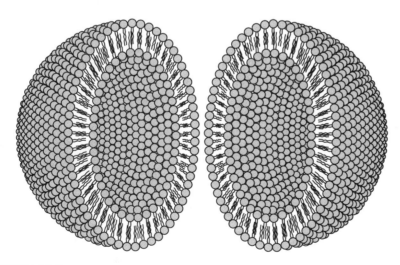

▲ FIGURE 17-5
When an extended bilayer folds on itself, it forms a vesicle, a floppy, spherical array (shown here cut in half so that the inside is visible). A vesicle contains an inner core of aqueous solution separated by the bilayer membrane from the bulk aqueous solution in which the vesicle is suspended.

Terpenes

All the lipids described so far have been derived from a linear arrangement of carbon chains. We now turn our attention to another class of lipids known as **terpenes,** three of which are shown in Figure 17-6. They can be distinguished from fatty acids because they are derived biochemically from a branched five-carbon unit, **isopentenyl pyrophosphate.** (Terpenes are sometimes called **isoprenoids** because they formally consist of skeletal oligomers of isoprene, 2-methyl-1,3-butadiene.) As a result, most terpenes have a carbon count that is a multiple of five (5, 10, 15, etc.) and have methyl groups as substituents, although others lack one or more of the expected number of carbons. Another distinction between the fats and waxes and the terpenes is that fats and waxes are universally composed of molecules joined by ester linkages, whereas terpenes are not. For this reason, fats and waxes are often referred to as **saponifiable lipids,** whereas terpenes are called **nonsaponifiable lipids.** Furthermore, terpenes often bear extensive oxygen substitution along the chain (as alcohols, carbonyls, or double bonds derived by the dehydration of alcohols). Such oxygen functionality is usually absent in fats or waxes.

The terpenes are an extraordinarily diverse group of natural compounds (there are more than 22,000 known structures), especially given their common origin from isopentenyl pyrophosphate. Their study is made all the more complex because the function of most of the terpenes present in plants is not known. It may be that some of these compounds at one time served as defensive agents against predators (for example, insects) that no longer exist, and the plants that synthesize them have not yet evolved to the point at which they no longer expend the energy consumed in the production of these chemical defense agents. It is also possible that other terpenes are simply chemical by-products of biochemical transformations in plants that are not as well tuned as those in mammals. We will return shortly to a discussion of the importance of several terpenes for which the biological functions have been well established.

Terpenes are subdivided into several classes according to the number of isoprene units that they contain (Figure 17-7). A majority of the terpenes that have been isolated have an even number of these units, although this finding may not correspond to the actual composition in the plant. Thus, **monoterpenes** are molecules with ten carbons, accounting for two isoprene units; **diterpenes** have twenty carbons (four isoprene units); **triterpenes** have thirty carbons (six isoprene units), and **tetraterpenes** have forty carbons (eight isoprene units). In addition, a significant number of terpenes with three isoprene units (fifteen carbons) have been isolated and form the class known as **sesquiterpenes.** A more-limited number of terpenes, known as **sesterterpenes,** are composed of five isoprene units.

There are a number of biologically important compounds that are closely related to isoprenoid terpenes. For example, lanosterol, a biosyn-

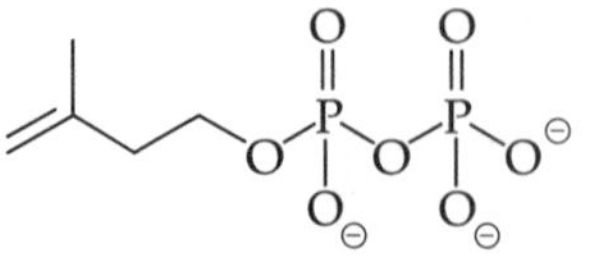

Isopentenyl pyrophosphate 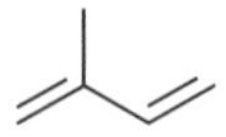   Isoprene 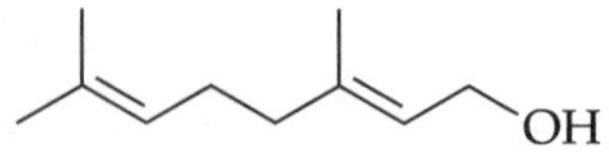   Geraniol

▲ FIGURE 17-6

The carbon skeletons of terpenes are oligomers of isopentenyl pyrophosphate. They can be conceptually related to a simple diene, isoprene (2-methyl-1,3-butadiene). A wide variety of functional groups are found in terpenes, including aldehydes, ketones, alcohols, and alkene groups.

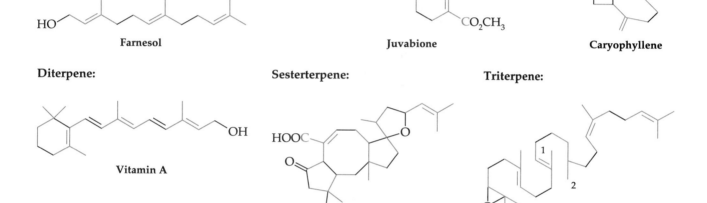

▲ FIGURE 17-7
Representative isoprenoid terpenes containing ten (monoterpenes), fifteen (sesquiterpenes), twenty (a diterpene), twenty-five (a sesterterpene), or thirty (a triterpene) carbon atoms. The isoprene units are indicated in color.

thetic precursor of cholesterol, is a triterpene that cannot be broken into un-rearranged isoprene units. This is because methyl groups migrate from one carbon atom to an adjacent cation in the isoprenoid structure that is its im-mediate biochemical precursor (Figure 17-8). This reaction is one step of a complex biosynthesis of cholesterol from isopentenyl pyrophosphate: sub-stantial contributions toward working out this sequence were made by the American chemist Konrad Bloch, who received the Nobel Prize in medicine in 1964, and by the British chemist John Cornforth, who received the Nobel Prize in chemistry in 1975. Cholesterol is derived from lanosterol by a process that removes three of the methyl groups. As such, cholesterol is not a triterpene because it is constituted of only twenty-seven carbons and is more properly referred to as a **nortriterpene.** The prefix **nor-** is used to refer to a substance whose structure closely resembles that of another but is lack-ing some small feature, usually a methyl group. Carbon degradation processes also exist for other terpenes, and all classes of terpenes have norterpene members.

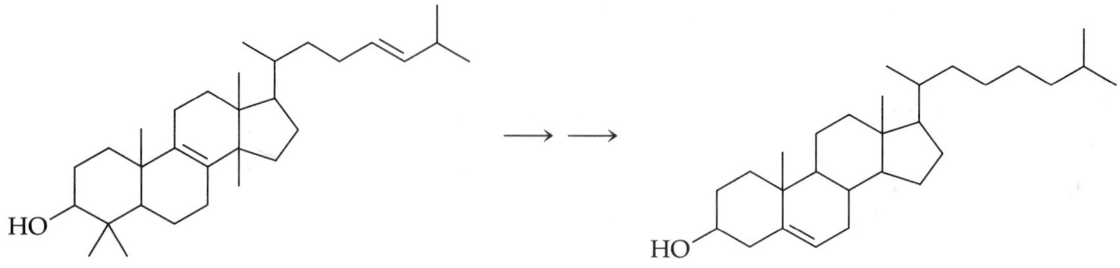

Lanosterol   **Cholesterol**

▲ FIGURE 17–8

Lanosterol is a triterpene that cannot be broken into isoprene units because it is formed through carbocationic rearrangements in which two methyl groups migrate from one carbon to the next. When lanosterol is converted into cholesterol, one of these methyl groups is lost, along with two others, making cholesterol a nortriterpene. (As a diversion, try to break lanosterol into isoprene units to discover which methyl groups are in the "wrong" places. Hint: Start at the upper right.)

Although the biological functions of the terpenes in plants are not generally known, a number of terpene and terpenelike compounds have pronounced biological effects in human beings and other mammals. For example, **nepetalactone** (Figure 17-7) is the constituent of catnip that is pleasing to cats although not all are affected. Indeed, many of the terpenes have what we consider to be pleasant odors and are used extensively in perfumery. In high concentrations, many terpenes such as camphor have a strong, yet not unpleasant, odor and were used as principal constituents of quack medicines and nostrums.

EXERCISE 17-E

Identify the five-carbon fragments corresponding to the isoprene unit in the following terpenes.

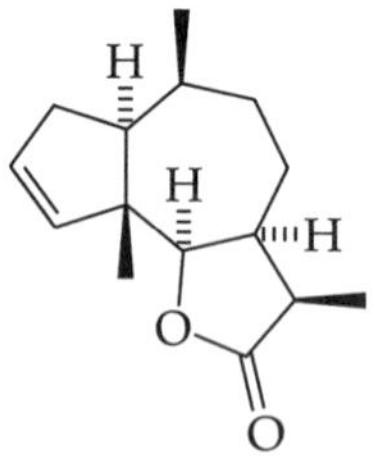

Ambrosin

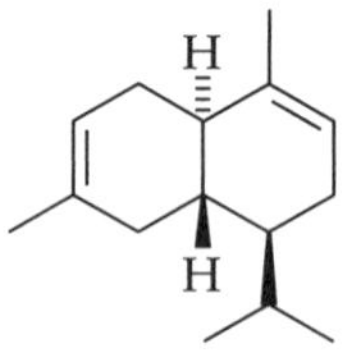

β-Cadinene
(one of many sesquiterpenes from cedar trees)

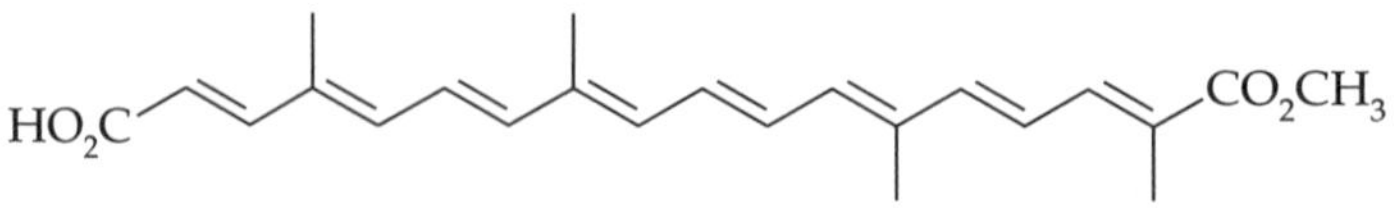

Bixen
(used as a golden yellow food coloring)

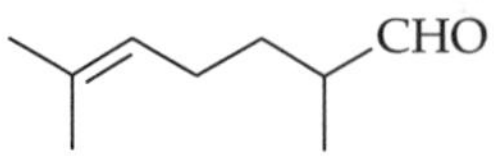

Citronellal
(used as insect repellent)

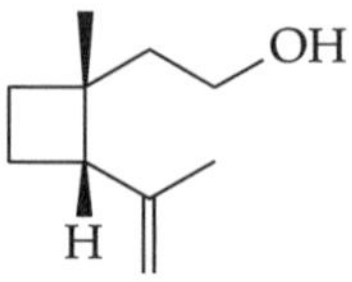

Grandisol
(sex attractant of the cotton boll weevil)

NATURAL INSECTICIDES

Many oxygenated compounds have been isolated from plants that are toxic to insects. Indeed, some of them, such as the pyrethrins (from pyrethrum flowers), are marketed commercially as insecticides. Others, such as azadirachtin (from the seeds *Neem*, a very common tree in India and Burma), are more difficult to isolate from the plant, too complicated to synthesize in quantities sufficient for large-scale use, and therefore quite expensive.

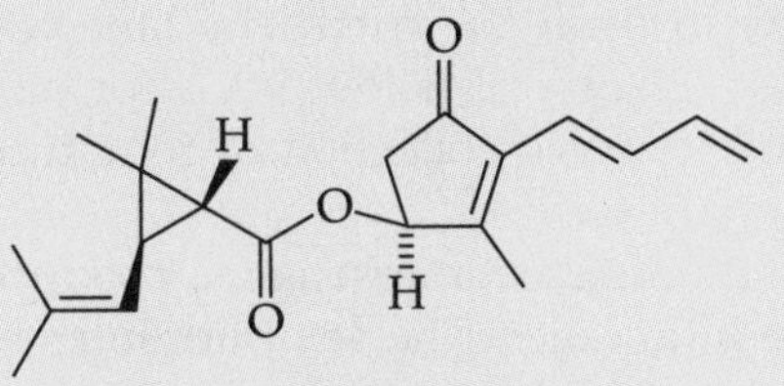

Pyrethrin I

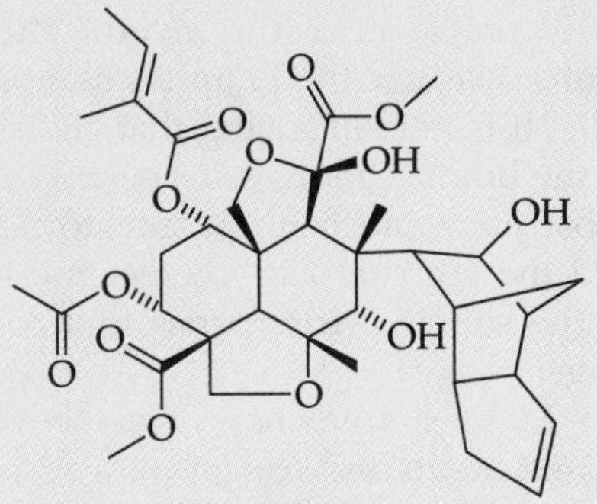

Azadirachtin

A number of terpenes, some much more complex than those shown in Figure 17-8, have been shown to have potent activity in inhibiting the division of cancerous cells. Most of them have proved to be too toxic to be of value in chemotherapy. A notable exception is **taxol** (Figure 17-9), currently the best chemical agent for suppressing the growth of hard tumors. Unfortunately, it is found in only very small amounts in the bark of the Pacific yew tree. Although this tree is not rare, especially in the western United States and Canada, the bark of three trees is required for the isolation of sufficient taxol for the treatment of one patient and its harvesting usually kills the tree. Clinical use of taxol is thus limited by the available supply, and some scientists are concerned that intensive demand for taxol may ultimately result in the extinction of the species.

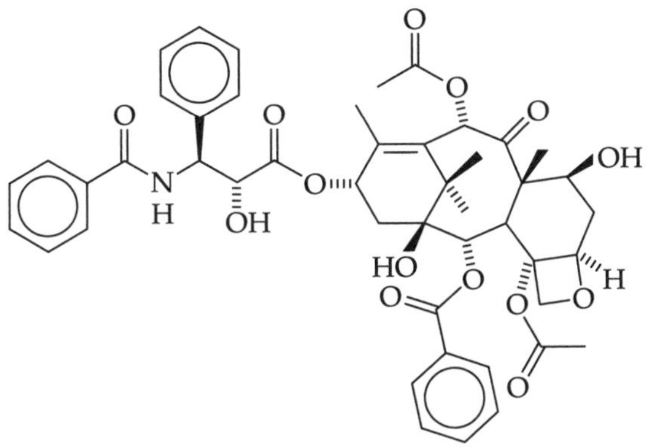

Taxol

▲ FIGURE 17-9
Taxol is a complex, highly functionalized triterpene with potent anticancer activity. It has so far eluded total synthesis from simple laboratory reagents.

Steroids

The term **steroid** is used to define a large group of naturally occurring compounds that have a fused-ring system of three 6-membered rings and one 5-membered ring, as shown in Figure 17-8 for lanosterol and cholesterol. In contrast with the uncertainty surrounding the functions of terpenes formed in plants, the biological functions of many of the terpenes found in animals have been established. For example, the steroid cholesterol, which has a negative connotation in the minds of many because of its high concentration in fatty deposits in arterial walls, is nonetheless an essential material for life. It is a constituent of cell membranes, or lipid bilayers, and is especially prevalent in the **myelin sheaths** that surround and insulate nerve axons. Because the brain is essentially a collection of interconnected nerve cells, it is not surprising that cholesterol constitutes 10% of the dry weight of the brain. The sesquiterpene farnesol undergoes dimerization and many subsequent biochemical transformations to form the triterpene lanosterol and the nortriterpene cholesterol. Indeed, lanosterol is the precursor of all of the steroidal triterpenes that contain this basic collection of four fused rings.

Many steroids have been shown to be chemical messengers, produced in one organ and triggering a response in another. The **sex hormones** are steroids that both determine sexual characteristics and regulate sexual functions (Figure 17-10). They are grouped into three classes: **estrogens** (female); **androgens** (male); and **progestins** (pregnancy hormones). **Estradiol,** an estrogen steroid, is produced by the ovaries and is responsible for secondary female sexual characteristics. The corresponding male sex hormone, **testosterone,** is produced by the testes and is responsible for the secondary male sexual characteristics. These steroids undergo, respectively, reduction and oxidation, before being excreted in the urine, with estradiol forming estrone and testosterone forming **androsterone.**

The progestins are hormones specific to pregnancy (Figure 17-11). For example, **progesterone** stimulates changes in the uterus lining necessary for the implantation of an egg and simultaneously suppresses ovulation. Synthetic steroids—for example, **norethynodrel (Enovid)**—have been devel-

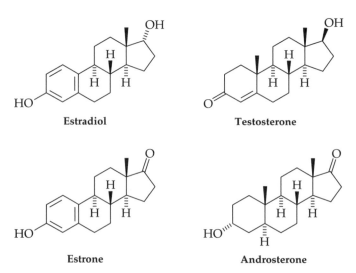

Estradiol

Testosterone

Estrone

Androsterone

▲ FIGURE 17-10

Structures of several steroids that function as sex hormones. Like all steroids, they contain a 6-6-6-5 fused-ring system.

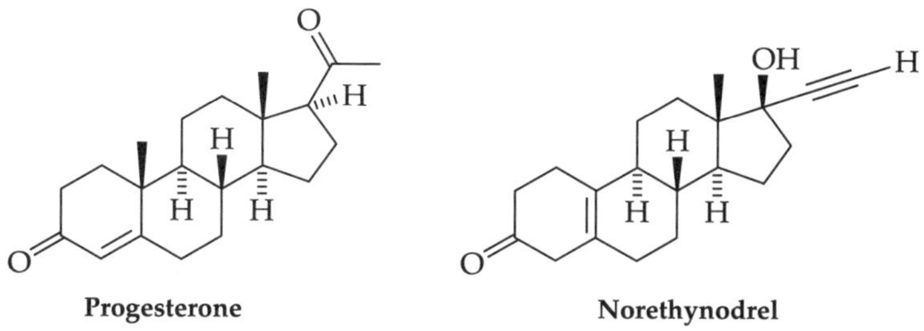

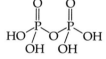

Progesterone   **Norethynodrel**

▲ FIGURE 17-11

The synthetic steroid norethynodrel mimics the hormone activity of the natural steroid progesterone, suppressing ovulation.

oped as effective contraceptives, acting in the same way on the ovaries as progesterone. (Progesterone itself cannot be used as an oral contraceptive because it is readily degraded in the stomach.)

17-2 ▸ Terpene Biosynthesis

Let us briefly consider how terpenes are formed biologically, concentrating on the key carbon–carbon bond-forming processes. We begin with the two major five-carbon building units, isopentenyl pyrophosphate and its isomer **dimethylallyl pyrophosphate.**

Isopentenyl pyrophosphate   **Dimethylallyl pyrophosphate**

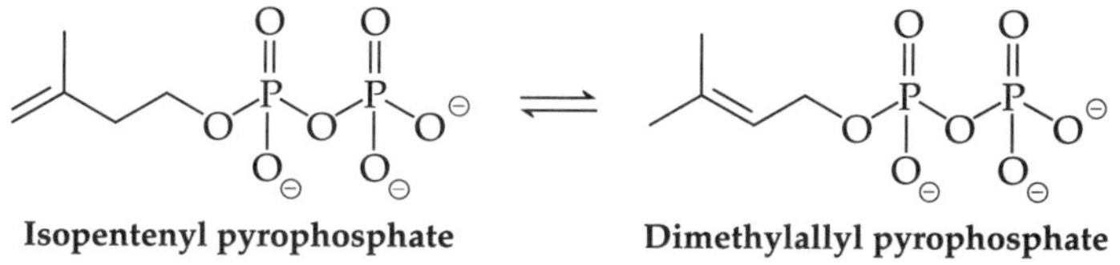

The pyrophosphate group, $-OP(O)(OH)-O-P(O)(OH)-O-$, is so named because phosphoric acid forms pyrophosphoric acid (as well as larger molecules) upon heating. This functionality plays several diverse roles in biochemical reactions. One very important function is the storage of energy; the anhydride linkage reacts with water in an exothermic hydrolysis process quite analogous to the hydrolysis of carboxylic acid anhydrides. (This topic is covered in detail in Chapter 22.) In the biosynthesis of terpenes, the pyrophosphate is a good leaving group.

Pyrophosphoric acid

Isopentenyl pyrophosphate and dimethylallyl pyrophosphate differ only in the position of the double bond. These two isomers are easily interconverted in biological systems, and we can mimic these transformations in the laboratory, using processes that include cationic intermediates (Figure 17-12). Protonation of either isopentenyl or dimethylallyl pyrophosphate leads to the same tertiary carbocation, as long as Markovnikov's Rule (Chapter 10) is followed in each case. In biological systems, an enzyme catalyzes the interconversion, providing a proton from an acidic site.

EXERCISE 17-F

Assuming that isopentenyl and dimethylallyl pyrophosphate are in equilibrium, which of these isomeric structures do you expect to be present in greater concentration? Explain.

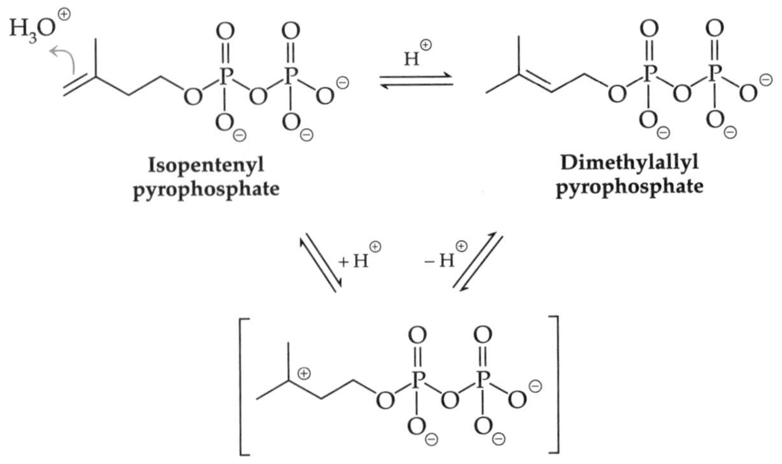

▲ FIGURE 17-12

The tertiary cation formed by protonation of isopentenyl pyrophosphate at C-1 is also formed by protonation of dimethylallyl pyrophosphate at C-3. Therefore, a reversible protonation-deprotonation sequence is a mechanism by which these compounds can be interconverted.

Isopentenyl and dimethylallyl pyrophosphate are bifunctional molecules, both having a pyrophosphate unit in addition to the double bond. Here, the phosphate group serves as a leaving group. The reactivity of the phosphate or pyrophosphate unit as leaving group derives from the stability of the phosphate anion as the conjugate base of a strong acid.

Let us compare the relative rates of hydrolysis of three kinds of esters: those derived from a carboxylic acid, from phosphoric acid, and from sulfuric acid (Figure 17-13). These acids differ greatly in their acidity as a direct consequence of the greater stability of the monoanion of sulfuric acid than those of phosphoric and acetic acids.

The chemistry of the esters derived from these acids also varies quite dramatically. We have already seen in Chapters 12 and 13 that nucleophiles react with carboxylic acid esters by nucleophilic acyl substitution through a tetrahedral intermediate formed by attack at the carbonyl carbon (Figure 17-14). In contrast, sulfate and sulfonate esters, such as tosylates, are derivatives of alcohols that react with nucleophiles at the alcoholic carbon rather

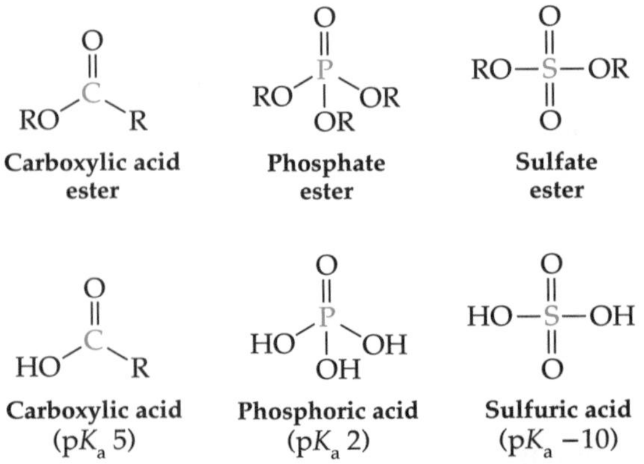

▲ FIGURE 17-13

The structures of carboxylic, phosphoric, and sulfuric acids and their esters are analogous, although the stabilities of their monoanions vary within the series.

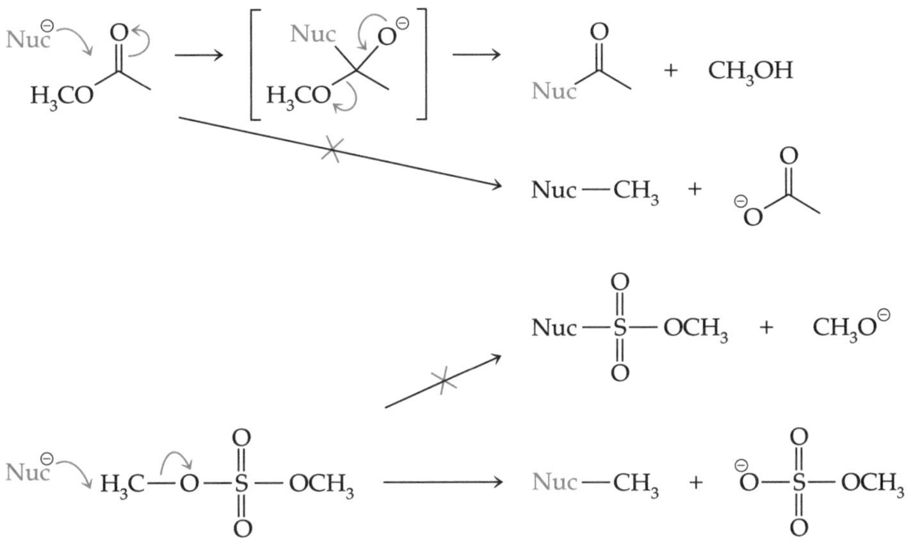

▲ FIGURE 17-14

Nucleophilic attack on a carboxylic acid ester (upper reaction) takes place at the carbonyl carbon, whereas nucleophilic attack on a sulfate ester (lower reaction) takes place on the ester alkyl group.

than at sulfur. This contrasting behavior can be directly attributed to the greater facility of the sulfate and sulfonate anions to act as leaving groups compared with the less-stable acetate anion. Because the acidity of phosphoric acid is midway between that of acetic acid and sulfuric acid, we can understand why phosphate esters undergo both types of reactions (Figure 17-15), leading to products resulting from nucleophilic attack at both phosphorus and carbon.

EXERCISE 17-G

Recalling the factors that affect acidity (Chapter 6), explain the qualitative difference in acidity (and therefore trends in stability of the corresponding conjugate base) of acetic acid, phosphoric acid, and sulfuric acid.

In terpene biosynthesis, the phosphate ester group serves the same role that sulfonate or sulfate esters do for the laboratory chemist in activating carbon toward substitution reactions. Dimethylallyl pyrophosphate is

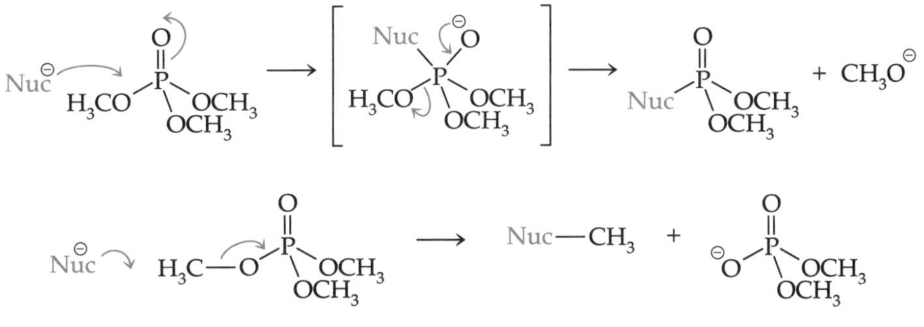

▲ FIGURE 17-15

Nucleophilic attack on a phosphoric acid ester takes place both at phosphorus (upper reaction) and at the alkyl group carbon of the ester (lower reaction).

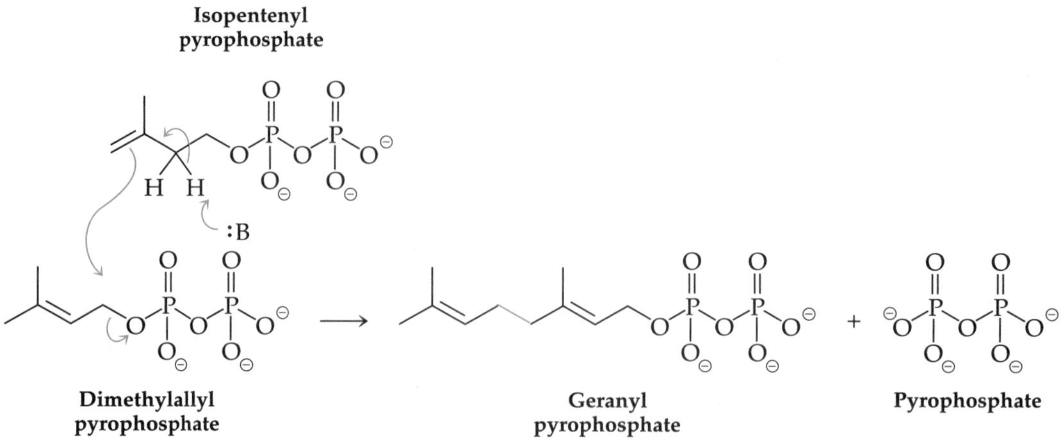

▲ FIGURE 17-16

In the nucleophilic substitution that produces geranyl pyrophosphate, deprotonation of isopentenyl pyrophosphate converts it into a nucleophile that attacks the allylic carbon of dimethylallyl pyrophosphate, displacing pyrophosphate as the leaving group.

especially activated toward substitution at C-1 because that position is allylic to the double bond. Thus, carbon–carbon bond formation results from the reaction of dimethylallyl pyrophosphate with the π bond of isopentenyl pyrophosphate, with simultaneous loss of a proton to a base, as shown in Figure 17-16.

A carbon–carbon bond is also formed between geranyl pyrophosphate and isopentenyl pyrophosphate, resulting in the construction of a fifteen-carbon terpene known as farnesyl pyrophosphate (Figure 17-17). The alcohols corresponding to geranyl pyrophosphate and farnesyl pyrophosphate, **geraniol** and **farnesol,** are found in plants and contribute significantly to the scent of flowers. For example, geraniol has the odor of roses, and both the alcohol and the acetate ester are important materials for the perfume industry.

The carbon–carbon bond-forming process based on the allylic activation of a pyrophosphate group is also important in the formation of cyclic terpenes. For example, nerol pyrophosphate (a geometric isomer of a ge-

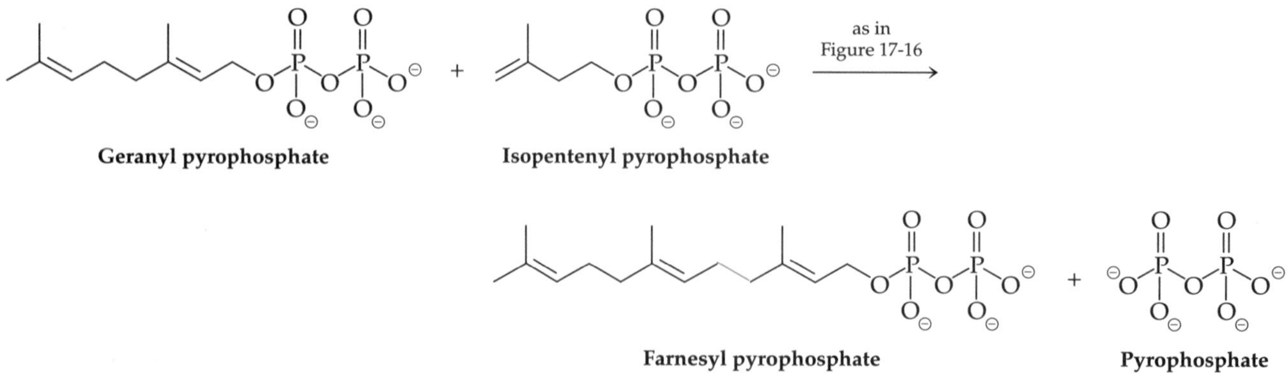

▲ FIGURE 17-17

A reaction similar to that shown in Figure 17-16 results in the addition of a second equivalent of isopentenyl pyrophosphate, displacing the pyrophosphate group from geranyl pyrophosphate and producing farnesyl pyrophosphate.

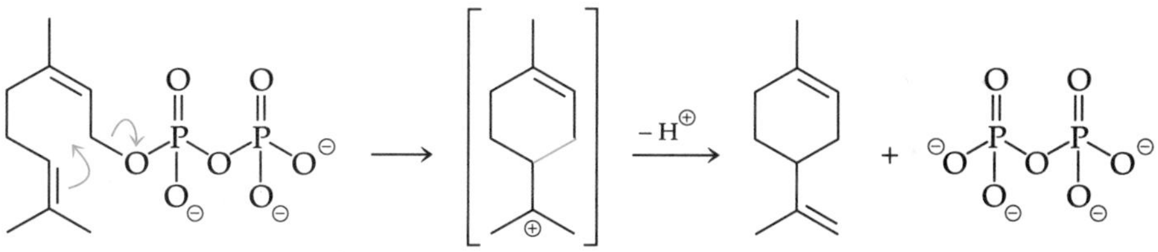

Nerol pyrophosphate   **Limonene**   **Pyrophosphate**

▲ FIGURE 17-18
The loss of pyrophosphate from nerol pyrophosphate is assisted
intramolecularly by the electrons in a remote π bond. A tertiary cation is then
formed. The driving force for the cyclization is the conversion of a C==C π
bond into a C—C σ bond.

ranyl pyrophosphate) undergoes cyclization with loss of the pyrophosphate
group to form an intermediate cation (Figure 17-18). Loss of a proton from
this cation then affords the monoterpene limonene.

EXERCISE 17-H

Write a detailed, step-by-step mechanism for the formation of farnesyl pyrophosphate from geranyl pyrophosphate and isopentenyl pyrophosphate.

EXERCISE 17-I

Limonene is a chiral molecule. First, identify the center(s) of chirality present in limonene. Generally, only one enantiomer of limonene is found in
natural sources. However, limonene can be racemized (the enantiomers interconverted) by strong acid. Explain. (Hint: Consider the chirality of
cations other than the one illustrated in Figure 17-18 that might be formed
by protonation of limonene.)

Carbon–carbon bond formation is also initiated with other functional
groups. For example, squalene monoepoxide undergoes cyclization to form
a six-membered ring at the same time as the three-membered ring of the
epoxide is opened, as shown in Figure 17-19. The resulting intermediate
cation undergoes further cyclization reactions in which all the remaining
alkene π bonds are replaced by new carbon–carbon σ bonds. In biological
systems, this process takes place within a single enzyme and no products
that would result from any of the intermediate cations have been isolated.
Although it was once believed that the transformation of an acyclic molecule into a tetracyclic one takes place in a single step and includes only one
transition state, it is now thought that it is more likely to happen in discrete
stages through a number of different cation intermediates. Many attempts
to effect the same transformation in the absence of enzymes have led to
products with, at most, three of the rings.

EXERCISE 17-J

Recalling the discussion of the thermodynamics of the polymerization of
ethylene to polyethylene in Chapter 16, predict $\Delta H°$ for the conversion of
the monocyclic cation illustrated in Figure 17-19 into the tetracyclic cation,
assuming that no ring strain is added.

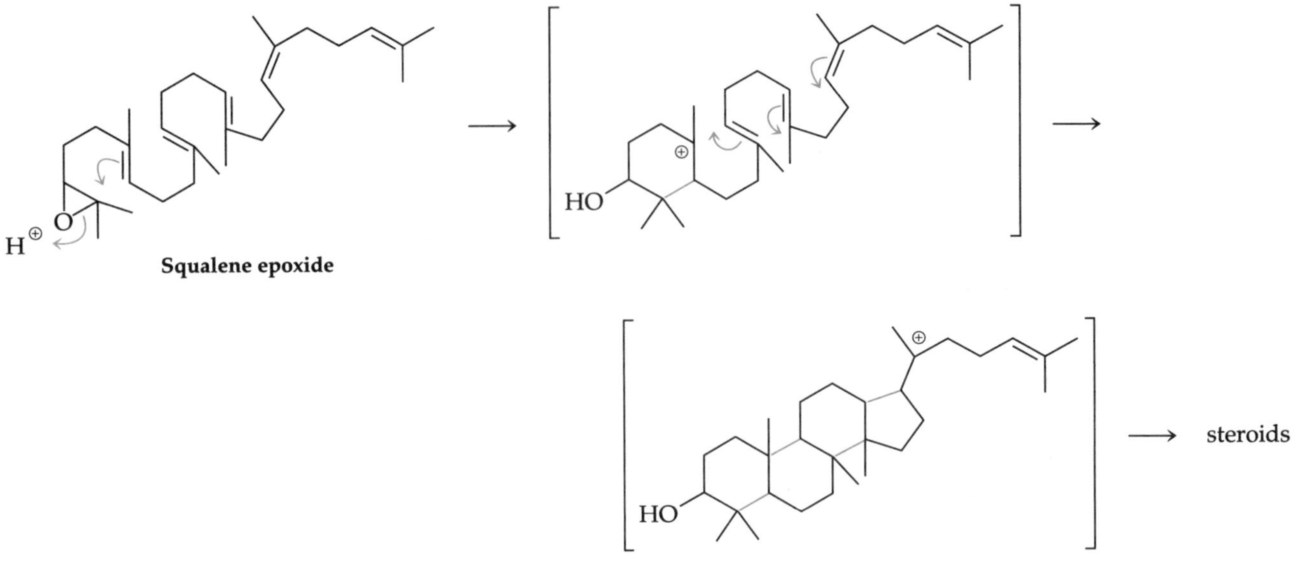

▲ **FIGURE 17-19**
The first cyclization of squalene monoepoxide requires the ring opening of a protonated epoxide assisted by simultaneous carbon–carbon bond formation with the electrons of a π bond. The resulting tertiary carbocation undergoes a further cyclization, again forming a carbon–carbon σ bond and breaking a π bond. Two further cyclizations result in a carbon cation with the steroid ring system (three 6- and one 5-membered rings).

17-3 ▸ Carbohydrates

$C_m(H_2O)_n$

Carbohydrate

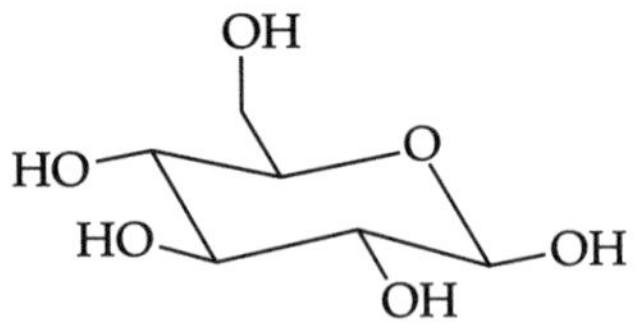

Glucose

$C_6H_{12}O_6 = C_6(H_2O)_6$

The natural materials considered so far have been mainly hydrocarbonlike, with few functional groups. In contrast, the *carbohydrates* generally have one oxygen for each carbon and, as mentioned in Chapter 16, their name implies that they are formally hydrates of carbon. For example, the formula of glucose, $C_6H_{12}O_6$, can alternatively be written as $C_6(H_2O)_6$. We have already encountered glucose as a constituent of the important biopolymers starch and cellulose. As noted in Chapter 16, glucose is stable in the cyclic hemiacetal form shown in the margin. It is appropriate to briefly examine this functionality of glucose because it is present in nearly all carbohydrates.

In Figure 17-20, a simple, cyclic hemiacetal undergoes ring opening to form a noncyclic hydroxyaldehyde. These structures are quite analogous to those available for glucose and, in fact, like the natural carbohydrate, this simpler system favors the cyclic form. This is generally the case for hydroxyaldehydes and hydroxyketones in which there is not undue strain in the cyclic hemiacetal (or hemiketal); that is, when the cyclic form is either a five- or a six-membered ring. This preference contrasts with what we would find for the combination of an aldehyde (or ketone) with a separate molecule containing a hydroxyl functional group.

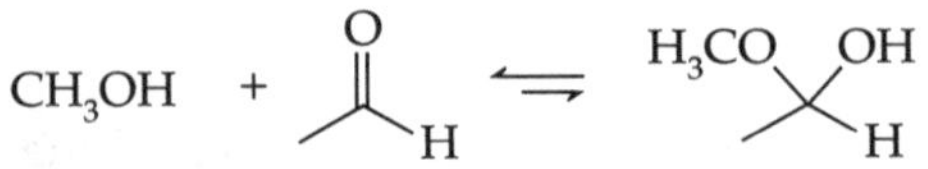

In this case, the separate molecules of alcohol and aldehyde are more stable than they are when combined into a hemiacetal. All of the reasons for the relatively greater stability of a cyclic hemiacetal than the corresponding hydroxyaldehyde and of the noncyclic hemiacetal than the separate alcohol

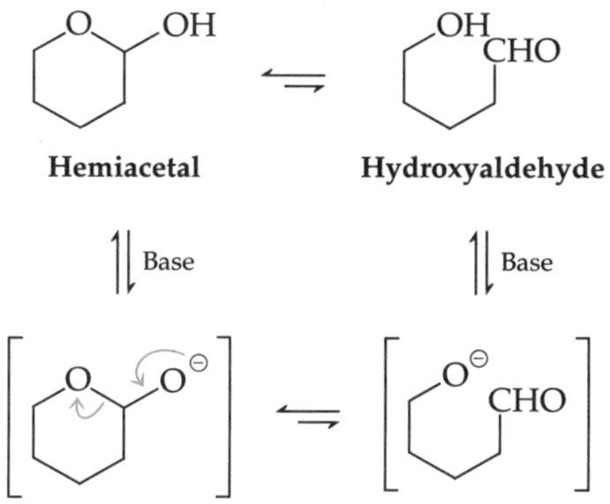

Hemiacetal   **Hydroxyaldehyde**

▲ FIGURE 17-20

The conversion of a hemiacetal to the ring-opened hydroxyaldehyde in basic solution involves deprotonation to form an alkoxy anion. This anion then undergoes cleavage of the carbon–oxygen bond of the ring with simultaneous formation of the carbon-oxygen π bond of an aldehyde. Protonation of the alkoxide produces the ring opened, hydroxyaldehyde.

and aldehyde are not clear, although entropic factors are undoubtedly important (two molecules must be joined to form a noncyclic hemiacetal).

EXERCISE 17-K

The interconversion of an acyclic hydroxyaldehyde and a hemiacetal, illustrated in Figure 17-20 as catalyzed by base, can also be effected by acid catalysis. Write a clear reaction mechanism for the acid-catalyzed conversion.

17-4 ▸ Classification of Sugars

Carbohydrates with three, four, five, and six carbons are called **trioses, tetroses, pentoses,** and **hexoses,** respectively, and representatives of each of these classes are commonly found in nature. Notice that all of the names of the simple carbohydrates end in **-ose,** a common ending for this class of natural materials.

Trioses

D-Glyceraldehyde and dihydroxyacetone are the simplest examples of trioses. The **stereochemical designator** D refers to an arrangement about a center of chirality that matches the three dimensional arrangement in D-glyceraldehyde (that is, *R*). Thus, this compound serves as the three-dimensional reference point for all similarly functionalized molecules. Glyceraldehyde, like other carbohydrates, can be depicted by a Fischer projection.

Trioses

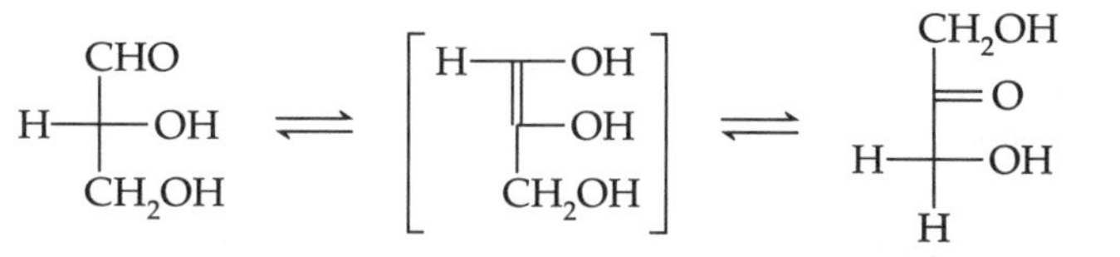

D-Glyceraldehyde   **Enediol**   **Dihydroxyacetone**
                          intermediate

Although Fischer projections do not adequately convey the three-dimensional shapes of molecules, they are a ready means for comparing stereochemistry. (Recall that, in a Fischer projection, the vertical bonds to a center of chirality recede away from the viewer, whereas the horizontal bonds come toward the viewer.)

These two simple carbohydrates are also the simplest members of two classes into which all carbohydrates are subdivided. Glyceraldehyde and other carbohydrates containing an aldehyde carbonyl group are referred to as **aldoses,** whereas those with a ketone are called **ketoses.** Thus, these two sugars are also referred to as an **aldotriose** and a **ketotriose.** Note that upon enolization by treatment with acid or base, both of these trioses form the same enediol. Although this enediol is less stable than either of the carbonyl forms, it does serve as an intermediate for the conversion of one carbohydrate into the other. This process can be readily catalyzed by base or acid, and we shall encounter this transformation again in Chapter 22.

EXERCISE 17-L

Using curved arrows to indicate the flow of electrons, write a clear, detailed reaction mechanism for the acid-catalyzed conversion of glyceraldehyde into dihydroxyacetone. Do the same for the base-catalyzed reaction.

Aldotetroses

Both D-erythrose and D-threose, common tetroses, contain two chiral centers.

Tetroses

$$
\begin{array}{cc}
\text{CHO} & \text{CHO} \\
\text{H}\!-\!\!-\!\text{OH} & \text{HO}\!-\!\!-\!\text{H} \\
\text{H}\!-\!\!-\!\text{OH} & \text{H}\!-\!\!-\!\text{OH} \\
\text{CH}_2\text{OH} & \text{CH}_2\text{OH} \\
\textbf{D-Erythrose} & \textbf{D-Threose}
\end{array}
$$

They differ only in the configuration at C-2 (counting from the aldehyde end of the molecule). Because there are two centers of chirality in these molecules, there are indeed a total of four possible stereoisomers. However, only the two shown are found in nature. Here, as in almost all naturally occurring carbohydrates, the stereochemistry at the chiral center most remote from the aldehyde end has the D configuration: that is, the same as in D-glyceraldehyde.

EXERCISE 17-M

Determine whether erythrose and threose can be interconverted by reversibly forming the enol. Why is or is not this possible?

Aldopentoses

Because only D-carbohydrates generally occur naturally, only half the number of possible stereoisomers are actually found. As a consequence, we would expect only four D-pentoses, despite the fact that three chiral centers would allow for a total of eight (2^3) possible stereoisomers. These pentoses, shown in Figure 17-21 are **D-ribose, D-arabinose, D-xylose,** and **D-lyxose.** Of

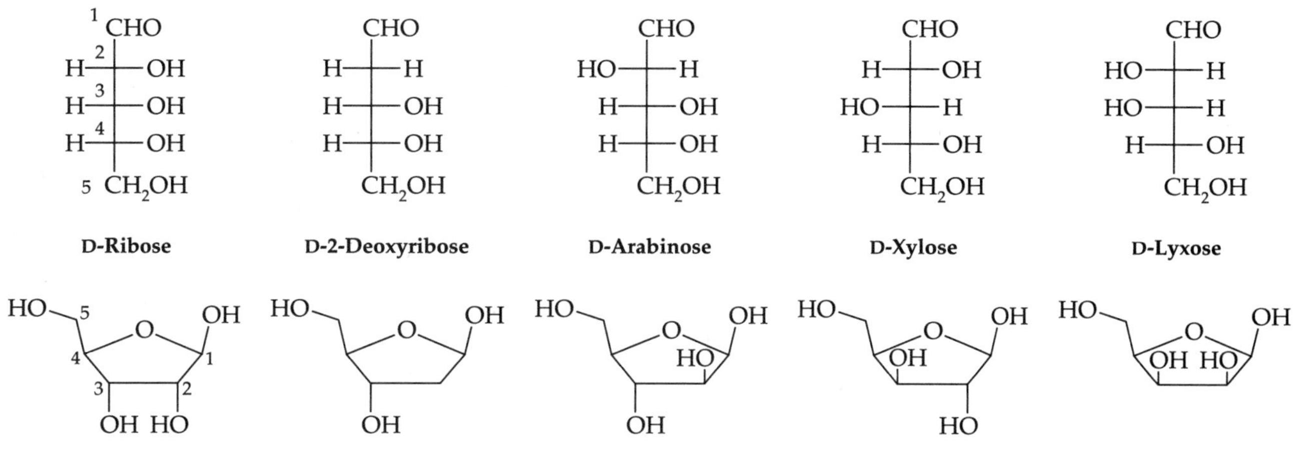

CHO — D-Ribose

^{1}CHO
H—^{2}—OH
H—^{3}—OH
H—^{4}—OH
$_5$ CH$_2$OH

D-Ribose

CHO
H——H
H——OH
H——OH
CH$_2$OH

D-2-Deoxyribose

CHO
HO——H
H——OH
H——OH
CH$_2$OH

D-Arabinose

CHO
H——OH
HO——H
H——OH
CH$_2$OH

D-Xylose

CHO
HO——H
HO——H
H——OH
CH$_2$OH

D-Lyxose

▲ FIGURE 17-21

The four possible pentoses are represented here both as Fischer projections and as ring-closed hemiacetal structures. Deoxyribose differs from ribose in lacking an OH group at C-2.

them, only the first three are found to any extent in nature. Xylose is the second most prevalent simple sugar, after glucose. It is found primarily in a polymer that is very similar to cellulose (Chapter 16). Arabinose is an exception to the general rule that naturally occurring carbohydrates are of the D-configuration, because significant amounts of the L form also occur in nature. A fifth pentose of great importance is D-2-deoxyribose; it is a significant constituent of the biopolymer **deoxyribonucleic acid (DNA),** as is ribose in the **ribonucleic acids (RNA).** The prefix **deoxy-** means that an oxygen functional group (in this case, an OH group) has been replaced by a C–H bond.

The five common pentoses in Figure 17-21 are shown both in their acyclic forms as Fischer projections and in their cyclic, hemiacetal forms. In each case, the cyclic hemiacetal has one more chiral center than is present in the noncyclic arrangement. Thus, there are two diastereomeric cyclic forms of each of the pentoses (only one of which is shown in Figure 17-21), just as for the hexoses. The **β form** drawn in Figure 17-21 is defined for ribose, as well as all carbohydrates, as that stereoisomer in which the hydroxyl group at C-1 is *cis* relative to the last carbon of the chain (in this case, C-5), as in Figure 17-22. Alternatively, the **α form** has the opposite configuration at

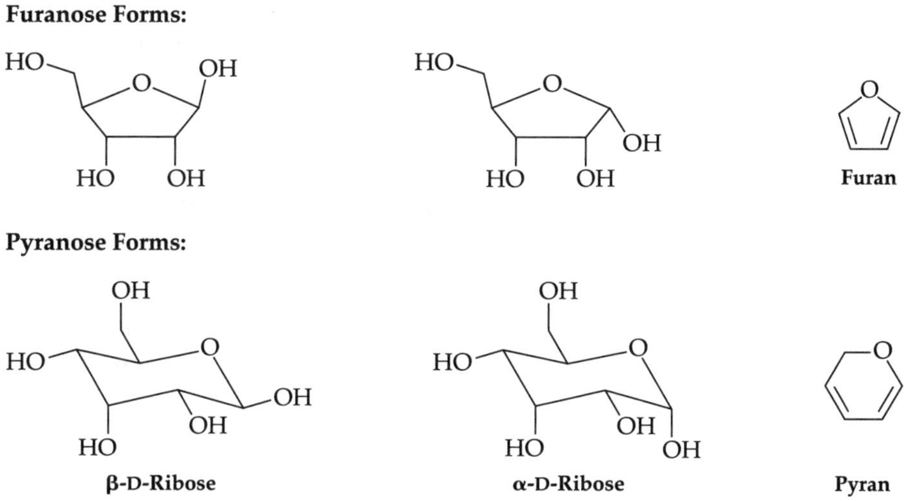

Furanose Forms:

Pyranose Forms:

β-D-Ribose   **α-D-Ribose**   **Pyran**

Furan

◄ FIGURE 17-22

The α and β forms of cyclic hemiacetals differ in configuration at C-1. In a furanose form, a five-membered ring is present, whereas a pyranose form includes a six-membered ring.

C-1, the hemiacetal carbon. Consequently, the α form is that in which the hydroxyl group at C-1 is *trans* to the last carbon of the chain. The β and the α forms of glucose are chiral diastereomeric molecules, and each occurs in nature as only a single enantiomer. These α and β cyclic forms differ in configuration only at C-1, and such isomers are called **anomers.** We might expect that each form would have a unique specific rotation and, indeed, this is the case. The specific rotation of the α form of glucose is +19° and that of the β form is +112°. When each of these pure substances is dissolved in water, it undergoes a relatively slow process of uncatalyzed interconversion with its anomer, resulting in the same equilibrium mixture from both isomers. Because this process produces a change in rotation from that of either pure substance to that of the equilibrium mixture, the transformation is known as **mutarotation.**

EXERCISE 17-N

Write a complete, detailed reaction mechanism for the mutarotation of glucose in acid solution. Do the same for the reaction catalyzed by base.

These cyclic, five-membered-ring hemiacetals are known as **furanoses,** a name derived from the simple compound furan. Alternatively, the oxygen on C-5 can be bound to C-1, forming a six-membered-ring hemiacetal, which also has α and β stereoisomers. Cyclic hemiacetals in the carbohydrate series having six-membered rings are named **pyranoses** after the simple parent pyran. Although ribose itself in aqueous solution is present mainly (76%) in the pyranose form, the free carbohydrate constitutes only a small fraction of the ribose found in nature. In the vast majority of cases, ribose is joined with other molecules and is in the β-furanose form.

Aldohexoses

Hexoses, with a total of six carbons, have four stereocenters. As with the smaller, four- and five-carbon carbohydrates, the only common, naturally occurring forms of the hexoses have the D configuration, thus limiting the total number of possible stereoisomers to eight. Of these, only the three illustrated in Figure 17-23—**glucose, mannose,** and **galactose**—are commonly found in nature. Glucose is by far the most prevalent: it is also the only hexose in which (in the β form) all the hydroxyl substituents on the six-membered ring are in equatorial positions. The other two hexoses found in nature, mannose and galactose, differ from glucose at only one chiral center. Thus, in their β forms, all substituents, but one, of these sugars are equatorial.

EXERCISE 17-O

It is convenient to use the D,L notation to describe the stereochemistry of the last stereocenter in carbohydrates. Unfortunately, no similarly convenient system exists for the specification of stereochemistry at the other stereocenters. For practice, assign Cahn, Ingold, Prelog (R, S) designations for stereochemistry at all of the centers of glucose, mannose, and galactose. (Hint: Keep in mind that interchanging the two horizontal substituents in a Fischer projection inverts the stereochemistry of that carbon. Thus, the problem is much simpler than it may at first appear.)

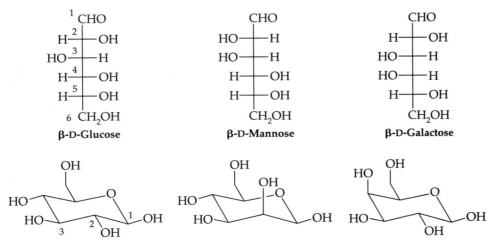

▲ FIGURE 17-23

The three hexoses found commonly in nature are shown here as Fischer projections and as the β anomers of the cyclic, hemiacetal forms.

The interconversion between the β and the α forms of glucose (and of all carbohydrates) is very rapid when catalyzed by either acid or base. Interconversion of these anomers can be accomplished by processes that cleave either the bond between C-1 and the oxygen in the ring or the bond between C-1 and the attached hydroxyl group oxygen. Although the β form, with the hydroxyl group at C-1 in the equatorial position, is the more-stable form, the difference in energy between the α and the β forms is very small, as can be seen by the equilibrium that favors the β form by a factor of only about 2:1.

β-D-Glucose   65 : 35   α-D-Glucose

Glucose can be converted into a derivative, known as a **glucoside,** by replacement of the C-1 hydroxyl by an alkoxy group derived from an alcohol.

β-D-Glucose   β-D-Methyl   35 : 65   α-D-Methyl
                     glucopyranoside   glucopyranoside

In the simple case, in which the alcohol is methanol, the products are known as methyl glucosides. The mechanism for their formation is parallel to the conversion of a hemiacetal into an acetal.

EXERCISE 17-P

Write a detailed mechanism for the formation of β-D-methyl glucoside from β-D-glucose in the presence of acid. At what point in the sequence is the stereochemistry at the anomeric carbon determined? From a consideration of your mechanism, do you think that there would be any difference if the reaction started from α-D-glucose?

At equilibrium, the α anomer of methyl glucoside predominates by a factor of approximately 2:1. The origins of this anomalous preference for the axial disposition of the substituent on the six-membered ring (as well as the relatively small preference for the β orientation of the hydroxyl group in glucose itself) are not well understood. This unusual favoring of the α, or axial, orientation (or, conversely, the disfavoring of the β anomer) is known as the **anomeric effect.**

Ketoses

All of the carbohydrates considered so far have the carbonyl group at the end of the chain and are thus aldehydes. One very important exception to this general structural feature of the carbohydrates is found in fructose, a ketose with the carbonyl group on the second carbon of the chain, as shown in Figure 17-24.

Certainly, we can expect some of the chemistry of fructose to differ from that of the aldoses in the same way that the reactions of ketones differ from those of aldehydes. For example, the aldoses can be oxidized to carboxylic acids, whereas fructose and other keto-sugars cannot, without simultaneous carbon–carbon bond cleavage. On the other hand, ketones and aldehydes have many reactions in common. Thus, fructose exists predominantly in a cyclic hemiketal form in solution. Both five- and six-membered hemiketals are present, with the latter being favored 4:1 over the former. Although only the β forms are depicted in Figure 17-24, fructose in solution exists as α and β mixtures of both 2-pyranose and 2-furanose arrangements.

EXERCISE 17-Q

To establish the structure and stereochemistry of glucose, Emil Fischer, who won a Nobel Prize in chemistry in 1902 for his characterization of the chemistry of carbohydrates, oxidized both the aldehyde carbon and C-6 to carboxylic acids. Is the resulting diacid a chiral molecule? What would your answer be if the starting material were mannose? Galactose?

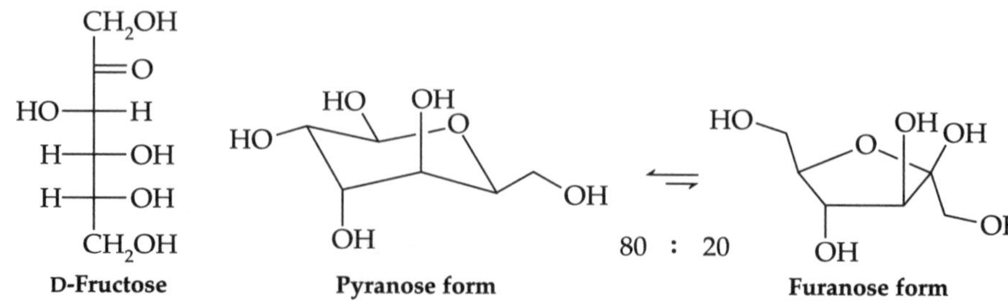

▲ FIGURE 17-24

In fructose, the six-membered-ring pyranose form is favored in the equilibrium with both the ring-opened and the five-membered-ring furanose forms.

17-5 ▸ Oligomeric Carbohydrates

Carbohydrates constitute a very significant fraction of the mass of all biological materials. Indeed, if all of the carbohydrates existing in plant and animal material in the world were converted into carbon dioxide (for example, by burning), the quantity of that gas in the atmosphere would increase by 50%. Most of the carbohydrates do not occur in nature in their free, monomeric form. They are most commonly found in the polymers that have already been described; namely, cellulose, starch, and related materials. They are also found as dimers, known as **disaccharides,** of which the three most prevalent are **sucrose, maltose,** and **lactose** (Figure 17-25).

Of these disaccharides, sucrose (ordinary table sugar) is the most common. It occurs naturally in many plants and is obtained commercially from both sugar cane and sugar beets. Sucrose is composed of the two simple sugars (glucose and fructose) bound to each other as illustrated in Figure 17-25 by an α linkage from glucose to a β linkage from fructose. This arrangement, a rather uncommon one in nature, is referred to as a **bis-acetal,** in which one oxygen is shared between two acetal functional groups.

The bis-acetal linkage between the carbohydrate subunits in sucrose is readily cleaved under acidic conditions, producing a mixture of glucose and fructose. The specific rotation of sucrose differs from that of the 1:1 mixture of glucose and fructose. In fact, this mixture has a rotation of −20°, opposite in sign to the specific rotation of sucrose. For this reason, the 1:1 mixture of glucose and fructose that is obtained upon cleavage of sucrose is known as **invert sugar,** and the biological catalyst, an enzyme, that catalyzes this process is known as **invertase.** Enzymes are often named by

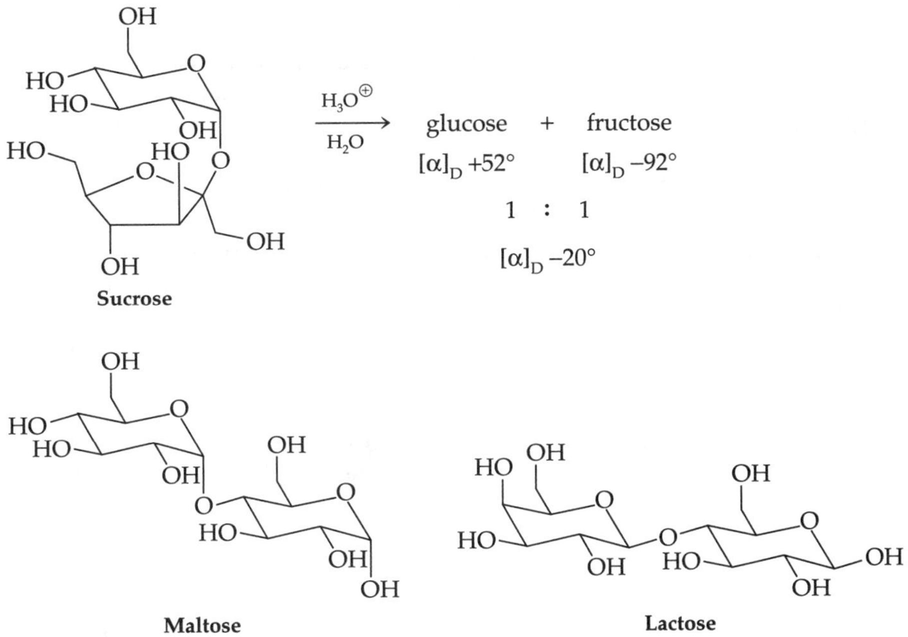

▲ FIGURE 17-25

Disaccharides are linked as acetals. The linkage can be of either the β form, as in lactose; or the α form, as in maltose; or both, as in the α linkage to glucose and the β linkage to fructose in sucrose. Hydrolysis (acid in water) of the acetal linkage of disaccharides forms the individual component sugars. For example, cleavage of sucrose forms a 1 : 1 mixture of glucose and fructose. The optical rotation of this mixture ([−20°]) is opposite in sign to that of sucrose ([+66°]), and the mixture is therefore often referred to as "invert" sugar.

adding the suffix **-ase** to a term descriptive of the transformation that is catalyzed by the enzyme.

Two other reasonably common disaccharides are maltose, formed from two α-linked glucose molecules, and lactose, in which a galactose is β linked to a glucose. As in the interconversion of the α and β anomers of glucose, the glucose unit of lactose is found in solution to be approximately 65% in the β form. Like sucrose, the disaccharides maltose and lactose can be cleaved both with acid in aqueous solution and by enzymes. These enzymes are specialized catalysts that are generally active only for a particular type of linkage. For example, **α-glucosidase** is capable of catalyzing the cleavage of only α-glucosidic linkages. Correspondingly, lactose is cleaved by the specific enzyme **β-galactosidase,** an enzyme that is capable of uniquely hydrolyzing a β linkage between two sugars. This chemical specificity of distinct enzymes is very important for complex living systems because many different chemical transformations must occur simultaneously without one interfering with each other. (We will return to this point in

---

FOLK MEDICINE AND MODERN PHARMACEUTICALS

As science progressed into the twentieth century, there was a strong feeling of elitist superiority among scientists. For example, "home" remedies for diseases were dismissed as "unscientific" and unworthy of study. Fortunately, the situation changed as many chemists in the pharmaceutical industry realized that, at least in some cases, there were sound bases for these medicines. In the early days of science, there were some who had the insight to realize the importance of folk medicine. For example, in 1795, William Withering, a physician, heard of a peasant woman who was famous for curing chronic heart problems. Upon investigation, Dr. Withering discovered that one of the ingredients of the cure was the herb foxglove and began using this plant to treat patients with congestive heart failure. It was not until much later that chemists isolated a number of different steroidal glycosides (for example, digitoxin) from foxglove and demonstrated their dramatic effects as stimulants of heart action. Digitoxin is an effective drug used today but it must be administered with care because the difference between the effective and lethal doses is comparatively small. Indeed, extracts of this plant have been used as an arrow poison by natives in various parts of the world, and both the ancient Egyptians and Romans used an extract from the sea onion containing similar compounds as both a heart tonic and a rat poison. The therapeutic maintenance dose of digitoxin is 0.1 mg (or 1 μg/kg); the lethal dose for cats is 200 μg/kg.

Digitoxin

Chapter 20.) In general, di- and polysaccharides such as cellulose are not digested by human beings because we do not have the enzymes that can cleave this β linkage.

Write a mechanism for the acid-catalyzed cleavage of sucrose to form glucose and fructose.

Conclusions

We have explored in this chapter some of the large number of molecules of nature that contain oxygen functional groups. The most simple among them are the fats and waxes: in a fat, long-chain fatty acids are connected by ester functional groups to a glycerol core; whereas, in a wax, a simple ester linkage connects a fatty acid to a fatty alcohol. Despite the presence of these ester linkages, fats and waxes consist mainly of straight-chain hydrocarbon units. The esters undergo hydrolysis in strong base (saponification) to form the free fatty acids. The presence of two sections of distinctly different polarity, one hydrocarbonlike (hydrophobic) and the other polar (hydrophilic), provides the driving force for the assembly of lipids into micelles, bilayers, and vesicles. Some large molecules spanning the bilayer thickness can transport small polar molecules across the membrane and can therefore elicit specific biological responses in cells.

Terpenes are nonsaponifiable, naturally occurring molecules that are mainly hydrocarbon in nature, although common oxygen-containing functional groups (such as carboxylic acids and esters, alcohols, epoxides, aldehydes, and ketones) also are present. Terpenes are constructed in living systems (by biosynthesis) through the oligomerization of isopentenyl pyrophosphate. This common source constitutes the chemical reason why terpenes generally contain multiples of five carbons and why their structures can often be subdivided into isoprene units. Many of the terpenes are biologically active, and the steroids (with four fused rings) play an important role as mammalian hormones, relaying information from one organ to another.

An important chemical function of pyrophosphate is as a leaving group. The chemical reactions of phosphate esters resemble those typically observed for both carboxylic acid esters and sulfate esters. The cation formed upon loss of a pyrophosphate group is an important intermediate that induces a multiring cyclization in the laboratory that resembles the enzymatic cyclization occurring in plants.

Carbohydrates are an important class of oxygenated natural products in which the presence of multiple functional groups greatly increases their solubility in water compared with fats, waxes, and terpenes. Carbohydrates are named by a prefix indicating the number of carbon atoms and the suffix **-ose.** In an aldose, an aldehyde oxidation state exists at C-1; whereas the ketone functionality of a ketose is at another position of the carbon backbone. An aldose can be converted into a ketose through an enediol. The designation D or L relates the sugar's absolute configuration to the naturally occurring triose D-glyceraldehyde. Containing either an aldehyde or a ketone group in the same molecule that bears an alcohol functional group, a carbohydrate undergoes intramolecular cyclization to become a hemiacetal. In this ring-closed form, an additional center of asymmetry is produced, permitting the existence of α- or β-anomers. Fischer projections are convenient tools for specifying configuration at nonanomeric carbon atoms. The cyclized structures can contain either five atoms (furanoses) or six atoms (pyranoses). Glucosides are formed when the hemiacetal OH group is converted into an OR group by interaction either with a simple alcohol or with the OH group of another sugar. Carbohydrates can exist as individual molecules; they can also be linked together by acetal groups (called a glycosidic linkage) to form dimers and polymers. Sucrose, composed of glucose and fructose, is an example of the former, whereas starch and cellulose (Chapter 16) are common examples of polymers of glucose.

Summary of New Reactions

Allylic Substitution of Pyrophosphate

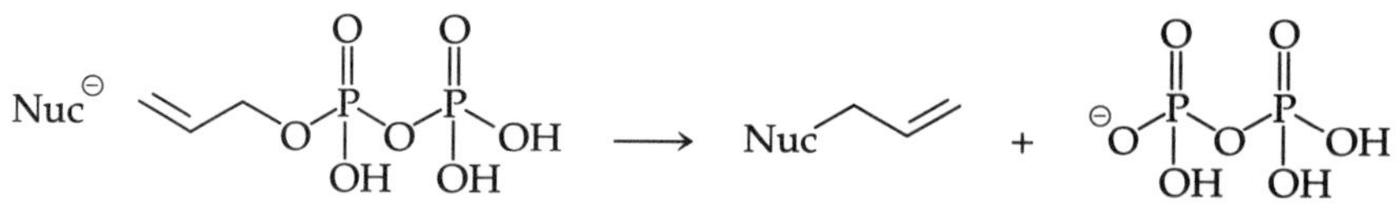

Cationic Cyclization

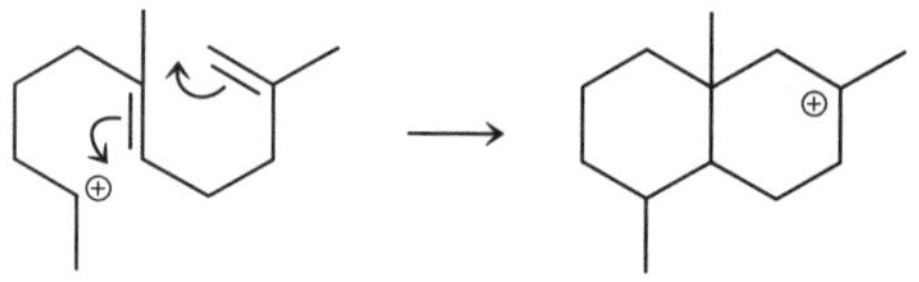

Anomerization

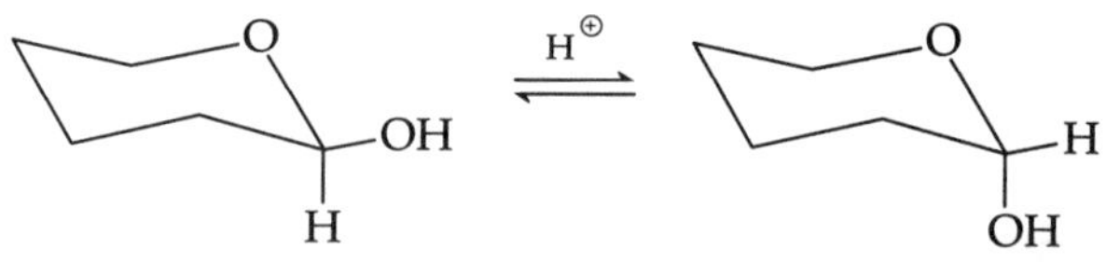

Review Problems

17-1 The melting points of *cis* unsaturated fatty acids are lower than those of the corresponding saturated acids. Would you expect the melting point of a *trans* unsaturated fatty acid to be closer to that of the *cis* isomer or that of the saturated fatty acid? Explain your reasoning clearly.

17-2 The phospholipids (illustrated in Figure 17-2), which are the fundamental building blocks of cell membranes, are mixed esters of glycerol and are chiral. Identify the center(s) of chirality in the phospholipid presented in the next column, and draw clear, three-dimensional representations of the possible stereoisomers.

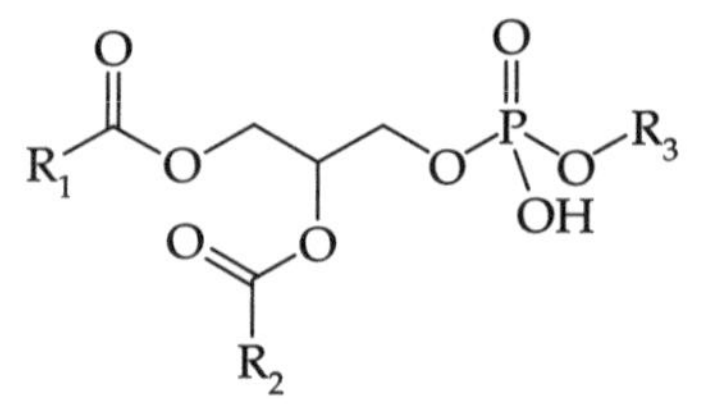

17-3 Although terpenes are not biochemically derived from isoprene, this diene does polymerize under radical and cationic conditions to form polyisoprenoids. Predict whether such polymerizations would have a propensity toward the joining of monomer units in a head-to-head or a head-to-tail (see next page) fashion. Explain your answer.

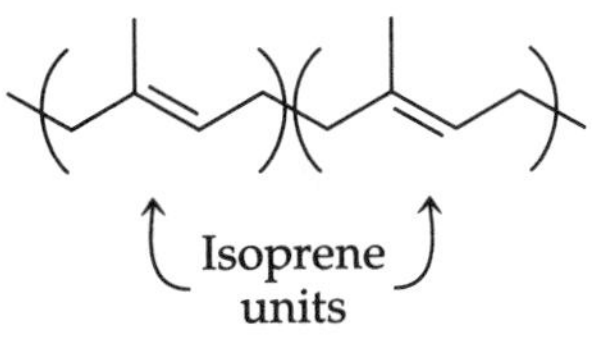

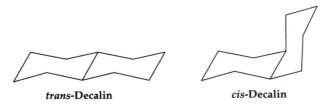

Head-to-head

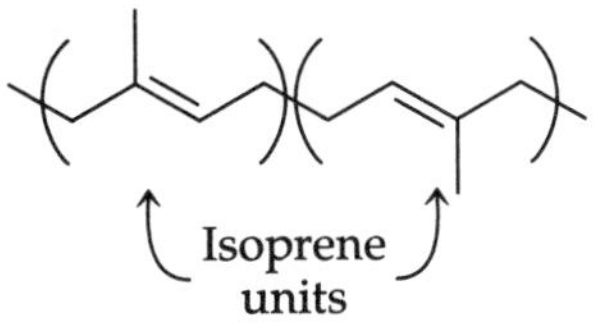

Head-to-tail

17-4 Recall that cyclohexyl ring systems are energetically favored in chair conformations and that *trans*-decalin is more stable than the *cis* isomer.

trans-Decalin   **cis-Decalin**

Notice that the stereochemistry at the bridgehead carbons of *trans*-decalin is such that both substituent hydrogens are axial and that, in *cis*-decalin, one is axial and the other is equatorial. Now consider the stereochemistry at the bridgehead carbons of the structure below, which includes three of the four rings of a steroid system. Decide whether the decalin systems formed from rings A and B and from B and C are *cis* or *trans* fused. Draw a three-dimensional representation of the structure below that is based on either the *trans*- or the *cis*-decalin conformations.

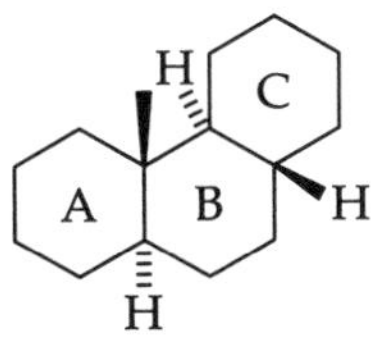

17-5 There are many different types of organic cleaning agents that are fundamentally similar to the salts of long-chain fatty acids. Two examples are:

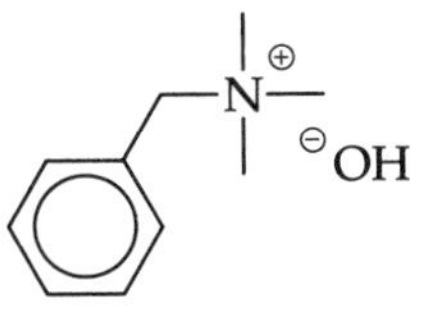

Triton B

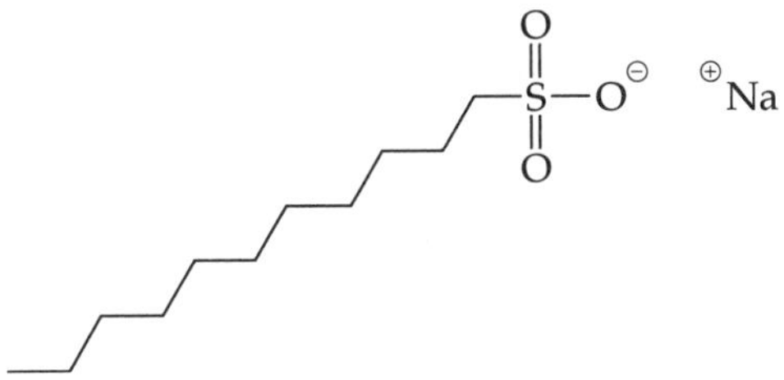

Sodium lauryl sulfonate

(You may find one or the other listed as an ingredient on the bottle of the shampoo that you use.) Identify in each case the hydrophobic and the hydrophilic parts of the molecule.

17-6 The three naturally occurring aldohexoses (glucose, mannose, and galactose) exist preferentially in solution as cyclic hemiacetal forms. We might expect that glucose is thermodynamically more stable than all other aldohexoses because (in the β form) all substituents are in equatorial positions. Mannose and galactose would be expected to be only slightly less stable because only one of the hydroxyl groups is axial in each (again in the β form). For practice, draw the other chair conformation of each of these sugars, noting the number of 1,3-diaxial interactions.

17-7 The acid-catalyzed hydrolysis of sucrose (refer to Exercise 17-R) can be initiated by cleavage of any of four different bonds. Indicate these bonds, using the abbreviated structure below, and write a detailed mechanism showing the electron flow for each bond-breaking step.

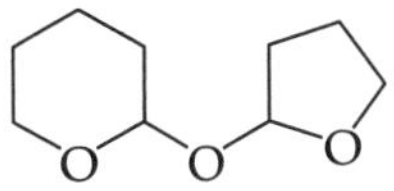

17-8 Vitamin C has a variety of natural functions, including acting as an antioxidant. It resembles carbohydrates in that each carbon bears at least one oxygen, although not all are at the oxidation level found in, for example, glucose. Identify which carbon(s) have higher and which have lower oxidation levels than a normal aldohexose. Notice that vitamin C has an enediol functionality; it is this subunit that undergoes oxidation with oxidants such as molecular oxygen, preventing oxidation of other important biological molecules.

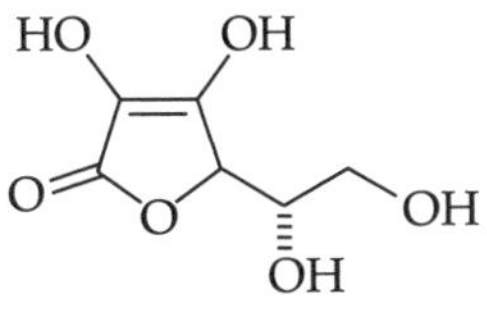

Vitamin C (ascorbic acid)

Write the two proton tautomers of vitamin C that result from isomerization of the enediol to ketoalcohols. Note the relation between carbonyl groups in these two isomers. Which of these isomers is more stable (based on the relation of the carbonyl groups to each other) and which carbon of the enediol system in vitamin C is more nucleophilic toward an electrophilic oxidant?

Identify the center of chirality in ascorbic acid that corresponds to that used for the D,L classification system in carbohydrates. Does vitamin C have the D or the L configuration?

17-9 Vitamin C is also called ascorbic acid because the pK_a of the first dissociation is 3.6. Compare this value with that of the individual functional groups, and suggest why the combination of functional groups present leads to a significantly stronger acid than would be expected for each group alone.

Chapter 18

Structures and Reactions of Naturally Occurring Compounds Containing Nitrogen Functional Groups

In Chapter 17, we examined several classes of natural compounds containing oxygen atoms in various functional groups. Nitrogen-containing functional groups also play an important role in biological chemistry, although the classification system for these materials is not as well defined. In this chapter, we consider amino acids in detail, as well as peptides and proteins—polymeric materials with multiple amide functional groups. We also examine the structures of the polynucleic acids—polymeric materials with sugar–phosphate backbones and heterocyclic amine pendent groups—as well as aminosaccharides and the alkaloids. This last category is an especially diverse one because it comprises all materials isolated from plants that contain a *basic, sp³*-hybridized nitrogen atom (for example, a simple amine). We will see in this chapter that the substitution of nitrogen for one of the hydroxyl groups in a carbohydrate produces an aminosaccharide; such compounds are essential chemical elements of a variety of biological recognition systems that will be discussed in Chapter 19.

We start by reviewing the methods for forming carbon–nitrogen bonds presented in earlier chapters, and we then explore how these transformations are used specifically in sequences that incorporate nitrogen into some biologically important organic molecules.

18-1 ▸ Methods for Forming Carbon–Nitrogen Bonds: A Review

We have learned several methods by which carbon–nitrogen bonds can be formed and will see that compounds containing C–N multiple bonds undergo some of the reactions of their carbonyl compound counterparts. We begin by reviewing bond formation by nucleophilic substitution at sp^3 carbon to form amines, nucleophilic acyl substitution to form amides, and imine formation by nucleophilic addition. We will then see that C–C bond formation with imines can take place by a pathway analogous to Grignard additions to carbonyl compounds and will learn a new condensation reaction that combines separate reactions described in the first half of the book. Finally, we will review nitrile reduction as a route to primary amines and the Beckmann rearrangement as a route to amides.

Amines

Simple amines can be prepared, as explained in Chapter 8, by S_N2 substitution on a carbon that bears a leaving group. For example, methylamine can be readily converted into ethylmethylamine by an S_N2 reaction with ethyl iodide.

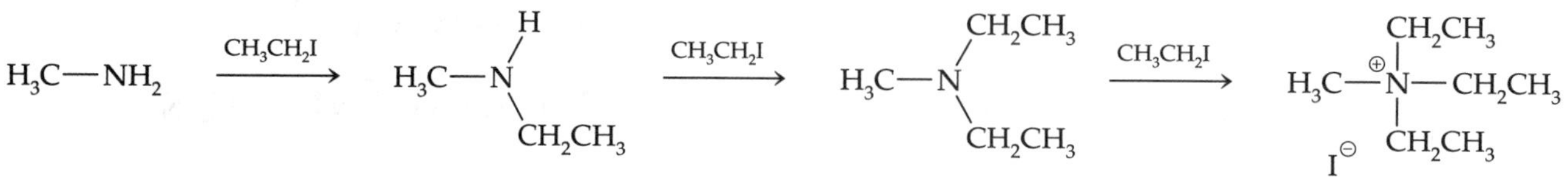

This method is seriously limited, however, because the products themselves are amines that undergo further reaction with the electrophilic partner. In this example, the product reacts further to form diethylmethylamine and then even further to form triethylmethylammonium iodide.

One exception to the generalization that the direct alkylation of amines does not result solely in a monoalkylated product is the nucleophilic ring opening of epoxides by amines.

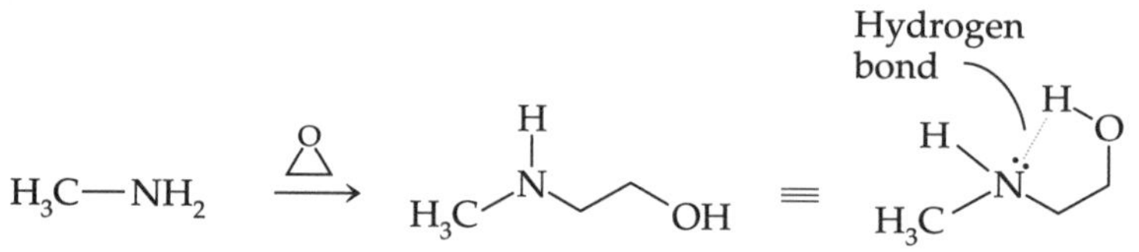

The reaction is considerably slower than that of a comparably substituted alkyl bromide and, except for monosubstituted epoxides, requires elevated temperatures. The product, a β-aminoalcohol, is considerably less reactive than the starting amine, probably because of stabilization by intramolecular hydrogen bonding between the lone pair of electrons on nitrogen and the hydrogen of the hydroxyl group. This hydrogen-bonding interaction also results in significantly lower basicity and nucleophilicity for nitrogen.

Amides

The reaction of primary and secondary amines (as well as ammonia) with a reactive carboxylic acid derivative results in the formation of amides by nucleophilic acyl substitution, as discussed in Chapter 12.

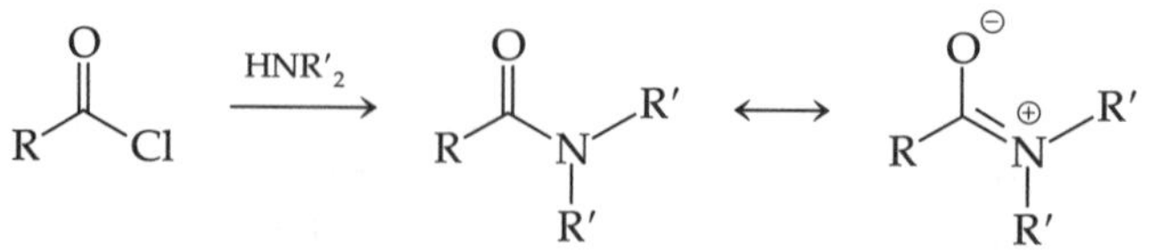

Because of resonance delocalization of the nitrogen lone pair, the nucleophilic character of the nitrogen atom in the product amide is substantially lower than that of the starting amine. Amides can be further transformed into amines by reduction, most conveniently in the laboratory with $LiAlH_4$.

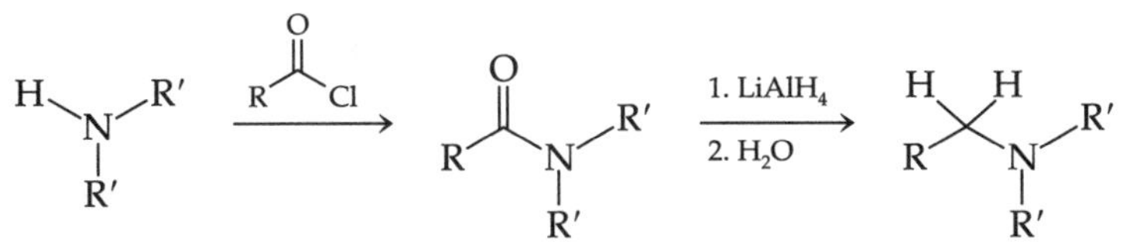

We can see that the overall process—amide formation followed by reduction—adds one substituent to the nitrogen that *must* be attached through a methylene group.

Imines

Amines also react quite selectively with carbonyl compounds to afford products resulting from a one-to-one pairing of the partners. The reaction of primary amines with ketones and aldehydes produces imines, with loss of water (Chapter 12).

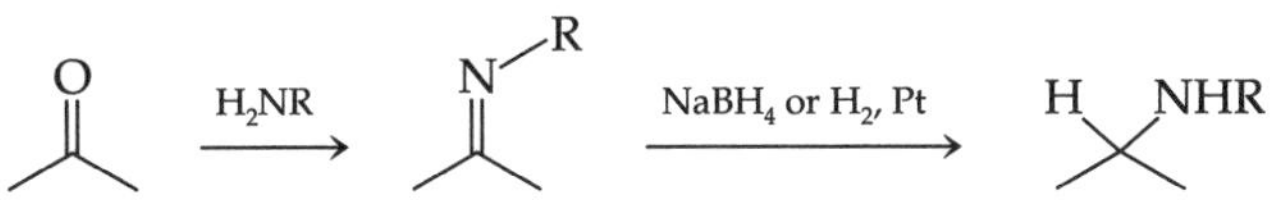

The imine is not highly favored in the equilibrium, but the reaction can be driven toward the product imine by removal of water (for example, by azeotropic removal or with molecular sieves). The resulting imine can then be reduced, either with complex hydride reagents such as $NaBH_4$ or with molecular hydrogen and a metal catalyst. The amine product is produced by reduction at both carbon and nitrogen, in contrast with the reduction of amides, in which only the carbonyl carbon is reduced. As a result, in the reduction of an imine from a primary amine, the product has one hydrogen and two carbon substituents on nitrogen, with one alkyl substituent derived from the starting amine and the other from the carbonyl compound.

Reductive amination can also be carried out with imines derived from ammonia. Generally, these imines are not sufficiently stable to allow their isolation. However, they can be trapped as they are formed by reduction with $NaBH_3CN$ (sodium cyanoborohydride), a hydride reagent that is stable in protic solvents (such as water and methanol) at a pH at which imine formation proceeds at an appreciable rate. Reductive amination by this process works equally well with ammonia (R = H) and primary amines (R = alkyl).

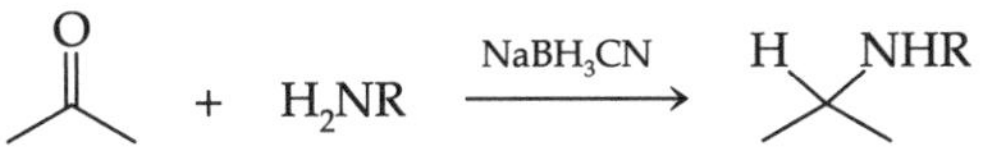

The reduction of an imine by $NaBH_4$ or $NaBH_3CN$ consists of the formal addition of a hydride ion as a nucleophile to the carbon and a proton to the nitrogen atom, as discussed in Chapter 12.

EXERCISE 18-A

Sodium cyanoborohydride is less active as a hydride donor than is sodium borohydride. As a result, $NaBH_3CN$ can be used in aqueous solution under acidic pH conditions in which $NaBH_4$ is rapidly destroyed. Explain this difference in reactivity.

Nucleophilic Addition to Imines

A carbon nucleophile such as a Grignard reagent or an alkyllithium can be added to the carbon of the imine functional group in a reaction analogous to the addition of these nucleophiles to ketones and aldehydes (Chapter 13).

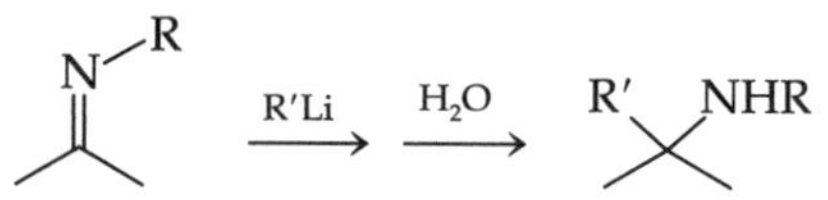

This mode of reaction, resulting in the formation of a new carbon–carbon bond, is quite general and a wide variety of Grignard reagents can be used. However, only secondary amines can be prepared by this method.

EXERCISE 18-B

Provide appropriate starting materials and reagents for the synthesis of the following amines.

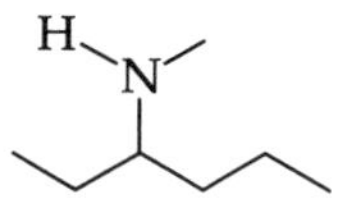

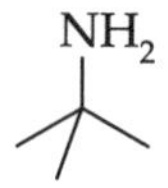

Mannich Condensation

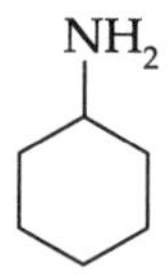

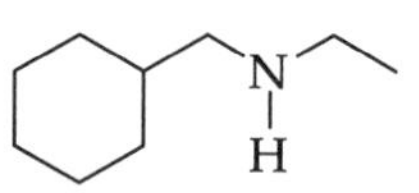

Carbon nucleophiles other than Grignard reagents and alkyllithiums also add to imines. An important process in which such a reaction takes place is the **Mannich condensation,** a combination of a simple aldehyde (often formaldehyde), a primary or secondary amine, and a ketone (Figure 18-1). Although we have not seen this reaction before, it is a combination of three reactions that we have seen: activation of the α-carbon of a ketone as a nucleophile by enol formation (Chapters 8 and 13); imine formation (Chapter 12); and nucleophilic addition to a π system (Chapter 13). First, the ketone is tautomerized to its enolic form. Then, formaldehyde and a primary amine react to form an imine that reacts as an electrophile with the ketone enol in a final step. Each of these steps is acid catalyzed; generally, a small amount of a weak acid, such as acetic acid, is added to the reaction mixture as a catalyst.

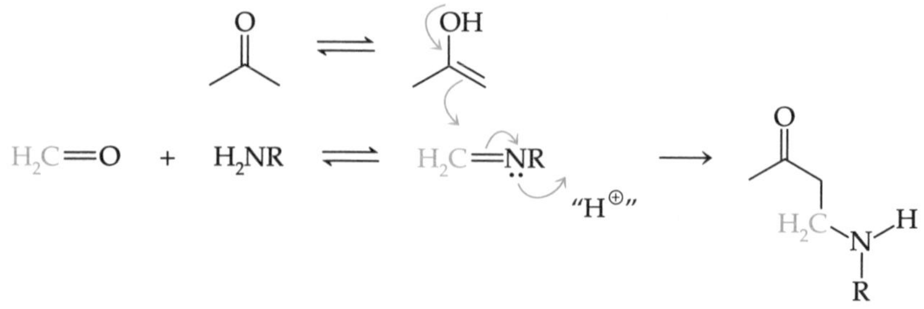

▲ FIGURE 18-1

In a Mannich condensation, formaldehyde is converted into an imine by condensation with an amine. The ketone coreactant is enolized to its enol form, which has nucleophilic character on the α carbon. Nucleophilic attack by this α carbon at the imine carbon forms a new carbon–carbon bond. In the resulting condensation product, a —CH$_2$—group (from formaldehyde) links the α carbon of the ketone with the amine nitrogen atom.

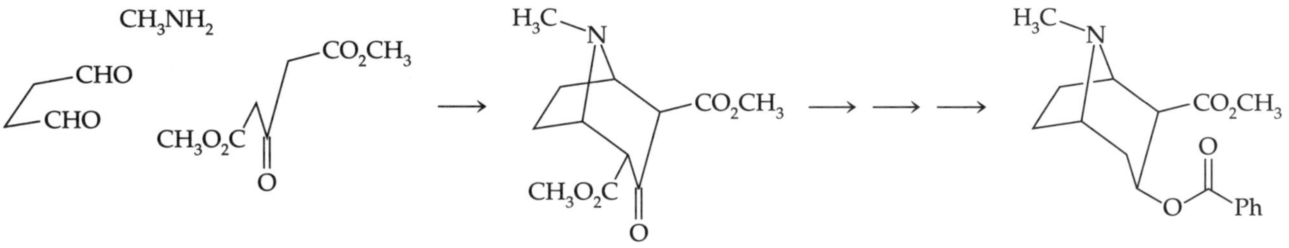

▲ FIGURE 18-2

A double Mannich reaction forms the ring system found in a number of natural products; for example, in cocaine.

The Mannich reaction is very useful in natural product synthesis. For example, the **tropane** ring system in cocaine can be formed in a single reaction by a double Mannich reaction. The dialdehyde (compound A) shown in Figure 18-2 is condensed with methylamine and dicarbomethoxyacetone (compound B) in the presence of an acid catalyst. This straightforward process provides the bicyclic product (compound C).

EXERCISE 18-C

Write a clear, detailed reaction mechanism for the formation of compound C in Figure 18-2.

EXERCISE 18-D

Suggest a sequence of reactions that could be used to convert intermediate C in Figure 18-2 into cocaine. (Hint: Compound C contains a β-ketoester functional group.)

Nitrile Reduction

The nitrogen atom of an amine functional group can also be introduced into an organic substrate by S_N2 substitution with cyanide ion, followed by reduction of the resulting nitrile to form a primary amine, as discussed in Chapter 12.

$$R—CH_2Br \;+\; K^{\oplus}\, {}^{\ominus}C\equiv N \;\xrightarrow{\text{DMSO}}\; R—CH_2—C\equiv N \;\xrightarrow[H^{\oplus}]{H_2,\,Pt}$$

$$R—CH_2—CH_2—\overset{\oplus}{N}H_3 \;\xrightarrow{Na_2CO_3}\; R—CH_2—CH_2—NH_2$$

This method works well only if the displacement is at a primary carbon. With more-substituted alkyl halides, cyanide ion also acts as a base, effecting elimination. The primary amines that can be produced are thus limited to those in which the nitrogen is attached to a —CH₂—CH₂— group. Reduction of nitriles can be accomplished either with $LiAlH_4$ or with H_2 and a metal catalyst. Because amines deactivate metal catalysts and decrease the rate of catalytic hydrogenation, an acid such as acetic acid is included in these reactions so that the amine product is converted into an ammonium ion. After reduction, neutralization of the acid with a base such as Na_2CO_3 frees the amine from the salt.

Beckmann Rearrangement of Oximes

An industrially very important method of carbon–nitrogen bond formation is the Beckmann rearrangement, a reaction that converts an oxime into an amide by a rearrangement reaction that was described in Chapter 14.

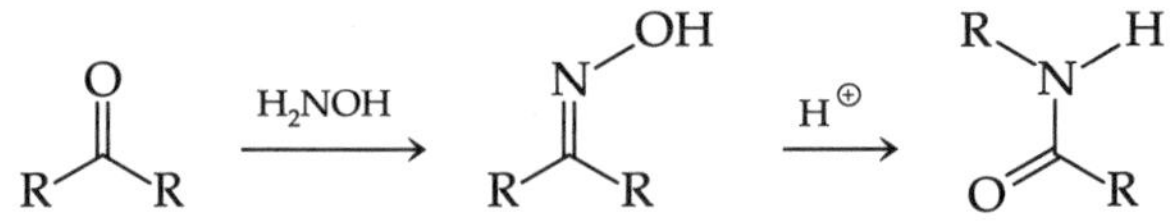

Oximes are formed from ketones by reaction with hydroxylamine; in this reaction, the equilibrium favors the oxime to a greater extent than the imine is favored in the analogous reaction between a ketone and an amine. Treatment of the oxime with concentrated acid initiates a series of steps that ultimately form the product amide by migration of one of the original carbon substituents of the ketone to nitrogen. The sequence is shown in Figure 18-3: a carbon substituent migrates from carbon to nitrogen simultaneously with the loss of water. Readdition of water to the carbocation and proton tautomerization (by protonation followed by deprotonation) affords the product amide.

Large quantities (approximately 5 million kilograms) of caprolactam are produced each year from cyclohexanone by the Beckmann rearrangement. (Recall from Chapter 16 that caprolactam is the precursor of nylon 6.)

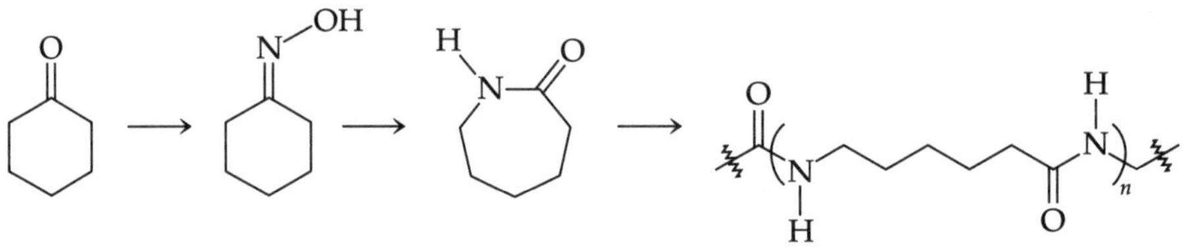

Caprolactam   Nylon 6

EXERCISE 18-E

Consider the structure of intermediate A in Figure 18-3. Are there other resonance structures for this cation? Would you expect this cation to be bent, as shown, or linear? Explain.

EXERCISE 18-F

Write a complete, detailed reaction mechanism for the conversion of the oxime of cyclohexanone into caprolactam. Would you expect the intermediate carbocation formed upon carbon migration in this reaction to be more or less stable than that formed in an acyclic system? Explain.

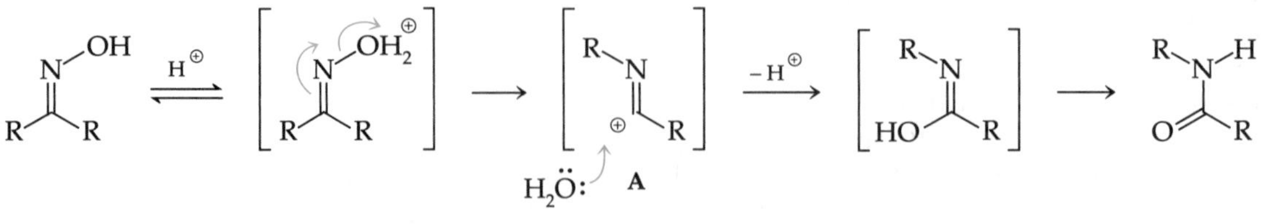

▲ FIGURE 18-3
In a Beckmann rearrangement, an oxime is converted into an amide.

18-2 ▸ Amino Acids

Amino acids constitute an important class of nitrogen-containing naturally occurring compounds. The biological function of amino acids is well established: they are the constituent "monomer" units from which the biopolymers based on the peptide bond are built (for example, peptides and proteins, including enzymes). We examined the structural properties of peptides in Chapter 16 but dealt there with only a few specific examples of amino acids. Here, we will examine all of the common amino acids found in nature; in the next section, we will consider their chemistry.

For convenience, amino acids are divided into four classes based on both the polarity and the acidity of the groups present in the side chains: (1) hydrophobic; (2) hydrophilic; (3) acidic; and (4) basic. The twenty amino acids commonly found in proteins in living systems are shown in Figures 18-4 and 18-5, subdivided into these four categories. Each of these amino

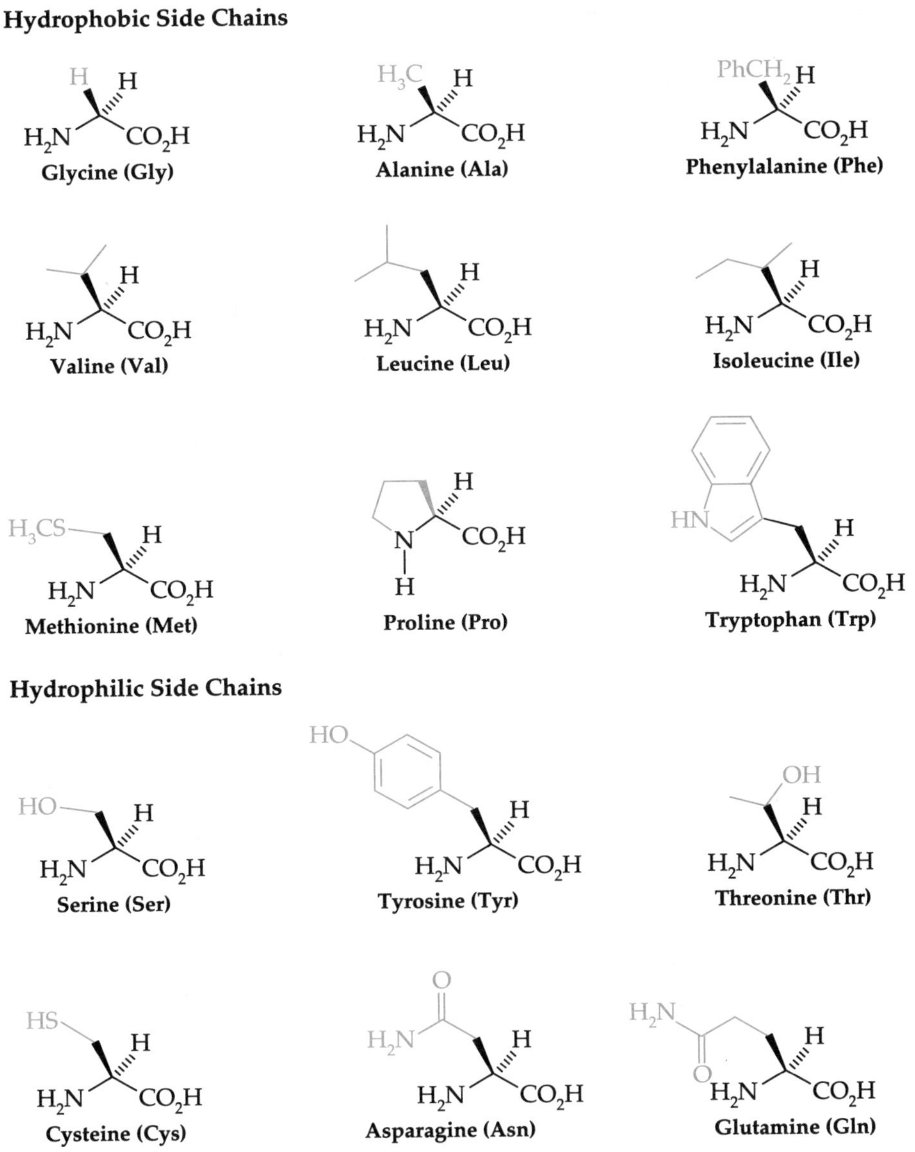

Hydrophobic Side Chains

Glycine (Gly)

Alanine (Ala)

Phenylalanine (Phe)

Valine (Val)

Leucine (Leu)

Isoleucine (Ile)

Methionine (Met)

Proline (Pro)

Tryptophan (Trp)

Hydrophilic Side Chains

Serine (Ser)

Tyrosine (Tyr)

Threonine (Thr)

Cysteine (Cys)

Asparagine (Asn)

Glutamine (Gln)

▲ FIGURE 18–4

The naturally occurring amino acids having neutral side chains. Their names are not systematic; abbreviations are given in parentheses.

Acidic Side Chains

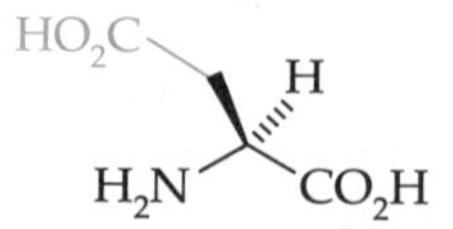

Aspartic acid (Asp) 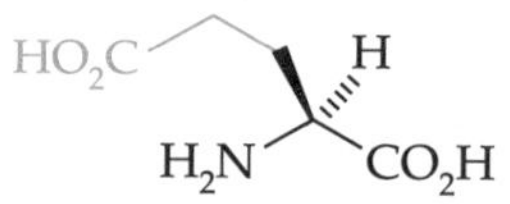   Glutamic acid (Glu)

Basic Side Chains

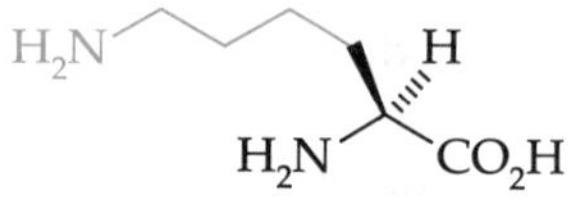

Lysine (Lys) 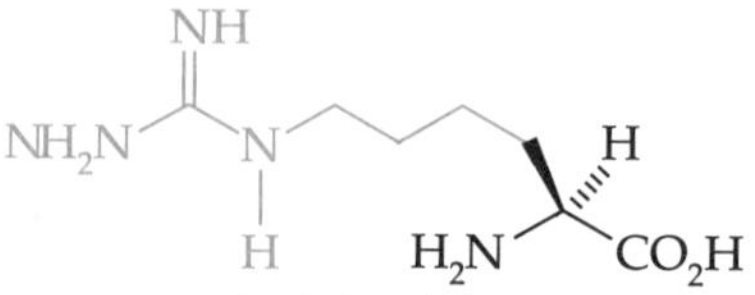   Arginine (Arg) 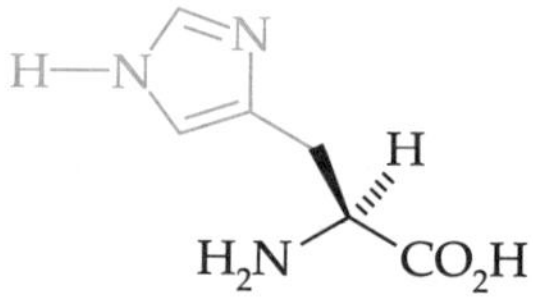   Histidine (His)

▲ FIGURE 18-5
The naturally occurring amino acids having acidic or basic side chains. As with amino acids with neutral side chains, their names are not systematic; abbreviations are given in parentheses.

acids contains a —CH(CO$_2$H)(NH$_2$) grouping; the side-chain residue for each is unique. Although their names are often rather arbitrary, their structures are comparable. For example, **glycine** is the parent amino acid and is the only one that is achiral. **Phenylalanine** derives both in name and in structure from **alanine,** whereas **tyrosine** is phenylalanine with a phenolic hydroxyl group. **Serine** and **cysteine** differ from alanine by the presence of a hydroxyl and a thiol group, respectively. **Glutamine** contains one more methylene group than **asparagine.** Just as glycine is unique in having no center of chirality, **isoleucine** and **threonine** are the only common amino acids that have two centers of chirality. **Proline** is unique among amino acids in that its side chain is joined in a ring with the α-amino nitrogen atom.

In Chapter 16, we learned that peptide chains prefer a more- or less-extended, zigzag arrangement (as in Figure 18-6). This conformation is not possible, however, if the chain includes a proline residue because of proline's five-membered ring. Furthermore, a proline residue in a peptide chain does not have a hydrogen on nitrogen and can thus participate only as a hydrogen-bond acceptor, not a donor. The hydrogen bonding that links one peptide chain to another is thus disrupted when proline is present. The "zig" instead of "zag" induced by the presence of a proline residue also disrupts hydrogen bonding in the α-helix.

Several other classification systems for amino acids that recognize their biological properties are commonly used. For example, amino acids are classified as either **glucogenic** or **ketogenic,** depending on how they are

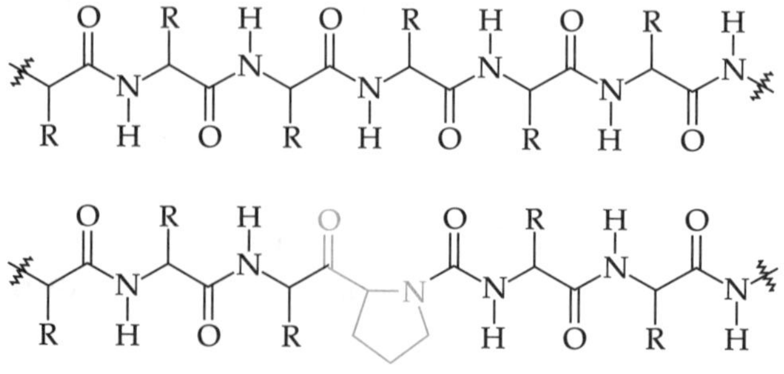

▶ FIGURE 18-6
The normal zigzag arrangement along a peptide backbone (upper structure) is interrupted by the presence of a proline residue (lower structure).

biodegraded. Alternatively, amino acids are grouped on the basis of the precursors from which they are synthesized in living systems. In another scheme, they are classified as either **essential** or **nonessential,** depending on whether a particular living system must obtain them from dietary sources or can synthesize them itself. In human beings, the following eight amino acids are essential and must be obtained from the diet: isoleucine, leucine, lysine, methionine, phenylalanine, threonine, tryptophan, and valine.

Hydrophobic and Hydrophilic Properties

In Figure 18-4, the amino acids with neutral side chains are divided into two groups: those with hydrocarbonlike (hydrophobic) units; and those with polar functional groups that impart hydrophilic character to the side chain. Others have functional groups that participate in important chemical reactions. For example, as already seen in Chapter 16, cysteine residues in separate peptide chains can be coupled by the formation of sulfur–sulfur bonds, serving as cross-links between separate peptide chains.

Because biochemists frequently discuss biopolymers derived from ten to hundreds or even more amino acid residues, they often use the three-letter-abbreviation codes given in Figures 18-4 and 18-5 to designate the individual amino acids, instead of writing their structures. We will not use these abbreviations in this chapter because our focus is on the unique chemistry imparted by the side chain of the amino acid.

Acidic and Basic Properties

Amino acids are also classified according to whether their side chains are acidic or basic (Figure 18-5). Because of the polarity associated with acidic and basic functional groups, all of these amino acids are hydrophilic. As with the neutral amino acids, these amino acids have related structures. For example, **glutamic acid** has one more methylene group than **aspartic acid,** and both are related to the hydrophilic neutral amino acids, glutamine and asparagine, respectively. **Arginine** can be viewed as a derivative of **lysine** in which the simple amino group of the latter has been converted into what is referred to as a **guanidine** unit, $-NHC=NH(NH)-$. This functional group is much more basic than is a simple amine; and, indeed, arginine is the most basic of the common amino acids. It is convenient to describe the acidity or basicity of the various functional groups present in amino acids in terms of the dissociation constant (pK_a) of each functionality in its acidic, protonated form. For example, the two dissociation constants of protonated alanine, $CH_3CH(NH_3^+)COOH$, are 2.35 and 9.87 (Figure 18-7). There are three dissociation constants for arginine, with pK_as of 2.18, 9.09, and 13.2. The first dissociation (the lowest number) converts the carboxylic acid into its carboxylate anion. In both cases, the protonated amino acid is substantially more acidic than acetic acid (pK_a 5.2). The second step is the loss of a proton from the α-ammonium ion. For arginine, the third deprotonation occurs at the guanidinium ion to form a neutral guanidine unit.

Only one of two possible proton tautomers for **histidine** is shown in Figure 18-5. An alternate arrangement is shown at the right of Figure 18-8. These two proton tautomers are in dynamic equilibrium in solution. Histidine residues in enzymes play an important role because of the facility with which a proton can be gained by one nitrogen at the same time that another is lost from the other nitrogen. We will address this point again in detail in Chapter 20, where catalysis by enzymes is treated in some depth.

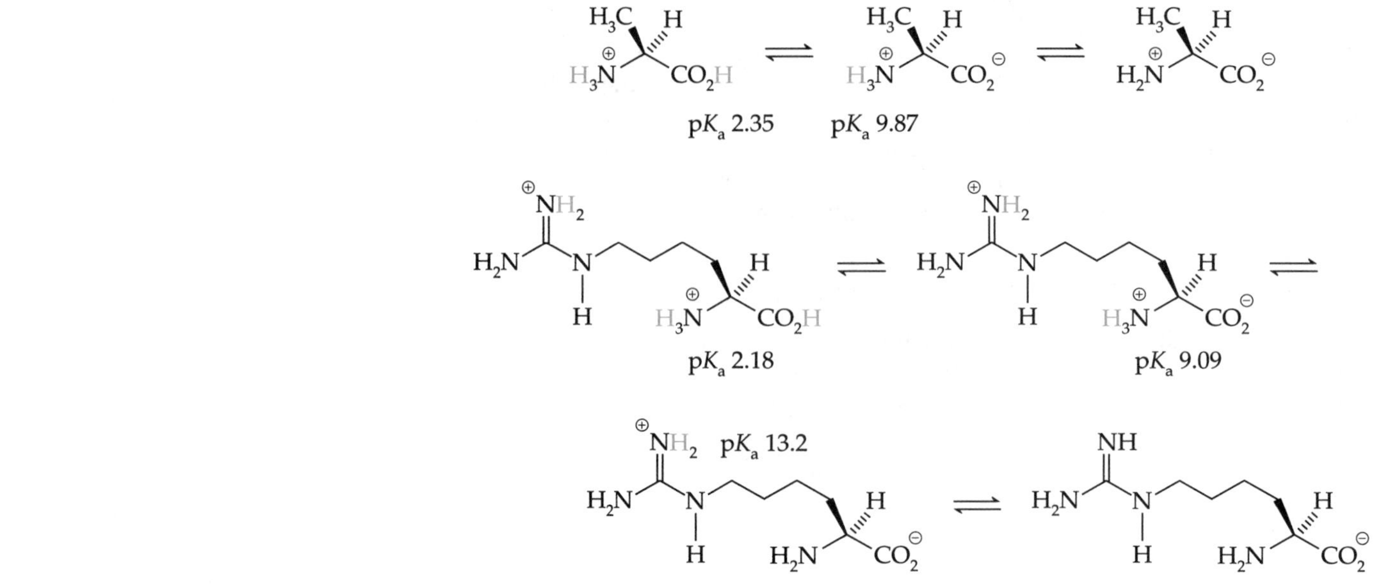

▲ FIGURE 18-7

Amino acids bear at least two sites with acidic and basic properties. The acid and base strength of these groups are characterized by pK_a values of the protonated forms. Those with acidic or basic side chains have more than two characteristic acid dissociation constants.

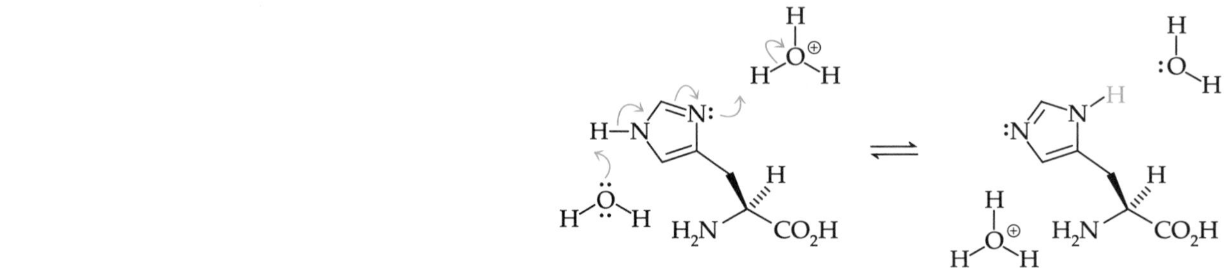

▲ FIGURE 18-8

Tautomerization of histidine results in the shift of a hydrogen atom from one ring nitrogen to the other. As in keto-enol or imine-enamine tautomerizations, this shift is accomplished by protonation at one site and deprotonation at the other.

EXERCISE 18-G

Explain why the carboxylic acid group of a nitrogen-protonated amino acid should be more acidic than a simple acid such as acetic acid. Also explain why the guanidine unit is more basic than a simple amine (pK_a 10–11).

Zwitterionic Character of Amino Acids

All amino acids discussed so far are **amphoteric:** they contain both an acidic carboxylic acid and a basic α-amino functional group. In aqueous solution at a specific pH (referred to as the **isoelectric point**) that varies with structure, amino acids exist mainly as zwitterions. In the zwitterionic form, the amino group has gained a proton and exists as a cation and the carboxylic acid has lost a proton to form a carboxylate anion (Figure 18-9). As the acid-

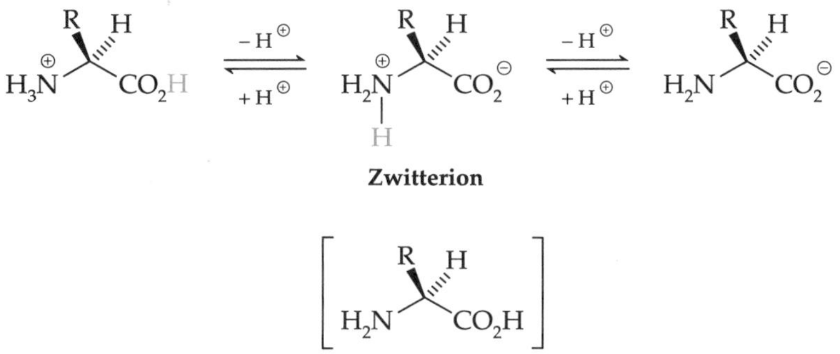

Zwitterion

▲ FIGURE 18-9

The neutral form of an α-amino acid, shown in brackets, is only a minor component of the mixture of protonated and deprotonated forms. At the isoelectric point, there are equal amounts of positively and negatively charged groups and the zwitterion shown dominates the equilibrium. As the pH decreases, the carboxylate is protonated, producing the cation at the left of the equilibrium. At higher pH, the ammonium ion is deprotonated, yielding the anion shown at the right.

ity of the solution is increased (at a lower pH), the concentration of carboxylic acid increases as the carboxylate is protonated and the equilibrium is displaced toward the cationic form, with an ammonium ion and carboxylic acid. Conversely, with an increase in pH to a more-basic solution, the equilibrium is displaced toward the anionic form, with a carboxylate and a neutral amine. In fact, the free amine and free carboxylic acid form of the amino acids depicted in Figures 18-4 and 18-5 exist to only a very small extent in aqueous solution at any pH.

Amino acids can be separated by *electrophoresis,* a technique described in Chapter 4, in which molecules are drawn through a medium by an applied electric field. Only charged compounds migrate, and the greater the charge (on a weight-adjusted basis), the faster they move. Electrophoresis is a valuable analytical tool for the study of biologically important compounds because many of them are ionic materials. Alternate techniques such as gas chromatography and solid-liquid chromatography, in which normal solid phases (for example, silica gel) are used, are not appropriate because the high polarity of these biomolecules results in excessively strong interactions with the solid phase.

18-3 ▸ Peptides

Almost invariably, the individual, free amino acids just discussed are not biologically important. On the other hand, they have great significance when they are joined together as biopolymers. As mentioned in Chapter 16, smaller biopolymers, containing fewer than a hundred amino acids, are referred to as peptides and polypeptides, and larger biopolymers are called proteins. The linking bond between two amino acids joined together to form an amide is referred to as a peptide bond. Peptides serve mainly as chemical messengers, or **hormones,** whereas the larger proteins are used for a variety of purposes: as structural materials, as the critical recognition elements of the immune system, and, most importantly, as enzymes, which are the catalysts for biochemical transformations.

Although there are only twenty common amino acids, the number of possible combinations, even for short peptides, is immense. For example,

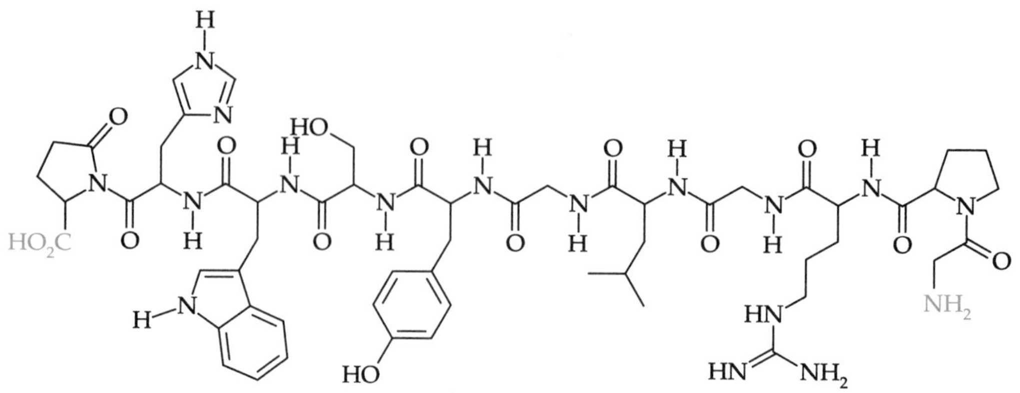

Leuteinizing hormone release hormone (LHRH)

▲ FIGURE 18-10

LHRH is a peptide hormone comprising ten amino acids.

consider the peptide hormone **LHRH (luteinizing hormone release hormone),** which has ten amino acids subunits (Figure 18-10). LHRH serves as a critical component of human fertility, communicating information from the hypothalamus gland to the pituitary gland and then to the ovaries at the time that an ovum is to ripen and be released. If we were to isolate this peptide and determine only that it was a decapeptide, then the number of possible sequences, derived from the twenty naturally occurring amino acids, would be $20^{10} = 1.0 \times 10^{13}$. By determining which amino acids are present, we reduce the number of possible structures tremendously. However, if all ten are different, there are still 3,628,800 (10!) different possible sequences. It is because of this tremendous potential for structural diversity that amino acid polymers find so many applications.

Except for the amino acid at either end, each amino acid of LHRH participates in two amide bonds, one from the carboxyl group to the amino group of the next amino acid (in one direction) and the other from the amino group to the carboxyl group of the next amino acid (in the other direction). However, the amino acid units at the two ends are different because each participates in only one amide bond. Thus, at one end of the peptide, there is a free amino group: this end is called the amino terminus, or simply the N terminus. At the other end, there is a carboxylic acid group: this end is referred to as the carboxyl terminus, or the C terminus.

18-4 ▸ Peptide Synthesis

The laboratory synthesis of small-to-medium-sized peptides is an important focus of the pharmaceutical industry, where the goal is to prepare useful quantities of peptide hormones, as well as analogs of these natural substances. In an analog, unnatural amino acids (such as the enantiomeric D form) substitute for those naturally present to enhance activity or protect the peptide from degradation in the body. The amidation reaction reviewed in Section 18-1 is important to this task.

Four features complicate the synthesis of polymers derived from amino acids. First, linking amino acids together requires the formation of an amide by the specific condensation between the carboxylic acid of one amino acid and the amine of another. Because each amino acid contains both functional groups, protection of the amino group of one and the carboxyl group of the

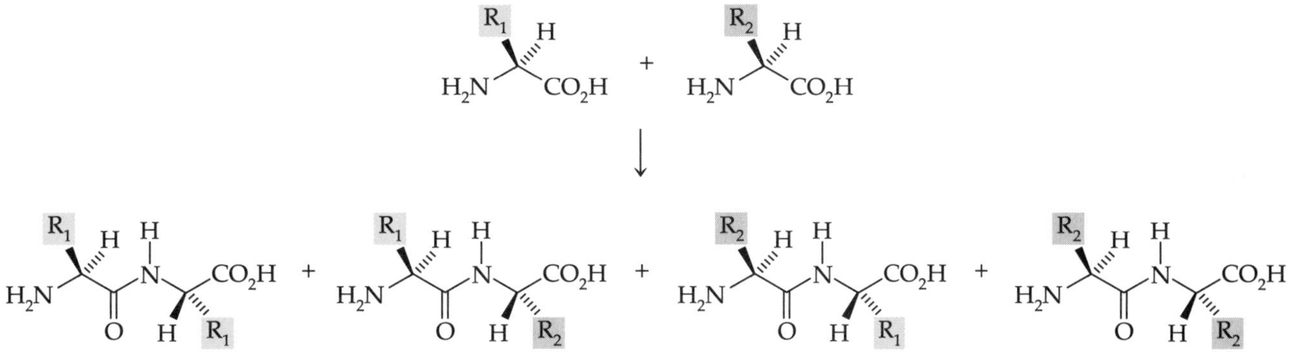

▲ FIGURE 18-11
There are four possible combinations of two amino acids bearing alkyl groups R_1 and R_2: the dipeptide with two R_1 groups, that with R_2 at the C terminus and R_1 at the N terminus, that with R_1 at the C terminus and R_2 at the N terminus, and that with two R_2 groups.

other is required. Otherwise, a complex mixture would result (Figure 18-11), consisting of two homogeneous and two heterogeneous dipeptides.

EXERCISE 18-H

How many unique tripeptides can be formed from any combination of two different amino acids? from three?

Second, amino acid derivatives are prone to epimerization of the α carbon by a deprotonation-reprotonation sequence. Loss of stereochemical purity as the multistep synthesis proceeds leads to a large number of diastereomeric peptide impurities with structures nonetheless very similar to the desired peptide.

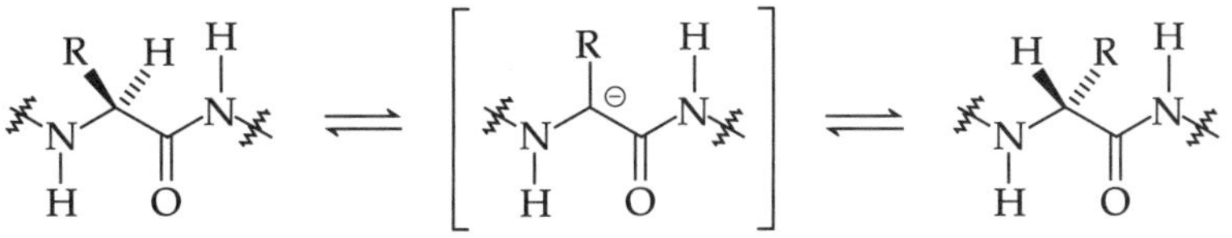

EXERCISE 18-I

The stereochemical center α to the carboxyl carbon of an amino acid derivative epimerizes more rapidly when the amino group is part of an amide than in the amino acid itself. Explain clearly why this is the case.

Third, to obtain pure peptide products, it is necessary that each growing peptide undergo every sequential condensation reaction. A peptide lacking even one of the amino acids from the total sequence constitutes an impurity that is difficult to remove.

Finally, many peptides are quite insoluble in solvents that are suitable for the condensation reactions used to form the amide linkages. For example, the simple peptide composed of six alanines linked together (hexaalanine) is almost completely insoluble in all solvents except water, and the nucleophilic reactivity of water precludes its use as a solvent for the reactions that form peptide bonds.

ARTIFICIAL SWEETENERS

Although the science of chemistry has advanced greatly in the twentieth century, many important discoveries are still made by accident. For example, the artificial sweetener, aspartame (NutraSweet) is a dipeptide that corresponds to the last two units of the peptide gastrin, a potent natural stimulant of gastric secretions. In the course of research on gastrin, James Schlatter, a chemist working at Searle, prepared the methyl ester of the dipeptide of aspartic acid and phenylalanine in 1965 and by chance tasted some of the crystalline powder that had gotten on his hands. Because the dipeptide is composed of two natural amino acids, Schlatter and other chemists at Searle realized that it was likely to be nontoxic and might therefore replace other, "less natural" artificial sweeteners such as cyclamate and saccharin. Indeed, five years later, cyclamates were banned by the Food and Drug Administration. Aspartame is 160 times as sweet as sucrose, whereas cyclamate is 30 times and saccharin is 500 times as sweet. Interestingly, neither of the constituent amino acids of aspartame is sweet: indeed, aspartic acid is bitter.

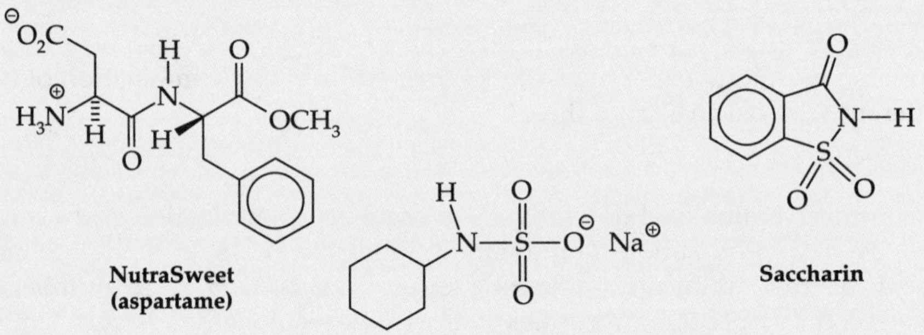

**NutraSweet
(aspartame)**

**Assugrin
(sodium cyclamate)**

Saccharin

Merrifield Solid–Phase Peptide Synthesis

The problem of insolubility was solved by the American chemist R. Bruce Merrifield, who was awarded the Nobel Prize in 1984 for his contribution to the advancement of peptide chemistry. In the **Merrifield synthesis,** the first amino acid of the peptide to be synthesized is chemically attached to porous beads of a polystyrene polymer support through a σ bond that is resistant to the reaction conditions required to form the peptide bonds. Additional amino acids are added step-by-step, with appropriate protection and deprotection steps. When all of the residues have been added, the complete peptide is released from the polystyrene support.

The polymer support is prepared by chloromethylating polystyrene by a Friedel-Crafts alkylation reaction (Chapter 11). In this process, a chloromethyl group is added to the aromatic nucleus (mainly at the position *para* to the point of attachment to the polymer chain) by reaction with chloromethyl methyl ether (Figure 18-12). This reaction almost certainly takes place through the methoxymethyl cation formed by loss of chloride ion by an S_N1 reaction. Reaction of this cation with the aromatic ring produces a methoxymethylene ($-CH_2OCH_3$) substituent that is further converted by reaction with HCl and $AlCl_3$ into the chloromethylene group.

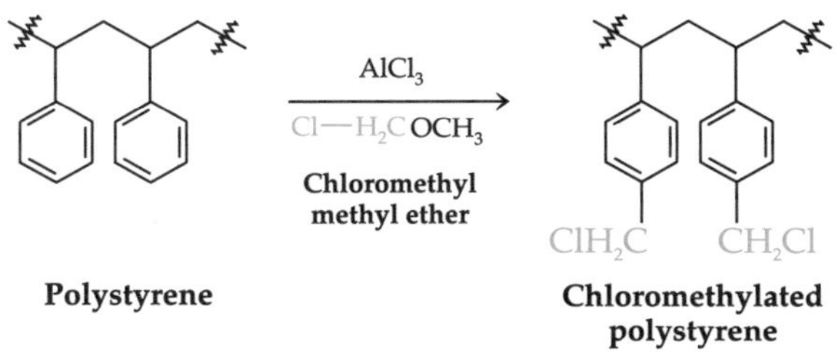

Polystyrene   →   **Chloromethylated polystyrene**

AlCl₃
Cl—H₂COCH₃
Chloromethyl methyl ether

◄ **FIGURE 18-12**
A Friedel-Crafts alkylation of polystyrene with chloromethyl methyl ether introduces a benzylic chloride that serves as a linking carbon for covalent attachment of an amino acid.

EXERCISE 18-J

Chloromethyl methyl ether is formed in situ from HCl, CH₃OH, and H₂C=O. Write a detailed, clear, and reasonable multistep mechanism for this conversion. This reaction also produces small amounts of bis-chloromethyl ether. (The potent carcinogenic activity of this compound was discovered, unfortunately, only through epidemiological studies of chemical plant workers.) Write a detailed mechanism for the formation of this dangerous by-product.

The amino acid units cannot be added uniquely to the growing peptide if both the amino and the carboxyl ends are free; thus, protection of one of these functional groups is required. Let us envision the process of synthesis schematically, in which a protected amino group is represented by —NHP (Figure 18-13). The first amino acid unit (starting from the carboxyl end of the peptide) is attached to the resin by reaction of the carboxylic acid group as a carboxylate ion with the benzylic chloride (an S_N2 reaction). A weak base such as triethylamine is used to convert the carboxylic acid into the carboxylate ion. The protecting group on nitrogen is then removed, exposing the free amino group for reaction with the carboxylic acid of the next amino acid to be added. What are the requirements that must be met by the protecting group? First, it must be stable to the reaction conditions used to form the amide bond during peptide synthesis. Second, conditions for removing this group from nitrogen must be sufficiently mild that the amide bonds already formed are not affected. The *t*-butyloxycarbonyl (*t*-BOC) unit discussed in Chapter 15 as a protecting group is used quite commonly in

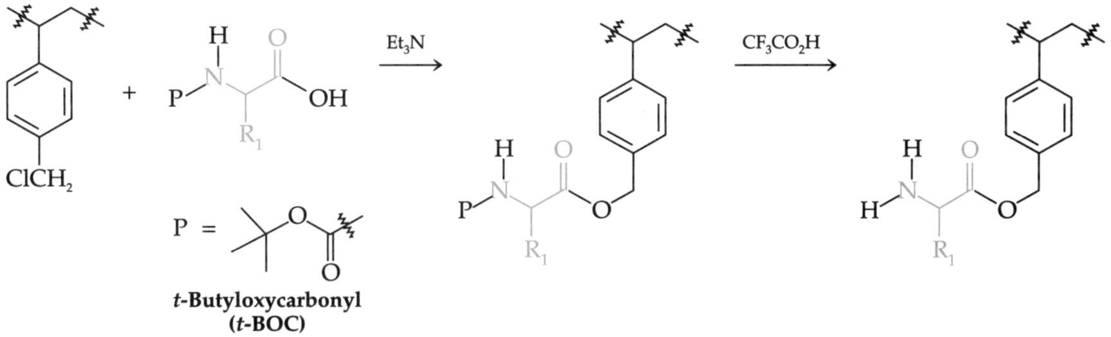

t-Butyloxycarbonyl (t-BOC)

▲ FIGURE 18-13
The amino group of the first amino acid of the peptide sequence is protected as a *t*-butyloxycarbonyl group. The free carboxylate functionality effects nucleophilic, S_N2 displacement of chloride ion from chloromethylated polystyrene, attaching the amino acid to the resin. Removal of the protecting group by treatment with trifluoroacetic acid produces a covalently attached amino acid with a free amino group.

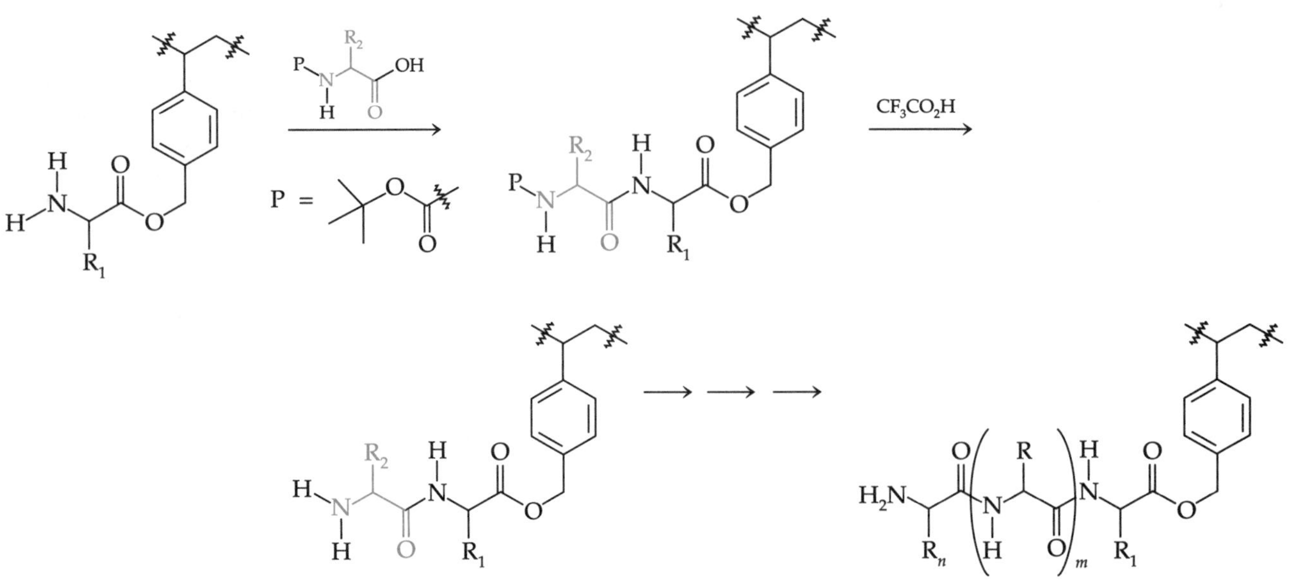

▲ FIGURE 18-14

The free amino end of an amino acid (or peptide) attached to the polystyrene resin reacts with the carboxylate end of an added N-protected amino acid. A new amide bond is thus formed, extending the peptide chain. Deprotection of the N-terminus by treatment with trifluoroacetic acid produces a covalently bound peptide with a free amino end that is ready for another amidation cycle.

solid-phase peptide synthesis. Deprotection of the amino group by removal of the *t*-BOC group is readily accomplished by treatment with trifluoroacetic acid. The synthesis requires two steps for the introduction of each amino acid unit: (1) formation of the peptide bond by reaction of the growing peptide with an N-protected amino acid; and (2) deprotection of the amino group in preparation for a repetition of the first step (Figure 18-14). The process is repeated until all of the peptide bonds have been formed. The product is then removed from the resin by cleavage of the benzylic carbon–oxygen bond connecting the peptide to the resin. A strong acid, either HF or HBr, in trifluoroacetic acid (Figure 18-15) effects an S_N2 displacement of the carboxylic acid.

EXERCISE 18-K

Write a clear, detailed reaction mechanism for the removal of the protecting group from a *t*-BOC-protected amino group. (Hint: Remember that a tertiary carbocation is exceptionally stable.)

The Merrifield synthesis of peptides usually proceeds in the direction illustrated in Figure 18-14, with attachment of each additional amino acid residue to the N-terminus of the growing polymer. This method is generally preferred to growth in the opposite direction (addition to the C-terminus) because of a side reaction that results when the carboxylic acid end of a peptide is converted into a more-reactive derivative. This process, illustrated for a carboxylic acid chloride (Figure 18-16), is an intramolecular reaction that forms an **azlactone.** (This cyclization takes place spontaneously at room temperature once the terminal carboxylic acid has been activated.) Although the azlactone group can react with the free amine of a

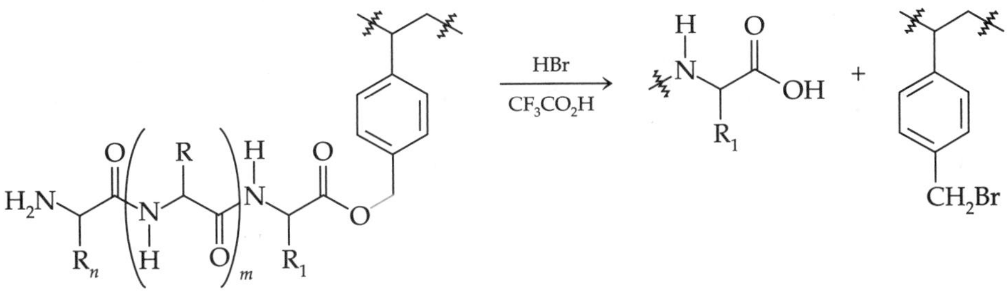

▲ FIGURE 18-15

After the cycle shown in Figure 18-14 has been repeated to sequentially attach all of the desired amino acids, the peptide is removed from the resin by treatment with HBr.

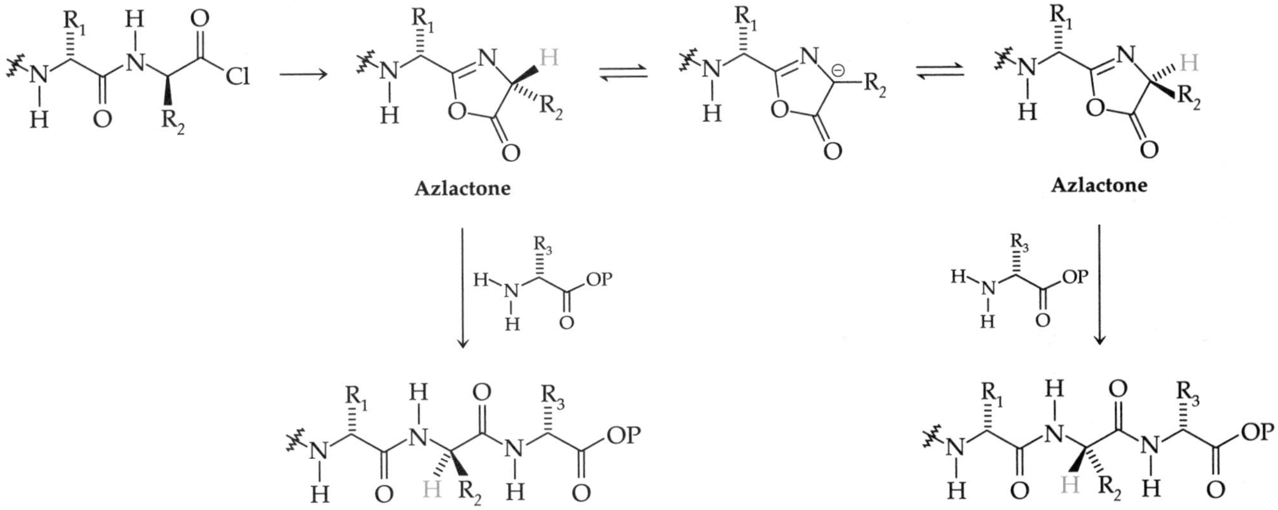

▲ FIGURE 18-16

The azlactone group is formed when a free carboxylic acid end of a peptide is converted into an acid chloride (or other reactive acyl derivative). Because of the high acidity of the hydrogen α to nitrogen in the azlactone, epimerization can take place, leading to the formation of disastereomeric contaminants.

carboxyl-protected amino acid to form an amide, it is especially prone to epimerization.

EXERCISE 18-L

Write a detailed reaction mechanism for the formation of an azlactone and for its reaction with an amine to form an amide, as illustrated in Figure 18-16. (Assume that azlactone is formed by nucleophilic acyl substitution.) What special feature of the anion of an azlactone makes it particularly stable?

The formation of an amide bond between amino acids can be accomplished in a number of ways, but all require that the carboxylic acid be converted into an activated acyl derivative that is more susceptible to nucleophilic acyl substitution. However, highly activated acyl derivatives of carboxylic acids are more susceptible to racemization at the α carbon because this center is acidified by the same electronic effect that enhances the reactivity of the carbonyl group. Thus, the choice of the activated acyl derivative becomes one of balancing the speed of the desired acylation against the rate of the undesirable epimerization.

A particularly good acyl derivative in this regard is one formed from a **carbodiimide** and the carboxylic acid. This addition product, which is quite similar to an anhydride, is not isolated but rather is generated in situ from the carboxylic acid in the presence of the amine.

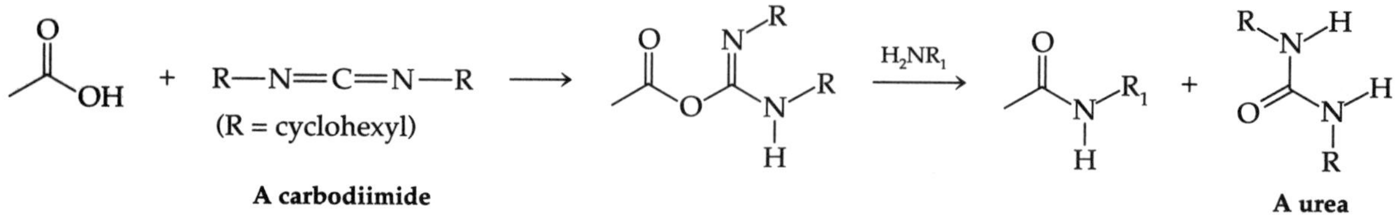

Nucleophilic acyl substitution takes place in the normal addition-elimination fashion and produces a urea (derived from the carbodiimide) in addition to the desired amide.

We can now construct a complete sequence for the synthesis of the simple dipeptide D-Ala–D-Ala, as shown in Figure 18-17. (We will see in Chap-

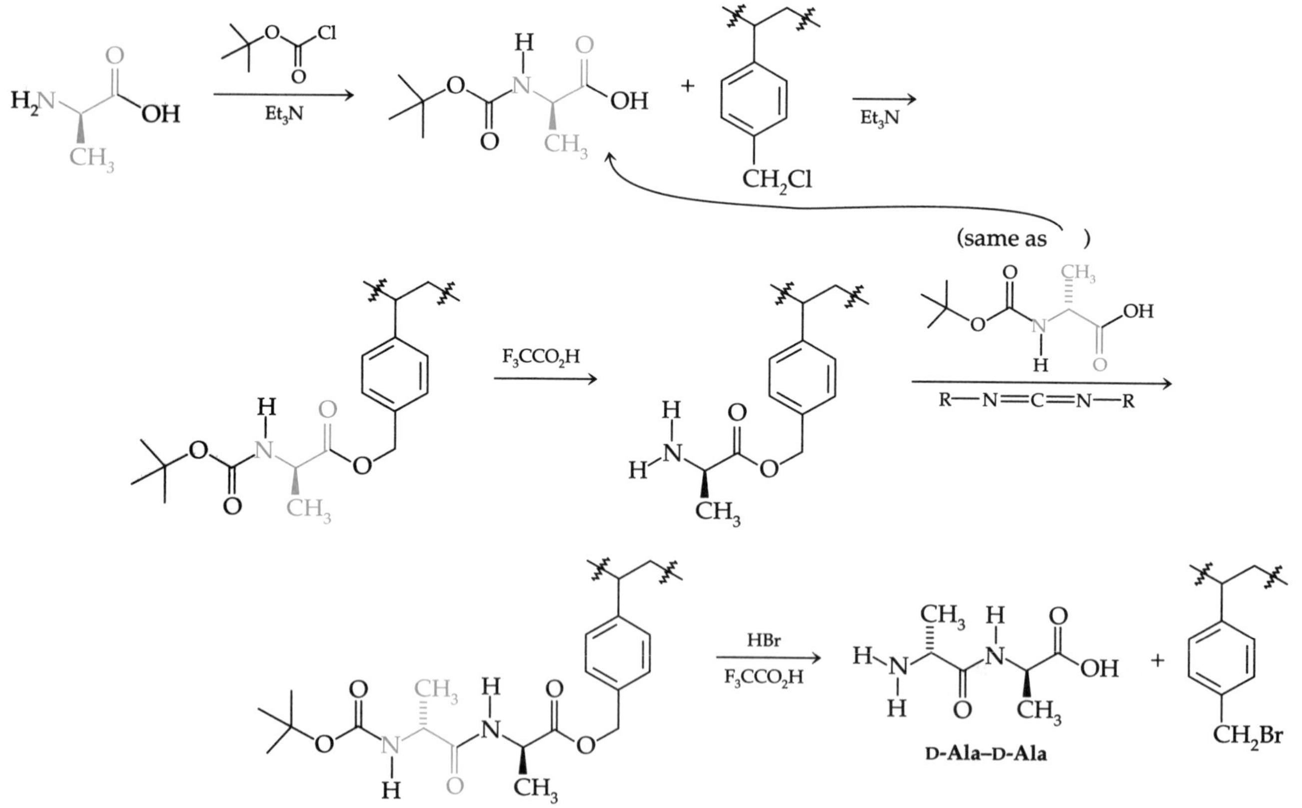

▲ FIGURE 18-17

Combining the individual steps shown in Figures 18-13 through 18-15 leads to a rational synthesis of a dipeptide. For the synthesis of D-Ala–D-Ala, the steps are: (1) protection of the amino group of the amino acid to be present as the C-terminus; (2) attachment to the polystyrene resin; (3) deprotection of the amino group of the bound amino acid; (4) amidation of the bound amino end with the carboxylate end of a second N-protected amino acid; and (5) deprotection of the N-terminus and removal of the peptide from the resin.

ter 23 how the presence of this two-amino-acid fragment plays a pivotal role in the destruction of bacterial cell walls by certain antibiotics.) The first step is the protection of the amine group of D-alanine by the *t*-BOC group. This protected amino acid is then attached to the Merrifield resin by an S_N2 reaction. The nitrogen-protecting group is then removed by treatment with trifluoroacetic acid. A second N-protected amino acid—in this case, another D-alanine—is then added, along with a carbodiimide. This treatment results in the formation of a peptide bond between the amino terminus of the first bound amino acid and the carboxylic acid of the second amino acid in solution. An excess of the amino acid in solution is used to insure an as-complete-as-possible conversion of the bound peptide into the more-extended product. After each step of the sequence, the resin is washed with an appropriate solvent to remove by-products and excess reagents. The completed peptide chain is removed from the resin by cleavage of the link between the carboxyl end and the resin with HBr in trifluoroacetic acid. A separate step to remove the nitrogen-protecting group from the last amino acid added is not required because the strongly acidic conditions used to cleave the bond to the resin are more than sufficient for removing the *t*-butoxycarbonyl group as well.

This peptide synthesis has now been automated, and several commercial units are available that automatically add the desired reagents according to a programmed sequence providing the required reaction times for each step. Even quite complex peptides containing a hundred amino acid residues have been prepared in this way.

EXERCISE 18-M

Show in detail the sequence of steps and the reagents required to synthesize Ala-Gly-Val, in which all the amino acids have the natural L configuration.

18-5 ▸ Alkaloids

Alkaloids, by definition, are naturally occurring materials that have a basic, nitrogen functional group; they are found primarily in plants. A number of structurally diverse examples of alkaloids are illustrated in Figure 18-18. Many higher plants contain a variety of alkaloids as complex mixtures: in most cases, we do not understand the biological importance of these materials to the plants in which they are produced. The only general explanation for the existence of this class of natural products (which comprises many tens of thousands of members) is that they serve to inhibit feeding on the plant material by predators. This supposition is based mainly on the observation that many have a bitter taste (to human beings). For example, the slightly bitter taste of **quinine** water is the result of only a few thousandths of a percent of that substance.

Nonetheless, alkaloids have received more attention than any other single class of naturally occurring compounds because a large fraction of them exhibit striking biological activity in insects and mammals, especially human beings. For example, nicotine is a potent insecticidal alkaloid, and the physiological effects of **nicotine, mescaline,** and **heroin** on people are well known.

Alkaloids have been isolated from plants for centuries—both as pure substances and as mixtures. The use of these materials as medicinal agents has historically formed the basis for the "wisdom" of the medicine man and

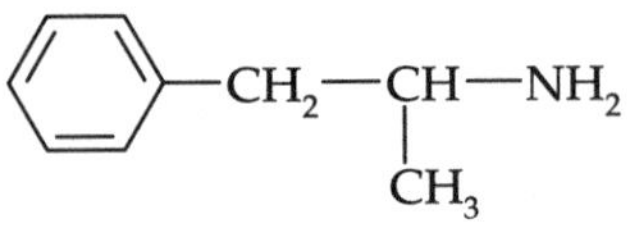

Amphetamine
(stimulant, LD_{50} 180 mg/kg, rats)

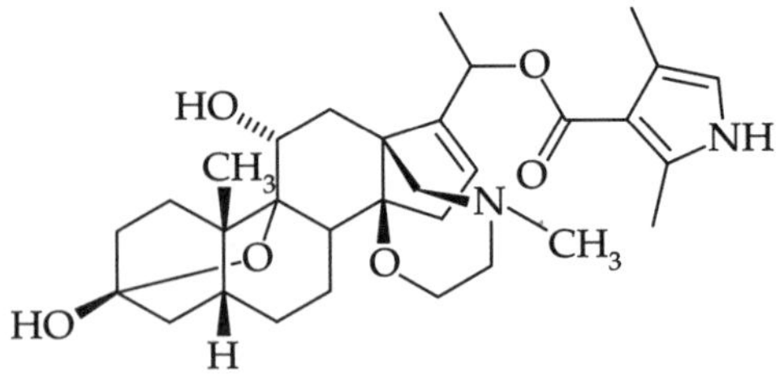

Batrachotoxin
(frog, poison, LD_{50} 2 μg/kg, mice)

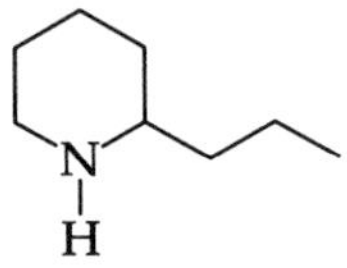

Coniine
(hemlock, poison, LD_{50} 10 mg/kg, human beings)

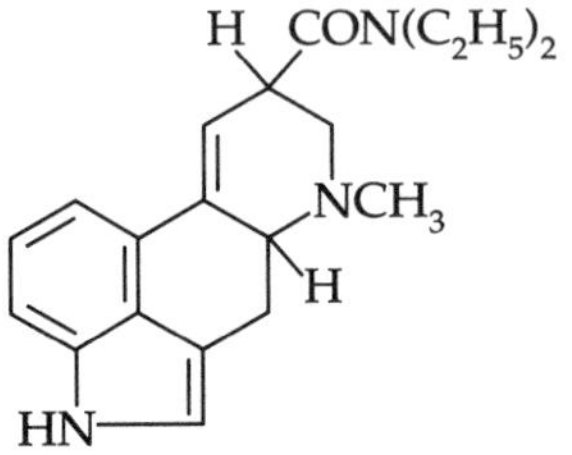

Lysergic acid diethylamide (LSD)
(ergot, hallucinogen)

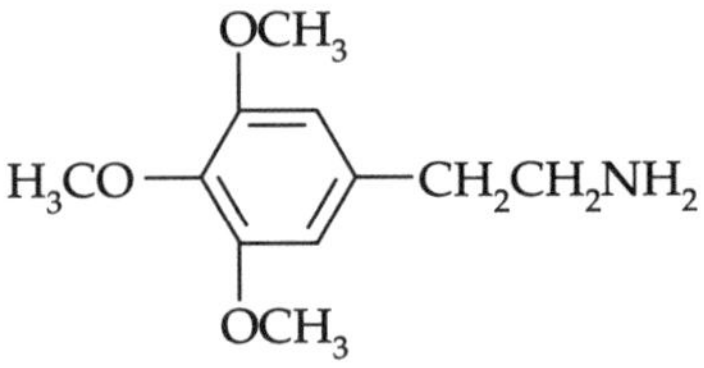

Mescaline
(peyote cactus, hallucinogen, LD_{50} 370 mg/kg, rats)

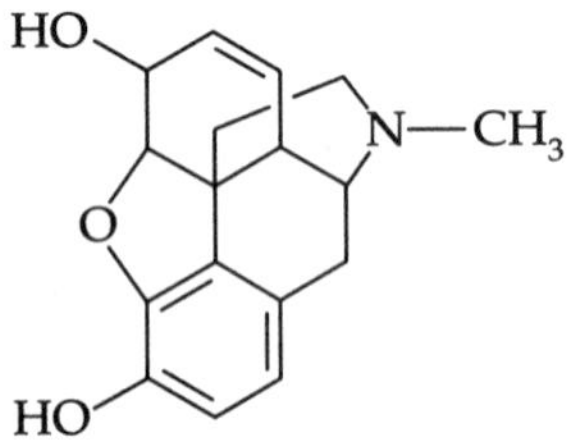

Morphine
(opium poppy, narcotic, analgesic)

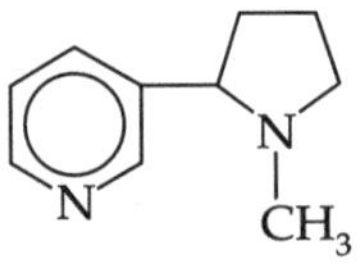

Nicotine
(tobacco, LD_{50} 230 mg/kg, rats)

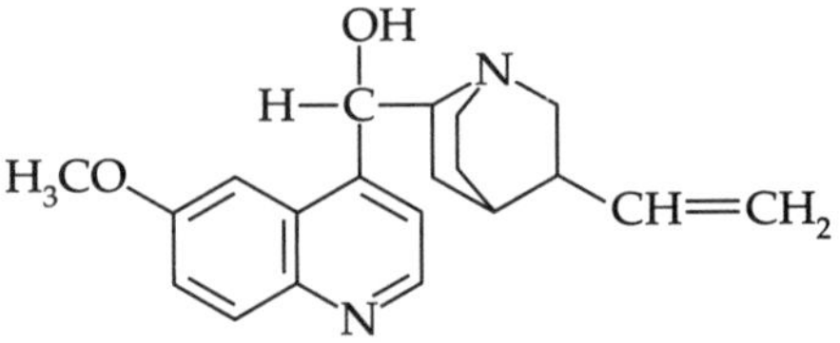

Quinine
(cinchona tree, antimalarial)

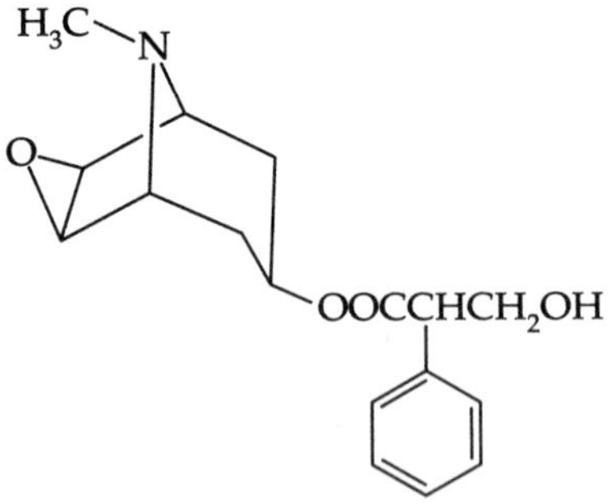

Scopolamine
(treatment of motion sickness, LD_{50}, 3.8 gm/kg, rats)

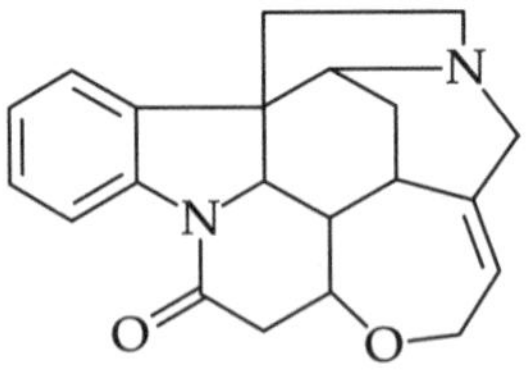

Strychnine
(strychos plant, poison, LD_{50} 1 mg/kg, human beings)

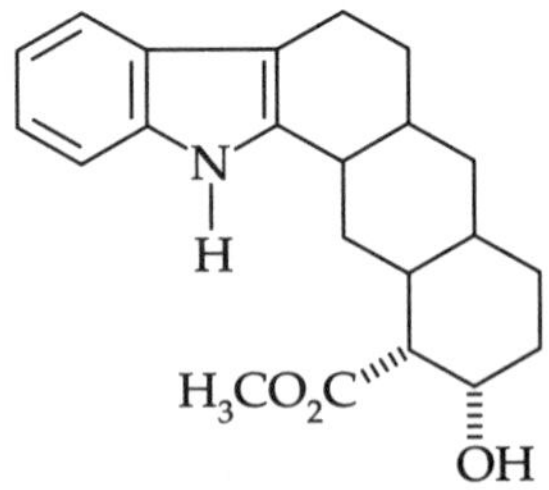

Yohimbine
(roots of *Rauwolfia serpentina* [mandrake], vasodilator, aphrodisiac?)

▲ FIGURE 18-18
Alkaloids vary in structure, but all contain nitrogen, are basic, and have natural origins in plants. The LD_{50} (lethal dose) is the amount required to kill 50% of a sample of the indicated animals.

the witch doctor in tribal societies and continues in today's homeopathy (more prevalent in Europe than in the United States). Indeed, the use of plant extracts persists today in the form of many home remedies used for a variety of ailments ranging from the common cold to cancer. In the past,

DOES NATURAL MEAN GOOD AND UNNATURAL MEAN BAD?

The idea that things that are "natural" must be good for us and that "unnatural chemicals" prepared in the laboratory are suspect has many examples that contradict both sides of this notion. Certainly, people would be much the worse without the synthetic "wonder drugs" such as antibiotics. Nonetheless, the idea persists, with reactions far out of proportion to reality of possible consequences. For example, the synthetic compound Alar was developed and marketed as an inhibitor of the enzymes that cause apples to ripen and ultimately to rot. Alar is degraded in the body to 1,1-dimethylhydrazine, which, in turn, is oxidized to a nitrosamine. Every nitrosamine tested in laboratory animals has been shown to be carcinogenic and this one is no exception. As a result, a campaign was mounted to ban Alar, and it succeeded, causing millions of dollars of apples treated with this compound to be destroyed. The action was based on only a limited knowledge of the facts: one mushroom contains ten times as much 1,1-dimethylhydrazine as two apples treated with Alar.

Alar   1,1-Dimethyl-   A nitrosamine
                hydrazine

physicians scoffed at the notion of such cures, but chemists working in the pharmaceutical industry have been able to isolate many useful alkaloids (and other compounds) from plants. The destruction of large areas of natural vegetation in developing countries and the resulting extinction of large numbers of plant species has raised concern over the possible loss of valuable organic compounds.

Many alkaloids are poisonous and some, such as **coniine,** the hemlock poison given to Socrates, are known specifically for this biological activity. Others have less-drastic biological properties when given in small doses. **Cocaine** is, unfortunately, a frequently used illicit drug that causes artificial mood elevation and euphoria; **atropine** is used to dilate pupils for eye examinations; and **yohimbine,** a reputed aphrodisiac, has been used as an antidepressant.

The names, if not the structures, of many of the alkaloids are familiar to us because of the very potent ability of these substances to disrupt the **central nervous system (CNS).** A general structural feature of this group of psychoactive compounds is a β-**phenethylamine** subunit. The chemical basis for the general CNS activity of molecules containing this unit is treated in Chapter 23.

Morphine

β-**Phenethylamine**

EXERCISE 18-N

Identify the β-phenethylamine moiety, if possible, for each of the alkaloids shown in Figure 18-18. Devise a synthesis of β-phenethylamine starting from benzene and any other reagents.

18-6 ▸ Nucleic Acids

Another important group of nitrogen-containing compounds are the **polynucleic acids,** very large and seemingly quite complex molecules. A short section of a polynucleic acid is illustrated in Figure 18-19. We will describe here only the constituent structure of these biopolymers: a detailed chemical description of how they serve to store and code information, specifically that which translates into sequences of amino acids in the peptides and enzymes critical to living systems, is treated in Chapter 19.

The polynucleic acid backbone is built on two different monomer units: a phosphate group and a ribose sugar unit (in the furanose form), the latter of which bears a heterocyclic base. This polymer is known as a **ribonucleic acid,** or RNA. Alternatively, in deoxyribonucleic acid, or DNA, the sugar unit in the backbone is deoxyribose.

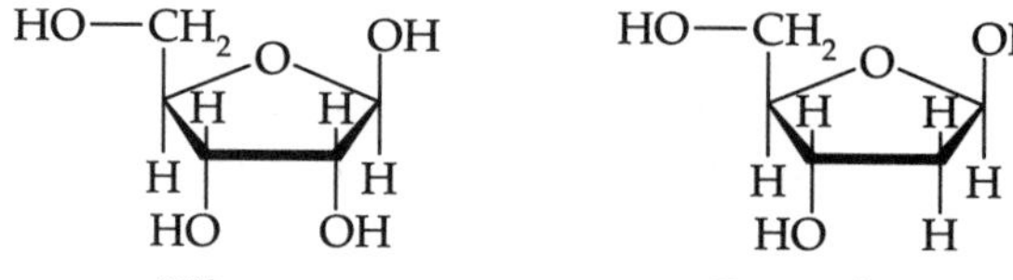

Ribose   Deoxyribose

First, let us consider the structures of the **nucleic acid bases** that are appended to the ribose (or deoxyribose) sugar units. As you will recall from

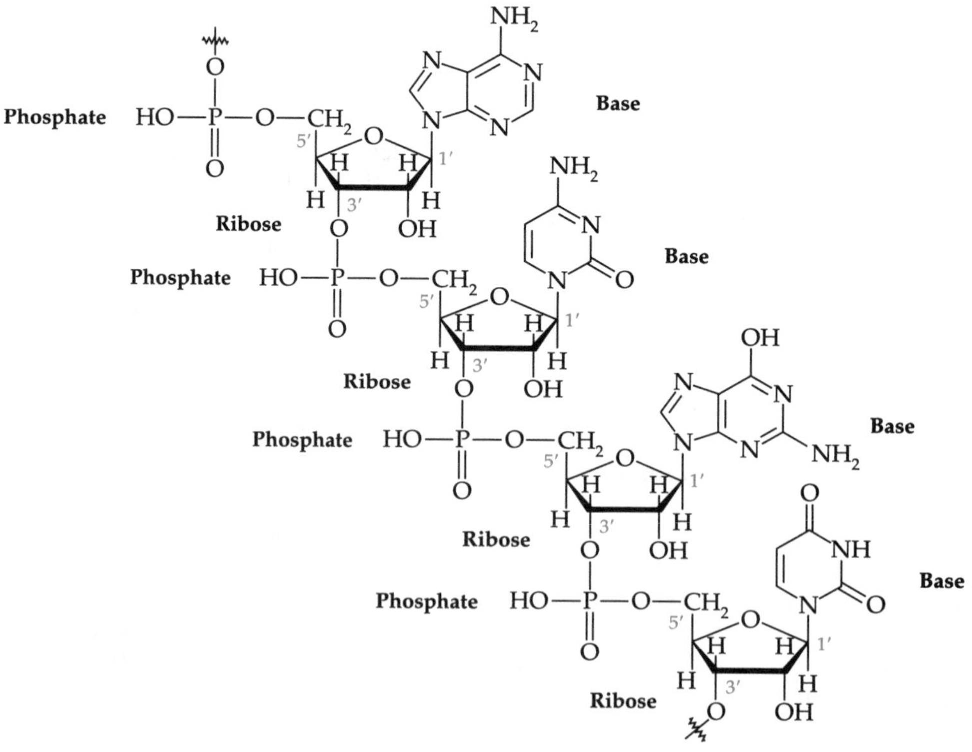

Polynucleic acid

▲ FIGURE 18-19
Polynucleic acids contain a backbone consisting of ribose units connected through a phosphate diester linkage between the C-5′ OH group of one sugar and a C-3′ OH group of another. C-1′ of each ribose unit is bound to a purine or pyrimidine base.

Chapter 3, these bases are heterocyclic amines. They are divided into two classes: the **purine bases** (adenine and guanine) and the **pyrimidine bases** (cytosine, uracil, and thymine).

Pyrimidine Bases

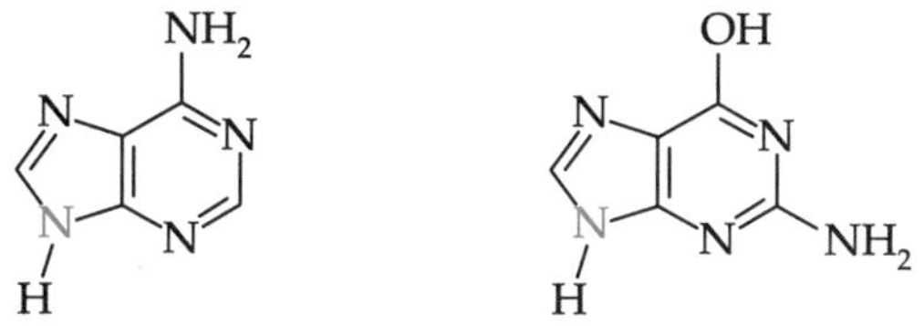

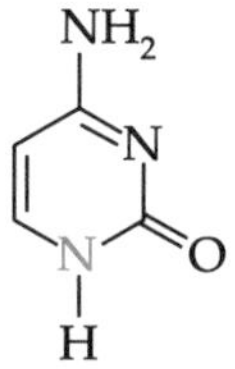

Cytosine

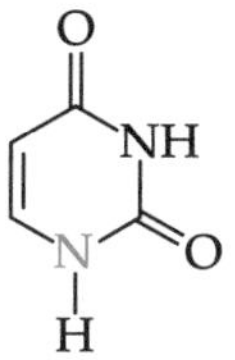

Uracil

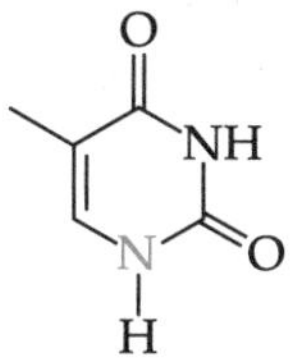

Thymine

Purine Bases

Adenine   Guanine

The names purine and pyrimidine derive from those of the parent, heterocyclic skeleton (Chapter 3), whereas the names of the individual bases are truly trivial (note that thymine is methyluracil). These bases are the critical units of polynucleic acids; they are the chemical basis for the storage and coding of information. Their precise sequence in a nucleic acid chain ultimately determines the specific sequences of amino acids in peptides and enzymes.

These naturally occurring aromatic bases contain several heteroatoms, both nitrogen and oxygen. We will see in Chapter 19 that hydrogen bonding between these functional groups serves as the key to information storage in the genetic code. One of the nitrogen atoms (that indicated in color in the margin) always serves to link the base to the ribose unit. Recall from Chapter 17 that carbohydrates such as glucose can be converted into glucosides by reaction with alcohols, following the normal reaction pathway for the formation of acetals. This transformation is illustrated for ribose in Figure 18-20, in which such a glucosidic linkage is formed with methanol. Use of a secondary amine in place of the alcohol results in the formation of an aminal, a structure quite similar to that of an acetal. A similar reaction between a sugar unit and a base leads to a **nucleoside** (Figure 18-21), in which the base is a substituent on C-1 of a ribose (or deoxyribose) sugar unit.

Each of the bases in a nucleic acid or a nucleoside has at least two nitrogen atoms and, when the reaction between one of these bases and ribose is carried out under laboratory conditions, a mixture results (Figure 18-22). The five-membered rings in adenine and guanine resemble the imidazole ring, a substituent in histidine (Section 18-2). Therefore, it should not be

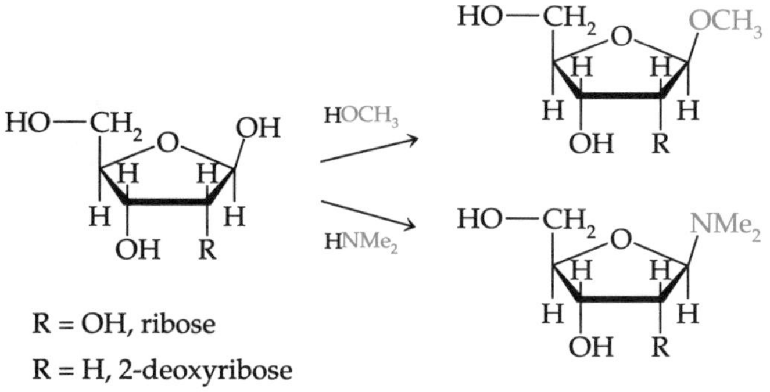

R = OH, ribose

R = H, 2-deoxyribose

▲ **FIGURE 18-20**

The attachment of a base to the furanose form of ribose or deoxyribose (lower reaction) to produce an aminal takes place through a pathway similar to the formation of glucosides (upper reaction).

Nucleosides

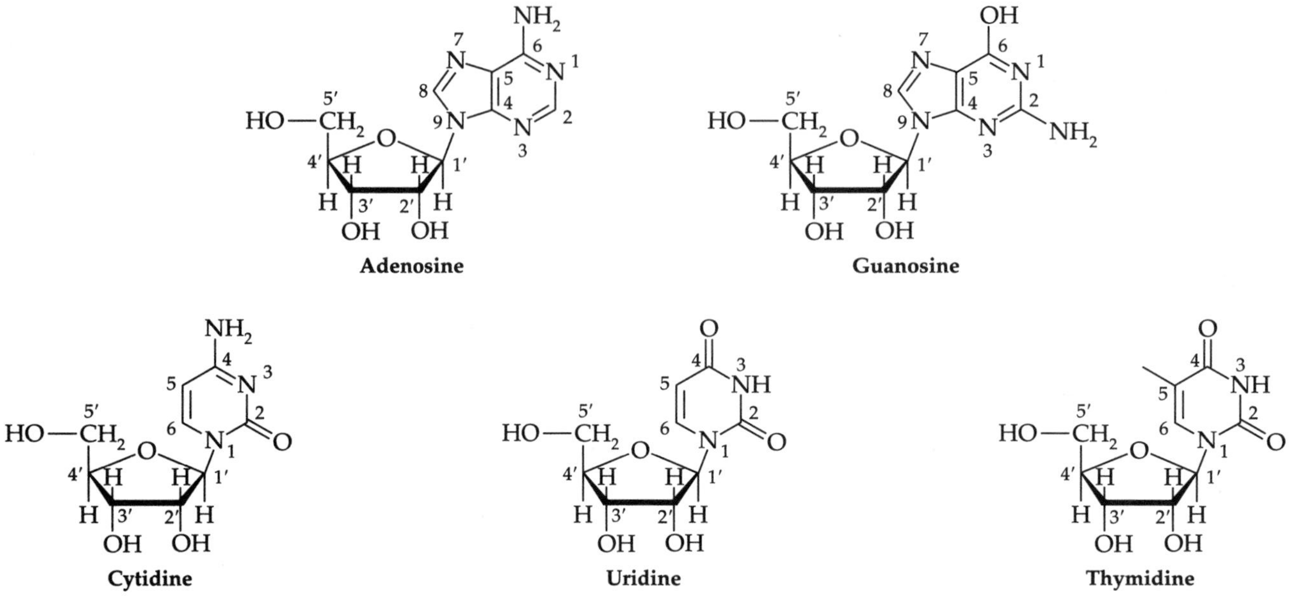

▲ FIGURE 18-21

In a nucleoside, a heterocyclic base is attached to C-1' of ribose or deoxyribose. (The two units that constitute a nucleoside, the base and the carbohydrate fragments, have separate numbering systems, with the numbers identifying atoms in the carbohydrate part primed.)

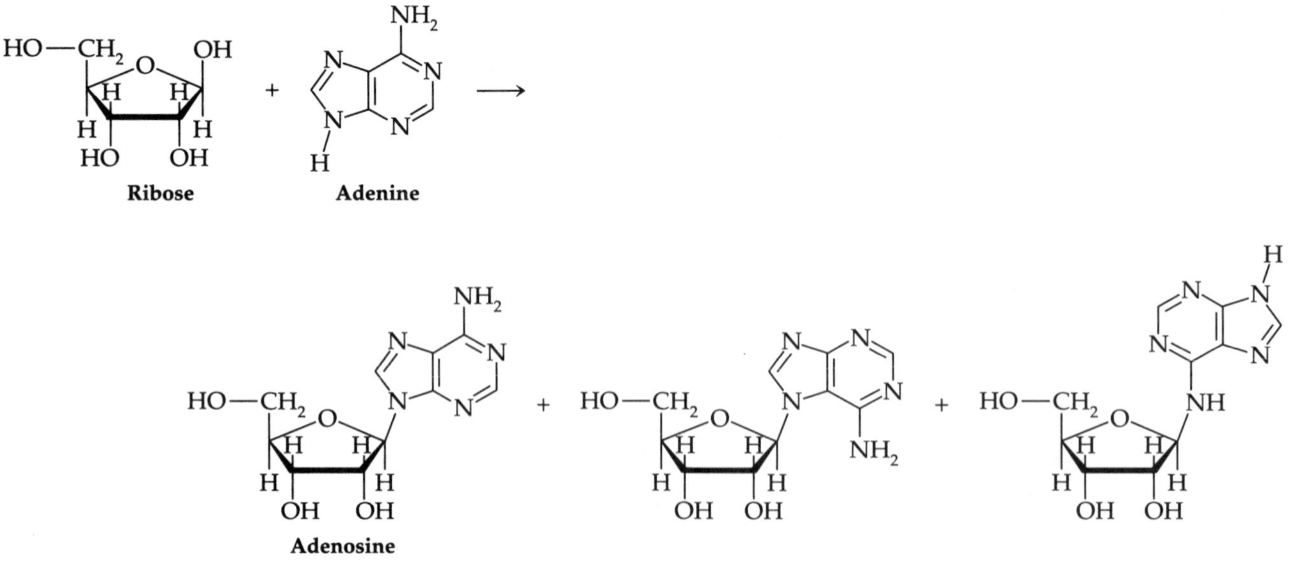

▲ FIGURE 18-22

The nucleic acid bases are in tautomeric equilibria resulting in several different structures, each of which has nucleophilic character on different ring nitrogen atoms. Consequently, several isomeric products (differing in the site of the heterocyclic base at which backbone sugar attaches) result from the reaction of ribose with one of the bases, illustrated here with adenine.

surprising that, in solution, adenine and guanine exist as mixtures of proton tautomers, differing in which nitrogen bears the proton. Other possible proton tautomers for these heterocyclic bases will be considered in Chapter 19, when we describe how these bases interact by hydrogen bonding for information storage. The biochemical synthesis of the nucleosides requires enzymes that specifically catalyze the formation of only one of the possible tautomeric structures.

EXERCISE 18-O

Draw both tautomers of adenine; write reaction mechanisms for the interconversion of these forms under both acidic and basic reaction conditions.

Let us now turn our attention to the polymer formed from the combination of these individual nucleosides with phosphoric acid. Figure 18-23 illustrates the structure of a ribonucleic acid in which the bases are represented diagrammatically. This illustration enables us to visualize the polynucleic acid as having been derived from a 1:1 mixture of two monomers, phosphoric acid and the nucleosides. Both of these monomers are polyfunctional and are thus capable, at least in theory, of forming highly branched polymeric networks. Nonetheless, nucleic acids are formed only as shown in Figure 18-23: that is, with the monomers serving as bifunctional links and with only the C-3' and C-5' hydroxyl groups of ribose participating in the binding.

We have represented the polymer as being composed of phosphate diesters in which the remaining oxygen is a free hydroxyl group. However, this group has a sufficiently high acidity [pH of $(RO)_2P-OH = 2$] that it is essentially completely deprotonated under the near-neutral conditions found in living systems. Thus, DNA and RNA exist as **polyanions** and are therefore **polyelectrolytes.** For this reason, electrophoresis is particularly valuable for the separation and identification of polynucleic acids.

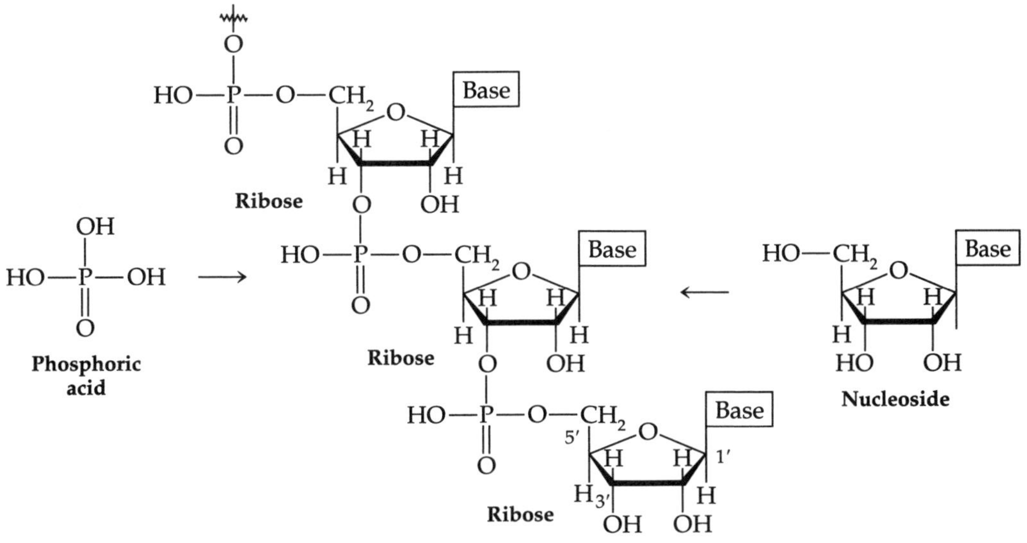

▲ FIGURE 18-23

This representation of a nucleic acid emphasizes the nucleoside-phosphate backbone, whereas it deemphasizes the structural contributions of the heterocyclic bases.

18-7 ▸ Aminocarbohydrates

We have seen in Chapter 17 that carbohydrates play a central role in energy storage. In addition, carbohydrates substituted with amino groups, **aminocarbohydrates** or **aminosugars,** serve a crucial role in molecular recognition. For example, different aminocarbohydrate units serve as the unique feature that distinguishes the various blood-group types. Here we will briefly discuss the chemical features that distinguish aminocarbohydrates from their all-oxygen analogs.

Many different aminocarbohydrates are found in nature. The structures of some of the more-important ones are shown in Figure 18-24, along with the structures of three antibiotics, **streptomycin, erythromycin,** and **kanamycin A,** all of which are multicyclic polyoxygenated antibiotics that have at least one appended aminocarbohydrate. These antibiotics are resistant to biological degradation that occurs by cleavage of the acetal linkages, in part because of the presence of the basic nitrogen atom(s).

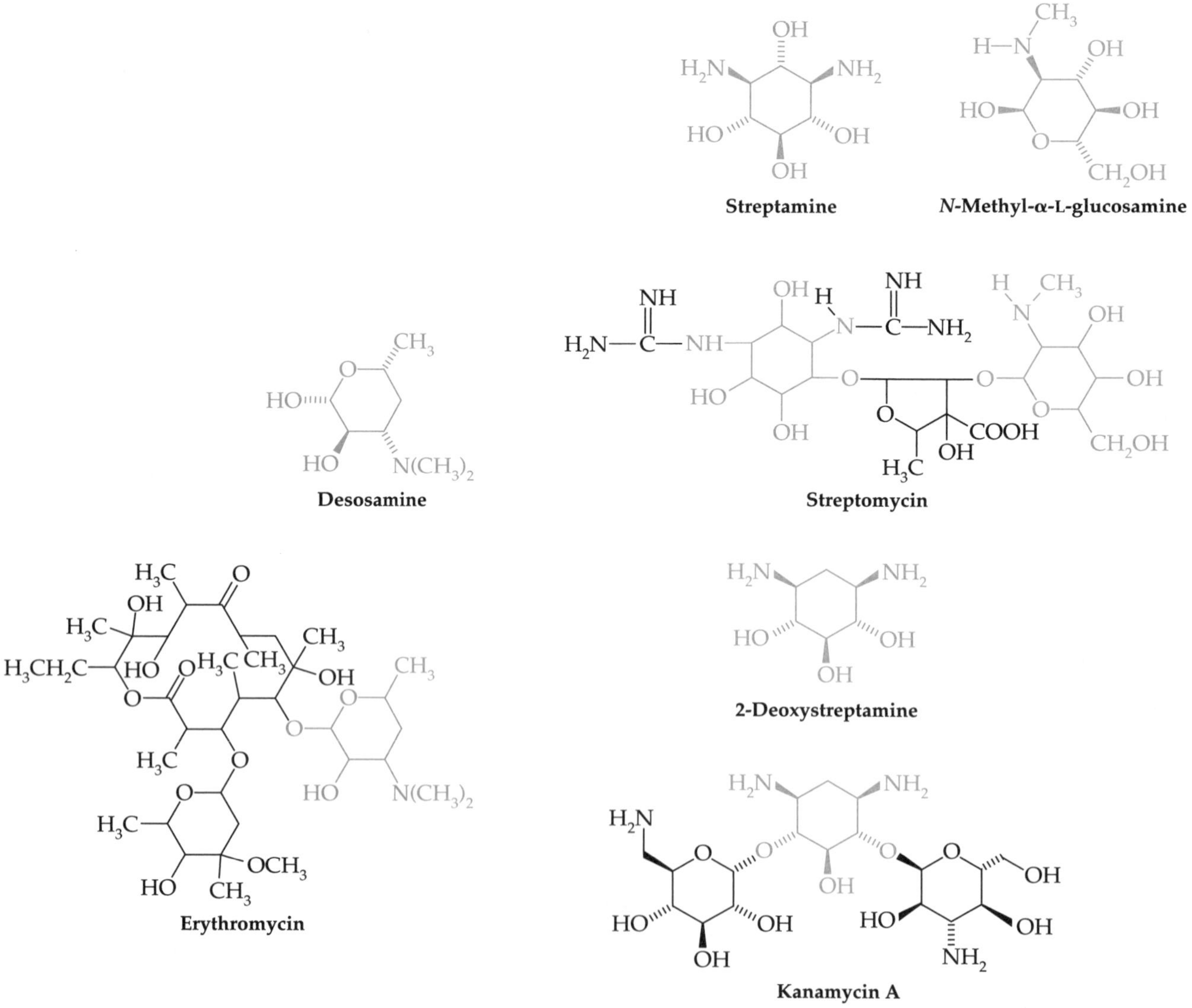

Streptamine   **N-Methyl-α-L-glucosamine**

Desosamine   **Streptomycin**

2-Deoxystreptamine

Erythromycin

Kanamycin A

▲ FIGURE 18-24

A number of active antibiotics contain subunits with six-membered rings (either carbocyclic or pyranose) bearing hydroxyl and amino groups.

NATURAL CHEMICAL DEFENSE SYSTEMS

Numerous toxic compounds are produced by animals, apparently as chemical defenses against predators. Frogs, especially those that live in the tropics, excrete alkaloids, many of which are fatal to mammals in very small doses. For example, bufotenine and bufotoxin have been isolated from the toad *Bufo marinus*. Both of these compounds increase the contractive power of weak heart muscles, and the latter has been reported to cause terrifying hallucinations. The dried skin of these animals has been used in zombie powder, leading to the suggestion (unsubstantiated) that these and perhaps other compounds present in the skin of the toad are responsible for inducing the trancelike state associated with zombies.

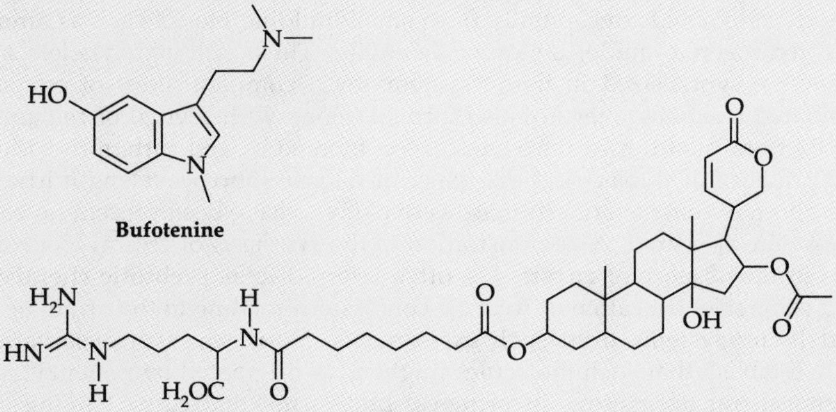

Bufotenine

Bufotoxin

A related alkaloid, homobatrachotoxin, has been isolated from birds of the genus *Pitohui* that are endemic to the New Guinea subregion. The same compound (along with batrachotoxin) is found in Colombian poison-dart frogs of the genus *Phyllobates*. Only 2 μg of these alkaloids will kill a 1 kg mouse (proportionally less for mice of normal size). It is probable that frogs and birds synthesize these alkaloids by modification of closely related substances obtained by eating plants.

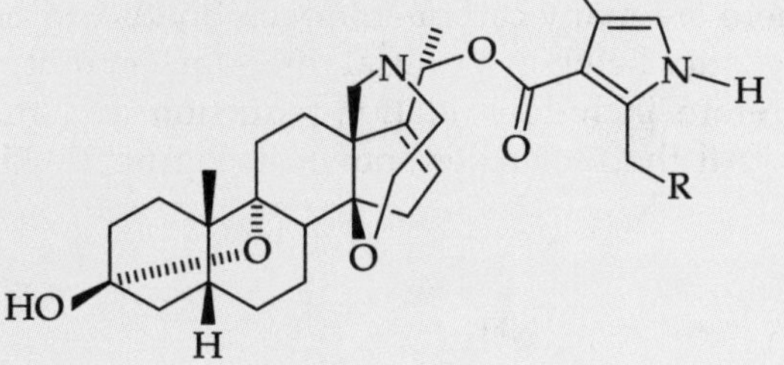

Batrachotoxin (R = H)
Homobatrachotoxin (R = CH₃)

18-8 ▶ Abiotic Synthesis

Throughout Chapters 17 and 18, we have seen many natural products, a large fraction of which appear to be structurally quite complex. For example, carbohydrates such as glucose and the nucleic acid bases have several heteroatom-based functional groups, and, in the former case, there are numerous stereocenters as well.

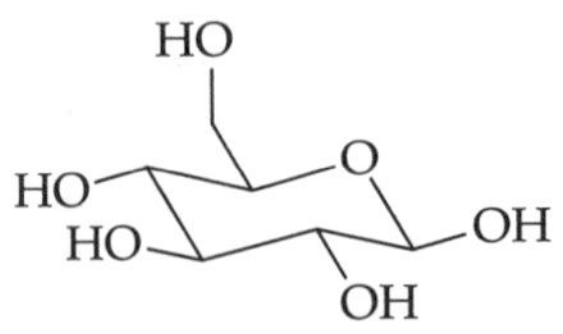

α-D-Glucose 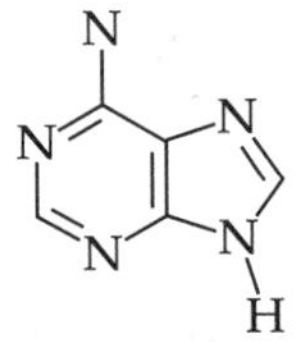   Adenine

In part, this functional complexity is necessary for their use as components of complicated biochemical systems because it establishes the basis for specific interactions between molecules. Were the molecules of nature less complicated, the distinction between one molecule and the next would be less.

Although we can consider the molecules of biochemical systems to be complex from a structural point of view, they constitute arrays that can be readily assembled conceptually from small building blocks such as ammonia, hydrogen cyanide, and formaldehyde. Thus, although nucleic acid bases are synthesized in living systems by a complex series of enzyme-mediated reactions, they are also formed (along with several of the amino acids) from mixtures of nitrogen, carbon monoxide, and carbon dioxide in the presence of an electrical discharge or intense short-wavelength irradiation. Both of these energy sources were likely to have been present on earth before life appeared. A demonstration of the synthesis of critical biochemicals in the absence of enzymes is often referred to as **prebiotic chemistry,** and some scientists attempt to draw conclusions relating to the origin of life and living systems from such experiments. However, such observations only establish that such molecules might have originated by nonenzymatic chemical transformations in primeval broths, not that current living systems have necessarily derived from them. The term **abiotic synthesis** is more appropriate because it does not have a connotation of time.

Adenine

Let us examine one possible abiotic synthesis of adenine for which the source of the atoms is illustrated in Figure 18-25. In fact, this abiotic synthesis closely follows the biosynthesis of adenine that takes place in cells. Using the lessons that we learned in Chapter 15 on retrosynthetic analysis, we can see that in adenine there are many carbon–nitrogen bonds but only two carbon–carbon bonds, the latter being in a central three-carbon unit. An effective synthesis must therefore include oxidation-reduction and functional-group transformations. All the molecules shown in Figure 18-25 as

▶ FIGURE 18-25

One possible retrosynthetic analysis of adenine breaks it into commonly encountered fragments.

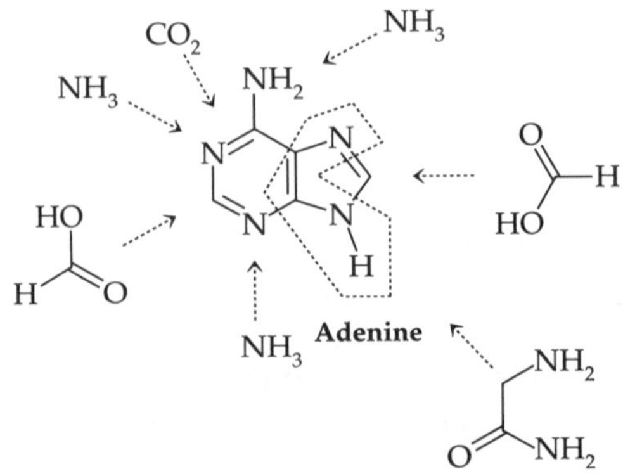

"starting materials," ammonia, formic acid, and the amide of the amino acid glycine, have been detected in interstellar space. The synthesis that is proposed here uses only those reactions covered in earlier chapters.

EXERCISE 18-P

Identify which of the starting sources of carbon for adenine in Figure 18-25 must undergo a net reduction, which must be oxidized, and which remain at the same oxidation level in adenine.

The synthesis starts with the formation of the formamide of α-amino-acetamide through condensation with formic acid (Figure 18-26). Indeed, amides can be formed from amines and carboxylic acids at high temperature, although the severity of the reaction conditions limits the usefulness of the method for complex molecules. Alternatively, this same product can be formed by the addition of the amine nitrogen to formaldehyde followed by oxidation.

In the second step (Figure 18-27, on page 642), condensation of ammonia with the acetamide carbonyl group forms a C=N double bond similar to that present in a simple imine. Intramolecular condensation then forms the five-membered ring. Dehydration followed by proton tautomerization establishes a six-π-electron aromatic ring, the heterocyclic system known as imidazole.

EXERCISE 18-Q

Write a detailed, step-by-step reaction mechanism for the condensation of the primary amide of glycine with formic acid (step 1, Figure 18-26).

We can now understand that the condensation reactions in this sequence are driven by the aromatic stabilization energy of the substituted imidazole. Looking ahead, we also realize that the final product (adenine) has an additional aromatic ring.

In the next stage of the synthesis, we begin to add the necessary atoms that constitute the six-membered ring. Although the aminoimidazole is stabilized by aromaticity, it does have an enamine functional group. Recall

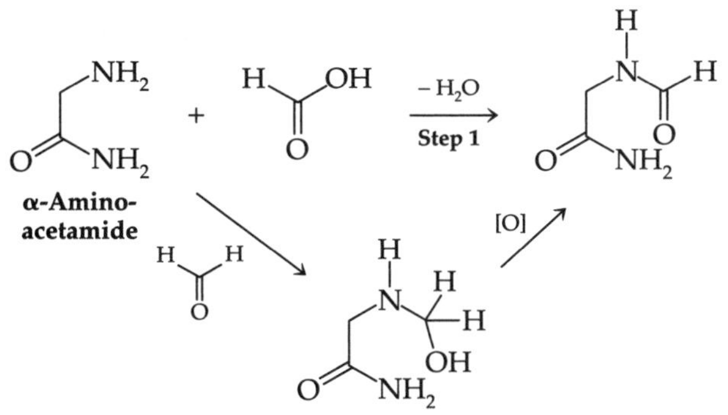

▲ FIGURE 18-26
The formamide of α-acetamide can be formed either by acylation of the primary amine (upper reaction) or by reaction of the primary amine with formaldehyde, followed by oxidation of the resulting aminol (lower reaction).

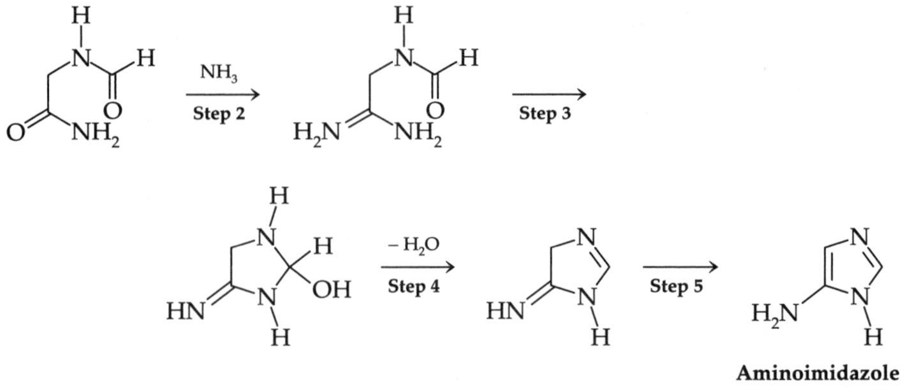

▲ FIGURE 18-27

Condensation of the primary amide group of the diamide product in Figure 18-26 with ammonia results in the formation of an iminoamide. The —NH2 group of an iminoamide is especially nucleophilic, and intramolecular cyclization by addition to the amide carbonyl takes place (analogous to that in imine formation), producing an endocyclic C=N bond. Proton tautomerization completes the synthesis of the aminoimidazole.

from Chapter 12 that enamines are moderately active nucleophiles and can, for example, react with electrophiles. In this case, the electrophile is carbon dioxide (Figure 18-28) and the carbon–carbon bond formed in this reaction (step 6 in the sequence) completes the three-carbon central unit of adenine.

We are now close enough to our target to see that what remains is the addition of one more carbon and two more nitrogen atoms. The next operation is the conversion of the just-formed carboxylic acid into the amide (Figure 18-29) by reaction with ammonia, in a process analogous to the first step

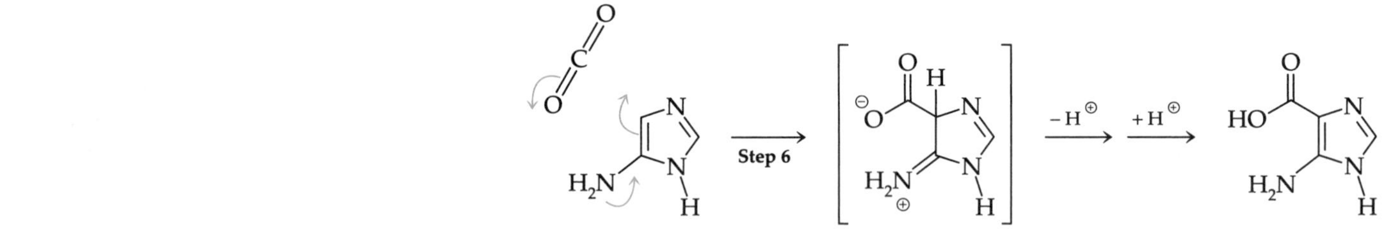

▲ FIGURE 18-28

Carboxylation of the enamine group in the aminoimidazole product in Figure 18-27 is analogous to the nucleophilic reaction of a Grignard reagent with carbon dioxide to form a carboxylic acid. Again, proton tautomerization aromatizes the imidazole ring.

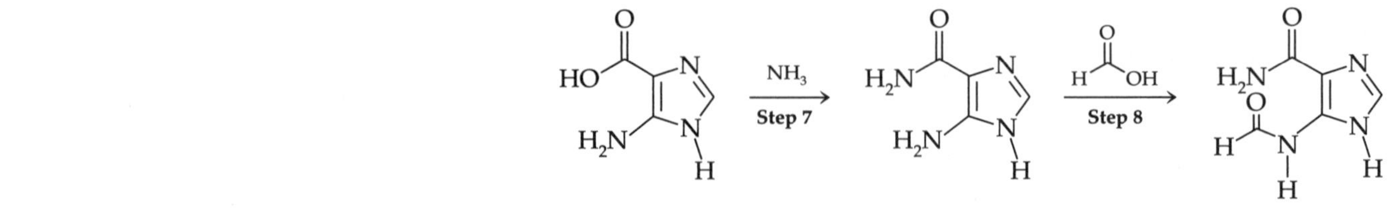

▲ FIGURE 18-29

Reaction of the carboxylic acid with ammonia forms an amide, as does the reaction of the amino group with formic acid. All of the atoms needed for the six-membered ring of adenine are now in place.

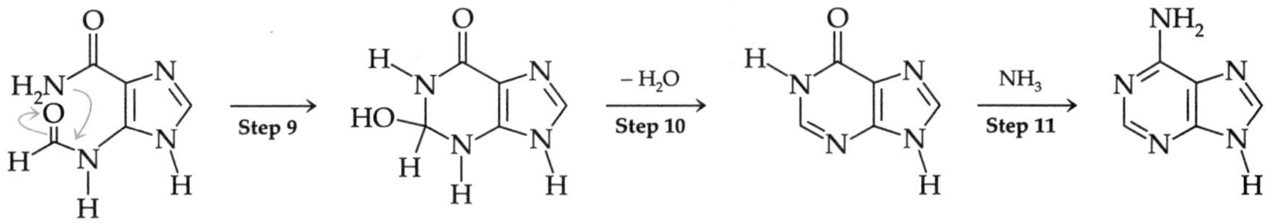

▲ FIGURE 18-30

Nucleophilic attack of the primary amide nitrogen on the formamide carbonyl closes the six-membered ring by forming a C–N bond. Dehydration, followed by tautomerization, aromatizes the ring, producing adenine.

in the sequence. A second amide group is then formed between formic acid and the primary enamino group in step 8.

EXERCISE 18-R

Would you expect the primary enamino group that undergoes acylation in step 8 of Figure 18-29 to be more or less reactive than a simple primary amine (such as dimethylamine) in a nucleophilic acyl substitution process?

The six-membered ring of adenine is formed through a condensation reaction analogous to steps 3 and 4 (Figure 18-27), forming the imidazole ring (Figure 18-30). The final step is the reaction of ammonia with the remaining carbonyl group.

EXERCISE 18-S

Write a clear detailed mechanism for the transformation shown in step 11 of Figure 18-30.

There are eleven steps in this sequence and, by the criterion of length, it is the most-complicated synthesis that we have examined so far. We can group the individual steps of this synthesis according to the type of transformation accomplished (Figure 18-31). From this perspective, the synthesis is quite straightforward because it consists of a repeated use of two functional-group transformations, the formation of amides (A) and imines (B),

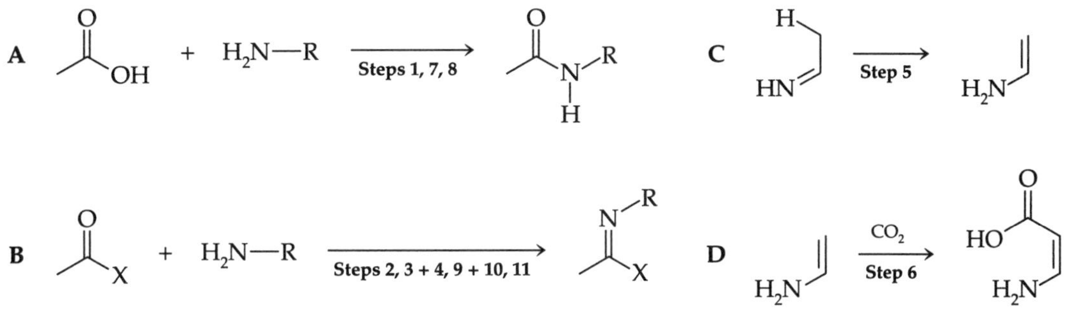

▲ FIGURE 18-31

In the adenine synthesis in Figures 18-26 through 18-30, three steps are conversions of carboxylic acids into amides by reaction with a primary amine (A); four steps are iminations of an acid derivative (B); one step is an imine–enamine tautomerization (C); and one step is a nucleophilic carboxylation (D).

a proton tautomerization (C), and the formation of one carbon–carbon bond (D). Furthermore, both functional-group transformations are accomplished through bond formation between nitrogen and a carbonyl carbon. Thus, although there is a relatively complex array of individual functionalities present in adenine, the molecule can be assembled using only a few elementary synthetic transformations. In a living system, it is important that relatively simple chemistry be required for the construction of complex molecules so as to maximize the overall chemical efficiency of the organism. From this perspective, adenine is a "natural."

EXERCISE 18-T

Analyze the structure of cytosine in the same way as that of adenine was analyzed in Figure 18-25. How many fundamental synthetic operations (analogous to the eleven shown in Figure 18-31 for the synthesis of adenine) would be required?

Carbohydrates (Ribose)

We now consider the abiotic synthesis of ribose, the carbohydrate that, in combination with bases such as adenine, forms nucleotides and nucleic acids. Ribose (and other carbohydrates) can be formed by a multistep synthesis requiring, in this case, only one reaction type. It can be synthesized by the treatment of formaldehyde with calcium hydroxide.

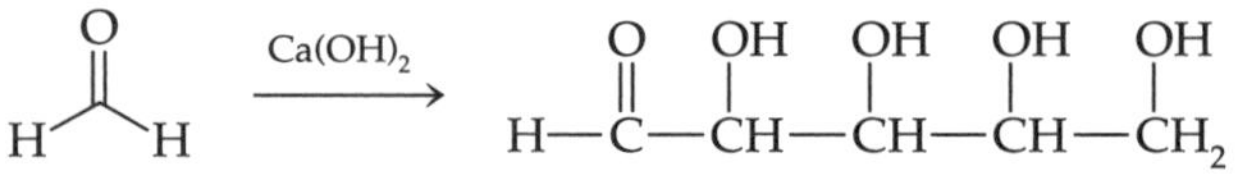

The carbon–carbon bonds are formed through nucleophilic addition to a carbonyl group by the anion derived by deprotonation of formaldehyde.

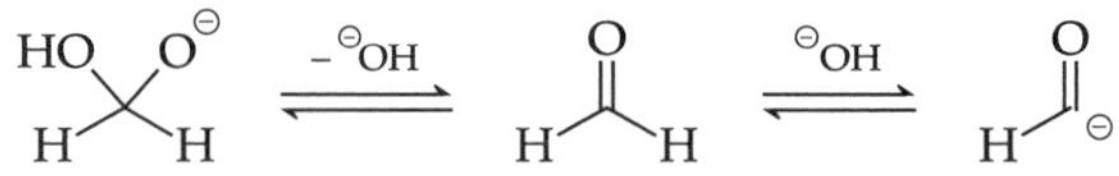

At first, this anion may seem unusual because, in all earlier examples in which carbonyl groups were converted into anionic nucleophiles, the proton was removed from the α carbon rather than from the carbonyl carbon itself. But, in formaldehyde and other aldehydes lacking α-C–H bonds, the interaction with base cannot follow the normal course. As in the Cannizarro reaction (Chapter 12), hydroxide can also act as a nucleophile, adding to the carbonyl group. In the present case, however, this process is reversible and nonproductive. (Recall the discussion of the interconversion of formaldehyde and paraformaldehyde in Chapter 16).

EXERCISE 18-U

Would you expect the anion obtained by deprotonation of formaldehyde to be more or less stable than a vinyl anion? Explain the reason for your answer.

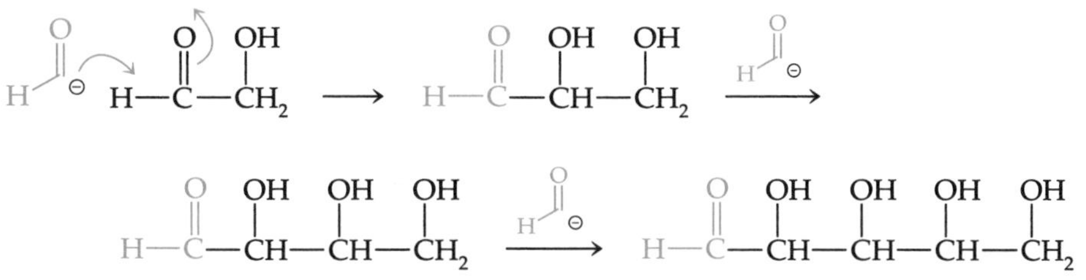

▲ FIGURE 18-32

The attack of the formaldehyde anion on the aldehyde of a carbohydrate extends the chain, providing a carbohydrate with one additional carbon and an aldehyde functional group that can undergo the same nucleophilic attack.

By default, then, hydroxide ion can function only as a strong base, by removal of the aldehydic proton, resulting in a relatively unstable but highly reactive anion. Nucleophilic addition of this formaldehyde anion to another molecule of formaldehyde produces the first and simplest of the carbohydrates.

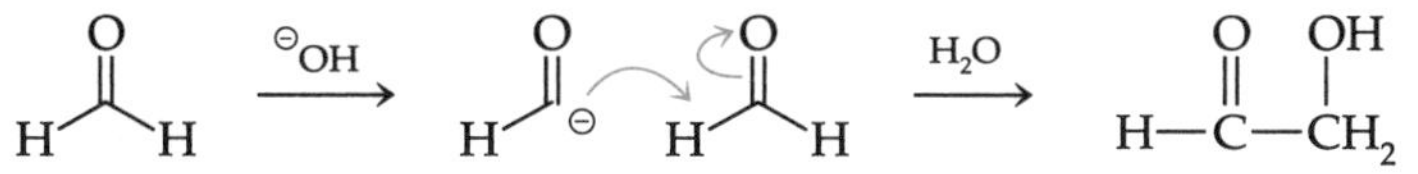

This can, in turn, be elaborated by further additions of the formaldehyde anion to the growing, carbonyl terminus of the molecule. Thus, we have a simple reaction that can be used repetitively to construct molecules that are structurally complex but synthetically accessible (Figure 18-32).

The reaction of formaldehyde anion (in which the carbonyl carbon acts as a nucleophile) is certainly not the first possibility that we might have imagined. After all, the vast majority of carbonyl transformations require the reaction of the carbonyl carbon with the opposite electronic character, as an electrophile. Such reactions, requiring reactive species with an abnormal, or inverted, charge density, are commonly referred to as examples of **umpolung,** a German expression meaning, literally, reversed polarity. We will see more examples of this type of reaction in Chapter 21.

EXERCISE 18-V

Recall that certain carbohydrates are more stable in cyclic, hemiacetal forms. Use this knowledge to explain why there might be a preference for the pentose and hexose carbohydrates from this abiotic synthesis from formaldehyde rather than shorter or longer chains.

18-9 ▶ Synthetic Methods for Preparing Nitrogen-containing Compounds

Many of the synthetic transformations described in this chapter are specific examples of reactions considered earlier. As an aid for study, however, they are listed in Table 18-1 in a format similar to that used in earlier chapters.

Table 18-1 ▸ How to Prepare Carbon–Nitrogen Bonds

In an amine:
  amine alkylation; or
  nucleophilic opening of an epoxide by an amine; or
  reductive amination of an aldehyde or ketone; or
  addition of a carbon nucleophile to an imine; or
  Mannich condensation; or
  nitrile reduction

In an amide:
  amidation of an acyl chloride (or other activated acid); or
  Beckmann rearrangement of an oxime to an amide; or
  amidation of a *t*-BOC-protected amino acid; or
  treatment of an azlactone with a carboxy-protected amino acid; or
  amination of an amino acid in the presence of a carbodiimide

In an imide:
  imidation of a formamide

In a nucleoside:
  amination of a hemiacetal to an aminal

Conclusions

Naturally occurring nitrogen-containing molecules range in complexity from the relatively simple amino acids through peptides and alkaloids to the apparently quite complicated polynucleic acids. In this chapter, we have concentrated mainly on the structural features of these molecules, dealing only briefly with those chemical transformations needed to understand their properties.

Carbon–nitrogen bonds can be formed by several possible routes: nucleophilic substitution of an alkyl halide, amidation of an acid derivative, reductive amination, nucleophilic addition to an imine, Mannich condensations, cyanide displacement followed by nitrile reduction, or oximation followed by a Beckmann rearrangement. Nucleophilic attack on an imine is similar to the attack of Grignard reagents on carbonyl groups, and the Mannich condensation is a new combination of simpler reactions considered earlier.

Amino acids exist as zwitterions and participate in acid-base equilibria in which a cation, the neutral zwitterion, and an anion are interconverted at a characteristic pH. Various amino acids differ as well in their hydrophobic or hydrophilic character because of differences in the attached group bound at the α position.

Amino acids are linked by amide bonds in peptides and proteins. The chemical synthesis of moderately sized peptides presents special challenges. The use of the Merrifield resin for solid-phase peptide synthesis solves most of these difficulties. This technique involves immobilization of an N-protected amino acid on a chloromethylated polystyrene resin, followed by deprotection to produce a surface-attached amino acid with a free amino group. Treatment with an N-protected amino acid in the presence of a carboxylic acid activator (such as carbodiimide) produces a surface-bound peptide. This sequence can be repeated, ultimately leading to a peptide that is cleaved from the resin by treatment with HBr or HF. The growth of a peptide from the C-terminus is more problematic because of epimerization of the azlactone intermediate.

Alkaloids are basic, nitrogen-containing compounds produced by plants. Many physiologically active alkaloids contain a β-phenethylamine subunit.

The seemingly complex structures of the nucleic acids are constituted from nucleotides (purine or pyrimidine aminals of ribose or deoxyribose) joined through a phosphate ester linkage between a C-3' OH group of one nucleotide and a C-5' OH group of another. The nucleic acids exist as anions at physiological pH because the phosphate acid backbone is deprotonated. They are therefore polyelectrolytes.

Abiotic routes to even such complex molecules as nucleic acid bases or sugars often employ simple combinations of the reactions covered in the first half of the book. In contrast, formaldehyde can be deprotonated to form an acyl anion. This species reacts as a nucleophile in carbon–carbon bond formation, providing a route to carbohydrates.

Summary of New Reactions

Mannich Condensation

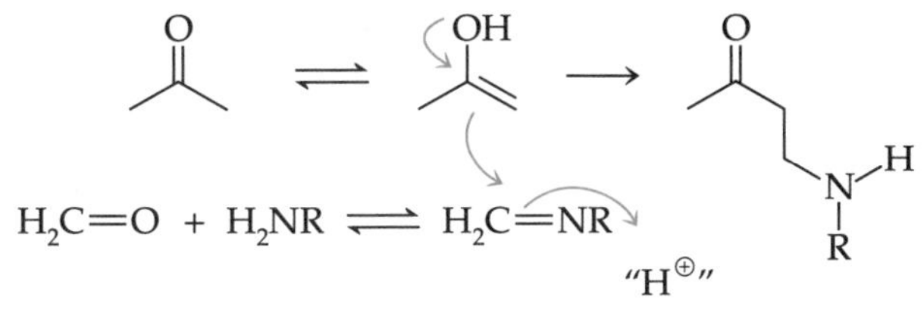

Azlactone Synthesis

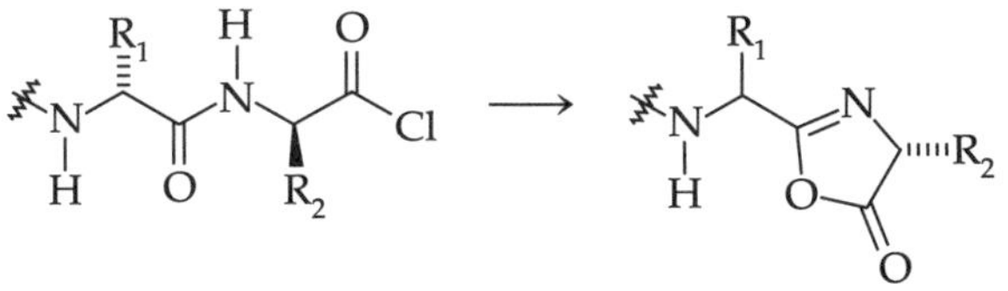

Azlactone

Formation of the Peptide Amide Bond with Carbodiimide

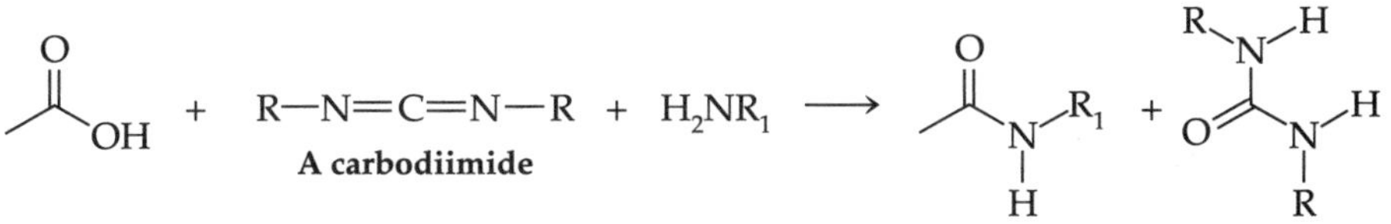

A carbodiimide

Aminol Formation

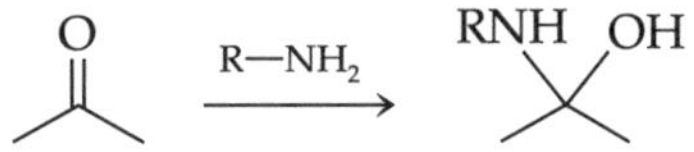

Nucleophilic Addition of an Acyl Anion

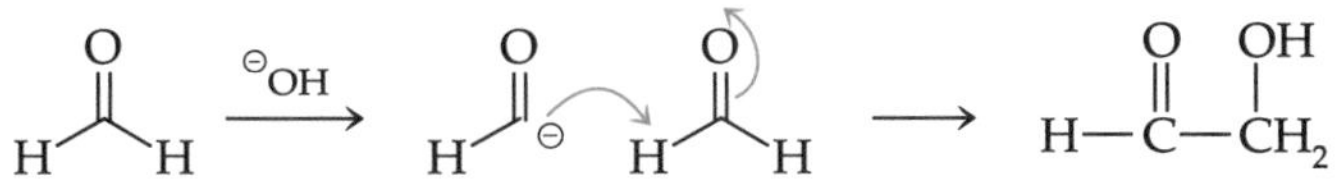

Review Problems

18-1 Each of the following amino alcohols can be prepared by the reaction of an amine with an epoxide. Suggest appropriate starting materials that can produce each of the following compounds through nucleophilic ring opening of an epoxide by nitrogen.

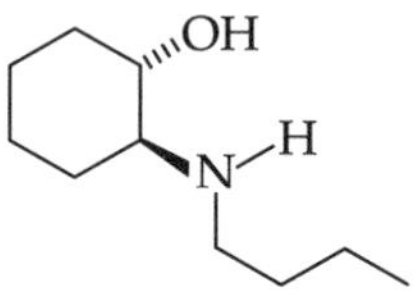

A

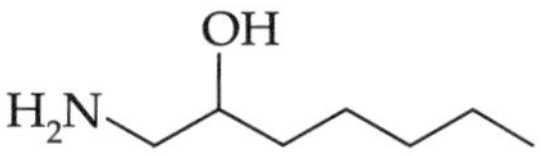

B

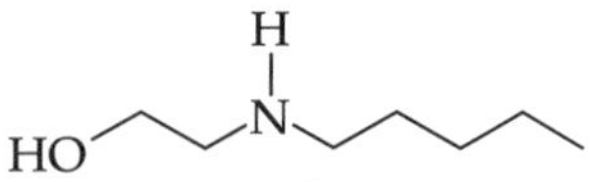

C

18-2 Reduction of an amide by $LiAlH_4$ can be used to prepare each amine shown below. Indicate the starting material required for each product.

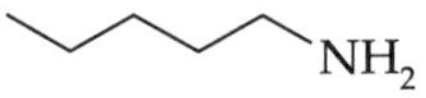

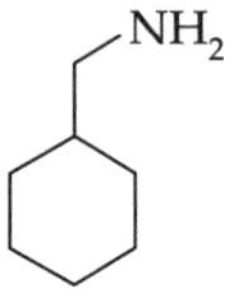

A   B

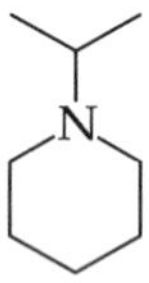

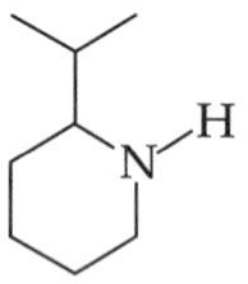

C   D

18-3 Determine what starting materials (both amine and carbonyl compound) are required, and the sequence by which they must be combined, for the synthesis of the following amines by reductive amination.

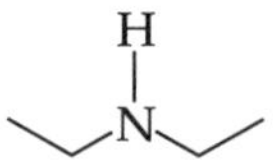

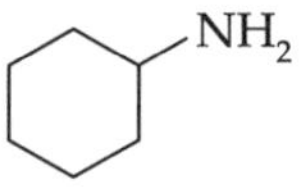

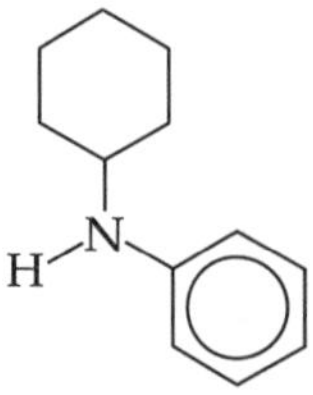

A   B   C

18-4 Which imine and which Grignard reagent react to form each of the following amines?

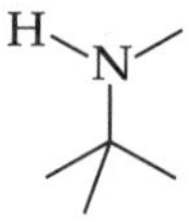

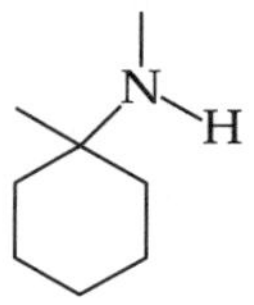

A   B

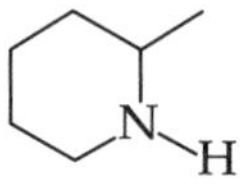

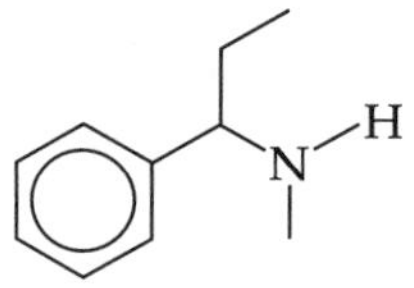

C   D

18-5 Indicate the necessary starting materials for the synthesis of each of the following aminoketones by a Mannich condensation.

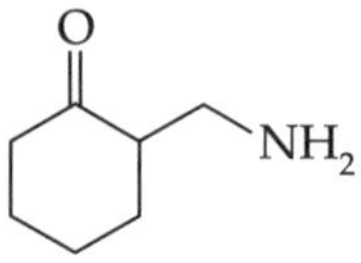

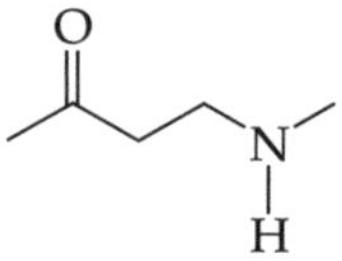

A   B

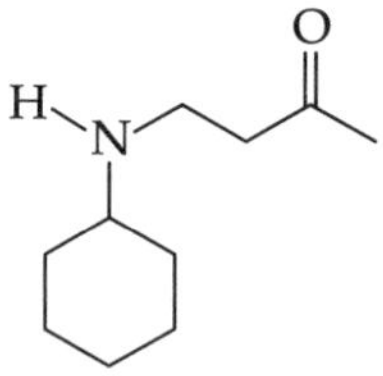

C

18-6 Which of the following amines can be synthesized by:

(a) reductive amination?
(b) Mannich condensation?
(c) reduction of a nitrile?
(d) reduction of an amide?

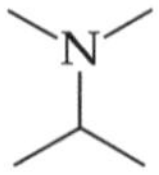

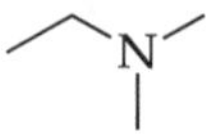

A   B

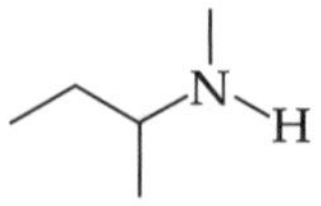

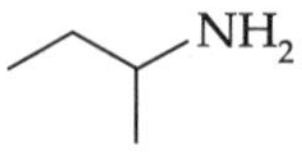

C   D

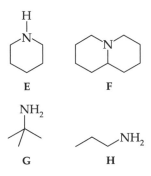

E   F

G   H

18-7 Show the structure of the product that would be obtained from the Beckmann rearrangement of the following oximes. Be sure to consider the question of regiochemistry.

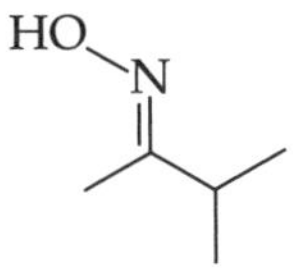

A 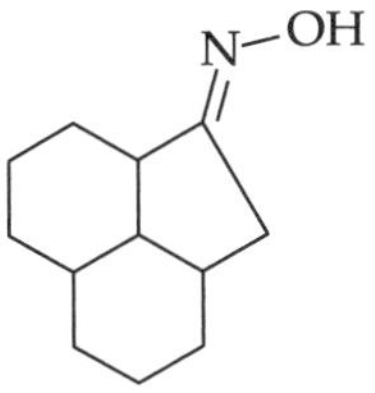   B

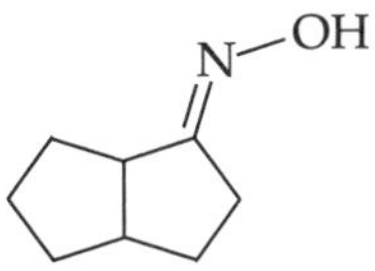

C

18-8 Show the structure of the ketone that could be used as starting material for a Beckmann rearrangement that would produce the following amides and lactams.

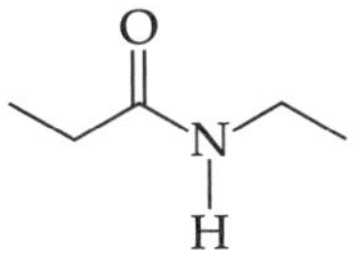

A 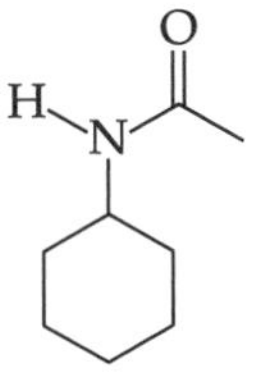   B

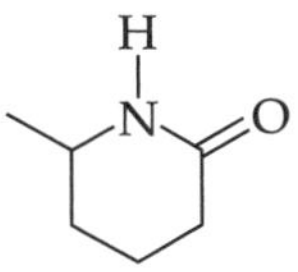

C 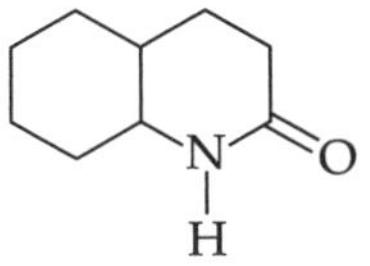   D

18-9 The synthesis of polypeptides by the sequential addition of one amino acid after another is a linear synthesis and, as such, has the drawbacks that were discussed in Chapter 15. Determine what the final yield would be of the complete and correct sequence of a decapeptide if the yield in each coupling reaction were 90%, 95%, or 99%.

18-10 It was noted that the more-reactive carboxylic acid derivatives (toward nucleophilic acyl substitution) undergo base-catalyzed epimerization more rapidly than do less-reactive ones. Explain why a carboxylic acid chloride is more readily deprotonated at the α position than is a carboxylic acid amide.

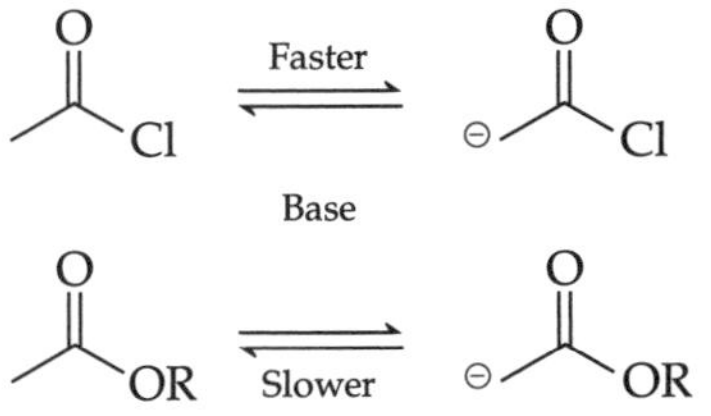

18-11 Acetal linkages formed from aminosugars are more resistant to acid-catalyzed cleavage than when the nitrogen is not present. Suggest a rationale for this reduced rate, indicating which step(s) in the multistep process leading to hydrolysis is most affected by the presence of the nitrogen.

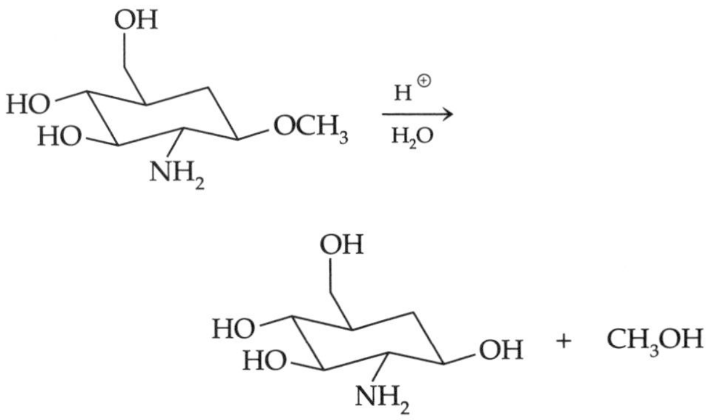

Chapter 19

Noncovalent Interactions
and Molecular Recognition

In virtually every chapter of this book, we have seen examples both of strong bonds, such as the covalent carbon–carbon bonds that hold molecules together, and of weaker interactions, such as hydrogen bonds that form additional links within and between molecules. For example, hydrogen bonds are responsible for linking peptide chains together in a β-pleated-sheet arrangement and, alternatively, for holding a single chain in an α-helical, secondary structure (Chapter 16). These dramatically different shapes are nonetheless fixed by the same basic force: the hydrogen bond. In this chapter, we will examine in detail how the hydrogen bond and other weak forces dramatically affect interactions between molecules. Hydrogen bonds provide one way by which molecules "recognize" those partners with which they form the strongest links. **Molecular recognition** is defined as weak, reversible, and *selective* binding between two reagents. Selectivity is the key to molecular recognition and, without such recognition, enzymes would not be able to selectively accelerate specific reactions. We will treat catalysis (especially, enzyme catalysis) in detail in the next chapter. Here, we present some of the details of the relatively weak, noncovalent interactions that hold specific pairs of molecules together.

19-1 ▸ Nonpolar (Hydrophobic) Interactions

Hydrogen is unique among the elements in that all of its electron density is used in bonding when it is combined with other elements. By sharing electron density with a neighboring atom, the hydrogen nucleus is relatively exposed on the side opposite the bond. Thus, as discussed in Chapters 1 and 2, hydrogen atoms interact with the electron density surrounding other atoms in the same or other molecules by weak, but significant, *electrostatic attractive interactions*. When this attraction is to the electrons of another σ bond, it is called a *van der Waals attractive interaction*. In contrast, the term *hydrogen bond* is used when the attraction is between a hydrogen atom participating in a polar covalent σ bond and an atom with one or more accessible

651

lone pairs of electrons. This latter interaction is generally stronger because the concentration of electron density in a lone pair is greater than that surrounding a covalently bound hydrogen atom.

Van der Waals
interaction

Hydrogen bond

We have already seen how weak interactions influence the physical properties of organic compounds. The effect of these interactions can be illustrated by comparing the melting points and boiling points of methane with those of mono-, di-, tri-, and tetrafluoromethane and of water (Table 19-1). Notice that both the melting points and the boiling points increase when one and then two fluorine atoms replace hydrogen atoms of methane (CH_2F_2 does not crystallize), but then both values decrease for CHF_3 and CF_4, even though the molecular weight increases uniformly along the series. As hydrogen atoms are progressively replaced with fluorine atoms, the van der Waals attractive forces between the hydrogen atoms are replaced with stronger attractive interactions between hydrogen atoms and fluorine nonbonded lone pairs. The strength of these attractive interactions reaches a maximum when the number of hydrogens and fluorines is equal. As the number of fluorines increases to three and then four, the repulsive interactions between the additional lone pairs of electrons dominate over the attractive forces introduced by intermolecular polar interactions. Notice that water, although lighter than all four of the fluoromethanes, has higher melting and boiling points than any of them, a direct result of strong hydrogen bonding between an unshared pair of electrons on the oxygen of one water molecule and a proton on oxygen of another.

Because the periphery of a typical molecule is composed mostly of hydrogen atoms (molecules such as the fluorocarbons being exceptions), the contribution of van der Waals interactions to intermolecular attractions does not differ greatly from one molecule to another. In contrast, the number and strength of hydrogen bonds can vary dramatically. Thus, the weak van der Waals forces play a much smaller role than do hydrogen bonds in molecular recognition, a process that by definition results in **differentiation** because of greater attractive interaction between one pairing of molecules than between other combinations.

Table 19-1 ▸ Melting and Boiling Points of Methane, Fluoromethanes, and Water			
Compound	Molecular Weight	mp (°C)	bp (°C)
CH_4	16	−182	−164
CH_3F	34	−142	−78
CH_2F_2	52		−52
CHF_3	70	−160	−83
CF_4	88	−184	−128
H_2O	18	0	100

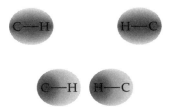

▲ FIGURE 19-1
When two C–H bonds are brought to within interaction distance, electron density shifts toward carbon in one molecule and toward hydrogen in the other. The dipole of each of the carbon–hydrogen bonds is thus changed in a small way and the result is referred to as an *induced dipole moment*. The result is an increased attraction between the electrons of one carbon–hydrogen bond and positive charge of the nucleus of the other. The magnitude of the change in dipole moment of carbon–hydrogen bonds when brought into proximity is quite small; as a result, there is not a large change in the magnitude of the attractive interaction.

Both van der Waals and dipole–dipole attractions derive from the same force: the electrostatic attraction between unlike charges. For a van der Waals interaction, the net electrostatic attraction of the electrons of one atom for the nucleus of another is increased by an induced polarization of the bonding electrons. Likewise, a dipole–dipole interaction is the result of electrostatic attraction of the electron density of a polarized bond for another nucleus that bears partial positive charge as a function of bond polarization.

The magnitude of a van der Waals attractive interaction is increased in compounds that have one or more bonds with **induced dipole moments.** When a hydrogen atom is brought near another hydrogen atom participating in a partially polarized C–H bond, the surrounding electron density of the two C–H bonds repel each other. If we were able to take a snapshot of the electron density at any particular instant, we would find that the electron density surrounding one hydrogen would be closer to the other hydrogen (and farther from its attached carbon) than it is in its normal distribution in an isolated molecule, whereas that on the neighboring hydrogen would be shifted toward its carbon (Figure 19-1).

19-2 ▸ Polar Interactions: Dipole–Dipole Interactions

The strength of intermolecular attractions rises dramatically as the magnitude of the bond dipoles present within the two interacting molecules increases. As an example, compare the boiling points of nitriles with those of alkynes of similar molecular weights (Figure 19-2). Clearly, the boiling point of the nitrile is consistently higher than that of the alkyne of the same weight. This higher boiling point is the result of strong dipolar interactions between pairs of nitrile molecules in the liquid phase, an association that must be broken as the molecules pass into the gaseous state. Because alkynes have only small dipole moments, dipole–dipole association is not a significant factor affecting the boiling points of these molecules.

In contrast, the melting points of alkynes and nitriles, especially those of higher molecular weight, are very similar (Figure 19-3). Although dipole–dipole interactions are important in the association of pairs and small groups of molecules in solution, the attractive and repulsive interactions nearly cancel in a three-dimensional solid matrix in which each molecule

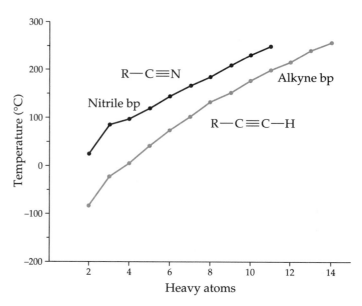

▲ FIGURE 19-2
The higher boiling points of nitriles than those of alkynes of the same weight are a consequence of strong intermolecular dipolar interactions in the more highly polarized C≡N triple bond.

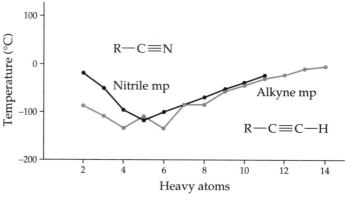

▲ FIGURE 19-3
The melting points of nitriles and the related alkynes are nearly equal because attractive and repulsive dipole–dipole interactions in the solid state nearly offset each other.

Hexagonal
closed pack

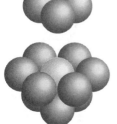

Top three spheres
lifted to reveal
central sphere
(in color)

is in close contact with many neighbors. Indeed, in the hexagonal close-packing arrangement shown in the margin, which is common for organic molecules (as well as for spheres such as marbles), each molecule is surrounded by six others in the same plane, as well as three above and three below for a total of twelve neighbors. (The central sphere in the representation is shown in color for clarity.)

The change with molecular weight of both the melting points and the boiling points of nitriles and alkynes can be represented by reasonably smooth curves (Figures 19-2 and 19-3). The reason for the zigzag in the curve for the melting point of alkynes is not well understood but is seen quite often in the melting behavior of molecules that are held together in the solid mainly by van der Waals attractive interactions. The same behavior is observed, for example, with straight-chain alkanes. The abnormally high melting boiling points of hydrogen cyanide, acetonitrile, and propio-

nitrile are due to the fact that each of these molecules has only one conformation, an effect that will be discussed shortly.

19-3 ▸ Polar Interactions: Hydrogen Bonds

The large differences in the boiling and melting points of water and
methane are a clear indication that the contribution of hydrogen bonds to
intermolecular attraction is much larger than that of van der Waals interactions. The hydrogen bond is an electrostatic, weak interaction between a
hydrogen atom and another electronegative atom bearing at least one lone
pair of electrons; that is, one of the heteroatoms N, O, P, or S, or a halogen
(F, Cl, Br, or I). The magnitude of this attractive interaction increases with
the electronegativity of the heteroatom. Typically, the hydrogen atom covalently bound to one of these heteroatoms is also associated with another
heteroatom. Because these heteroatoms are more electronegative than carbon, the hydrogen–heteroatom bond is even more polarized and the hydrogen even less electron rich than if it were attached to carbon. Thus, the
hydrogen bond includes a total of three atoms: the hydrogen atom and two
heteroatoms (typically), both of which are attracted to this single proton.

EXERCISE 19-A

Suggest an order of boiling points for propane, ethylamine, and ethyl alcohol, all of which have approximately the same molecular weight. Check
your answer (by consulting the data in Chapter 3, Table 3-1, or by using the
Handbook of Chemistry and Physics) and then explain this order.

It is conceivable that the hydrogen atom could be located in a symmetrical fashion, midway between the heteroatoms. However, this arrangement
represents the transition state for the transfer of the proton from one heteroatom to the other and is higher in energy than either of the arrangements
in which it is closer to one of the heteroatoms (at an optimal σ-bonding distance). Furthermore, in simple systems, such a transfer is generally accompanied by a change in valency for both heteroatoms, as shown in Figure 19-4

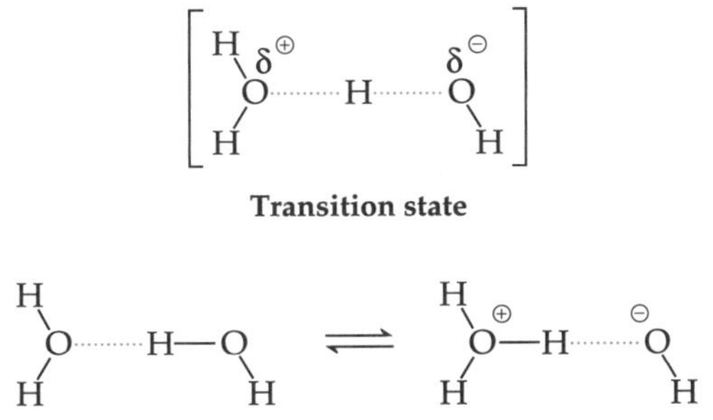

Transition state

▲ FIGURE 19-4
A hydrogen bond between water molecules is formed when a proton of one
molecule is attracted to the nonbonded lone pair on another. As the molecules
become associated and the intermolecular attraction increases, the covalent
O–H bond lengthens and the weak hydrogen bond shortens. A proton-
transfer reaction can occur through an intermediate transition state in which
the lengths (and strengths) of the two O–H bonds are equal. In this
symmetrical transition state, it is no longer possible to refer to one of these
bonds as covalent and the other as a hydrogen bond.

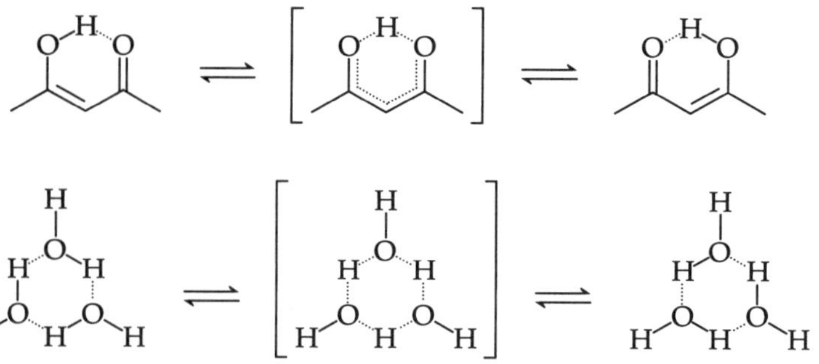

▲ FIGURE 19-5

In the upper example, the proton of the O—H group of the enolic form of a β-diketone is associated with the ketone carbonyl oxygen in a six-membered arrangement. Proton transfer in a symmetrical transition state converts the 4-keto-2-enol into the 2-keto-4-enol. In the lower example, three water molecules associate as a trimer. Proton transfer exchanges the specific hydrogens bound to a given oxygen atom but does not change the trimer overall.

for the transfer of a proton from one water molecule to another. In such cases, one of the heteroatoms is considerably more effective in attracting the proton—another argument for an unsymmetrical arrangement.

However, there are situations in which proton transfer results in an arrangement identical (or nearly so) in energy. Two examples are shown in Figure 19-5. The upper example in Figure 19-5 is the *intra*molecular proton transfer in the enol of a β-diketone; the lower example in the same figure is an *inter*molecular transfer of three protons among three associated water molecules. In both cases, the starting and ending states are identically constituted and the transition state consists of a structure in which each hydrogen atom is simultaneously and equally associated with two oxygen atoms, but in which these two bonds are weaker in total than when bonding is between hydrogen and only one oxygen atom. We have encountered a symmetrical transition state before, in Chapter 8 in the S_N2 displacement of bromide from methyl bromide.

Although organic chemists tend to focus on the carbon-based part of a reaction, the S_N2 reaction could be viewed alternatively as the transfer of the carbon from the leaving group to the nucleophile (as, for example, in the margin, where a methyl group is transferred from one bromine atom to another). In a like fashion, the symmetrical positioning of a hydrogen atom between two heteroatoms represents the transition state for the transfer of the proton from one heteroatom to the other. As in nucleophilic substitution, the arrangement of lowest energy for the three atoms taking part in a proton transfer is linear, with maximum separation of the two pairs of electrons associated with the proton. Furthermore, both unsymmetrical ground states leading to this transition state are lower in energy with this alignment. As discussed in Chapter 16, this linear relation exists in the hydrogen bonding in peptides, both in the β-pleated sheet and in the α-helical secondary structures (Figure 19-6).

This linear arrangement is favored both as the ground state for hydrogen bonding and as the transition state for proton transfer, but other alignments also are possible. It is only necessary that the balance between attractive and repulsive interactions provide for net bonding. In both examples given in Figure 19-5, the alignments of hydrogen bonds deviate substantially from the ideal, linear arrangement. Nonetheless, the enolic forms of β-diketones and β-ketoesters are substantially stabilized by intramolecular hydrogen bonding.

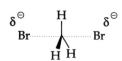

◄ FIGURE 19-6
Hydrogen bonds are responsible for the precise intermolecular interactions in both the β-pleated sheet and the α-helical structures of peptides.

β-Pleated sheet   **α-Helix**

─── Axis of helix

EXERCISE 19-B

The enol content of three dicarbonyl compounds is provided below:

acetylacetone ($CH_3COCH_2COCH_3$):   76%

ethyl acetoacetate ($CH_3COCH_2COOCH_2CH_3$):   8%

diethyl malonate ($CH_3CH_2OCOCH_2COOCH_2CH_3$):   0.01%

Explain these differences, keeping in mind the resonance stabilization of the ester group.

A simple way to examine the strong influence that hydrogen bonding has on bi- and multimolecular interactions is to compare the solubilities of various compounds in water, a solvent that is an excellent hydrogen-bond acceptor and donor. For example, cyclopentane is virtually insoluble in water, whereas tetrahydrofuran is miscible with water. In this case, water acts solely as a hydrogen-bond donor because tetrahydrofuran does not have a hydrogen bound to a heteroatom.

The basic nitrogen atom of an amine serves as a hydrogen-bond acceptor, with similar effects on solubility. Pyridine, a somewhat weaker base than a typical aliphatic amine, is miscible with water, but the analogous hydrocarbon, benzene, dissolves only to the extent of 0.01% (Figure 19-7).

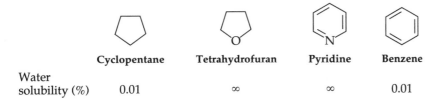

	Cyclopentane	Tetrahydrofuran	Pyridine	Benzene
Water solubility (%)	0.01	∞	∞	0.01

▲ FIGURE 19-7
The water solubility of a common solvent molecule depends on that molecule's ability to hydrogen bond with water.

Table 19-2 ▸ Melting and Boiling Points of Isomeric Butanols		
Compound	**bp (°C)**	**mp (°C)**
n-Butanol	117	−90
i-Butanol	108	−108
s-Butanol	99	−115
t-Butanol	82	26

Hydrogen bonding also affects the boiling points and melting points of isomeric alcohols (Table 19-2). Notice that the boiling points of the isomeric butanols drop as the degree of steric congestion about oxygen increases (and hence the ability to participate in linear hydrogen bonding decreases). A similar effect is seen in the melting points, except for *t*-butanol, for which the value is more than 100 °C higher than that for the other three butanols.

This dramatic difference in the melting points of isomers is an example of the effect of *symmetry* and *rigidity* on the physical properties of molecules. An energetic factor that must also be considered in a comparison of the gaseous, liquid, and solid states is entropy; that is, the degree of randomness. The degree of order in crystalline materials is quite high; that in liquids is considerably lower; and separated molecules in the gas phase are totally randomly arranged. Furthermore, gas-phase molecules are completely free of intermolecular interactions and thus can tumble at will. Those in liquids are more restricted as the result of attractive interactions with neighboring molecules. The solid state is the most rigid of the three phases. Because long-range order is entropically unfavorable, there is always a favorable energetic contribution from entropy to the transformation from the crystal to the liquid and then to the gaseous state.

The symmetric arrangement of molecules in a crystal is lost when the solid is allowed to melt or evaporate. Among the four isomeric butanols, *t*-butanol is the most-symmetrical molecule: because much symmetry is preserved in gaseous *t-butanol*, the loss of symmetry in the progression of *t*-butanol from solid to liquid is less energetically costly than it is for the other butanols. Hence, entropy plays a considerably less important role in the energetics of these phase changes for *t*-butanol than for its less-symmetrical isomers.

EXERCISE 19-C

The effect of symmetry on the melting points of crystalline compounds is quite general. Predict, on the basis of symmetry, which molecule (A or B) in the following pairs has the higher melting point.

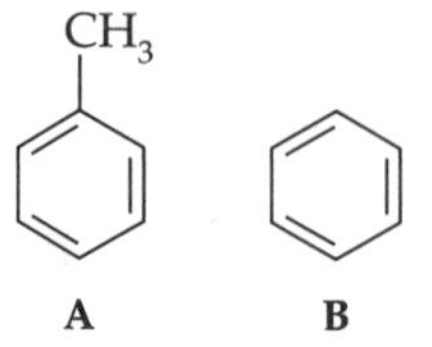

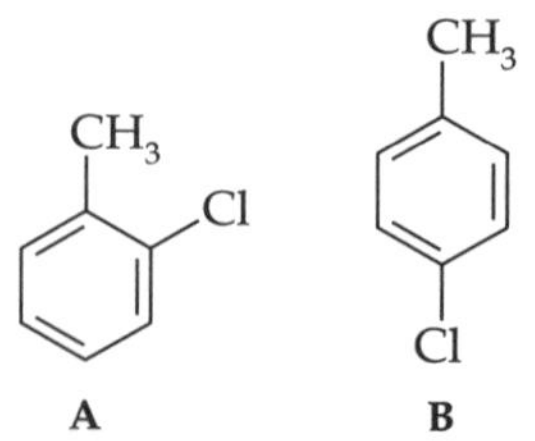

Unfortunately, and despite the large body of information available, a precise energy for a "typical" hydrogen bond cannot be given. Each one is only a relatively minor contributor to the total energy of a molecule and its arrangements, and it is not yet possible to factor out the contribution of hydrogen bonding from other factors such as entropy. Generally, a single hydrogen bond is believed to contribute between 1 and 5 kcal/mole to attractive interactions between molecules, with the strongest contribution being from the interaction between relatively basic heteroatoms and relatively acidic hydrogens such as that shown in the margin. In the extreme, when these groups are sufficiently basic on the one hand and acidic on the other, the proton is transferred in an acid-base reaction, becoming covalently bound to the basic site.

Hydrogen bonds are quite important contributors to molecular interactions, especially when the arrangement of atoms is such that two or more hydrogen bonds can simultaneously hold a pair of molecules together. We will examine multiple hydrogen bonds in Section 19-5, but first we consider an interaction that is similar to the hydrogen bond—that between metal cations and lone pairs of electrons of heteroatoms.

19-4 ▸ Polar Interactions: Metal–Heteroatom Bonds

Many small cations, especially alkali metal ions, are attracted to the lone-pair electron density on a heteroatom in much the same way as are hydrogen atoms. (Heteroatoms attracted to metal cations are referred to as **ligands.**) These interactions are one of the two major contributors to the water solubility of salts such as sodium chloride. The other contribution is the converse interaction of the hydrogens in water molecules with the negatively charged counterions (chloride ion in this case). Often, these anions have four nonbonded lone pairs of electrons. Because both of the species in contact with water are ions, these interactions are generally stronger associations than simple hydrogen bonds with neutral heteroatoms.

The metal–lone-pair interactions are especially favorable when more than one heteroatom is present in a single molecule such that multiple and simultaneous contacts between several heteroatoms and the same metal ion are possible. Such species are called **ionophores.** When two heteroatoms are present, as in ethylenediamine (Figure 19-8), such ligands for metals are referred to as being **bidentate** (meaning "two teeth"). A ligand with three

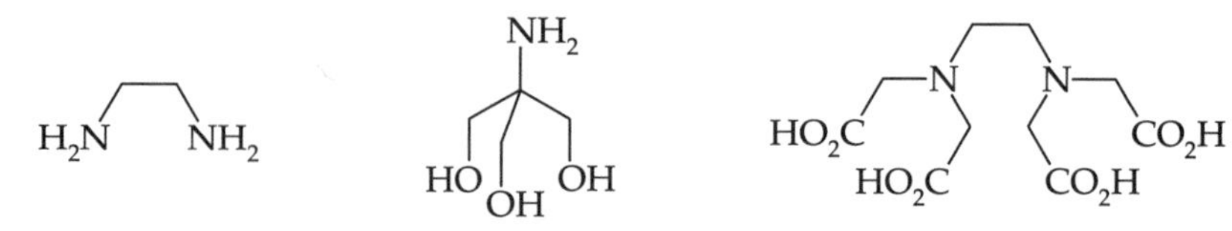

Ethylenediamine **Tris(hydroxymethyl)-aminomethane (TRIS)** **Ethylenediamine tetraacetic acid (EDTA)**

▲ FIGURE 19-8
Molecules containing several heteroatoms that can coordinate with metals and other cations are called multidentate. Ethylene diamine has two coordinating nitrogen atoms and is a bidentate ligand. TRIS and EDTA have several possible coordination sites and are called polydentate ligands.

heteroatoms is **tridentate;** and one with even more heteroatoms, as in TRIS and EDTA, is **polydentate.**

Cyclic ionophores are a special class of polydentate ligands for metals. There are many naturally occurring ionophores (both cyclic and acyclic), many of which play critical biological roles. For example, **vitamin B_{12}**, **heme** (the oxygen carrier in mammalian blood), and **chlorophyll** (a compound that is essential for photosynthesis) are tetradentate nitrogen-based ligands (Figure 19-9). Recently, several naturally occurring polydentate ionophores have been identified in which several oxygen atoms coordinate

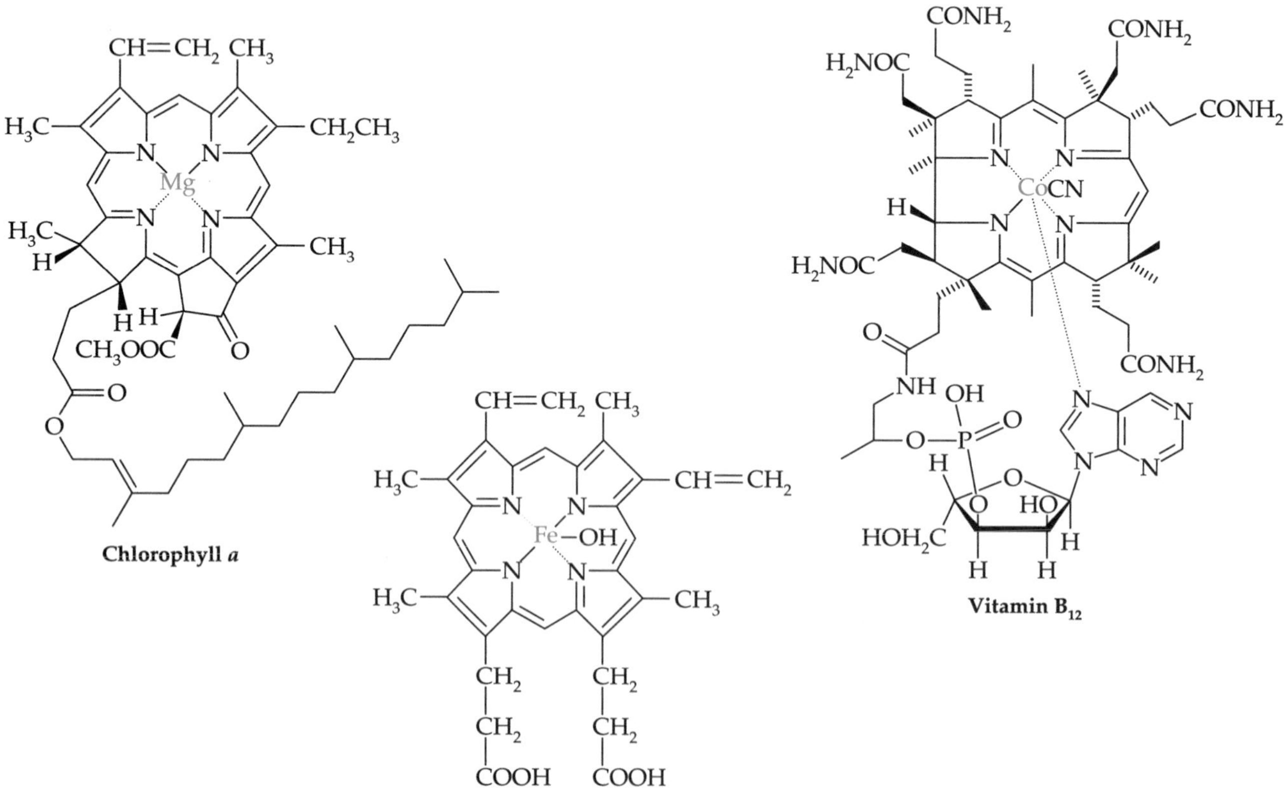

Chlorophyll *a* **Vitamin B_{12}** **Heme**

▲ FIGURE 19-9
Chlorophyll *a*, heme, and vitamin B_{12} are porphyrins, compounds in which four nitrogen atoms (present as four pyrrole nitrogens in a large ring) coordinate to a metal ion. In vitamin B_{12}, the metal ion (cobalt) is also coordinated to a covalently attached nucleotide base.

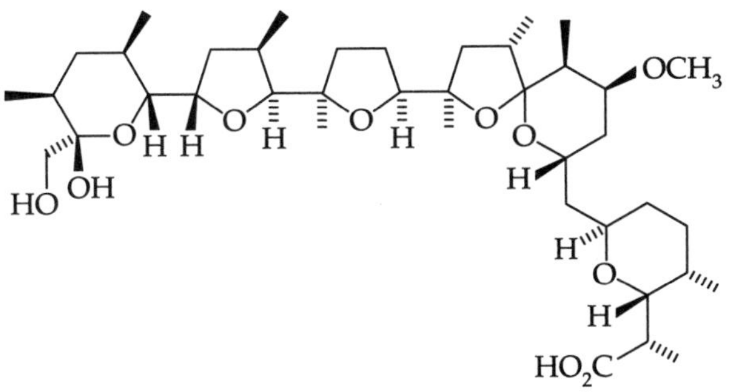

Nigericin

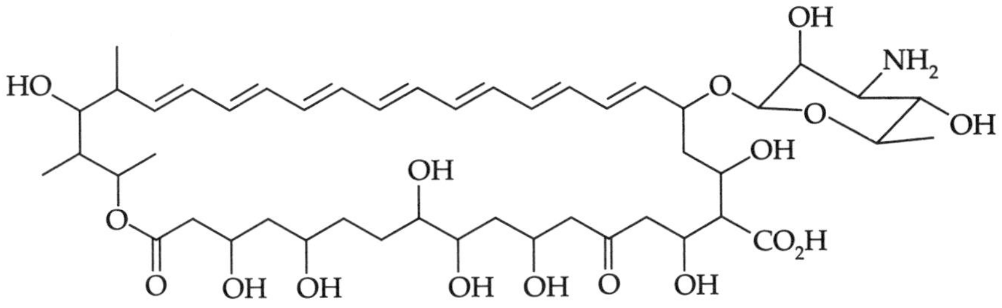

Amphotericin B

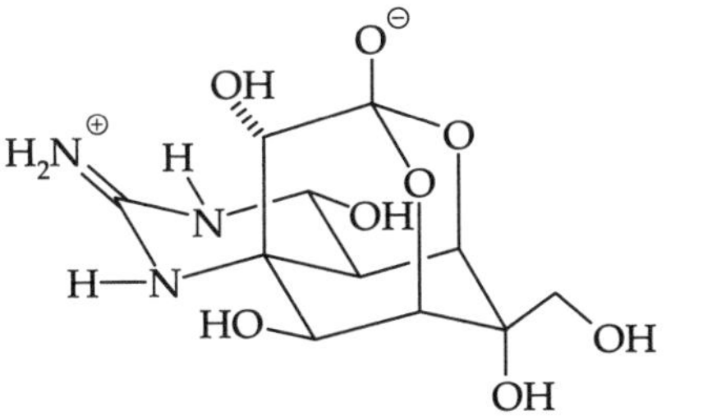

Tetrodotoxin

▲ FIGURE 19-10

Naturally occurring oxygen ionophores have many coordinating oxygen atoms, enabling them to assist in metal ion transport across membranes. Because this transport significantly affects nerve transmission, some (like tetrodotoxin) are deadly poisons, whereas others are medicinal agents.

with a metal ion (Figure 19-10). Some of these ionophores have antibacterial activity (and sometimes, unfortunately, toxic effects in mammals) by virtue of their ability to encase metal ions within an exterior that is mostly hydrocarbon in nature. Wrapped in this hydrophobic shell, ions can permeate through cell membranes, thus upsetting the normal ion balance. An extreme example of the toxic effect of ion transport is produced by **tetrodotoxin,** a highly oxygenated compound resembling a carbohydrate that is found (mainly) in the liver and sexual organs of the puffer fish. Quantities as small as a few micrograms of this toxin cause total disruption of nerve function in mammals, which is dependent on the maintenance of unequal concentrations of both sodium and potassium ions on the two sides of a nerve-cell membrane. Gourmets in Japan and other countries around the world consider the puffer fish a delicacy; and, for some, the ultimate experience is to sense a slight numbing effect on the lips caused by traces of tetrodotoxin in the flesh.

Other cyclic ionophores have been developed by chemists led by the Americans Donald Cram and Charles Pedersen and by Jean-Marie Lehn, a French scientist, all of whom shared the Nobel Prize in chemistry in 1987 for their pioneering research on synthetic ionophores. These compounds, like other cyclic polydentate ligands, have unusually high specificity for particular ions. This selectivity can be measured by determining the ability of the ionophore to "transport" a metal cation from an aqueous medium to a hydrocarbon solvent. Essentially, the ionophore replaces the solvation interaction of the oxygen of water. To do so effectively, however, the ionophore must surround the cation and provide a sufficiently large number of heteroatoms to meet the ligation requirements of the metal. This ability, in turn, depends on the "fit" of the metal ion within the hole that is formed within the cavity of the ligand. Cations that are too large cannot penetrate deeply enough to contact more than two or three of the heteroatoms. For cations that are too small, the distance from side to side of the ionophore is too large for the cation to interact effectively with all of the heteroatoms simultaneously. For example, the ionophore shown in Figure 19-11 has a cavity that is nearly the ideal size for a sodium ion and, as a result, the interactions between this ion and the ionophore are quite strong. In contrast, both the lithium ion, which is smaller than the sodium ion, and the potassium ion, which is larger, bind much more poorly to the ligand. We thus have our first example of molecular recognition—between an ionophore and a specific ion.

EXERCISE 19-D

Suggest an experimental method for determining the relative ability of an ionophore to solubilize an inorganic salt in a nonpolar, organic solvent such as benzene.

Entropy also influences the magnitude of interactions between metal ions and otherwise comparable cyclic and noncyclic polydentate ligands. For example, compare the cyclic and noncyclic polyethers shown in Figure 19-12. These two polyethers are very similar and, indeed, differ in formula by only two hydrogens. Nonetheless, they differ by four orders of magnitude in their ability to associate with a potassium ion. This dramatic difference is due to the effect of entropy: the organization required to wrap the noncyclic ligand around the potassium ion deprives this ligand of significant conformational freedom. Indeed, there are fifteen σ bonds about which the molecule is free to rotate when it is not associated with a cation. Conversely, the interaction with the potassium ion deprives this ligand of these

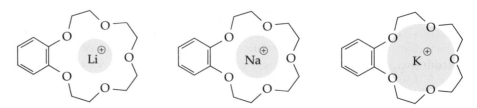

▲ FIGURE 19-11

The ability of an ionophore to bind cations depends on how well the cavity size of the cyclic ether matches the radius of the cation. The binding constant of benzo[15-crown-5] ether, the ligand shown here, is much higher with a sodium ion (Na^+), which fits well, than with Li^+ (too small) or K^+ (too large). (Benzo refers to the presence of the fused benzene ring, 15 to the number of atoms in the ring, and 5 to the number of oxygen atoms available in the ring to associate with a metal ion.)

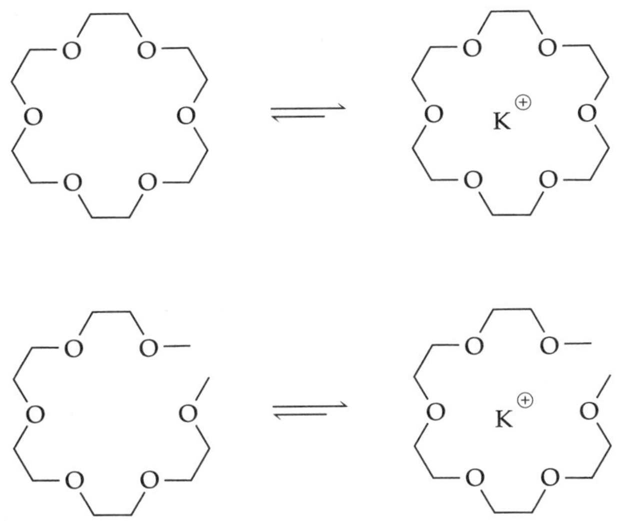

◄ **FIGURE 19-12**
The binding constant of a metal ion by a cyclic ether (upper equilibrium) is much higher than that of the analogous open-chain ether because of the entropy cost required to bring the open-chain ether into a conformation for binding.

rotational degrees of freedom. The cyclic ligand has far less conformational freedom in its uncomplexed form; therefore, association with the potassium ion (although still entropically unfavorable) is much less energetically costly than for the acyclic ligand because a smaller change in entropy takes place with the former. A large fraction of the entropy cost associated with binding has been paid earlier in preparing the cyclic ligand, which has many fewer degrees of freedom than does the acyclic analog.

Our treatment of conformations of cyclic systems in Chapter 5 were mainly concerned with differences in enthalpy; that is, differences originating from such factors as bond length or angle distortion. However, even such a conformationally "ideal" arrangement as chair cyclohexane can be relatively unstable from an entropic point of view when compared with its acyclic counterparts.

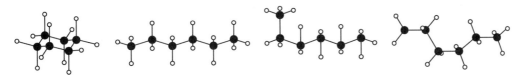

For example, in a comparison between cyclohexane and *n*-hexane, there is little difference in conformational enthalpy between the chair conformation of the former and that conformation of the latter in which all bonds are oriented in an *anti* fashion. On the other hand, there is only one stable conformation of cyclohexane (the chair), whereas, in addition to the all-*anti* conformation, there are four unique conformations of *n*-hexane with only one *gauche* bond: two enantiomeric conformers in which the *gauche* bond is between carbons 2 and 3; and two more in which the *gauche* bond is between carbons 3 and 4. Each of these conformers is about 0.8 kcal/mole less stable than the all-*anti* conformer because each has a *gauche* interaction similar to that in butane. However, because of the number of these conformers, the total of their populations is greater than that for the all-*anti* form.

Thus, cyclic systems can be significantly less favorable from an entropic point of view than their acyclic counterparts. Furthermore, as the number of atoms in the cyclic arrangement increases, the influence of entropy (disfavoring the cyclic form) increases. Yet, we have also seen how this entropic disadvantage inherent in cyclic systems is an advantage in molecular recognition, precisely because the cyclic system has already incorporated the entropic disadvantage associated with the interactions required for the recognition event. As a result, the interaction of the cyclic system to form the complex is more favorable than that of its open-chain analog.

19-5 ▸ Multiple Hydrogen Bonds in Two Dimensions

Let us now consider what happens when pairs of molecules are held in close contact by two or more hydrogen bonds, a situation that differs in several ways from the single hydrogen bonds thus far considered. Perhaps the most-obvious difference is that the strength of the association is greater because the effects of the hydrogen bonds are additive. However, this is true only if the orientation of the three participating atoms (the two heteroatoms and the hydrogen) for each hydrogen bond is ideal: thus, a unique feature of a system containing multiple hydrogen bonds is that the relative positioning of the participating atoms is critical. Minor deviations from an ideal geometry can cause a substantial decrease in the strength of the bonding interactions and a situation in which several nonideal hydrogen bonds are no better (or even worse) than a single bond with an optimal spatial orientation. Keep in mind that even the best of hydrogen bonds are weak; as a result, they are in dynamic equilibrium, rapidly forming and breaking, under most normal conditions. For this reason, molecules associate through more than one hydrogen bond only if the energies of these interactions in total exceed other pairings linked by fewer bonds or only one bond.

Multiple contacts are often necessary to bind a substrate *reversibly* within an enzyme cavity. If the hydrogen bonding is too weak, the complex does not form and no acceleration of the reaction occurs. On the other hand, if the bonding is too strong, the enzyme (present in minor concentration) is tied up irreversibly and is not available for repeated cycles, so that, again, no acceleration takes place.

The carboxylic acid functional group can form two hydrogen bonds between a pair of molecules.

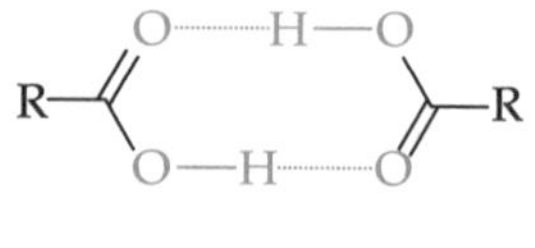

Indeed, carboxylic acids exist as dimers in solution except in solvents that are themselves very good hydrogen-bond acceptors and donors. Notice that the dimer structure is symmetrical, with each carboxylic acid group donating one hydrogen and one lone pair to the two hydrogen bonds. The hydrogen atoms are located on a line between the pairs of oxygen atoms. The magnitude of this hydrogen-bonding interaction can be judged by comparing the boiling points of a straight-chain carboxylic acid with that of the methyl ester of the same molecular weight (the methyl ester of the carboxylic acid with one carbon atom fewer), shown graphically in Figure 19-13. In each case, the acid has a considerably higher boiling point as a result of the two hydrogen bonds between a pair of molecules. These bonds are sufficiently strong that they are not broken in going from the liquid to the gaseous state. Thus, distillation of a carboxylic acid actually requires the vaporization of the dimer, a complex that is twice as heavy as the carboxylic acid itself.

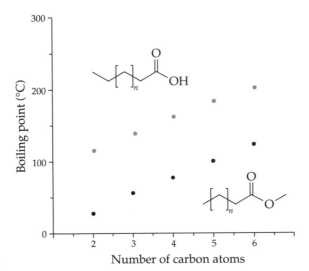

The boiling points of carboxylic acids are consistently higher than those of methyl esters of the same molecular weight because of strong, intramolecular hydrogen bonding in the acid.

19-6 ► Genetic Coding, Reading, and Misreading

Strong hydrogen bonds are formed to hydrogen atoms attached to nitrogen, as, for example, in the intermolecular bonding of peptides (Chapter 16). Among many other biologically important pairs held together through hydrogen bonding to nitrogen, the most important are the bases in DNA and RNA. These bases form the fundamental units for the storage and transmission of information in biological systems.

We were introduced to these bases in Chapter 18, although we did not explore their chemistry in any detail there. As the name of the class to which they belong implies, these substances have one or more nitrogen atoms (although not all of the "bases" are strongly basic) and complementary carbonyl groups that can serve as hydrogen-bond acceptors. In addition, each N—H group can function as a hydrogen-bond donor. Two sets of base pairings are illustrated in Figure 19-14, in which thymine is paired with adenine and cytosine is paired with guanine. The distance between each pair of heteroatoms taking part in a hydrogen bond is nearly the same for all the hydrogen bonds; that is, each of the two hydrogen bonds joining thymine and adenine and the three joining cytosine and guanine contributes substantially to the total interactions.

▲ FIGURE 19-14
Complementary base pairing between thymine and adenine is accomplished through two strong hydrogen bonds. That between cytosine and guanine takes place through three hydrogen bonds.

These hydrogen-bonding patterns form a simple system of recognition between molecules, and the appropriate pairings of guanine with cytosine (G–C) and adenine with thymine (A–T) are said to be **complementary.** Alternate base pairings—for example, between adenine and guanine or adenine and cytosine—would have fewer hydrogen-bonding interactions than the preferred number shown in Figure 19-14. This specificity of the bases for their appropriate partners forms the essence of the duplication and reading of the **genetic code,** the system by which information is stored and used in living systems.

EXERCISE 19-E

Draw pictures representing hydrogen bonding in the pairing between adenine and guanine and between adenine and cytosine, trying to maximize the number of hydrogen bonds. (It might be helpful to draw the structure of adenine on tracing paper so that you can readily overlay and move this base relative to cytosine and guanine.) Using your drawings, explain why adenine prefers to base-pair with thymine.

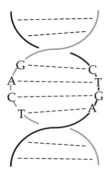

Recall from Chapter 18 that these bases are found chemically bound to C-1 of a carbohydrate in the biopolymers DNA and RNA. For the bases in two polymer chains to associate in a hydrogen-bonding sense, the two molecules must twist about each other and be arranged in a head-to-tail fashion, as shown in the margin. (The handedness of the twisting of the double helix comes from the chirality of the carbohydrate. However, visualizing how the twist comes about is quite difficult without molecular models.) Imagine, then, a sequence of bases in one DNA chain such as G-A-C-T. For a second chain to effectively associate with the first, it must have the complementary sequence C-T-G-A. This sequence represents information that, as we will see shortly, is translated into a specific sequence of amino acids in the biosynthesis of peptides and proteins, including enzymes.

Before we proceed, let us review the nomenclature of the nucleic acid subunits presented in Chapter 18. Each of the bases (guanine, adenine, cytosine, and uracil or thymine) is attached to either ribose (for RNA) or deoxyribose (for DNA). These subunits are called **nucleosides,** and the names of the bases are appropriately modified (Figure 19-15).

Base	Nucleoside	Nucleotide
Adenine	Adenosine	Adenylic acid

▲ FIGURE 19-15

Adenine derivatives: as a free nucleic acid base; as adenosine, a nucleoside (bound to ribose); and as adenylic acid, a nucleotide (a phosphorylated nucleoside). The corresponding compounds derived from cytosine are cytidine and cytidylic acid; from guanine; guanosine and guanylic acid; and from uracil, uridine and uridylic acid.

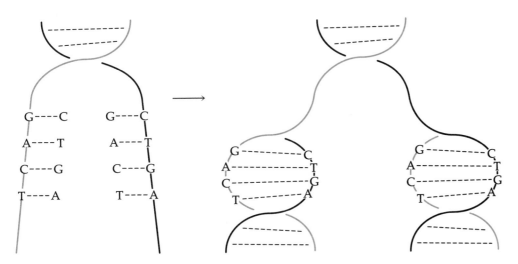

Base complementarity pairs C with G and A with T.

All living systems depend on the translation of the information contained in specific sequences of bases in DNA into peptide sequences. Thus, DNA represents stored information, that, when read, provides the primary sequence of amino acids for all peptides, proteins, and enzymes. It is equally important that this information be in a form that can be reproduced and passed on from cell to "daughters" and from organism to offspring. Imagine that the two strands of a double helix unwind in a solution rich in the various nucleotides. Each of the separated strands can then associate by hydrogen bonding with free bases in solution, and the proper, complementary pairings will be favored (Figure 19-16). Enzyme-catalyzed formation of a phosphate ester bond between the nucleotides then links the bases together and completes the synthesis of a complementary strand for each of the originals, in essence duplicating the original pair.

This process would be highly unfavorable energetically if the strands were to fully separate before associations with the nucleotides in the solution replaced the hydrogen bonds lost by unwinding of the original double helix (DNA can be as long as 3×10^9 bases). Instead, the complementary bases are replaced and the phosphate bonds of the polymer are formed at the same time that the strands uncoil from each other (Figure 19-17). In this fashion, the information stored in a specific sequence of bases within a

▲ FIGURE 19-17

Only partial unwinding is needed for replication by complementary base pairing. As a complementary sequence of paired bases is formed, polymerization produces a strand that is bound to the dissociated single strand of DNA or RNA. This sequence specificity accomplishes information transfer through replication.

DNA strand can be precisely duplicated in a process critical to the propagation (reproduction) of living systems.

The sequence of bases along a DNA strand determines the sequence of amino acids in a peptide. Because there are only four different bases, more than one base is needed to encode each of the twenty common amino acids. If a code unit consisted of two bases, there would be $4^2 = 16$ different permutations—also too few. Indeed, each amino acid is encoded by a specific three-base sequence, and thus only some of the $4^3 = 64$ combinations are needed to encode all of the amino acids. Some of the other combinations are used as special markers; for example, to indicate the beginning or end of a sequence; some of the possible combinations are not used; and some of the amino acids are encoded by more than one sequence of three bases.

EXERCISE 19-F

Write the sixteen possible coding combinations of pairs made from C, T, A, and G to convince yourself of the validity of the foregoing calculation.

In order for the genetic code for the synthesis of peptides to be read, the information from a segment of one strand of double-stranded DNA is converted into a complementary strand of RNA by a process called **transcription.** Recall that RNA is distinguished from DNA mainly by the presence of ribose rather than the deoxyribose units of DNA. There is another, more-subtle difference: thymine is found only in DNA and uracil is found only in RNA. These bases differ from each another only by the presence of a methyl group in thymine: their ability to base-pair with adenine is essentially the same (Figure 19-18).

The transcribed strand of RNA, known as **messenger RNA** (mRNA) is complementary to the segment of DNA from which it is transcribed. It thus contains the same sequence of bases as the other strand of the double-stranded DNA (except in the substitution of uracil for thymine), as shown in Figure 19-19. In this process, only a short segment of the DNA is transcribed into RNA, whose sequence carries the code for a single peptide. The mRNA moves out of the cell nucleus, where it was synthesized, to a **ribosome**—a body in the cell cytoplasm.

The ribosome contains all of the biological reagents needed to synthesize peptides—that is, to **translate** mRNA into protein. Critical to this process is **transfer RNA** (tRNA), a small nucleic acid molecule ranging in size from seventy to eighty nucleotides. Most of these nucleotides are the same in all tRNAs, except for two unique regions: one that has a sequence of three bases (an **anticodon**) that complements the three-base **codon** in mRNA for a particular amino acid; and another that provides a site for the attachment of a specific amino acid. The amino acid is chemically bound to

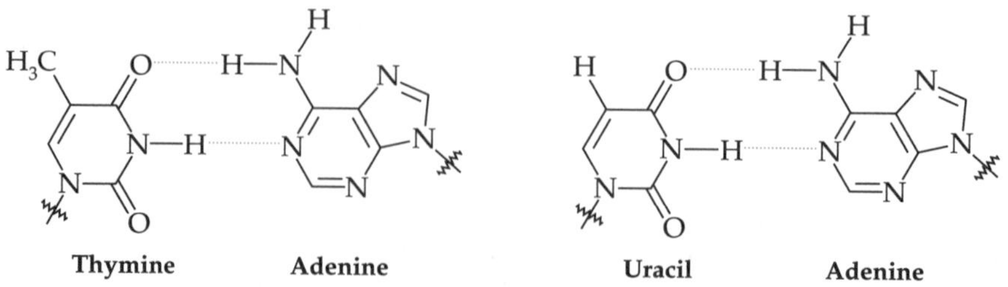

Thymine   Adenine   Uracil   Adenine

▲ FIGURE 19-18

Thymine and uracil base-pair with adenine by the same pattern of hydrogen bonds. Thymine is found only in DNA and uracil is found only in RNA.

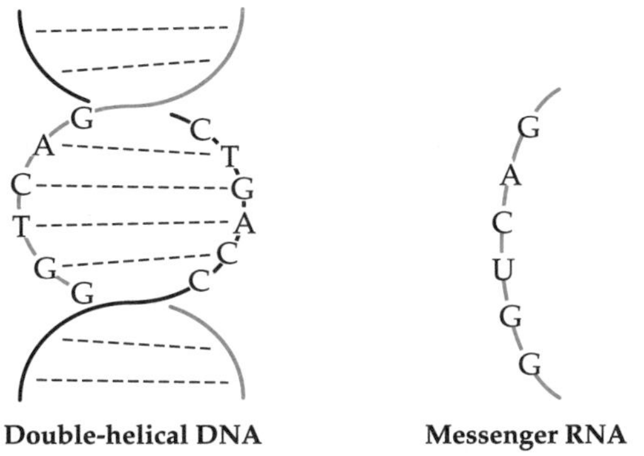

Double-helical DNA   **Messenger RNA**

a specific terminal nucleotide through a mixed anhydride (Figure 19-20) and is thus activated for nucleophilic acyl substitution. A tRNA bearing its specified amino acid associates with a three-base codon of mRNA (Figure 19-21). With each incoming aminoacyl-tRNA, the amino acid at the other

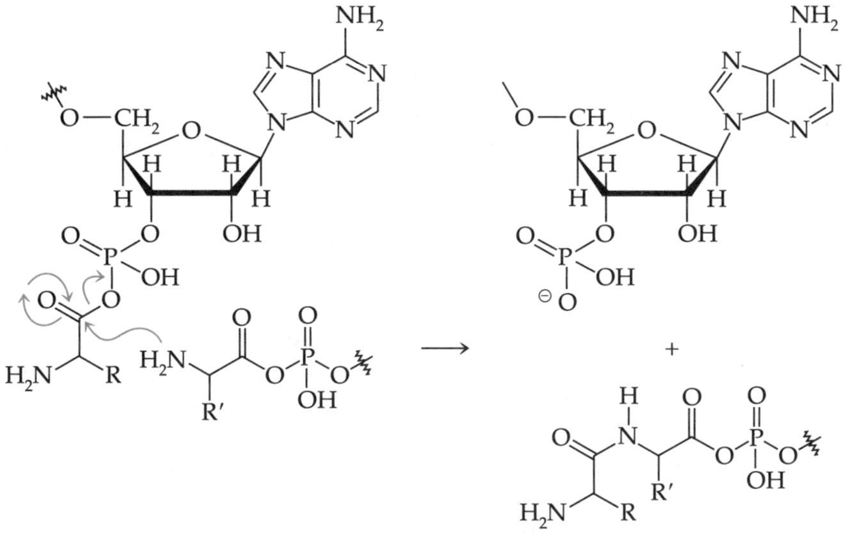

▲ FIGURE 19-20
The terminal amino group of an amino acid in which the acid group is phosphorylated to a mixed anhydride is activated toward nucleophilic attack by the free amino group of another phosphorylated amino acid.

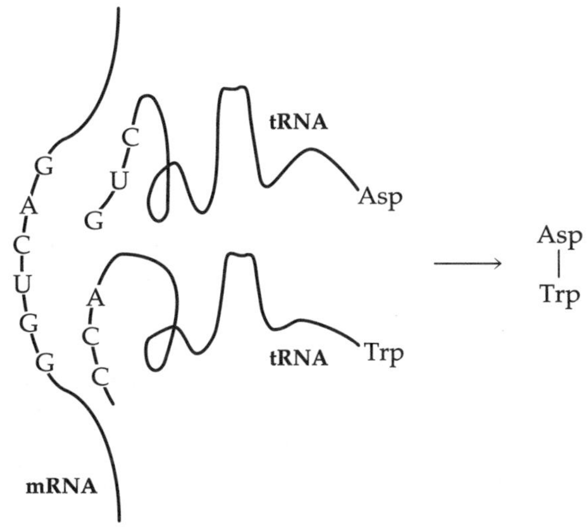

◀ FIGURE 19-21
A three-base-pair codon selects for a specific amino acid that is delivered in sequence to the growing peptide chain.

end of the tRNA becomes linked to the growing peptide chain by the formation of an amide (peptide) bond. In the short example in Figure 19-21, the G-A-C sequence in mRNA is complementary to the C-U-G sequence in the tRNA coding for aspartic acid (Asp), and the next three bases, U-G-G, are complementary to the set in tRNA for tryptophan (Trp). The two amino acids bound to the tRNAs are brought into proximity and are then linked together by an enzyme in a process that is similar to the replication of the DNA. A large number of codes are required for the synthesis of a large peptide.

EXERCISE 19-G

A coding system using four bases per code unit would provide far more codes than necessary. (How many?) What advantage might such a system offer in transcription of the code? What disadvantage would there be for living systems?

This system of information storage, transcription, and translation might seem foolproof, but it is not. There are several chemical changes that occur in the nucleotide bases that can lead to the misreading of the code for a particular amino acid. For example, simple proton tautomerization of adenine converts it into a form that can form two hydrogen bonds to cytosine and thereby mimic the chemical activity of a guanine residue (Figure 19-22).

More-permanent changes also can alter the hydrogen-bonding pattern of a nucleotide base. Many chemicals taken in through the diet are capable of altering the genetic code: nitrite ion (NO_2^-) is perhaps the most notorious because of its widespread use as a meat preservative in the form of potassium (or sodium) nitrite. It is an effective preservative, functioning as an antibacterial and antifungal agent and as an antioxidant. Primary amines react readily with nitrous acid (HNO_2), formed from nitrite ion in slightly acidic solution, to form diazonium ions, which are excellent leaving groups. In aqueous systems, reaction of amines with nitrous acid can lead to net overall replacement of the amino nitrogen by a hydroxyl group. In this way, a cytosine residue can be converted into a uracil group, thus coding for adenine instead of guanine.

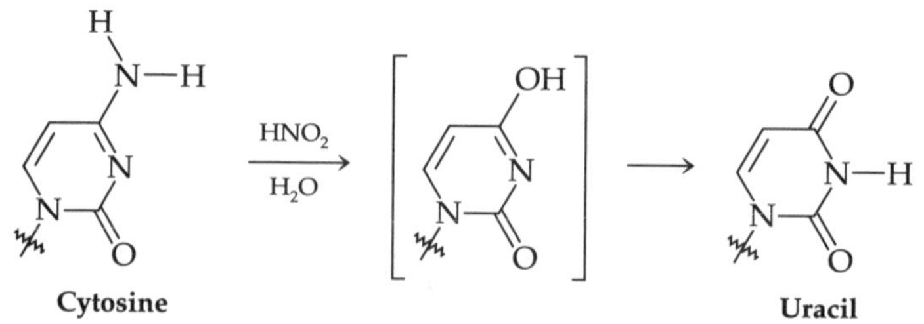

Cytosine   Uracil

In many cases, such a change has little consequence, especially if it alters a three-base code sequence to one that has no meaning (recall that not all of the sixty-four possible combinations are interpreted). However, in some cases, such an alteration leads to the substitution of one amino acid for another. For example, the sequence T-C-G in DNA produces mRNA with the sequence A-G-C, the code for tRNA carrying serine. Chemical modification that transforms a cytosine into uracil in DNA produces the sequence T-T-G, which codes for A-A-C in mRNA, a sequence complementary to that in the tRNA carrying asparagine. The end result is the synthesis

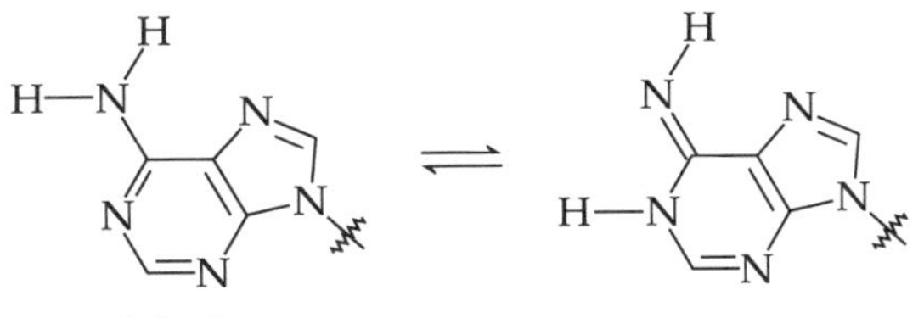

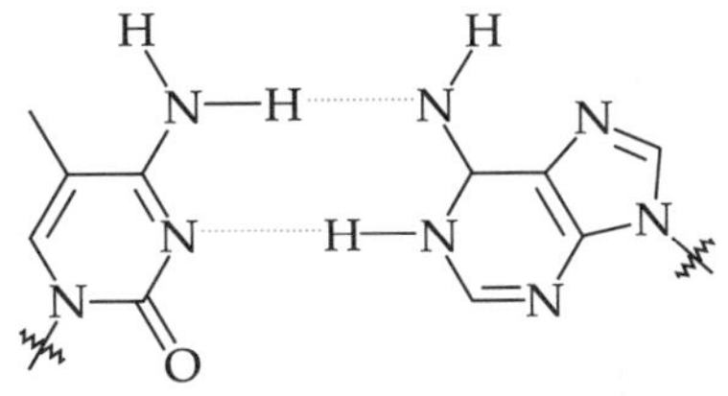

◄ FIGURE 19-22
◄ FIGURE 19-22
Tautomerization of adenine converts it into an isomer that bases-pairs with cytosine in much the same way as guanine does.

of a peptide with asparagine in place of serine. Even when such changes do affect the amino acid sequence in a peptide, it is rare that they have a major effect on the organism because the modified peptide most likely has little or no activity as a hormone or as an enzyme.

EXERCISE 19-H

Suggest a reasonable mechanism for the conversion of cytosine into uracil with a strong acid and water as the only reagents.

19-7 ▸ Molecular Recognition of Chiral Molecules

Many of the molecules of nature are chiral, almost always as the result of carbon centers of chirality with four different groups attached. In addition, almost all of these chiral molecules are present as single enantiomers, as is the case for the carbohydrates and the amino acids (Chapters 17 and 18). In this section, we examine some of the underlying reasons for this handedness of nature. In so doing, we set the stage for the discussions to follow in the next chapter that deal with the intimate associations at the transition states of most biochemical transformations.

Three-Point Contacts Are Necessary for Chiral Recognition

For simplicity, we represent a chiral molecule as a tetrahedron with four numbered groups attached (Figure 19-23). This depiction is just like that used to describe the Cahn-Ingold-Prelog rules for the assignment of the chirality specifiers *R* and *S* in Chapter 5. In the present context, however, we use such pictures to represent an entire chiral molecule in which the four regions differ in the types and strengths of possible weak interactions with other molecules. Thus, one group might be a large, hydrophobic hydrocarbon region, another a polar, hydrogen-bonding functional group such as a hydroxyl group, and so forth. Nonetheless, we use the *R,S* notation in referring to one of the enantiomers as if the numbered groups were substituents on a chiral center in the Cahn-Ingold-Prelog sense.

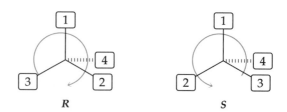

▲ FIGURE 19-23

In assigning absolute configuration at a center of chirality, we assign priorities to the four unique groups according to the Cahn-Ingold-Prelog rules. With the group of lowest priority directed away from the observer, the progression from highest to lowest priority for the remaining substituents is clockwise in an *R* configuration (left) and counter-clockwise in an *S* configuration (right).

In Figure 19-24, these two enantiomeric representations have been rotated so that three of the four groups of the *R* enantiomer are oriented toward the same groups of its mirror image. We can see that, in this orientation, there is close proximity between the same groups of each enantiomer (1↔1, 2↔2, 3↔3).

Let us now compare this pairing of an *R* with an *S* enantiomer with that in which both partners are of the same configuration (*R* with *R* and *S* with *S*), as in Figure 19-25. (Recall that interchanging any two substituents

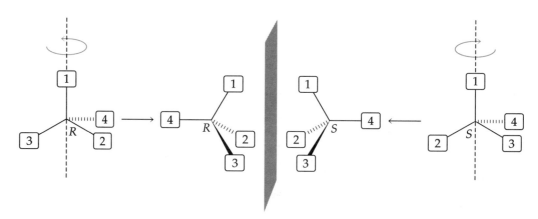

▲ FIGURE 19-24

Because enantiomers are mirror images, identical groups can be directed toward the same regions in space, permitting three molecular contacts.

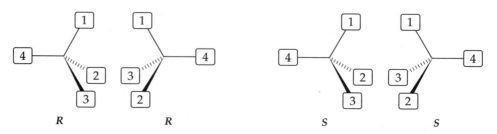

▲ FIGURE 19-25

Two groups are mismatched when an *R* (or an *S*) enantiomer associates with itself.

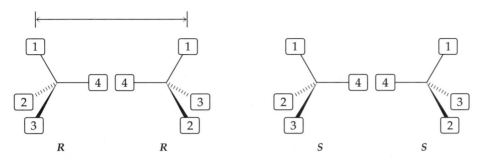

▲ FIGURE 19-26

With only a single contact between compounds bearing a center of asymmetry, recognition of one enantiomer over another is not possible because the other groups are too remote from each other for there to be a significant difference in the magnitude of interactions.

at a chiral center effects the conversion of one enantiomer into its mirror image.) Notice that now only one of the interactions is between like groups (1↔1), whereas the other two are between unlike groups (2↔3 and 3↔2). Clearly, then, the association of a chiral molecule with its enantiomer as shown in Figure 19-24 is not identical with that obtained by pairing the enantiomer with itself as in Figure 19-25.

This *three-point contact* between chiral molecules forms the basis for **chiral recognition.** Before we consider the consequences of these interactions, let us see why three contacts are required for chiral recognition by examining alternate associations of pairs of molecules in which only one or two of the four groups are in close contact. Figure 19-26 illustrates a one-point pairing in which the groups of lowest priority (4) are in proximity. Notice that there appears to be little difference between the unlike (R with S) and like (R with R) pairings. Indeed, to the extent that groups 1, 2, and 3 in one molecule are distant from those in the other, there is little or no difference in energy between these alternate interactions. Even when two of the groups are near one another, as in Figure 19-27, the energy difference between the R,R and the R,S pairings is still small because groups 2 and 3 are directed away from each other in both pairs. Thus, significant differentiation based on chirality can occur only when there is substantial interaction between three groups in a three-point contact.

Simple physical analogies may help us understand and visualize the differences between matched and unmatched pairings. Hands and feet (as well as many other common objects) are chiral: that is, the object is not superimposable upon its mirror image. As discussed in Chapter 5, a glove made for one hand does not readily fit the other. In fact, the misfit of a glove

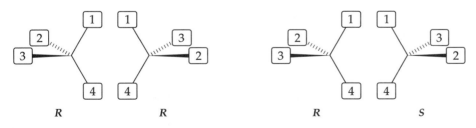

▲ FIGURE 19-27

Even when there are two contacts between compounds bearing centers of chirality, the difference between the magnitudes of interactions for the two possible diastereomeric pairings is small.

CHIRALITY IN EVERYDAY LIFE

The concept of chirality has many implications in our lives. Many of the objects with which we deal daily are chiral, although most people pay little attention to this subtlety. For example, books (we read left to right, top to bottom), scissors (special "enantiomeric" versions are available for left-handed people), and cars (steering wheel on the left, except in England, Jamaica, and some fifty other countries) are chiral objects—yet by-and-large we do not notice their handedness. The human brain also appears to be chiral. Each half, the left and right hemispheres, seems to have well-defined tasks. For example, logical analysis (among many other operations) is normally carried out on the left side of the brain, whereas abstract thinking and artistic interpretation is the responsibility of the right side. In some persons, the connection between the hemispheres of the brain allows for an abnormally high level of communication. These people often have difficulty in seemingly simple operations such as spelling, often interchanging the letters d, q, p, and b, all of which differ only because of a defined perspective (a "p" is a backward "q" and an upside-down "b"). Often, such people are gifted in ways that are said to compensate for this handicap; having artistic ability, for example, perhaps derives in part, from having a greater facility for three-dimensional perceptions.

---

with the wrong hand is very similar to the mismatching of chiral stereocenters. If we maintain the orientation of the palm of the glove, then the thumb and fingers are on the wrong side. Conversely, if we turn the glove over so that the thumb is on the correct side, then the palm is not.

If the smallest finger and the thumb were the same shape and size, the hand would be symmetrical from left to right, and the glove could be worn with either face oriented toward the palm. In a like fashion, if there were no difference between the palm and the backside of the glove, then one shape would fit either hand equally well (as is the case for most children's mittens).

Resolution

We can see energetic consequences of chiral recognition in the separation of a racemic mixture into individual enantiomers. This process, known as *resolution* (Chapter 5), consists of the interaction of the racemic mixture with a single enantiomer of another chiral compound referred to as the **resolving agent.** There are two different possible pairings in a resolution: with a resolving agent of the *R* configuration, *R* can pair with *R* and *R* with *S* (Figure 19-28). These alternate possible pairings are diastereomers and thus have different physical properties such as solubility. However, there must be some chemical reason why the two different molecules combine in a single crystal. Indeed, most resolutions require either the combination of a racemic acid with a basic resolving agent such as an amine or the opposite combination of a racemic amine with the single enantiomer of a chiral acid. In either case, the acid-base reaction is a driving force for the formation of the crystalline salt.

In many cases, the acidic or basic resolving agent is one that is obtained from natural sources as a single enantiomer. As an example, let us consider

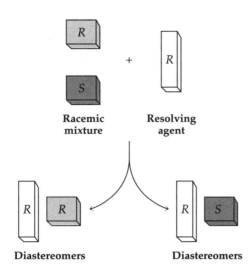

◄ FIGURE 19-28
When a racemate forms a bond
to a chiral reagent, a pair of
diastereomers is produced.
Because diastereomers have
different chemical and physical
properties, they can be
separated by several possible
methods; for example, by
column chromatography,
selective crystallization,
distillation, or extraction.

the resolution of mandelic acid with the alkaloid cinchonine (pronounced
as sing'ke-nen'), shown in Figure 19-29. This alkaloid, like the structurally
related antimalarial alkaloid quinine (Chapter 18), is obtained by extraction
of the bark of the cinchona tree.

(S)-Mandelic acid

Cinchonine   **(R)-Mandelic acid**

mp 80 °C   mp 165 °C

▲ FIGURE 19-29
Resolution of racemic mandelic acid into separate enantiomers can be
accomplished by separation of the diastereomeric salts formed by its acid-base
reaction with cinchonine, a tertiary amine. Neutralization of each of the salts
with NaOH will reform the neutral amine that can be easily separated from
the carboxylate ion of mandelic acid.

Combining racemic mandelic acid with cinchonine results in the formation of two diastereomeric salts that differ dramatically in their melting points. Furthermore, the lower-melting diastereomer is twice as soluble in water (20 g/l versus 10 g/l); thus, these diastereomers are readily separated by recrystallization. After this resolution process, the individual enantiomers of mandelic acid and the resolving agent are easily separated by an acid-base extraction.

EXERCISE 19-1

Suggest an appropriate sequence of extraction operations that would result in the separation of mandelic acid and cinchonine from the salt.

Biological Significance of Chirality

Handedness has profound consequences in living systems. In most cases, the interactions that invoke a biological response take place at receptor sites in the body that are chiral and exist with only one handedness. Thus, each of the two enantiomers of a biologically active material would not be expected to elicit the same response; indeed, the differences are often dramatic. For example, each of the enantiomeric terpenes (−)-carvone and (+)-carvone has a significant odor: the first is a major contributor to the odor of spearmint and the second is responsible for that of caraway seeds.

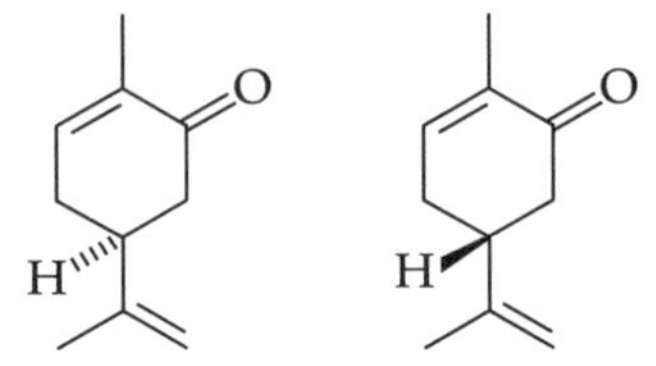

(−)-Carvone   (+)-Carvone

Each of these plants produces only one enantiomer, whereas, interestingly, racemic carvone is found in gingergrass. Likewise, the naturally occurring, L enantiomers of leucine, phenylalanine, tyrosine, and tryptophan have a bitter taste, whereas each of the nonnatural D isomers is sweet. In this case, the two enantiomers must interact with different receptors because the sensations of sweetness and bitterness result from stimulation of different areas of the tongue.

A tragic example of the interaction of each enantiomer with a different receptor site was uncovered when the antidepressant **thalidomide** was sold as the racemic mixture in the early 1960s, even though only the R enantiomer has the desired activity.

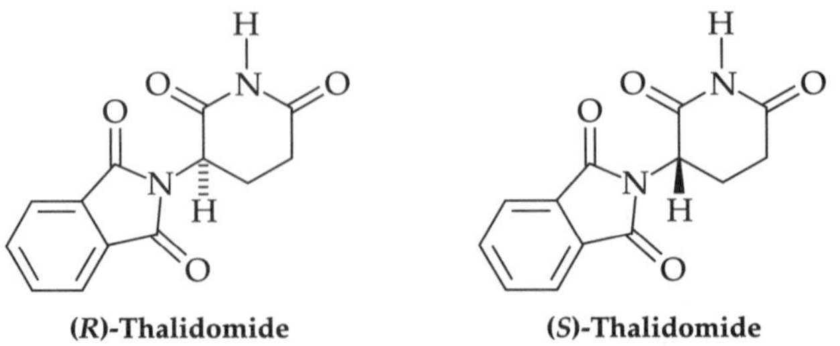

(R)-Thalidomide   (S)-Thalidomide

RESOLUTION AND THE PHARMACEUTICAL INDUSTRY

Despite the lesson to be learned from the thalidomide experience, the majority of chiral pharmaceutical agents that are marketed today (> 85%) are sold as racemic mixtures. The reason is simple: economics. Although we can relatively easily understand the concept behind the process of resolution, it is often quite difficult for the industrial chemist to devise methods for the separation of enantiomers that do not add substantially to the cost of manufacture. In some cases, the three-dimensional specificity of enzymes is used to separate a racemic mixture. For example, there are a variety of enzymes that catalyze the hydrolysis of esters. In some cases, when these enzymes are used as catalysts for the hydrolysis of racemic mixtures, only one of the enantiomeric esters undergoes hydrolysis. Separation of the remaining ester from the alcohol and acid hydrolysis products (on the basis of polarity differences) results in separation of the enantiomers.

Because of the keen insight of Frances Kelsey, a researcher at the Federal Drug Administration (FDA), thalidomide was never approved for use in the United States. However, this prescription drug was already in use in Canada and in European countries and, despite strong warnings against its use by pregnant women or even women likely to become pregnant, thalidomide was being prescribed for the treatment of "morning sickness." The antidepressant activity of thalidomide is due to one enantiomer; the other was found to be mutagenic and antiabortive. With this combination of activities, this enantiomer not only initiated genetic alterations that resulted in deformed fetuses, but also contributed to their retention, preventing the natural expulsion of damaged fetuses owing to other causes. The result of the use of thalidomide was the birth of many very seriously deformed children, often having underdeveloped arms and legs. Curiously, the observation that Kelsey had used to hold back approval of thalidomide was that it caused abortions at high doses in rats.

Conclusions

In this chapter, we have explored the consequences of weak interactions between molecules. In addition to van der Waals and dipole–dipole interactions, hydrogen bonds play a major role in holding molecules together. Such interactions significantly influence physical properties such as melting points, boiling points, and solubility. In a similar fashion, metal cations such as those derived from the alkali earth metals associate by relatively weak, but still quite significant, electrostatic interactions with lone pairs of electrons. Natural ionophores and synthetic ionophores bind metal ions most effectively when multiple ligand sites coordinate with the cation. Entropy considerations, including the effect of molecular symmetry, influence the equilibrium binding constant between ligating hosts and included guests.

Although a linear arrangement of the three atoms in a hydrogen bond is optimal, nonlinear hydrogen bonds are often found. Hydrogen-bond energies range from about 1 to 5 kcal/mole. Strong hydrogen-bonding interaction can take place between molecules in two- or three-dimensions. Complementary base pairing in two dimensions is responsible for information storage and transcription and is based on the specific recognition of cytosine by guanine and of adenine by thymine in DNA. Uracil fills the role of thymine in RNA, mRNA, and tRNA. A three-letter code specifies the sequence of amino acids delivered to a given peptide chain.

In combination, three-point hydrogen-bonding contacts can lead to three-dimensional recognition of centers of asymmetry. Although these forces are significantly weaker than normal chemical covalent bonds, they constitute the strongest reversible, noncovalent interactions generally found between molecules. They therefore form the basis for selective interactions and transformations. Chiral recognition results only from three-point contact; one or two-point contacts are inadequate to differentiate chiral centers. The resulting subtle, yet selective, interactions have wide-ranging consequences on the chemistry of living systems.

Review Problems

19-1 Indicate which substance (A or B) in the following pairs can be expected to have the higher boiling point. Briefly explain your answer, indicating the key molecular difference in each case that leads to a difference in the degree of association of molecules in the liquid phase.

(a)

A

B

(b)

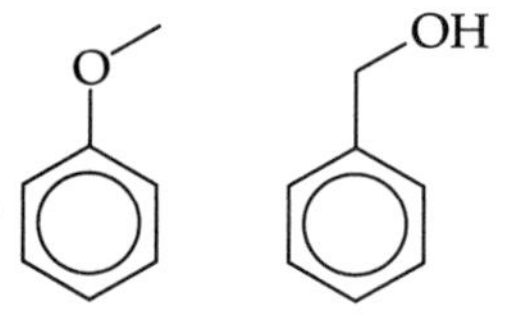

A   B

(c)

A   B

(d) H—C≡N   H—C≡C—H

A   B

(e)

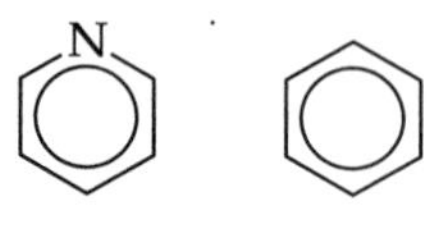

A   B

(f)

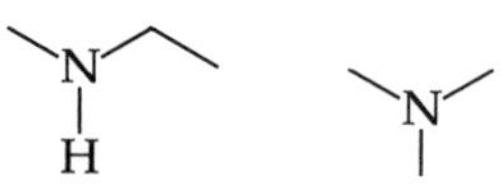

A   B

19-2 Several heterocyclic systems exist in equilibrium between a hydroxyaromatic and a ketonic form. Suggest a reasonable explanation for the observation that the equilibrium between 2-hydroxypyridine and the ketonic form known as 2-pyridone favors the latter, whereas the corresponding equilibrium for hydroxybenzene (phenol) favors the aromatic form. Write a reasonable mechanism for these interconversions as catalyzed by acid and by base.

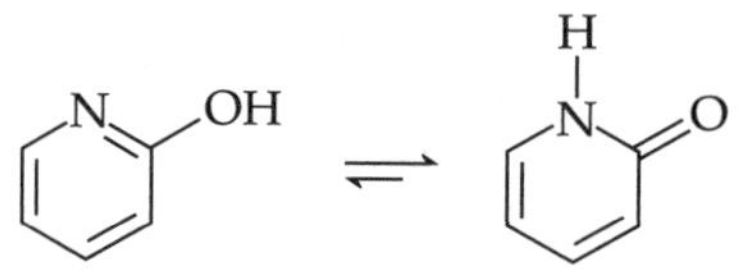

2-Hydroxy-   2-Pyridone
pyridine

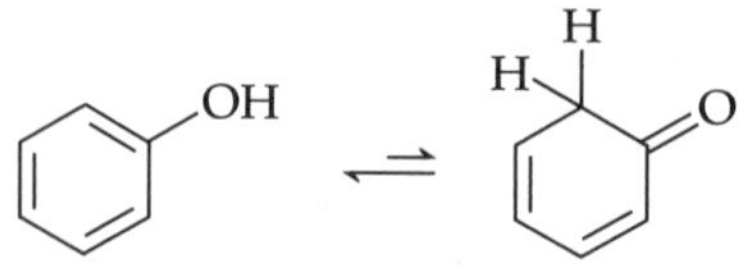

Phenol

19-3 Suggest an explanation for the observation that the reaction of 2-aminopyridine with strong base can lead to replacement of nitrogen by oxygen.

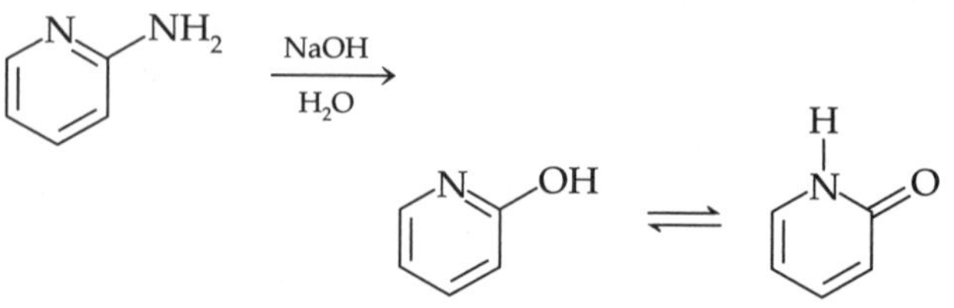

$$\xrightarrow[\text{H}_2\text{O}]{\text{NaOH}}$$

Yet, under similar reaction conditions, aniline is unreactive. Write a mechanism for the reaction leading from 2-aminopyridine to 2-pyridone.

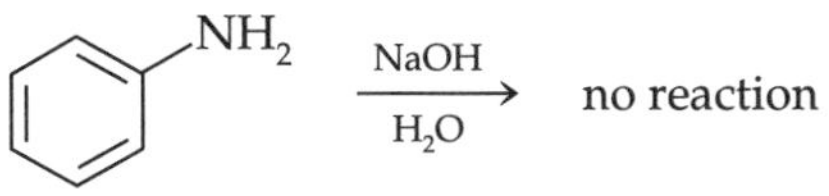

19-4 Dimethylsulfoxide (DMSO), $(CH_3)_2SO$, and acetone are very similar in structure because each has a polarized π bond to oxygen. Although DMSO is only 20 mass units heavier than acetone (molecular weight 68 versus 48), it has a very much higher boiling point (189 versus 57 °C). Can you suggest a reason for this difference? Dimethyl sulfone, $(CH_3)_2SO_2$, has a boiling point of 238 °C, which is 49 degrees higher than that of DMSO, but a melting point that is substantially higher (109 versus 18 °C). What molecular feature of dimethyl sulfone contributes to its higher melting point?

19-5 A common technique for extracting polar organic molecules from aqueous solution into an organic solvent is to add large quantities of sodium chloride to the water. Explain why adding salt decreases the solubility of organic compounds in water.

19-6 How would you expect the boiling point of amides to change as the number of methyl groups (0, 1, and 2) on nitrogen increases? Check your prediction with experimental data (for example, in the *Handbook of Chemistry and Physics*) and then explain the observed order.

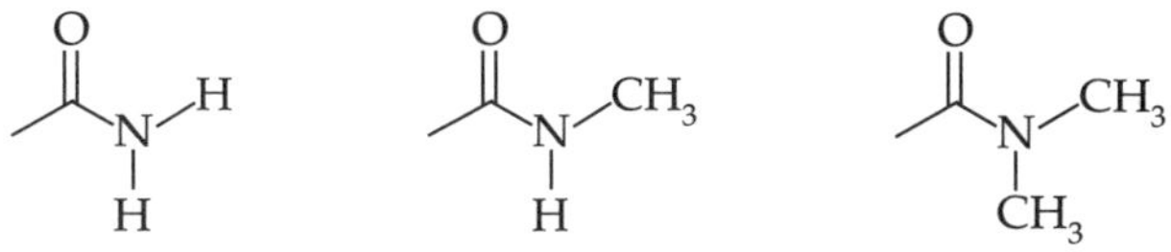

19-7 Large peptide and protein molecules generally have both hydrophobic and hydrophilic side chains. When these polymers form three-dimensional, globular structures, the amino acid residues with hydrophobic groups tend to be together, as do those with hydrophilic groups. Which would you expect to be on the outside and which on the inside of the three-dimensional arrangement of a biologically important protein?

Chapter 20

Catalyzed Reactions

Many of the chemical reactions discussed in earlier chapters take place under different reaction conditions, with very different rates and efficiencies. For example, various carboxylic acid derivatives have widely different reactivities: carboxylic acid chlorides are sufficiently reactive to be hydrolyzed within minutes in water (as long as they are somewhat soluble) at room temperature, whereas amides are stable toward hydrolysis in neutral water for years.

Such large differences in reactivity can be used to advantage as one method of controlling reaction rates.

Alternatively, the reactivity for a given derivative can be influenced by altering the concentrations of the reagents. For example, the rate of hydrolysis of an amide is pH dependent and increases dramatically as the concentration of hydroxide ion is increased. Other reactions, such as the dehydration of alcohols, are slow in the absence of hydronium ion and are accelerated as the solution pH is decreased.

Living systems do not have the flexibility that the chemist has in the laboratory: of the great range of chemical transformations that take place in living systems, each must proceed at a reasonable rate under the same conditions of temperature, pH, and solvent. Regardless of the criterion that one chooses to define life, all such characteristics refer to a change in the system from one time to the next. All living systems must be in a state of constant

---

THE EFFECT OF pH ON ENZYME FUNCTION

Many different reactions are catalyzed by enzymes, and each enzyme operates at maximum rate only within a fairly narrow range of experimental conditions. For example, although reaction rates generally increase with temperature, most enzymes denature at elevated temperatures and thus lose their catalytic properties. Acidity is also important: for example, the digestive enzymes in the stomach (where the pH is roughly 1.5) function best under acidic conditions, whereas digestive enzymes in the small intestine (where the pH is approximately 8) prefer alkaline conditions. The enzyme that causes cut apples to turn brown on exposure to oxygen does not perform well below pH 3.5; for this reason, lemon juice applied to the surface of freshly cut apples increases the acidity and keeps them from turning brown.

---

chemical flux, at which most of the chemical reactions taking place are not at their individual equilibrium states. Further, all living systems depend on the compartmentalization of reactions into regions (for example, cells) to achieve this far-from-equilibrium state. Conversely, equilibrium for a living system is equivalent to death.

There must be a special feature common to this collection of reactions, therefore, such that the rate of each reaction can be controlled by nonthermodynamic factors. In fact, virtually all biochemical transformations are greatly accelerated by the presence of biological catalysts, or **enzymes.** Without these catalysts, it would be impossible for all of the molecules important to living systems to be present simultaneously and for their rates of reaction to be controlled as demanded by the metabolism of the organism.

To understand the reactions that take place in living systems, it is necessary to have a good understanding of the concepts of catalysis in general, as well as the specific features that are unique to enzymes. In this chapter, we will explore general concepts of reaction catalysis and then see how these ideas can be used to explain the catalysis achieved by enzymes. The importance of organization into macromolecular units (such as cells) is more appropriately discussed in courses in biochemistry, microbiology, and molecular biology and is not covered here.

20-1 ▸ General Concepts of Catalysis

A catalyst accelerates the rate of a chemical reaction without being consumed. Many catalytic cycles can therefore be induced by a single catalyst molecule, and stoichiometric quantities of the catalyst are not needed. Catalysis is accomplished in the laboratory by stabilizing the transition state of the rate-determining step of a chemical reaction, either by providing electrostatic stabilization of partially charged centers that develop as the transition state is formed or by prearranging the reactants into a geometry appropriate for a desired bimolecular reaction, or both. This section compares the important features of specific catalysts that are effective either in the laboratory or in living systems.

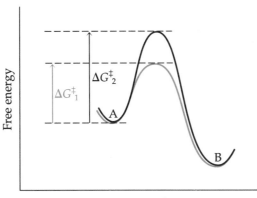

◄ FIGURE 20-1
A chemical catalyst accelerates a reaction by lowering the activation energy $\Delta G^{\ddagger}_1$ from that needed in the absence of the catalyst $\Delta G^{\ddagger}_2$.

Reaction coordinate

Transition-State Stabilization

There are two fundamentally distinct ways by which the rate of a chemical reaction can be increased. The first, and more common, is the use of a reaction pathway that is very similar to or even identical with the uncatalyzed process except for some special feature, such as solvent polarity, that leads to a decrease in the energy of the transition state (or, conversely, an increase in the energy of the starting materials). With such a catalyst, the activation energy ($\Delta G^{\ddagger}$) is lowered and the reaction rate is increased (Figure 20-1). Even though the catalyzed reaction is accelerated, the nature and sequence of the bond-making and -breaking steps remain essentially unchanged. The second way is to catalyze a reaction by making available a pathway that is entirely different: for example, one in which bond breaking and making are homolytic rather than heterolytic.

To understand catalysis, it is critical that we have a firm understanding of transition-state theory and how it relates to the rates of chemical reactions. Recall from Chapter 6 the shape of an energy diagram for a simple one-step, exothermic transformation, as in Figure 20-2. We can gain direct information only for species that lie in energy wells, or local minima—in this case, the starting material, A, and the product, B. All other aspects of the diagram are inferred. For example, the energetic position of the transition state is inferred from the activation energies of the forward or reverse reactions, which, in turn, are derived from the rates of these transformations. Certainly, we can obtain much indirect evidence on the transition state by examining such parameters as the effect of solvent polarity on the rate of reaction. For example, transition states that have considerably

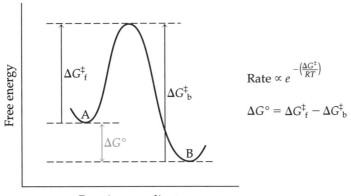

$$\text{Rate} \propto e^{-\left(\frac{\Delta G^{\ddagger}}{RT}\right)}$$

$$\Delta G^{\circ} = \Delta G^{\ddagger}_f - \Delta G^{\ddagger}_b$$

Reaction coordinate

◄ FIGURE 20-2
Assuming that entropy contributions can be neglected, the free energy for a concerted reaction (one that proceeds without one or more intermediates) is equivalent to the difference in activation energies for the forward (A to B) and reverse (B to A) conversions.

greater polar character than do the starting materials are favored to a greater extent by polar solvents. In turn, this stabilization lowers the energy of the transition state relative to that of the starting material and results in a faster rate than that of the same reaction carried out in a less-polar or non-polar solvent.

Effect of Solvation on S_N2 Reactions

Figure 20-3 illustrates energy curves for a simple S_N2 reaction of a neutral amine (a nucleophile) with a neutral substrate (an alkyl halide) in water (a polar protic solvent) and in ethyl acetate (a less polar, aprotic reaction medium). This reaction starts with an uncharged species but proceeds to charged products through a transition state in which partial charge is developed. Of the two solvents, water is much better at stabilizing ions, by both donating and accepting hydrogen bonds. Thus, although water stabilizes the starting materials, its effect on the transition state is even greater (the greatest stabilization is that for the product ions—species with full charges). Because the energy of the transition state is lowered by solvation to a greater extent than that of the reactants, the activation energy ($\Delta G^{\ddagger}$) is lower than that for the same transformation carried out in ethyl acetate. For those reactions that develop charged products from neutral starting materials, the rate of reaction is increased by polar (and especially polar, protic) solvents. We can see the effect of polar solvents on the rates of reactions in which charged species combine to form neutral products by considering the reaction in Figure 20-3 in the reverse direction. In this case, the greater stabiliza-

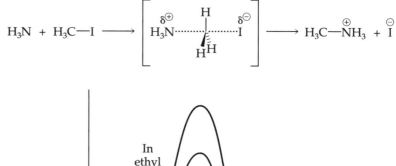

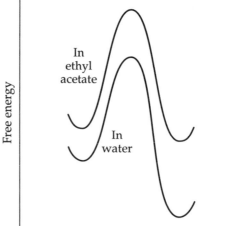

Reaction coordinate

▲ FIGURE 20-3
Because of enhanced dipole–dipole interactions with the solvent, a polar solvent such as water stabilizes the reactant, the transition state, and the product in this concerted S_N2 reaction. Because the transition state is stabilized to a greater extent than the reactant and product, the reaction is accelerated in water compared with the rate observed in ethyl acetate, a less-polar solvent.

tion of the ions by a polar, protic solvent such as water compared with that of the transition state leads to an increase in the activation energy ($\Delta G^{\ddagger}$) and a corresponding decrease in the rate of the reverse reaction.

Although the rate of the reaction in the forward direction is accelerated by the presence of water, water is not a catalyst because it binds to the product through solvation. A true catalyst does not affect (dramatically) the stability of either the starting material or the product; as a consequence, it does not significantly perturb the equilibrium position. In our example, water stabilizes the product to a greater extent than it does either the starting material or the transition state. Were we to carry out this reaction with a "catalytic amount" (less than one equivalent) of water, we would find that the reaction would be rapid at first but would then slow down dramatically because water would bind tightly to the products and would not be available to stabilize the transition state for reaction of additional starting material. Nonetheless, when water is used as the reaction solvent, its effect is quite similar to that of a catalyst in that the energy of the transition state is lowered relative to that of the starting materials, resulting in a decrease in the reaction's activation energy. We conclude that a compound functioning as a catalyst cannot become more tightly bound to either starting material or product than it is to the transition state of a given reaction.

Let us now consider an alternate influence of solvation on the rates of reaction in the S_N2 displacement of bromide ion in methyl bromide by hydroxide ion (Figure 20-4). This is a displacement reaction in which the

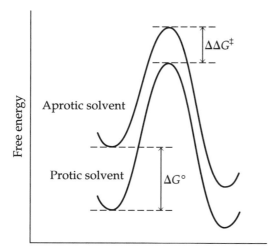

▲ FIGURE 20-4
In a nucleophilic substitution reaction employing a charged nucleophile, the ionic reactant is significantly stabilized by a polar solvent. In the transition state, the highly localized charges present in the reactant and product ions are dispersed; although a polar solvent also stabilizes this charge-diffuse transition state, it does so less effectively than it stabilizes the reactants or products. The activation energy for the forward reaction is therefore increased, and the reaction proceeds at a slower rate in the polar than the nonpolar solvent.

concentrated negative charge in the reactant hydroxide ion is more widely dispersed at the transition state. Reactions of this type are slower in polar, protic solvents because the stabilization from solvation is greater for the starting material than it is for the transition state. Conversely, the reaction is greatly accelerated in an aprotic solvent because the starting hydroxide ion is less effectively stabilized than is the transition state in this medium compared with the same species in the absence of solvation. However, most metal hydroxides are only poorly soluble in aprotic solvents and, conversely, organic molecules such as methyl bromide dissolve to only a very limited extent in protic solvents such as water. There is, then, a dilemma here: the lowest activation energy for this substitution reaction is attained in a solvent in which only limited amounts of one of the key reagents can dissolve.

There are two methods used in laboratory reactions to solubilize such ions as hydroxide in relatively nonpolar solvents: both techniques rely upon decreasing the degree of association of the counterion with the hydroxide ion (Figure 20-5). We have already seen in Chapter 19 how macrocyclic ("crown") ethers can surround a small cation such as lithium or potassium with oxygen ligands, leaving the exterior of the complex very hydrocarbon-like and thereby increasing its solubility in lipophilic solvents. A second method exchanges a tetraalkylammonium cation for Na^+ as a counterion for the hydroxide ion. The tetraalkylammonium cation resembles a crown ether metal ion complex in that a cation is surrounded by hydrocarbon chains with a high affinity for hydrocarbonlike solvents. These methods differ in that the cation in the second case is covalently bound to its lipophilic partner; whereas, in the former, the association is the result of electrostatic attraction between the cation and lone pairs of electrons on the ether oxygens.

Tetraalkylammonium ions can be used as carriers to transport anions such as hydroxide between an aqueous solution and a hydrocarbon solvent such as benzene. The ammonium ion need not be used in a stoichiometric amount and, indeed, serves as a true catalyst for the reaction, increasing the

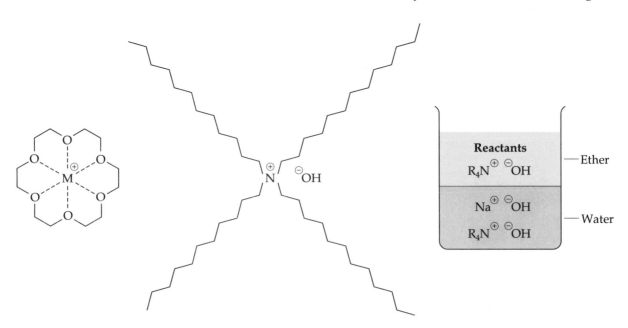

▲ **FIGURE 20-5**
Tight electrostatic association between a tetraalkylammonium cation and a hydroxide ion makes the pair sufficiently hydrophobic to dissolve in ether. This allows the hydroxide ion to be transferred from the aqueous phase to the ether layer. The movement of a reactant from one phase to another in which a reaction takes place is called phase-transfer catalysis.

rate of the reaction by increasing concentration of hydroxide (Figure 20-5) without itself being consumed. Because the reactive nucleophile has been transferred from one phase to another in which the reaction proceeds, this association leads to rate acceleration by a process called **phase-transfer catalysis.**

EXERCISE 20-A

How would the change from a nonpolar to a polar protic solvent affect the rate of an S_N2 reaction in which the nucleophile is neutral and the substrate positively charged? Provide a specific example of a reaction corresponding to this type.

20-2 ▸ Avoiding Charge Separation in Multistep Reactions

We now examine a slightly more complex reaction in which catalysis can operate. Figure 20-6 is a simple energy diagram describing the two-stage process that converts a carboxylic acid ester and water into the corresponding carboxylic acid and alcohol. Recall from Chapter 12 that this reaction proceeds through a tetrahedral intermediate, represented here as a simple uncharged species. Let us consider, however, what must take place in the bond-forming and bond-breaking steps that lead to this intermediate. As the oxygen atom of water approaches and begins to form a bond with the carbonyl carbon, it loses electron density, ultimately to become trivalent and therefore positively charged (Figure 20-7). Simultaneously, the π bond of the carbonyl group must be broken to an extent at least as great as that to which the new C–O bond is being formed to prevent a buildup of more than eight electrons around carbon. Upon completion of these bond changes, the original carbonyl oxygen is monovalent and thus bears a full negative charge.

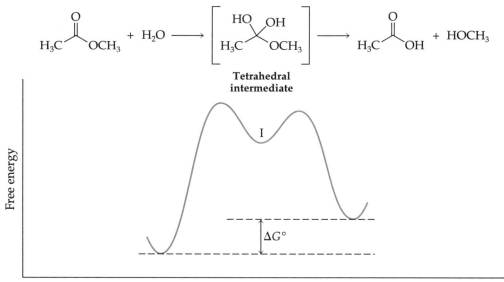

Tetrahedral intermediate

▲ FIGURE 20-6
Transesterification is a two-step process that proceeds through a tetrahedral intermediate formed by nucleophilic addition in the initial step.

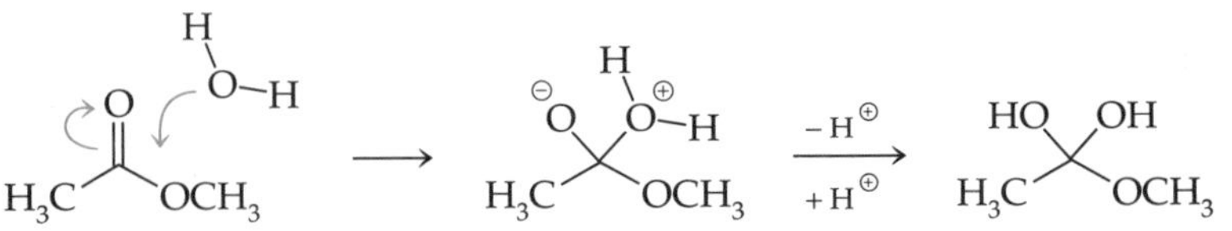

▲ FIGURE 20-7
Under neutral conditions, nucleophilic attack by water on methyl acetate is slow. The intermediate zwitterion contains one negatively charged and one positively charged oxygen atom. Deprotonation and reprotonation would produce a neutral tetrahedral intermediate.

The first tetrahedral intermediate formed in Figure 20-7 differs from that in Figure 20-6 most significantly in that there is complete charge separation in the latter. Such charge separation is highly unfavorable energetically and, indeed, a zwitterionic tetrahedral intermediate would very rapidly undergo proton exchange to form the neutral intermediate. We will return to the nature of this proton transfer shortly.

Now, let us see how the reaction differs when it is conducted in an acidic medium. Under these conditions, there is a significant concentration of the protonated form of the ester (A) in which the carbonyl oxygen bears a positive charge (Figure 20-8). As a bond is formed between the carbonyl carbon and the oxygen of water, a positive charge develops, as before, on oxygen as it becomes trivalent. However, as the π bond of the carbonyl system is broken, the positive charge on that oxygen decreases until it is formally neutral in the intermediate, B. Throughout this sequence, there is no charge separation; instead, there is only a transfer of positive charge from the protonated ester's carbonyl oxygen to the oxygen of the attacking water molecule. This feature is responsible for a dramatic energetic difference between the uncatalyzed and catalyzed processes illustrated in Figures 20-7 and 20-8. The transition state for oxygen–carbon bond formation in the latter is considerably more stable than that in the reaction taking place without prior association with a proton. Because the proton consumed at the beginning is ultimately released at the end of the scheme in Figure 20-8, this reaction constitutes an example of true catalysis.

Let us compare the energy diagram for the catalyzed (c) and uncatalyzed (uc) processes (Figure 20-9). The most-important difference is that the activation energy for the catalyzed process is considerably lower than that for the uncatalyzed reaction (the black curve). Therefore, the catalyzed reaction is faster. In addition, the energy diagram for the catalyzed reaction has three transition states, whereas that for the uncatalyzed reaction has only two. In the uncatalyzed pathway, the first transition state corresponds

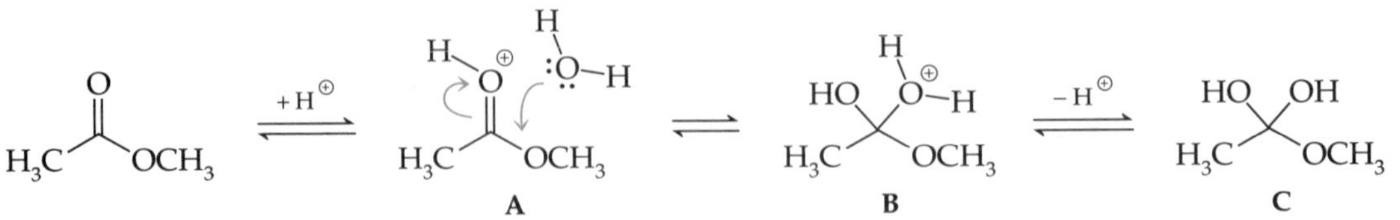

▲ FIGURE 20-8
In the acid-catalyzed addition of water to methyl acetate, the carbonyl oxygen is protonated in the reactant. When water attacks, no further charge is created: a cationic reactant is converted into a cationic intermediate with charge on the protonated carbonyl oxygen being replaced by positive charge developed on the attacking nucleophile. Deprotonation of this tetrahedral intermediate produces the same diol as that produced in the reaction under neutral conditions depicted in Figure 20-7.

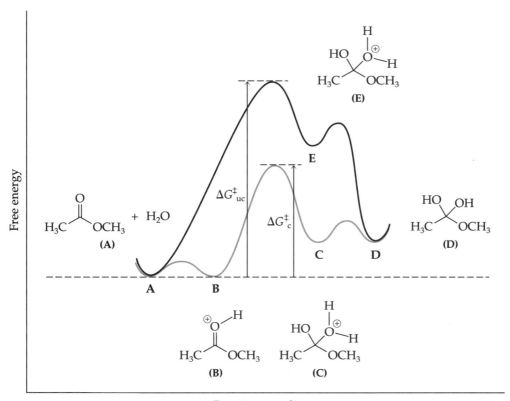

▲ FIGURE 20-9

In the acid-catalyzed addition of water to methyl acetate, the neutral ester (A) is in equilibrium with its carbonyl-oxygen-protonated form (B). The protonated ester is more reactive toward nucleophilic addition, as is evidenced by the lower activation energy required for the catalyzed pathway compared with the uncatalyzed pathway. A cationic tetrahedral intermediate (C) is formed in this nucleophilic addition. This cation C equilibrates with its neutral form (D) in an acid-base reaction. In the uncatalyzed pathway, the neutral ester is transformed into a high-energy, zwitterionic intermediate (E), which is converted into the hydrate (D) by protonation and deprotonation on the two charged oxygen atoms. The higher energy of the transition state leading to intermediate E is responsible for the slower rate observed in the absence of the acid catalyst.

to carbon–oxygen bond formation as the nucleophile bonds to the carbonyl carbon, whereas the second corresponds to the proton reorganization required to neutralize the separated charges produced in the first step. For the catalyzed reaction (the orange curve), the rate-limiting step (that with the highest activation energy) is preceded and followed by very rapid proton-transfer steps, neither of which results in an appreciable change in energy. The catalyzed and uncatalyzed reaction pathways differ at all points along the two reaction profiles in Figure 20-9 except for the beginning ester in water and the neutral tetrahedral intermediate (D).

_____

EXERCISE 20-B

Draw schematic representations for the five transition states in the catalyzed and uncatalyzed reaction pathways of Figure 20-9. Use dashed lines to represent partial bonds that are undergoing either formation or cleavage.

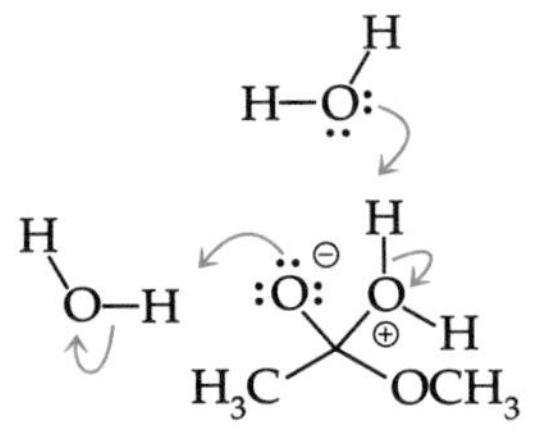

Let us now return to the question of how the zwitterionic tetrahedral intermediate undergoes proton transfers so as to arrive at the neutral tetrahedral intermediate. Indeed, bimolecular proton transfers between heteroatoms often occur with very small energies of activation. As a consequence, it is quite difficult to uncover the details of these reactions. Nonetheless, we can make reasonable presumptions on the basis of what is known about proton-transfer reactions and about geometric constraints on hydrogen bonding. We can rule out an intramolecular transfer through the transition state at the top of Figure 20-10 because of the very serious distortion of this transition state from the ideal, linear arrangement for proton transfer discussed in connection with hydrogen bonding in Chapter 19. Instead, intermolecular proton transfer forms the neutral tetrahedral intermediate, the negatively charged oxygen gaining a proton from the medium and the proton on the positively charged oxygen being transferred to a base such as water or methanol. These two proton transfers can be simultaneous, but this is unlikely for two reasons. First, each bond-forming and bond-breaking process contributes its own activation energy to any reaction in which there are simultaneous, multiple transformations. Second, in addition to this enthalpy factor, each proton transfer requires its own organization of molecules into the entropically unfavorable arrangement necessary for the transformation. Thus, for two proton transfers to and from the zwitterionic tetrahedral intermediate to be simultaneous, three molecules (the proton donor, the proton acceptor, and the intermediate) would have to come together simultaneously in the proper arrangement. Such three-body collisions are improbable.

By a process of elimination, then, we have arrived at two alternate multistep processes that differ only in the order in which the protons are added and removed. They can be expected to have very similar activation energies and thus to contribute approximately equally to the overall process.

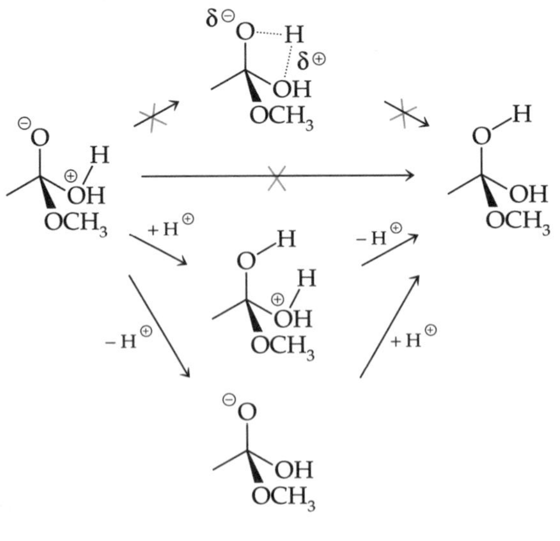

▲ FIGURE 20-10

There are three possible routes by which charge can be neutralized in the tetrahedral, zwitterionic intermediate produced by hydration of acetone under neutral conditions. They are: an intramolecular proton transfer through a four-membered transition state (upper reaction); protonation of the alkoxide oxygen, followed by deprotonation at the oxonium ion (center reaction); and deprotonation at the oxonium ion, followed by protonation on the alkoxide oxygen (lower reaction). The transition state for the intramolecular proton transfer shown at the top is too strained to render that pathway energetically acceptable when compared with the other two, intermolecular routes.

There is a final possibility for proton transfer: a mechanism that includes bimolecular transfer but nonetheless has some of the features of the intramolecular transfer shown in Figure 20-10. The two-step sequences for intermolecular proton transfer in Figure 20-10 consist of the removal of a proton by one water molecule and the addition of a proton from another one. Let us consider how both functions could be served by a single molecule of water.

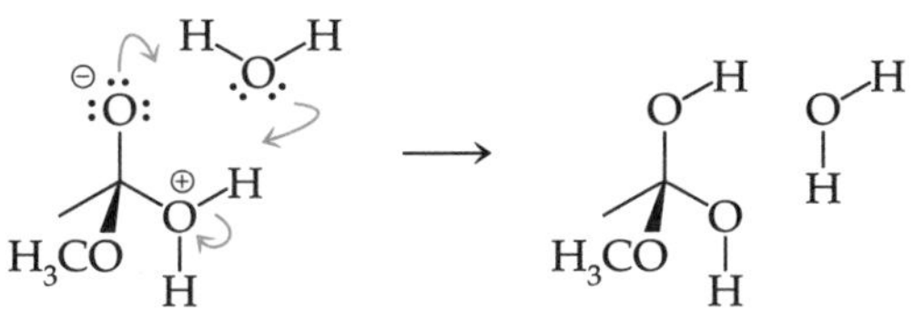

With a single water molecule serving both functions simultaneously—on the one hand, gaining and, on the other, losing a proton—this molecule undergoes no net change. In this transition-state arrangement, a total of six atoms take part in the intramolecular proton transfer instead of the four atoms participating in the intramolecular transfer shown in the upper reaction of Figure 20-10, which was deemed to be too high in energy. Although the preferred linear relation of hydrogen associated with two oxygen atoms in a hydrogen bond is not attained in this transition state, the OHO and OHO angles are larger than those of the four-atom intramolecular transfer shown in the upper reaction of Figure 20-10. This improvement in bond angles is achieved at a cost, however: there are now two bond-breaking and two bond-making steps taking place simultaneously, each of which adds to the energy of the transition state.

EXERCISE 20-C

Write mechanisms for all four of the proton-addition/proton-removal stages shown in Figure 20-10. Be sure to include arrows to indicate the flow of electrons. Carefully examine the species present as starting materials and as products. (Be sure to include the water and its product.) Determine whether you would expect any large changes in energy as each of these four reactions proceeds.

The expression **charge relay mechanism** is used to refer to this type of reaction, in which the function of the water molecule is to transfer a proton, and therefore the charge, from one charged atom in a molecule to an uncharged one that is not spatially oriented correctly for direct transfer between these centers.

One might wonder whether water could relay charge at the same time that the bond between the oxygen of the attacking water molecule and the carbonyl carbon is formed.

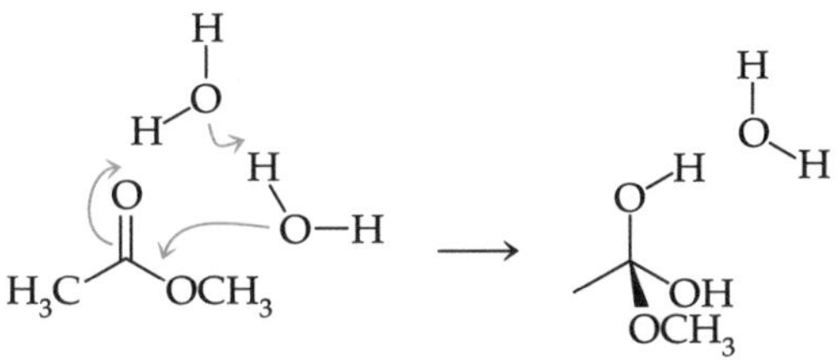

This mechanism has many favorable features, the most important of which is that no significant charge separation develops at the transition state. However, it is disfavored by entropy because it requires the simultaneous interaction of three molecules, here an ester and two water molecules.

Such highly organized arrangements, however, can be readily achieved within enzymes because they themselves are already highly organized. In essence, the cost of the entropic requirement of a highly organized transition state has already been paid in the construction of the enzyme. Because enzymes are catalysts and serve to transform thousands of molecules of starting material into product, the energetic cost of their synthesis is not a significant factor overall for the organism.

20-3 ▸ Distinction between Catalysis and Induction

We have seen how an acidic medium can prevent charge separation in the transition state for the hydrolysis of a carboxylic acid ester. Let us now consider whether a change to a more-alkaline pH (increasing hydroxide ion concentration) also can prevent this energetically unfavorable separation of charge. Indeed, with a sufficiently high concentration of sodium hydroxide, the negatively charged hydroxide ion becomes an effective nucleophile, forming a bond with the carbonyl carbon of the carboxylic acid ester (Figure 20-11). The nature of the charge in the starting materials and that in this tetrahedral intermediate is essentially identical: in both cases, negative charge is localized on a single oxygen atom. Thus, nucleophilic addition of hydroxide consists only of the transfer of charge: no additional charge separation in addition to that provided by NaOH develops in the transition state.

The neutral tetrahedral intermediate is then formed by transfer of a proton from water, reforming the hydroxide ion that was consumed in the first step. Again, the transition state for this step does not include the development of charge separation; it involves only transfer of negative charge from one oxygen to another. Thus, the ester is converted into the neutral tetrahedral intermediate in two steps, neither of which develops additional charge separation; in this sequence, hydroxide ion is a true catalyst. As a result, the formation of the tetrahedral intermediate is much more rapid when hydroxide ion, rather than water, is the nucleophile.

EXERCISE 20-D

Draw an energy diagram for the two steps shown in Figure 20-11, keeping in mind that the transfer of a proton from one heteroatom to another is generally a reaction with a very low energy of activation.

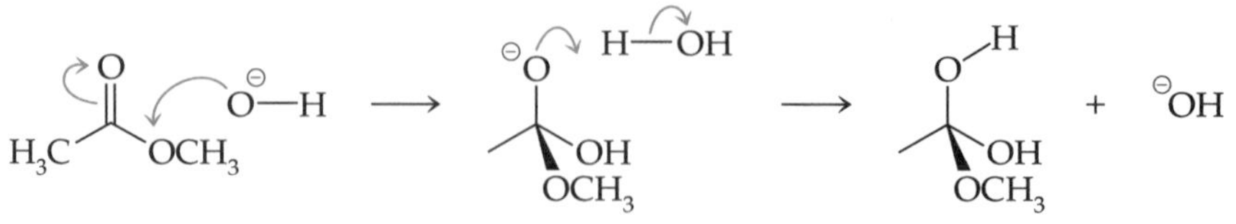

▲ FIGURE 20-11
Hydroxide ion is a reactive nucleophile that attacks an ester, producing an anionic, tetrahedral intermediate by addition to the carbonyl carbon. Protonation of this anion produces a neutral tetrahedral intermediate.

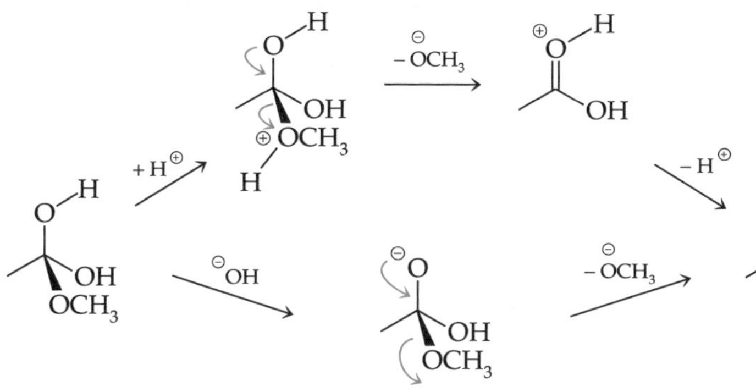

▲ FIGURE 20-12

The neutral tetrahedral intermediate from Figure 20-11 is transformed into a carboxylic acid in either acid or base. The acid-catalyzed reaction, shown in the upper reaction sequence, is initiated by protonation of the oxygen of the OCH_3 group. Methanol is a better leaving group than methoxide ion ($^-OCH_3$) and the loss of methanol takes place readily, with the assistance of lone-pair electron density of the OH groups. Deprotonation produces the carboxylic acid. In base, a proton is removed from the most-acidic site, one of the identical OH groups, as shown in the lower sequence. The electron density from the anionic oxygen atom assists in the expulsion of $^-OCH_3$.

Let us now continue our examination of the reaction as it proceeds to the final product from the tetrahedral intermediate in both acidic and basic media. The sequences differ according to the pH in much the same fashion as do the reactions leading to the tetrahedral intermediate (Figure 20-12). In the upper reaction, in acid, a proton is gained in the first step and later lost. Therefore, this stage of the reaction is acid-catalyzed. In the lower reaction, base present in the medium removes a proton, reforming the same anionic tetrahedral intermediate as that in Figure 20-11. Reformation of the carbonyl π bond with loss of methoxide ion leads to the carboxylic acid. Because methanol is a weaker acid than acetic acid (Chapter 5), these two species react, as shown in Figure 20-13, to form methanol and the carboxylate ion. Indeed, the equilibrium favors acetate and methanol by some eleven orders of magnitude.

EXERCISE 20-E

Verify the role of the acid as a catalyst in the sequence shown in Figure 20-12 (leading from the neutral tetrahedral intermediate to the carboxylic acid) by comparing how the reaction sequence might differ if the reaction were to take place at neutral pH.

In the net conversion of methyl acetate and water into acetate ion and methanol, the hydroxide ion is *not* a catalyst because it is not regenerated.

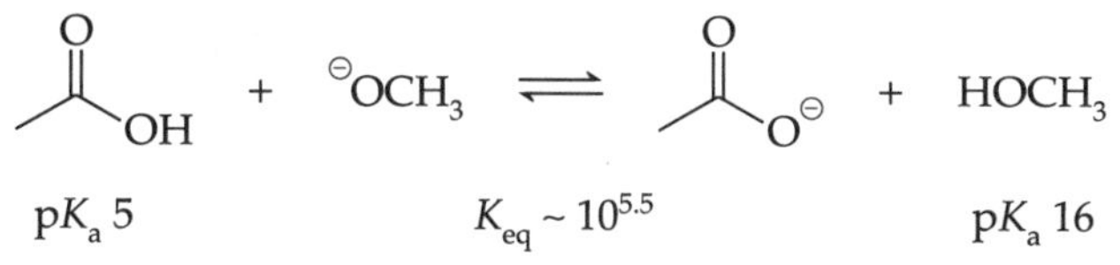

▲ FIGURE 20-13

Carboxylic acids are more acidic than simple alcohols, and thus the equilibrium between a carboxylic acid and methoxide ion lies far to the side of the carboxylate ion and methanol at equilibrium.

Thus, there is a fundamental difference between the hydrolysis of an ester in the presence of acid and in the presence of base. In the first case, the reaction is catalyzed, whereas, in the second, it is not. We say that the second process is **induced by base** to distinguish it from a truly catalytic sequence. Nonetheless, both reactions have many features in common with regard to how the reaction is accelerated.

Many enzymatic processes resemble hydrolysis with base in that a cofactor serves an essential role and is consumed in the process. Such reactions can still be correctly referred to as being catalyzed by the enzyme because, unlike the cofactor, the enzyme remains unchanged at the end of the reaction.

There is a way in which we can adjust the reaction conditions so that hydroxide does participate as a catalyst in the hydrolysis of an ester. Until the last (neutralization) stage shown in Figure 20-13, hydrolysis under basic conditions is catalytic because the methoxide generated can, and indeed does, react with water to form hydroxide ion. Recall that a buffered solution is one that contains a mixture of an acid and its conjugate base in rapid equilibrium. Small amounts of additional acid or base do not significantly affect the pH of the solution, only the position of the equilibrium. By performing the hydrolysis of an ester in a buffered, basic solution, the hydroxide concentration remains relatively unchanged as long as there is significantly more buffer present than acid generated by the hydrolysis. Under these conditions, the hydroxide ion is a true catalyst: it is the conjugate base of the buffer that is consumed, rather than the hydroxide ion.

The distinction between an acid-catalyzed and a base-induced hydrolysis of an ester is an important one for the laboratory chemist. In the former case, the amount of acid can be adjusted so as to provide a practical rate of conversion. However, there must be at least one mole-equivalent (referred to as a stoichiometric amount) of base present for the reaction to proceed to completion. In fact, even an equimolar amount is not sufficient because, as the reaction nears completion, the concentration of the hydroxide ion and, correspondingly, the rate of the reaction is diminished. In contrast, the distinction between a catalyzed and an induced reaction is far less important in biological systems where small amounts of material are converted in a large volume of buffered medium.

20-4 ▸ Base Catalysis

In the foregoing discussions, the base that accelerated the reaction (for example, hydroxide ion) was also the nucleophile. Although this situation is frequently encountered, some other species can often act as a base to generate the active nucleophile by acid-base reaction with a protonated precursor. For example, consider the hydrolysis of an ester by sodium carbonate in methanol containing a small amount of water. Here, sodium carbonate, the base, generates methoxide ion (and ultimately hydroxide ion) as nucleophile. (The use of a mild base such as carbonate is often chosen when there are other functional groups present in the molecules that might also undergo a base-induced or based-catalyzed transformation.)

In the first step, methoxide ion effects transesterification, converting the cyclohexyl ester of acetic acid into methyl acetate; in the second step, methyl acetate is hydrolyzed to acetate ion by attack of hydroxide ion. Notice that the first step does not consume base; it is thus base catalyzed. In the second step, the acetic acid initially produced reacts irreversibly with base, making this step a base-induced, not base-catalyzed, one.

Because sodium carbonate is only slightly soluble in methanol, the effective pH of the solution is only mildly basic. Nonetheless, the methoxide and hydroxide ions are produced in sufficient quantity for their reaction as nucleophiles. In essence, the reaction contains a sufficient quantity of base (some that is not even in solution) to effect the desired hydrolysis without the high pH that would result if the required equivalent of methoxide and hydroxide ions were present.

Because sodium carbonate is not as basic as sodium hydroxide, the presence of a full equivalent does not lead to as basic a solution as would be obtained with sodium methoxide. Yet, the equilibrium between sodium carbonate and sodium methoxide is quite rapid and, even though the equilibrium favors the former by some eight orders of magnitude, sufficient methoxide is present that the reaction can proceed.

$$\underset{pK_a \sim 16}{\overset{O}{\underset{\ominus O}{\overset{\|}{C}}\underset{}{O^\ominus}}} + CH_3OH \;\rightleftharpoons\; \underset{pK_a \sim 8}{\overset{O}{\underset{\ominus O}{\overset{\|}{C}}\underset{}{OH}}} + CH_3O^\ominus$$

In this reaction, the chemist seeks a balance between the concentration of the required nucleophile and the pH of the solution. Clearly, when the pH is very low, the rate of the reaction is retarded to a point at which it is too slow to be usable. Conversely, at higher pH, the rate of reaction increases but so do other, undesirable base-catalyzed and base-induced reactions. We will see shortly how a similar balance is important for the controlled function of the enzyme chymotrypsin.

20-5 ▸ Intermolecular and Intramolecular Reactions Compared

Most of the reactions considered in the first half of the book are bimolecular; that is, two reactants come together in the rate-determining transition state. For example, the hydrolysis of an ester, whether carried out under acidic or basic conditions, has a transition state in which a bond is formed between the carbonyl carbon and a nucleophilic oxygen of either water or hydroxide ion. Several energetic difficulties must be overcome in reaching such a transition state. First, two separate molecules must be brought together in a process that is, therefore, entropically unfavorable. Furthermore, these two reacting species cannot be brought together in a random fashion but must be oriented in an arrangement favorable for bond formation, thus adding additional entropic cost to the formation of the transition state. In addition, each reactant molecule is originally surrounded by (and stabilized by) solvent molecules. For a reaction to occur, some of the solvent molecules associated with each reactant must be removed in order that the two species can approach to within the distance necessary for bond formation in the transition state, as illustrated schematically in Figure 20-14.

Both entropy and solvation are far less important when the two reacting partners are already connected by bonds that are not affected by the

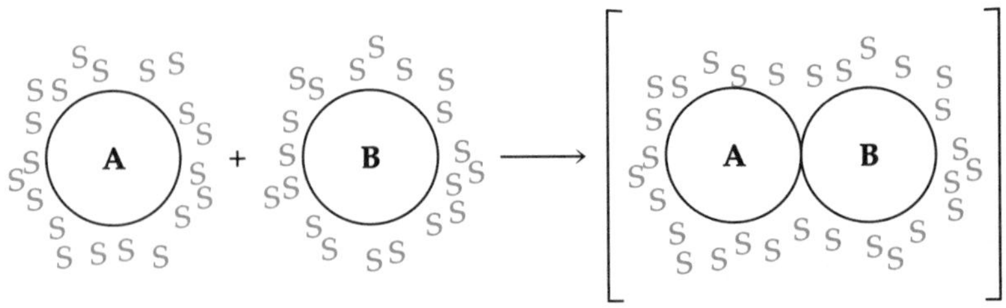

▶ FIGURE 20-14
When two starting materials, A and B, are stabilized by interaction with solvent molecules (S), desolvation is required before they can be brought into contact for chemical reaction.

reaction. For example, the rate of acid-catalyzed esterification of a carboxylic acid by an alcohol (equation 1) is much lower than those of intramolecular ring closure of hydroxyacids to form lactones (equations 2 and 3). (Recall that lactone is the name for a cyclic ester.)

$$RCO_2H \quad + \quad HOR \quad \xrightarrow{H^\oplus} \quad RCO_2R \tag{1}$$

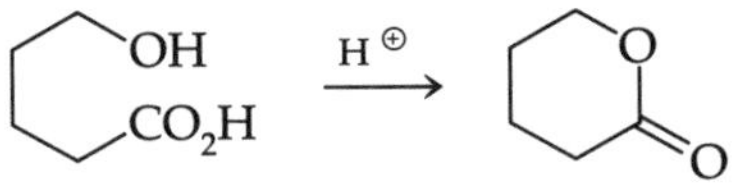

$$\tag{2}$$

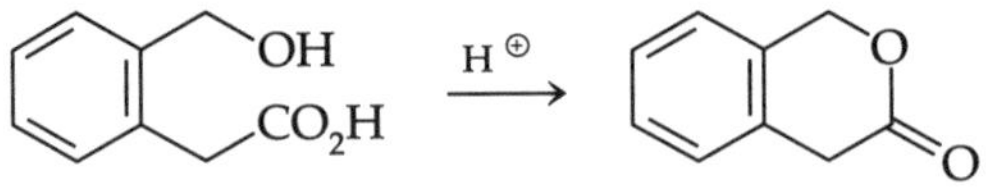

$$\tag{3}$$

For the first reaction (equation 1), the two separate reagents must become associated to arrive at the transition state for the reaction. In contrast, the reactions that form lactones (equations 2 and 3) are unimolecular and have less entropically unfavorable transition states; as a result, they have higher rates of reaction. Not only are the two reactive functional groups (the —OH and —CO$_2$H groups) already associated by covalent bonds, but less desolvation is required to reach the transition state.

Two intramolecular examples of ester (lactone) formation are shown: one in which the hydroxyl group is attached at the end of a flexible alkyl chain (equation 2) and the other in which it is bound to a more-rigid, aromatic spacer (equation 3). A difference is also expected between the rates of formation of these two lactones. In equation 3, the reacting groups are held more closely together than they are in equation 2, and less additional organization is required to reach the transition state for lactone formation. Furthermore, as a result of the proximity of these two groups in the starting hydroxy acid, fewer intervening solvent molecules must be removed from the region between the reacting functional groups for bond formation to occur. The effects of entropy and solvation are even more important in enzyme-catalyzed reactions.

20-6 ▶ Transition–Metal Catalysis

As mentioned in Chapter 2, transition metals such as palladium and platinum catalyze the addition of molecular hydrogen to an alkene, a reaction that otherwise does not proceed at a measurable rate.

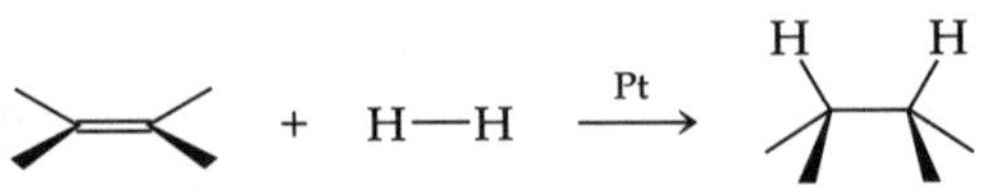

To understand why the metal catalyst is required, let us examine conceivable mechanisms for this process that follow homolytic or heterolytic pathways.

First, we can rule out the concerted pathway illustrated here.

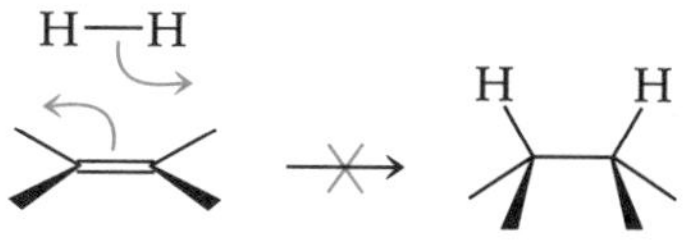

As noted in Chapter 13, a simple rule applies in most situations for concerted, cyclic processes: they take place most readily when the number of electrons is the same as that in an aromatic system defined according to Hückel's Rule. For example, the Diels-Alder reaction, requiring a combination of four electrons in a diene with two electrons in a dienophile, proceeds by a concerted pathway because the transition state has $4n + 2$ delocalized electrons. On the other hand, the direct addition of a molecule of hydrogen to an alkene, with only four electrons (two from the σ bond in H_2 and two from the π bond of the alkene) is not concerted because such a reaction would involve only $4n$ electrons.

There are two possible multistep processes for accomplishing the bond making and breaking required for this transformation: homolytic or heterolytic. However, simple bond cleavage in either mode is energetically prohibitive because each of these routes requires the input of at least as much energy as is needed for the dissociation of the hydrogen–hydrogen bond: 104 kcal/mole.

$$H\text{—}H \quad \longrightarrow \quad H^{\cdot} + H^{\cdot} \qquad \Delta H^{\circ} = 104 \text{ kcal/mole}$$

$$\longrightarrow \quad H^{\oplus} + H^{\ominus} \qquad \Delta H^{\circ} = 104 \text{ kcal/mole} + ?$$

In fact, heterolytic rupture of the hydrogen–hydrogen bond would require even more energy because of the additional energy needed for charge separation.

Alternatively, we can consider the possible interaction of a molecule of hydrogen with an alkene by the homolytic and heterolytic processes shown in Figure 20-15. Again, we can arrive at a minimum energy requirement for

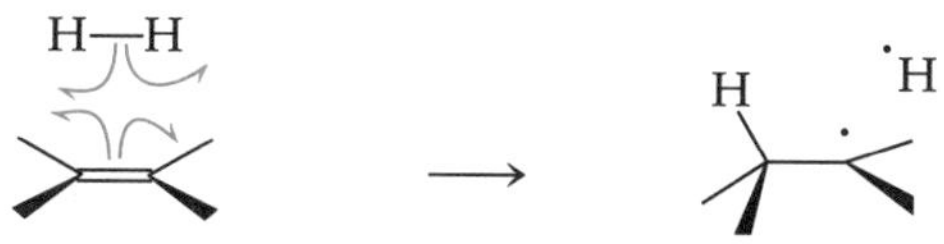

Broken   **Made**

H—H	104 kcal/mole		C—H 99 kcal/mole		$\Delta H^{\circ} = +68$ kcal/mole $(167 - 99)$
C=C π	63 kcal/mole				
	167 kcal/mole				

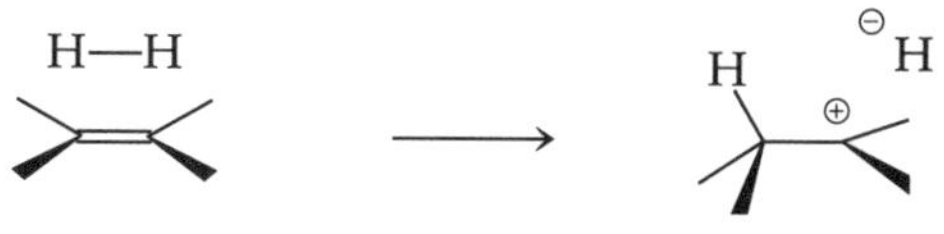

▲ **FIGURE 20-15**

Without a metal catalyst, the first step in both homolytic and heterolytic pathways for the reduction of an alkene by hydrogen gas is highly endothermic.

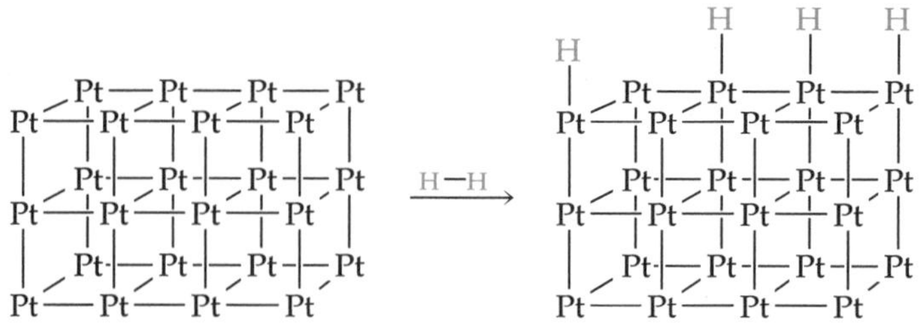

▲ FIGURE 20-16

Catalytic hydrogenation proceeds by replacing the hydrogen–hydrogen σ bond in H_2 with two Pt–H covalent bonds.

both of these possibilities by evaluating the bond breaking and making that takes place in the homolytic pathway. Two bonds are broken (the hydrogen–hydrogen bond and the carbon–carbon π bond) and a single carbon–hydrogen bond is formed. The energy released upon formation of the carbon–hydrogen bond does not nearly compensate for that consumed in the bond-breaking processes, and the homolytic pathway would require at least 68 kcal/mole, a prohibitively large value. (Recall that the activation energy for an endothermic process must be equal to or greater than $\Delta H°$.) The endothermicity of the heterolytic alternative shown in Figure 20-15 is even greater, as a result of charge separation. Thus, there appears to be no direct way by which an alkene and molecular hydrogen can react to form an alkane.

Indeed, the hydrogenation of alkenes requires a metal catalyst. As discussed in Chapters 2 and 10, molecular hydrogen is absorbed by a metal such as platinum in a process in which the hydrogen–hydrogen bond is broken and replaced by two new platinum–hydrogen bonds. The energy released in Pt–H bond making more than compensates for that required to cleave the hydrogen molecule (Figure 20-16). Although the hydrogen atoms are shown in Figure 20-16 as being bound only to the metal atoms on the top surface, hydrogen atoms in fact migrate freely throughout the metal. (Hydrogen is the only element small enough to do this.) This process, and others that will be considered shortly, is facilitated greatly by the ability of a transition metal to expand and contract its valence shell and number of associated ligands. This characteristic of transition metals is the result of unfilled valence shells where the addition or subtraction of an electron does not greatly alter the energy. In contrast, elements in the second row of the periodic table have normal valency requirements that result in filled electron shells. With atoms such as carbon and nitrogen, addition or subtraction of electron density results in an electronic state of much higher energy.

The next step consists of the interaction of the π-electron density of the alkene with a platinum atom not bound to hydrogen.

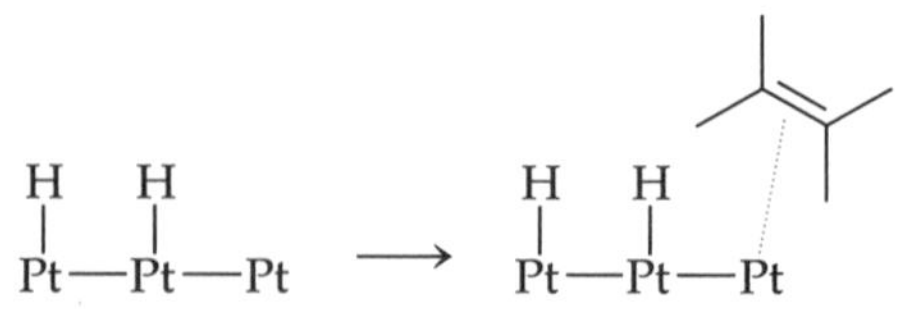

Again, empty orbitals of the metal are available to accept this additional electron density. With both hydrogen atoms and the alkene now associated with the platinum surface, hydrogen is transferred from platinum to carbon, forming a discrete platinum–carbon bond.

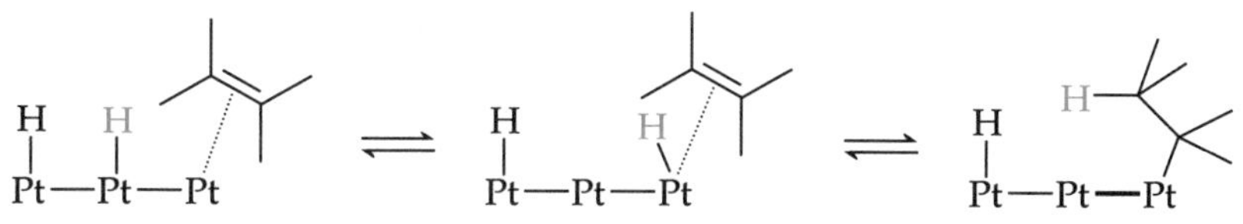

This step is neither highly exothermic nor endothermic; in addition, it has a low energy of activation. The energy consumed in the cleavage of the platinum–hydrogen bond and the carbon–carbon π bond is compensated by that released in the formation of the platinum–carbon and carbon–hydrogen bonds.

The hydrogen atoms associated with the platinum are very mobile and can readily move from one platinum atom to another on the metal surface. The first step shown in Figure 20-17 is the migration of a hydrogen from one platinum to that already bound to the alkyl fragment. In the final step of the reaction mechanism, both the platinum–hydrogen and the platinum–carbon bonds are cleaved in a process that simultaneously forms the second carbon–hydrogen bond in the product. This step is known as a **deinsertion** reaction because the reverse process, an **insertion** reaction, can be viewed simply as the interjection of platinum between hydrogen and carbon of an alkane. Unlike the starting alkene, the product alkane does not have readily accessible electron density for interaction with the platinum surface and thus returns to the solution, exposing the metal surface for another cycle of hydrogenation. Although the reduction is exothermic, all of the steps in the addition of hydrogen to an alkene in the presence of a noble metal catalyst are reversible. The reverse reaction is favored by entropy (two molecules formed from one) and, because the contribution of entropy varies with temperature, $\Delta G° = \Delta H° - T\Delta S°$, the alkene and hydrogen are favored at high temperature.

This reverse hydrogenation (oxidation) is an important industrial reaction. Alkanes burn much more rapidly than do alkenes, and too rapid a rate of oxidation results in knocking (pinging), in the cylinder of an internal combustion engine. Because of this rapid oxidation, pressure builds up before the piston reaches the top of its stroke in the cylinder. When crude oil feed-stocks are obtained from oil wells too rich in alkanes, high-temperature treatment with noble metal catalysts is used to effect partial dehydrogenation, resulting in the conversion of some of the alkanes into alkenes and cyclohexanes into aromatics, as well as the formation of molecular hydrogen, a valuable by-product.

EXERCISE 20-F

Clearly explain why the mechanism detailed herein for catalytic hydrogenation results in *cis* addition of hydrogen.

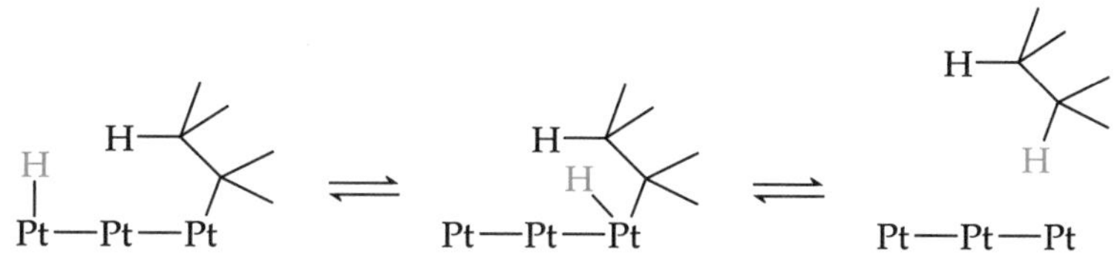

▲ FIGURE 20-17
Hydrogen atoms can migrate easily within and along the surface of platinum. When a hydrogen atom moves close to the site at which an alkyl group is attached, it can shift from platinum to carbon. The alkane thus formed then desorbs from the platinum surface, freeing the metal for another catalytic cycle.

20-7 ▸ Catalysis by Enzymes

Chemical reactions in living systems are much more varied and complex than any chemical experiment yet devised and carried out in the laboratory. For the many reactions of nature to take place simultaneously in water near pH 7 requires that each one be accelerated by a separate catalyst specific for that process. This is the role of enzymes, large polypeptide molecules that have molecular weights as high as 100,000 and, in some cases, even higher. Their size allows for a virtually unlimited number of possible permutations and variations. Indeed, for a peptide consisting of sixty-three amino acids that has been constructed from the twenty common amino acids, there are $20^{63} = 10^{82}$ possible permutations, a number larger than the estimated number of all elementary particles in the universe.

Enzyme–Substrate Binding

As we learned in Chapter 16, large polypeptides adopt a folded and twisted three-dimensional structure that is uniquely determined by the sequence of amino acids present. Nonbonded intermolecular interactions control, in part, whether the chain adopts a helical or β-pleated-sheet arrangement along any particular segment and how these segments fold back on each other so as to maximize van der Waals, hydrogen-bonding, and electrostatic interactions between the amino acid side chains. These attractive interactions often result in a large, three-dimensional mass and, although much of the volume is filled by the backbone and residues of the polypeptide, there are often hydrophobic and hydrophilic pockets on the surface and holes inside. The three-dimensional structure of an enzyme is chiral, and only one sense of handedness is present because the constituent amino acids are generally only L-enantiomers. The pockets and holes are chiral, and it is in these sites that catalysis occurs in most enzymes. The relatively small part of an enzyme in which catalysis occurs is referred to as the **active site.**

The binding of a substrate molecule at the active site often requires a close matching in shape and electrostatic properties of the reactant and pocket. As a result, the molecule undergoing reaction, called the **substrate,** often binds rather tightly in the active site before reaction. This process is represented schematically as:

$$\text{E} + \text{S} \longrightarrow \text{E·S} \longrightarrow \text{E·P} \longrightarrow \text{E} + \text{P}$$

in which E is the enzyme, S the substrate, P the product, and E·S and E·P are the enzyme bound to either the substrate or the product. In many cases, the first step, forming the E·S complex, is very fast and is limited only by the rate of diffusion of the substrate through the solution to the active site. The overall reaction rate is then determined by the rate of the second step, which includes the actual bond changes that convert the substrate into the product.

The binding between the product and the enzyme should not be too strong. Otherwise, the reaction rate would decrease in direct proportion to the amount of product until the catalysis would all but stop at the point at which the concentration of the product equaled that of the enzyme. When this does happen, the enzyme is said to be **product inhibited.** Relatively weak favorable interactions between the enzyme and the product, however, can serve a useful function in complex biological systems. In such cases, the product serves to regulate its own rate of formation in a **feedback** process. An important and developing area of pharmaceutical research is based on the concept of moderating the rate of biological transformations that over-produce product. Structural mimics of the product or the starting material can act as **inhibitors** by binding tightly to the enzyme (without themselves undergoing reaction) and blocking the natural substrate. We will examine the structural features of one such inhibitor in Chapter 23.

Catalysis by the Enzyme Chymotrypsin

With this background, we can now consider the mechanism of catalysis by a specific enzyme, **chymotrypsin,** whose structure has been determined by single-crystal, x-ray crystallographic techniques. This enzyme catalyzes the hydrolysis of peptide bonds. Such enzymes are referred to as **proteases,** emphasizing their role in cleaving proteins. Recall from Section 20-2 that the hydrolysis of esters (and, indeed, other carboxylic acid derivatives such as amides) at or near pH 7 is extremely slow because the addition of water as a nucleophile to the carbonyl carbon results in charge separation in the transition state for the reaction. Conversely, at high hydroxide ion concentrations, the pH is too basic for many other functional groups to be stable. We learned in Section 20-4, however, that a balance between reaction rate and pH can be established in base-catalyzed reactions.

When catalyzed by chymotrypsin, amide hydrolysis is achieved with a carboxylate ion acting as a base. Because the pK_a of a typical carboxylic acid is approximately 5, the equilibrium favors the carboxylate anion at pH 7. Furthermore, because a carboxylate is present in the enzyme cavity quite near the site of active site, its effective concentration is much higher than if a base such as sodium acetate were merely added to the solution.

Let us specifically consider the hydrolysis of the amide bond to a phenylalanine residue in a peptide. The entire process leading from E·S to E·P is shown in Figure 20-18 (on page 702), in which all amino acid residues but those taking part in the reaction are represented by wavy lines. In this sequence, the substrate becomes temporarily bound through an ester linkage to a serine residue in the enzyme: as a result, this process is called **covalent catalysis.** In the first step, the hydroxyl-group oxygen of serine serves as a nucleophile, adding to the carbonyl carbon of the amide group. Simultaneously, the proton of the hydroxyl group is transferred to one nitrogen of the imidazole group of a nearby histidine residue, and the imidazole loses a proton to the carboxylate ion of an aspartate residue. No charge separation develops as a result of these proton transfers; this first step results only in the transfer of negative charge from the carboxylate ion to the tetrahedral intermediate. The imidazole group serves to relay a proton from serine to the carboxylate ion and, in so doing, merely undergoes proton tautomerization. The histidine residue does not directly take part in the net reaction and functions as a catalyst.

In the second step, the carbonyl-group π bond is reformed from the tetrahedral intermediate, with the loss of the nitrogen substituent. A second series of proton transfers, in which the histidine residue serves as intermediary, results in the simultaneous protonation of nitrogen and deprotonation of oxygen of the aspartate residue. Thus, the leaving group is effectively the free amine, not its anion. In step 3, a water molecule acts as a nucleophile, adding to the ester carbonyl group formed in step 2, resulting in a new tetrahedral intermediate. Again, a series of proton transfers, ultimately to the carboxylate ion, ensures that this transformation results only in charge transfer, not in charge separation. Collapse of the second tetrahedral intermediate (step 4) results again in formation of the carbonyl π bond and rupture of the carbon–oxygen bond linking the substrate to the serine

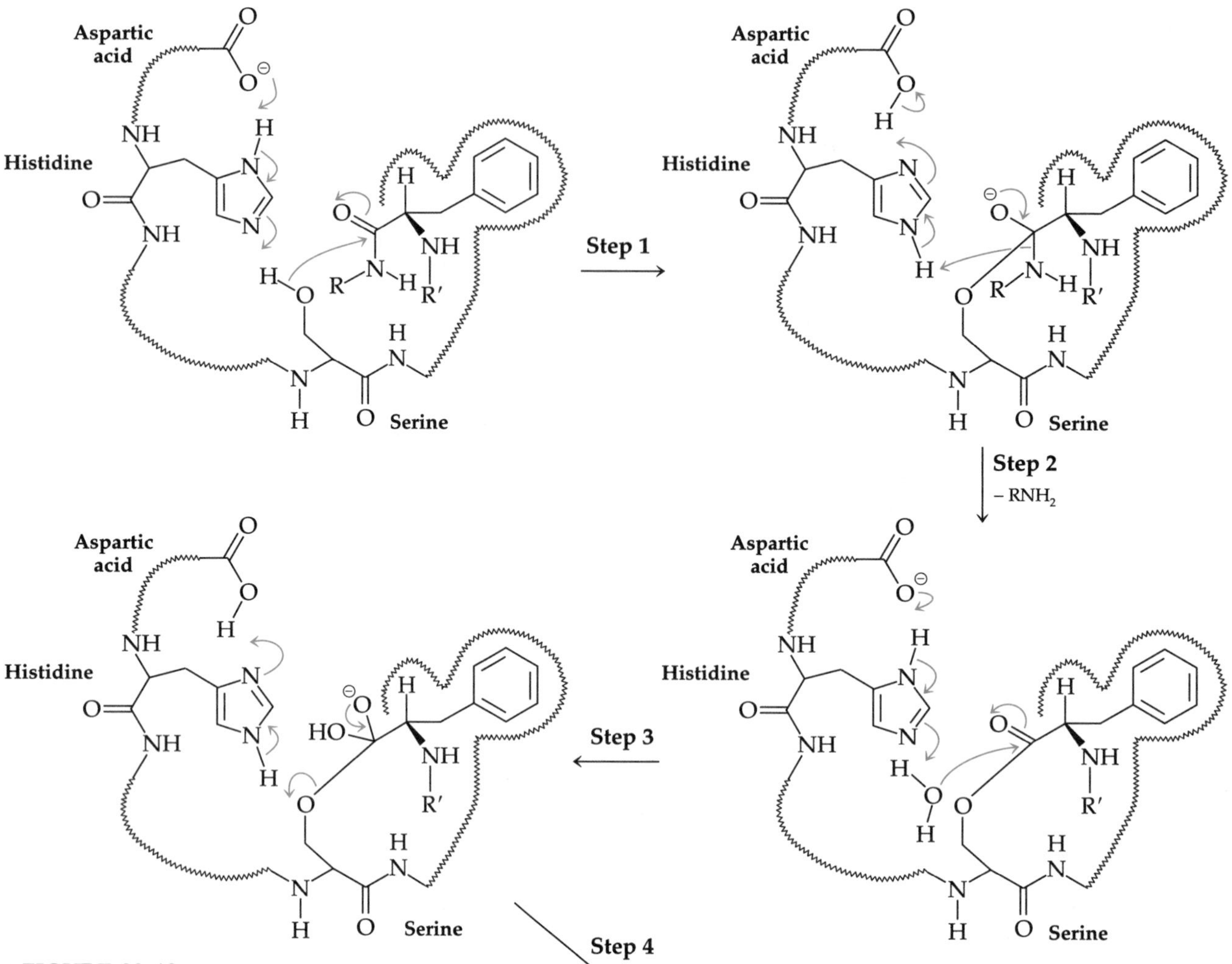

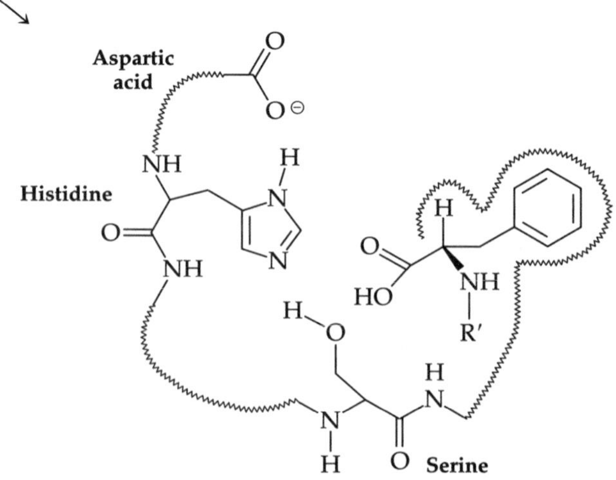

▲ FIGURE 20-18

The enzymatically catalyzed hydrolysis of a peptide bond takes place by bringing together the amino acid to be hydrolyzed, the acid catalyst, and the nucleophile within a spatially well organized pocket. In step 1, a carboxylate residue acts as a general base to deprotonate histidine which, as its conjugate base, deprotonates the OH group of serine, greatly enhancing serine's nucleophilic character. Nucleophilic attack on the amide carbonyl carbon by the deprotonated serine residue covalently bound to the enzyme produces a tetrahedral intermediate. The electron density on the negatively charged oxygen atom in this intermediate is used to reform the C=O bond in step 2 as the amine RNH_2 is lost and the general base is regenerated. At this stage, the resulting ester is covalently bound to the enzyme through the serine linkage. In step 3, a tautomer of histidine is formed by deprotonation by an enzyme-bound aspartate carboxylate anion, permitting reprotonation of histidine by a water molecule present near the active site. This activates the water molecule as a nucleophile, permitting attack on the covalently attached ester and producing a second tetrahedral intermediate. Reformation of the C=O double bond, with expulsion of the serine OH group, in step 4 produces the free hydrolyzed carboxylic acid, as the

remote carboxylic acid reprotonates the histidine tautomer. At this point, the enzyme has also been restored to its initial condition, leaving a free serine residue and a histidine in a tautomeric form ready to activate another nucleophile by general base catalysis by a carboxylate residue within the enzymatic active site.

residue. The product acid is now free of covalent bonds to the enzyme and diffuses into the solution. Overall in this catalytic process, the rate is accelerated because charge separation does not develop in these pathways.

The rate acceleration of amide hydrolysis by chymotrypsin is achieved through a reaction pathway that is quite different from that which is followed (albeit very much more slowly) without the enzyme. This is not the only way that enzymes provide for catalysis. Many enzymes operate in the simple way explored in Section 20-1 by stabilizing the transition state through favorable interactions with amino acid residues present in the active site. For example, zinc and other metal cations that are present near the active site in many enzymes effect the hydrolysis of carboxylic and phosphoric acid derivatives. These metal ions effectively lower the transition-state energy through favorable electrostatic interactions with developing negative charge (Figure 20-19). Thus, the activation energy for the reaction is lowered as a direct result of the stabilization of the transition state (Figure 20-20).

Notice that in this energy diagram, the starting material in the catalyzed reaction is also stabilized to a certain extent. Indeed, this is often the case for enzymatic reactions in which the E·S complex is more stable than the separate substrate and enzyme. It is important, however, that this stabilization of the complex be *less* than that of the transition state. Otherwise, the activation energy would be raised instead of lowered.

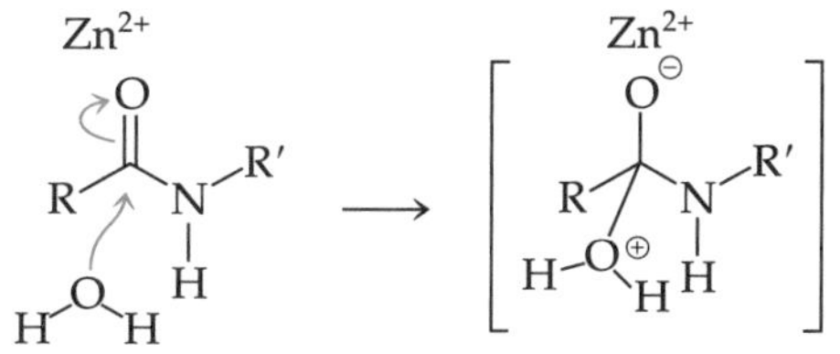

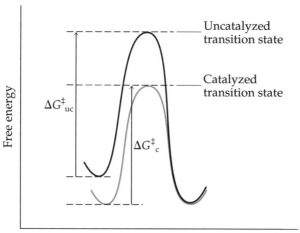

▲ FIGURE 20-19
Multivalent metal cations stabilize charged tetrahedral intermediates by associating electrostatically with centers of negative charge.

▲ FIGURE 20-20
A reaction is accelerated by complexation with a stabilizing reagent if the transition state is stabilized to a greater extent than the reactants.

EXERCISE 20-G

Which amino acid residues in an enzyme could provide a positive charge in the vicinity of the developing negative charge on the oxygen of the carbonyl group of an amide or ester substrate?

20-8 ▸ Enzymes and Chiral Recognition

Chymotrypsin is an effective catalyst for the hydrolysis of only certain peptide bonds; namely, those in which the carboxyl group is that of phenylalanine. Even minor structural variations in this amino acid residue result in a vastly lower rate or even no reaction at all. Because enzymes are chiral, it should not be surprising that even the seemingly subtle change from L- to D-phenylalanine has a dramatic effect. Indeed, the active site of chymotrypsin is a "pocket" (left side of Figure 20-21) that has regions that are the proper size and shape for each of the three groups attached to the center of chirality of phenylalanine: a hydrogen, a nitrogen, and a phenyl group. The enzyme is thus matched to this particular substrate, the phenylalanine residue, in what is generally referred to as a **lock-and-key** manner. Just as only keys of just the right shape and size fit a given lock, only a limited number of quite similar substrates fit in the active site of a given enzyme and undergo catalyzed reaction.

An increase in the size of any one of the three groups results in a substrate that can no longer fit into the active site, often preventing any reaction from taking place. We can see from Table 20-1 that replacement of a hydrogen by a methyl group, as well as the interchange of two of the substituents (which produces the enantiomeric substrate), results in no reaction. In both cases, a relatively large substituent would have to fit in the small pocket that normally accommodates hydrogen in order for the enzyme to accept the substrate (Figure 20-22).

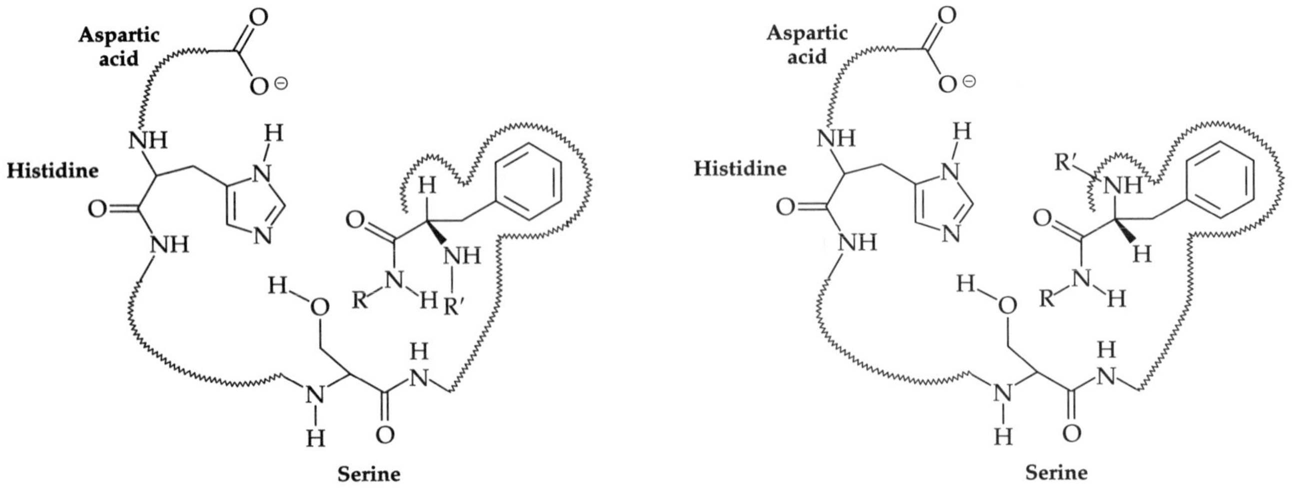

▲ FIGURE 20-21

The "fit" of a given peptide into the chiral cavity at the active site of chymotrypsin is enantiospecific: there is a much better match between the enantiomer shown at the left and this cavity than that shown at the right.

Table 20-1 ▸ Relative Rates of Hydrolysis of Various Amides by Chymotrypsin

Substrate	Relative Rate of Hydrolysis
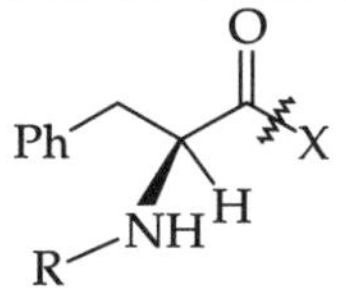	3×10^4
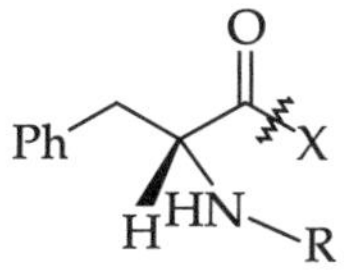	0
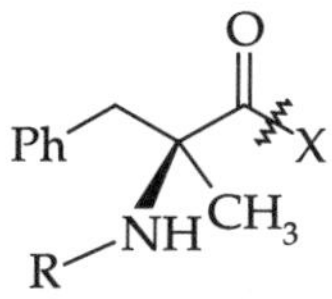	0
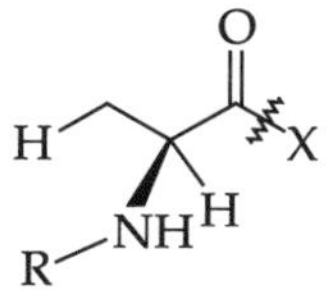	1
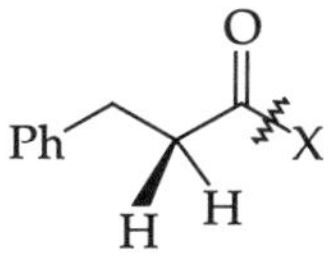	7

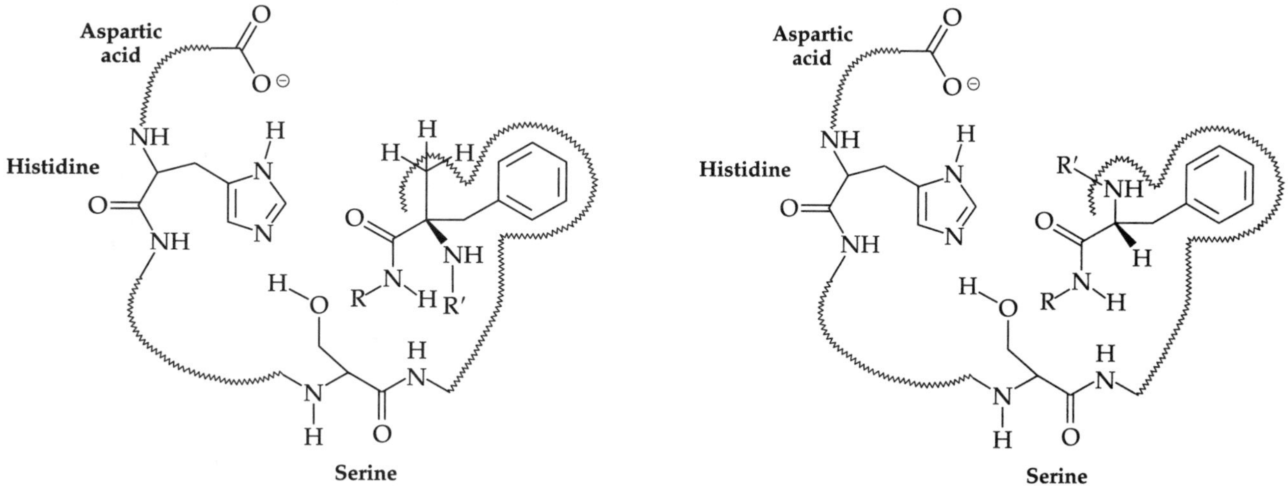

▲ FIGURE 20-22

A bulky substituent at the α position of a peptide to be hydrolyzed interferes with binding to the enzyme. Poorer binding equilibrium constants are observed with an α-alkylated substrate (as shown in the complex on the left) or with the "wrong" enantiomer (as shown in the complex on the right).

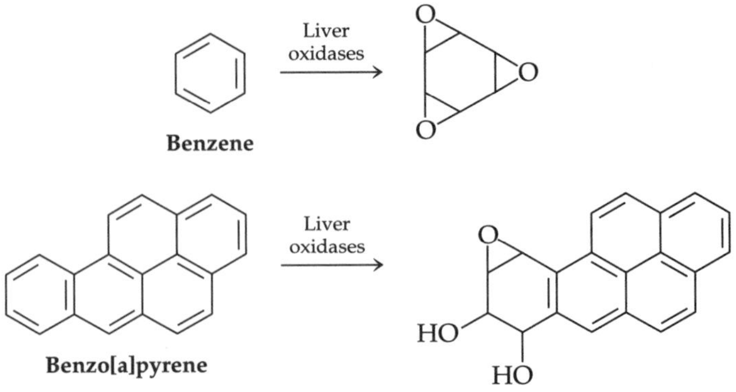

▲ FIGURE 20-23

Liver oxidases induce epoxidation of double bonds present in some fused aromatic hydrocarbons. The epoxide products are highly activated toward nucleophilic attack, making covalent attachment to DNA possible, with disastrous effects on information replication.

Furthermore, attractive interactions with an enzyme's amino acid residues hold the substrate relatively tightly in the active site. Changing the nature of the substrate results in a dramatic decrease in the rate acceleration induced by the enzyme. For example, without the phenyl group (that is, with alanine), the rate drops by more than four orders of magnitude; and, without the nitrogen, by more than three. It is clear, then, that chymotrypsin is a selective enzyme that is an effective catalyst for the hydrolysis of only certain peptide bonds. This specificity is the result of the very well defined structure of the active site. The large number of different enzymes, each with its own set of specific substrates (and reactions), is possible only because of the structural complexity afforded enzymes by their large size.

Many enzymes are just as selective as chymotrypsin, whereas others accept a wide range of substrates. For example, there are oxidizing enzymes in the liver responsible for increasing the water solubility of unwanted materials by the addition of oxygen substituents. It is logical that these enzymes would be relatively nonselective because their function is to remove an undefined variety of unwanted, and possibly even toxic, substances. Unfortunately, these enzymes operate not only on "natural" waste materials but also on external contaminants. Benzene and polycyclic aromatic hydrocarbons such as benzo[a]pyrene are oxidized to polyepoxides that are sometimes further transformed into diols (Figure 20-23). In general, these oxygen functional groups are quite reactive toward nucleophilic addition and substitution. It is believed that some of these polyepoxides/glycols react with nucleotides on both strands of DNA, linking the double helix permanently together by strong chemical bonds. This association dramatically alters cell replication and can lead, in turn, to the cell becoming cancerous. For example, exposure to benzo[a]pyrene leads to cancer, apparently after this polycyclic aromatic hydrocarbon is oxidized to the epoxydiol that chemically attaches to DNA. (Benzo[a]pyrene has been identified as one of the hydrocarbons in tobacco smoke.)

20-9 ▸ Artificial Enzymes: Catalytic Antibodies

An exciting, emerging field of organic chemistry deals with the design and construction of artificial enzymes: simple molecules that have some of the features of natural enzymes and are thus able to catalyze organic reactions.

The process of developing artificial enzymes is complicated because few of the necessary design criteria are known and thus it is necessary to fine tune the structure in an almost trial-and-error fashion. Such catalysts must have at least some of the structural complexity of enzymes and thus are difficult to construct by planned sequences in the laboratory. Although some systems that have been designed show high levels of catalytic activity, they are generally those for which the substrate has been chosen to match the catalyst.

Researchers working at the interface between organic chemistry and molecular biology have recently devised a way of inducing animal and bacterial cells to produce catalysts for selected reactions. This process takes advantage of the natural defense system of **antibodies,** moderate-size peptide complexes that are responsible for alerting the **immune system** to the presence of foreign substances. Antibodies are composed of different regions: some regions are the same for all antibodies produced within a given cell and others, called **variable regions,** are produced with nearly unlimited diversity (Figure 20-24). When a foreign substance (antigen) by chance binds into a cavity of a variable region of an antibody, the immune system is activated to produce more of this specific antibody, and to destroy the "captured" molecule.

Chemists have developed methods by which cells can be stimulated to produce antibodies that bind a specific molecule. Complementary techniques also have been perfected for the isolation of these specific antibodies. When the molecules used to stimulate antibody production closely resemble the transition state of a particular reaction, antibodies that bind this transition-state mimic are isolated. In some cases, these antibodies also bind the substrate(s) for the reaction and, because the antibody has been selected

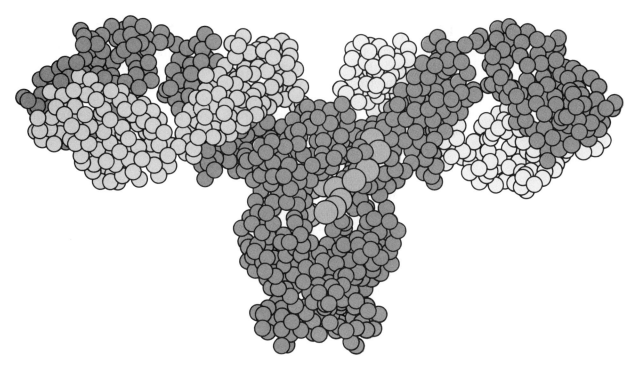

▲ FIGURE 20-24

Large sections of all catalytic antibodies appear to be the same. The variable region found within the antigen-binding fragment is responsible for the chemical selectivity of the antibody. (Courtesy of Brent Iverson, University of Texas)

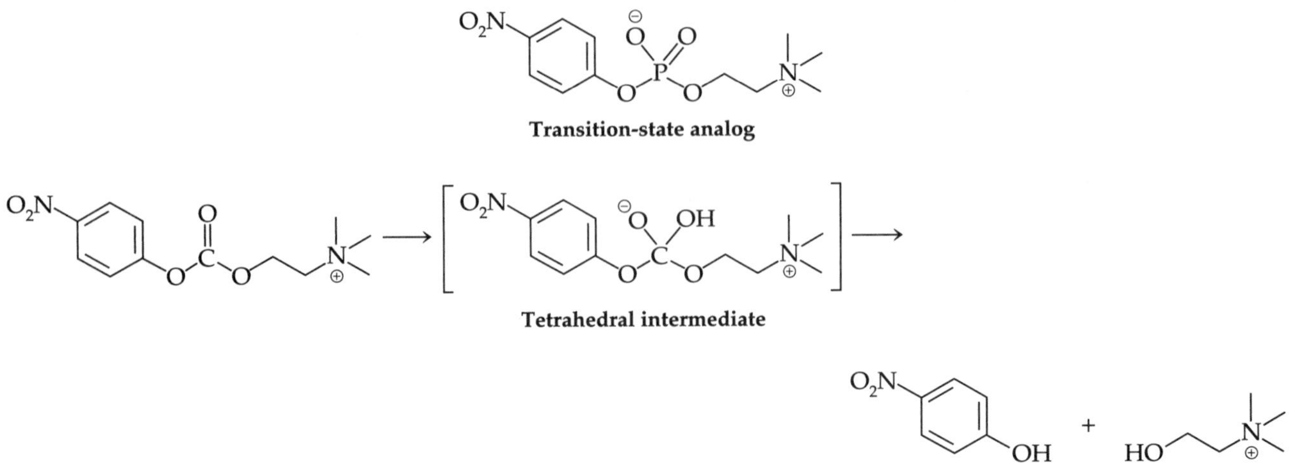

Transition-state analog

Tetrahedral intermediate

▲ FIGURE 20-25
The phosphate ester shown at the top is structurally analogous to the
tetrahedral intermediate formed in the base-induced hydrolysis of the
carbonate ester shown in the reaction. Both the charge demand and the
geometry are similar in the phosphate ester and in the tetrahedral
intermediate. Antibodies expressed in response to exposure to the phosphate
ester at the top catalyze the hydrolysis reaction.

to match the transition-state mimic, binding to the true transition state is
stronger than to the starting material. As we have already seen, a species
that interacts more strongly with the transition state than with the starting
material is a catalyst for the reaction. These naturally produced catalysts are
called, logically enough, **catalytic antibodies.** An example of such a **transition-state analog** is shown in Figure 20-25, along with the reaction that is
catalyzed by the antibodies that bind to the analog. The phosphate ester
exists at neutral pH mainly as the anion shown, and provides a charge distribution and three-dimensional shape that maps closely to those in the
tetrahedral intermediate in carbonate ester hydrolysis. Because the first step
in hydrolysis is endothermic, the transition state resembles the tetrahedral
intermediate and hence the phosphate anion. Presumably, the antibody has
amino acid residues in close proximity to the phosphate group that stabilize
the negative charge through electrostatic interaction in much the same fashion as do enzymes with metal ions (Figure 20-19).

The development of catalytic antibodies for a particular reaction
requires that an appropriate mimic for the transition state be available.
Although this may not be possible for all reactions, catalytic antibodies
capable of accelerating a large number of useful transformations are likely
to emerge in the near future. One especially promising use for catalytic antibodies is the synthesis of single enantiomers of pharmaceutical agents
because, as discussed in Chapter 19, it is generally true that only one enantiomer of a racemate has the desired effect. We also saw how, in some cases,
administration of both enantiomers (as the racemate) can have disastrous
consequences.

EXERCISE 20-H

What is the energetic consequence of an antibody developed to mimic the
product of a reaction? Would such an antibody accelerate the rate of the
reaction? Would it be a catalyst?

Conclusions

The rates of chemical reactions are accelerated by catalysts, species that affect the free energy of activation of a reaction without themselves being consumed in the process. Catalysis is accomplished by stabilizing the transition state of the rate-determining step in a chemical transformation. When a greater separation of charge develops in the transition state of a given reaction than in the reactant, polar solvents or polar local environments (for example, inside a polar enzymatic cavity) preferentially stabilize the transition state and enhance the rate of reaction. When a reaction takes place through a transition state with a lower degree of charge separation than in the reactant, nonpolar (hydrophobic) environments enhance chemical reactivity. A catalyst cannot become more tightly bound to either starting material or product than it is to the transition state of a given reaction. It can, however, act through a charge relay mechanism.

Acid and base catalysts accelerate the rates of reactions by causing the reactants to bear substantial amounts of localized charge so that, in contrast, charge is more highly dispersed in the transition state than in the reactant. For example, an acid accelerates a nucleophilic acyl substitution by protonation of the carboxyl oxygen, producing a cationic reactant, whereas bases accelerate the same transformations by producing a highly activated anionic nucleophile. These catalysts thereby minimize additional separation of charge along the reaction pathway.

An intramolecular reaction usually takes place at a faster rate than does the intermolecular version of the same reaction unless the intramolecular pathway includes an especially strained transition state. Preassociation of a reagent with a substrate therefore usually accelerates a given reaction. Reversible association of a reagent with a complexant with matched hydrophobic or hydrophilic character allows for the movement of the reagent to a site where enhanced reactivity can take place, as happens in phase-transfer catalysis.

When a reagent (for example, an acid or base) is fully regenerated after completing an accelerated cycle, it is said to act as a catalyst; when it is consumed, the reaction is said to be induced by the reagent to distinguish it from a truly catalytic sequence.

The coordination of a transition-metal cation with a carboxyl oxygen atom induces similar acceleration of nucleophilic acyl substitution as that produced by complexation with an acid. This complexation relies on the metal's ability to readily expand and contract its valence shell. Analogous bonding changes with zero-valent transition-metal surfaces account for their utility as heterogeneous hydrogenation catalysts. Metals such as platinum or nickel provide a reactive surface for the catalyzed addition of hydrogen to alkenes, a process that brings together two species in an exothermic reaction that would not proceed without this catalytic association and activation.

The structural complexity of enzymes is responsible for both the catalytic activity and the substrate specificity necessary for the biochemical transformations critical to living systems. Enzymes bind both a specific substrate and the required reagent within a common cavity (or active site), thereby reducing the requirements of desolvation in approaching the transition state by preorganizing the reactants. The equilibrium constant for binding of a given substrate to an enzyme is controlled by the size of the available enzymatic cavity and to the charge distribution within the active site. Enzymatic binding is sufficiently specific to permit chiral recognition, with one enantiomer becoming associated with the enzyme with a much higher binding constant than the other. Both steric and electrostatic factors influence the arrangement of the reagent, substrate, and acid- or base-catalysts within the enzymatic pocket.

Chymotrypsin catalyzes hydrolysis reactions by covalent catalysis, a process in which an intermediate is reversibly covalently bound to a residue present within the enzymatic cavity. The preferential stabilization of a given transition state within the active site accounts for the enhanced reactivity induced by the enzyme catalyst. Tight binding of a product molecule within the active site inhibits the catalyzed reaction; weak, reversible binding of a product is a mechanism for feedback inhibition.

New catalysts developed from antibodies are expressed by binding with molecules (mimics or antigens) that structurally and electrostatically resemble the transition state of the reaction being accelerated. When a living organism produces an immune response to the antigen, a catalytic antibody capable of binding preferentially to a transition state of a desired reaction is produced. Catalytic antibodies thus provide selectivity, as well as rate acceleration, for selected organic transformations.

Review Problems

20-1 Each of the following transformations requires the presence of either acid or base, as indicated above the reaction arrow. Indicate for each transformation whether the reaction is acid (or base) induced or catalyzed; that is, is the reagent required to be present in stoichiometric amounts or will a catalytic amount suffice?

(a)

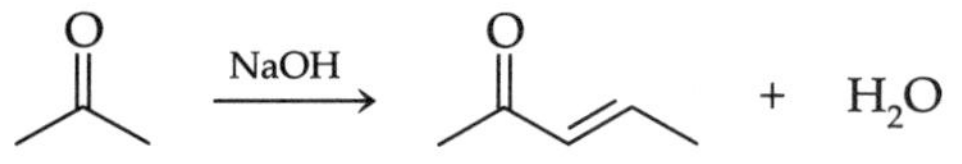

(b)

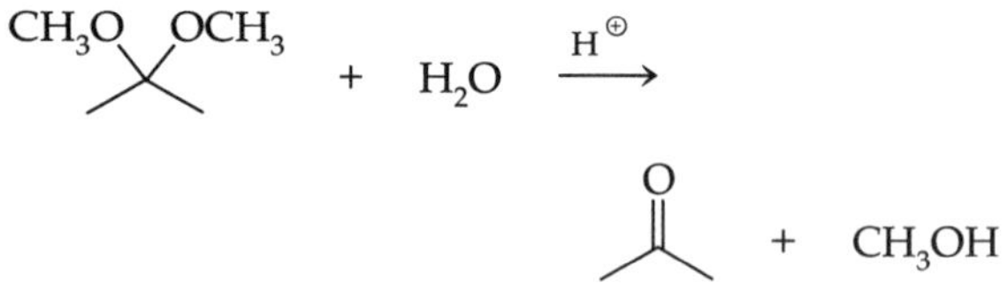

(c)

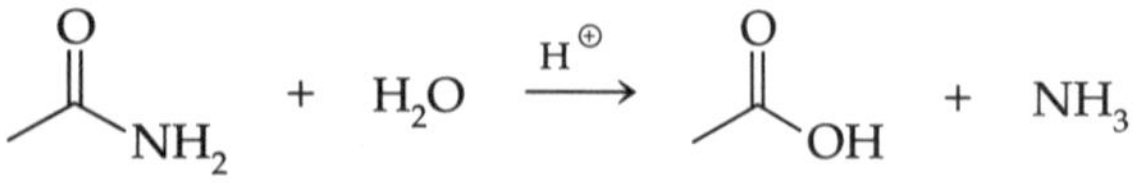

(d)

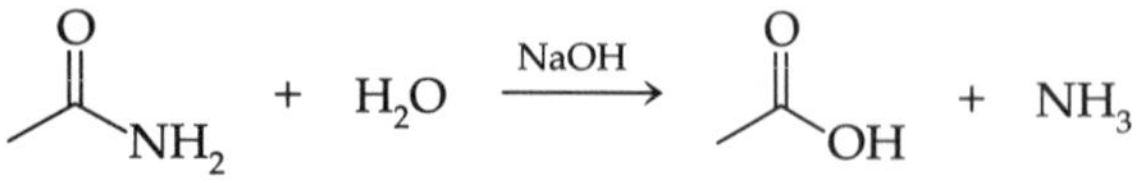

20-2 Solvent can alter the activation energy and therefore change the rate of a reaction. How would you expect the rate of the following reaction to differ in a polar, protic solvent such as water compared with a nonpolar, aprotic solvent such as dimethyl ether? Be sure to explain your answer clearly.

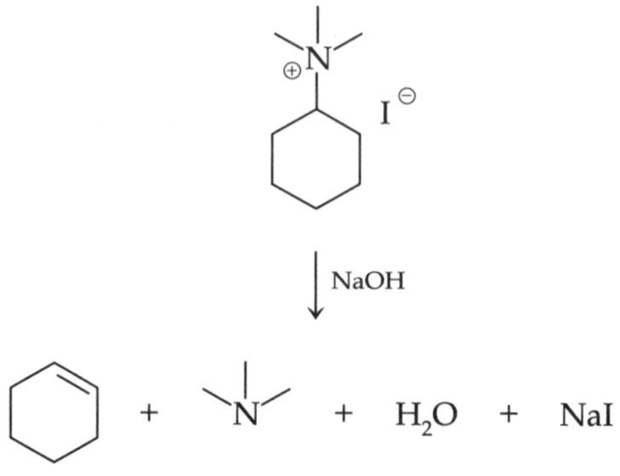

20-3 Isomerization of alkenes is sometimes detected in catalytic reductions when D_2 instead of H_2 is used. In such cases, the deuterium is found in unexpected places, as shown in the example below.

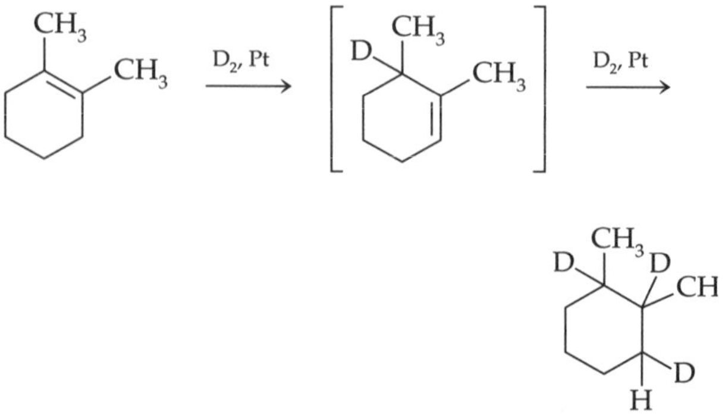

This anomalous behavior can be explained on the basis of the known observation that all steps in the catalytic reduction of an alkene to an alkane are reversible. However, the sequence need not be carried through to completion before reversal to effect alkene isomerization. What is the first point in the mechanistic sequence detailed in Figure 20-17 in which reversal to alkene could lead to isomerization? Write a mechanistic sequence for this isomerization and for the formation of the D_3 product shown here.

20-4 Phospholipase C is an important zinc-containing enzyme that is responsible for disassembling cell-wall material by catalyzing the hydrolysis of phospholipids. In the reaction, a phosphate ester functionality is cleaved, as shown below, resulting in the loss of the ionic head group that is critical to the formation of the lipid bilayer structure of the cell (Chapter 17). First, write a simple mechanism that is parallel to nucleophilic acyl substitution, using hydroxide as a base for this hydrolysis. Then show how zinc ions, properly positioned, could stabilize the transition state leading to the intermediate in this substitution reaction.

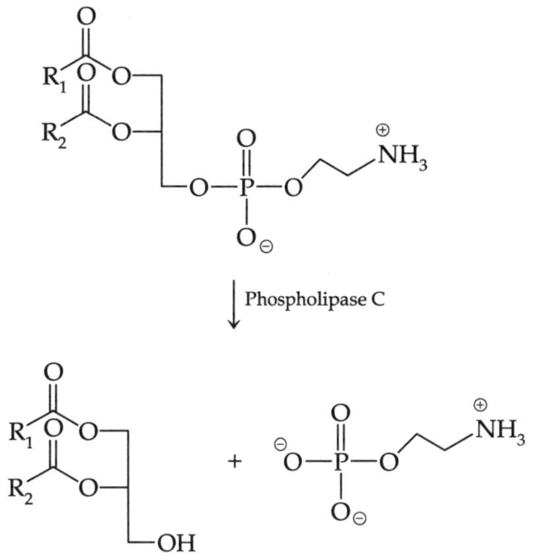

Chapter 21

Cofactors for Biological Redox Reactions

Many biologically important molecules have more-complicated structures than the laboratory reagents studied in the first half of this course. For example, in the reduction of simple ketones and aldehydes to alcohols, $LiAlH_4$ and $NaBH_4$ are quite effective reagents, whereas a much larger reducing agent, the reduced form of nicotinamide adenine dinucleotide (NADH) with a molecular weight greater than 500, is used to accomplish these same reactions within several living systems. Nonetheless, even though many of the reagents employed in the transformations of large biological molecules are themselves quite elaborate, the chemistry taking place in biologically relevant reactions is often no more complicated than that occurring in the simple examples presented in earlier chapters.

In this chapter, we will learn some of the reasons that more-complex reagents are required for seemingly simple chemical reactions taking place in living organisms: to function within a biological system, a reagent, or **cofactor,** must have the ability to recognize a target molecule from the complex mixture present in biological fluids and must be able to be recycled so that the product derived from the cofactor can be used again by the same organism. We will also learn the rather complex structures and functions of some of the reagents that take part in many biochemical transformations. We will see, for example, how several specific complex redox reagents (pyridoxamine phosphate, nicotinamide adenine dinucleotide, and flavin adenine dinucleotide) can induce reactions that result in the shift of a chemical equilibrium between the oxidized and reduced forms of an organic compound. We will also see how a thiol ester represents an activated derivative of a carboxylic acid for nucleophilic acyl substitution in biological reactions.

In considering the reactivity of thiamine pyrophosphate in the presence of lipoic acid, we encounter an example of a biological reversal of normal chemical reactivity (umpolung), a concept first mentioned in Chapter 18. This cofactor functions by inducing in an α-ketoacid the chemical reactivity more typical of a β-ketoacid. To understand this reaction more fully, we will learn how to analyze a given multifunctional compound retrosynthetically to determine alternating chemical reactivity patterns and to use laboratory reagents to overcome limitations imposed by normal chemical reactivity patterns. We will learn how to convert an aldehyde into a dithioacetal, which upon deprotonation forms an acyl anion equivalent.

Finally, mechanisms are given for removing one carbon from an amino acid (by interaction with thiamine pyrophosphate) and for transferring one carbon atom from a reagent to a growing carbon chain (by interaction with tetrahydrofolic acid). These latter reactions require reagents with reverse polarity.

21-1 ▸ Molecular Recognition

Biological reagents generally have higher molecular weights and more functional groups than do the rather simple oxidation and reduction reagents presented so far. Why are the reagents employed in biological transformations so complex? Simply because they must function within the diverse mixture of materials that are always present in living systems. Biological molecules must be able to pass through or be excluded by membranes at different times. Even after a reagent reaches the site where a needed reaction is to take place (inside or outside a given cell), it must interact only with its specific reaction partner: the chemical reagent must be able to distinguish a target reactant from all other molecules, even those having similar functional groups.

An enzyme achieves this matching of a substrate with a reagent by providing a defined environment that brings the substrate and the reagent (cofactor) together in a well-defined geometry (Figure 21-1). The enzyme has an irregular cavity into which the specific reagent and its target substrate can bind in a way that enables a chemical reaction to proceed rapidly. Only those reactions that are accelerated by the close association of a substrate with a reagent within the enzyme take place with high chemical efficiency. This molecular recognition is critical to the operation of complex living systems. Without it, oxidations and reductions could not take place simultaneously (as is necessary for energy conversion).

Without specific molecular recognition by common chemical association, all possible reactions between all reagents and substrates would take place. If the required reagents simply reacted with each other, equilibrium would soon be established: the oxidizing reagents would react to destroy the reducing reagents, and the chemical balance of a living system would be disrupted. In living systems, equilibrium is equivalent to death. Thus,

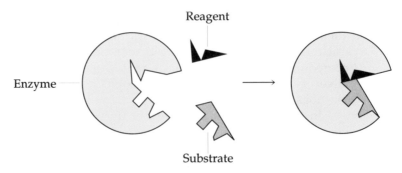

▲ FIGURE 21-1

The shape of an enzyme cavity matches closely in a complementary fashion the shape of the substrate and any required biochemical reagent (cofactor).

molecular recognition (recall Chapter 19) not only accelerates a desired reaction, but also inhibits free chemical reaction between homogeneously dispersed reagents. In this chapter, we focus on the reactions taking place between a given biological reagent and a specific substrate.

21-2 ▸ Recycling of Biological Reagents

Despite the structural complexity of a biological reagent, we can simplify our study by focusing on that part of the molecule, "the working end," that effects the transformation. We must keep in mind, however, that the remainder of the molecule is critical for molecular recognition and for delivery of the reagent to the site at which the desired reaction is to take place.

There is another significant distinction between laboratory reagents and their biochemical counterparts. Many of the reagents that we have studied so far (such as lithium aluminum hydride) are highly reactive, and the reactions in which they take part are quite exothermic and, thus, essentially irreversible. For the laboratory chemist, this exothermicity is highly desirable because it ensures that the reaction can be carried to completion within a convenient length of time. The reagent itself is fully consumed and is converted into a form from which it would be difficult to derive the original reagent. For example, the reduction of carbonyl groups with sodium borohydride ultimately affords the borate ion, and reforming the reagent (borohydride) from the product (borate) can be accomplished only with great difficulty. In the laboratory, this by-product is discarded rather than reused.

It would be highly inefficient for living systems to construct reagents having the complexity required for molecular recognition for only a single use. Rather, they are "recycled," often in a place within the cell or organism that is quite distant from the site where they are used. However, for effective and efficient recycling, the energy difference between the active and "spent" forms of the reagent must not be large. As a consequence, these reagents do not provide a large "driving force" for biochemical transformations; often, these reactions have equilibrium constants near one. Clearly, this would not be desirable in a laboratory transformation where the goal is to obtain the complete conversion of a starting material into a product. In biological systems, the reactions are driven toward product by the constant influx of starting material, produced by another reaction, and are pulled toward product by the further transformation of the first product into yet another material. This system of "driving" biochemical sequences is thus an application of Le Châtelier's Principle. The dynamic nature of living systems is the key feature that distinguishes biological chemistry from that which takes place in a closed reaction flask, where change stops either when all the starting material is consumed (for essentially irreversible processes) or when equilibrium is reached.

In the following representative, multistep sequence, the reaction starts with compound A and ultimately arrives at compound D through intermediates B and C.

$$A \rightleftharpoons B \rightleftharpoons C \rightleftharpoons D$$

The thermodynamics of the conversion of A into D does not require each of the steps (A → B, B → C, and C → D) to be exothermic: it is only necessary that the overall process be energetically favorable. We will see in Chapter 22 (on energy storage and utilization) that there are two sources of driving force for biochemical transformations.

EXERCISE 21-A

From the energy diagram below, determine which steps are endothermic and which are exothermic. Is the conversion of B into C faster or slower than the reformation of A from B? Is the conversion of C into D faster or slower than the reformation of B from C?

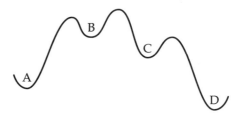

21-3 ▸ Cofactors: Chemical Reagents for Biological Transformations

Before we turn to the roles of several cofactors in biological systems, it is worthwhile to reiterate some important points made in the preceding sections. Cofactors are the biochemical equivalents of laboratory reagents, except that cofactors are always recycled. Cofactors are complex molecules that function in combination with enzymes. Enzymes act as catalysts to control the rate of the transformation of one biomolecule into another, whereas cofactors act as reagents to accomplish a net chemical conversion. The chemical reactions induced by a laboratory reagent and by a cofactor are often parallel: the reaction mechanisms of a cofactor are certainly no more complicated than those that we have already learned, especially if we focus our attention solely on that part of the cofactor molecule that undergoes reaction. This "myopic" view of a cofactor is not unreasonable because most of its structural complexity serves mainly to enable it to recognize the target molecule and to enhance binding to the enzyme.

We deal first with three common cofactors that function as biochemical reducing agents: **pyridoxamine phosphate, nicotinamide adenine dinucleotide (NADH)**, and **flavin adenine dinucleotide (FADH$_2$)**, shown in Figure 21-2. In their reduced forms, these cofactors act formally as hydride transfer agents. Despite their considerably greater structural complexity, these cofactors behave chemically like the complex metal hydrides discussed in Chapter 12.

Each cofactor shown in Figure 21-2 is an active reducing agent derived from a simpler molecule. Pyridoxamine phosphate is a derivative of **pyridoxol (vitamin B$_6$)**, and NADH and NADPH are biosynthetically related to **nicotinamide (vitamin PP)**. Likewise, **vitamin B$_2$ (riboflavin)** is the biochemical precursor of FADH$_2$. Many precursor molecules, although essential to living systems, cannot be synthesized within an organism: instead, they must be obtained from the diet. Molecules that are required to sustain life but that cannot be synthesized by an organism are referred to as **vitamins.** Because vitamins are recycled in most biochemical transformations, they are usually needed in only small amounts.

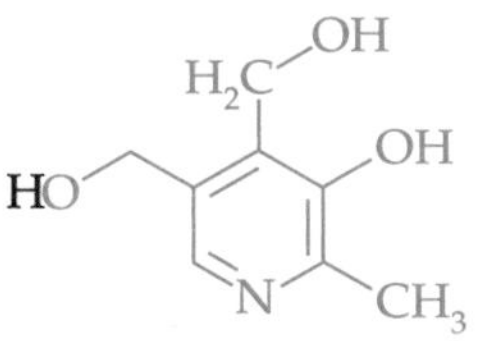

**Pyridoxol
(vitamin B$_6$)**

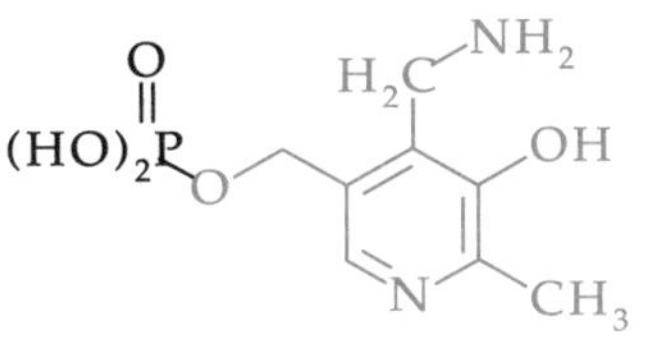

**Pyridoxamine phosphate
(active form)**

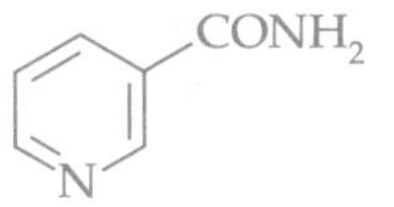

**Nicotinamide
(vitamin PP)**

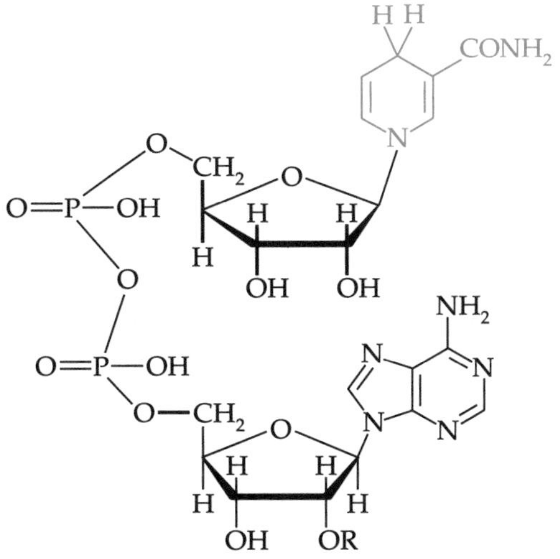

**Nicotinamide adenine dinucleotide phosphate (NADPH) R = PO(OH)$_2$
Nicotinamide adenine dinucleotide (NADH)   R = H**

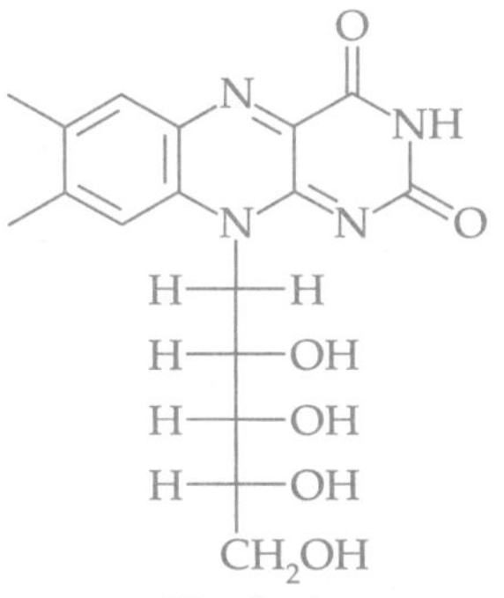

**Riboflavin
(vitamin B$_2$)**

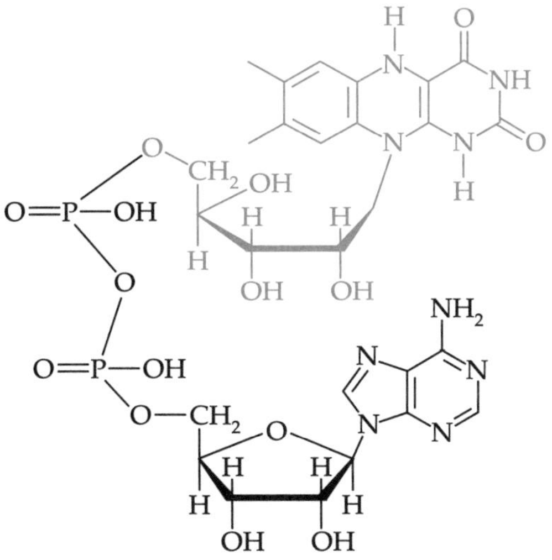

Flavin adenine dinucleotide (FADH$_2$)

▲ FIGURE 21-2
The four cofactors shown at the right are derivatives of the vitamins shown at the left. These cofactors function as biological reducing agents.

21-4 ▸ Pyridoxamine Phosphate: Reductive Amination of Alpha-Ketoacids as a Route to Alpha-Amino Acids

Pyridoxamine phosphate is the reductive amination product of **pyridoxal phosphate** (Figure 21-3). It serves both as a reducing agent and as a source of nitrogen for the production of α-amino acids. The starting material for the α-amino acid is the corresponding α-ketoacid.

This reaction requires not only the α-ketoacid and pyridoxamine phosphate, but also an enzyme that functions as a catalyst for the reaction. The enzyme allows for molecular recognition of the cofactor and the substrate (in this case, pyridoxamine phosphate and the α-ketoacid), and it provides a defined environment with a lower activation energy for the reaction.

The key transformations in this *reductive amination* are shown in Figure 21-4. The first step consists of the interaction of the amino group of pyridoxamine phosphate with the carbonyl group of the α-ketoacid, producing a **ketimine.** This ketimine is transformed into an **aldimine** through an intermediate **quinoid** form. Both the transformation of the ketimine into the quinoid form and the transformation of the latter into the aldimine are simply proton tautomerizations. The overall result of this sequence of proton transfers is a shift of a proton from the sp^3-hybridized carbon attached to the imine nitrogen in pyridoxamine phosphate to the imine carbon, converting the ketimine into an aldimine. The sp^3-hybridized carbon attached to nitrogen in the resulting aldimine was the ketone carbon in the original α-ketoacid.

All naturally occurring enzymes are peptides composed of L-amino acids; that is, those with the *S* configuration. They are therefore themselves chiral. As such, they can, and almost always do, impart "handedness" to reactions in which substrates or products or both are chiral. In the reaction that produces α-amino acids, the enzyme controls the approach of the cofactor to the carbonyl group of the ketoacid so that only *(S)*-enantiomers are produced.

EXERCISE 21-B

Recalling that imines bearing α-C–H bonds can tautomerize to form enamines, consider the specific ketimine in Figure 21-4 in which R = CH₃. Write a complete mechanism for an imine–enamine tautomerization, clearly showing the structures of all relevant intermediates. Explain why this tautomerization is less likely than the ketimine-aldimine tautomerization shown in the figure.

The ketimine–aldimine tautomerization in Figure 21-4 results in the loss of a proton from one carbon and its replacement on another. This overall exchange can take place either in acid (by the addition of one proton, followed by the loss of the other) or in base (by the reverse sequence); that is, the removal of one proton followed by the addition of the other. Either of these sequences results in a charged intermediate of substantially higher energy than the neutral reactant; namely, a cation in acid or an anion in base. Such charged intermediates are avoided when the addition and removal of the protons is simultaneous. This can be accomplished only if an acidic site is near the atom to be protonated and a basic site is near the proton to be removed. The enzyme provides an environment in which acidic

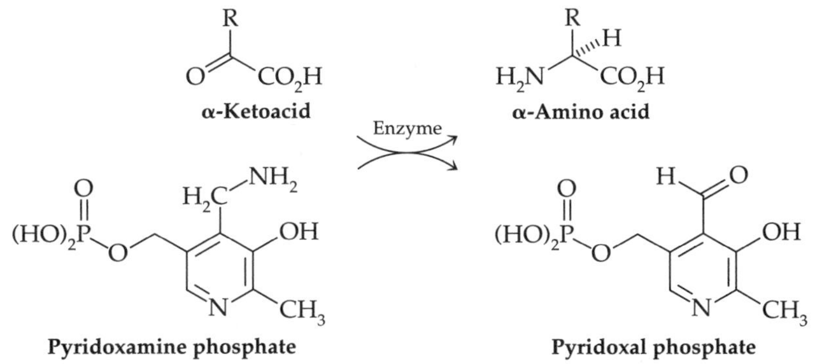

▲ FIGURE 21-3
In this coupled reaction, an α-ketoacid is reductively aminated (to an amino acid) as pyridoxamine phosphate is oxidatively deaminated (to pyridoxal phosphate). Pyridoxamine phosphate functions both as the source of nitrogen and as the reducing agent. In biological systems, an enzyme greatly accelerates the rate of this reaction.

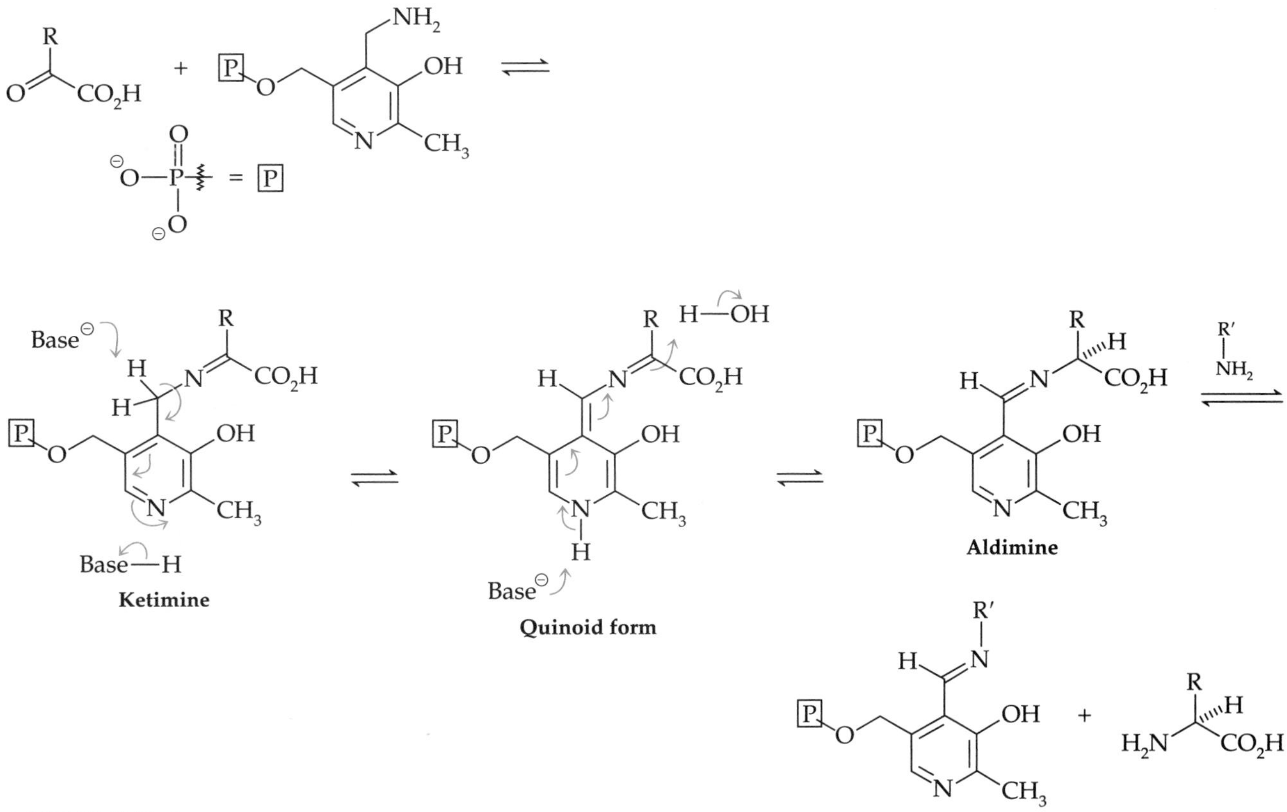

▲ FIGURE 21-4
The coupled reductive amination of an α-ketoacid begins with a nucleophilic attack of the amino group of pyridoxamine phosphate on the ketone group of the α-ketoacid, producing an imine. Deprotonation at the position adjacent to the imine nitrogen and reprotonation on the ring nitrogen at the other end of the conjugated π system produces a quinoid intermediate. The ketimine exists in equilibrium with the quinoid form and the aldimine. Because the deprotonation of the amine and the reprotonation on the imine carbon take place within a chiral pocket of an enzyme, the chiral center formed in the aldimine is produced exclusively as the L-enantiomer. Treatment of the aldimine with the amino group of a lysine residue present on the enzyme near the site of reaction releases the free α-amino acid.

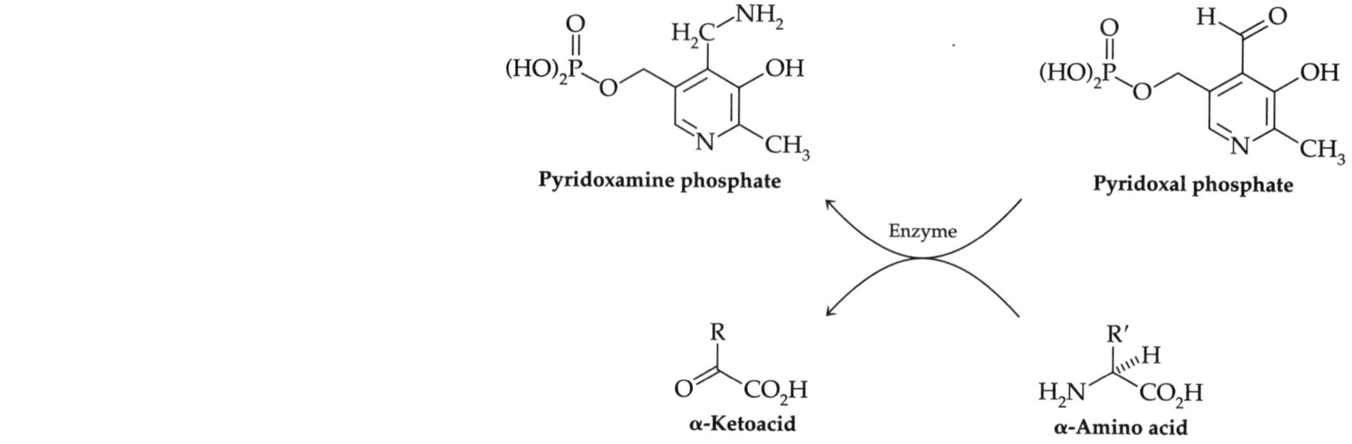

▲ FIGURE 21-5

In an oxidative deamination, a $CH(NH_2)$ group is converted into a $C=O$ group; reductive amination is the reverse process, the conversion of a $C=O$ group to a $CH(NH_2)$ group. Here, pyridoxal phosphate is reductively aminated as an α-amino acid is oxidatively deaminated. These reactions constitute the inverse of the reactions shown in Figure 21-4.

and basic sites are precisely positioned so that the pyridine ring nitrogen can be protonated at the same time as the C–H proton is removed, thus catalyzing the reaction. Like the imidazole ring of histidine (Chapter 18), the pyridine nitrogen accomplishes the transfer of a proton to a more-remote site. In addition, the aromatic ring of pyridine serves alternately first as an electron sink and then as an electron source. The α-amino acid is released from the aldimine in the last step by transamination, a process in which an imine is formed at the primary amino group of a lysine present in the enzyme catalyzing this overall transformation.

All steps illustrated in Figure 21-4 are readily reversible because they are neither highly exothermic nor highly endothermic. The pyridoxamine phosphate that is consumed in the overall production of one amino acid in Figure 21-4 is regenerated by the biodegradation of another amino acid to form a ketoacid by oxidative deamination (Figure 21-5). In this way, nitrogen is obtained from one or more amino acids brought into the living system through the diet. The unneeded α-ketoacids that are simultaneously produced are degraded, ultimately to carbon dioxide, in a process that also releases energy. At the same time, specifically required α-amino acids are produced by reductive amination of the corresponding α-ketoacids. Thus, the random mixture of α-amino acids taken in by an organism through the diet is converted into the specific amino acid mixture needed for a given biological function.

At first, this process of degrading amino acids to provide a source of nitrogen and energy and then rebuilding the carbon framework, followed by reductive amination, seems wasteful. However, this process has the advantage that the amount of each amino acid available to a living system is controlled by the rate of production of the required ketoacid and is not governed by the ratio of amino acids ingested from the diet.

EXERCISE 21-C

Write a complete mechanism for the last step in the reaction shown in Figure 21-4 in which the aldimine is converted into a different one as the α-amino acid is released. (For simplicity, abbreviate the structures; for example, use RCH₂NH₂ to represent pyridoxamine phosphate.)

21-5 ▸ NADPH: Hydride Reduction of Beta–Ketoacids

In the reduction of a carbonyl group, NADPH functions very much like the laboratory hydride reagents lithium aluminum hydride and sodium borohydride. All three compounds serve to transfer a hydride equivalent: a proton plus the two electrons needed to form a carbon–hydrogen bond at a carbonyl carbon group. In Figure 21-6, the part of the NADPH molecule not directly participating in the reduction is represented by a circle so that attention is drawn to the part of the molecule responsible for the reduction. (The complete structure of NADPH is shown in Figure 21-2.) In this example, NADPH effects the reduction of a β-ketoester to a β-hydroxyester, one of several steps in the biosynthesis of the straight-chain fatty acids. The presence of an electron pair on nitrogen of the dihydropyridine enhances the electron density of the ring, making NADPH a more-active hydride transfer agent.

The formal loss of hydride (H⁻) leaves the cofactor NADP with a positive charge. Although the oxidized form of the reagent, NADP, contains an aromatic pyridine nucleus, it is quite electron deficient because the nitrogen is positively charged. Again, there is little change in energy in this reaction, and the NADP formed can (and does) serve as a hydride acceptor in other reactions that result in the oxidation of another substrate. In this way, the redox reagent pair NADPH and NADP can be recycled, with one acting as an oxidant and the other as a reductant. From the overall view of NADP and NADPH inducing both oxidation and reduction reactions, the cofactor functions as a catalyst because it is recycled and is not consumed.

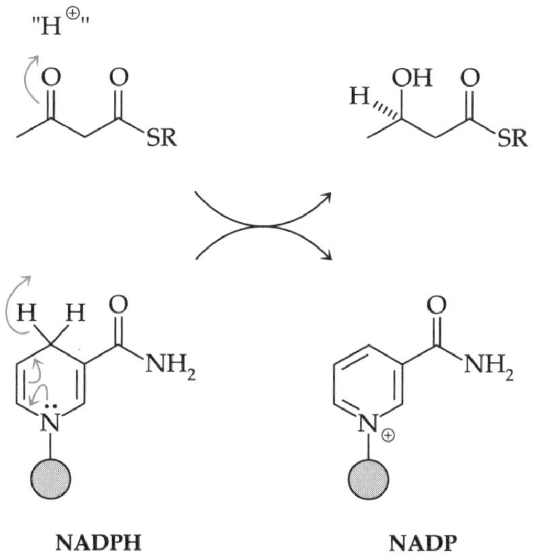

▲ FIGURE 21-6

Nicotinamide adenine dinucleotide phosphate (NADPH) is a biological reducing agent that acts as a formal source of hydride. As in a complex metal hydride reduction, NADPH transfers a hydride equivalent by nucleophilic addition to an activated carbonyl carbon.

EXERCISE 21-D

NADP is a biologically important pyridinium ion. Draw alternate resonance structures for a simple pyridinium ion that do not place positive charge on nitrogen. Consider the reaction of a pyridinium ion with a nucleophile to form a new bond. Which of the atoms of the pyridinium ring are most electrophilic and therefore most easily attacked by the nucleophile?

It might be reasonably expected that the oxidative degradation of fatty acids would take place precisely through the reverse of the reactions illustrated in Figure 21-6, but this is not the case. There are two important differences. First, the oxidizing agent in fatty acid degradation is not NADP but a dephosphorylated analog known as NAD. Second, the oxidation specifically involves the *R* configuration of the hydroxythiol ester, whereas the biosynthesis produces the *S* alcohol (Figure 21-7). Because the absolute configuration at the center of chirality differs, the forward and reverse reactions must take place at different enzymatic sites, a point discussed more thoroughly in Chapter 22.

As before, an important distinction between this biological redox conversion and those induced by laboratory reagents is that the energy difference between the oxidized and reduced forms is much smaller. The reverse process (the biological regeneration of the reduced form) is therefore also easier and is accomplished by the transfer of a proton and two electrons from some other biological species that, as a consequence, is oxidized.

EXERCISE 21-E

It is now possible to obtain nuclear magnetic resonance information directly from living animals. The technique has been used to follow the course of

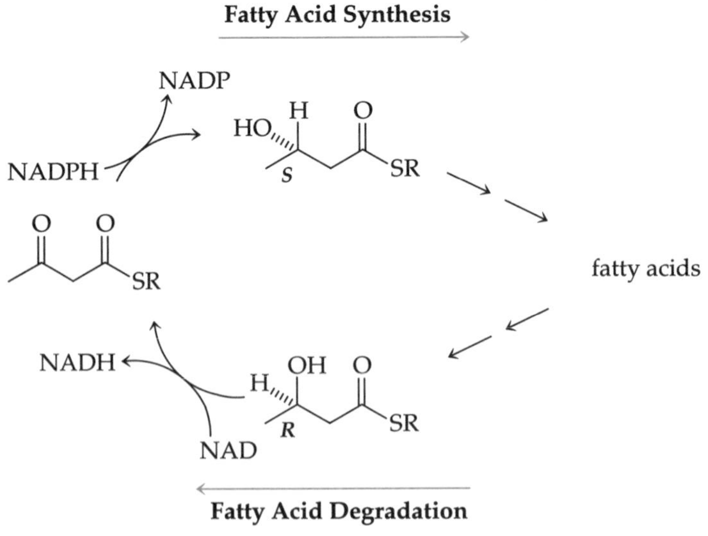

▲ FIGURE 21-7
A key step of fatty acid synthesis is the conversion of a β-ketoester group into a β-hydroxyester. Equally important in fatty acid degradation is the oxidation of the hydroxyl group of a β-hydroxyester to the corresponding β-ketoester. The former conversion, shown in the upper half of the cycle, is mediated by NADPH, whereas the latter conversion, shown in the lower half of the cycle, is induced by NAD. Each reaction is accelerated by a different enzyme.

biological transformations by "feeding" to an animal samples of biochemical intermediates that have been enriched in carbon-13. As the material is consumed and transformed, changes in the chemical shift of the carbon can be followed by NMR spectroscopy. How might this technique be used to establish that the L-enantiomer of a β-hydroxyester is an intermediate in the synthesis of fatty acids, whereas the D-enantiomer is part of the degradation sequence? Which carbon would you choose to enrich so that the differences in shifts would be greatest?

21-6 ▸ FADH$_2$: Electron-Transfer Reduction of an Alpha,Beta-Unsaturated Thiol Ester

The reducing part of flavin adenine dinucleotide (FADH$_2$) is the flavin subunit set apart by color in Figure 21-2. This unit serves as the source of hydrogen when the cofactor effects a biological reduction. In some circumstances, FADH$_2$ functions as a source of hydride in a manner analogous to that which we have seen for NADPH. In other reactions, FADH$_2$ serves as a source of electrons. In this case, FADH$_2$ reduction resembles dissolving-metal reductions (Chapter 12) in which an alkali metal such as lithium or sodium provides a source of electrons, with hydrogen being supplied in a second step as one or more protons. This mechanism is sometimes called **single electron transfer** because the individual steps consist of the transfer of one electron at a time. The reduction of a substrate by FADH$_2$ results in the oxidation of the cofactor to FAD. Just as we have seen for the NADH-NAD pair, FAD later functions as an oxidant, becoming reduced to reform FADH$_2$.

An example of a biological reduction by FADH$_2$ is illustrated in Figure 21-8. Here, an unsaturated thiol ester is reduced to a saturated one, a transformation that also is part of fatty acid biosynthesis. The reductions by both FADH$_2$ (alkene to alkane) and NADPH (ketone to alcohol) afford means of "storing" energy, which is later released when the fatty acids are consumed in the reverse, oxidation reactions.

Overall, the mechanism of this reduction includes the transfer of two electrons and two protons to the unsaturated ester (Figure 21-9). In the first

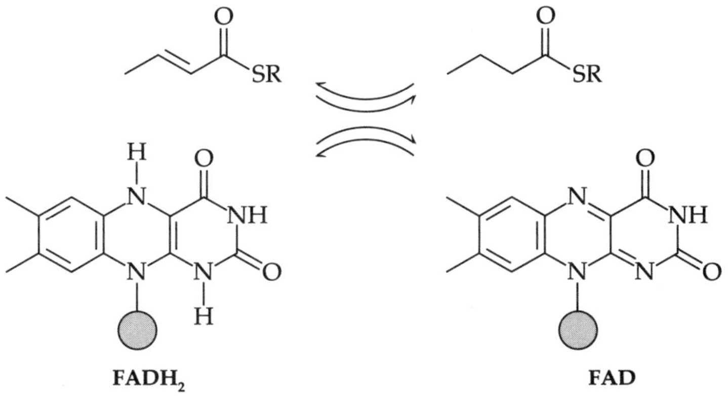

▲ **FIGURE 21-8**

The reduced form of flavin adenine dinucleotide (FADH$_2$) is a reducing agent that effects the reduction of a carbon–carbon double bond in an α,β-unsaturated thiol ester to the saturated thiol ester.

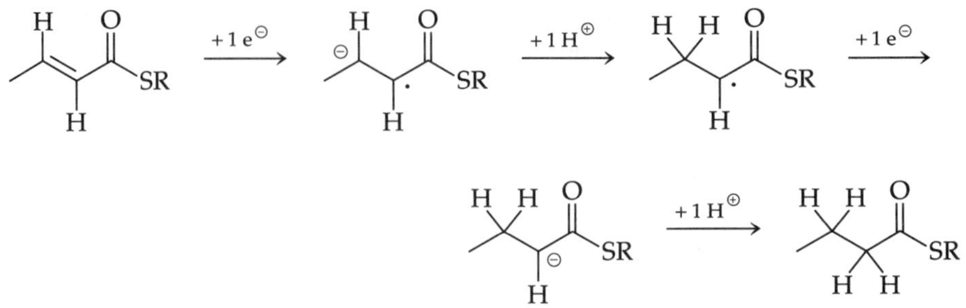

▲ FIGURE 21-9

The addition of an electron to an α,β-unsaturated thiol ester produces a resonance-stabilized radical anion. Protonation on carbon produces a resonance-stabilized thiol ester enolate radical that is converted into a thiol ester enolate anion by the addition of another electron. Protonation of this anion on carbon completes the reduction.

step, an electron is transferred from $FADH_2$ to the unsaturated ester, populating the lowest-lying antibonding orbital and resulting in a radical anion, for which a resonance contributor is shown. (Recall from Chapter 6 that the π bond of the neutral molecule is weakened upon forming a radical anion.)

Protonation of this anion radical on the β carbon in the second step affords the most-stable radical possible, one that is resonance stabilized by conjugation with the carbonyl group in a fashion analogous to that in enolates derived from aldehydes, ketones, and esters. Indeed, the third step of the reduction is the transfer of a second electron to the radical to produce the enolate anion of the thiol ester. Protonation of this enolate on carbon affords the saturated product as a thiol ester.

EXERCISE 21-F

There are many additional resonance structures for the radical anion shown in Figure 21-9. These can be subdivided into four groups differing in which atom (the carbonyl oxygen, carbonyl carbon, α carbon, or β carbon) bears the negative charge. Draw one resonance structure for each of the four groups. Which group is likely to contribute the most to the hybrid, based solely on the position of the negative charge?

EXERCISE 21-G

Write an alternate mechanism for the reduction shown in Figure 21-9, in which $FADH_2$ acts as a hydride donor in a Michael sense. How could this alternative mechanism be differentiated from that shown in Figure 21-9?

21-7 ▸ Acetyl CoA: Activation of Carboxylic Acids (as Thiol Esters) toward Nucleophilic Attack

Thiol esters were the substrates in each of the preceding biological redox reactions. This functional group plays an essential role in carboxylic acid chemistry in living systems. In many circumstances, the thiol ester serves as a carboxylic acid derivative that is more reactive toward nucleophilic attack (for example, toward hydrolysis) than is the corresponding oxygen ester. We have seen in Chapter 12 that the enhanced reactivity of thiolesters is the result of decreased resonance donation of sulfur's lone pair of electrons into the carbonyl π system compared with that from oxygen.

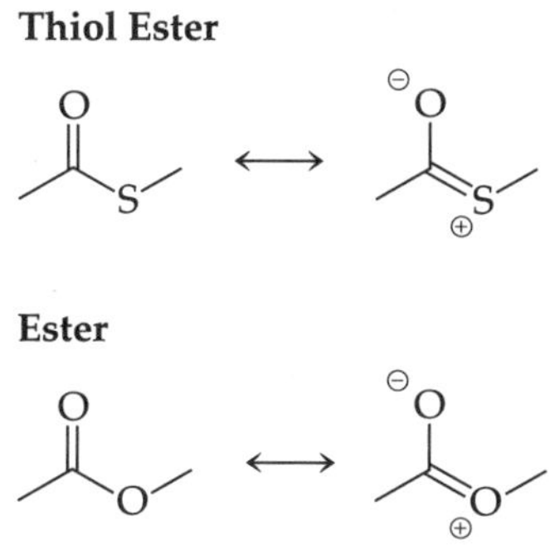

Thiol Ester

Ester

Comparing the two zwitterionic, charged resonance structures at the right, we see positive charge either on sulfur or on oxygen and a π bond between carbon and either sulfur or oxygen. Although the more-electronegative oxygen is less stable when positively charged than is sulfur, the delocalization of lone-pair electron density from sulfur into the carbonyl π system requires overlap between a $3p$ orbital on sulfur and the π system of the carbonyl group, which is composed of $2p$ orbitals. Because of the large difference in size between these orbitals, such overlap plays only a minor role in thiol esters. Addition of a nucleophile to the carbonyl carbon of an ester results in the breaking of the π bond and thus also results in loss of any stabilization resulting from delocalization. Because thiol esters are less stabilized by this delocalization, they undergo addition of a nucleophile to the carbonyl carbon more rapidly than do carboxylic acid esters.

Thiol esters are comparable in reactivity to carboxylic acid anhydrides and can be thought of as "activated" derivatives of carboxylic acids. Because the activation energies for the nucleophilic acyl substitution of thiol esters are substantially lower than those of the corresponding oxygen esters, thiol esters represent an ideal balance between stability and reactivity in an aqueous medium. Carboxylic acid chlorides react too rapidly with water, whereas carboxylic acid esters based on oxygen require a substantially greater activation energy in order to undergo nucleophilic acyl substitution.

At times, thiol esters also serve the function of temporarily holding a carboxylic acid unit attached to an enzyme. For example, the sulfur of a cysteine residue in a peptide can form a covalent linkage as a thiol ester with a carboxylic acid. The most-important thiol compound is **coenzyme A (CoA)**, shown in Figure 21-10. As we might expect, coenzyme A often participates in reactions that require the activation of a carboxylic acid. We will encounter the thiol ester of coenzyme A and acetic acid, known as **acetyl coenzyme A (acetylCoA)**, again in Chapter 22 where it is used in fatty acid biosynthesis.

Some of the pieces of which the relatively complex structure of this cofactor is composed are familiar. For example, in NADPH, FADH$_2$, and coenzyme A, the active part of the molecule is bound through a diphosphate unit to an adenine nucleotide that differs in these three molecules only by the presence and placement of an additional phosphate group. Critical to the function of coenzyme A is the presence of a thiol group, derived from the very simple molecule 2-aminoethane thiol (known as **cysteamine**), which is linked through an amide bond to the vitamin **pantothenic acid**.

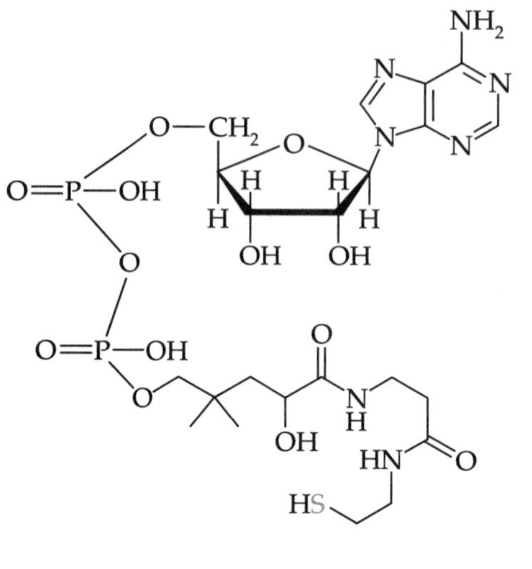

Coenzyme A (CoA)

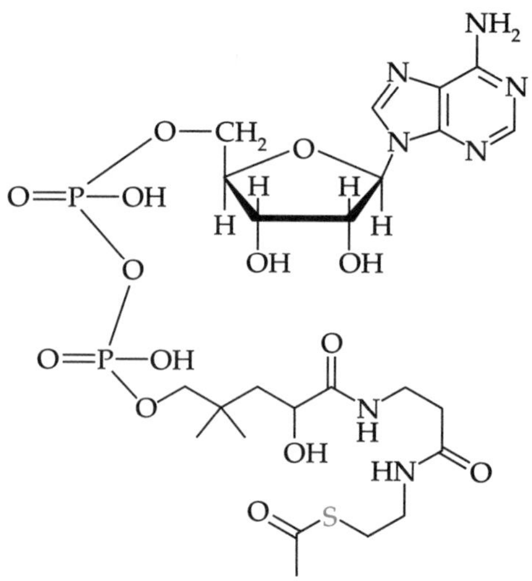

Acetyl coenzyme A (Acetyl CoA)

▲ FIGURE 21-10
The acetylation of the cofactor coenzyme A on the terminal thiol group produces a thiol ester. Acetyl CoA is a derivative of acetic acid that is activated toward nucleophilic acyl substitution.

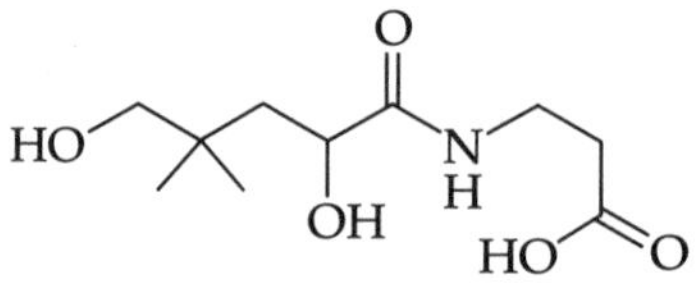

Pantothenic acid

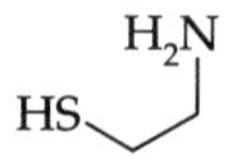

2-Aminoethane thiol

EXERCISE 21-H

Write a mechanism for acetyl CoA hydrolysis in both acidic and basic media. Explain why the thiol ester rather than the amide is hydrolyzed under these conditions.

21-8 ▸ Thiamine Pyrophosphate and Lipoic Acid: Decarboxylation of Alpha-Ketoacids

Thiamine pyrophosphate is a cofactor important in the degradation of amino acids. As discussed in Section 21-4, the cofactor pyridoxamine induces the oxidative deamination of amino acids. The resulting α-ketoacid is further transformed, in the presence of thiamine pyrophosphate and lipoic acid, by a process known as **oxidative decarboxylation** to a carboxylic acid that is one carbon shorter.

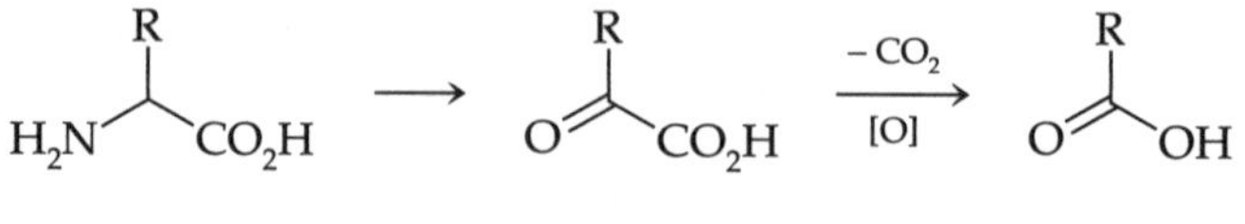

This process is reminiscent of the decarboxylation of β-ketoacids discussed in Chapter 8. However, there is an important electronic difference

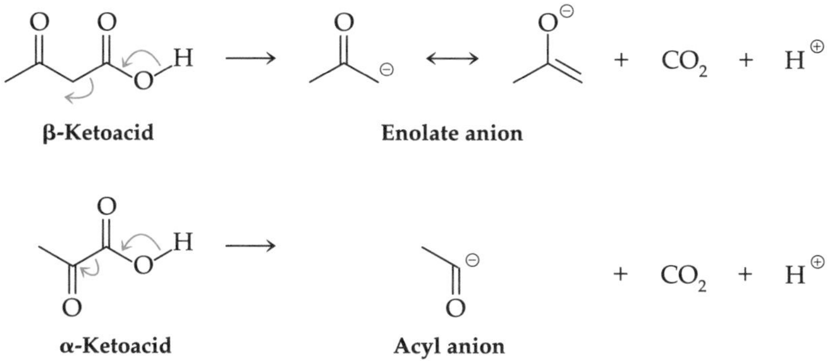

▲ FIGURE 21-11
Decarboxylation of a β-ketoacid produces a resonance-stabilized enolate anion, whereas the analogous reaction of an α-ketoacid would lead to an unstable acyl anion.

between the decarboxylations of β-ketoacids and α-ketoacids that is the direct result of the difference in the relative positions of the two carbonyl groups in these two systems. These two processes are illustrated in Figure 21-11. Recall that loss of carbon dioxide from the salt of a β-ketoacid results in the formation of a resonance-stabilized enolate anion. However, decarboxylation of an α-ketoacid by a parallel route leads not to a resonance-stabilized anion but to the very unstable acyl anion. (Recall the discussion of the acyl anion derived from formaldehyde in Chapter 18.) Because the anion that would be formed is unstable, α-ketoacids do not readily undergo decarboxylation. Instead, decarboxylation of α-ketoacids in biological systems requires that these ketoacids first be modified so that the anion resulting from the loss of carbon dioxide is stabilized.

EXERCISE 21-I

Compare and contrast the structures and stabilities of an acyl anion and an enolate anion, explaining why the latter is more stable.

To understand how this process is accomplished biochemically, we first need to become familiar with the structures of cofactors **thiamine pyrophosphate** (Figure 21-12) and **lipoic acid** (Figure 21-14). Thiamine

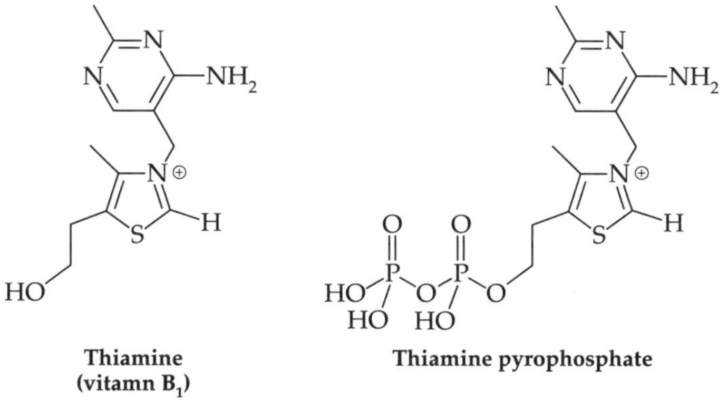

Thiamine
(vitamn B₁)

Thiamine pyrophosphate

▲ FIGURE 21-12
The vitamin thiamine (B₁) takes part in the decarboxylation of α-ketoacids as thiamine pyrophosphate.

pyrophosphate is derived from **vitamin B$_1$** (also known as thiamine) by phosphorylation of the free hydroxyl group. Notice that the pyrophosphate (also called, more correctly, a diphosphate) shown here differs from the phosphate group shown in Figure 21-3 in that an additional phosphate linkage is present in a form that is analogous to a carboxylic acid anhydride. For convenience, we have represented the pyrophosphate unit in the acid form but keep in mind that, in water near pH 7, the fully deprotonated form dominates.

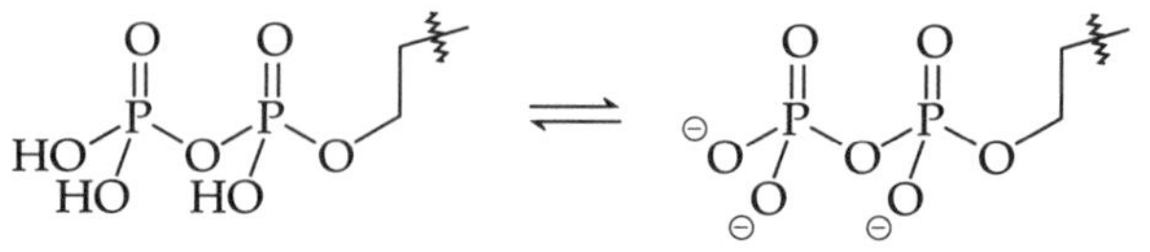

The unique features of these phosphoric acid anhydride linkages are discussed again in Chapter 22.

The chemical reactivity of thiamine derives from the relatively high acidity of the single hydrogen of the thiazole ring (pK_a = 17.6). Although anions derived from the deprotonation of sp^2-hybridized carbons are uncommon, this example is a special case in which stability is imparted to the vinyl carbanion by the adjacent positively charged nitrogen (Figure 21-13). As stated in Chapter 8, molecules containing centers of opposite charge on adjacent atoms are called ylides. It is often possible to write an alternate resonance structure for an ylide, in which electron density has been shifted from the negatively charged carbon to the atom bearing positive charge. In this case, moving the π electrons to nitrogen results in the neutral valence-bond structure at the right in Figure 21-13. However, this structure is not dominant in the hybrid because it leaves the carbon that lost the proton with access to only six valence electrons, the electronic structure of a carbene.

In contrast with the complex cofactor structures presented so far, that of lipoic acid is quite simple. In its oxidized disulfide form shown at the right in Figure 21-14, it serves as an oxidizing agent. Thiamine and lipoic

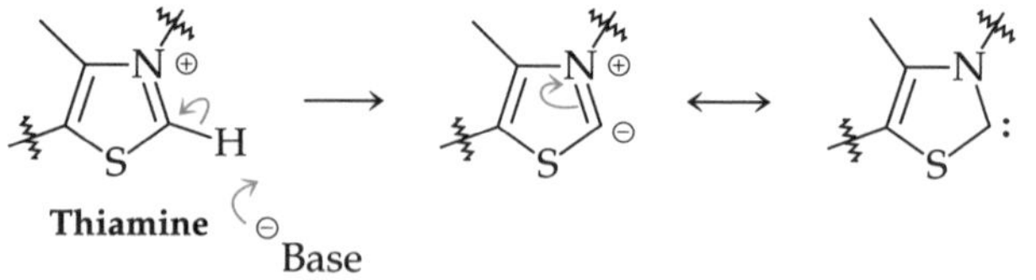

▲ FIGURE 21-13
Deprotonation of thiamine at the site adjacent to the positively charged nitrogen atom produces an ylide. A resonance hybrid for this ylide in which the electrons in the C–N π bond are localized onto nitrogen emphasizes the similarity of the electronic structure of this ylide to that of a carbene.

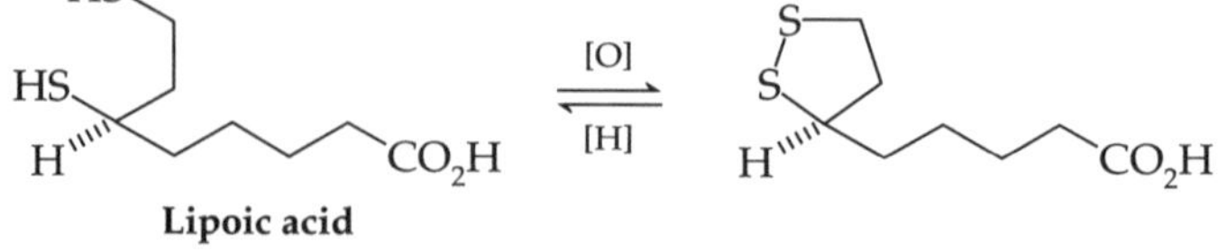

▲ FIGURE 21-14
Lipoic acid, a dithiol, is readily interconverted with its cyclic disulfide (—S—S—) form by oxidation-reduction.

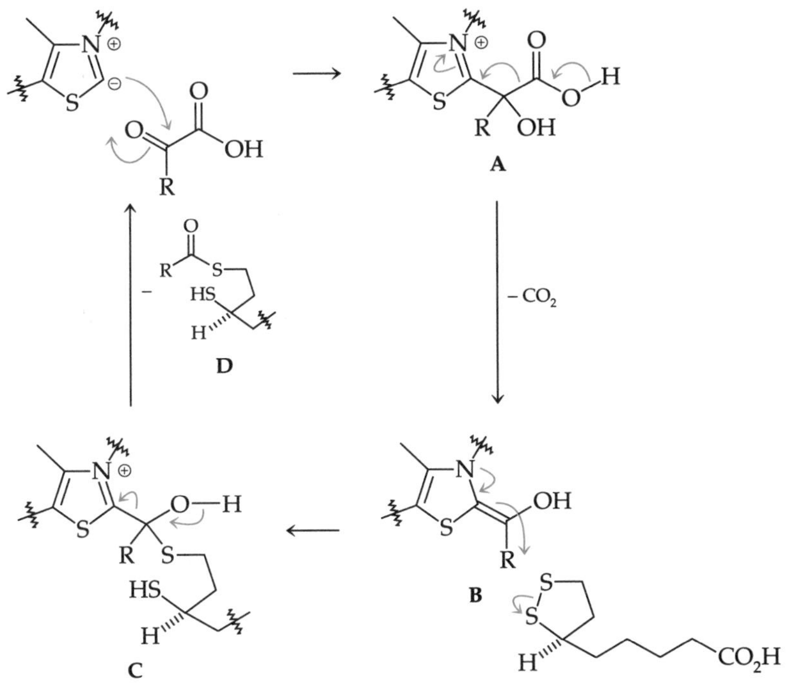

▲ **FIGURE 21-15**
The nucleophilic carbon atom in the ylide formed by deprotonation of
thiamine adds to an α-ketoacid, producing an adduct (A) in which a C=N
bond is positioned β to the carboxylic acid (analogous to the β-ketone group
in a β-ketoacid). Therefore, decarboxylation takes place readily, forming an
activated enol (B). Reaction of this intermediate as a nucleophile at sulfur of
the cyclic disulfide form of lipoic acid effects oxidation of this carbon.
Formation of a carbonyl group from intermediate C produces an activated
thiol ester (D) and regenerates thiamine pyrophosphate: thus, this cofactor is a
catalyst for the decarboxylation of α-ketoacids.

acid, in combination, effect the oxidative decarboxylation of an α-ketoacid,
a process that is very difficult to accomplish directly. This reaction begins
with the anion of thiamine pyrophosphate (Figure 21-13) acting as a nucleophile, adding to the carbonyl group of the α-ketoacid, as shown in Figure
21-15, to form adduct A. In this process, the α-keto group is converted into
an alcohol, and a carbon–heteroatom π bond has been attached at a position
β to the carboxylic acid carbon, an arrangement that resembles the relation
of the two π systems in a β-ketoacid.

Facile decarboxylation of adduct A is now possible by a route similar
to that described earlier for a β-ketoacid. The thiamine nitrogen is already
positively charged in adduct A, and loss of carbon dioxide and a proton
results in a neutral intermediate (B). Thus, this process is unlike the decarboxylation of a β-ketoacid that consists of the intramolecular transfer of the
proton from the carboxylic acid to the keto-oxygen through a cyclic transition state (Chapter 8).

The carbon residue of the starting α-ketoacid is now covalently
attached to thiamine pyrophosphate in intermediate B, and the remaining
steps are necessary to disconnect these two pieces, forming the decarboxylated acid and regenerating thiamine pyrophosphate. In the next step, intermediate B acts as a nucleophile, attacking the oxidized form of lipoic acid

and forming a new carbon–sulfur bond while simultaneously effecting cleavage of the sulfur–sulfur bond in the cofactor. This reaction employs the electron pair of the ring nitrogen (an enamine) to displace a thiol from the disulfide by an S_N2 reaction. If we compare the resulting intermediate with adduct A, we see that one of the sulfur atoms of lipoic acid has replaced the carboxylic acid carbon of the original α-ketoacid. Reforming the carbonyl group by the loss of thiamine pyrophosphate produces compound D, the hydrolysis of which reforms lipoic acid and the decarboxylated acid.

EXERCISE 21-J

Show how the decomposition of intermediate C in Figure 21-15 is similar to the nucleophilic substitution of a simple thiol ester.

21-9 ▸ Mimicking Biological Activation with Reverse Polarity Reagents

In effecting the decarboxylation of an α-ketoacid, thiamine induces chemistry normally observed only with β-ketoacids. This cofactor exemplifies the use of reagents to specifically invert normal chemical reactivity, a methodology that is very important for the design of new and unusual chemical reactions. In Chapter 18, we learned about a reverse polarity (umpolung) reagent (the formyl anion) that permitted the typically characteristic electrophilic reactivity of a carbonyl carbon to be reversed. In a like fashion, we can view thiamine as a nucleophile at one stage in the oxidative decarboxylation of an α-ketoacid and as a leaving group at a later stage. These two functions have opposite electronic requirements: nucleophiles readily donate electron density to form a new bond; and leaving groups readily accept the electron density of a bond undergoing cleavage. To understand normal polarity for chemical reactions, we must develop a method for the analysis of a given structure according to the positions of various appended functional groups.

The concept of reverse polarity reagents has become important in synthetic chemistry because it affords a graphic way of visualizing reagents that have unusual reactivity. In the same way, it is also useful to analyze the relative reactivity of functional groups and neighboring atoms in a target molecule to see if they exhibit normal reactivity patterns. If not, a reverse polarity reagent may be useful for the synthesis of such molecules. For example, let us examine the reactivities imparted by the presence of two carbonyl groups in a β-ketoester and in an α-ketoester.

Carbonyl carbons typically react as electrophiles: that is, because they are electron deficient, they typically react with nucleophiles. Conversely, the α carbon can be readily converted into an enolate anion, which is a reactive nucleophile. This reactivity is represented in a pictorial fashion by pluses (for electrophilic sites that react with nucleophiles) or minuses (for nucleophilic sites that react with electrophiles) within parentheses, (+) or (–), assigned to the relevant atoms, as shown for acetone and methylvinyl ketone, an α,β-unsaturated ketone in Figure 21-16. Be aware that these symbols represent reactivity, not charges, although it is often the case that this reactivity parallels the partial charge of an atom resulting from the presence of an electron-withdrawing or electron-donating group.

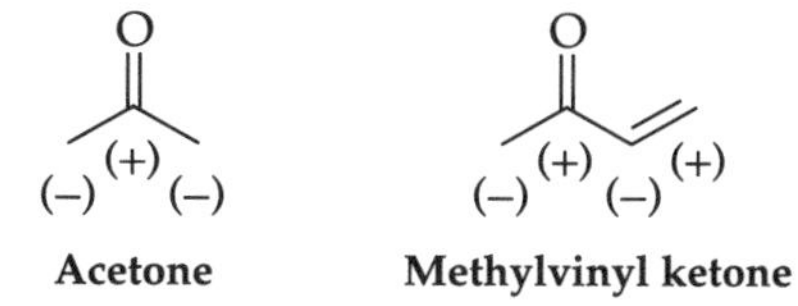

O
‖
Acetone
(–) (+) (–)

O
‖
Methylvinyl ketone
(–) (+) (–) (+)

▲ FIGURE 21-16

The normal polarity of a carbonyl group results in substantial electrophilic character of the carbonyl carbon. The removal of a proton from either of the α carbons results in the formation of a nucleophilic enolate at this site. Thus, the reactivity of a ketone can be characterized as electrophilic (+) at the carbonyl carbon and, after deprotonation, nucleophilic (–) at the adjacent carbons. Unsaturated ketones react with some nucleophilic species at the β carbon (conjugate addition); thus, this site has electrophilic character (+). The characteristic reactivities of the carbon atoms of ketones thus alternate in a regular fashion.

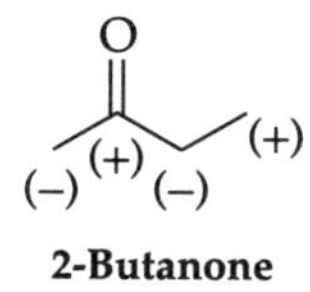

It is also useful to consider that the β carbon of saturated ketones is a site of potential electrophilic reactivity (as illustrated in the margin for 2-butanone). Just as the α carbon of a neutral ketone is a potential rather than a real nucleophile, the β carbon of a ketone is a potential electrophile because its conversion into an α,β-unsaturated ketone does produce this reactivity. It is important to note that reactivity alternates in a regular fashion along the chain. This alternation is quite reasonable, because the strong electron-withdrawing nature of the carbonyl group imparts electrophilic reactivity to the carbonyl carbon and stabilizes a carbanion at the α carbon.

O
‖
2-Butanone
(–) (+) (–) (+)

EXERCISE 21-K

Explain the reactivities shown for methylvinyl ketone in Figure 21-16 by providing an example of a likely chemical reaction at each of the carbon atoms in the indicated sense.

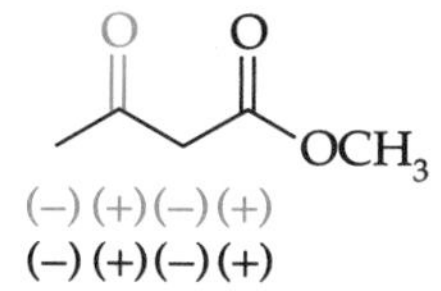

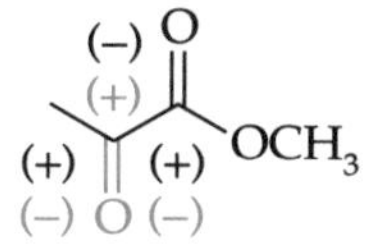

When more than one functionality is present, it is useful to use color to distinguish the reactivities imparted by each to adjacent atoms, as in the α- and β-ketoesters shown in the margin. The sequence of reactivities shown in black is that imparted by the ester, whereas that shown in color represents the effect of the keto group. In the β-ketoester, the reactivity imparted by both of the carbonyl groups to each of the carbon atoms of the chain is the same. As a consequence, the alternation of reactivities imparted by each of the carbonyl groups is preserved in the β-ketoester. Recall from Chapter 6 that the carbon between the two carbonyl groups is especially acidic (pK_a 11 for a β-ketoester). Indeed, the superimposition of the reactivities provides us with a sense of the additional stability imparted to the anion at this position. On the other hand, the reactivities imparted to each carbon of an α-ketoester by the two carbonyl groups do not match. We will see the consequence of this mismatch shortly.

A consideration of potential reactivity can help in the retrosynthetic analysis of a target molecule because the characteristic reactivity at a specific site does not generally change from starting material to product as long as the fundamental nature of the functional group remains unaltered. For example, let us analyze the transformation of two molecules of methyl acetate into methyl acetoacetate, a reaction that we know as the Claisen condensation.

O   O
‖   ‖
        OCH₃
(–) (+) (–) (+)
(–) (+) (–) (+)

(–) O
(+) ‖
        OCH₃
(+) ‖ (+)
(–) O (–)

Claisen Condensation

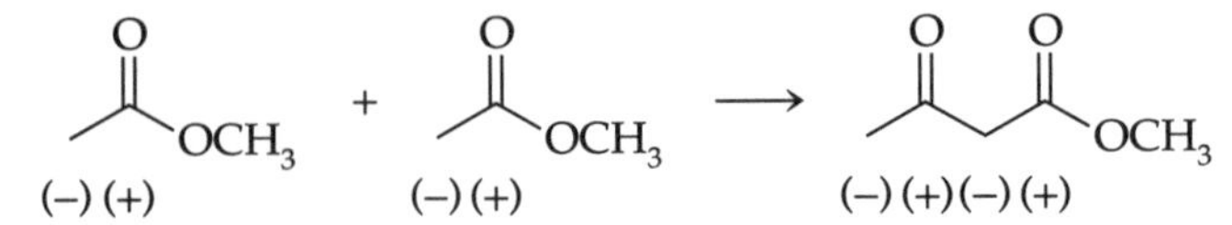

Recall that the Claisen condensation consists of the interaction of the α carbon of one ester as an enolate carbanion in a nucleophilic acyl substitution on the carbonyl group of a second ester molecule: the reactivity of each carbon of the starting materials remains the same in the product β-ketoester. Let us imagine, however, that we had not yet learned this specific reaction but had, nonetheless, the task of constructing methyl acetoacetate from starting materials with fewer carbon atoms. Such a synthesis requires carbon–carbon bond formation: as we learned in Chapter 14, we should try to incorporate this reaction late in a proposed synthetic sequence (that is, early in the retrosynthetic analysis). Rather than consulting the list of carbon–carbon bond-forming reactions in Chapter 14, let us proceed by considering, in turn, the formation of each of the carbon–carbon bonds by the combination of a nucleophile with an electrophile. Our analysis is simplified because there are only three such bonds in our example: the three possible nucleophile-electrophile combinations are illustrated in Figure 21-17. Notice that, in each set of fragments, a pair of electrons is available on an atom that we have determined to have potential nucleophilic character in the product. Because the reactivities alternate along the chain, the positively charged carbon atom is naturally one that has electrophilic character in the β-ketoester.

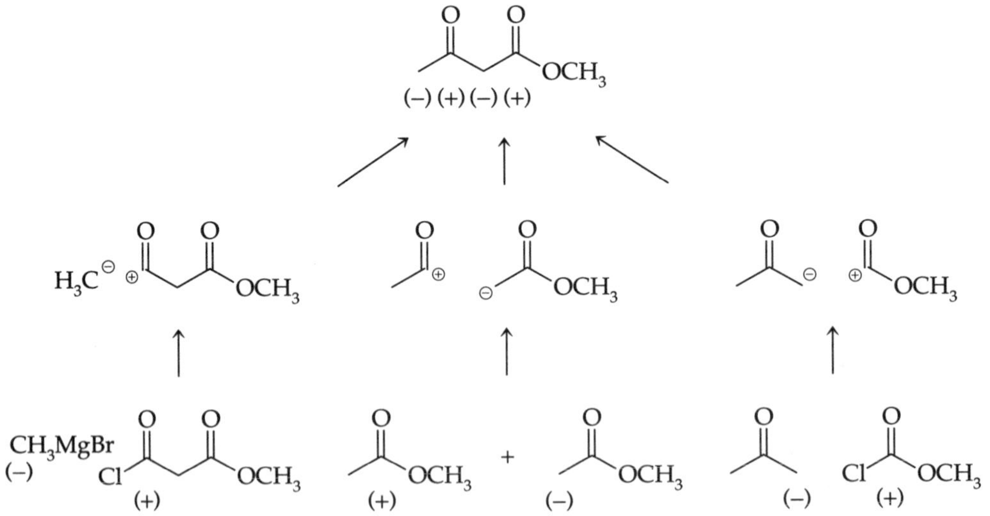

▲ FIGURE 21-17
Because the characteristic reactivities of the atoms of a β-ketoester alternate in a regular fashion, it is possible to construct this species by three separate routes, each forming a different bond. Each route combines an electrophilic with a nucleophilic starting material whose characteristic reactivity is preserved and matches that found in the corresponding atoms of the product. For example, in the left-hand route, a carbon–carbon bond is formed by the reaction of a nucleophilic carbon of a Grignard reagent with the electrophilic carbonyl carbon of an acid chloride. In the product β-ketoester, the added methyl group is adjacent to a carbonyl group and thus has (potential) nucleophilic character. The carbonyl group of the acid chloride remains in the product and has electrophilic character both before and after the reaction.

The analysis to this point has led us to ionic fragments, requiring the implausible simultaneous formation of both a carbanion and a carbocation. However, in each case we can envision stable reactants whose reactivity would enable the bond to be formed by a reaction in which specific atoms act as nucleophiles or electrophiles, as appropriate, to correspond to our analysis. For example, we can envision the CH_3^- fragment shown at the left of Figure 21-17 as being derived from an organometallic reagent, such as methyllithium or methyl magnesium bromide. Each of these reagents has substantial negative charge on the methyl carbon as a result of the highly polarized carbon–metal bond. The reaction of a Grignard reagent with the half-ester, half-acid chloride of malonic acid shown at the lower left might well be expected to result in methyl acetoacetate by nucleophilic acyl substitution on the more reactive of the two carboxylic acid derivatives (the acid chloride). Although we have not seen this specific reaction before, it poses a chemically reasonable combination of reagents.

In addition, this analysis has uncovered two other possibilities involving enolate anions of an ester (center) and of a ketone (right). The analysis of the center reaction in Figure 21-17 has led us to "rediscover" the Claisen condensation, as a combination of functional groups that should readily form a β-ketoester. Similarly, nucleophilic acyl substitution on methyl chloroformate by the enolate anion produced by deprotonation of acetone (the reaction at the right) should also produce the desired product. Although we have concentrated on the chemistry of only one of these syntheses of a β-ketoester, all three indeed work in the laboratory.

In this example, the chemical reactivity of each carbon atom in the product matches that of the corresponding atom in the proposed starting materials in all three combinations. This matching is a consequence of the alternation in reactivity imparted by carbonyl groups and of the positioning of the two functional groups of the product so that their reactivities correspond. In other words, by constructing a chart of electron demand, we can use our knowledge of organic chemistry to predict new synthetic transformations.

EXERCISE 21-L

Although the preceding analysis has led us to three routes for constructing a new carbon–carbon bond in the synthesis of methyl acetoacetate, it does not evaluate them from a practical point of view. Suppose that, instead of using methyl chloroformate in the third reaction, we had chosen to react the enolate of acetone with dimethyl carbonate ($CH_3OCO_2CH_3$). What problem(s) might be encountered with this proposed combination of reagents? Might there be similar difficulties with methyl chloroformate? (Hint: Consider the relative acidities and electrophilic reactivities of each of the starting materials and the product.)

Let us now undertake a similar analysis for an α-ketoester (Figure 21-18). Here, the relation of the carbonyl groups is such that these functional groups impart opposite senses of reactivity to each of the carbons of the chain. (The alternating reactivity imparted by the ester is shown in black, whereas that imparted by the ketone group is shown in color.) Because the potential reactivities of the two carbonyl groups in the product do not match, an analysis parallel to that used in Figure 21-17 does not lead to an unambiguous, straightforward reaction between a nucleophile and an electrophile: we cannot find a unique pairing of starting materials that leads

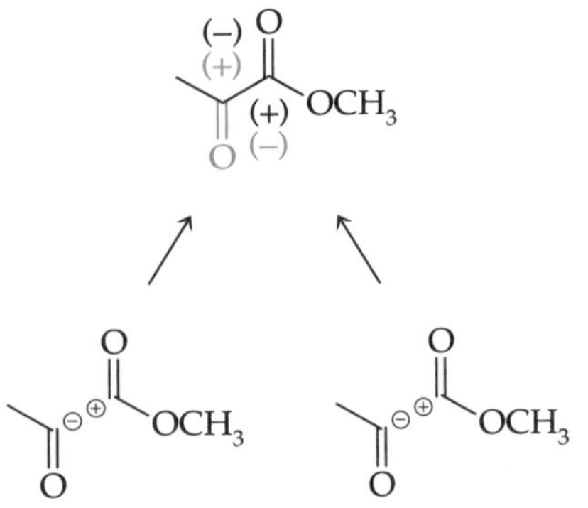

▲ FIGURE 21-18

With the reactivities expected from an analysis focusing on the ester (shown in black), the fragments shown at the lower right are predicted; whereas, with the reactivities expected from an analysis focusing on the ketone group (shown in color), those shown at the lower left are expected. Although we know a carbonyl carbon to be electrophilic, both routes include an acyl anion—of a ketone at the lower left and of an ester at the lower right. Neither of these approaches accomplishes the construction of the α-ketoester because of the instability of the acyl anion. This result can be predicted without detailed analysis simply by noting that the two carbonyl groups of the β-ketoester impart opposite characteristic reactivities.

naturally to the product. Because the analysis predicts opposite reactivity when starting from the ester and ketone groups, we do not find an unambiguous clue to the reactivities that might be required in the starting materials: each of the two possible nucleophile-electrophile combinations shown in Figure 21-18 seem equally inappropriate because both require the use of an unstable, acyl anion as a nucleophile.

How then can we form an α-ketoester if the potential reactivities imparted by functional groups do not match? This analysis suggests that at some point we must either: (1) use a reaction in which a functional group reacts in an "unnatural" fashion; or (2) interconvert normal functional groups such that their natural reactivity is inverted. An example of the first option was presented in Chapter 18, where the acyl anion of formaldehyde was used as a nucleophile. However, such anions are not particularly stable and are restricted to aldehydes lacking hydrogens on the α carbon. As a consequence, they are only rarely used in carbon–carbon bond-forming reactions.

On the other hand, we have already seen several examples in which the characteristic reactivity of a carbon in a reagent can be reversed by the transformation of one functional group into another. The conversion of an alkyl halide into an organometallic reagent such as a Grignard reagent is a simple example of reactivity inversion that we have encountered several times.

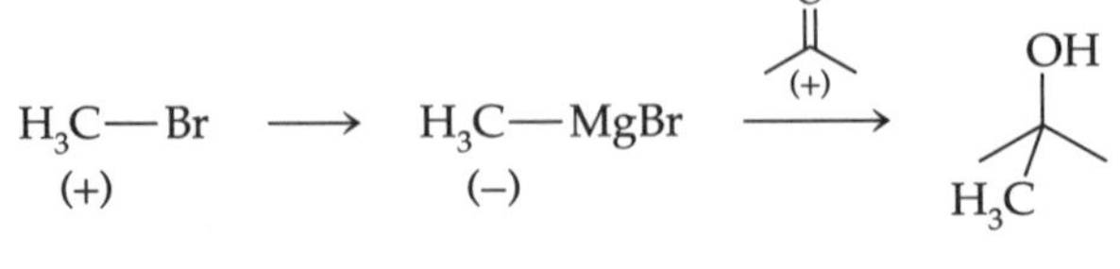

Because of the polarization of the carbon–halogen bond, the carbon of an alkyl bromide characteristically reacts as an electrophile. Conversely, in the derived organomagnesium reagent, the replacement of the more-electronegative halogen by the less-electronegative metal results in polarization of the carbon–metal bond that is the reverse of that in an alkyl halide. Thus, the carbon of a Grignard reagent reacts as a nucleophile.

In a like fashion, the conversion of an alkyl halide into a phosphonium salt and then into an ylide (a Wittig reagent) also produces a nucleophilic carbon, as illustrated by its reaction with acetone.

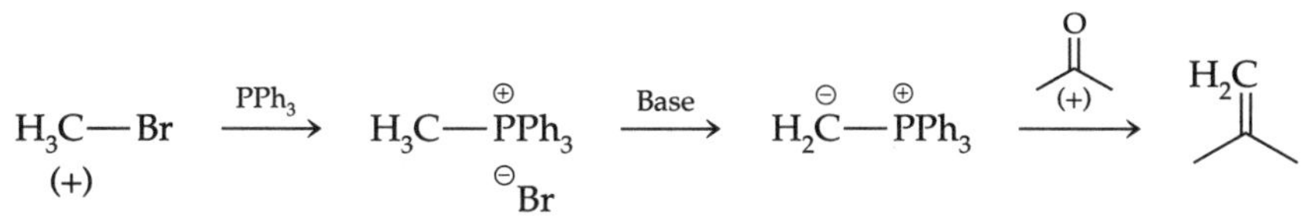

In both cases, the nature of the transformation leaves no clue in the reactivities of the carbons in the product to the original, electrophilic reactivity of the added carbon because the original functionality, an alkyl halide, is not present in the product.

Chemists have formulated a number of reactions that temporarily mask the normal reactivity of a functional group while imparting the opposite reactivity. For example, consider the conversion of an aldehyde into a dithioacetal (Figure 21-19). This transformation proceeds by a mechanism essentially identical with that given in Chapter 12 for the formation of the corresponding oxygen acetal. In the dithioacetal, the normal electrophilic reactivity of the carbonyl carbon of the aldehyde is no longer present. More-

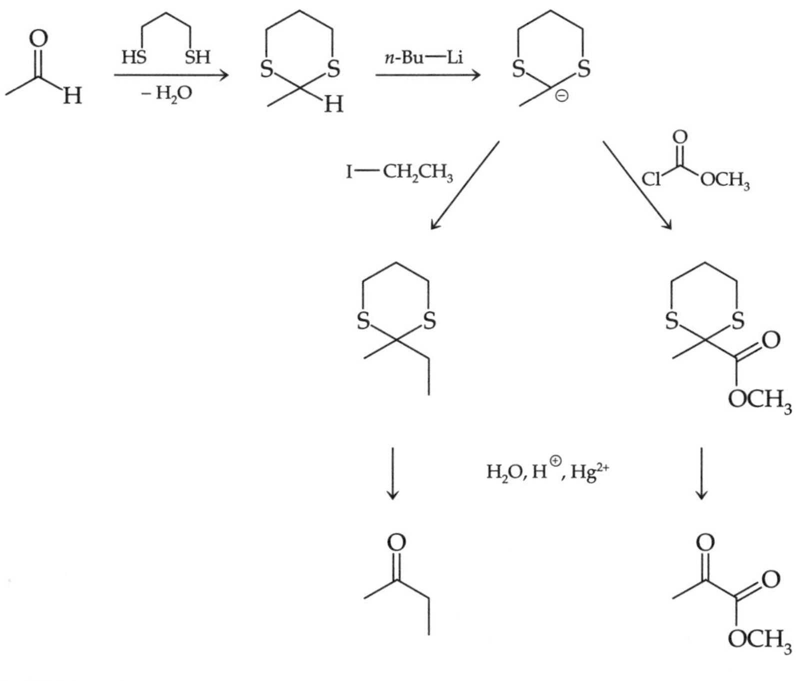

▲ **FIGURE 21-19**
The dithioacetal derivative of an aldehyde can be converted into an anion; thus, the dithioacetal can be considered to have (potential) nucleophilic character at what, in the aldehyde, is an electrophilic, carbonyl carbon. Deprotonation of the dithioacetal produces an anion that can function as a nucleophile for reaction with an alkyl iodide by S_N2 displacement (left) and with an acid chloride by nucleophilic acyl substitution on the acid chloride (right) to form new carbon–carbon bond. Hydrolysis of the dithioacetal group forms the carbonyl group of the ketone product.

over, the proton on the carbon bearing the two sulfur atoms is now moderately acidic and can be removed by a strong base such as a lithium dialkylamide. (Recall that sulfur and other third-row elements, though not strongly electronegative, are polarizable and can effectively stabilize the negative charge on an adjacent carbon atom.)

This anion reacts as a nucleophile with many of the normal, electrophilic carbon species such as alkyl halides and acid chlorides. Once the carbon–carbon bond is formed, the thioketal in the product is hydrolyzed to regenerate the carbonyl group. The overall process forms a carbon–carbon bond to the carbonyl carbon in a reaction employing the carbonyl carbon as a nucleophile. As shown in Figure 21-19, this sequence makes it possible to convert an aldehyde into a ketone or into an α-ketoester.

As a result, the dithioacetal anion is referred to as an **acyl anion equivalent.** Many such equivalents have been developed through the years by synthetic chemists.

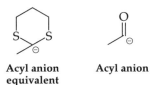

Acyl anion   **Acyl anion**
equivalent

<hr />

EXERCISE 21-M

Write a step-by-step mechanism for the formation of a dithioacetal as in the first step of Figure 21-19.

<hr />

EXERCISE 21-N

Show how you could use the acyl anion equivalent shown above to prepare 2,3-hexanedione and 2,5-hexanedione.

<hr />

21-10 ▸ Tetrahydrofolic Acid: A One-Carbon Transfer Cofactor for the Methylation of Nucleic Acids

Thiamine provides an energetically reasonable pathway for the loss of one carbon from α-ketoacids, resulting in the formation of a shorter carboxylic acid and carbon dioxide. This process releases energy, as heat, as a result of the production of carbon dioxide. On the other hand, other biochemical syntheses require the addition of a one-carbon unit for their biological function. **Tetrahydrofolic acid** (Figure 21-20) serves as a carbon carrier, transferring one carbon obtained from biodegradation to an intermediate in a biosynthetic pathway. (Tetrahydrofolic acid is sometimes referred to by the abbreviation THF. Because this abbreviation is also used by organic chemists for the common solvent tetrahydrofuran, we do not use it here so as to avoid confusion.) **Folic acid,** the biological precursor of tetrahydrofolic acid, is also known as **vitamin B$_c$** (or as **vitamin M**) and must be obtained by human beings from the diet, although it is synthesized by bacteria. We will return to this difference between mammals and bacteria in Chapter 23 where we will see the chemical mode of action of the group of antibiotics known as sulfa drugs. These compounds selectively kill bacteria by interfering with the synthesis of folic acid.

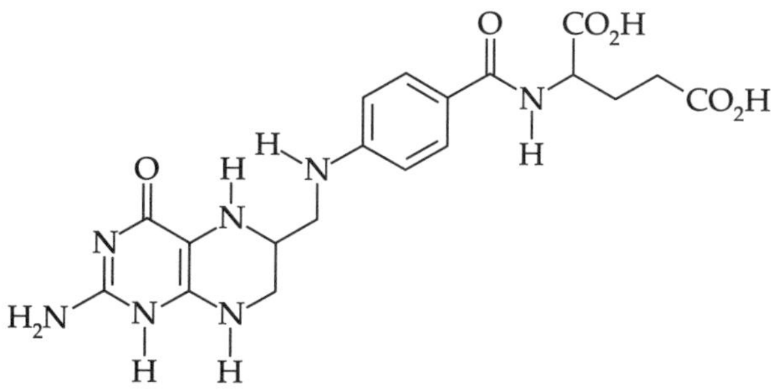

▲ FIGURE 21-20

Tetrahydrofolic acid and its oxidized derivative, folic acid, are used to transfer one carbon atom to growing carbon chains in biosynthesis.

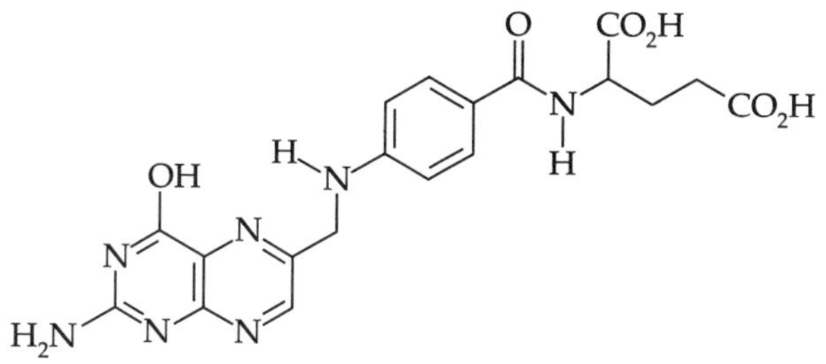

EXERCISE 21-O

Identify the atoms that have been reduced when tetrahydrofolic acid is formed from folic acid. How many equivalents of hydrogen would be required if this process were to be effected by catalytic hydrogenation? Notice that there are other differences between the structures of folic acid and tetrahydrofolic acid besides the presence of additional hydrogen. What special feature(s) of each of these structures provides unique stabilization?

The carbon that is transferred by tetrahydrofolic acid is derived from serine in the synthesis of glycine. This carbon is then added, for example, to **deoxyuridylic acid,** forming **deoxythymidylic acid,** one of the monomer units of DNA (Figure 21-21). (Deoxythymidylic acid is often referred to simply as thymidylic acid.) In this sequence, methylene-tetrahydrofolic acid is the carrier of the carbon (shown in color in Figure 21-21). The added carbon bridges two of the nitrogen atoms of tetrahydrofolic acid and is at the oxidation level of formaldehyde. To emphasize bond formation and cleavage in subsequent reactions, we abbreviate the structures of tetrahydrofolic acid and methylene-tetrahydrofolic acid as shown in Figure 21-22. The methylene carbon of methylene-tetrahydrofolic acid is functionally similar to that in an acetal in that it is attached to two heteroatoms. Indeed, this functionality, known as an **aminal,** undergoes quite similar reactions to those of an acetal. An aminal is formed by the reaction of two equivalents of an amine (or, as in this case, a diamine) with an aldehyde. Aminals react rapidly with water under both acidic and basic conditions to regenerate the original carbonyl compound (Figure 21-23).

EXERCISE 21-P

Write a mechanism for the formation of an aminal from formaldehyde and ethylenediamine, $H_2N—CH_2CH_2—NH_2$.

EXERCISE 21-Q

Write a clear and complete step-by-step reaction mechanism for the sequence shown in Figure 21-23. Which would you expect to react more rapidly with water: aminals or acetals? Explain your answer clearly.

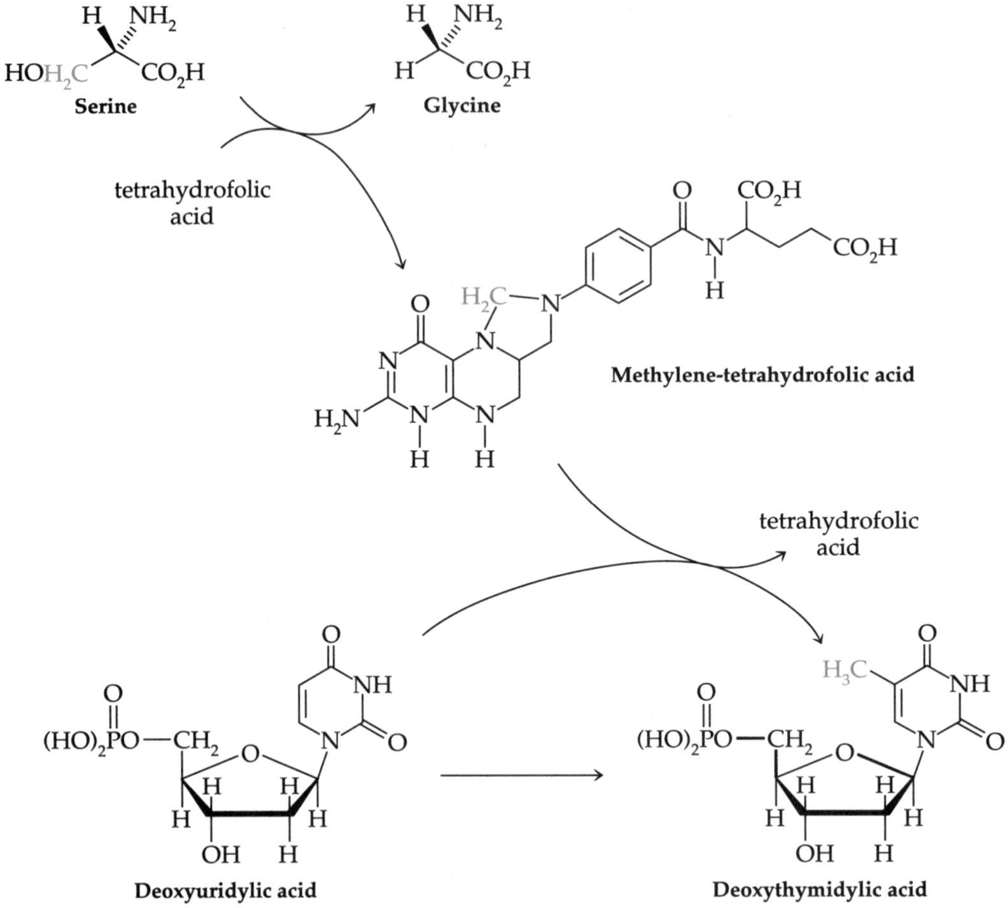

▲ FIGURE 21-21
Serine transfers a single carbon atom as a methylene group to tetrahydrofolic acid. Methylene-tetrahydrofolic acid then relays the —CH₂— unit to deoxyuridylic acid, producing deoxythymidylic acid, which is incorporated directly into DNA.

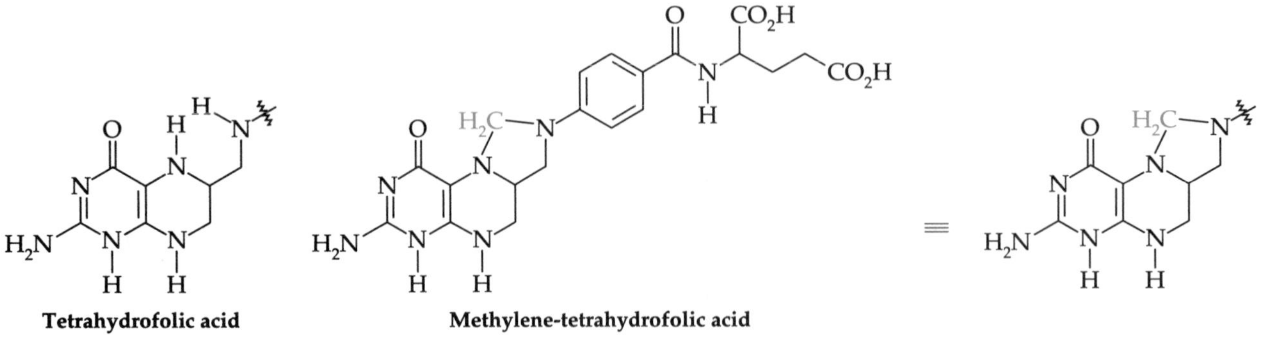

▲ FIGURE 21-22
Abbreviated structures of tetrahydrofolic acid and methylene-tetrahydrofolic acid are used to emphasize the carbon–carbon bond-making and -breaking reactions through which a methylene unit (—CH₂—) is transferred.

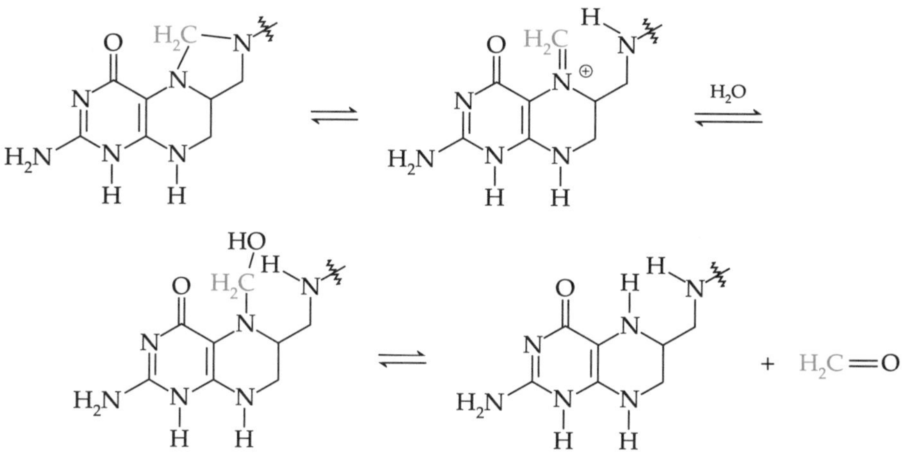

◄ FIGURE 21-23
The aminal shown at the upper left exists in equilibrium with the iminium on and a free amine shown at the upper right. Hydrolysis of the iminium ion releases formaldehyde, while forming a free amine (lower right).

Let us now speculate on how one carbon might be transferred from serine to tetrahydrofolic acid to form methylene-tetrahydrofolic acid. (The details of the actual biological sequence are not yet known.) Overall, two of the bonds to the hydroxymethylene carbon in serine must be replaced by two new bonds to nitrogen atoms of tetrahydrofolic acid. This sequence is initiated by the loss of water, followed by the addition of a ring nitrogen of tetrahydrofolic acid in an elimination-addition sequence that effects a net substitution on serine (Figure 21-24).

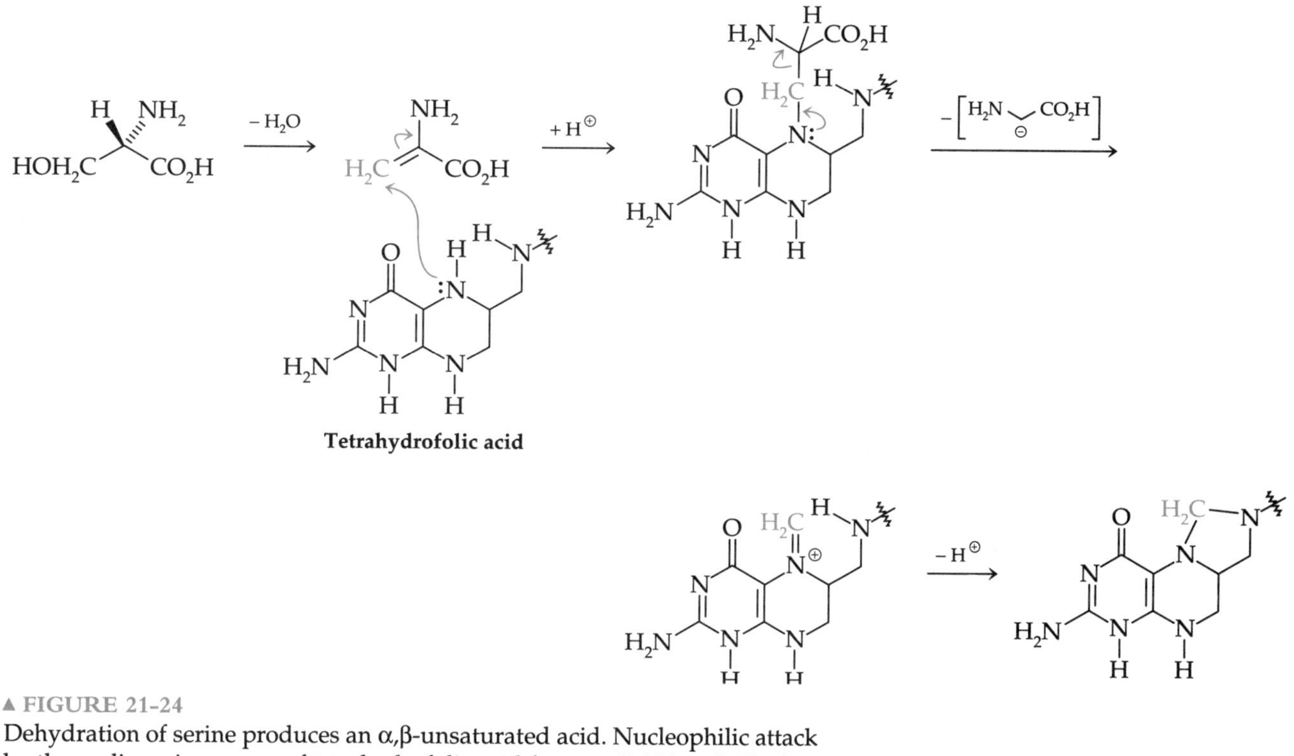

Tetrahydrofolic acid

▲ FIGURE 21-24

Dehydration of serine produces an α,β-unsaturated acid. Nucleophilic attack by the cyclic amino group of tetrahydrofolic acid forms a C–N bond. The lone pair of electrons on nitrogen assists in the loss of the α-amino acid anion equivalent, producing an iminium ion. Nucleophilic attack by the second amino group of tetrahydrofolic acid on the iminium carbon forms the cyclic aminal. A comparison of this product with the starting material (tetrahydrofolic acid) reveals that a methylene unit has been transferred.

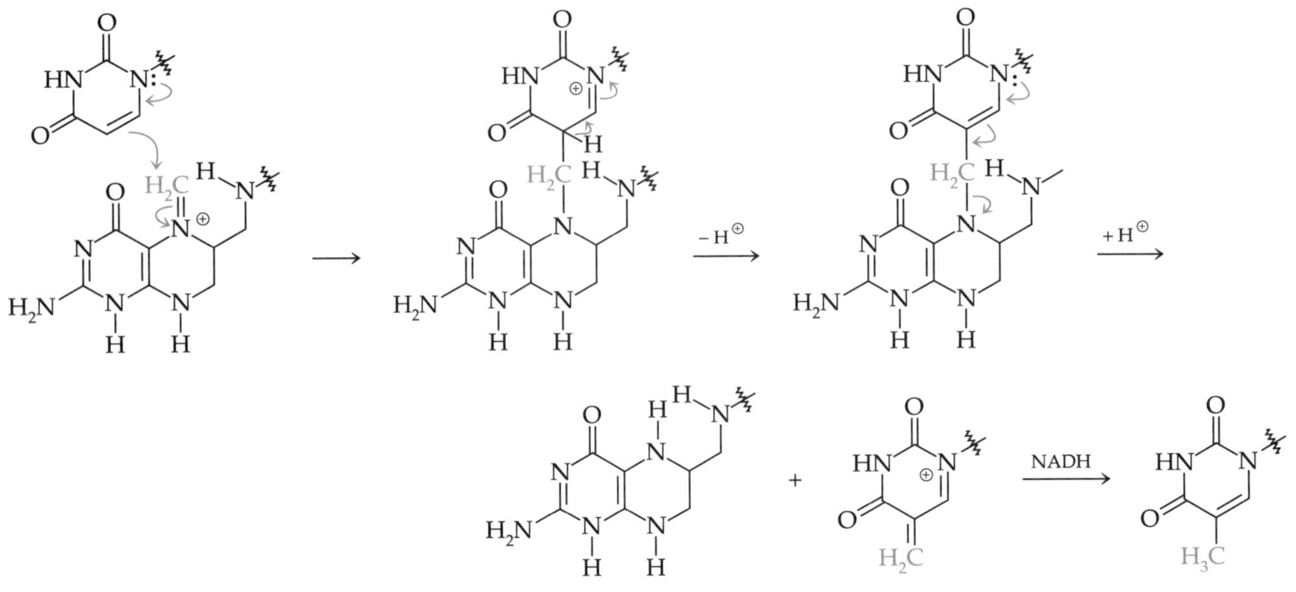

▲ FIGURE 21-25
Intermolecular nucleophilic attack on the iminium ion produced by
methylenation of tetrahydrofolic acid produces a new carbon–carbon bond.
Deprotonation of the resulting iminium salt restores a lone pair on nitrogen
that is then used to displace the tetrahydrofolic acid. Tautomerization of the
resulting salt leads to deoxythymidylic acid.

The carbon–carbon bond can now be cleaved, with the formal loss of
the carbonyl stabilized carbanion. Notice that the iminium ion intermediate
formed at this point is the same functionality as that in the Mannich reaction in Chapter 18. Nucleophilic addition of the second nitrogen to the
methylene carbon, with loss of a proton, completes the sequence.

We can readily reverse the sequence of steps in Figure 21-24 to transfer
the methylene carbon to another site. With deoxyuridylic acid, nucleophilic
addition to the intermediate iminium ion introduces the necessary
carbon–carbon bond for deoxythymidylic acid (Figure 21-25). After deprotonation, the tetrahydrofolic acid residue is lost, and the formal addition of
a hydride equivalent (for example, from NADH) completes the transfer of
carbon and the formation of deoxythymidylic acid. Tetrahydrofolic acid is
responsible for the transfer of one carbon in a number of biological
sequences that result in the introduction of CH_3, CH_2, $CH=$, and CHO
groups. Indeed, glycine is also synthesized (in the liver of vertebrates) from
ammonia, methylene-tetrahydrofolic acid, carbon dioxide, and NADH.

EXERCISE 21-R

An alternative route for forming methylene-tetrahydrofolic acid would
require an S_N2 displacement of H_2N—CH^-—CO_2H by the second nitrogen. Write a detailed reaction mechanism for this pathway. Which factors
favor the route shown in Figure 21-24? Which favor the S_N2 pathway?

Conclusions

Cofactors are important reagents that are essential components of several key biochemical transformations. Although much of the chemistry of cofactors is analogous to that observed for classical laboratory reagents, cofactors are more structurally complex than the reagents presented so far in this book. This complexity is necessary because cofactors must have parts that enable them to recognize their target molecules and themselves to be recognized by enzymes. Their reactions also differ significantly in that they are close to thermoneutral in order that the "spent" cofactor can be converted into its active form for reuse.

The cofactors treated in this chapter often react by pathways similar to those presented in earlier chapters with simpler reagents. Pyridoxamine pyrophosphate accomplishes reductive amination by a route similar to the nucleophilic addition and reduction that we have seen before, and NADH and NADPH are biological reagents that react by pathways quite similar to those followed by the hydride donors LiAlH$_4$ and NaBH$_4$. FADH$_2$ is a biological reducing agent that functions either as a hydride donor or by single electron transfer, the latter pathway bearing similarity to that of a dissolving-metal reduction. Acetyl CoA activates carboxylic acid derivatives by forming a thiol ester by methods parallel to those learned earlier with acid chlorides or anhydrides.

The cofactor thiamine pyrophosphate is responsible for the removal of carbon dioxide in the biodegradation of α-ketoacids. The special function of thiamine pyrophosphate is to invert the normal reactivity pattern of an α-ketoacid to mimic the decarboxylation reaction usually observed with a β-ketoacid. Thiamine's function in this decarboxylation thus places it in the special class of reverse-polarity reagents. Lipoic acid, in its oxidized, disulfide form, reacts in combination with thiamine to effect the overall oxidative decarboxylation of an α-amino acid to a simple carboxylic acid with one fewer carbon. A dithioacetal is a simple laboratory umpolung reagent that similarly inverts the normal chemical reactivity of the carbonyl carbon atom of an aldehyde. Deprotonation of a dithioacetal produces an acyl anion equivalent, which can be trapped by alkylation reactions to produce a dithioketal that can be hydrolyzed to a ketone.

The transfer of a one-carbon unit in the course of biodegradation and biosynthesis can be accomplished through the reversible binding of a methylene group to tetrahydrofolic acid. In an intermediate step, the one-carbon unit is transferred from serine by reversible aminal formation.

Summary of New Reactions

Alcohols

β-Hydroxythiol ester by reduction of a β-ketothiol ester with NADPH in fatty acid biosynthesis

Hydration of an α,β-unsaturated thiol ester in fatty acid degradation

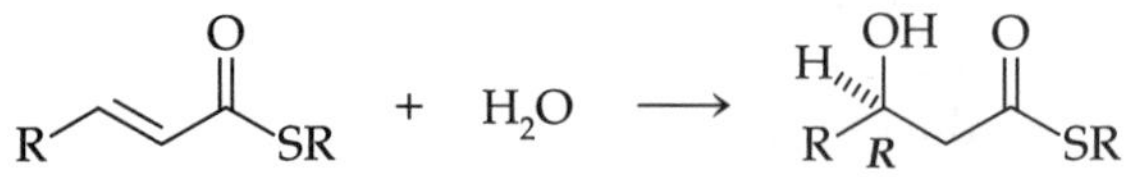

Aldehydes

Decarboxylation of an α-ketoacid catalyzed by thiamine

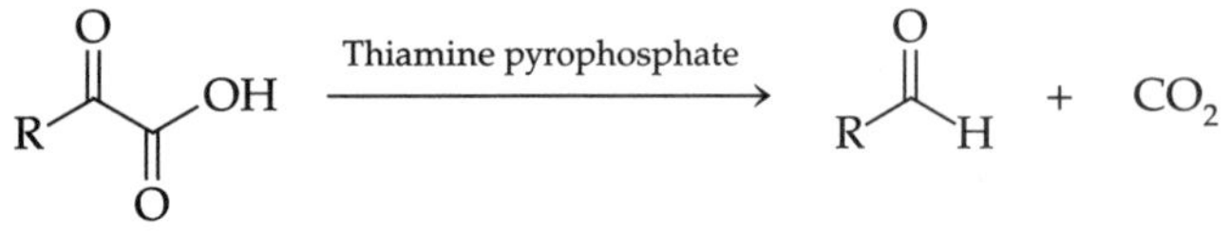

Amines

α-Amino acid by the pyridoxamine-pyrophosphate–mediated reductive amination of an α-ketoacid

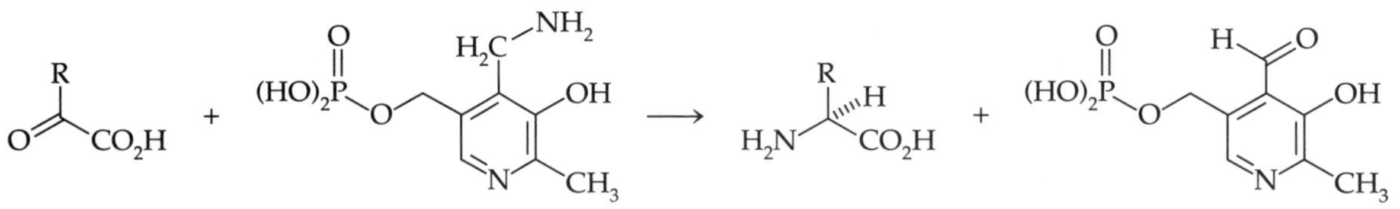

α-Ketoacid   **Pyridoxamine phosphate**   **α-Amino acid**   **Pyridoxal phosphate**

Carboxylic acids

Thiamine-pyrophosphate–catalyzed oxidative decarboxylation of an α-ketoacid with oxidized form of lipoic acid

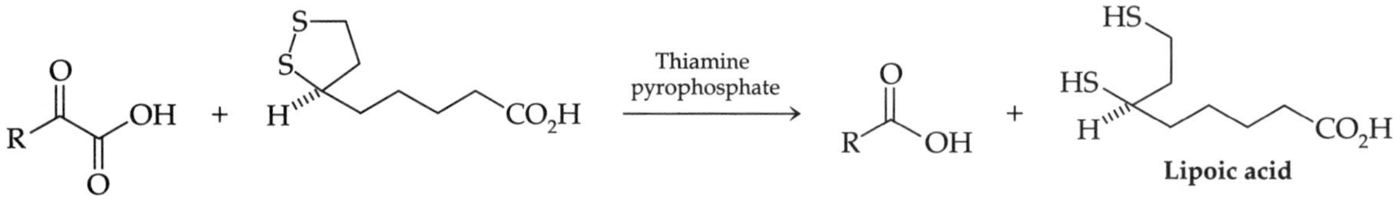

Lipoic acid

Glycine

Dehydroxymethylenation of serine with transfer of CH₂ group to tetrahydrofolic acid, forming methylene-tetrahydrofolic acid

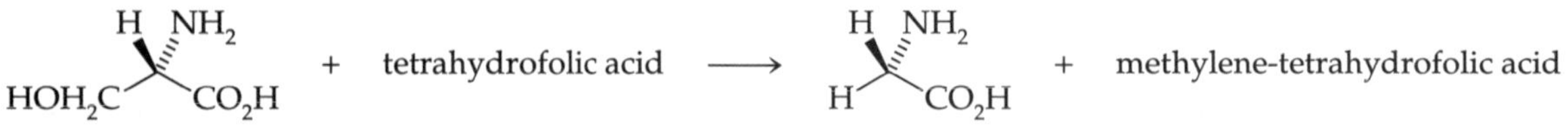

Serine   **Glycine**

Ketones

α-Ketoacid by oxidative deamination of an α-amino acid with pyridoxal phosphate

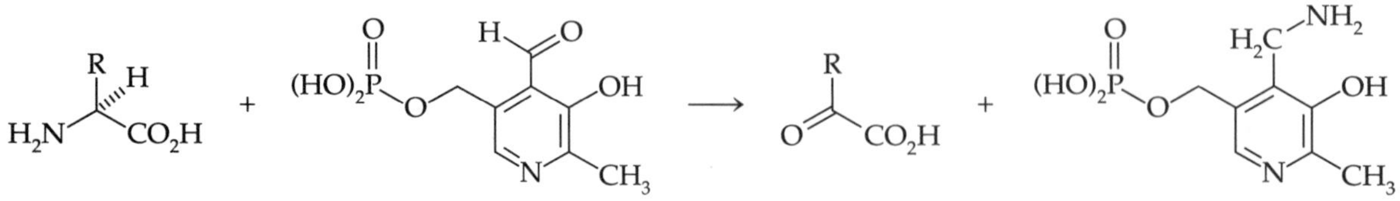

α-Amino acid   **Pyridoxal phosphate**   **α-Ketoacid**   **Pyridoxamine phosphate**

β-Ketothiol ester by oxidation of β-hydroxythiol ester with NAD

Alkylation of a thioacetal anion followed by hydrolysis

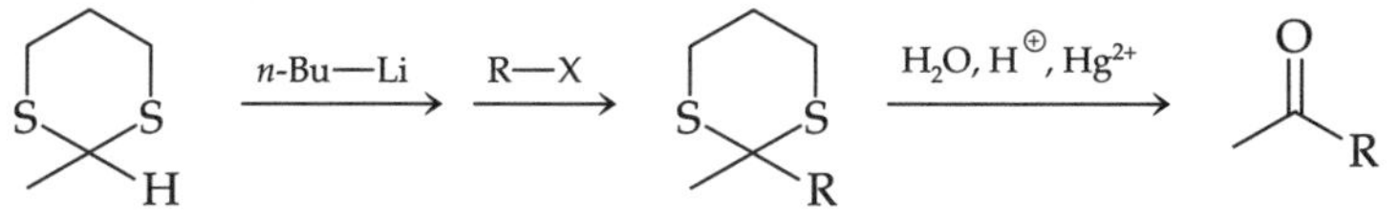

α-Ketoesters by acylation of a thioacetal anion followed by hydrolysis

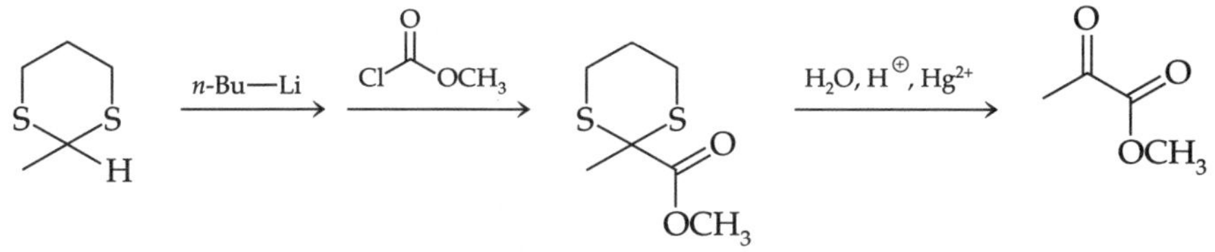

Thiol esters

FADH₂–mediated reduction of an α,β-unsaturated thiol ester

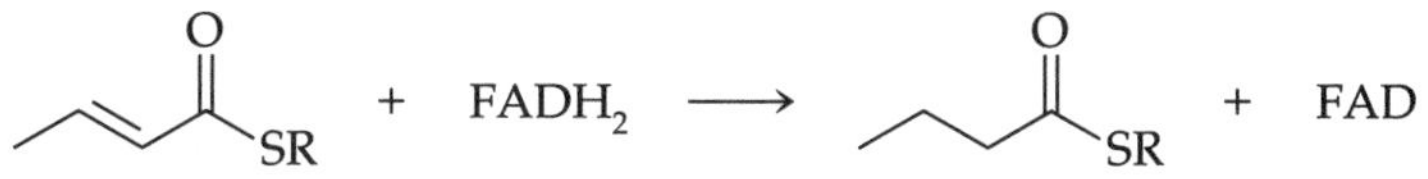

α,β-Unsaturated thiol ester by oxidation of a thiol ester with FAD

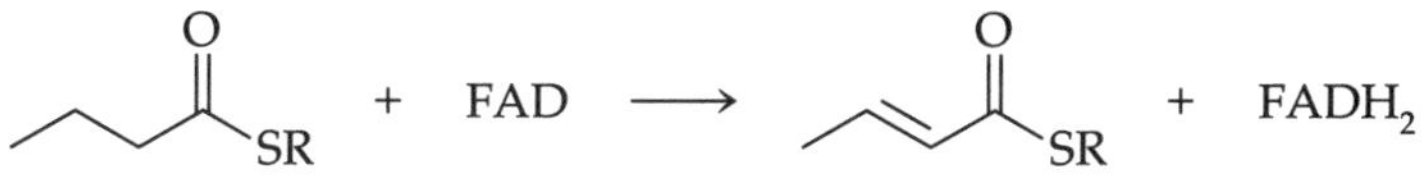

Review Problems

21-1 Classify each of the following reactions as written as: (1) highly exothermic; (2) having an equilibrium near 1; or (3) an endothermic reaction favoring starting materials.

(a)

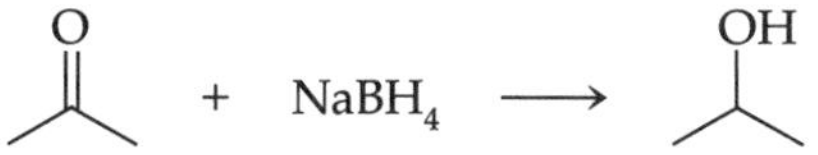

(b)

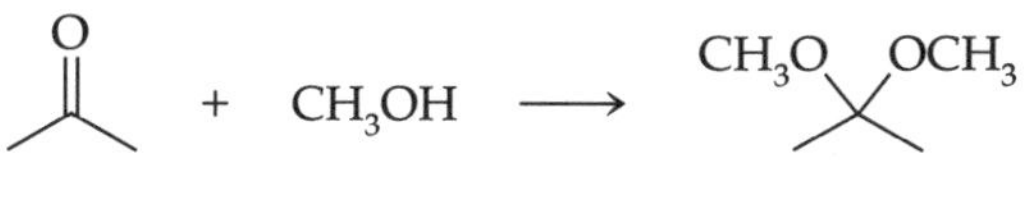

(c)

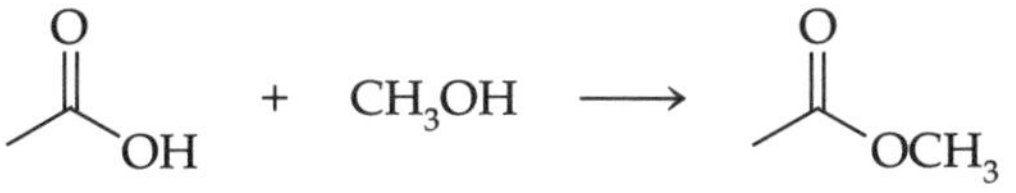

(d)

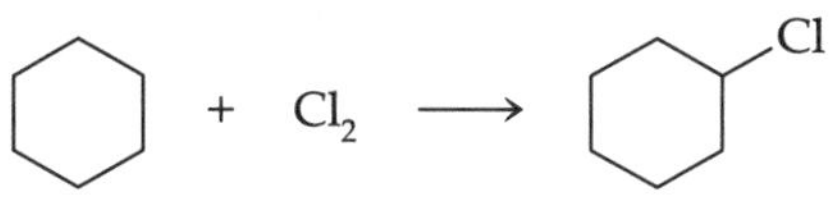

(e)

(f)

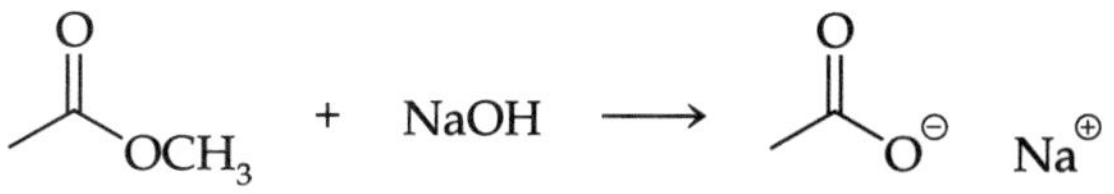

21-2 Hydroquinone is an important antioxidant that readily traps oxidizing reagents, especially radicals, by undergoing oxidation to form quinones. Use a radical (R·) to remove a hydrogen atom from one oxygen of hydroquinone and analyze with resonance structures how this radical is highly delocalized and therefore stabilized. Now remove the other hydroxyl hydrogen, providing a reasonable set of arrows (remember, half-headed) that produces quinone. (See chemical structures at the top of the next page.)

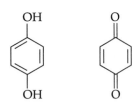

Hydroquinone   **Quinone**

21-3 Butylated hydroxytoluene (BHT) is an important antioxidant for the food industry that has been approved by the Federal Drug Administration for human consumption. Compare the structure of this phenol with that of hydroquinone, noting the similarities and the differences.

**Butylated hydroxytoluene
(BHT)**

21-4 The oxidation of FADH$_2$ to FAD can take place through radical intermediates. Use the ideas that you developed in solving Problem 21-2 to write a mechanism for this oxidation, assuming that the two hydrogens (shown in color) removed in forming FAD from FADH$_2$ are sequentially abstracted by two radicals, R$^{\cdot}$.

FADH$_2$

$\downarrow$ 2 R$^{\cdot}$

FAD

21-5 Compare the structures of toluene and *N*-methylpyridinium ion, paying particular attention to the arrangement of electrons. Draw resonance structures for the ion that do not have positive charge on nitrogen, noting which carbons of the nucleus bear charge in the hybrid structure.

Toluene   ***N*-Methylpyridinium ion**

21-6 Thiol esters are more reactive than their oxygen analogs because of reduced resonance overlap between the sulfur atom and the carbonyl π system. Thione esters differ from thiol esters in the position of the sulfur atom. How would you expect thionesters to differ in stability and reactivity from thiol esters and from "normal" carboxylic acid esters?

Thiol ester   **Thione ester**

21-7 In some ways, the reactions of tetrahydrofolic acid in the transfer of one carbon resemble the steps in the Mannich reaction (Chapter 18). Review the Mannich reaction and then compare it with the steps in methylene transfer described in this chapter. Which of the steps are comparable?

21-8 Analyze the patterns of reactivity imparted by each of the functional groups in the following bifunctional molecules. In which cases would it be possible to form the molecules shown by a connection of nucleophilic and electrophilic carbons if this reactivity were consistent with that imparted by the functional groups?

(a)

(b)

(c)

(d)

Chapter 22

Energy Storage
in Organic Molecules

In this chapter, we will deal specifically with the chemical properties of lipids and carbohydrates (whose structures were discussed in Chapter 17), especially as they relate to how these natural materials store chemical energy. We will see how these compounds are broken into smaller and smaller pieces, ultimately forming carbon dioxide, water, and heat. In addition, some of these reactions are also coupled to the conversion of a diphosphate into a triphosphate, a process that by itself is approximately 7 kcal/mole endothermic. Thus, we will find that the degradation of lipids and carbohydrates serves two functions: (1) to produce thermal energy; and (2) to supply stored, chemical potential energy. Throughout the chapter, we will find a close correspondence between biochemical transformations and the laboratory equivalents that we have studied in earlier chapters.

22-1 ▸ Reaction Energetics

We begin by reviewing some basic concepts of reaction kinetics and thermodynamics presented in Chapter 6 so that we can apply these simple concepts to the more-complicated, multistep processes of biochemical transformations. However, because we will be dealing with relatively small energy differences in this chapter, we will include the effects of entropy, using free energy, $G°$, rather than enthalpy, $H°$ (recall that $G° = H° - TS°$). Consider an energy diagram relating two species, A and B, in which A is higher in energy (and therefore less stable) than B (Figure 22-1). The state of equilibrium is defined as that condition in which the rate of conversion of A into B is precisely balanced by an equal rate of conversion of B into A. In this example, the activation energy for the forward reaction (A → B), $\Delta G^{\ddagger}_{f}$, is less than that for the backward reaction (B → A), $\Delta G^{\ddagger}_{b}$. Because the rate is proportional to $\exp(-\Delta G^{\ddagger}/RT)$, the intrinsic rate of the process that converts A into B is higher than the rate of conversion of B into A. To maintain equilibrium, the concentration of species A must be lower than that of B.

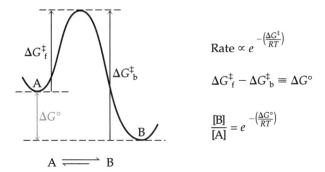

$$\text{Rate} \propto e^{-\left(\frac{\Delta G^{\ddagger}}{RT}\right)}$$

$$\Delta G^{\ddagger}_f - \Delta G^{\ddagger}_b \equiv \Delta G^{\circ}$$

$$\frac{[B]}{[A]} = e^{-\left(\frac{\Delta G^{\circ}}{RT}\right)}$$

▲ FIGURE 22-1

In all reactions, the rate is controlled by the magnitude of the activation energy, $\Delta G^{\ddagger}$. In a reaction in which the number of reactant and product molecules is the same, and in which entropy effects (ΔS°) are assumed to be negligible, the enthalpy of reaction (ΔH°) is given by the difference in activation energies for the forward and reverse reactions. The equilibrium constant, which describes the ratio of the concentrations of the product B to the reactant A at steady state, is dependent on the magnitude of the free-energy difference ΔG° ($\Delta G^{\circ} = \Delta H^{\circ} - T\Delta S^{\circ}$).

Indeed, because

$$K = \frac{[B]}{[A]}$$

and

$$\Delta G^{\circ} = -RT \ln K$$

the ratio of concentrations of A and B can be expressed as $\exp -(\Delta G^{\circ}/RT)$ in which $\Delta G^{\circ} = \Delta G^{\ddagger}_f - \Delta G^{\ddagger}_b$ and in which ΔG° is always uniquely identical with the difference in the free energies of activation for the forward and backward reactions.

Although some simple organic transformations take place with a single starting material being converted into a single product, most transformations require a separate reagent and result in more than one product. For example, the oxidation of an alcohol to a ketone requires an oxidizing agent such as Cr^{6+}, which itself is ultimately transformed in the reaction into Cr^{3+}. It then becomes impossible to speak of the thermodynamics of the transformation of an alcohol into a ketone solely on the basis of the stabilities of the two organic species, the reactant and the product.

Indeed, an alcohol and a ketone cannot be directly related because they are not isomeric: the ketone lacks two of the hydrogens present in the alcohol. At best, we can compare the relative stabilities of the reactants—the alcohol combined with the oxidizing agent—and the products—the ketone and the reduced form of the oxidant. We cannot speak of the thermodynamics of the A-into-B conversion as being independent of the C-into-D conversion: the energetics of the reaction requires a coupling of the two transformations. In Figure 22-2 (on page 748), these additional species have been added to the energy diagram. The concentrations of the two products (B and D) and the two starting materials (A and C) are related by the same equation as that used for the simpler case (Figure 22-1).

ORGANIC POLYMERS IN BATTERIES

The energy that we use in many areas of daily life is stored in various materials in the form of reduction potential that is released as substantial quantities of heat upon combustion with molecular oxygen. Both natural gas (methane) and gasoline (a complex mixture of organic hydrocarbons) release relatively large quantities of energy on burning.

$$CH_4 + 2 O_2 \longrightarrow CO_2 + 2 H_2O \qquad \Delta H° = -213 \text{ kcal/mole}$$

There are other ways in which organic molecules can participate in an energy-storage system. For example, the recently developed lithium batteries are composed of three layers: an oxidized form of polyaniline; an intervening polymer layer rich in water; and a layer of lithium metal. When the first layer is connected with the third by an external circuit, electrons flow from lithium through the circuit to the polyaniline layer, effecting oxidation of the lithium and reduction of the polymer. Concurrently, lithium ions and the anion associated with the polymer migrate into the intervening water-rich layer. Electric current will flow until all of the lithium has been oxidized and all of the polymer reduced. The process can be reversed by applying a potential in the opposite direction: the lithium ions are reduced to the metal and the polyaniline layer is oxidized.

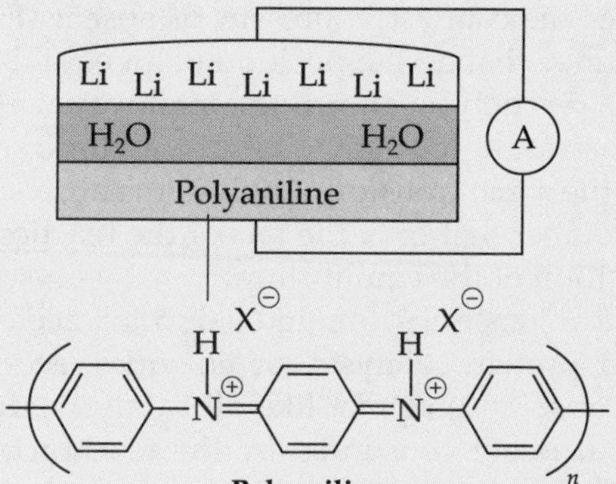

Polyaniline

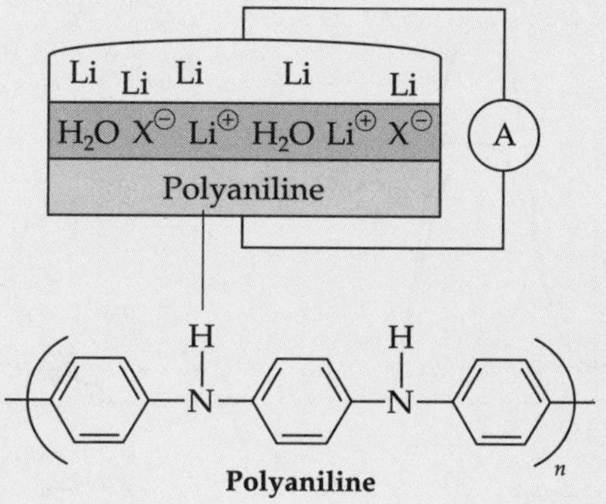

Polyaniline

Thus, a reaction profile provides important information about kinetics and thermodynamics. Because biological systems efficiently accelerate specific chemical reactions, we now consider how a reaction rate can be influenced by the addition of a catalyst and compare the efficiency of a multistep process with a concerted, single-step conversion.

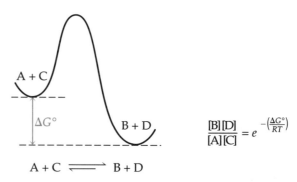

▲ **FIGURE 22-2**

In a bimolecular reaction between A and C to produce B and D, the equilibrium constant can be expressed as a product of the concentrations of the products to those of the reactants. As in the unimolecular reaction depicted in Figure 22-1, the position of the equilibrium is dependent on the magnitude of the difference in free energy of the reactants and products $\Delta G°$.

Catalysis

What is the effect of a catalyst on the energetics of a reaction? Recall that, by definition, a catalyst is a species that increases the rate of a reaction but does not itself undergo a net chemical transformation. Because the catalyst is not changed, the difference in energy between starting materials and products is the same, whether the catalyst is present or absent (Figure 22-3). Therefore, the catalyst cannot change the position of an equilibrium: nonetheless, it does influence the reaction by affording a pathway with a transition state of lower energy. As a result, the activation energy for a catalyzed process is lower than that for the same transformation in the absence of the catalyst. It is important to remember that only the *rate* of the reaction is affected by the catalyst, not the position of the equilibrium.

Most biochemical reactions are indeed catalyzed processes; a large number of different protein catalysts, or enzymes, accelerate the rates of these reactions. Because an enzyme, like all catalysts, does not undergo a net transformation, it is not consumed by the reaction and therefore need not be present in stoichiometric amounts. Nonetheless, the rate of the reac-

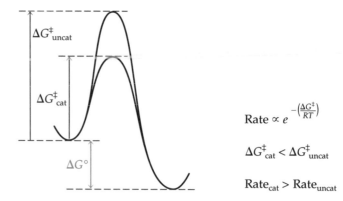

▲ **FIGURE 22-3**

A catalyst influences the magnitude of the activation energies for the forward and reverse reactions but does not affect the potential energies of the reactants or products. A catalyst therefore influences the rate at which an equilibrium is established but not its position. In lowering the activation energy barrier for the forward reaction, a catalyst accelerates the rate of formation of product.

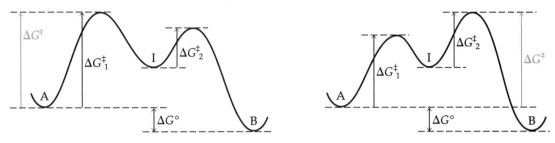

▲ FIGURE 22-4
In a two-step sequence taking place through an intermediate, I, either
the first step (shown in the reaction profile at the left) or the second step
(shown at the right) can be rate determining.

tion is proportional to the concentration of the catalyst (as well as each of
the reactants), and the rate of reaction decreases as the amount of the catalyst is reduced.

Enzymes are important not only because they can accelerate biochemical reactions, but also because the rate at which product is produced is controlled directly by changes in the concentration of the enzyme. The effect of
the enzyme, like all catalysts, is to provide a transition state with a lower
energy: this has an identical effect on the energies of activation (and therefore the rates) of both the forward and the backward reactions. This is an
important point to keep in mind because, in biochemical reactions, some of
the pathways for the synthesis of complex molecules are the same as those
used to degrade these compounds.

Multistep Transformations

The second important point relating to reaction energetics concerns the consequence on reaction rates of multiple steps. Figure 22-4 shows two possible
scenarios for a two-step reaction that takes place through the formation of
an intermediate that is higher in energy than either the starting material or
the product. Although there are two discrete steps in each of these reac-
tions, and therefore two activation energies ($\Delta G^{\ddagger}_1$ and $\Delta G^{\ddagger}_2$), the rate of the
reaction is determined solely by the amount of energy required ($\Delta G^{\ddagger}$) to
proceed from the starting material(s) to the *highest* point on the energy diagram. This energy of activation is equal to $\Delta G^{\ddagger}_1$ in the first example but is
equal to neither $\Delta G^{\ddagger}_1$ nor $\Delta G^{\ddagger}_2$ in the second. In both examples, the equilibrium positions of the overall reactions are determined solely by $\Delta G°$ and are
not influenced by the shapes of the energy curves or by the presence or
absence of relatively high energy intermediates.

In multistep biochemical sequences, one or more endothermic steps are
often tolerated in a process that is favorable overall because of one or more
exothermic steps. The last step must be exothermic if the sequence is to
yield significant quantities of the ultimate product.

22-2 ▸ Complex Reaction Cycles

Most reactions discussed so far have been written as in equation 1, in which
the required reagent, C, is placed above the arrow.

$$A \xrightarrow{C} B \qquad\qquad (1)$$

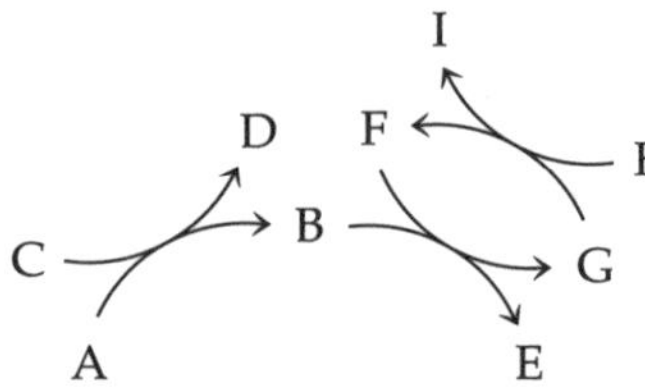

▲ FIGURE 22-5

In many biochemical transformations, several reactions are coupled, with the product of one reaction being used as a reactant in another. Here, for example, H is converted into I with the simultaneous conversion of G into F. The production of G from F requires a coupled reaction of B into E, with B having been produced from A when coupled with the conversion of C into D. If any of these reactions is interrupted, the cycle stops.

$$A + C \longrightarrow B + D \tag{2}$$

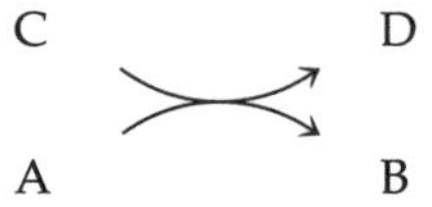

$$(3)$$

However, when the reagent is itself consumed, a more-accurate representation is that in equation 2, in which both the reagent, C, and the product, D, derived from it are explicitly included. In depicting biochemical reactions, it is often more convenient to use the method shown in equation 3, in which the proximity of the two arrows indicates that the transformation of C into D is intimately coupled with (and required for) that of A into B. This representation is particularly convenient for showing reactions in which several transformations are coupled together, as illustrated in Figure 22-5. As in Chapter 12, the curved arrows used to indicate coupled reactions have a different meaning from those used to indicate electron flow in the mechanisms considered in the first half of the book. Here, the curved arrows are used to emphasize the link between coupled reactions. For example, the net reaction of A into E in Figure 22-5 takes place in two steps, the first simultaneously converting C into D as A is converted into B; and the second converting F into G as B is transformed into E. Species G is converted back into F, for another cycle, as H is transformed into I. With such complex sequences, it is important to keep in mind that it is the overall energetics of *all* of the transformations taken together that determines the position of the equilibrium. In the example shown in Figure 22-5, a highly exothermic conversion of G and H into F and I can compensate for the endothermicity of all of the preceding steps.

Many biosynthetic sequences are used reversibly either to prepare or to degrade biological materials. Clearly, the forward and backward reactions cannot both be exothermic. The endothermic process must be coupled with one or more other transformations that are sufficiently exothermic to compensate for the endothermicity of the desired reaction step.

22-3 ▶ Energy Storage in Anhydrides

Three phosphorylated species are critical in **biochemical energy storage** (Figure 22-6). Each of these molecules is based on the adenine-ribose unit present in the cofactors NAD and NADP (Chapter 18) and contains one, two, or three phosphoric acid units bound to the 5′ carbon of ribose. In

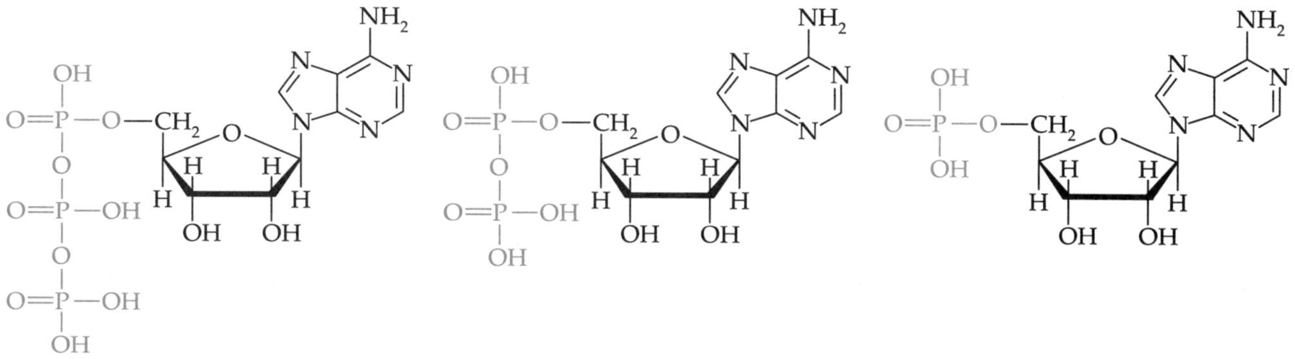

Adenosine triphosphate (ATP)

Adenosine diphosphate (ADP)

Adenosine monophosphate (AMP)

▲ FIGURE 22-6

The mono-, di-, and triphosphate esters of adenine are principal energy-storage molecules for all biological systems.

adenosine diphosphate (ADP) and **adenosine triphosphate (ATP),** the additional phosphate groups are linked by anhydride bonds that are effective stores of chemical potential energy.

The **phosphoric acid anhydride** functionality present in di- and triphosphates is quite similar to the carboxylic acid anhydride in that they are hydrolytically unstable relative to two separate acid units. In effect, this means that both of the hydrolysis reactions illustrated in Figure 22-7 are exothermic. Thus, we would expect the hydrolysis of the phosphoric acid anhydride units in ADP and ATP also to be exothermic. Indeed, the hydrolysis of ATP to ADP (and phosphate) is exothermic by 7.0 kcal/mole and the hydrolysis of both anhydride bonds in the conversion of ATP into AMP produces 16.3 kcal/mole of energy.

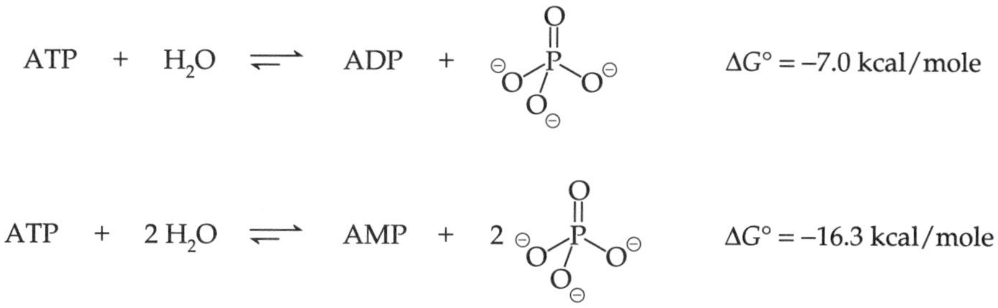

$$\text{ATP} + \text{H}_2\text{O} \rightleftharpoons \text{ADP} + \text{phosphate} \qquad \Delta G° = -7.0 \text{ kcal/mole}$$

$$\text{ATP} + 2\,\text{H}_2\text{O} \rightleftharpoons \text{AMP} + 2\,\text{phosphate} \qquad \Delta G° = -16.3 \text{ kcal/mole}$$

For this reason, these phosphoric acid anhydride units are known as **high-energy phosphate bonds.**

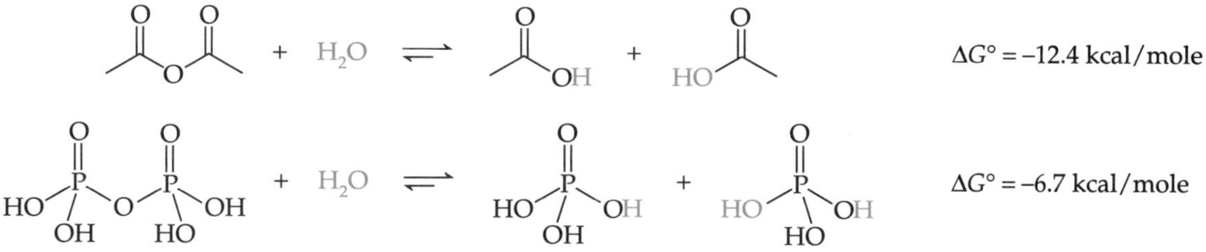

$$\Delta G° = -12.4 \text{ kcal/mole}$$

$$\Delta G° = -6.7 \text{ kcal/mole}$$

▲ FIGURE 22-7

The hydrolysis of a diphosphate (lower reaction) bears close chemical similarity to that of a carboxylic acid anhydride (upper reaction).

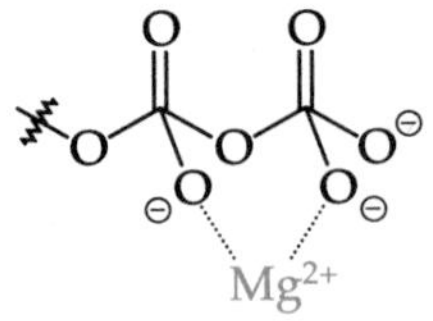

The exact amount of energy released upon hydrolysis of ATP to either ADP or AMP depends on the experimental conditions used. Such factors as the acidity of the solution and the presence of cations (especially divalent ones such as magnesium) have small, but very real, effects on the equilibrium position. For example, although all three species are represented as fully protonated in Figure 22-6, the phosphoric acid units of ATP, ADP, and AMP are essentially completely deprotonated at pH 7. With monovalent cations such as sodium and potassium as counterions for the phosphate units, the energy of the phosphoric anhydride units is increased by repulsion between the negatively charged oxygen anions. This is one reason why the value for the conversion of ATP into ADP (–7.0 kcal/mole) is larger than that for the conversion of pyrophosphate into phosphoric acid and why the hydrolysis of both anhydride linkages (ATP ⟶ AMP) releases more than twice the energy of one (16.3 versus 7.0 kcal/mole). On the other hand, in the presence of Mg^{2+}, the anhydride linkage is slightly stabilized by the bridge formed by the metal dication between the oxygen anion of one phosphate subunit and that of another. In no case, however, do these factors reverse the energetics: the process is always exothermic in the direction of cleavage of the phosphoric acid anhydride units. We will see throughout this chapter that the conversion of ATP into ADP is a source of chemical energy used to compensate for the endothermicity of other transformations to which it is coupled.

EXERCISE 22-A

The interconversion of various derivatives of phosphoric acid is believed to take place by means of a nucleophilic addition-elimination mechanism. Write such a mechanism for the hydrolysis of trimethylphosphate in aqueous acid, forming first dimethylphosphate and then methylphosphate. Now, do the same for hydrolysis with aqueous sodium hydroxide. Be careful to consider the state of protonation of each intermediate and product as it would exist in an alkaline medium.

Trimethyl-phosphate **Dimethyl hydrogen phosphate**

Methyl dihydrogen phosphate

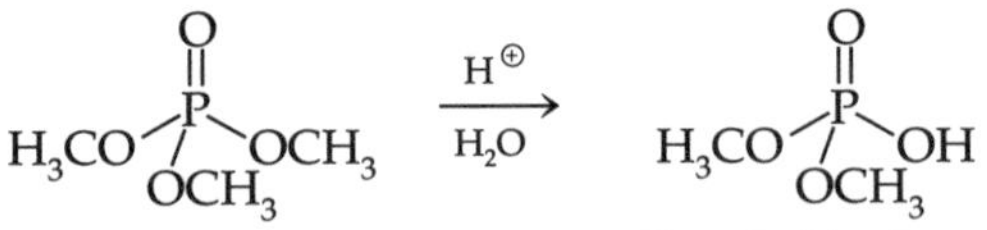

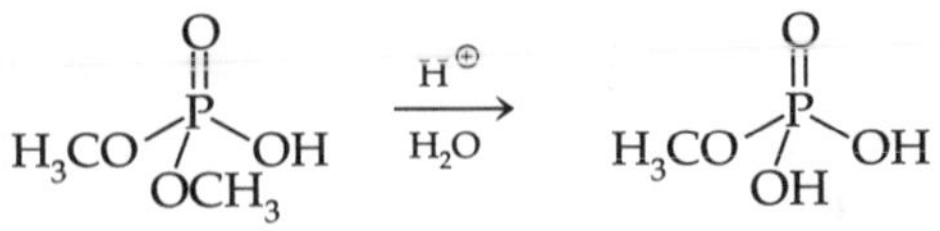

22-4 ▸ Energy Storage in Redox Reactions

A second form of potential energy is stored whenever NAD (or NADP) is reduced to NADH (or NADPH). Nature frequently uses oxidation-reduction couples such as NAD-NADH to store potential energy. For example, green plants effect the conversion of carbon dioxide and water into glucose (and other carbohydrates) and molecular oxygen under the influence of light.

$$6\,CO_2 \;+\; 6\,H_2O \quad \underset{\text{Respiration}}{\overset{\text{Photosynthesis}}{\rightleftharpoons}} \quad C_2H_{12}O_6 \;+\; 6\,O_2 \qquad \Delta G° = +686 \text{ kcal/mole}$$

Carbon dioxide is the highest possible oxidation level for carbon and thus the reaction as written is a reduction of the carbon of carbon dioxide to those of glucose in the process known as **photosynthesis.** The overall reaction achieves the storage of solar energy in carbon–hydrogen and carbon–carbon chemical bonds.

In the formation of glucose (the forward reaction), this complex transformation, which proceeds by many steps, is very endothermic. Conversely, the reverse reaction (oxidation of glucose to carbon dioxide by bond cleavage, a process called **respiration**) is exothermic and constitutes a major source of both chemical and simple thermal energy for living systems. Here, we can think of CO_2 reduction (photosynthesis) as a reaction that leads to a high-energy product (a sugar), which later releases its energy when the reverse process of oxidation (respiration) is carried out.

The conversion of a ketone into an alcohol by reduction with NADPH is shown in Figure 22-8 as a coupled reaction. Typically, the oxidation of NADPH to NADP coupled to the reduction of a ketone to an alcohol is, overall, an exothermic reaction. We cannot say that the exothermicity is the result of the conversion either of the ketone into the alcohol or of NADPH into NADP: it is the result of both transformations and requires that all four of the species take part. The reverse reaction, the reduction of NADP with the concomitant oxidation of the alcohol to the ketone, must be an endothermic reaction. However, the conversion of an alcohol into a ketone, when effected by Cr^{6+}, is an exothermic reaction. The energetics of the interconversion of a ketone and an alcohol depend not only on these two species, but also on the reagent(s) that effects the redox reaction.

Biological systems do not use Cr^{6+} as an oxidant: in fact, Cr^{6+} compounds are generally quite toxic because they interfere with enzymes that require other metal ions for their activity. Rather, biological systems ultimately depend on molecular oxygen, O_2, to achieve such oxidations. The reaction of molecular oxygen as a biochemical oxidant is very complicated, especially in its early stages. Although many of the mechanistic features of

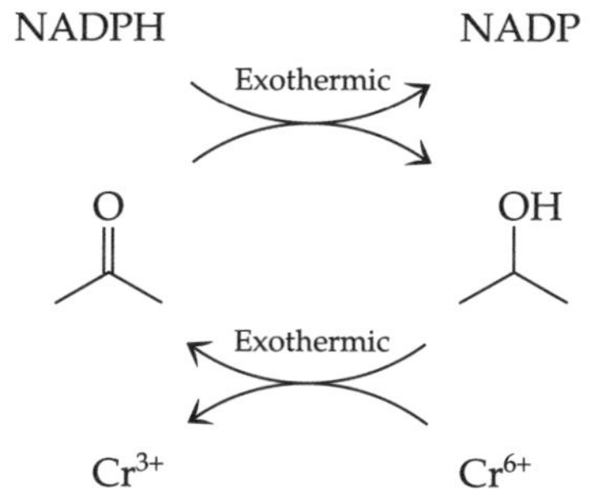

▲ FIGURE 22-8

The redox interconversion of acetone and 2-propanol can be coupled to a complementary reaction of a reducing agent. For example, the reduction of acetone can be driven by the oxidation of NADPH or the oxidation of 2-propanol can be driven by the reduction of Cr^{6+}. Either reaction is exothermic because of the difference in redox potential of the coupled reagent.

$$\text{NADH} \;+\; \text{H}^{\oplus} \;+\; \tfrac{1}{2}\,\text{O}_2 \;\longrightarrow\; \text{NAD} \;+\; \text{H}_2\text{O} \qquad \Delta G^\circ = -52.4\ \text{kcal/mole}$$

$$\text{FADH}_2 \;+\; \tfrac{1}{2}\,\text{O}_2 \;\longrightarrow\; \text{FAD} \;+\; \text{H}_2\text{O} \qquad \Delta G^\circ = -36.2\ \text{kcal/mole}$$

▲ **FIGURE 22-9**

Both NADH and FADH$_2$ undergo an exothermic oxidation with molecular oxygen.

this process have been established, a detailed description of these reactions is more appropriate to a course in biochemistry. Nonetheless, we can abbreviate the process to the very simple scheme in Figure 22-9, in which NADH is oxidized by molecular oxygen to NAD in a process that is exothermic by 52.4 kcal/mole. Likewise, the oxidation of FADH$_2$ by molecular oxygen releases 36.2 kcal/mole of energy.

22-5 ▶ Energy Storage in Fatty Acid Biosynthesis

Let us now apply what we have learned about simple biochemical energetics to see how the synthesis of fatty acids achieves the storage of potential energy. The direct oxidation of fatty acids is an exothermic process: for example, the burning of **palmitic acid** in air (Figure 22-10) releases a large quantity of energy. Conversely, the conversion of carbon dioxide into a fatty acid must be endothermic. We now examine in some detail how this process occurs in nature.

Fatty acid biosynthesis is a process that builds long-chain fatty acids from acetic acid. The formation of carbon–carbon bonds is an intrinsic part of the process, as are the reductions required to replace most of the oxygens present in the acetic acid building blocks with hydrogens.

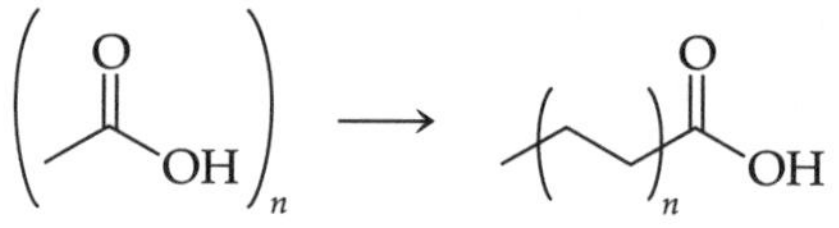

Carbon–Carbon Bond Formation

Carbon–carbon bond formation in fatty acid biosynthesis is very similar to the Claisen condensation discussed in Chapter 13 (Figure 22-11). However,

$$\text{CH}_3(\text{CH}_2)_{14}\text{CO}_2\text{H} \;+\; 23\ \text{O}_2 \;\longrightarrow\; 16\ \text{CO}_2 \;+\; 16\ \text{H}_2\text{O} \qquad \Delta G^\circ = -2386\ \text{kcal/mole}$$
Palmitic acid

▲ **FIGURE 22-10**

The combustion of a fatty acid in air is highly exothermic.

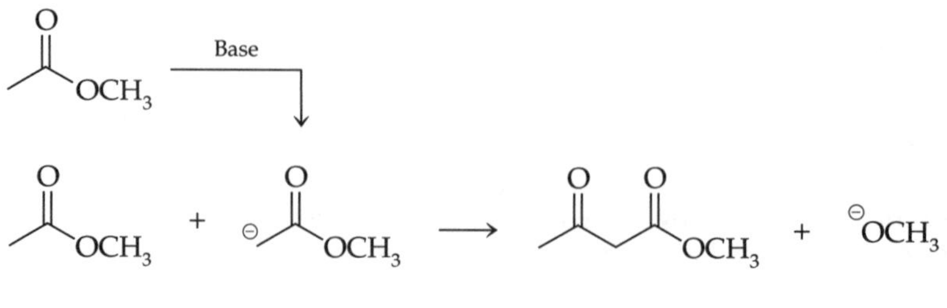

▶ **FIGURE 22-11**

A Claisen condensation is accomplished when an ester enolate reacts as a nucleophile with a neutral ester, effecting nucleophilic acyl substitution and producing a β-ketoester.

there are several differences between the biosynthetic pathway for carbon–carbon bond formation and the laboratory-based condensation reaction. The first is that, instead of oxygen esters, **thiol esters** are involved, presumably because thiol esters are significantly more reactive toward nucleophilic acyl substitution than are their oxygen analogs, as we learned in Chapter 21.

EXERCISE 22-B

Explain clearly the chemical rationale for the higher reactivity of a thiol ester compared with a carboxylic acid ester toward nucleophilic acyl substitution.

The reaction of a thiol with a carboxylic acid to produce a thiol ester and water, as illustrated in Figure 22-12, is endothermic and must therefore be coupled with an exothermic reaction if it is to take place. The reaction of water with ATP to produce AMP (and the pyrophosphate ion) results in the cleavage of a high-energy bond. Because the latter process is more exothermic than the former is endothermic, the combination to produce a thiol ester and AMP is energetically favorable.

The second difference from the classical Claisen reaction is that the nucleophilic partner in this nucleophilic acyl substitution is a species derived from a thiol ester of acetic acid by carboxylation at the α carbon (Figure 22-13). This carboxylation is not well understood, and so we do not deal with it in detail. It is known, however, that **biotin (vitamin H)** participates as a temporary carrier of carbon dioxide and that the overall process consumes one molecule of ATP. The condensation of this nucleophile (perhaps in its enol form) with the CoA ester of acetic acid is illustrated in Figure 22-13. This reaction initially forms a β-carboxy-β-ketoester. Loss of carbon dioxide from this tricarbonyl compound is a facile, exothermic process

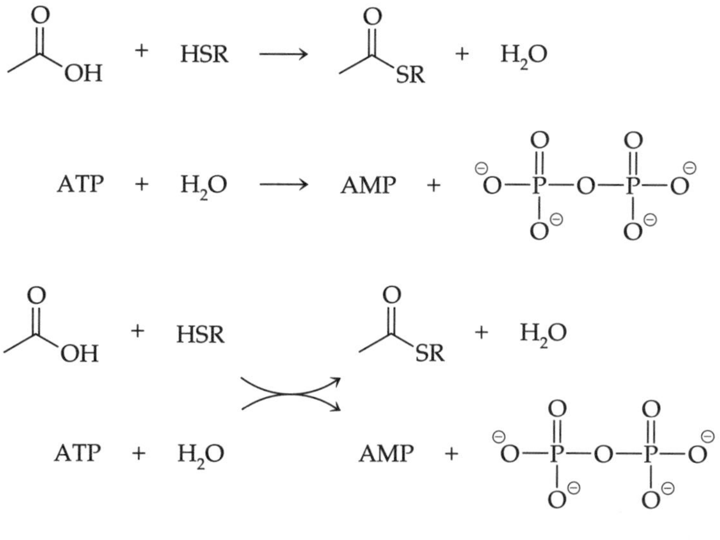

$$\Delta G° = (-10 \text{ kcal/mole}) + (+7.5 \text{ kcal/mole}) = -2.5 \text{ kcal/mole}$$

▲ **FIGURE 22-12**
The endothermicity of the formation of a thiol ester from a carboxylic acid is compensated by the exothermicity of the hydrolysis of ATP into AMP when these reactions are coupled.

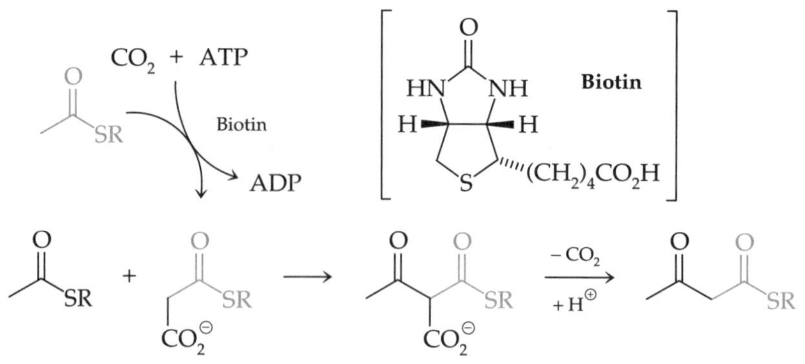

▲ FIGURE 22-13

Carboxylation of a thiol ester by biotin, installs a second carboxylic acid functional group into the thiol ester. By analogy with a malonic ester, this compound is activated toward the formation of either an enol or an enolate anion, which initiates the Claisen condensation. The new carbon–carbon bond of the Claisen condensation having been formed, the —CO_2^- group is no longer needed and is lost as carbon dioxide. This last decarboxylation step takes advantage of the presence of two β carbonyl groups that assist in the loss of CO_2.

that helps to make the overall process more exothermic than it otherwise would be. Thus, the chemical energy of ATP is stored temporarily in the carboxylated thiol ester of acetic acid and is released in a subsequent step by loss of carbon dioxide.

We can think of carbon dioxide and biotin as catalysts: overall, two molecules of acetic acid thiol ester and one molecule of ATP are consumed. This process can be repeated to build more-complex chains, with two carbons being added in each cycle.

EXERCISE 22-C

Write a mechanism for the formation of the intermediate β-carboxy-β-ketoester shown in Figure 22-13. What is the chemical function of the carboxylation step? (Hint: Recall the relative acidities of simple esters and malonic esters given in Chapters 6 and 8.)

EXERCISE 22-D

Explain why the decarboxylation of the intermediate β-carboxy-β-ketothiol ester in Figure 22-13 is relatively fast, and write a mechanism for that conversion. (Hint: Again, recall the mechanism of decarboxylation of the malonic esters and acetoacetic esters discussed in Chapter 8.)

EXERCISE 22-E

Refresh your memory of nucleophilic acyl substitution reaction mechanisms by writing a detailed reaction mechanism for the process illustrated below as induced by strong base. Why cannot this reaction be catalyzed by strong base? (Hint: Reexamine your answer to Exercise 22-C.)

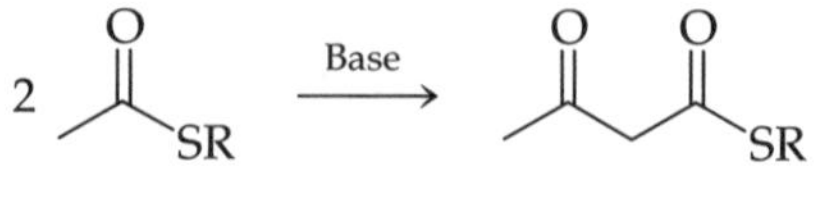

Reduction

The sequence described in the preceding section accomplishes the formation of a carbon–carbon bond between two acetic acid molecules. To proceed to fatty acids, the β-ketothiol ester must be reduced, replacing the oxygen of the ketone group with two hydrogens. This is accomplished by the three steps shown in equations 4 through 6, each of which was considered in some detail in Chapter 21.

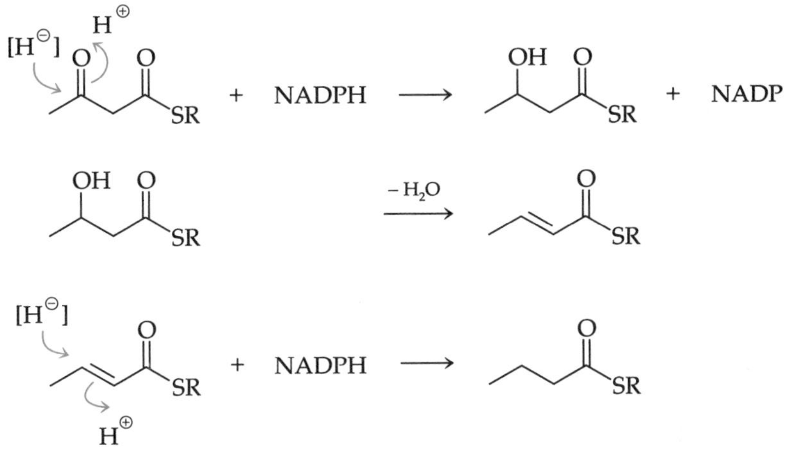

(4)

(5)

(6)

First, the keto group of the β-ketothiol ester is reduced by NADPH, forming a β-hydroxythiol ester (equation 4). Next, the hydroxyl group is lost by dehydration in a process that does not require a cofactor but nonetheless is enzyme catalyzed (equation 5). Last, further reduction of the unsaturated ester is effected by NADPH (equation 6) by a process similar to a complex metal hydride reduction or a Michael addition.

Overall, then, the chemical potential energy of one ATP and two NADPH molecules has been stored temporarily in the thiolester of butyric acid by means of the formation of a carbon–carbon bond and the chemical reactions presented in equations 4 through 6.

EXERCISE 22-F

What chemical alternatives do you know for the reduction of a carbonyl to a methylene group?

EXERCISE 22-G

Loss of water from a β-hydroxyester is highly favored in the direction of the conjugated regioisomer shown at the right below. Explain this result, assuming that the reaction is induced by base. Now do the same for the reaction catalyzed by acid.

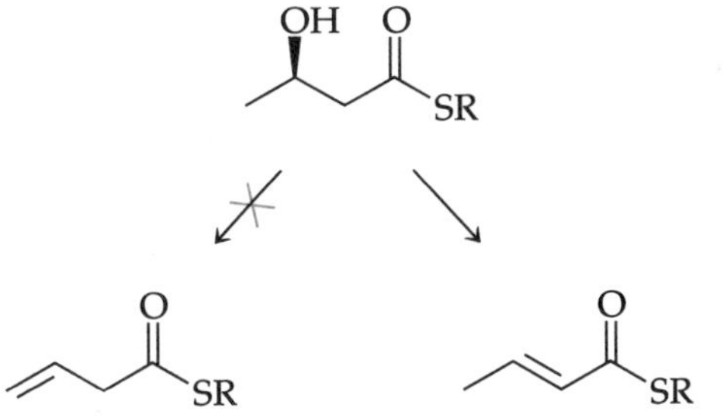

EXERCISE 22-H

Assign oxidation levels to the α and β carbons of the hydroxythiol ester in equation 5. Is the dehydration a redox reaction? That is, is there a net change in oxidation level in the molecule?

Synthesis of Longer Chains

Clearly, we would not consider butanoic acid to be a long-chain fatty acid. However, repetition of carbon–carbon bond formation and the steps in equations 4 through 6 can be used to build longer chains. We again use the carboxylated thiol acetic acid ester shown in Figure 22-13, but this time it is condensed with the thiol butanoic acid ester produced in equation 6, rather than with an acetate ester (Figure 22-14). The steps are exactly analogous to those illustrated in Figure 22-13 except that one of the two components that are to be joined by the new carbon–carbon bond now has four carbons. The resulting β-ketoester, then, has six carbons. In accord with the steps shown in equations 4 through 6, the keto group of this β-ketoester is reduced, the product alcohol dehydrated, and the unsaturated ester reduced to form a six-carbon fatty acid chain (as a thiol ester).

Each cycle of the sequence adds two carbons (from the malonic acid thiol ester unit) to the growing, aliphatic straight-chain thiol ester. As noted earlier, the third carbon is lost, as carbon dioxide, thus providing a driving force to carry this step in the sequence to completion. The original thiol ester of acetic acid gained two carbons to become a butanoic acid ester to which two more carbons were added, forming the thiol ester of hexanoic acid. Each time this process is repeated, two carbons are added, which explains why fatty acids have an even number of carbon atoms. The process

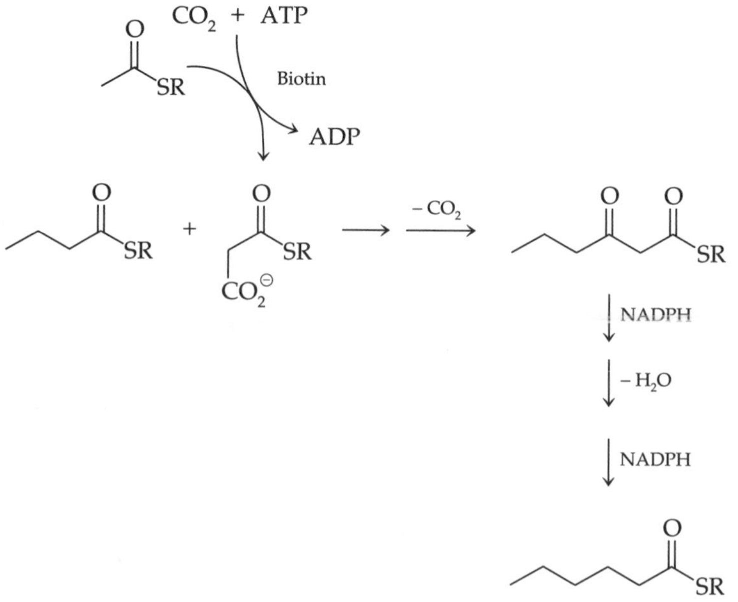

▲ FIGURE 22-14

The thiol ester of butyric acid (produced as shown in equations 4 through 6) is elaborated to a six-carbon thiol ester by the same reactions by which it was formed. The equivalent of the Claisen condensation with carboxylated thiol acetate forms the carbon–carbon bond. Reduction is effected again by the reactions in equations 4, 5, and 6.

of fatty acid biosynthesis is regulated by the "rate limiting" step of conversion of the thiol ester of acetic acid into the malonic acid thiol ester. This step consumes ATP, and the biosynthesis of fatty acids stops in its absence.

EXERCISE 22-1

Suppose that a source of acetic acid in which the carboxylate carbon is enriched in ^{13}C were available to an organism conducting fatty acid biosynthesis.

(a) Indicate the position of ^{13}C-enriched carbons in octanoic acid if it were formed in this way.

(b) How might $CH_3{}^{13}CO_2H$ be prepared in the laboratory? $^{13}CH_3CO_2H$?

22-6 ▸ Energy Release in Fatty Acid Degradation

We have just seen how fatty acids are built progressively from acetic acid, in the course of which each two-carbon chain-lengthening step consumes one ATP and two NADPH molecules. A significant fraction of this energy can be released later, when biological demands require it, in fatty acid degradation. In this process, except for minor details, the reactions of the steps in fatty acid biosynthesis are reversed. When an organism has a low demand for energy, fatty acid degradation stops and the fatty acids accumulate as lipids. Unfortunately, this accumulation of lipids has well-known consequences in human beings.

The steps in fatty acid degradation are illustrated in Figure 22-15. We have arbitrarily selected hexanoic acid to illustrate this process and have highlighted the two carbons that are removed from this six-carbon acid, forming acetic and butyric acid. In the first step, two hydrogens are lost to

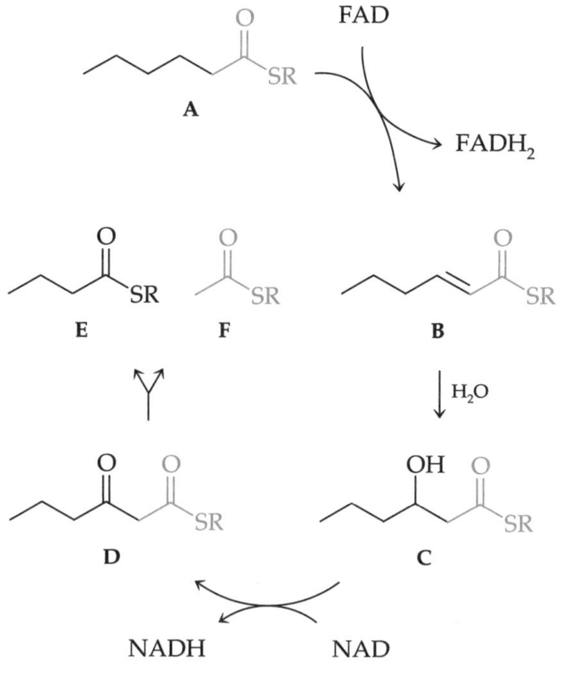

◀ FIGURE 22-15
The steps by which a fatty acid is degraded are the reverse of the reactions in its construction. The FAD-mediated oxidation of a fatty acid thiol ester (A) to the α,β-unsaturated thiol ester (B), followed by hydrolysis, produces a β-hydroxythiol ester (C). Oxidation of the —OH group, coupled with the reduction of NAD, leads to a β-ketothiol ester (D) from which a retro-Claisen condensation produces two shorter thiol esters (E and F).

form an α,β-unsaturated thiol ester. This oxidation is accomplished with FAD, not with NADP (which would exactly reverse the fatty acid biosynthesis step), although the reasons for this are not clear. It is likely that this oxidation includes radical intermediates; in contrast, the reduction in the final stage of fatty acid biosynthesis is usually represented as the transfer of a hydride ion.

In the next stage, the double bond is hydrated by the addition of water. As in fatty acid synthesis, this step requires an enzyme (as catalyst) but no cofactor. Again, a chiral enzyme directs the hydration, and only one stereoisomer of this alcohol is formed; its configuration is opposite that produced in the biosynthesis.

EXERCISE 22-J

What intermediate would be formed if the conversion of compound A into compound B in Figure 22-15 were accomplished by single-electron transfer, as discussed for the $FADH_2$ reductions in Chapter 21? Show the electron flow needed to convert this intermediate into the observed product.

Oxidation of the alcohol to the β-ketothiol ester is effected by the consumption of NAD, producing NADH. Finally, the β-ketothiol ester undergoes a reverse (retro-) Claisen condensation. This last process cleaves the carbon–carbon bond between the second and third carbons in the six-carbon chain, producing a two-carbon fragment (a thiol ester of acetic acid) and the thiol ester of butanoic acid.

Overall in this sequence, two biochemical reducing agents, $FADH_2$ and NADH, are produced, transferring some of the chemical potential energy stored in the fatty acid. In the simplest energetic sense, either of these molecules can reduce molecular oxygen in an exothermic process. Furthermore, either reducing agent provides sufficient driving force for the reduction of a ketone. Thus, fatty acids serve as a storage medium for reducing equivalents in a structurally quite simple form. Because fatty acids contain only CH and CO_2H functional groups, they are chemically quite stable in the absence of the enzymes that initiate the beginning stages of fatty acid degradation.

By the repeated use of these degradation steps, saturated fatty acids are broken down into thiol esters of acetic acid.

For example, the eighteen-carbon fatty acid stearic acid is degraded ultimately to nine acetate units by this pathway. This process constitutes an oxidation because half of the carbons in the acetate products are at the carboxylic acid oxidation level, compared with only one in the original fatty acid.

22-7 ▸ The Krebs Cycle

The acetate units produced in the degradation of fatty acids are further oxidized in a fascinating sequence of steps known as the **tricarboxylic acid (TCA) cycle,** or **Krebs cycle,** for which its discoverer, Sir Hans Adolf Krebs shared the Nobel Prize in 1953 for physiology or medicine with Fritz Albert Lipmann who discovered coenzyme A. This sequence of reactions results ultimately in the decarboxylation of acetic acid and produces additional equivalents of hydride reduction potential, stored in NADH and FADH$_2$. First, recall that acetic acid cannot readily lose carbon dioxide chemically (equation 9) in the same way that β-ketoacids do (equation 7). Also recall that special reactivity must be induced by thiamine pyrophosphate (Chapter 21) in order to decarboxylate an α-ketoacid: this reaction fails (equation 8) under conditions in which CO$_2$ is lost easily from a β-ketoacid.

$$\text{(7)}$$

$$\text{(8)}$$

$$\text{(9)}$$

The ketone carbonyl group in a β-ketoacid provides a means by which the negative charge produced in the loss of carbon dioxide is stabilized as an enolate anion (equation 7). The corresponding α-ketoacid, even though it does have a ketone carbonyl group, is not properly constituted so as to generate a stable anion upon loss of carbon dioxide (equation 8) and, instead, requires a sequence of several steps involving thiamine for decarboxylation, as in the biochemical degradation of amino acids (Chapter 21).

The situation with simple carboxylic acids is similar to that of α-ketoacids. Direct loss of carbon dioxide would lead to the highly unstable, charge-intensive methyl anion (equation 9), a process that does not take place under any normal reaction conditions. Therefore, oxidative energy release through decarboxylation of acetic acid requires the formation of a derivative with functionality properly positioned to provide anion stabilization after loss of CO$_2$.

Decarboxylation is an integral part of the TCA cycle, an overview of which is given in Figure 22-16. Thiol acetate feeds into the TCA cycle by reaction with a key starting material, oxaloacetic acid. As this cycle proceeds, two of the carbons of oxaloacetic acid are lost as carbon dioxide and replaced by the two carbons of another acetate unit. Thus, oxaloacetic acid is a catalyst in the TCA cycle and is not consumed as acetate is degraded to two molecules of carbon dioxide. Several oxidation steps in the cycle take place with the simultaneous conversions of NAD into NADH and of FAD into FADH$_2$, and there are other similarities with the steps in fatty acid degradation. Indeed, the cycle begins with the reaction of a thiol ester of acetic acid with a ketone carbonyl group of oxaloacetic acid in an aldollike reaction, a process similar to the Claisen condensation that initiates fatty acid biosynthesis. The product, **citric acid,** has a new carbon–carbon bond and a tertiary alcohol group. (In Figure 22-16, the acetic acid carbons are identified by black dots so that we may see how they are transformed in the progression of the cycle.)

Both the TCA cycle and fatty acid degradation take place completely within mitochondria, small structures within the cell that are responsible

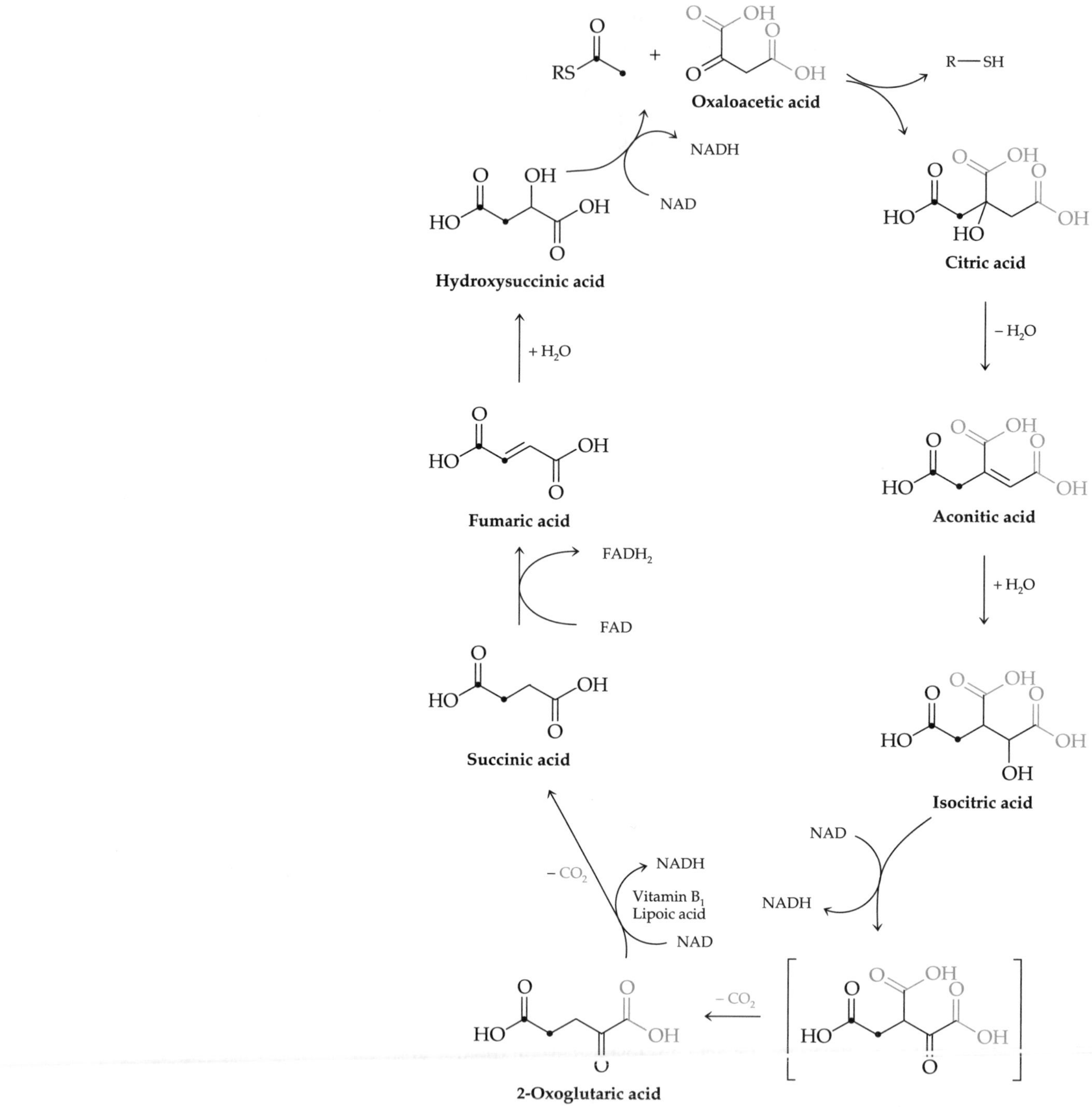

▲ FIGURE 22-16

In the tricarboxylic acid (TCA) cycle, citric acid undergoes sequential dehydration, rehydration, and oxidation. These steps introduce a ketone carbonyl group β to a carboxylic acid (lower right) that undergoes ready decarboxylation to produce 2-oxoglutaric acid. Decarboxylation of 2-oxoglutaric acid (an α-ketoacid) takes place through the routes described in Chapter 18, with succinic acid being produced. The succinic acid is oxidized to fumaric acid, which undergoes hydration followed by oxidation to form oxaloacetic acid. Condensation of a thiol ester with this α-ketoester reforms citric acid and permits the cycle to be repeated. One complete cycle converts an acetate unit (entering the cycle at the step shown at the top left) into two equivalents of carbon dioxide (leaving the cycle at the steps shown at the bottom and bottom left), while releasing stored chemical energy.

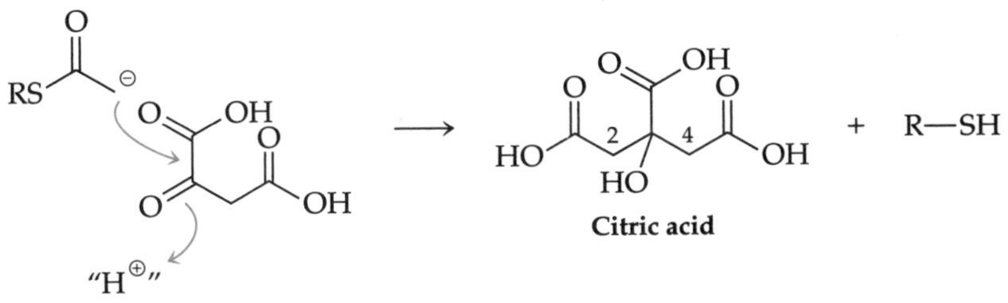

▲ FIGURE 22-17
Nucleophilic addition of a thiol acetate enolate anion to the highly
electrophilic α-keto group of oxaloacetic acid produces citric acid.

for the production of ATP from ADP. The NADH produced in the TCA
cycle does not leave the mitochondria but is oxidized (indirectly) by molecular oxygen with the production of one molecule of NAD and three molecules of ATP. Thus, the chemical potential energy produced in this cycle is
available to the cell only as ATP and not as the reduction potential of
NADH.

Citric acid is formed by the nucleophilic addition of the α carbon of thi-
olacetate to the α-keto group of oxaloacetic acid, followed by hydrolysis of
the thiol ester (Figure 22-17).

EXERCISE 22-K

Write a detailed, step-by-step mechanism for the formation of citric acid as
shown in Figure 22-17.

At first, you might think that C-4 is identical with C-2 in citric acid,
because it appears that the left- and right-hand sides of the molecule are the
same. However, this is not the case. A tetrahedral carbon atom for which
three of its four substituents are different (as for C-3) is a **prochiral** center.
Although the molecule is superimposable on its mirror image and therefore
achiral by definition, C-3 becomes a center of chirality by any transformation
that changes one of the two identical groups. The handedness of this new
center of chirality depends on which of the identical groups has been
changed, and thus, from this point of view, these two groups are really not
identical. Transformation of one group leads to the *R* configuration in the
new center and, for this reason, the group is called **pro-R**. Correspondingly,
the other group is termed **pro-S**. For example, if the C-1 carboxylic acid group
of citric acid is converted into a methyl ester (at the left in Figure 22-18), the

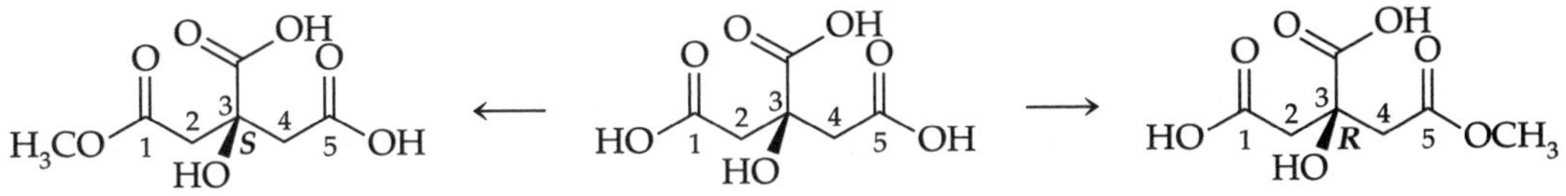

▲ FIGURE 22-18
The conversion of the carboxylic acid functionality at either C-1 or C-5 to an
ester differentiates two identical substituents on C-3 and converts this atom
into a center of chirality. Esterification of the C-1 carboxylic acid produces a
monoester with the *S* configuration; esterification of C-5 affords the *R*
configuration.

A prochiral molecule can interact with a chiral surface (for example, at the active site of an enzyme) in different ways, depending on which of the two identical (prochiral) substituents approaches the surface.

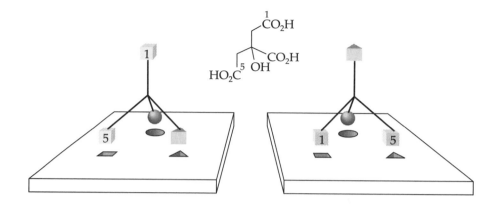

two groups bound to C-3 are no longer identical and C-3 becomes a center of chirality. Thus, we can distinguish between these two groups on the basis of the configuration of the product formed upon selective transformation of one of them. Such groups are called **enantiotopic** because they lie on opposite sides of the plane of symmetry of an achiral molecule. In this case, C-2 has a higher priority than C-4, and the resulting ester has the *S* configuration.

Conversely, when the carboxylic acid at C-5 is converted into a methyl ester, C-3 has the *R* configuration, as shown at the right in Figure 22-18. As noted earlier, enzymes almost always control processes that generate stereochemical centers so that only one configuration is formed with a given enzyme.

We can visualize this process of **chiral recognition** with the help of Figure 22-19, in which geometric shapes are used to represent the structure of citric acid, exemplifying the general case of a prochiral center for which three of the four groups are different. Imagine that a "stick" structure having groups corresponding to two cubes, a sphere, and a wedge is placed on a board containing three holes (square, circular, and triangular) so that the cube labeled 5 (to correspond to C-5 of citric acid) is aligned with a square hole on the surface. We see that both the wedge and the sphere also align with their corresponding geometric holes, as at the left in Figure 22-19. If we rotate the structure so that cube 1 aligns with the square hole in the board, then we have no choice but to attempt to match either cube 5, as at the right, or the sphere with the triangular hole. This analogy contains no real chemical details, but it serves to describe how an enzyme is capable of recognizing the details of the stereochemistry of molecules. Instead of a board having holes of various geometric shapes, the pocket of an enzyme has cavities that differ from each other in polarity, size, and shape. Each cavity is tuned to a part of a substrate molecule, affording recognition of that particular part. As a result of this precise matching, however, many enzymes serve as catalysts for only a limited number of substrates, and some are effective for only one specific molecule.

The next two steps of the TCA cycle accomplish an isomerization of citric acid to **isocitric acid** by dehydration and then rehydration. Note that, in the conversion of citric acid into isocitric acid, the hydroxyl group is moved from the center of the molecule to C-4.

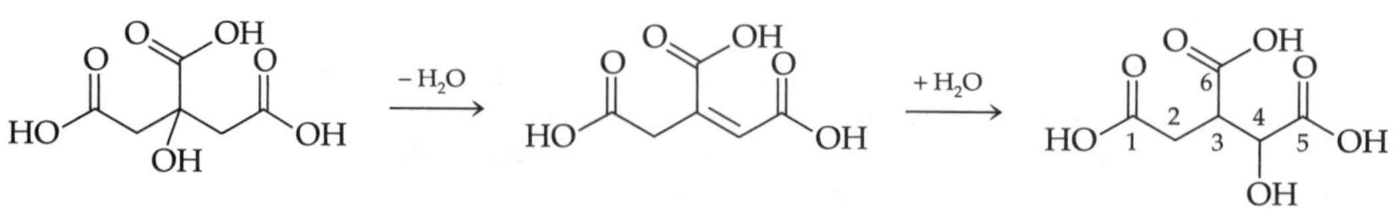

Citric acid → (− H₂O) → → (+ H₂O) → **Isocitric acid**

FLUOROACETIC ACID: A SIMPLE BUT DEADLY MOLECULE

Fluoroacetic acid is very toxic, as was first observed in animals that ate the leaves of the South African plant "Gifblaar" (*Dichapetalum cymosum* Hook), which contains this unusual, natural compound. The toxic effect of fluoroacetic acid is the result of its incorporation into the TCA cycle to form fluorocitric acid. This compound binds tightly to the enzyme responsible for the isomerization of citric acid to isocitric acid and blocks it from performing this critical role. As a result, the level of citric acid increases dramatically and, deprived of the chemical potential energy and reduction potential produced in the TCA cycle, the animals often die. Only from about 2 to 5 mg per kilogram of body weight is required to kill an average-sized man (less than 0.5 g).

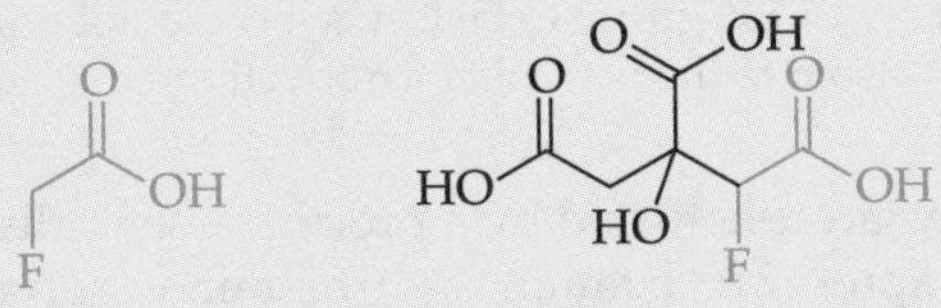

Fluoroacetic acid   **Fluorocitric acid**

In the first step, an enzyme discriminates between the pro-*R* and the pro-*S* groups so that dehydration occurs only with the loss of hydrogen from C-4. This selective process has no stereochemical consequence because there are no centers of chirality in the product alkene. However, the removal of a hydrogen atom specifically from C-4 means that neither of the carbon atoms of the acetate unit introduced in the first step participates in the double bond. This has the consequence that neither of these carbons is lost as carbon dioxide in subsequent reactions.

EXERCISE 22-L

All of the individual steps in the TCA cycle are catalyzed by enzymes. Thus, the outcome need not correspond to what we might expect under laboratory conditions. Under acid-catalyzed hydration conditions, which isomer—citric acid or isocitric acid—would you expect to be formed more rapidly from the intermediate alkene? Write a mechanism for these hydration reactions.

Isocitric acid is then oxidized by NAD (producing NADH) to a molecule that is both an α-ketoacid (relative to the C-5 carboxylate) and a β-ketoacid (relative to the C-6 carboxylate), as shown in Figure 22-20.

EXERCISE 22-M

Draw the transition state for the decarboxylation shown in Figure 22-20 and indicate the electron flow that accompanies bond breaking and making.

This ketotriacid now readily undergoes decarboxylation; the carbon lost in this step was present in oxaloacetic acid and is *not* derived from the added acetic acid residue. The resulting **2-oxoglutaric acid** undergoes

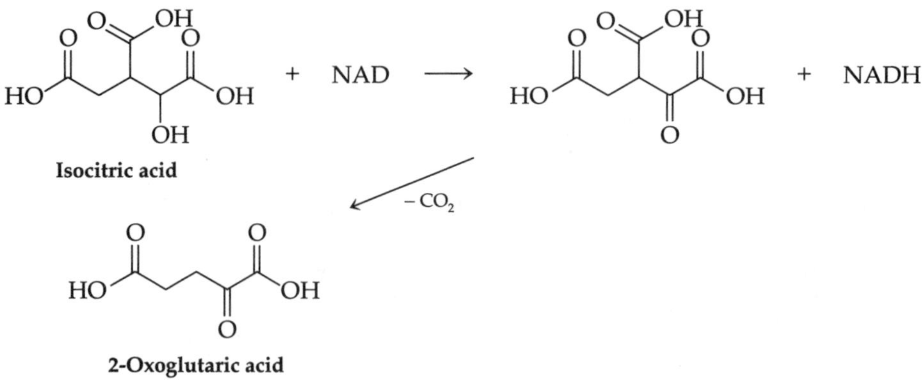

Isocitric acid + NAD ⟶ + NADH

− CO₂

2-Oxoglutaric acid

▲ FIGURE 22-20
Decarboxylation of the ketoacid formed by oxidation of isocitric acid proceeds readily because the —CO₂H group to be lost is located not only α to the C-5 carboxylic acid, but also β to the two other carboxylic acid groups.

decarboxylation as an α-ketoacid by a process essentially identical with that described in Chapter 21: vitamin B₁ (thiamine) serves as cofactor and another molecule of NAD is reduced to NADH in the process.

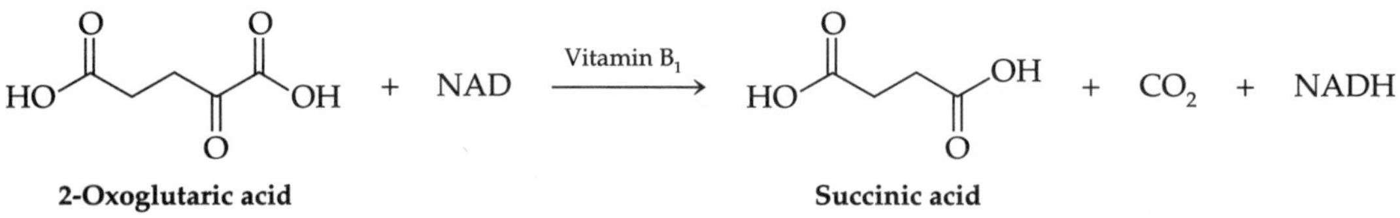

2-Oxoglutaric acid + NAD $\xrightarrow{\text{Vitamin B}_1}$ + CO₂ + NADH

Succinic acid

The resulting symmetrical diacid, **succinic acid,** is then oxidized by FAD to the unsaturated diacid, **fumaric acid,** in a fashion analogous to that in fatty acid degradation (Figure 22-21). In analogy with fatty acid degradation, fumaric acid undergoes hydration, producing an alcohol that is oxidized to a ketone with NAD. This product ketodiacid is oxaloacetic acid, and the cycle has returned to its starting point. In this cycle, carbons 5 and 6 of citric acid, which are lost in the two decarboxylation steps, are replaced by the two carbons from a molecule of acetic acid. Oxaloacetic acid is a β-ketoacid from which decarboxylation takes place readily: the TCA cycle allows this reaction to occur through a route in which a thiol ester, which cannot itself undergo this reaction, is converted into a derivatized oxaloacetic acid that does undergo facile decarboxylation.

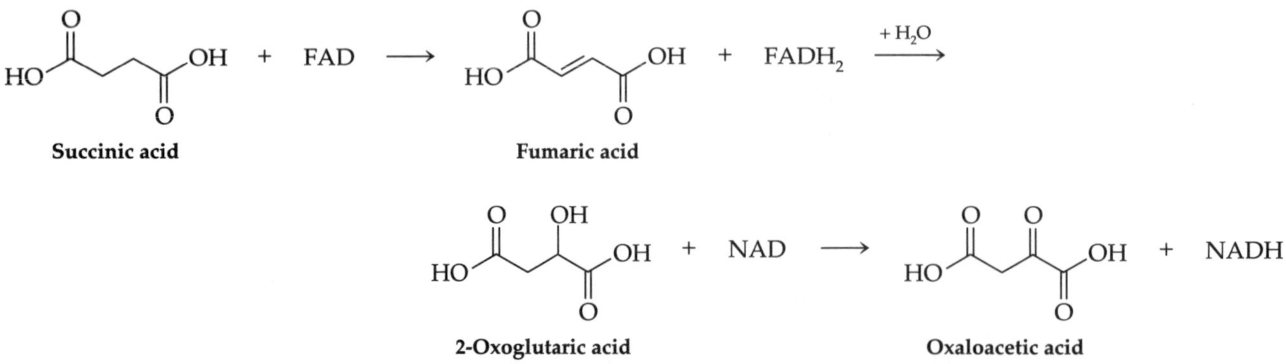

Succinic acid + FAD ⟶ **Fumaric acid** + FADH₂ $\xrightarrow{+ \text{H}_2\text{O}}$

2-Oxoglutaric acid + NAD ⟶ **Oxaloacetic acid** + NADH

▲ FIGURE 22-21
Succinic acid is converted into oxaloacetic acid by a sequence resembling the reverse of fatty acid biosynthesis: oxidation; hydration; oxidation.

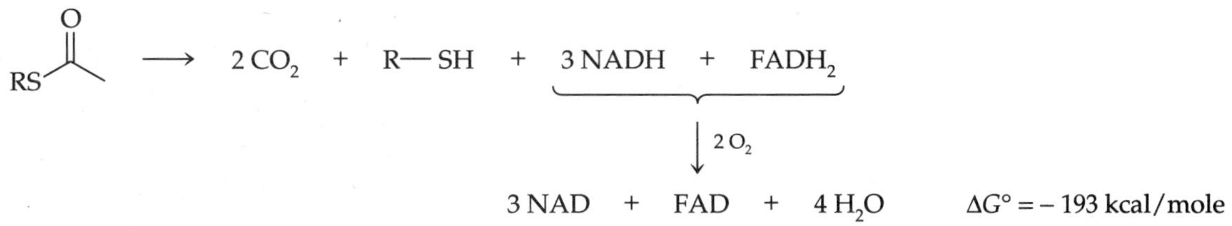

3 NAD + FAD + 4 H$_2$O $\Delta G° = -193$ kcal/mole

▲ FIGURE 22-22

The net conversion of a thiol acetate into carbon dioxide in the TCA cycle is highly exothermic.

Figure 22-22 summarizes the net conversion accomplished in the TCA cycle, in which one molecule of acetate is consumed, producing two molecules of carbon dioxide, three of NADH, and one each of FADH$_2$ and coenzyme A (written as R—SH). If all of the reduction potential generated in the NADH and FADH$_2$ formed in the cycle were consumed in the reduction of molecular oxygen to water, a total of 193 kcal/mole (Figure 22-22) would be released as a consequence of this oxidation of a thiol ester of acetic acid to carbon dioxide. This is an enormous amount of energy to be released at once: it is thus important to the organism that it be released in carefully controlled fashion.

22-8 ▸ Controlling Heat Release

The energy released by the simple combustion of palmitic acid is 2386 kcal/mole (Figure 22-10). The biological breakdown of palmitic acid into eight thiol acetic acid units (Figure 22-23) produces two NADPH molecules for each acetate (thus, a total of sixteen). Degradation through the TCA cycle of the eight acetic acid equivalents produced from palmitic acid produces twenty-four NADH and eight FADH$_2$ molecules. In total, then (Figure 22-24), the biological degradation of palmitic acid to sixteen carbon

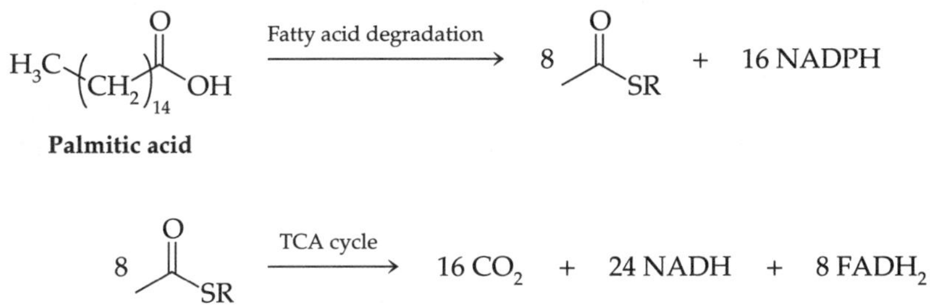

▲ FIGURE 22-23

The energy released in the biodegradation of a fatty acid is the sum of the energies released in cleaving to thiol acetates and in degrading the thiol acetates in the TCA cycle.

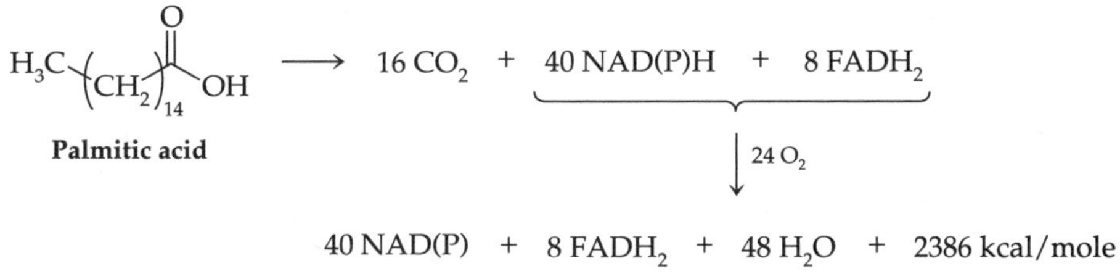

◀ FIGURE 22-24

The energy released in the biological degradation of palmitic acid is exactly equivalent to that produced by the combustion of this fatty acid in air.

ENERGY CONSERVATION

The process of making rum from molasses begins with fermentation of sugars into ethanol. However, the microorganisms that are responsible for this reduction process cannot survive alcohol concentrations above 13% (26 proof). Thus, to obtain 80 proof rum, the crude mixture of alcohol, water, and other volatile compounds is separated by the use of large distillation columns. At the Bacardi production facility outside San Juan, Puerto Rico, 70% of the energy required to operate the stills that produce 100,000 gallons of rum daily, and indeed the entire operation, is obtained by converting organic waste materials into methane gas, using microbes that effect this reduction efficiently when deprived of oxygen.

dioxide molecules produces forty NADH (or NADPH) molecules and eight $FADH_2$ molecules.

If all of these reduced cofactors were reoxidized by molecular oxygen, 2386 kcal of energy would be released. This value is precisely the same as that obtained when palmitic acid is converted by direct oxidation with molecular oxygen into carbon dioxide and water. This is exactly as we would expect, because thermodynamics does not depend on the reaction pathway. However, were such a direct oxidation to be the process followed biologically, all of the energy stored in the fatty acid would be released as heat and none would be available for other biochemical transformations. Furthermore, the heat released would be dangerous to the organism. Instead, this potential energy is much more efficiently utilized by being broken into smaller amounts that can be incorporated into subsequent biochemical transformations.

We have previously encountered situations in which it is advantageous to have a source of chemical energy, or driving force, that can be incorporated into a reaction that would otherwise be unfavorable. In the first stages of fatty acid biosynthesis (Figure 22-13), ATP is consumed in a process that provides ultimately for the loss of carbon dioxide, making an otherwise marginally favorable carbon–carbon bond-forming process quite exothermic. By coupling chemical reactions with oxidation-reduction agents, biological systems effect the oxidation of NADH (as well as NADPH and $FADH_2$) and the simultaneous conversion of ADP into ATP.

$$\frac{1}{2} O_2 \ + \ NADH \ + \ 3\,ADP \ \xrightarrow{\text{Enzymes}} \ NAD \ + \ 2\,H_2O \ + \ 3\,ATP \qquad \Delta G^\circ = -31.4 \text{ kcal/mole}$$

$$\frac{1}{2} O_2 \ + \ FADH_2 \ + \ 2\,ADP \ \xrightarrow{\text{Enzymes}} \ FAD \ + \ H_2O \ + \ 2\,ATP \qquad \Delta G^\circ = -22.2 \text{ kcal/mole}$$

Unlike the direct oxidation of NADH (and $FADH_2$) with molecular oxygen (which releases all of the stored energy as heat), the biologically mediated process stores some of this energy in the high-energy phosphate bond of ATP.

Recall that the simple hydrolysis of ATP to ADP releases 7.0 kcal/mole of energy. Thus, when one NADH molecule is consumed to produce one NAD molecule and three ATP molecules, 21.0 (3×7.0) kcal/mole of energy is stored in the ATP. Likewise, the exothermicity of the combination of molecular oxygen with $FADH_2$ is coupled with the conversion of two ADP molecules into two ATP molecules. Here, direct oxidation would release 36.2 kcal/mole (Figure 22-9) but, when the process is mediated biologically, 14.0 kcal/mole (2 × 7.0) is stored in the two molecules of ATP and only 22.2 kcal/mole is released as heat. Once again, the process is still exothermic, but not all of the energy stored in $FADH_2$ is released as thermal energy in its conversion into FAD.

We can thus now look at the overall energetics of the conversion of palmitic acid into carbon dioxide (Figure 22-23) from the perspective of how much ATP could be produced. Let us assume that, instead of a direct oxidation of the NADH and $FADH_2$, these materials are oxidized in processes biologically coupled with the conversion of ADP into ATP. In this case, the 40 molecules of NADPH produce 120 molecules of ATP, and the 8 molecules of $FADH_2$ produce 16 molecules of ATP for a total of 136 ATPs.

$$\begin{array}{ccccc} & 24\ O_2 & & 48\ H_2O & \\ 40\ NAD(P)H & +\ 8\ FADH_2 & \rightleftarrows & 40\ NAD(P) & +\quad 8\ FAD \qquad \Delta G^\circ = -952\ \text{kcal/mole} \\ & 136\ ADP & & 136\ ATP & \end{array}$$

Because each molecule of ATP represents 7.0 kcal/mole of stored chemical energy, the total energy produced from one mole of palmitic acid is 952 kcal. This amount comprises 40% of the total of 2386 kcal/mole released on direct oxidation. The efficiency of the conversion of the stored energy of palmitic acid into ATP is quite impressive. Analogous efficiencies for energy conversion outside the biological realm are rare. For instance, in the conversion of thermal energy into electricity in power plants, 40% is the best efficiency yet obtained.

22-9 ▸ Energy Release from Carbohydrates through Glycolysis

Let us compare the biological sequence for fatty acid degradation with that which releases the energy stored in carbohydrates in photosynthesis. The breakdown of carbohydrates, called **glycolysis,** also produces acetic acid (as a thiol ester), which, as we have seen, enters the TCA cycle, a major pathway for energy release. We will now see how glucose is converted into acetic acid. Like fatty acid degradation and the TCA cycle, this transformation takes place through reactions analogous to standard organic reactions.

Glycolysis takes place in a sequence of four stages, as outlined in Figure 22-25:

1. The isomerization of glucose to fructose
2. The cleavage of fructose into three-carbon fragments through a retro-aldol reaction
3. The functionalization of the three-carbon fragments (to form pyruvic acid) to set the stage for decarboxylation
4. Decarboxylation to form a thiol ester of acetic acid

There are many chemical steps in this process, although not much stored chemical energy is produced. Nonetheless, glucose (and other carbohydrates) contribute significantly to energy storage because the acetic acid produced by glycolysis enters the tricarboxylic acid cycle and is further degraded to carbon dioxide, releasing considerable stored potential energy

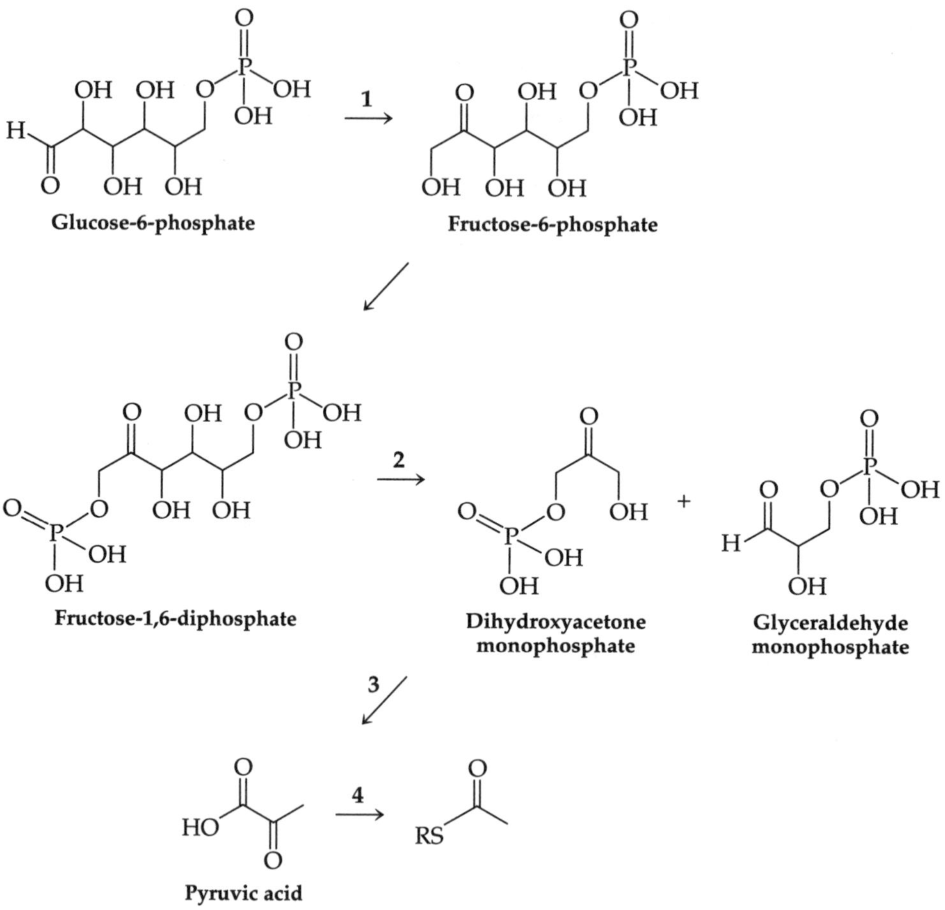

▲ FIGURE 22-25
The degradation of carbohydrates begins with the isomerization of glucose-6-phosphate (an aldose) to fructose-6-phosphate (a ketose) in step 1. Phosphorylation of the C-1 alcohol produces fructose-1,6-diphosphate, which reacts by a retro-aldol reaction in step 2 to form two three-carbon fragments; namely, dihydroxyacetone monophosphate and glyceraldehyde monophosphate. After the oxidation levels of the three carbons of dihydroxyacetone monophosphate are adjusted to produce pyruvic acid in step 3, decarboxylation can take place in step 4 by routes similar to those discussed for other α-ketoacids in Chapter 18.

in the form of the reducing species NADPH and FADH$_2$. Let us consider the first three stages in sequence. (We have already covered decarboxylation of α-ketoacids in Chapter 21 and the TCA cycle earlier in this chapter.)

Isomerization of Glucose to Fructose

As glucose enters the mitochondria, it is phosphorylated by ATP to form glucose-6-phosphate. Thus, glycolysis starts by *consuming* chemical potential energy. The phosphorylated derivative next undergoes isomerization to fructose-6-phosphate. We might expect the ketone fructose to be more stable than the aldehyde glucose, but this is not the case and the isomerization illustrated in Figure 22-26 is endothermic by 0.4 kcal/mole. Recall from Chapter 17, however, that both glucose and fructose exist in solution in cyclic forms, not as the free carbonyl groups shown for clarity in Figure 22-26. It is in the cyclic hemiacetal form that glucose is more stable than the hemiketal form of fructose. [Recall that anomerization of the sugars takes

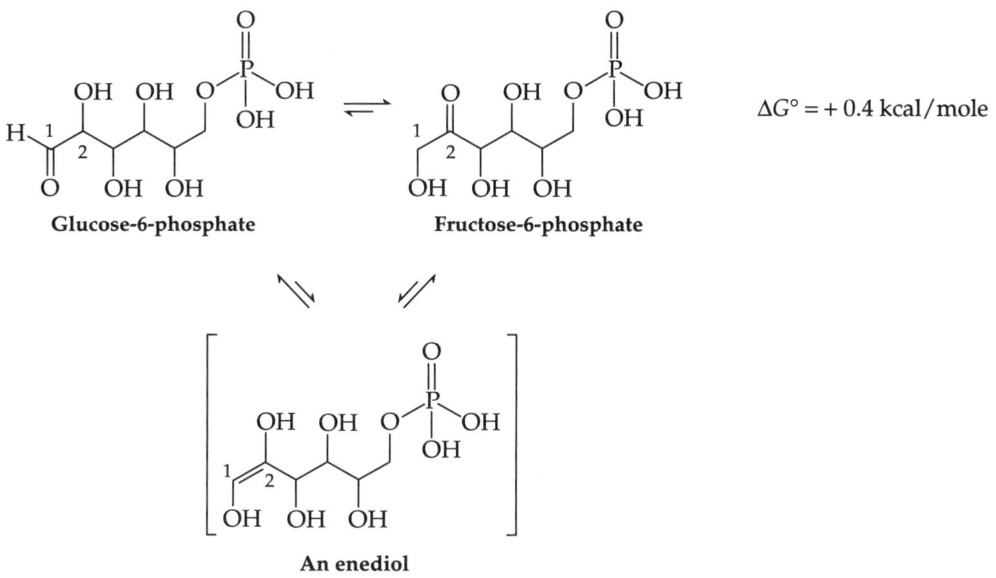

$\Delta G° = + 0.4$ kcal/mole

Glucose-6-phosphate   **Fructose-6-phosphate**

An enediol

▲ FIGURE 22-26

Although glucose-6-phosphate is more stable than fructose-6-phosphate, an equilibrium can be established by the tautomerization of a C-2 hydrogen through an intermediate enediol.

place through their noncyclized (open-chain) forms, so that these species are quite accessible as chemical intermediates.]

The isomerization of glucose to fructose is accomplished through a sequence of proton tautomerizations analogous to the reversible conversion of a carbonyl group into an enol. The open-chain form of glucose has an aldehyde functional group at C-1 from which an enol can be formed by tautomerization of a proton from C-2 to the aldehyde oxygen. The resulting enol differs from those considered in earlier chapters in that it bears two hydroxyl groups on the double bond and is therefore referred to as an **enediol.** The proton on C-2 can thus as easily tautomerize to the doubly bound oxygen on C-1.

We can examine the chemistry of enediols in more detail by considering a simpler system, the interconversion of glyceraldehyde and dihydroxyacetone (Figure 22-27), that nonetheless includes the relevant functional transformation taking place in the conversion of glucose into fructose. Earlier, we considered enols to be tautomers of ketones in which the same carbon bears the hydroxyl in the enol and the carbonyl oxygen in the ketone. However, in the enediol intermediate in Figure 22-27, each carbon of the double bond (C-1 and C-2) has a hydroxyl group. Thus, it is not possible to determine from the structure of the enediol whether it derived from an

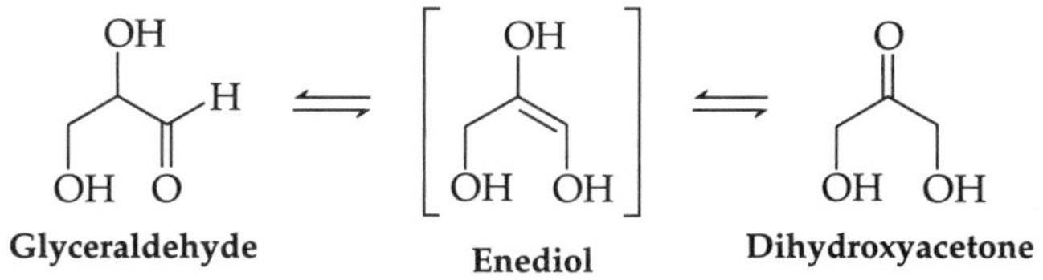

Glyceraldehyde   **Enediol**   **Dihydroxyacetone**

▲ FIGURE 22-27

Isomerization of glyceraldehyde to dihydroxyacetone requires two tautomerization steps and proceeds through an enediol intermediate. Overall, two hydrogens, one from the C-2 hydroxyl group and the other from C-2 have migrated to C-1.

aldehyde with a carbonyl group at C-1 (glyceraldehyde) or from a ketone group at C-2 (dihydroxyacetone); that is, this intermediate can serve as a precursor for either of the carbonyl compounds. Protonation at C-2 forms glyceraldehyde after the loss of a proton from the C-1 OH group, whereas protonation at C-1 results in the formation of dihydroxyacetone after deprotonation of the C-2 OH group. We will encounter this isomerization again, later in the glycolysis sequence.

The pathway for the interconversion of glucose and fructose is analogous to this isomerization of glyceraldehyde to dihydroxyacetone (both as monophosphate esters). Although the details of the sequence of proton removals and replacements in this isomerization are not known with certainty, a species analogous to the enediol illustrated must be formed as an intermediate. We do know that glucose (which is the major form in which carbohydrates are stored in living beings) is converted into fructose as its diphosphate ester, although isomerization in fact takes place on the monophosphate ester. This phosphate group is critical in enabling the enzyme that catalyzes this conversion to recognize glucose-6-phosphate, but it does not otherwise directly take part in the isomerization.

EXERCISE 22-N

Write a mechanism for the transformation of glucose into fructose under both acidic and basic conditions. Suggest a reason why the carbonyl group of fructose might not be as stable as that of a simple ketone such as acetone.

Cleavage of Fructose into Three–Carbon Fragments

After glucose has been converted into fructose, glycolysis proceeds with cleavage of the C-3—C-4 bond by the equivalent of a retro-aldol reaction. Let us start by reviewing the aldol reaction, a transformation discussed in detail in Chapter 13 that is similar to the carbonyl addition reactions in fatty acid biosynthesis and in the tricarboxylic acid cycle. The aldol reaction produces a carbon–carbon bond when the nucleophilic carbon of an enolate or an enol (as illustrated in Figure 22-28) reacts with an electrophilic carbonyl carbon. The reverse of this transformation, a process known as the **retro-aldol reaction,** has close analogy with the loss of carbon dioxide from a β-ketoacid (Figure 22-29). Unlike the decarboxylation step, however, the retro-aldol reaction does not have the large energetic driving force of the formation of the very stable carbon dioxide molecule. Indeed, the equilibrium favors the aldol product only slightly.

By drawing fructose as an extended zigzag ketone rather than as a cyclic hemiketal, as in the margin, we can see more readily that this molecule is also a β-hydroxyketone. Like the simpler β-hydroxyketone shown in Figure 22-29, fructose can undergo a retro-aldol reaction. In the biological

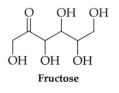

Fructose

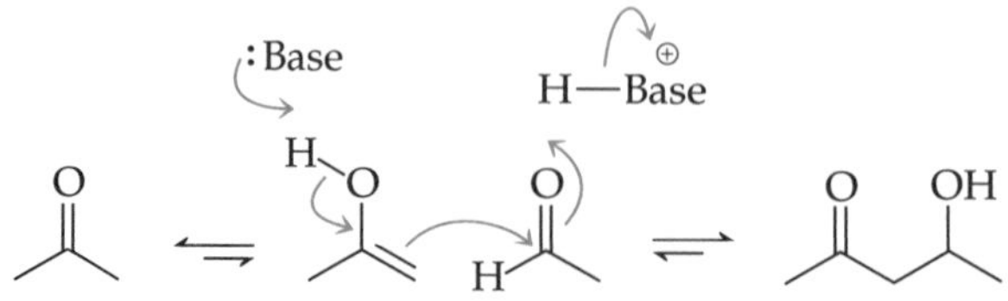

▲ FIGURE 22-28
In an aldol condensation, an enol adds as a nucleophile to a carbonyl group, producing a β-hydroxycarbonyl compound.

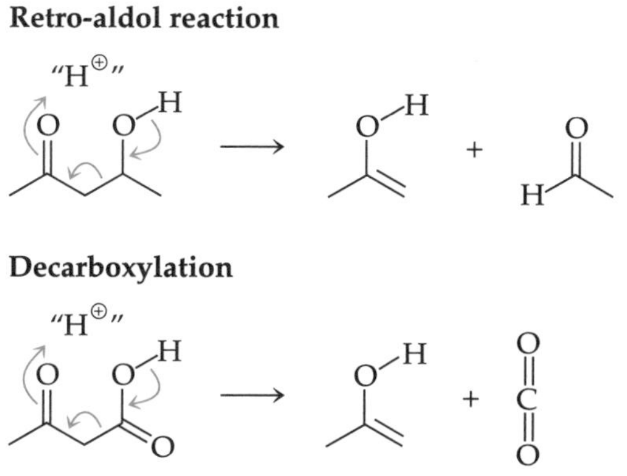

Retro-aldol reaction

Decarboxylation

▲ FIGURE 22-29
In the decarboxylation of a β-ketoacid (lower reaction), an enol and a $C=O$ bond (one of the bonds in CO_2) are produced. This reaction bears close mechanistic resemblance to the retro-aldol reaction (upper reaction) that is the reverse of the aldol condensation shown in Figure 22-28.

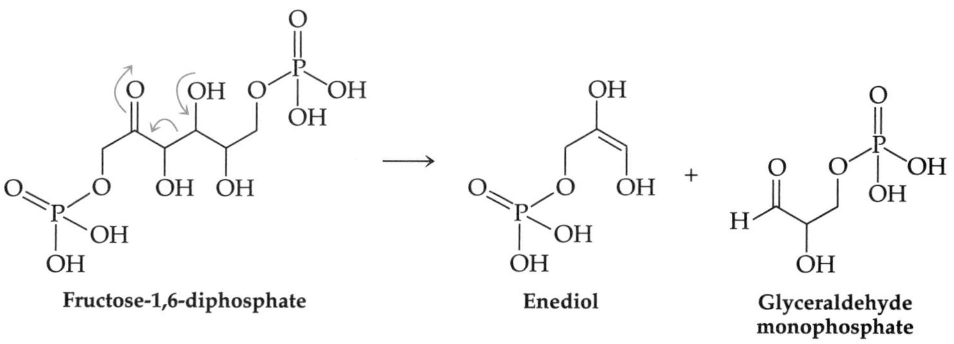

Fructose-1,6-diphosphate   Enediol   Glyceraldehyde monophosphate

▲ FIGURE 22-30
A retro-aldol reaction of fructose-1,6-diphosphate cleaves the C-3—C-4 bond, producing two 3-carbon fragments.

process, fructose-6-phosphate is converted into fructose-1,6-diphosphate (consuming yet another equivalent of ATP), which then undergoes a retro-aldol reaction, producing two 3-carbon fragments by cleavage of the central carbon–carbon bond of fructose (Figure 22-30). One of the fragments is a familiar molecule, glyceraldehyde monophosphate; the other is an enediol.

EXERCISE 22-O

It is conceivable that glucose (in its aldehyde form) could undergo a retro-aldol reaction in the presence of base, although this reaction is not known to occur naturally. What products would be expected from this retro-aldol reaction? Write a detailed mechanism accounting for the bond breaking and making in this proposed transformation.

Indeed, isomerization of the enediol formed in the retro-aldol reaction of fructose diphosphate can form either dihydroxyacetone monophosphate or glyceraldehyde monophosphate, as shown in Figure 22-31. The isomerization of dihydroxyacetone monophosphate to glyceraldehyde monophosphate converts an achiral molecule into a chiral one. When affected by the enzyme isomerase, this transformation forms only the D enantiomer. We can

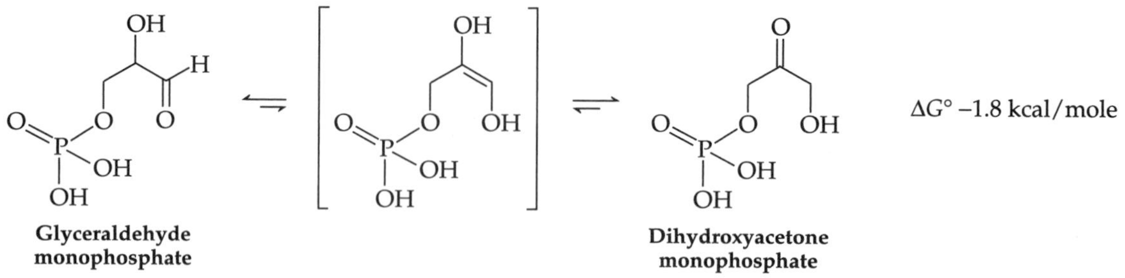

▲ FIGURE 22-31
An enediol is an intermediate in the tautomeric interconversion of an aldose to a ketose.

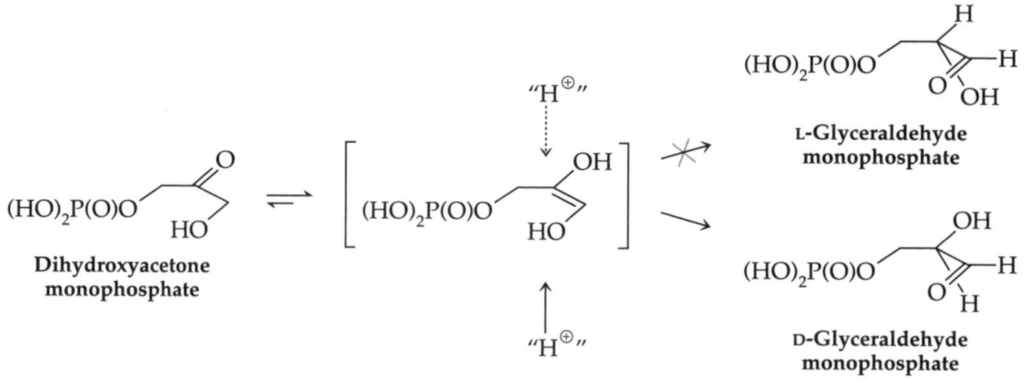

▲ FIGURE 22-32
The planar enediol intermediate can be protonated from either the top or the bottom face. When the enzyme catalyzing the isomerization of dihydroxyacetone to glyceraldehyde delivers a proton exclusively to the bottom face, the naturally occurring D enantiomer is formed.

see in Figure 22-32 how the addition of a proton from the bottom face of the enediol results in the observed D configuration. The alternate process, in which a proton is added from the top to produce L-glyceraldehyde, is not observed in biological systems where the enzyme **isomerase** catalyzes the protonation. Once again we see that enzymes are capable of making a stereo-chemical distinction. Because the addition of the proton results in the formation of a center of chirality, we refer to the precursor sp^2-hybridized alkene carbon as prochiral in the same sense that C-3 of citric acid is prochiral. Similar stereochemical control is also observed in both fatty acid biosynthesis and the tricarboxylic acid cycle.

Our knowledge of simple carbonyl-group chemistry tells us that it is reasonable for the ketone to be the more-stable product derivable from the enediol, although the energy difference is not large (1.8 kcal/mole). Whether only one or both of these carbonyl compounds is produced directly from the enediol of fructose diphosphate is not known (and may never be) because **isomerase** effects rapid interconversion of dihydroxyace-tone monophosphate and glyceraldehyde monophosphate. Notice that the name of the enzyme speaks directly to the type of chemistry that it catalyzes: an isomerase catalyzes isomerization. Correspondingly, the enzyme that catalyzes the retro-aldol reaction of fructose diphosphate is known as **aldolase**.

It is instructive to consider the reverse of the cleavage reaction of fructose diphosphate in Figure 22-30 because both the forward and the reverse reactions have the same transition state. The reverse carbon–carbon bond-

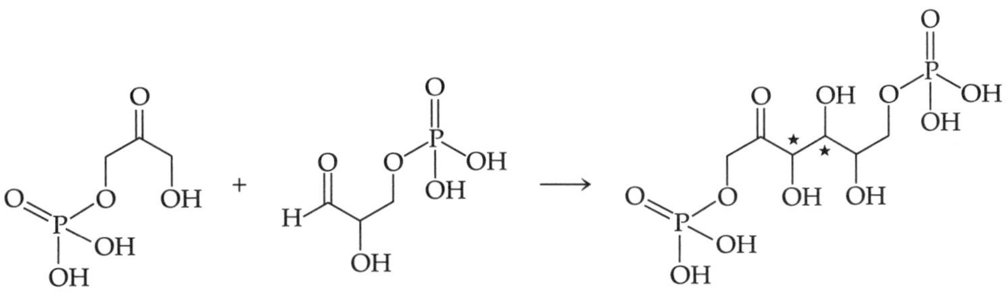

Dihydroxyacetone monophosphate **Glyceraldehyde monophosphate** **Fructose-1,6-diphosphate**

▲ FIGURE 22-33
Two chiral centers are produced in the aldol condensation of dihydroxyacetone monophosphate and glyceraldehyde monophosphate.

forming reaction, in which fructose diphosphate is built from dihydroxyacetone monophosphate and glyceraldehyde monophosphate, also produces two centers of chirality (marked with asterisks in Figure 22-33). This process can be accomplished with glyceraldehyde itself (that is, lacking a phosphate group) in the laboratory. Barium hydroxide acts as a base to catalyze the interconversion of glyceraldehyde and dihydroxyacetone. Then the enolate of dihydroxyacetone undergoes an aldol reaction to produce the carbon skeleton of fructose (Figure 22-34).

Of the six carbons of fructose, three are centers of chirality (carbons 3, 4, and 5). Only one of these stereocenters (C-5) is present in the starting materials, coming from C-2 of glyceraldehyde. Thus, the carbons linked in this process, C-3 and C-4, are new stereocenters formed in the reaction. Because each of them can have either the *R* or the *S* configuration, there are four stereoisomers possible: *RR*, *RS*, *SR*, and *SS*.

EXERCISE 22-P

Draw clear three-dimensional representations and Fischer representations of the four possible stereoisomers that result from the aldol reaction shown in Figure 22-34.

In this simple laboratory transformation, only two of the four possible stereoisomers are formed, one of which is fructose. The presence of the divalent cation barium in the reaction is responsible for this selectivity, functioning to bridge the enolate oxygen and the carbonyl group of the reacting aldehyde, as illustrated in the margin. In general, the energetic difference between competing, cyclic transition states is greater than that for noncyclic transition states that accomplish the same transformation. Six atoms participate directly in this cyclic process, and the transition state can

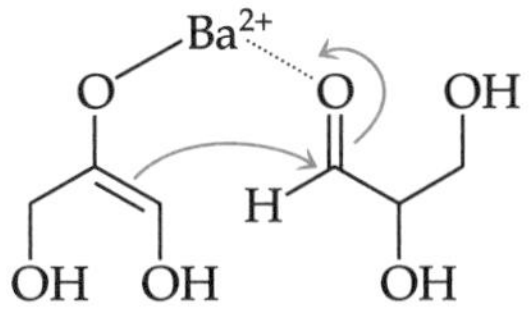

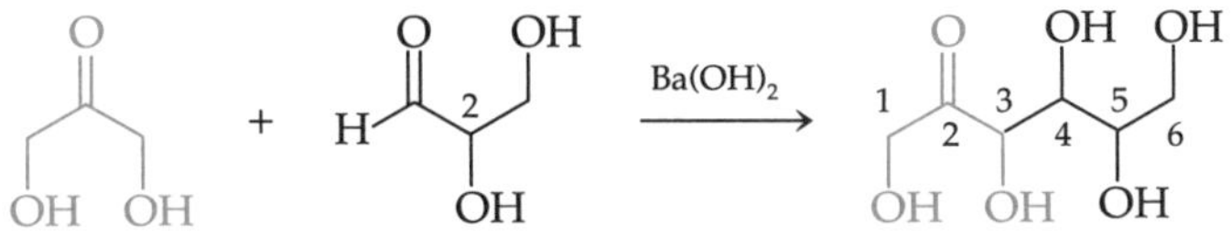

▲ FIGURE 22-34
Barium hydroxide induces enolatelike reactivity in dihydroxyacetone, producing an intermediate that condenses with glyceraldehyde to form fructose.

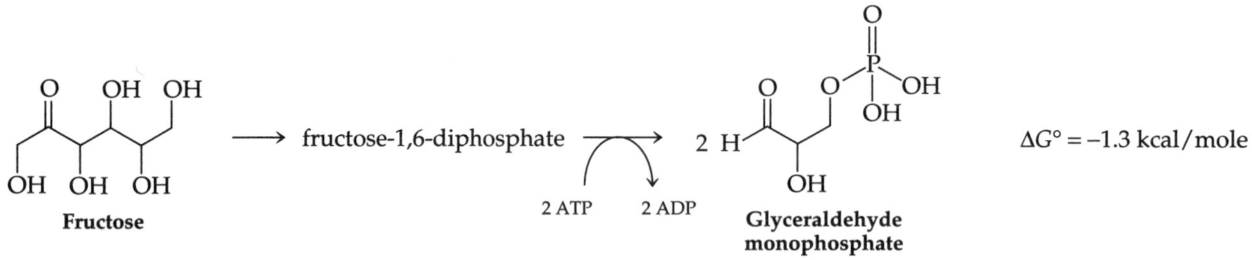

The endothermicity inherent in the cleavage of a carbon–carbon bond in fructose-1,6-diphosphate is balanced by coupling with the exothermic hydrolysis of ATP to ADP and phosphate.

be conformationally defined roughly as a chair. As a result of this conformation, only one of the four stereoisomers is formed.

The sequence of steps to this point in glycolysis converts glucose through fructose diphosphate into two molecules of glyceraldehyde monophosphate. Although glucose constitutes a major vehicle by which chemical potential energy is stored, the part of glycolysis summarized in Figure 22-35 in fact consumes two molecules of ATP in order to form the two phosphate bonds present in the two molecules of glyceraldehyde monophosphate. Even with the added driving force of the ATP-into-ADP conversion, the transformation of glucose into two molecules of glyceraldehyde monophosphate is exothermic by only 1.3 kcal/mole.

Conversion of the Three–Carbon Fragments into Acetic Acid Derivatives

Glyceraldehyde monophosphate is then further transformed into carbon dioxide, acetic acid, and stored chemical potential energy in the form of NADH and ATP. To begin, inorganic phosphate adds to the aldehyde of glyceraldehyde phosphate in the same way that other nucleophilic species, such as water and cyanide, react to make hydrates and cyanohydrins (Figure 22-36). The alcohol formed by the addition of phosphate undergoes oxidation by NAD to form an interesting **mixed anhydride** between phos-

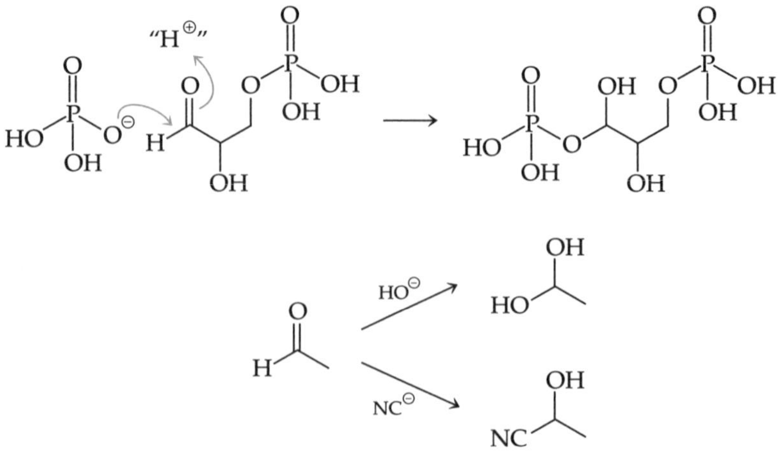

▲ FIGURE 22-36
Nucleophilic attack by phosphate on glyceraldehyde-3-phosphate in the upper reaction is mechanistically similar to hydrate formation and cyanohydrin formation shown in the lower reaction.

THE CHEMISTRY OF ACHES AND PAINS

Yogurt is made by treating milk with the organism *Lactobacillus bulgaricus*, which, among other things, produces lactic acid through the sequence of glycolysis. In this process, the reactions are similar to those described for the production of pyruvic acid except that the oxidation shown in Figure 22-37 is not carried out. In animals, lactic acid is not normally produced by glycolysis because sufficient oxidation potential is available.

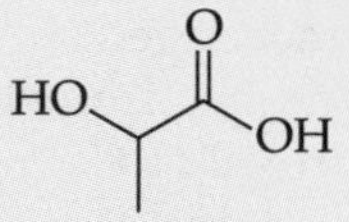

Lactic acid

However, upon extended exertion and without adequate supplies of oxygen, lactic acid is produced instead of pyruvic acid. The concentration of lactic acid in muscle tissue increases as glycolysis is called upon to provide a rapid increase in ATP as a source of energy, and it is the lactic acid that causes tired muscles to ache. Lactic acid contributes the gamey flavor that we attribute to meat from animals that have been killed on the run by hunters.

phoric acid and a substituted carboxylic acid (Figure 22-37). The reaction of this anhydride with ADP transfers a phosphate group, forming ATP and releasing the carboxylic acid. This sequence thus achieves oxidation of the

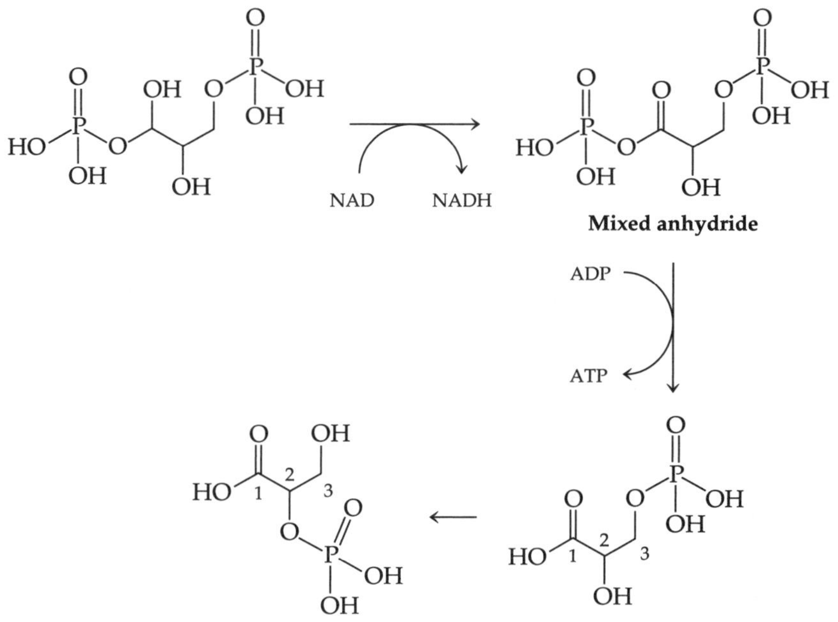

▲ FIGURE 22-37
The oxidation of glyceraldehyde diphosphate by NAD changes C-1 to the oxidation level of a carboxylic acid. The resulting linkage is a mixed anhydride between a carboxylic acid and a phosphoric acid. The energy released upon hydrolysis of this high-energy bond provides a driving force for the phosphorylation of ADP, converting it into ATP. In a final step, the phosphate group remaining on the C-3 OH group is transferred to one on C-2.

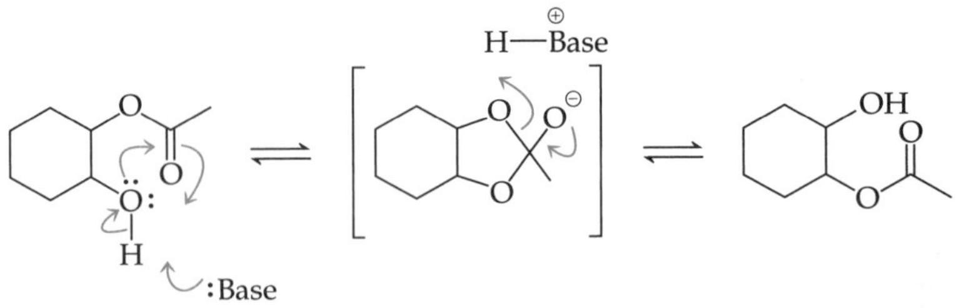

► FIGURE 22-38
An acid group can be transferred from one OH group to another within the same molecule by nucleophilic acyl substitution analogous to an intermolecular transesterification.

aldehyde of D-glyceraldehyde monophosphate, while leaving the oxidation state of the secondary alcohol unchanged.

The next reaction is yet another isomerization, one that transfers the phosphate group from the hydroxyl at C-3 of glyceraldehyde to one at C-2 (Figure 22-37). Let us step away from glycolysis for a moment to consider how this conversion takes place. Such isomerizations are very common occurrences not only with phosphate esters, but also with carboxylic acid esters. It may be easier to visualize this process with a carboxylic acid ester group, as illustrated in the isomerization of a monoacetate ester of *cis*-1,2-cyclohexane diol in Figure 22-38.

This transformation is similar to other nucleophilic acyl substitution reactions of carboxylic acid derivatives (Chapter 12). It proceeds through a tetrahedral intermediate that, lacking a carbonyl group, is less stable than either the starting material or product. Reactions that transfer carboxyl (or phosphoryl) groups intramolecularly, as in Figure 22-38, are often very fast when they proceed through five- or six-membered-ring transition states or intermediates. The transfer of the phosphate group from C-3 to C-2 illustrated in Figure 22-37 proceeds through an analogous cyclic intermediate and is very fast. Enzymatic catalysis accelerates the reaction even further.

Returning to glycolysis, we find that the next step in the sequence is the loss of water from the monophosphate ester to yield an enol phosphate carboxylic acid. In this step, the oxidation level of the terminal carbon in the product has been decreased, whereas that of the carbon bearing the phosphate group has been increased.

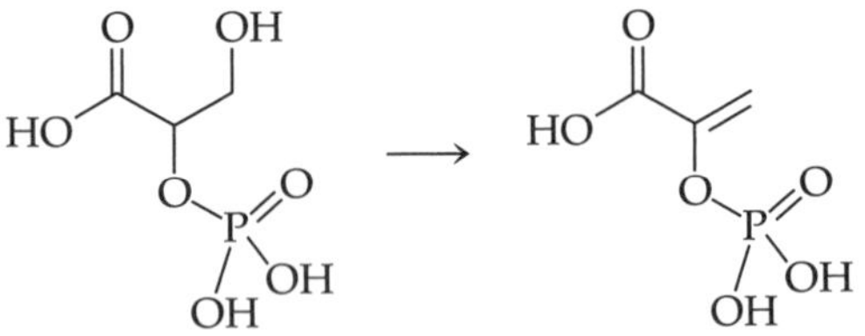

This is similar to one sequence in the tricarboxylic acid cycle in which the oxidation level of a carbon atom bound to a hydroxyl group is transferred to an adjacent carbon by dehydration, followed by rehydration. This sequence is illustrated by the conversion of 1-propanol into 2-propanol.

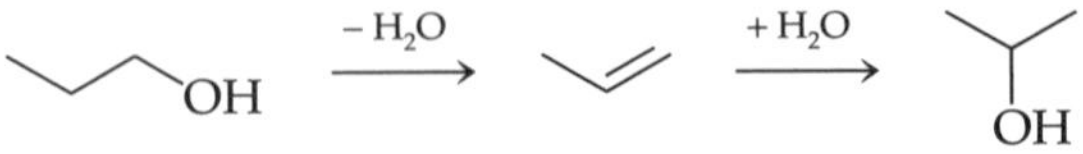

Although the overall conversion does not result in net oxidation or reduction, the oxidation level of C-1 decreases and that of C-2 increases. This oxidation and reduction occurs in two stages; in the intermediate alkene, 1-propene, the oxidation level of C-1 has decreased, whereas that of C-2 has increased. The same change happens again as the alkene is rehydrated to produce 2-propanol.

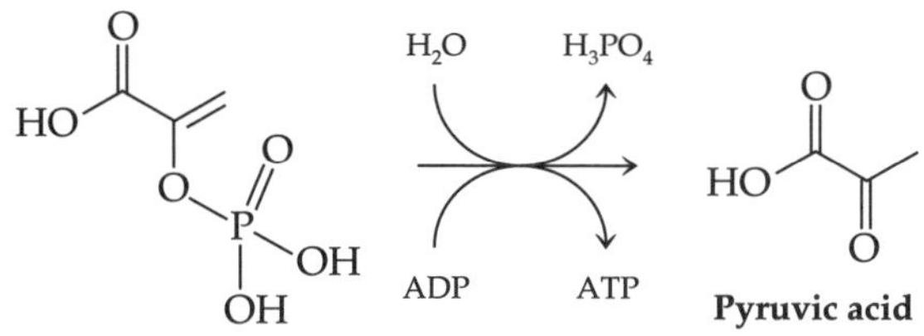

Hydrolysis of an enol phosphate produces a ketone group, thereby converting the reactant into pyruvic acid.

The enol phosphate formed in glycolysis by dehydration can be viewed, as its name implies, as a derivative of a carbonyl compound. Indeed, acid-catalyzed hydrolysis of this functional group releases pyruvic acid as well as inorganic phosphate (Figure 22-39). This simple process releases a large amount of thermal energy. Enolic derivatives of ketones are considerably less stable than are the ketones themselves because of the greater strength of the π bond of a carbonyl group compared with that of an alkene (93 versus 63 kcal/mole). In biological systems, this energy is not released fully as heat; some of it is captured by the simultaneous conversion of ADP into ATP.

We have encountered pyruvic acid before in Chapter 21, where we learned how it is degraded by decarboxylation (with the use of thiamine) to produce acetic acid. After conversion into a thiol ester, acetic acid enters the tricarboxylic acid cycle and is itself converted into two molecules of carbon dioxide. Thus, respiration, which in total constitutes the complete degradation of a hexose to carbon dioxide, is now complete. The net transformation accomplished in glycolysis is summarized in Figure 22-40 (on page 780). Each molecule of glucose produces 2 pyruvate molecules and 8 ATPs (two directly and six indirectly from 2 NADHs). (Remember that 2 ATPs are consumed in making fructose-1,6-diphosphate.) These 2 pyruvate molecules produce 2 acetates and 2 NADHs by decarboxylation. The oxidation of these 2 acetates, producing CO_2 in the TCA cycle, leads to another 22 ATPs. Again, counting each NADH as the equivalent of 3 ATPs, we produce a total of 38 ATPs by respiration in the conversion of glucose into carbon dioxide. Because each ATP stores 7.3 kcal/mole of chemical potential energy, each mole of glucose is converted into 277 kcal/mole (38 × 7.3) of potential energy. This value is 38% of that obtained when glucose is burned in air to form carbon dioxide and water, an efficiency that is almost the same as that for fatty acid degradation.

$$C_6H_{12}O_6 \ + \ 6\,O_2 \ \xrightarrow{\text{Respiration}} \ 6\,CO_2 \ + \ 6\,H_2O \ + \ 38\,\text{ATP} \qquad \Delta G° = -277 \text{ kcal/mole}$$

$$C_6H_{12}O_6 \ + \ 6\,O_2 \ \xrightarrow{\text{Combustion}} \ 6\,CO_2 \ + \ 6\,H_2O \qquad \Delta G° = -686 \text{ kcal/mole}$$

EXERCISE 22-Q

Representing ADP by the abbreviated structure shown below, write a reasonable reaction mechanism for the conversion of the enol phosphate into pyruvic acid concurrently with the transfer of the phosphate group to produce ATP.

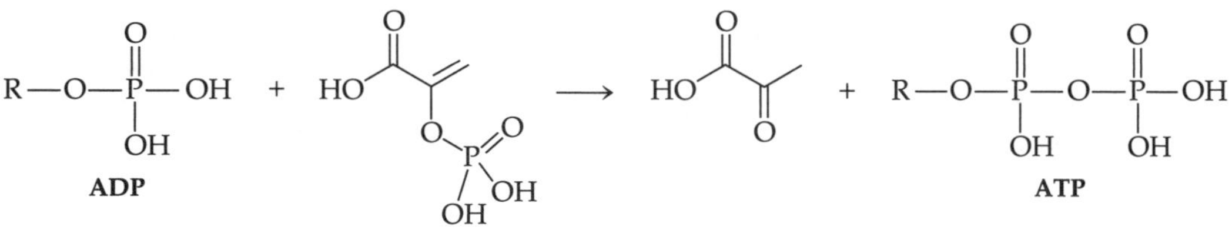

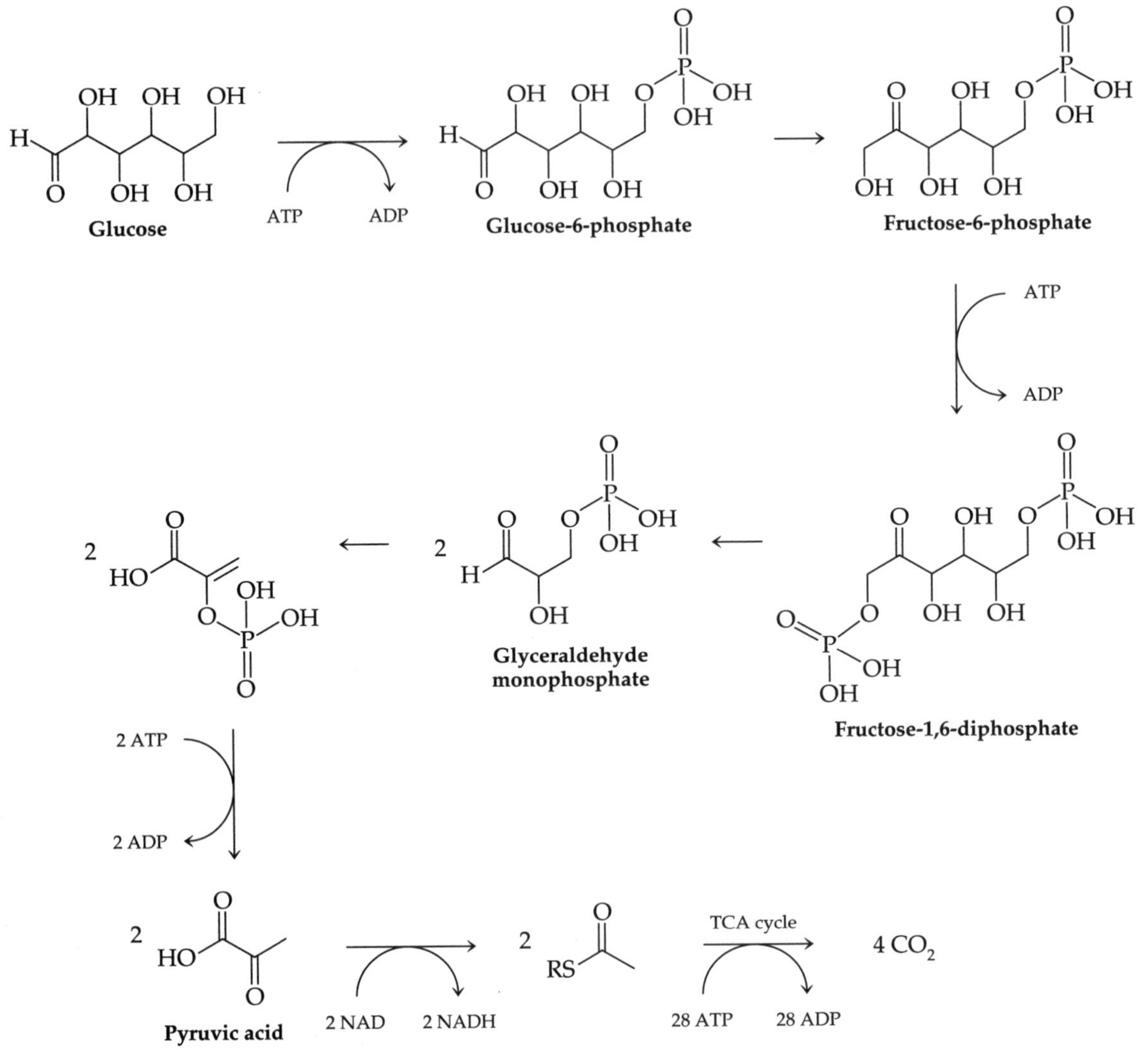

▲ FIGURE 22-40

Metabolic breakdown of sugars (glycolysis) in animals takes place in several steps: phosphorylation of glucose; isomerization to fructose-6-phosphate; a second phosphorylation to fructose-1,6-diphosphate; a retro-aldol reaction producing glyceraldehyde monophosphate; dehydration and intramolecular transesterification of the phosphate ester to an enol phosphate; hydrolysis to pyruvic acid; decarboxylation to an acetate, which is converted into a thiol ester and degraded in the TCA cycle to CO_2.

22-10 ▸ Biological Reactions in Energy Storage and Utilization

In this chapter, we have encountered a number of biochemical transformations that are involved in the storage and retrieval of energy. These reactions form an essential part of the chemistry of living organisms, including both plants and animals. Nonetheless, each of these reactions belongs to one of the basic categories discussed in earlier chapters. Examples of most of the types of reactions that we have encountered previously are represented, including: carbon–carbon bond formation; oxidation-reduction; and functional-group transformations. These biologically important reactions are summarized in Table 22-1.

Table 22-1 ▸ Biological Reactions Involved in Energy Storage and Utilization

Fatty Acid Biosynthesis

Claisen condensation:

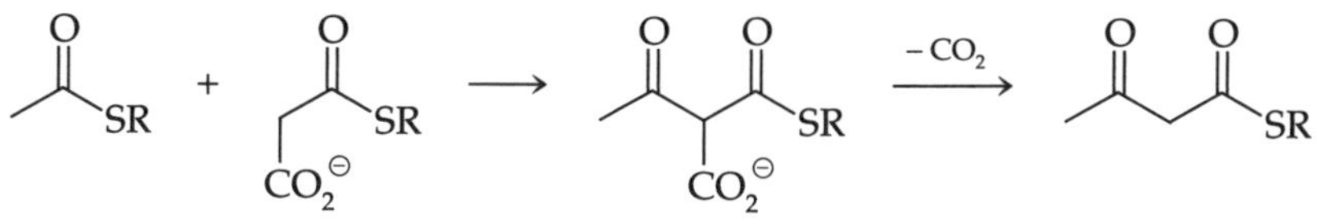

Reduction, dehydration, and reduction:

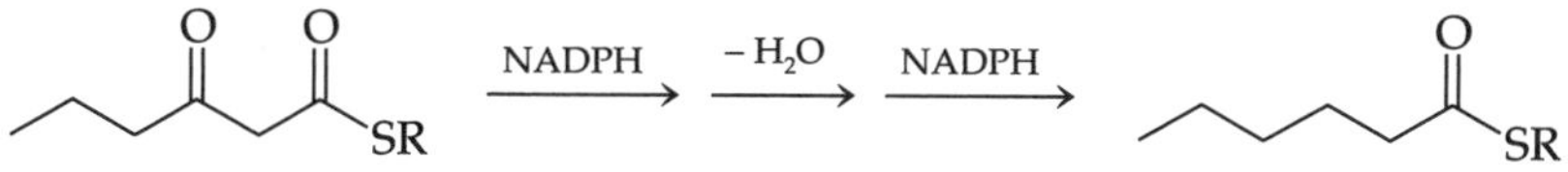

Fatty Acid Degradation

Oxidation (alkane to alkene):

Hydration:

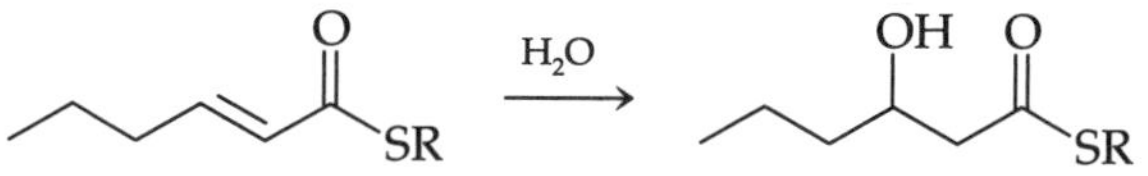

Oxidation (alcohol to ketone):

Retro-Claisen condensation:

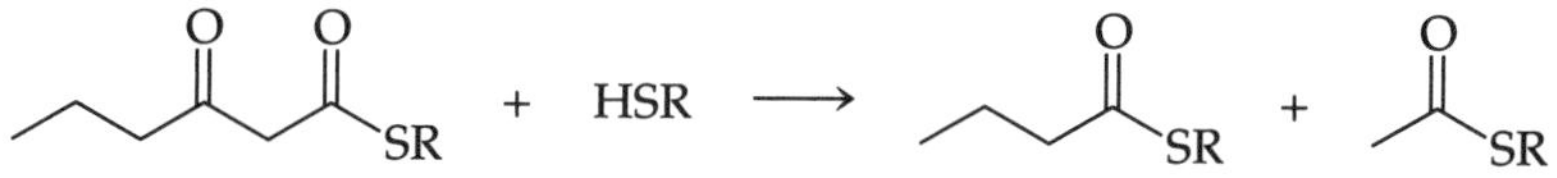

Tricarboxylic Acid Cycle

Aldollike reaction:

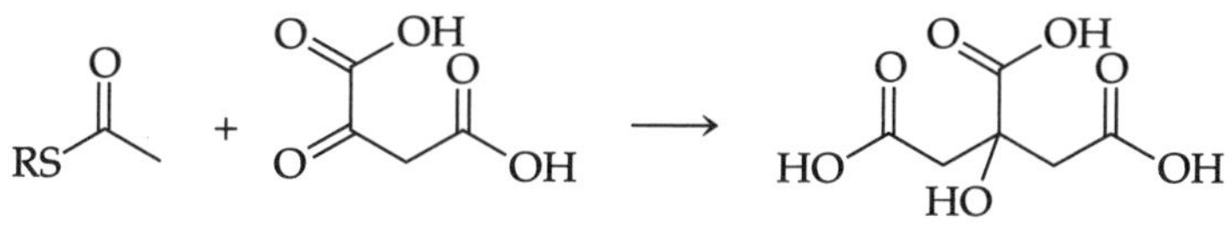

Oxaloacetic acid   Citric acid

Dehydration:

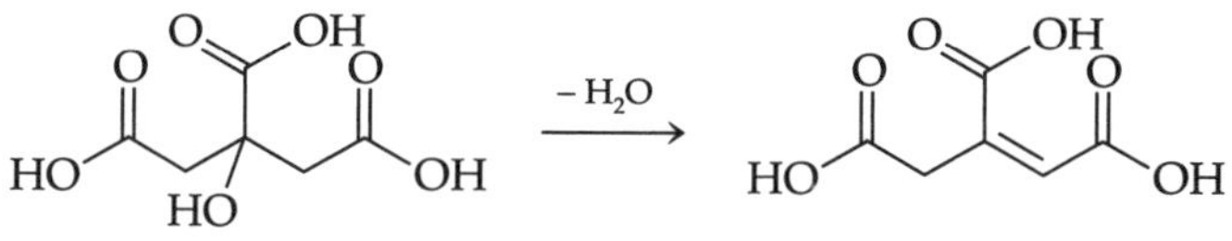

Citric acid   Aconitic acid

Table 22-1 ▸ Biological Reactions Involved in Energy Storage and Utilization *(continued)*

Hydration of an alkene:

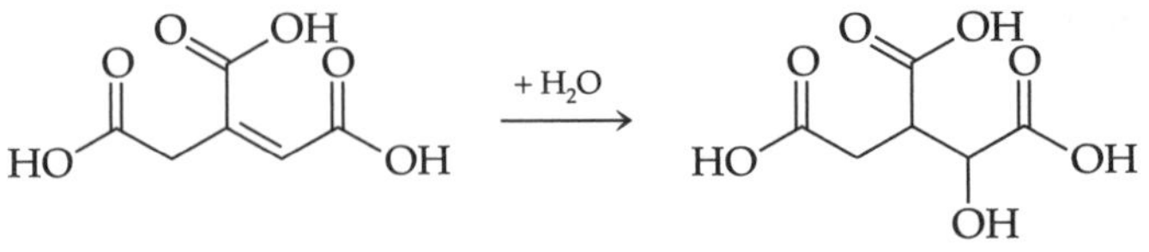

Aconitic acid → Isocitric acid

Oxidation of an alcohol to a ketone:

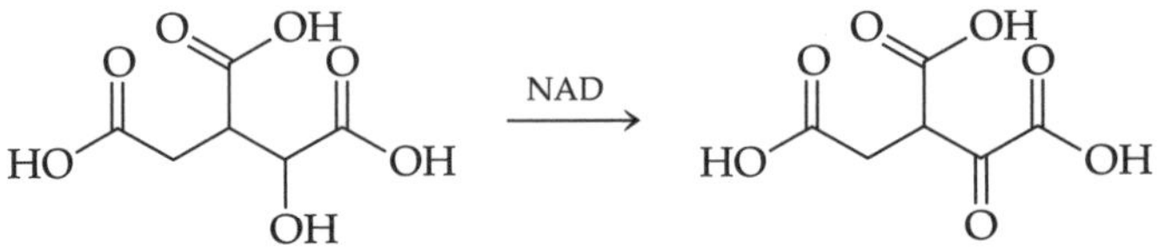

Isocitric acid → (NAD)

Decarboxylation of a β-diacid:

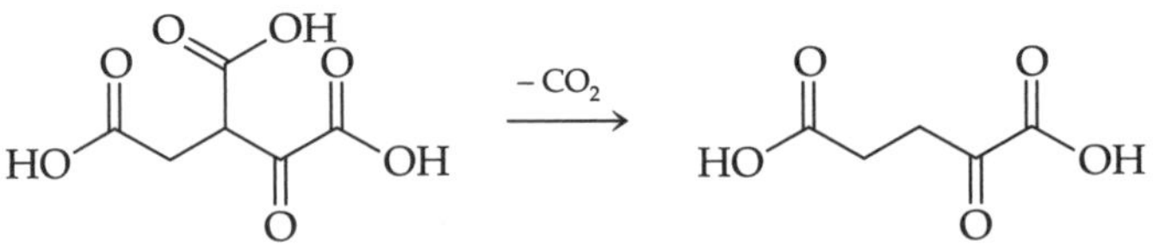

−CO₂ → 2-Oxoglutaric acid

Oxidative decarboxylation of an α-ketoacid:

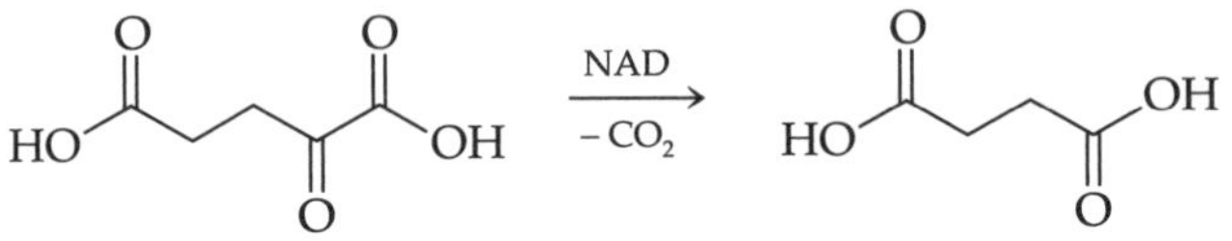

2-Oxoglutaric acid → (NAD, −CO₂) → Succinic acid

Oxidation of an alkane to an alkene:

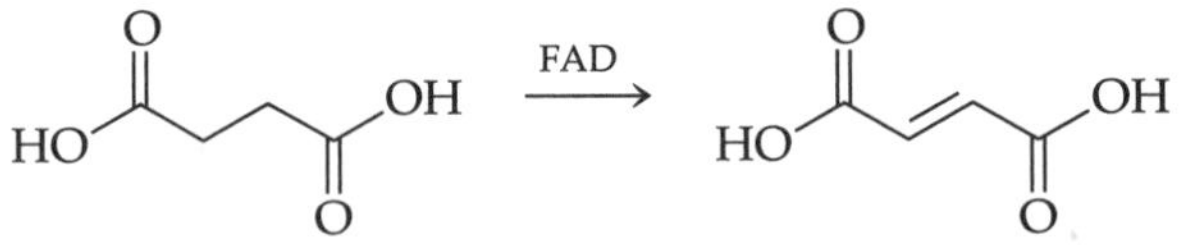

Succinic acid → (FAD) → Fumaric acid

Hydration of an alkene:

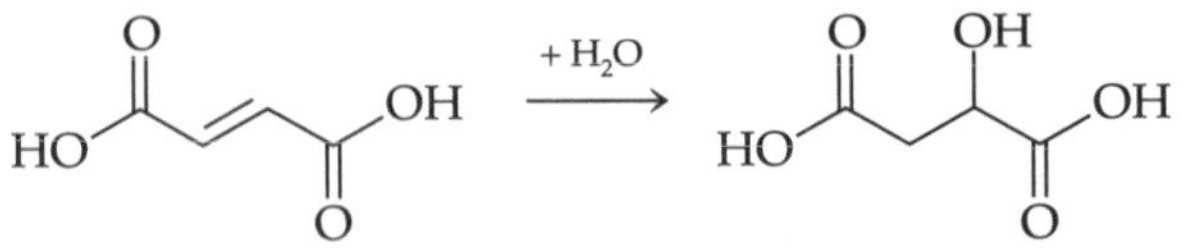

Fumaric acid → (+H₂O) → Hydroxysuccinic acid

Oxidation of an alcohol to a ketone:

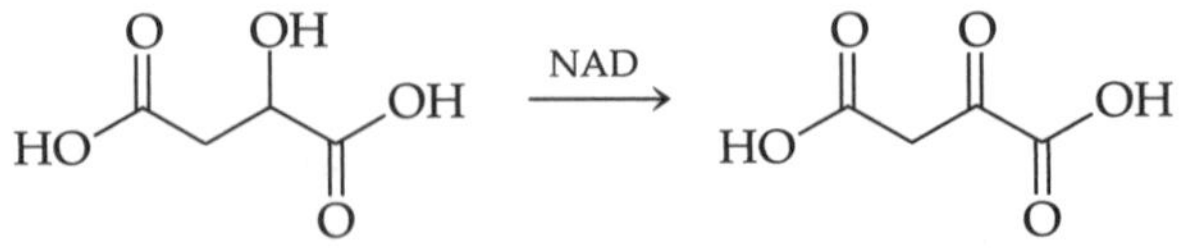

Hydroxysuccinic acid → (NAD) → Oxaloacetic acid

Table 22-1 ▸ Biological Reactions Involved in Energy Storage and Utilization *(continued)*

Glycolysis

Retro-aldol reaction:

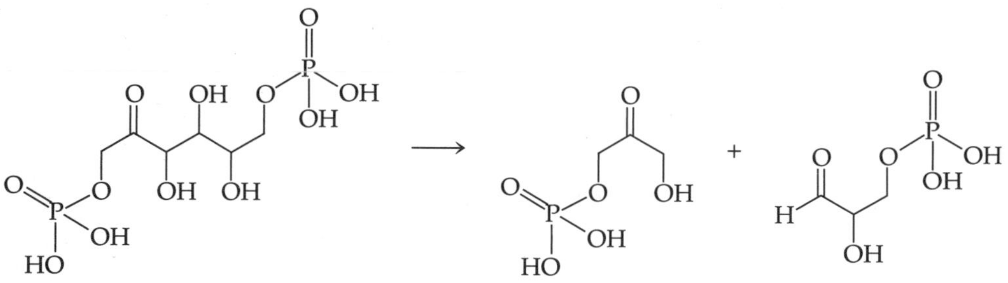

Fructose-1,6-diphosphate ⟶ Dihydroxyacetone monophosphate + Glyceraldehyde monophosphate

Keto-enol tautomerization:

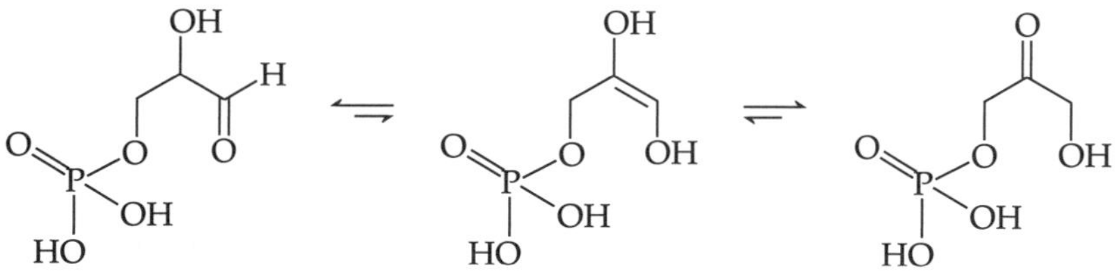

Glyceraldehyde monophosphate ⇌ ⇌ Dihydroxyacetone monophosphate

Oxidation of an aldehyde to a carboxylic acid:

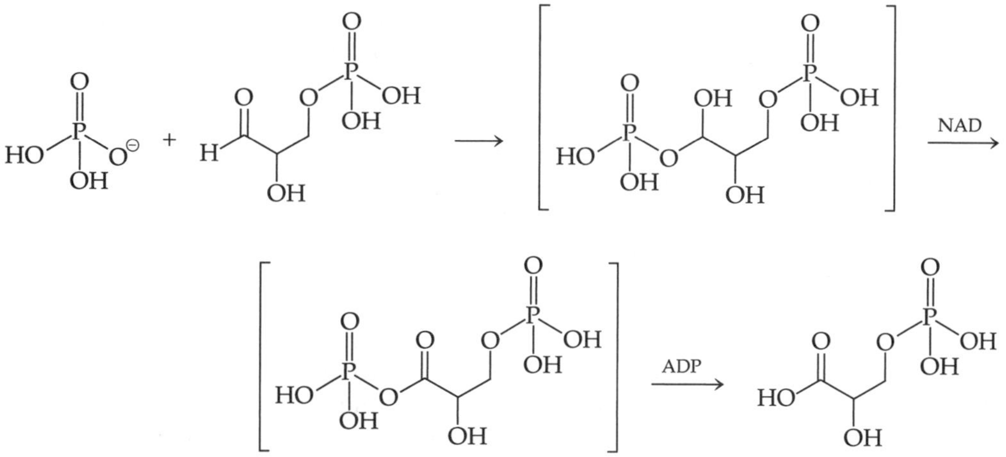

Intramolecular transesterification:

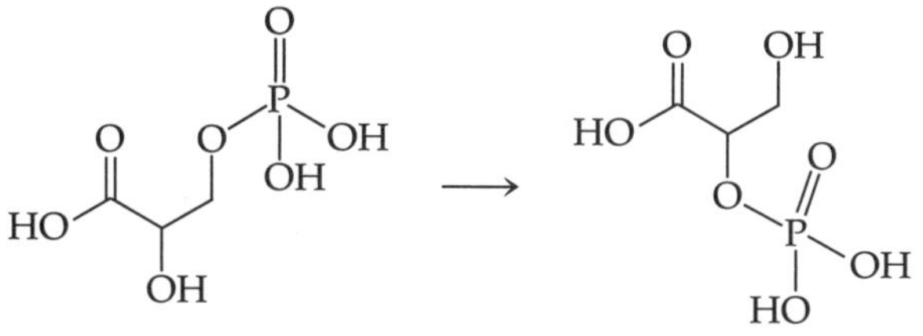

> **Table 22-1 ▸ Biological Reactions Involved in Energy Storage and Utilization** *(continued)*

Dehydration:

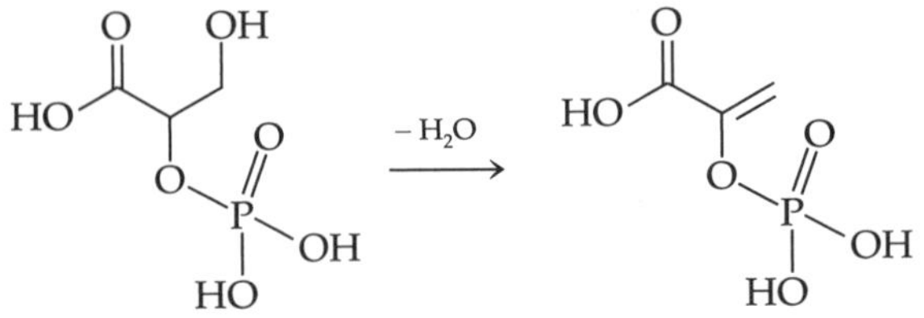

Hydrolysis of an enol derivative to a ketone:

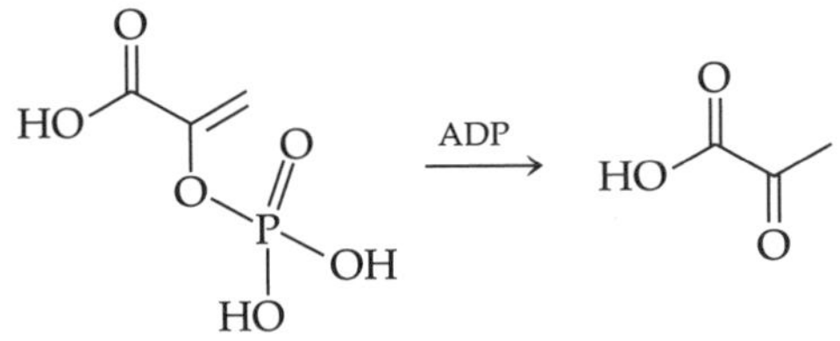

Conclusions

In the storage of chemical potential energy in biological systems, a key recurring theme is the coupling of several reactions in which one or more endothermic steps are driven by an exothermic partner. The biological transformations taking place in these reactions are closely analogous to those in the elementary reactions of organic chemistry. In some cases, the chemistry seems complex because the structures themselves are large and have a variety of functional groups. However, usually only one of these groups undergoes reaction. Among the most-common energy-releasing reactions in biological systems are the hydrolyses of adenosine tri- and diphosphates, the oxidation of NADPH to NADP, the oxidation of $FADH_2$ to FAD, and decarboxylation of α- and β-ketoesters and thiol esters. The reverse of these reactions constitute energy-storage steps when coupled with highly exothermic steps.

Energy stored in biomolecules, especially in fats and carbohydrates, is released both as heat and as chemical potential energy. The latter can be called upon to drive an otherwise unfavorable reaction toward product in a controlled fashion. The breakdown of both lipids and carbohydrates is quite efficient in converting the heats of these reactions into stored chemical energy, with approximately 40% of available energy being ultimately stored in ATP.

Fatty acid biosynthesis begins with the conversion of acetic acid to acetyl CoA, a thiolester. Carboxylation of acetyl CoA, catalyzed by biotin, introduces a carboxylic acid equivalent into the thiolester, activating it toward a Claisenlike condensation. The resulting β-ketothiol ester is reduced, dehydrated, and reduced again to provide a two-carbon chain-extended thiol ester. Repetition of this sequence produces even-numbered fatty acids after hydrolysis.

Fatty acid degradation is accomplished by the reverse sequence, initiated by FAD-mediated oxidation of a fatty acid thiol ester to an α,β-unsaturated thiol ester. Hydrolysis, followed by oxidation produces a β-ketothiol ester, from which two shorter thiol esters are formed in a retro-Claisen-like condensation.

The tricarboxylic acid (TCA or Krebs) cycle is a series of biochemical transformations that are responsible for the degradation of acetate to carbon dioxide. In this cycle, citric acid is dehydrated, rehydrated, oxidized, and decarboxylated to produce 2-oxoglutaric acid. Decarboxylation of this α-ketoacid leads to succinic acid, which is oxidized to fumaric acid. Upon hydration and oxidation, oxaloacetic acid is formed. Condensation with acetyl CoA reforms citric acid and permits the cycle to be repeated. One complete cycle converts an acetate unit into two equivalents of carbon dioxide, while releasing stored chemical energy. Some of the energy stored in acetic acid is thus converted into chemical potential energy in NADH and $FADH_2$, in a form that can be later used in other complex chemical transformations.

The breakdown of carbohydrates takes place in stages. First, glucose (a hexose) is split into two 3-carbon fragments that undergo oxidative loss of carbon dioxide to form acetic acid. This sequence begins with the isomerization of glucose-6-phosphate (an aldose) to fructose-6-phosphate (a ketose). Phosphorylation is needed for molecular recognition by an enzyme, which induces a retro-aldol reaction to form two 3-carbon fragments (dihydroxyacetone monophosphate and glyceraldehyde monophosphate). Oxidation produces pyruvic acid, which is decarboxylated to acetic acid. As a thiol ester derivative (acetyl CoA), the acetic acid then enters the TCA cycle and is itself oxidatively degraded to carbon dioxide. Most of the sugar's chemical potential energy is released in this last stage.

Reactions in which a center of chirality is formed are controlled by enzymes so that only one absolute configuration is produced. This control, which is a natural consequence of the chirality of the enzymes themselves, has additional significance in that enantiomers are treated as unique molecules in living systems. The concept of prochirality is significant in the transformation of molecules with mirror symmetry into those having chiral centers: prochiral groups take part both in the TCA cycle and in glycolysis.

Review Problems

22-1 In some of the following reactions, the reagent indicated above the arrow is required to be present in a stoichiometric amount; in others, the reagent is a catalyst and is not consumed. Indicate for each reaction whether or not the reagent is a catalyst.

(a)

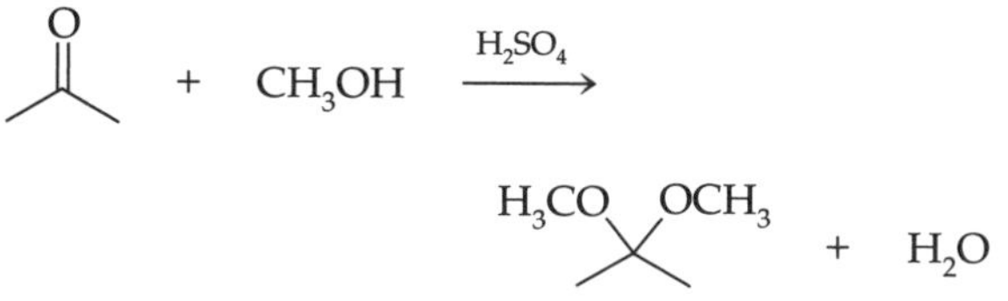

(b)

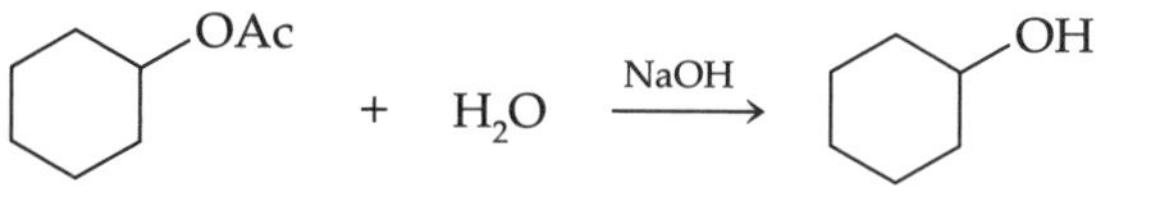

(c)

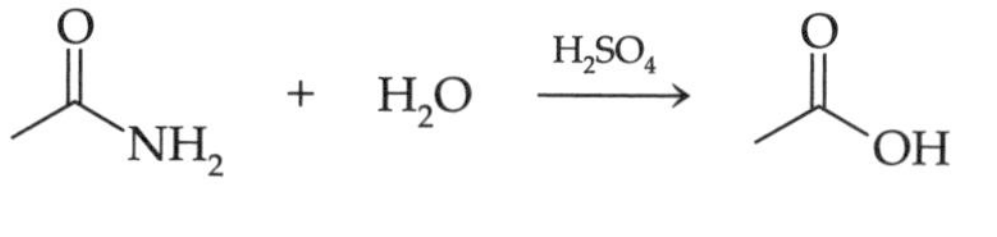

(d)

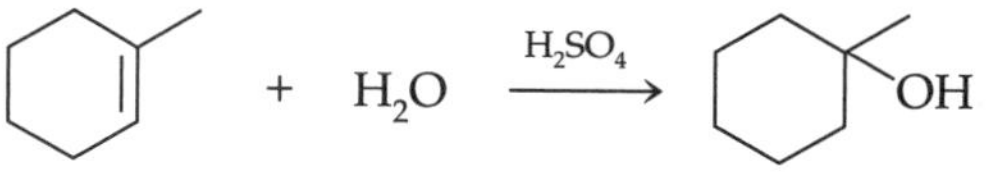

(e) $CH_3Br \xrightarrow{NaOH} CH_3OH$

22-2 Some of the reactions in Problem 22-1 take place by a pathway that includes a single transition state with no intermediates, whereas others are multistep reactions with intermediates (and more than one transition state). Classify these reactions as either single step (SS) or multistep (MS) and draw an energy diagram for each. For multistep reactions, indicate which transition state is of highest energy and is therefore the rate-limiting step in the overall reaction.

22-3 In the first energy diagram below, identify the transition state of highest energy. Decide if the equilibrium between the intermediate, I, and the reactant, A, favors I or the starting material. Do the same for the conversion of I into the product, B. Identify the transition state of second highest energy. Would the overall reaction proceed more rapidly if this transition state were lower in energy?

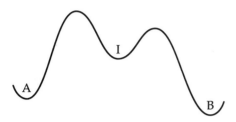

Carry out the same analysis for the following alternative energy diagram.

22-4 The rate of nucleophilic acyl substitution is usually limited by the addition of the nucleophile to form a tetrahedral intermediate rather than by the collapse of this species to product with expulsion of the leaving group. Typically, the energy profile for acyl substitution resembles the first curve in Problem 22-3, in which the first step has the higher activation energy. Under this condition, does the stability of the leaving group directly affect the rate of reaction? Would your answer be different if the reaction followed the second energy diagram in Problem 22-3?

22-5 We have seen that anhydrides react with water to form carboxylic acid in an exothermic process; they can therefore be viewed as a chemical means for the storage of thermal energy released upon reaction with water. First, explain why an anhydride is a less-stable and more-reactive carbonyl unit than the acid. Would you expect the following anhydrides to react more or less exothermically with water than acetic acid anhydride?

(a)

$$F_3C \overset{\displaystyle O}{\underset{}{\parallel}} C \underset{O}{} C \overset{\displaystyle O}{\underset{}{\parallel}} CF_3$$

(b)

$$H \overset{\displaystyle O}{\underset{}{\parallel}} C \underset{O}{} C \overset{\displaystyle O}{\underset{}{\parallel}} H$$

(c)

$$H_3CO \overset{\displaystyle O}{\underset{}{\parallel}} C \underset{O}{} C \overset{\displaystyle O}{\underset{}{\parallel}} OCH_3$$

22-6 Consider the consequence of a propanoic thiol ester taking part in the biosynthesis of a fatty acid. What saturated fatty acid would be produced from one acetate and one propanoic thiol ester unit if the key nucleophilic acyl substitution were on the acetate unit? On the propionate unit?

22-7 Apply the sequence detailed in this chapter for fatty acid biosynthesis to prepare the following fatty acid. Is there any step at which the normal course of the sequence cannot occur as a result of the presence of the extra methyl group? Explain your answer clearly.

22-8 We learned in Chapter 17 that there are unsaturated fatty acids with *cis* double bonds at specific positions along the carbon chain. Review the structures of these fatty acids and decide whether or not the double bond is in the correct regiochemical position to have resulted from a synthesis in which the α,β-unsaturated ester intermediate *did not* undergo reduction of the double bond. (Note that, even if the position were correct, the geometry would not be because the intermediate in the synthesis has the *trans* geometry.)

22-9 At one stage of fatty acid degradation, an α,β-unsaturated ester is hydrated to produce a β-hydroxyester. This process formally represents the addition of water across the double bond and can occur, in principle, with the opposite regiochemistry of that observed; that is, with the introduction of the hydroxyl group at the α carbon.

Write a reaction mechanism for this hydration under both basic and acidic conditions, and explain why the formation of the β-hydroxyester is faster than that of the α-regioisomer.

22-10 In the Krebs cycle, two seemingly identical hydrogens on a methylene group are in fact unique. Specific replacement of one leads to one enantiomer, whereas replacement of the other results in the other. (For example, the formation of (*R*)- and (*S*)-2-butanol from butane, as shown below.)

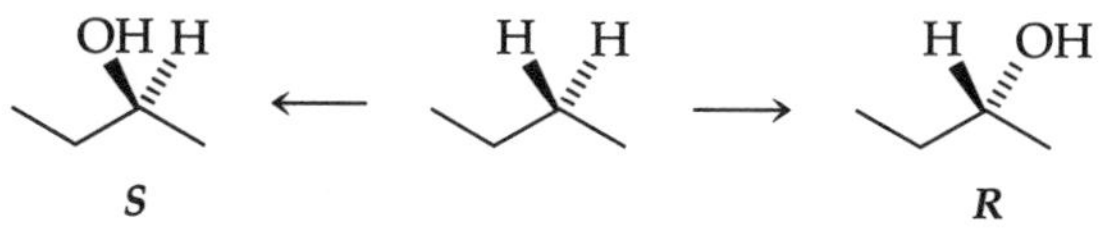

Each molecule illustrated below has several methylene groups. For each such CH_2 group, determine if replacement of one of the hydrogens generates a new center of chirality. If so, are enantiomers or diastereomers generated?

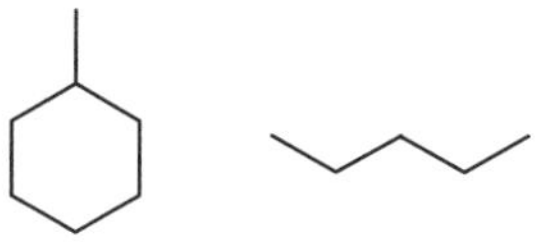

22-11 Ascorbic acid contains a stable enediol functionality. Draw structures for two alternate

hydroxy carbonyl forms (tautomers) and write a mechanism for the acid-catalyzed interconversion of these two ketonic forms that proceeds through vitamin C.

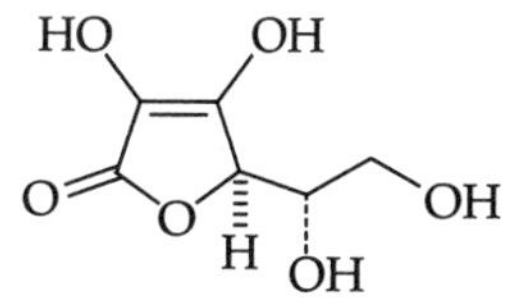

Vitamin C
(ascorbic acid)

22-12 The formula of vitamin C ($C_6H_8O_6$) is close to that of hexose carbohydrates ($C_6H_{12}O_6$). Is vitamin C at a higher or lower oxidation state than carbohydrates? How many centers of chirality are there in vitamin C (refer to the structure in problem 22-11)? Identify the end of the carbon chain having the lower oxidation level and classify vitamin C as belonging to either the D or the L series. The stereochemistry of vitamin C is as shown above.

Chapter 23

Molecular Basis
for
Drug Action

As we approach the end of our study of organic chemistry, it is appropriate to briefly summarize the substance of the course as it relates to practical problems. One important area for the practice of organic chemistry is in the design and synthesis of new drugs in the pharmaceutical industry. The development of an effective pharmaceutical agent is a very complicated endeavor. Despite tremendous advances in the understanding of the chemical basis of disease states, the success rate obtained in the progression from initial concept to final product is low. It has been estimated that, of every fifty compounds tested in pharmaceutical laboratories, only one shows promising biological activity. Structural modification of this lead compound produces an agent worthy of further study only one time in 100. Of these agents, only one in ten passes successfully through clinical trials with human patients and becomes marketable.

There are many reasons for this low rate of success. Effective pharmaceutical agents must have not only structures that can elicit the desired response, but many other features as well. They must have some water solubility so that they can be transported through the blood; they must be at least somewhat resistant to chemical degradation in the body; and, perhaps most importantly, they must not interfere elsewhere in the complicated biochemistry that is essential to life.

Organic chemistry has played a major role in the development of new drugs. Our knowledge of disease states and their causes and of the details of biochemistry is expanding at a tremendous rate. Using this knowledge, organic chemists play an ever-increasing role in the design and development of new and effective pharmaceutical agents.

The choice and synthesis of a medicinally active molecule in many ways exercises what you have learned from this book. We began this study by reviewing basic concepts of bonding and structure that are part of a first-year course in chemistry. We extended these basic features to the relatively large collections of atoms found in organic substances. In particular, stereochemistry, both configurational and conformational, accounts for some of the exceptional diversity found among organic compounds. We then began a detailed analysis of the various reactions common to organic chemistry, developing an organization based on the mechanisms of these transformations. With a fundamental knowledge of structure and reactivity in hand, we progressed to polymers and learned that some of the macroscopic properties of these materials can be predicted from a knowledge of the functional groups present, whereas other features derive from their polymeric nature.

After a study of the structures and properties of various naturally occurring organic substances, we learned that the structural and functional complexity of many biochemicals plays an important role in specific molecular recognition, by which each species can perform a particular role without taking part in other transformations. We then learned the special features of catalyzed reactions, both for laboratory transformations and for biologically important processes in which enzymes serve the role of catalyst.

We are now in a position to use this accumulated knowledge to gain an understanding of the molecular basis for the action of modern, as well as older, pharmaceutical agents. The use of chemicals for "healing" is generally referred to as chemotherapy, although that term often takes on a more specific connotation in the treatment of cancer.

23-1 ▸ Chemical Basis of Disease States

In a general sense, we can classify most disease states as resulting from one of the following conditions:

1. *Under*production of a critical biochemical

2. *Over*production of a critical biochemical

3. The invasion of an alien living species that produces substances that are toxic to the host or consumes materials required by the host

4. The too-rapid growth of part of the organism

Of these conditions, the first is generally the easiest to control because all that is required is a supplement from an external source to make up for the deficiency. For example, diabetics lack sufficient production of insulin, a peptide hormone fifty-one amino acids in length that triggers the cleavage of glycogen to glucose, initiating the process of energy release through glycolysis. Because the enzyme pepsin (present in the stomach), like chymotrypsin, cleaves peptide bonds, insulin cannot be taken orally. However, subcutaneous injection of insulin leads to its rapid transfer to the blood stream, where it has its physiological effect.

Many other disease states result from a deficiency of essential biochemicals. Indeed, the importance of vitamins was first discovered because of the diseases that resulted from a dietary deficiency: scurvy (vitamin C); rickets (vitamin D); and beriberi (vitamin B_6, or thiamine). There are also diseases that result from a deficiency of inorganic ions: for example, goiter is caused by too little iodide ion. Dietary deficiencies are generally easy to correct, but other diseases resulting from the lack of enzymes caused by genetic faults are much more difficult to treat. For example, sickle-cell anemia is caused by alterations in the amino acid sequence of hemoglobin, reducing its ability to carry oxygen.

The overproduction of acid in the stomach and intestines is not only painful but can lead to the destruction of the lining of the digestive tract, producing ulcers. We will see in Section 23-3 how the modern drug cimetidine (Tagamet) blocks acid production by mimicking the structure of the natural material histamine, which triggers acid production.

Not all areas of the body can be reached so readily, however. For example, transport across the membrane surrounding a nerve cell is quite selective, and generally only small polar molecules are carried into the interior of these cells. We will see in Section 23-4 how morphine and other psychoactive compounds mimic the actions of naturally occurring peptides in the brain, leading to similar, but often greatly enhanced responses.

Living systems, though similar to one another, are not identical. Some of these differences derive from the precise structure of certain biomolecules that are thus characteristic of that particular organism. The immune system present in animals, including human beings, is a natural defense that relies on subtle differences in the composition of the exterior of cell walls. This system depends on the production of "random" antibodies and the tight association of a specific type of antibody with a specific alien material: for example, the specific binding to one carbohydrate on the surface of an invading cell. Production of this particular antibody is then increased and special cells (called T cells) consume the antibodies along with their attached alien material.

The difference in the carbohydrates present on the surface of blood cells is the chemical basis for the classification of different human **blood**

STEREOCHEMISTRY AND PHARMACEUTICALS

There are approximately 1850 different compounds that are marketed worldwide as pharmaceutical agents. From the following breakdown, we can see that most are prepared in the laboratory (semisynthetic means that a compound from nature is modified in the laboratory).

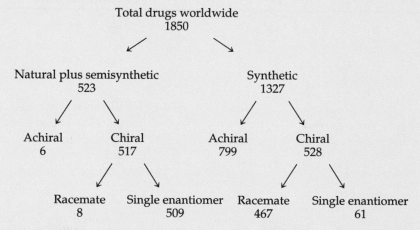

Notice that almost all of the natural and semisynthetic drugs are chiral and that, further, most of these are marketed as single enantiomers. On the other hand, the majority of the synthetic drugs are achiral, and most of the chiral synthetic drugs are sold as racemates. It is almost invariably the case that only one of the enantiomers of the racemic mixture has the desired effect. Because it is possible that the other enantiomer can have undesirable effects (recall the story of thalidomide in Chapter 5, page 183), there is an increasing desire in the pharmaceutical industry to market single enantiomers.

types. Figure 23-1 shows the oligosaccharides responsible for the classification of blood as types A, O, AB, and B. Although these structures are complex, they differ *only* in the region shown in color. When blood of a different type is administered to a patient, antibodies associate with these foreign sequences of carbohydrates and, as a result, the alien cells are destroyed. One might wonder why a person lacks antibodies for his or her own blood cells. In total, there are approximately 10^{12} different antibodies owing to different permutations of amino acids in that part of an antibody referred to as the variable region, which typically contains about sixty-three amino acids.

Type O

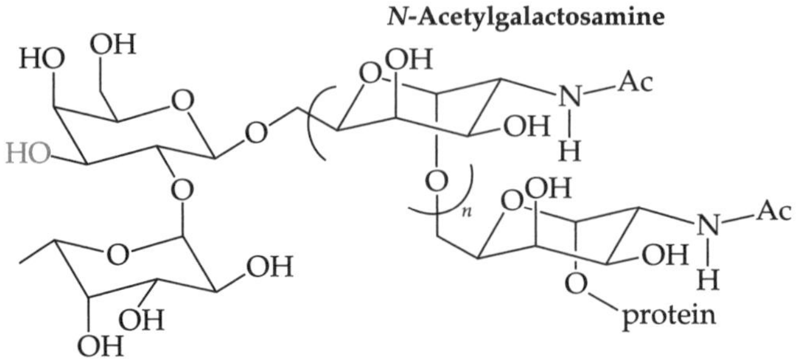

Type A

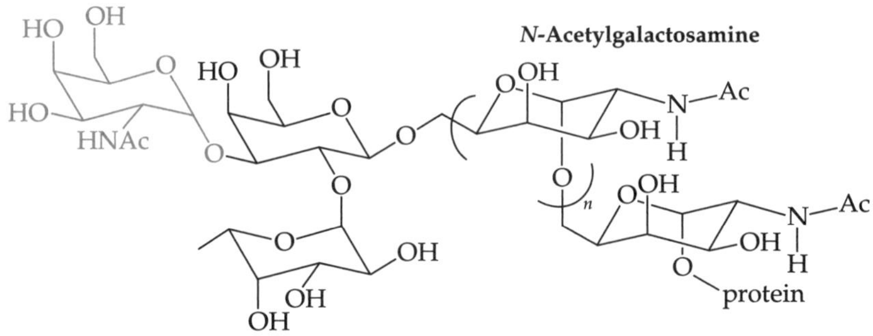

Type B

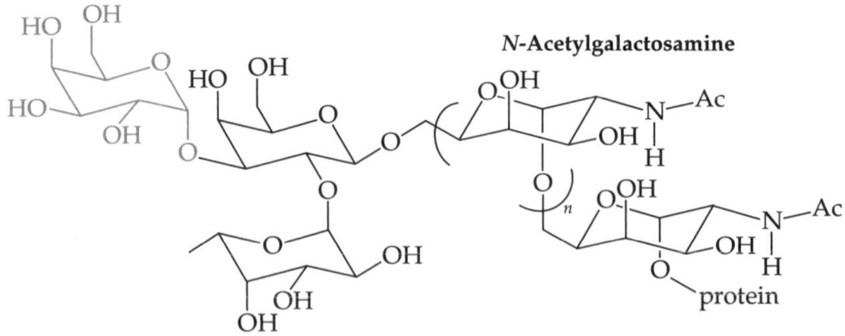

▲ FIGURE 23-1

The differences between the various human blood types are due to the presence of slightly different saccharides on the surface of blood cells. Although these differences are not large, they are sufficient to trigger the immune system when, for example, type B blood is given to a person who has type A. Type O blood lacks the sugar substituent and thus does not trigger the immune system; type O blood can be given to patients having any of the blood types (the universal donor).

In early stages of fetal development, those antibodies that would trigger a person's immune system against itself are removed.

There are even greater differences between human beings and other living systems. For example, bacteria synthesize tetrahydrofolic acid, the cofactor that transfers single carbon units (Chapter 21), but mammals do not and must obtain this essential compound from the diet. A drug that interferes with the synthesis of tetrahydrofolic acid can therefore kill bacteria, but does not affect human beings. In Section 23-6 we will see that the sulfa drugs function as antibacterial agents in exactly this fashion.

Antibiotics belonging to the cephalosporin and penicillin classes function in a different fashion, by interfering with the construction of bacterial cell walls. The selective, deleterious effect of these chemicals on bacteria is possible because of a fundamental difference between the cell walls of bacteria and animals. We will examine in Section 23-7 how this difference in the "construction" of bacterial cell walls and those of mammals provides an opportunity for the selective destruction of the bacteria.

Cancer is a most-difficult disease state to control. For reasons not yet understood, an aberration in normal cells accelerates the natural rate of cell multiplication. The most-apparent difference between cancerous and normal cells is this enhanced rate of division, and many anticancer chemicals target this difference, some of the details of which are presented in Section 23-8.

23-2 ▸ Intact Biological Systems as Chemical Factories

In Chapter 20, we equated equilibrium with death: with only a few exceptions (such as viruses), living systems are in a constant state of flux. New raw materials constantly enter a system, while products and by-products are excreted. It is difficult to imagine how this dynamic state could be maintained without compartmentalization. Indeed, the basic building block of living systems, the cell, is itself divided into smaller units such as the nucleus and ribosomes. We learned in Chapter 17 how cell membranes are constructed of lipid bilayers containing important special substructural features that facilitate the flow of inorganic ions and small organic molecules into and out of the cell. These membranes serve a dual role in isolating the contents of the cell and its substructures and in permitting selective transport. We will learn in Section 23-6 that many antibacterial agents are effective because they interfere with one or the other of these two critical functions of membranes.

It is one thing to destroy foreign organisms in a laboratory environment (*in vitro;* that is, "in glass") and quite another to do so *in vivo* (in a living organism) without simultaneously destroying the host. Life is a delicate state that lies quite far from equilibrium. There are many simple ways to destroy this state because all living things are sensitive to such environmental features as temperature and pH. There are no known life forms that survive temperatures in excess of 150 °C or treatment with concentrated sulfuric acid, which denatures the critical biopolymers (peptides, enzymes, nucleic acids). As we shall see, pharmaceutical agents that are to be used to destroy an invader organism, such as a virus or a bacterium, or a cancer cell must be not only effective at the assigned task, but selective as well.

23-3 ▸ Beta–Blockers: Modern Antacids

For many years, the standard treatment for the overproduction of acid in the stomach was the administration of antacids such as sodium bicarbonate or a mixture of aluminum and magnesium hydroxides. These treatments are still used: check for yourself the ingredient statement on a commercial antacid such as Maalox or Alka-Seltzer. Excessive bicarbonate, however, can make the stomach alkaline, triggering the production of even more stomach acid. The metal hydroxides constitute a better treatment because they are not soluble in water and do not increase the pH above neutrality. However, these techniques address only the symptoms of excessive acidity, not the cause.

A major breakthrough was the discovery that **histamine,** a potent vasodilator, also stimulates the secretion of pepsin and acid into the stomach. Compounds such as bromopheniramine and terfenadine, called **antihistamines** (Figure 23-2), had been used to inhibit vasodilation by competing with histamine for binding at the site at which histamine exerts its effect (sometimes resulting in the excess release of fluid into the nasal passageways). It only remained for chemists at Smith-Kline-Beecham in England to develop another mimic of histamine that would act at the receptor sites where histamine triggers the release of acid and pepsin. (These are distinctly different receptors: those molecular features that favor tight binding to one most likely would not be optimal for binding to the other.) Their search was successful and produced cimetidine (Figure 23-3), known as Tagamet when sold by Smith-Kline-Beecham and by other names when sold by other pharmaceutical companies. The sale of beta-blockers worldwide for the treatment of ulcers exceeds $2 billion.

EXERCISE 23-A

In some cases, the structural analogy between a substance such as an antihistamine and the natural substance with which it competes for binding is clear; whereas, in other cases, the relation is less obvious. Compare the structure of histamine with those of brompheniramine, terfenadine, and cimetidine, and identify the structural features that the three synthetic compounds have in common with histamine.

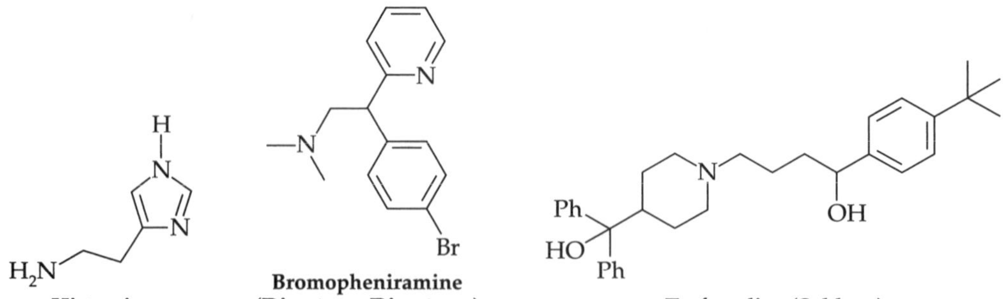

Histamine

**Bromopheniramine
(Dimetapp, Dimetame)**

Terfenadine (Seldane)

▲ FIGURE 23-2

Bromopheniramine (marketed under the trade names Dimetapp and Dimetane) and terfenadine (sold as Seldane) are synthetic compounds that bind to the same receptor site as histamine but do not trigger the same response. Because they interfere with the natural action of histamine, they are called antihistamines and are used to relieve the symptoms of common colds and allergic reactions to pollen.

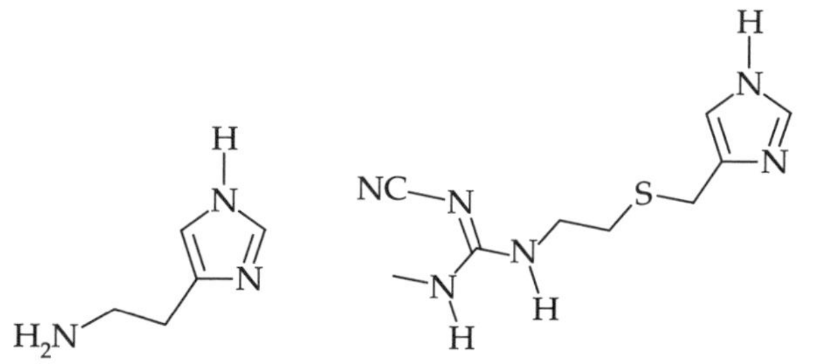

▲ FIGURE 23-3

Cimetidine (sold as the antiulcer drug Tagamet) structurally resembles histamine. It binds to receptors that trigger the release of acid into the stomach, thus reducing the irritation to inflamed stomach lining, permitting the ulceration to heal.

The process of drug discovery is both challenging and fascinating because there are many requirements that must be met by a pharmaceutical agent: it must have the desired effect (in this case, binding to, and thus blocking, the receptor site for histamine without triggering the same response); it must be able to move through the body to the target, and preferably through the stomach wall without being destroyed so that it can be administered orally; it must not elicit other responses by taking part in other biochemical transformations; and it must ultimately be excreted, either intact or after degradation in the liver, after a reasonable period of time. We can imagine how some of these issues can be addressed at the molecular level. For example, transport throughout the body requires some water solubility. Others can be dealt with at the macroscopic level. For example, the drug can be coated with a substance that is resistant to the action of stomach acid and pepsin but is ultimately degraded in the intestines. This latter aspect of drug therapy and pharmaceutical research is called **formulation** and is just as important as the molecular aspects of drug design. Unfortunately, finding the right balance among all the necessary features and properties for a drug still substantially requires trial and error.

23-4 ▸ Beta-Phenethylamines: Peptide Mimics

The discovery of cimetidine (Tagamet) is an example of a modern approach to the development of new pharmaceutical agents that requires a reasonable level of understanding of the biochemical processes of a disease state. Using this knowledge, the skilled synthetic chemist can then prepare molecules designed to elicit a desired response. However, some very useful drugs have been developed and used for many years before this knowledge became available. Indeed, the majority of the "miracle drugs" were uncovered by chance. This section and the next three sections deal with several biologically active and useful therapeutic agents that were uncovered by chance, although many details of their modes of action are now available.

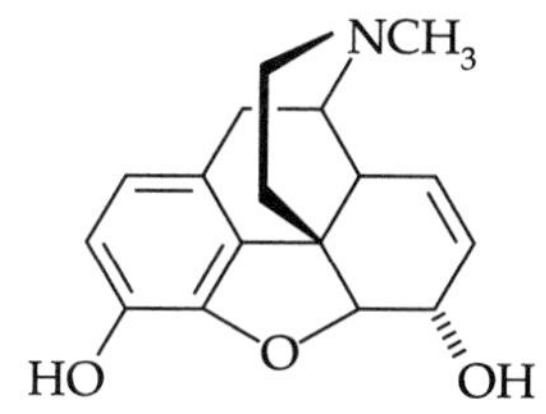

Morphine and heroin have been used for many centuries as **psychoactive drugs** to achieve a state of euphoria. Heroin is hydrolyzed inside brain cells to form morphine and thus produces an identical physiological effect. It is preferred by drug abusers, however, because it is less polar than morphine and crosses from the blood into the brain more readily, resulting in a more-rapid and more-elevated "high." More recently, similar drug-induced neurological activity has been used medicinally to relieve severe pain—for

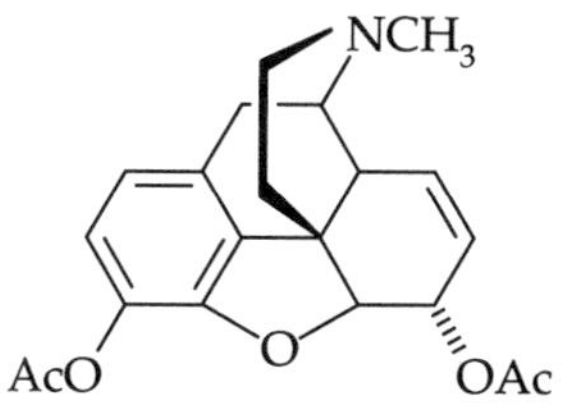

Morphine

Heroin

example, in terminally ill patients who have cancer. Morphine, the major active component of the poppy, rightly derives its name from the Greek god of dreams, Morpheus. The veil that morphine hangs upon the mind is unsurpassed in clouding the sensation of both physical and mental pain. Most unfortunately, the pain-abating character of morphine is accompanied by an equally virulent power to cause physical dependence.

Morphine is also an "old" compound in that it was the first of the naturally occurring bases to be isolated in pure form. A German apothecary named Friedrich Sertürner reported his discovery in 1805, but it was not until 1835 that the French scientist Jean Baptiste André Dumas coined the term alkaloid for this group of compounds. Although morphine was the first alkaloid isolated (Chapter 1), its structure was not established until 1925. Because it has such a long history, one might expect that its mode of action would have been described in detail many years ago. However, this is not the case, and only when details of how it affects the brain were uncovered did it become clear why this had been difficult: morphine mimics the natural action of small pentapeptides called **endorphins,** a name derived from *endo*genous m*orphine,* that are present in the brain in extraordinarily low concentrations. Endorphins serve as natural pain relievers that function in the brain to change or remove the perception of nerve signals. It is believed that their concentration is increased under conditions of high stress, as, for example, when long-distance runners push themselves to their limits and beyond (described as "hitting the wall").

Two endorphins known specifically as **enkephalins,** Met-enkephalin and Leu-enkephalin, are shown here. These peptides are distinguished by the presence of either a methionine or a leucine residue at the carboxyl terminus. (Notice that these residues are quite similar in molecular shape.)

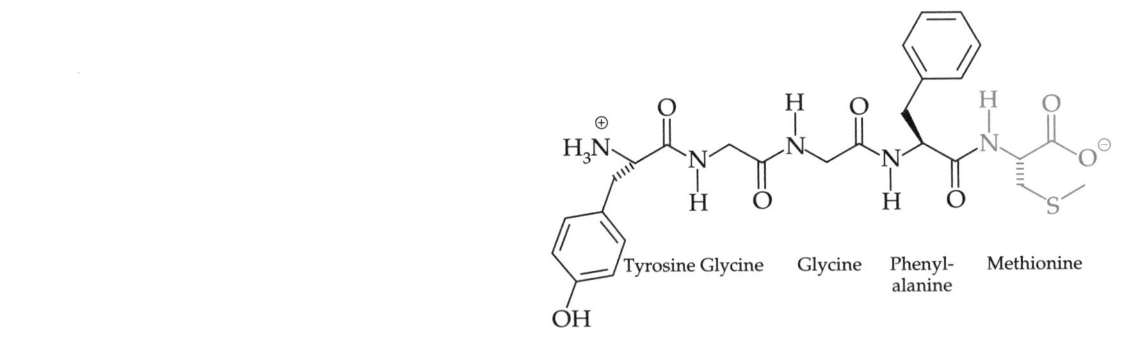

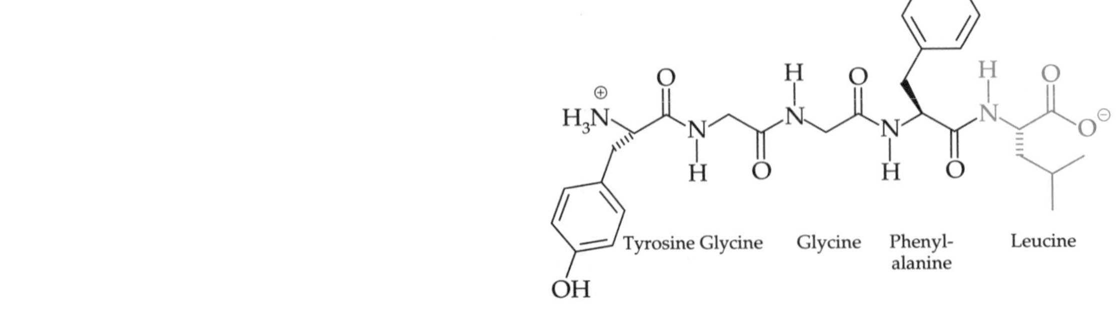

The endorphins are believed to be naturally produced pain killers, selectively intervening with the perception of pain but not with nerve signals from other senses. Morphine has the same effect because its three-dimensional shape mimics that of the enkephalins and produces the same

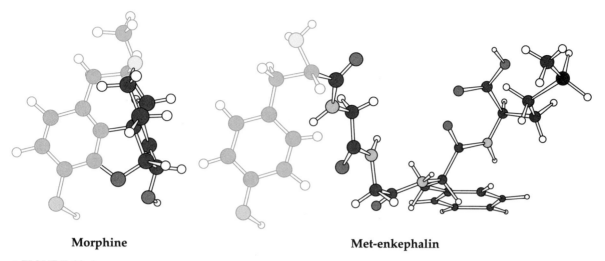

Morphine   **Met-enkephalin**

▲ FIGURE 23-4

In these three-dimensional representations of morphine and Met-enkephalin, the region of each molecule shown in color is believed to be the part that conforms to a common binding site in the brain.

response. Shown in Figure 23-4 is a representation of Met-enkephalin in a conformation similar to that of morphine. Through a detailed study of the biological effect of many different peptides resembling the natural enkephalins, it has been possible to demonstrate that the phenolic ring and amino group of tyrosine, as well as the phenyl ring of phenylalanine, are essential for pain-relieving activity.

Many other compounds that greatly affect the brain have a structural feature in common: the presence of a β-phenethylamine moiety. Many of those shown in Figure 23-5 (on page 798) have similar effects on the perception of pain but influence other brain functions as well.

EXERCISE 23-B

Identify the β-phenethylamine subunit present in each of the compounds shown in Figure 23-5.

23-5 ▶ Blocking Tetrahydrofolic Acid Synthesis

As the science of chemistry progressed in the nineteenth and twentieth centuries, the idea that chemicals can adversely affect invading bacteria but not a host came to the front. For example, the German chemist Paul Ehrlich considered that drugs could interact differently with the host and the parasite and explored the possibility that altering the structure of a toxic substance would change its relative toxicity, described as the **chemotherapeutic index,** the ratio of toxic dose for the host to that for the invading organism. Ehrlich's initial studies led to the synthesis of atoxyl, which he prepared by heating the extremely toxic metal arsenic with aniline in air (Figure 23-6). The chemotherapeutic index of atoxyl was only about 10, and long-term use led to serious toxicity to the host as well. Further modifications of the structure led in 1912 to arsenphenolamine, known as Salvarsan, the first really effective treatment for syphilis. Although Salvarsan is toxic to human beings, its effect on the spirochete that causes syphilis is much greater.

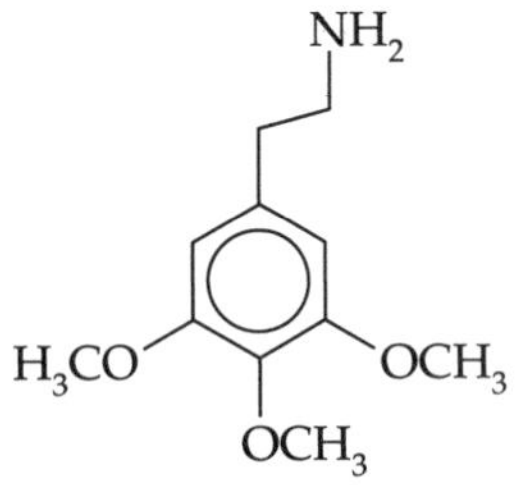

Mescaline
(causes serious
psychological
disturbances)

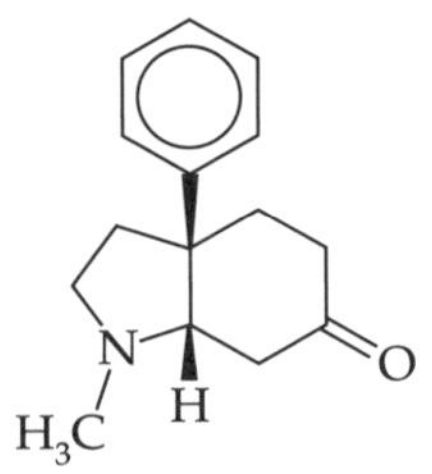

Mesembrine

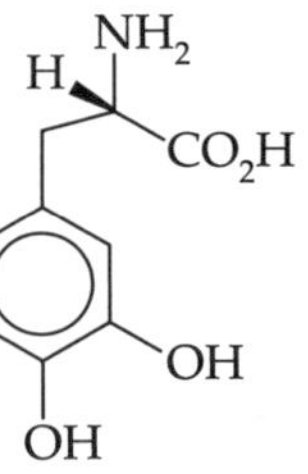

Levodopa
(L-dopa, for treatment
of Parkinson's disease)

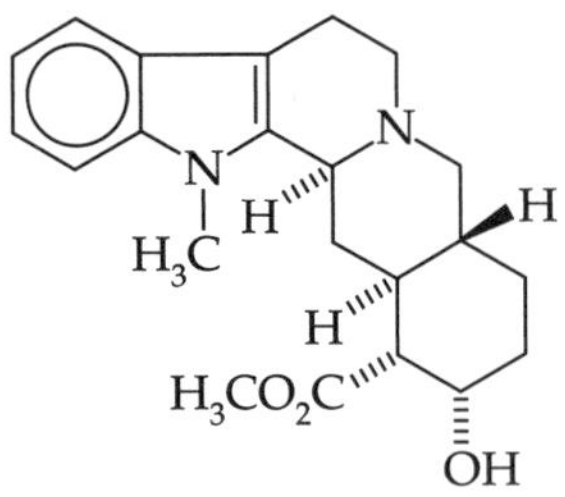

Yohimbine
(has been used
as an aphrodisiac)

Lysergide
(lysergic acid
diethyl amide, or LSD,
causes serious
psychological
disturbances)

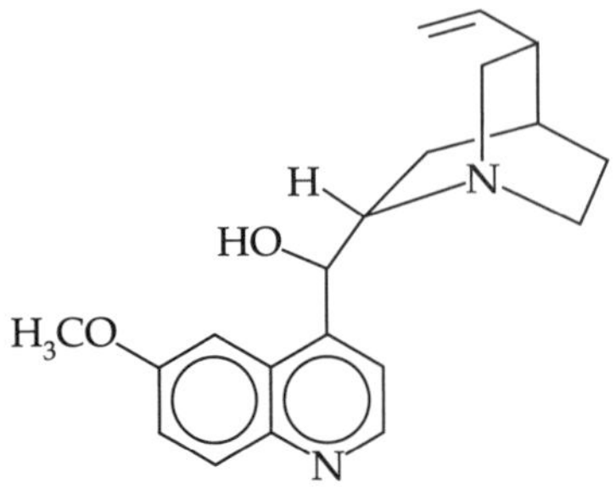

Quinine
(antimalarial, has
also been used as
an analgesic in
animals)

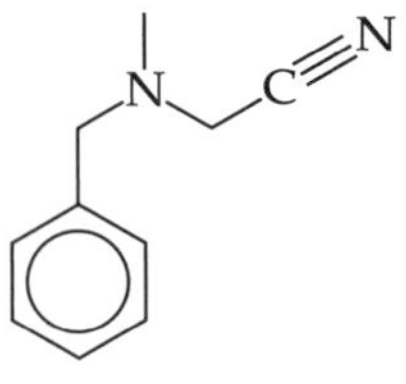

***l*-Deprenyl**
(used for treatment
of depression)

▲ **FIGURE 23-5**

All of the compounds shown here have two features in common. One is the presence of a β-phenethylamine substructural unit. The other is that they affect the central nervous system when administered to human beings and other animals.

► **FIGURE 23-6**

Atoxyl (*p*-aminophenylarsonic acid) was one of the first drugs to be prepared by chemists. Although it did have the desired activity against syphilis, it was too toxic to human beings to be of use for treatment of this disease.

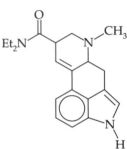

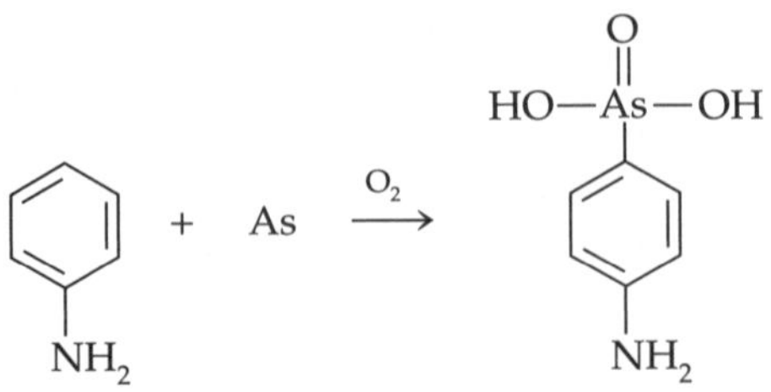

p-**Aminophenylarsonic acid**
(atoxyl)

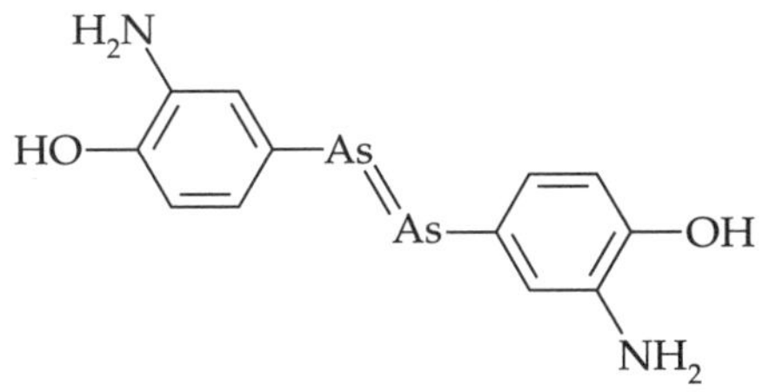

Salvarsan

Ehrlich continued his studies, concentrating on dyes produced by chemical companies. (Indeed, the large-scale production of chemicals was started in order to produce coloring agents. The new chemical industry was located in Basel, Switzerland, on the Rhine River just as it left that country, becoming the border between France and Germany, taking the industrial waste away from Switzerland.) Many dyes are produced by the coupling reaction between an amine and an aryl diazonium salt (Figure 23-7). The extended conjugation in the product is responsible for the strong absorption of light in the visible region of the spectrum. The structural similarity between these dyes and Salvarsan is clear, with the substitution of nitrogen for arsenic. However, the primary reason that Ehrlich chose to study dyes was in the hope that he could find compounds that would selectively bind to bacteria. Any preference was easily seen with highly colored materials. The first effective antibacterial agent developed from this investigation was the sulfonamide known as Prontosil.

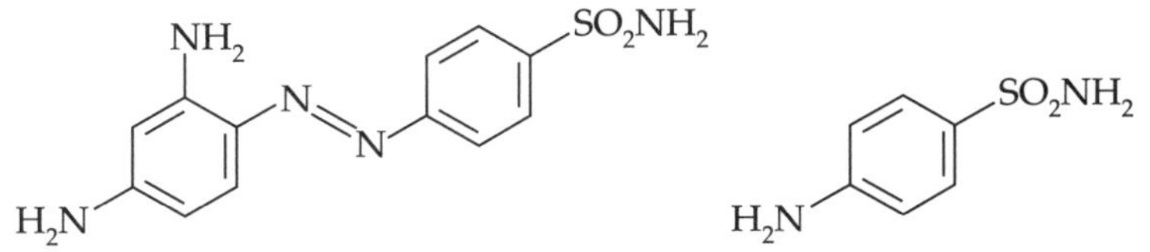

Prontosil   **Sulfanilamide**

Chemists then began a study of the relation between structure and activity to ascertain what molecular features were necessary for antibacterial properties. It was soon discovered that just the right-hand part of Prontosil, in the form of *p*-aminobenzenesulfonamide (**sulfanilamide**), has the antibacterial activity, and the era of the sulfa drugs was born.

EXERCISE 23-C

There are two ways in which Prontosil can be prepared by using the reaction of a diazonium salt with an aromatic ring. Show both ways and suggest which is the better method. (Hint: This reaction can be viewed as an electrophilic aromatic substitution reaction: see Chapter 11.)

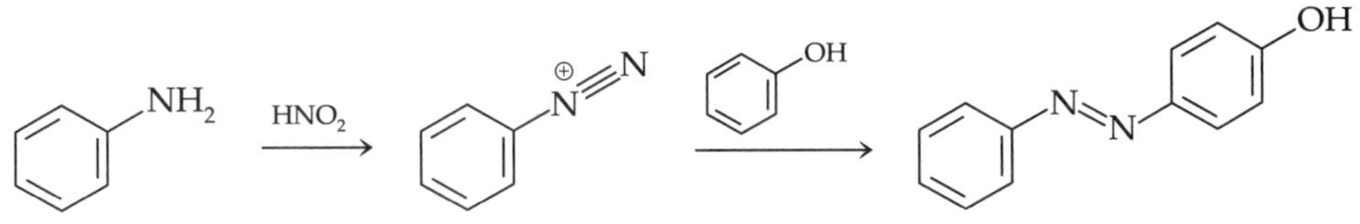

▲ FIGURE 23-7

Nitrosation of aniline produces a diazonium salt from which azo compounds can be produced by coupling with electronic-rich, aromatic compounds such as phenol, as discussed in Chapter 10.

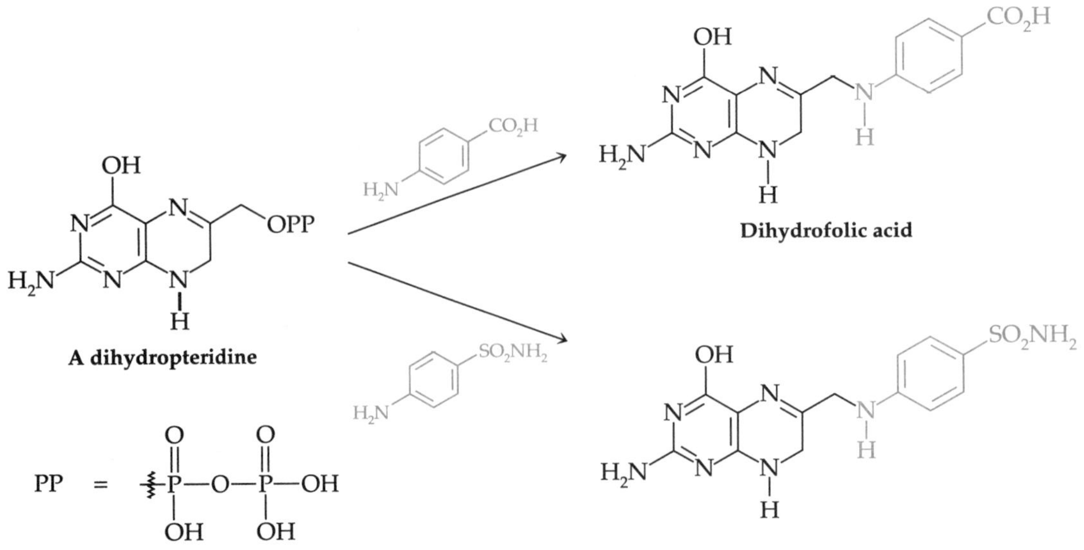

▲ FIGURE 23-8

The essential cofactor dihydrofolic acid is synthesized in bacteria by the combination of a dihydropteridine with *p*-aminobenzoic acid (upper route). In the presence of sulfanilamide, a bacterium incorporates this molecule into a structure that is similar to the natural cofactor. However, this unnatural material blocks the active site of one or more enzymes that use dihydrofolic acid as a one-carbon transfer agent, leading to a breakdown in critical biochemical transformations and the death of the bacterium.

How can such a simple molecule kill bacteria? Furthermore, why does it not have a toxic effect on human beings? The answer to both questions lies in the necessity for bacteria to synthesize tetrahydrofolic acid from *p*-aminobenzoic acid and a pteridine pyrophosphate (Figure 23-8). Mammals lack this pathway and must obtain this cofactor from the diet.

Sulfanilamide (and other sulfa drugs) are sufficiently similar in structure and reactivity to *p*-aminobenzoic acid that they can take part in the biosynthesis. Apparently, the modified cofactor produced in this way consumes pteridine, reducing the amount of the "real" cofactor and possibly interfering with the enzyme that uses tetrahydrofolate as a one-carbon transfer agent. Although it is easy to recognize the structural similarity between sulfanilamide and *p*-aminobenzoic acid, unraveling the details of *how* the antibiotic interferes with bacteria is quite complex.

In addition to the hypothesis advanced herein for the mode of action of sulfa drugs (for which there is substantial experimental verification), there

TETRAHYDROFOLIC ACID: A COFACTOR SUPPLIED BY BACTERIA

Taking sulfa drugs for infections is effective in killing not only harmful bacteria, but also beneficial bacteria that are part of the "fauna and flora" in our intestinal systems. These bacteria are normally a major supplier of tetrahydrofolic acid. Thus, orally administered sulfa drug for an extended period (2–3 months), as in, for example, the treatment of chronic acne, can lead to a deficiency in this essential cofactor. It is for this reason that doctors often recommend that sulfa drug therapy be accompanied by eating yogurt with active culture.

are several other possibilities. For example, the antibiotic could bind tightly to the enzyme responsible for the synthesis of *p*-aminobenzoic acid or to the enzyme responsible for its incorporation into tetrahydrofolate. Alternatively, the bogus tetrahydrofolate itself could block the enzyme responsible for its synthesis (as well as that of tetrahydrofolic acid). Unraveling these details is an important role that chemists play in developing an understanding of the molecular basis for drug action. This information, in turn, is an invaluable resource for those chemists who design and synthesize new pharmaceutical agents.

23-6 ▸ Antibiotics Affecting Membrane Structure and Ion Balance across Membranes

It has been known for some time that surfactants (surface active agents) have an adverse effect on bacteria. For example, many common household cleaning agents contain tetraalkylammonium salts. (Your knowledge of organic chemistry should help you to understand the function of various ingredients in commercial products.) These salts appear to have antibacterial activity because they are adsorbed into and therefore disrupt the bacterial cell membrane.

Physical disruption of the cell membrane is not the only mechanism for the action of antibiotics. One class of antibiotics, the ionophores (discussed in Chapter 19), destroy bacteria by disrupting the normal ionic balance across cell membranes. Some of the ionophores are effective antibacterial agents because they encapsulate metal ions in a hydrophobic exterior, and thus carry them through a membrane. Other ionophores function by becoming part of the cell membrane and forming new channels through which ions can move relatively freely. Recall that the unique chemical property of ionophores is their ability to act as multidentate ligands for metal ions. Simple ionophores, such as the cyclic polyether 18-crown-6, bind tightly to alkali metal cations and can carry the cation, with its associated anion(s), from water into an organic solvent.

There are several naturally occurring antibiotics that bind Na^+, K^+, and Li^+ within a substantially hydrocarbonlike exterior. These encapsulated ions can then be transported through the lipid bilayer of cell membranes, producing a dramatic change from the natural ionic balance. Valinomycin (Figure 23-9) is a cyclic dodecadepsipeptide that can carry ions across

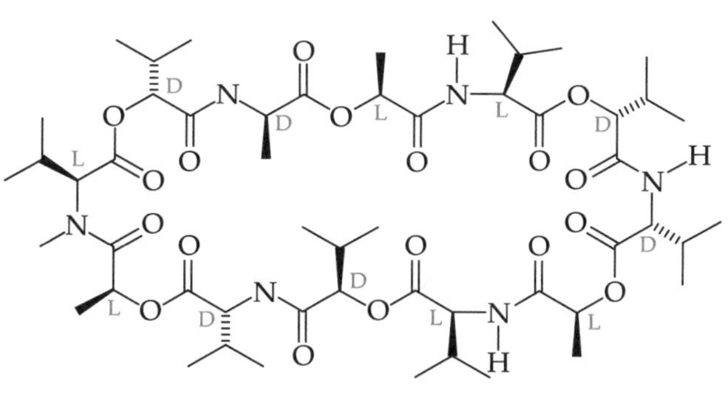

Valinomycin

▲ FIGURE 23-9
Valinomycin is a naturally occurring, cyclic dodecadepsipeptide. In addition to normal peptide bonds, valinomycin contains several ester linkages. Both L and D configurations of amino acids are found in this peptide.

▶ FIGURE 23-10

Gramicidin A is a linear pentadecapeptide containing amino acids having either the L or the D configuration.

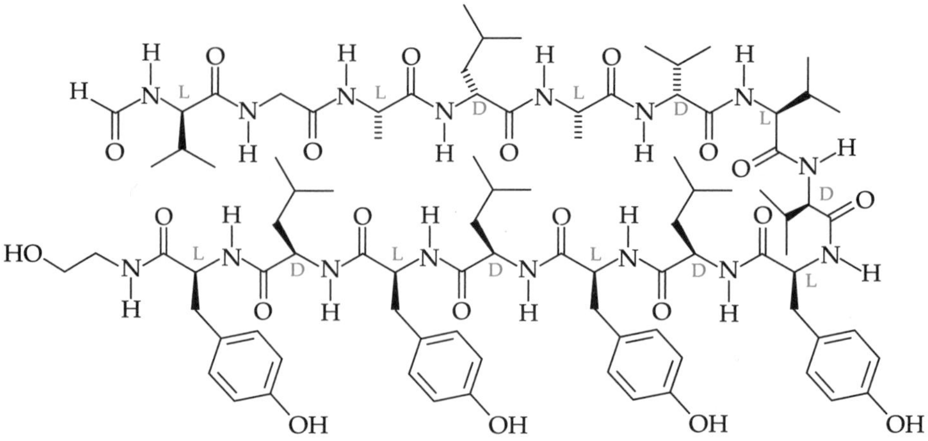

Gramicidin A

membranes, disrupting the normal balance with catastrophic effects on the cell. (**Depsipeptides** are those that have ester linkages in addition to amide, or peptide, bonds.) There are many other antibacterial agents with apparently the same mode of action as valinomycin. However, most of them are also quite toxic to mammals and have a low therapeutic index. As a result, they are not of clinical importance.

Gramicidins are linear pentadecapeptides produced by *Bacillus brevis*, a common bacterium found in soil. The major peptide (88%) is gramicidin A, shown in Figure 23-10. This peptide contains both D- and L-amino acids that alternate except for the presence of glycine, an achiral residue at the second unit from the amino end. The substitution of a single D-amino acid in a linear peptide chain prevents the formation of the α-helix for several residues at either side. However, the alternation of configuration found in gramicidin affords the opportunity to form a *left-handed helix* with 6.3 amino acid residues per turn (Figure 23-11). This helix is markedly different from the α-helix with 3.6 residues per turn presented in Chapter 15. The diameter of

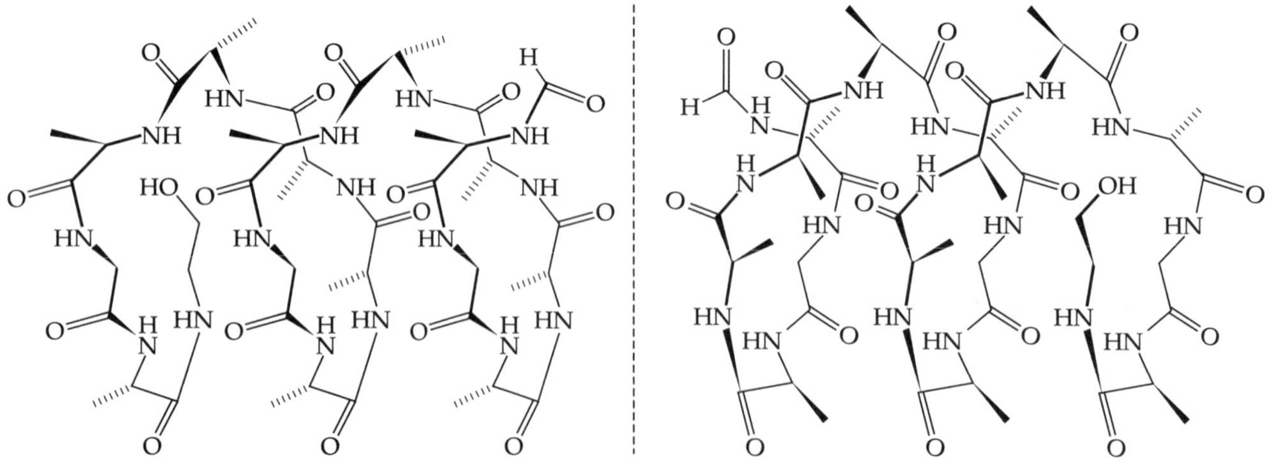

▲ FIGURE 23-11

Because of the presence of a number of amino acids with the unnatural D configuration, gramicidin A forms an unusual secondary structure with a left-handed helix. This helix is considerably larger than that normally formed from peptides composed entirely of L-amino acids, and it has 6.3 instead of 3.4 amino acids per turn.

the gramicidin helix is also larger than that of a normal helix, with a central hole of about 4 Å. Thus, it is possible for ions as large as Cs$^+$ to be contained inside and move freely from one end of this "tube" to the other. It is believed that two of these helices join together to form a conduit for ions across the cell membrane.

The **polyene antibiotics** also appear to function by creating additional ion channels through cell membranes. Amphotericin B is produced by a soil bacterium, *Streptomycetes nodosus*, apparently to protect itself from attack by fungi.

Amphotericin B

Not all bacteria are harmful and, indeed, some are beneficial. Bacteria in the intestines are a major source of folic acid for mammals and aid in our constant battle with fungus infections. The cell membranes of fungi more closely resemble the mammalian cell membrane than do those of bacteria; thus, developing a selective **antifungal agent** is generally much more difficult than finding an antibiotic. Outbreaks of infections by the yeast *Candida albicans* can occur in the mouth (oral thrush) and in the vagina, especially when the beneficial bacteria that control these yeasts are destroyed.

23-7 ▸ Disruption of Bacterial Cell Walls

Because bacteria must survive in dramatically different media (for example, tap water and "salty" blood) in order to be transferred from one person to another, the bacterial cell wall must be able to resist the swelling that occurs when the ionic strength inside the bacterium is several orders of magnitude greater than that outside. In this situation, water migrates into the cell faster than it leaves (osmosis), resulting in an increase in pressure inside the bacterium that may be as high as 25 atmospheres. An ordinary lipid bilayer membrane is very weak and does not remain intact under even much lower pressure differentials. In a bacterium, the cell membrane has a protective outside wall composed of peptidoglycans, long chains of carbohydrates that are cross-linked together by short (7–12 amino acids) peptides (Figure 23-12). Interestingly and very importantly, several "unnatural" D-amino acids are incorporated into these peptides, providing the opportunity for a chemical to selectively affect bacteria because these D-amino acids are not present in animals.

Although a relatively rigid, protective wall is essential to the survival of the bacterium, it must be able to expand as the cell divides. To provide for this expansion, bacteria have enzymes that can cleave these peptide bonds, thereby removing the cross-links, and other enzymes that can repair this damage to the outer wall by forming new peptide bonds. As the cell

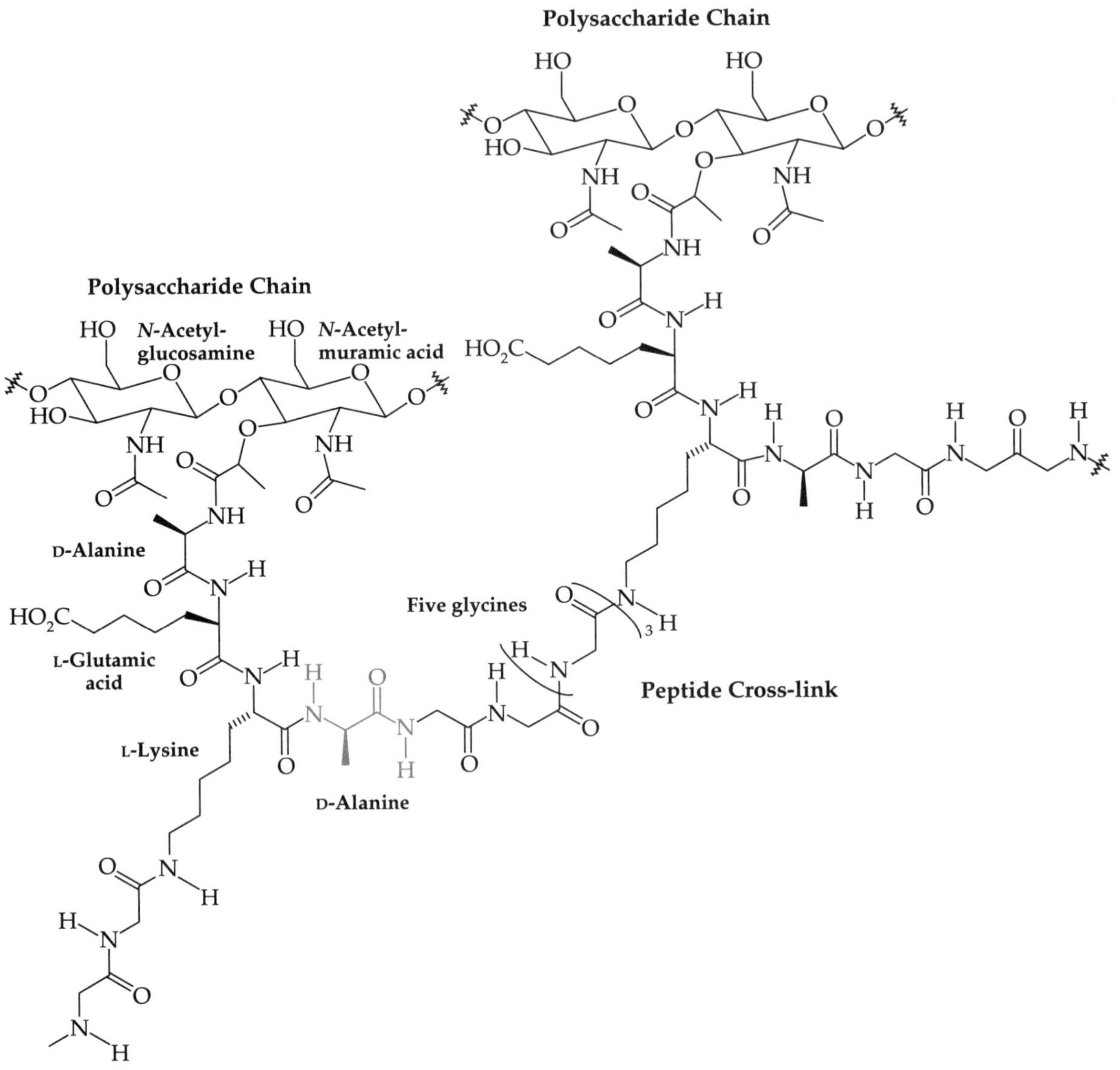

▲ FIGURE 23-12
The cell walls of many bacteria are composed of polysaccharide chains that are cross-linked by short peptide chains to form a rigid polymer matrix. The β-lactam antibiotics greatly weaken bacterial cell walls by interfering with the construction of these cross-links at the point containing D-alanine. Without complete cross-linking, the cell wall cannot withstand the large internal pressure developed by a bacterium when exposed to water of low ionic strength; the cell wall then ruptures and the bacterium dies.

divides, there is a net addition of new material as the wall is unstitched and then restitched until there is enough for two cells. Clearly, interfering with the ability to remove the cross-links prevents bacteria from multiplying, and blocking the formation of the cross-links results in rupture of the cell wall.

In a critical stage of cross-linking, the terminal residue is removed from a branch consisting of two D-alanines (for *Staphylococcus aureus*). The remaining D-alanine is then linked by a peptide bond to the amino group of the terminal glycine of a neighboring chain, as illustrated in Figure 23-13.

The discovery by Alexander Fleming of the antibacterial properties of a *Penicillium* fungus in 1928 dramatically altered the course of history, yet his

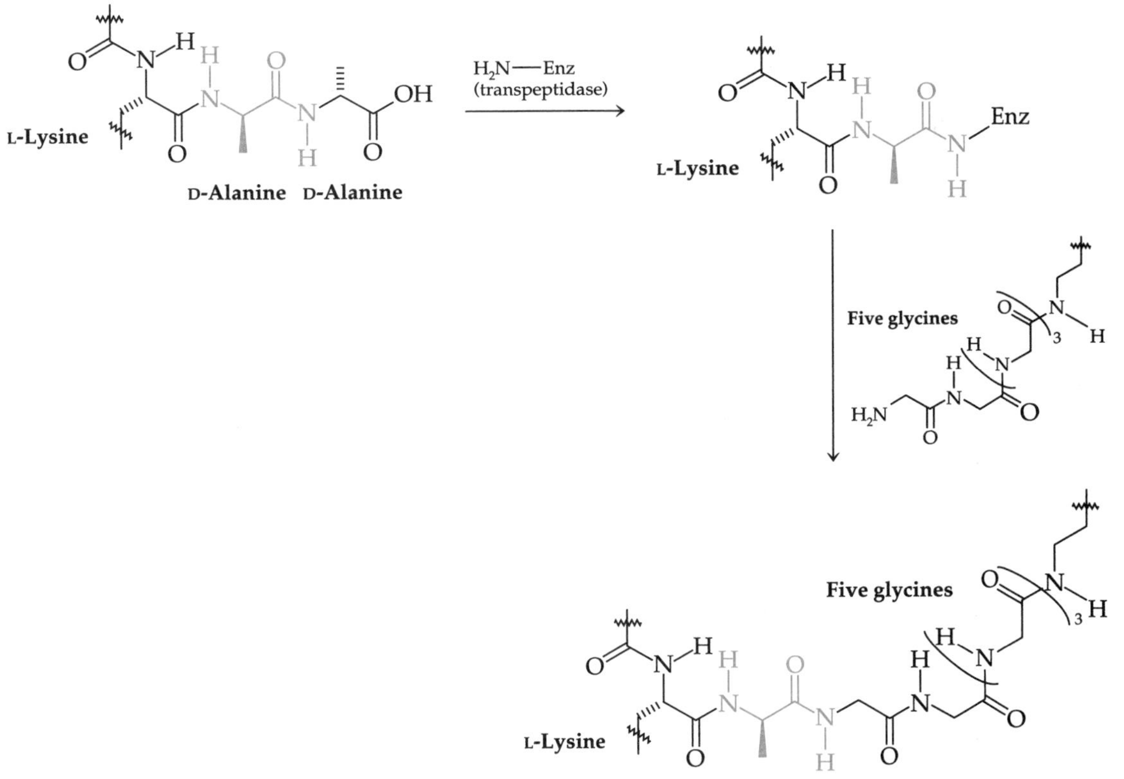

▲ FIGURE 23-13

The construction of the cross-link between polysaccharide chains in bacterial cell walls requires the formation of peptide bonds. A short (tripeptide) chain attached to one polysaccharide chain is first covalently bound by an amide bond to the enzyme transpeptidase. By transamidation, the carboxyl functional group is transferred to the amino group of a terminal glycine of a pentapeptide attached to another polysaccharide chain. The formation of this amide bond covalently links one polysaccharide to the other.

observations constitute one of the most-amazing cases of serendipity in science. Fleming, an English bacteriologist, was making more or less routine studies, growing various bacteria known to be pathogenic (that is, capable of causing disease), such as *Staphylococcus aureus,* at different temperatures for part of his contribution to a textbook on bacteriology. In another laboratory, other scientists were studying molds, and it is suspected that the contamination of Fleming's Petri dish came from this source. He left this dish at room temperature and departed for a week of vacation, during which the fungi grew rapidly, producing penicillin and retarding the growth of the bacteria. Had Fleming put this particular plate in an incubator at 37 °C, the course of history might well have been quite different. At that temperature, the *Staphylococcus* bacteria grow much more rapidly than does the mold and, once a bacterial colony is mature, it is unaffected by penicillin. That Fleming did not discard this plate marks him as a true scientist, because the process of scientific discovery includes both discovering and recognizing the unexpected, not executing the expected.

Modifications of the structure of the active component ultimately led (after thirteen years) to the first penicillin to be used by people, Penicillin G (Figure 23-14). Many modifications of this basic structure have been made, all of which affect the side-chain group attached to nitrogen, resulting in

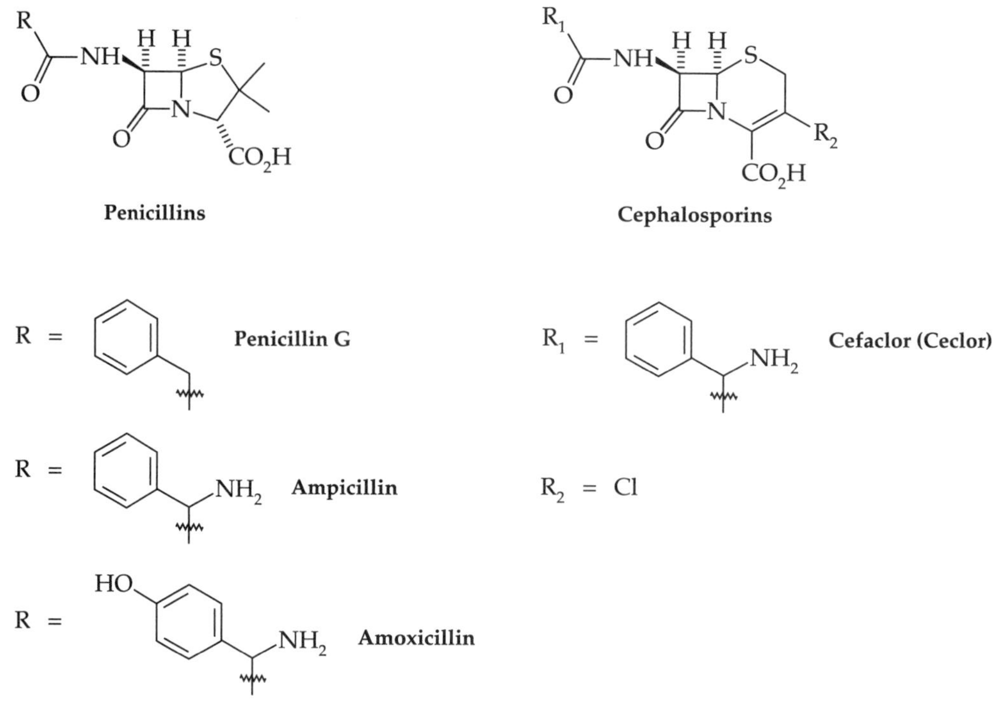

▲ FIGURE 23-14

All penicillins and cephalosporins have a four-membered lactam ring fused at nitrogen to another ring that contains sulfur. Shown here are a few of the many β-lactam antibiotics that are sold as antibacterial agents by the trade names given.

such familiar antibiotics as Ampicillin and Amoxicillin. Interestingly, the simple penicillin structure lacking a substituent on nitrogen has no antibacterial activity.

All of the penicillin antibiotics, as well as the structurally related cephalosporins, function in similar ways. Notice that they have in common a four-membered ring lactam (β-lactam), with an amino group attached to the α carbon. They are thus α-amino acid derivatives and are of the L configuration. They also have a carboxylic acid functional group, also with an α-amine substituent, but of the D configuration. Three-dimensional representations of the penicillin nucleus and the D-alanine–D-alanine branch that is critical to cross-linking in bacterial cell walls are shown in Figure 23-15. Because of the close similarity in shape of these two fragments, the bacterial enzyme responsible for cross-linking instead incorporates the antibiotic into the cell wall, preventing the formation of the necessary cross-link. It has been suggested that it requires only one such "nick" in the stitching of the cell wall to lead to the rupture of a bacterium. As a result, β-lactam antibiotics are effective at very low concentrations. For example, Penicillin G can kill bacteria at dilutions of 1:50,000,000, which corresponds to only 20 μg per liter. Figure 23-16 shows the dramatic change in the cell wall of *Staphylococcus* bacteria after treatment with penicillin.

It may well be that this mechanism of action for the β-lactam antibiotics is only one of several ways in which the synthesis of bacterial cell walls can be impeded. Another explanation is that the β-lactam becomes covalently attached to the enzyme, thus permanently blocking the reaction site and killing the enzyme. This mode of inactivation of enzymes is referred to as **suicide inhibition** because the enzyme undergoes an irreversible transformation that blocks its catalytic function.

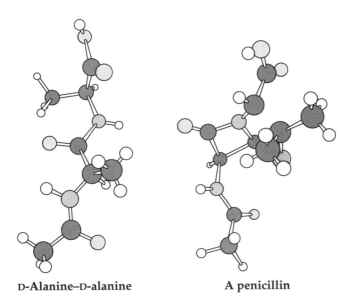

D-Alanine–D-alanine   **A penicillin**

▲ FIGURE 23-15
These ball-and-stick molecular representations show the overall similarity
between the dipeptide D-alanine–D-alanine and a representative penicillin,
β-lactam antibiotic. Because of this similarity in structure, penicillins become
attached to and thus block the action of the transpeptidase enzyme that is
critical to the construction of the cross-links in bacterial cell walls.

The strain in the four-membered ring of penicillin and cephalosporin
antibiotics makes these cyclic amides unusually reactive. Indeed, the 25
kcal/mole of strain energy exceeds the resonance-stabilization energy (18
kcal/mole) of an amide. As a result, some of these antibiotics are rapidly
hydrolyzed in the stomach under acidic conditions and cannot be administered orally.

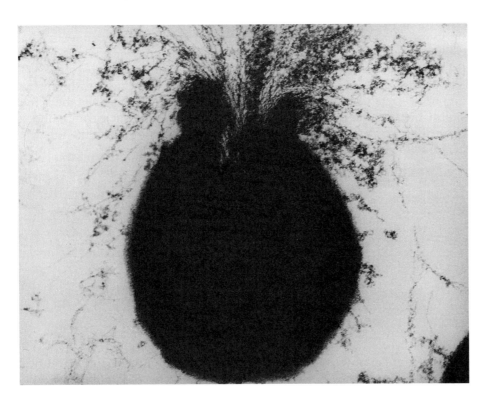

◄ FIGURE 23-16
This photomicrograph
catches a *Staphylococcus aureus*
bacterium in the act of literally
exploding as a result of defects
in the cell membrane caused by
exposure to penicillin.
(Courtesy of Victor Lorian,
M.D., Bronx Lebanon Hospital
Center.)

A PHARMACEUTICAL AGENT ACTIVE AGAINST
MULTICELL ORGANISMS

Carbomycin is a macrolide aminosugar isolated from fermentation broths of the mold *Streptomyces halstedii.* Carbomycin, marketed under the name Magnamycin, is a potent antibiotic effective against many of the same bacteria as penicillin but, in addition, is active against *Rickettsiae,* parasitic organisms that are responsible for such diseases as typhus and Rocky Mountain spotted fever.

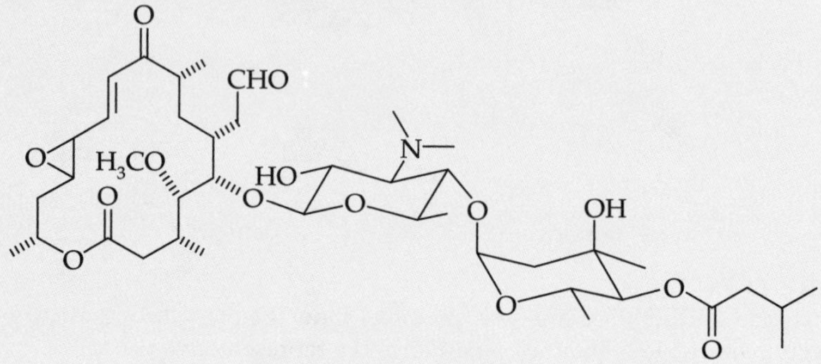

**Carbomycin
(Magnamycin)**

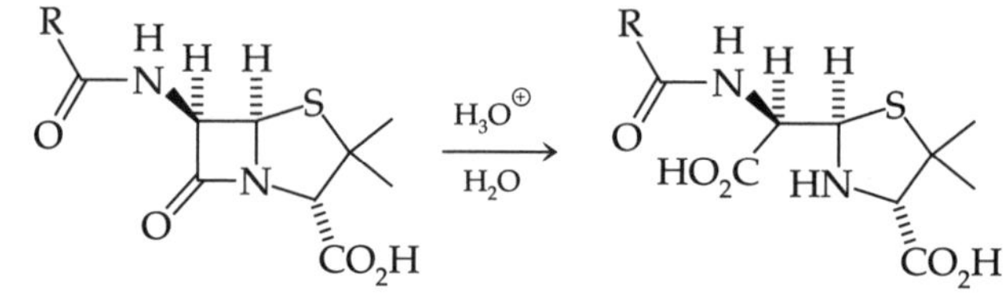

Furthermore, mutations of bacteria sometimes produce new enzymes that catalyze the cleavage of the β-lactam ring. These enzymes, referred to as **β-lactamases,** destroy antibiotics before they can disrupt bacterial cell membranes. This effect can be partially overcome through the administration of a β-lactamase inhibitor, a compound that binds strongly at the active site of the enzyme, blocking its normal function. One such compound is cloxacillin. Mixtures with Ampicillin have been used in which the function of cloxacillin is to protect Ampicillin from enzymatic cleavage.

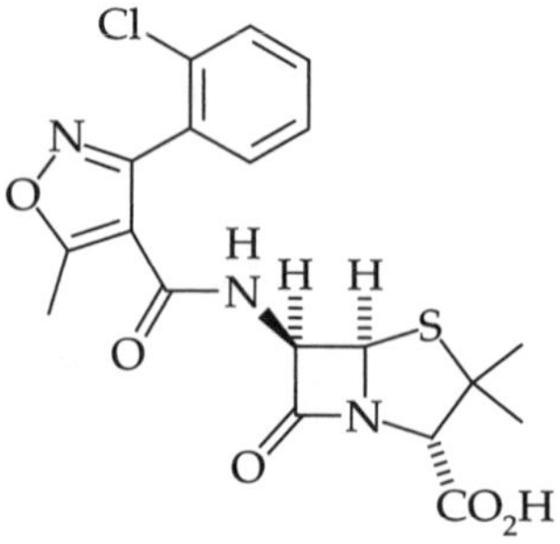

Cloxacillin

There are many criteria that determine the suitability of pharmaceutical agents for the chemical treatment of a disease state. For antibiotics, especially those that must be injected to avoid degradation in the stomach, resistance to enzymatic degradation in the blood is a very important characteristic. In fact, the cephalosporin marketed as Ceftriaxone is the leading injectable antibiotic precisely because it need only be administered once each day (it is effective against Lyme disease and drug-resistant gonorrhea, as well as many other bacterial infections).

Ceftriaxone

23-8 ▸ Drugs Affecting Nucleic Acid Synthesis

All living systems use polynucleic acids for the storage of genetic information. Thus, compounds that interfere with the synthesis of these vital materials are potentially toxic to all life forms. Recall from Chapter 18 that we followed a sequence for the synthesis of adenine that was parallel to its biosynthesis and included stages that sequentially added carbon and nitrogen atoms to glycine. Both methotrexate and hadacidin inhibit the biosynthesis of the purine-based nucleotides, but in different ways.

Tetrahydrofolic acid is responsible for the addition of single carbon atoms in the biosynthesis of a number of other important biochemicals. For example, this cofactor is the source of the two carbons of adenosine shown in color in the margin (recall the discussion of the abiotic synthesis of adenine in Chapter 18). Methotrexate is structurally quite similar to tetrahydrofolic acid (Figure 23-17) and interferes with the reductase that is responsible for its formation from folic acid.

Hadacidin acts in a different way; namely, by interfering with the transfer of nitrogen. Normally, aspartic acid is a source of nitrogen for the synthesis of a nucleic acid, as shown in Figure 23-18. This process is a transamination quite similar to those described in Chapter 21 for the formation of α-amino acids from their α-ketoacid precursors. The enzyme responsible for the covalent bonding of the nitrogen from aspartic acid to the base is competitively inhibited by hadacidin, but apparently it has no effect on other enzymes that act on aspartic acid. Because nucleotide synthesis is critical to all living systems, compounds such as methotrexate and hadacidin are quite toxic to mammals, as well as bacteria. However, they do have application as antineoplastic (or anticancer) agents. Because cancer cells undergo much more rapid growth and multiplication than normal cells, their requirements for polynucleic acids are greater, resulting in a favorable chemotherapeutic index for compounds that interfere with nucleotide synthesis, such as methotrexate and hadacidin.

Adenosine

EXERCISE 23-D

Suggest plausible reaction mechanisms for both steps shown in Figure 23-18 as they might occur in the presence of acid.

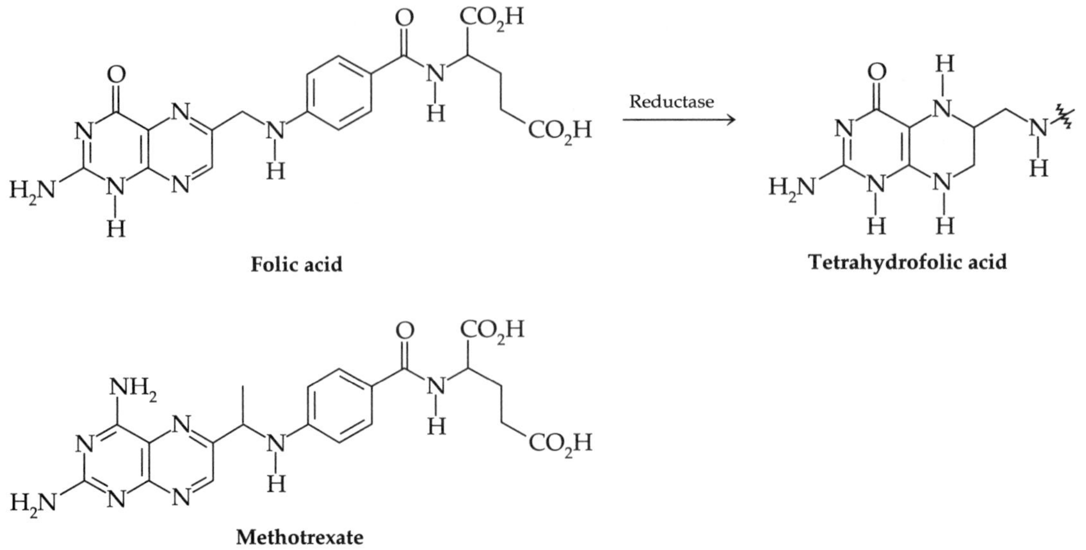

Folic acid → Reductase → **Tetrahydrofolic acid**

Methotrexate

▲ FIGURE 23-17

Methotrexate is an antibacterial agent that interferes with the synthesis of the cofactor tetrahydrofolic acid. In the synthesis of this cofactor, folic acid is reduced by a reductase enzyme. Methotrexate is sufficiently similar to folic acid that it binds to this enzyme and inhibits its role as a catalyst for the production of the cofactor.

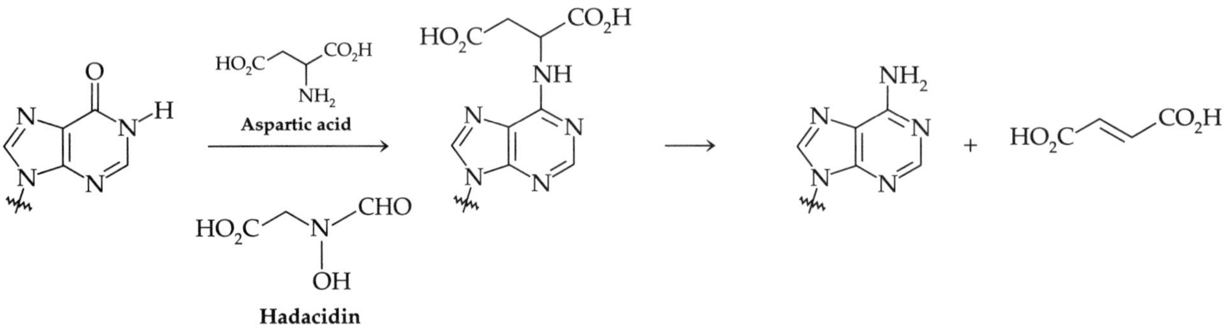

Aspartic acid

Hadacidin

▲ FIGURE 23-18

Hadacidin is an effective anticancer compound because it interferes with the synthesis of the nucleoside adenosine. Adenosine is biosynthesized by transfer of nitrogen from aspartic acid in a process that, like most biological reactions, is catalyzed by an enzyme. This enzyme is inhibited by hadacidin, which binds tightly in the active site.

Conclusions

The design of medicinal agents has progressed dramatically from the early development of Salvarsan to the design of effective beta-blockers for the treatment of hyperacidity. Many of the concepts explored in this course are important for understanding the mode of action of modern pharmaceutical agents.

Many disease states are caused by chemical imbalance from over- or underproduction or from deficiencies resulting from an inadequate diet. Many of these diseases can be effectively treated through chemical means.

ANTICANCER COMPOUNDS FROM MOLDS

Many unusual compounds that inhibit the growth of cancer cells have been isolated from plants and other natural sources. For example, dynemicin A (produced by a mold and isolated from a Texas soil sample) contains an unusual ene-diyne (one alkene and two alkynes) conjugated system that undergoes the cyclization shown below, forming a biradical. This process is energetically favorable because of the formation of the aromatic ring, yet the product biradical is nonetheless quite reactive. Each of the radical carbon centers reacts with one of the chains of a DNA duplex, linking them together and thus preventing replication of the genetic material, a process that requires that the two strands separate. Because cancer cells divide much more frequently than do normal cells, they are affected to a greater extent. Dynemicin A is toxic to normal cells as well and thus cannot be used for the treatment of cancer. Chemists are currently synthesizing other molecules containing the ene-diyne system that it is hoped will be less toxic to normal cells and yet retain the anticancer activity of the natural compound.

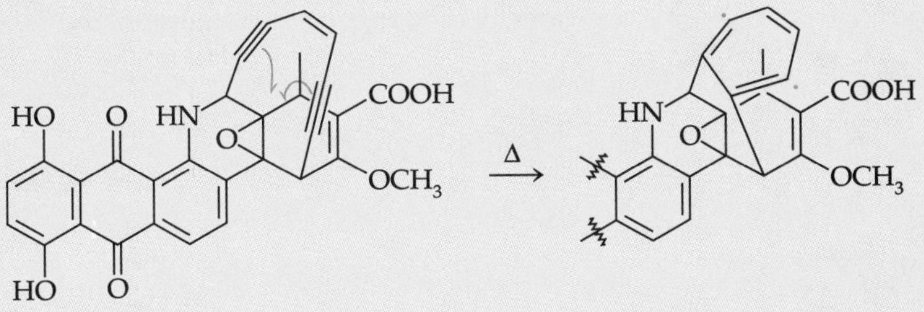

Dynemicin A

A deficiency can be corrected by administering the needed chemical, and an enzyme inhibitor can be used to reduce the amount of a substance that is being overproduced by a natural pathway.

A biological response can also sometimes result from administering compounds that only vaguely resemble the natural material. For example, the alkaloid morphine relieves pain because its shape mimics that of the endorphins, naturally occurring pain killers. Many biologically active agents share with morphine the β-phenethylamine unit.

Disease states resulting from the invasion of another living entity (such as a bacterium) also can be treated with chemicals. Modern, quite-effective antibacterial agents have been developed that are destructive to the microbe while being relatively harmless to human beings and other animals. This selective toxicity results from the distinctive chemistry of bacteria and animals. Bacterial cell walls are substantially stronger than those of mammalian cells because bacteria must be protected from the harsh environment. β-Lactam antibiotics destroy cells by interfering with the construction of bacterial cell walls at the critical stage of cross-linking, at which much of the strength is imparted. Bacteria must manufacture tetrahydrofolic acid, a critical cofactor responsible for the transfer of one carbon in biosynthesis and biodegradation. Sulfa drugs are effective antibiotics because they block the synthesis of tetrahydrofolic acid.

Review Problems

23-1 Many molecules with a β-phenethylamine moiety can have pronounced effects on the mind. For review, list all of the ways that you can think of that could be used to synthesize the parent, starting with benzene and any other needed organic and inorganic reagents.

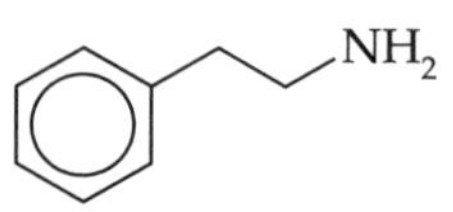

β-Phenethylamine

23-2 The reaction of an aryl diazonium ion with another aryl ring bearing an electron-donating substituent G produces a diazo linkage between the benzene rings. Write a detailed and clear reaction mechanism for this transformation, keeping in mind what you learned about electrophilic, aromatic substitution.

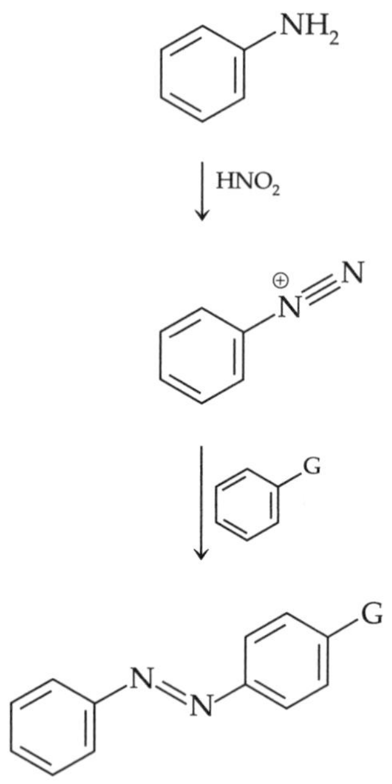

23-3 Write a detailed reaction mechanism for the conversion of aniline into its diazonium ion with nitrous acid.

23-4 Although aromatic diazonium ions prepared from aryl amines are sufficiently stable to be isolated, the treatment of aliphatic amines with nitrous acid leads to extensive further reactions and generally to a number of products. Explain this dramatic difference in reactivity between aryl and aliphatic diazonium ions.

23-5 Examine the structure of valinomycin (Figure 23-9) to determine if there are any sequences of amino acids that repeat (ignore stereochemistry at this point). What are the largest of such segments and how many are there? Calculate the number of different stereoisomers possible from the presence of twelve amino acid residues, all of which are chiral. Can any of them be meso stereoisomers?

Appendix

....

Summary of Methods for the Preparation of Functional Groups

C—H

1. Catalytic hydrogenation of alkenes and alkynes:

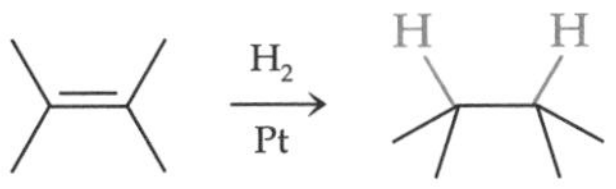

$$\text{(alkene)} \xrightarrow[\text{Pt}]{\text{H}_2} \text{(alkane with 2 H added)}$$

2. Hydrolysis of Grignard reagents:

$$\text{R—MgBr} \xrightarrow{\text{H}_2\text{O}} \text{R—H}$$

3. Clemmensen reduction of an aldehyde or ketone:

$$\text{(ketone)} \xrightarrow{\text{Zn, HCl}} \text{(alkane, 2 H added)}$$

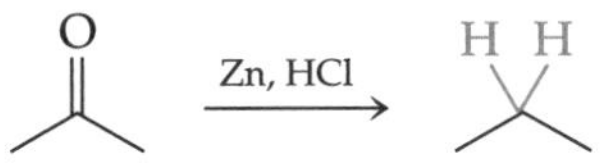

4. Wolff-Kishner reduction of an aldehyde or ketone:

$$\text{(ketone)} \xrightarrow[\text{KOH}]{\text{H}_2\text{NNH}_2} \text{(alkane, 2 H added)}$$

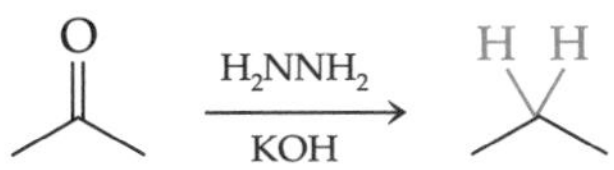

5. Decarboxylation of β-ketoacids:

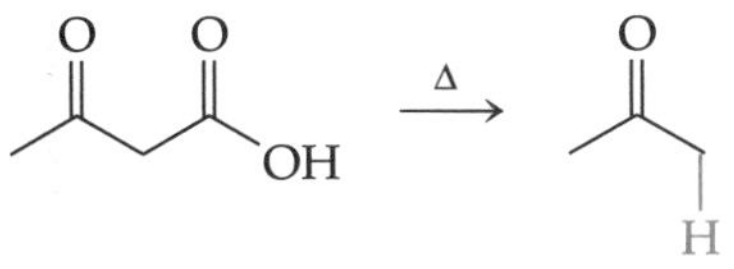

$$\text{(β-ketoacid)} \xrightarrow{\Delta} \text{(ketone)}$$

C—C

1. S$_N$2 displacement by cyanide:

$$\text{R—Br} \xrightarrow{{}^{\ominus}\text{C}\equiv\text{N}} \text{R—C}\equiv\text{N}$$

2. S$_N$2 displacement by acetylide anions:

$$\text{R—Br} + {}^{\ominus}\text{=}\text{—R}' \longrightarrow \text{R}\text{=}\text{—R}'$$

3. Grignard addition:

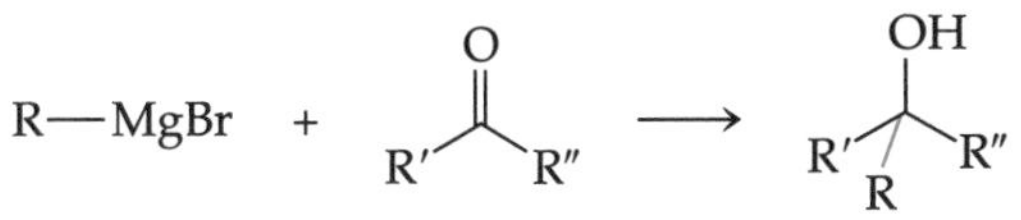

$$\text{R—MgBr} + \text{(ketone)} \longrightarrow \text{(tertiary alcohol)}$$

R' and R" = H, alkyl, aryl
R = alkyl, aryl

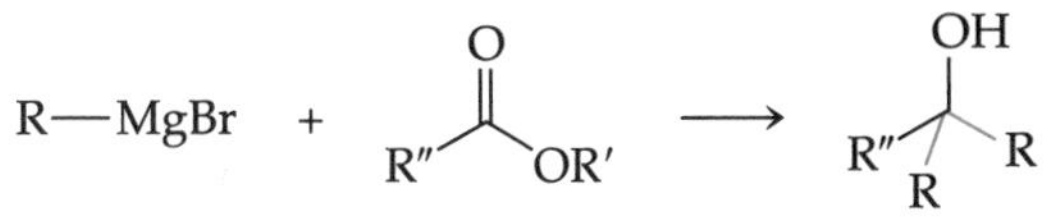

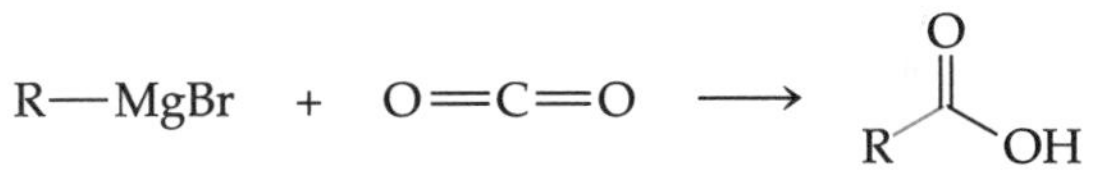

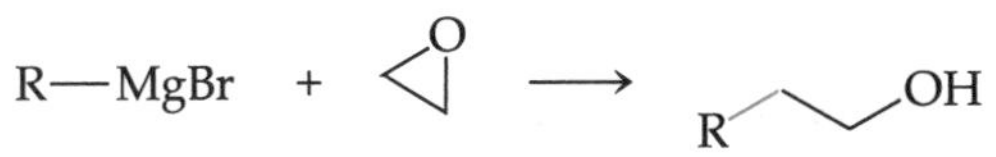

$$\text{R—MgBr} + \text{(ester)} \longrightarrow \text{(tertiary alcohol)}$$

$$\text{R—MgBr} + \text{O}=\text{C}=\text{O} \longrightarrow \text{(carboxylic acid)}$$

$$\text{R—MgBr} + \text{(epoxide)} \longrightarrow \text{R}\diagup\diagup\text{OH}$$

(continued)

813

Summary of Methods for the Preparation of Functional Groups (*continued*)

4. Friedel-Crafts acylation:

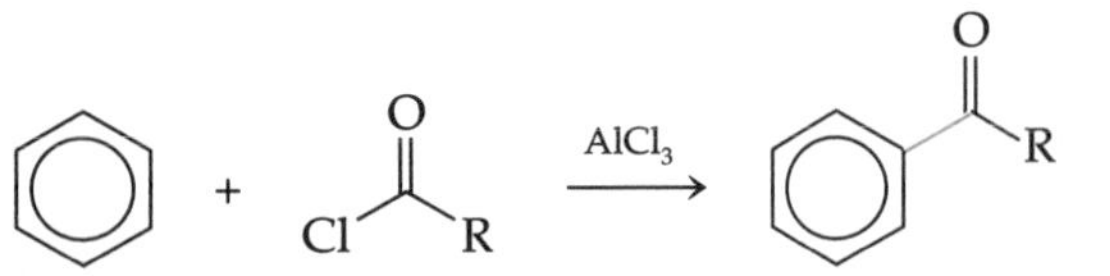

5. Friedel-Crafts alkylation:

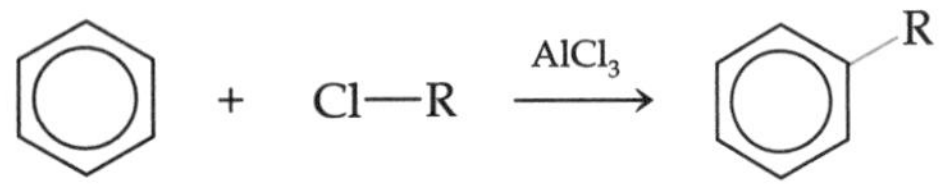

6. Diels-Alder reaction:

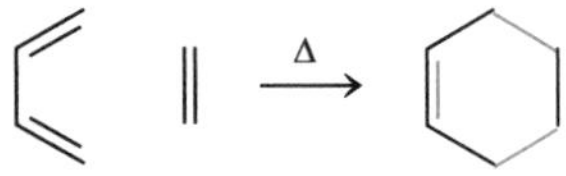

7. Conjugate addition to α,β-unsaturated carbonyl groups:

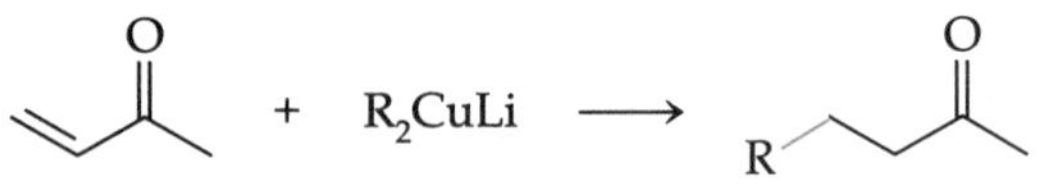

8. Michael reaction:

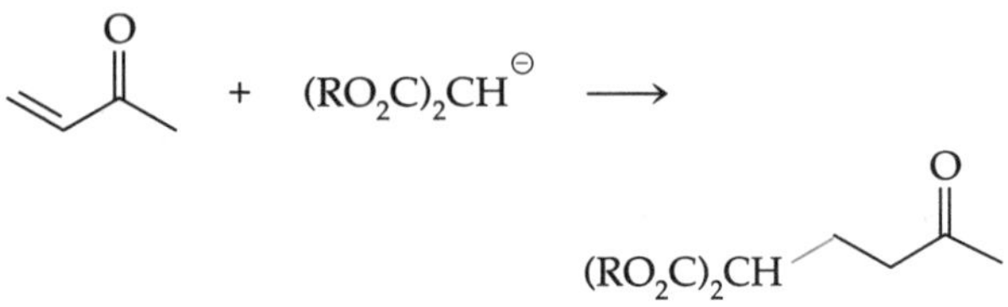

9. Aldol reaction:

10. Acetoacetic ester synthesis:

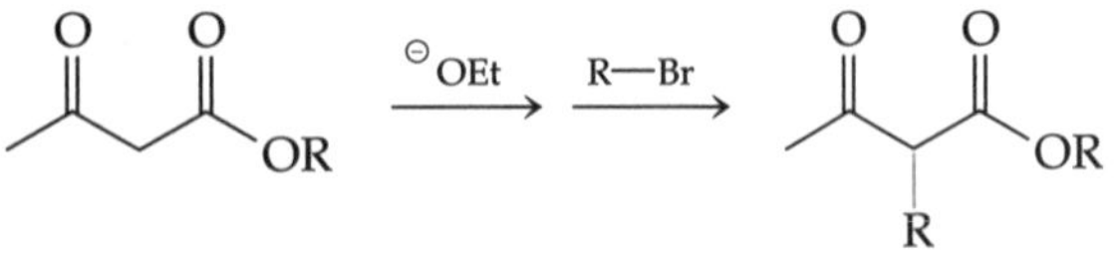

11. Malonic ester synthesis:

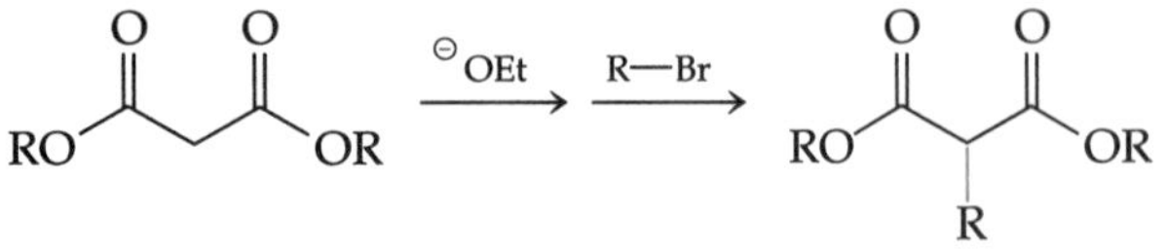

12. Claisen condensation:

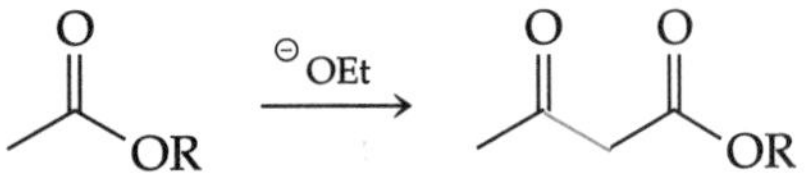

13. Cope rearrangement:

14. Claisen rearrangement:

C=C

1. Dehydrohalogenation:

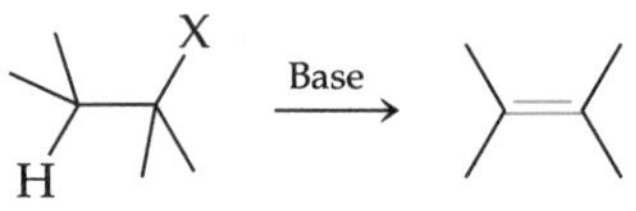

2. Dehydration:

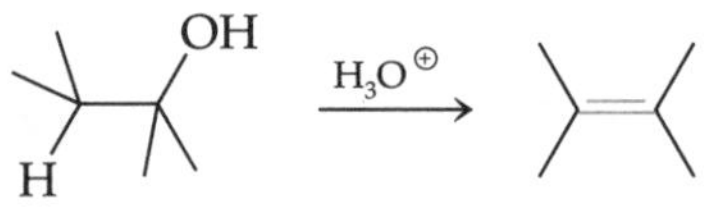

3. Hofmann elimination:

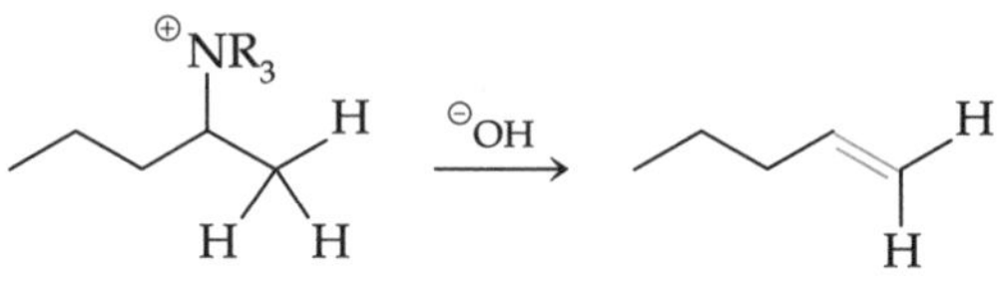

4. Catalytic hydrogenation of alkynes:

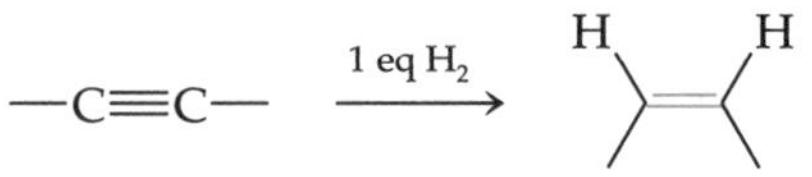

5. Dissolving metal reduction of alkynes:

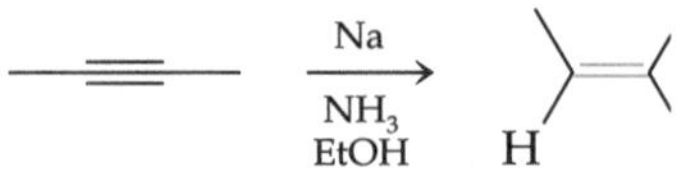

6. Reductive elimination:

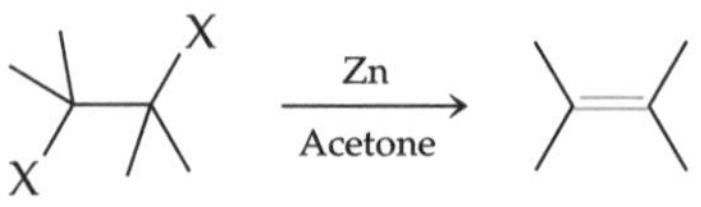

7. Wittig reaction:

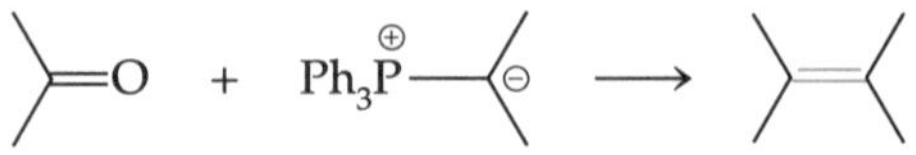

8. Aldol condensation:

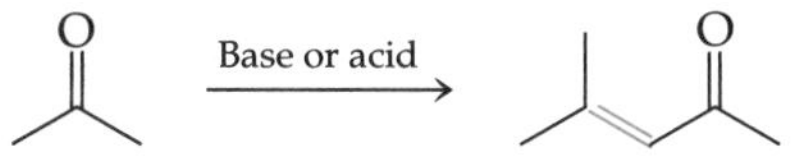

C≡C

1. Dehydrohalogenation:

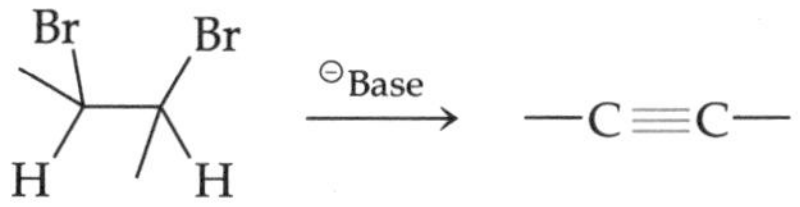

2. S$_N$2 displacement by acetylide anions:

$$-C\equiv C^{\ominus} \;+\; R-Br \longrightarrow \;-C\equiv C-R$$

C—X

1. Free radical halogenation:

$$R-H \xrightarrow[h\nu]{X_2} R-X$$

2. Addition of HX:

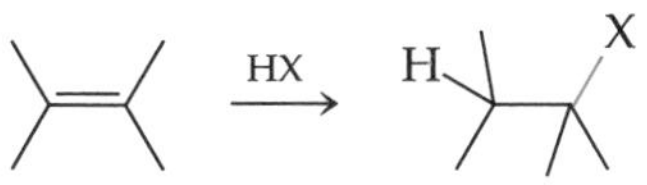

3. Conversion of alcohols into alkyl halides:

$$R-OH \xrightarrow{PX_3,\ POX_3,\ HX,\ or\ SOX_2} R-X$$

4. Electrophilic aromatic substitution:

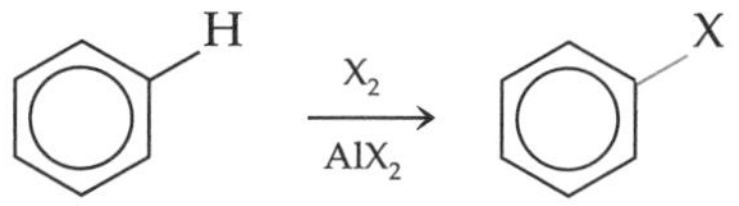

5. α-Halogenation of ketones:

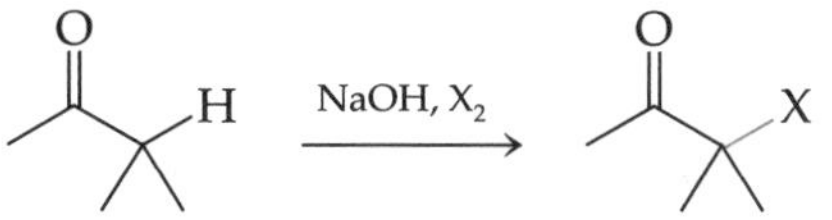

6. Hell-Volhard-Zelinsky reaction:

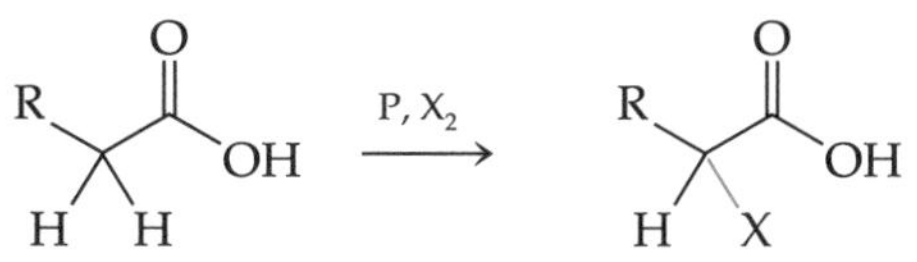

C—OH

1. Hydrolysis of alkyl halides:

$$R-X \xrightarrow{\ominus OH} R-OH$$

2. Hydration of alkenes:

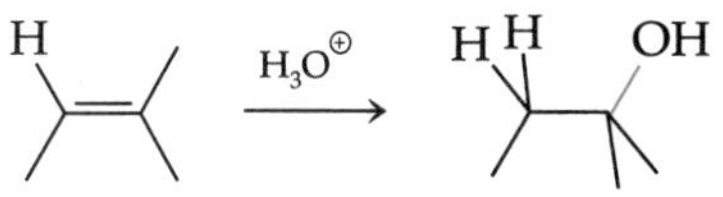

3. Oxymercuration-demercuration:

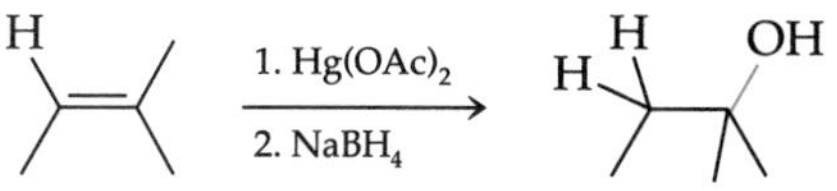

4. Hydroboration-oxidation:

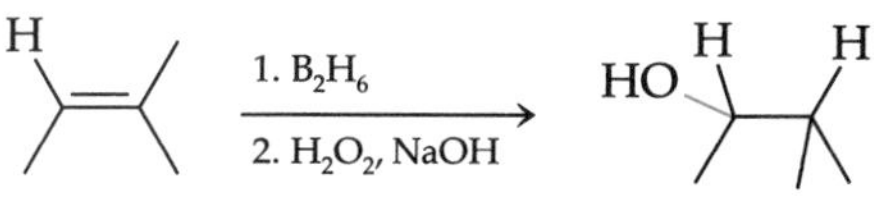

5. Hydration of benzyne:

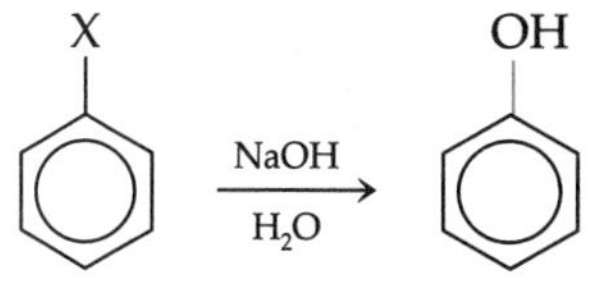

6. Nucleophilic opening of epoxides:

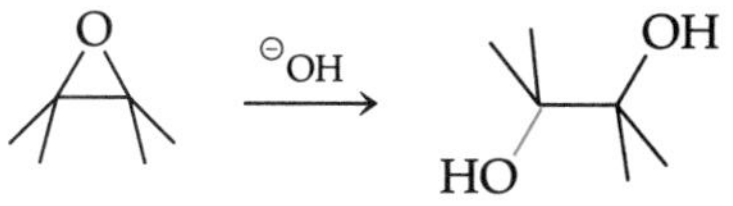

7. Grignard reaction of aldehydes and ketones:

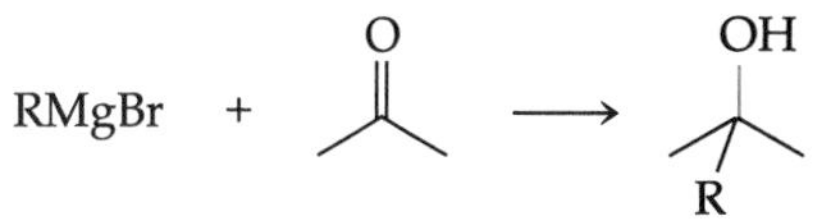

8. Grignard reaction of esters:

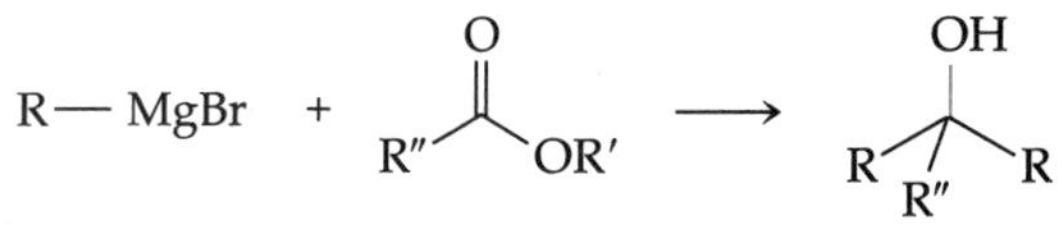

9. Metal hydride reductions of aldehydes and ketones:

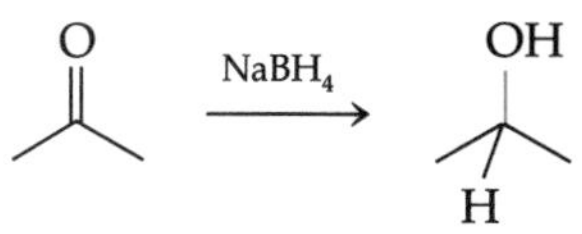

10. Metal hydride reduction of esters:

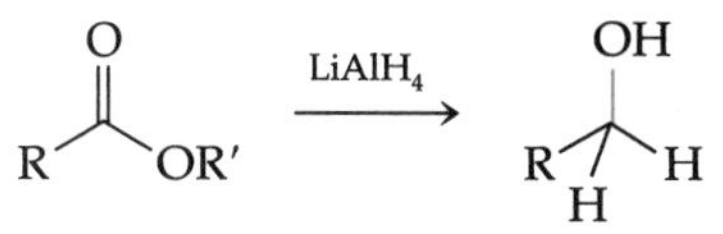

(continued)

Summary of Methods for the Preparation of Functional Groups (*continued*)

11. Aldol reaction:

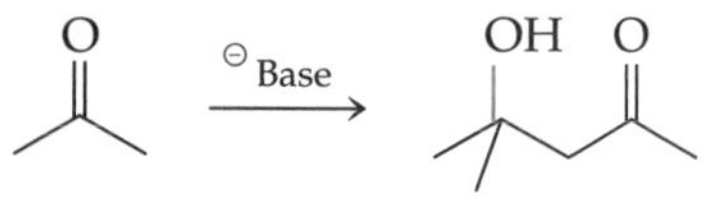

12. Cannizzaro reaction:

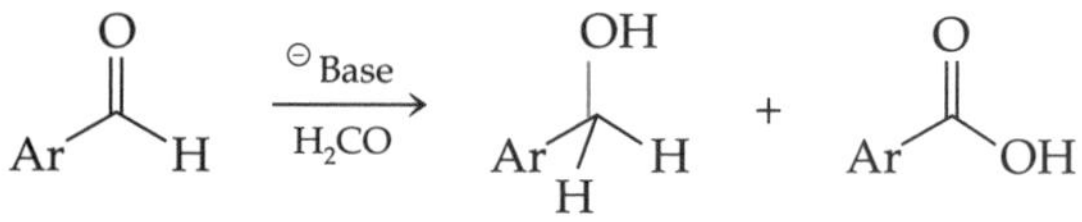

R—C≡N

1. S_N2 displacement by cyanide:

$$R—Br \quad + \quad {}^{\ominus}C≡N \quad \longrightarrow \quad R—C≡N$$

2. Cyanohydrin formation:

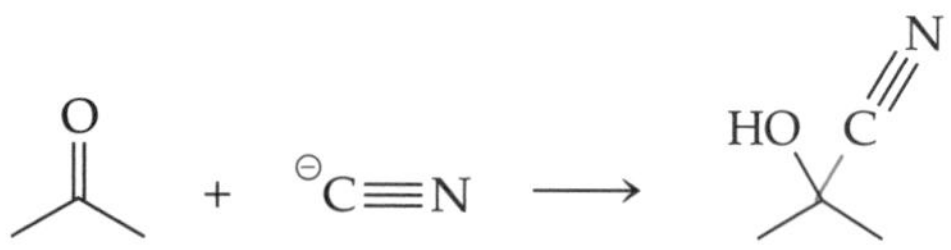

3. Dehydration of amides:

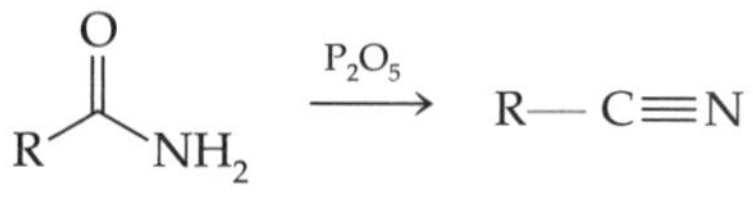

R—NH₂

1. Aminolysis of alkyl halides:

$$R—X \quad \xrightarrow{NH_3} \quad R—NH_2$$

2. Reduction of aromatic nitro compounds:

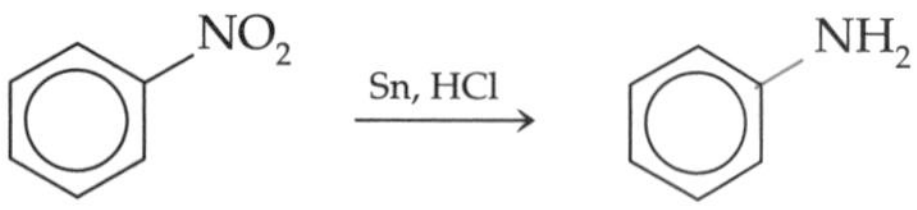

3. Amination of benzyne:

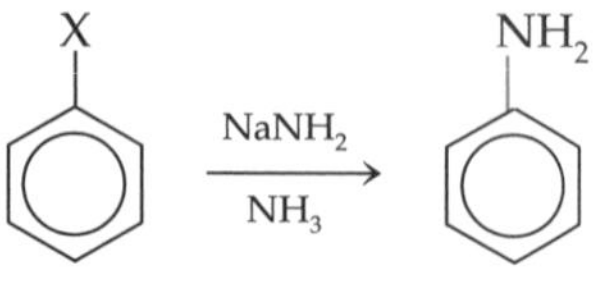

4. Reductive amination of ketones:

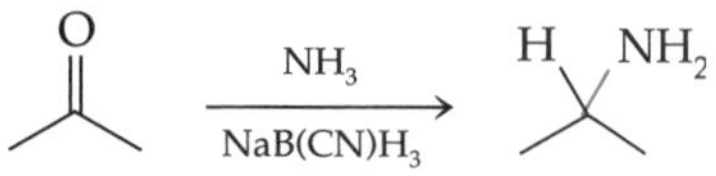

5. Lithium aluminum hydride reduction of amides:

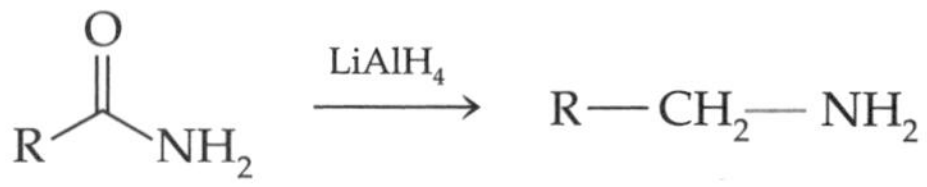

6. Hofmann rearrangement:

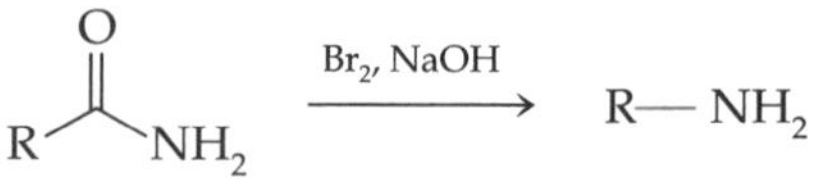

7. Catalytic hydrogenation of nitriles:

$$R—C≡N \quad \xrightarrow{H_2, Ni(Ra)} \quad R—CH_2—NH_2$$

8. Gabriel synthesis:

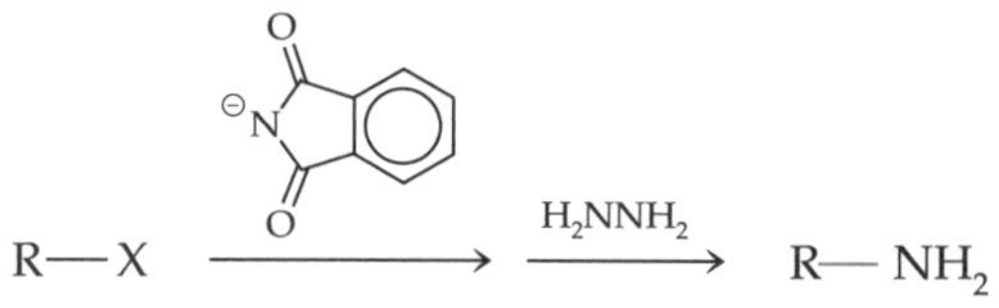

9. Catalytic hydrogenation of azides:

$$R—X \quad \xrightarrow{NaN_3} \quad R—N_3 \quad \xrightarrow[Pt]{H_2} \quad R—NH_2$$

R—O—R′

1. Williamson ether synthesis:

$$R—O^{\ominus} \quad + \quad Br—R′ \quad \longrightarrow \quad R—O—R′$$

2. Peracid oxidation of alkenes:

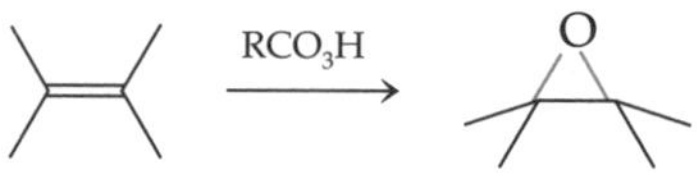

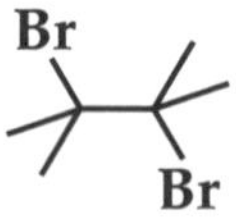

Br / Br (structure)

1. Bromination of alkenes:

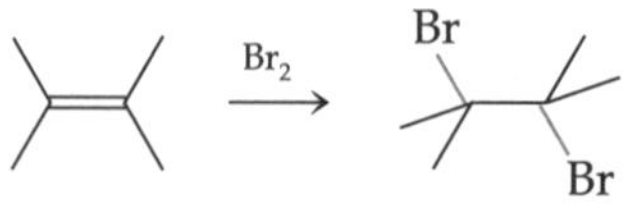

1. Oxidation of primary alcohols:

$$\underset{\text{H}}{\overset{\text{OH}}{\diagup}}\xrightarrow[\text{Pyridine}]{\text{Cr}^{6+}}\overset{\text{O}}{\diagup}\text{H}$$

2. Aldol reaction:

$$\overset{\text{O}}{\diagup}\text{H}\xrightarrow{\overset{\ominus}{}\text{Base}}\underset{\text{H}}{\overset{\text{OH}\quad\text{O}}{\diagup}}\text{H}$$

3. Ozonolysis of alkenes:

$$\underset{\text{H}}{\overset{\text{H}}{\diagup}}\xrightarrow[\text{2. Zn, HOAc}]{\text{1. O}_3}2\overset{\text{O}}{\diagup}\text{H}$$

4. Acetal hydrolysis:

$$\underset{\text{H}}{\overset{\text{RO}\quad\text{OR}}{\diagup}}\xrightarrow{\text{H}_3\text{O}^{\oplus}}\overset{\text{O}}{\diagup}\text{H}$$

1. Ketal hydrolysis:

$$\underset{}{\overset{\text{RO}\quad\text{OR}}{\diagup}}\xrightarrow{\text{H}^+, \text{H}_2\text{O}}\overset{\text{O}}{\diagup}$$

2. Hydrolysis of terminal alkynes:

$$\text{H}\!\!-\!\!\equiv\!\!-\xrightarrow{\text{Hg}^{2+}, \text{H}_2\text{O, H}^+}\overset{\text{O}}{\diagup}$$

3. Chromate oxidation of secondary alcohol:

$$\underset{}{\overset{\text{H}\quad\text{OH}}{\diagup}}\xrightarrow{\text{Cr}^{6+}}\overset{\text{O}}{\diagup}$$

4. Friedel-Crafts acylation:

5. Claisen condensation:

$$\overset{\text{O}}{\underset{\text{OR}}{\diagup}}\xrightarrow{\overset{\ominus}{}\text{OEt}}\overset{\text{O}\quad\quad\text{O}}{\underset{\text{OR}}{\diagup}}$$

6. Decarboxylation of β-ketoacids:

$$\overset{\text{O}\quad\text{O}}{\underset{\text{OH}}{\diagup}}\xrightarrow{\Delta}\overset{\text{O}}{\diagup}$$

7. Pinacol rearrangement:

$$\underset{}{\overset{\text{HO}\quad\text{OH}}{\diagup}}\xrightarrow{\text{H}^{\oplus}}\overset{\text{O}}{\diagup}$$

8. Claisen rearrangement:

$$\overset{\text{O}}{\diagup}\!\!\diagdown\longrightarrow\overset{\text{O}}{\diagup}$$

$$\overset{\text{O}}{\underset{\text{OH}}{\diagup}}$$

1. Hydrolysis of carboxylic acid derivatives:

$$\overset{\text{O}}{\underset{\text{X}}{\diagup}}\xrightarrow[\text{(or NaOH)}]{\text{H}^+, \text{H}_2\text{O}}\overset{\text{O}}{\underset{\text{OH}}{\diagup}}$$

2. Nitrile hydrolysis:

$$-\text{C}\!\equiv\!\text{N}\xrightarrow[\text{(or NaOH)}]{\text{H}^+, \text{H}_2\text{O}}\overset{\text{O}}{\underset{\text{OH}}{\diagup}}$$

3. Oxidation of primary alcohols:

$$-\text{CH}_2-\text{OH}\xrightarrow[\text{H}_2\text{O}]{\text{Cr}^{6+}}\overset{\text{O}}{\underset{\text{OH}}{\diagup}}$$

4. Oxidation of aldehydes:

$$\overset{\text{O}}{\underset{\text{H}}{\diagup}}\xrightarrow[\text{H}_2\text{O}]{\text{Cr}^{6+}}\overset{\text{O}}{\underset{\text{OH}}{\diagup}}$$

5. Permanganate oxidation of alkyl side chains of arenes:

$$\underset{}{\overset{\text{CH}_2\diagdown\text{R}}{\bigcirc}}\xrightarrow{\text{KMnO}_4}\underset{}{\overset{\text{O}}{\bigcirc}\text{OH}}$$

6. Iodoform reaction:

$$\overset{\text{O}}{\underset{\text{CH}_3}{\diagup}}\xrightarrow[]{\text{I}_2, \text{NaOH}}\xrightarrow{\text{H}^+}\overset{\text{O}}{\underset{\text{OH}}{\diagup}}$$

(continued)

Summary of Methods for the Preparation of Functional Groups (*continued*)

7. Carboxylation of Grignard reagents:

$$R-Br \xrightarrow[\text{2. CO}_2]{\text{1. Mg}} \xrightarrow{\text{H}^+} \overset{O}{\underset{OH}{\|}}$$

8. Decarboxylation of β-diacids:

$$HO \overset{O}{\|} \overset{O}{\|} OH \xrightarrow{\Delta} \overset{O}{\|} OH$$

9. Benzilic acid rearrangement:

$$Ar \overset{O}{\underset{O}{\|}} Ar \xrightarrow{{}^{\ominus}OH} HO \overset{O}{\underset{Ar}{\underset{Ar}{\|}}} OH$$

1. Treatment of acid chloride with thionyl chloride:

$$\overset{O}{\underset{OH}{\|}} \xrightarrow[\text{(or PCl}_3 \text{ or PCl}_5)]{\text{SOCl}_2} \overset{O}{\underset{Cl}{\|}}$$

$$\overset{O}{\|} \overset{O}{\underset{O}{\|}}$$

1. Acid dehydration:

$$\overset{O}{\underset{OH}{\|}} \xrightarrow{\Delta \text{ or } P_2O_5} \overset{O}{\|} \overset{O}{\underset{O}{\|}}$$

$$\overset{O}{\underset{NR_2}{\|}}$$

1. Amidation of carboxylic acid derivatives:

$$\overset{O}{\underset{X}{\|}} \xrightarrow{\text{HNR}_2} \overset{O}{\underset{NR_2}{\|}}$$

$$X = Cl, OR, OAc$$

2. Beckmann rearrangement:

$$\overset{O}{\|} \xrightarrow[\text{2. H}_2\text{SO}_4]{\text{1. H}_2\text{NOH}} \overset{O}{\underset{\underset{H}{N}}{\|}}$$

$$\overset{O}{\underset{OR}{\|}}$$

1. Esterification of carboxylic acids:

$$\overset{O}{\underset{OH}{\|}} + HOR \xrightarrow{\text{H}^{\oplus}} \overset{O}{\underset{OR}{\|}}$$

2. Transesterification:

$$\overset{O}{\underset{OR'}{\|}} + HOR \xrightarrow{\text{H}^{\oplus}} \overset{O}{\underset{OR}{\|}} + HOR'$$

3. Baeyer-Villiger oxidation:

$$\overset{O}{\|} \xrightarrow{\text{RCO}_3\text{H}} \overset{O}{\underset{O}{\|}}$$

$$RO \overset{}{\underset{}{\bigwedge}} OR$$

1. Acetal (ketal) formation:

$$\overset{O}{\underset{H(R)}{\|}} \xrightarrow{\text{H}^+, \text{ROH}} RO \overset{}{\underset{H(R)}{\bigwedge}} OR$$

$$\overset{N^{\diagup R}}{\|}$$

1. Imine formation:

$$\overset{O}{\|} + H_2N-R \longrightarrow \overset{N^{\diagup R}}{\|}$$

Glossary

....

The parenthetical insertion at the end of each entry is a citation to the chapter(s) and section(s) in which the term is discussed more completely.

▼

Abiotic synthesis: preparation of a compound, often of biological relevance, without the use of biological agents such as enzymes or nucleic acids (18-8, 18-9)

Absolute configuration: the three-dimensional structure of a molecule that has one or more centers of chirality (5-6)

Absolute stereochemistry: unambiguous specification of all spatial positions about a center of chirality (5-6)

Absorption spectroscopy: measurement of the dependence of the intensity of absorbed light on wavelength for light in the visible and ultraviolet regions (4-3)

Acetal: a functional group bearing an alkyl group, a hydrogen atom, and two alkoxy groups on one carbon atom [$RCH(OR)_2$]; produced in the acid-catalyzed alcoholysis of an aldehyde or a hemiacetal; a protecting group for an aldehyde (12-6, 15-8)

Acetoacetic ester: α-acetylated derivative of an ester; $CH_3(CO)CH_2CO_2R$ (8-4)

Acetoacetic ester synthesis: method for preparing an α mono- or dialkylated derivative of a methyl ketone by the sequential alkylation of an acetoacetic ester anion, hydrolysis of the alkylated ester, and decarboxylation of the resulting β-ketoacid (8-4)

Acetyl CoA: thiol ester of CoA and acetic acid; critical intermediate in fatty acid biosynthesis and degradation, in the citric acid cycle, and in glycolysis (21-7, 22-5, 22-7, 22-9, 22-10)

Acetylene: *See* **Ethyne** (2-4)

Acetyl group: —$COCH_3$ (8-4)

Acetylide anion: *See* **Alkynide anion** (8-3)

Achiral: descriptor of a molecule in which at least one of its conformations has a mirror plane of symmetry; lacking handedness (5-5)

Acid catalyzed: descriptor of a reaction that is accelerated in the presence of an acid, but the acid is not consumed in forming the product (10-2, 20-2)

Acid chloride: RCOCl; functional group in which a carbonyl carbon bears an alkyl or aryl group and a chlorine atom (3-10)

Acid-induced reaction: a reaction in which acid is required and is not regenerated at the end of the sequence (20-3)

Activation energy (E_{act}): the energy difference between a ground-state reactant and the transition state (5-1)

Activation energy barrier: *See* **Activation energy** (6-1)

Active electrophile: a more-active-than-normal form of an electrophilic reagent often prepared by interaction of an electrophilic reagent with a Lewis or Brønsted acid (11-2)

Active site: the relatively small part of an enzyme at which catalysis occurs (20-7)

Acyclic: lacking rings (1-6)

Acyl anion: R—C≡O⁻; an unstable anion, for which chemical equivalents are available for synthesis (18-8)

Acyl anion equivalent: a reagent that provides a nucleophilic equivalent of [RC≡O]⁻ (21-9)

Acylation: replacement of H by an acyl group (11-2)

Acylium ion: —RC≡O⁺; a resonance-stabilized cation in which positive charge is distributed between carbon and oxygen (11-2)

Addition polymer: macromolecule produced in a polymerization in which all atoms present in the monomer are retained in the polymeric product (16-2, 16-3)

Addition reaction: a chemical conversion in which two simple molecules combine to form a product of higher molecular weight (7-1)

Adenosine diphosphate (ADP): diphosphate ester of adenine; a principal energy storage molecule in biological systems (22-3)

Adenosine monophosphate (AMP): monophosphate ester of adenine (22-3)

Adenosine triphosphate (ATP): triphosphate ester of adenine; a principal energy storage molecule in biological systems (22-3)

ADP: *See* **Adenosine diphosphate** (22-3)

Adsorption: association with a solid surface, often reversible (4-2)

Alcohol: a compound bearing the OH functional group (3-7)

Alcoholysis: a reaction in which an alcohol displaces a leaving group or is added across a multiple bond (12-6)

Aldehyde: RCHO; functional group in which a carbonyl carbon bears a hydrogen and an alkyl or aryl group (3-9)

Aldimine: RCH=NR′; the imine of an aldehyde (21-4)

Aldol: a β-hydroxyalcohol; a molecule containing both an aldehyde and an alcohol functional group (13-2)

Aldolase: an enzyme that catalyzes a retro-aldol reaction; as in the degradation of fructose diphosphate in glycolysis (22-9)

Aldol condensation: production of a more-complex α,β-unsaturated aldehyde (or ketone), with the elimination of water, upon treatment of two equivalents of an aldehyde (or ketone) with acid or base (7-1, 13-2)

Aldol reaction: formation of a β-hydroxyaldehyde (or ketone) from two molecules of an aldehyde (or ketone) (13-2)

Aldose: a sugar with an aldehyde at C-1 (17-4)

Aliphatic hydrocarbons: a family of compounds containing hydrogen and carbon atoms but lacking aromatic rings (2-1)

Alkaloids: a structurally diverse set of natural products isolated from plants and consisting of compounds that contain a basic, sp^3-hybridized nitrogen atom (18-5)

Alkanes: a family of saturated hydrocarbons with an empirical formula C_nH_{2n+2} for acyclic members (1-5)

Alkenes: a family of unsaturated hydrocarbons containing one or more double bonds; compound with an empirical formula C_nH_{2n} for acyclic members with one double bond (2-1)

Alkoxide: anion obtained by deprotonation of the —OH group of an alcohol (8-2)

Alkylation: replacement of H by an alkyl group (8-4, 11-2)

Alkylborane: functional group in which carbon is attached to a trivalent boron atom (10-3)

Alkyl group: a fragment derived from an alkane by removal of one hydrogen (1-7)

Alkyl halide: functional group in which carbon is bound to a halogen atom (3-13)

Alkynes: a family of hydrocarbons containing a triple bond; compounds with an empirical formula C_nH_{2n-2} for acyclic members with one triple bond (2-4)

Alkynide anion: an anion formed by deprotonation of a terminal alkyne; RC≡C⁻ (8-3)

Allene: an unsaturated hydrocarbon containing two orthogonal double bonds emanating in opposite directions from a common sp-hybridized carbon atom (2-4)

Allyl cation: a resonance-stabilized carbocation in which the vacant p orbital is adjacent to a π bond (3-7)

Allyl group: an alkyl substituent in which the point of attachment is adjacent to a double bond; —CH₂CH=CH₂ (2-3)

α-amino acid: compound in which an amino group and a carboxylic acid are attached to the same carbon atom (16-4, 18-1)

α anomer: stereoisomer of the cyclic form of a carbohydrate in which the hydroxyl group at C-1 is *trans* to the last carbon of the chain (axial in six-membered-ring carbohydrates) (17-4)

α-glucosidase: an enzyme that catalyzes the cleavage of α-glucosidic linkages (17-4)

α-helix: a right-handed spiraling structure imposed by intramolecular hydrogen bonding between groups along a single peptide chain (16-6, 19-3)

α-ketoacid: a carboxylic acid bearing a ketone group at the α position (21-8)

Aluminate: a species containing an O–Al bond (12-2)

Ambiphilicity: tendency of an —XH group to act as both an acid and a base (3-4)

Amide: RCONR₂′; functional group in which a carbonyl carbon bears an alkyl or aryl group and an amino group (3-10, 18-1)

Aminal: a functional group bearing one hydrogen, an alkyl group, and two amino groups on one carbon atom [RCH(NR′₂)₂]; produced in the reaction of a secondary amine with an aldehyde (21-11)

Amine: alkyl or aryl derivative of ammonia (3-1, 8-2, 18-1)

Amino acids: *See* **α-amino acid**

Aminocarbohydrates: carbohydrates substituted with amino groups (18-7)

Aminosugars: *See* **Aminocarbohydrates** (18-7)

Ammonia: NH₃; the simplest compound containing sp^3-hybridized nitrogen (3-1)

Ammonolysis: formal addition of ammonia across a double bond, as in benzyne (11-3)

AMP: *See* **Adenosine monophosphate** (22-3)

Amphoteric: a compound that contains both an acidic and a basic site; for example, α-amino acids (18-2)

Amylopectin: a highly branched water-insoluble starch (16-4)

Amylose: *See* **Starch** (16-4)

Androgen: a male hormone (17-1)

Angle strain: destabilization caused by deformation from normal bonding angles for atoms in a cyclic compound (1-6)

Anhydride: RCO₂COR; functional group in which two carbonyl carbons bearing alkyl or aryl groups are linked through an oxygen atom (3-10)

Aniline: C₆H₅NH₂; amino-substituted benzene (3-12)

Anion: a negatively charged ion (3-2)

Anionic polymerization: formation of a polymer by a process in which the growing end is a carbanion (16-3)

Annulation: formation of a ring on an existent ring (13-2)

Anomeric effect: unusual favoring of the α, or axial, orientation (or, conversely, the disfavoring of the β anomer) in an anomeric equilibrium (17-4)

Anomerization: interconversion between the α- and β-anomers of a carbohydrate (17-4, 22-9)

Anomers: stereoisomers of cyclic hemiacetals, usually carbohydrates, that differ in configuration at the hemiacetal carbon (17-4)

***Anti* addition:** the formation of an addition product by delivery of an electrophile and nucleophile to opposite faces of a double (or triple) bond (10-2)

Antiaromatic hydrocarbon: a planar, conjugated, cyclic, unsaturated hydrocarbon consisting of sp^2-hybridized carbons, lacking the chemical stability of a Hückel aromatic; most such systems contain $4n$ π electrons and are said to be conjugatively destabilized, although the choice of an appropriate model with which to compare them is not absolutely clear (2-1)

Antibiotics: chemicals that attack microorganisms (18-7)

Antibodies: peptide complexes of moderate size that are responsible for alerting the immune system to the presence of foreign substances (20-8)

Antibonding molecular orbital: a molecular orbital that, when occupied by electrons, destabilizes a molecule relative to the separated atoms (2-1)

Anticodon: a sequence of three bases that complements the three-base codon in mRNA and selects for a particular amino acid for protein synthesis (19-6)

***Anti* conformer:** conformational isomer in which two large groups on adjacent atoms are separated by a 180° dihedral angle (5-2)

Antifungal agent: a pharmaceutical agent that selectively attacks and destroys fungi (23-6)

Antihistamine: a compound used to inhibit vasodilation by competing with histamine for binding at a physiologically active site (23-3)

Anti-Markovnikov regiochemistry: that taking place in the opposite sense from that predicted by Markovnikov's Rule; an addition in which a proton is delivered to the more-substituted carbon and the nucleophile to the less-substituted carbon of an alkene (10-3)

***Anti*-periplanar:** descriptor of the geometric relation in which the bonds to substituents on adjacent atoms of a σ bond are coplanar, with a dihedral angle of 180°; the preferred geometry for an E_2 elimination (9-2)

Applied field (H_{app}): external magnetic field applied to a sample in a nuclear magnetic resonance spectrometer (4-3)

Aprotic solvent: a solvent molecule lacking a polar X–H bond (3-2)

Arene: an aromatic hydrocarbon or derivative (2-3)

Aromatic hydrocarbons: a family of planar, sp^2-hybridized, conjugated, cyclic, unsaturated hydrocarbons with unusual chemical stability; according to Hückel's Rule, such compounds contain $4n + 2$ electrons in their π systems (2-3)

Aromaticity: special stability afforded by a planar cyclic array of p orbitals containing a Hückel number ($4n + 2$) of electrons (2-3, 6-7)

Arrhenius equation: $k = Ae^{-\Delta G^{\ddagger}/RT}$; mathematical correlation of the rate of a reaction with its activation energy (5-2)

Arrow notation: the use of half- and full-headed curved arrows to indicate electron motion in a chemical-reaction mechanism (7-2)

Arrow pushing: the use of curved arrows to describe the movement of electrons as a reaction proceeds (7-4)

Aryl group: an arene fragment lacking one substituent from a ring carbon (2-3)

Aryl halide: functional group in which a halogen is attached to an arene ring (9-4)

-ase: suffix descriptive of an enzyme that catalyzes a specific transformation (17-4)

Atactic: stereochemical designator of a polymer with random orientation of groups at centers of chirality (16-6)

Atomic orbitals: probability surfaces, associated with an atom, within which an electron is likely to be found (1-1)

ATP: *See* **Adenosine triphosphate** (22-3)

Average bond energy: typical energy of a specific type of bond; obtained from heats of formation (for example, the average bond energy of a C–H bond is obtained as 1/4 of the heat required to convert methane into carbon and hydrogen—that is, of $CH_4 \rightarrow C + 4$ H, $\Delta H°/4 = 99$ kcal/mole; and that for C–C is obtained by measuring the heat of formation of ethane and subtracting the bond energies of the six C–H bonds) (3-7)

Axial: descriptor of a group pointing roughly orthogonally from the pseudoplane of a chair conformation (5-4)

Azlactone: activated, cyclized derivative of the carboxy terminus of a peptide or amino acid (18-4)

Azo dyes: highly colored compounds containing an —N═N— linkage; among the first synthetic colorfast agents (11-3, 23-5)

▼

Backside attack: the approach of a nucleophilic reagent from the side opposite that from which the leaving group is displaced (7-5)

Baeyer-Villiger oxidation: transformation of a ketone into an ester by reaction with a peracid; net change is the insertion of an oxygen atom between the carbonyl carbon and an adjacent carbon of the ketone (14-3)

Base-induced reaction: a reaction in which base is required but in which it is not regenerated at the end of the sequence (12-7, 20-3)

Base-line separation: an efficient separation of two compounds in which the peaks detected as representative of elution of the component molecules do not overlap; that is, the detector response returns to the base line between peaks (4-2)

Base pairing: a system of recognition based on optimal hydrogen-bonding patterns between nucleic acid bases; for example, guanine with cytosine (G–C) and adenine with thymine (A–T) (19-6)

Base peak: the most-intense peak in a mass spectrum (4-3)

Basicity: the affinity of an atom or ion for a proton (12-5)

Beckmann rearrangement: transformation of the oxime of a ketone into an amide by reaction with a strong acid; net change from the ketone is the insertion of an NH group between the carbonyl carbon and an adjacent carbon (7-1, 14-2, 16-4, 18-1)

Benzenoid ring: a six-membered ring in a polycyclic aromatic compound that retains three formal double bonds (11-5)

Benzilic acid rearrangement: anionic skeletal rearrangement of an α-diketone to an α-hydroxyacid induced by treatment with aqueous hydroxide (14-1)

Benzyl cation: a resonance-stabilized carbocation in which the vacant *p* orbital is adjacent to an aryl ring (3-7)

Benzyl ether: a protecting group for an alcohol (15-8)

Benzyne: C_6H_4; an unstable species with a triple bond in a benzene ring; a highly reactive ring compound related to benzene in having two hydrogen atoms removed from adjacent ring positions (9-4)

β anomer: stereoisomer of the cyclic form of a carbohydrate in which the hydroxyl group at C-1 is *cis* to the last carbon of the chain (equatorial in six-membered-ring carbohydrates) (17-4)

Beta-blocker: a compound that interferes with binding to certain receptor sites; for example, that which stimulates acid production in the stomach (23-4)

β-dicarbonyl compound: functional group containing two carbonyl groups attached to a common atom (8-4)

β-galactosidase: an enzyme that catalyzes the hydrolysis of the β linkage between carbohydrates, one of which is galactose (17-4)

β-helix: a left-handed spiraling structure imposed by intramolecular hydrogen bonding between groups along a single peptide chain; not found in naturally occurring α-amino acids (16-6, 23-6)

Betaine: a zwitterionic species with a negatively charged atom and a positively charged atom that are separated by two atoms; name derived from a compound called betaine [$^-O_2CCH_2N^+(CH_3)_3$] isolated from beets; an intermediate in the Wittig reaction (13-1)

β-ketoacid: functional group containing a keto group and a carboxylic acid attached to a common atom (8-4)

β-phenethylamine: Ar—C—C—N; subunit present in many psychoactive compounds (18-5, 23-4)

β-pleated sheet: a folded sheetlike structure imposed by intermolecular hydrogen bonding between peptide chains (16-6, 19-3)

Bidentate: a compound with two ligating heteroatoms; from Latin, meaning "two teeth" (19-4)

Bilayer: a structure comprising two layers of lipids or surfactants in which the hydrocarbon tails point toward the interior and the polar groups are on the two surfaces, solvated by water (or another polar protic solvent); held together by the van der Waals attractive interactions between the hydrocarbon tails (17-1, 23-6)

Bimolecular reaction: a reaction that requires a collision between two reactants in the rate-determining step (6-9)

Biochemical energy storage: reservoirs of energy in chemical bonds; often stored as anhydrides of phosphoric acid and reduced forms of redox cofactors (22-3, 22-4)

Biochemical reducing agent: a cofactor that acts as a reducing agent, either through transfer of a hydride equivalent or by providing electrons (21-3)

Biopolymer: macromolecule occurring in nature (16-4)

Biosynthesis: natural pathway for the construction of biologically important molecules (13-1)

Biotin: vitamin H; a biological carrier of carbon dioxide (22-5)

Biradical: a chemical species bearing two noninteracting radical centers (2-1, 10-4)

Bis-acetal: functional group in which one oxygen is shared between two acetal functional groups (17-4)

Blood type: one of four classifications of human blood cells differentiated by the identity of carbohydrate residues present on the cell surface (23-1)

Boat conformation: the eclipsed conformation of cyclohexane or an analogous six-atom cyclic compound in which the spatial placement of C-1 and C-4 roughly resembles the bow and stern of a boat (5-3)

Boiling point: temperature at which a given liquid is converted into a gas at standard pressure; temperature at which the effect of entropy overcomes associative intermolecular forces in a liquid phase (1-9, 2-5, 19-1)

Bond alternation: a repeating sequence of short and long (single and double) bonds in an extended π system (2-2, 2-3)

Bond angle: the angle formed by two bonds intersecting at an atom; varies with hybridization: typically about 109.5° at an sp^3-hybridized atom; 120° at an sp^2-hybridized atom; and 180° at an sp-hybridized atom (1-2, 2-1, 2-4)

Bond-dissociation energy: the quantity of heat consumed when a covalent bond is homolytically cleaved (3-7)

Bonding molecular orbital: a molecular orbital that, when occupied by electrons, stabilizes a molecule relative to the separated atoms (2-1)

Bond length: equilibrium distance between two covalently bonded atoms; varies with hybridization: 1.54 Å between sp^3-hybridized carbon atoms in ethane; 1.33 Å between sp^2-hybridized carbon atoms in ethene; and 1.06 Å between sp-hybridized carbon atoms in ethyne (1-2, 2-1, 2-4)

Borate: a species containing one or more B–O bonds (12-2)

Branched polymer: a macromolecule in which chemical bonds interconnect chains so that a complex, three-dimensional network results (16-1)

Bridgehead atom: one that is common to both rings in a bicyclic (or multicyclic) compound (5-4)

Bromonium ion: three-membered cyclic cationic intermediate in which bromine bears formal positive charge; formed by the addition of Br^+ (or a source of this species) to an alkene (10-4)

Brønsted acid: a proton (H^+) donor (3-4)

Brønsted base: a proton (H^+) acceptor (3-4)

▼

Cahn-Ingold-Prelog rules: used in specifying absolute stereochemistry (5-6)

Calorimeter: a device used to measure the heat released or consumed in a chemical reaction (1-8)

Cannizzaro reaction: conversion of an aldehyde lacking α hydrogens into equal amounts of the corresponding carboxylic acid and alcohol upon treatment with sodium hydroxide or potassium hydroxide (12-6)

Carbamate: *See* **Urethane** (15-8)

Carbanion: negatively charged trivalent carbon bearing an unshared electron pair (6-4)

Carbene: a neutral reactive intermediate in which a carbon atom bears two σ bonds and two unshared electrons; contains only six electrons in its outer shell (6-4, 10-4)

Carbocation: a positively charged trivalent carbon atom containing only six electrons in its outer shell (3-7)

Carbodiimide: R—N═C═N—R; reagent used to activate a carboxylic acid toward amide formation (18-4)

Carbohydrate: polyhydroxylated aldehyde or ketone with a molecular formula $C_n(H_2O)_n$ (16-4, 17-3, 17-4)

Carbon–carbon bond-forming reaction: chemical transformation in which two previously unconnected carbon atoms become covalently bound (15-1)

Carbonium ion: *See* **Carbocation** (3-7)

Carbonyl group: C═O; functional group containing a carbon–oxygen double bond (3-9)

Carbowax: a synthetic poly(ethyleneglycol) (16-3)

Carboxylic acid: RCO_2H; functional group in which a carbonyl carbon bears an alkyl or an aryl group and an OH group (3-10)

Carboxylic acid anhydride: *See* **Anhydride** (3-10)

Carcinogens: cancer-inducing agents (2-3)

Catalysis: acceleration of a reaction by a catalyst (20-1, 22-1)

Catalyst: a species that is not involved in the overall stoichiometry of a reaction and is recovered unchanged after the reaction but is needed for the reaction to proceed at a reasonable rate; a reagent that facilitates a reaction without itself ultimately forming chemical bonds in the product or appearing in the stoichiometric equation describing the reaction (2-1, 7-1, 12-4)

Catalytic antibody: an enzyme expressed by the immune system of an organism in response to an injected transition-state analog of a desired reaction (20-8)

Catalytic cycle: the complete sequence of steps by which a chemical transformation is accelerated in the presence of a catalyst (20-7)

Catalytic hydrogenation: a reaction catalyzed by a heterogeneous catalyst (usually a noble metal) in which hydrogen is added across one or more multiple bonds (2-1, 7-1, 12-4, 20-6)

Cation: a positively charged ion (3-2)

Cationic polymerization: formation of a polymer by a process in which the growing end is a carbocation (16-3)

Cellulose: water-soluble biopolymer containing from 3000 to 5000 glucose units connected exclusively by β linkages (16-4)

Cellulose acetate: optically transparent polymer obtained by treating cellulose with acetic anhydride, thus converting many of the polysaccharide hydroxyl groups into acetate esters (16-4)

Center of chirality: a tetrahedral atom (usually carbon) bearing four different groups (5-5)

Chain reaction: a chemical conversion in which the one of the products is a reactive species that initiates another cycle of the reaction; a reaction that, after initiation, repeats a cycle of propagation steps until one of the reactants is consumed (7-7)

Chair conformation: the staggered conformation of cyclohexane or an analogous six-atom cyclic compound roughly resembling the back, seat, and footrest of a chair (5-3)

Charge-relay mechanism: a reaction in which a molecule, often water, transfers a proton (and, therefore, charge) from one position to another in the same molecule or in another, in which direct transfer is impossible because of the spatial orientation and the distance separating the two sites (20-2)

Charge separation: development of centers of positive and negative charge upon interaction of neutral reagents (20-2)

Chemical bond: an energetically favorable interaction between two atoms that is induced by a pair of electrons mutually attracted to both nuclei or by electrostatic attraction between two ions (1-3)

Chemical shift: the magnitude of the change of the observed resonance energy for a given nucleus relative to that observed for a standard (usually tetramethylsilane); the position on an NMR spectrum at which a given nucleus absorbs (4-3)

Chemotherapeutic index: the ratio of toxic dose to the host to that for the invading organism (23-6)

Chiral: the property of handedness; when applied to molecules, lacking a mirror plane through any conformation (5-5)

Chiral center: *See* **Center of chirality** (5-5)

Chirality: handedness; the property of an object (in this context, a molecule) whereby the object is not superimposable on its mirror image (5-5)

Chiral molecule: a molecule lacking an internal plane of symmetry; a molecule that is not superimposable on its mirror image; the most-common indicator of chirality is the presence of a carbon atom bonded to four different groups (5-1, 5-5)

Chiral recognition: specific, reversible interaction between two chiral molecules based on three-point contact that is different for the different diastereomeric pairings (19-7, 22-7)

Chloromethyl polystyrene: solid polymeric support used in the Merrifield peptide synthesis (18-4)

Chloronium ion: three-membered cyclic cationic intermediate in which chlorine bears formal positive charge; formed by the addition of Cl^+ (or a source of this species) to an alkene (10-4)

Chromate oxidation: oxidation with Cr^{6+}, often of alcohols to aldehydes, ketones, or carboxylic acids, which is accompanied by a color change of the inorganic reagent from red-orange to green (Cr^{3+}) (9-7)

Chromatogram: a plot of a detector response as a function either of the volume of effluent flowing through the column or of time (4-2)

Chromatographic resolution: degree of separation of a mixture of compounds (4-2)

Chromatographic separation: isolation of individual components of a mixture through a chromatographic technique (4-2)

Chromatography: technique by which components of a mixture are partitioned between two different phases, attaining separation because of a difference in solubility of the component molecules in each phase (4-1)

***Cis*-isomer:** a geometric isomer in which the largest groups are on the same side of a double bond or ring (2-1)

Citric acid cycle: *See* **Tricarboxylic acid cycle** (22-7)

Claisen condensation: reaction producing a β-ketoester upon treatment of an ester with base (13-3)

Claisen rearrangement: a [3,3]-sigmatropic shift of a substituted allyl vinyl ether; pericyclic reaction in which allyl vinyl ether is converted into a rearranged β,γ-enone; sometimes called an oxa-Cope rearrangement (14-3)

Clemmensen reduction: reduction of a ketone to a —CH_2— group by treatment with zinc in HCl (11-2)

Codon: a three-base sequence on messenger RNA that specifies the amino acid to be used in protein synthesis; complementary to the anticodon on tRNA (19-6)

Coenzyme A (CoA): a complex thiol that, as an thiol ester derivative, accelerates nucleophilic acyl substitution in several biochemical transformations (21-7)

Cofactor: a recyclable biological reagent (21-3)

Column chromatography: liquid chromatography conducted with an open chromatography column through which the eluent flows in response to gravity (4-2)

Combustion: burning in air (1-8)

Complementary: descriptor of a favorable hydrogen-bonding interaction between bases (19-6)

Complex metal hydride: a reagent in which hydride is bound to boron or aluminum and which is soluble in organic solvents, providing the equivalent of the hydride ion in nucleophilic reactions; most-common members of this group are $NaBH_4$, $LiAlH_4$, and $NaBH_3(CN)$ (12-2)

Complex metal hydride reduction: the use of a complex metal hydride to convert an aldehyde into the corresponding primary alcohol, a ketone into a secondary alcohol, an ester into a primary alcohol, an imine into an amine, or an amide into an amine (12-2)

Concerted reaction: a chemical transformation that proceeds directly from reactant to product through a single transition state and without a reactive intermediate (6-1, 14-1)

Condensation polymer: macromolecule produced in a polymerization in which a small-molecule by-product is formed (16-2, 16-4)

Condensation reaction: a chemical conversion in which two molecules combine to form a more-complex product, with the loss of a small molecule, usually water or an alcohol (7-1)

Configurational isomers: stereoisomers that can be interconverted only by the breaking and reforming of a covalent bond (5-1)

Conformational analysis: energetic description of conformational interconversion; relates the relative atomic positions to the changes in potential energy during rotation about a σ bond (5-2)

Conformational anchor: a substituent (usually large) that so strongly prefers the equatorial position that it blocks conformational flipping of the six-membered ring to which it is attached (5-4)

Conformational isomers: stereoisomers that can be interconverted by rotation about a σ bond (5-2)

Conformational lock: *See* **Conformational anchor** (5-4)

Conformer: a conformational isomer (5-2)

Conjugate acid: cation obtained by the addition of a proton to a Brønsted base (6-6)

Conjugate addition: the addition of a reagent across a four-carbon conjugated π system, producing a 1,4-adduct and a double bond between C-2 and C-3 (10-2)

Conjugate base: anion obtained by the removal of a proton from a Brønsted acid (6-6)

Conjugated diene: a diene with an array of *p* orbitals on adjacent atoms; that is, one in which the double bonds constituting the π system interact directly without interruption by an intervening sp^3-hybridized atom (2-2)

Conjugation: a series of alternating single and double bonds along a carbon chain with adjacent *p* orbitals (2-2)

Connectivity: the sequence of attachment of atoms along a chain in a molecule (2-1)

Constitutional isomers: isomers having the same molecular formula in which the atoms are connected in different sequences (5-1)

Convergent synthesis: a branched synthesis in which two or more synthetic intermediates react with each other (15-6)

Cope rearrangement: a [3,3] sigmatropic shift; process by which a new carbon–carbon σ bond is formed between C-1 and C-6 in a substituted 1,5-hexadiene at the same time as the bond between C-3 and C-4 is broken, with both π bonds shifting to take up new positions between different carbon atoms (14-1)

Coupled reactions: two simultaneous reactions in which the second provides sufficient energy to overcome the endothermicity of the first (12-8, 13-1, 22-2)

Coupling: the spin-spin interaction of the magnetic spin of a nucleus with one or more neighboring nuclei in nuclear magnetic resonance spectroscopy, causing a signal to be split into a characteristic pattern reflecting the number of magnetically active neighboring nuclei (4-3)

Coupling constant: the magnitude of spin-spin splitting of an NMR signal by magnetically active neighboring nuclei; *J*, usually expressed in Hz (4-3)

Covalent bond: a chemical bond in which an electron pair is shared by two atoms (1-3)

Covalent catalysis: an accelerated chemical reaction in which the substrate becomes temporarily bound through a covalent linkage to an active site on the catalyst (20-7)

Crossed aldol condensation: aldol condensation between two different carbonyl compounds (13-2)

Crossed Claisen condensation: Claisen condensation between two different esters (13-3)

Cross-linking: covalent interconnections between polymer chains that result in a three-dimensional network; process in which a bifunctional molecule is incorporated into two separate polymer chains (16-1, 16-3)

Crown ether: synthetic cyclic molecule containing several (usually from four to ten) ether groups that can assume a conformation in which a metal cation is effectively complexed with the ethereal oxygen atoms at the center of the ring cavity (8-4, 19-4)

Cumulated diene: a diene in which the two orthogonal double bonds share a common carbon atom (2-2)

Curved full-headed arrows: used in mechanistic organic chemistry to indicate the motion of an electron pair (6-5)

Cyanide: ^-CN (8-4, 13-1)

Cyano group: $R\text{---}C\equiv N$; functional group with a carbon–nitrogen triple bond; also called a nitrile (3-6)

Cyanohydrin: α-cyanoalcohol, the product of HCN addition to a carbonyl compound (13-1)

Cyclic: containing one or more rings (1-6)

Cyclic bromonium ion: *See* **Bromonium ion** (10-4)

Cyclic chloronium ion: *See* **Chloronium ion** (10-4)

Cyclic halonium ion: *See* **Halonium ion** (10-4)

Cycloaddition reaction: a concerted pericyclic reaction resulting from the combination of two separate π systems into a cyclic product (14-1)

Cycloalkanes: saturated hydrocarbons containing one or more rings with the empirical formula in which two hydrogens per ring are subtracted from that of an acyclic alkane (C_nH_{2n+2}) (1-6)

Cyclopropane: C_3H_6; the simplest cycloalkane (1-6)

Cycloreversion: a pericyclic reaction (the inverse of a cycloaddition) in which a cyclic molecule fragments into two or more smaller π systems (14-1)

▼

d: relative stereochemical designator of a molecule with a positive (dextrorotatory) specific rotation: from the Greek for "right rotating"; counterpart of *l* (5-8)

d,l: indicator of an optically inactive racemic modification (5-8)

D: absolute stereochemical descriptor that relates substituent disposition at a given center of chirality to that in natural D-glyceraldehyde; counterpart of L (5-8, 17-4)

D-glyceraldehyde: triose carbohydrate, serves as a reference compound for stereochemical designation of sugars; 2-(R)-propanal-2,3-diol (17-4)

D,L: absolute stereochemical descriptors that relate substituent disposition at a center of chirality to that in D- and L-glyceraldehyde; refers to a racemic mixture when used together as D,L (17-4)

Dacron: a commercial polyester produced by linking dimethyl terephthalate with ethylene glycol (16-4)

Decalin: bicyclo[4.4.0]decane; two fused six-membered rings (5-4)

Decarboxylation: loss of CO_2, usually from a carboxylic acid; particularly easy from a β-ketocarboxylic acid (8-4)

Degenerate rearrangement: a skeletal rearrangement in which the breaking and forming of bonds leads to a product that is chemically identical with the reactant (14-1)

Dehalogenation: formal loss of X_2 from a dihalide (9-6)

Dehydration: formal loss of water, usually from an alcohol (3-7, 7-1, 9-5)

Dehydrobromination: loss of HBr from an alkyl bromide (9-2)

Dehydrohalogenation: formal loss of HX from an alkyl halide (7-1)

Deinsertion: the opposite of an insertion reaction; one in which an atom (often a transition metal), covalently associated with two groups, is removed as the two groups become covalently bound to each other (20-6)

Delocalization: spreading of π-electron density over an entire π system (2-3)

Denaturation: disruption of the secondary and tertiary structure of a protein (16-6)

Deoxy-: prefix indicating that an oxygen functional group (often OH) has been replaced by a C–H bond (17-4)

Deoxyribonucleic acid: *See* **DNA** (17-4)

Deoxyribose: sugar unit found in the backbone of DNA (17-4)

Depsipeptide: a reagent with both ester and peptide linkages (23-6)

Detector: a device that produces a signal in response to the presence of a compound of interest (4-2)

Dextrorotatory: *See d* (5-8)

Diastereomers: nonmirror-image stereoisomers (5-8)

Diazo coupling: connection of two aromatic rings through an azo linkage, usually by electrophilic attack on one ring by an aryl diazonium salt (11-3)

Diazonium salt: $[Ar—N\equiv N]^+ X^-$; prepared by treatment of a primary aniline with nitrous acid, HNO_2 (11-3, 23-5)

Diazotization: conversion of a primary amine into a diazonium salt (11-3)

Dieckmann condensation: an intramolecular variant of the Claisen condensation (13-3)

Diels-Alder reaction: concerted cyclization of a conjugated diene and an alkene (called a dienophile) to produce a cyclohexene; the most frequently encountered [4 + 2] cycloaddition (6-5)

Dienes: compounds containing two double bonds (2-2)

Dienophile: the alkene component in a Diels-Alder reaction that reacts with a diene (6-5)

Digonal: a carbon atom with only two bonds (6-4)

Dimer: compound containing most or all of the atoms of two molecules of a starting material (10-4, 16-1)

Dimethylallyl pyrophosphate: structural isomer of isopentenyl pyrophosphate; $(CH_3)_2CH\equiv CHCH_2OPO_3PO_3^{3-}$ (17-2)

Dipolar aprotic solvent: *See* **Polar protic solvent** (3-2)

Dipole–dipole interaction: intermolecular attraction or repulsion deriving from the electrostatic forces between bond dipoles in the two interacting molecules (19-2)

Dipole moment: the vector pointing from the weighted center of positive to the center of negative charge in a collection of atoms—typically, a molecule (3-2)

Directive effect: a substituent effect that influences the regiochemistry of a reaction (11-3)

Disaccharide: two carbohydrate molecules joined by an acetal or a ketal linkage (17-4)

Disease state: unnatural condition of an organism caused by: under- or overproduction of a critical biochemical; invasion of an alien living species that produces substances that are toxic to the host; or too-rapid growth of part of the organism (23-1)

Disproportionation: a reaction in which a species of intermediate oxidation level is converted into equal amounts of a more-oxidized product and a more-reduced product (12-6)

Dissolving-metal reduction: a chemical reduction accomplished with a zero-valent metal (usually sodium or potassium in liquid ammonia containing small quantities of ethanol) by protonation of radical anion or dianion intermediates, or both, produced by reductive electron transfer (so-called because the metal appears to dissolve as it is converted into a soluble cation) (12-4)

Diterpene: terpene containing twenty carbons; derived from four isoprene units (17-1)

DNA: deoxyribonucleic acid; the principal genetic information storage unit; found in cell nuclei; biopolymer composed of deoxyribonucleotide units linked through a sugar–phosphate backbone (17-4)

Double bond: a σ and a π bond between sp^2-hybridized atoms (2-1)

Doublet: a two-line multiplet (4-3)

Downfield: chemical shift of a nucleus that resonates at a higher δ value than a reference nucleus; that is, one shifted to a lower frequency; deshielded; left-hand part of an NMR chart (4-3)

▲
Early transition state: one that is reactantlike (6-2)

Eclipsed conformation: spatial arrangement in which each σ bond at one carbon is coplanar with one on an adjacent atom (dihedral angle = 0°); when viewed end-on in a Newman projection, conformation with exactly aligned bonds on adjacent atoms (5-2)

Effective collision: collision between two reactants with the correct orientation and sufficient energy to overcome the activation energy barrier and form product (6-9)

Effective field (H_{eff}): magnetic field "felt" at a nucleus of interest in an NMR scan; differs from the applied field by the tiny local magnetic field (H_{loc}) induced by the electron cloud surrounding the nucleus (4-3)

(*E*)-isomer: a geometric isomer in which the groups of highest priority are on opposite sides of a double bond; from the German, *entgegen*, opposite (2-1)

Electrocyclic reaction: a concerted pericyclic intramolecular ring-forming reaction (14-1)

Electromagnetic radiation: a particle (called a photon) or a wave traveling at the speed of light; includes infrared, visible, ultraviolet, and x-ray ranges. When regarded as a wave, light is described by its wavelength (λ) or its frequency (v) (4-3)

Electron acceptor: a group that withdraws electron density from an attached atom (11-4)

Electron affinity: tendency of an atom to gain an electron, thus forming an anion (1-4)

Electron configuration: an atomic-orbital description of an electron associated with a given atom, listing the electron's principal quantum number, its hydrogenic orbital type, and the number of electrons occupying the suborbital (1-1)

Electron donor: a group that releases electron density to an attached atom (11-4)

Electronegativity: the tendency of an atom to attract electrons, thus polarizing a covalent bond (1-4, 6-7)

Electronic effect: perturbation of molecular properties by shifts in electron density by a substituent (6-7, 11-4)

Electron transfer: the release of one or more electrons from one atom, ion, or compound to another atom, ion, or compound (1-4, 21-5)

Electrophile: an electron-deficient reagent that attacks centers of electron density; from the Greek *electros*, electron, and *philos*, loving (3-2)

Electrophilic addition: a chemical reaction in which a $C\!=\!C$ π bond is replaced by two σ bonds upon reaction with an electrophilic reagent, often a proton, followed by addition of a nucleophile (7-6, 10-1)

Electrophilic aromatic substitution: *See* **Electrophilic substitution** (11-1)

Electrophilicity: tendency of an atom, an ion, or a group of atoms to accept electron density from a carbon center (3-2)

Electrophilic substitution: replacement of a substituent (usually hydrogen) on an aromatic ring upon interaction of an aromatic π system with an active electrophile (11-1)

Electrophoresis: the migration of a charged molecule under the influence of an electric field; used to separate charged organic species, often proteins, nucleic acids, and other polyelectrolytes (4-2, 18-2)

Electrostatic attraction: favorable interaction between two species of opposite charge (1-4, 19-1)

Electrostatic repulsion: unfavorable interaction between two species of like charge (1-4)

Elimination reaction: a reaction (the inverse of an addition reaction) in which a single complex molecule splits into two simpler products; chemical reaction in which two groups on adjacent atoms are lost as a double bond is formed (7-1, 9-1)

Eluent: the mobile phase in liquid chromatography (4-2)

Elution: the motion of solute and solvent through the stationary phase in a chromatography column (4-2)

Elution time: the time required for a given compound to pass through a chromatography column (4-2)

Empirical formula: quantitative description of the relative proportion of elements present in a compound in smallest whole numbers (1-6)

Enamine: the nitrogen analog of an enol, in which an amino substituent is attached to an alkenyl carbon (8-3)

Enantiomers: stereoisomers related to each other as nonsuperimposable mirror images; stereoisomers with opposite configuration at each center of chirality (5-5)

Enantiotopic: a descriptor of identical groups that lie on opposite sides of the plane of symmetry of an achiral molecule (22-7)

Endergonic reaction: a chemical transformation in which free energy input is needed; one in which the free energy content of the products is higher than that of the reactants (*see also* **Endothermic reaction**) (6-1)

Endorphin: a natural pentapeptide found in the brain in extraordinary low concentration that induces euphoria or blocks pain; name derived from *endo*genous m*orphine* (23-5)

Endothermic reaction: a conversion with a positive enthalpy change (*see also* **Endergonic reaction**) (6-1)

Enediol: functional group bearing two hydroxyl groups on the double bond (22-9)

Energy barrier: the amount of energy required to reach the most-unfavorable point along the path followed in the conversion of one species into another (5-1)

Energy diagram: a graphic representation of the change in free energy (or enthalpy) encountered in the course of a reaction (6-1)

Energy of activation: *See* **Activation energy** (5-1)

Enol: functional group in which a hydroxyl group is attached to an alkenyl carbon (6-5, 10-2)

Enolate anion: a resonance-stabilized anionic intermediate obtained by removal of a proton from the α position of a carbonyl compound or the OH group of an enol (6-5, 6-7, 8-4, 13-2)

Enolization: keto to enol tautomerization; conversion of a ketone or aldehyde into its enol form (13-2, 17-4)

Entgegen: *See* **(*E*)-isomer** (2-1)

Enthalpy: heat of reaction (5-1)

Enthalpy change: ($\Delta H°$); heat of reaction; difference between the bond energies of the reactants and products (5-1)

Entropy: disorder; free motion (5-1)

Entropy change: ($\Delta S°$); difference in disorder between reactants and products (5-1)

Enyne: an organic compound containing a double bond and a triple bond (2-4)

Enzyme: a protein that functions as a biological catalyst (17-2, 17-5, 20-7, 21-3)

Enzyme catalysis: acceleration of a chemical reaction by reversible association of a substrate with the active site of an enzyme (20-7, 21-3)

E1-CB reaction: mechanism for a unimolecular heterolytic elimination, taking place by the loss of the leaving group, L, from the deprotonated form (anionic conjugate base) of the neutral substrate in the rate-determining step (9-1)

E1 reaction: mechanism for a unimolecular heterolytic elimination, taking place by breaking of the carbon–leaving-group σ bond, with the formation of a carbocation, in the rate-determining step (9-1)

Epoxidation: preparation of an epoxide from an alkene (10-4)

Epoxide: a three-membered-ring functional group containing oxygen (10-4)

Epoxy resin: structurally rigid material obtained by cross-linking a diol with epichlorohydrin (16-4)

Equatorial: descriptor of a group whose orientation is roughly parallel with the pseudoplane of a chair conformation (5-4)

Equilibrium: state in which the forward rate of an ideally reversible reaction is equal to the reverse rate (5-2)

Equilibrium constant: ($K = [C][D]/[A][B]$; measure of the equilibrium position of reaction A + B = C + D; ratio of the forward and reverse rate constants of a reversible reaction at equilibrium (6-6)

Essential: descriptor of an amino acid that must be obtained from dietary sources for a particular living system (18-2)

Ester: RCO_2R'; functional group in which a carbonyl carbon bears an OR group (3-10, 8-2)

Ester enolate anion: resonance-stabilized anionic species obtained by removal of a proton from the α position of an ester (8-4, 13-3)

Estrogen: a female hormone (17-1)

Ethane: C_2H_6; the simplest saturated hydrocarbon containing a carbon–carbon bond (1-5)

Ethene (also called ethylene): C_2H_4; the simplest unsaturated hydrocarbon containing a double bond between sp^2-hybridized carbon atoms (2-1)

Ether: a functional group in which two alkyl or aryl groups are attached to an sp^3-hybridized oxygen atom (3-8)

Ethylene glycol: $HOCH_2CH_2OH$ (16-3)

Ethylene oxide: the simplest epoxide (C_2H_4O) (8-4)

Ethyne (also called acetylene): $HC\equiv CH$; the simplest alkyne containing a triple bond between sp-hybridized carbon atoms (2-4)

E2 reaction: mechanism for a bimolecular concerted elimination in which bonds to both the proton and the leaving group are broken in the rate-determining step (9-1)

Excited state: electronic configuration with a higher energy content than that of the ground state; often produced by absorption of a photon, promoting an electron from a bonding molecular orbital or from a nonbonding one to an antibonding molecular orbital (4-3)

Exergonic reaction: a reaction in which free energy is released; one in which the total free energy content of the products is lower than that of the reactants (6-1)

Exothermic reaction: a conversion with a negative enthalpy change (6-1)

Extraction: selective partitioning of a compound between two immiscible liquids, often a nonpolar organic phase and an aqueous or alcoholic phase (4-2)

▼
FAD: flavin adenine dinucleotide; cofactor used for electron-transfer oxidation— for example, in the oxidation of a saturated to an α,β-unsaturated thiol ester in fatty acid degradation (21-6)

FADH₂: flavin adenine dinucleotide; cofactor used for the electron-transfer reduction of an α,β-unsaturated thiol ester in fatty acid synthesis and degradation (21-6)

Fat: fatty acid ester of glycerol (17-1) ·

Fatty acid: a long, straight-chain carboxylic acid containing an even number of carbon atoms (17-1)

Fatty acid biosynthesis: biological pathway by which acetate (through acetyl CoA) is converted into a long-chain, unbranched carboxylic acid through a series of Claisen-like condensations (21-5, 22-5)

Fatty acid degradation: biosynthetic pathway by which long-chain, unbranched carboxylic acids are converted into acetyl CoA through a series of retro-Claisen-like condensations (22-6)

Fatty acids: long, straight-chain carboxylic acids (17-1)

Feedback: process by which a product serves to regulate its own rate of formation; often accomplished through partial product inhibition of enzyme catalysis (20-7)

Fingerprint region: the region in the infrared (from 400 to about 1100 cm^{-1}) that usually exhibits a series of complex, low-energy bands that are characteristic of a specific molecule (rather than a functional group) (4-3)

Fischer projection: a line notation used to indicate absolute configuration in which the intersection of two lines indicates the position of a chiral carbon, with horizontal lines indicating substituents directed toward the observer and vertical lines indicating substituents directed away from the observer (5-8, 17-4)

Flame ionization detector: a gas chromatography detector that senses the presence of ions that are generated as the effluent from the column is burned in a hydrogen flame (4-2)

Fluid mosaic: an expression used to describe the mobile nature of lipid bilayers (17-1)

Formal charge: a construct used to describe electron distribution in a molecule by comparing the number of valence electrons in a neutral atom with the sum of the number of unshared electrons plus half the number of shared electrons available to that atom; difference between the number of electrons accessed by an atom in a molecule and that in its elemental state (1-2)

Formulation: final preparation of a pharmaceutical in a form acceptable for delivery to the target organ or organism (23-3)

Formyl anion: $[HC{=}O]^-$; anion produced by deprotonation of formaldehyde (21-9)

Fragmentation pattern: a molecule-specific set of fragment ions obtained by bombarding a neutral molecule with high-energy electrons in a mass spectrometer (4-3)

Free energy: property of a system; has contributions from both enthalpy ($H°$) and entropy ($S°$); measure of the potential energy of a molecule or group of molecules (6-1)

Free energy change: a measure ($\Delta G°$) of the potential energy change in a chemical reaction; includes enthalpy ($\Delta H°$) and entropy ($\Delta S°$) components; $\Delta G° = \Delta H° - T\Delta S°$; related to the equilibrium constant as $\Delta G° = -RT \ln K$ (6-1, 6-6)

Free radical halogenation: a homolytic substitution of halogen for hydrogen, often in an alkane (7-7)

Free rotation: motion attained when orbital overlap is unaffected by rotation about the internuclear axis of a σ bond (1-5)

Friedel-Crafts acylation: reaction of a carboxylic acid chloride with an aromatic compound in the presence of a Lewis acid, resulting in the replacement of a hydrogen by an acyl substituent (11-2)

Friedel-Crafts alkylation: reaction of an alkyl halide with an aromatic compound in the presence of a Lewis acid, resulting in the replacement of a hydrogen by an alkyl substituent (11-2)

Functional group: a site in a molecule at which it undergoes characteristic and selective chemical reactions (1-8)

Functional-group compatibility: descriptor of a reagent or reaction that is sufficiently chemically selective so that only the desired functional group (of the several present in the molecule) interacts with the reagent (15-7)

Functional-group transformation: a chemical reaction in which one functional group is changed into another (15-1)

Furan: C_4H_4O; five-atom oxygen-containing heteroaromatic molecule (3-12)

Furanose: a carbohydrate containing a cyclic, five-membered-ring hemiacetal (17-4)

▼

Gabriel synthesis: synthesis of a primary amine by alkylation of phthalimide anion, followed by treatment of the resulting *N*-alkylphthalimide with basic hydrazine (8-2)

Gas chromatography: chromatographic technique in which a vaporized sample is carried by a gaseous mobile phase over a stationary phase (usually either a solid or a solid coated with a nonvolatile liquid) (4-2)

Gauche **conformer:** conformational isomer in which two large groups on adjacent atoms are separated by a 60° dihedral angle (5-2)

Gel electrophoresis: a separation technique that uses an electric field to induce movement of polyelectrolytes through a gel (*see also* **Electrophoresis**) (4-2)

Geminal dihalide: a compound in which two halogen atoms are attached to the same carbon atom (8-3, 9-3, 10-2)

Geminal diol: a functional group bearing two ——OH substituents on the same carbon atom (*see also* **Hydrate**) (12-6)

General acid catalysis: a reaction accelerated by any base capable of generating the specific acid (often H^+) required for the reaction (20-4)

General base catalysis: a reaction accelerated by any base capable of generating the specific base (often OH^-) required for the reaction (20-4)

Genetic code: system of information storage, transcription, and translation based on complementary base pairing in DNA; interpretation of the sequence of three base pairings in DNA–RNA transcription (19-6)

Geometric isomer: an isomer with the same connectivity along the backbone but with different spatial disposition of one or more groups around a bond with restricted rotation; *cis-trans* isomers (2-1, 5-1)

Geometric isomerization: a chemical conversion in which the relative positions of groups bound to a functional group with restricted rotation are reversed (7-1)

Glass: a polymer based on a three-dimensional network of tetrahedrally arranged silicon atoms linked by oxygen (16-3)

Glucoside: a cyclic derivative of glucose in which the C-1 hemiacetal hydroxyl group has been replaced by an alkoxy group (17-4)

Glycerol: 1,2,3-propanetriol (17-1)

Glycine: $H_2NCH_2CO_2H$; the simplest α-amino acid (18-2)

Glycol: a 1,2- or 1,3-diol (16-3)

Glycolysis: breakdown of carbohydrates in which a retro-aldollike reaction is a key step (22-9)

Grignard reagent: a reagent in which carbon is directly bound to magnesium (8-3, 13-1)

Ground state: the most-stable, lowest-energy electronic configuration (4-3)

Guanidine: ——NHC≡$NH(NH)$—— (18-2)

▼

Half-chair: high-energy conformation that represents the transition state obtained upon converting a chair into a boat conformation; has all but one atom of the ring in the same plane (5-3)

Haloform reaction: conversion of a methyl ketone into the corresponding carboxylic acid and haloform (CHX_3) upon treatment with aqueous base and dihalogen (8-4)

Halogenation: addition or substitution of halogen; as in the α-halogenation of methyl ketones in the haloform reaction (8-4), the formal addition of dihalogen to an alkene (10-4), and electrophilic aromatic substitution (11-2)

Halonium ion: three-membered cyclic cationic intermediate in which a halogen bears formal positive charge; formed by the reaction of X^+ with an alkene; important for chlorine and bromine (10-4)

Hammond postulate: an assertion that a transition state most closely resembles the stable species that lies closest to it in energy (6-2)

Hard: descriptor of a charge-intensive reagent; often applied in the description of nucleophiles, electrophiles, acids, and bases (13-1)

Hatched lines: graphic representation in three-dimensional structures indicating a group positioned away from the observer (1-5)

Heat of combustion: heat released when one mole of a compound is completely oxidized to CO_2 and H_2O (1-8, 2-1)

Heat of formation: a theoretical description of the energy that would be released if a molecule were formed from its component elemental atoms in their standard states (1-8)

Heat of hydrogenation: heat released when one mole of an unsaturated compound is completely hydrogenated to a saturated compound (2-1)

Heat of reaction: the energy difference between a reactant and a product (5-1)

Helix: *See* **α-helix; β-helix** (16-6, 19-3, 23-6)

Hell-Volhard-Zelinski reaction: method for the monobromination α to a carboxylic acid by treatment of a carboxylic acid bearing α hydrogens with bromine in the presence of phosphorus tribromide (8-4)

Hemiacetal: [RCH(OR)(OH)]; a functional group bearing an alkyl group, a hydrogen atom, an alkoxy group, and a hydroxy group on one carbon atom; the product of the nucleophilic addition of an alcohol to an aldehyde (12-6)

Hemiketal: [RR'C(OR)(OH)]; a functional group bearing two alkyl groups, an alkoxy group, and a hydroxy group on one carbon atom; the product of the nucleophilic addition of an alcohol to a ketone (12-6)

Heteroaromatic molecule: an aromatic molecule containing a ring heteroatom (3-12)

Heteroatom: any atom other than carbon and hydrogen (3-1)

Heterocycle: a cyclic molecule in which the ring contains one or more heteroatoms (3-12)

Heterocyclic aromatic: *See* **Heteroaromatic molecule** (3-12)

Heterolysis: *See* **Heterocyclic cleavage** (3-2, 3-7, 7-2)

Heterolytic cleavage: cleavage of a bond in which both electrons are shifted to one of the atoms of the bond (3-2, 3-7, 7-2)

Hexose: a six-carbon sugar (17-4)

High-energy phosphate bonds: phosphoric acid anhydride units critical in biological energy storage (22-3)

High performance liquid chromatography: *See* **High pressure liquid chromatography** (4-2)

High pressure liquid chromatography (HPLC): liquid chromatography in which the mobile phase is driven through a sealed chromatography column by a mechanical pump (4-2)

Hofmann elimination: a kinetically controlled elimination reaction in which the less-substituted alkene is preferentially formed (9-2)

Hofmann rearrangement: conversion of an amide into an amine containing one carbon fewer upon treatment with bromine in aqueous base (14-2)

HOMO: highest occupied molecular orbital (4-3)

Homolysis: *See* **Homolytic cleavage** (7-2)

Homolytic cleavage: cleavage of a bond in which the electrons are shifted, one to each of the atoms of the bond (3-2, 3-7, 7-2)

Hormone: a compound that controls essential biological functions, playing an important regulatory role in controlling key biochemical pathways (17-1, 18-3)

Hückel's Rule: an empirical generalization that any planar, cyclic, conjugated system containing $4n + 2$ π electrons (in which n is an integer) experiences unusual aromatic stabilization, whereas those containing $4n$ π electrons do not (2-3)

Hund's Rule: when possible, electrons singly occupy orbitals of identical energy (1-1)

Hybridization effect: influence of mixing of s and p orbitals; the greater the fraction of s character (50% in an sp-hybrid, 33% in an sp^2-hybrid, 25% in an sp^3-hybrid) of the hybrid orbital, the more electronegative is the atom (6-7)

Hybrid orbitals: those formed by mathematically mixing hydrogenic (s, p, d, etc.) atomic orbitals (1-2, 2-1, 2-4)

Hydrate: product of nucleophilic addition of water to an aldehyde or ketone (*see also* **Geminal diol**) (12-6)

Hydration: formal addition of water (10-2, 11-3)

Hydrazone: condensation product of hydrazine (NH_2NH_2) with an aldehyde or ketone; $R_2C{=}NNH_2$; often a highly colored solid used as a diagnostic test for the presence of a carbonyl group (12-7)

Hydroboration: addition of a carbon–boron and a carbon–hydrogen bond to an alkene (10-3)

Hydroboration-oxidation: a reaction sequence used to achieve anti-Markovnikov hydration of an alkene; initiated by concerted *syn* addition of borane, followed by oxidation with basic hydroperoxide (10-3)

Hydrocarbons: compounds that contain only carbon and hydrogen (1-1)

Hydrochlorination: addition of HCl to a multiple bond (7-1)

Hydrogenation: addition of H_2 (2-1)

Hydrogen bond: the weak association of a hydrogen atom attached to one electronegative heteroatom with a nonbonded electron pair on a second electronegative atom in the same or another molecule (X——H···Y, in which X and Y are electronegative heteroatoms) (3-2, 16-4, 19-3)

Hydrogenic atomic orbitals: atomic orbitals calculated precisely for hydrogen, including spherical *s* orbitals, propeller-shaped *p* orbitals, dumbbell-shaped *d* orbitals, more complexly shaped *f* orbitals, and so forth (1-1)

Hydrogen peroxide: H_2O_2 (10-3)

Hydrohalogenation: formal addition of HX (X = halide) to a multiple bond (10-2)

Hydrolysis: a reaction in which water displaces a leaving group or is added across a multiple bond (7-6, 12-6, 12-7)

Hydronium ion: H_3O^+ (10-2)

Hydrophilic: preference for association with an aqueous environment; property of polar molecules (17-1, 18-1)

Hydrophobic: preference for association with a nonaqueous environment; property of nonpolar molecules (17-1, 18-1, 19-1)

Hyperconjugation: an orbital description of the stabilizing effect derived by interaction of an aligned σ bond with an adjacent *p* orbital (2-1)

▼

Imide: RCONHCOR; functional group in which two carbonyl carbons bearing an alkyl or aryl group are linked through a nitrogen atom (3-10)

Imine: a compound containing a C═N double bond (3-5, 12-7, 18-1)

Imine-enamine tautomerization: process by which a proton is shifted from the α carbon of an imine to the imine nitrogen, or from the NH group of an enamine to the adjacent alkenyl carbon; a 1,3 shift of a proton in an imine or enamine (12-7)

Iminium ion: cation formed by protonation or alkylation of an imine on nitrogen; $(R_2C═NR_2)^+$ (8-4, 12-2, 12-7)

Immune system: biological system that produces antibodies to a foreign substance and effects its destruction or excretion (20-8)

Index of hydrogen deficiency: half the difference between the number of hydrogens present in a hydrocarbon and that expected for a straight-chain alkane ($2n + 2$); indicative of the number of multiple bonds or rings or both present (2-1)

Inductive effect: charge polarization through a series of σ bonds, causing a shift of electron density from or to a charged or polar site (6-7)

Inert gas: an atom that does not easily enter into chemical bonding with other atoms because the valence electron shell is filled; found at the right-hand column of the periodic table (1-1)

Infrared spectroscopy: technique that measures the absorption light of energies ranging from about 4000 to 400 cm^{-1} (4-3)

Initiation step: the first step of a radical reaction in which the number of radicals produced is greater than the number of radicals present in the reactant (7-7)

Initiator: a substance with an easily broken covalent bond that fragments to radicals that can induce a radical chain reaction (7-7)

Insertion: a reaction in which an atom, often a transition metal, becomes covalently associated with two atoms that were originally covalently bound to each other (20-6)

Integration: measurement of the relative area under each peak of a spectrum (4-2)

Integration curve: a measure of the area under each peak of a spectrum or chromatogram (4-3)

Intermediate: *See* **Reactive intermediate** (6-1, 6-4)

Intermolecular: descriptor of a phenomenon taking place between molecules (1-9)

Intermolecular hydrogen bond: a hydrogen bond connecting electronegative atoms in separate molecules (3-2, 20-5)

Intermolecular proton transfer: movement of H^+ from a position in a molecule to a bonded position in another molecule (19-3)

Intramolecular: descriptor of a phenomenon taking place within one molecule (1-9)

Intramolecular hydrogen bond: a hydrogen bond connecting electronegative atoms within the same molecule (3-2, 20-5)

Intramolecular proton transfer: movement of H^+ from one position to another in the same molecule (19-3)

Inversion of configuration: reversal of configuration at a center of chirality attained by the formation of a new bond on the face opposite the site at which a bond is broken (7-5)

Invertase: enzyme that catalyzes the cleavage of sucrose to a 1:1 mixture of glucose and fructose (17-5)

Invert sugar: 1:1 mixture of glucose and fructose obtained upon cleavage of sucrose (17-5)

In vitro: a reaction conducted in a laboratory environment; from the Latin for "in glass" (23-2)

In vivo: a reaction conducted within a living organism; from the Latin "in life" (23-2)

Iodoform test: chemical color test for the presence of an R(CO)CH$_3$ functionality by treatment with aqueous base and iodine, evidenced by the formation of a yellow precipitate of CHI$_3$ (8-4)

Ion channels: compounds, often proteins, with hydrophobic exteriors and hydrophilic internal regions, sometimes containing several charged amino acid residues, that dissolve readily in the interior of a phospholipid bilayer, thus spanning the bilayer, through which polar or charged molecules can move from one side of the membrane to the other (17-1)

Ionic bond: attractive electrostatic association between two oppositely charged ions (1-4)

Ionic polymerization: formation of a polymer by a process in which the growing end is an ion (16-3)

Ionophore: compound containing several heteroatoms arranged so that multiple, simultaneous contacts with a metal ion or other highly polar molecule are possible (19-4, 23-6)

Ion pairing: electrostatic association between oppositely charged ions (6-4)

Irreversible reaction: an exothermic reaction in which the activation energy for the reverse reaction is sufficiently large that the reaction proceeds only in the forward direction under practical conditions (6-1)

Isocyanate: RN$=$C$=$O; an intermediate in the Hofmann rearrangement (14-2)

Isoelectric point: the pH in aqueous solution at which an amphoteric molecule exists as a neutral entity with an equal number of positive and negative charges; in a simple α-amino acid, the pH at which it exists as a zwitterion (18-2)

Isoelectronic: two atoms with the same electronic configuration (6-4)

Isolated diene: a diene in which the double bonds do not interact directly with each other because of one or more intervening sp^3-hybridized atoms (2-2)

Isomerase: an enzyme that catalyzes isomerization (22-9)

Isomerization: a chemical conversion in which compounds with the same molecular formula, but different structures, are interconverted (7-1)

Isomers: different structural arrangements constituted from the same atoms (1-5)

Isopentenyl pyrophosphate: CH$_2$$=$C(CH$_3$)CH$_2CH_2OPO_3PO_3^{3-}$; branched five-carbon derivative of 2-methylbutadiene that is the biochemical precursor of the terpenes (17-1, 17-2)

Isoprene: 2-methylbutadiene (16-3)

Isoprenoids: derivatized skeletal oligomers of isoprene (17-1)

Isopropyl group: $-$CH(CH$_3$)$_2$; attached through the secondary carbon (1-7)

Isotactic: stereochemical designator of a polymer in which all groups at centers of chirality along the chain point in the same direction (16-6)

Isotopic labeling: the replacement of an isotope of highest natural abundance with another isotope at a specific position in a molecule; for example, replacement of ^{1}H by ^{2}H (D) or of ^{12}C by ^{13}C (8-3)

IUPAC rules: a set of guidelines for naming organic compounds in which a root gives the number of backbone carbon atoms and a suffix defines the functional group, with each substituent and its number being given by a prefix (1-7)

K_a: the acid-dissociation equilibrium constant; $K_a = K[H_2O] = ([A^-][H_3O^+])/[HA]$ (6-6)

Kekulé structures: historically, cyclic C$_6$ structures suggested by August Kekulé to depict benzene as having localized double bonds; now, the term is used to describe valence bond resonance representations of benzene (2-3)

Ketal: a functional group bearing two alkyl groups and two alkoxy groups on one carbon atom [RR'C(OR)$_2$]; produced in the acid-catalyzed alcoholysis of a ketone or a hemiketal; used as a protecting group for a ketone (12-6, 15-8)

Ketimine: the imine of a ketone (21-4)

Ketimine–aldimine tautomerization: a 1,3 shift of a proton to or from the α position of an imine, catalyzed by acid or base, that interconverts an aldimine and a ketimine (21-4)

Keto–enol tautomerization: process by which a proton is shifted from the α carbon of a ketone to the carbonyl oxygen, or from the O–H group of an enol to the remote alkenyl carbon; a 1,3 shift of a proton in an aldehyde or ketone (6-5, 17-4)

Ketone: R_2CO; functional group in which a carbonyl carbon bears alkyl or aryl groups or both (3-9)

Ketose: a sugar with a ketone functional group, usually at C-2 (17-4)

Ketyl: a radical anion obtained when an electron is added is the carbonyl group of a ketone (6-4)

Kinetic control: descriptor of a chemical reaction in which the reverse reaction takes place slowly or not at all, so that the relative concentration of products directly correlates with the relative rates of their formation rather than their relative stabilities (6-8)

Kinetics: a description of factors influencing the rate at which a reaction proceeds (6-5)

Krebs cycle: *See* **Tricarboxylic acid cycle** (22-7)

▼

l: relative stereochemical designator of a molecule with a negative (levorotatory) specific rotation: from the Greek for "left rotating"; counterpart of *d* (5-8)

L: absolute stereochemical descriptor that relates substituent disposition at a given center of chirality to that in natural L-glyceraldehyde; counterpart of D (5-8)

Lactam: a cyclic amide (16-4)

Lactone: a cyclic ester (14-3)

Late transition state: one that is productlike (6-2)

Leaving group: group displaced from a reactant in a substitution or elimination reaction (7-5, 8-3, 9-1, 12-2)

Le Chatelier's Principle: observation that the position of an equilibrium A + B = C + D can be shifted to the right either by increasing the concentrations of A or B or both or by decreasing the concentrations of C or D or both (7-1, 12-7)

Levorotatory: *See l* (5-8)

Lewis acid: an electron-pair acceptor (3-4)

Lewis base: an electron-pair donor (3-4)

Lewis dot structure: a representation in which electrons available to a given atom are indicated either as a nonbonded lone pair (by a pair of dots) or as a shared bonding pair (as a pair of dots between two atoms) (1-2)

LHRH: luteinizing hormone release hormone; a peptide hormone (18-3)

Linear polymer: a macromolecule in which the monomer units are attached end-to-end (16-1)

Linear synthesis: a plan that effects a sequence of transformations in which the product of one reaction is the reactant in the next (15-6)

Lipid bilayer: *See* **Bilayer** (17-1)

Lipids: group of simple, naturally occurring molecules that are soluble in nonpolar solvents; composed mostly of carbon, hydrogen, and oxygen atoms (17-1)

Lipoic acid: a dithiol cofactor important in the oxidative degradation of α-ketoacids (21-8)

Lipophilic: hydrophobic (17-1)

Liquid chromatography: chromatographic technique in which a solid or liquid sample is carried by a liquid mobile phase over a stationary phase (usually a solid composed of small particles around which the liquid phase can flow) (4-2)

Living polymer: a macromolecule in which the end of the chain is chemically reactive but in which two such ends will not react with each other; often applied to anionic, cationic, and organometallic polymerizations (16-3)

Lock-and-key: descriptor of the highly specific, tight association between a substrate and the active site of an enzyme (20-7)

Lone pair: two nonbonded electrons of opposite spin accommodated in an atomic or hybrid atomic orbital (3-1)

Lucas reagent: a mixture of Brønsted and Lewis acids that induces the conversion of an alcohol into the corresponding alkyl chloride (3-7)

Lucas test: chemical means for distinguishing tertiary, secondary, and primary alcohols by the rate of formation of the corresponding alkyl chloride from an alcohol upon treatment with the Lucas reagent (3-7)

LUMO: lowest unoccupied molecular orbital (4-3)

▼

Macromolecule: *See* **Polymer** (16-1)

Magnetic resonance imaging: a three-dimensional map of water concentration in an object; often used in medical applications for visualizing organs or anomalous growths (4-3)

Malonic ester: a diester in which both ester groups are bound to the same carbon atom; $CH_2(CO_2R)_2$ (8-4)

Malonic ester synthesis: method for preparing mono- and dialkylated carboxylic acids by sequential alkylation of a malonic ester anion, hydrolysis of the alkylated diester, and decarboxylation of the resulting β-diacid (8-4)

Mannich reaction: a condensation in which a simple aldehyde (often formaldehyde), a primary or secondary amine, and a ketone are combined (18-1)

Markovnikov's Rule: empirical prediction that the regiochemistry of the addition of HX to an unsymmetrical alkene takes place so as to place the proton on the less-substituted carbon atom of the multiple bond (7-6, 10-2)

Mass spectroscopy: technique that determines the mass of ions formed when molecules are bombarded with high-energy electrons (4-3)

Mechanism: *See* **Reaction mechanism** (7-1, 7-3)

Mechanistic organic chemistry: subarea of organic chemistry that focuses on the study of how reactions take place (7-3)

Melting point: temperature at which a solid is converted into a liquid at standard pressure; temperature at which the effect of entropy overcomes associative intermolecular crystal-packing forces (1-9, 2-5, 19-1)

Membrane: two-dimensional lipid bilayer; found on the outer surface of a cell, separating it from the external aqueous medium (17-1)

Merrifield synthesis: solid-state peptide synthesis on a porous polystyrene support (18-4)

Meso compound: an optically inactive molecule that contains a plane of symmetry that mirrors internally each center of chirality with another (5-8)

Messenger RNA (mRNA): a transcribed strand of RNA complementary to a segment of DNA; a replicated sequence of complementary bases of a DNA strand, except in the substitution of uracil for thymine (19-6)

Meta **director:** a functional group that directs electrophilic aromatic substitution to the *meta* position (11-4)

Meta **substitution:** the relation of substituents that are 1,3 to each other on an aromatic ring (2-3)

Methane: CH_4; the simplest hydrocarbon, composed of carbon surrounded by four hydrogens (1-2)

Micelle: a roughly spherical aggregation of many soaplike molecules with hydrophobic and hydrophilic parts; the polar or ionic head groups at the surface surround a hydrocarbonlike core (*see also* **Bilayer**) (17-1)

Michael addition: reaction in which a resonance-stabilized carbanion reacts with an α,β-enone in a conjugate addition (13-1)

Microscopic reversibility: requirement that the same transition state is encountered in the forward and backward direction in any reversible chemical reaction (6-5)

(−): levorotatory (*see l*) (5-8)

Mirror image: a reflected projection of an object (5-5)

Mirror plane: a plane through which each part of an object on one side of the plane is reflected to an identical part on the opposite side (5-5)

Mixed anhydride: anhydride with two different carboxylic acid subunits; often used to describe an anhydride derived from a carboxylic acid and a phosphoric acid (22-9)

Mobile phase: flowing medium used in chromatography to carry the components of a mixture through the stationary phase; flow can be induced by gravity, pressure, or capillary action (4-2)

Mobility: measure of the ease with which a given compound can move (for example, through a chromatography column) (4-2)

Molecular formula: description of the number of each type of atom present in a molecule (1-6)

Molecular ion: an unfragmented (parent) ion formed by loss of an electron from a molecule; has the same mass as that of the sample being analyzed by mass spectrometry (4-3)

Molecular orbitals: probability surfaces in a molecule within which an electron is likely to be found; constructed by the overlap of atomic orbitals (2-1)

Molecular recognition: selective, weak, reversible binding between two reagents (19-1, 21-1)

Molozonide: five-membered ring containing three oxygen atoms; produced by the direct addition of O_3 to an alkene (10-4)

Monomer: the chemical precursor of a polymer (16-1)

Monoterpenes: terpene containing ten carbons; derived from two isoprene units (17-1)

Multiplet: pattern obtained by the splitting of the signal for a magnetically active nucleus into several lines (4-3)

Multiplicity: the number of peaks into which a signal is split (4-3)

Mutarotation: process that produces a change in optical rotation of a solution of two or more equilibrating species from that of a pure substance to that of the equilibrium mixture; change in optical rotation that takes place when a pure sugar anomer is dissolved (17-4)

Myelin sheath: lipid bilayer that surrounds and insulates nerve axons (17-1)

▼

NADH: nicotinamide adenine dinucleotide; cofactor that effects the reduction of α-ketoacids in fatty acid biosynthesis (21-4, 22-4)

NADPH: phosphorylated derivative of NADH, with many of the same functions (21-4, 22-5)

Natural amino acid: an amino acid found in nature; of the L configuration (18-2)

N-bromosuccinimide: a reagent useful for regioselective allylic or benzylic bromination (7-7)

Neoprene: poly(2-chlorobutadiene) (16-3)

Newman projection: representation used to indicate stereochemical relations between groups bound to adjacent carbon atoms; conformational descriptor in which a triad juncture inscribed within a circle is used to represent dihedral angles between σ bonds on one carbon and those attached to the adjacent atom (5-2, 9-2)

Nicotinamide adenine dinucleotide (NADH): a biological reducing agent that provides a hydride equivalent (12-3)

Nitration: replacement of H by an NO_2 group (11-2)

Nitrile: $R\text{—}C\equiv N$; functional group in which an sp-hybridized nitrogen atom is triply bound to carbon (also called a **Cyano group**) (3-6, 8-4, 18-1)

Nitrilium cation: resonance-stabilized alkylated nitrile cation, $R\text{—}C\equiv N\text{—}R'^{+}$; an intermediate in the Beckmann rearrangement (14-2)

Nitrogen inversion: rapid redisposition of the nonbonding lone electron pair of an amine to the opposite side of the molecule, converting the starting amine into its mirror image (5-10)

Noble metal: typically a coinage metal that is highly stable in the zero-valent state; used as a catalyst for catalytic hydrogenation; common metals used in this way are gold, platinum, and nickel (2-1, 12-4)

Node: a position in an atomic or molecular orbital at which electron density is zero (1-1)

Nonessential: descriptor of an amino acid that can be synthesized by the organism itself (18-1)

Nonpolar covalent bond: a chemical bond characterized by the absence of appreciable partial charge separation because of nearly equal sharing of the electrons constituting the bond by the two bonded atoms (1-3)

Nonsaponifiable lipids: those that cannot by hydrolyzed by aqueous base to soaps; terpenes (17-1)

Nor–: prefix used to identify a substance that closely resembles the structure of another but is lacking some small feature, usually a methyl group (17-1)

Normal alkane: a straight-chain alkane (1-7)

Normal-phase chromatography: liquid chromatographic technique in which less-polar compounds elute first through a polar stationary phase, often unmodified silica gel or alumina (4-2)

n,π* transition: an electronic transition of an electron from one of the nonbonded lone pairs of electrons to a π* (antibonding) orbital (4-3)

Nuclear magnetic resonance (NMR) spectroscopy: spectroscopic technique for measuring the amount of energy needed to bring a spinning nucleus (most commonly ^{1}H or ^{13}C in organic molecules) into resonance when a molecule is placed in a strong magnetic field and is irradiated with radio-frequency waves (4-3)

Nucleic acid: polymer composed of alternating sugar (ribose or deoxyribose) and phosphate units along a backbone and with one of several heterocyclic bases appended to the sugar unit joined through a phosphate ester linkage between a C-3' OH group of one nucleoside and a C-5' OH group of another; used to store and code information, specifically that which translates into sequences of amino acids in the peptides and enzymes critical to living systems (18-6)

Nucleic acid base: a purine or pyrimidine base found in DNA or RNA: adenine, guanine, cytosine, uracil, and thymine (18-6)

Nucleophile: an electron-rich reagent that attacks centers of positive charge; from the Greek *nucleo*, nucleus, and *philos*, loving (3-2)

Nucleophilic acyl substitution (at an *sp²*-hybridized center): a substitution reaction initiated by attack by an electron-rich reagent (a nucleophile) on a carboxylic acid derivative (12-7)

Nucleophilic addition: an addition reaction initiated by attack by an electron-rich reagent (a nucleophile) on a carbonyl compound or derivative (12-1)

Nucleophilic substitution: a chemical conversion in which a leaving group is displaced by an electron-rich (nucleophilic) reagent (7-5, 8-1)

Nucleophilicity: tendency of an atom, an ion, or a group of atoms to release electron density to form a bond with a carbon atom (3-2, 12-5)

Nucleoside: component of RNA and DNA with purine or pyrimidine base attached to C-1 of a ribose or deoxyribose sugar unit (18-6, 19-6)

Nucleotide: a phosphate ester derivative of a nucleoside (18-6, 19-6)

Nylon 6: polyamide formed by the ring-opening polymerization of caprolactam (16-4)

Nylon 66: polyamide formed in the cross reaction between adipic acid and 1,6-diaminohexane (16-4)

▼

Olefin: an alkene (2-1)

1,2-addition: mode of addition in which two groups are bound to adjacent carbons in the product (10-2)

1,4-addition: mode of addition in which two groups are bound to the ends of a four-atom system in the product. *See also* **Conjugate addition** (10-2)

1,3-diaxial interaction: steric interaction between axial substituents bound to carbon atoms in a six-membered ring resulting in a steric destabilization (5-4)

Optically active: a sample that rotates the plane of polarized light; a sample containing an excess of one enantiomeric chiral molecule (5-7)

Optically inactive: descriptor of a compound or mixture that does not rotate a plane of polarized light (5-7)

Optical purity: the excess of one enantiomer over the other in a mixture as determined by comparison of the optical rotation of the sample with that of one presumed to be a single enantiomer (5-7)

Orbital: probability surface describing the volume in which an electron is likely to be found (1-1)

Orbital overlap: spatial intersection of atomic or hybrid atomic orbitals required for forming a chemical bond (1-3)

Orbital phasing: description of the relative wave property of electrons in orbitals that results in either favorable or unfavorable interaction; like phasing results in bonding and unlike phasing results in antibonding interactions (2-1)

Organic chemistry: the chemistry of carbon compounds (1-1)

Organic synthesis: subarea of organic chemistry that focuses on the construction of interesting new molecules or complex existing molecules (for example, natural products) (7-3)

Organocadmium: a reagent in which carbon is directly bound to cadmium (8-3)

Organolithium: a reagent in which carbon is directly bound to lithium (8-3, 13-1)

Organomercury: a reagent in which carbon is directly bound to mercury (8-3)

Organometallic compound: functional group in which carbon is bound to a metal atom (8-3)

Organozinc: a reagent in which carbon is directly bound to zinc (8-3)

***Ortho, para* director:** a functional group that directs electrophilic aromatic substitution to the *ortho* and *para* positions (11-4)

***Ortho* substitution:** a compound bearing substituents at the 1 and 2 positions of an aromatic ring (2-3)

Osmosis: migration of a liquid or gas (usually water) across a membrane, from the side containing the lower concentration of a molecule or salt to the side containing the higher concentration (23-7)

Oxidation: a chemical transformation resulting in the loss of electrons and hydrogen atoms or in the addition of oxygen atoms or other electronegative heteroatoms or in both (3-5, 9-7)

Oxidation potential: energetic change, measured as an electrochemical potential, associated with the loss of an electron from a chemical species (7-2)

Oxidation-reduction reaction: chemical transformation in which the oxidation level of a reactant and its reaction partner are equivalently changed, with one substrate gaining electrons and the other losing them; also used to refer to a reaction in which a substrate undergoes both oxidation at one atom and reduction at another (15-1)

Oxidative decarboxylation: conversion of a functionalized carboxylic acid (often an α-ketoacid or an α-amino acid) into a carboxylic acid containing one carbon fewer by the loss of CO_2 (21-8)

Oxidative degradation: cleavage of a carbon skeleton (often at a $C\!=\!\!=\!C$ double bond) with the introduction of new carbon–oxygen bonds (10-4, 21-5)

Oxidizing agent: one that effects an oxidation (3-5)

Oxime: condensation product of hydroxylamine (NH_2OH) with an aldehyde or ketone; $R_2C\!=\!\!=\!NOH$; often a highly colored solid used as a diagnostic test for the presence of a carbonyl group; an intermediate in the Beckmann rearrangement (12-7, 13-2)

Oxirane: *See* **Epoxide** (10-4, 18-1)

Oxonium ion: cation produced when oxygen bears three σ bonds (3-7, 10-2)

Oxymercuration-demercuration: a reaction sequence used to achieve Markovnikov hydration of an alkene without accompanying skeletal rearrangements; initiated by treatment with mercuric acetate in aqueous acid, followed by $NaBH_4$ (10-3)

Ozonation: addition of O_3 to an alkene (10-4)

Ozone: an electrophilic allotrope of oxygen that exists in a zwitterionic form in which the central oxygen formally bears positive charge; O_3 (10-4)

Ozonide: five-membered ring containing three oxygen atoms; produced by rearrangement of a molozonide in the addition of O_3 to an alkene (10-4)

Ozonolysis: sequence in which a $C\!=\!\!=\!C$ double bond is oxidatively converted into two carbonyl groups through sequential treatment with O_3, followed by Zn in acetic acid (10-4)

▼

Paper chromatography: chromatographic technique in which a mixture of compounds is separated by elution by the liquid phase passing by capillary action through a sheet of chromatographic paper (4-2)

Para **substitution:**   a compound bearing substituents at the 1 and 4 positions of an aromatic ring (2-3)

Pauli Exclusion Principle: a theoretical statement that each electron must be unique—that is, each must have a distinct set of principal, secondary, azimuthal, and spin quantum numbers; statement that no more than two electrons, which must have opposite spins, can occupy the same orbital (1-1)

Pentose: a five-carbon sugar (17-4)

Peptidase: an enzyme that catalyzes the cleavage of peptide bonds (17-1)

Peptide: polyamide composed of one hundred or fewer linked α-amino acids (16-4, 16-6, 18-3)

Peptide bond: an amide linkage (16-6)

Peptide mimic: a pharmaceutical agent, structurally and functionally similar to a small peptide hormone, that mimics the physiological function of the natural compound (23-5)

Peptidoglycan: a constituent of a bacterial cell wall consisting of long chains of carbohydrates cross-linked by short peptides (23-7)

Peracid: RCO_3H; an oxygenated relative of a carboxylic acid (10-4)

Pericyclic reaction: a concerted chemical conversion taking place through a transition state that can be described as a cyclic array of interacting orbitals (14-1)

Pericyclic rearrangement: a skeletal rearrangement proceeding through a concerted, pericyclic transition state (14-1)

Periodic table: an orderly arrangement of the elements grouped according to their atomic number and electronic configuration (1-1)

Periplanar: *See Anti*-**periplanar;** *Syn*-**periplanar** (9-2)

Peroxide: functional group containing an oxygen–oxygen σ bond; ROOR (10-3)

Pharmaceuticals: biologically active compounds sold by drug companies; may be synthetic, semisynthetic, or obtained from natural sources (15-7)

Phase-transfer catalyst: a compound that provides enhanced solubility in organic solvents to a reagent through reversible binding, providing for greatly increased concentrations of the reagent in a nonaqueous phase (20-1)

Phenol: C_6H_5OH; OH-substituted benzene (3-12)

Phenyl anion: highly unstable anion formed by deprotonation of benzene (9-3)

Phenyl group: a C_6H_5 fragment with one hydrogen fewer than benzene (2-3)

Phenylhydrazone: condensation product of phenylhydrazine ($PhNHNH_2$) with an aldehyde or ketone; $R_2C\!=\!\!=\!NNHPh$; often a highly colored solid used as a diagnostic test for the presence of a carbonyl group (12-7)

Phosphate ester: ester of phosphoric acid; $(RO)_3P\!=\!\!=\!O$ (17-2)

Phosphine: functional group containing trivalent phosphorus (PR_3) (8-2)

Phospholipids: dicarboxylate, monophosphate esters of glycerol (17-1)

Phosphonium salt: a tetravalent phosphorus cation ($^+PR_4$); obtained by protonation or alkylation of a phosphine (8-2)

Phosphonium ylide: an α-deprotonated phosphonium salt; R_3P^+—$(CR_2)^-$ (8-2)

Phosphoric acid anhydride: condensation product obtained by dehydration of two equivalents of phosphoric acid (22-3)

Phosphoric acid derivatives: a family of compounds containing the —$PO(OR)_3$ group (12-10)

Photoexcitation: process by which a photon ($h\nu$) is absorbed by a molecule, causing the promotion of one of the electrons from a bonding to an antibonding orbital (4-3)

Photosynthesis: complex biological process by which carbon dioxide is converted into carbohydrates in a series of reactions initiated by the absorption of light energy (22-4)

Physical organic chemistry: subarea of organic chemistry that relates structure to reactivity in explaining reaction mechanisms (7-3)

Pi (π) bond: a covalent bond in which electron density is symmetrically arranged above and below the axis connecting the two bonded atoms; results from the sideways overlap of p orbitals (2-1)

Pinacol rearrangement: acid-catalyzed conversion of a 1,2-diol into a ketone with migration of a carbon–carbon bond (7-1, 14-1)

π,π^* transition: an electronic transition taking place through the promotion of an electron in a π (bonding) orbital to a π^* (antibonding) orbital (4-3)

pK_a: the negative logarithm of K_a; a larger value is indicative of a weaker acid (6-6)

Plane of symmetry: a symmetry element that bisects a molecule such that half of the molecule is the mirror image of the other half (5-5)

Plane-polarized light: light that has the electric vectors of all photons aligned in a single plane; obtained by passing ordinary light though a polarizer (5-7)

Plastics: polymers that can be heated and molded while relatively soft; from the Greek *plastikos,* fit to be molded (16-3)

Pleating: deviation from a planar arrangement in the hydrogen-bonded structure of two intermolecularly associated peptide chains to avoid steric interaction of the alkyl groups at the α position (16-6)

Plexiglas: poly(methyl methacrylate); —$[CH_2C(CH_3)(CO_2CH_3)]_n$— (16-3)

(+): dextrorotatory (*see d*) (5-8)

Poison: a species that binds irreversibly to the surface of a catalyst, obviating its catalytic activity (12-4)

Polar aprotic solvent: one that lacks an acidic proton on a heteroatom (3-2)

Polar covalent bond: a chemical bond characterized by appreciable charge separation because of unequal sharing of the electrons constituting the bond between two atoms (3-1)

Polarimeter: an instrument used in quantitatively measuring optical rotation (5-7)

Polarizability: measure of the ease with which the electron distribution in a molecule can shift in response to a change in electric field; the ability of an atom to accommodate electron density (3-7)

Polarization: partial charge separation induced by a difference in electronegativity between carbon and a heteroatom (1-4)

Polarized light: *See* **Plane-polarized light** (5-7)

Polar protic solvent: one that has an acidic proton on a heteroatom (3-2)

Polyacetal: —$(CHRO)_n$— (16-3)

Polyamide: polymer in which the repeat units are joined by an amide linkage (16-4)

Polycarbonate: —$(ROCO_2)_n$— (16-4)

Polycyclic aromatic hydrocarbon: an aromatic compound containing fused rings (2-3, 11-5)

Polydentate: compound with many ligating heteroatoms; from Latin, meaning "many teeth" (19-4)

Polyelectrolyte: a high-molecular-weight molecule that readily ionizes to form a multiply charged species when dissolved in water or other polar solvents (4-2, 18-6)

Polyene: an unsaturated hydrocarbon or derivative containing more than two double bonds (2-2)

Polyene antibiotic: a long-chain, multiply unsaturated compound that functions by creating additional ion channels through cell membranes (23-6)

Polyester: polymer in which the repeat units are joined by an ester linkage (16-1, 16-4)

Polyether: polymer in which the repeat units are joined by an ether linkage (16-3)

Poly(ethylene glycol): —$(CH_2CH_2O)_n$— (16-3)

Polymer: large molecule composed of many repeating subunits; from the Greek *polumeres,* having many parts (16-1)

Polymerization: process of linking monomer units into a polymeric matrix (16-1)

Polynucleic acid: *See* **Nucleic acid** (18-6)

Polypeptide: polyamide derived from α-amino acids, specifically composed of one hundred or more α-amino acids (16-4, 16-6, 18-1)

Polysaccharide: polyacetal formed by condensation of a hemiacetal group of one sugar unit with an alcohol group of another, taking place with the loss of water (16-4)

Polystyrene: —$(CH_2CH(Ph)_n$— (16-3)

Polyurethane: —$(OCONH)_2$— polymer in which the repeat units are joined by a urethane (carbamate) linkage (16-4)

Poly (vinyl alcohol): —$(CH_2CH(OH)_n$— (16-3)

Poly (vinyl chloride): —$(CH_2CH(Cl)_n$— (16-3)

Positional isomerization: a chemical conversion in which the position of a functional group is altered (7-1)

Positional isomers: isomers in which the sequence of atoms along a chain differs in the position of one or more functional groups (2-2)

Potential energy surface: a plot of the changes in potential energy taking place as a reaction proceeds (6-1)

Potential energy well: an energy minimum in a potential energy diagram representing a molecule or intermediate with a real-time existence (5-2)

Prebiotic chemistry: the synthesis of critical biochemicals in the absence of chemicals formed by living systems, as might have occurred before life was present on earth (18-8)

Primary alcohol: RCH_2—OH; an alcohol in which the OH group is attached to a primary carbon

Primary amine: RNH_2; an amine in which nitrogen is attached to one carbon substituent (3-1)

Primary carbon: a carbon atom chemically bonded to only one other carbon atom (1-7)

Primary structure of a peptide or protein: the sequence of amino acid units in a peptide or protein chain (16-6)

Prochiral: an achiral center that can become a center of chirality either by replacement of one of two identical groups or by addition to a π system (10-2, 22-7)

Product inhibition: binding of a product to a catalyst, inhibiting further catalytic cycles (20-7)

Propagation step(s): the principal product-forming sequence in a free radical chain reaction in which a reactant radical is converted into product and a different radical; the number of product radicals in a propagation step is equal to the number of reactant radicals; step in a free radical chain that carries on the chain (7-7)

Protease: an enzyme that catalyzes the hydrolysis of peptide bonds (20-7)

Protecting group: a functional group that can be formed reversibly and lacks the reactivity of another part of the molecule. Useful protecting groups must have three features: the reactions used to form them must be readily reversible; both the protection and deprotection steps must proceed in high yield; and the characteristic reactivity of the starting functional group must not be present in the protected product (10-4, 15-8, 18-4)

Protein: poly(α-amino acid) composed of more than one hundred α-amino acids (16-6, 18-3)

Protic solvent: a solvent molecule incorporating a polar X–H bond (3-2)

Protonated alcohol: cationic species produced upon association of a proton with a nonbonded lone pair of the oxygen atom of an alcohol (3-7)

Protonation: covalent attachment of a proton (H^+) to an atom bearing either a nonbonded lone pair of electrons or a π bond (7-6)

Proton decoupling: simplification of a nuclear magnetic resonance spectrum by irradiation of the sample with radio frequencies either at a specific region or over the entire chemical-shift range at which protons absorb; results in saturation of the populations in the high-spin state and loss of coupling to the irradiated nuclei; technique used routinely to simplify ^{13}C NMR spectra (4-3)

Psychoactive drugs: pharmaceutical agents used to achieve a state of euphoria or to block intense pain (23-4)

Puckered: a descriptor of a nonplanar cycloalkane that has fewer eclipsing C–H interactions and lower torsional strain than its planar analog (5-3)

Purines: a family of bicyclic heteroaromatic molecules composed of a five-membered ring fused to a six-membered ring and containing two nitrogens in a 1,3 relation in each ring; two members of the family (adenine and guanine) are nucleic acid bases (18-6)

Pyramidal: spatial arrangement in which a central atom and three attached groups are located at the corners of a pyramid (3-1)

Pyranose: a carbohydrate containing a cyclic, six-membered-ring hemiacetal (17-4)

Pyridine: C_5H_5N; six-membered-ring, nitrogen-containing heteroaromatic molecule (3-12)

Pyridoxamine phosphate: the reductive amination product of pyridoxal phosphate; a cofactor that serves both as a reducing agent and as a source of nitrogen for the production of α-amino acids (21-4)

Pyrimidines: a family of monocyclic heteroaromatic molecules composed of a six-membered ring containing two nitrogens in a 1,3 relation; three members of the family (cytosine, uracil, and thymine) are nucleic acid bases (18-6)

Pyrophosphate group: [—OP(O)(OH)—O—P(O)(OH)—O—]; monoanhydride of phosphoric acid (17-2)

Pyrrole: C_4H_4NH; five-membered-ring, nitrogen-containing heteroaromatic molecule (3-12)

▼

Quartet: a four-line multiplet (4-3)

Quaternary ammonium ion: positively charged ion in which nitrogen is attached to four carbon substituents (3-1)

Quaternary carbon: a carbon atom chemically bonded to four other carbon atoms (1-7)

Quaternary structure of a peptide or protein: clusters formed as several large polypeptide or protein units join together to form a functional object (16-6)

Quinoid form: a six-membered ring with one or more exocyclic double bonds resembling quinone (21-4)

▼

R: absolute stereochemical designator employed in the Cahn-Ingold-Prelog rules; used to describe the stereoisomer in which a clockwise rotation is required to move from the group of highest priority attached to a chiral tetravalent atom to that of lowest priority when the substituent of lowest priority is directed away from the observer; counterpart of *S* (5-8)

Racemate: *See* **Racemic mixture** (5-7, 19-7)

Racemic mixture: an optically inactive mixture composed of equal amounts of enantiomers (5-7)

Racemic modification: *See* **Racemic mixture** (5-7)

Racemization: loss of optical activity when one enantiomer is converted into a 50:50 mixture of enantiomers (7-6)

Radical: a chemical species bearing a single unpaired electron on an atom; a chemical species with an odd number of electrons (2-2)

Radical anion: a reactive intermediate with one more electron than needed for the electron configuration of a stable neutral molecule (6-4)

Radical cation: a reactive intermediate lacking one electron from the complement needed for a stable neutral molecule (6-4)

Radical chain reaction: a chain reaction in which a free radical is produced in the initiation and propagation steps and is consumed in the termination steps (7-7)

Radical hydrobromination: an anti-Markovnikov hydrobromination of an alkene taking place through the radical addition of a bromine atom; initiated by peroxide decomposition (10-3)

Radical polymerization: polymerization, initiated by a radical, in which the chain-carrying step is a radical (16-3)

Rate-determining step: the step in a multistep sequence whose transition state lies at highest energy (6-3)

Rate-limiting step: *See* **Rate-determining step** (6-3)

Reaction coordinate: variation of a specific structural feature (for example, a bond length or angle) that measures how far a reaction has proceeded (6-1)

Reaction mechanism: the sequence of bond-making and bond-breaking by which a reactant is converted into a product; a detailed description of the electron flow, including the identity of any intermediate(s) formed, that takes place in a chemical reaction (7-1, 7-3)

Reaction profile: *See* **Energy diagram** (6-1)

Reactive intermediate: a metastable species with a high energy relative to that of reactant and product; lies at an energy minimum (in a potential energy well) along a reaction coordinate (5-2, 6-1, 6-4)

Rearrangement reaction: a chemical conversion in which the molecular skeleton is altered so that the sequence in which atoms are attached is changed (7-1, 10-2, 14-1)

Redox reagent: one that can induce an oxidation or a reduction (3-5)

Reducing agent: one that effects a reduction (3-5)

Reduction: a chemical transformation induced by the addition of electrons or hydrogen atoms or by the removal of oxygen or other electronegative atoms or by both (3-5)

Reduction potential: energetic change, measured as an electrochemical potential, associated with the addition of an electron to a chemical species (7-2)

Reductive amination: the conversion of a carbonyl group into an amine by means of an imine; process by which a carbonyl group is converted into an amine through reduction of an intermediate imine (12-7, 18-1, 21-4)

Reformatsky reaction: Claisen-like condensation of a preformed zinc ester enolate with a ketone or an aldehyde (13-3)

Refractive index: the ratio of the speed of light in a vacuum to that in a material; the path of light is bent upon passing from one medium to another of different refractive index (4-2)

Refractive index detector: a device that produces an electrical signal in response to the difference in refractive index of a solvent with and without a solute; often used in conjunction with high pressure liquid chromatography (4-2)

Regiochemistry: orientation of a chemical reaction on an unsymmetrical substrate (7-7)

Regiocontrol: the formation of one regioisomer to a greater extent than others in a chemical reaction (7-7)

Regioselective: descriptor of a reaction in which there is a clear preference for one of two or more possible regioisomers (7-7)

Relative stereochemistry: specification of the stereochemical relation between two molecules (5-6)

Repeat unit: the segment of atoms or groups that is sequentially repeated in a polymer chain (16-1)

Resin: a highly viscous polymeric glass (16-5)

Resolution: method of separating a racemic mixture into two pure enantiomers; often accomplished by forming and then separating diastereomers, followed by regeneration of the original reactant (5-8, 19-7)

Resolving agent: single enantiomer of a chiral compound that reversibly binds with both enantiomers of a racemic mixture, producing separable diastereomers (5-8, 19-7)

Resonance (in nuclear magnetic resonance): condition in which the applied radio-frequency energy matches the energy difference between the parallel and antiparallel spin states of the nucleus, so that the energy is absorbed, causing its spin to "flip" from the lower-energy parallel state to the higher-energy antiparallel state (4-3)

Resonance contributors: *See* **Resonance structures** (2-3)

Resonance effect: stabilization by delocalization of π electrons; donation or withdrawal of electron density by overlap with a neighboring π system (6-7)

Resonance hybrid: an energetically weighted composite of contributing resonance structures (2-3)

Resonance structures: valence-bond representations of possible distributions of electrons in a molecule, differing only in positions of electrons and *not* in positions of atoms (2-3)

Respiration: biological oxidation of glucose to carbon dioxide (22-4)

Restricted rotation: inhibition of rotation about a σ bond (3-10)

Retention: descriptor of the stereochemical course of a reaction in which the stereochemistry of all bonds in the product duplicate those of the reactant (10-4)

Retention time: interval required for a molecule to elute from a chromatography column; influenced by the magnitude of noncovalent interactions between the compounds being separated and the stationary phase (4-2)

Retro-aldol reaction: the reverse of an aldol reaction by which a β-hydroxycarbonyl compound is cleaved to two carbonyl derivatives (22-9)

Retro-Diels-Alder reaction: concerted fragmentation of a cyclohexene to a butadiene and an alkene; the reverse of a Diels-Alder reaction (6-5)

Retrosynthetic analysis: method for planning an organic synthesis by working backward, step by step from the product to possible starting materials (7-2, 15-1, 15-2)

Reverse phase chromatography: liquid chromatographic technique in which more-polar compounds elute first through a highly nonpolar stationary phase, often silica-gel coated with a long-chain alkylsilane (4-2)

Reverse polarity reagent: *See* **Umpolung** (18-8, 21-9)

Reversible reaction: one that can proceed backward or forward with similar ease many times (6-1)

R_f **value:** the ratio of the distance migrated by a substance compared with the solvent front (4-2)

Ribonucleic acid: *See* **RNA** (17-4)

Ribose: sugar unit found in the backbone of RNA (17-4)

Ribosome: a body in the cell cytoplasm containing the biological reagents needed to synthesize peptides (19-6)

Rigidity: stiffness (19-3)

Ring annelation: the attachment of a new ring to an existing cyclic structure (*see also* **Annulation**) (13-2)

Ring-opening polymerization: polymerization reaction in which the driving force for bond formation between repeat units is supplied by relief of ring strain in a monomer (16-3)

Ring strain: destabilization caused by angle strain and eclipsing interactions in a cyclic compound (1-6)

RNA: ribonucleic acid; biopolymer composed of ribonucleotide units linked through a sugar–phosphate backbone; transcribes the genetic information stored in DNA and directs protein synthesis; found in cell nuclei (17-4)

Robinson ring annulation: the use of an intramolecular aldol reaction to construct a six-membered ring (13-2)

Rotational isomers: stereoisomers that are interconverted by rotation about covalent bonds (1-5)

Rubber: naturally occurring poly(2-methylbutadiene) (16-3)

▼

S: absolute stereochemical designator employed in the Cahn-Ingold-Prelog rules; used to describe the stereoisomer in which a counterclockwise rotation is required to move from the group of highest priority attached to a chiral tetravalent atom to that of lowest priority when the substituent of lowest priority is directed away from the observer; counterpart of *R* (5-8, 19-7)

Saccharide: *See* **Carbohydrate** (16-4)

Saponifiable lipids: those that can by hydrolyzed by aqueous base to fatty acids; fats and waxes (17-1)

Saponification: making of soaps by hydrolysis of fatty acid esters with aqueous hydroxide; one of the oldest known chemical reactions (17-1)

Saturation: condition of a compound containing only sp^3-hybridized atoms (1-1)

Sawhorse representation: a depiction of a molecule by the use of solid wedges and hatched lines to represent three-dimensional shape (1-5)

Schiff base: an *N*-alkylated imine ($R_2C=NR'$) (12-7)

Secondary alcohol: $R_2CH—OH$; an alcohol in which the OH group is attached to a secondary carbon atom (3-7)

Secondary amine: R_2NH; an amine in which nitrogen is attached to two carbon substituents and one hydrogen atom (3-1)

Secondary carbon: a carbon atom chemically bonded to only two other carbon atoms (1-7)

Secondary structure of a peptide or protein: complex three-dimensional structure describing local organization of chain segments such as α-helices and β-pleated sheets (16-6)

Selectivity: the formation of one product in preference to other possible products (8-2)

Self-exchange: a substitution reaction in which the incoming group and the leaving group are identical (7-5)

Semicarbazone: condensation product of semicarbazide [$H_2NNHC(O)NH_2$] with an aldehyde or ketone; $R_2C=NNHC(O)NH_2$; often a highly colored solid used as a diagnostic test for the presence of a carbonyl group (12-7)

Semisynthetic: a naturally occurring (or cultured) material that is chemically altered, sometimes in a relatively minor way, in the laboratory (15-7)

Sesquiterpene: terpene containing fifteen carbons; derived from three isoprene units (17-1)

Sesterterpene: terpene containing twenty-five carbons; derived from five isoprene units (17-1)

Sex hormones: steroids that determine sexual characteristics and regulate sexual function (17-1)

Shielding: the shift of a nuclear magnetic resonance signal from that expected from the applied field caused by donation of electron density to the observed nucleus (4-3)

Side-chain oxidation: the conversion of an alkyl or acyl side chain on an aromatic ring into a —CO_2H group upon treatment with hot aqueous $KMnO_4$ (11-3)

Sigma (σ) bond: a covalent chemical bond in which electron density is arranged symmetrically along the axis connecting the two bonding atoms; results from direct overlap of hybrid orbitals having some *s* character (1-5)

Sigmatropic rearrangement: skeletal rearrangement accomplished through the shift of a σ bond to the opposite end of a π system—for example, as in the Cope rearrangement; entails the migration of a group from one end of a π system to the other 14-1)

Sigmatropic shift: a pericyclic reaction in which a σ-bound substituent migrates from one end of a π system to the other (14-1)

Silk: naturally occurring polymer of glycine and alanine (16-4)

Simmons-Smith reaction: formation of a cyclopropane through stereospecific carbene addition to an alkene through treatment of a vicinal dihalide with a zinc-copper couple in the presence of an alkene (10-4)

Simmons-Smith reagent: I—CH_2—ZnI; a reagent with carbene character prepared by the treatment of methylene iodide with a zinc-copper couple in ether (8-3, 10-4)

Single-electron transfer: chemical reaction in which the key step consists of the exchange of one electron (21-6)

Singlet: a molecule in which all electrons are paired, generally with two electrons of opposite spin in each molecular orbital (6-4)

S_N1 reaction: a step-by-step unimolecular nucleophilic substitution that proceeds through an intermediate carbocation (7-6, 8-1)

S_N2 reaction: a concerted bimolecular nucleophilic substitution that takes place by backside attack of a nucleophile and leads to a substitution product with inverted configuration at the substituted carbon (7-5, 8-1)

Soap: mixture of salts of long-chain fatty acids obtained by base hydrolysis of fats (17-1)

Soft: descriptor of a charge-diffuse reagent; often applied in the description of nucleophiles, electrophiles, acids, and bases (13-1)

Solid wedges: graphic representation in three-dimensional structure indicating a group positioned near the observer (1-5)

Solvation: association of solvent molecules about a solute (3-2, 20-1)

Solvent front: the farthest point reached by the solvent in chromatography (4-2)

Solvolysis: a reaction in which solvent displaces a leaving group or is added across a multiple bond (7-6)

sp-hybrid orbitals: hybrid orbitals formed by mathematically mixing one *s* and one *p* atomic orbital; the two hybrid orbitals formed by this mixing are dispersed in a linear array separated by 180° (2-4)

sp²-hybrid orbitals: hybrid orbitals formed by mathematically mixing one *s* and two *p* atomic orbitals; the three hybrid orbitals formed by this mixing are dispersed in a plane and are separated by 120° (2-1)

sp³-hybrid orbitals: hybrid orbitals formed by mathematically mixing one *s* and three *p* atomic orbitals; the four hybrid orbitals formed by this mixing are tetrahedrally dispersed (separated by 109.5°) from the atom's nucleus (1-2)

Specific acid catalysis: a reaction in which only a specific acid can effect a rate acceleration (20-4)

Specific base catalysis: a reaction in which only a specific base can effect a rate acceleration (20-4)

Specific rotation ([α]):   the extent to which a given molecule (on a weight basis) rotates a plane of polarized light. The observed rotation is the product of the specific rotation, the concentration in the sample compartment, and the path length of the sample cell; $\alpha = [\alpha] \times l \times c$, in which l = path length (in dm) and c = concentration (in g/ml) (5-7)

Spectroscopy: a set of techniques that measure the response of a molecule to the input of energy (4-1)

Spectrum: a display of peak intensity detected by a given spectroscopic method as a function of incident energy (4-3)

Staggered conformation: spatial arrangement in which each σ bond at one carbon is fixed at a 60° dihedral angle from one on an adjacent atom; when viewed end-on in a Newman projection, the conformation in which the bonds on one atom exactly bisect those on the adjacent atom (5-2)

Starch: water-soluble biopolymer containing as many as 4000 glucose units connected by α linkages (16-4)

Stationary phase: immobile medium (usually a solid or highly viscous liquid) through which a mixture passes in chromatography (4-2)

Stereoelectronic control: requirement for precise orbital alignment for a proposed reaction (9-2)

Stereoisomers: isomers that differ only in the position of atoms in space (1-5, 2-1, 5-1)

Stereorandom: descriptor of a reaction without any stereochemical preference (10-2)

Stereoselective: descriptor of a reaction in which there is a clear preference for one possible stereoisomer (7-7)

Steric effect: destabilization resulting from van der Waals repulsion between groups that are too close to each other (5-2, 6-7, 11-4)

Steric strain: *See* **Steric effect** (5-2, 6-7, 11-4)

Steroid: a member of a group of naturally occurring, often oxygenated, tetracyclic compounds that have a fused-ring system with three 6-membered rings and one 5-membered ring; as a class, steroids often have important hormonal functions (17-1)

Strecker synthesis: method for the preparation of an α-amino acid by treatment of an aldehyde with ammonium chloride in the presence of potassium cyanide to form an α-amino nitrile, which is then hydrolyzed to the α-amino acid (13-1)

Structural isomers: isomers in which the carbon backbones differ (1-5)

Substituent effect: altered reactivity induced by the presence of a substituent group on a reactant, often affecting rates, stereochemistry, or regiochemistry (or all three) of a reaction (11-4)

Substitution reaction: a chemical conversion in which one atom or group of atoms in a molecule is replaced by another (7-1)

Substrate: a molecule undergoing reaction under the influence of one or more external reagents (20.7)

Sugar: *See* **Carbohydrate** (16-4)

Suicide inhibition: an irreversible transformation in which the catalytic function of an enzyme is blocked (23-7)

Sulfa drugs: sulfonamide antibiotics (21-10, 23-5)

Sulfanilamide: a *p*-aminobenzenesulfonamide antibiotic; a sulfonamide derivative of aniline (23-6)

Sulfate ester: ester of sulfuric acid; $(RO)_2SO_2$ (17-2)

Sulfonamide: RSO_2NH_2; a functional group in which an $—SO_2NH_2$ group is attached to an alkyl or aryl group (3-11, 12-9)

Sulfonation: replacement of H by an $—SO_3H$ group (11-2)

Sulfonic acid: RSO_3H; a functional group in which an $—SO_3H$ group is attached to an alkyl or aryl group (3-11, 12-9)

Superimposable: descriptor of the relation of two molecules for which a conformation exists so that each of the four substituents at a center of chirality can be placed upon each other and are thus oriented in exactly the same direction in space (5-5)

Symmetry: structural correspondence of constituent parts on opposite sides of a plane, center, or axis of symmetry (5-5, 19-3)

***Syn* addition:** the formation of product by delivery of an electrophile and a nucleophile to the same face of a multiple bond (10-2, 12-4)

Syndiotactic: stereochemical designator of a polymer in which the alkyl groups at centers of chirality along a polymer chain point alternately in one direction and then the opposite along the polymer chain (16-6)

***Syn* eclipsed conformer:** a conformational isomer in which two large groups on adjacent carbons are disposed with a dihedral angle of 0° (5-2)

***Syn*-periplanar:** descriptor of the geometric relation in which the bonds to substituents on adjacent atoms of a σ bond are coplanar, with a dihedral angle of 180°; a possible geometry for an E_2 elimination, although less preferred than the *anti*-periplanar alignment (9-2)

Synthetic: prepared in the laboratory (15-7)

Synthetic efficiency: evaluation of the utility of a proposed synthesis; depends on the number of steps, the yield of each step, the ease and safety of the reaction conditions, the ease of purification of intermediates, and the cost of starting materials, reagents, and personnel time (15-6)

▼

Tautomerization: a change from one structure to another in which the only changes are the position of attachment of a hydrogen atom and the position of π bond(s); typically, a 1,3 (or 1,5) shift of a proton to or from a heteroatom in a three-atom system containing a double bond; catalyzed by acid or base (6-5, 8-3, 17-4)

Tautomers: constitutional isomers that differ only in the position of an acidic hydrogen along a three-atom segment containing a heteroatom and a double bond (6-5, 8-3)

Teflon: —$(CF_2)_n$—, a polymer produced by polymerization of tetrafluoroethylene (16-3)

Termination reaction: one that stops a chain reaction by consuming a reactive intermediate without producing another or by converting two reactive intermediates into one stable product (7-7)

Termolecular reaction: a reaction that requires a collision between three reactants in the rate-determining step; termolecular reactions are rare (6-9)

Terpene biosynthesis: condensation of isopentenyl pyrophosphate (17-2)

Terpenes: family of relatively nonpolar natural products (lipids) derived biochemically from isopentenyl pyrophosphate and thus usually containing $5n$ (n = integer) carbon atoms (17-1)

Tertiary alcohol: R_3C—OH; an alcohol in which the —OH group is attached to a tertiary carbon atom (3-7)

Tertiary amine: R_3N; an amine in which nitrogen is attached to three carbon substituents (3-1)

t-**butoxycarbonyl:** group used as an amine protecting group; abbreviated *t*-BOC (18-4)

t-**butyl ester:** a protecting group for a carboxylic acid (15-8)

t-**butyl group:** —$C(CH_3)_3$, attached through the tertiary carbon (1-7)

Tertiary carbon: a carbon atom chemically bonded to only three other carbon atoms (1-7)

Tertiary structure of a peptide or protein: three-dimensional description of how the β-pleated sheets and α-helices are spatially dispersed; describes protein folding (16-6)

Tetrahedral carbon: an sp^3-hybridized carbon atom bearing four substituents directed at 109.5° from each other (1-2)

Tetrahedral intermediate: intermediate in nucleophilic addition and nucleophilic acyl substitution obtained upon covalent bond formation between an attacking nucleophile and a carbonyl carbon (12-2, 20-2)

Tetrahydrofolic acid: cofactor that effects the methylation of nucleic acids by a one-carbon transfer from serine (21-10)

Tetraterpene: terpene containing forty carbons; derived from eight isoprene units (17-1)

Tetravalent carbon: an sp^3-hybridized carbon atom bearing four substituents (1-2)

Tetravalent intermediate: *See* **Tetrahedral intermediate** (12-2, 20-2)

Tetrose: a four-carbon sugar (17-4)

Thermal conductivity detector: a gas chromatography detector that measures the difference in thermal conductivity between the carrier gas alone and that observed as a sample elutes from the column (4-2)

Thermodynamic control: descriptor of a chemical reaction in which the reverse reaction takes place at a rate not substantially different from that of the forward reaction, establishing equilibrium; describes a reaction in which the relative stabilities of the possible products, rather than the activation energies for their formation, defines the course of the reaction; reaction that preferentially forms the most-stable product (6-8)

Thermodynamics: a description of the relative energies of the reactants and products and the equilibrium established between them (6-5)

Thermoneutral reaction: a conversion in which the reactants and products have the same energy content (6-1)

Thiamine pyrophosphate: a cofactor important in the degradation of amino acids (21-8)

Thin-layer chromatography: a chromatographic technique in which a mixture of compounds is separated by elution by the liquid phase by capillary action through a flat solid support such as a sheet of glass, plastic, or aluminum foil coated with a thin layer of silica gel or alumina (4-2)

Thioacetal: $(RS)_2CRH$; protonated form of an acyl anion equivalent (18-9, 21-9)

Thioether: a functional group in which two alkyl or aryl groups are attached to an sp^3-hybridized sulfur atom, R—S—R′ (3-11, 8-2)

Thiol: compounds bearing the —SH functional group (3-11)

Thiol ester: RCOSR'; functional group in which a carbonyl carbon bears an alkyl or aryl group and an SR group (3-11, 21-7, 22-5)

Thiophene: C_4H_4S; five-membered-ring, sulfur-containing heteroaromatic molecule (3-12)

Three-point contact: association between pairs of molecules in which three of the four groups are in close contact (19-7)

Through bond: transmission of some effect through the electron density connecting atoms in covalent bonds (6-7)

Torsional strain: the increased electron–electron repulsion upon rotation about a σ bond from a staggered to an eclipsed conformation (5-2)

Tosylate: a *p*-toluenesulfonate ester (8-2, 12-9)

Transcription: reading of the encoded stored information in DNA in RNA synthesis (19-6)

Transesterification: the interconversion of one carboxylic acid ester into another (12-7, 22-9)

Transfer RNA (tRNA): a small nucleic acid containing from seventy to eighty nucleotides with two unique regions, one containing an anticodon and the other a site for the attachment of a specific amino acid (19-6)

***Trans*-isomer:** a geometric isomer in which the largest groups are on opposite sides of a double bond or ring (2-1)

Transition metal: an element with an incomplete inner electron shell; metallic elements broadly found in the center of the third, fourth, and fifth rows of the periodic table that exist in multiple stable valence states (20-6)

Transition state: the least-favorable arrangement (highest-energy state) through which molecules must pass as a reaction proceeds; an activated complex at the top of a potential energy curve, having only a fleeting existence (5-1, 5-2)

Transition-state analog: a stable species that closely mimics the geometry and charge distribution of the transition state of a reaction (20-8)

Transition-state stabilization: a decrease in the energy of the highest-energy configuration along a reaction coordinate; attained by altering the environment of the interacting reagents (20-1)

Transition-state theory: a theory that asserts that the rate of a reaction varies exponentially with the energy required to reach the transition state (6-9)

Transmetallation: an exchange of metals between an organometallic compound and either a metal or a different organometallic compound (8-3)

Tricarboxylic acid (TCA) cycle: sequence of reactions by which acetate units as acetyl CoA are degraded to carbon dioxide. Some of the energy released in this oxidation is stored as chemical reduction potential in the cofactors FADH2, NADH, and NADPH (22-7)

Tridentate: compound with three ligating heteroatoms; from Latin, meaning "three teeth" (19-4)

Triglycerides: a triester between glycerol and three fatty acids (*see also* **Fats**) (17-1)

Trigonal: a carbon atom with three σ bonds (6-4)

Trimer: compound containing most of or all of the atoms of three molecules of a starting material (10-4, 16-1)

Triose: a three-carbon sugar (17-4)

Triple bond: one σ and two π bonds between adjacent *sp*-hybridized atoms (2-4)

Triplet: a molecule in which not all electrons are spin paired, with two electrons of the same spin being accommodated in two different orbitals (6-4)

Triplet (in NMR): a three-line multiplet (4-3)

Triterpene: terpene containing thirty carbons; derived from six isoprene units (17-1)

Twist-boat conformation: a twisted boatlike conformation attained as chair conformations are interconverted (5-4)

▼

Ultraviolet spectroscopy: technique that measures a molecule's tendency to absorb light of wavelengths between 200 and 400 nm (a region of energy just higher than that detectable by the human eye) (4-3)

Umpolung: descriptor of a reagent with reversed polarity; a reaction employing an umpolung reagent (18-8, 21-9)

Unimolecular reaction: a reaction in which only a single species takes part in the rate-determining step (6-9)

Unnatural amino acid: one not found in nature, either bearing a different side chain from those of naturally occurring amino acids or of the opposite (D) configuration (23-7)

Unsaturated fatty acid: long, straight-chain carboxylic acid containing at least one double bond along the chain (17-1)

Unsaturation: condition of a compound containing non-sp^3-hybridized atoms; consequently, a description of a molecule containing one or more multiple bonds (2-1)

Upfield: chemical shift of a nucleus that resonates at a lower δ value than does a reference nucleus—that is, for most uncharged molecules, at a higher frequency and thus closer to that of tetramethylsilane; right-hand part of an NMR chart (4-3)

Urethane: a carbonyl group bound on one side to the nitrogen of an amine and on the other to the oxygen of an alcohol; used as a protecting group for an amine (15-8, 16-4)

Valence electrons: electrons occupying an incompletely filled quantum level (1-1)

Valence shell: the outermost atomic shell that typically contains electrons (1-1)

van der Waals attraction: energetically favorable force resulting from the interaction of the electrons of one molecule and the nuclei of another; intermolecular dipole–dipole interaction (1-4, 19-1)

van der Waals repulsion: energetically unfavorable force resulting from the interaction of the bonded electrons of one molecule and those of another or of the nuclei of one molecule with those of another; repulsive intermolecular dipole–dipole interaction (1-4)

Variable region: the part of an antibody that is unique for a particular foreign invader; responsible for antibody specificity (20-8)

Vesicle: a bilayer extended to form a closed surface (17-1)

Vicinal dihalide: a compound that has two halogen atoms bound to adjacent carbon atoms (8-3, 9-3, 10-4)

Vinyl group: an alkene substituent connected by a bond to one of the sp^2-hybridized carbon atoms (2-3)

Vinyl halide: a functional group in which a halogen is attached to an alkenyl carbon (9-3)

Viscosity: resistance to flow; stiffness (17-1)

Visible spectroscopy: a physical technique that measures absorption of light of wavelengths between about 400 and 800 nm (the region of energy detectable by the human eye) (4-3)

Vitamin: a cofactor that is required to sustain life and that cannot be synthesized by a host animal; it must therefore be obtained from the diet (21-3)

Vulcanization: cross-linking of rubber by heating it with sulfur (16-3)

Wagner-Meerwein rearrangement: a cationic rearrangement in which a carbon substituent participates in a 1,2 shift (14-1)

Water: H_2O; the simplest compound containing sp^3-hybridized oxygen (3-7)

Wavelength: distance from peak to peak of a wave; describes the energy of electromagnetic radiation (4-3)

Wax: fatty acid ester of a long, straight-chain alcohol (17-1)

Williamson ether synthesis: production of an ether by reaction of an alkoxide ion with an alkyl halide (or sulfonate ester) (8-2)

Wittig reaction: reaction by which an aldehyde or ketone is converted into an alkene by condensation with a phosphonium ylide (13-1)

Wolff-Kishner reduction: reduction of a ketone to a —CH_2— group by treatment with sodium hydroxide and hydrazine (NH_2NH_2) (11-2)

Wool: a structurally complex, naturally occurring protein heavily cross-linked with sulfur–sulfur bonds (16-4)

Ylide: a zwitterion bearing opposite charges on adjacent atoms (8-2, 13-1)

Zaitsev's Rule: empirically derived prediction of preferential formation of the thermodynamically more stable, more highly substituted regioisomer in an elimination reaction (9-2)

Ziegler-Natta catalyst: an organometallic polymerization initiator that produces isotactic polypropylene (16-6)

(Z)-isomer: a geometric isomer in which the groups of highest priority are on the same side of a double bond (from the German, *zusammen*, together) (2-1)

Zusammen: *See* **(Z)-isomer** (2-1)

Zwitterions: neutral species that contain equal numbers of positively and negatively charged centers (2-3, 17-1)

Index

Abiotic synthesis, definition of, 640
Absolute configuration
    assigning, 177
    definition of, 177
Absorption maxima characteristic of
    unsaturated molecules, 144 (Table 4-4)
Absorption spectroscopy, 140
    conjugation effects in, 143
Acetaldehyde, 101, 421, 462
    Grignard addition to, 513
Acetal(s)
    formula of, 422
    preparation of, from aldehydes, 421
    as protecting groups for aldehydes, 527
    structure of, 421
Acetal formation, 500 (Table 14-2)
    acid catalysis in, 421
    alcoholysis of aldehydes, 500
    aldehyde protecting group, 423
    from aldehydes, mechanism of, 421
Acetal hydrolysis
    for aldehyde formation, 498
    mechanism of, 421
Acetic acid, 101
Acetic anhydride, 437
Acetoacetic ester, structure of, 290
Acetoacetic ester synthesis
    for C–C bond formation, 496
    mechanism of, 290
Acetone, 101, 404, 414, 424, 452, 459, 465
    conversion of, into 2-methylpropanol,
        509
    formation of, from 2-propanol by
        oxidation, 746
    reduction of, 715
Acetone enolate anion, 461
    energy diagram for protonation, 218
Acetophenone, 98, 286, 288
Acetyl CoA. See Acetyl coenzyme A
Acetyl coenzyme A, 724
    in fatty acid biosynthesis, 755
    structure of, 726
Acetyl group, 101
    chemical test for, 286
Acetylcyclohexane, 488

Acetylene, 52
    as a precursor of benzene, 47
Acetylide anions
    addition to carbonyl compounds,
        mechanism of, 456
    as carbon nucleophiles, 277
    displacement of alkyl halides, 495
Acetylsalicylic acid, 391
Acid anhydrides
    formation of, 499 (Table 14-2)
    nomenclature of, 92
    structure of, 91
Acid-base equilibria, 207
Acid-base reaction, 71
Acid catalysis
    in alkene hydration, 336
    in amide hydrolysis, 433
    in ester hydrolysis, 431, 688
    in nitrile hydrolysis, 438
    in thiol ester hydrolysis, 431
    in transesterification, 433
Acid chlorides
    formation of, 499 (Table 14-2)
    formation of, from carboxylic acids
        with thionyl chloride, 499
    hydrolysis of, 434
    nomenclature of, 92
    preparation from carboxylic acids with
        $SOCl_2$, 434
    preparation from carboxylic acids,
        mechanism of, 435
    rate of hydrolysis of, 681
    structure of, 91
Acid dissociation constant, 207
Acidity
    effect of aromaticity on, 215
    effect of bond energy on, 209
    effect of electronegativity on, 209
    effect of hybridization on, 212
    effect of inductive effects on, 211 (Table
        6-2)
    effect of resonance on, 212
    pK_a as a measure of, 208
Aconitic acid, structure of, 762
Acrylonitrile, structure of, 549

Activation energy, 745
    definition of, 157
    from energy diagram, 195
Activation energy barrier, 195
Active site of an enzyme, definition of, 700
Acyl anion
    as product of decarboxylation of α-
        ketoester, 727
    carbon–carbon bond formation with, 644
Acyl anion equivalent, 736
Acylation, definition of, 376
Acylium ion, structure of, 376
Addition, 1,2 vs.1,4
    electrophilic, 338
    nucleophilic, 457
1,4-Addition, nucleophiles for, 456
Addition polymer, definition of, 545
Addition polymerization, 545
Addition reactions
    definition of, 228
    for preparation of functional groups,
        362 (Table 10-1)
1,2-Adduct in diene hydrohalogenation,
    338
1,4-Adduct in diene hydrohalogenation,
    338
Adenine, 97
    abiotic synthesis of, 640
    structure of, 635
    tautomerization of, 671
Adenine–thymine base pairing, 665
Adenosine
    inhibition of biosynthesis of, 809
    preparation of, from ribose and
        adenine, 636
    structure of, 636, 666
Adenosine diphosphate, 751
Adenosine monophosphate, structure of,
    751
Adenosine triphosphate, structure of, 751
Adenylic acid, structure of, 666
Adipic acid, 560
ADP hydrolysis, energetics of, 768
ADP. See Adenosine diphosphate
Adsorption, 111

851

Agent Orange, 373
Alanine
    acidic sites in, 622
    pK_as of, 621
    structure of, 562, 619
D-Alanine–D-alanine, 807
    synthesis of, 630
Alar, structure of, 633
Alcohols
    acidity of, 208
    acidity order of, 212
    boiling points of, 64
    conversion of, into alkyl halide by S_N1
        reaction, 250
    conversion of, into bromide by
        treatment with PBr$_3$, 533
    dehydration of, for C=C bond
        formation, 496
    designation of subclasses of, 78
    laboratory classification of, 86
    mechanism of dehydration of, 320
    oxidation of, 89, 322
    oxidation of, by NADP, 753
    protecting groups for, 529
    structures and names of long chain,
        583
    structures of, 78
Alcohols, preparation of
    from aldehydes and a Grignard reagent,
        453
    from aldehydes by complex metal
        reduction, 404
    from aldehydes or ketones by addition
        of organolithiums, 456
    from aldehydes or ketones by addition
        of acetylide anions, 456
    from alkenes by hydration, 336
    from alkenes by hydroboration-
        oxidation, 347
    from alkenes by oxymercuration-
        demercuration, 349
    by disproportionation of aldehydes by
        the Cannizzaro reaction, 497
    from α,β-enones by 1,2-addition, 457
    from epoxides by nucleophilic ring
        opening, 614
    from esters and a Grignard reagent, 454
    from esters by complex metal
        reduction, 406
    from ethylene oxide and a Grignard, 454
    from formaldehyde and a Grignard, 453
    by Grignard addition to aldehydes, 495
    by Grignard addition to esters, 495
    by Grignard addition to ethylene oxide,
        495
    by Grignard addition to ketones, 495
    from Grignard opening of ethylene
        oxide, 285
    by Grignard reaction of carbonyl
        compounds, 497
    by hydration of alkenes, 497
    by hydroboration-oxidation of alkenes,
        497
    from ketones and a Grignard, 453
    by oxymercuration-demercuration of
        alkenes, 497
Alcoholysis
    of aldehydes for acetal formation, 500
    of ketones for ketal formation, 500
Aldehydes
    acetal formation with, 421
    acidity of, 214
    complex metal reduction of, 404
    nomenclature of, 90
    protecting groups for, 527

    reaction of, in base-catalyzed aldol
        reaction, 462
    structure of, 89
Aldehydes, conversion of, into
    alkenes by the Wittig reaction, 459
    amides through oximes and the
        Beckmann rearrangement, 487
    cyanohydrins, 452
    β-hydroxyaldehydes by the aldol
        reaction, 462
    imines, 424
    oximes, 487
Aldehydes, preparation of, 498 (Table 14-2)
    by acetal hydrolysis, 498
    by acid-catalyzed aldol condensation,
        463
    by aldol reaction, 498
    from alkenes by ozonolysis, 359
    by chromate oxidation of primary alco-
        hols in the presence of pyridine, 498
    by ozonolysis of alkenes, 498
    from primary alcohols by oxidation, 323
Aldimine, structure of, 718
Aldohexose, definition of, 604
Aldol
    definition of, 462
    structure of, 462
Aldol condensation, 230, 460
    for C=C bond formation, 496
    definition of, 462
    intramolecular, 467
    mechanism of the acid-catalyzed, 463
    mechanism of reaction of ketones by
        the acid-catalyzed, 464
    mechanism of the base-catalyzed, 462
Aldol dehydration
    in acid, 464
    in base, 462
Aldol reaction
    for aldehyde formation, 498
    for C–C bond formation, 496
    for C–OH bond formation, 497
    definition of, 462
    mechanism of reaction of ketones by
        the base-catalyzed, 464
    mechanism of the base-catalyzed, 462
Aldolase, function of, 774
Aldopentose, 602
Aldose, definition of, 602
Aldotetrose, definition of, 602
Aldotriose, definition of, 602
Alka-Seltzer, as antacid, 794
Alkali metal cations, transport of, 801
Alkali metals, molten, 549
Alkaloid, definition of, 613
Alkanes
    boiling points of, 24
    definition of, 13
    formula of, 17
    heat of combustion of, 23
    melting points of, 24
    nomenclature of, 18
Alkanes, preparation of
    by catalytic hydrogenation of alkenes,
        411, 495
    by catalytic hydrogenation of alkynes,
        495
    by Clemmensen reduction of aldehydes,
        495
    by Clemmensen reduction of ketones,
        495
    by coupling of organometallic reagents,
        285
    by hydrolysis of Grignard reagents, 495
    from ketones by chemical reduction, 377

    by Wolff-Kishner reduction of
        aldehydes, 495
    by Wolff-Kishner reduction of ketones,
        495
Alkenes, 27
    bromination of, for vicinal dibromide
        formation, 498
    carbene addition to, 355
    catalytic hydrogenation of, 72, 411
    epoxidation of, 356
    formula of, 29
    heat of combustion of, 37
    heat of formation of, 41
    heats of hydrogenation of, 38
    hydration of, 94, 336
    hydration of, by hydroboration-
        oxidation, 347
    hydration of, by oxymercuration-
        demercuration, 349
    hydrohalogenation of, 333
    nomenclature of, 34
    ozonation of, 359
    radical addition to, 344
    stereochemistry of bromination of,
        353
cis-Alkenes, preparation of, by catalytic
    hydrogenation of alkynes, 411, 496
trans-Alkenes, preparation of, by
    dissolving-metal reduction of
    alkynes, 414, 496
Alkenes, preparation of
    from alcohols by dehydration, 320
    from alkyl halides by
        dehydrohalogenation, 306
    from alkynes by catalytic
        hydrogenation, 411
    from carbonyl compounds by the Wittig
        reaction, 459
    by dehydration of alcohols, 496
    by dehydrohalogenation of alkyl
        halides, 496
    by elimination from quaternary
        ammonium salts, 496
    by reductive elimination of vicinal
        dihalides, 496
    from vicinal dihalides by
        dehalogenation, 321
    by Wittig reaction, 496
Alkoxy radical, 345
Alkylarenes, side chain oxidation of, 379
α-Alkylated acetoacetic esters,
        preparation of, by acetoacetic ester
        synthesis, 496
β-Alkylated ketones
    preparation of, by conjugate addition to
        α,β-enones, 495
    preparation of, by Michael addition to
        α,β-enones, 496
α-Alkylated malonic esters, preparation
        of, by malonic ester synthesis, 496
Alkylation
    definition of, 372
    of enamines, mechanism of, 288
    of enolate ions, limitations of, 288
    of enolate ions, mechanism of, 288
Alkyl azides
    preparation of, from alkyl halides, 276
    reduction of, 276
Alkyl bromides, preparation of, from
        alcohols, 270, 440
Alkyl chlorides
    formation with phosphoryl chloride,
        mechanism of, 440
    preparation of, from alcohols, 86, 270,
        440

Alkyl fluorides, hybridization in, 99
Alkyl halides
  amination of, 273
  conversion of, into alkenes by Wittig reaction, 460
  conversion of, into alkynes for C–C bond formation, 495
  conversion of, into nitriles for C–C bond formation, 495
  dehydrohalogenation of, for C=C bond formation, 306, 496
  hydrolysis, reaction profile for, 250
  relative rates of heterolytic cleavage of, 100
  structure of, 100
Alkyl halides, preparation of
  from alkenes by hydrohalogenation, 333
  by free radical halogenation of alkanes, 496
  by hydrohalogenation of alkenes, 496
  by treatment of alcohols with phosphoryl or sulfonyl halides, 497
Alkyllithium, addition of, to imines, 616
Alkyl substituents, structure of, 20 (Table 1-3)
Alkyl tosylates, preparation from alcohols, 271
Alkyne hydration, mechanism of, 343
Alkynes
  acidity of terminal, 277
  chemical test for terminal, 278
  definition of, 51
  dissolving-metal reduction of, 414
  electrophilic addition to, 341
  formula of, 52
  melting points of, 653
  nomenclature of, 53
  structure of, 51
Alkynes, preparation of,
  by acetylide anion displacement of alkyl halides, 495
  from acetylide anions, 285
  by acetylide ion alkylation, 496
  from amines by bromination-dehydrobromination, 318
  by dehydrohalogenation of vicinal dihalides, 496
  from vinyl halides by dehydrohalogenation, 316
Allenes, 54
  optical activity in, 186
Allyl cation, 87
Allyl group, definition of, 50
Allylic alcohol, 484
Allylic bromination, 259
Allylic ester enolate ions, preparation of, from β,γ-unsaturated esters by Claisen rearrangement, 494
Allyl phenyl ethers
  Claisen rearrangement of, 493
  conversion of, into ortho-allylphenols by a Claisen rearrangement, 493
Allyl radical, 87
Allyl vinyl ethers
  Claisen rearrangement of, 493
  conversion of, into γ,δ-enones by a Claisen rearrangement, 493
Alprazolam, structure of, 523
Alternation, 44
Alumina, 112
Aluminate, 405
Aluminum hydroxide, as antacid, 794
Ambiphilic reactivity, 200
Ambiphilicity, 71

Ambrosin, structure of, 592
Amide linkages, planarity of, 569
Amides
  complex metal reduction of, 406
  dehydration of, for nitrile formation, 497
  formation of, 500 (Table 14-2)
  geometric isomers of monosubstituted, 569
  nomenclature of, 92
  partial double-bond character in the C–N bond in, 614
  rate of hydrolysis of, 681
  relative rates of enzymatic hydrolysis, 706
  resonance in, 569
  resonance stabilization of, 614
  restricted rotation in, 92
Amides, hydrolysis of
  acid catalysis in, 433
  in base, mechanism of, 433
  mediation by chymotrypsin, 701
Amides, preparation of
  from acid chlorides by amidation, 434
  from acid chlorides, 614
  from aldehydes through oximes by the Beckmann rearrangement, 487
  by amidation of carboxylic acid derivatives, 500
  from amines by acylation, 614
  by Beckmann rearrangement, 500
  from carboxylic acid chlorides by amidation, 429
  from esters by amidation, 433
  from ketones through oximes by the Beckmann rearrangement, 487
  from ketones through the Beckmann rearrangement, 618
  from oximes by the Beckmann rearrangement, 487, 618
Aminals
  definition of, 635
  structure of, 739
Amination of benzyne for C–N bond formation, 498
Amines
  alkylation of, 614
  boiling points of, 64
  designation of subclasses of, 61
  formal charge calculation in, 62
  nomenclature of, 60
  polar covalent bonding in, 62
  protecting groups for, 530
  structure of, 60
Amines, preparation of
  from alkyl halides, 273, 614
  from amides by complex metal reduction, 406
  from amides by metal hydride reduction, 614
  from amides by the Hofmann rearrangement, 489
  by aminolysis of alkyl halides, 498
  from carbonyl compounds by reductive amination, 424, 615
  by catalytic hydrogenation of amides, 498
  by catalytic hydrogenation of nitriles, 498
  by the Gabriel synthesis, 498
  by Hofmann rearrangement of amides, 498
  from imines by catalytic hydrogenation, 413
  from imines by sodium cyanoborohydride reduction, 423

  by lithium aluminum hydride reduction of amides, 498
  from nitriles by reduction, 617
  by nucleophilic addition to imines, 616
  from primary, synthesis of, 275
  by reduction of nitro compounds, 498
  by reductive amination of ketones, 498
α-Aminoacetamide, 641
α-Amino acids
  acidic, 620
  basic, 620
  as components of polypeptides, 561
  degradation of, 720, 726
  essential, 621
  glucogenic, 620
  hydrophilic, 619
  hydrophobic, 619
  ketogenic, 620
  natural (S) configuration of, 571
  nonessential, 621
  protection of, 627
  structures of, 562, 619
  unnatural, 624
  zwitterionic form of, 623
α-Amino acids, preparation of
  from α-aminonitriles by hydrolysis, 453
  from α-ketoacids, 718
  from aldehydes by a Strecker synthesis, 453
p-Aminobenzenesulfonamide. See Sulfanilamide
Aminocarbohydrates, definition of, 638
2-Aminoethane thiol, structure of, 726
1-Aminohexane, 61
2-Aminohexane, 61
Aminoimidazole, structure of, 642
Aminol, definition of, 641
Aminolysis of alkyl halides, for C–N bond formation, 498
α-Aminonitriles
  hydrolysis of, 453
  preparation of, from aldehydes by treatment with KCN in aqueous NH₄Cl, 453
  structure of, 453
Aminosugars, structures of, 638
Ammonia, 488
  acidity of, 209
  alkylation of, solvent effects on, 684
  hybridization in, 59
  hydrogen bonding in, 64
  industrial synthesis of, 60
Ammonium salts, structure of, 274
Ammonium sulfate, 488
  as by-product of caprolactam synthesis, 560
Amoxicillin, structure of, 523, 806
Amoxil, structure of, 523
AMP. See Adenosine monophosphate
Amphetamine, structure of, 632
Amphoteric, definition of, 622
Amphotericin B
  antibacterial activity of, 803
  structure of, 661, 803
Ampicillin, structure of, 806
Amygdalin, 452
Amylopectin, structure of, 558, 559
Analgesic, morphine as, 632
Androgens, 594
Androsterone, structure of, 594
-ane, suffix for alkanes, 19
Anesthetics, 272
Anhydrides
  hydrolysis of, 437

Anhydrides (*cont.*)
  preparation of, from carboxylic acids by dehydration, 436
Aniline, 98, 378, 383, 491
  amide formation with, 535
  bromination of, 383
  diazotization of, 378
Anilines, preparation of
  by amination of benzyne , 498
  from benzyne, 319
  from nitrobenzene by reduction, 378
  by reduction of nitro compounds, 498
Anionic polymerization, 547
  initiators for, 548
  mechanism of, 548
Anionic rearrangement, 481
Anisole, 98
Anomer, definition of, 604
Anomeric effect, definition of, 606
Anthocyanins, 146
Anti addition
  in alkene bromination, 352
  definition of, 340
Antianxiety agent, alprazolam as, 523
Antiaromatic, 48
Antibacterial activity, 801
Antibacterial agent, nitrite ions as, 670
Antibiotic, Amoxil as, 523
Antibiotics, 440
Antibodies, 791
  definition of, 708
Antibonding molecular orbital, 31
Anticodon, 668
Anti conformation in *n*-alkanes, 663
Anti conformer, 162
Antifungal agent, 803
  nitrite ions as, 670
Antigen, 708
Antihistamine, 794
  trefenadine as, 523
Antimalarial agent, quinine as, 632
Anti-Markovnikov addition, 344
Anti-Markovnikov sense, 346
Antimenopausal agent, conjugated estrogen as, 523
Antioxidant agent, nitrite ions as, 670
Anti-periplanar arrangement, 310
Antistatic agents, 554
Antiulcerative agent, Zantac as, 523
Aphrodisiacs, 403
  yohimbine, 632
Applied magnetic field, 120
Aprotic solvent, definition of, 66
D-Arabinose, structure of, 603
Arenes
  alkylation of, 374
  bromination of, 372
  chlorination of, 372
  definition of, 48
  nitration of, 374
  sulfonation of, 374
Arene sulfonic acids, preparation of, from arenes by electrophilic sulfonation, 374
Arginine
  acidic sites in, 622
  pK_as of, 621
  structure of, 620
Aromatic hydrocarbon, definition of, 43
Aromatic spacers, 696
Aromaticity
  definition of, 46
  effect on acidity, 215
Arrhenius equation, 164
  activation energy from, 220

Arrow poisons, 608
Arrow pushing, 238
  definition of, 69
Artificial enzymes, 707
Artificial sweeteners, structures of, 626
Artistic ability, 674
Aryl alkanes, preparation of, from arenes by Friedel-Crafts alkylation, 375, 495
Aryl bromides, preparation of, from arenes by electrophilic bromination, 372
Aryl chlorides, preparation of, from arenes by electrophilic chlorination, 372
Aryl diazonium salt, reaction of, with phenol, 799
Aryl halides, dehydrohalogenation of, 319
Aryl ketones, preparation of, from arenes by Friedel-Crafts acylation, 377, 495
-ase, definition of, as a suffix, 608
Asparagine, structure of, 619
Aspartame, structure of, 626
Aspartic acid
  inhibition of biosynthesis of, by hadacidin, 809
  structure of, 620
Aspirin, 391
Assugrin, structure of, 626
Atactic, definition of, 567
Atomic orbitals
  definition of, 6
  shapes of, 6
Atomic structure, 7
Atoxyl
  structure of, 798
  synthesis of, 798
ATP. *See* Adenosine triphosphate
ATP hydrolysis
  energetics of, 768
  energy release from, 752
Atropine, use of, 633
Axial substituents, 169
Azadirachtin, structure of, 593
Azeotropic distillation, removal of water by, 615
Azlactone
  epimerization of, 629
  structure of, 629
Azo compounds, structure of, 379

Bacterial cell wall, disruption of, 803
Bacterium, explosion of, 807
Baeyer-Villiger oxidation
  for ester formation, 500
  mechanism of, 492
Bakelite, structure of, 564
Barbiturates, 294
Barium hydroxide, in abiotic carbohydrate formation, 775
Base, definition of, 207
Base catalysis of transesterification, 694
Base complementarity, 667
Base induction, definition of, 693
Base-line separation, 118
Base peak, 150
Base size in regioselectivity in E2 elimination, 309
Basicity
  relation to electronegativity, 415
  relation to nucleophilicity, 416
Batrachotoxin, structure of, 632, 639
Batteries, energy storage in, 747
Beckmann rearrangement, 230, 618
  for amide formation, 500

  mechanism of, 487
  relative rates of, 489
  for the synthesis of caprolactam, 560
Benzaldehyde, 98, 420, 455, 465
  mass spectrum of, 150
Benzamide, 491
Benzene, 391, 392, 412
  heat of hydrogenation, 45
  oxidation of, in the liver, 707
  structure of, 43
  water solubility of, 657
Benzenediazonium chloride, 378
Benzenesulfonic acid, 374
Benzenesulfonyl chloride, 438
Benzil, 481
Benzilic acid rearrangement, 481
  for carboxylic acid formation, 499
Benzoate esters, structure of, 469
Benzoic acids, 98, 392, 491
  preparation of, from alkyl arenes by oxidation, 379
Benzophenone, 488
Benzophenone oxime, 488
Benzoyl chloride, as an acylating agent, 535
Benzo[15-crown-5]ether, 662
Benzo[a]pyrene, 50
  carcinogenicity of, 51, 707
Benzyl bromides
  formation of benzyl ethers with, 529
  preparation of, from toluenes by radical bromination, 379
Benzyl cation, 87
Benzyl ethers
  formation of, 529
  hydrogenolysis of, 529
  hydrolysis of, 529
  as protecting group for alcohols, 529, 539
Benzylic bromination, 379
Benzyl radical, 87
Benzyne
  preparation of, from aryl halides by dehydrobromination, 319
  structure of, 319
Beriberi, 790
Berzelius, Johns Jakob, 541
Beta blockers, 794
Betaine, definition of, 459
Bidentate ligand, definition of, 659
Bimolecular nuclear substitution, 240
Bimolecular reactions, 222
Biological reducing agent, 410
Biosynthesis
  definition of, 451
  terpenes, 595
Biotin, in fatty acid biosynthesis, 755
Biradicals
  definition of, 30
  from triplet carbenes, 355
Birth control pills, 468
Bis-acetal, definition of, 607
Bis-isocyanate, structure of, 563
Bis-lactone, 542
Bisphenol-A, structure of, 557, 565
Bitrex, 275
Bixen, structure of, 592
Bloch, Konrad, 591
Blood types, chemical basis for, 792
Boat cyclohexane, 168
t-BOC. *See t*-Butyloxycarbonyl
Boiling points, 24, 64 (Table 3-1)
  of alkanes, 24 (Table 1-5)
  of carboxylic acids, 665
  of hydrocarbons, 55
  of isomeric butanols (Table 19-2), 658

of methyl esters, 665
Boltzmann distribution, temperature dependence of, 221
Boltzmann, Ludwig, 221
Bombardier beetle, 254
Bond alternation, 44
Bond cleavage
heterolytic, 67
homolytic, 67
Bond-dissociation energies, 80 (Table 3-6)
definition of, 67
Bond energies
average, 72 (Table 3-4)
typical, 68 (Table 3-3)
Bond energy, effect of, on acidity, 209
Bonding molecular orbital, 31
Bond lengths
double bonds, 34
in hydrocarbons, 53
single bonds, 34
typical, 68 (Table 3-3)
Bond strengths, carbon–fluorine bonds, 99
Boric acid, 404
Boron trifluoride, 69
Branched polymer, definition of, 544
Breathalyzer test, 323
Bridged cation, 343, 350
Bridged ions, 354
Bridgehead positions, 172
α-Bromacids, preparation of, by Hell-Volhard-Zelinski reaction, 497
Bromination of alkenes
anti addition, 352
mechanism of, 351
Bromine, 489
Bromine decolorization as chemical test for alkenes, 351
N-Bromoamide, 490
Bromobenzene, 358, 392
structure of, 48
2-Bromobutane, 177, 308, 314
2-Bromo-3-chlorobutane, 182
1-Bromo-2,2-dimethylpropane, 478
α-Bromoesters, preparation of, 287
2-Bromo-3-methylbutane, 251
1-Bromo-1-methylcyclohexane, 307
Bromonium ion, structure of, 352
Brompheniramine, structure of, 794
2-Bromopropane, 235
N-Bromosuccinimide, 259
o-Bromotoluene, 392
p-Bromotoluene, 392
Brønsted acid, definition of, 69
Brønsted base, definition of, 69
Brown, Herbert C., 347
Bufotenine, structure of, 639
Bufotoxin, structure of, 639
Building insulation, polyurethanes for, 564
1,3-Butadiene, 338, 482
mechanism of polymerization, 551
molecular orbitals in, 142
polymerization of, 550
Butadiyne, in interstellar space, 138
Butane
conformational analysis of, 161
structure of, 15
Butanoic acid, acidity of, 209
1-Butanol, 85
2-Butanol, 85
structure of, 426
synthesis of, from 2-butanone, 513
synthesis of, from 2-propanol, 512
t-Butanol, high melting point of, 658
Butanols
boiling points of, 658 (Table 19-2)

melting points of, 658 (Table 19-2)
2-Butanone, 465
alkylation of, 514
enolate anion from, 514
metal hydride reduction of, 513
regioselectivity in enolate anion formation from, 515
synthesis of, from 2-propanol, 505
2-Butenal, 463
1-Butene, 34, 231, 337
2-Butene, 231, 354, 355
allylic bromination of, 259
barrier to rotation in, 155
cis-2-Butene, 34, 315
energy diagram for geometric isomerization of, 156
trans-2-Butene, 34, 315
energy diagram for geometric isomerization of, 156
Butenes
catalytic hydrogenation of, 39
heat of hydrogenation of, 39
hyperconjugation in, 41
Butylated hydroxytoluene (BHT), 325
n-Butyl bromide, 231
s-Butyl bromide, 231
t-Butyl bromide, 235
hydrolysis of, 248
n-Butyl butyrate, as cleaning solvent, 586
t-Butyl cation, 85
t-Butylchloroformate, formation of urethanes with, 530
t-Butylesters
cleavage of, by trifluoracetic acid, 530
formation of, 530
hydrolysis of, 530
as protecting groups for carboxylic acids, 529
t-Butyl group as a conformational anchor, 170
n-Butyllithium, 279, 455, 459
n-Butylmagnesium bromide, 280
t-Butyloxycarbonyl group as a protecting group, 627
t-Butyl radical, 83
Butyne, 52
2-Butyne, 341
By-products
difficulties in separation of, 521
effect of, on synthetic efficiency, 521

β-Cadinene, structure of, 592
Caffeine, 97
Cahn-Ingold-Prelog rules, 177
Calciferol, 486
Calcium channel blocker, 523
Calcium hydroxide for the synthesis of carbohydrates from formaldehyde, 644
Calorie, definition of, 119
Calorimetry, 22
Camphor, 176
structure of, 591
Candida albicans, control of, by bacteria, 803
Cannizzaro reaction, 644
contrasted with the benzilic acid rearrangement, 481
disproportionation of aldehydes, 419
for C–OH bond formation, 497
mechanism of, 420
Capillary action, 115
Caprolactam, 488
conversion of, into nylon 6, 561
synthesis of, from cyclohexanone, 618
Caraway seeds as source of (+)-carvone, 676

Carbamates. *See* Urethanes
Carbamic acids, decarboxylation of, 490, 530
Carbanions
hybridization in, 199
rearrangement of, 481
structure of, 199
Carbene addition to alkenes, stereospecificity in, 355
Carbenes
ambiphilic reactivity in, 200
structure of, 200
Carbenoid, 356
Carbocation rearrangement, 251
absence of, in Friedel-Crafts acylation, 377
absence of, in oxymercuration-demercuration, 350
by alkyl shifts in, 478
in electrophilic addition, 340
in E1 elimination, 308
in Friedel-Crafts alkylation, 375
by hydrogen shifts in, 477
Carbocation stability
effect of an adjacent oxygen atom on, 480
in regiochemical control of additions, 338
Carbocations, 84
conjugation in, 86
designation of subclasses of, 85
hyperconjugative stabilization in, 85
hypervalent, 335
order of stability of, 479
stability order of subclasses of, 85, 87
structure of, 199
rearrangement. *See* Carbocation rearrangement
Carbodiimide, structure of, 630
Carbohydrate oligomers, 607
Carbohydrates
acetal formation in, 558
coiling in, 575
cyclic hemiacetal forms of, 600
definition of, 557
degradation of, 769
sheetlike structure of, 575
structure of, as cyclic hemiacetals, 558
water solubility of, 575
4-Carbomethoxycyclohexanol, conversion of, into 4-hydroxymethylcyclohexanone, 525
Carbomycin, structure of, 808
Carbon
allotropes of, 4
electronic structure of, 8
Carbon black as additive for rubber, 553
Carbon–bromine bond formation, Hell-Volhard-Zelinski reaction, 497
Carbon–carbon bond formation. *See also* Carbon–carbon double-bond formation; Carbon–carbon single-bond formation; Carbon–carbon triple-bond formation
classification of, 506
reactions for, 507 (Table 15-1)
Carbon–carbon double-bond formation, 496 (Table 14-2)
Carbon–carbon single-bond formation, 495 (Table 14-2)
acetoacetic ester synthesis, 496
aldol reaction, 496
Claisen condensation, 496
Claisen rearrangement, 496
conjugate addition to α,β-enones, 495

Carbon–carbon single-bond formation,
  (*cont.*)
  conversion of alkyl halides into alkynes,
    495
  conversion of alkyl halides into nitriles,
    495
  Cope rearrangement, 496
  Diels-Alder reaction, 495
  Friedel-Crafts acylation, 495
  Friedel-Crafts alkylation, 495
  Grignard addition to aldehydes, carbon
    dioxide, esters, ethylene oxide,
    ketones, 495
  malonic ester synthesis, 496
  Michael addition, 496
Carbon–carbon triple-bond formation, 496
    (Table 14-2)
  dehydrohalogenation of vicinal
    dihalides, 496
Carbon carrier, tetrahydrofolic acid as, 736
Carbon dioxide, conversion of, into
    carbohydrates, 753
Carbon esters
  acylation with, 534
  structure of, 469
Carbon fixation, 147
Carbon–fluorine bonds, strengths of, 99
Carbon–halogen bond formation, 496
    (Table 14-2)
  electrophilic aromatic halogenation, 497
  free radical halogenation of alkanes, 496
  α-halogenation of ketones, 497
  hydrohalogenation of alkenes, 496
  treatment of alcohols with phosphoryl
    or sulfonyl halides, 497
Carbon–hydrogen bond formation, 495
    (Table 14-2)
  catalytic hydrogenation of alkenes, 495
  catalytic hydrogenation of alkynes, 495
  Clemmensen reduction of aldehydes, 495
  Clemmensen reduction of ketones, 495
  hydrolysis of Grignard reagents, 495
  β-ketocarboxylic acid decarboxylation,
    495
  Wolff-Kishner reduction of aldehydes,
    495
  Wolff-Kishner reduction of ketones, 495
Carbon–hydroxyl group bond formation,
    497 (Table 14-2)
  aldol reaction, 497
  Cannizzaro reaction, 497
  Grignard reaction of carbonyl
    compounds, 497
  Grignard reaction of esters, 497
  hydration of alkenes, 497
  hydration of benzyne, 497
  hydroboration-oxidation of alkenes, 497
  hydrolysis of alkyl halides, 497
  metal hydride reduction of aldehydes,
    esters, ketones, 497
  nucleophilic opening of epoxides, 497
  oxymercuration-demercuration of
    alkenes, 497
Carbonium ion. *See* Carbocation, 84
Carbonless paper, 305
Carbon–nitrogen single-bond formation,
    498 (Table 14-2)
  amination of benzyne, 498
  aminolysis of alkyl halides, 498
  catalytic hydrogenation of amides, 498
  catalytic hydrogenation of nitriles, 498
  Gabriel synthesis, 498
  Hofmann rearrangement of amides, 498
  lithium aluminum hydride reduction of
    amides, 498

reduction of nitro compounds, 498
reductive amination of ketones, 498
synthesis of, 614, 646 (Table 18-1)
Carbon nucleophiles, preparation of, from
    alkyl halides, 277
Carbon transfer from serine, mechanisim
    of, 739
Carbonyl compounds
  addition of acetylide ions to, 456
  addition of cyanide ion to, 452
  addition of organolithiums to, 456
  conversion of, into 2,4-
    dinitrophenylhydrazones, 426
  conversion of, into hydrazones, 426
  conversion of, into oximes, 426
  conversion of, into semicarbazones, 426
  equilibrium with hydrates, 419 (Table
    12-1)
  preparation of, by the aldol
    condensation, 496
  nucleophilic attack on, 456
  π bonding in, 90
  reactivity order of, toward hydride
    reducing agents, 409
  reactivity order of, toward nucleophilic
    attack, 428
  stability order, of, 429
  structure of, 89
Carbonyl group
  electrophilic attack on, 90
  insertion of —NH— adjacent to, 488
  insertion of —O— adjacent to, 491
  nucleophilic attack on, 90
  polarization of, 90
  relative reactivity of, toward
    nucleophiles, 725
Carbowax, structure of, 554
Carboxylates, solubility of calcium salts
    of, 586
Carboxylic acid anhydride. *See* Anhydride
Carboxylic acid bromide, 287
Carboxylic acid chlorides. *See* Acid
    chlorides
Carboxylic acid derivatives
  with amines, for amide formation, 500
  esterification, for ester formation, 500
  structure of, 91
Carboxylic acids
  acidity of, 208
  boiling points of, 665
  dehydration, of, for acid anhydride
    formation, 499
  dimer of, 542
  intramolecular transfer to a hydroxy
    group, 778
  mechanism of reaction with $SOCl_2$, 435
  nomenclature of, 91
  protecting groups for, 529
  structure of, 91
N-Carboxylic acids
  decarboxylation of, 490
  structure of, 490
Carboxylic acids, preparation of, 499
    (Table 14-2)
  from acid anhydrides by hydrolysis,
    429
  from acid chlorides by hydrolysis, 434
  from alkenes by permanganate
    oxidation, 359
  from alkynes by ozonolysis, 359
  from amides by hydrolysis, 433
  from anhydrides by hydrolysis, 437
  by benzilic acid rearrangement, 499
  from CO_2 and a Grignard reagent, 454
  by decarboxylation of β-diacids, 499

by disproportionation of aldehydes by
    the Cannizzaro reaction, 497
from esters by hydrolysis, 429
by Grignard reagent carboxylation, 495,
    499
by hydrolysis of acid derivatives, 499
by iodoform reaction, 499
from malonic ester, 293, 496
from methyl ketones by iodoform
    reaction, 286
by nitrile hydrolysis, 437, 499
from organonitriles by hydrolysis, 284
by oxidation of aldehydes, 499
by oxidation of primary alcohols, 323,
    499
by permanganate oxidation of aryl
    alkanes, 499
from thiol esters by hydrolysis, 429
Carboxylic acid treatment with thionyl
    chloride, acid chloride formation, 499
4-Carboxymethylcyclohexanone
  acetal of, 527
  lithium aluminum hydride reduction
    of, 525
  sodium borohydride reduction of, 525
Carcinogen, 50
  1,1-dimethylhydrazine as, 633
Cardizem, structure of, 523
β-Carotene, 42, 144
  absorption spectrum of, 145
Carrier gas, 111
Carvone, 176
  structure of, 676
Caryophyllene, structure of, 591
Catalysis, 748
  avoiding charge separation in, 687
  differentiation from induction, 692
  by enzymes, 692, 700
  general concepts of, 682
  intermolecularity, 695
  by transition metals, 696
  transition-state stabilization, 683
Catalyst, 228
  definition of, 38
  effect of, on position of equilibria, 419
Catalyst poisons, definition of, 412
Catalytic antibodies, artificial enzymes, 707
Catalytic hydrogenation, 697
  of alkenes for C–H bond formation, 495
  of alkynes for C≡C bond formation, 496
  of alkynes for C–H bond formation, 495
  catalyst preparation for, 410
  definition of, 38
  difficulties with aromatic compounds,
    412
  difficulties with tetrasubstituted
    alkenes, 412
  of imines, 72
  mechanism of, 411
  of nitriles for C–N bond formation, 498
  reactivity order of unsaturated
    compounds toward, 413
  selectivity for functional groups, 413
  syn addition in, 411
  thermodynamics for, 412
Cationic polymerization, 547
  mechanism of, 360, 548
Cationic rearrangement. *See* Carbocation
    rearrangement
Cation stability, substituent effects on, 381
Catnip, nepetalactone in, 592
Ceclor, structure of, 523, 806
Cefaclor, structure of, 523, 806
Ceftriaxone, structure of, 809
Cell membrane, disruption of, 801

Cellulose, 607
 structure of, 558, 575
Cellulose acetate
 for photographic film, 559
 structure of, 559
Center of chirality, definition of, 174
Central nervous system, 633
Cephalosporins, structure of, 806
Chair cyclohexane, 168
Charge-diffuse nucleophiles, 457
Charge-intensive nucleophiles, 457
Charge relay in action of chymotrypsin, 702
Charge relay mechanism, definition of, 691
Chemical bond, definition of, 9
Chemical flux in living systems, 682
Chemical shifts
 delta scale for describing, 122
 downfield, 121
 upfield, 121
Chemical tests
 for acetyl group, iodoform test, 286
 for alcohol or aldehyde, chromate oxidation, 323
 for alcohols, Lucas test, 86
 for aldehydes and ketones, hydrazone formation, 426
 for alkenes, bromine decolorization, 351
 for oxidizable groups, chromate oxidation, 77
 for oxidizable groups, permanganate oxidation, 77, 359
Chemical yield, effect of, on synthetic efficiency, 520
Chemistry, definition of, 1
Chemotherapeutic index, definition of, 797
Chemotherapy 593
Chiral atom. *See* Center of chirality
Chiral center. *See* Center of chirality
Chiral recognition, 671
 by enzymes, 705
 example of, 764
Chirality, definition of, 173
Chlorination of alkenes, 354
Chlorination of arenes, mechanism of, 372
Chloroacetic acid, acidity of, 210
4-Chloroaniline, acylation of, 535
Chlorobenzene, 372, 388, 391
m-Chlorobenzoic acid, 392
2-Chloro-1,3-butadiene, polymerization of, 550
2-Chlorobutanoic acid, acidity of, 209
3-Chlorobutanoic acid, acidity of, 209
4-Chlorobutanoic acid, acidity of, 209
Chloroethylene, polymerization into poly(vinylchloride), 548
Chlorofluorocarbons, 257
Chloromethylated polystyrene for Merrifield synthesis, 626
Chloronium ion, structure of, 354
m-Chloroperbenzoic acid, 356, 492
Chlorophyll, 147
Chlorophyll *a*, 511
 structure of, 660
2-Chloropropanoic acid, 183
Chloroprene, structure of, 550
o-Chlorotoluene, 391
p-Chlorotoluene, 391
Chlorotrifluoroethylene, structure of, 549
Cholesterol
 biosynthesis of, 591
 structure of, 592
Chromate, as a color indicator, 77
Chromate ester, structure of, 322
Chromate oxidation

chemical test for alcohols or aldehydes, 323
 mechanism of, 322
 of 3-methyl-2-butanol, 515
 of 2-propanol, 506
 of 2-propanol into acetone, 746
 of secondary alcohols, for ketone formation, 499
Chromatogram, 114
Chromatographic separation, 113
Chromatography, definition of, 110
Chromic acid oxidations, reagents for, 322
Chymotrypsin, function of, 701
Cimetidine, structure of, 795
Cinchona tree, as source of quinine, 632
Cinchonine
 as a resolving agent for acids, 675
 structure of, 675
Circularly polarized light, 179
Citral, structure of, 591
Citric acid
 as the basis of the TCA cycle, 761
 conversion of, into isocitric acid, 764
 structure of, 762
Citronellal, structure of, 592
Claisen condensation, 732
 for C–C bond formation, 496
 definition of, 468
 in fatty acid biosynthesis, 754
 intramolecular, 470
 for ketone formation, 499
 mechanism of, 468
Claisen rearrangement
 of allyl vinyl ethers, ketone formation, 499
 for C–C bond formation, 496
 contrasted with Friedel-Crafts alkylation, 493
 mechanism of, 493
Clemmensen reduction, 377
 of aldehydes for C–H bond formation, 495
 of ketones for C–H bond formation, 495
Clorsulfuron, 440
Cloxacillin, structure of, 808
^{13}C NMR chemical shifts, substituent effects on, 124 (Table 4-1)
 alcohol dehydration, 496
 aldol condensation, 496
 alkyl halide dehydrohalogenation, 496
 catalytic hydrogenation of alkynes, 496
 dissolving-metal reduction of alkynes, 496
 Hofmann elimination from a quaternary ammonium salt, 496
 reductive elimination of vicinal dihalide, 496
 Wittig reaction, 496
CoA. *See* Coenzyme A
Coal, 541
Cocaine, structure of, 617
Cockroach, 336
Codeine, 432
Codon, 668
Coenzyme A
 function of, 725
 structure of, 726
Cofactor(s)
 definition of, 713
 recycling of, 715
Coiling of proteins, 575
Column chromatography, 112
Combustion, 22
Complementary base pairing, 666
Complex reaction cycles, 749

Concentration effect in bimolecular reactions, 222
Concerted reaction, definition of, 195, 482
Condensation polymer, definition of, 545
Condensation reaction, definition of, 229
Configuration, IUPAC rules for designating, 181
Configurational isomers, definition of, 155
Conformational analysis, 159
Conformational anchor, 170
Conformational isomers, 159
Conformers. *See* Conformational isomers
Coniine, structure of, 632
Conjugate acid, 70
Conjugate addition
 of amines, 458
 definition of, 339
 to α,β-enones for C–C bond formation, 495
Conjugate addition to enones, 456
 dialkylcuprates for, 456
 mechanism of, 457
Conjugate base, definition of, 207
Conjugated double bond, 42
Constitutional isomers, definition of, 155
Convergent synthesis, definition of, 520
Cope, Arthur C., 483
Cope rearrangement
 for C–C bond formation, 496
 mechanism of, 483
Cornforth, John, 591
Coronary vasodilator, Diltiazem as, 523
Cotton boll weevil, grandisol as sex attractant for, 592
Coumarin, 405
Coupled reactions, 459, 750, 753
 in acid chloride formation, 437
 in biological reductive amination, 720
 definition of, 436
 in fatty acid biosynthesis, 721
 in fatty acid degradation, 723
 in one-carbon transfer reactions, 738
Coupling, 126
Coupling constant, 130
Covalent bond, definition of, 9
Covalent catalysis, definition of, 701
Cracking towers, 236
Crafts, James, 374
Cram, Donald, 661
Creams, polymers in, 554
Criteria for evaluating synthetic efficiency, 519
Critical micelle concentration, 585
Crossed aldol condensations, definition of, 465
Crossed aldol reactions, difficulties in, 465
Crossed Claisen condensation
 difficulties in, 469
 mechanism of, 470
Cross-linked polymer, definition of, 544
Cross-linking, definition of, 552
18-Crown-6-ether, 279, 801
Crown ethers, structure of, 662
Crude oil, 541
 refining of, 699
Cumulated double bond, 42
Curved-arrow notation, 233
Cyanide ion, 77
 addition of, to aldehydes and ketones in formation of cyanohydrins, 497
 as a carbon nucleophile, 277, 284
 displacement on alkyl halides for nitrile formation, 495, 497
 nucleophilic addition of, to carbonyl compounds, 452

Cyanidine, 146
Cyano group, 77
Cyanoacetylene, in interstellar space, 138
α-Cyanoalcohol. *See* Cyanohydrin
α-Cyanoester enolate ion, structure of, 282
Cyanohydrin
    formation, mechanism of, 452
    preparation by cyanide addition to aldehydes and ketones, 497
    preparation from aldehydes or ketones by treatment with cyanide ion, 452
Cyanuric acid, 440
Cyclic halonium ions, 354
Cyclic hemiacetal, 600
Cyclization
    of carbohydrates, mechanism of, 601
    diester enolate anions, 471
    by an electrocyclic reaction, 485
    polyisoprenes, 600
Cycloaddition reaction, definition of, 482
Cycloalkanes
    conformations, torsional and angle strain in, 166
    definition of, 16
    formula of, 17
    geometric isomerism in, 21
    heats of combustion of, 167 (Table 5-3)
    ring strain in, 167
    strain energy in, 18
    three-dimensional structure of, 165
Cyclobutadiene, 46
Cyclobutane, 17
    conformation of, 166
    puckering in, 166
Cyclobutylmethanol, 479
Cycloheptatriene, acidity of, 215
Cyclohexadiene, 412, 485
Cyclohexane, 325
    chair-to-chair ring flipping in, 168
    conformational analysis of multiply substituted, 171
    entropy of, 663
    radical bromination of, 229
    three-dimensional structure of, 167
Cyclohexanol
    conversion of, into cyclohexene, 681
    dehydration of, 229
Cyclohexanone, 425, 426, 492
    alkylation of, 532
    deprotonation of, 532
    enolate anion, 532
    oxime, 488
    synthesis of caprolactam from, 618
Cyclohexatriene, 44
Cyclohexene(s), 339, 352, 354, 412, 482
    catalytic hydrogenation of, 228
    hydration of, 228
    hydrochlorination of, 245
    preparation of, by Diels-Alder reaction, 495
    reaction profile for hydrochlorination of, 245
    synthesis of, from cyclohexanol, 681
Cyclohexenone, 413, 456
N-Cyclohexylacetamide, 488
Cyclohexyl acetate, hydrolysis of, 694
Cyclohexyl bromide, dehydrobromination of, 228
Cyclohexyl iodide, S$_N$2 displacement by bromide ion, 229
Cyclohexylmethanol
    alkylation of, 532
    deprotonation of, 532

Cyclohexylmethylketone, 488
Cyclohexylmethylmethylether, 532
Cyclooctane, 176
Cyclooctatetraene, 46
Cyclopentadiene, acidity of, 215
Cyclopentadienyl anion, 47
Cyclopentadienyl cation, 48
Cyclopentane
    conformation of, 166
    water solubility of, 657
Cyclopentyl cation, 479
Cyclopropane(s), 16
    preparation of, from alkenes by carbene addition, 355
    synthesis of, with Simmons-Smith reagent, 280
Cyclopropenyl anion, 48
Cyclopropenyl cation, 47
Cycloreversion, definition of, 483
Cysteamine, structure of, 725
Cysteine
    cross-linking in polypeptides through linkages between, 562
    structure of, 562, 619
Cytosine, 97
    conversion of, into uracil by nitrite, 670
    structure of, 635
Cytosine–guanine base pairing, 665

D, as stereochemical designator, 601
d,l-, representation for designating stereoisomers, 182
Dacron, structure of, 556
Dalton, John, 2
Davy, Humphry, 272
Dean-Stark trap, 432
cis-Decalin, 173
trans-Decalin, 172
Decarboxylation, 773
    comparison of β-ketoacids and α-ketoacids, 727
    definition of, 291
    of β-diacids, for carboxylic acid formation, 499
    of α-ketoacids, mechanism of, 729
    of β-ketoacids, 291
    ketone formation, 499
    mechanism of, 291
    of oxaloacetic acid, 766
Decoupled NMR spectra, 129
Degenerate rearrangement, definition of, 483
Degree of unsaturation, 29
Dehalogenation, mechanism of, 321
Dehydrating agent, 437
Dehydration, definition of, 84
Dehydration of alcohols, mechanism of, 320
Dehydration of aldol
    in acid, 464
    in base, 462
Dehydration of β-ketoalcohols, 321
Dehydrobromination, definition of, 306
Dehydrohalogenation, 306
    of aryl halides, 319
    of vicinal dihalides, for C≡C bond formation, 496
    of vinyl halides, 316
Deinsertion, definition of, 699
Deodorants, polymers in, 554
Deoxy-, definition of, as a prefix, 603
Deoxyribonucleic acid. *See* DNA
D-Deoxyribose, structure of, 603, 634
2-Deoxystreptamine, structure of, 638

Deoxythymidylic acid, synthesis of, 737
Deoxyuridylic acid, conversion of, into deoxythymidylic acid, 737
l-Deprenyl
    biological effect of, 798
    structure of, 798
Deprotonation, definition of, 199
Depsipeptides, definition of, 802
Desosamine, structure of, 638
Dessicant, 437
Detector, 114
Deuterium labeling, 283
Dextrorotatory, 182
Diabetes, treatment of, with insulin, 790
Dialkylzinc, preparation of, from alkyl halides, 280
1,6-Diaminohexane, conversion into nylon, 560
Diamond, structure of, 3, 544
Diastereomers
    definition of, 182
    separation of, 675
1,3-Diaxial interactions, 170
    energetic cost of, 171 (Table 5-4)
Diazonium salts
    preparation of, from aniline by diazotization, 378
    structure of, 378
Diazotization, definition of, 378
2,3-Dibromobutane, 184
Dicarbonyl compounds, conversion of, into cyclized products, 467
Dichloroacetic acid, acidity of, 210
Dichlorocarbene, 355
Dicyanogen in interstellar space, 138
Dieckmann condensation, as intramolecular variant of the Claisen condensation, 470
Diels-Alder reaction, 205, 482
    for C–C bond formation, 495
    entropy effects in, 205
Dienes
    definition of, 42
    hydrohalogenation of, 338
Diester enolate anions, cyclization of, 471
Digitoxin, structure of, 608
Digoxin, 522
    structure of, 523
Dihydrofolic acid
    biosynthesis of, 800
    structure of, 800
Dihydropteridine, structure of, 800
Dihydroxyacetone, structure of, 601
Dihydroxyacetone monophosphate, 770, 774
Diimines, preparation of, from divinyl hydrazines by Claisen rearrangement, 493
β-Diketone enolate ion, structure of, 282
1,3-Diketones, acidity of, 214
α-Diketones, 482
β-Diketones, hydrogen bonding in, 656
Diltiazem, structure of, 523
Dimer
    carboxylic acid, 664
    structure of carboxylic acid, 542
Dimetane. *See* Bromopheniramine
Dimetapp. *See* Bromopheniramine
N,N-Dimethylacetamide, 406, 407
Dimethylallylpyrophosphate, structure of, 595
Dimethylamine, 61
p-(Dimethylamino)azobenzene, 379
2,3-Dimethyl-2,3-butanediol, 480
3,3-Dimethyl-1-butene, 350

Dimethylcyclohexane, conformational analysis of, 172

2,2-Dimethyl-*trans*-6-deuterocyclohexyl bromide, 312

3,4-Dimethyl-1,5-hexadiene, 484

1,1-Dimethylhydrazine, in mushrooms, 633

Dimethyl sulfoxide, 94, 279

Dimethyl terephthalate, use of, in the synthesis of dacron, 556

2,4-Dinitrophenylhydrazine, 426

2,4-Dinitrophenylhydrazone, preparation of, from carbonyl compounds, 426

1,2-Diols
  dehydration of, 480
  preparation of, by nucleophilic opening of epoxides by hydroxide ion, 497
  rearrangement of, 480

Dipeptide, structure of, 562, 625

Dipole-dipole
  attractions, 653
  interactions, 684

Dipole moment, 63

Disaccharides, structures of, 607

Disease states, definition of, 790

Disproportionation, definition of, 420

Dissolving-metal reductions
  of alkynes for alkene bond formation, 496
  of alkynes, mechanism of, 415
  definition of, 414
  dianions in, 413
  of ketones, mechanism of, 414
  radical anions in, 413
  reagents for, 413

Diterpenes, definition of, 590

Dithioacetal, structure of, 735

Dithioacetal anion
  alkylation of, 735
  umpolung in the preparation of, 735

Diuretic, triamterene, 523

p-Divinylbenzene
  as a cross-linking agent, 552
  structure of, 552

Divinylhydrazine, Claisen rearrangement of, 493

Diyne, 54

DMSO. *See* Dimethylsulfoxide

DNA
  information storage in, 667
  strand linking in, by reaction with polyepoxides, 707
  structure of, 634

DNP. *See* 2,4-Dinitrophenylhydrazone

Doering, William v. E., 73

L-Dopa, structure of, 798

Double bond, definition of, 29

Double helix, structure of, 667

Dumas, Jean Baptiste André, 796

Dyazide, structure of, 523

Dynamicin A, 55

E1-CB elimination
  in dehydrohalogenation of vinyl halides, 317
  mechanism of, 305

E1 elimination reaction
  acceleration of, by stabilization of cationic intermediate, 246
  effect of leaving group on, 312
  mechanism of, 245, 304
  rate-determining step in, 245
  regioselectivity in, 307
  stereochemistry of, 314

E2 elimination
  effect of leaving group on, 312
  mechanism of, 305
  requirement for *anti*-periplanar transition state in, 310
  stereochemistry of, 315
  stereoelectronic control in, 311

Early transition state, 197
  S_N2 reaction, 242

Eclipsed conformation, definition of, 159

EDTA (ethylenediaminetetraacetic acid), 660

Effective magnetic field, 120

Ehrlich, Paul, 797

Electrocyclic reaction
  definition of, 485
  reverse, 486

Electromagnetic radiation, energetic order of, 119

Electromagnetic spectrum, 119

Electron configuration, 7

Electron repulsion, 78

Electron transfer, 11

Electron-transfer reduction of α,β-unsaturated thiol esters, mechanism of, 724

Electronegativity
  definition of, 9
  effect of, on acidity, 209
  progression of, in the periodic table, 63

Electronic transitions, 141

Electrophiles, 67 (Table 3-2)
  activation of, for electrophilic aromatic substitution, 371
  definition of, 66

Electrophilic addition
  to alkenes, mechanism of, 332
  to alkynes, mechanism of, 341
  cationic rearrangement in, 340
  definition of, 245
  effect of phase on, 334
  energetic analysis of, 245
  stereorandomness in, 340
  thermodynamics of, 332

Electrophilic aromatic halogenation for C–X bond formation, 497

Electrophilic aromatic substitution
  definition of, 371
  mechanism of, 370
  meta direction by electron acceptors, 385
  ortho,para direction by electron donors, 382
  ortho,para direction by halogens, 386
  of polycyclic aromatic hydrocarbons, 393
  rate acceleration by electron donors, 382
  rate retardation by electron acceptors, 385
  rate retardation by halogens, 387
  regiochemical preference in, 380
  regiochemical preference in, of polycyclic aromatics, 393
  substituent effects in, 387 (Table 11-1)
  using substituent effects in synthesis, 390

Electrophilic aromatic substitution, directive effects in, 382
  of halogens in, 386
  of multiple substituents, 389

Electrophilic aromatic substitution for preparation of functional groups, 395 (Table 11-2)

Electrophoresis
  definition of, 116
  for separation of amino acids, 623

Electrostatic attraction in a crystal lattice, 12

Electrostatic forces, in polymer association, 543

Electrostatic interaction, 4, 651
  definition of, 66

Electrostatic repulsion, 5

Elimination reactions
  definition of, 228
  mechanistic possibilities for, 303
  for preparation of functional groups, 326 (Table 9-1)

Eluent, 112

Elution, 112

Elution time, 112

α,β-Enals, preparation of, from β-hydroxyaldehydes by dehydration
  in acid, 464
  in base, 462

Enamines
  alkylation of, 288
  preparation of, from carbonyl compounds, 425
  preparation of, from secondary amines, 425
  structure of, 282

Enanthotoxin, a conjugated polyene, 551

Enantiomers, definition of, 175

Enantiospecificity in enzymatic reactions, 705

Encapsulation within vesicles, 589

Endocyclic double bond, definition of, 307

Endoergic reaction, definition of, 194

Endorphins, definition of, 796

Endothermic reaction
  definition of, 194
  energy diagram for, 194

-ene, suffix for alkenes, 19

Enediol
  geometry of, 774
  as intermediate in the interconversion of glucose and fructose, 771
  as intermediate in triose isomerization, 601
  isomerization of, 773

Energy barrier, definition of, 157

Energy diagram, 193, 746

Energy of activation. *See* Activation energy

Energy storage, in anhydrides, 750

Enkephalins
  function of, 796
  structure of, 796

Enolate anions
  of aldehydes in aldol condensation, 460
  alkylation of, 288, 506
  C vs. O protonation of, 204
  definition of, 461
  of ketones in the aldol condensation, 460
  as nucleophile in base-catalyzed aldol reactions and condensations, 462
  preparation of, from aldehydes, ketones, or esters, 281
  relative stability of, 214, 289
  structure of, 203

Enol phosphate, hydrolysis of, 779

Enols
  of aldehydes in the aldol condensation, 460
  bond strengths in, 204
  of ketones in the aldol condensation, 460
  as nucleophiles in acid-catalyzed aldol condensations, 463
  preparation of, from enolate protonation, 281

Enones, preparation of, from β-ketoalcohols by dehydration, 321

α,β-Enones, preparation of, β-ketoalcohols
    by dehydration
    in acid, 464
    in base, 462
γ,δ-Enones
    preparation of, by Claisen
      rearrangement, 496
    preparation of, from allyl vinyl ethers
      by a Claisen rearrangement, 493
Enovid, structure of, 595
Enthalpy, 745
    definition of, 156
Enthalpy changes, calculation in free
    radical halogenation, 252
Entropy, 745
    definition of, 156
    effect of, on complexation, 662
Enyne, 54
Enzymatic chiral recognition, 705
Enzymatic molecular recognition, 714
Enzymes, 467
    active sites of, 700
    binding of, with substrates, 700
    as catalysts, 692
    definition of, 682
    denaturation of, 682
Epichlorohydrin
    structure of, 565
    synthesis of epoxy resins from, 566
Epimerization of glucose, mechanism of,
    601
Epoxidation of alkenes, mechanism of, 357
Epoxides
    nucleophilic opening of, 357, 614
    preparation of, by peracid oxidation of
      alkenes, 356, 498
Epoxy glues, 564
Epoxy resins, 564
Equatorial substituents, 169
Equilibrium constant, 746
    definition of, 164
    relation of, to free energy, 206
Equilibrium distribution, dependence of,
    on energy difference, 165 (Table 5-2)
Ergosterol, 486
Ergot alkaloid, structure of, 632
Erythromycin, structure of, 638
D-Erythrose, structure of, 602
Eschenmoser, Albert, 511
Ester enolate ions
    alkylation, limitations of, 289
    alkylation, mechanism of, 289
    as nucleophile in base-catalyzed
      Claisen condensations, 468
    structure of, 461
Esterification, acid-catalyzed, 230
Esters
    acidity of, 214
    comparison of rates of hydrolysis of
      carboxylic acid, phosphate, and
      sulfate esters, 596
    comparison of structures of carboxylic
      acid, phosphate, and sulfate esters,
      596
    complex metal reduction of, 406
    conversion of, into β-hydroxyesters by
      Reformatsky reaction, 471
    conversion of, into β-ketoesters by
      Claisen condensation, 468
    nomenclature of, 92
    structure of, 91
Esters, hydrolysis of
    acid catalysis in, 431, 433, 694
    base-induced, 694

equilibrium position in, 431
    pH dependence, 695
Esters, preparation of, 500 (Table 14-2)
    from acid chlorides by alcoholysis, 434
    by Baeyer-Villiger oxidation of ketones,
      500
    from carboxylic acid chlorides by
      esterification, 429
    by ester enolate ion alkylation, 289
    by esterification of carboxylic acid
      derivatives, 500
    from esters by transesterification, 433
    from ketones by the Baeyer-Villiger
      oxidation, 492
    by transesterification, 500
Estradiol, 468
    structure of, 594
Estrogen replacement by conjugated
    estrogen, 523
Estrogen, structure of conjugated, 523
Estrogens, 594
Ethanal, oxidation level of, 93
Ethane
    acidity of, 212
    conformational analysis of, 159
    energy diagram for rotation about C–C
      bond in, 161
    free radical chlorination of, 252
    in interstellar space, 138
    oxidation level of, 75
Ethanol, 582
    ^{13}C NMR spectrum of, 122
    ^{1}H NMR spectrum of, 123
    oxidation level of, 75
    production by fermentation, 768
Ethene, 52, 358, 482, 543
    acidity of, 212
    catalytic hydrogenation of, 74
    in interstellar space, 138
    molecular orbitals in, 143
    structure of, 29
Ethers
    boiling points of, 64
    formation of, 498 (Table 14-2)
    preparation of, by Williamson ether
      synthesis, 498
    preparation of, from alkyl halides, 271
    as solvents, 273
    structure of, 88
Ethyl acetate, 407
Ethylamine, oxidation level of, 75
Ethylbenzene, benzilic bromination of, 259
Ethyl bromide, ^{13}C NMR spectrum of, 129
Ethyl cation, 85
Ethyl chloride, oxidation level of, 75
Ethylene. *See* Ethene
Ethylenediamine, as bidentate ligand, 659
Ethylenediaminetetraacetic acid
    as a polydentate ligand, 660
    structure of, 660
Ethylene glycol, 357
    use of, in the synthesis of dacron, 556
Ethylene oxide, 285, 357
Ethyl iodide, as alkylating agent for
    amines, 614
Ethyl magnesium bromide, 513
Ethyl radical, structure of, 82
Ethylmethylamine, synthesis of, 614
Ethyltriphenylphosphonium bromide, 459
Ethyne, 47
    acidity of, 212
    in interstellar space, 138
    oxidation level of, 75, 93
Exocyclic double bond, definition of, 307

Exoergic reaction, definition of, 194
Exothermic reaction
    definition of, 194
    energy diagram for, 194
Extraction, 111

FADH$_2$. *See* Flavin adenine dinucleotide
Farnesol
    sources of, 598
    structure of, 591
Farnesyl pyrophosphate, mechanism of
    formation of, 598
Fats, structure of, 581
Fatty acid biosynthesis, 721, 754
    carbon–carbon bond formation in, 756
    conversion of β-keto group into a
      methylene group, 757
    energetics of, 768
    NADPH in, 722
    steps in, 781
Fatty acid degradation, 722
    energy release in, 759
    NADH in, 722
    steps in, 781
Fatty acids
    combustion of, 754
    definition of, 581
    energy storage in, 769
    melting points of, 583
    names of, 583
    structures of, 583
Feedback control in enzyme catalysis, 700
Fermentation, conversion of
    carbohydrates into alcohol, 582, 768
Fertilizer, 440, 488
Fingerprint region of the infrared
    spectrum, 138
Firestone, Charles, 553
Fischer, Edmond, 761
Fischer projections, 601
    drawing of, 185
Flame ionization detector, 117
Flavin adenine dinucleotide
    structure of, 717
    oxidation of, 754
Fleming, Alexander, 804
Flower pigments, 146
Fluid mosaic, definition of, 587
Fluoroacetic acid, incorporation of, into
    citric acid, 765
1-Fluorobutane, 100
2-Fluorobutane, 100
Fluorocitric acid, as toxic enzyme
    inhibitor, 765
Fluoromethanes, melting points of, 652
    (Table 19-1)
2-Fluoro-2-methylpropane, 100
Foaming, 586
    of polyurethane, 564
Folding of proteins, 575
Folic acid
    conversion of, into tetrahydrofolic acid,
      810
    function of, 736
Food coloring, 592
Food preservatives, 325
α-Form of cyclic hemiacetals, 603
β-Form of cyclic hemiacetals, 603
Formal charge
    calculation of, 11
    definition of, 10
Formaldehyde, 5, 89
    in Mannich condensation, 616
    molecular orbitals in, 146

polymerization of, 564
synthesis of carbohydrates from, 644
Formaldehyde imine, 71
Formate esters, 469
Fragmentation pattern, 148
Free energy, 745
definition of, 193
relation of, to entropy and enthalpy, 156
relation of, to equilibrium constant, 206
Free radical halogenation, 252
of alkanes for C–X bond formation, 496
dependence of energetics on halogen, 256
reaction profile for, 255
regioselectivity in, 257
relative reactivity of halogen atoms in, 255
Free rotation, 13
Frequency, 119
Friedel, Charles, 374
Friedel-Crafts acylation
for C–C bond formation, 495
of 4-chloroaniline, 535
directive effects in, 537
of isobutylbenzene, 533
for ketone formation, 499
mechanism of, 377
Friedel-Crafts alkylation, 374
for C–C bond formation, 495
chloromethylation of polystyrene, 627
mechanism of, 375
Frontside attack, 241
D-Fructose
structure of, 606
carbon–carbon bond cleavage of, 772
Fructose-1,6-diphosphate, structure of, 770
Fructose-6-phosphate, structure of, 770
Full-headed arrows, 69
Fullerenes, 4
Fumaric acid
formation of, from succinic acid, 766
structure of, 762
Functional groups
classification of transformations, 506
compatibility, definition of, 524
conversions of, by S_N2 reactions, 295 (Table 7-1)
definition of, 59
preparation of, by addition reactions, 362 (Table 10-1)
preparation of, by elimination reactions, 326 (Table 9-1)
transformations of, 508 (Table 15-3)
Furan, 95
structure of, 603
Furanose, definition of, 603
Fused rings, 172
preparation of, from dicarbonyl compounds, 467

Gabriel synthesis
for C–N bond formation, 498
mechanism of, 275
β-D-Galactose, structure of, 605
β-Galactosidase, activity of, 608
Gas chromatography, 111, 117
Gasohol, 582
Gasoline, 22
Gauche conformer, 162
Gel electrophoresis, 116
Geminal dihalide, definition of, 280
Geminal diol, definition of, 417
Generic descriptors of pharmaceuticals, 522
Genetic code, definition of, 666

Geometric isomerization, 155
definition of, 231
Geometric isomers, definition of, 34, 155
Geraniol
sources of, 598
structure of, 590, 591
Geranyl pyrophosphate, mechanism of formation of, 598
Gingergrass, as source of racemic carvone, 676
Glass replacements, 549
Glass, structure of, 551
Glucose
cyclic hemiacetal form of, 600
formation of, in photosynthesis, 753
mechanism of isomerization of, to fructose, 771
oxidation of, 753
structure of, 557, 600
β-D-glucose
anomeric equilibrium of, 605
epimerization, mechanism of, 601
structure of, 605
Glucose-6-phosphate, structure of, 770
α-Glucosidase, activity of, 608
Glucoside, definition of, 605
Glutamic acid, structure of, 620
Glutamine, structure of, 619
D-Glyceraldehyde, structure of, 601
Glyceraldehyde monophosphate
energetics of degradation of, 776
structure of, 770, 774
Glycerol, structure of, 581
Glycine, structure of, 562, 619
Glycolysis
definition of, 769
energetics of, 779
steps in, 769, 780
Goiter, 791
Gorter, E., 587
Gramicidin A
antibacterial activity of, 802
structure of, 802
Grandisol, structure of, 592
Graphite, 4
structure of, 544
Grendel, F., 587
Grignard reactions for preparation of functional groups, 455 (Table 13-1)
Grignard reagents
additions, for C–C bond formation, 495
carboxylation, for carboxylic acid formation, 499, 642
as examples of umpolung in the preparation of, 734
preparation of, from alkyl or aryl halides, 280
protonation of, 283
reaction of, with ethylene oxide, mechanism of, 285
Grignard reagents, nucleophilic addition to
acetaldehyde, 515
aldehydes for C–OH bond formation, 497
carbonyl compounds, mechanism of, 453
carbonyl groups, 453
CO_2, 454
CO_2, mechanism of, 454
esters, for C–OH bond formation, 454, 497
esters, mechanism of, 454
imines, 616

ketones, for C–OH bond formation, 497
2-methylpropanal, 515
Grignard, Victor, 280
Guanidine unit, structure of, 621
Guanine, 97
structure of, 635
Guanine-cytosine base pairing, 665
Guanosine, structure of, 636

Haber-Bosch process, 60
Hadacidin
function of, 809
structure of, 810
Hair sprays, polymers in, 554
Half-chair cyclohexane, 169
Half-headed arrows, 69
Halides, rationale for failure in nucleophilic addition, 416
Hallucinogen, mescaline as, 632
Haloform reaction, mechanism of, 286
α-Halogenation of ketones for C–X bond formation, 497
Halogen reactivity in free radical halogenation, 255
α-Haloketones, preparation of, by halogenation of enolate anions, 497
Halonium ions, 354
Hammond Postulate, 197
Hard, definition of, 313
Hard nucleophiles, definition of, 457
Heartbeat regulation, Digoxin for, 523
Heat capacity, 22
Heat of combustion, definition of, 22
of alkanes, 23 (Table 1-4)
of alkenes, 37
Heat of formation, 23
of alkenes, 41
Heat of hydrogenation
of alkenes, 38
of benzene, 45
Heat of reaction ($\Delta H°$), definition of, 156
Helical peptides, as membrane channels, 588
α-Helix
hydrogen bonding in, 656
intramolecular hydrogen bonding in, 573
structure of, 573, 657
β-Helix, definition of, 574
Helix unwinding, 667
Hell-Volhard Zelinski reaction, 287, 470
for C–X bond formation, 497
Heme, structure of, 660
Hemiacetal
formula of, 422
structure of, 421
Hemiacetal
formation from aldehydes, mechanism of, 421
hydrolysis, mechanism of, 421
Hemiketal, 722
structure of, 422
Hemlock, as source of coniine, 632
Hemoglobin, 791
Herbicide, 440
Heroin, 432
structure of, 631
Hertz, 119
Heteroaromatic molecules, definition of, 95
Hückel aromaticity in, 95
Heteroatoms
definition of, 59
stereochemical inversion at, 188
stereoisomerism at, 187

Heterolysis, 233
Heterolytic cleavage, 233
Heterolytic reactions, energy changes in, 234
Hexaalanine, low solubility of, 625
Hexachloroethane, 176
Hexachlorophene, 373
1,5-Hexadiene, 483
1,5-Hexadien-3-ol, 484
Hexagonal close packing, definition of, 654
n-Hexane
    conformations and entropy of, 663
    mass spectrum of, 149
Hexanoic acid thiol ester, synthesis of, 758
1-Hexanol, 78
2-Hexanol, 78
1,3,5-Hexatriene, 485
    molecular orbitals in, 143
5-Hexenal, 484
trans- 2-Hexene, 359
trans-3-Hexene, 359
Hexenes, structures of isomeric, 39
Hexoses, structures of, 601
Hexynes, structures of isomeric, 53
High-energy phosphate bonds, 751
High performance liquid chromatography. *See* High pressure liquid chromatography
High pressure liquid chromatography, 112
High-resolution mass spectroscopy, 148
Histamine, 794
    receptor for, 795
Histidine
    proton tautomerism in, 622
    structure of, 620
Hoffmann, Roald, 484
Hofmann elimination
    definition of, 309
    from a quaternary ammonium salt for C=C bond formation, 496
Hofmann orientation in E2 elimination, 308
Hofmann rearrangement, 489
    of amides for C–N bond formation, 498
    mechanism of, 490
Home remedies, 608
Homobatrachotoxin, structure of, 639
Homolysis, 233
Homolytic cleavage, 67, 233
Homolytic reactions, energy changes in, 233
Homolytic substitution, 252
Hormones, peptides as, 623
¹H NMR chemical shifts, representative, 128 (Table 4-2)
HPLC. *See* High pressure liquid chromatography
Hückel aromaticity in anionic conjugate bases, 215
Hückel's Rule, 46
Human body, MRI imaging of, 136
Hund's Rule, 8
Hybrid orbitals, 9
    fractional *s* character of, 9
Hybridization
    effect of, on acidity, 212
    sp, 51
    *sp*², 27
    *sp*³, 8
Hydrate(s)
    definition of, 417
    equilibrium with carbonyl compounds, 419 (Table 12-1)
    structure of, 417
Hydrate formation
    under acidic conditions, 418

under basic conditions, 417
    mechanism of, 417
Hydration
    of alkenes for C–OH bond formation, 497
    of alkynes, 342
    of benzyne for C–OH bond formation, 497
    mechanism of, 336
Hydrazine, 426
    in the Gabriel synthesis, 275
    as reducing agent in Wolff-Kishner reduction, 377
Hydrazones
    in chemical test for aldehydes and ketones, 426
    preparation of, from carbonyl compounds, 426
    structure of, 426
Hydride transfer, from NADPH, 721
Hydroboration-oxidation
    of alkenes for C–OH bond formation, 497
    mechanism of, 347
    regiocontrol in, 348
    stereocontrol in, 348
Hydrobromic acid, 628
    acidity of, 209
Hydrocarbons
    boiling points of, 55
    bond lengths in, 53
    definition of, 3
    melting points of, 55
    oxidation of, 325
    preparation of, from alkyl halides by zinc reduction , 284
    preparation of, from alkyl halides by Grignard reagent protonation, 283
Hydrochloric acid, acidity of, 209
Hydrochlorination of an alkene, reaction profile for, 245
Hydrofluoric acid
    acidity of, 209
    for deprotection, 628
Hydrogel
    structure of, 563
    use of, in contact lenses, 563
Hydrogen, electronic structure of, 7
Hydrogen bonds, 651
    definition of, 64
    energy of, 570, 659
    intermolecular, 65, 571, 656
    intramolecular, 65, 571, 656
    intramolecular, in β-aminoalcohols, 614
    linear, 656
    optimal geometry for, 655
Hydrogen peroxide, 492
    as a disinfectant, 252
Hydrohalogenation
    of alkenes for C–X bond formation, 496
    mechanism of, 333
    regiochemistry of, 337
Hydroiodic acid, acidity of, 209
Hydrolysis
    of acid derivatives, for carboxylic acid formation, 499
    of alkyl halides for C–OH bond formation, 497
    definition of, 248
    of Grignard reagents for C–H bond formation, 495
    of terminal alkynes for ketone formation, 498
Hydronium ion, 418
Hydroperoxides, structure of, 344
Hydrophilic, definition of, 585
Hydrophobic, definition of, 585

Hydroquinone, 325
Hydroxide ion, 418
    as a nucleophile, 481
    reaction of, with methyl bromide 685
β-Hydroxyaldehydes
    conversion of, into α,β-enals, 462, 464
    preparation of, from aldehydes by the aldol reaction, 462
β-Hydoxycarbonyl compounds, preparation of, by the aldol reaction, 496
β-Hydroxyesters, preparation of, from ketones and esters by the Reformatsky reaction, 470
β-Hydroxyketones, preparation of, from ketones by the aldol reaction, 462
Hydroxylamine, 426, 487
4-Hydroxymethylcyclohexanone, 532
    acetal of, 527
    2-alkylation of, 531
    benzylation of, 532
    synthesis of, from 4-carboxymethyl-cyclohexanone, 525
5-Hydroxypentanoic acid, lactone formation from, 696
3-Hydroxypropanal, 462
Hydroxysuccinic acid, structure of, 762
Hyperconjugation
    definition of, 40
    geometry required for, 40
Hypervalent carbocations, 335
Hypothyroidism, treatment of, with levothyroxine, 523

Ibuprofen, synthesis of, 533
σ-Inductive electron withdrawal, 386
Imidazole, 96
    structure of, 635
Imides, structure of, 91
Imine-enamine tautomerization, 424
Imine formation, 500 (Table 14-2)
    amination of ketones, 500
    mechanism of, 424
Imines
    catalytic hydrogenation of, 72, 413
    hybridization in, 71
    preparation of, from carbonyl compounds, 423, 615
    reduction of, 615
    structure of, 71
Iminium ion, 407, 423
Immune system, 708
Index of hydrogen deficiency, 29
Induced dipole moment, 653
Induction, differentiation from catalysis, 692
Inductive effects
    definition of, 209
    influence of, on acidity, 211 (Table 6-2)
Inert gas, 7
Information storage in DNA, 667
Infrared spectroscopy, 135
    characteristic of functional groups, 137 (Table 4-3)
Inhibitors, 701
Initiation
    of free radical halogenation, 253
    of radical addition, 345
Insect repellent, 592
Insecticides, 17
    terpenes as natural, 593
Insects, moisture-barrier coating of, 582
Insertion
    into a C–X bond, 278
    definition of, 699

of —NH— adjacent to carbonyl group, 488
of —O— adjacent to carbonyl group, 491
Instant glue, structure of, 549
Insulin, 589
    hydrolysis of, in the stomach, 790
Integration curve, 130
Intramolecular Claisen condensation, 470
Intramolecular cyclization, 485
Invertase, 607
Invert sugar, definition of, 607
Iodoenolate anion, structure of, 286
Iodoform, 286
Iodoform reaction, 286
    for carboxylic acid formation, 499
Iodoform test, 286
Ion channels, 803
Ionic bond, 10
Ionophores
    as antibacterial agents, 801
    definition of, 659
Ion pairing, 199
    in alkyllithiums, 279
Ion transport
    by polydentate ligands, 662
    by tetraalkylammonium ions, 686
Irreversible reaction, 195
Iso-, prefix for alkyl groups, 20
Isobutylbenzene, Friedel-Crafts acylation of, 533
Isocitric acid
    formation of, from citric acid, 764
    structure of, 762
Isocyanates
    nucleophilic addition to, 490
    structure of, 490
Isoelectric point, definition of, 622
Isolated double bond, 42
Isoleucine, structure of, 619
cis-Isomer, 36
trans-Isomer, 36
Isomerase, function of, 774
Isomerism, 14
Isomerization, 478
cis-trans Isomerization, 157
Isomerization reaction, definition of, 231
(*E*)-Isomers, 36
(*E,Z*)-Isomers, rules for assigning structure to, 36
Isopentenyl pyrophosphate
    carbocation from protonation of, 596
    structure of, 590, 595
Isoprene, structure of, 550, 590
Isoprenoids, definition of, 590
Isopropyl bromide, hydrolysis of, 248
Isopropyl cation, 85
Isopropyl radical, 83
Isotactic, definition of, 567
Isothiocyanate, 187
Isotopic abundance from mass spectroscopy, 149
Isotopic labeling of carbonyl oxygen, 418

Jackson, Charles, 272
Juvabione, structure of, 591

Kanamycin A, structure of, 638
Kekulé structures, 44
Kel-F, structure of, 549
Kelsey, Frances, 183, 677
Ketals
    formation of, 500 (Table 14-2)
    formula of, 422

hydrolysis of, for ketone formation, 498
as protecting groups for ketones, 423, 527
structure of, 422
Ketimine, structure of, 718
Ketimine-aldimine tautomerization, in biological reductive amination, 718
α-Ketoacid, decarboxylation of, 761
β-Ketoacid
    for C–H bond formation, 495
    decarboxylation of, 291, 761
β-Ketoalcohols
    conversion of, into α,β-enones, 462, 464
    preparation of, from ketones by the aldol reaction, 462
9-Ketodecenoic acid, 403
Keto-enol tautomerism, 457
    definition of, 204
β-Ketoester enolate anion, alkylation of, 290
α-Ketoesters, decarboxylation of, 727
β-Ketoesters
    acidity of, 469, 471
    decarboxylation of, 727
    hydrogen bonding in, 656
    preparation of, by Claisen condensation, 499
    preparation of, from esters by Claisen condensation, 468
    reduction of, 721
    structure of, 468
Ketone enolate ion, structure of, 461
Ketones
    acidity of, 214
    amination of, for imine formation, 500
    aminomethylation of, by Mannich condensation, 616
    complex metal reduction of, 404
    nomenclature of, 90
    protecting groups for, 527
    reduction of, by NADPH, 753, 757
    reduction of, by sodium borohydride, 533
Ketones, conversion of, into
    alkenes by Wittig reaction, 459
    amides through oximes and Beckmann rearrangement, 487
    cyanohydrins, 452
    esters by Baeyer-Villiger oxidation, 492
    β-hydroxyesters by Reformatsky reaction, 471
    β-hydroxyketones by aldol reaction, 462
    imines, 424
    oximes, 487
Ketones, preparation of, 498 (Table 14-2)
    by acetoacetic ester synthesis, 290, 496
    from alkenes by ozonolysis, 359
    from arenes by Friedel-Crafts acylation, 377
    by chromate oxidation of secondary alcohols, 499
    by Claisen condensation, 499
    by Claisen rearrangement of allyl vinyl ethers, 499
    by decarboxylation of β-ketocarboxylic acids, 495, 499
    from 1,2-diols by pinacol rearrangement, 480
    by enolate alkylation, 288
    from α,β-enones by conjugate addition, 456
    from α,β-enones by Michael addition, 457
    by Friedel-Crafts acylation, 499
    by hydrolysis of terminal alkynes, 498
    by ketal hydrolysis, 498

by pinacol rearrangement, 499
from secondary alcohols by oxidation, 323
β-Ketonitrile enolate ion, structure of, 282
Ketoprofen
    structure of, 532
    synthesis of, 534
Ketose, definition of, 602, 606
Ketotriose, definition of, 602
Ketyl radical anion, 414
    electron distribution in, 202
Kinetic control in enolate ion protonation, 217
Kinetics, definition of, 203
Knocking, 22
    definition of, 699
Krebs cycle, 760
Krebs, Edwin, 760

β-Lactam
    definition of, 806
    resonance stabilization of, 807
β-Lactamase, 808
Lactic acid, 777
    structure of, 542
Lactide, structure of, 542
Lactone
    formation of, 696
    structure of, 492
Lactose, structure of, 607
Lanosterol, 591
    structure of, 592
Lanoxin, structure of, 523
Late transition state, 197
    in S$_N$2 reaction, 242
Latex rubber, structure of, 550
Leaving group, 240
    effect of, on elimination efficiency, 312
LeChatelier's Principle, 230, 422, 431
    application of, to complex systems, 715
Lehn, Jean-Marie, 661
Leucine, structure of, 619
Leu-enkephalin, structure of, 797
Leutenizing hormone release hormone, structure of, 624
L-dopa
    biological effect of, 798
    structure of, 798
Levorotatory, 182
Levothyroxine, structure of, 523
Lewis acid
    activation of alkyl chlorides for electrophilic aromatic substitution, 375
    activation of dihalogen for electrophilic aromatic substitution, 371
    definition of, 66
    production of acylium ion from carboxylic acid chlorides, 376
Lewis acidity, 69
Lewis base, definition of, 66
Lewis basicity, 69
Lewis dot structure, 9
LHRH. *See* Leutenizing hormone release hormone
Ligands, definition of, 659
Limonene, formation of, from nerol pyrophosphate, 599
Linalool
    ^{1}H NMR spectrum (90 MHz) of, 126
    ^{1}H NMR spectrum (360 MHz) of, 127
Linear carbon, 5
Linear polymer, definition of, 543
Linear synthesis, definition of, 520

Line notation, 14
Lipid bilayer, structure of, 588
Lipids
  definition of, 581
  solubility of, 584
Lipoic acid
  function of, 727
  oxidative cyclization of, 728
Liquid chromatography, 111
Lithium
  electronic structure of, 7
  as an energy source, 747
Lithium acteylide, 455
Lithium aluminum hydride, 404
  hydrogen evolution from, 406
Lithium aluminum hydride reduction
  of amides for C–N bond formation, 498
  of esters, mechanism of, 407
  of ketones, mechanism of, 405
  of nitriles, 617
  of secondary amides, mechanism of, 408
  of tertiary amides, mechanism of, 407
Lithium dialkylamide, 459
  acetone enolate formation with, 506
  for deprotonation of ketones, 532
Lithium dialkylcuprate, 281
Lithium diisopropylamide, 461
Lithium dimethylcuprate, 457
Lithium hydride, 404
Lithium ion
  complexation with dimethylsulfoxide, 279
  complexation with 18-crown-6-ether, 279
  complexation with tetrahydrofuran, 279
  size of, 662
Liver oxidase, degradation of waste materials by, 707
Living polymer, definition of, 548
Local magnetic field, 120
Lock, conformational, 170
Lock-and-key, definition of, 705
Lotions, polymers in, 554
LSD. *See* Lysergic acid diethyl amide
Lubricants, 554
Lucas test, alcohol classification by, 86
Lucite, structure of, 549
Lycopene, 144
Lysergic acid, structure of, 632
Lysergic acid diethyl amide, structure of, 798
Lysergide
  biological effect of, 798
  structure of, 798
Lysine, structure of, 620
D-Lyxose, structure of, 603

Maalox, as antacid, 794
Magnamycin, structure of, 808
Magnetic field, 120
Magnetic resonance imaging, 135
Malonic acid thiol ester in fatty acid synthesis, 759
Malonic ester, structure of, 292
Malonic ester synthesis
  for C–C bond formation, 496
  mechanism of, 293
Malononitrile anion, 457
Maltol, 465
Maltose, structure of, 607
Mammalian vision, 158
Mandelic acid, resolution of, 675
Mannich condensation, mechanism of, 616
β-D-Mannose, structure of, 605
Margarine colorant, 379
Markovnikov's Rule, 247, 337, 342, 343, 360, 480, 595

Markovnikov, Vladimir, 247
Mass spectroscopy, 147
Matrix isolation, 201
MCPBA. *See meta*-Chloroperbenzoic acid
Meat preservative with nitrite ions, 670
Mechanistic organic chemistry, 237
Melting points, 24
  of alkanes, 24 (Table 1-5)
  of alkynes, 653
  of fluoromethanes, 652 (Table 19-1)
  of hydrocarbons, 55
  of isomeric butanols, 658 (Table 19-2)
  of nitriles, 653
Membrane disruption of ion balance by polydentate ligands, 661
Membrane transport, 791
Membranes, structure of, 587
Menthol, structure of, 591
Mercuronium ion, structure of, 343
Merrifield, R. Bruce, 626
Merrifield resin, 627
Merrifield synthesis, 626
  sequence of steps in, 630
Mescaline
  biological effect of, 798
  structure of, 631, 632, 798
Mesembrine
  biological effect of, 798
  structure of, 798
Meso compound, definition of, 184
Messenger RNA, 668
Met-enkephalin, structure of, 797
Meta isomer, 49
Metal hydride reduction
  of aldehydes for C–OH bond formation, 497
  of esters for C–OH bond formation, 497
  of ketones for C–OH bond formation, 497
Metal insertion into a C–X bond, 278
Methane, 5
  acidity of, 209
  as an energy source, 747
  three-dimensional structure of, 9
Methanol, 78
  hydrogen bonding in, 79
Methionine, structure of, 619
Methotrexate
  inhibition of biosynthesis of tetrahydrofolic acid by, 809
  structure of, 810
N-Methylacetamide, 408
Methyl acetate, hydrolysis of, 687
Methylamine, 60, 614
  bond cleavage in, 68
  solvation of, 65
Methylation, dimethylsulfate in, 535
Methyl benzoate, 385
Methyl bromide, 239
trans-2-Methylbromocyclohexane, 310
2-Methyl-1,3-butadiene, polymerization of, 550
3-Methyl-2-butanol, chromate oxidation of, 515
3-Methyl-2-butanone, synthesis from 2-propanol, 514, 519
2-Methyl-2-butene, 459
Methyl cation, 84
Methyl cyanoacrylate, structure of, 549
Methylcyclohexane
  conformation of, 169
  flagpole interactions in, 170
N-Methylcyclohexanone imine, 413, 425
1-Methylcyclohexene, 346
  reaction profile for hydrochlorination of, 247

Methylcyclopropane, 17
o-Methyl-*N,N*-dimethylaniline, bromination of, 389
Methylene-tetrahydrofolic acid, 738
Methyl esters, boiling points of, 665
Methyl ether, 88
Methyl fluoride, 99
Methyl formate, 469
α-D-Methylglucopyranoside, structure of, 605
β-D-Methylglucopyranoside, structure of, 605
N-Methyl-α-L-glucosamine, structure of, 638
Methyl Grignard, 457
Methyl halides, bond-dissociation energies of, 100
Methyl iodide, 240, 532
  alkylation of acetone with, 506
Methyl ketones
  chemical test for, 286
  preparation of, by hydrolysis of terminal alkynes, 498
  preparation of, from alkynes by hydration, 343
Methylmagnesium bromide, 278
Methyl methacrylate, structure of, 549
3-Methyl-2-pentene, 36
4-Methyl-1-pentene, 35
2-Methylpropane
  free radical bromination of, 259
  free radical chlorination of, 259
2-Methyl-2-propanol, 78, 85, 478
  synthesis from acetone, 509
2-Methylpropene, 344, 388
Methyl radical, 67
  structure of, 82
Methyl sulfide, 94
Methyl vinyl ketone, 457
Micelles
  hydrophobic core of, 585
  interface of, 585
  polar head region in, 585
  structure of, 585
Michael addition
  conjugate addition to α,β-enones for C–C bond formation, 496
  definition of, 457
Microscopic reversibility, 203
Miracle drugs, 795
Mirror plane, 174
Mixed anhydride, 776
Mobile phase, 111
Mobility, 111
Molasses, use of, in the production of rum, 768
Molecular ion, 148
Molecular orbitals, 31
  antibonding, 140
  bonding, 140
  phasing of, 32
Molecular recognition
  definition of, 651
  importance of, in biological reactions, 714
  multiple contacts for, 664
  reversible binding for, 664
Molecular sieves, removal of water by, 615
Molozonide, 358
Monoterpenes, definition of, 590
Morphine, 2, 432, 791
  effect of, on central nervous system, 633
  structure of, 632
Morton, William, 272
Motion sickness, scopolamine for, 632

mRNA. *See* Messenger RNA
Multiplet, definition of, 126
Multiplicity, definition of, 130
Multistep reactions, 749
Mushrooms, 1,1-dimethylhydrazine in, 633
Mustard gas, 269
Mutarotation, definition of, 604
Myelin sheaths, cholesterol in, 594
Mylar, structure of, 556

n-, prefix for unbranched alkyl groups, 20
NADH. *See* Nicotinamide adenine dinucleotide
NADPH. *See* Nicotinamide adenine dinucleotide phosphate
Naphthalene, 393
Narcotic, morphine as, 632
Natta, Giulio, 567
Natural isotopic abundance of carbon-13, 123
Neo, prefix for alkyl groups, 20
Neoprene, structure of, 550
Nepetalactone, structure of, 591
Nerol pyrophosphate, mechanism of cyclization of, 599
Newman projections
    drawing of, 159
    for visualizing E2 transition states, 311
    for visualizing stereocontrol in E1 eliminations, 314
    for visualizing stereocontrol in E2 eliminations, 315
Nicotinamide, structure of, 717
Nicotinamide adenine dinucleotide, 410
    as biological oxidant, 324
    reduction of, 752
    structure of, 717
Nicotinamide adenine dinucleotide phosphate
    reduction of β-ketoesters by, 721
    structure of, 717
Nicotine, structure of, 631
Nigericin, structure of, 661
Nitration, definition of, 371
Nitration of arenes, mechanism of, 373
Nitriles
    hybridization in, 77
    melting points of, 653
    reduction of, 617
    structure of, 77
Nitriles, hydrolysis of, 437, 488, 533, 534
    by acid catalysis, 438
    for carboxylic acid formation, 499
    mechanism of, 438
Nitriles, preparation of, 497 (Table 14-2)
    from alkyl halides by S_N2 displacement, 437, 617
    by amide dehydration, 437, 497
    by cyanide displacement from alkyl halides, 495, 497, 533, 534
Nitrilium ion, structure of, 488
Nitrite ion, as source of nitrous acid, 670
Nitroacetic acid, acidity of, 210
Nitroarenes, preparation of, from arenes by electrophilic nitration, 374
Nitrobenzene, 374, 378, 385
    bromination of, 386
    reduction of, 378
Nitrogen inversion, 188
1-Nitronaphthalene, 394
Nitronium ion, structure of, 373
Nitrosamine, toxicity of, 633
Nitrous acid, 378
    conversion of uracil into cytosine by, 670

NMR spectroscopy, medical applications of, 134
NMR spectrum, 121
Noble metal, definition of, 410
Node, 6
Nomenclature
    for alkenes, 34
    for alkynes, 53
    for heteroatom-containing groups, 101 (Table 3-7)
    IUPAC rules for alkanes, 18 (Table 1-2)
    IUPAC rules for hydrocarbons, 35
Nonpolar covalent bonding, 10
Nonsaponifiable lipid, definition of, 590
Nor-, definition of, as a prefix, 591
Norethynodrel, structure of, 595
Normal phase chromatography, 116
Nortriterpenes, definition of, 591
n,π* transition, 141
Nuclear magnetic resonance (NMR), 119
Nuclear spin, alignment of, 120
Nucleic acid
    base pairing, 665
    structure of, 637
    synthesis of, 809
Nucleophiles, 67 (Table 3-2)
    definition of, 66
Nucleophilic acyl substitution
    acceleration in acid, 428
    acceleration in base, 429
    definition of, 427
    mechanism of, 428
Nucleophilic addition
    acceleration in acid, 402
    acceleration in base, 402
    to carbonyl groups, definition of, 402
    of hydroxide ion, 481
    thermodynamics of, 416
Nucleophilic addition and substitution
    carbon nucleophiles in, 451
    for preparation of functional groups, 442 (Table 12-2)
Nucleophilic attack, preferred sites for
    in carboxylic acid esters, 597
    in phosphate esters, 597
    in sulfate esters, 597
Nucleophilic attack, relative reactivity of carbonyl compounds toward, 93
Nucleophilic carbon–carbon bond formation in functional-group preparation, 473 (Table 13-2)
Nucleophilic opening of epoxides for C–OH bond formation, 497
Nucleophilicity
    dependence of, on state of protonation, 418
    factors affecting, 416
    relation of, to basicity, 313
Nucleoside
    definition of, 635
    structure of, 666
Nucleotide, structure of, 666
Number of steps, effect of, on synthetic efficiency, 520
Nutrasweet, structure of, 626
Nylon 6, 488, 618
    structure of, 560, 561
    synthesis by Beckmann rearrangement, 560
Nylon 66, structure of, 560
Nylons, nomenclature for, 560

(*E,E*)-2,6-Octadiene, 484
Octane rating
    of alkenes, 40

definition of, 22
Octanoic acid, IR spectrum of, 140
2-Octanol, 110, 138
    IR spectrum of, 139
2-Octanone, 110, 138
    IR spectrum of, 139
Odor recognition, 176
Oil of caraway, 176
Oil of spearmint, 176
Olah, George, 335
Ophiobolin, structure of, 591
Opium, 2
Opsin, 158
Optical inactivity, racemic mixture as a cause of, 181
Optically active, definition of, 179
Optical purity, calculation of, 180
Optical stereoisomers, definition of, 155
Oral thrush, control of, by bacteria, 803
Orbital overlap, 9
Orbital size mismatch, 95
Organic chemistry, definition of, 3
Organic reactions, study techniques for, 236, 295
Organic synthesis, 237
Organocadmium, 281
Organolithiation, mechanism of, 279
Organolithiums, 455
    addition to carbonyl compounds, mechanism of, 456
    nucleophilic addition to carbonyl compounds, 456
    preparation from alkyl halides, 278
Organomercury, 281
Organometallic reagents, preparation of, from alkyl halides, 278
Organonitriles
    hydrolysis of, 284
    preparation of, from alkyl halides, 284
Organozincs, 281
    as intermediates in reduction of alkyl halides, 284
Orlon, structure of, 549
Ortho isomer, 49
Overall yield, calculation of, 521
Oxa-Cope rearrangement. *See* Claisen rearrangement
Oxalate esters, structure of, 469
Oxaloacetic acid
    formation of, from fumaric acid, 766
    structure of, 762
    in the TCA cycle, 761
Oxidation
    of aldehydes, for carboxylic acid formation, 499
    definition of, 74
    of primary alcohols for aldehyde formation, 498
    of primary alcohols for carboxylic acid formation, 499
Oxidation level, means of establishing, 74
Oxidation reaction, 231
Oxidation-reduction reactions, 507 (Table 15-2)
    classification of, 506
Oxidative decarboxylation, 726
Oxidative degradation, definition of, 358
Oxidizing agents, 76 (Table 3-5)
Oximes
    conversion of, into amides by Beckmann rearrangement, 487
    as intermediates in Beckmann rearrangement, 618
    preparation of, from carbonyl compounds, 426

Oximes (*cont.*)
structure of, 426
2-Oxoglutaric acid, structure of, 762
Oxonium ion
definition of, 84
deprotonation of, 249
trapping of carbocations to produce, 248
Oxygen
singlet, 142
as a stable triplet, 142
Oxygen atoms, as electron donors, 480
Oxymercuration-demercuration
of alkenes for C–OH bond formation, 497
mechanism of, 349
Ozonation of alkenes, mechanism of, 359
Ozone, structure of, 358
Ozone hole, 257
Ozonide, 358
Ozonolysis, 358
of alkenes for aldehyde formation, 498
contrasted with Wittig reaction, 459

Pacific yew tree, as source of taxol, 593
Palmitic acid
biodegradation of, 767
combustion of, 767
degradation of, to carbon dioxide, 754
Pantothenic acid, structure of, 726
Paper chromatography, 115
Paracelsus, Phillipus, 2
Paraformaldehyde
as source of formaldehyde, 644
structure of, 555, 644
Para isomer, 49
Parent ion, 148
Parkinson's disease, 798
Partial charge separation, means for indicating, 63
Partitioning, 110
Pasteur, Louis, 180
Pauli Exclusion Principle, 6
PBr$_3$, 441
PCl$_3$, 441
Pedersen, Charles, 661
PEG. *See* Poly(ethylene glycol)
Penicillin G, structure of, 806
Penicillin V, synthesis of, 526
Penicillins, structure of, 806
Penicillium fungus (source of penicillin), 804
1,3-Pentadiene, acidity of, 213
Penta-4-en-2-one, 413
3-Pentanol, 131
Pentoses, structures of, 601
Pepsin, 790, 794
Peptidase, definition of, 589
Peptide, definition of, 561
Peptide bond, structure of, 569
Peptide hormones, 589
Peptide hydrolysis, mediation of, by chymotrypsin, 703
Peptide mimics, 795
Peptides
C terminus of, 624
definition of, 623
epimerization of, 625
hydrogen bonding in, 620
naturally occurring, 718
N terminus of, 624
preparation of, from amino acids by the Merrifield synthesis, 626
solubility of, 625
zig-zag conformation of, 620
Peptidoglycans, 803
Peracetic acid, 356, 492

Peracid oxidation of alkenes for epoxide formation, 498
Peracids, structure of, 356, 492
Perfumes, 405
terpenes in, 598
Pericyclic reaction, definition of, 482
Pericyclic rearrangement, definition of, 483
Periodic table, 3
Perkin, William, 394
Perkin's mauve, 394
Permanganate, as a color indicator, 77
Permanganate oxidation
of alkyl aromatics, 379
of aryl alkanes for carboxylic acid formation, 499
as chemical test for oxidizable groups, 359
Peroxides, structure of, 344
Petroleum
cracking of, 236
crude, 541
Peyote cactus, as source of mescaline, 632
Pharmaceutical agents, 789
Pharmaceuticals
chiral, 791
generic descriptors of, 522
naming of, 524
psychoactive, 535
semisynthetic, 526, 791
Phase-transfer catalysis, 687
Phasing of molecular orbitals, 32
β-Phenethylamines, 183
as peptide mimics, 795
structure of, 633
Phenol oxidation, 325
Phenols, 98
acidity of, 208
directive effect of —OH group on ring alkylation, 565
preparation of, from benzyne, 319
preparation of, by hydration of benzyne, 497
preparation of, from phenylallylether by a Claisen rearrangement, 493
2-Phenoxyethanol, synthesis of penicillin V from, 526
Phenyl cation, 319
Phenyl group, definition of, 50
N-Phenylacetamide, 383
bromination of, 384
Phenylacetate, bromination of, 384
Phenylacetic acid, 287
Phenylalanine
structure of, 619
taste of, 676
N-Phenylbenzamide, 488
2-Phenyl-1-butene, 339
p-Phenylenediamine
structure of, 561
synthesis of *p*-phenyleneterephthalamide from, 561
p-Phenyleneterephthalamide, structure of, 561
Phenylhydrazine, 426
Phenylhydrazone, structure of, 426
3-Phenyl-1-propene, 340
Pheromones, 403
Phosphate esters
hydrolysis of, 709
role in terpene biosynthesis, 597
Phosphatidylcholines, structures of, 587
Phosphatidylethanolamines, structure of, 587
Phosphatidylserines, structures of, 587

Phosphines, structure of, 276
Phosphite ester
mechanism of formation, 441
structure of, 441
Phospholipids, definition of, 587
Phosphonium salts, structure of, 276
Phosphonium ylides, 459
preparation of, from alkyl halides, 276
Phosphoric acid, 440
Phosphoric acid anhydride
hydrolysis of, 751
structure of, 728, 751
Phosphorus–oxygen double bond, 459
Phosphoryl halides, 440
Photoexcitation, 141
Photographic development, 325
Photographic film, 559
Photosynthesis, 753
Phthalimide, in Gabriel synthesis, 275
Physical organic chemistry, 237
Physical properties, effect of symmetry on, 658
Pi (π) bond, definition of, 28
π-electron donation, 386
Pinacol rearrangement, 230, 480
for ketone formation, 499
α-Pinene, 480
Pinging, 22
π,π * transition, 141
Pivalate esters, structure of, 469
pK_as
characteristic of functional groups, 210 (Table 6-1)
comparison of carboxylic, phosphoric, and sulfuric acids, 596
definition of, 207
Planck's constant, 119
Plastic bags, poly(vinylchloride) in, 548
Plastics, definition of, 547
Platinum as a heterogeneous catalyst, 698
β-Pleated sheet
hydrogen bonding in, 656
intermolecular hydrogen bonding in, 573
structure of, 572, 657
Plexiglas, structure of, 549
Plunkett, R.J., 549
+/−, representation for designating stereoisomers, 182
POBr$_3$, 440
POCl$_3$, 440
Polar covalent bonding, 10
Polarimetry, 179, 180
Polar interactions, 655
Polarizer, 179
Pollen, 794
Polyacetals
end groups for, 555
structure of, 555
Polyacrylamide, 116
Polyamides, structure of, 560
Polyaniline, in batteries, 747
Polyanions, 637
Polybutadiene
isomerism in, 551
repeat units in, 551
Polycarbonates, structure of, 557
Polycyclic aromatic hydrocarbons, 49
electrophilic aromatic substitution in, 393
Polydentate ligand, definition of, 660
Polyelectrolytes, 116, 637
Polyene antibiotics, 803
Polyenes, 551
definition of, 42

Polyepoxides, formation of, in the liver by liver oxidase, 707
Polyesters
    as condensation polymers, 545
    structure of, 541, 556
Polyethers, structure of, 554
Polyethylene
    as an addition polymer, 545
    molecular weight of, 546
    structure of, 543
Poly(ethylene glycol), structure of, 554
Polyisoprenes, cyclization of, 600
Polyketals, structure of, 555
Polylactic acid, 542
Polymerization
    anionic, definition of, 547
    cationic, definition of, 547
    definition of, 541
    radical, 546
    ring-opening, 554
    types of, 545
Polymers
    definition of, 112, 541
    flexibility in, 547
    head-to-head, 549
    head-to-tail, 549
    linear, structure of, 543
    1 : 1 pairing of monomers, 543
    regular, structure of, 543
    resistance to oils, 550
    solubility of, 567
    stereoregularity in, 567
    three-dimensional, 544
    two-dimensional, 544
Polynucleic acids
    definition of, 613
    structure of, 634
Polyols, structure of, 553
Polypeptides
    definition of, 561, 623
    structure of, 562, 568
Polypropylene, structure of, 567
Polysaccharides
    in bacterial cell-wall structure, 804
    cross-linking of, in bacteria cell wall, 805
    difficulties in digestion of, 609
    structure of, 557
Polystyrene, 360
    chloromethylated, for Merrifield synthesis, 626
    NMR imaging of, 137
    structure of, 548
Polyurethanes
    definition of, 562
    structures of, 563
Poly(vinyl acetate), structure of, 554
Poly(vinyl alcohol), structure of, 553, 554
Poly(vinyl chloride), structure of, 548
Porphyrins, structure of, 660
Positional isomer, definition of, 34
Positional isomerization, definition of, 231
Potassium ion, size of, 662
Potential energy surface, 193
Potential energy well, 161
PPTA. *See p*-Phenyleneterephthalamide
Prebiotic chemistry, 640
Precalciferol, 486
Premarin, structure of, 523
Priestley, Joseph, 272
Primary alcohols, preparation of
    by Grignard reaction of esters, 497
    by metal hydride reduction of aldehydes, 497
    by metal hydride reduction of esters, 497

Primary amides
    conversion of, into amines by Hofmann rearrangement, 489
    structure of, 91
Primary amines, 61
    synthesis of, 275
Primary carbon, 20
Primary structure of protein, 574
Prochiral, definition of, 339, 764
Product inhibition, definition of, 700
Progesterone, structure of, 594
Progestins, 594
Proline, structure of, 619
Prontosil, structure of, 799
1,2-Propadiene, 54
Propagation of radical addition, 345
Propagation steps in free radical halogenation, 253
Propanal
    enolate anion formation from, 517
    formation of, from 1-propanol, 518
    Grignard addition to, 513
Propane
    acidity of, 213
    free radical bromination of, 258
    free radical chlorination of, 258
    in interstellar space, 138
    structure of, 15
1-Propanol, oxidation of, 518
2-Propanol, 405
    conversion of, into 2-butanol, 512
    conversion of, into 2-butanone, 505
    conversion of, into 3-methyl-2-butanone, 514, 519
    dehydration of, 518
    formation of acetone from, 746
Propene, 34
    acidity of, 213
    geometry required for hyperconjugation in, 40
    hydroboration-oxidation of, 518
n-Propylbenzene, 380
Propyne, in interstellar space, 138
Protease, definition of, 701
Protected alkene, 354
Protecting groups
    for alcohols, 529
    for aldehydes, 527
    from amines, 530, 627
    for carboxylates, 529
    characteristics of, 528
    definition of, 354
    for ketones, 527
    use of, in practical synthesis, 531
Protein folding, 575
Proteins
    definition of, 561, 574, 623
    denaturation of, 586
    as membrane channels, 588
    primary structure of, 574
    quaternary structure of, 575
    secondary structure of, 574
    tertiary structure of, 574
Protic solvent, definition of, 66
Protonated alcohol, structure of, 84
Protonated carbonyl compounds, as electrophiles in acid-catalyzed aldol condensations, 463
Proton transfer in ester hydrolysis, 690
Pteridine, 97
Puffer fish, as source of tetrodotoxin, 661
Purine, 97
Purine bases, structure of, 635
Putrescine, 61

Pyran, structure of, 603
Pyranose, definition of, 603
Pyranose-furanose equilibration, 606
Pyrazine, 96
Pyrethrin I, structure of, 593
Pyrethrin II, 17
Pyridazine, 96
Pyridine, 96, 434
    Kekulé structure in, 96
    water solubility of, 657
Pyridoxamine phosphate, structure of, 717
Pyridoxol, structure of, 717
Pyridoxol phosphate in reductive amination, 718
Pyrimidine, 96
Pyrimidine bases, structure of, 635
Pyrophosphoric acid, structure of, 595
Pyrrole, 95
Pyrrolidine enamine, 288
Pyruvic acid, 777, 779
    structure of, 770

Quack medicines, terpenes in, 592
Quantum numbers, 6
Quartet multiplet, 127
Quaternary ammonium salt, 61
Quinine, 73
    biological effect of, 798
    structure of, 631, 798
Quinoid form, definition of, 718
Quinoline, 97
Quinone, 325

R, stereochemical designator, 177
Racemate. *See* Racemic mixture
Racemic mixture, 181
Racemic modification. *See* Racemic mixture
Radical
    definition of, 30
    stability order of subclasses of, 81
Radical addition to alkenes, 344
    mechanism of, 345
Radical anions
    electrochemical generation of, 414
    electron distribution in, 202
Radical cation, electron distribution in, 201
Radical chain reaction, definition of, 253
Radical cyclization of alkenes, 361
Radical polymerization
    definition of, 546
    initiation of, 547
    mechanism of, 361, 546
    termination of, 547
Radicals
    conjugation in, 86
    designation of subclasses of, 82
    hybridization in, 82
    hyperconjugative stabilization in, 82
    order of stability in, 345
    stability order of subclasses of, 87
    structure of, 82, 199
Radio frequency radiation, 121
Rantidine, structure of, 523
Rate-determining step, definition of, 198
Reaction coordinate, 194
Reaction mechanism, definition of, 238
Reaction profile. *See* Energy diagram
Reactions, classification of, 227
Reactive intermediate, recognition of, in an energy diagram, 161, 196
Reactivity order, radicals, 345
Rearrangement reactions
    definition of, 230

Rearrangement reactions (*cont.*)
  in functional-group preparation, 494
    (Table 14-1)
Red No. 2, 378
Redox reactions, color indicators for, 77
Redox reagents, definition of, 74
Reducing agents, 76 (Table 3-5)
Reduction reaction, 231
  definition of, 74
  of nitro compounds for C–N bond
    formation, 498
Reductive amination, 615
  biosynthesis of amino acids by, 718
  definition of, 424
  of ketones for C–N bond formation, 498
Reductive elimination of vicinal dihalide,
  for C=C bond formation, 496
Reformatsky reaction, 470
Refractive index, definition of, 114
Refractive index detector, 114
Refrigerants, 257
Regiochemistry for protonation of
  1-methylcyclohexene, 247
Regiocontrol, definition of, 258
Regioselectivity
  definition of, 258
  in E1 elimination, 307
  in E2 elimination, 309
  free radical halogenation, 257
  in free radical halogenation, Hammond
    Postulate, 258
Relative rates, Beckmann rearrangement,
  489
Relative stereochemistry, definition of, 177
Renegade molecules, 197
Repeat unit, definition of, 542
Resin, definition of, 564
Resolution
  definition of, 112
  of enantiomeric mixtures, 183
  separation of enantiomers by, 674
Resolving agent, 674
Resonance, effect of, on acidity, 212
Resonance stabilization of a β-lactam, 807
Resonance structures, definition of, 44
Respiration (conversion of carbohydrates
  into carbon dioxide and energy), 753
Restricted rotation, 92
Retention times, 118
11-*cis*-Retinal, 158
Retinal damage by ultraviolet light, 257
Retro-aldol reaction, 772
Retro-Diels-Alder reaction, 206
Retrosynthetic analysis, 509, 640, 731
  definition of, 237
Reverse-phase chromatography, 116
Reverse-polarity reagent, definition of,
  645, 730
Reversible reactions, 421
R_f value, 116
Rhodopsin, 158
Riboflavin, structure of, 717
Ribonucleic acid. *See* RNA
D-Ribose, structure of, 603
α-D-Ribose
  abiotic synthesis of, 644
  structure of, 603, 634
β-D-Ribose, structure of, 603
Ribosome, definition of, 668
Rickets, 790
Rickettsiae, 808
Rigidity, effect of
  on physical properties, 658
  on rates of cyclization reactions, 696

Ring annulation, definition of, 468
Ring flipping, 168
Ring-opening polymerization, 554, 563
Ring strain, 16
Ripening, inhibition of, by
  1,1-dimethylhydrazine, 633
Ripening agent, 30
RNA, structure of, 634
Robinson ring annulation, 467
Rocky Mountain spotted fever, 808
Rotational isomers, definition of, 14
Royal purple, 394
Rum, production of, in Puerto Rico, 768

s-, as prefix for alkyl groups, 20
(*S*), stereochemical designator, 177
Saccharides. *See* Sugars, 557
Saccharin, structure of, 626
Salicin, 391
Salicylic acid, 391
Salvarsan, 797, 799
Saponifiable lipid, definition of, 590
Saponification, definition of, 584
Saturation, 29
Sawhorse representation, 14
Schiff base
  definition of, 425
  mechanism of formation of, 425
  preparation of, from carbonyl
    compounds, 425
  preparation of, from secondary amines,
    425
Schlatter, James, 626
Schrödinger, Erwin, 197
Scopolamine, structure of, 632
Scurvy, 790
Secondary alcohols, preparation of
  by Grignard addition to a ketone, 509
  by metal hydride reduction of ketones,
    497
Secondary amine, 61
Secondary carbon, 20
Sedatives, 535
Seldane. *See* Terfenadine
Selectivity, definition of, 274
Self-exchange, 239
  reaction profile for, 241
Semicarbazide, 426
Semicarbazones
  preparation of, from carbonyl
    compounds, 426
  structure of, 426
Serendipity, in the discovery of Teflon, 549
Serine
  as a source of one carbon in
    biosynthesis, 738
  structure of, 619
Sertürner, Friedrich, 796
Sesquiterpenes, definition of, 590
Sesterterpenes, definition of, 590
Sex attractant, 592
Sex hormones, 468, 594
Shielding, 121
Sickle-cell anemia, 791
Sigma (σ) bond
  definition of, 13
  lengths, 34
Sigma (σ) bonding, 13
Sigmatropic rearrangement, definition of,
  483
Sigmatropic shift, 483
  of hydrogen, 486
Silica gel, 112
Silk, structure of, 562

Simmons-Smith reaction, 356
Simmons-Smith reagent, 280
Single electron transfer, in reductions by
  $FADH_2$, 723
Singlet carbene, electronic structure of, 200
Site electrophilicity, assignment of, 731
Site nucleophilicity, assignment of, 731
Skeletal-rearrangement reactions in
  functional-group preparation, 494
  (Table 14-1)
S_N1 reaction
  conversion of alcohol into alkyl halide
    by, 250
  mechanism of, 48, 268
  rate dependence on concentration in,
    249
  rate-determining step in, 248
  reactivity order in, 249
S_N2 reaction
  backside attack in, 241
  effect of substrate structure, 243
  mechanism of, 268
  opening of bromonium ions, 353
  preparation of functional groups by, 244
    (Table 7-1)
  reactivity order in, 242
  solvent effects on, 684
  steric effects in, 242
Socrates, 633
Sodium alkoxides, preparation of, 414
Sodium alkylbenzene sulfonate, structure
  of, 586
Sodium alkylsulfonate, structure of, 586
Sodium bicarbonate, as antacid, 794
Sodium borohydride, 404
  reduction of acetone by, 715
  for reduction of imines, 615
  mechanism of reduction with, 404
Sodium carbonate, for hydrolysis of
  esters, 694
Sodium cyanoborohydride, for reduction
  of imines, 615
Sodium cyclamate, structure of, 626
Sodium D-line, 179
Sodium hydride, 404
Sodium ion, size of, 662
Sodium nitrite, as source of nitrous acid,
  670
Sodium pentothal, 294
SO_2 evolution, 435
Soft, definition of, 313
Soft nucleophiles, definition of, 457
Solar energy, storage of, in carbohydrates,
  753
Solubility, 24
  control of, by hydrogen bonding, 657
  as a factor affecting chemical equilibria,
    417
Solubilization of salts by
  tetraalkylammonium ions, 686
Solvation
  definition of, 65
  effect of, on reaction rate, 683
Solvent
  aprotic, 66
  characteristic of an ideal, 89
  protic, 66
Solvent front, 116
Solvent polarity, effect of, on reaction rate,
  683
Solvolysis, definition of, 250
sp^2-hybridized atom, electronegativity of, 39
Space-filling models, 546
Spearmint, as source of (-)-carvone, 676

Specific rotation, 179
Spectroscopy, definition of, 110
Speed of light, 119
Spermine, 61
Spin-spin decoupling, 130
Squalene epoxide, mechanism of cationic
    cyclization of, 600
Squalene monoepoxide, 599
Squalene oxide, structure of, 591
Square planar carbon, 197
Staggered conformation, definition of, 159
Staphylococcus aureus, 804, 807
Starch, 607
    structure of, 557, 575
Stationary phase, 111
Stereochemistry of alkene bromination, 353
Stereoelectronic control, 311
Stereoisomers
    calculation of number of possible, 184
    definition of, 14, 155
Stereorandom, definition of, 339
Stereorandomness in electrophilic
    addition, 340
Stereospecificity in singlet carbene
    addition, 355
Steric effect
    definition of, 162
    on solvation, 212
Steric interactions, energetic costs of, 163
    (Table 5-1)
Steric strain, definition of, 162
Sterilant, 357
Steroid, 468
    definition of, 594
cis-Stilbene, 158
trans-Stilbene, 158
Stimulant(s), 97
    amphetamine as, 632
Strain energy, 18 (Table 1-1)
Strecker synthesis, 453
Streptamine, structure of, 638
Streptomyces halstedii, 808
Streptomycetes nodosus, 803
Streptomycin, structure of, 638
Strychnine, structure of, 632
Styling gels, polymers in, 554
Styrene, 338, 360, 361
    industrial production of, 519
    structure of, 548
Styrofoam, 548
Substituent effects on cation stability, 381
Substitution reaction, definition of, 229
Substrate, definition of, 700
Succinic acid
    conversion of, into fumaric acid, 766
    structure of, 762
Sucrose, structure of, 607
Sugars. *See* Carbohydrates
Suicide inhibition, definition of, 806
Sulfa drugs, definition of, 800
Sulfanilamide, structure of, 799
Sulfonamides, 440
    mechanism of formation of, 439
    preparation of, from amines and
        sulfonyl chloride, 439
    structure of, 95
Sulfonate ester. *See* Sulfonic acid ester
Sulfonation, definition of, 371
Sulfonic acid ester
    mechanism of formation of, 438
    preparation of, from sulfonic acids by
        esterification, 438
    preparation of, from sulfonyl chlorides
        by alcoholysis, 438

Sulfonic acids
    preparation of, from arenes by
        electrophilic sulfonation, 374
    structure of, 95
Sulforaphane, 187
Sulfur, expanded valence shell in, 95, 438
Super acid, 335
Supercritical carbon dioxide, 97
Symmetry, effect of, on physical
    properties, 658
Syn addition
    definition of, 340
    in hydroboration, 347
Syndiotactic, definition of, 567
Syn eclipsed conformer, 162
Syn-periplanar arrangement, 310
Synthesis, large scale, 519
Synthetic efficiency
    in abiotic chemistry, 643
    criteria for evaluating, 519
Synthetic routes, criteria for choosing
    among, 517
Synthetic rubber, structure of, 550
Synthoid, structure of, 523
Syphilis, 797

t-, prefix for alkyl groups, 20
Tagamet, function of, 795
Tartaric acid, 180
Tautomerism, 424
Tautomerization, 456, 484
    of adenine, 671
    in biological reductive amination, 718
    definition of, 204
    in interconversion of glucose and
        fructose, 771
Tautomers, 204
Taxol, structure of, 593
TCA cycle. *See* Tricarboxylic acid cycle
T cells, 791
Teflon, 99
    structure of, 549
Terephthalic acid, 379
    structure of, 561
Terfenadine, structure of, 794
Termination of free radical halogenation
    reactions, 254
Termination of radical addition, 346
Termolecular reaction, 220
Terpenes
    biosynthesis of, 595
    definition of, 590
Tertiary amides, acidity of, 214
Tertiary amine, 61
Tertiary carbon, 21
Testosterone, 468
    structure of, 594
Tetraalkylammonium ion, as phase-
    transfer agent, 686
2,3,7,8-Tetrachlorodibenzodioxin, 373
1,1,6,6-Tetradeutero-1,5-hexadiene, 483
Tetraethylborate, 404
Tetrafluorethylene, structure of, 549
Tetrahedral carbon, 5
Tetrahedral intermediate(s)
    in covalent catalysis, 702
    in ester hydrolysis, 690
Tetrahydrofolic acid
    biosynthesis of, from folic acid, 810
    blocking physiological synthesis of,
        797
    function of, 736
    inhibition of biosynthesis of, 809

    structure of, 737
    synthesis inhibition by sulfa drugs, 800
Tetrahydrofuran, 279
    water solubility of, 657
Tetramethylammonium cation, 61
Tetramethylsilane, 121
Tetraterpenes, definition of, 590
Tetrodotoxin, structure of, 661
Tetroses, structures of, 601
Thalidomide, 183
    physiological activity of enantiomers,
        676
Theobromine, 97
Thermal conductivity detector, 117
Thermodynamic control in enolate ion
    protonation, 217
Thermodynamics, definition of, 203
Thermoneutral reaction, definition of, 194
THF. *See* Tetrahydrofuran
Thiamine
    decarboxylation by, 766
    deprotonation of, 728
Thiamine pyrophosphate, 761
    function of, 726, 727
    function of, in the TCA cycle, 762
    structure of, 727
Thin-layer chromatography, 115
Thioethers, structure of, 94
Thiol acetate, 761
Thiol ester hydrolysis
    acid catalysis in, 431
    mechanism of, in acid, 430
    reaction profile for, 430
Thiol esters
    hydrolysis of, in base, 429
    preparation (multistep) of, from
        amides, 435
    relative reactivity of, 755
    relative reactivity of carbonyl groups
        toward nucleophiles, 725
    structure of, 91, 94
Thiols, structure of, 94
Thionyl chloride, 434
Thiophene, 95
Thiourea, 294
Three-base-pair codon, 669
Three-dimensional structure, drawing of,
    14
Three-point contact for chiral recognition,
    671
Threonine, structure of, 619
D-Threose, structure of, 602
Through-bond effects, 210
Thymidine, structure of, 636
Thymine, 97
    base pairing of, with adenine, 665
    structure of, 635
Tobacco, as source of nicotine, 632
Toluene, 49, 392
    bromination of, 381
p-Toluenesulfonyl chloride. *See* Tosyl
    chloride
Torsional strain, definition of, 160
Tosyl chloride, 439
    alcoholysis of, 439
    ammonolysis of, 439
Tosylate esters, preparation of, from
    alcohols by treatment with tosyl
    chloride, 439
Transcription, definition of, 668
Transesterification, 687
    definition of, 433
Transfer RNA, 668
Transition metal catalysis, 696

Transition state
　　definition of, 157
　　recognition of, in an energy diagram, 161
Transition-state analog, 709
Transition-state mimic, 708
Transition-state stabilization, 749
　　multivalent metal cations in, 704
　　solvent effects on, 683
Transition-state theory, 220
　　in catalysis, 683
Transmetallation, as a route to organometallic reagents, 280
Transpeptidase, 805
Treatment of alcohols with phosphoryl or sulfonyl halides, for C–X bond formation, 497
Trefenadine, structure of, 523
Triamterene, structure of, 523
Tricarboxylic acid cycle
　　energy conversion in, 767
　　mechanism of, 762
　　steps in, 781
　　stereoselectivity in, 763
Trichloroacetic acid, acidity of, 210
Trichloroethane, as cleaning solvent, 586
2,4,5-Trichlorophenoxyacetic acid, 373
Tridentate ligand, definition of, 660
Triglycerides, definition of, 581
Trigonal carbon, 5
　　geometry at, 28
Trimethylamine, 61
Trioses, structures of, 601
1,1,2-Triphenylethene, 232
Triphenylphosphine oxide, 459
Triple bond, bond length of, 52
Triplet carbene, 200
　　electronic structure of, 200
　　formation of, 356
Triplet multiplet, 127
Tris(hydroxymethyl)aminomethane
　　structure of, 660
　　as tridentate ligand, 660
TRIS. *See* Tris(hydroxymethyl)aminomethane
Triterpenes, definition of, 590
tRNA. *See* Transfer RNA
Tropane, structure of, 617
Tryptophan
　　structure of, 619
　　taste of, 676
Tswett, Mikhail, 111
Twist-boat cyclohexane, 168
Typhus, 808
Tyrosine
　　structure of, 619
　　taste of, 676

Umpolung, definition of, 645
Unimolecular reaction, definition of, 219
Unnatural amino acids, 624
Unsaturated compounds, reactivity order toward catalytic hydrogenation, 413
Unsaturated fatty acids
　　geometric configuration of, 584
　　names of, 583
　　structures of, 583
Unsaturated molecules, characteristic absorption bands in, 144 (Table 4-4)
α,β-Unsaturated thiol esters, reduction of, 723
Unsaturation, 29
Unsymmetrical ketones

complications in the aldol reaction and condensation, 465
deprotonation of, 465
regiopreference for migration in the Baeyer-Villiger oxidation of, 493
regiopreference for migration in the Beckmann rearrangement of, 488
Uracil, 97
　　structure of, 635
Urea, 2, 440
　　crystal packing of, 570
　　structure of, 570, 630
Urethanes
　　formation of, 530
　　hydrolysis of, 530
　　as protecting group for amines, 530
　　structures of, 530
Uridine, structure of, 636

Vaginal yeast, control of, by bacteria, 803
Valence electrons, 7
Valence shell, 7
Valine, structure of, 619
Valinomycin, structure of, 801
Valium
　　structure of, 535
　　synthesis of, 535
van der Waals attractions, 12, 651
　　in membrane structure, 587
　　in polymer association, 543
　　between polymer chains, 546
van't Hoff, Jacobus Henricus, 541
Variable region in a catalytic antibody, 708
Vasodilation, 794
Vasodilator, yohimbine as, 632
Vesicle, structure of, 589
Vicinal dibromides
　　as alkene protecting group, 354
　　formation of, 498 (Table 14-2)
　　preparation of, by bromination of alkenes, 351, 498
Vicinal dichlorides, preparation of, from alkenes by chlorination, 354
Vicinal dihalides,
　　definition of, 280
　　dehalogenation of, 321
Vinyl acetate, radical polymerization of, 553
Vinyl anion, structure of, 317
Vinyl cation
　　stability of, 341
　　structure of, 317
Vinyl chloride, structure of, 548
Vinyl group, definition of, 50
Vinyl halides
　　dehydrohalogenation of, 316
　　preparation of, from alkynes by electrophilic addition, 341
　　preparation of, from vicinal dihalides, 318
5-Vinylnorbornene
　　as a cross-linking agent, 552
　　preparation of, by a Diels-Alder reaction, 552
　　structure of, 552
Viscosity, definition of, 584
Vitamin A
　　as a conjugated polyene, 551
　　structure of, 591
Vitamin B$_1$. *See* Thiamine pyrophospate
Vitamin B$_{12}$, 511
　　structure of, 660
Vitamin B$_2$, structure of, 717
Vitamin B$_6$

structure of, 717
　　in treatment of beriberi, 790
Vitamin B$_c$. *See* Folic acid
Vitamin C, in treatment of scurvy, 790
Vitamin D, in treatment of rickets, 790
Vitamin D$_3$, 486
Vitamin H. *See* Biotin
Vitamin M. *See* Folic acid
Vitamin PP, structure of, 717
von Leeuwenhoek, Anton, 61
Vulcanization, allylic C–S cross-links in, 553
Vulcanized rubber, 553

Wagner-Meerwein rearrangement, 478
Water, 77
　　acidity of, 209
　　hybridization in, 77
　　ion product of, 418
　　solvation by, 685
　　structure of, 78
Water pipes, poly(vinylchloride) in, 548
Wavelength, 119
Wave number scale, 135
Waxes, structure of, 581
Wells, Horace, 272
Whole body NMR spectroscopy, 134
Williamson ether synthesis, 271
　　for ether formation, 498
　　in formation of benzyl ethers, 529
　　mechanism of, 272
Withering, William, 608
Wittig, Georg, 459
Wittig reaction, 459
　　for C=C bond formation, 496
　　mechanism of, 459
　　net conversion, 460
Wöhler, Friedrich, 2
Wolff-Kishner reduction, 377
　　of aldehydes for C–H bond formation, 495
　　of ketones for C–H bond formation, 495
Woodward, Robert B., 73, 484, 511
Woodward-Hoffmann rules, 484
Wool, structure of, 562
Working backward, 509

Xanax, structure of, 523
Xylene, 49
Xylomollin, 403
D-Xylose, structure of, 603

-yl, suffix for alkyl group, 19
Ylide
　　definition of, 276
　　of thiamine, 728
-yne, suffix for alkynes, 19
Yogurt, 777
Yohimbine
　　biological effect of, 798
　　structure of, 632, 798

(Z)-isomer, 36
Zaitsev orientation in E2 elimination, 308
Zaitsev's Rule, 307
Zantac, structure of, 523
Ziegler, Karl, 567
Ziegler-Natta catalyst, 567
Zinc-copper couple, 356
Zinc enolate, 470
Zombies, 639
Zwitterion, 459
　　definition of, 45

Average Bond Energies (kcal/mole)

Example: $CH_4 \rightarrow C + 4\,H$, $\Delta H°/4 = 99$ kcal/mole

C—H 99	C—C 83	C$=$C 146	C$\equiv$C 200	
N—H 93	C—N 73	C$=$N 147	C$\equiv$N 213	
O—H 111	C—O 86	C$=$O 179	C$\equiv$O 257	O$=$C$=$O 225 (each)
H—H 104	N—N 39	N$=$N 100	N$\equiv$N 226	
H—F 135	O—O 35	^{3}(O$=$O) 119		
H—Cl 103	C—Cl 81	Cl—Cl 58		
H—Br 87	C—Br 68	Br—Br 46		
H—I 71	C—I 51	I—I 36		

Bond Dissociation Energies (kcal/mole)

Bond	X =	H	F	Cl	Br	I	OH	NH$_2$	CH$_3$
Ph—X		111	126	96	81	65	111	102	101
CH$_3$—X		105	108	85	70	57	92	85	90
CH$_3$CH$_2$—X		100	108	80	68	53	94	84	88
(CH$_3$)$_2$CH—X		96	107	81	68	54	94	84	86
(CH$_3$)$_3$C—X		93		82	68	51	93	82	84
PhCH$_2$—X		88		72	58	48	81		75
H$_2$C$=$CH—CH$_2$—X		86		68	54	41	78		74
H—X		104	136	103	87	71	119	107	105
X—X		104	38	59	46	36	51	66	90